Units

D0217686

Symbol	Name	Measure of	Conversion Factors
A	ampere	electrical current	1 C/sec
Å	Ångström	length	10^{-10} m, 0.1 nm
Bq	becquerel	radioactivity	1 disintegration/sec, 60 dpm
C	coulomb	electrical charge	1 A sec
°C	centigrade degree	temperature	°K − 273
Ci	curie	radioactivity	3.7×10^{10} Bq, 2.22×10^{12} dpm
cm	centimeter	length	10^{-2} m, 10^7 nm
cpm	counts/min	radioactivity	dpm × counting efficiency[a]
d	dalton	molecular mass	1.66×10^{-24} g ($^1/_{12}$ mass of a carbon atom)
dpm	disintegrations/min	radioactivity	0.016 Bq, cpm/counting efficiency[a]
g	gram	mass	6.02×10^{23} daltons
°K	Kelvin	temperature	°C + 273
kb	kilobase	nucleotides	1000 bases or base pairs
kcal	kilocalorie	energy	4.18 kilojoules
kd	kilodalton	molecular mass	1000 d
kJ	kilojoule	energy	0.24 kilocalories
L	liter	volume	1000 mL
m	meter	length	100 cm, 10^9 nm
M	molar	concentration	moles solute per liter of solution
μg	microgram	mass	10^{-6} g
min	minute	time	60 sec
mL	milliliter	volume	1 cm^3
mole	mole	number	6.02×10^{23} molecules
mV	millivolt	electrical potential	10^{-3} volts
nm	nanometer	length	10^{-9} m, 10 Å
s	siemen	conductance	A/V
sec	second	time	3600 sec/hour; 86,400 sec/day
V	volt	electrical potential	1000 mV

[a] See table of radioactive isotopes (inside back cover) for efficiency of counting of specific isotopes.

A PROBLEMS APPROACH

MOLECULAR BIOLOGY OF

THE CELL
FOURTH EDITION

A PROBLEMS APPROACH

JOHN WILSON & TIM HUNT

GS Garland Science
Taylor & Francis Group
NEW YORK AND LONDON

Garland
Vice President: Denise Schanck
Managing Editor: Sarah Gibbs
Senior Editorial Assistant: Kirsten Jenner
Editorial Assistants: Kate Harper and Sbita Reandi
Production Editor: Angela Bennett
Layout: Angela Bennett
Copy Editors: Harriet Stewart-Jones and Angela Bennett
Design: Blink Studio, London
Indexer: Merrall Ross International Ltd.
Manufacturing: Nigel Eyre

Back cover This photograph was taken on 10 December 2001 in front of one of the mosaics of the Golden Hall of Stockholm's City Hall. Both John Wilson (left) and Tim Hunt (right) are more formally attired than usual. Actually, neither had worn 'white tie and tails' ever before.

Front cover Human Genome: Reprinted by permission from *Nature*, International Human Genome Sequencing Consortium, 409:860–921, 2001 © Macmillan Magazines Ltd. Adapted from an image by Francis Collins, NHGRI; Jim Kent, UCSC; Ewan Birney, EBI; and Darryl Leja, NHGRI; showing a portion of Chromosome 1 from the initial sequencing of the human genome.

Chapter motif reprinted with permission from MD Adams et al. *Science* 287:2185–2195, 2000 © American Association for the Advancement of Science.

John Wilson received his Ph.D. from the California Institute of Technology and pursued his postdoctoral work at Stanford University. He is currently Professor of Biochemistry and Molecular Biology at Baylor College of Medicine in Houston. His research interests include genetic recombination, genome stability, and gene therapy. He has taught medical and graduate students for many years, coauthored books on immunology, molecular biology, and biochemistry, and received numerous teaching honors, including the Distinguished Faculty Award and Robertson Presidential Award for excellence in education.

Tim Hunt received his Ph.D. in biochemistry from The University of Cambridge where he supervised undergraduates in cell biology for more than 20 years. He spent many summers in the late 1970s and early 1980s teaching molecular biology at the Marine Biological Laboratories in Woods Hole, Massachusetts. In 1990, he moved to a position at ICRF Clare Hall Laboratories just outside London, where he works on the control of the cell cycle. A Fellow of the Royal Society of London and foreign associate of the US National Academy of Sciences, he shared the 2001 Nobel Prize in Physiology or Medicine with Leland Hartwell and Paul Nurse.

© 2002, 1994, 1989 by John Wilson and Tim Hunt

All rights reserved. No part of this book covered by the copyright hereon may be reproduced or used in any format in any form or by any means—graphic, electronic, or mechanical, including photocopying, recording, taping, or information storage and retrieval systems—without permission of the publisher.

Library of Congress Cataloging-in-Publication Data

Wilson, John.
  Molecular biology of the cell : a problems approach / John Wilson & Tim Hunt. --4th ed.
    p. cm.
  Completely rewritten and greatly expanded to correspond to the structure of the fourth
  edition of Molecular biology of the cell by Bruce Alberts ... [et al.]
  Includes bibliographical references and index.
  ISBN 0-8153-3577-6 (pbk.)
   1. Cytology--Problems, exercises, etc. 2. Molecular biology--Problems, exercises, etc.
 I. Hunt, Tim, 1943- II. Title

QH581.2 .M64 2002 Suppl.
571.6'076--dc21

                                                         2002069639

Published by Garland Science, a member of the Taylor & Francis Group, 29 West 35th Street, New York, NY 10001-2299, and 11 New Fetter Lane, London, EC4P 4EE.

Printed in the United States of America

15 14 13 12 11 10 9 8 7 6 5 4 3 2 1

For Lynda, Mary,
Celia, and Agnes

Preface

"You know, the proper method for inquiring after the properties of things is to deduce them from experiments." Isaac Newton, 1672.

This book of Problems is our third attempt, now deliberately titled 'An Approach.' It is intended as a companion to the Fourth Edition of *Molecular Biology of the Cell* (MBoC) by Bruce Alberts, Julian Lewis, Sandy Johnson, Martin Raff, Keith Roberts, and Peter Walter. Our aim has been to explore the questions that ought ideally to surface during the reading of a textbook, but rarely do. How do scientists know that? And how did they find it out? Such things are rarely obvious, especially to people who have never worked in a real research laboratory, but the process of discovery is in fact the whole point. Science is a way of knowing, not a dull catalogue of facts. To be understood properly, principles must be examined, questioned, and played with. We would like to encourage a more questioning attitude toward cell biology than is possible in a textbook.

The other problem with textbooks is that, in their quest for clear, elegant, and manageable presentation, they leave out most of the experimental evidence that supports the current view of a topic. Probably our most important task is to introduce readers to the experimental foundations of cell and molecular biology and show how the links between the behavior of molecules and the biology of cells have been revealed. In the real world of research biology, half the battle is knowing what question to ask; the other half is finding a way to answer it. In biology, new knowledge never comes from just sitting back and thinking: experiments are the only way we have to find out how nature works. Unfortunately, no book can replace the actual doing of experiments, but we have tried to capture some of the flavor of the process, especially the thinking that leads from observation to interpretation.

Although our intentions remain the same in this edition, we have made some significant changes in style and content. The book is heavily revised, and we now cover many more topics than in previous editions. For example, you will now find problems dealing with the opening chapters of MBoC, discussing the nature of life, its evolution, simple chemistry, radioactivity, pH, biochemistry, and protein chemistry. We ourselves learned and relearned a great deal in this process, and hope that some of this rubs off on our readers. We discovered to our consternation and frequent puzzlement, that what is supposed to be simple—elementary—often is not. We learned anew how fruitful and illuminating it is to struggle with simple concepts in different surroundings and new contexts.

We hope that the new sections will make our book more generally useful and complete. Especially in these early chapters, we have sometimes elaborated very considerably on concepts that are summarized in a few sentences in MBoC. In addition, we would strongly recommend two little books (both of them still in print) for further study: Howard C Berg's brilliant introduction to diffusion, entitled *Random Walks in Biology* and Irwin H Segel's excellent *Biochemical Calculations: How to Solve Mathematical Problems in General Biochemistry*. Both provided us with enlightenment and ideas for problems.

In preparing this new edition, we also decided to modify two features of the previous book. Biology is notoriously and unavoidably full of unfamiliar terms, which we previously dealt with as a series of definitions with blanks to be filled in. Now, we supply the terms without the definitions under the heading of 'Terms to Learn,' to emphasize that a certain amount of new vocabulary must be learned: there's no escape. These terms are in bold in MBoC and in its glossary, where they are defined. Second, we have scattered true/false questions throughout the text as a kind of leavening, rather than, as we did in previous editions, present them in a block. In many cases, these questions will also require a little more thought than their predecessors.

We have borrowed and learned from Peter Walter, who composed a large set of wonderful questions for *Essential Cell Biology* (the little sister of MBoC). We have greatly increased the number of these 'intermediate-level' questions, which were largely absent before but now comprise roughly a quarter of the questions in this edition. We hope that they will prove popular and useful.

The core of this Problems book, however, remains the research-based questions. Science does not stand still, and we have added more than 200 new problems to cover the gap between 1994 and 2002, the time since the last edition. These additions, along with selected problems based on older literature, have more than doubled the number of such questions. To provide students and teachers ready access to the research upon which these questions are based, as well as a fuller discussion of the issues involved, we have, as before, given references with the answers. As a result, our lists of References (p. 671) and Cited Researchers (p. 683) have grown characteristically with this edition.

Although this revision has created a book whose size may seem intimidating, the positive aspect is that there should be something here for everyone. Thus the reader will find examples of practically all the ways there are of finding things out in biology— or so we believe. Advances in electronic publication have allowed us to include many more 'real' data, where previously we would have used a sketch. But we have been highly selective, and have often edited out distracting features, enhanced contrast, and subtracted background in our efforts to make the heart of the matter as clear as possible.

This Problems book, then, provides a kind of running commentary on MBoC. As we wrote in a previous Preface, our secret aim has been to try to provoke readers to ask questions, as well as answer them. The goal of a textbook is to make things as plain as possible, and the better the book, the less the reader is interrogated or made to stop and think. Yet, if the text does not challenge the reader, and the reader does not challenge the text, much of the richness and depth of understanding is lost. It all seems so clear and so obvious. In real live research, understanding is born out of puzzlement, doubt, criticism, and debate. Groping one's way through the fog of uncertainty is a slow and often discouraging process; eureka moments (even if one is lucky) are few and far between. Nevertheless, those moments catch the essence of the drama, and we have tended to focus on them, where we have been able to cast them in the form of a problem. In this way, for student and teacher alike, we hope to encourage a questioning approach to cell and molecular biology. For without curiosity, there would be neither science nor scientists.

How can this book be used? It is not intended for bedtime reading. We composed it, as before, by means of constant dialogue and discussion, and we suspect that the most fruitful use of the problems will be to stimulate discussions in class, or between students. Tackling selected problems as homework will also surely help. Teachers will find ideas for exam questions here, and as before we have deliberately left half the questions unanswered. Answers to these questions are available to instructors on the WWW via www.classwire.com/garlandscience/, on application to Garland for an account and a password.

As always, we would like to hear from our readers, especially in cases where we have gotten things wrong or have stated them badly. You can email your comments and suggestions to John Wilson at jwilson@bcm.tmc.edu or Tim Hunt at tim.hunt@cancer.org.uk.

Acknowledgements

If an army marches on its stomach, authors work best well fed in pleasant surroundings. We started to plan this revision of The Problems Book towards the end of the last century (in February 1999) in a beautiful old house overlooking the Pacific Ocean in Santa Cruz, California. Karen, Darrell, and Denise Darling took care of us and the other authors of MBoC on this and subsequent meetings held at their home. Most of the work was done in London, however, and we are specially grateful to Betty and Bruce Alberts for letting us use their splendid pied-à-terre in London, and to Emily Preece and her staff for their support. A special thanks, too, to the girls at Café Roma for their lunches during the final stages of this work.

More important than creature comforts, however, has been various help from a long list of people. First and foremost we thank Alastair Ewing of the Open University for working through almost all the problems, thereby saving us from innumerable errors and infelicities. Nobody else comes close to Alastair as an editor, although the authors of MBoC, plus David Morgan and Julie Theriot, were very helpful with their particular chapters and Kathy Suprenant read the mitosis chapter. Peter Walter showed us how to construct short, simple, entertaining problems, and Keith Roberts and Nigel Orme were friendly, patient and inspiring guides to illustration. The following people answered specific queries, contributed problems, provided photographs and figures, or read drafts of chapters: David Allis, Susan Berget, Mark Carrington, Xiu-Bao Chang, Pietro Decamilli, Ramon Diaz-Arrastia, Al Edwards, Steve Freeland, Susan Fullilove, Yukio Fujiki, Steve Halford, Bob Goldman, Jim Haber, J Allie C Hajian, Lee Hartwell, Ming-Derg Lai, Juliet Lee, Peter Lund, Kathryn Meier, Mark Mooseker, Deborah Mowshowitz, Andrew Murray, Kate Nobes, Fumio Oosawa, Hamida Qavi, Jun Qin, Tom Pollard, Scottie Robinson, David Roth, Robert Sabatini, Thomas Shea, George N. Somero, Paul Treichel, Roger Tsien, Steve West, Rick Young, and Anthony Weiss. We are particularly grateful to Martin Rees for his lucid guide to estimating the mass of the universe, Russ Doolittle for explaining how to construct phylogenetic trees, and Ray Deshaies for his illuminating recollections of a key turning point during his career as a graduate student.

The support from Garland Science Publishing was outstanding throughout and very much appreciated. Denise Schanck was quietly but effectively encouraging right from the start. Sarah Gibbs was a paragon of organization and made sure that we always had everything we needed; her positive attitude was a great solace during the dark moments when the task seemed impossible. Kirsten Jenner translated the text into first pages, which took very careful fiddling, and she together with Kate Harper and Sbita Reandi took care of permissions. Conversations with Nasreen Arain and Adam Sendroff gave us a much better feel for the marketing perspective. Emma Hunt converted all our old figures into Adobe Illustrator™ format, and Angela Bennett became a great friend during the stages of page layout and proof-reading. Angela was terrific to work with: firm but flexible in discussions of layout and editorial nuances, and with an eagle eye for detail. The look and feel of the book is largely her doing.

Finally, we should apologize to our families and colleagues for our all too frequent absences, and thank them for their indulgence. We hope that the cause was good and just.

A Couple of Things to Know

AVOGADRO'S NUMBER (6.02 × 10^{23} MOLECULES/MOLE)

Avogadro's number (N) is perhaps the most important constant in molecular sciences, and it appears again and again in this book. Do you know how it was determined? We didn't, or had forgotten if we ever knew. How can one measure the number of molecules in a mole? And who did it first? You will not find this information in modern biology books, partly because it is ancient history, and partly because it was the business of physicists; some pretty good physicists too, as we shall see.

Amadeo Avogadro had no idea how many molecules there were in 22.4 L of a gas. His hypothesis, presented in 1811, was simply that equal volumes of all gases contained the same number of molecules, irrespective of their size or density. Not until much later, when the reality of molecules was more widely accepted and the microscopic basis for the properties of gases was being worked out, were the first estimates attempted. An Austrian high school teacher called Josef Loschmidt used James Clerk Maxwell's recently developed kinetic theory of gases to estimate how many molecules there were in a cubic centimeter of air. Maxwell had derived an expression for the viscosity of a gas, which is proportional to the density of the gas, to the mean velocity of the molecules, and to their mean free path. The latter could be estimated if one knew the size and number of the molecules. Loschmidt simply made the assumption that when a gas was condensed into a liquid, its molecules were packed as closely as they could be, like oranges in a display on a fruit stand, and from this he was able to get a pretty accurate value for Avogadro's number. Not surprisingly, in Austria they often refer to N as 'Loschmidt's number.' In fact, it wasn't until 1909 that the term 'Avogadro's number' was suggested by Jean Perrin, who won the 1926 Nobel prize for physics (his lecture is available on the Nobel web site, and his book on Atoms [Les Atomes, 1913, translated from the original French by D. Ll. Hammick, reprinted in 1990 by Ox Bow Press] is highly recommended—and accessible—reading. It has been called the finest book on physics of the 20th century).

You may be surprised to discover, as we were, that estimating Avogadro's number was an important topic of Albert Einstein's Ph.D. thesis. Abraham Pais's wonderful biography of Einstein, *Subtle is the Lord* (subtitled *The Science and the Life of Albert Einstein*, 1982 Oxford University Press) devotes Chapter 5, 'The Reality of Molecules,' to this period of the great physicist's life and work. Einstein found three independent ways to estimate N: from the viscosity of dilute sucrose solutions; from his analysis of Brownian motion, and from light scattering by gases near the critical point, including the blueness of the sky. Because the sky is five million times less bright than direct sunlight, Avogadro's number is 6×10^{23}. Isn't that romantic?

But Einstein's was not the last word on the subject. Indeed, according to Pais, he made an "elementary but nontrivial mistake" in his thesis that was later corrected, and it was really Perrin who brought the whole field together with his experiments on Brownian motion. The Nobel presentation speech contains this line:

"His [Perrin's] measurements on the Brownian movement showed that Einstein's theory was in perfect agreement with reality. Through these measurements a new determination of Avogadro's number was obtained."

For most methods of counting molecules, neither the physics nor the math is easy to follow, but two are simple to understand. The first comes from radioactive decay, and another Nobel prize-winning physicist, Ernest Rutherford. When radium decays, it emits alpha particles, which are helium nuclei. If you can count the radioactive decay events with a Geiger counter and measure the volume of helium emitted, you can estimate Avogadro's number. The second way is much more modern. You can see large proteins and nucleic acids with the aid of an electron microscope.

CALCULATIONS AND UNIT ANALYSIS

Many of the problems in this book involve calculations. Where the calculations are based on an equation (for example, the Nernst equation or the equation for volume of a sphere), we provide the equation along with a brief explanation of symbols, and often their values. Many calculations, however, involve the conversion of information from one form into another, equivalent form. For example, if the concentration of a protein is 10^{-9} M, how many molecules of it would be present in a mammalian nucleus with a volume of 500 μm^3? Here, a concentration is given as M (moles/L), whereas the desired answer is molecules/nucleus; both values are expressed as 'number/volume' and the problem is to convert one into the other.

Both kinds of calculation use constants and conversion factors that may or may not be included in the problem. The Nernst equation, for example, uses the gas constant R (2.0×10^{-3} kcal/°K mole) and the Faraday constant F (23 kcal/V mole). And conversion of moles/L to molecules/nucleus requires Avogadro's number N (6.0×10^{23} molecules/mole). All of the constants, symbols, and conversion factors that are used in this book are listed inside the book covers (along with the standard genetic code, the one-letter amino acid code, useful geometric formulas, and data on common radioisotopes used in biology).

For each type of calculation, we strongly recommend the powerful general strategy known as unit analysis (or dimensional analysis). If units (for example, moles/L) are included along with the numbers in the calculations, they provide an internal check on whether the numbers have been combined correctly. If you've made a mistake in your math, the units will not help, but if you've divided where you should have multiplied, for example, the units of the answer will be nonsensical: they will shout 'error.' Consider the conversion of 10^{-9} M (moles/L) to molecules/nucleus. In the conversion of moles to molecules, do you multiply 10^{-9} by 6×10^{23} (Avogadro's number) or do you divide by it? If units are included, the answer is clear.

$$\frac{10^{-9}\,\text{moles}}{L} \times \frac{6 \times 10^{23}\,\text{molecules}}{\text{mole}} = \frac{6 \times 10^{14}\,\text{molecules}}{L} \quad \textbf{YES}$$

$$\frac{10^{-9}\,\text{moles}}{L} \times \frac{\text{mole}}{6 \times 10^{23}\,\text{molecules}} = \frac{1.7 \times 10^{-33}\,\text{mole}^2}{\text{molecules L}} \quad \textbf{NO}$$

Similarly, in the conversion of liters to nuclei, the goal is to organize the conversion factors to transform the units to the desired form.

$$\frac{6 \times 10^{14}\,\text{molecules}}{L} \times \frac{1\,L}{1000\,\text{mL}} \times \frac{\text{mL}}{\text{cm}^3} \times \frac{\text{cm}^3}{(10^4\,\mu m)^3} \times \frac{500\,\mu m^3}{\text{nucleus}} = \frac{300\,\text{molecules}}{\text{nucleus}}$$

If you do this calculation with pure numbers, you must worry at each step whether to divide or multiply. If you attach the units, however, the decision is obvious. It is important to realize that any set of (correct) conversion factors will give the same answer. If you are more comfortable converting liters to ounces, that's fine, so long as you know a string of conversion factors that will ultimately transform ounces to μm^3.

There are a few simple rules for handling units in calculations.
1. Quantities with different units cannot be added or subtracted. (You cannot subtract 3 meters from 10 kcal.)
2. Quantities with different units can be multiplied or divided; just multiply or divide the units along with the numbers. (You can multiply 3 meters times 10 kcal; the answer is 30 kcal meters.)
3. All exponents are unitless. (You can't use $10^{6\,\text{mL}}$.)
4. You cannot take the logarithm of a quantity with units.

Throughout this book, we have included the units for each element in every calculation. If the units are arranged so that they cancel to give the correct units for the answer, the numbers will take care of themselves.

Credits

CHAPTER 1

Figure 1–1. Data courtesy of Steve Freedland.

CHAPTER 2

Figure 2–23. From HC Berg, Random Walks in Biology. Princeton: Princeton University Press © 1993.
Figure 2–28. Courtesy of Zermatt Tourism.
Figure 2–41. Adapted from AL Lehinger, DL Nelson & MM Cox, Principles of Biochemistry. New York: Worth Publishers, 2000.

CHAPTER 3

Figure 3–14. From FA Smit, *Svenska Vetensk. Acad. Handl.* 24:1–40, 1898.
Figure 3–14 (top). From GA Boulenger, Pisces. Report on the collections of natural history made in the Antarctic regions during the voyage of the "Southern Cross". *Brit. Mus. (Nat. Hist.)* 174–189, 1902.
Figure 3–17 (bottom). Courtesy of Eli Lilly and Company. Reprinted from JE Thomas, M Smith, B Rubinfeld, M Gutowski, RP Beckmann & P Polakis *J. Biol. Chem.* 271:28630–28635, 1996. © American Society for Biochemistry and Molecular Biology.
Figure 3–19. From AW Karzai, MM Susskind & RT Sauer, *EMBO J.* 18:3793–3799, 1999 by permission of Oxford University Press.
Figure 3–33. From NR Brown, ME Noble, AM Lawrie, MC Morris, P Tunnah, G Divita, LN Johnson & JA Endicott, *J. Biol. Chem.* 274:8746–8756, 1999. © American Society for Biochemistry and Molecular Biology.
Figure 3–34. From AJ Flint, T Tiganis, D Barford & NK Tonks, *PNAS* 94:1680–1685, 1997. © National Academy of Sciences, USA.

CHAPTER 4

Figure 4–5. From D Vollrath & RW Davis, *Nucleic Acids Res.* 15:7865–7876, 1987 by permission of Oxford University Press.
Figure 4–6. Courtesy of Laurent J Beauregard.
Figure 4–27. Reprinted from K Shibahara & B Stillman, *Cell* 96:575–585. © 1999, with permission from Elsevier Science.
Figure 4–30. Reprinted from K Kimura & T Hirano, *Cell* 90:625–634. © 1997, with permission from Elsevier Science.

CHAPTER 5

Figure 5–16. Courtesy of Victoria Foe.
Figure 5–23 and 5–67. Reprinted from RD Klemm, RJ Austin & SP Bell, *Cell* 88:493–502. © 1997, with permission from Elsevier Science.
Figure 5–25. Reprinted from K Shibahara & B Stillman, *Cell* 96:575–585. © 1999, with permission from Elsevier Science.
Figure 5–26. Reprinted with permission from TM Nakamura, GB Morin, KB Chapman, SL Weinrich, WH Andrews, J Lingner, CB Harley & TR Cech, *Science* 277:955–959, 1997. © American Association for the Advancement of Science.
Figure 5–31. From C Masutani, M Araki, A Yamada, R Kusumoto, T Nogimori, T Maekawa, S Iwai & F Hanaoka, *EMBO J.* 18:3491–3501, 1999 by permission of Oxford University Press.
Figure 5–37. Reprinted with permission from E Van Dyck, AZ Stasiak, A Stasiak & SC West, *Nature* 398:728–73, 1999. © Macmillan Magazines Limited.
Figure 5–46. From R Shah, R Cosstick & SC West, *EMBO J.* 16:1464–1472, 1997 by permission of Oxford University Press.

CHAPTER 6

Figure 6–1. Courtesy of Ulrich Sheer.

Figure 6–2. Reprinted from MJ Thomas, J Platas & DK Hawley, *Cell* 93:627–637. © 1998, with permission from Elsevier Science.

Figure 6–5. Reprinted with permission from SL McKnight & R Kingsbury, *Science* 217:316–324, 1982. © American Association for the Advancement of Science.

Figure 6–7. © Fred Stevens.

Figure 6–11C. Reprinted from E Nudler, A Mustaev, E Lukhtanov & A Goldfarb, *Cell* 89:33–41. © 1997, with permission from Elsevier Science.

Figure 6–13. Reprinted with permission from A Dugaiczyk, SL Woo, EC Lai, ML Mace Jr, L McReynolds & BW O'Malley, *Nature* 274:328–33, 1978. © Macmillan Magazines Limited.

Figure 6–25B. From B Jady & T Kiss, *EMBO J.* 20:541–551, 2001 by permission of Oxford University Press.

Figure 6–34. From AW Karzai, MM Susskind & RT Sauer, *EMBO J.* 18:3793–3799, 1999 by permission of Oxford University Press.

Figure 6–35. From S Horowitz and MA Gorovsky *PNAS* 82:2452–2455, 1985. © National Academy of Sciences, USA.

Figure 6–37. Reprinted from SA Teter, WA Houry, D Ang, T Tradler, D Rockabrand, G Fischer, P Blum, C Georgopoulos & FU Hartl, *Cell* 97:755–765. © 1999, with permission from Elsevier Science.

Figure 6–38. Reprinted with permission from E Deuerling, A Schulze-Specking, T Tomoyasu, A Mogk & B Bukau, *Nature* 400:693–696, 1999. © Macmillan Magazines Limited.

Figure 6–41A. Reprinted with permission from U Schubert, LC Anton, J Gibbs, CC Norbury, JW Yewdell & JR Bennink, *Nature* 404:770–774, 2000. © Macmillan Magazines Limited.

Figure 6–42. From H Yamano, C Tsurumi, J Gannon & T Hunt, *EMBO J.* 17:5670–5678, 1998 by permission of Oxford University Press.

Figure 6–44 and 6–45. Reprinted with permission from A Bachmair, D Finley & A Varshavsky, *Science* 234:179–186, 1986. © American Association for the Advancement of Science.

CHAPTER 7

Figure 7–1. Reprinted with permission from I Wilmut, AE Schnieke, J McWhir, AJ Kind & KHS Campbell, *Nature* 385:810–813, 1997. © Macmillan Magazines Limited.

Figure 7–3 and 7–57. Courtesy of Tim Myers and Leigh Anderson, Large Scale Biology Corporation

Figure 7–8B. Reprinted from H Kabata, O Kurosawa, I Arai, M Washizu, SA Margarson, RE Glass & N Shimamoto, *Science* 262:1561–1563, 1993. © American Association for the Advancement of Science.

Figure 7–10. Adapted from MS Lee, GP Gippert, KV Soman, DA Case & PE Wright, *Science* 245:635–637, 1989.

Figure 7–12B. From LR Patel, T Curran & TK Kerppola, *PNAS* 91:7360–7364, 1994. © National Academy of Sciences, USA.

Figure 7–22. Adapted from M Ptashne, A Genetic Switch: Phage λ and Higher Organisms. Cambridge, MA: Cell Press, 1992.

Figure 7–26B. From E Larshan & F Winston, *Genes & Development* 15:1946–1956, 2001. © Cold Spring Harbor Press.

Figure 7–30B. From X Bi, M Braunstein, GJ Shei & JR Broach, *PNAS* 96:11934–11939, 1999. © National Academy of Sciences, USA.

Figure 7–31. Reprinted with permission from J Zieg, M Silverman,

M Hilmen & M Simon, *Science* 196:170–172, 1977. © American Association for the Advancement of Science.

Figure 7–35. Reprinted with permission from D Whitmore, NS Foulkes & P Sassone-Corsi, *Nature* 404:87–91, 2000. © Macmillan Magazine Limited.

Figure 7–39. Reprinted with permission from GD Penny, GF Kay, SA Sheardown, S Rastan & N Brockdorff, *Nature* 379:131–137, 1996. © Macmillan Magazine Limited.

Figure 7–52. From SR Thompson, EB Goodwin & M Wickens, *Mol. Cell Biol.* 20:2129–2137, 2000. © American Society for Microbiology.

Figure 7–55. Reprinted with permission from CY Chen & P Charnow, *Science* 268:415–417, 1995. © American Association for the Advancement of Science.

CHAPTER 8

Figure 8–3. From M Meselson & FW Stahl, *PNAS* 44:671–682, 1958.

Figure 8–7. All attempts have been made to contact the copyright holder and we would be pleased to hear from them.

Figure 8–9. Courtesy of JC Revy/Science Photo Library.

Figure 8–26. Courtesy of Leander Lauffer & Peter Walter.

Figure 8–33. From AY Ting, KH Kain, RL Klemke & RY Tsien, *PNAS* 98:15003–15008, 2001. © National Academy of Sciences, USA.

Figure 8–34. From AY Ting, KH Kain, RL Klemke & RY Tsien, *PNAS* 98:15003–15008, 2001. © National Academy of Sciences, USA.

Figure 8–35. Reprinted from T Tugal, XH Zou-Yang, K Gavin, D Pappin, B Canas, R Kobayashi, T Hunt & B Stillman, *J. Biol. Chem.* 273:32421–32429, 1996. © American Society for Biochemistry and Molecular Biology.

Figure 8–40. Courtesy of Michele Sawadogo & Robert Roeder.

Figure 8–44. Reprinted from AM MacNicol, AJ Muslin & LT Williams, *Cell* 73:571–583. © 1993, with permission from Elsevier Science.

CHAPTER 11

Figure 11–7. Courtesy of DW Fawcett, from A Textbook of Histology, 12th edn. New York: Chapman and Hall, 1994.

Figure 11–10A. Reprinted with permission from T Hoshi, WN Zagotta & RW Aldrich, *Science* 250:568–571, 1990. © American Association for the Advancement of Science.

CHAPTER 12

Figure 12–7. From K Ribbeck, G Lipowsky, HM Kent, M Stewart & D Gorlich, *EMBO J.* 17:6587–6598, 1998 by permission of Oxford University Press.

Figure 12–8. Reprinted from B Wolff, JJ Sanglier & Y Wang, *Chemistry & Biology* 4:139–147. © 1997, with permission from Elsevier Science.

Figure 12–9. From N Kudo, N Matsumori, H Taoka, D Fujiwara, EP Schreiner, B Wolff, M Yoshida & S Horinouchi, *PNAS* 96:9112–9117, 1999. © National Academy of Sciences, USA.

Figure 12–10. From N Kudo, N Matsumori, H Taoka, D Fujiwara, EP Schreiner, B Wolff, M Yoshida & S Horinouchi, *PNAS* 96:9112–9117, 1999. © National Academy of Sciences, USA.

Figure 12–11. Reprinted from M Fornerod, M Ohno, M Yoshida & IW Mattaj, *Cell* 90:1051–1060. © 1997, with permission from Elsevier Science.

Figure 12–12. From MV Nachury & K Weis, *PNAS* 96:9622–9627, 1999. © National Academy of Sciences, USA.

Figure 12–16B. Reprinted from M Donzeau, K Kaldi, A Adam, S Paschen, G Wanner, B Guiard, MF Bauer, W Neupert & M Brunner, *Cell* 101:401–412. © 2000, with permission from Elsevier Science.

Figure 12–20. From N Kinoshita, K Ghaedi, N Shimozawa, RJ Wanders, Y Matsuzono, T Imanaka, K Okumoto, Y Suzuki, N Kondo & Y Fujiki, *J. Biol. Chem.* 273:24122–24130, 1998. © American Society for Biochemistry and Molecular Biology.

Figure 12–22. Adapted from V Dammai & S Subramani, *Cell* 105:187–196, 2001.

CHAPTER 13

Figure 13–2B. Courtesy of BMF Pearse, adapted from CJ Smith, N Grigorieff & BM Pearse, *EMBO J.* 17:4943–4953, 1998.

Figure 13–2C. Courtesy of Wyn Locke.

Figure 13–6. Reprinted from H Tokumaru, K Umayahara, LL Pellegrini, T Ishizuka, H Saisu, H Betz, GJ Augustine & T Abe, *Cell* 104:421–432. © 2001, with permission from Elsevier Science.

Figure 13–7. Reprinted from H Tokumaru, K Umayahara, LL Pellegrini, T Ishizuka, H Saisu, H Betz, GJ Augustine & T Abe, *Cell* 104:421–432. © 2001, with permission from Elsevier Science.

Figure 13–14. Courtesy of Margit Burmeister. Reproduced from AA Peden, RE Rudge, WW Lui & MS Robinson, *J. Cell Biol.* 156:327–336, 2002. © The Rockefeller University Press.

Figure 13–15. From SM Wilson, R Yip, DA Swing, TN O'Sullivan, Y Zhang, EK Novak, RT Swank, LB Russell, NG Copeland & NA Jenkins, *PNAS* 97:7933–7938, 2000. © National Academy of Sciences, USA.

Figure 13–17A. From MM Perry & AB Gilbert *J.Cell Sci.*39:257–272, 1979. © The Company of Biologists Limited.

Figure 13–22. Reproduced from M Matteoli, K Takei, MS Perin, TC Sudhof & P De Camilli, *J. Cell Biol.* 117:849–861, 1992. © The Rockefeller University Press.

Figure 13–23. From JH Koenig amd K Ikeda, *J. Neuroscience* 9:3844–3960, 1989. © The Society of Neuroscience.

Figure 13–24B and C. From T Xu, U Ashery, RD Burgoyne & E Neher, *EMBO J.* 18:3293–304, 1999 by permission of Oxford University Press.

CHAPTER 14

Figure 14–3B. Reprinted from R Yasuda, H Noji, K Kinosita Jr, & M Yoshida, *Cell* 93:1117–1124. © 1998, with permission from Elsevier Science.

Figure 14–8. Adapted from David Keilin, The History of Cell Respiration and Cytochrome. Cambridge: Cambridge University Press, 1966.

Figure 14–9. Adapted from David Keilin, The History of Cell Respiration and Cytochrome. Cambridge: Cambridge University Press, 1966.

CHAPTER 15

Figure 15–29. Courtesy of Helfrid Hochegger.

Figure 15–30. Reprinted with permission from JE Ferrell Jr & EM Machleder, *Science* 280:895–898, 1998. © American Association for the Advancement of Science.

Figure 15–32. From A Toker & A Newton, *Journal of Biological Chemistry* 275:8271–8274, 2000. © American Society for Biochemistry & Molecular Biology.

Figure 15–35. From H Aberle, A Bauer, J Stappert, A Kispert & R Kemler, *EMBO J.*16:3797–3804, 1997 by permission of Oxford University Press.

Figure 15–36. Reprinted from CG Winter, B Wang, A Ballew, A Royou, R Karess, JD Axelrod & L Luo, *Cell* 105:81–91, 2001. © with permission from Elsevier Science.

Figure 15–37. Reprinted with permission from JA Porter, DP von Kessler, SC Ekker, KE Young, JJ Lee, K Moses & PA Beachy, *Nature* 374:363–366, 1995. © Macmillan Magazine Limited.

Figure 15–38B and C. Reprinted with permission from JA Porter, DP von Kessler, SC Ekker, KE Young, JJ Lee, K Moses & PA Beachy, *Nature* 374:363–366, 1995. © Macmillan Magazine Limited.

Figure 15–39. Reprinted with permission from JA Porter, DP von Kessler, SC Ekker, KE Young, JJ Lee, K Moses & PA Beachy, *Nature* 374:363–366, 1995. © Macmillan Magazines Limited.

CHAPTER 16

Figure 16–4. Courtesy of Thomas D Pollard.

Figure 16–8. Reproduced from EM Mandelkow, E Mandelkow & RA Milligan, *J. Cell Biol.*114:977–991, 1991. © The Rockefeller University Press.

Figure 16–10. Reproduced from M Yoon, RD Moir, V Prahlad & RD Goldman, *J. Cell Biol.*143:147–157, 1998. © The Rockefeller University Press.

Figure 16–14A. From MR Bubb, I Spector, AD Bershadsky & ED Korn, *J. Biol. Chem.*275:21975–21980, 2000. © American Society for Biochemistry and Molecular Biology.

Figure 16–14B and C. From MO Steinmetz, D Stoffler, A Hoenger, A Bremer & U Aebi, *J. Struct Biol.* 119:295–320, 1997. © Academic Press.

Figure 16–21. From R Leguy, R Melki, D Pantaloni & MF Carlier, *J. Biol.Chem.* 275:21975–21980, 2000. © American Society for Biochemistry & Molecular Biology.

Figure 16–22. Reprinted with permission from Y Zheng, ML Wong, B Alberts & T Mitchison, *Nature* 378:578–583, 1995. © Macmillan Magazines Limited.

Figure 16–23A. Reprinted from JA Theriot & DC Fung, *Methods Enzymol.* 298:114–122, 1998. © Academic Press Limited.

Figure 16–23B. Courtesy of Margaret Coughlin, Tim Mitchison & Julie Theriot.

Figure 16–28. Reprinted from J Fan, AD Griffiths, A Lockhart, RA Cross & LA Amos, *J. Mol. Biol.* 259:325–330, 1996. © Academic Press Limited.

Figure 16–29. Reprinted from J Fan, AD Griffiths, A Lockhart, RA Cross & LA Amos, *J. Mol. Biol.* 259:325–330, 1996. © Academic Press Limited.

Figure 16–33C. Reproduced from EL Reese & LT Haimo, *J. Cell Biol.* 151:155–166, 2000. © The Rockefeller University Press.

Figure 16–34. Courtesy of HE Huxley.

Figure 16–38. Courtesy of Lewis Tilney.

Figure 16–39. Courtesy of Keith Roberts.

Figure 16–42A. Reproduced from G Piperno, B Huang, Z Ramanis & DJ Luck, *J. Cell Biol.* 88:73–79, 1981. © The Rockefeller Univerity Press.

Figure 16–42B. Reproduced from G Piperno, B Huang, Z Ramanis & DJ Luck, *J. Cell Biol.* 88:73–79, 1981. © The Rockefeller Univerity Press.

Figure 16–44. Courtesy of Kate Nobes.

Figure 16–46A. Reproduced from A Mallavarapu & T Mitchison, *J. Cell Biol.* 146:1097–1106, 1999. © The Rockefeller University Press.

Figure 16–52. From CL Ho, JL Martys, A Mikhailov, GG Gundersen, RK Liem, *J. Cell Sci.* 111:1767–1778, 1998. © The Company of Biologists Limited.

Figure 16–53. From MO Steinmetz, D Stoffler, A Hoenger, A Bremer & U Aebi, *J. Struct Biol.* 119:295–320, 1997. © Academic Press.

CHAPTER 17

Figure 17–16. Reprinted from C Michaelis, R Ciosk & K Nasmyth, *Cell* 91:35–41. © 1997, with permission from Elsevier Science.

Figure 17–17. Reprinted with permission from F Uhlmann, F Lottspeich & K Nasmyth, *Nature* 400:37–42, 1999. © Macmillan Magazines Limited.

Figure 17–22. From SA Datar, HW Jacobs, AF de la Cruz, CF Lehner & BA Edgar, *EMBO J.* 19:4543–4554, 2000 by permission of Oxford University Press.

Figure 17–25. Courtesy of Julia Burne.

Figure 17–27. Reprinted with permission from JC Goldstein, NJ Waterhouse, P Juin, G Evanl & DR Green, *Nature Cell Biology* 2:156–162, 2000. © Macmillan Magazines Limited.

Figure 17–32. Reprinted from H Yoshida, YY Kong, R Yoshida, AJ Elia, A Hakem, R Hakem, JM Penninger & TW Mak, *Cell* 94:739–750. © 1998, with permission from Elsevier Science.

Figure 17–32. Reprinted from K Kuida, TF Haydar, CY Kuan, Y Gu, C Taya, H Karasuyama, MS Su, P Rakic & RA Flavell, *Cell* 94:325–337. © 1998, with permission from Elsevier Science.

Figure 17–34. From S D'Atri, L Tentori, PM Lacal, G Graziani, E Pagani, E Benincasa, G Zambruno, E Bonmassar & J Jiricny, *Mol. Pharm.* 54:334–341, 1998. © The American Society for Pharmacology and Experimental Therapeutics.

Figure 17–38. From M Zecca, K Basler & G Struhl, *Development* 121:2265–2278, 1995. © The Company of Biologists Limited.

CHAPTER 18

Figure 18–2. Courtesy of Terry D Allen.

Figure 18–3. Reprinted from A Losada & T Hirano, *Curr. Biol.* 11:268–272. © 2001, with permission from Elsevier Science.

Figure 18–07. Courtesy of Conly L Rieder, Wadsworth Center.

Figure 18–18. Reprinted from C Antonio, I Ferby, H Wilhelm, M Jones, E Karsenti, AR Nebreda & I Vernos, *Cell* 102:425–435. © 2000, with permission from Elsevier Science.

Figure 18–19. Reprinted from H Funabiki & A Murray, *Cell* 102:411–424. © 2000, with permission from Elsevier Science.

Figure 18–20. Reproduced from B de Saint Phalle & W Sullivan, *J. Cell Biol.* 1141:1383–139, 1998. © The Rockefeller University Press.

Figure 18–22. Reprinted from AR Skop & JG White, *Curr. Biol.* 11:735–746. © 2001, with permission from Elsevier Science.

Figure 18–24. From A Bassini, S Pierpaoli, E Falcieri, M Vitale, L Guidotti, S Capitani & G Zauli, *Br. J. Haematol.* 104:820–828. © 1999, with permission from Blackwell Science.

Contents

Problems

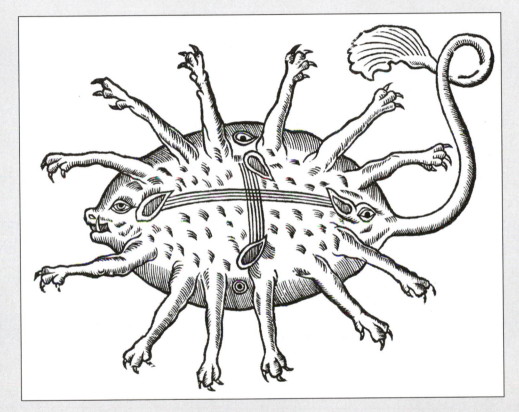

Sea creature sighted between Antibes and Nice in 1562. From Curious Woodcuts of Fanciful and Real Beasts: A selection of 190 sixteenth-century woodcuts from Gesner's and Topsell's Natural Histories, edited by Edmund V Gillon Jr. New York: Dover Publications Inc., 1971.

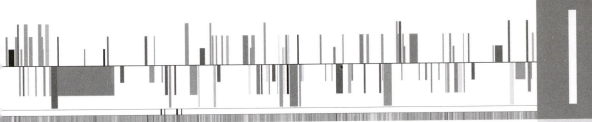

CELLS AND GENOMES

THE UNIVERSAL FEATURES OF CELLS ON EARTH

TERMS TO LEARN

amino acid	nucleotide	ribonucleic acid (RNA)
enzyme	phenotype	ribosomal RNA (rRNA)
gene	plasma membrane	transcription
genome	polypeptide	transfer RNA (tRNA)
genotype	protein	translation
messenger RNA (mRNA)		

1–1 'Life' is easy to recognize but difficult to define. The dictionary defines life as "The state or quality that distinguishes living beings or organisms from dead ones and from inorganic matter, characterized chiefly by metabolism, growth, and the ability to reproduce and respond to stimuli." Biology text-books usually elaborate slightly; for example, according to a popular text, living things:

1. Are highly organized compared to natural inanimate objects.
2. Display homeostasis, maintaining a relatively constant internal environ-ment.
3. Reproduce themselves.
4. Grow and develop from simple beginnings.
5. Take energy and matter from the environment and transform it.
6. Respond to stimuli.
7. Show adaptation to their environment.

Score yourself, a car, and a cactus with respect to these characteristics.

1–2* NASA has asked you to design a module that will seek signs of life on Mars. What will your module look for?

1–3 An adult human is composed of about 10^{13} cells, all of which are derived by cell division from a single fertilized egg.

   A. Assuming that all cells continue to divide (like bacteria in rich media), how many generations of cell divisions would be required to produce 10^{13} cells?

*Answers to these problems are available to Instructors at www.classwire.com/garlandscience/

B. Human cells in culture divide about once per day. Assuming all cells continue to divide at this rate during development, how long would it take to generate an adult organism?

C. Why is it, do you think, that adult humans take longer to develop than these calculations might suggest?

1–4* You have embarked on an ambitious research project: to create life in a test tube. You boil up a rich mixture of yeast extract and amino acids in a flask along with a sprinkling of the inorganic salts known to be essential for life. You seal the flask and allow it to cool. After several months, the liquid is as clear as ever, and there are no signs of life. A friend suggests that excluding the air was a mistake, since most life as we know it requires oxygen. You repeat the experiment, but this time you leave the flask open to the atmosphere. To your great delight, the liquid becomes cloudy after a few days and under the microscope you see beautiful small cells that are clearly growing and dividing. Does this experiment prove that you managed to generate a novel life form? How might you redesign your experiment to allow air into the flask, yet eliminate the possibility that contamination is the explanation for the results?

1–5 The genetic code (see inside back cover) specifies the entire set of codons that relate the nucleotide sequence of mRNA to the amino acid sequence of encoded proteins. Ever since the code was deciphered nearly four decades ago, some have claimed it must be a frozen accident, while others have argued that it was shaped by natural selection.

A striking feature of the genetic code is its inherent resistance to the effects of mutation. For example, a change in the third position of a codon often specifies the same amino acid or one with similar chemical properties. But is the natural code more resistant to mutation (less susceptible to error) than other possible versions? The answer is an emphatic "Yes," as illustrated in Figure 1–1. Only one in a million computer-generated 'random' codes is more error-resistant than the natural genetic code.

Does the extraordinary mutation resistance of the genetic code argue in favor of its origin as a frozen accident or as a result of natural selection? Explain your reasoning.

1–6* You have begun to characterize a sample obtained from the depths of the oceans on Europa, one of Jupiter's moons. Much to your surprise, the sample contains a life-form that grows well in a rich broth. Your preliminary analysis shows that it is cellular and contains DNA, RNA, and protein. When you show your results to a colleague, she suggests that your sample was contaminated with an organism from Earth. What approaches might you try to distinguish between contamination and a novel cellular life-form based on DNA, RNA, and protein, and why?

1–7 In the 1940s Erwin Chargaff made the remarkable observation that in samples of DNA from a wide range of organisms the mole percent of G [G/(A+T+C+G)] was equal to the mole percent of C, and the mole percents of A and T were equal. This was an essential clue to the structure of DNA. Nevertheless, Chargaff's 'rules' were not universal. For example, in DNA from the virus ϕX174 the mole percents are A = 24, C = 22, G = 23, and T = 31. What is the structural basis for Chargaff's rules, and how is it that DNA from ϕX174 doesn't obey the rules?

1–8* In 1944, at the beginning of his book, *What is Life*, the great physicist Erwin Schrödinger (of cat fame) asked the following question: "How can the events *in time and space* which take place within the spatial boundary of a living organism be accounted for by physics and chemistry?" What would be your answer today? Do you think there are peculiar properties of living systems that disobey the laws of physics and chemistry?

1–9 (**True/False**) Genes and their encoded proteins are co-linear; that is, the

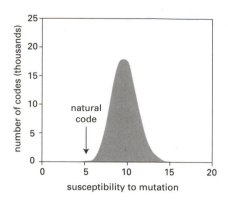

Figure 1–1 Susceptibility to mutation of the natural code shown relative to that of millions of other computer-generated codes (Problem 1–5). Susceptibility measures the average change in amino acid properties caused by random mutations in a genetic code. A small value indicates that mutations tend to cause only minor changes.

order of amino acids in proteins is the same as the order of the codons in the RNA and DNA. Explain your answer.

1–10*   There are about 35,000 genes in the human genome. If you wanted to use a stretch of the DNA of each gene as a unique identification tag, roughly what minimum length of DNA sequence would you need? What length of DNA in nucleotides would have a diversity (the number of different possible sequences) equivalent to at least 35,000 and would be expected to be present once in the haploid human genome (3.2×10^9 nucleotides)? (Assume that A, T, C, and G are present in equal amounts in the human genome.)

1–11   Cell growth depends on nutrient uptake and waste disposal. You might imagine, therefore, that the rate of movement of nutrients and waste products across the cell membrane would be an important determinant of the rate of cell growth. Is there a correlation between a cell's growth rate and its surface-to-volume ratio? Assuming that the cells are spheres, compare a bacterium (radius 1 μm), which divides every 20 minutes, with a human cell (radius 10 μm), which divides every 24 hours. Is there a match between the surface-to-volume ratios and the doubling times for these cells? (The volume of a sphere = $4/3\pi r^3$; the surface area = $4\pi r^2$.)

1–12*   (True/False) The disease of cystic fibrosis is caused by mutations in an ABC transporter that mediates chloride ion transport in human cells. Explain your answer.

1–13   Which of the following correctly describe the direct coding relationships for replication, transcription, and translation?
  A. DNA → DNA
  B. DNA → RNA
  C. DNA → Protein
  D. RNA → DNA
  E. RNA → RNA
  F. RNA → Protein
  G. Protein → DNA
  H. Protein → RNA
  I. Protein → Protein

THE DIVERSITY OF GENOMES AND THE TREE OF LIFE

TERMS TO LEARN

archaea	homolog	procaryote
bacteria	model organism	ortholog
eucaryote	paralog	virus

1–14*   It is not so difficult to imagine what it means to feed on the organic molecules that living things produce. That is, after all, what we do. But what does it mean to 'feed' on sunlight, as phototrophs do? Or, even stranger, to 'feed' on rocks, as lithotrophs do? Where is the 'food,' for example, in the mixture of chemicals (H_2S, H_2, CO, Mn^+, Fe^{2+}, Ni^{2+}, CH_4, and NH_4^+) spewed forth from a hydrothermal vent?

1–15   At the bottom of the seas where hydrothermal vents pour their chemicals into the ocean, there is no light and little oxygen, yet giant (two-meter long) tube worms live there happily. These remarkable creatures have no mouth and no anus, living instead off the excretory products and dead cells of their symbiotic lithotrophic bacteria. These tube worms are bright red because they contain large amounts of hemoglobin, which is critical to the survival of their symbiotic bacteria, and, hence, the worms. This specialized hemoglobin carries O_2 and H_2S. In addition to providing O_2 for its own oxidative metabolism, what role might this specialized hemoglobin play in the symbiotic relationship that is crucial for life in this hostile environment?

1–16* (**True/False**) The only major way in which carbon is newly incorporated into organic compounds on Earth is via photosynthesis. Explain your answer.

1–17 The overall reaction for the production of glucose ($C_6H_{12}O_6$) by oxygenic (oxygen-generating) photosynthesis,

$$6 \, CO_2 + 6 \, H_2O + light \rightarrow C_6H_{12}O_6 + 6 \, O_2 \qquad \text{(Equation 1)}$$

was widely interpreted as meaning that light split CO_2 to generate O_2, and that the carbon was joined with water to generate glucose. In the 1930s a graduate student at Stanford University, C.B. van Neil, showed that the stoichiometry for photosynthesis by purple sulfur bacteria was

$$6 \, CO_2 + 12 \, H_2S + light \rightarrow C_6H_{12}O_6 + 6 \, H_2O + 12 \, S \qquad \text{(Equation 2)}$$

On the basis of this stoichiometry, he suggested that the oxygen generated during oxygenic photosynthesis derived from water, not CO_2. His hypothesis was confirmed two decades later using isotopically labeled water. Yet how is it that the 6 H_2O in Equation 1 can give rise to 6 O_2? Can you suggest how Equation 1 might be modified to clarify exactly how the products are derived from the reactants?

1–18* How many possible different trees (branching patterns) can be drawn for eubacteria, archaea, and eucaryotes, assuming that they all arose from a common ancestor?

1–19 Highly conserved genes such as those for ribosomal RNA are present as clearly recognizable relatives in all organisms on Earth; thus, they have evolved very slowly over time. Were such genes 'born' perfect?

1–20* Several procaryotic genomes have been completely sequenced and their genes have been counted. But how is it exactly that one recognizes a gene in a string of Ts, As, Cs, and Gs? Is the problem any different for genes in eucaryotic genomes?

1–21 Which one of the processes listed below is NOT thought to contribute significantly to the evolution of new genes?
  A. Duplication of genes to create extra copies that can acquire new function.
  B. Formation of new genes *de novo* from noncoding DNA in the genome.
  C. Horizontal transfer of DNA between cells of different species.
  D. Mutation of existing genes to create new functions.
  E. Shuffling of domains of genes by gene rearrangement.

1–22* (**True/False**) The human hemoglobin genes, which are arranged in two clusters on two chromosomes provide a good example of an orthologous set of genes. Explain your answer.

1–23 Genes participating in informational processes such as replication, transcription, and translation are transferred between species much less often than are genes involved in metabolism. The basis for this inequality is unclear at present, but one suggestion is that it relates to the underlying complexity of the two types of processes. Informational processes tend to involve large aggregates of different gene products, whereas metabolic reactions are usually catalyzed by enzymes composed of a single protein.
  A. Archaea are more closely related to eubacteria in their metabolic genes, but are more similar to eucaryotes in the genes involved in informational processes. In terms of evolutionary descent, do you think archaea separated more recently from eubacteria or eucaryotes?
  B. Why would the complexity of the underlying process—informational or metabolic—have any effect on the rate of horizontal gene transfer?

1–24* Why do you suppose that horizontal gene transfer is more prevalent in single-celled organisms than multicellular organisms?

1–25 Natural selection is such a powerful force in evolution because cells with even a small growth advantage quickly outgrow their competitors. To illustrate this process, consider a cell culture that contains 10^6 bacterial cells that double every 20 minutes. A single cell in this culture acquires a mutation that allows it to divide with a generation time of only 15 minutes. Assuming that there is an unlimited food supply and no cell death, how long would it take before the progeny of the mutated cell became predominant in the culture? The number of cells N in the culture at time t is described by the equation $N = N_0 \times 2^{t/G}$, where N_0 is the number of cells at zero time and G is the generation time. (Before you go through the calculation, make a guess: do you think it would take about a day, a week, a month, or a year?)

1–26* You are interested in finding out the function of a particular gene in the mouse genome. You have sequenced the gene, defined the portion that codes for its protein product, and searched the database; however, neither the gene nor the encoded protein resembles anything in the database. What (various) types of information about the gene or the encoded protein would you like to know in order to narrow down the possible functions, and why? Focus on the information you want, rather than on the techniques you might use to get that information.

GENETIC INFORMATION IN EUCARYOTES

TERMS TO LEARN
genetic redundancy

1–27 (**True/False**) Eucaryotic cells contain either mitochondria or chloroplasts, but not both. Explain your answer.

1–28* Animal cells do not have cell walls or chloroplasts, whereas plant cells have cell walls and chloroplasts. Fungal cells are somewhere in between; they have cell walls but lack chloroplasts. Are fungal cells more likely to be animal cells that gained the ability to make cell walls, or plant cells that lost their chloroplasts? This question represented a difficult issue for early investigators who sought to assign evolutionary relationships based solely on cell characteristics and morphology. How do you suppose that this question was eventually decided?

1–29 (**True/False**) Most of the DNA sequences in a bacterial genome code for proteins, whereas most of the sequences in the human genome do not. Explain your answer.

1–30* Giardiasis is an acute form of gastroenteritis caused by the protozoan parasite, *Giardia lamblia. Giardia* is a fascinating eucaryote, which contains a nucleus but no mitochondria and no recognizable endoplasmic reticulum or Golgi apparatus—one of the very rare examples of such a cellular organization among eucaryotes. This might be because it represents an ancient lineage that separated from the rest of eucaryotes before mitochondria were acquired and internal membranes were developed. Or it might be a stripped-down version of a more standard eucaryote that has lost these structures because they are not necessary in the parasitic lifestyle it has adopted. How might you use nucleotide-sequence comparisons to distinguish between these alternatives?

1–31 (**True/False**) The only horizontal gene transfer that has occurred in animals is from the mitochondrial genome to the nuclear genome. Explain your answer.

1–32* It is difficult to obtain information about the process of gene transfer from the mitochondrial to the nuclear genome in animals because there are few

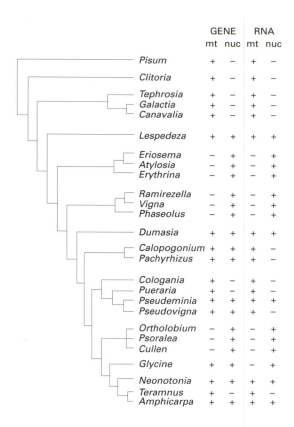

	GENE		RNA	
	mt	nuc	mt	nuc
Pisum	+	−	+	−
Clitoria	+	−	+	−
Tephrosia	+	−	+	−
Galactia	+	−	+	−
Canavalia	+	−	+	−
Lespedeza	+	+	+	+
Eriosema	−	+	−	+
Atylosia	−	+	−	+
Erythrina	−	+	−	+
Ramirezella	−	+	−	+
Vigna	−	+	−	+
Phaseolus	−	+	−	+
Dumasia	+	+	+	+
Calopogonium	+	+	+	−
Pachyrhizus	+	+	+	−
Cologania	+	−	+	−
Pueraria	+	−	+	−
Pseudeminia	+	+	+	+
Pseudovigna	+	+	+	−
Ortholobium	−	+	−	+
Psoralea	−	+	−	+
Cullen	−	+	−	+
Glycine	+	+	−	+
Neonotonia	+	+	+	+
Teramnus	+	−	+	−
Amphicarpa	+	+	+	+

Figure 1–2 Summary of *Cox2* gene distribution and transcript data in a phylogenetic context (Problem 1–32). The presence of the intact gene or a functional transcript is indicated by a (+); the absence of the intact gene or a functional transcript is indicated by (−).

differences among animal mitochondrial genomes. The same set of 13 (or occasionally 12) protein genes is encoded in all the numerous animal mitochondrial genomes that have been sequenced. In plants, however, the situation is different with quite a bit more variability in the set of proteins encoded in the mitochondrial genome. Analysis of plants, thus, can provide valuable information on the process of gene transfer.

The respiratory gene *Cox2*, which encodes subunit 2 of cytochrome oxidase, was functionally transferred to the nucleus during flowering plant evolution. Extensive analyses of plant genera have pinpointed the time of appearance of the nuclear form of the gene and identified several likely intermediates in ultimate loss from the mitochondrial genome. A summary of *Cox2* gene distributions between mitochondria and nuclei, along with data on their transcription, is shown in a phylogenetic context in Figure 1–2.

A. Assuming that transfer of the mitochondrial gene to the nucleus occurred only once (an assumption supported by the structures of the nuclear genes), indicate the point in the phylogenetic tree where the transfer occurred.

B. Are there any examples of genera in which the transferred gene and the mitochondrial gene both appear functional? Indicate them.

C. What is the minimal number of times that the mitochondrial gene has been inactivated or lost? Indicate those events on the phylogenetic tree.

D. What is the minimal number of times that the nuclear gene has been inactivated or lost? Indicate those events on the phylogenetic tree.

E. Based on this information, propose a general scheme for transfer of mitochondrial genes to the nuclear genome.

1–33   Although stages in the process of mitochondrial gene transfer can be deduced from studies such as the one in the previous question, there is much less information on the mechanism by which the gene is transferred from mitochondria to the nucleus. Does a fragment of DNA escape the mitochondria and enter the nucleus? Or does the transfer somehow involve an

TABLE 1–1 The difference matrix for the first 30 amino acids of the hemoglobin α chains from five species (Problem 1–34).

	HUMAN	FROG	CHICKEN	WHALE	FISH
Human	0	?	11	8	17
Frog		0	?	17	20
Chicken			0	?	20
Whale				0	?
Fish					0

Human	VLSPADKTNVKAAWGKVGAHAGEYGAEALE
Frog	LLSADDKKHIKAIMPAIAAHGDKFGGEALY
Chicken	VLSAADKNNVKGIFTKIAGHAEEYGAETLE
Whale	VLSPTDKSNVKATWAKIGNHGAEYGAEALE
Fish	SLSDKDKAAVRALWSKIGKSADAIGNDALS

Figure 1–3 Alignment of the first 30 amino acids of the hemoglobin α chains from five species (Problem 1–34). Amino acids are represented by the one-letter code (see inside back cover).

RNA transcript of the gene as the intermediary? The *Cox2* gene provides a unique window on this question. The initial transcript of mitochondrial *Cox2* genes is modified by RNA editing, a process that changes several specific Cs to Us. How might this observation allow you to decide whether the informational intermediary in transfer was DNA or RNA? What do you think the answer is?

1–34* Nucleotide sequence comparisons are fundamental to our current conception of the tree of life, to our understanding of how mitochondria and chloroplasts were acquired and their subsequent evolution, to the importance and magnitude of horizontal gene transfer, and to the notion that by focusing on a few model organisms we can gain valid insights into all of biology. For these reasons we've designed this problem and the following ones to introduce the common methods and assumptions that underlie the art of nucleotide sequence comparison.

A phylogenetic tree represents the history of divergence of species from common ancestors. Construction of such trees from DNA or protein sequences can really only be done with computers: the data sets are enormous and the algorithms are subtle. Nevertheless, some of the fundamental principles of tree construction can be illustrated with a simple example. Consider the first 30 amino acids of the hemoglobin α chains for the five species shown in Figure 1–3.

A. In a common approach, known as the distance-matrix method, the first step is to construct a table of all pairwise differences between the sequences. A partially filled-in example is shown in Table 1–1. Complete the table by filling in the blanks indicated by question marks.

B. According to the information in the completed Table 1–1, which pair of species is most closely related? What is the assumption that underlies your choice?

C. The information in Table 1–1 can be used to arrange species on the phylogenetic tree shown in Figure 1–4. The branching order is determined using a simple kind of cluster analysis. The two most similar species are placed on the adjacent branches at the upper left in Figure 1–4. The species with the fewest average differences relative to this pair is placed on the next branch. In the next step these three species are combined and the average differences from the remaining species are calculated and used to fill in the next branch, and so on. Use this method to arrange the species on the tree in Figure 1–4.

D. Is the branching order you determined in part C the same as you would get by simply using the number of differences relative to human to place the other species on the tree? Why is the method of cluster analysis superior?

1–35 In the previous question the branching order, or topology, of the tree was established, but actual distances (number of differences) were not assigned to the line segments that make up the tree (Figure 1–5). To calculate distances for line segments is, once again, tedious by hand but easy by computer. The following exercise gives a feeling for how such calculations are done.

A. Using the numbers from the completed distance matrix in Table 1–1 and the

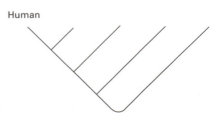

Human

Figure 1–4 A general phylogenetic tree for five species (Problem 1–34).

branching order determined in the previous problem, write down all the equations for the differences between species in terms of the line segments that make up the tree (Figure 1–5). Are there enough equations to solve for the lengths of the seven line segments?

B. One straightforward, not too exhausting method for solving these equations is to consider them three at a time; for example,

$$I \rightarrow II = a + b$$
$$I \rightarrow III = a + c + d$$
$$II \rightarrow III = b + c + d$$

There are 10 such three-at-a-time equations for five species. Using the information in the distance-matrix table (see Table 1–1), solve two of these sets of equations—Human/Whale/Chicken and Human/Whale/Frog—for a and b. Are the values for a and b the same in the two solutions?

1–36* Some genes evolve rapidly, whereas others are highly conserved. But how can we tell whether a gene has evolved rapidly, or simply had a long time to diverge from it relatives? The most reliable approach is to compare several genes from the same two species, as shown for rat and human in Table 1–2. Two measures of rates of nucleotide substitution are indicated in the table. Nonsynonymous changes refer to single nucleotide changes in the DNA sequence that alter the encoded amino acid (ATC → TTC, which is I → F, for example). Synonymous changes refer to those that do not alter the encoded amino acid (ATC → ATT, which is I → I, for example). (As is apparent in the genetic code, inside back cover, there are many cases where several codons correspond to the same amino acid.)

A. Why are there such large differences between the synonymous and nonsynonymous rates of nucleotide substitution?

B. Considering that the rates of synonymous changes are about the same for all three genes, how is it possible for the histone H3 gene to resist so effectively those nucleotide changes that alter the amino acid sequence?

C. In principle, a gene might be highly conserved because it exists in a 'privileged' site in the genome that is subject to very low mutation rates. What feature of the data in Table 1–2 argues against this possibility for the histone H3 gene?

1–37 Plant hemoglobins were found initially in legumes, where they function in root nodules to lower the oxygen concentration so that the resident bacteria can fix nitrogen. These hemoglobins impart a characteristic pink color to the root nodules. When these genes were first discovered, it was so surprising to find a gene typical of animal blood that it was hypothesized that the plant gene arose by horizontal transfer from some animal. Many more hemoglobin genes have now been sequenced, and a phylogenetic tree based on some of these sequences is shown in Figure 1–6.

A. Does this tree support or refute the hypothesis that the plant hemoglobins arose by horizontal gene transfer?

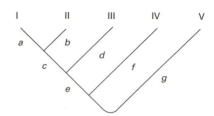

Figure 1–5 A general phylogenetic tree for five species with line segments indicated (Problem 1–35).

TABLE 1–2 Rates of nucleotide substitutions in three genes from rat and human (Problem 1–36).

GENE	AMINO ACIDS	RATES OF CHANGE	
		NONSYNONYMOUS	SYNONYMOUS
Histone H3	135	0.0	4.5
Hemoglobin α	141	0.6	4.4
Interferon γ	136	3.1	5.5

Rates are expressed as nucleotide changes per site per 10^9 years. The average rate of nonsynonymous changes for several dozen rat and human genes is about 0.8.

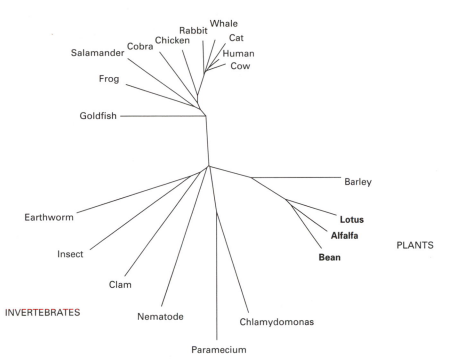

VERTEBRATES

Whale
Rabbit
Chicken   Cat
Cobra   Human
Salamander   Cow

Frog

Goldfish

Barley

Earthworm   **Lotus**
                             Alfalfa
                                        PLANTS
Insect   **Bean**

Clam

INVERTEBRATES
          Nematode   Chlamydomonas

Paramecium

PROTOZOA

Figure 1–6 Phylogenetic tree for hemoglobin genes from a variety of species (Problem 1–37). The legumes are shown in *bold*.

B. Supposing that the plant hemoglobin genes were originally derived from a parasitic nematode, for example, what would you expect the phylogenetic tree to look like?

1–38*   Rates of evolution appear to vary in different lineages. For example, the rate of evolution in the rat lineage is significantly higher than in the human lineage. These rate differences are apparent whether one looks at changes in protein sequences that are subject to selective pressure or at changes in noncoding nucleotide sequences, which are not under obvious selection pressure. Can you offer one or more possible explanations for the slower rate of evolutionary change in the human lineage versus the rat lineage?

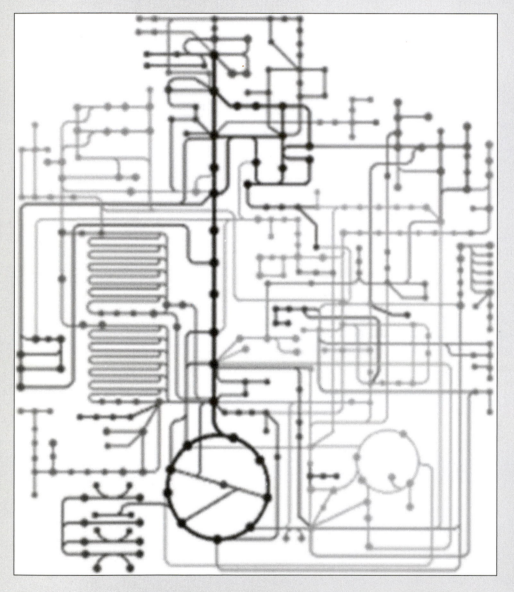

A map of metabolism. The aim of classical biochemistry was to explain how food and drink was transformed into flesh and blood. The transformations of simple precursors into more complex molecules, and the breakdown of complex molecules to simpler ones occupied thousands of researchers for at least 50 years. The fruits of this prodigious labor, now augmented by genome sequencing projects, are summarized by diagrams such as these, based on the Biochemical Pathways Wallcharts published by the firm of Boehringer Mannheim (now Roche Biochemicals). These maps, edited by Dr. Gerhard Michal, have a long tradition on the walls of life science laboratories. Adapted from a version developed by the Kyoto Encyclopaedia of Genes and Genomes (KEGG) located at http://www.genome.ad.jp/kegg/.

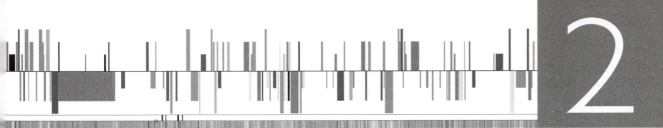

CELL CHEMISTRY AND BIOSYNTHESIS

THE CHEMICAL COMPONENTS OF A CELL

TERMS TO LEARN

acid	electron	molecule
adenosine triphosphate (ATP)	hydrogen bond	neutron
atom	hydrolysis	nucleotide
atomic weight	hydronium ion (H_3O^+)	pH scale
Avogadro's number	hydrophilic	polar
base	hydrophobic	protein
chemical bond	hydrophobic force	proton (H^+)
chemical group	ion	ribonucleic acid (RNA)
condensation reaction	ionic bond	sugar
covalent bond	molecular weight	van der Waals attraction
deoxyribonucleic acid (DNA)		

2–1 The organic chemistry of living cells is said to be special for two reasons: it occurs in an aqueous environment and it accomplishes some very complex reactions. But do you suppose it's really all that much different from the organic chemistry carried out in the top laboratories in the world? Why or why not?

2–2* The mass of a hydrogen atom—and thus of a proton—is almost exactly 1 dalton. If protons and neutrons have virtually identical masses, and the mass of an electron is negligible, shouldn't all elements have atomic weights that are nearly integers? A perusal of the periodic table shows that this simple expectation is not true. Chlorine, for example, has an atomic weight of 35.5. How is it that elements can have atomic weights that are not integers?

*Answers to these problems are available to Instructors at www.classwire.com/garlandscience/

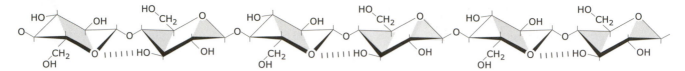

2–3 A carbon atom contains six protons and six neutrons.
A. What are its atomic number and atomic weight?
B. How many electrons does it have?
C. How many additional electrons must it add to fill its outermost shell? How does this affect carbon's chemical behavior?
D. Carbon with an atomic weight of 14 is radioactive. How does it differ in structure from nonradioactive carbon? How does this difference affect its chemical behavior?

Figure 2–1 Structure of the polysaccharide cellulose (Problem 2–4).

2–4* To gain a better feeling for atomic dimensions, assume that the page on which this question is printed is made entirely of the polysaccharide cellulose (Figure 2–1). Cellulose is described by the formula $(C_6H_{12}O_6)_n$, where n is a large number that varies from one molecule to another. The atomic weights of carbon, hydrogen, and oxygen are 12, 1, and 16, respectively, and this page weighs 5 grams.
A. How many carbon atoms are there in this page?
B. In paper made of pure cellulose, how many carbon atoms would be stacked on top of each other to span the thickness of this page (the page is 21 cm × 27.5 cm × 0.07 mm). (Rather than solving three simultaneous equations for the carbon atoms in each dimension, you might try a shortcut. Determine the linear density of carbon atoms by calculating the number of carbon atoms on the edge of a cube with the same volume as this page, and then adjust that number to the thickness of the page.)
C. Now consider the problem from a different angle. Assume that the page is composed only of carbon atoms, which have a van der Waals radius of 0.2 nm. How many carbon atoms stacked end to end at their van der Waals contact distance would it take to span the thickness of the page?
D. Compare your answers from parts B and C and explain any differences.

2–5 In the United States the concentration of glucose in blood is commonly reported in milligrams per deciliter (dL = 100 mL). Over the course of a day in a normal individual the circulating levels of glucose vary around a mean of about 90 mg/dL (Figure 2–2). What would this value be if it were expressed as a molar concentration of glucose in blood, which is the way it is typically reported in the rest of the world?

2–6* A few of the radioactive isotopes that are commonly used in biological experiments are listed in Table 2–1, along with some of their properties.
A. How do each of these unstable isotopes differ in atomic structure from the most common isotope for that element; that is, ^{12}C, ^{1}H, ^{32}S, and ^{31}P?
B. ^{32}P decays to a stable structure by emitting a β particle—an electron—according to the equation: $^{32}P \rightarrow {}^{32}S + e^-$. The product sulfur atom has the same atomic weight as the radioactive phosphorus atom. What has happened?

(A) STRUCTURE OF GLUCOSE   (B) GLUCOSE LEVELS IN BLOOD

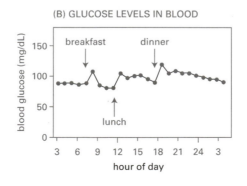

Figure 2–2 Circulating blood glucose (Problem 2–5). (A) Structure of glucose. (B) Typical variation in blood glucose over the course of a day.

TABLE 2–1 Radioactive isotopes and some of their properties (Problem 2–6).

RADIOACTIVE ISOTOPE	EMISSION	HALF-LIFE	MAXIMUM SPECIFIC ACTIVITY (Ci/mmol)
^{14}C	β particle	5730 years	0.062
^3H	β particle	12.3 years	29
^{35}S	β particle	87.4 days	1490
^{32}P	β particle	14.3 days	9120

C. ^{14}C, ^3H, and ^{35}S also decay by emitting an electron. (The electron can be readily detected, which is one reason why these isotopes are so useful in biology.) Write the decay equations for each of these radioactive isotopes and indicate whether the product atom is the most common isotope of the element generated by the decay.

D. Would you expect the product atom in each of these reactions to be charged or uncharged? Explain your answer.

2–7 Imagine that a ^{32}P-phosphate has been incorporated into the backbone of DNA. When the atom decays, would you expect the DNA backbone to remain intact? Why or why not?

2–8* As indicated in Table 2–1, the time it takes for half of a population of radioactive atoms to decay (their half-lives) ranges from about two weeks for ^{32}P to more than 5000 years for ^{14}C. Imagine that you created two atoms of ^{32}P today. When would you expect the first atom to decay? When would the second atom decay?

2–9 Radioactive decay is a first-order process (Figure 2–3). Thus, the number of radioactive atoms, N, that remain at time t is related to the number of radioactive atoms present initially, N_0, by the equation

$$N = N_0 \, e^{-\lambda t}$$

or rearranging and taking the log of both sides,

$$2.303 \log_{10} \frac{N}{N_0} = -\lambda t$$

where λ is a decay constant, which is different for each radioactive isotope.

The half-life ($t_{1/2}$) of a radioactive isotope is the time required for half the original number of atoms to decay; that is, when N/N_0 equals 0.5. For each of the isotopes in Table 2–1, calculate the decay constant λ in the same units as the half-life and then in minutes.

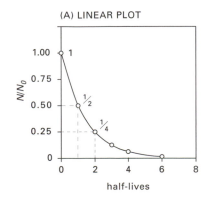

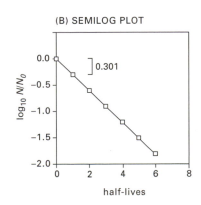

Figure 2–3 First order decay of a radioactive isotope (Problem 2–9). (A) A linear plot. (B) A semilog plot. The log of 2 is 0.301.

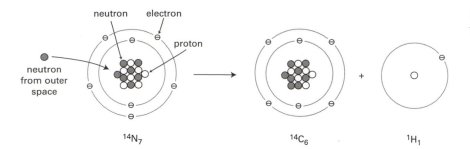

2–10* Neutrons of cosmic radiation constantly bombard Earth's upper atmosphere, converting a fairly constant fraction of $^{14}N_7$ to $^{14}C_6$, as shown in Figure 2–4. This ^{14}C enters the biosphere as CO_2, being incorporated first into plants via photosynthesis and then via the food chain into animals. As a result, all living plants and animals contain the same fraction of ^{14}C as in atmospheric CO_2, which is sufficient to yield 15.3 disintegrations per minute (dpm) per gram of carbon. The number of dpm equals the number of radioactive atoms, N, times the decay constant λ, expressed in min^{-1} (see Problem 2–9).

A. What fraction of the carbon atoms in living organisms does ^{14}C constitute?

B. The curie (Ci), which is the standard unit of radioactive decay, corresponds to 2.22×10^{12} dpm. How many curies of ^{14}C are in a 70 kg person? Humans are about 18.5% carbon by weight.

C. When an organism dies, it no longer incorporates carbon from the environment so that the quantity of ^{14}C decays with a half-life of 5730 years. What is the age of a biological sample that contains 3.0 dpm/g of carbon?

2–11 (**True/False**) After 10 half-lives, only about 1/1000 of the original radioactivity remains. Explain your answer.

2–12* Specific activity refers to the amount of radioactivity per unit amount of substance, most commonly in biology expressed on a molar basis, for example, as Ci/mmol. If you examine Table 2–1 (see Problem 2–6), you will see that there seems to be an inverse relationship between maximum specific activity and half-life. Do you suppose this is just a coincidence or is there an underlying reason? Explain your answer.

2–13 If both carbon atoms in every molecule of glycine were ^{14}C, what would its specific activity be in Ci/mmol? What proportion of glycine molecules is labeled in a preparation that has a specific activity of 200 µCi/mmol?

2–14* C, H, and O account for 95% of the elements in living organisms (Figure 2–5). These atoms are in the ratio of CH_2O, which is the general formula for carbohydrates. Does this mean we are mostly sugar? Why or why not?

2–15 The chemical properties of elements are determined by the behavior of electrons in their outer shell.

A. How many electrons can be accommodated in the first, second, and third electron shells of an atom?

B. How many electrons would atoms of H, C, N, O, P, and S preferentially gain or lose in order to obtain a completely filled outer electron shell?

C. What are the valences of H, C, N, O, P, and S?

2–16* Distinguish between a salt and a molecule.

2–17 Discuss whether the following statement is correct: "An ionic bond can, in principle, be thought of as a very polar covalent bond. Polar covalent bonds, then, fall somewhere in between ionic bonds at one end of the spectrum and nonpolar covalent bonds at the other end."

2–18* Order the following list of processes in terms of their energy content from smallest to largest.

A. ATP hydrolysis in cells.

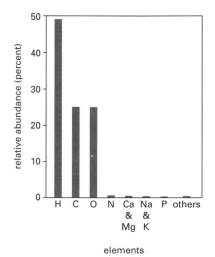

Figure 2–5 Abundance of elements in living organisms (Problem 2–14).

water (H₂O)

hydrogen sulfide (H₂S)

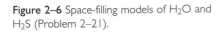

Figure 2–6 Space-filling models of H_2O and H_2S (Problem 2–21).

   B. Average thermal motions.
   C. C–C bond.
   D. Complete oxidation of glucose.
   E. Noncovalent bond in water.

2–19   Why are polar covalent bonds and the resulting permanent dipoles so important in biology?

2–20*   Hydrogen bonds and van der Waals attractions are important in the interactions between molecules in biology.
   A. Describe the differences and similarities between van der Waals attractions and hydrogen bonds.
   B. Which of the two types of interactions would form (1) between two hydrogens bound to carbon atoms, (2) between a nitrogen atom and a hydrogen bound to a carbon atom, and (3) between a nitrogen atom and a hydrogen bound to an oxygen atom?

2–21   Oxygen and sulfur have similar chemical properties because both elements have six electrons in their outermost electron shells. Indeed, both oxygen and sulfur form molecules with two hydrogen atoms: water (H_2O) and hydrogen sulfide (H_2S) (Figure 2–6). Surprisingly, water is a liquid, yet H_2S is a gas, even though sulfur is much larger and heavier than oxygen. Propose an explanation for this striking difference.

2–22*   The molecular weight of ethanol (CH_3CH_2OH) is 46 and its density is 0.789 g/cm³.
   A. What is the molarity of ethanol in beer that is 5% ethanol by volume? [Alcohol content of beer varies from about 4% (lite beer) to 8% (stout beer).]
   B. The legal limit varies for blood alcohol content of a driver, but 80 mg of ethanol per 100 mL of blood (usually referred to as a blood alcohol level of 0.08) is typical. What is the molarity of ethanol in a person at this legal limit?
   C. How many 12-oz (355-mL) bottles of 5% beer could a 70-kg person drink and remain under the legal limit? A 70-kg person contains about 40 liters of water. Ignore the metabolism of ethanol, and assume that the water content of the person remains constant.
   D. Ethanol is metabolized at a constant rate of about 120 mg per hour per kg body weight, regardless of its concentration. If a 70-kg person were at twice the legal limit (160 mg/100 mL), how long would it take for their blood alcohol level to fall below the legal limit?

2–23   What does the 'p' in pH stand for?

2–24*   Imagine that you have a beaker of pure water at neutral pH (pH 7.0).
   A. What is the concentration of H_3O^+ ions and how were they formed?
   B. What is the molarity of pure water? (Hint: 1 liter of water weighs 1 kg.)
   C. What is the ratio of H_3O^+ ions to H_2O molecules?

2–25   By a convenient coincidence the ion product of water, $K_w = [H^+][OH^-]$, is a nice round number: $1.0 \times 10^{-14} \ M^2$.
   A. Why is a solution at pH 7.0 said to be neutral?
   B. What is the H^+ concentration and pH of a 1 mM solution of NaOH?
   C. If the pH of a solution is 5.0, what is the concentration of OH^- ions?

2–26*   **(True/False)** A 10^{-8} M solution of HCl has a pH of 8. Explain your answer.

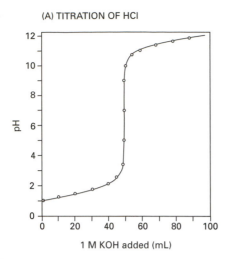

(A) TITRATION OF HCl

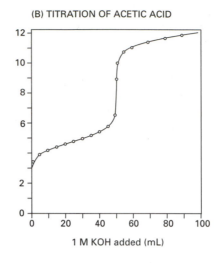

(B) TITRATION OF ACETIC ACID

Figure 2–7 Titration curves for solutions of HCl and acetic acid (Problem 2–28).

2–27   Imagine that you put some crystals of sodium chloride, potassium acetate, and ammonium chloride into separate beakers of water. Predict whether the pHs of the resulting solutions would be acidic, neutral, or basic. Explain your reasoning.

2–28*   Solutions containing 500 mL of 0.1 M HCl or 500 mL of 0.1 M acetic acid were titrated by addition of increasing amounts of 1 M KOH as shown in Figure 2–7.

   A. How many mL of KOH were added to neutralize all of the protons derived from HCl or acetic acid? The point at which neutralization occurs is termed the equivalence point.

   B. For each solution estimate the pH at the equivalence point.

   C. Estimate the pK for acetic acid.

2–29   The Henderson–Hasselbalch equation

$$pH = pK + \log_{10}\frac{[A^-]}{[HA]}$$

is a useful transformation of the equation for dissociation of a weak acid, HA

$$K = \frac{[H^+][A^-]}{[HA]}$$

   A. It is instructive to use the Henderson–Hasselbalch equation to determine the extent of dissociation of an acid at pHs above and below the pK. For the pH values listed in Table 2–2, fill in the values for $\log_{10}[A^-]/[HA]$ and $[A^-]/[HA]$, and indicate the percentage of the acid that has dissociated.

TABLE 2–2 Dissociation of a weak acid at pH values above and below the pK (Problem 2–29).

pH	$\log_{10}\dfrac{[A^-]}{[HA]}$	$\dfrac{[A^-]}{[HA]}$	% DISSOCIATION
pK+4			
pK+3			
pK+2			
pK+1			
pK			
pK–1			
pK–2			
pK–3			
pK–4			

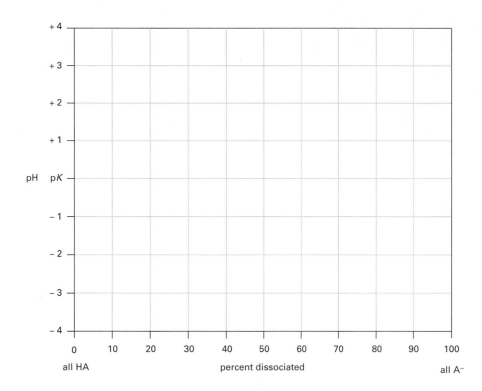

Figure 2–8 Graph for plotting values from Table 2–2 (Problem 2–29).

B. Using the graph in Figure 2–8, sketch the relationship between pH of the solution and the fractional dissociation of a weak acid. Will the shape of this curve be the same for all weak acids?

2–30* Cells maintain their cytosolic pH in a narrow range around pH 7.0 by using a variety of weak acids to buffer against changes in pH. This is essential because a large number of processes in cells generate or consume H^+ ions. Weak acids resist changes in pH—the definition of a buffer—most effectively within about one pH unit on either side of their pK values, as can be seen from the titration curve for acetic acid (see Figure 2–7B). Ionization of phosphoric acid provides an important buffering system in cells. Phosphoric acid has three ionizable protons, each with a unique pK.

Problem 11–26 considers the use of fluorescent probes to measure intracellular pH.

$$H_3PO_4 \xrightleftharpoons{pK=2.1} H^+ + H_2PO_4^- \xrightleftharpoons{pK=6.9} H^+ + HPO_4^{2-} \xrightleftharpoons{pK=12.4} H^+ + PO_4^{3-}$$

A. Using the values derived in Problem 2–29, estimate how much of each of the four forms of phosphate (H_3PO_4, $H_2PO_4^-$, HPO_4^{2-}, or PO_4^{3-}), as percentage of the total, are present in the cytosol of cells at pH 7. (No calculators permitted.)

B. What is the ratio of $[HPO_4^{2-}]$ to $[H_2PO_4^-]$ ($[A^-]/[HA]$) in the cytosol at pH 7? If the cytosol is 1 mM phosphate (sum of all forms), what is the concentration of $H_2PO_4^-$ and HPO_4^{2-} in the cytosol? (Calculators permitted.)

2–31 Describe in a general way how you might prepare 1 liter of 1 mM phosphate buffer at pH 6.9. You have available 0.1 M solutions of H_3PO_4, KH_2PO_4, K_2HPO_4, and K_3PO_4, 1.0 M solutions of HCl and KOH, and a pH meter. There are several possible (correct) answers. Be creative. Do all methods lead to the same final solution?

2–32* The amino acid glycine (H_2NCH_2COOH) has two ionizable groups: the carboxylic acid group (–COOH) and the basic amine group (–NH_2). Adding NaOH to a solution of glycine at pH = 1 gives the titration curve shown in Figure 2–9.

A. Write the expressions ($HA \rightleftharpoons H^+ + A^-$) for dissociation of the carboxylic acid and amine groups.

B. Estimate the pK values for the carboxylate and amine groups of glycine.

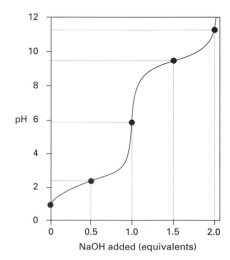

Figure 2–9 Titration of a solution of glycine (Problem 2–32). One equivalent of OH^- is the amount required to completely neutralize one acidic group.

TABLE 2–3 Values for the ionizable groups on several amino acids (Problem 2–34).

AMINO ACID	pK VALUES		
	–COOH	–NH₃⁺	R GROUP
Leucine	2.4	9.6	
Proline	2.0	10.6	
Glutamate	2.2	9.7	4.3 (carboxyl)
Histidine	1.8	9.2	6.0 (imidazole)
Cysteine	1.8	10.8	8.3 (sulfhydryl)
Arginine	1.8	9.0	12.5 (guanidino)
Lysine	2.2	9.2	10.8 (amino)

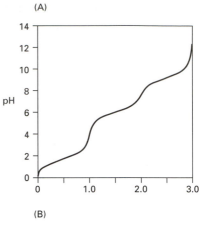

(A)

(B)

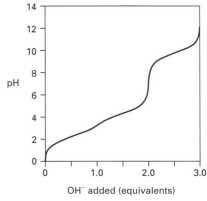

Figure 2–10 Titration curves for two amino acids (Problem 2–34).

C. Indicate the predominant ionic species of glycine at each point shown on the curve in Figure 2–9.

D. The isoelectric point of a solute is the pH at which it carries no net charge. Estimate the isoelectric point for glycine from the curve in Figure 2–9.

2–33 If you want to order glycine from a chemical supplier, you have three choices: glycine, glycine sodium salt, and glycine hydrochloride. Write the structures of these three compounds.

2–34* From the pK values for the amino acids listed in Table 2–3 decide which ones correspond to the titration curves shown in Figure 2–10.

2–35 The ionizable groups in amino acids can influence one another, as shown by the pK values for the carboxyl and amino groups of alanine and various oligomers of alanine (Figure 2–11). Suggest an explanation for why the pK of the carboxyl group increases with oligomer size, while that of the amino group decreases.

2–36* Suggest a rank order for the pK values (from lowest to highest) for the carboxyl group on the aspartate side chain in the following environments on a protein. Explain your ranking.
1. An aspartate side chain on the surface of a protein with no other ionizable groups nearby.
2. An aspartate side chain buried in a hydrophobic pocket on the surface of a protein.
3. An aspartate side chain in a hydrophobic pocket adjacent to a glutamate side chain.
4. An aspartate side chain in a hydrophobic pocket adjacent to a lysine side chain.

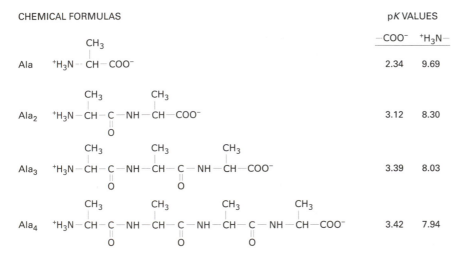

CHEMICAL FORMULAS

	pK VALUES	
	–COO⁻	⁺H₃N–
Ala	2.34	9.69
Ala₂	3.12	8.30
Ala₃	3.39	8.03
Ala₄	3.42	7.94

Figure 2–11 pK values for the carboxyl and amino groups in oligomers of alanine (Problem 2–35).

2–37   A histidine side chain is known to play an important role in the catalytic mechanism of an enzyme; however, it is not clear whether histidine is required in its protonated or unprotonated state. To answer this question you measure enzyme activity over a range of pH, with the results shown in Figure 2–12. Which form of histidine is necessary for the active enzyme?

Problem 3–69 explores the role of ionizable side chains in the active site of lysozyme.

2–38*   Inside cells the two most important buffer systems are provided by phosphate and proteins. The quantitative aspects of a buffer system pertain to both the effective buffering range (how near the pH is to the pK for the buffer) and the overall concentration of the buffering species (which determines the number of protons that can be handled). As discussed in Problem 2–30, $H_2PO_4^- \rightleftharpoons HPO_4^{2-}$ has a pK of 6.9 with an overall intracellular phosphate concentration of about 1 mM. In red blood cells the concentration of globin chains (molecular weight = 15,000) is about 100 mg/mL and each has 10 histidines, with pK values between 6.5 and 7.0. Which of these two buffering systems is quantitatively the more important in red blood cells?

2–39   The most important buffer in the bloodstream is the bicarbonate/CO_2 system. It is much more important than might be anticipated from its pK because it is an open system in which the CO_2 is maintained at a relatively constant value by exchange with the atmosphere. (By contrast, the buffering systems described in Problem 2–38 are closed systems with no exchange.) The equilibria involved in the bicarbonate/CO_2 buffering system are

$$CO_2(gas) \rightleftharpoons CO_2(dis) \overset{pK_1 =}{\underset{2.3}{\rightleftharpoons}} H_2CO_3 \overset{pK_2 =}{\underset{3.8}{\rightleftharpoons}} H^+ + HCO_3^- \overset{pK_3 =}{\underset{10.3}{\rightleftharpoons}} H^+ + CO_3^{2-}$$

pK_3 ($HCO_3^- \rightleftharpoons H^+ + CO_3^{2-}$) is so high that it never comes into play in biological systems. pK_2 ($H_2CO_3 \rightleftharpoons H^+ + HCO_3^-$) seems much too low to be useful, but it is influenced by the dissolved CO_2, which is directly proportional to the partial pressure of CO_2 in the gas phase. The dissolved CO_2 in turn is kept in equilibrium with H_2CO_3 by the enzyme carbonic anhydrase.

$$K_1 = \frac{[H_2CO_3]}{[CO_2(dis)]} = 5 \times 10^{-3}, \text{ or } pK_1 = 2.3$$

The equilibrium for hydration of CO_2(dis) can be combined with the equilibrium for dissociation of H_2CO_3 to give a pK' for CO_2(dis) $\rightleftharpoons H^+ + HCO_3^-$

$$K' = \frac{[H^+][HCO_3^-]}{[CO_2(dis)]} = K_1 \times K_2$$

$$pK' = pK_1 + pK_2 = 2.3 + 3.8 = 6.1$$

Even the pK' of 6.1 seems too low to maintain the blood pH around 7.4, yet this open system is very effective, as can be illustrated by a few calculations. The total concentration of carbonate in its various forms, but almost entirely CO_2(dis) and HCO_3^-, is about 25 mM.

A. Using the Henderson–Hasselbalch equation, calculate the ratio of HCO_3^- to CO_2(dis) at pH 7.4? What are the concentrations of HCO_3^- and CO_2(dis)?
B. What would the pH be if 5 mM H^+ were added under conditions where CO_2 was not allowed to leave the system; that is, if the concentration of CO_2(dis) was not maintained at a constant value?
C. What would the pH be if 5 mM H^+ were added under conditions where CO_2 was permitted to leave the system; that is, if the concentration of CO_2(dis) was maintained at a constant value?

2–40*   During an all-out sprint, muscles metabolize glucose anaerobically, producing a high concentration of lactic acid, which lowers the pH of the blood and of the cytosol. It is thought that the lower pH inside the cell decreases the efficiency of certain glycolytic enzymes, which reduces the rate of ATP production and leads to the fatigue sprinters experience well before their

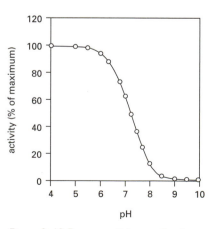

Figure 2–12 Enzyme activity as a function of pH (Problem 2–37).

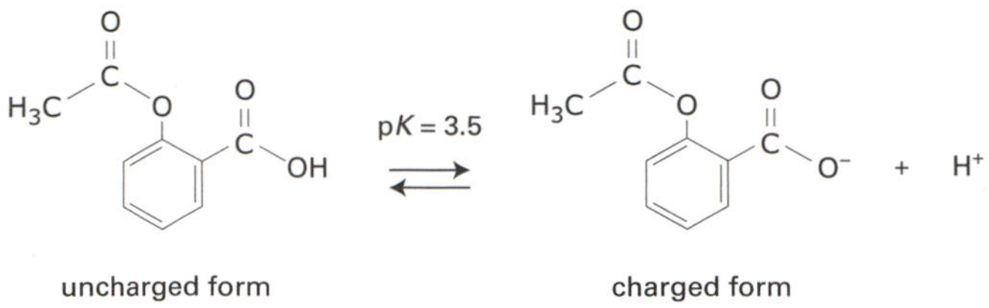

Figure 2–13 Aspirin (Problem 2–42).

1, 3 bisphosphoglycerate

pyruvate

cysteine

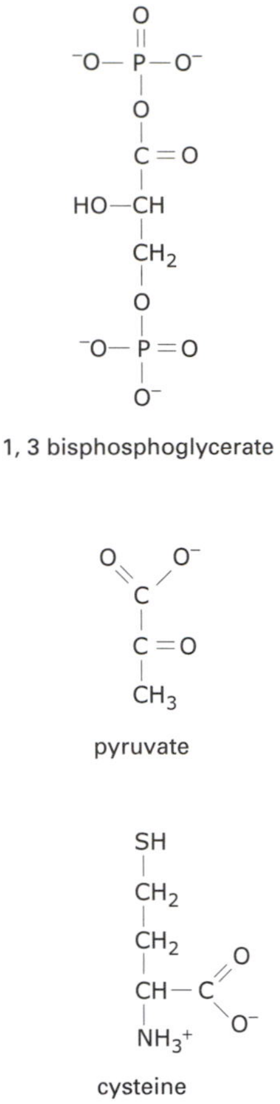

Figure 2–14 Three molecules that illustrate the seven most common functional groups in biology (Problem 2–45). 1,3-Bisphosphoglycerate and pyruvate are intermediates in glycolysis and cysteine is an amino acid.

fuel reserves are exhausted. With this as a background, would you advise sprinters to hold their breath or to breathe rapidly for a minute immediately before the race? Explain your answer.

2–41   (**True/False**) Strong acids bind protons strongly. Explain your answer.

2–42*   Aspirin is a weak acid (Figure 2–13) that is taken up into the bloodstream by diffusion through cells in the lining of the stomach and the small intestine. Aspirin crosses the plasma membrane of a cell most effectively in its uncharged form; in its charged form it cannot cross the hydrophobic lipid bilayer of the membrane. The pH of the stomach is about 1.5 and that of the lumen of the small intestine is about 6.0. Is the majority of the aspirin absorbed in the stomach or in the intestine? Explain your reasoning.

2–43   What, if anything, is wrong with the following statement: "When NaCl is dissolved in water, the water molecules closest to the ions will tend to orient themselves so that their oxygen atoms point toward the sodium ions and away from the chloride ions." Explain your answer.

2–44*   If noncovalent interactions are so weak in a water environment, how can they possibly be important for holding molecules together in cells?

2–45   The three molecules in Figure 2–14 contain the seven most common reactive groups in biology. Most molecules in the cell are built from these functional groups. Indicate and name the functional groups in these molecules.

2–46*   Ball-and-stick and space-filling models of glucose are shown in Figure 2–15. In both illustrations there are two different sizes of hydrogen atoms (identified by arrows). Is this accurate or a mistake? Explain your answer.

2–47   In solution, linear D-glucose forms a ring by reaction of the hydroxyl oxygen on carbon 5 with the carbon of the aldehyde at position 1 (Figure 2–16A). Depending on which side carbon 1 is attacked, the resulting hydroxyl group can be located either above the ring (β) or below the ring (α). The α form is represented in Figure 2–16A. Shown along with the Fischer projection of linear D-glucose and the Haworth projection of circular α-D-glucose, are two more realistic representations of the energetically favorable chair conformation of α-D-glucose.

Problems 3–27, 3–28, and 3–30 look at the noncovalent interactions that hold proteins together.

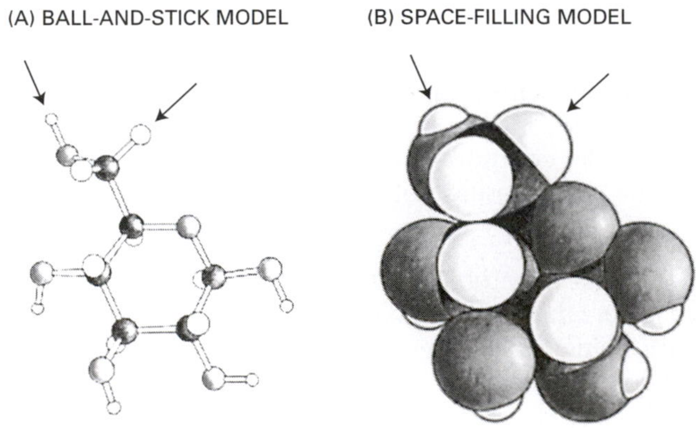

(A) BALL-AND-STICK MODEL   (B) SPACE-FILLING MODEL

Figure 2–15 Ball-and-stick and space-filling models of a glucose molecule (Problem 2–46). *Arrows* identify a 'small' and a 'large' hydrogen atom in each model.

(A) D-GLUCOSE

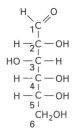

Fischer projection

Haworth projection

chair conformation

space-filling model

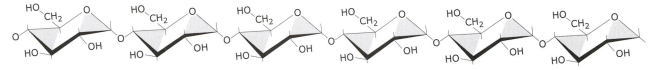

(B) AMYLOSE

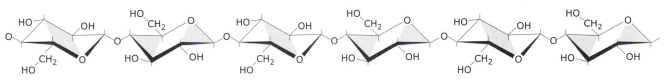

(C) CELLULOSE

Polysaccharides are formed by linkage of sugar monomers via glycosidic bonds. Amylose and cellulose, for example, are polysaccharides composed entirely of hexoses (six-carbon sugars) that are linked together by glycosidic bonds between the number 1 and 4 carbons of adjacent monomers. Structures of amylose and cellulose with the hexoses in the chair conformation are shown in Figure 2–16B and C. Which, if either, of these polysaccharides is composed entirely of D-glucose? What are the similarities and differences between the structures of amylose and cellulose?

2–48* The drug thalidomide was once prescribed as a sedative to help with nausea during the early stages of pregnancy. One of its optical isomers, (*R*)-thalidomide (Figure 2–17), is the active agent responsible for its sedative effects. It was synthesized, however, as a mixture of both optical isomers—a not uncommon practice that usually causes no problems. Unfortunately, the other optical isomer is a teratogen that led to horrific series of birth defects characterized by malformed or absent limbs. On the structural formula in Figure 2–17A identify the carbon that is responsible for its optical activity (its chiral center) and sketch the structure of the teratogenic form of thalidomide.

Figure 2–16 Structures of carbohydrates (Problem 2–47). (A) Several structural representations of D-glucose. (B) Amylose. (C) Cellulose. The glycosidic bonds that connect the hexose monomers in the polysaccharide chains are all 1→4, which indicates the numbers of the two carbons that are linked.

(A) THALIDOMIDE CHEMICAL FORMULA

(B) THALIDOMIDE SPACE-FILLING MODEL

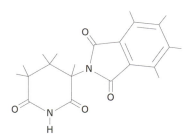

Figure 2–17 The structure of the sedative, (*R*)-thalidomide (Problem 2–48).

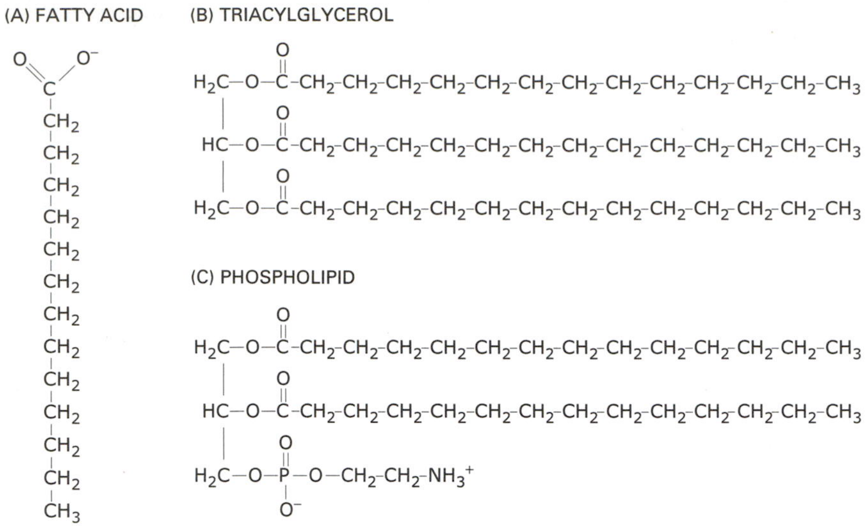

Figure 2–18 A fatty acid, a triacylglycerol, and a phospholipid (Problem 2–49).

2–49 What does the term 'amphipathic' mean? Figure 2–18 shows a fatty acid, a triacylglycerol, and a phospholipid. Indicate which of these molecules are amphipathic and illustrate the reason why. How do their amphipathic characteristics account for the typical structures these molecules form in cells?

2–50* A short polypeptide is shown in Figure 2–19. Identify the N-terminus, the C-terminus, the α carbons, and side chains. Mark the atoms involved in the peptide bonds and those that form each amino acid.

2–51 Why do you suppose that only L-amino acids and not a random mixture of L- and D-amino acids are used to make proteins?

2–52* Two short chains of nucleic acids are shown in Figure 2–20. Identify the 5′ and 3′ ends of each oligonucleotide, name the component bases and sugars, and indicate the components that make up a nucleoside and a nucleotide. Which one is RNA and which is DNA? How can you tell?

2–53 The proteins in a mammalian cell account for 18% of its net weight. If the density of a typical mammalian cell is about 1.1 g/mL and the volume of the cell is 4×10^{-9} mL, what is the concentration of protein in mg/mL?

2–54* There are many different, chemically diverse ways in which small molecules can be linked to form polymers. For example, ethene ($CH_2=CH_2$) is used commercially to make the plastic polymer polyethylene (...–CH_2–CH_2–CH_2–CH_2–...). The individual subunits of the three major classes of biological macromolecules, however, are all linked by similar reaction mechanisms, namely, by condensation reactions that eliminate water. Can you think of any benefits that this chemistry offers and why it might have been selected in evolution?

2–55 (**True/False**) Covalent bonds could be used in place of noncovalent bonds to mediate most of the interactions of macromolecules. Explain your answer.

Problems 3–10 and 3–13 deal with amphipathic α helices and β sheets.

Problem 3–19 asks how many different proteins of a given size can be made with 20 amino acids.

Problem 3–65 explores the specificity of an enzyme composed entirely of D-amino acids.

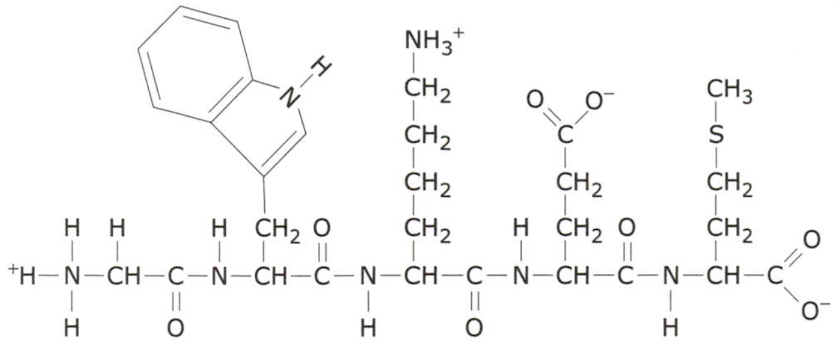

Figure 2–19 A polypeptide (Problem 2–50).

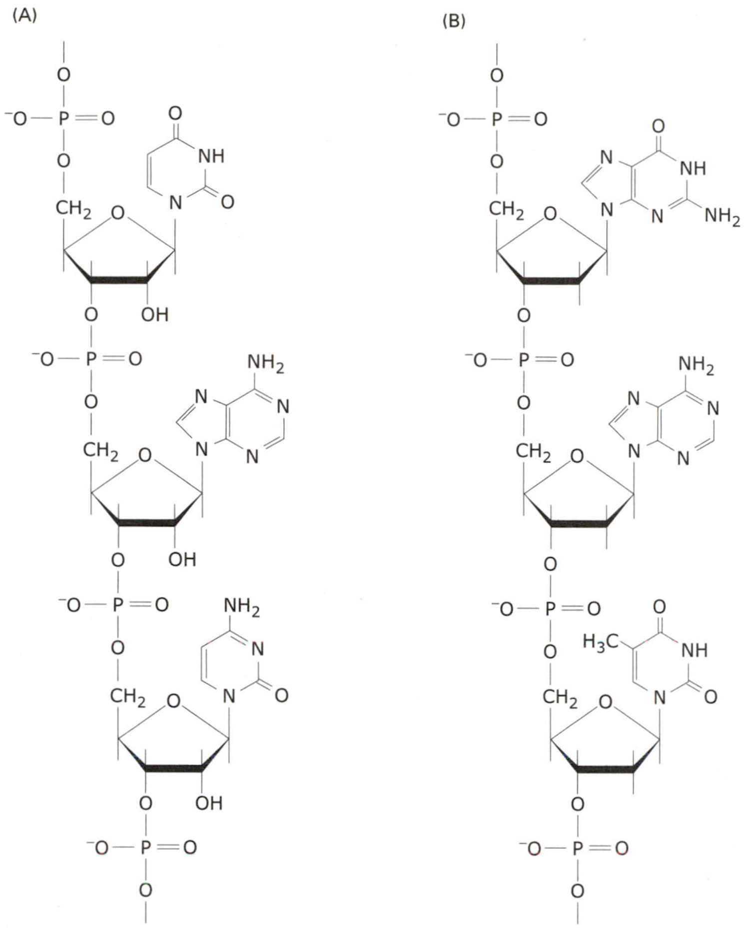

(A)

(B)

Figure 2–20 Two oligonucleotides (Problem 2–52).

CATALYSIS AND THE USE OF ENERGY BY CELLS

TERMS TO LEARN

acetyl CoA	entropy	NADP$^+$/NADPH
activated carrier	enzyme	oxidation
activation energy	equilibrium	photosynthesis
ADP	equilibrium constant (K)	reduction
ATP	free energy (G)	respiration
catalyst	free energy change (ΔG)	standard free energy change ($\Delta G°$)
coupled reaction	metabolism	substrate
diffusion	NAD$^+$/NADH	

2–56* Distinguish between catabolic and anabolic pathways of metabolism, and indicate in a general way how such pathways are linked to one another in cells.

2–57 The second law of thermodynamics states that systems will change spontaneously toward arrangements with greater entropy (disorder). Living systems are so intricately ordered, however, it seems they surely must violate the second law. Explain briefly—and in a way your parents could understand—how life is fully compatible with the laws of thermodynamics.

2–58* The equation for photosynthesis in green plants is

$$\text{light energy} + CO_2 + H_2O \rightarrow \text{sugars} + O_2 + \text{heat energy}$$

Would you expect this reaction to be carried out by a single enzyme? Why is heat energy also produced along with sugars and O_2?

(A) OXIDATION STATES OF CARBON

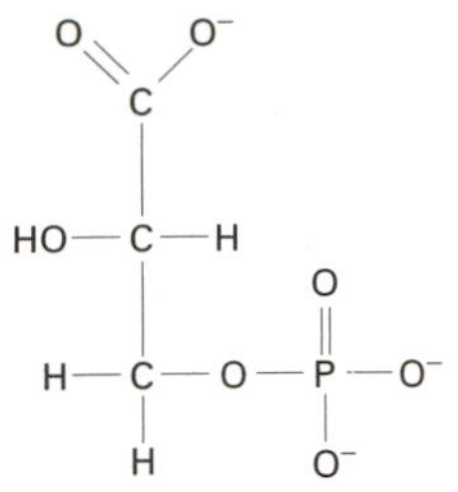

carbon 1: $2(O) + C$
$-4 + C = -1, \therefore C = +3$

carbon 2: $2(H) + O + C$
$+2 \quad -2 + C = 0, \therefore C = 0$

carbon 3: $2(H) + 4(O) + P + C$
$+2 \quad -8 \quad +5 + C = -2, \therefore C = -1$

overall sum: $4(H) + 7(O) + P + 3C$
$+4 \quad -14 \quad +5 \quad +2 = -3$

3-phosphoglycerate

(B) A SERIES OF TWO-CARBON MOLECULES

ethane	H_3C——CH_3
ethene	H_2C=CH_2
ethanol	H_3C——CH_2OH
acetaldehyde	H_3C——CHO
acetate	H_3C——COO^-
acetamide	H_3C——$CONH_2$
ethylamine	H_3C——$CH_2NH_3^+$
phosphoethanol	H_3C——$CH_2PO_4^{2-}$
thioethane	H_3C——CH_2SH

Figure 2–21 Oxidation states of carbon atoms in molecules (Problem 2–62). (A) Method for assigning oxidation states to individual carbon atoms in a molecule. The sum of the oxidation states of the attached atoms plus the carbon atom equals the charge, if any, on those atoms. (B) A list of two-carbon molecules on which to try the method.

2–59 (**True/False**) Animals and plants use oxidation to extract energy from food molecules. Explain your answer.

2–60* (**True/False**) If an oxidation occurs in a reaction, it must be accompanied by a reduction. Explain your answer.

2–61 In the reaction $2\,Na + Cl_2 \rightarrow 2\,Na^+ + 2\,Cl^-$, what is being oxidized and what is being reduced? How can you tell?

2–62* It is not nearly so easy to follow oxidations and reductions (redox reactions) in organic molecules. In biological redox reactions involving organic molecules in solution, it is the carbon atoms that are usually oxidized or reduced. (That's not always the case, but it's a good place to start.) Three rules make it possible to follow what happens.
1. The oxidation states of all the atoms in the molecule—except carbon—are designated as follows: H is +1, O is –2, N is –3, S is –2, and P is +5.
2. The oxidation state of each carbon atom is calculated as the overall charge (if any) on the attached atoms minus the sum of the oxidation states of all the attached atoms. For this calculation attached carbon atoms—and all atoms connected through such carbon atoms—are ignored.
3. As a check, the sum of the oxidation states on all the atoms should equal the overall charge on the molecule.
   An example of this method of assessing the oxidation states of carbon atoms in 3-phosphoglycerate is shown in Figure 2–21A.
   A. Use this method to assign oxidation states to each of the carbon atoms in the molecules listed in Figure 2–21B.
   B. Using the sum of the oxidation states of the carbon atoms as a measure of the overall oxidation state of a molecule, order the molecules in Figure 2–21B from most reduced to most oxidized.
   C. Are the differences in oxidation state between the molecules equal to one electron or two electrons? Do you think there is any particular significance to your answer?
   D. Which molecules are at the same overall oxidation state? If you can remember which functional groups are at the same oxidation state, then you can usually tell at a glance whether a redox reaction has occurred—without going through this method of assigning oxidation states.

2–63 By comparing the oxidation states before and after a reaction, one can decide whether a redox reaction has occurred. A short segment of the pathway of reactions that occur in the citric acid cycle is shown in Figure 2–22.
   A. Which of the reactions are redox reactions? Have the molecules involved in the redox reactions been oxidized or have they been reduced? How can you tell? Which carbons have lost electrons and which have gained them?
   B. The redox reactions involve electron carriers that are not shown. Have the electron carriers been reduced or have they been oxidized?

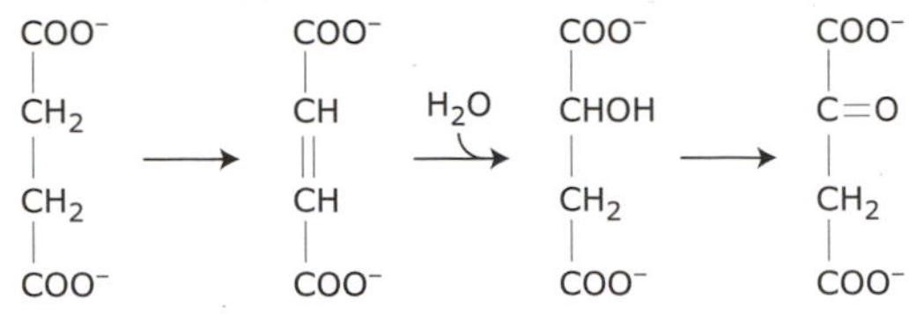

$$\text{succinate} \xrightarrow{\quad} \text{fumarate} \xrightarrow{\ H_2O\ } \text{malate} \xrightarrow{\quad} \text{oxaloacetate}$$

Figure 2–22 A segment of the citric acid cycle (Problem 2–63).

2–64* If an uncatalyzed reaction occurs at the rate 1 event per century and if an enzyme speeds up the rate by a factor of 10^{14}, how long in seconds does it take the enzyme to catalyze one event?

2–65 'Diffusion' sounds slow—and over everyday distances it is—but on the scale of a cell it is very fast. The average instantaneous velocity of a particle in solution, that is, the velocity between collisions, is

$$v = (kT/m)^{\frac{1}{2}}$$

where $k = 1.38 \times 10^{-16}$ g cm^2/°K sec^2, T = temperature in °K (37°C is 310°K), m = mass in g/molecule.

Calculate the instantaneous velocity of a water molecule (molecular mass = 18 daltons), a glucose molecule (molecular mass = 180 daltons), and a globin molecule (molecular mass = 15,000 daltons) at 37°C. Just for fun, convert these numbers into kilometers/hour. Before you do any calculations, try to guess whether the molecules are moving at slow crawl (<1 km/hr), an easy walk (5 km/hr), or a record-setting sprint (40 km/hr).

2–66* The instantaneous velocity tells you little about the time it takes for a molecule to move cellular distances because its trajectory is constantly altered by collisions with other molecules in solution (Figure 2–23). The average time it takes for a molecule to travel x cm by diffusion in three dimensions is

$$t = x^2/6D$$

where t is the time in seconds and D is the diffusion coefficient, which is a constant that depends on the size and shape of the particle. Glucose and globin, for example, have diffusion coefficients of about 5×10^{-6} cm^2/sec and 5×10^{-7} cm^2/sec, respectively. Calculate the average time it would take for glucose and globin to diffuse a distance of 20 μm, which is approximately the width of a mammalian cell.

Problems 16–21 and 16–95 deal with diffusion as a means for distributing cellular constituents.

2–67 If a cell in mitosis is cooled to 0°C, the microtubules in the spindle depolymerize into tubulin subunits. The same is true for microtubules made from pure tubulin in a test tube; they assemble readily at 37°C, but disassemble at low temperature. In fact, many protein assemblies that are held together by noncovalent bonds show the same behavior: they disassemble when cooled. This behavior is governed by the basic thermodynamic equation

$$\Delta G = \Delta H - T\Delta S$$

where ΔH is the change in enthalpy (chemical bond energy), ΔS is the change in entropy (disorder of the system), and T is the absolute temperature.

A. The change in free energy (ΔG) must be negative for the reaction (tubulin subunits → microtubules) to proceed at high temperature. At low temperature, ΔG must be positive to permit disassembly; that is, to favor the reverse reaction. Decide what the signs (positive or negative) of ΔH and ΔS must be, and show how your choices account for polymerization of tubulin at high temperature and its depolymerization at low temperature (Assume that the ΔH and ΔS values themselves do not change with temperature.)

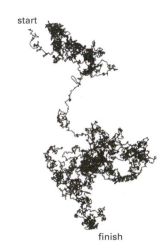

start

finish

Figure 2–23 A two-dimensional, simulated random walk of a molecule in solution (Problem 2–66).

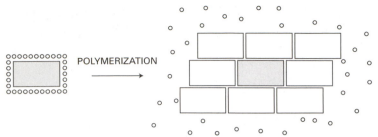

Figure 2–24
Polymerization of tubulin subunits into a microtubule (Problem 2–67). The fates of one subunit (*shaded*) and its associated water molecules (*small spheres*) are shown.

B. Polymerization of tubulin subunits into microtubules at body temperature clearly occurs with an increase in the orderliness of the subunits (Figure 2–24). Yet tubulin polymerization occurs with an increase in entropy (decrease in order). How can that be?

2–68* The polymerization of subunits into a pentameric ring is shown in Figure 2–25. The equilibrium constants for association of a subunit at each step in the assembly of the tetramer (that is, K_1, K_2, and K_3) are approximately equal at 10^6 M^{-1}. The equilibrium constant for association of the final subunit in the ring (K_4), however, is $>10^{12}$ M^{-1}. Why is association of the final subunit so much more highly favored than association of the initial subunits? What would you have expected the equilibrium constant for the association of the final subunit to be?

2–69 The enzyme carbonic anhydrase is one of the speediest enzymes known. It catalyzes the hydration of CO_2 to H_2CO_3, which then rapidly dissociates as described in Problem 2–39. Carbonic anhydrase accelerates the reaction by 10^7-fold, hydrating 10^5 CO_2 molecules per second at its maximal speed.

A. What are the factors that might limit the speed of the enzyme?

B. The curve in Figure 2–26 shows the distribution of energy of molecules in solution, and illustrates schematically the proportions that have sufficient energy to undergo an uncatalyzed reaction (to the right of threshold B, Figure 2–26) or a catalyzed reaction (to the right of threshold A, Figure 2–26). On such a curve, to what does the 10^7-fold rate enhancement by carbonic anhydrase correspond?

2–70* (**True/False**) Linking the energetically unfavorable reaction A → B to a second, favorable reaction B → C will shift the equilibrium constant for the first reaction. Explain your answer.

2–71 Discuss the statement: "The criterion for whether a reaction proceeds spontaneously is ΔG not $\Delta G°$, because ΔG takes into account the concentrations of the substrates and products."

2–72* At a particular concentration of substrates and products the reaction below has a negative ΔG.

$$A + B \rightarrow C + D \qquad \Delta G = -4.5 \text{ kcal/mole}$$

At the same concentrations, what is ΔG for the reverse reaction?

$$C + D \rightarrow A + B$$

2–73 Phosphoglucose isomerase catalyzes the interconversion of glucose 6-phosphate (G6P) and fructose 6-phosphate (F6P).

$$G6P \rightleftharpoons F6P$$

The ΔG for this reaction is given by the equation

$$\Delta G = \Delta G° + 2.3 \, RT \log_{10} \frac{[\text{F6P}]}{[\text{G6P}]}$$

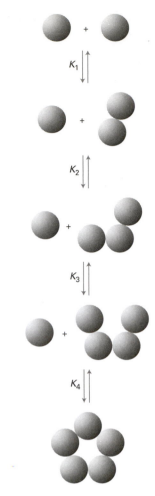

Figure 2–25 Polymerization of subunits into a pentameric ring (Problem 2–68).

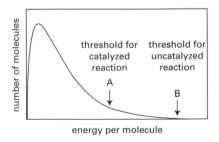

Figure 2–26 Energy distribution of molecules in solution (Problem 2–69). Molecules with sufficient energy to undergo a catalyzed reaction have energies above (to the *right* of) threshold A. Molecules with sufficient energy to undergo an uncatalyzed reaction have energies above (to the *right* of) threshold B.

where $R = 1.98 \times 10^{-3}$ kcal/°K mole and $T = 310$°K. A useful number to remember is that 2.3 $RT = 1.41$ kcal/mole at 37°C, which is body temperature.

Problem 14–26 examines the relationship between concentration and ΔG for ATP hydrolysis.

A. At equilibrium, [F6P]/[G6P] is equal to the equilibrium constant (K) for the reaction. Rewrite the above equation for the reaction at equilibrium.

B. At equilibrium, the ratio of [F6P] to [G6P] is observed to be 0.5. At this equilibrium ratio, what are the values of ΔG and $\Delta G°$?

C. Inside a cell the value of ΔG for this reaction is –0.6 kcal/mole. What is the ratio of [F6P] to [G6P]? What is $\Delta G°$?

2–74* The values for $\Delta G°$ and for ΔG in cells have been determined for many different metabolic reactions. What information do these values provide about the rates of these reactions?

2–75 Phosphorylation of glucose (GLC) by ATP to produce glucose 6-phosphate (G6P) and ADP is the first step in glucose metabolism after entry into cells. It can be written as the sum of two reactions

$$
\begin{array}{lll}
(1) & \text{GLC} + \text{P}_i \rightarrow \text{G6P} + \text{H}_2\text{O} & \Delta G° = 3.3 \text{ kcal/mole} \\
(2) & \underline{\text{ATP} + \text{H}_2\text{O} \rightarrow \text{ADP} + \text{P}_i} & \Delta G° = -7.3 \text{ kcal/mole} \\
\text{NET:} & \text{GLC} + \text{ATP} \rightarrow \text{G6P} + \text{ADP} &
\end{array}
$$

A. What is the value of the equilibrium constant K for reaction (1)?

B. In a liver cell [GLC] and [P_i] are both maintained at about 5 mM. What would the equilibrium concentration of G6P be under these conditions if reaction (1) were the sole source of G6P? Does this concentration of G6P seem reasonable for the initial step in glucose metabolism in cells? Why or why not?

C. In cells the phosphorylation of glucose is accomplished by addition of phosphate from ATP; that is, by the sum of reactions (1) and (2). What is $\Delta G°$ for the net reaction? What is the equilibrium constant K for the net reaction?

D. In liver cells [ATP] is maintained at about 3 mM and [ADP] at about 1 mM. Under these conditions what would be the equilibrium concentration of G6P? Does this concentration of G6P seem reasonable for the initial step in glucose metabolism in cells? Why or why not?

E. The concentration of G6P in liver cells is around 200 µM. Why is this concentration different than the one you calculated in part C? What is ΔG for the phosphorylation of glucose as it occurs under cellular conditions?

2–76* Thermodynamically, it is perfectly valid to consider the cellular phosphorylation of glucose as the sum of two reactions.

$$
\begin{array}{lll}
(1) & \text{GLC} + \text{P}_i \rightarrow \text{G6P} + \text{H}_2\text{O} & \Delta G° = 3.3 \text{ kcal/mole} \\
(2) & \underline{\text{ATP} + \text{H}_2\text{O} \rightarrow \text{ADP} + \text{P}_i} & \Delta G° = -7.3 \text{ kcal/mole} \\
\text{NET:} & \text{GLC} + \text{ATP} \rightarrow \text{G6P} + \text{ADP} &
\end{array}
$$

But biologically it makes no sense at all. Hydrolysis of ATP (reaction 2) in one part of the cell can have no effect on phosphorylation of glucose (reaction 1) elsewhere in the cell, given that [ATP], [ADP], and [P_i] are maintained within narrow limits. How does the cell manage to link these two reactions?

2–77 Successive steps in a metabolic pathway such as glycolysis, which converts glucose to pyruvate in 10 steps, are connected by metabolic intermediates. The product of the first reaction in the pathway provides a substrate for the second reaction, which in turn generates a product that is a substrate for the third step, and so on. Such pathways are not at equilibrium. Instead, intermediates flow along pathways so that net conversion, for example, of glucose to pyruvate can occur. Flow along such pathways is what distinguishes living cells from dead ones (which are at equilibrium).

The flow of metabolites through a metabolic pathway can occur *only* when the ΔG value for *each* step is negative. This is a true statement. Convince yourself of its validity by considering a segment of a pathway D → E → F, which has an overall (D → F) ΔG that is negative under cellular conditions.

When there is a flow through the pathway, D is constantly added and F is constantly removed. Consider what will happen if the concentration of the intermediates were such that the ΔG for D → E was momentarily positive and the ΔG for E → F was negative.

2–78*   Red blood cells obtain energy in the form of ATP by converting glucose to pyruvate via the glycolytic pathway (Figure 2–27). The values of $\Delta G°$ and ΔG have been calculated for each of the steps in glycolysis in red blood cells that are actively metabolizing glucose, and these are shown in Table 2–4. The $\Delta G°$ values are based on the known equilibrium constants for the reactions; the ΔG values are calculated from the $\Delta G°$ values and actual measurements of concentrations of the intermediates in red cells.

Despite the assertion in Problem 2–77 that *all* ΔG values must be negative, three reactions in red-cell glycolysis have slightly positive ΔG values. What do you suppose is the explanation for the results in the table?

2–79   Each phosphoanhydride bond between the phosphate groups in ATP is a high-energy linkage with a $\Delta G°$ value of –7.3 kcal/mole. Hydrolysis of this bond in cells normally liberates usable energy in the range of 11 to 13 kcal/mole. Why do you think a range of values for released energy is given for ΔG, rather than a precise number, as for $\Delta G°$?

2–80*   A 70-kg adult human (154 lbs) could meet his or her entire energy needs for one day by eating 3 moles of glucose (540 g). (We don't recommend this.) Each molecule of glucose generates 30 ATP when it is oxidized to CO_2. The concentration of ATP is maintained in cells at about 2 mM, and a 70-kg adult has about 25 L of intracellular fluid. Given that the ATP concentration remains constant in cells, calculate how many times per day, on average, each ATP molecule in the body is hydrolyzed and resynthesized.

2–81   Consider the effects of two enzymes. Enzyme A catalyzes the reaction

$$ATP + GDP \rightleftharpoons ADP + GTP$$

whereas enzyme B catalyzes the reaction

$$NADH + NADP^+ \rightleftharpoons NAD^+ + NADPH$$

Discuss whether the enzymes would be beneficial or detrimental to cells.

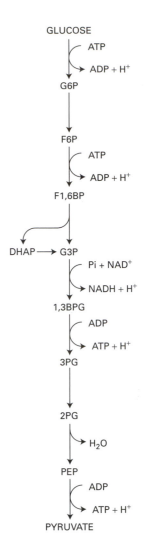

Figure 2–27 The glycolytic pathway (Problem 2–78). See Table 2–4 for the key to abbreviations.

TABLE 2–4 The reactions of glycolysis in red blood cells and their associated $\Delta G°$ and ΔG values (Problem 2–78).

STEP	REACTION	$\Delta G°$	ΔG
1	GLC + ATP → G6P + ADP + H$^+$	–4.0	–8.0
2	G6P → F6P	+0.4	–0.6
3	F6P + ATP → F1,6BP + ADP + H$^+$	–3.4	–5.3
4	F1,6BP → DHAP + G3P	+5.7	–0.3
5	DHAP → G3P	+1.8	+0.6
6	G3P + P$_i$ + NAD$^+$ → 1,3BPG + NADH + H$^+$	+1.5	–0.4
7	1,3BPG + ADP → 3PG + ATP	–4.5	+0.3
8	3PG → 2PG	+1.1	+0.2
9	2PG → PEP + H$_2$O	+0.4	–0.8
10	PEP + ADP + H$^+$ → PYR + ATP	–7.5	–4.0

GLC = glucose, G6P = glucose 6-phosphate, F6P = fructose 6-phosphate, F1,6BP = fructose 1,6-bisphosphate, DHAP = dihydroxyacetone phosphate, G3P = glyceraldehyde 3-phosphate, 1,3BPG = 1,3-bisphosphoglycerate, 3PG = 3-phosphoglycerate, 2PG = 2-phosphoglycerate, PEP = phosphoenolpyruvate, PYR = pyruvate.

2–82* Match the activated carriers below with the group carried in high-energy linkage.

A. Acetyl CoA   1. acetyl group
B. *S*-Adenosylmethionine   2. carboxyl group
C. ATP   3. electrons and hydrogens
D. Carboxylated biotin   4. glucose
E. NADH, NADPH, FADH$_2$ 5. methyl group
F. Uridine diphosphate glucose   6. phosphate

2–83 Which of the following reactions will occur only if coupled to a second, energetically favorable reaction?

A. glucose + $O_2 \rightarrow CO_2 + H_2O$
B. $CO_2 + H_2O \rightarrow$ glucose + O_2
C. nucleoside triphosphate + $DNA_n \rightarrow DNA_{n+1} + 2\ P_i$
D. nucleosides $\rightarrow$ nucleoside triphosphates
E. ADP + $P_i \rightarrow$ ATP

HOW CELLS OBTAIN ENERGY FROM FOOD

TERMS TO LEARN

citric acid cycle	fermentation	nitrogen fixation
electron-transport chain	glycogen	oxidative phosphorylation
FAD/FADH$_2$	glycolysis	starch
fat	GTP	

2–84* Food from the diet is broken down in three stages of metabolism beginning with digestion in the gut. A portion of this food is stored for use between meals as glycogen—a polymer of glucose—in liver and as triacylglycerols in fat cells. When the supply of digested food from the gut is exhausted, these stored fuels are released into the circulation as glucose and fatty acids. Describe the three stages of catabolism of the fuel reserves stored in liver and fat cells.

2–85 If a cell contains 10^9 ATP molecules that it hydrolyzes and replaces once per minute, how long will it take for a cell to consume its own volume (1000 μm^3) of oxygen? Roughly 90% of the ATP in the cell is regenerated by oxidative phosphorylation (the remainder is regenerated by substrate-level phosphorylation, which is described in Problem 2–91). Assume that 5 molecules of ATP are regenerated by each molecule of oxygen (O_2) that is converted to water. Recall that a mole of a gas occupies 22.4 L.

2–86* Assuming that there are 5×10^{13} cells in the human body and that ATP is turning over at a rate of 10^9 ATP per minute in each cell, how many watts is the human body consuming? (A watt is a Joule per second, and there are 4.18 Joules/calorie.) Assume that hydrolysis of ATP yields 12 kcal/mole.

2–87 Does a Snickers™ candy bar (65 g, 325 kcal) provide enough energy to climb from Zermatt (elevation 1660 m) to the top of the Matterhorn (4478 m, Figure 2–28), or might you need to stop at Hörnli Hut (3260 m) to eat another one? Imagine that you and your gear have a mass of 75 kg, and that all of your work is done against gravity (that is, you're just climbing straight up).

$$\text{work (J)} = \text{mass (kg)} \times g\ (\text{m/sec}^2) \times \text{height gained (m)}$$

where g is acceleration due to gravity (9.8 m/sec^2). One Joule is 1 kg m^2/sec^2 and there are 4.18 kJ per kcal.

2–88* From a chemical perspective, the glycolytic pathway (see Figure 2–27) can be thought of as occurring in two stages. The first stage from glucose to glyceraldehyde 3-phosphate (G3P) prepares glucose so its cleavage yields two equivalent three-carbon fragments that are converted into G3P. The second

Figure 2–28 The Matterhorn (Problem 2–87).

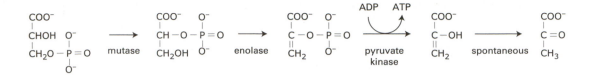

Figure 2–29 Conversion of 3-phosphoglycerate to pyruvate during glycolysis (Problem 2–90).

stage—from G3P to pyruvate—harvests energy in the form of ATP and NADH.

A. Write the balanced equation for the first stage of glycolysis (glucose → G3P).

B. Write the balanced equation for the second stage of glycolysis (G3P → pyruvate).

C. Write the balanced equation for the overall pathway (glucose → pyruvate).

2–89 The same ten reaction steps that make up the glycolytic pathway in Figure 2–27 are found in most living cells, from bacteria to humans. One could envision countless alternative chemical reaction mechanisms for sugar oxidation that could, in principle, have evolved to take the place of glycolysis. Discuss the near universality of glycolysis in cells in the context of evolution.

2–90* One of the two ATP generating steps in glycolysis is outlined in Figure 2–29. This sequence of reactions yields ATP and produces pyruvate, which is subsequently converted to acetyl CoA and oxidized to CO_2 in the citric acid cycle. Under anaerobic conditions ATP production from phosphoenolpyruvate (PEP) accounts for half of a cell's ATP supply. These 'substrate-level' phosphorylation events (so named to distinguish them from oxidative phosphorylation in mitochondria) were the first to be understood. Consider the conversion of 3-phosphoglycerate (3PG) to PEP, which constitutes the first two reactions in Figure 2–29. Recall that

$$\Delta G° = -2.3 \, RT \log_{10} K$$
$$= -1.41 \text{ kcal/mole } \log_{10} K$$

where K is the equilibrium ratio of the products over the reactants.

A. If 10 mM 3PG is mixed with phosphoglycerate mutase, which catalyzes its conversion to 2-phosphoglycerate (2PG) (Figure 2–29), the equilibrium concentrations at 37°C are 8.3 mM 3PG and 1.7 mM 2PG. How would the ratio of equilibrium concentrations change if 1 M 3PG had been added initially? What is the equilibrium constant K for the conversion of 3PG into 2PG, and what is the $\Delta G°$ for the reaction?

B. If 10 mM PEP is mixed with enolase, which catalyzes its conversion to 2PG, the equilibrium concentrations at 37°C are 2.9 mM 2PG and 7.1 mM PEP. What is the $\Delta G°$ for conversion of PEP to 2PG? What is the $\Delta G°$ for the reverse reaction?

C. What is the $\Delta G°$ for the conversion of 3PG to PEP?

2–91 The study of substrate-level phosphorylation events such as that in Figure 2–29 led to the concept of the 'high-energy' phosphate bond. The term 'high-energy' bond is somewhat misleading since it refers not to the strength of the bond, but rather, to the free-energy change (ΔG) upon hydrolysis. In the conversion of 3PG to PEP (the first two reactions in Figure 2–29), a low-energy phosphate bond is turned into a high-energy phosphate bond. The standard free-energy change ($\Delta G°$) for hydrolysis of the phosphate group in 3PG is about –3.3 kcal/mole, whereas hydrolysis of the phosphate group in PEP (converting it to pyruvate) has a $\Delta G°$ of –14.8 kcal/mole. How is it that moving the phosphate to the 2 position of glycerate and removing water, which has an overall $\Delta G°$ of 0.4 kcal/mole, can have such an enormous effect on the free energy for the subsequent hydrolysis of the phosphate bond?

Consider the set of reactions shown in Figure 2–30.

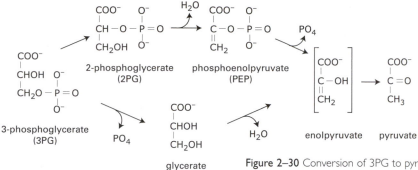

Figure 2–30 Conversion of 3PG to pyruvate and phosphate by two routes (Problem 2–91). The *bracketed* compound, enolpyruvate, is a transient intermediate.

A. What is $\Delta G°$ for conversion of 3PG to pyruvate by way of PEP?

B. What is $\Delta G°$ for conversion of 3PG to pyruvate by way of glycerate? What is $\Delta G°$ for conversion of glycerate to pyruvate? (Recall that thermodynamic quantities are state functions; that is, they describe differences between the initial and final states—they are independent of the pathway between the states.)

C. Propose an explanation for why the phosphate bond in PEP is a high-energy bond, whereas that in 3PG is a low-energy bond. (Assume that removal of water from glycerate has a $\Delta G°$ of –0.5 kcal/mole.)

D. In cells the conversion of PEP to pyruvate is linked to the synthesis of ATP as shown in Figure 2–29. What is $\Delta G°$ for the linked reactions?

2–92* (**True/False**) Because glycolysis is only a prelude to the oxidation of glucose in mitochondria, which yields 15-fold more ATP, glycolysis is not really important for human cells. Explain your answer.

2–93 At first glance, fermentation of pyruvate to lactate appears to be an optional add-on reaction to glycolysis (Figure 2–31). After all, couldn't cells growing in the absence of oxygen simply discard pyruvate as a waste product? In the absence of fermentation which products derived from glycolysis would accumulate in cells under anaerobic conditions? Could the flow of glucose through the glycolytic pathway continue in the absence of oxygen in cells that cannot carry out fermentation? Why or why not?

2–94* In the absence of oxygen, cells consume glucose at a high, steady rate. When oxygen is added, glucose consumption drops precipitously and is then maintained at the lower rate. Why is glucose consumed at a high rate in the absence of oxygen and at a low rate in its presence?

2–95 Arsenate (AsO_4^{3-}) is chemically very similar to phosphate (PO_4^{3-}) and is used as an alternative substrate by many phosphate-requiring enzymes. In contrast to phosphate, however, the anhydride bond between arsenate and a carboxylic acid group is very quickly hydrolyzed in water. Knowing this, suggest why arsenate is a compound of choice for murderers, but not for cells. Formulate your explanation in terms of the step in glycolysis at which 1,3-bisphosphoglycerate is converted to 3-phosphoglycerate, generating ATP (Figure 2–32).

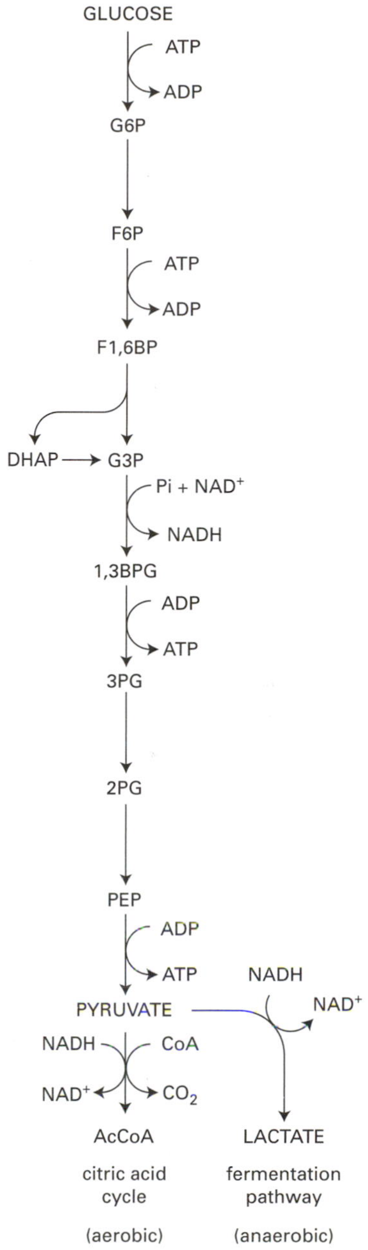

Figure 2–31 Fermentation of pyruvate to lactate (Problem 2–93). H^+ ions are not shown.

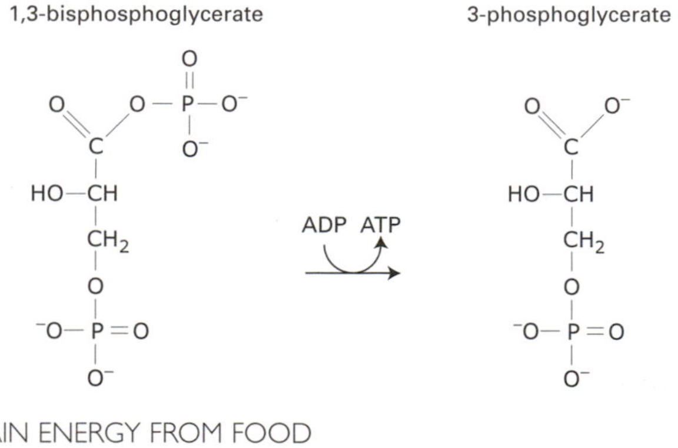

Figure 2–32 Conversion of 1,3-bisphosphoglycerate and ADP to 3-phosphoglycerate and ATP (Problem 2–95).

2–96* Muscles contain creatine phosphate (CP) as an energy buffer to maintain the levels of ATP in the initial stages of exercise. Creatine phosphate can transfer its phosphate to ADP to generate creatine (C) and ATP, with a $\Delta G°$ of -3.3 kcal/mole.

$$CP + ADP \rightarrow C + ATP \qquad \Delta G° = -3.3 \text{ kcal/mole}$$

A. In a resting muscle [ATP] = 4 mM, [ADP] = 0.013 mM, [CP] = 25 mM, and [C] = 13 mM. What is the ΔG for this reaction in resting muscle? Does this value make sense to you? Why or why not?

B. Consider an initial stage in vigorous exercise, when 25% of the ATP has been converted to ADP. Assuming that no other concentrations have changed, what is the ΔG for the reaction at this stage in exercising muscle? Does this value make sense?

C. If the ATP in muscle could be completely hydrolyzed (in reality it never is), it would power an all-out sprint for about 1 second. If creatine phosphate could be completely hydrolyzed to regenerate ATP, how long could a sprint be powered? Where do you suppose the energy comes from to allow a runner to finish a 200-meter sprint?

2–97 The liver provides glucose to the rest of the body between meals. It does so by breaking down glycogen, forming glucose 6-phosphate in the penultimate step. Glucose 6-phosphate is converted to glucose by splitting off the phosphate ($\Delta G° = -3.3$ kcal/mole). Why do you suppose the liver employs this mechanism rather than simply reversing the reaction by which glucose 6-phosphate was formed from glucose (GLC + ATP $\rightarrow$ G6P + ADP, $\Delta G° = -4.0$ kcal/mole)? By reversing this reaction the liver could generate glucose *and* make ATP.

2–98* In 1904 Franz Knoop performed what was probably the first successful labeling experiment to study metabolic pathways. He fed many different fatty acids labeled with a terminal benzene ring to dogs and analyzed their urine for excreted benzene derivatives. Whenever the fatty acid had an even number of carbon atoms, phenylacetate was excreted (Figure 2–33A). However, whenever the fatty acid had an odd number of carbon atoms, benzoate was excreted (Figure 2–33B).

From these experiments Knoop deduced that oxidation of fatty acids to CO_2 and H_2O involved removal of two-carbon fragments from the carboxylic acid end of the chain. Can you explain the reasoning that led him to conclude that two-carbon fragments, as opposed to any other number, were removed, and that degradation was from the carboxylic acid end, as opposed to the other end?

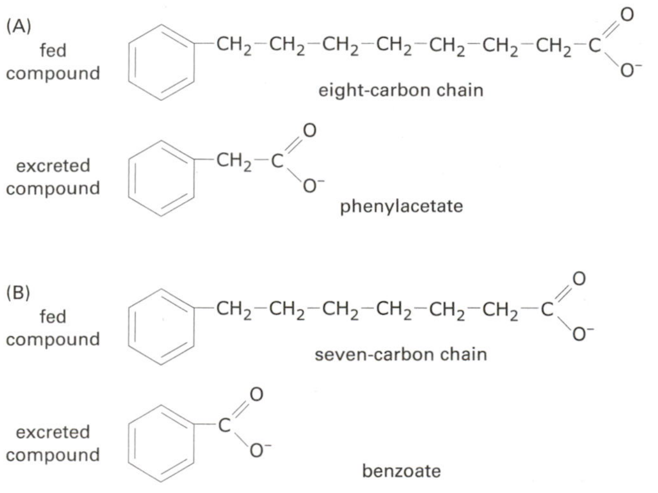

Figure 2–33 The original labeling experiment to analyze fatty acid oxidation (Problem 2–98). (A) Fed and excreted derivatives of an even-chain fatty acid. (B) Fed and excreted derivatives of an odd-chain fatty acid.

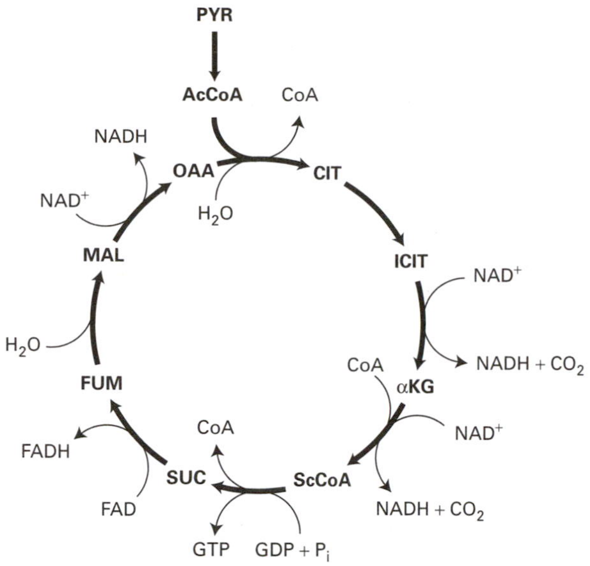

Figure 2–34 The citric acid cycle (Problem 2–99). PYR = pyruvate, AcCoA = acetyl coenzyme A, CIT = citrate, ICIT = isocitrate, αKG = α-ketoglutarate, ScCoA = succinyl coenzyme A, SUC = succinate, FUM = fumarate, MAL = malate, and OAA = oxaloacetate.

2–99 In 1937 Hans Krebs deduced the operation of the citric acid cycle (Figure 2–34) from careful observations on the oxidation of carbon compounds in minced preparations of pigeon flight muscle. (Pigeon breast is a rich source of mitochondria, but the function of mitochondria was unknown at the time.) The consumption of O_2 and the production of CO_2 were monitored with a manometer, which measures changes in volume of a closed system at constant pressure and temperature. Standard chemical methods were used to determine the concentrations of key metabolites. (Remember, radioactive isotopes were not available then.)

In one set of experiments Krebs measured the rate of consumption of O_2 during the oxidation of endogenous carbohydrates in the presence or absence of citrate. As shown in Table 2–5, addition of a small amount of citrate resulted in a large increase in the consumption of oxygen. Szent-Györgyi (1924, 1937) and Stare and Baumann (1936) had previously shown that fumarate, oxaloacetate, and succinate also stimulated respiration in extracts of pigeon breast muscle.

When metabolic poisons, such as arsenite or malonate (whose modes of action were undefined), were added to the minced muscles, the results were much different. In the presence of arsenite, 5.5 mmol of citrate were converted into about 5 mmol of α-ketoglutarate. In the presence of malonate an equivalent conversion of citrate into succinate occurred. Furthermore, in the presence of malonate roughly 5 mmol of oxygen were consumed (above background levels in the absence of citrate), which was twice as much as in the presence of arsenite.

Finally, Krebs showed that the minced muscles were capable of synthesizing citrate if oxaloacetate was added and all traces of oxygen were excluded. None of the other intermediates in the cycle led to a net synthesis of citrate in the absence of oxygen.

TABLE 2–5 Respiration in minced pigeon breast in the presence and absence of citrate (Problem 2–99).

TIME (minutes)	OXYGEN CONSUMPTION (mmol)		
	NO CITRATE	3 mmol CITRATE	DIFFERENCE
30	29	31	2
60	47	68	21
90	51	87	36
150	53	93	40

A. If citrate ($C_6H_8O_7$) were completely oxidized to CO_2 and H_2O, how many molecules of O_2 would be consumed per molecule of citrate? What is it about the results in Table 2–5 that caught Krebs's attention?

B. Why is the consumption of oxygen so low in the presence of arsenite or malonate? If citrate were oxidized to α-ketoglutarate ($C_5H_6O_5$), how much oxygen would be consumed per molecule of citrate? If citrate were oxidized to succinate ($C_4H_6O_4$), how much oxygen would be consumed per molecule of citrate? Does the observed stoichiometry agree with the expectations based on these calculations?

C. Why, in the absence of oxygen, does oxaloacetate alone cause an accumulation of citrate? Would any of the other intermediates in the cycle cause an accumulation of citrate in the presence of oxygen?

D. Toward the end of the paper Krebs states, "While the citric acid cycle thus seems to occur generally in animal tissues, it does not exist in yeast or in *E. coli*, for yeast and *E. coli* do not oxidize citric acid at an appreciable rate." Why do you suppose Krebs got this point wrong?

2–100* (**True/False**) The reactions of the citric acid cycle do not directly require the presence of oxygen. Explain your answer.

2–101 The last reaction of the citric acid cycle, which regenerates oxaloacetate (OAA) from malate (MAL), has a very positive $\Delta G° = 7.1$ kcal/mole.

$$\text{MAL} + \text{NAD}^+ \rightarrow \text{OAA} + \text{NADH} \qquad \Delta G° = 7.1 \text{ kcal/mole}$$

Despite its unfavorable equilibrium position, material must flow through this reaction quite readily in mitochondria—otherwise the cycle could not turn.

A. How is flow through the cycle accomplished in the face of such an overwhelmingly positive $\Delta G°$?

B. If the ratio of [NAD$^+$]/[NADH] is maintained at about 10 in mitochondria, what is the minimum ratio of [MAL]/[OAA] when the cycle is turning?

2–102* What, if anything, is wrong with the following statement: "The oxygen consumed during the oxidation of glucose in animal cells is returned as CO_2 to the atmosphere." How might you support your answer experimentally?

2–103 Compare the structures of the sugar glucose and the fatty acid palmitate in Figure 2–35. Can you give an intuitive explanation as to why oxidation of a sugar yields only about half the energy as the oxidation of the same weight of a fatty acid?

2–104* Humans are unable to synthesize all 20 amino acids. In the course of a day, when do you suppose you can make net amounts of new protein? Explain your answer.

2–105 Pathways for synthesis of amino acids in microorganisms were worked out in part by cross-feeding experiments among mutant organisms that were defective for individual steps in the pathway. Results of cross-feeding experiments for three mutants defective in the tryptophan pathway—*trpB*$^-$, *trpD*$^-$, and *trpE*$^-$—are shown in Figure 2–36A. The mutants were allowed to grow briefly in the presence of a very small amount of tryptophan, producing a pale streak. Heavier growth was observed at points where some streaks were close to other streaks. These spots of heavier growth indicate that one mutant can cross-feed (supply an intermediate) to the other one.

A. From the pattern of cross-feeding shown in Figure 2–36A, deduce the order of the steps controlled by the products of the *trpB*, *trpD*, and *trpE* genes. Explain your reasoning.

B. If accumulated intermediates at the block are responsible for the cross-feeding phenomenon, it should be possible to grow individual mutants on some intermediates. The three mutants were tested for growth on tryptophan and intermediates in the pathway (Figure 2–36B), with the results shown in Table 2–6. Use this information to arrange the defective genes relative to the tryptophan pathway.

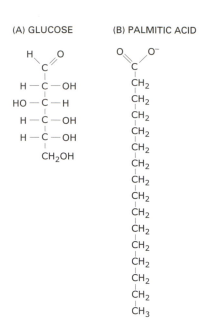

Figure 2–35 Structures of glucose and palmitate (Problem 2–103).

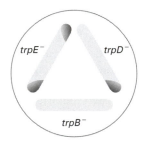

(A) CROSS-FEEDING RESULT

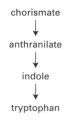

(B) TRYPTOPHAN SYNTHETIC PATHWAY

chorismate
↓
anthranilate
↓
indole
↓
tryptophan

Figure 2–36 Determination of pathway for tryptophan synthesis using cross-feeding experiments (Problem 2–105). (A) Results of a cross-feeding experiment among mutants defective for steps in the tryptophan biosynthetic pathway. *Dark areas* show regions of cell growth. (B) The tryptophan biosynthetic pathway. Several steps precede chorismate in the pathway and there are several steps between anthranilate and indole.

TABLE 2–6 Growth of mutants on intermediates in the pathway for tryptophan biosynthesis (Problem 2–105).

STRAIN	GROWTH ON MINIMAL MEDIUM SUPPLEMENTED WITH				
	NONE	CHORISMATE	ANTHRANILATE	INDOLE	TRYPTOPHAN
Wild type	+	+	+	+	+
trpB⁻	–	–	–	–	+
trpD⁻	–	–	–	+	+
trpE⁻	–	–	+	+	+

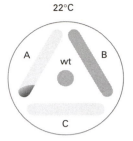

22°C

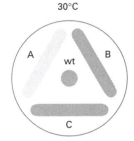

30°C

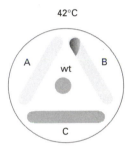

42°C

Figure 2–37 Results of cross-feeding experiments with three strains defective in proline biosynthesis (Problem 2–106). *Dark areas* show regions of cell growth.

2–106* You have isolated three different strains of bacteria—*proA⁻*, *proB⁻*, and *proC⁻*—that require added proline for growth. One is cold-sensitive, one is temperature-sensitive (heat-sensitive), and one has a deletion of the gene. You carry out cross-feeding experiments by streaking the strains out on agar plates containing minimal medium supplemented with a very low level of proline. After growth at three temperatures, you observe the results shown in Figure 2–37.

  A. Identify the types of mutations—cold-sensitive, temperature-sensitive, or deletion—in each of the strains.
  B. Deduce the order in which the gene products act in the pathway for proline biosynthesis.
  C. Does the identification of three different genes that affect proline biosynthesis mean that there are three steps in the biosynthetic pathway? Explain your answer.
  D. Why do you suppose there was no cross-feeding of *proA⁻* by strain *proC⁻* at 30°C or 42°C, or between the wild-type bacteria and the mutant strains under any conditions?

2–107 A cyclic reaction pathway requires that the starting material is regenerated and available at the end of each cycle. Intermediates in the citric acid cycle, however, are siphoned off for use in a variety of other metabolic reactions. Why does the citric acid cycle not quickly cease to exist?

2–108* An exceedingly sensitive instrument (yet to be devised) shows that one of the carbon atoms in Charles Darwin's last breath is resident in your bloodstream, where it forms part of a hemoglobin molecule. Suggest how this carbon atom might have traveled from Darwin to you, and list some of the molecules it could have passed through.

EFTu, cysteinyl tRNA, and guanine nucleotides. Hydrolysis of the bound GTP causes a conformational change that delivers the tRNA to the ribosome and frees the elongation factor for another delivery round. Coordinates from 1B23 determined by P Nissen, M Kjeldgaard, S Thirup & J Nyborg. *Struct. Fold. Des.* 7:143–156, 1999. Coordinates from from 1TUI determined by G Polekhina, S Thirup, M Kjeldgaard, P Nissen, C Lippmann & J Nyborg. *Structure* 4: 1141–1151, 1996.

PROTEINS

3

THE SHAPE AND STRUCTURE OF PROTEINS

TERMS TO LEARN

α helix	polypeptide backbone	quaternary structure
β sheet	primary structure	secondary structure
binding site	protein	side chain
coiled-coil	protein domain	subunit
conformation	protein module	tertiary structure

3–1   Amino acids are commonly grouped as nonpolar (hydrophobic) or as polar (hydrophilic), based on the properties of the amino acid side chains. Quantification of these properties is important for a variety of protein structure predictions, but they cannot readily be determined from the free amino acids themselves. Why are the properties of the side chains difficult to measure using the free amino acids? How might you determine hydrophilicity or hydrophobicity of amino acid side chains?

3–2*   What are the four weak (noncovalent) interactions that determine the conformation of a protein?

3–3   Typical proteins have a stability ranging from 7 to 15 kcal/mole at 37°C. Stability is a measure of the equilibrium between the folded and unfolded forms of the protein.

Folded [F] ⇌ Unfolded [U], $K = [U]/[F]$

For a protein with a stability of 9.9 kcal/mole, calculate the fraction of unfolded protein that would exist at equilibrium at 37°C. At equilibrium,

$$\Delta G° = - RT\ln K = - 2.3RT\log K,$$

where $R = 1.98 \times 10^{-3}$ kcal/°K mole and T is temperature in °K (37°C = 310°K).

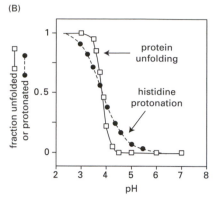

(A)

(B)

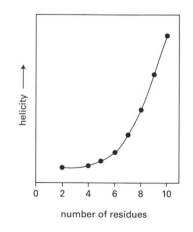

Figure 3–1 Denaturation of proteins (Problem 3–6). (A) Histone protonation. (B) Titration curves for protein unfolding and histone protonation.

3–4* You wish to try your hand at predicting the structure of lysozyme. Because lysozyme has several hundred weak interactions that contribute to its folded conformation, you have decided first to measure their overall contribution, so that you will know how much leeway there is when it comes time to assign values to individual interactions. Since the stability of lysozyme is

$$\Delta G° = G°_{\text{unfolded}} - G°_{\text{folded}}$$

and

$$G° = H° - TS°$$

you determine the standard enthalpy, $H°$, a measure of bond strength, and the standard entropy, $S°$, a measure of disorder, for both the folded and unfolded states. Your values (kcal/mole) are shown below for measurements made at 37°C.

	$H°$	$TS°$
Unfolded state	128	119
Folded state	75	76

A. What is the stability of lysozyme at 37°C?
B. Does this calculation give you hope that you will be able to predict the structure of lysozyme? Why or why not?

3–5 When egg white is heated, the proteins in it denature and the egg white hardens. The denaturation is irreversible, but hard-boiled egg white can be dissolved by heating it in a solution containing a strong detergent (such as sodium dodecyl sulfate) together with a reducing agent, like 2-mercaptoethanol. Neither reagent alone has any effect.
A. Why does boiling an egg white cause it to harden?
B. Why does it require both a detergent and a reducing agent to dissolve the hard-boiled egg white?

3–6* Most proteins denature at both high and low pH. At high pH, the ionization of internal tyrosines is thought to be the main destabilizing influence, whereas at low pH, the protonation of buried histidines (Figure 3–1A) is the likely culprit. A titration curve for the unfolding of the enzyme ribonuclease is shown in Figure 3–1B. Superimposed on it is the expected titration curve for the ionization of a histidine side chain with a pK of about 4, which is typical for a buried histidine (the pK for the side chain of the free amino acid is 6). The titration curve for denaturation is clearly much steeper than that for the side chain. Given the discrepancy between the titration curves for protein unfolding and histidine protonation, how can it be true that protonation of histidine causes protein unfolding?

3–7 (**True/False**) A protein is at a near entropy minimum (point of lowest disorder, or greatest order) when it is completely stretched out like a string and when it is properly folded up. Explain your answer.

3–8* Although α helices are common components of polypeptide chains, they need to be of a certain minimum length. To find out how chain length affects α-helix formation, you measure the circular dichroism (a measure of helicity) for a series of peptides of increasing length (Figure 3–2). Why is there essentially no helix formation until the chain is at least six amino acids long?

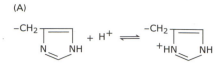

Figure 3–2 Helicity of various peptides of increasing length (Problem 3–8).

(A) HELIX WHEEL

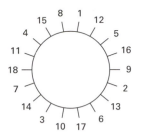

(B) PEPTIDE SEQUENCES

S L I K S V I E M V D E W F R T F L

F L I R V L R K V F R V L T R I L S

R L F R S R V L K I A V I R F L L I

Figure 3–3 Helix-wheel projection (Problem 3–10). (A) Helix wheel. The *circle* (wheel) represents the helix as viewed from one end. *Numbers* show the positions of the amino acid side chains, as projected on the wheel. The positions of the first 18 amino acids are shown; amino acid 19 would occupy the same position as amino acid 1. Amino acid 1 is closest to the reader; amino acid 18 is farthest away. (B) Peptide sequences. The N-termini are shown at the *left*; hydrophobic amino acids are *shaded*; hydrophilic amino acids are *unmarked*. (See inside back cover for one-letter amino acid code.)

3–9   The uniform arrangement of the backbone carbonyl oxygens and amide nitrogens in an α helix gives the helix a net dipole, so that it carries a partial positive charge at the amino end and a partial negative charge at the carboxyl end. Where would you expect the ends of α helices to be located in a protein? Why?

3–10*   α Helices are often embedded in a protein so that one side faces the surface and the other side faces the interior. Such helices are often termed amphipathic because the surface side is hydrophilic and the interior side is hydrophobic. A simple way to decide whether a sequence of amino acids might form an amphipathic helix is to arrange the amino acids around what is known as a 'helix-wheel projection' (Figure 3–3). If the hydrophobic and hydrophilic amino acids are segregated on opposite sides of the wheel, the helix is amphipathic. Using the helix-wheel projection, decide which of the three peptides in Figure 3–3 might form an amphipathic helix. (The mnemonic 'FAMILY VW' will help you recognize hydrophobic amino acids.)

3–11   Examine the segment of β sheet shown in Figure 3–4. For each strand of the sheet decide whether it is parallel or antiparallel to each of its neighbors.

3–12*   (**True/False**) Each strand in a β sheet is a helix with two amino acids per turn. Explain your answer.

3–13   Like α helices, β sheets often have one side facing the surface of the protein and one side facing the interior, giving rise to an amphipathic sheet with one hydrophobic surface and one hydrophilic surface. From the sequences listed below pick the one that could form a strand in an amphipathic β sheet. (See inside back cover for one-letter amino acid code.)

A. A L S C D V E T Y W L I
B. D K L V T S I A R E F M
C. D S E T K N A V F L I L
D. T L N I S F Q M E L D V
E. V L E F M D I A S V L D

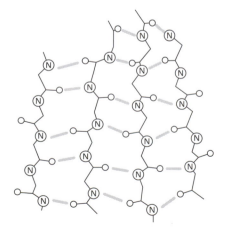

Figure 3–4 A segment of β sheet from the interior of thioredoxin (Problem 3–11). Amide nitrogens are indicated by *circled Ns*; hydrogen bonds are shown as *gray lines*.

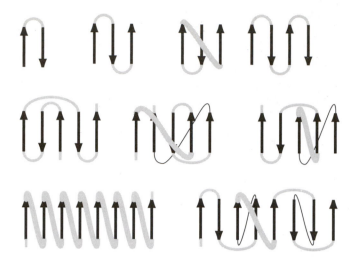

Figure 3–5 Topological representations of several protein folds (Problem 3–14). *Vertical arrows* represent strands in β sheets; *gray connectors* may be loops or helices. *Thick gray diagonal lines* are above the plane of the page; *thin black lines* lie below the plane of the page.

3–14* Several different protein folds are represented in schematic form in Figure 3–5. These diagrams preserve the topology of the protein and allow one to decide, for example, whether a protein is folded in a new way or is an example of a protein fold that is already known. These diagrams also permit a ready demonstration of a fundamental principle of protein folding. For each of these folds, imagine that you could grasp the N and C termini and pull them apart. Would any of the illustrated folds produce a knot when fully stretched out?

3–15 It is a common observation that antiparallel strands in a β sheet are connected by short loops, but that parallel strands are connected by α helices. Why do you think this is?

3–16* Small proteins may have only one or two amino acid side chains that are totally inaccessible to solvent. Even in large proteins, only about 15% of the amino acids are fully buried. A list of buried side chains from a study of twelve proteins is shown in Table 3–1. The list is ordered by the proportion of amino acids of each type that are fully buried. What types of amino acids are most commonly buried? Least commonly buried? Are there any surprises on this list?

3–17 (**True/False**) Loops of polypeptide that protrude from the surface of a protein often form the binding sites for other molecules. Explain your answer.

3–18* In 1968 Cyrus Levinthal pointed out a complication in protein folding that is widely known as the Levinthal paradox. He argued that because there are astronomical numbers of conformations open to a protein in the denatured state, it would take a very long time for a protein to search through all the possibilities to find the correct one, even if it tested each possible conformation exceedingly rapidly. Yet denatured proteins typically take less than a second to fold inside the cell or in the test tube. How do you suppose that proteins manage to fold so quickly?

3–19 "To produce one molecule of each possible kind of polypeptide chain, 300 amino acids in length, would require more atoms than exist in the universe." But the universe is very large. Can this statement really be correct? Since counting atoms is a tricky business, let's consider the problem from the standpoint of mass. The mass of the observable universe is estimated to be about 10^{80} grams, give or take an order of magnitude or so. If the average mass of each amino acid is 110 daltons, what would be the mass of one molecule of each possible kind of polypeptide chain 300 amino acids in length? Is this greater than the mass of the universe?

TABLE 3–1 Proportions of amino acids that are inaccessible to solvent in a study of twelve proteins (Problem 3–16). See inside back cover for one-letter amino acid code.

AMINO ACID SIDE CHAIN	PROPORTION BURIED
I	0.60
V	0.54
C	0.50
F	0.50
L	0.45
M	0.40
A	0.38
G	0.36
W	0.27
T	0.23
S	0.22
E	0.18
P	0.18
H	0.17
D	0.15
Y	0.15
N	0.12
Q	0.07
K	0.03
R	0.01

3–20* Comparison of a homeodomain protein from yeast and *Drosophila* shows that only 17 of 60 amino acids are identical. How is it possible for a protein to change over 70% of its amino acids and still fold in the same way?

3–21 Neither yeast nor *Drosophila* has been around for more than a few hundred million years, yet they are separated by more than a billion years of evolution. How can that possibly be true?

3–22* Often, the hard part of protein structure determination by x-ray diffraction is getting good crystals. In difficult cases, there are two common approaches for obtaining crystals: (1) using fragments of the protein and (2) trying homologous proteins from different species.

A. Examine the protein in Figure 3–6. Where would you cleave this protein to obtain fragments that might be expected to fold properly and perhaps form crystals?

B. Why do you suppose that homologous proteins from different species might differ in their ability to form crystals?

3–23 Some 1000 different protein folds are now known, and it is estimated that there may only be around 2000 in total. Protein folds seem to stay the same as genes evolve, giving rise to large families of similarly folded proteins with related functions. Does this mean that the last common ancestor to all life on Earth had 1000–2000 different genes?

3–24* A common strategy for identifying distantly related homologous proteins is to search the database using a short signature sequence indicative of the particular protein function. Why is it better to search with a short sequence than with a long sequence? Don't you have more chances for a 'hit' in the database with a long sequence?

3–25 (**True/False**) The computational method known as threading allows one to obtain an approximate three-dimensional structure for a protein as soon as its amino acid sequence is known. Explain your answer.

3–26* The so-called kelch motif consists of a four-stranded β sheet, which forms what is known as a β propeller. It is usually found to be repeated four to seven times, forming a kelch-repeat domain in a multidomain protein. One such kelch repeat domain is shown in Figure 3–7. Would you classify this domain as an 'in-line' or 'plug-in' type domain?

3–27 Examine the three protein monomers in Figure 3–8. From the arrangement of complementary binding surfaces, which are indicated by similarly shaped protrusions and invaginations, decide which monomer would assemble into a ring, which would assemble into a chain, and which would assemble into a sheet.

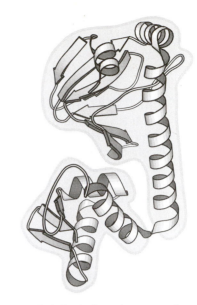

Figure 3–6 Catabolite activator protein from *E. coli* (Problem 3–22). *Shading indicates its domain structure.*

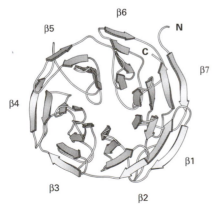

Figure 3–7 The kelch repeat domain of galactose oxidase from *D. dendroides* (Problem 3–26). The seven individual β propellers are indicated. The N- and C-termini are indicated by N and C.

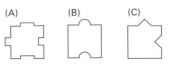

Figure 3–8 Three protein monomers (Problem 3–27).

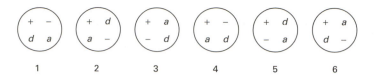

Figure 3–9 Binding surfaces for six different proteins (Problem 3–30). In each case the bulk of the protein is below the plane of the page.

3–28* Cro is a bacterial gene regulatory protein that binds to DNA to turn genes off. It is a symmetrical 'head-to-head' dimer. Each of the two subunits of the dimer recognizes a particular short sequence of nucleotides in DNA. If the sequence of nucleotides recognized by one subunit is represented as an arrow (→), so that the 'head' of the arrow corresponds to DNA recognized by the 'head' of the subunit, which of the following sequences in DNA represents the binding site for the Cro dimer?

A. →→
B. →←
C. ←←
D. ←→
E. Could be more than one of the above.

3–29 Why is it that there are numerous examples of 'head-to-head' and 'tail-to-tail' dimers, but few, if any, examples of 'head-to-tail' dimers?

3–30* Proteins bind to one another via weak interactions across complementary surfaces. Oppositely charged amino acids are apposed, as are hydrogen-bond donors and acceptors, and protrusions match invaginations so that van der Waals contacts can be optimized. When two copies of a protein bind to form a 'head-to-head' dimer, they use the same binding surface. Examine the binding surfaces of the six proteins shown in Figure 3–9, where charged amino acids are indicated by + and –, and hydrogen-bond donors and acceptors are indicated by *d* and *a*. (Protrusions and invaginations—three-dimensional shapes—are not represented in the binding surfaces in Figure 3–9 just because it is difficult to do so, but their absence does not change the general principles derived from this problem.) In which cases could two copies of one protein form a 'head-to-head' dimer in which the charges and hydrogen-bonding groups are appropriately matched? Can you spot any common feature of the surfaces that allows such dimers to form?

3–31 Nuclear lamin C is a member of the intermediate filament family. Thus, it should show regions of the coiled-coil heptad repeat motif AbcDefg, where A and D are hydrophobic amino acids and b, c, e, f, and g can be almost any amino acid. The sequence of nuclear lamin C is shown in Figure 3–10 with potential coiled-coil regions highlighted. Examine the segment marked 'coil 1A.' Does it conform to the heptad repeat? (Don't forget the mnemonic, FAMILY VW.)

coil 1A
METPSQRRATRSGAQASSTPLSPTRITRLQEKED**LQELNDRLAVYIDRVRSLETENA**
coil 1B
GLRLRITESEEVVSREVSGIKAA**YEAELGDARKTLDSVAKERARLQLELSKVREEFK**

ELKARNTKKEGDLIAAQARLKDLEALLNSKEAALSTALSEKRTLEGELHDLRGQVAK

LEAALGEAKKQLQDEMLRRVDAENRLQTMKEELDFQKNIYSEELRETKRRHETRLVE

coil 2
IDNGKQREFESRLAD**ALQQLRAQHEDQVEQYKKELEKTYSAKLDNARQSAERNSNLV**

GAAHEELQQSRIRIDSLSAQLSQLQKQLAAKEAKLRDLEDSLARERDTSRRLLAEKE

REMAEMRARMQQQLDEYQQLLDIKLALDMQIHAYRKLLEGEEERLRLSPSPTSQRSR

GRASSHSSQTQGGGSVTKKRKLESTESRSSPSQHARTSGRVAVEEVDEEGKFVRLRN

KSNEDQSMGNWQIKRQNGDDPLLTYRFPPKFTLKAGQVVTIWAAGAGATHSPPTDLV

WKAQNTWGCGNSLRTALINSTGEEVAMRKLVRSVTVVEDDEDEDGDDLLHHHHVSGS

RR

Figure 3–10 The amino acid sequence of nuclear lamin C (Problem 3–31).

(A)

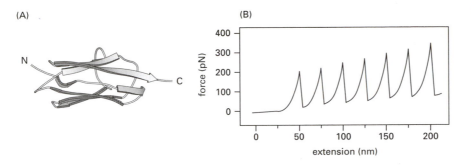

(B)

Figure 3–11 Springlike behavior of titin (Problem 3–32). (A) The structure of an individual Ig domain. (B) Force in piconewtons versus extension in nanometers obtained by atomic force microscopy.

3–32* Titin, which has a molecular weight of 3×10^6, is the largest polypeptide yet described. Stringlike titin molecules extend from muscle thick filaments to the Z disc; they are thought to act as springs to keep the thick filaments centered in the sarcomere. Titin is composed of a large number of repeated immunoglobulin (Ig) sequences of 89 amino acids, each of which is folded into a domain about 4 nm in length (Figure 3–11A).

Problem 16–75 discusses the location of titin in striated muscle.

You suspect that the springlike behavior of titin is caused by the sequential unfolding (and refolding) of individual Ig domains. You test this hypothesis using the atomic force microscope, which allows you to pick up one end of a protein molecule and pull with an accurately measured force. For a fragment of titin containing seven repeats of the Ig domain, this experiment gives the sawtooth force versus extension curve shown in Figure 3–11B. If the experiment is repeated in solution of 8M urea (a protein denaturant) the peaks disappear and the measured extension becomes much longer for a given force. If the experiment is repeated after the protein has been cross-linked by treatment with glutaraldehyde, once again the peaks disappear but the extension becomes much smaller for a given force.

A. Are the data consistent with your hypothesis that titin's springlike behavior is due to the sequential unfolding of individual Ig domains? Explain your reasoning.
B. Is the extension for each putative domain-unfolding event the magnitude you would expect? (In an extended polypeptide chain, amino acids are spaced at intervals of 0.34 nm.)
C. Why is each successive peak in Figure 3–11B a little higher than the one before?
D. Why does the force collapse so abruptly after each peak?

3–33 You are skeptical of the blanket statement that cysteines in intracellular proteins are not involved in disulfide bonds, while in extracellular proteins they are. To test this statement you carry out the following experiment. As a source of intracellular protein you use reticulocytes, which have no internal membranes and, thus, no proteins from the ER or other membrane-enclosed compartments. As examples of extracellular proteins, you use bovine serum albumin (BSA), which has 37 cysteines, and insulin, which has 6. You denature the soluble proteins from a reticulocyte lysate and the two extracellular proteins so that all cysteines are exposed. To probe the status of cysteines, you treat the proteins with *N*-ethylmaleimide (NEM), which reacts covalently with the thiol groups of free cysteines, but not with sulfur atoms in disulfide bonds. In the first experiment you treat the denatured proteins with radiolabeled NEM, then break any disulfide bonds with dithiothreitol (DTT) and react a second time with unlabeled NEM. In the second experiment you do the reverse: you first treat the denatured proteins with unlabeled NEM, then break disulfide bonds with DTT and treat with radiolabeled NEM. The proteins are separated according to size by electrophoresis on a polyacrylamide gel (Figure 3–12).

A. Do any cytosolic proteins have disulfide bonds?
B. Do the extracellular proteins have any free sulfhydryl groups?
C. How do you suppose the results might differ if you used lysates of cells that have internal membrane-enclosed compartments?

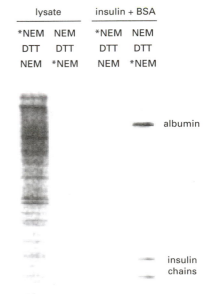

lysate		insulin + BSA	
*NEM	NEM	*NEM	NEM
DTT	DTT	DTT	DTT
NEM	*NEM	NEM	*NEM

albumin

insulin chains

Figure 3–12 Test for disulfide bonds in cytosolic and extracellular proteins (Problem 3–33). The order of treatment with NEM and DTT is indicated at the top of each lane; *NEM indicates radiolabeled NEM.

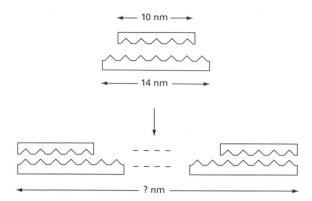

Figure 3–13 Vernier assembly of two proteins into a fiber of defined length (Problem 3–36).

3–34* The error rate for protein synthesis is estimated to be about 1/10,000; that is, the synthesis machinery incorporates one incorrect amino acid for each 10,000 it inserts. At this error rate, what fraction of proteins will be synthesized correctly for proteins 1000 amino acids, 10,000 amino acids, and 100,000 amino acids in length? [The probability of a correct sequence, P_C, equals the fraction correct for each operation, f_C, raised to a power equal to the number of operations, n. $P_C = (f_C)^n$. For an error rate of 1/10,000, $f_C = 0.9999$.]

3–35 It is often said that protein complexes are made from subunits (that is, individually synthesized proteins) rather than as one long protein because it is more likely to give a correct final structure.

   A. Assuming the same error rate as in the previous problem, what fraction of bacterial ribosomes would be constituted correctly if the proteins were synthesized as one large protein versus assembled from individual proteins? For the sake of calculation assume that the ribosome is composed of 50 proteins, each 200 amino acids in length, and that the subunits—correct and incorrect—are assembled with equal likelihood into the complete ribosome.

   B. Is the assumption that correct and incorrect subunits assemble equally well likely to be true? Why or why not? How would a change in that assumption affect the calculation in part A?

3–36* In a vernier type of assembly, rodlike proteins of different length form a staggered complex that grows until their ends exactly match.

   A. Consider the assembly of a long fibrous complex from two sets of rodlike proteins, one 10 nm in length and the other 14 nm (Figure 3–13). How long will the final fiber be if assembly proceeds until both ends are flush?

   B. Initiation of growth of such fibers usually begins with the binding of one kind of subunit to a subunit of the other type. Does the 'stagger' in the very first complex formed make any difference in the length of the final fiber?

PROTEIN FUNCTION

TERMS TO LEARN

active site	equilibrium constant (K)	motor protein
allosteric protein	feedback inhibition	protein kinase
antibody	GTP-binding protein	protein phosphatase
antigen	ligand	regulatory site
Ca^{2+} ATPase	linkage	substrate
catalyst	lysozyme	transition state
enzyme		

3–37 Antarctic notothenioid fish (Figure 3–14) avoid freezing in their perpetually icy environment because of an antifreeze protein that circulates in their blood. This evolutionary adaptation has allowed the *Notothenioidei* suborder to rise to dominance in the freezing Southern Ocean. It is said that all

Dissostichus eleginoides, the Chilean sea bass

Pagothenia borchgrevinki

Figure 3–14 Two notothenioid fish (Problem 3–37). The notothenioid family now dominates Antarctica's continental shelf, accounting for 50% of the species and 95% of the biomass of fish. The Chilean sea bass is commonly served in restaurants.

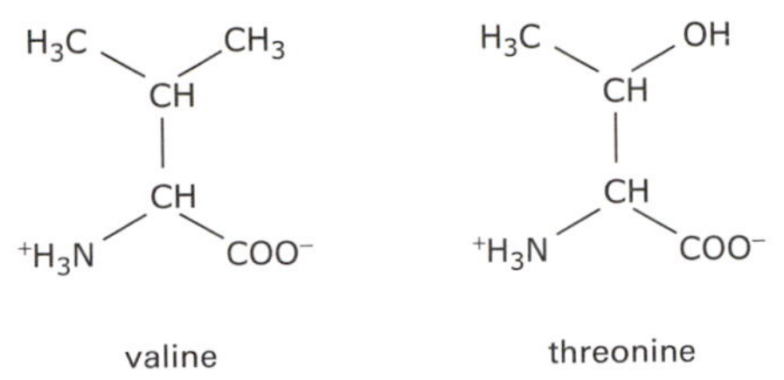

Figure 3–15 Structures of valine and threonine (Problem 3–38).

proteins function by binding to other molecules. To what ligand do you suppose antifreeze proteins bind to keep the fish from freezing? Or do you think this might be an example of a protein that functions in the absence of any molecular interaction?

3–38* Aminoacyl-tRNA synthetases attach specific amino acids to their appropriate tRNAs in preparation for protein synthesis. The synthetase that attaches valine to tRNA^{Val} must be able to discriminate valine from threonine, which differ very slightly in structure: valine has a methyl group where threonine has a hydroxyl group (Figure 3–15). Valyl-tRNA synthetase achieves this discrimination in two steps. In the first it uses a binding pocket whose contours allow valine or threonine (but not other amino acids) to bind, but the binding of valine is preferred. This site is responsible for coupling the amino acid to the tRNA. In the second step the enzyme checks the newly made aminoacyl tRNA using a second binding site that is very specific for threonine and hydrolyzes it from the tRNA. How do you suppose it is that the second binding site can be very specific for threonine, whereas the first binding site has apparently been able to evolve to be only moderately specific for valine?

3–39 (**True/False**) The tendency for an amino acid side-chain group such as –COOH to release a proton, its pK, is the same for the amino acid in solution and for the amino acid in a protein. Explain your answer.

3–40* (**True/False**) For a family of related genes that do not match genes of known function in the sequence database, it should be possible to deduce their function using 'evolutionary tracing' to see where conserved amino acids cluster on their surfaces. Explain your answer.

3–41 The binding of platelet-derived growth factor (PDGF) to the PDGF receptor stimulates phosphorylation of 8 tyrosines in the receptor's cytoplasmic domain. The enzyme phosphatidylinositol 3′-kinase (PI3-kinase) binds to one or more of the phosphotyrosines through its SH2 domains and is thereby activated. To identify the activating phosphotyrosines, you synthesize 8 pentapeptides that contain the critical tyrosines (at the N-terminus) in their phosphorylated or unphosphorylated forms. You then mix an excess of each of the various pentapeptides with phosphorylated PDGF receptor and PI3-kinase. Immunoprecipitation of the PDGF receptor will bring down any bound PI3-kinase, which can be assayed by its ability to add ^{32}P-phosphate to its substrate (Figure 3–16).

A. Do your results support the notion that PI3-kinase binds to phosphotyrosines in the activated PDGF receptor? Why or why not?

B. The amino acid sequences of the PDGF pentapeptides tested above (numbered according to the position of tyrosine in the PDGF receptor) and of peptide segments that are known to bind to PI3-kinase in other activated receptors are shown below.

684	YSNAL	YMMMR	(FGF receptor)
708	YMDMS	YTHMN	(insulin receptor)
719	YVPML	YEVML	(hepatocyte growth factor receptor)
731	YADIE	YMDMK	(steel factor receptor)
739	YMAPY	YVEMR	(CSF-1 receptor)
743	YDNYE		
746	YEPSA		
755	YRATL		

pentapeptides with tyrosine

pentapeptides with phosphotyrosine

684   708   719   731   739   743   746   755

Figure 3–16 Assay for PI3-kinase in immunoprecipitates of the PDGF receptor (Problem 3–41). Pentapeptides are indicated by numbers that refer to their position in the PDGF receptor. *Black circles* indicate incorporation of ^{32}P-phosphate into the substrate for PI3-kinase.

What are the common features of peptide segments that form binding sites for PI3-kinase?

C. Which of the three common types of protein–protein interaction—string–surface, helix–helix, or surface–surface—does the binding of PI3-kinase with the PDGF receptor most likely illustrate?

3–42* Which pair(s) of proteins in Figure 3–9 (see Problem 3–30) could bind to one another to form a heterodimer in such a way that all binding groups are satisfied? (Charged amino acids are indicated by + or – and hydrogen-bond donors and acceptors are indicated by d and a.)

3–43 Protein A binds to protein B to form a complex, AB. A cell contains an equilibrium mixture of protein A at a concentration of 1 μM, protein B at a concentration of 1 μM, and the complex AB also at 1 μM.

A. What is the equilibrium constant, K, for the reaction A + B → AB?

B. What would the equilibrium constant be if A, B, and AB were each present in equilibrium at a concentration of 1 nM?

C. At this lower concentration, about how many extra hydrogen bonds would be needed to hold A and B together tightly enough to form the same proportion of the AB complex? (Free-energy change is related to the equilibrium constant by the equation: $\Delta G° = -2.3\,RT\log_{10}K$, where R is 1.98×10^{-3} kcal/°K mole and T is 310°K. Assume that the formation of one hydrogen bond is accompanied by a favorable free-energy change of about 1 kcal/mole.)

3–44* An antibody binds to another protein with an equilibrium constant, K, of 5×10^9 M^{-1}. When it binds to a second, related protein, it forms three fewer hydrogen bonds, reducing its binding affinity by 2.8 kcal/mole. What is the K for its binding to the second protein? (Free-energy change is related to the equilibrium constant by the equation: $\Delta G° = -2.3\,RT\log_{10}K$, where R is 1.98×10^{-3} kcal/°K mole and T is 310°K.)

3–45 Antibodies are often used to identify the location of a protein within a cell to gain clues to its function. But one must be careful in interpreting the results. A case in point is the *Brca1* gene, which was identified as the mutated gene in one type of familial breast cancer. Its sequence failed to identify a homolog with a known function. Two groups of scientists raised antibodies to segments of the Brca1 protein (predicted from the sequence to be about 220 kd) and then reacted them with breast cancer cells using a method that allowed the bound antibodies to be visualized by fluorescence microscopy. One group used an antibody (C20) raised against a 20-amino acid peptide in the C-terminal region, and reported that Brca1 was located in secretory vesicles and on the plasma membrane. A second group, which used antibodies raised against an N-terminal fragment (BPA1) and antibodies raised against a C-terminal fragment (BPA2), reported that Brca1 was localized to the nucleus. To clear up this contradiction, an immunoblot was performed using all three antibodies against the proteins in lysates from three different breast cancer cell lines (Figure 3–17). (An immunoblot is performed by separating the proteins according to their molecular mass by SDS polyacrylamide-gel electrophoresis, blotting them onto a filter paper, and then reacting them with antibodies in a way that allows the places where antibodies bind to be readily visualized.)

Based on the results of the immunoblot, propose an explanation for the

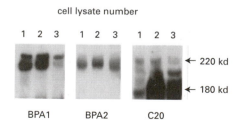

cell lysate number

BPA1   BPA2   C20

← 220 kd
← 180 kd

Figure 3–17 Immunoblot of three antibodies raised against Brca1 (Problem 3–45).

contradictory results obtained by the two groups. Where do you think Brca1 is located in cells?

3–46* The equilibrium constant for a reaction like that of antibody (Ab) binding to a protein (Pr) to form an antibody–protein complex (Ab–Pr) is equal to the ratio of the association rate constant, k_{on}, to the dissociation rate constant, k_{off} ($K = $ [Ab–Pr]/([Ab][Pr]) $= k_{on}/k_{off}$). Recall that the association rate (k_{on}[Ab][Pr]) equals the dissociation rate (k_{off}[Ab–Pr]) at equilibrium. Consider two such reactions. The first has an on rate constant of 10^5/M sec and an off rate constant of 10^{-3}/sec at 37°C. The second has an on rate constant of 10^3/M sec and an off rate of 10^{-5}/sec at 37°C.

A. What are the equilibrium constants for these two reactions?
B. At equal concentrations of antibody and protein, which of these reactions will reach its equilibrium point more quickly?
C. You wish to use an antibody to purify the protein you are studying. You are concerned that the complex may fall apart in the time it takes you to isolate it, and you are unsure how the off rates relate to the half-time for dissociation; that is, the time at which half the complex will have dissociated. It can be shown the fraction of complex remaining at time t ([Ab–Pr]$_t$) relative to that present initially ([Ab–Pr]$_0$) is

$$\frac{[\text{Ab–Pr}]_t}{[\text{Ab–Pr}]_0} = e^{-k_{off}t}$$

an equation more easily dealt with in its logarithmic form,

$$2.3 \log_{10} \frac{[\text{Ab–Pr}]_t}{[\text{Ab–Pr}]_0} = -k_{off}t$$

Using this relationship decide how long it will take for half of the Ab–Pr complexes in each of the above two reactions to dissociate. Neglect any contribution from new complex being formed in the association reaction.

3–47 Equilibrium dialysis provides a simple method for determining the equilibrium constant for binding of a ligand by a protein. The protein is confined inside a dialysis sac, formed by an artificial membrane with pores through which the protein cannot pass, but which the much smaller ligand can freely permeate. The ligand, usually radiolabeled, is added to the solution surrounding the dialysis sac and after equilibrium has been established the concentration of the ligand is measured in both compartments. The concentration in the external compartment is the concentration of free (unbound) ligand and the concentration in the dialysis sac is the sum of the bound plus free ligand. By measuring these values in the presence of various initial ligand concentrations, the value of the equilibrium constant can be determined. By convention, the equilibrium is usually considered for the dissociation reaction (Pr–L $\rightarrow$ Pr + L) rather than the association reaction (Pr + L $\rightarrow$ Pr–L) and the equilibrium constant is referred to as the dissociation constant K_d.

It is useful to look at the transformation of the standard equilibrium relationship into the form commonly used to analyze the data. At equilibrium,

$$K_d = \frac{[\text{Pr}][\text{L}]}{[\text{Pr–L}]}$$

Substituting [Pr]$_{TOT}$ – [Pr–L] for [Pr] and rearranging gives

$$[\text{Pr–L}] = \frac{[\text{Pr}]_{TOT}[\text{L}]}{K_d + [\text{L}]}$$

This is an equation for a rectangular hyperbole. It can be rearranged to give a linear form, as was first done by George Scatchard in 1947 (hence, graphs of such data are commonly known as Scatchard plots).

$$\frac{[\text{Pr–L}]}{[\text{L}]} = \frac{-[\text{Pr–L}]}{K_d} + \frac{[\text{Pr}]_{\text{TOT}}}{K_d}$$

At constant protein concentration and a variety of ligand concentrations, a plot of bound over free ligand ([Pr–L]/[L]) against bound ligand ([Pr–L]) gives a line with slope equal to $-1/K_d$ and an x-intercept equal to $[\text{Pr}]_{\text{TOT}}$, which is the total concentration of binding sites.

In the early 1960s, when the nature of the genetically identified repressors of bacterial gene expression had not been defined, Walter Gilbert and Benno Muller-Hill used equilibrium dialysis to measure the binding of an inducer of gene expression (IPTG) to the lactose repressor protein. They used radio-labeled IPTG and two partially purified preparations of lactose repressor protein: one from wild-type cells and the other from mutant cells in which induction of the lactose operon occurred at lower concentrations of IPTG (assumed to carry a lactose repressor that bound IPTG more tightly). Their data are shown as a Scatchard plot in Figure 3–18.

A. What are the K_ds for the two lines shown in the figure?
B. Which line corresponds to the wild-type lactose repressor and which to the mutant (tighter IPTG-binding) repressor?

3–48* Another common method for determining K_d is to use gel electrophoresis to separate the bound and free forms of the ligand. (Note that this method requires that the off rate be slow enough that the complex will not dissociate during the time it takes to carry out the electrophoresis, usually a few hours.) This method was used to show that the protein, SmpB, binds specifically to a special species of tRNA (tmRNA) that is used in bacteria to eliminate the incomplete proteins made from truncated mRNAs. In this experiment tmRNA was labeled and included in the binding reactions at a concentration of 0.1 nM (10^{-10} M). Purified SmpB protein was included at a range of concentrations and the mixture was incubated until the binding reaction was at equilibrium. Free and bound tmRNA were then separated by electrophoresis and made visible by autoradiography (Figure 3–19).

A. Examine the equation for K_d in the previous problem. When the concentrations of bound and free ligand are equal, what is the relationship between the concentration of free protein and the K_d?
B. By visual inspection of Figure 3–19, estimate the K_d. Do you have to worry about the concentration of bound protein in this experiment? Why or why not?
C. The concentration of labeled tmRNA in these experiments was 100 pM. Would the results have been the same if tmRNA were used at 100 nM? At 100 μM?

3–49 If the data in Figure 3–19 are plotted as fraction tmRNA bound versus SmpB concentration, one obtains a symmetrical S-shaped curve as shown in Figure 3–20. This curve is a visual display of a very useful relationship between K_d and concentration, which has broad applicability. The expression for fraction of ligand bound is derived from the equation for K_d by substituting ($[\text{L}]_{\text{TOT}} - [\text{L}]$) for [Pr–L] and rearranging. Because the total concentration of ligand ($[\text{L}]_{\text{TOT}}$) is equal to the free ligand ([L]) plus bound ligand ([Pr–L]),

$$\text{Fraction Bound} = \frac{[\text{L}]}{[\text{L}]_{\text{TOT}}} = \frac{[\text{Pr}]}{[\text{Pr}] + K_d}$$

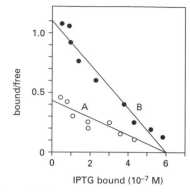

Figure 3–18 Scatchard plot of equilibrium dialysis data for the binding of IPTG to the lactose repressor (Problem 3–47).

Problem 6–65 looks at the mechanism of action of SmpB.

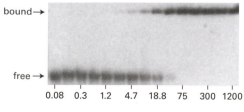

bound →
free →
0.08 0.3 1.2 4.7 18.8 75 300 1200
concentration of SmpB (nM)

Figure 3–19 Assay of the binding of purified SmpB protein to ^{32}P-labeled tmRNA (Problem 3–48). From left to right across the gel, the experiment in each successive lane used a 2-fold increase in concentration of SmpB protein; concentrations in every other lane are indicated.

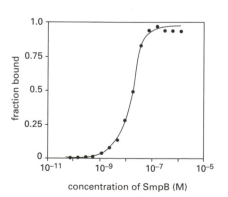

Figure 3–20 Fraction of tmRNA bound versus SmpB concentration (Problem 3–49).

(An equivalent relationship in terms of the fraction of protein bound can be derived in an analogous way; it is $[Pr]/[Pr]_{TOT} = [L]/([L] + K_d)$.)

Using this relationship, calculate the fraction of ligand bound for protein concentrations expressed in terms of K_d, using Table 3–2.

3–50* The lactose repressor regulates expression of a set of genes for lactose metabolism, which are adjacent to its binding site on the bacterial chromosome. In the absence of lactose in the medium, the binding of the repressor turns the genes off. When lactose is added, an inducer is generated that binds to the repressor, preventing it from binding to its DNA target, thereby turning on gene expression.

Inside *E. coli* there are about 10 molecules of lactose repressor (10^{-8} M) and 1 binding site (10^{-9} M) on the bacterial genome. The dissociation constant, K_d, for binding of the repressor to its binding site is 10^{-13} M. When an inducer of gene expression binds to the repressor, the K_d for repressor binding to its DNA binding sites increases to 10^{-10} M. Use the relationships developed in the previous problem to answer the questions below.

A. In a population of bacteria growing in the absence of lactose, what fraction of the binding sites would you expect to be bound by repressor?
B. In bacteria growing in the presence of lactose, what fraction of binding sites would you expect to be bound by the repressor?
C. Given the information in this problem, would you expect the inducer to turn on gene expression? Why or why not?
D. The repressor binds nonspecifically to any sequence of DNA with a K_d of about 10^{-6} M, which represents a very low affinity. Can you suggest in a qualitative way how such low affinity, nonspecific binding would alter the calculations in parts A and B and your conclusion in part C?

3–51 You have raised a specific, high-affinity monoclonal antibody against the enzyme you are working on, and have identified its interaction site as a stretch of six amino acids in the enzyme. Your advisor suggests that you could use the antibody to purify the enzyme by affinity chromatography. This technique would involve attaching the antibody to the inert matrix of a column, passing a crude cell lysate over the column, allowing the antibody to bind your enzyme but not other proteins, and finally eluting your enzyme by washing the column with a solution containing the six amino acid peptide corresponding to the binding site. The principal advantage of affinity chromatography is that it allows a rapid, one-step purification under mild conditions that retain enzyme activity.

In a preliminary experiment you show that if you incubate the antibody with the peptide corresponding to the binding site it will no longer bind to your enzyme, demonstrating that the antibody binds the peptide. Thus encouraged, you bind the antibody to the column and show that it completely removes your enzyme from the crude cell lysate, and does not bind any other proteins. When you try to elute your enzyme with a solution containing a high concentration of the peptide, however, you find that essentially none of your enzyme comes off the column. What could have gone wrong? (Think about what must happen for the enzyme to come off the column.)

3–52* Binding of fragments and competition for binding can be used to identify the portion of a larger ligand that is critical for binding. Fibronectin, which is a large glycoprotein component of the extracellular matrix, binds to fibronectin receptors on cell surfaces. It can stick cells to the surface of a plastic dish, to which they would otherwise not bind, forming the basis of a simple binding assay. By attaching small fragments of fibronectin to dishes, it was possible to identify the cell-binding domain as a 108-amino acid segment about three-quarters of the way from the N-terminus.

Synthetic peptides corresponding to different portions of the 108-amino acid segment were then tested in the cell-binding assay to localize the active region precisely. Two experiments were conducted. In the first, peptides

TABLE 3–2 Fraction of ligand bound versus protein concentration (Problem 3–49).

PROTEIN CONCENTRATION	FRACTION BOUND (%)
$10^4\ K_d$	
$10^3\ K_d$	
$10^2\ K_d$	
$10^1\ K_d$	
K_d	
$10^{-1}\ K_d$	
$10^{-2}\ K_d$	
$10^{-3}\ K_d$	
$10^{-4}\ K_d$	

TABLE 3–3 Fibronectin-related peptides tested for their ability to promote cell sticking (Problem 3–52).

PEPTIDE	SEQUENCE	CONCENTRATION REQUIRED FOR 50% CELL ATTACHMENT
Fibronectin		0.10
Peptide 1	YAVTGRGDSPASSKPISINYRTEIDKPSQM(C) *	0.25
Peptide 2	VTGRGDSPASSKPI(C)	1.6
Peptide 3	SINYRTEIDKPSQM(C)	>100
Peptide 4	VTGRGDSPA(C)	2.5
Peptide 5	SPASSKPIS(C)	>100
Peptide 6	VTGRGD(C)	10
Peptide 7	GRGDS(C)	3.0
Peptide 8	RGDSPA(C)	6.0
Peptide 9	RVDSPA(C)	>100

*The (C) at the C-terminus indicates the cysteine linkage to the carrier protein.

TABLE 3–4 Fibronectin-related peptides tested for their ability to block cell sticking (Problem 3–52).

PEPTIDE	PERCENT OF INPUT CELLS STICKING
GRGDSPC	2.0
GRGDAPC	1.9
GKGDSPC	48
GRADSPC	49
GRGESPC	44
None	47

were attached covalently to protein-coated plastic via a cysteine and tested for their ability to promote cell sticking (Table 3–3). In the second experiment, plastic dishes were coated with native fibronectin, and cells that stuck to the dishes in the presence of the synthetic peptides were counted (Table 3–4).

A. The two experiments use different assays to detect the cell-binding segment of fibronectin. Does the sticking of cells to the dishes mean the same thing in both assays? Explain the difference between the assays.

B. From the results in Tables 3–3 and 3–4, deduce the amino acid sequence in fibronectin that is recognized by the fibronectin receptor.

C. How might you make use of these results to design a method for isolating the fibronectin receptor?

3–53 The attachment of bacteriophage T4 to *E. coli* K illustrates the value of multiple weak interactions, which allow relative motion until fixed connections are made (Figure 3–21). During infection, T4 first attaches to the surface of *E. coli* by the tips of its six tail fibers. It then wanders around the surface until it finds an appropriate place for attachment of its baseplate. When the baseplate is securely fastened, the tail sheath contracts, injecting the phage DNA into the bacterium (Figure 3–21). The initial tail-fiber cell-surface interaction is critical for infection: phages that lack tail fibers are totally noninfectious.

Analysis of T4 attachment is greatly simplified by the ease with which resistant bacteria and defective phages can be obtained. Bacterial mutants resistant to T4 infection fall into two classes: one lacks a major outer membrane protein called OmpC (outer membrane protein C); the other contains alterations in the long polysaccharide chain normally associated with bacterial lipopolysaccharide (LPS). The infectivity of T4 on wild-type and

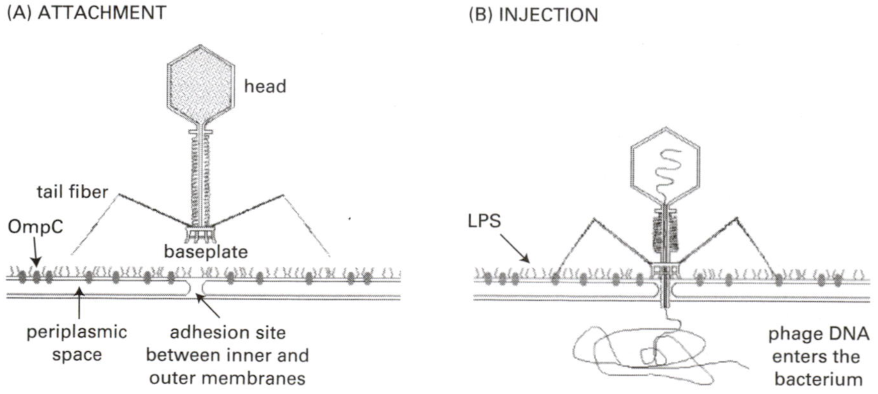

(A) ATTACHMENT

head
tail fiber
OmpC
baseplate
periplasmic space
adhesion site between inner and outer membranes

(B) INJECTION

LPS
phage DNA enters the bacterium

Figure 3–21 Infection by bacteriophage T4 (Problem 3–53). (A) Attachment to bacterial surface. (B) Injection of its DNA.

TABLE 3–5 Infectivity of phage T4 on various bacterial mutants (Problem 3–53).

BACTERIAL STRAIN	PHAGE T4 INFECTIVITY RELATIVE TO NONMUTANT BACTERIA
$ompC^+$ LPS$^+$	1
$ompC^-$ LPS$^+$	10^{-3}
$ompC^+$ LPS$^-$	10^{-3}
$ompC^-$ LPS$^-$	10^{-7}

mutant cells is indicated in Table 3–5. These results suggest that each T4 tail fiber has two binding sites: one for LPS and one for OmpC. Electron micrographs showing the interaction between isolated tail fibers and LPS suggest that individual associations are not very strong, since only about 50% of the fibers are seen bound to LPS.

A. Assume that at any instant each of the six tail fibers has a 0.5 probability of being bound to LPS and the same probability of being bound to OmpC. With this assumption, the fraction of the phage population on the bacterial surface that will have none of its six tail fibers attached in a given instant is $(0.5)^{12}$ (which is the probability of a given binding site being unbound, 0.5, raised to the number of binding sites, two on each of six tail fibers). In light of these considerations, what fraction of the phage population will be attached at any one instant by at least one tail fiber? (The attached fraction is equal to one minus the unattached fraction.) Suppose that the bacteria were missing OmpC. What fraction of the phage population would now be attached by at least one tail fiber at any one instant?

B. Surprisingly, the above comparison of wild-type and $ompC^-$ bacteria suggests only a very small difference in the attached fraction of the phage population at any one instant. As shown in Table 3–5, however, phage infectivities on these two strains differ by a factor of 1000. Can you suggest an explanation that might resolve this apparent paradox?

3–54* Consider an uncatalyzed reaction A $\rightleftharpoons$ B. The rate constants for the forward and reverse reactions are $k_f = 10^{-4}$/sec and $k_r = 10^{-7}$/sec. Thus, the rates or velocities (v) of the forward and reverse reactions are

$$v_f = k_f \, [A] \qquad \text{and} \qquad v_r = k_r \, [B]$$

The overall reaction rate is

$$v = v_f - v_r = k_f \, [A] - k_r \, [B]$$

A. What is the overall reaction rate at equilibrium?
B. What is the value of the equilibrium constant, K?
C. You now add an enzyme that increases k_f by a factor of 10^9. What will the value of the equilibrium constant be with the enzyme present? What will the value of k_r be?

3–55 Examine Figure 3–22, which compares the energetics of a catalyzed and uncatalyzed reaction during the progress of the reaction from substrate (S) to product (P). The highest peak in such a diagram corresponds to the transition state, which is an unstable, high-energy arrangement of substrate that is intermediate between substrate and product. The free energy required to surmount this barrier to the reaction is termed the activation energy. Enzymes function by lowering the activation energy, thereby allowing a more rapid approach to equilibrium.

With this diagram in mind, consider the following question. Suppose the enzyme in the diagram was mutated in such a way that its affinity for the substrate increased by a factor of 100. Assume there was no other effect beyond increasing the depth of the trough labeled ES (enzyme–substrate complex) in Figure 3–22. Would you expect the rate of the reaction catalyzed

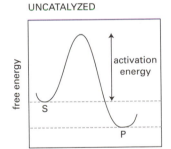

UNCATALYZED

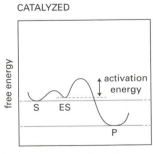

CATALYZED

Figure 3–22 Catalyzed and uncatalyzed reactions showing the free energy at various stages in the progress of the reaction (Problem 3–55).

by the altered enzyme to be higher than, lower than, or equal to the reaction rate catalyzed by the normal enzyme?

3–56* Many enzymes obey simple Michaelis–Menten kinetics, which are summarized by the equation:

$$\text{rate} = \frac{V_{max}\,[S]}{[S] + K_M}$$

where V_{max} = maximum velocity, [S] = concentration of substrate, and K_M = the Michaelis constant.

It is instructive to plug a few values of [S] into the equation to see how rate is affected. What are the rates for [S] equal to zero, equal to K_M, and equal to infinite concentration?

3–57 The rates of production of product, P, from substrate, S, catalyzed by enzyme, E, were measured under conditions where very little product was formed. The results are summarized in Table 3–6.

A. Why is it important to measure rates of product formation under conditions where very little product is formed?

B. Plot these data as rate versus substrate concentration. Is this plot a rectangular hyperbola as expected for an enzyme that obeys Michaelis–Menten kinetics? What would you estimate as the K_M and V_{max} values for this enzyme?

C. To obtain more accurate values for the kinetic constants, the Lineweaver–Burke transformation of the Michaelis–Menten equation is often used so that the data can be plotted as a straight line.

Michaelis–Menten equation:

$$\text{rate} = \frac{V_{max}\,[S]}{[S] + K_M}$$

Lineweaver–Burke equation:

$$\frac{1}{\text{rate}} = \left(\frac{K_M}{V_{max}}\right)\frac{1}{[S]} + \frac{1}{V_{max}}$$

This equation has the form of a straight line, $y = ax + b$. Thus, when $1/\text{rate}$ (y) is plotted versus $1/[S]$ (x), the slope of the line equals K_M/V_{max} (a) and the y intercept is $1/V_{max}$ (b). Furthermore, it can be shown that the x intercept is equal to $-1/K_M$.

Plot $1/\text{rate}$ versus $1/[S]$ and determine the kinetic parameters K_M and V_{max}. (The values for $1/\text{rate}$ and $1/[S]$ are shown in Table 3–6.)

3–58* Suppose that the enzyme in Problem 3–57 is regulated by phosphorylation such that the K_M for its substrate is increased by a factor of 3, but V_{max} is unaltered. At a concentration of substrate equal to the K_M for the unphosphorylated enzyme, decide whether phosphorylation activates the enzyme or inhibits it? Explain your reasoning.

3–59 Which one of the following properties of an enzyme is responsible for its saturation behavior; that is, a maximum rate insensitive to increasing substrate concentration?

A. The enzyme does not change the overall equilibrium constant for a reaction.

B. The enzyme lowers the activation energy of a chemical reaction.

C. The enzyme is a catalyst that is not consumed by the reaction.

D. The enzyme has a fixed number of active sites where substrate binds.

E. The product of the enzyme reaction usually inhibits the enzyme.

3–60* For an enzyme that follows Michaelis–Menten kinetics, by what factor does the substrate concentration have to increase to change the rate of the reaction from 20% to 80% V_{max}?

TABLE 3–6 Initial rates of product formation at various substrate concentrations (Problem 3–57).

RATE (µmol/min)	[S] (µM)
0.15	0.08
0.21	0.12
0.7	0.54
1.1	1.23
1.3	1.82
1.5	2.72
1.7	4.94
1.8	10.00

1/RATE (min/µmol)	1/[S] (1/µM)
6.7	12.5
4.8	8.3
1.4	1.9
0.91	0.81
0.77	0.55
0.67	0.37
0.59	0.20
0.56	0.10

A. A factor of 2
B. A factor of 4
C. A factor of 8
D. A factor of 16
E. The factor required cannot be calculated without knowing K_M.

3–61   The Michaelis constant K_M is often spoken of as if it were a measure of the affinity of the enzyme for the substrate: the lower the K_M, the higher the binding affinity. This would be true if K_M was the same as K_d, but it is not. For an enzyme-catalyzed reaction

$$E + S \overset{k_1}{\underset{k_2}{\rightleftharpoons}} ES \overset{k_3}{\rightarrow} E + P$$

$$K_M = \frac{(k_2 + k_3)}{k_1}$$

A. In terms of these rate constants, what is K_d for dissociation of the ES complex to E + S?
B. Under what conditions is K_M approximately equal to K_d?
C. Does K_M consistently overestimate or underestimate the binding affinity? Or does it sometimes overestimate and sometimes underestimate the binding affinity?

3–62*   The 'turnover number,' or k_{cat}, for an enzyme is the number of substrate molecules converted into product by an enzyme molecule in a unit time when the enzyme is fully saturated with substrate. It is equal to k_3 in the simple representation of an enzyme-catalyzed reaction shown in Problem 3–61. The maximum rate of a reaction, V_{max}, equals k_3 times the concentration of enzyme. (Remember that the maximum rate occurs when all of the enzyme is present as the ES complex.) Carbonic anhydrase catalyzes the hydration of CO_2 to form H_2CO_3. Operating at its maximum rate, 10 µg of pure carbonic anhydrase (M_r 30,000) in 1 mL hydrates 0.90 g of CO_2 in 1 minute. What is the turnover number for carbonic anhydrase?

3–63   (**True/False**) Higher concentrations of enzyme give rise to a higher turnover number. Explain your answer.

3–64*   You are trying to determine whether it is better to purify an enzyme from its natural source or to express the gene in bacteria and then purify it. You purify the enzyme in the same way from both sources and show that each preparation gives a single band by denaturing gel electrophoresis, a common measure of purity. When you compare the kinetic parameters, however, you find that both enzymes have the same K_M but the enzyme from bacteria has a 10-fold lower V_{max}. Propose possible explanations for this result.

3–65   The enzyme hexokinase adds a phosphate to D-glucose, but ignores its mirror image, L-glucose. Suppose that you were able to synthesize hexokinase entirely from mirror-image D-amino acids, instead of from the normal L-amino acids.
A. Assuming that the 'D' enzyme would fold to a stable conformation, what relationship would you expect it to bear to the normal 'L' enzyme?
B. Do you suppose the 'D' enzyme would add a phosphate to L-glucose, and ignore D-glucose?

3–66*   In 1948 Linus Pauling proposed what is now considered to be a key aspect of enzyme function.

"I believe that an enzyme has a structure closely similar to that found for antibodies, but with one important difference, namely, that the surface configuration of the enzyme in not so closely complementary to its specific substrate as is that on an antibody, but is instead complementary to an

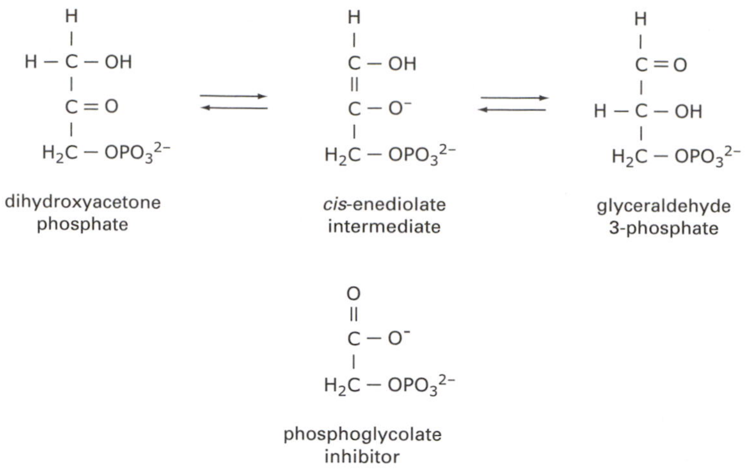

Figure 3–23 The reaction catalyzed by triosephosphate isomerase, and the enzyme inhibitor, phosphoglycolate (Problem 3–66).

unstable molecule with only transient existence—namely, the 'activated complex' [transition state, in modern parlance] for the reaction that is catalyzed by the enzyme. The mode of action of an enzyme would then be the following: the enzyme would show a small power of attraction for the substrate molecule or molecules, which would become attached to it in its active surface region. This substrate molecule, or these molecules, would then be strained by the forces of attraction to the enzyme, which would tend to deform it into the configuration for the activated complex, for which the power of attraction by the enzyme is the greatest....The assumption made above that the enzyme has a configuration complementary to the activated complex, and accordingly has the strongest power of attraction for the activated complex, means that the activation energy for the reaction is less in the presence of the enzyme than in its absence, and accordingly that the reaction would be speeded up by the enzyme."

The enzyme triosephosphate isomerase catalyzes the interconversion of glyceraldehyde 3-phosphate and dihydroxyacetone phosphate through a *cis*-enediolate intermediate (Figure 3–23). Phosphoglycolate (Figure 3–23) is a competitive inhibitor of triosephosphate isomerase with a K_d of 7 µM. The normal substrates for the enzyme have K_ds of about 100 µM. Do you think that phosphoglycolate is a transition-state analog? Why or why not?

3–67   It is sometimes difficult to decide whether a molecule is a transition-state analog, especially in reactions with multiple substrates. *N*-(phosphonacetyl)-L-aspartate (PALA) is a very effective inhibitor of the reaction catalyzed by aspartate transcarbamoylase (Figure 3–24). PALA resembles the intermediate in the reaction and it binds with a K_d of 27 nM. The K_d for carbamoyl phosphate is 27 µM and that for aspartate is 11 mM. Although PALA binds more tightly than either substrate, it would be expected to because it combines elements of both substrates into the same molecule and therefore benefits from all the binding contacts made by either substrate separately. One quick way to evaluate the binding of the substrates relative to the inhibitor is by comparing the numerical value for the product of the K_ds for the substrates with that for the K_d for the inhibitor (all K_ds must be in the same units). If the K_d for the inhibitor is less than the product of the K_ds for the two substrates, the inhibitor is likely to be a transition-state analog. By this criterion is PALA a transition-state analog or a 'bisubstrate' analog?

3–68*   Aspartate transcarbamoylase catalyzes the second step in the synthesis of the pyrimidine nucleotides, UMP and CMP. When PALA is added to the growth medium at 2.7 µM (100-fold more than its K_d for binding to the enzyme), cultured cells are very effectively killed, as might be expected from its ability to inhibit this critical enzyme. At a low frequency, however, cells arise that grow in the presence of PALA. Analysis of these resistant cells shows that they make an aspartate transcarbamoylase identical to the one in normal cells, but that they synthesize 100-times more of it.

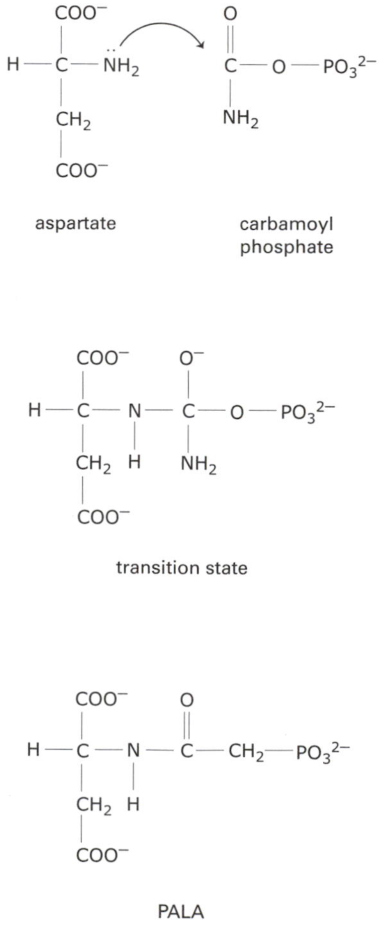

aspartate   carbamoyl phosphate

transition state

PALA

Figure 3–24 Reaction catalyzed by aspartate transcarbamoylase and the inhibitor, PALA (Problem 3–67).

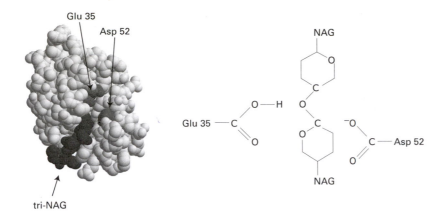

Figure 3–25 Forms of Glu 35 and Asp 52 required for polysaccharide cleavage by lysozyme (Problem 3–69). In the space-filling model the positions of Glu 35 and Asp 52 are shown relative to the trisaccharide, tri-NAG, which is not quite long enough to be cleaved. In the schematic diagram the positions of Glu 35 and Asp 52 are shown relative to the glycosidic bond to be cleaved in a polysaccharide composed of NAG (*N*-acetylglucosamine) residues.

A. How does producing more enzyme allow resistant cells to grow in the presence of PALA?

B. Many enzymes become resistant to the effects of an inhibitor by mutational changes that reduce the binding of the inhibitor, and therefore reduce inhibition. Why do you suppose cells with this type of resistance to PALA were never found?

3–69   The mechanism for lysozyme cleavage of its polysaccharide substrate requires Glu 35 in its nonionized form, whereas the nearby Asp 52 must be ionized (Figure 3–25). The pKs for the side chain carboxyl groups on the two amino acids in solution are virtually identical.

A. How can one carboxyl group be charged and the other uncharged in the active site of lysozyme?

B. The pH optimum for lysozyme is about 5? Why do you suppose that the activity decreases above and below this optimum?

3–70*   Lysozyme achieves its antibacterial effect by cleaving the polysaccharide chains that form the bacterial cell wall. In the absence of this rigid mechanical support, the bacterial cell literally explodes due to its high internal osmotic pressure. The cell wall polysaccharide is made up of alternating sugars, *N*-acetylglucosamine (NAG) and *N*-acetylmuramate (NAM), linked together by glycosidic bonds (Figure 3–26). Lysozyme normally cleaves after

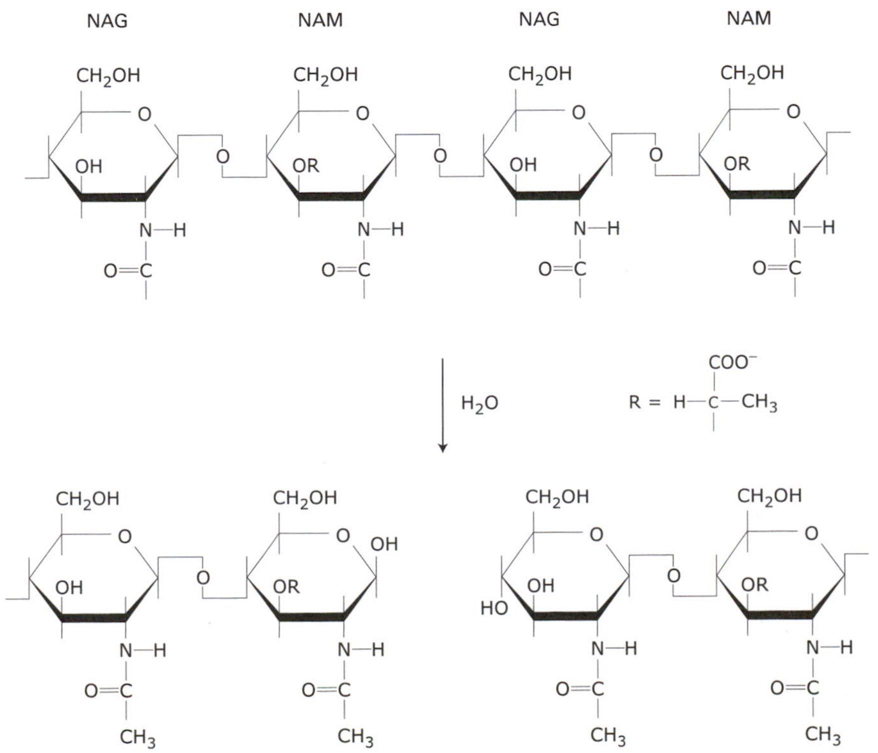

Figure 3–26 Arrangement of NAM and NAG in the bacterial cell-wall polysaccharide (Problem 3–70).

NAM units in the chain (that is, between NAM and NAG), but will also cleave artificial substrates composed entirely of NAG units. When the crystal structure of lysozyme bound to tri-NAG was solved, it was deduced that the binding cleft comprised six binding sites, A through F, for sugars and that tri-NAG filled the first three sites. From the crystal structure it was not apparent, however, which of the five bonds between the six sugars was the one that was normally cleaved. Tri-NAG is not cleaved by lysozyme, although longer NAG polymers are. It was clear from modeling studies that NAM is too large to fit into site C. Where are the catalytic groups responsible for cleavage located relative to the six sugar binding sites?

A. Between sites A and B

B. Between sites B and C

C. Between sites C and D

D. Between sites D and E

E. Between sites E and F

3–71 Egg whites, a rich source of nutrients, can be left out at room temperature for days or weeks and nothing grows in them. Egg white provides three main defenses against microorganisms. One is lysozyme, which cleaves bacterial cell walls. A second is a protein called avidin, which binds the essential vitamin biotin with extremely high affinity, making it unavailable to microorganisms. A clue to the third is the observation that washing eggs in water from rusty pipes can cause them to go bad. What necessity for the life of microorganisms might be provided by water from rusty pipes, and how might the egg 'defend' against it?

3–72* How is it that hemoglobin, which is designed to carry oxygen, can bind it efficiently in the lungs, yet release it efficiently in the tissues?

3–73 Assume that two enzymes catalyze successive steps in a metabolic pathway (that is, the product of one enzyme is the substrate for the next) and that the rate of reaction of each enzyme is limited by the rate of diffusion of its substrate. The rate of diffusion is a physical property of the medium and cannot be altered by any changes to the enzymes. Why is it then that linking the two enzymes into a complex results in an increase in the metabolic flux through the linked enzymes relative to the unassociated enzymes?

3–74* If you were in charge of enzyme design for a cell, for what circumstances might you design an enzyme that had a K_M much, much lower than the prevailing substrate concentration ($[S] >> K_M$)? A K_M around the prevailing substrate concentration ($[S] \cong K_M$)? A K_M much, much higher than the prevailing substrate concentration ($[S] << K_M$)?

3–75 Rous sarcoma virus (RSV) carries an oncogene called *v-src*, which encodes a continuously active protein tyrosine kinase that leads to unchecked cell proliferation. Normally, v-Src carries an attached fatty acid (myristoylate) group that allows it to bind to the cytoplasmic side of the plasma membrane. A mutant version of v-Src that does not allow attachment of myristoylate does not bind to the membrane. Infection of cells with RSV encoding either the normal or the mutant form of v-Src leads to the same high level of protein tyrosine kinase activity, but the mutant v-Src does not cause cell proliferation.

A. Assuming that the normal v-Src is all bound to the plasma membrane and that the mutant v-Src is distributed throughout the cytoplasm, calculate their relative concentrations in the neighborhood of the plasma membrane. For the purposes of this calculation, assume that the cell is a sphere with a radius of 10 µm and that the mutant v-Src is distributed throughout, whereas the normal v-Src is confined to a 4-nm-thick layer immediately beneath the membrane. (For this problem, assume the membrane has no thickness. The volume of a sphere is $4/3\ \pi r^3$.)

B. The target (X) for phosphorylation by *v-src* resides in the membrane. Explain why the mutant v-Src does not cause cell proliferation.

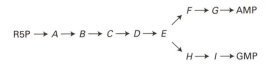

Figure 3–27 Schematic diagram of the metabolic pathway for synthesis of AMP and GMP from R5P (Problem 3–77).

3–76* Which of the following does NOT describe a mechanism cells use to regulate enzyme activities?
A. Cells control enzyme activity by phosphorylation and dephosphorylation.
B. Cells control enzyme activity by the binding of small molecules.
C. Cells control the rates of diffusion of substrates to enzymes
D. Cells control the rates of enzyme degradation.
E. Cells control the rates of enzyme synthesis.
F. Cells control the targeting of enzymes to specific organelles.

3–77 Synthesis of the purine nucleotides, AMP and GMP, proceeds by a branched pathway starting with ribose 5-phosphate (R5P), as shown schematically in Figure 3–27. Using the principles of feedback inhibition, propose a regulatory strategy for this pathway that ensures an adequate supply of both AMP and GMP and minimizes the build-up of the intermediates (*A–I*) when supplies of AMP and GMP are adequate.

3–78* Pathways devoted to the synthesis of specific bioproducts such as purines are commonly regulated via feedback inhibition by the final product. By contrast, the flow of metabolites through the web of pathways devoted to overall energy metabolism—production and utilization of ATP as well as the build-up and breakdown of internal fuel reserves—is regulated by metabolites whose concentrations reflect the energy status of the cell. ATP-like signal metabolites (ATP, NADH, etc.) tend to accumulate when the cell is slowly consuming ATP to meet its energy needs; AMP-like signal metabolites (ADP, AMP, P_i, NAD^+, etc.) tend to accumulate when the cell is rapidly using ATP.

Consider the pathways for the synthesis and breakdown of glycogen, the main fuel reserve in muscle cells (Figure 3–28). The synthetic pathway is controlled by glycogen synthase, whereas the breakdown pathway is controlled by glycogen phosphorylase. In resting muscle, which type of signal metabolite would be expected to accumulate and how would they be expected to affect the activity of the two regulatory enzymes of glycogen metabolism? What about in exercising muscle?

3–79 (**True/False**) Enzymes that undergo cooperative allosteric transitions invariably contain multiple identical subunits. Explain your answer.

3–80* The enzyme glycogen phosphorylase uses phosphate as a substrate to split off glucose 1-phosphate from glycogen, which is a polymer of glucose. Glycogen phosphorylase in the absence of any ligands is a dimer that exists in two conformations: a predominant one with low enzymatic activity and a more rare one with high activity. Both phosphate, a substrate that binds to the active site, and AMP, an activator that binds to an allosteric site, alter the conformational equilibrium by binding preferentially to one conformation. To which conformation of the enzyme would you expect phosphate to bind?

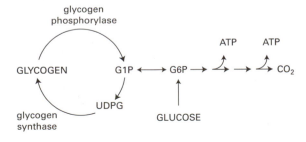

Figure 3–28 Pathways for glycogen synthesis and its breakdown to glucose 6-phosphate, which is an intermediate along the pathway for glucose metabolism to CO_2 (Problem 3–78). G6P stands for glucose 6-phosphate, G1P for glucose 1-phosphate, and UDPG for uridine diphosphoglucose.

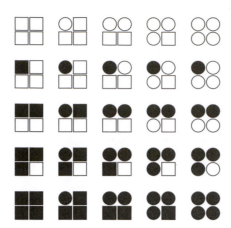

Figure 3–29 A subset of all the possible arrangements of a tetramer, composed of subunits with either of two conformations, and different numbers of bound ligands (Problem 3–81). *Circles* and *squares* represent the two conformations of the subunits; *black* indicates a subunit with a bound ligand.

To which conformation would you expect AMP to bind? How does the binding of either molecule alter the activity of the enzyme?

3–81   Monod, Wyman, and Changeux (MWC) originally explained the kinetic behavior of allosteric enzymes using four postulates, as summarized below.
1. All subunits are identical and are arranged symmetrically in the protein.
2. Each subunit carries a binding site for each ligand.
3. The protein can exist in at least two conformations that conserve the symmetry of the protein. The different conformations may have very different affinities for the ligands. (Ligands may be asymmetrically distributed among the subunits.)
4. The binding affinity of a ligand depends only on the conformational state of the enzyme and not on the occupancy of neighboring sites.

A subset of all the possible arrangements of subunits in a cooperative enzyme composed of four identical subunits, each with two conformations, and with different numbers of bound ligands is shown in Figure 3–29. Assuming that the ligand binds much more tightly to one conformation of subunit (circle), decide which of the tetrameric species are consistent with the MWC postulates. Black subunits in this diagram indicate those with a bound ligand. What would your answer be if the ligand bound equally well to the two conformations of subunit?

3–82*   Aspartate transcarbamoylase (ATCase) is an allosteric enzyme with six catalytic and six regulatory subunits. It exists in two conformations: one with low enzymatic activity and the other with high activity. In the absence of any ligands the low-activity conformation predominates. Malate is an inhibitor of ATCase that binds in the active site at the position where the substrate aspartate normally binds. If the activity of ATCase is measured at low aspartate concentrations, a very peculiar effect of malate is observed: at very low malate concentrations there is an *increase* in ATCase activity (Figure 3–30).
A. How is it that malate, a bona fide inhibitor, can increase ATCase activity under these conditions?
B. Would you expect malate to have the same peculiar effect if the measurements were made in the presence of a high concentration of aspartate? Why or why not?

3–83   Using a chemically modified version of ATCase, it is possible to monitor binding at active sites due to local changes that occur upon ligand binding. It is also possible to measure global changes in ATCase conformation by effects on the protein's sedimentation rate in a centrifugal field. You want to know how binding at the active sites of ATCase relates to global conformational changes. You incubate the modified ATCase with increasing concentrations of succinate (a substrate analog that can bind to the active site in each subunit). You measure succinate binding spectrally and global

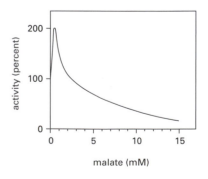

Figure 3–30 The activity of ATCase with increasing concentrations of the inhibitor malate (Problem 3–82). These measurements were made at an aspartate concentration well below its K_M.

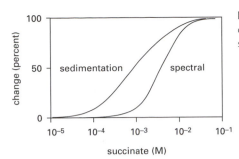

change (percent)

Figure 3–31 Changes in binding and global conformation of ATCase with increasing succinate (Problem 3–83).

succinate (M)

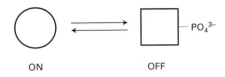

ON   OFF

Figure 3–32 Phosphorylated and nonphosphorylated states of a metabolic enzyme (Problem 3–84).

conformational changes by sedimentation. Your results are shown in Figure 3–31. Are these results what you might expect for an allosteric protein like ATCase? Why or why not?

3–84* Many proteins inside cells are regulated by phosphorylation and dephosphorylation. Imagine a metabolic reaction for which the enzyme is completely active when not phosphorylated and completely inactive when phosphorylated (Figure 3–32). Inside the cell the rate of this metabolic reaction can vary continuously between very fast and very slow (or zero) at a given substrate concentration. How is it that the reaction in the cell can proceed at any rate between very fast and zero, whereas an individual enzyme molecule is either on or off?

3–85 **(True/False)** Continual addition and removal of phosphates by protein kinases and protein phosphatases is wasteful of energy—since their combined action consumes one molecule of ATP—but it is a necessary consequence of effective regulation by phosphorylation. Explain your answer.

3–86* Cyclin-dependent protein kinase 2 (Cdk2) regulates critical events in the progression of the cell cycle in mammalian cells. Cdk2 can form a complex with cyclin A and can be phosphorylated by another protein kinase, Civ1, to produce P-Cdk2. To determine the roles of cyclin A and phosphorylation in the function of Cdk2, you purify nonphosphorylated and Civ1-phosphorylated Cdk2. You mix these two forms of Cdk2 and cyclin A in various combinations with ^{32}P-ATP and assay for phosphorylation of histone H1 (Figure 3–33). You also measure the binding affinity of various forms of Cdk2 for ATP, ADP, cyclin A, and histone H1 (Table 3–7).

A. From Figure 3–33, what is required for Cdk2 to phosphorylate histone H1 efficiently?

B. How do the requirements identified in part A specifically affect the function of Cdk2 relative to its target, histone H1 (Table 3–7 and Figure 3–33)?

C. The usual intracellular concentrations of ATP and ADP are in the range of 0.1 to 1 mM. Assume that the binding of cyclin A to Cdk2 or P-Cdk2 does not alter the affinities of either form of Cdk2 for ATP and ADP. Is it likely that the observed changes in affinity for ATP and ADP are important for Cdk2 function? Why or why not?

TABLE 3–7 The observed dissociation constants (K_ds) of Cdk2 for ATP, ADP, Cyclin A, and Histone H1 (Problem 3–86).

	K_d (µM)			
	ATP	ADP	CYCLIN A	HISTONE H1
Cdk2	0.25	1.4	0.05	not detected
P-Cdk2	0.12	6.7	0.05	100
Cdk2 + cyclin A				1.0
P-Cdk2 + cyclin A				0.7

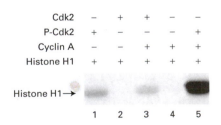

Cdk2	–	+	+	–	–
P-Cdk2	+	–	–	–	+
Cyclin A	–	–	+	+	+
Histone H1	+	+	+	+	+

Histone H1→

1   2   3   4   5

Figure 3–33 Phosphorylation of histone H1 by various combinations of Cdk2 and cyclin A (Problem 3–86). The amount of radioactive phosphate attached to histone H1 in lanes 1 and 3 is 0.3% and 0.2%, respectively, of that in lane 5.

TABLE 3–8 Kinetic parameters of purified PTP1B mutants (Problem 3–87).

ENZYME	V_{max} (nmol/min/mg)	K_M (nm)	k_{cat} (min⁻¹)
Wild type	60,200	102	2244
Tyr-46 → Leu	4,160	1700	155
Glu-115 → Ala	5,700	45	212
Lys-120 → Ala	19,000	80	708
Asp-181 → Ala	0.61	126	0.023
His-214 → Ala	700	20	26
Cys-215 → Ser	no activity		
Arg-221 → Lys	11	80	0.41
Arg-221 → Met	3.3	1060	0.12

3–87 Genome sequencing has revealed a surprisingly large number of protein tyrosine phosphatases (PTPs), very few of which have known roles in the life of a cell. PTP1B was the very first member of the PTP family to be discovered, and its three-dimensional structure and catalytic mechanism are well defined. PTP1B will dephosphorylate almost any phosphotyrosine-containing protein or peptide in the test tube, but in the cell it is likely to have better-defined targets. One strategy to identify a binding target is to incubate the known protein with a cell lysate, and then isolate the known protein and identify any proteins that are bound to it. An enzyme such as PTP1B, however, binds and releases its target as part of its catalytic cycle. You reason that if you interfere with catalysis you might be able to increase the dwell time of the substrate, making it stable enough for isolation. To this end, you make several mutants of PTP1B by changing specific amino acids in the active site, and measure their kinetic parameters (Table 3–8).

You pick mutant Cys-215 → Ser (C215S) for study because the sulfhydryl of Cys-215 initiates a nucleophilic attack on the phosphotyrosine, releasing the phosphate. You pick a second mutant—mutant 2—because it has promising kinetic properties. You use genetic engineering tricks to fuse these two mutants and the wild-type enzyme to glutathione S-transferase (GST), so that each of the three proteins can be rapidly purified by precipitation with glutathione-Sepharose. You express the GST-tagged proteins, precipitate them from cell lysates, separate the precipitated proteins by gel electrophoresis, and identify phosphotyrosine-containing proteins using anti-phosphotyrosine antibodies. As shown in Figure 3–34, GST-mutant 2 bound two phosphotyrosine-containing proteins, whereas GST-C215S bound none.

A. Which mutant PTP1B in Table 3–8 is likely to correspond to GST-mutant 2? Why do you think this protein gave a successful result?

B. Why might GST-C215S have failed to precipitate any phosphotyrosine-containing proteins?

3–88* The Ras protein is a GTPase that functions in many growth-factor signaling pathways. In its active form, with GTP bound, it transmits a downstream signal that leads to cell proliferation; in its inactive form, with GDP bound, the signal is not transmitted. Mutations in the gene for Ras are found in many cancers. Of the choices below, which alteration of Ras activity is most likely to contribute to the uncontrolled growth of cancer cells?

A. A mutation that prevents Ras from being made.

B. A mutation that increases the affinity of Ras for GDP.

C. A mutation that decreases the affinity of Ras for GTP.

D. A mutation that decreases the affinity of Ras for its downstream targets.

E. A mutation that decreases the rate of hydrolysis of GTP by Ras.

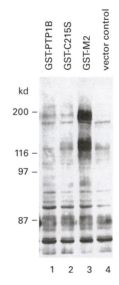

Figure 3–34 Phosphotyrosine-containing proteins precipitated with GST-tagged PTP1B enzymes (Problem 3–87). GST-PTP1B is the fused wild-type protein; GST-M2 is the fused mutant-2 protein. The positions of 'marker' proteins of known molecular mass are indicated on the *left*.

3–89   The activity of Ras is carefully regulated by two other proteins, a guanine nucleotide-exchange factor (GEF) that stimulates uptake of GTP by Ras, and a GTPase-activating protein (GAP) that stimulates hydrolysis of GTP by Ras. The activities of these regulatory proteins are in turn also regulated. Which of the following changes in GAP and GEF proteins might cause a cell to proliferate excessively?

A. A nonfunctional GAP

B. A permanently active GAP

C. A nonfunctional GEF

D. A permanently active GEF

3–90*   (**True/False**) Conformational changes in proteins never exceed a few tenths of a nanometer. Explain your answer.

3–91   Motor proteins generally require ATP (or GTP) hydrolysis to ensure unidirectional movement.

A. In the absence of ATP would you expect a motor protein to stop moving, to wander back and forth, to move in reverse, or to continue moving forward but more slowly?

B. Assume that the concentrations of ATP, ADP, and phosphate were adjusted so that the free-energy change for ATP hydrolysis by the motor protein was equal to zero (instead of very negative, as it is normally). Under these conditions would you expect a motor protein to stop moving, to wander back and forth, to move in reverse, or to continue moving forward but more slowly?

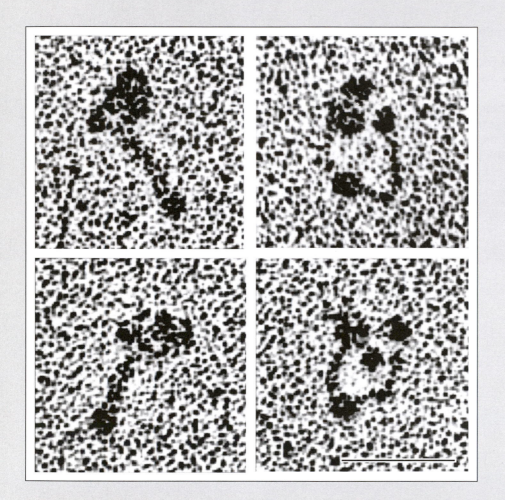

Condensins and cohesins. These are the molecules responsible for condensing eucaryotic chromosomes preparatory to mitosis and for holding sister chromatids together until the metaphase to anaphase transition. These previously unpublished micrographs of rotary-shadowed human condensin (left pair) and cohesin (right pair) were taken by David Anderson and Harold Erickson (Duke University), Ana Losada and Tatsuya Hirano (Cold Spring Harbor Laboratories). Scale bar is 50 nm,

DNA AND CHROMOSOMES

THE STRUCTURE AND FUNCTION OF DNA

TERMS TO LEARN

antiparallel	deoxyribonucleic acid (DNA)	genome
base pair	double helix	nuclear lamina
complementary	gene	template

4–1 The start of the coding region for the human β-globin gene reads 5′-ATGGT-GCAC-3′. What is the sequence of the complementary strand for this segment of DNA?

4–2* Upon returning from a recent trip abroad, you explain to the customs agent that you are bringing in a sample of DNA, deoxyribonucleic acid. He is aghast that you want to bring an acid into his country. What is the acid in DNA? Should the customs agent be wary?

4–3 Examine the space-filling models of the base pairs shown in Figure 4–1. Each base pair includes two bases, two deoxyribose sugars, and two phosphates. Can you identify the locations of the purine base, the pyrimidine base, the sugars, and the phosphates? Identify each base pair as CG, GC, TA, or AT.

*Answers to these problems are available to Instructors at www.classwire.com/garlandscience/

(A) 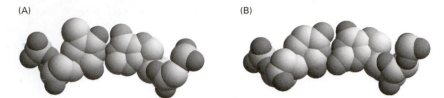 (B)

Figure 4–1 Space-filling models of two base pairs (Problem 4–3). Carbon and phosphorus atoms are *light gray*, nitrogen atoms are *intermediate*, and oxygen atoms are *dark gray*. No hydrogen atoms are shown.

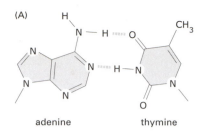

adenine   thymine

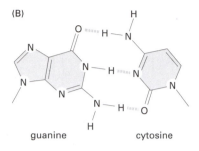

guanine   cytosine

Figure 4–2 An AT and a GC base pair, as viewed along the helix axis (Problem 4–4). Each base is attached to its deoxyribose sugar via the line extending from the ring nitrogen.

4–4* The chemical structures for an AT and a GC base pair are shown in Figure 4–2, along with their points of attachment to the sugar-phosphate backbones.

    A. Indicate the positions of the major and minor grooves of the DNA helix on these representations.

    B. Draw the structure of a TA base pair in the same way as in Figure 4–2.

    C. Do the same chemical moieties (for example, the methyl group of thymine) always project into the same groove?

    D. With a sufficiently powerful microscope, do you think it would be possible to read directly the sequence of DNA? Why or why not?

4–5 Human DNA contains 20% C on a molar basis. What are the mole percents of A, G, and T?

4–6* DNA isolated from the bacterial virus M13 contains 25% A, 33% T, 22% C, and 20% G. Do these results strike you as peculiar? Why or why not? How might you explain these values?

4–7 A segment of DNA from the interior of a single strand is shown in Figure 4–3. What is the polarity of this DNA from top to bottom?

4–8* Bacteriophage T4 attaches to its bacterial host and injects its DNA to initiate an infection that ultimately releases hundreds of progeny virus. In 1952 Alfred Hershey and Martha Chase radiolabeled the DNA of bacteriophage T4 with $^{32}PO_4^{3-}$ and the proteins with ^{35}S-methionine. They then mixed the labeled bacteriophage with bacteria and after a brief time agitated the mixture vigorously in a blender to detach T4 from the bacteria. They then separated the phage from the bacteria by centrifugation. They demonstrated that the bacteria contained 30% of the ^{32}P label but virtually none of the ^{35}S label. When new bacteriophage were released from these bacteria, they were also found to contain ^{32}P but no ^{35}S. How does this experiment demonstrate that DNA rather than protein is the genetic material? (Note that bacteriophage T4 contains only protein and DNA.)

4–9 The diploid human genome comprises 6.4×10^9 bp and fits into a nucleus that is 6 μm in diameter.

    A. If base pairs occur at intervals of 0.34 nm along the DNA helix, what is the length of DNA in a human cell?

    B. If the diameter of the DNA helix is 2.4 nm, what fraction of the volume of the nucleus is occupied by DNA? (Volume of a sphere is $4/3\,\pi r^3$ and volume of a cylinder is $\pi r^2 h$.)

4–10* (**True/False**) Human cells do not contain any circular DNA molecules. Explain your answer.

4–11 DNA forms a right-handed helix. Pick out the right-handed helix from those shown in Figure 4–4.

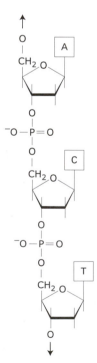

Figure 4–3 Three nucleotides from the interior of a single strand of DNA (Problem 4–7). *Arrows* at the ends of the DNA strand indicate that the structure continues in both directions.

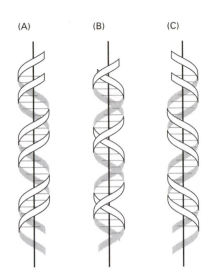

(A)   (B)   (C)

Figure 4–4 Three 'DNA' helices (Problem 4–11).

4–12* One gram of cultured human cells contains about 10^9 cells and occupies roughly 1 mL. If the average molecular mass of a base pair is 660 daltons and each cell contains 6.4×10^9 bp, what mass of DNA is present in this sample? If all the DNA molecules in the sample were laid end to end to form a single thread, would it be long enough to reach from the Earth to the Moon (385,000 kilometers)?

CHROMOSOMAL DNA AND ITS PACKAGING IN THE CHROMATIN FIBER

TERMS TO LEARN

artificial chromosome	gene	kinetochore
cell-division cycle (cell cycle)	histone	linker DNA
centromere	histone octamer	nucleosome
chromatin	homologous chromosome (homolog)	replication origin
chromosome	intron	telomere
conserved synteny	karyotype	transcription unit
exon		

4–13 In the 1950s the techniques for isolating DNA from cells all yielded molecules of about 10,000 to 20,000 base pairs. We now know that the DNA molecules in all cells are very much longer. Why do you suppose such short pieces were originally isolated?

4–14* One way to demonstrate that a chromosome has a single DNA molecule uses a technique called pulsed-field gel electrophoresis, which can separate DNA molecules up to 10^7 bp in length. Ordinary gel electrophoresis cannot separate such long molecules because the steady electric field stretches them out so they travel end-first through the gel matrix at a rate that is independent of their length. If the electric field is changed periodically, however, the DNA molecules are forced to reorient to the new field before continuing their snakelike movement through the gel. The time for reorientation is dependent on length, so that longer molecules move more slowly through the gel.

The results of pulsed-field gel electrophoresis of the DNA from the yeast, *Saccharomyces cerevisiae*, are shown in Figure 4–5. How many chromosomes does *S. cerevisiae* have?

4–15 (**True/False**) Human females have 23 different chromosomes, whereas human males have 24. Explain your answer.

4–16* Consider the following statement: A human cell contains 46 molecules of DNA in its nucleus. Do you agree with it? Why or why not?

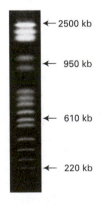

← 2500 kb

← 950 kb

← 610 kb

← 220 kb

Figure 4–5 Pulsed-field gel electrophoresis of *S. cerevisiae* chromosomes (Problem 4–14). To minimize the handling of DNA, which would surely break it, the cells themselves are placed at the top of the gel and gently opened by addition of a lysis buffer. The DNA molecules in this gel have been exposed to the dye ethidium bromide, which fluoresces under ultraviolet light when it is bound to DNA. This treatment allows the DNA—otherwise invisible—to be seen.

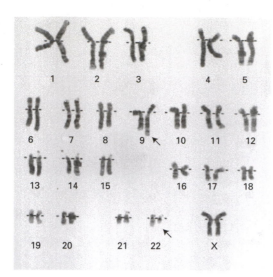

Figure 4–6 Karyotype illustrating the typical alteration seen in chronic myelogenous leukemia (Problem 4–18).

4–17 Human chromosome 1 contains about 2.8×10^8 bp. At mitosis this chromosome measures 10 μm in length. Relative to its fully extended length, how compacted is the DNA molecule in chromosome 1 at mitosis? (Recall that a DNA bp is 0.34 nm in length.)

4–18* An abnormal human karyotype is shown in Figure 4–6. This particular karyotype is found in the cancer cells of more than 90% of patients with chronic myelogenous leukemia. Arrows indicate two abnormal chromosomes. Describe the event that led to this abnormal karyotype. Is this patient male or female?

4–19 The human U2 small nuclear RNA (U2 snRNA), which is present at thousands of copies per nucleus, plays an important role in mRNA processing. You have made a bacteriophage lambda clone that is 43 kb long and carries two copies of the *U2* gene, 6 kb apart. The restriction map of this clone is shown in Figure 4–7. When you cut human genomic DNA to completion with HindIII, HincII (H2), or KpnI (K) and analyze the restriction digest by blot hybridization against the *U2* gene, you detect a single intense band at 6 kb (Figure 4–8, lanes 9 to 11). If you cut with BglII (B), EcoRI (R), or XbaI (X), which do not cut the cloned segment (Figure 4–7), you also detect a single intense band, but of a size greater than 50 kb (Figure 4–8, lanes 1 to 3). If you incubate the genomic DNA with HindIII and remove samples at various times, you see a ladder of bands (lanes 4 to 9). If you cut 2 ng of the cloned DNA with KpnI and run it alongside 10 μg of the genomic KpnI digest, two bands are visible—each of equal intensity to the 6-kb band from the genomic digest (compare lanes 11 and 12).

A. What is the organization of the *U2* genes in the human genome. Explain how the restriction digests define it.

B. Why are two bands visible in the digest of the cloned DNA (lane 12), whereas only one is visible in the digest of genomic DNA (lane 11)?

C. Given that 2 ng of cloned DNA produces a band of equal intensity to that from 10 μg of genomic DNA (lanes 11 and 12), calculate the number of

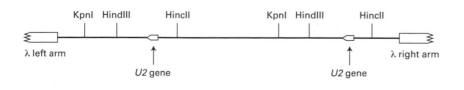

Figure 4–7 Restriction map of a bacteriophage lambda clone carrying two *U2* genes (Problem 4–19). There are no restriction sites for EcoRI, BglII or XbaI in the stretch of DNA shown. The bacteriophage lambda arms are not shown to scale.

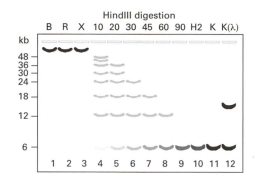

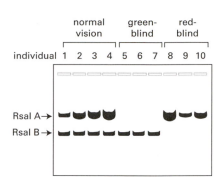

Figure 4–8 Autoradiograph of various restriction digests of human genomic DNA probed with a radiolabeled *U2* gene (Problem 4–19). Numbers under HindIII indicate time of digestion in minutes. B = BglII; R = EcoRI; X = XbaI; H2 = HincII; K = KpnI. K(λ) indicates a KpnI digest of the bacteriophage DNA carrying the two *U2* genes.

U2 genes in the human genome. (The bacteriophage lambda clone is 43 kb, and the haploid human genome is 3.2×10^6 kb.)

D. The sequence of the human genome in this region identifies only 3 genes encoding U2 snRNA. What do you suppose is the basis for the difference in the number of *U2* genes identified by sequencing and by the calculation in part C?

4–20* About 5% of the human genome consists of duplicated segments of chromosomes, many of which are highly homologous, indicating a relatively recent origin. The high degree of homology occasionally allows inappropriate recombination events to occur between the duplications, which can decrease or increase the number of duplicated segments. Such events are responsible for several human diseases, including the red–green color blindness that affects 8% of the male population. The genes for the red and green visual pigments lie near one another on the X chromosome, one in each copy of the duplicated segment. They are 98% identical throughout most of their length, in both exons and introns; however, the genes can be distinguished by a RsaI restriction fragment length polymorphism (RFLP). The RsaI-A gene gives a longer RsaI fragment than the RsaI-B gene (Figure 4–9).

A. To determine which gene encodes which pigment, several normal, red-blind, and green-blind males were screened using a hybridization probe specific for the RsaI RFLP (Figure 4–9). Which gene encodes the red visual pigment, and which encodes the green visual pigment?

B. The intensity of hybridization in normal individuals was constant for the RsaI-B gene, but surprisingly variable for the RsaI-A gene. This anomaly was investigated by digesting the DNA from selected individuals with NotI, which cleaves once within the RsaI-A gene but does not cleave the RsaI-B gene. The restriction fragments were separated by pulsed-field gel electrophoresis and hybridized with a probe that recognizes both genes (Figure 4–10). What is the basis for the variable intensity of hybridization of RsaI-A genes in males with normal color vision? Can your explanation account for the high frequency of color blindness?

C. What is the size of the duplicated chromosomal segment at this site in the human genome?

4–21 The total number of protein-coding genes in the human genome can be calculated in several ways. It is important to remember that all such numbers are estimates at present, because it is still difficult to identify a gene from the DNA sequence. Chromosome 22 has about 700 genes in 48 Mb of sequence, which represents 1.5% of the estimated 3200 Mb in the haploid genome. Using these numbers, how many genes would you estimate for the haploid human genome? If your estimate is significantly larger or smaller than the accepted value of approximately 30,000 genes, suggest possible explanations for the discrepancy.

4–22* On chromosome 22 (48 Mb) there are about 700 genes, which average 19,000 bp in length and contain an average of 5.4 exons, each of which

Figure 4–9 RsaI RFLPs in normal, green-blind, and red-blind males (Problem 4–20). RsaI A refers to the RFLP characteristic of the RsaI-A gene; RsaI B refers to the RFLP characteristic of the RsaI-B gene. Individual males are indicated by number.

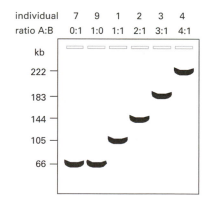

Figure 4–10 NotI digests of DNA from selected normal and color-blind individuals (Problem 4–20). Numbers for individuals correspond to the numbers in Figure 4–9. The ratio of RsaI-A genes to RsaI-B genes (A:B) is estimated from the intensity of hybridization in Figure 4–9. The sizes of NotI fragments are indicated in kb.

Figure 4–11 Transposable elements and genes in 1 Mb regions of chromosomes 2 and 22 (Problem 4–24). Lines that project *upward* indicate exons of known genes. Lines that project *downward* indicate transposable elements; they are so numerous (constituting nearly 50% of the human genome) they merge into nearly a solid block outside the Hox clusters.

averages 266 bp. On average what fraction of a gene is present in mRNA? What fraction of the chromosome do genes occupy?

4–23   (**True/False**) The majority of human DNA is thought to be unimportant junk. Explain your answer.

4–24* Transposable elements of four types—long interspersed elements (LINES), short interspersed elements (SINES), LTR retrotransposons, and DNA transposons—are inserted more or less randomly throughout the human genome. However, at the four homeobox gene clusters, HoxA, HoxB, HoxC, and HoxD, these elements are rare, as illustrated for HoxD in Figure 4–11. Each Hox cluster is about 100 kb in length and contains 9 to 11 genes, whose differential expression along the anteroposterior axis of the developing embryo establishes the basic body plan for humans (and for other animals). Why do you suppose that transposable elements are so rare in the Hox clusters?

4–25   (**True/False**) In a comparison between the DNAs of related organisms such as humans and mice, conserved sequences represent functionally important exons and regulatory regions, and nonconserved sequences generally represent noncoding DNA. Explain your answer.

4–26* The earliest graphical method for comparing nucleotide sequences—the so-called diagon plot—still yields one of the best visual comparisons of sequence relatedness. An example is illustrated in Figure 4–12, where the human β-globin gene is compared to the human cDNA for β globin (Figure 4–12A) and to the mouse β-globin gene (Figure 4–12B). Diagon plots are generated by comparing blocks of sequence, in this case blocks of 11 nucleotides at a time. If 9 or more of the nucleotides match, a dot is placed on the diagram at the coordinates corresponding to the blocks being compared. A comparison of all possible blocks generates diagrams such as the ones shown in Figure 4–12, in which sequence homologies show up as diagonal lines.

A.   From the comparison of the human β-globin gene with the human β-globin cDNA (Figure 4–12A), deduce the positions of exons and introns in the β-globin gene.

Figure 4–12 Diagon plots (Problem 4–26). (A) Human β-globin cDNA compared with the human β-globin gene. The β-globin cDNA is a complementary DNA copy of the β-globin mRNA. (B) Mouse β-globin gene compared with the human β-globin gene. The positions of the exons in the human β-globin gene are indicated by *shading* in (B). The 5′ and 3′ ends of the sequences are indicated. The human gene sequence is identical in the two plots. To accommodate the short β-globin cDNA sequence (549 nucleotides) and the sequence of the β-globin gene (2052 nucleotides) in similar spaces, while maintaining proportional scales within each plot, the scale of (A) is about three times that of (B).

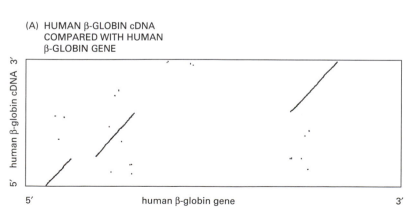

(A) HUMAN β-GLOBIN cDNA COMPARED WITH HUMAN β-GLOBIN GENE

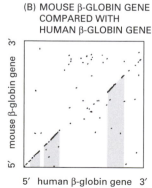

(B) MOUSE β-GLOBIN GENE COMPARED WITH HUMAN β-GLOBIN GENE

B. Are the entire exons of the human β-globin gene (indicated by shading in Figure 4–12B) homologous to the mouse β-globin gene? Identify and explain any discrepancies.

C. Is there any homology between the human and mouse β-globin genes that is outside the exons? If so, identify its location and offer an explanation for its preservation during evolution.

D. Have either of the genes undergone a change of intron length during their evolutionary divergence? How can you tell?

4–27 (**True/False**) Sequences of cellular RNAs (obtained from cDNA libraries) provide the most reliable way to identify genes in human chromosomal DNA. Explain your answer.

4–28* Your advisor, the brilliant bioinformatician, has a high regard for your intellect and industry. She suggests that you write a computer program that will identify the exons of protein-encoding genes directly from the sequence of the human genome. In preparation for that task, you decide to write down a list of the features that might distinguish coding sequences from intronic DNA and sequences outside of genes. What features would you list? (You may wish to review basic aspects of gene expression in MBoC Chapter 6.)

4–29 Why do you expect to encounter a STOP codon about every 20 codons, or so, on average in a random sequence of DNA?

4–30* Chromosome 3 in orangutans differs from chromosome 3 in humans by two inversion events (Figure 4–13). Draw the intermediate chromosome that resulted from the first inversion and explicitly indicate the segments included in both inversions.

4–31 Define a 'gene.'

4–32* List the three specialized DNA sequences and their functions that allow chromosomes to be maintained during the cell cycle.

4–33 A classic experiment linked telomeres from *Tetrahymena* to a linearized yeast plasmid, allowing the plasmid to grow as a linear molecule—that is, as an artificial chromosome (Figure 4–14). A circular 9-kb plasmid was constructed to contain a yeast origin of replication (ARS1) and the yeast *LEU2* gene. Cells that are missing the chromosomal *LEU2* gene, but have taken up the plasmid, can grow without leucine added to the medium. The plasmid was linearized with BglII, which cuts once (Figure 4–14), and then mixed with 1.5-kb *Tetrahymena* telomere fragments generated by cleavage with BamHI (Figure 4–14). The mixture was incubated with DNA ligase and the two restriction nucleases, BglII and BamHI. The ligation products included molecules of 10.5 kb and 12 kb in addition to the original components. The 12-kb band was purified and transformed into yeast, which were then selected for growth in the absence of leucine. Samples of DNA from one transformant were digested with HpaI, PvuII, or PvuI and the fragments

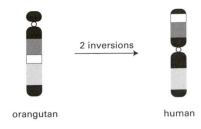

Figure 4–13 Chromosome 3 in orangutans and humans (Problem 4–30). Differently *shaded blocks* indicate segments of the chromosomes that were derived by previous fusions.

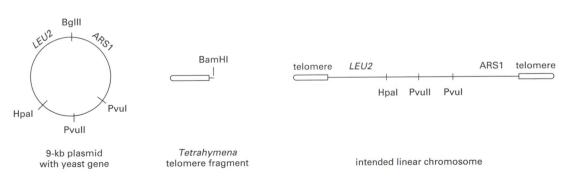

Figure 4–14 Structure of 9-kb plasmid, telomere fragment, and the intended linear chromosome with *Tetrahymena* telomeres (Problem 4–33). The sites of unique cutting by restriction enzymes are indicated.

were separated by gel electrophoresis and visualized after hybridization to a plasmid-specific probe (Figure 4–15).

A. How do the results of the analysis in Figure 4–15 distinguish between a linear and a circular form of the plasmid in the transformed yeast?

B. Explain how ligation of the DNA fragments in the presence of the restriction nucleases BamHI and BglII ensures that you get predominantly the construct you want. The recognition site for BglII is 5′-A*GATCT-3′ and for BamHI is 5′-G*GATCC-3′, where the asterisk (*) is the site of cutting.

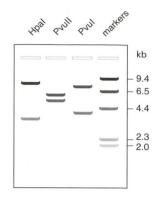

Figure 4–15 Autoradiograph of restriction analysis of plasmid structure (Problem 4–33). Marker DNAs of known sizes are shown on the *right*.

4–34* Describe the consequences that would arise if a eucaryotic chromosome (150 Mb in length) had one of the following features:

A. A single replication origin located in the middle of the chromosome. (DNA replication in animal cells proceeds at about 150 nucleotide pairs per second.)

B. A telomere at only one end of the chromosome.

C. No centromere.

4–35 Early in development, most human cells turn off expression of an essential component of telomerase, the enzyme responsible for addition of telomere repeat sequences (5′-TTAGGG) to the ends of chromosomes. Thus, as our cells proliferate their telomeres get shorter and shorter, but are normally not lost over the course of a lifetime. If cells are removed from the body and grown in culture, however, they ultimately enter a state of replicative senescence and stop dividing when their telomeres get too short. By contrast, most human tumor cells express active telomerase, allowing them to maintain their telomeres and grow beyond the normal limit imposed by senescence—good for them, bad for us.

Anticipating a universal cure for cancer, you set up a company to screen chemical 'libraries' for telomerase inhibitors. The company share price takes a tumble, however, when a rival group generates a strain of telomerase-knockout mice. These mice breed happily for several generations, but by the sixth-generation (when their telomeres are much shorter than normal) they have a greatly increased tendency to die of tumors, compared to their telomerase-plus littermates. The tumors tend to arise in tissues that show high proliferation rates, such as testis, skin, and blood. Why has this observation shaken the confidence of your investors? Is there a flaw in your hypothesis?

4–36* (**True/False**) In the living cell chromatin usually adopts the extended 'beads-on-a-string' form. Explain your answer.

4–37 A single nucleosome is 11 nm long and contains 146 bp of DNA (0.34 nm/bp). What packing ratio (DNA length to nucleosome length) has been achieved by wrapping DNA around the histone octamer? Assuming that there are an additional 54 bp of extended DNA in the linker between nucleosomes, how condensed is 'beads-on-a-string' DNA relative to fully extended DNA? What fraction of the 10,000-fold condensation that occurs at mitosis does this first level of packing represent?

4–38* You are studying chromatin structure in rat liver DNA. When you digest rat liver nuclei briefly with micrococcal nuclease, extract the DNA, and run it on an agarose gel, it forms a ladder of broad bands spaced at about 200-nucleotide intervals. If you use the enzyme DNase I instead, there is a much more continuous smear of DNA on the gels with only the haziest suggestion of a 200-nucleotide repeat. If you denature the DNase I-treated DNA before fractionating it by gel electrophoresis, however, you find a new ladder of bands with a regular spacing of about 10 nucleotides.

You are puzzled by the different results with these two enzymes. When you describe the experiments to the rest of your research group, one colleague suggests that the difference derives from the steric properties of the DNA-binding sites on the two enzymes: micrococcal nuclease can only bind and cleave DNA that is free, whereas DNase I can bind and cut free DNA and DNA that is bound to the surface of a nucleosome. Your colleague predicts that if DNA is bound to any surface and digested with DNase I, it will generate a 10-nucleotide ladder. You test this prediction by binding DNA to polylysine-coated plastic dishes and digesting with the two enzymes. Micrococcal nuclease causes minimal digestion, but DNase I generates a 10-nucleotide ladder, verifying your friend's prediction.

A. Why does brief digestion of nuclei with micrococcal nuclease yield a ladder of bands spaced at intervals of about 200 nucleotides?

B. If you digested nuclei extensively with micrococcal nuclease, what pattern would you expect to see after fractionation of the DNA by gel electrophoresis?

C. Explain how your colleague's suggestion accounts for the generation of a 10-nucleotide ladder when nuclei are digested with DNase I.

4–39 Assuming that the histone octamer forms a cylinder 9 nm in diameter and 5 nm in height and that the human genome forms 32 million nucleosomes, what volume of the nucleus (6 μm in diameter) is occupied by histone octamers? (Volume of a cylinder is $\pi r^2 h$; volume of a sphere is $4/3\,\pi r^3$.) What fraction of the nuclear volume do the DNA and the histone octamers occupy (see Problem 4–9)?

4–40* Nucleosomes can be assembled onto defined DNA segments. When a particular 225-bp segment of human DNA was used to assemble nucleosomes, and then digested with micrococcal nuclease, a quantitative yield of fragments 146-bp in length resulted. Subsequent digestion of these fragments with a restriction enzyme that cuts once within the original 225-bp sequence produced two well-defined bands at 37 and 109 bp. Why do you suppose two well-defined fragments were generated by restriction digestion, rather than a range of fragments of different sizes? How would you interpret this result?

4–41 You have been sent the first samples of a newly discovered Martian microorganism for analysis of its chromatin. The cells resemble Earthly eucaryotes and are composed of similar molecules, including DNA, which is located within a nucleuslike structure in the cell. One member of your team has identified two basic histonelike proteins associated with the DNA in roughly an equal mass ratio with the DNA. You isolate nuclei from the cells and treat them with micrococcal nuclease for various times. You then extract the DNA and run it on an agarose gel alongside rat-liver nuclei that were briefly digested with micrococcal nuclease. As shown in Figure 4–16, the digest of rat-liver nuclei gives a standard ladder of nucleosomes, but the Martian organism gives a smear of digestion products with a nuclease-resistant limit of about 300 nucleotides. As a control, you isolate the Martian DNA free of all protein and digest it with micrococcal nuclease: it is completely susceptible, giving predominantly mono- and dinucleotides as the limit product.

Do these results suggest that the Martian organism has nucleosomelike structures in its chromatin? If so, how are they spaced along the DNA?

4–42* (**True/False**) The four core histones are relatively small proteins with a very high proportion of positively charged amino acids; the positive charge helps the histones bind tightly to DNA, regardless of its nucleotide sequence. Explain your answer.

4–43 Histone proteins are among the most highly conserved proteins in eucaryotes. Histone H4 proteins from a pea and a cow, for example, differ in only 2 of 102 amino acids. Comparison of the gene sequences shows many more differences, but only two that change encoded amino acids. These observations indicate that mutations that change amino acids must be selected against. Why do you suppose that amino acid-altering mutations in histone genes are deleterious?

4–44* Duplex DNA composed entirely of CTG/CAG trinucleotide repeats 5′-CTG in one strand and 5′-CAG in the other strand) is unusually flexible.

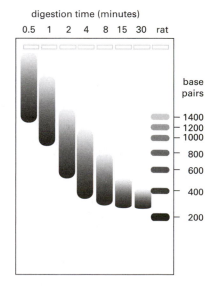

digestion time (minutes)

0.5 1 2 4 8 15 30 rat

base pairs

— 1400
— 1200
— 1000

— 800

— 600

— 400

— 200

Figure 4–16 Micrococcal digest of chromatin from a Martian organism (Problem 4–41). The results of digestion of rat-liver chromatin are shown on the *right*.

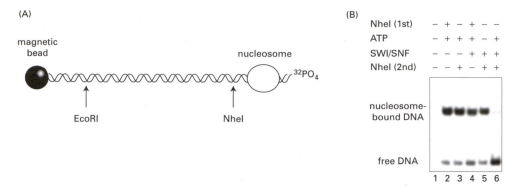

Figure 4–17 The action of the SWI/SNF complex on a nucleosome (Problem 4–47). (A) Nucleosome-containing substrate tethered to a magnetic bead. Sites of cleavage by the restriction enzymes NheI and EcoRI are indicated. (B) Results of incubation with the SWI/SNF complex before (NheI 2nd) or after (NheI 1st) cleavage with NheI. In the absence of cleavage (lane 1) the labeled substrate does not enter the gel.

If 75 CTG/CAG repeats are incorporated into a much longer DNA molecule and mixed with histones octamers, the first nucleosome that assembles nearly always includes the CTG/CAG repeat region. Can you suggest a reason why CTG/CAG repeats might be such effective elements for positioning nucleosomes? (Consider what the energy of the binding interaction between the histone octamers and the DNA must accomplish.)

4–45   Assuming that the 30-nm chromatin fiber contains about 20 nucleosomes (200 bp/nucleosome) per 50 nm of length, calculate the degree of compaction of DNA associated with this level of chromatin structure. What fraction of the 10,000-fold condensation that occurs at mitosis does this level of DNA packing represent?

4–46*   (**True/False**) Nucleosomes bind DNA so tightly that they cannot move from the positions where they are first assembled. Explain your answer.

4–47   Moving nucleosomes out of the way is important for turning genes on. In yeast the 11-subunit SWI/SNF complex, which is the founding member of the ATP-dependent chromatin remodeling complexes, is required for both activating and repressing gene transcription. How does it work? In principle, it could slide nucleosomes along the chromosome, knock them off the DNA, or transfer them from one duplex to another.

To investigate this problem, you assemble a nucleosome on a 189-bp segment of labeled DNA, which you then ligate to a longer piece of DNA tethered to a magnetic bead, as shown in Figure 4–17A. You incubate this substrate with the SWI/SNF complex either before or after you cut the DNA with NheI, which cleaves near the nucleosome, or with EcoRI, which cleaves near the bead. You use a magnet to separate the DNA that is still attached to the bead from the released DNA. If the nucleosome is present, the released DNA will run with a slower mobility on an agarose gel than if it is absent. Incubation with the SWI/SNF complex in the presence of ATP, followed by NheI digestion (NheI 2nd), releases most of the label as the free DNA fragment (Figure 4–17B, lane 6). By contrast, incubation with SWI/SNF after cleavage with NheI (NheI 1st) releases most of the label as nucleosome-bound DNA (lane 4). Similar experiments with EcoRI digestion showed that incubation with SWI/SNF released most of the label as nucleosome-bound DNA regardless of whether EcoRI cleavage preceded or followed the incubation.

Do your results distinguish between the three possible mechanisms for SWI/SNF action—nucleosome sliding, release, or transfer? Explain how your results argue for or against each of these mechanisms.

4–48*   Phosphorylation of serines and methylation and acetylation of lysines in histone tails affect the stability of chromatin structure above the nucleosome level and have important consequences for gene expression. Draw the structures for serine and phosphoserines, as well as those for lysine, monomethylated lysine, and acetylated lysines. Which modifications would alter the net charge on a histone tail? Would the changes in charge be expected to increase or decrease the ability of the tails to interact with DNA?

TABLE 4–1 Protection of globin cDNA by untreated and nuclease-treated chromatin samples (Problem 4–50).

SOURCE OF DNA	PROTECTED GLOBIN cDNA AFTER TREATMENT		
	NONE	MICROCOCCAL	DNase I
Fibroblast nuclei	93%	91%	91%
Red cell nuclei	95%	92%	25%
Red cell nucleosomes		91%	25%
Red cell nucleosomes (trypsin)		25%	

4–49 (**True/False**) Deacetylation of histone tails allows nucleosomes to pack together into tighter arrays, which usually reduces gene expression. Explain your answer.

4–50* The first paper to demonstrate different chromatin structures in active and inactive genes used nucleases to probe the globin loci in chicken red blood cells, which express globin mRNAs, and in chicken fibroblasts, which do not. Isolated nuclei from these cells were treated with either micrococcal nuclease or DNase I, and then DNA was prepared and hybridized in vast excess to a ^{3}H-thymidine-labeled globin cDNA. If the nuclear DNA has not been degraded, it will hybridize to the globin cDNA and protect it from digestion by S1 nuclease, which is specific for single strands of DNA.

Digestion of red cell nuclei or fibroblast nuclei with micrococcal nuclease (so that about 50% of the DNA is degraded) yielded DNA samples that still protected greater than 90% of the cDNA from subsequent digestion with S1 nuclease. Similarly, digestion of fibroblast nuclei with DNase I (so that less than 20% is degraded) yielded DNA that protected greater than 90% of the cDNA. An identical digestion of red cell nuclei with DNase I, however, yielded DNA that protected only about 25% of the cDNA. These results are summarized in Table 4–1.

When nucleosome monomers were isolated from red blood cells by digestion with micrococcal nuclease, their DNA protected more than 90% of globin cDNA. If the monomers were first treated with DNase I, however, the isolated DNA protected only 25% of globin cDNA. If the monomers were briefly treated with trypsin to remove 20 to 30 amino acids from the N-terminus of each histone, digestion of the modified nucleosomes with *micrococcal nuclease* yielded DNA that protected only 25% of globin cDNA (Table 4–1).

A. Which nuclease—micrococcal nuclease or DNase I—digests chromatin that is being expressed (active chromatin)? How can you tell?

B. Does trypsin treatment of nucleosome monomers appear to render a random population or a specific population of nucleosomes sensitive to micrococcal nuclease? How can you tell?

C. Is the alteration that distinguishes active chromatin from bulk chromatin a property of individual nucleosomes, or is it related to the way nucleosome monomers are packaged into higher-order structures within the cell nucleus?

THE GLOBAL STRUCTURE OF CHROMOSOMES

TERMS TO LEARN

euchromatin	mitotic chromosome	position effect
heterochromatin	nuclear matrix	position effect variegation
interphase chromosome	polyploid	sister chromatids
lampbrush chromosome	polytene chromosome	

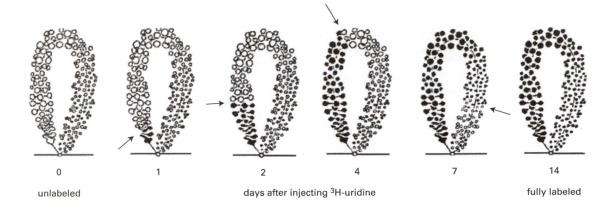

0 — unlabeled

1

2 — days after injecting ^{3}H-uridine — 4

7

14 — fully labeled

4–51 (**True/False**) In lampbrush chromosomes of amphibian oocytes, most of the DNA is in the loops, which are actively transcribed, while the remainder remains highly condensed in the chromomeres, which are not transcribed. Explain your answer.

4–52* One of the earliest studies of transcription in lampbrush chromosomes used oocytes from the newt *Triturus*. ^{3}H-uridine was injected into the oocytes, and after various times radioactive RNA was detected by autoradiography. Small loops, which are the most common type, were labeled throughout, even at the shortest times of labeling. By contrast, giant loops, which are rare, incorporated label progressively around the loop, beginning after about 1 day and continuing for 14 days before the entire loop was labeled (Figure 4–18).

If loops in lampbrush chromosomes represent single transcription units, in which RNA polymerase initiates and terminates synthesis at the base of the loop, as shown in Figure 4–19, which pattern of loop labeling would you expect: uniform, as in the small loops, or progressive, as in the giant loops?

4–53 Imagine that a human interphase chromosome could be transferred intact into an amphibian oocyte and that it could form a lampbrush chromosome. What might be learned from knowing the DNA and RNA sequences in the loops, and how might you determine their identity?

4–54* Although mammalian chromosomes, as yet, do not form lampbrush chromosomes in amphibian oocytes, chromosomes from different amphibians do. When demembranated sperm heads are injected into oocytes, the sperm chromosomes gradually swell and take on the general appearance of typical lampbrush chromosomes. When *Xenopus laevis* sperm heads were injected into *Xenopus laevis* oocytes, they formed lampbrush chromosomes

Figure 4–18 Autoradiographs of a giant chromatin loop from a lampbrush chromosome of the newt (Problem 4–52). *Arrows* show forward progress of labeled regions around the loop at various times after injection of labeled uridine into the oocytes. These giant loops have identical partners on the other side of the chromosome axis, but they are not shown for clarity.

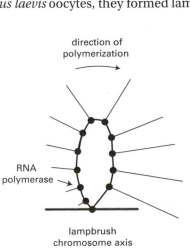

direction of polymerization

RNA polymerase

lampbrush chromosome axis

Figure 4–19 Diagrammatic representation of a chromatin loop that is a single transcription unit (Problem 4–52). The progress of RNA polymerase (shown as black circles) around the loop is illustrated along with the attendant growth of the nascent RNA chain associated with each polymerase.

like those in the oocyte, except that they had unpaired loops—as expected for the single chromatids of sperm—instead of the paired loops formed by the sister chromatids in the oocyte chromosomes.

When sperm heads from *Rana pipiens*, which forms large loops in its own oocyte chromosomes, were injected in *X. laevis* oocytes, the resulting lampbrush chromosomes had the small loops typical of those in *X. laevis* oocytes. Similarly, when sperm heads from *X. laevis* were injected into *Notophthalmus viridescens* oocytes, the resulting lampbrush chromosomes had the very large loop structure typical of *N. viridescens*.

Do these heterologous injection experiments support the idea that loop structure is an intrinsic property of a chromosome? Why or why not? What do these results imply about experiments designed to map the natural loop domains in mammalian chromosomes by forming them into lampbrush chromosomes in amphibians?

4–55 (**True/False**) Classical genetic studies coupled with more recent analysis of the sequence of the *Drosophila melanogaster* genome indicate that each band in a polytene chromosome corresponds to a single gene. Explain your answer.

4–56* The characteristic banding patterns of the giant polytene chromosomes of *Drosophila melanogaster* provide a visible map of the genome that has proven an invaluable aid in genetic studies for decades. The molecular basis for the more intense staining of bands relative to interbands, however, remains a puzzle. In principle, bands might stain more darkly because they contain more DNA than interbands due to overreplication, or the amount of DNA may be the same in the bands and interbands, but the DNA stains more prominently in the bands because it is more condensed or contains more proteins. These two possibilities—differential replication or differential staining—were distinguished by the experiments described below.

A series of radiolabeled segments spanning 315 kb of a *Drosophila* chromosome were used as hybridization probes to estimate the amount of corresponding DNA present in normal diploid tissues versus DNA from salivary glands, which contain polytene chromosomes. DNA from diploid and polytene chromosomes was digested with combinations of restriction enzymes. The fragments were then separated by gel electrophoresis and transferred to nitrocellulose filters for hybridization analysis. In every case the restriction pattern was the same for the DNA from diploid chromosomes and polytene chromosomes, as illustrated for two examples in Figure 4–20. The intensities of many specific restriction fragments were measured and expressed as the ratio of the intensity of the fragment from polytene

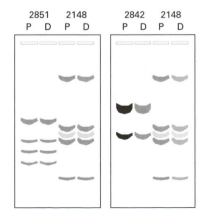

Figure 4–20 Autoradiograph of blot-hybridization analysis of polytene and diploid chromosomes (Problem 4–56). P and D refer to polytene and diploid, respectively. Numbers at the top refer to cloned DNA segments used as probes: 2851 and 2842 are from the 315-kb region under analysis (see Figure 4–21); 2148 is from elsewhere in the genome and was used in all hybridizations to calibrate the amount of DNA added to the gels.

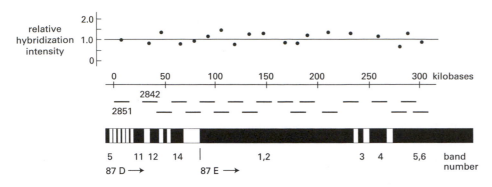

Figure 4–21 Relative amounts of DNA in diploid and polytene chromosomes at different points along the chromosome (Problem 4–56). The chromosomal segment covered by the cloned restriction fragments is shown at the *bottom*, along with the cytological designations for the chromosome regions and bands. The cloned fragments are shown *above* the chromosomes, and the positions of 2851 and 2842 are indicated. The ratio of hybridization of each restriction fragment to DNA from polytene chromosomes versus diploid chromosomes is plotted above each fragment.

chromosomes to the intensity of the corresponding fragment from diploid chromosomes (Figure 4–21).

How do these results distinguish between differential replication or differential staining as the basis for the difference between bands and interbands? Explain your reasoning.

4–57 In *Drosophila* polytene chromosomes, most puffs arise from the decondensation of a single chromosome band. The chromatin in a puff is arranged in a loop. Each loop contains a single gene. There are three times as many genes as bands. Is there a contradiction in these statements? Explain why or why not.

4–58* Look at the two yeast colonies in Figure 4–22. Each of these colonies contains about 100,000 cells descended from a single yeast cell, originally somewhere in the middle of the clump. When the *ADE2* gene is located at its normal chromosome location, its encoded protein is expressed and the colonies are white. When it is located near a telomere, however, the *ADE2* gene is inactivated, giving rise to red colonies with white sectors. The *ADE2* gene is still located near telomeres in both the red and white sectors. Explain why white sectors have formed near the rim of the colony. Based on the existence of these white sectors, what can you conclude about the propagation of the transcriptional state of the *ADE2* gene from mother to daughter cells?

4–59 High-density DNA microarrays can be used to analyze changes in expression of all the genes in the yeast genome in response to various perturbations. The effects of depletion of histone H4 and deletion of the *SIR3* gene on expression of all yeast genes were analyzed in this way, as summarized on the yeast chromosomes illustrated in Figure 4–23. Depletion of histone H4 was achieved in a strain in which the gene was engineered to respond to galactose. In the absence of galactose the gene is turned off and depletion of histone H4 is evident within 6 hours, leading to a decreased density of nucleosomes throughout the genome. Deletion of *SIR3* removes a critical component of the Sir protein complex that binds to telomeres and is responsible for deacetylation of telomeric nucleosomes.

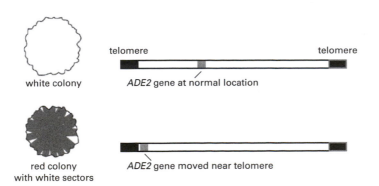

Figure 4–22 Position effect on expression of the yeast *ADE2* gene (Problem 4–58).

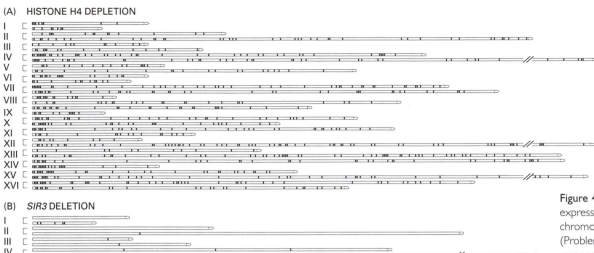

(A) HISTONE H4 DEPLETION

I
II
III
IV
V
VI
VII
VIII
IX
X
XI
XII
XIII
XIV
XV
XVI

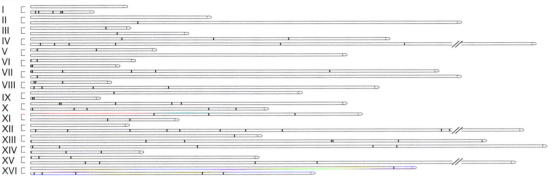

(B) *SIR3* DELETION

I
II
III
IV
V
VI
VII
VIII
IX
X
XI
XII
XIII
XIV
XV
XVI

Figure 4–23 Changes in expression of the genes on chromosomes I-XVI of yeast (Problem 4–59). (A) In response to depletion of histone H4. (B) In response to deletion of *SIR3*. *Black bars* indicate genes whose expression was increased relative to wild-type yeast by 3-fold or more. The very light *gray bars* show genes whose expression was decreased by 3-fold or more; they are not relevant to this problem but have been added for the sake of completeness. Chromosomes were split at their centromeres so that all their telomeres could be aligned at the left; *brackets* indicate the pairs of arms that make up individual chromosomes. Three chromosome arms have been shortened to fit into the figure, as indicated by diagonal lines.

A. Depletion of histone H4 significantly increased expression of 15% of all yeast genes (Figure 4–23A, black bars). Does loss of histone H4 increase expression of a greater fraction of genes near telomeres than in the rest of the genome? Explain how you arrived at your conclusion.

B. Deletion of the *SIR3* gene significantly increased expression of 1.5% of all yeast genes (Figure 4–23B, black bars). Does the absence of Sir3 protein increase expression of a greater fraction of genes near telomeres than in the rest of the genome? Explain how you arrived at your conclusion.

C. If you concluded that either one, or both, preferentially increased expression of genes near telomeres, propose a mechanism for how that might happen.

4–60* A classic paper examined the arrangement of nucleosomes around the centromere (CEN3) of yeast chromosome III. Because centromeres are the chromosome attachment sites for microtubules, it was unclear whether they would have the usual arrangement of nucleosomes. This study used plasmids into which were cloned various lengths of the native chromosomal DNA around the centromere (Figure 4–24). Chromatin from native yeasts and from yeasts that carried individual plasmids was treated briefly with micrococcal nuclease, and then the DNA was deproteinized and digested with the restriction enzyme BamHI, which cuts the DNA only once (Figure 4–24). The digested DNA was fractionated by gel electrophoresis and analyzed by Southern blotting using a segment of radiolabeled centromeric DNA as a hybridization probe (Figure 4–24). This procedure (called indirect end

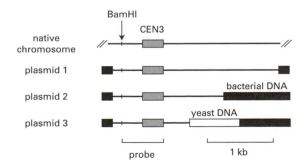

Figure 4–24 Diagrams of the structures of the native chromosome and three plasmids (Problem 4–60). The native chromosome is linear; its true ends extend well beyond the positions marked by the diagonal lines. The plasmids, which are circular, are shown here as linears for ease of comparison. The native yeast sequences around the centromere are shown as *thin lines*. Bacterial DNA sequences in the plasmids are shown as *black rectangles*. The yeast DNA in plasmid 3, which is shown as a *white rectangle*, is a segment of yeast chromosomal DNA far removed from the centromere.

labeling) allows visualization of all the DNA fragments that include the DNA immediately to the right of the BamHI-cleavage site. As a control, a sample of naked DNA from the same region was treated with micrococcal nuclease and subjected to the same analysis. An autoradiogram of the results is shown in Figure 4–25.

A. When the digestion with BamHI was omitted, a regular, though much less distinct, set of dark bands was apparent. Why does digestion with BamHI make the pattern so much clearer and easier to interpret?

B. Draw a diagram showing the micrococcal-nuclease-sensitive sites on the chromosomal DNA and the arrangement of nucleosomes along the chromosome. What is special about the centromeric region?

C. What is the purpose of including a naked DNA control in the experiment?

D. The autoradiogram in Figure 4–25 shows that the native chromosomal DNA yields a regularly spaced pattern of bands beyond the centromere; that is, the bands at 600 nucleotides and above are spaced at 160-nucleotide intervals. Does this regularity result from the lining up of nucleosomes at the centromere, like cars at a stoplight? Or, is the regularity an intrinsic property of the DNA sequence itself? Explain how the results with plasmids 1, 2, and 3 decide the issue.

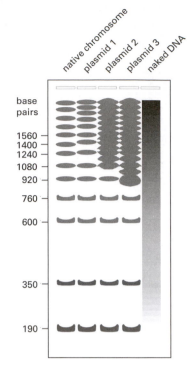

Figure 4–25 Results of micrococcal-nuclease digestion of DNA around CEN3 (Problem 4–60). Approximate lengths of DNA fragments in nucleotide pairs are indicated on the left of the autoradiogram.

4–61 Why is a chromosome with two centromeres (a dicentric chromosome) said to be unstable? Wouldn't a back-up centromere be a good thing for a chromosome, giving it two chances to form a kinetochore and attach to microtubules during mitosis? Wouldn't that ensure that the chromosome didn't get left behind at mitosis—sort of like using a belt and braces to keep your pants up?

4–62* **(True/False)** At the final level of condensation each chromatid of a mitotic chromosome is organized into loops of chromatin that emanate from a central axis. Explain your answer.

4–63 The typical coiled phone cord provides an everyday example of the phenomenon of supercoiling. Invariably, the cord becomes coiled about itself forming a tangled mess. These coiled coils are supercoils. Dangling the receiver and letting it spin until it stops can remove them. Similarly, supercoils can be reintroduced by twisting the receiver, which of course is how they get there in the first place.

DNA is coiled into a double helix that exhibits the same phenomenon of supercoiling (Figure 4–26). A relaxed circular DNA, with 10.5 bp per turn of the helix, will assume a more or less circular form when laid onto a surface. If one strand of the DNA is broken and wound around its partner two extra times (overwound—an increase in linking number of +2) and then rejoined, the molecule will twist on itself, forming two *positive* supercoils. If one strand in a relaxed circular DNA is broken and rejoined with two fewer turns (underwound—a decrease in linking number of –2), the molecule will twist to form two *negative* supercoils. Positive and negative supercoils each can assume two forms termed plectonemic and solenoidal, although plectonemic supercoils are the only ones that are stable in naked DNA. The effect of supercoiling is to preserve the preferred local winding of DNA at 10.5 bp per

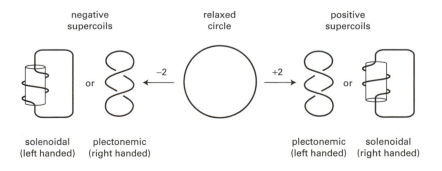

Figure 4–26 Relaxed and supercoiled circular DNA molecules (Problem 4–63). Duplexes of DNA are indicated by single lines. These DNA molecules differ only in the number of times one strand is wound around the other, a quantity known as the linking number. Solenoidal supercoils are shown as wrapped around a cylinder for illustrative purposes.

turn. In cells the degree of supercoiling of DNA is carefully controlled by special enzymes called topoisomerases that break and rejoin strands of DNA.

Circular plasmid DNA isolated from *E. coli* is highly supercoiled as evident when the DNA is separated by electrophoresis on an agarose gel (Figure 4–27, lane 1). When incubated for increasing times with *E. coli* topoisomerase I (which breaks and reseals a single DNA strand in negatively supercoiled DNA, but not in positively supercoiled DNA), several new bands appear between the supercoiled DNA and the relaxed DNA (lanes 2–5).

A. In the untreated DNA sample isolated from bacteria, why do you suppose a small fraction of the plasmid molecules is relaxed?

B. What are the discrete bands between the highly supercoiled and relaxed bands in Figure 4–27 that appear with increasing times of incubation with topoisomerase I? Why do they move at rates intermediate between relaxed and highly supercoiled DNA?

C. Estimate the number of supercoils that were present in the original plasmid molecules.

D. Did the bacterial plasmid originally contain positive or negative supercoils? Explain your answer.

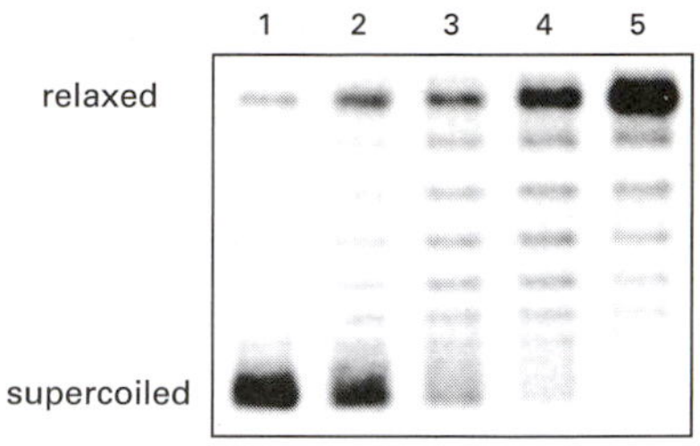

4–64* No source of energy (ATP, for example) needs to be included in the incubation for *E. coli* or other topoisomerase I enzymes. What does this imply about the equilibrium between relaxed and supercoiled molecules? Do supercoiled molecules represent a higher or a lower energy state than relaxed molecules?

4–65 Imagine that you assemble a single nucleosome on a closed circular, relaxed DNA molecule; that is, a circular duplex DNA with no breaks in either strand and zero supercoiling. Wrapping the DNA molecule around the histone octamer forms solenoidal supercoils, which are compensated for by plectonemic supercoils in another part of the molecule.

Of the four possible arrangements of solenoidal and plectonemic supercoils shown in Figure 4–28, which have a net supercoiling of zero? (Since no breaks were introduced into the DNA in the process of forming the nucleosome, it must retain an overall supercoiling of zero.) Indicate the sign of the supercoiling (positive or negative) on the structures you select.

4–66* Which of the two alternative arrangements of compensating solenoidal and plectonemic supercoils, generated by formation of a nucleosome (see Problem 4–65), represents the true biological situation? These alternatives were distinguished by incubating the nucleosome-bound DNA with either *E. coli* topoisomerase I, which can remove only negative plectonemic supercoils, or with calf thymus topoisomerase I, which can remove both negative and positive plectonemic supercoils. Histones were removed after the incubation with a topoisomerase, and the presence of supercoils in the naked DNA was assayed by gel electrophoresis (see Figure 4–27, Problem 4–63). (The sign of the plectonemic supercoils in the naked DNA can be determined by subsequent incubation with *E. coli* topoisomerase I, which relaxes negative supercoils, but not positive ones.)

Figure 4–27 Plasmid DNA treated with *E. coli* topoisomerase I for increasing times (Problem 4–63).

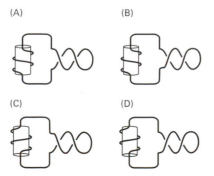

(A) (B)

(C) (D)

Figure 4–28 Four possible arrangements of circular DNA molecules with two solenoidal and two plectonemic supercoils (Problem 4–65). The position of the nucleosome is indicated by the cylinder.

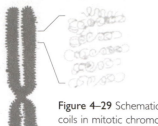

Figure 4–29 Schematic diagram of the large coils in mitotic chromosomes (Problem 4–67).

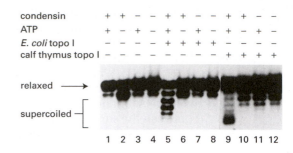

It was found that incubation of nucleosomal DNA with *E. coli* topoisomerase I gave DNA molecules with zero supercoils. By contrast, incubation with calf thymus topoisomerase I gave DNA molecules with two negative supercoils. Are the solenoidal supercoils around biological nucleosomes positive (right handed) or negative (left handed)? What results would you have expected for the other alternative?

Figure 4–30 Experiment to test for condensin-mediated coiling of DNA (Problem 4–67). Relaxed, closed circular DNA was incubated with condensin and ATP, as indicated, and then the deproteinized DNA was incubated with topoisomerase I from *E. coli* or calf thymus.

Problem 18–4 compares the activities of condensins and cohesins.

4–67   Condensins use the energy of ATP hydrolysis to drive the coiling of interphase chromosomes into the highly condensed chromosomes visible at mitosis (Figure 4–29). You realize that if condensins operate this way they may change the supercoiling of the DNA, since the chromosomal coils are just large solenoids. To test this hypothesis, you incubate relaxed, closed circular DNA with condensin and ATP. Then you incubate them with either *E. coli* or calf thymus topoisomerase I, remove condensin and assay for supercoils, as shown in Figure 4–30.

   A. Do you expect that supercoils present after topoisomerase treatment (Figure 4–30, lanes 5 and 9) will be positive or negative? Why? If the supercoils were generated as a result of condensin-mediated formation of solenoids, were the solenoids right handed (positive) or left handed (negative)?

   B. How do you know that condensin isn't simply an ATP-driven topoisomerase; that is, a topoisomerase that uses the energy of ATP to underwind or overwind the DNA, to introduce plectonemic supercoils? What would have been the outcome of the experiment if condensin acted in this way to introduce positive plectonemic supercoils? What would be the outcome if it introduced negative plectonemic supercoils?

4–68*   Why are the dye-stained bands in human mitotic chromosomes not thought to be the same as the bands observed in insect polytene chromosomes?

4–69   Each interphase chromosome tends to occupy a discrete and relatively small area within the nucleus. Does this mean that the particular site a chromosome occupies is critical for cell function? Why or why not?

5

DNA REPLICATION, REPAIR, AND RECOMBINATION

THE MAINTENANCE OF DNA SEQUENCES

TERMS TO LEARN

mutation	mutation rate

5–1   Mutations are introduced into the *E. coli* genome at the rate of 1 mutation per 10^9 base pairs per generation. Imagine that you start with a population of 10^6 *E. coli*, none of which carry any mutations in your gene of interest, which is 1000 nucleotides in length and not essential for bacterial growth and survival. In the next generation, after the population doubles in number, what fraction of the cells, on average, would you expect to carry a mutation in your gene? After the population doubles again, what would you expect the frequency of mutants in the population to be? What would the frequency be after a third doubling?

5–2*   To determine how reproducible mutation frequencies are, you do the following experiment. You inoculate each of 10 cultures with a single *E. coli* bacterium, allow the cultures to grow until each contains 10^6 cells, and then measure the number of cells in each culture that carry a mutation in your gene of interest. You were so surprised by the initial results that you repeated the experiment to confirm them. Both sets of results have the same surprising features, as shown in Table 5–1. If the rate of mutation is constant, why is there so much variation in the frequencies of mutant cells in different cultures?

TABLE 5–1 Frequencies of mutant cells in multiple cultures (Problem 5–2).

EXPERIMENT	CULTURE (mutant cells/10^6 cells)									
	1	2	3	4	5	6	7	8	9	10
1	4	0	257	1	2	32	0	0	2	1
2	128	0	1	4	0	0	66	5	0	2

5–3 In a genetics lab, Kim and Maria infected a sample from an *E. coli* culture with a virulent bacteriophage. They noticed that most of the cells were lysed, but a few survived. The survival in their sample was about 1×10^{-4}. Kim was sure the bacteriophage induced the resistance in the cells, while Maria thought that resistant mutants probably already existed in the sample of cells they used. Earlier, for a different experiment, they had spread a dilute suspension of *E. coli* onto solid medium in a large Petri dish, and, after seeing that about 10^5 colonies were growing up, they had replica plated that plate onto three other plates. Kim and Maria decided to use these plates to test their hypotheses. They pipette a suspension of the bacteriophage onto each of the three replica plates. What should they see if Kim is right? What should they see if Maria is right?

5–4* Evaluate the following statement: "No two cells in your body have the identical nucleotide sequence." Do you agree or disagree? Explain your reasoning.

5–5 Why do calculations based on amino acid differences in comparisons of the same protein in different species tend to underestimate the actual mutation rate? Is this true for estimates of rates based on the fibrinopeptides? Why or why not?

5–6* What, if anything, is wrong with the following statement: "Both germ-cell DNA stability and somatic-cell DNA stability are essential for the survival of the species." Explain your answer.

5–7 The incidence of cancer tends to increase with age, as shown in Figure 5–1, where the number of newly diagnosed cases of colon cancer in women in one year is plotted as a function of age at diagnosis. Studies of other types of cancer show the same sort of age dependence. Assuming that the rate of mutation is constant throughout life, why do you think it is that the incidence of cancer increases so dramatically with age?

5–8* It is easy to see how deleterious mutations in bacteria, which have a single copy of each gene, are eliminated by natural selection; the affected bacteria die and the mutation is thereby lost from the population. Eucaryotes, however, have two copies of most genes because they are diploid. It is often the case that an individual with two normal copies of the gene (homozygous, normal) is indistinguishable in phenotype from an individual with one normal copy and one defective copy of the gene (heterozygous). In such cases, natural selection can operate only on an individual with two copies of the defective gene (homozygous, defective). Imagine the situation in which a defective form of the gene is lethal when homozygous, but without effect when heterozygous. Can such a mutation ever be eliminated from the population by natural selection? Why or why not?

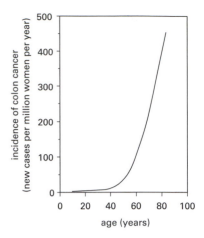

Figure 5–1 Colon cancer incidence as a function of age (Problem 5–7).

DNA REPLICATION MECHANISMS

5–9 The nucleotide sequence of one DNA strand of a DNA double helix is 5′-GGATTTTTGTCCACAATCA-3′. What is the sequence of the complementary strand?

5–10* The DNA fragment in Figure 5–2 is double stranded at each end but single stranded in the middle. The polarity of the top strand is indicated. Is the phosphate (PO_4^-) indicated on the bottom strand at the 5′ end or the 3′ end of the fragment to which it is attached?

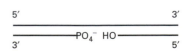

Figure 5–2 A DNA fragment with a single-stranded gap on the bottom strand (Problem 5–10).

5–11 (True/False) When read in the same direction (5′-to-3′), the sequence of nucleotides in a newly synthesized DNA strand is the same as in the parental template strand. Explain your answer.

5–12* In the electron microscope, it is possible to observe the replication fork directly and, for small DNA molecules, to see the entire replicating structure. In addition, by using appropriate techniques of sample preparation, one can distinguish double-stranded DNA from single-stranded DNA.

A series of hypothetical replicating molecules is illustrated schematically in Figure 5–3, with regions of single-stranded DNA shown as thin lines. In an important early electron microscopic study of bacteriophage lambda replication, some of these structures were observed commonly and others were never observed.

A. Draw a diagram of a replication structure with two forks moving in opposite directions. Label the ends of all strands (5′ or 3′), and indicate the leading and lagging strands at each replication fork.

B. Based on your knowledge of DNA replication, indicate the structure in Figure 5–3 you would expect to be observed most commonly. In the actual experiment, four of these structures were seen. Which ones do you think they were?

5–13 (True/False) Each time the genome is replicated, half the newly synthesized DNA is stitched together from Okazaki fragments. Explain your answer.

5–14* A born skeptic, you plan to confirm for yourself the results of a classic experiment originally performed in the 1960s by Meselson and Stahl. They concluded that each daughter cell inherits one and only one strand of its mother's DNA. To check their results, you 'synchronize' a culture of growing cells, so that virtually all cells begin and then complete DNA synthesis at the same time. You first grow the cells in a medium that contains nutrients highly enriched in heavy isotopes of nitrogen and carbon (^{15}N and ^{13}C in place of the naturally abundant ^{14}N and ^{12}C). Cells growing in this 'heavy' medium use the heavy isotopes to build all of their macromolecules, including nucleotides and nucleic acids. You then transfer the cells to a normal, 'light' medium containing ^{14}N and ^{12}C nutrients. Finally, you isolate DNA from cells that have grown for different numbers of generations in the light medium and determine the density of their DNA by density-gradient centrifugation. Your data, plotting the amount of DNA isolated versus its density, are shown in Figure 5–4. Are these results in agreement with your expectations? Explain the results.

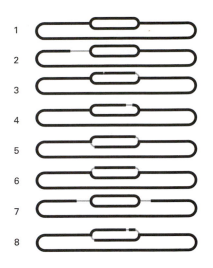

Figure 5–3 Hypothetical structures of replicating DNA molecules (Problem 5–12).

Problem 8–14 presents data from the original Meselson–Stahl experiment.

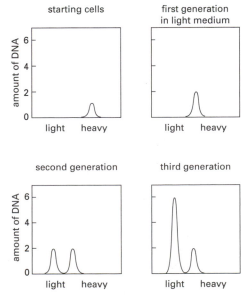

Figure 5–4 Density of DNAs isolated from cells that were grown for different times in 'light' medium after initial growth in medium enriched for heavy isotopes of nitrogen and carbon (Problem 5–14). Equal culture volumes were analyzed for each time point. Amount of DNA is in arbitrary units, with the peak amount of DNA in the sample containing starting cells set equal to 1.

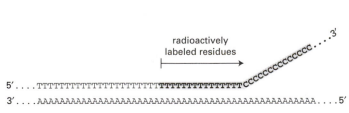

(A) ddCTP

(B) ddCMP

Figure 5–5 Potential replication substrates (Problem 5–15). (A) Dideoxycytidine triphosphate (ddCTP). (B) Dideoxycytidine monophosphate (ddCMP).

5–15 Look carefully at the structures of the molecules in Figure 5–5.
  A. What would you expect if dideoxycytidine triphosphate (ddCTP) were added to a DNA replication reaction in large excess over the concentration of deoxycytidine triphosphate (dCTP)?
  B. What would happen if ddCTP were added at 10% the concentration of dCTP?
  C. What effects would you expect if dideoxycytidine monophosphate (ddCMP) were added to a DNA replication reaction in large excess, or at 10% the concentration of dCTP?

5–16* How would you expect the loss of the 3′-to-5′ proofreading exonuclease activity of DNA polymerase in *E. coli* to affect the fidelity of DNA synthesis? How would its loss affect the rate of DNA synthesis? Explain your reasoning.

5–17 To study the 3′-to-5′ proofreading exonuclease activity of DNA polymerase I of *E. coli*, you prepare an artificial substrate with a poly(dA) strand as template and a poly(dT) strand as primer. The poly(dT) strand contains a few ^{32}P-labeled dT nucleotides followed by a few ^{3}H-labeled dC nucleotides at its 3′ end, as shown in Figure 5–6. You measure the loss of the labeled dTs and dCs either without any dTTP present, so that no DNA synthesis is possible, or with dTTP present, so that DNA synthesis can occur. The results are shown in Figure 5–7.
  A. Why were the Ts and Cs labeled with different isotopes?
  B. Why did it take longer for the Ts to be removed in the absence of dTTP than the Cs ?
  C. Why were none of the Ts removed in the presence of dTTP, whereas the Cs were lost regardless of whether or not dTTP was present?
  D. Would you expect different results in Figure 5–7B if you had included dCTP along with the dTTP?

5–18* You have discovered a novel organism that thrives in the ocean depths in the hostile environment of hydrothermal vents. In characterizing its replication, you are astounded to discover that it replicates both strands continuously, using two DNA polymerases: one that synthesizes DNA in the usual 5′-to-3′ direction and a second that synthesizes DNA in the 3′-to-5′ direction. Both polymerases use the standard nucleoside 5′-triphosphates for addition of nucleotides to growing DNA chains. You are most surprised, however, to find that *both* newly synthesized DNA strands are made with the same high degree of fidelity that characterizes DNA synthesis in *E. coli*.
  A. List the four processes that contribute to the high fidelity of DNA replication in *E. coli*.

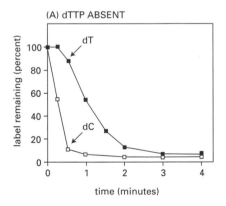

(A) dTTP ABSENT

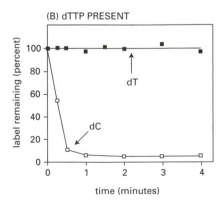

(B) dTTP PRESENT

Figure 5–7 Proofreading by DNA polymerase I (Problem 5–17). (A) In the absence of dTTP. (B) In the presence of dTTP.

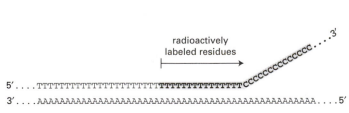

radioactively labeled residues

5′ . . . . TTTTTTTTTTTTTTTTTTTTTTT**TTTTTTTTTTTTTTTTT**CCCCCCCCCCCCC . . . . 3′
3′ . . . . AAAAAAAAAAAAAAAAAAAAAAAAAAAAAAAAAAAAAAAAAAAAAAAAAAAAAAAA . . . . 5′

Figure 5–6 Artificial substrate for studying proofreading by DNA polymerase I of *E. coli* (Problem 5–17). *Shaded* letters indicate nucleotides that are radioactively labeled.

site	M13 template sequences		DNA sequences linked to RNA primer
	5′ 3′		5′ 3′
1	A T C C T T G C G T T G A A A T		A G G A T
2	T C T T G T T T G C T C C A G A		C A A G A
3	A T T C T C T T G T T T G C T C		A G A A T
4	A C A T G C T A G T T T T A C G		C A T G T
5	A T T G A C A T G C T A G T T T		T C A A T
6	A T C T T C C T G T T T T T G G		A A G A T
7	A A A T A T T T G C T T A T A C		T A T T T
8	C T A G A A C G G T T A C C C T		T C T A G

Figure 5–8 RNA priming during M13 replication (Problem 5–20). M13 template sequences at several sites of RNA priming are shown adjacent to the DNA sequences that were found to be linked to RNA primers at each site. The RNA primers were removed from these DNA chains prior to sequencing.

B. Explain why it is surprising that both strands are replicated with high fidelity.

C. Suggest at least two ways by which high fidelity might be accomplished. If you need to invent additional enzymes to accomplish the task, describe their activities.

5–19   Discuss the following statement: "Primase is a sloppy enzyme that makes many mistakes. Eventually, the RNA primers it makes are replaced with DNA made by a polymerase with higher fidelity. This is wasteful. It would be more energy efficient if a DNA polymerase made an accurate copy in the first place."

5–20*   Does RNA priming occur at specific sites or at random sites on the template? The M13 viral DNA, which is a circular single strand, is an ideal template for studying this question. The M13 circle was copied in the presence of DNA polymerase, primosome (a complex of a helicase and an RNA primase), rNTPs, and dNTPs to make a double-stranded circle that still retained RNA primers. Under these conditions the 5′ ends of the RNA primers are not linked to the upstream DNA, leaving nicks in the newly synthesized strand that correspond to the beginning of each RNA primer. The circles were then digested with a restriction nuclease that makes a double-strand cut at a unique site. When the products were denatured and separated by gel electrophoresis, many discrete bands were observed. If the products were treated with RNase before electrophoresis, they all became five nucleotides shorter as judged by their faster electrophoretic migration.

The sequences of the newly synthesized DNA immediately adjacent to the RNA primers in several fragments were determined and aligned with the template. The template sequences and the corresponding DNA sequences linked to the RNA primers are shown in Figure 5–8.

From these data, deduce the start site for the RNA primer on each template sequence. What is the likely signal for starting the RNA primase reaction?

5–21   (**True/False**) In *E. coli*, where the replication fork travels at 500 nucleotide pairs per second, the DNA ahead of the fork must rotate at nearly 3000 revolutions per minute. Explain your answer.

5–22*   The *dnaB* gene of *E. coli* encodes a helicase (DnaB) that unwinds DNA at the replication fork. Its properties have been studied using artificial substrates like those shown in Figure 5–9. In such substrates DnaB binds preferentially to the longest single strand (the largest target) available. The experimental approach is to incubate the substrates under a variety of conditions and then subject a sample to electrophoresis on agarose gels. The short single strand will move slowly if it is still annealed to the longer DNA strand, but it will move much faster if it has been unwound and detached. The migration of the short single strand can be followed selectively by making it radioactive and examining its position in the gel by autoradiography. The migration of

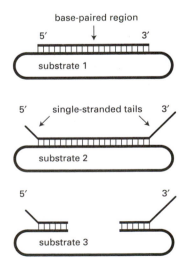

Figure 5–9 Substrates used to test the properties of DnaB (Problem 5–22).

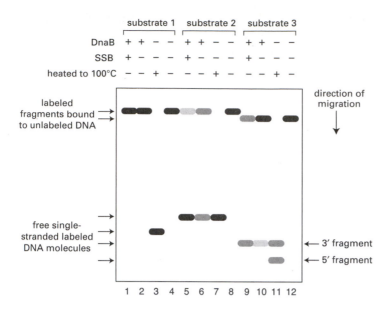

Figure 5–10 Results of several experiments to measure unwinding by DnaB (Problem 5–22). Only the single-stranded fragments were radioactively labeled. Their positions are shown by the bands in this schematic diagram. ATP was included with DnaB in all incubations, which were carried out at 37°C.

the labeled single strands in three different substrates is shown in Figure 5–10: in the absence of any treatment the labeled strands move slowly (lanes 4, 8, and 12); when the substrates are heated to 100°C, the labeled strands are detached and migrate more rapidly (lanes 3, 7, and 11).

The results of several experiments are shown in Figure 5–10. Substrate 1, the hybrid without tails, was not unwound by DnaB and ATP at 37°C (Figure 5–10, lanes 1 and 2). However, when either substrate with tails was incubated at 37°C with DnaB and ATP, a significant amount of small fragment was released by unwinding (lanes 6 and 10). For substrate 3 only the 3′ fragment was unwound (lane 10). All unwinding was absolutely dependent on ATP hydrolysis.

Unwinding was considerably enhanced by adding single-stranded DNA-binding protein (SSB) (compare Figure 5–10, lanes 5 and 6 and lanes 9 and 10). Interestingly, SSB had to be added about 3 minutes after DnaB; otherwise it inhibited unwinding.

A. Why is ATP hydrolysis required for unwinding?

B. In what direction does DnaB move along the long single-stranded DNA? Is this direction more consistent with its movement on the template for the leading strand or on the template for the lagging strand at the replication fork?

C. Why might SSB inhibit unwinding if it is added before DnaB but stimulate unwinding if added after DnaB?

5–23 SSB proteins bind to single-stranded DNA at the replication fork and prevent formation of short hairpin helices that would otherwise impede DNA synthesis. What sorts of sequences in single-stranded DNA might be able to form a hairpin helix? Write out an example of a sequence that could form a 5-nucleotide hairpin helix, and show the helix.

5–24* Like all organisms, bacteriophage T4 encodes an SSB protein that is important for removing secondary structure in the single-stranded DNA ahead of the replication fork. The T4 SSB protein is an elongated monomeric protein with a molecular weight of 35,000. It binds tightly to single-stranded, but not double-stranded, DNA. Binding saturates at a 1:12 weight ratio of DNA to protein. The binding of SSB protein to DNA shows a peculiar property that is illustrated in Figure 5–11. In the presence of excess single-stranded DNA (10 μg), virtually no binding is detectable at 0.5 μg SSB protein (Figure 5–11A), whereas almost quantitative binding is seen at 7.0 μg SSB protein (Figure 5–11B).

A. At saturation, what is the ratio of nucleotides of single-stranded DNA to molecules of SSB protein? (The average mass of a single nucleotide is 330 daltons.)

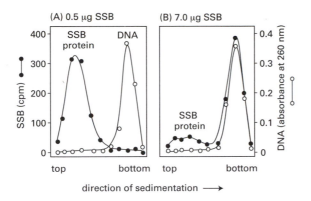

Figure 5–11 Binding of T4 SSB protein to single-stranded DNA (Problem 5–24). The binding of SSB protein to DNA was analyzed by centrifugation through sucrose gradients, on which the much more massive DNA sediments more rapidly than protein and is consequently found closer to the bottom of the gradient.

B. When the binding of SSB protein to DNA reaches saturation, are adjacent monomers of SSB protein likely to be in contact? Assume that a monomer of SSB protein extends for 12 nm along the DNA upon binding and that the spacing of bases in single-stranded DNA after binding is the same as in double-stranded DNA (that is, 10 nucleotides per 3.4 nm).

C. Why do you think that the binding of SSB protein to single-stranded DNA depends so strongly on the amount of SSB protein?

5–25 Conditional lethal mutations are very useful in genetic and biochemical analyses of complex processes such as DNA replication. Temperature-sensitive (ts) mutations, which are one form of conditional lethal mutation, allow growth at one temperature (for example, 30°C) but not at a higher temperature (for example, 42°C).

A large number of temperature-sensitive replication mutants have been isolated in *E. coli*. These mutant bacteria are defective in DNA replication at 42°C but not at 30°C. If the temperature of the medium is raised from 30°C to 42°C, these mutants stop making DNA in one of two characteristic ways. The 'quick-stop' mutants halt DNA synthesis immediately, whereas the 'slow-stop' mutants stop DNA synthesis only after many minutes.

A. Predict which of the following proteins, if temperature sensitive, would display a quick-stop phenotype and which would display a slow-stop phenotype. In each case explain your prediction.
   1. DNA topoisomerase I
   2. A replication initiator protein
   3. Single-strand binding protein
   4. DNA helicase
   5. DNA primase
   6. DNA ligase

B. Cell-free extracts of the mutants show essentially the same patterns of replication as the intact cells. Extracts from quick-stop mutants halt DNA synthesis immediately at 42°C, whereas extracts from slow-stop mutants do not stop DNA synthesis for several minutes after a shift to 42°C. Suppose extracts from a temperature-sensitive DNA helicase mutant and a temperature-sensitive DNA ligase mutant were mixed together at 42°C. Would you expect the mixture to exhibit a quick-stop phenotype, a slow-stop phenotype, or a nonmutant phenotype?

5–26* (**True/False**) The mismatch proofreading system in *E. coli* can distinguish the parental strand from the progeny strand as long as one or both are methylated, but not if both strands are unmethylated. Explain your answer.

5–27 DNA repair enzymes preferentially repair mismatched bases on the newly synthesized DNA strand, using the old DNA strand as a template. If mismatches were simply repaired without regard for which strand served as template, would mismatch repair reduce replication errors? Would such an indiscriminate mismatch repair be better than, worse than, or the same as no repair at all? Explain your answers.

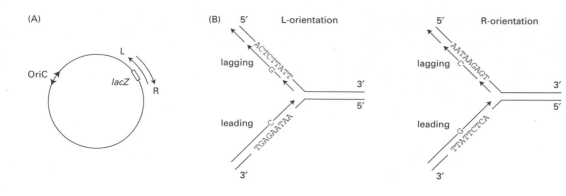

Figure 5–12 Fidelity of synthesis of the leading and lagging strands (Problem 5–29). (A) Site of insertion of *lacZ* into the *E. coli* chromosome. Orientations are denoted R and L. The *arrows* at OriC represent the two forks initiated at that site. (B) Arrangement of sequences relative to the leading and lagging strands in the L- and R-orientations. The G and C misincorporations that could lead to reversion are indicated.

5–28* If DNA polymerase requires a perfectly paired primer in order to add the next nucleotide, how is it that any mismatched nucleotides 'escape' the polymerase so that they become substrates for mismatch repair enzymes?

5–29 The different ways in which DNA synthesis occurs on the leading and lagging strands raises the question as to whether synthesis occurs with equal fidelity on the two strands. One clever approach used reversion of specific mutations in the *E. coli lacZ* gene to address this question. *E. coli* is a good choice for such a study because the same polymerase (DNA pol III) synthesizes both the leading and the lagging strands.

The *lacZ* CC106 allele can regain its function (revert) by converting the mutant AT base pair to the normal GC base pair. This allele was inserted in both orientations on one side of the normal origin of replication (Figure 5–12A). As shown in Figure 5–12B, misincorporation of G opposite T on one strand, or of C opposite A on the other strand, could lead to reversion. Previous studies had shown that C is only very rarely misincorporated opposite A, and when it is, DNA polymerase is inefficient at extending the chain. Thus, the most common source of reversion is from misincorporation of G opposite T.

To eliminate the complicating effects of mismatch repair, the experiments were done in two mutant strains of bacteria. One was defective for mismatch repair, which eliminates it from consideration; the other was defective in the proofreading exonuclease, and introduces so many mismatches that it overwhelms the mismatch-repair machinery.

Accurate frequencies of *lacZ* reversion were measured in the two strains, along with the frequencies of mutation at the *rif* gene, whose orientation in the chromosome was constant (Table 5–2).

A. On which strand, leading or lagging, does DNA synthesis appear to be more accurate? Explain your reasoning.

B. Can you suggest a reason why DNA synthesis might be more accurate on the strand you have chosen?

5–30* DNA damage can interfere with DNA replication. X-rays, for example, generate highly reactive hydroxyl radicals that can break one or both strands of

TABLE 5–2 Frequencies (per 10^6 cells) of revertants of *lacZ* and mutants of *rif* in mismatch-repair and proofreading deficient strains of *E. coli* (Problem 5–29).

lacZ ALLELE (mutation measured)	*lacZ* ORIENTATION	MISMATCH-REPAIR DEFICIENT		PROOFREADING DEFICIENT	
		$lac^- \rightarrow lac^+$	$rif^s \rightarrow rif^r$	$lac^- \rightarrow lac^+$	$rif^s \rightarrow rif^r$
CC106 (AT → GC)	L	0.27	7.0	2.7	82
CC106 (AT → GC)	R	0.51	6.4	6.7	80

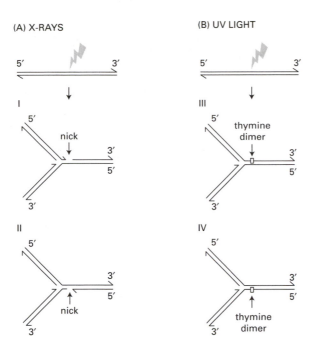

(A) X-RAYS

(B) UV LIGHT

Figure 5–13 Damage to the templates for the leading and lagging strands (Problem 5–30). (A) X-ray-induced nicks. (B) UV-light-induced thymine dimers. Structures I and III have damage in the template for lagging-strand synthesis. Structures II and IV have damage in the template for leading-strand synthesis.

DNA. UV light commonly generates cyclobutane dimers between adjacent T bases in the same DNA strand, which blocks progression of DNA polymerase. If such damage is not repaired, it can have serious consequences when a replication fork encounters it. See if you can predict the appearance of the replication fork after it encounters a nick or a thymine-dimer block in the templates for the leading and lagging strands. Replication forks just before they come to the damage are shown in Figure 5–13.

5–31 (**True/False**) Topoisomerase I does not require ATP to break and rejoin DNA strands because the energy of the phosphodiester bond is stored transiently in a phosphotyrosine linkage in the enzyme's active site. Explain your answer.

5–32* Approximately how many high-energy bonds are used to replicate the *E. coli* chromosome? How many molecules of glucose would *E. coli* need to consume to provide enough energy to copy its DNA once? How does this mass of glucose compare to the mass of *E. coli*, which is about 10^{-12} g? There are 4.6×10^6 base pairs in the *E. coli* genome. Oxidation of one glucose molecule yields about 30 high-energy phosphate bonds. Glucose has a molecular mass of 180 daltons, and there are 6×10^{23} daltons/g.

5–33 At the completion of replication of the circular genome of the animal virus SV40, the two daughter circles are interlocked like links in a chain. How are such interlinked molecules separated?

THE INITIATION AND COMPLETION OF DNA REPLICATION IN CHROMOSOMES

TERMS TO LEARN

origin recognition complex (ORC)	S phase	telomerase
replication origin		

5–34* (**True/False**) In a replication bubble, a single parental DNA strand serves as the template strand for leading-strand synthesis in one replication fork and as the template for lagging-strand synthesis in the other fork. Explain your answer.

5–35 The laboratory you joined is studying the life cycle of an animal virus that uses a circular, double-stranded DNA as its genome. Your project is to define the location of the origin(s) of replication and to determine whether replication proceeds in one or both directions away from an origin (unidirectional or bidirectional replication). To accomplish your goal, you isolated replicating molecules, cleaved them with a restriction nuclease that cuts the viral genome at one site to produce a linear molecule from the circle, and examined the resulting molecules in the electron microscope. Some of the molecules you observed are illustrated schematically in Figure 5–14. (Note that it is impossible to tell one end of a DNA molecule from the other in the electron microscope.)

You must present your conclusions to the rest of the lab tomorrow. How will you answer the questions that your advisor has posed for you? (1) Is there a single, unique origin of replication or several origins? (2) Is replication unidirectional or bidirectional?

5–36* (**True/False**) When bidirectional replication forks from adjacent origins meet, a leading strand always runs into a lagging strand. Explain your answer.

5–37 You are investigating DNA synthesis in a line of tissue culture cells using a classic protocol. In this procedure ³H-thymidine is added to the cells, which incorporate it at replication forks. Then the cells are gently lysed in a dialysis bag to release the DNA. When the bag is punctured and the solution slowly drained, some of the DNA strands adhere to the walls and are stretched in the general direction of drainage. This method allows very long DNA strands to be isolated intact and examined; however, the stretching collapses replication bubbles so that daughter duplexes lie side by side. The support with its adhered DNA is fixed to a glass slide, overlaid with a photographic emulsion, and exposed for 3 to 6 months. The labeled DNA shows up as tracks of silver grains.

You pretreat the cells to synchronize them at the beginning of S phase. In one experiment you release the synchronizing block and add ³H-thymidine immediately. After 30 minutes you wash the cells and change the medium so that the label is present at a third of its initial concentration. After an additional 15 minutes you prepare DNA for autoradiography. The results of this experiment are shown in Figure 5–15A. In the second experiment you release the synchronizing block and then wait 30 minutes before adding ³H-thymidine. After 30 minutes in the presence of ³H-thymidine, you once again change the medium to reduce the concentration of labeled thymidine and incubate the cells for an additional 15 minutes. The results of the second experiment are shown in Figure 5–15B.

A. Explain why in both experiments some regions of the tracks are dense with silver grains (dark), whereas others are less dense (light).

B. In the first experiment each track has a central dark section with light sections at each end. In the second experiment the dark section of each track has a light section at only one end. Explain the reason for the difference between the results in the two experiments.

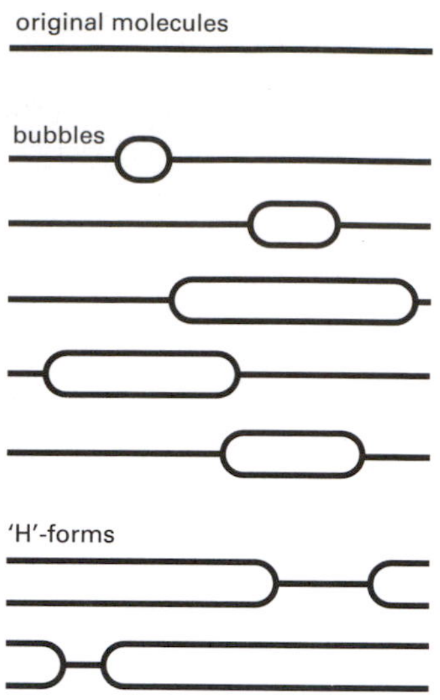

Figure 5–14 Parental and replicating forms of an animal virus (Problem 5–35).

(A)

(B)

50 μm

Figure 5–15 Autoradiographic investigation of DNA replication in cultured cells (Problem 5–37). (A) Addition of labeled thymidine immediately after release from the synchronizing block. (B) Addition of labeled thymidine 30 minutes after release from the synchronizing block.

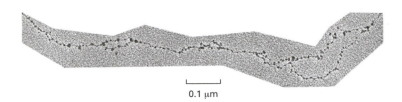

Figure 5–16 Electron micrograph showing multiple replication bubbles in a chromosome from the early embryo of *Drosophila* (Problem 5–38).

0.1 μm

C. Estimate the rate of fork movement (μm/min) in these experiments. Do the estimates from the two experiments agree? Can you use this information to gauge how long it would take to replicate the entire genome?

5–38* In the early embryo of *Drosophila* many replication origins are active so that several can be observed in a single electron micrograph, as shown in Figure 5–16.

A. Identify the four replication bubbles in Figure 5–16. Indicate the approximate locations of the origins at which each replication bubble was initiated, and label the replication forks 1 through 8 from left to right across the figure.

B. Estimate how long it will take until forks #4 and #5, and forks #7 and #8, respectively, collide with each other. The distance between nucleotides in DNA is 0.34 nm, and eucaryotic replication forks move at about 50 nucleotides/second. For this problem disregard the nucleosomes evident in Figure 5–16 and assume that the DNA is fully extended.

5–39 Assuming that there were no time constraints on replication of the genome of a human cell, what would be the minimum number of origins that would be required? If replication had to be accomplished in an 8-hour S phase and replication forks moved at 50 nucleotides/second, what would be the minimum number of origins required to replicate the human genome? (Recall that the human genome comprises a total of 6.4×10^9 nucleotides on 46 chromosomes.)

5–40* One important rule for eucaryotic DNA replication is that no chromosome or part of a chromosome should be replicated more than once per cell cycle. Eucaryotic viruses must evade or break this rule if they are to produce multiple copies of themselves during a single cell cycle. The animal virus SV40, for example, generates 100,000 copies of its genome during a single cycle of infection. In order to accomplish this feat, it synthesizes a special protein, termed T-antigen (because it was first detected immunologically). T-antigen binds to the SV40 origin of replication and in some way triggers initiation of DNA replication.

The mechanism by which T-antigen initiates replication has been investigated *in vitro*. When purified T-antigen, ATP, and SSB protein were incubated with a circular plasmid DNA carrying the SV40 origin of replication, partially unwound structures, like the one shown in Figure 5–17, were observed in the electron microscope. In the absence of any one of these components, no unwound structures were seen. Furthermore, no unwinding occurred in an otherwise identical plasmid that carried a six-nucleotide deletion at the origin of replication. If care was taken in the isolation of the plasmid so that it contained no nicks (that is, it was a covalently closed circular DNA), no unwound structures were observed unless topoisomerase I was also present in the mixture.

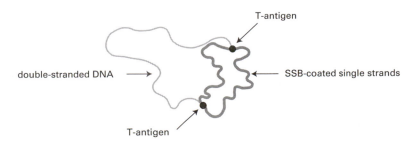

T-antigen

double-stranded DNA →

← SSB-coated single strands

T-antigen

Figure 5–17 A typical example of a plasmid molecule carrying an SV40 origin of replication after incubation with T-antigen, SSB protein, and ATP (Problem 5–40).

A. What activity in addition to site-specific DNA binding must T-antigen possess? How might this activity lead to initiation of DNA synthesis?

B. These experiments suggest that the structures observed in the electron microscope are unwound at the SV40 origin of replication. The location of the origin on the SV40 genome is precisely known. How might you use restriction nucleases in addition to the electron microscope to prove this point and to determine whether unwinding occurs in one or both directions away from the origin?

C. Why is there a requirement for topoisomerase I when the plasmid DNA is a covalently closed circle, but not when the plasmid DNA is linear? (Topoisomerase I introduces single-strand breaks into duplex DNA and then rapidly recloses them, so that the breaks have only a transient existence.)

D. Draw an example of the kind of structure that would result if T-antigen repeatedly initiated replication at an SV40 origin that was integrated into a chromosome.

5–41 Which one of the following statements about the newly synthesized strand of a human chromosome is correct?
A. It was synthesized from a single origin solely by continuous DNA synthesis.
B. It was synthesized from a single origin solely by discontinuous DNA synthesis.
C. It was synthesized from a single origin by a mixture of continuous and discontinuous DNA synthesis.
D. It was synthesized from multiple origins solely by continuous DNA synthesis.
E. It was synthesized from multiple origins solely by discontinuous DNA synthesis.
F. It was synthesized from multiple origins by a mixture of continuous and discontinuous DNA synthesis.
G. It was synthesized from multiple origins by either continuous or discontinuous DNA synthesis, depending on which specific daughter chromosome is being examined.

5–42* Fertilized frog eggs are very useful for studying the cell-cycle regulation of DNA synthesis. Foreign DNA can be injected into the eggs and followed independently of chromosomal DNA replication. For example, in one study ^{3}H-labeled viral DNA was injected. The eggs were then incubated in a medium supplemented with ^{32}P-dCTP and nonradioactive bromodeoxyuridine triphosphate (BrdUTP), which is a thymidine analog that increases the density of DNA into which it is incorporated. Incubation was continued for long enough to allow one or two cell cycles to occur; then the viral DNA was extracted from the eggs and analyzed on CsCl density gradients, which can separate DNA with 0, 1, or 2 BrdU-containing strands. Figure 5–18A and B show the density distribution of viral DNA after incubation for one and two cell cycles, respectively. If the eggs are bathed in cycloheximide (an inhibitor of protein synthesis) during the incubation, the results after incubation for one cycle or for two cycles are like those in Figure 5–18A. (The ability to look specifically at the viral DNA depends on a technical trick: the eggs were heavily irradiated with UV light before the injection to block chromosome replication.)

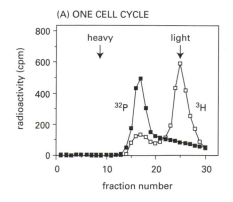

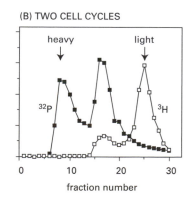

Figure 5–18 Density distribution of viral DNA after injection into fertilized frog eggs (Problem 5–42). (A) Results after one cell cycle. (B) Results after two cell cycles. The more dense end of the gradient is shown to the *left* (heavy), and the less dense end of the gradient is shown to the *right* (light).

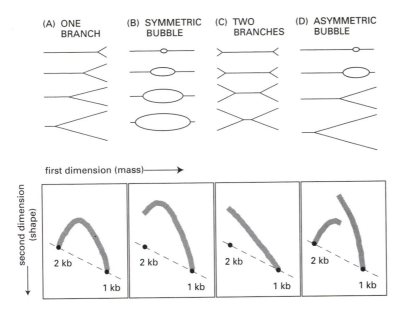

Figure 5–19 Expected patterns on two-dimensional gels for various DNA molecules (Problem 5–44). (A) Molecules with a single branch. (B) Molecules with a symmetrically located replication bubble. (C) Molecules with two branches. (D) Molecules with an asymmetrically located replication bubble. At the top 1-kb molecules are shown at progressive stages of replication to 2-kb molecules. At the bottom are shown the corresponding gel patterns that would result from a continuum of such intermediates.

(A) ONE BRANCH   (B) SYMMETRIC BUBBLE   (C) TWO BRANCHES   (D) ASYMMETRIC BUBBLE

first dimension (mass) ⟶

second dimension (shape)

2 kb / 1 kb   2 kb / 1 kb   2 kb / 1 kb   2 kb / 1 kb

A. Explain how the three density peaks in Figure 5–18 are related to replication of the injected DNA. Why was no ^{32}P radioactivity associated with the light peak, and why was no ^{3}H radioactivity associated with the heavy peak?

B. Does the injected DNA mimic the behavior that you would expect for the chromosomal DNA?

C. Why do you think that cycloheximide prevents the appearance of the most dense peak of DNA?

5–43   (**True/False**) Regardless of whether a gene is expressed in a given cell type, it will replicate at the same, characteristic time during S phase. Explain your answer.

5–44*   Autonomous replication sequences (ARSs), which confer stability on plasmids in yeast, are thought to be origins of replication. Proving that an ARS is an origin of replication, however, is difficult, mainly because it is very hard to obtain enough well-defined replicating DNA to analyze. This problem can be addressed using a two-dimensional gel-electrophoretic analysis that separates DNA molecules by mass in the first dimension and by shape in the second dimension. Because they have branches, replicating molecules migrate more slowly in the second dimension than do linear molecules of equal mass. By cutting replicating molecules with restriction nucleases, it is possible to generate a continuum of different branched forms that together give characteristic patterns on two-dimensional gels (Figure 5–19).

You apply this technique to the replication of a plasmid that contains *ARS1*. To maximize the fraction of plasmid molecules that are replicating, you synchronize a yeast culture and isolate DNA from cells in S phase. You then digest the DNA with BglII or PvuI, which cut the plasmid as indicated in Figure 5–20A. You separate the DNA fragments by two-dimensional electrophoresis and visualize the plasmid sequences by autoradiography after blot hydridization to radioactive plasmid DNA (Figure 5–20B).

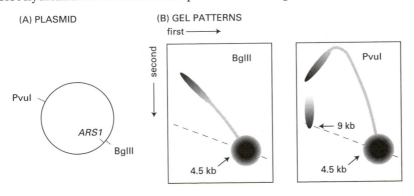

(A) PLASMID   (B) GEL PATTERNS   first ⟶   second   BglII   PvuI   PvuI   ARS1   BglII   ← 9 kb   4.5 kb   4.5 kb

Figure 5–20 Analysis of replication of an *ARS1*-containing plasmid (Problem 5–44). (A) Structure of the *ARS1* plasmid. (B) The two-dimensional gel patterns resulting from cleavage of replicating plasmids with BglII or PvuI (Problem 5–44).

A. What is the source of the intense spot of hybridization at the 4.5-kb position in both gels in Figure 5–20B?

B. Do the results of this experiment indicate that ARS1 is an origin of replication? Explain your answer.

C. There is a gap in the arc of hybridization in the PvuI gel pattern in Figure 5–20B. What is the basis for this discontinuity?

5–45   The shell of a *Drosophila* egg is made from more than 15 different chorion proteins, which are synthesized at a late stage in egg development by follicle cells surrounding the egg. The various chorion genes are grouped in two clusters, one on chromosome 3 and the other on the X chromosome. In each cluster the genes are closely spaced with only a few hundred nucleotides separating adjacent genes. During egg development the number of copies of the chorion genes increases by overreplication of a segment of the surrounding chromosome. This amplification can be detected by preparing DNA from eggs at different stages of development, digesting the DNA samples with a restriction nuclease, and analyzing them by Southern blotting using the cDNA from one chorion gene as a probe. As shown in Figure 5–21, the number of copies of this chorion gene increases substantially between stages 8 and 12, whereas the copy number of a control gene that is far removed from the chorion gene clusters stays constant.

The level of amplification around a chorion gene cluster can be determined using cloned probes covering the entire region. Measurements of the relative intensities of bands on autoradiographs such as the one in Figure 5–21 show that amplification of the chorion cluster on chromosome 3 is maximal in the region of the chorion genes but extends for nearly 50 kb on each side (Figure 5–22).

The DNA sequence responsible for amplification of the chorion cluster on chromosome 3 has been narrowed to a 510-nucleotide segment immediately upstream of one of the chorion genes. When this segment is moved to different places in the genome, those new sites are also amplified in follicle cells. No RNA or protein product seems to be synthesized from this amplification-control element.

A. Sketch what you think the DNA from an amplified cluster would look like in the electron microscope.

B. How many rounds of replication would be required to achieve a 60-fold amplification?

C. How do you think the 510-nucleotide amplification-control element promotes the overreplication of the chorion gene cluster?

5–46*   (**True/False**) If an origin of replication is deleted from a chromosome, the DNA on either side will ultimately be lost, as well, because it cannot be replicated. Explain your answer.

5–47   In yeast, origin selection is initiated by the origin recognition complex (ORC). ORC is a six-protein DNA-binding complex that recognizes DNA sequences within the yeast origin of replication. One of the components of ORC, Orc1, contains a protein motif (the Walker motif) that is commonly associated with binding and hydrolysis of ATP.

To determine whether binding of ATP, its hydrolysis, or both are required for the recognition of origin DNA by ORC, you carry out the following set of experiments. You first generate a mutation in the *ORC1* gene that introduces an amino acid change into the Walker motif of Orc1. You then isolate two versions of ORC: a wild-type form with normal Orc1 and a second form with mutant Orc1. Finally, you measure the binding of these two forms of ORC to origin DNA in the presence of different concentrations of ATP and a nonhydrolyzable analog, ATPγS. Binding under these various conditions, as revealed by a DNase I protection assay (DNase footprinting), is shown in Figure 5–23 for ATP. Exactly the same results were obtained when ATPγS was used.

A. Indicate the location of ORC binding on the origin DNA in Figure 5–23.

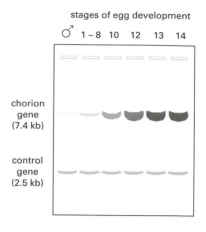

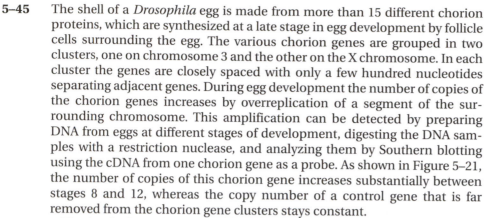

Figure 5–21 Blot hybridization of a chorion gene and a control gene in the male and at various stages of egg development in the female (Problem 5–45).

Problem 8–72 uses monoclonal antibodies to analyze members of the *Xenopus* ORC complex.

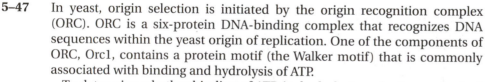

Figure 5–22 Levels of amplification in the region of the chromosome surrounding the chorion gene cluster (Problem 5–45).

B. Is ATP required for ORC to bind to origin DNA? How can you tell?

C. Is ATP hydrolysis required for ORC binding to origin DNA? How can you tell?

D. Is the Walker motif important to the function of ORC? Explain your answer.

5–48*   Origins of replication in mammalian cells have been difficult to define because the sequences required for proper origin function appear to extend over very large stretches of DNA. It may be that replication begins, not at fixed positions, but anywhere within an 'initiation zone.' You wish to characterize origin function within the initiation zone known to exist near the *DHFR* gene in hamster cells.

You synchronize a population of cells at the G1/S boundary and then release them from the block in the presence of bromodeoxyuridine (BrdU). This nucleotide analog is incorporated into the newly synthesized DNA, tagging it for isolation. Your goal is to isolate nascent leading strands. You harvest the DNA 15 minutes after release of the block in order to enrich for nascent strands near origins (Figure 5–24A). To avoid Okazaki fragments, which are ~100 nucleotides in length, you select BrdU-tagged single strands that are ~800 nucleotides long. You then quantify the amount of nascent strands at many sites in the 120-kb region that includes the initiation zone. You find that nascent strands are concentrated in a 10-kb region, as shown in Figure 5–24B.

A. What is your interpretation of the results shown in Figure 5–24B? Why are there peaks, and what does the double peak indicate?

B. Do you agree with the statement that replication begins anywhere within an initiation zone? Why or why not?

5–49   You have developed an assay for assembly of nucleosomes onto DNA in order to define the role of chromosome assembly factor 1 (CAF-1). You replicate SV40 DNA in a cell-free system in the presence of a labeled nucleotide to tag the replicated molecules, so you can follow them specifically in the assembly assay. After separating the DNA from soluble components in the cell-free replication system, you incubate it with purified CAF-1 and a source of histones. You then assay for nucleosome assembly by its effects on the supercoiling status of the circular SV40 genome. Genomes without nucleosomes remain relaxed, whereas genomes with nucleosomes become supercoiled. You separate different supercoiled forms of the DNA by electrophoresis on an agarose gel, which is stained to reveal total DNA and subjected to autoradiography to identify replicated DNA (Figure 5–25).

A. Is most of the SV40 DNA replicated or unreplicated? How can you tell?

B. Does CAF-1 assemble nucleosomes on replicated DNA, unreplicated DNA, or both? Explain your answer.

C. How do you suppose that CAF-1 recognizes the DNA it assembles into nucleosomes?

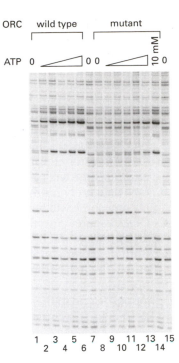

Figure 5–23 DNase footprinting assay to detect ORC binding to origin DNA (Problem 5–47). *Wedges* indicate increasing concentration of ATP in factors of ten, from 10 nM to 100 μM. In the footprinting assay one strand of the origin-containing DNA was labeled at one end. After ORC was allowed to bind, the complex was treated briefly with DNase I, which breaks the DNA at characteristic places except where it is protected by ORC.

(A)

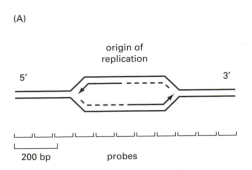

(B)

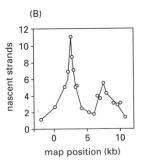

Figure 5–24 Analysis of an origin in the *DHFR* gene (Problem 5–48). (A) Arrangement of PCR probes around an origin of replication. The distribution of PCR probes—each about 100 bp long—is shown schematically below the origin. (B) Quantification of nascent strands in the initiation zone.

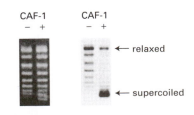

Figure 5–25 CAF-1 nucleosome assembly assay (Problem 5–49). The presence or absence of CAF-1 in the assembly assay is indicated by + or –, respectively.

5–50* The mechanism of DNA replication gives rise to the 'end-replication problem' for linear chromosomes. Over time, this problem leads to loss of DNA from the ends of chromosomes. In cells such as yeast loss of nucleotides during replication is balanced by addition of nucleotides by telomerase. In humans, however, telomerase is turned off in most somatic cells early in development, so that chromosomes become shorter with increasing rounds of replication. Consider one round of replication in a human somatic cell. Which one of the following statements correctly describes the status of the two daughter chromosomes relative to the parent chromosome?

A. One daughter chromosome will be shorter at one end; the other daughter chromosome will be normal at both ends.
B. One daughter chromosome will be shorter at both ends; the other daughter chromosome will be normal at both ends.
C. One daughter chromosome will be shorter at both ends; the other daughter chromosome will be shorter at only one end.
D. Both daughter chromosomes will be shorter at one end, which is the same end in the two chromosomes.
E. Both daughter chromosomes will be shorter at one end, which is the lopposite end in the two chromosomes.
F. Both daughter chromosomes will be shorter at both ends.

5–51 You have recently purified and partially sequenced a protein from a ciliated protozoan that seems to be the catalytic subunit of telomerase. You identify the homologous gene in fission yeast, which makes it possible to perform genetic studies that are impossible in the protozoan. You make a targeted deletion of the gene in a diploid strain and then induce sporulation to produce haploid organisms. All four spores germinate perfectly, and you are able to grow colonies on nutrient agar plates. Every 3 days, you re-streak colonies onto fresh plates. After four such serial transfers, the descendants of two of the original four spores grow poorly, if at all. You take cells from the 3-, 6-, and 9-day master plates, prepare DNA from them, and cleave the samples at a site about 35 nucleotides away from the start of the telomere repeats on two of the three chromosomes. You separate the fragments by gel electrophoresis, and hybridize them to a radioactive telomere-specific probe (Figure 5–26).

A. What is the average length of telomeres in fission yeast?
B. Do the data support the idea that you have identified yeast telomerase? If so, which spores lack telomerase?
C. Assuming that the generation time of this yeast is about 6 hours when growing on plates, by how much do the chromosomes shorten each generation in the absence of telomerase?
D. If you were to examine the yeast cells that stop dividing, do you suppose they would be longer, shorter, or about the same size as normal yeast cells?

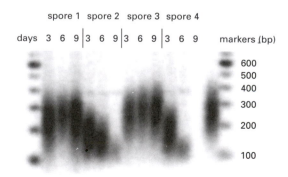

spore 1 spore 2 spore 3 spore 4

days 3 6 9│3 6 9│3 6 9│3 6 9   markers (bp)

600
500
400

300

200

100

Figure 5–26 Analysis of telomeres from four fission-yeast spores (Problem 5–51). The results from normal diploid yeast are shown at the *right*, adjacent to the markers.

DNA REPAIR

TERMS TO LEARN

base excision repair	nonhomologous end joining	SOS response
DNA repair	nucleotide excision repair	

5–52* Suppose that you are unable to repair the damage to DNA caused by the loss of purine bases. This defect causes the accumulation of about 5000 mutations per day in the DNA of each of your cells. As the average difference in DNA sequences between humans and chimpanzees is about 1%, it is only a matter of time until you turn into a chimp. What is wrong with this argument?

5–53 Your first foray into archaeological DNA studies ended in embarrassment. The dinosaur DNA sequences that you so proudly announced to the world later proved to be derived from contaminating modern human cells—probably your own. Setting your sights slightly lower, you decide to try to amplify residual mitochondrial DNA from a well-preserved Neanderthal skeleton. You also redesign your laboratory to minimize the possibility of stray contamination. You carefully prepare three different samples (A, B, and C) of bone from the femur and perform separate PCR reactions on them, one in the laboratory of a foreign collaborator. Sure enough, clear products of the expected size are seen in all three reactions. Cloned products from each PCR reaction are individually sequenced with the results shown in Figure 5–27.

```
Human  ACAGCAATCAACCCTCAACTATCACACATCAACTGCAACTCCAAAGCCACCCCT-CACCCAC
A1     .............T......-...T.........A...........A.GTT.T.A......
A2     .............T......G...T.........A...........A.G...T.G......
A3     .............T......G...T.........A...........A.G...T.A......
A4     .............T......G...T.T.......A...........A.G...T.A......
A5     .............T......G...T.........A...........A.G...T.A......
A6     .............T......G...T.........A...........A.G...T.A......
A7     .......T....T......G...T......G.A...........A.G...T.A......
A8     .............T......G...T.........A...........A.G...T.A......
A9     .............T......G...T.........A...........A.G...T.A......
A10    .............T......G...T.........A...........A.G...T.A......
A11    .............T......G...T.T.......A...........A.G...T.A......
A12    .............T......G...T.........A...........A.G...T.A......
A13    .............T....T.G...T.T.......A...........A.G...T.A......
A14    .............T..........T.........A...........A.G...T.A......
A15    .............T......G...T.........A...........A.G...T.A......
A16    .............T..........T.........A...........A.G...T.A......
A17    .............T......G...T.........A...........A.G...T.A......
A18    ...........................................G.........-......

B1     .............T......G...T.........A...........A.G...T.A......
B2     .............T......G...T.........A...........A.G...T.A......
B3     .T...........T......G...T.........A...........A.G...T.A......
B4     .............T......G...T.........A...........A.G...T.A......
B5     .............T......G...T.........T.AT.......A.G...T.A......
B6     .............T......G...T.........A...........A.G...T.A......
B7     .............T......G...T.........A...........A.G...T.A......
B8     .............T......G...T.........A...........A.G...T.A......
B9     .............T....T.G...T.........A...........A.G...T.A..T....
B10    .............T......G...T.........A.....T.....A.G...T.A......
B11    ...............................................-......
B12    ...............................................-......

C1     .............T......G...T.........A...........A.G...T.A......
C2     .............T......G...T.........A...........A.G...T.A......
C3     .............T......G...T.........A...........A.G...T.A......
C4     .T.....T...T......G...T...C.....A...........A.G...T.A......
C5     .T.....T...T......G...T.........A...........A.G...T.A..T....
C6     .T.....T...T......G...T.........A...........A.G...T.A..T....
C7     .............T......G...T.........A...........A.G...T.A..T....
C8     .T.....CT...T......G...T.........A...........A.G...T.A......
C9     .T.....T...T......G...T.........A...........A.G...T.A......
C10    .............T......G...T.........A...........A.G...T.A..T....
C11    ......C.............................................-......
C12    ...................................................-......
C13    ...................................................-......
C14    ...............T...................................-......
```

Figure 5–27 Sequences of mitochondrial DNA derived from Neanderthal samples (Problem 5–53). The sequence across the top is a reference human sequence. *Dots* indicate matches to the human sequence; *dashes* indicate missing nucleotides.

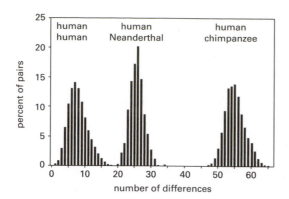

Figure 5–28 Pairwise comparisons of DNA sequences (Problem 5–53). The human–human distribution compared 994 individual, and distinct, human sequences with one another. These 994 sequences represent contemporary human mitochondrial lineages; that is, distinct sequences occurring in one or more individuals. The fraction of pairs with a given number of differences is plotted. The human–Neanderthal comparison is one Neanderthal sequence against the 994 contemporary human sequences. The human–chimpanzee comparison involved 986 contemporary human lineages with 16 contemporary chimpanzee lineages.

The sequence of the corresponding region of mitochondrial DNA from a human is shown at the top. Dots indicate matches to the human sequence; dashes indicate missing DNA.

To determine whether the common sequence differences you observe could be due to normal variation within the human population, you make pairwise comparisons of your consensus (most common) Neanderthal sequence with a large number of individual human sequences. You do the same for individual human sequences versus one another and versus chimpanzee sequences. Your pairwise comparisons are shown in Figure 5–28.

A. Have you successfully identified a Neanderthal mitochondrial DNA sequence? Explain your reasoning.

B. Did your extensive precautions in handling the sample eliminate human contamination?

C. What is the reason for choosing mitochondrial DNA for archaeological DNA studies? Wouldn't nuclear DNA sequences be more informative?

D. What would you consider to be the most important way to confirm or refute your findings?

5–54* (**True/False**) The variety of DNA repair mechanisms all depend on the existence of two copies of the genetic information: one in each of the two homologous chromosomes. Explain your answer.

5–55 Discuss the following statement: "The DNA repair enzymes that correct damage introduced by deamination and depurination must preferentially recognize such defects on newly synthesized DNA strands."

5–56* If you compare the frequency of the sixteen possible dinucleotide sequences in *E. coli* and human cells, there are no striking differences except for one dinucleotide, 5′-CG-3′. The frequency of CG dinucleotides in the human genome is significantly lower than in *E. coli* and significantly lower than expected by chance. Why do you suppose that CG dinucleotides are underrepresented in the human genome?

5–57 (**True/False**) Spontaneous depurination and the removal of a deaminated C by uracil DNA glycosylase both leave an identical intermediate, which is the substrate recognized by AP endonuclease. Explain your answer.

5–58* Several genes in *E. coli*, including *uvrA, uvrB, uvrC,* and *recA*, are involved in repair of UV damage. Strains of *E. coli* that are defective in any one of these genes are much more sensitive to killing by UV light than are nonmutant (wild-type) cells, as shown for *uvrA* and *recA* strains in Figure 5–29A. The sensitivity of cells that are mutant in two of these genes varies much more than that of the single mutants. The combinations of *uvr* mutations with one another show little increase in sensitivity relative to the *uvr* single mutants. The combination of *recA* with any of the *uvr* mutations, however, gives a strain that is exquisitely sensitive to UV light, as shown for *uvrArecA* on an expanded scale in Figure 5–29B.

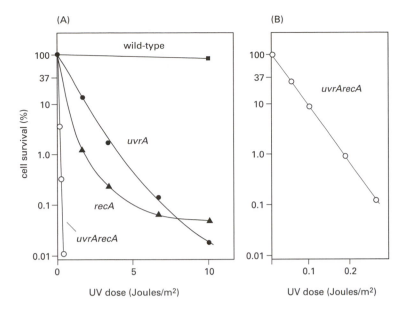

Figure 5–29 Cell survival as a function of UV dose (Problem 5–58). (A) Survival of wild-type cells, a *uvrA* mutant, a *recA* mutant, and a *uvrArecA* double mutant. (B) An expanded scale for *uvrArecA* survival.

A. Why do combinations of a *recA* mutation with a *uvr* mutation give an extremely UV-sensitive strain of bacteria, whereas combinations of mutations in different *uvr* genes are no more UV sensitive than the individual mutants?

B. According to the Poisson distribution, when a population of bacteria receives an overall average of one lethal 'hit,' 37% (e^{-1}) will survive because they actually receive no hits. For the double mutant *uvrArecA*, a dose of 0.04 Joules/m^2 gives 37% survival (Figure 5–29B). Calculate how many pyrimidine dimers constitute a lethal hit for the *uvrArecA* strain. *E. coli* has 4.6×10^6 base pairs in its genome (assume 50% GC). Exposure of DNA to UV light at 400 Joules/m^2 converts 1% of the total pyrimidine pairs (TT, TC, CT, plus CC) to pyrimidine dimers.

5–59 In addition to killing bacteria, UV light causes mutations. You have measured the UV-induced mutation frequency in wild-type *E. coli* and in strains defective in either the *uvrA* gene or the *recA* gene. The results are shown in Table 5–3. Surprisingly, these strains differ dramatically in their mutability by UV.

A. Assuming that the *recA* and *uvrA* gene products participate in different pathways for repair of UV damage, decide which pathway is more error prone. Which pathway predominates in wild-type cells?

B. The error-prone pathway results from misincorporation of nucleotides opposite a site of unrepaired damage using specialized DNA polymerases. One of these tends to incorporate an adenine nucleotide opposite a pyrimidine dimer. Is this a good strategy for dealing with UV damage? Calculate the frequency of base changes (mutations) using A insertion versus random incorporation (each nucleotide with equal probability) for *E. coli,* where pyrimidine dimers are approximately 60% TT, 30% TC and CT, and 10% CC.

5–60* (**True/False**) Only the first step in DNA repair is catalyzed by enzymes that are unique to the repair process; the later steps are catalyzed by enzymes that play more general roles in DNA metabolism. Explain your answer.

TABLE 5–3 Frequency of UV-induced mutations in various strains of *E. coli* (Problem 5–59).

STRAIN	SURVIVAL (%)	MUTATIONS/10^{10} SURVIVORS
Wild type	100	400
recA	10	1
uvrA	10	40,000

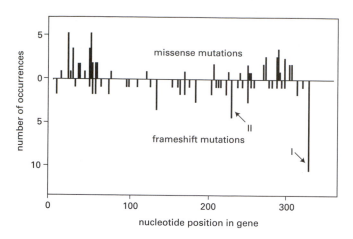

Figure 5–30 Pattern of UV-induced mutations in the *E. coli lacI* gene (Problem 5–61).

5–61 The patterns of UV-induced mutations occurring in the *E. coli lacI* gene have been extensively analyzed. Figure 5–30 shows the total number of independently isolated missense (amino acid substitution) mutations (above the line) and frameshift (change in the reading frame) mutations (below the line). There are almost equal numbers of mutations in each category. Missense mutations were identified by scoring for loss of function of the lac repressor, the protein specified by the *lacI* gene; frameshift mutations were scored by a gene-fusion assay, which is independent of the function of the lac repressor.

A. Why do you think that there are so many more missense mutations near the ends of the gene than in the middle? Why do you think that the frameshift mutations are more or less evenly distributed across the gene (except for one or two 'hot spots')?

B. DNA sequence analysis of the hot spot labeled I in Figure 5–30 revealed that its wild-type sequence was TTTTTC and that the mutated sequence was TTTTC. The next most common mutation (labeled II in Figure 5–30) was the change from GTTTTC to GTTTC. Analysis of other frameshifts indicated that they resulted most commonly from the loss of one base: no insertions were found. Based on what you know of the nature of UV damage, can you suggest a molecular mechanism for the loss of a single base pair?

5–62* Humans with the rare genetic disease xeroderma pigmentosum (XP) are extremely sensitive to sunlight and are prone to develop malignant skin cancers. Defects in any one of eight genes can cause XP. Seven XP genes encode proteins involved in nucleotide excision repair (NER). The eighth gene is associated with the XP variant (XP-V) form of the disease. Cells from XP-V patients are fine for NER but are deficient in their ability to replicate UV-damaged DNA. Using a clever assay, you manage to purify from a normal cell extract the enzyme that is missing from XP-V cells. You test its ability to synthesize DNA from the simple template in Figure 5–31A, which contains a TT sequence. This template can be modified to contain either a cyclobutane thymine dimer or the somewhat rarer 6-4 photoproduct (two Ts linked in a different way). You compare the ability of DNA polymerase α (a normal replicative polymerase) and the XP-V enzyme to synthesize DNA from the undamaged template, the template with a cyclobutane dimer, and the template with a 6-4 photoproduct. By labeling the primer at its 5′ end and denaturing the reaction products, it is possible to determine whether DNA synthesis has occurred (Figure 5–31B).

A. Is the XP-V enzyme a DNA polymerase? Why or why not?

B. How does the XP-V enzyme differ from DNA polymerase α on an undamaged template? On a template with a cyclobutane dimer? On a template with a 6-4 photoproduct?

C. How accurately do you suppose that the XP-V enzyme copies normal DNA? Would you guess it to be error-prone or faithful?

D. If NER is normal in patients with XP-V, why are they sensitive to sunlight and prone to skin cancers?

(A) SUBSTRATE

```
5' [³²P]-CACTGACTGTATG
              GTGACTGACATACTACTTCTACGACTGCTC - 5'
```

(B) RESULTS

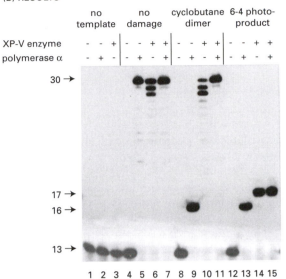

Figure 5–31 Comparison of DNA polymerase α and the XP-V enzyme (Problem 5–62). (A) Substrate for DNA polymerization. The adjacent Ts indicated by the *bracket* can be specifically modified to produce a cyclobutane dimer or a 6-4 photoproduct. (B) Results obtained with DNA polymerase α and the XP-V enzyme on damaged and undamaged templates. Numbers on the *left* indicate lengths of single-stranded DNA in nucleotides.

5–63 Mutagens such as *N*-methyl-*N'*-nitro-*N*-nitrosoguanidine (MNNG) and methyl nitrosourea (MNU) are potent DNA-methylating agents and are extremely toxic to cells. Nitrosoguanidines are used in research as mutagens and clinically as drugs in cancer chemotherapy because they preferentially kill cells in the act of replication.

The original experiment that led to the discovery of the alkylation repair system in bacteria was designed to assess the long-term effects of exposure to low doses of MNNG (as in chemotherapy), in contrast to brief exposure to large doses (as in mutagenesis). Bacteria were placed first in a low concentration of MNNG (1 μg/mL) for 1.5 hours and then in fresh medium lacking MNNG. At various times during and after exposure to the low dose of MNNG, samples of the culture were treated with a high concentration (100 μg/mL) of MNNG for 5 minutes and then tested for viability and the frequency of mutants. As shown in Figure 5–32, exposure to low doses of MNNG temporarily increased the number of survivors and decreased the frequency of mutants among the survivors. As shown in Figure 5–33, this adaptive response to low doses of MNNG was prevented if chloramphenicol (an inhibitor of protein synthesis) was included in the incubation.

A. Does the adaptive response of *E. coli* to low levels of MNNG require activation of preexisting protein or the synthesis of new protein?

B. Why do you think the adaptive response to a low dose of MNNG is so short-lived?

5–64* The nature of the mutagenic lesion introduced by MNNG and the mechanism

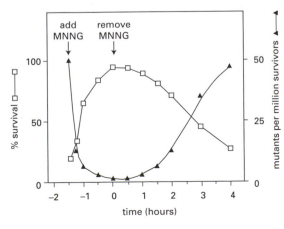

Figure 5–32 Adaptive response of *E. coli* to low doses of MNNG (Problem 5–63). MNNG at 1 μg/mL was present from −1.5 to 0 hours. Samples were removed at various times and treated briefly with 100 μg/mL MNNG to assess the number of survivors and the frequency of mutants.

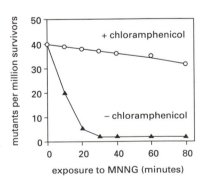

Figure 5–33 Effects of chloramphenicol on the adaptive response to low doses of MNNG (Problem 5–63). After different times of exposure to 1 μg/mL MNNG, samples were removed and treated with 100 μg/mL MNNG to measure susceptibility to mutagenesis.

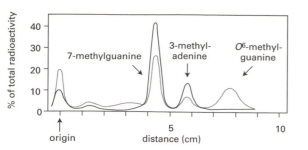

Figure 5–34 Chromatographic separation of labeled methylated purines in the DNA of untreated bacteria and bacteria treated with low doses of MNNG (Problem 5–64). *Solid line* indicates methylated purines from untreated bacteria; *dashed line* indicates those from MNNG-treated bacteria.

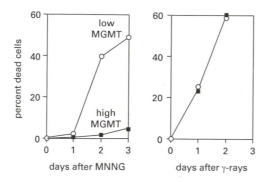

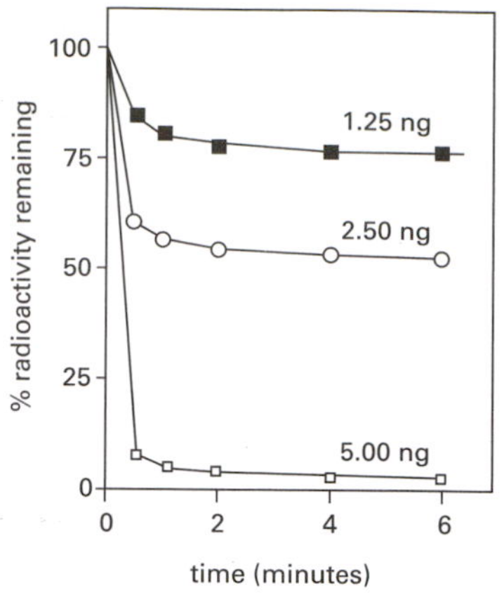

Figure 5–35 Removal of ^{3}H-labeled methyl groups from DNA by purified methyltransferase enzyme (Problem 5–64). The quantities of purified enzyme are indicated.

of its removal from DNA were identified in the following experiments. To determine the nature of the mutagenic lesion, untreated bacteria and bacteria that had been exposed to low doses of MNNG were incubated with 50 μg/mL ^{3}H-MNNG for 10 minutes. Their DNA was isolated and hydrolyzed to nucleotides, and the radioactive purines were then analyzed by paper chromatography as shown in Figure 5–34.

To examine the mechanism of removal of the mutagenic lesion, the enzyme responsible for removal was first purified. The kinetics of removal were studied by incubating different amounts of the enzyme (molecular weight 19,000) with DNA containing 0.26 pmol of the mutagenic base, which was radioactively labeled with ^{3}H. At various times samples were taken, and the DNA was analyzed to determine how much of the mutagenic base remained (Figure 5–35). When the experiment was repeated at 5°C instead of 37°C, the initial rates of removal were slower, but exactly the same end points were achieved.

A. Which methylated purine is responsible for the mutagenic action of MNNG?
B. What is peculiar about the kinetics of removal of the methyl group from the mutagenic base? Is this peculiarity due to an unstable enzyme?
C. Calculate the number of methyl groups that are removed by each enzyme molecule. Does this calculation help to explain the peculiar kinetics?

5–65 Alkylating agents, which are highly cytotoxic, are commonly used as anticancer drugs. The mechanism by which they kill cells, however, is not clear. Do they kill cells by damaging DNA, or by reacting with protein or lipid components of the cell? You know that O^6-methylguanine is highly mutagenic and you wonder whether it might also be responsible for cell killing. To test this possibility you construct a cell line that expresses high levels of O^6-methylguanine methyltransferase (MGMT), which efficiently removes such groups from the DNA. You then compare the sensitivity of these cells, and of control cells that express very little MGMT, to killing by the alkylating agent MNNG. As a control, you compare sensitivity to γ-irradiation, which kills cells by introducing double-strand breaks, and thus should not be dependent on cellular levels of MGMT. The results of your experiments are shown in Figure 5–36. What is responsible for the cytotoxic effects of alkylating agents?

5–66* Budding yeast cells preferentially repair double-strand breaks by homologous recombination. Haploid cells are especially sensitive to agents that cause double-strand breaks in DNA. If the breaks occur in the G_1 phase of the cell cycle, most of the yeast cells die; however, if the breaks occur in the G_2 phase, a much higher fraction of cells survive. Explain these results.

Figure 5–36 Killing by MNNG and γ-rays of cells that express high or low levels of MGMT (Problem 5–65). Results for cells expressing low levels of MGMT are shown in *open circles*; results for cells expressing high levels of MGMT are shown in *filled squares*.

(A) ELECTRON MICROGRAPH

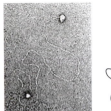

(B) NUCLEASE DIGESTION

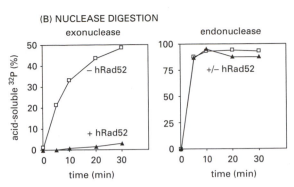

Figure 5–37 Analysis of role of hRad52 in repair of double-strand breaks (Problem 5–67). (A) Binding of hRad52 to linear DNA. A schematic representation is shown adjacent to the micrograph. (B) Nuclease digestion of DNA in the presence (+) and absence (−) of bound hRad52.

5–67 Most eucaryotic cells use two different mechanisms to repair double-strand breaks in DNA: homologous recombination and nonhomologous end joining. The recombination pathway is favored by a homolog of the Rad52 protein, first defined in yeast. The end-joining pathway is stimulated by relatives of the protein known as Ku.

You have purified human Rad52 protein (hRad52) and are studying its properties. When you mix hRad52 with linear DNA that has short (300-nucleotide) single-strand tails and examine the mixture in the electron microscope, you see structures like the one shown in Figure 5–37A. Such structures are much rarer when the linear DNA is blunt ended. You then expose hRad52-bound, uniformly radiolabeled DNA to nucleases and measure digestion of DNA by release of soluble fragments. As shown in Figure 5–37B, hRad52 protects the DNA from an exonuclease but not from an endonuclease.

A. Where does hRad52 bind on linear DNA? What features of the DNA are important for hRad52 binding?

B. How does hRad52 protect against digestion by the exonuclease, but not against digestion by the endonuclease?

C. How do the properties of hRad52 revealed by these observations fit into its role in recombinational repair of double-strand breaks?

5–68* The SOS pathway in *E. coli* represents an emergency response to DNA damage. As illustrated in Figure 5–38, under normal conditions the SOS set of damage-inducible genes is turned off by the LexA repressor, which also

Figure 5–38 The SOS response in *E. coli* (Problem 5–68).

UNINDUCED

damage-induced genes

lexA

recA

repression — repression | low basal rate   repression | low basal rate

LexA — RecA

activated RecA

RecA is activated when it binds to ssDNA formed by DNA-damaging agents

activated RecA mediates LexA cleavage

SOS response

genes expressed

cleaved LexA is inactive

LexA

RecA

high derepressed synthesis

damage-induced genes

lexA

recA

INDUCED

partially represses its own synthesis and that of RecA. In response to DNA damage, a signal (thought to be single-stranded DNA) activates RecA, which then mediates the cleavage of LexA. In the absence of LexA, all genes are maximally expressed. The SOS response increases cell survival in the face of DNA damage and transiently increases the mutation rate, hence variability, in the bacterial population. Although indispensable during an emergency, the constant expression of the SOS genes would be deleterious.

One aspect of the regulation of the SOS response appears paradoxical: the expression of LexA (the repressor of the SOS response) is substantially increased during an SOS response. If the object of the response is maximal expression of the damage-induced genes, why should the repressor be expressed at high levels? Put another way, why is LexA not expressed at a low rate all the time? Do you see any advantage that the induced expression of LexA offers for the regulation of the SOS response?

5–69   In addition to UvrABC endonuclease repair, recombinational repair, and SOS repair, bacteria have an even more potent repair system for dealing with pyrimidine dimers. This phenomenon was discovered by careful observation to define an uncontrolled variable in an investigation of the effect of UV light on bacteria—not unlike the scenario described below.

You and your advisor are trying to isolate mutants in *E. coli* using UV light as the mutagenic agent. To get plenty of mutants, you find it necessary to use a dose of irradiation that kills 99.99% of the bacteria. You have been getting much more consistent results than your advisor, who also requires 10-fold to a 100-fold higher doses of irradiation to achieve the same degree of killing. He wonders about the validity of your results since you always do your experiments at night after he has left. When he insists that you come in the morning to do the experiments in parallel, both of you are surprised when you get exactly the same results. You are a bit chagrined because the results are more like your advisor's, and you had confidently assumed you were better at doing lab work than your advisor, who is rarely seen in a lab coat these days. When you repeat the experiments in parallel at night, however, your advisor is surprised at the results, which are exactly as you had described them.

Now that you believe one another's observations, you make rapid progress. You find that you need a higher dose of UV light in the afternoon than in the morning to get the same degree of killing. Even higher doses are required on sunny days than on overcast days. Your laboratory faces west. What is the variable that has been plaguing your experiments?

GENERAL RECOMBINATION

TERMS TO LEARN

allele	genetic recombination	RecA protein
gene conversion	Holliday junction	synapsis
general recombination	hybridization	

5–70*   **(True/False)** General recombination requires long regions of homologous DNA on both partners in the exchange, whereas site-specific recombination requires short, specific nucleotide sequences, which in some cases need be present on only one of the exchanging partners. Explain your answer.

5–71   Using Figure 5–39A as a guide, draw the products of a crossover recombination event between the homologous regions of the molecules represented in Figure 5–39B. In the figure the DNA duplex is represented schematically by a single line and the targets for homologous recombination are represented by arrows.

5–72*   When plasmid DNA is extracted from *E. coli* and examined by electron microscopy, the majority of plasmids are monomeric circles, but there are a

(A)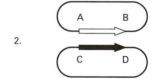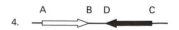

(B)
1.
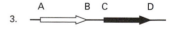
2.
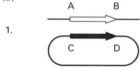
3.
4.
5.
6.
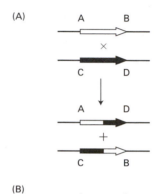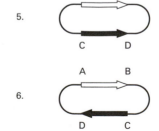

Figure 5–39 A variety of recombination substrates (Problem 5–71).

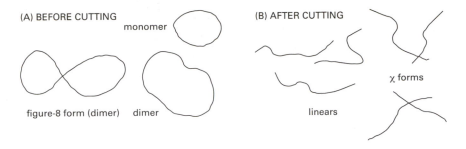

(A) BEFORE CUTTING

monomer

figure-8 form (dimer)   dimer

(B) AFTER CUTTING

χ forms

linears

Figure 5–40 Structures of plasmid molecules (Problem 5–72). (A) Before digestion. (B) After digestion with a single-cut restriction nuclease.

variety of other forms, including dimeric and trimeric circles. In addition, about 1% of the molecules appear as figure-8 forms, in which the two loops are equal (Figure 5–40A).

You suspect that the figure 8s are recombination intermediates in the formation of a dimer from two monomers (or two monomers from a dimer). However, to rule out the possibility that they represent twisted dimers or touching monomers, you cut the DNA sample with a restriction nuclease, which cuts at a single site in the monomer, and then examine the molecules. After cutting, only two forms are seen: 99% of the DNA molecules are linear monomers, and 1% are χ (Chi) forms (Figure 5–40B). You note that the χ forms have an interesting property: the two longer arms are the same length, as are the two shorter arms. In addition, the sum of the lengths of a long arm and a short arm is equal to the length of the monomer plasmid. The position of the crossover point, however, is completely random.

Unsure of yourself and feeling you are probably missing some hidden artifact, you show your pictures to a friend. She points out that your observations prove you are looking at recombination intermediates that arose by random pairing at homologous sites.

A. Is your friend correct? What is her reasoning?

B. How would you expect your observations to differ if you repeated the experiments in a strain of *E. coli* that carried a nonfunctional *recA* gene?

C. What would the χ forms have looked like if the figure 8s were intermediates in a site-specific recombination between the monomers?

D. What would the χ forms have looked like if the figure 8s were intermediates in a totally random, nonhomologous recombination between the monomers?

5–73   Specific DNA sequences known as Chi sites locally stimulate RecBCD-mediated homologous recombination in *E. coli*. Interaction between the RecBCD protein and the Chi site stimulates a rate-limiting step in the recombination pathway. To study this interaction in detail, RecBCD was purified and incubated with a linear, double-stranded DNA fragment containing a Chi site (Figure 5–41).

Different samples of the linear DNA were labeled specifically at the 5′ end on the left (5′L), at the 5′ end on the right (5′R), at the 3′ end on the left (3′L), and at the 3′ end on the right (3′R). Each sample was then incubated in a reaction buffer containing RecBCD. As a control, a separate aliquot of labeled DNA was incubated in the reaction buffer without RecBCD. After 1 hour of reaction the DNA was denatured by boiling, and the resulting single strands were separated by electrophoresis through a polyacrylamide gel. The pattern of radioactively labeled DNA fragments is shown in Figure 5–42. As a control, a sample of 3′R that had been incubated with RecBCD was run on the gel without first denaturing it (Figure 5–42, lane 6).

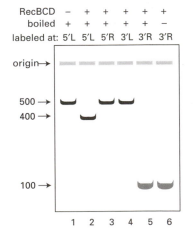

Figure 5–42 Results of incubating RecBCD with Chi-containing DNA fragments labeled at the ends of defined strands (Problem 5–73). Numbers adjacent to bands indicate the length of the labeled fragment in nucleotides. All samples were denatured before electrophoresis, except the unboiled sample in lane 6.

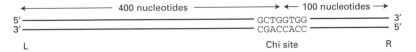

Figure 5–41 A linear DNA fragment containing a Chi site (Problem 5–73). The sequence of the Chi site is shown. L and R indicate the left and right ends of the fragment.

A. What is the evidence that RecBCD cuts the DNA at the Chi site? Does it cut one or both strands? If you decide it cuts only one strand, specify the strand and indicate your reasoning.
B. What is the evidence that RecBCD can act as a helicase—that is, what is the evidence that it can separate the strands of duplex DNA?
C. How might the action of RecBCD stimulate homologous recombination in the neighborhood of a Chi site?

5–74* RecA protein catalyzes both the initial pairing step of recombination and subsequent branch migration in *E. coli*. It promotes recombination by binding to single-stranded DNA and catalyzing the pairing of such coated single strands to homologous double-stranded DNA. One assay for the action of RecA is the formation of double-stranded DNA circles from a mixture of double-stranded linear molecules and homologous single-stranded circles, as illustrated in Figure 5–43. This reaction proceeds in two steps: circles pair with linears at an end and then they branch migrate until a single-stranded linear DNA is displaced.

One important question about the RecA reaction is whether branch migration is directional. This question has been studied in the following way. Single-stranded circles, which were uniformly labeled with ^{32}P, were mixed with unlabeled double-stranded linear molecules in the presence of RecA. As the single-stranded DNA pairs with the linear DNA, it becomes sensitive to cutting by restriction nucleases, which do not cut single-stranded DNA. By sampling the reaction at various times, digesting the DNA with a restriction nuclease, and separating the labeled fragments by electrophoresis, you obtain the pattern shown in Figure 5–44A.

A. By comparing the time of appearance of labeled fragments with the restriction map of the circular DNA in Figure 5–44, deduce which end (5' or 3') of the minus strand of the linear DNA the circular plus strand pairs with initially. Also deduce the direction of branch migration along the minus strand. (The linear double-stranded DNA was cut at the boundary between fragments *a* and *c* on the restriction map.)
B. Estimate the rate of branch migration, given that the length of this DNA is 7 kb.
C. What would you expect to happen if the linear, double-stranded DNA carried an insertion of 500 nonhomologous nucleotides between restriction fragments *e* and *a*?

5–75 Discuss the following statement: "The cross-strand exchange contains two distinct pairs of strands (crossing strands and noncrossing strands), which cannot be interconverted without breaking the phosphodiester backbone of at least one strand."

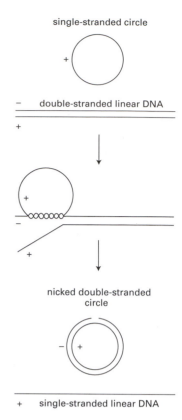

Figure 5–43 Strand assimilation assay for RecA (Problem 5–74). The single-stranded (+) circle is complementary to the minus (–) strand and identical to the plus strand of the duplex.

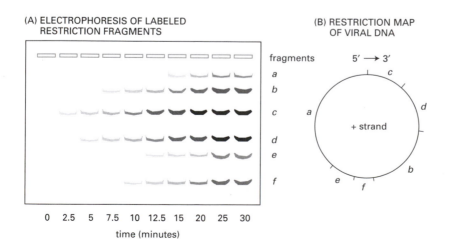

Figure 5–44 Analysis of branch migration catalyzed by RecA (Problem 5–74). (A) Electrophoretic separation of labeled restriction fragments as a function of time of incubation with RecA. (B) Sites of cleavage represented on the single-stranded DNA circle. Clockwise around the circle is 5'-to-3'.

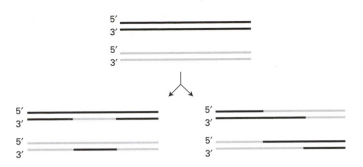

Figure 5–45 Parental and recombinant duplexes (Problem 5–76).

5–76* Two homologous parental duplexes and two sets of potential recombination products are illustrated in Figure 5–45. Diagram a Holliday junction between the parental duplexes that could generate the indicated recombination products. Label the left end of each strand in the Holliday junction 5′ or 3′ so that the relationship to the parental and recombinant duplexes is clear. Indicate which strands need to be cut to generate each set of recombination products. Finally, draw the recombinants as they would look after one round of replication.

5–77 The mechanism of Holliday-junction cleavage by RuvC has been investigated using artificial junctions created by annealing oligonucleotides together (Figure 5–46A). Each of the duplexes involved in these junctions has unique sequences at its ends, which allowed the oligonucleotides to anneal to form the indicated Holliday junctions; they also possess a core of 11 nucleotides that are homologous. The Holliday junction can branch migrate within the core region. A dimer of RuvC binds to Holliday junctions and cleaves a pair of adjacent strands between nucleotides 3 and 4 in the sequences 5′-ATTG, 5′-ATTC, 5′-TTTG, or 5′-TTTC (Figure 5–46A). These four sequences are represented by the consensus sequence 5′-^{A}/$_T$TT^G/$_C$.

You want to know whether the two subunits of the RuvC dimer coordinate the cleavages on the two strands or can act independently. To investigate this you make the three Holliday junctions shown in Figure 5–46A: one with two cleavable sequences, one with two uncleavable sequences, and one with one cleavable and one uncleavable sequence. You label the 5′ end of one strand in each, incubate them with RuvC, and analyze cleavage by electrophoresis on denaturing gels, so that the oligonucleotides separate from one another. The results are shown in Figure 5–46.

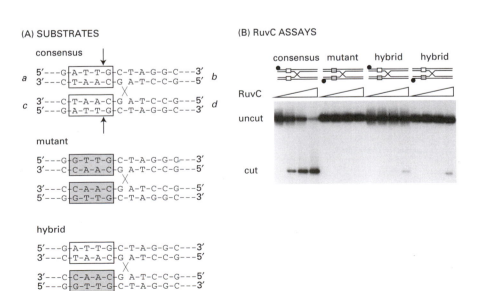

Figure 5–46 Analysis of Holliday junction cleavage (Problem 5–77). (A) Effects of DNA sequence on resolution of Holliday junctions. Cleavable sequences are shown in *open boxes*; uncleavable sequences are shown in *shaded boxes*. (B) Resolution of Holliday junctions by RuvC. Increasing concentrations of RuvC in each lane are indicated schematically by *triangles*. The location of the labeled 5′ end is indicated by *black dots*.

A. What fraction of all possible four-nucleotide sequences are cleaved by RuvC?

B. Does the requirement for resolution at a limited set of specific sequences unduly restrict the sites in the *E. coli* genome at which Holliday junctions can be resolved?

C. Do the two subunits of the RuvC dimer coordinate their cleavage of the two strands or can they act independently? Explain your reasoning.

D. Draw out the duplex products generated by RuvC resolution of the 'consensus' Holliday junction in Figure 5–46A, the one with its ends marked by letters.

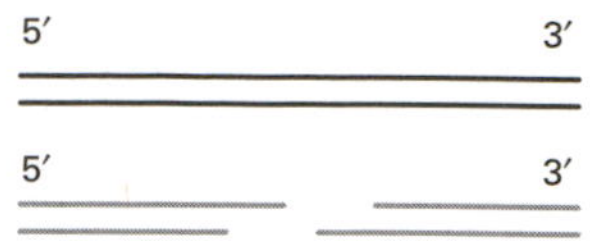

5–78* Draw the structure of the double Holliday junction that would result from strand invasion by both ends of the broken duplex into the intact homologous duplex shown in Figure 5–47. Indicate how DNA synthesis would be used to fill in any single-strand gaps in your double Holliday junction.

5–79 (**True/False**) All known mechanisms of gene conversion require a limited amount of DNA synthesis. Explain your answer.

5–80* Why is recombination between similar, but nonidentical, repeated sequences a problem in human cells? How does the mismatch-repair system protect against such recombination events?

Figure 5–47 A broken duplex with single-strand tails ready to invade an intact homologous duplex (Problem 5–78).

SITE-SPECIFIC RECOMBINATION

TERMS TO LEARN

bacteriophage	nonretroviral retrotransposon	transposable element
conservative site-specific recombination	retroviral-like retrotransposon	transpositional site-specific recombination
cut-and-paste transposition	retrovirus	transposon
DNA-only transposon	reverse transcriptase	
	site-specific recombination	

5–81 Discuss the following statement: "Transposable elements move around the genome, but rarely integrate into the middle of a gene, for gene disruption could be lethal to the cell. Such indiscriminate movement is selected against by evolution because it would kill not only the cell but the transposable element, as well."

5–82* You are studying the procaryotic transposon Tn10 and have just figured out an elegant way to determine whether Tn10 replicates during transposition or moves directly without intervening DNA replication. Your idea is based on the key difference between these two mechanisms: both parental strands of the Tn10 will move if transposition is nonreplicative, whereas only one parental strand will move if transposition is replicative (Figure 5–48).

You plan to mark the individual strands by annealing strands from two different Tn10s. Both Tn10s contain a gene for tetracycline resistance and a gene for lactose metabolism (*lacZ*), but in one, the *lacZ* gene is inactivated by a mutation. This difference provides a convenient way to follow the two Tn10s since *lacZ*⁺ bacterial colonies (when incubated with an appropriate substrate) turn blue, but *lacZ*⁻ colonies remain white. You denature and

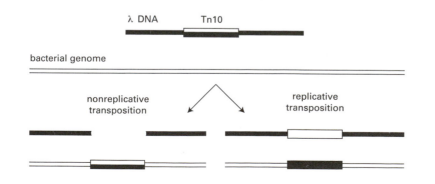

Figure 5–48 Replicative and nonreplicative transposition of a transposable element (Problem 5–82). The transposable element is shown as a heteroduplex, which is composed of two genetically different strands—one black and one white. During replicative transposition, one strand stays with the donor DNA and one strand is transferred to the bacterial genome. In nonreplicative transposition, the transposable element is cut out of the donor DNA and transferred entirely to the bacterial genome.

reanneal a mixture of the two transposon DNAs, which produces an equal mixture of heteroduplexes and homoduplexes (Figure 5–49). You introduce them into *lacZ⁻* bacteria, and spread the bacteria onto Petri dishes that contain tetracycline and the color-generating substrate.

Once inside a bacterium, the transposon will move (at very low frequency) into the bacterial genome, where it confers tetracycline resistance on the bacterium. The rare bacterium that gains a Tn10 survives the selective conditions and forms a colony. When you score a large number of such colonies, you find that roughly 25% are white, 25% are blue, and 50% are mixed, with one blue sector and one white sector.

A. Explain the source of each kind of bacterial colony and decide whether the results support a replicative or a nonreplicative mechanism for Tn10 transposition.

B. You performed these experiments using a recipient strain of bacteria that was incapable of repairing mismatches in DNA. How would you expect the results to differ if you used a bacterial strain that could repair mismatches?

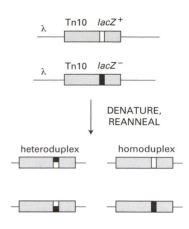

Figure 5–49 Formation of a mixture of heteroduplexes and homoduplexes by denaturing and reannealing two different Tn10 genomes (Problem 5–82). The Tn10 DNA is shown as a *box*. The sequence differences between the two Tn10s are indicated by the *black* and *white segments*.

5–83   The Ty elements of the yeast *Saccharomyces cerevisiae* move to new locations in the genome by transposition through an RNA intermediate. Normally, the Ty-encoded reverse transcriptase is expressed at such a low level that transposition is very rare. To study the transposition process, you engineer a cloned version of the Ty element so that the gene for reverse transcriptase is linked to the galactose control elements. You also 'mark' the element with a segment of bacterial DNA so that you can detect it specifically and thus distinguish it from other Ty elements in the genome. As a target gene to detect transposition, you use a defective histidine (*his⁻*) gene whose expression is dependent on the insertion of a Ty element near its 5′ end. You show that yeast cells carrying a plasmid with your modified Ty element generate *HIS⁺* colonies at a frequency of 5×10^{-8} when grown on glucose. When the same cells are grown on galactose, however, the frequency of *HIS⁺* colonies is 10^{-6}: a 20-fold increase.

You notice that cultures of cells with the Ty-bearing plasmid grow normally on glucose but very slowly on galactose. To investigate this phenomenon, you isolate individual colonies that arise under three different conditions: *his⁻* colonies grown in the presence of glucose, *his⁻* colonies grown in the presence of galactose, and *HIS⁺* colonies grown in the presence of galactose. You eliminate the plasmid from each colony, isolate DNA from each culture, and analyze it by gel electrophoresis and blot hybridization, using the bacterial marker DNA as a probe. Your results are shown in Figure 5–50.

A. Why does transposition occur so much more frequently in cells grown on galactose than it does in cells grown on glucose?

B. As shown in Figure 5–50, *his⁻* cells isolated after growth on galactose have about the same number of marked Ty elements in their chromosomes as *HIS⁺* cells that were isolated after growth on galactose. If transposition is independent of histidine selection, why is the frequency of Ty-induced *HIS⁺* colonies so low (10^{-6})?

C. Why do you think it is that cells with the Ty-bearing plasmid grow so slowly on galactose?

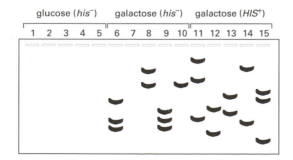

Figure 5–50 Analysis of cells that harbored a plasmid carrying a modified Ty element (Problem 5–83). Cells were initially grown on glucose or galactose as a carbon source. *his⁻* cells were grown in the presence of histidine; *HIS⁺* cells were grown in its absence. Bands indicate restriction fragments that hybridized to the marker DNA originally present in the Ty element carried by the plasmid.

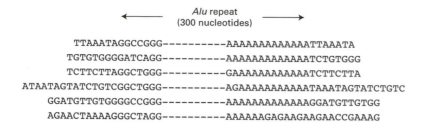

Figure 5–51 Nucleotide sequences of the six *Alu* inserts in the human albumin-gene family (Problem 5–84). *Dashed lines* indicate nucleotides in the internal part of the *Alu* sequences.

5–84* *Alu* sequences are present at six sites in the introns of the human serum-albumin and α-fetoprotein genes. (These genes are evolutionary relatives that are located side by side in mammalian genomes.) The same pair of genes in the rat contains no *Alu* sequences. The lineages of rats and humans diverged more than 85 million years ago at the time of the mammalian radiation. Does the presence of *Alu* sequences in the human genes and their absence from the corresponding rat genes mean that *Alu* sequences invaded the human genes only recently, or does it mean that the *Alu* sequences have been removed in some way from the rat genes?

To examine this question, you have sequenced all six of the *Alu* sequences in the human albumin-gene family. The sequences around the ends of the inserted *Alu* elements are shown in Figure 5–51.

A. *Alu* sequences create target-site duplications of a few nucleotides when they insert. Mark the left and right boundaries of the inserted *Alu* sequences and underline the nucleotides in the flanking chromosomal DNA that have been altered by mutation.

B. The rate of nucleotide substitution in introns has been measured at about 3×10^{-3} mutations per million years at each site. Assuming the same rate of substitution into the intron sequences that were duplicated by these *Alu* sequences, calculate how long ago the *Alu* sequences inserted into these genes. (Lump all the *Alu* sequences together to make this calculation; that is, treat them as if they inserted at about the same time.)

C. Why are these particular flanking sequences used in the calculation? Why were larger segments of the intron not included? Why were the mutations in the *Alu* sequences themselves not used?

D. Did these *Alu* sequences invade the human genes recently (after the time of the mammalian radiation), or have they been removed from the rat genes?

5–85 Cre recombinase is a site-specific enzyme that catalyzes recombination between two LoxP DNA recognition sequences. Cre recombinase pairs two LoxP sites in the same orientation, breaks both duplexes at the same point in the LoxP sites, and joins the ends with new partners so that each LoxP site is regenerated, as shown schematically in Figure 5–52A. Based on this mechanism, predict the arrangement of sequences that will be generated by Cre-mediated site-specific recombination for each of the two DNAs shown in Figure 5–52B.

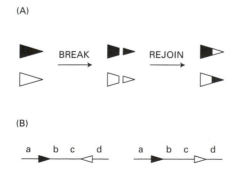

Figure 5–52 Cre recombinase-mediated site-specific recombination (Problem 5–85). (A) Schematic representation of Cre/LoxP site-specific recombination. The LoxP sequences in the DNA are represented by *triangles* that are colored so that the site-specific recombination event can be followed more readily. (B) DNA substrates containing two arrangements of LoxP sites.

5–86* You are investigating the properties of two site-specific recombination systems, Cre/LoxP and FLP/FRT, to determine which might be better for the chromosome engineering project you have in mind. You make the two sets of substrates shown in Figure 5–53, one with LoxP sites and one with FRT sites. You compare the ability of the respective recombinases to promote recombination between adjacent directly repeated recognition sites on a chromosome and recombination between a site on a plasmid and a site on a chromosome (Figure 5–53). You are surprised and puzzled by the results. When you introduce Cre into cells, you get a very high frequency of chromosomal recombination (10^{-1}) but a very low frequency of plasmid integration (10^{-6}). When you introduce FLP recombinase, you get a lower frequency of chromosomal recombination (10^{-2}) but a higher frequency of plasmid integration (10^{-5}). You had expected that chromosomal recombination might be more efficient than plasmid integration because the recognition sites are adjacent on the chromosome; however, that consideration alone cannot account for your results. Can you offer an explanation for the outcome of your experiments?

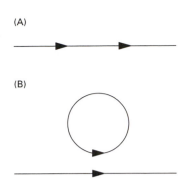

Figure 5–53 Comparison of Cre/LoxP and FLP/FRT site-specific recombination (Problem 5–86). (A) Chromosomal site-specific recombination between directly repeated sites. (B) Site-specific plasmid integration into a chromosome.

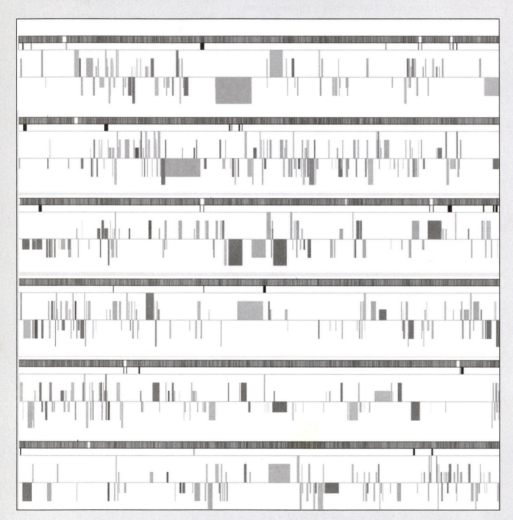

A portion of the *Drosophila* genome: "Often the most effective way to describe, explore, and summarize a set of numbers—even a very large set—is to look at pictures of those numbers. Furthermore, of all methods for analyzing and communicating statistical information, well-designed data graphics are usually the simplest and…the most powerful." Edward R. Tufte: The Visual Display of Quantitative Information. The jacket of this book presents a portion of the human genome using a different coding scheme. Arguably, of course, the clearest and most striking summary of any genome sequence is the organism to which it belongs. From MD Adams et al. *Science* 287:2185–2195, 2000. © AAAS.

HOW CELLS READ THE GENOME: FROM DNA TO PROTEIN

FROM DNA TO RNA

TERMS TO LEARN

DNA supercoiling	RNA polymerase	spliceosome
exon	RNA splicing	TATA box
general transcription factor	rRNA gene	terminator
intron	rRNA (ribosomal RNA)	transcription
mRNA (messenger RNA)	snoRNA (small nucleolar RNA)	transcription unit
nuclear pore complex	snRNA (small nuclear RNA)	trans-splicing
promoter		

6–1 Consider the expression 'central dogma,' which refers to the proposition that genetic information flows from DNA to RNA to protein. Is the word 'dogma' appropriate in this scientific context?

6–2* (**True/False**) The consequences of errors in transcription are less than those of errors in DNA replication. Explain your answer.

6–3 In the electron micrograph in Figure 6–1, are the RNA polymerase molecules moving from right to left or from left to right? How can you tell? Why are the RNA transcripts so much shorter than the length of DNA that encodes them?

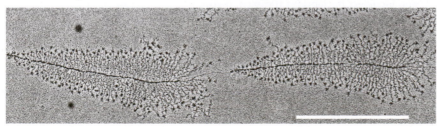

Figure 6–1 Transcription of two adjacent rRNA genes (Problem 6–3). The scale bar is 1 micrometer.

6–4* If RNA polymerase proofreads its product in a manner analogous to DNA polymerase, it will slow its rate of nucleotide incorporation after adding an incorrect base to the end of the growing RNA chain. This will allow time for removal of the mismatched nucleotide before the next one is added.

To investigate this possibility, scientists devised a clever technique to measure the rate of nucleotide incorporation by RNA polymerase at a defined point in a DNA template. The template contained a promoter and was covalently attached to agarose beads, so that it (and the attached RNA polymerase) could be removed from solution and washed, and then reincubated in a new mixture of nucleotides (Figure 6–2A). By using solutions with only one or a couple of nucleotides at a time, the RNA polymerase was 'walked' along the template to the site shown in the figure. At that point the RNA polymerase and template were resuspended in a solution containing one nucleotide, and the rate of incorporation of that nucleotide was measured as shown in Figure 6–2B. The rates of incorporation of A and G at position +44 in the RNA strand are shown in Table 6–1. In addition, the incorporation of the following C at position +45 was measured after A or G were incorporated at position +45 (Table 6–1).

A. Imagine that the RNA polymerase was stopped so that the last nucleotide in the RNA chain was the C at position +34. Describe how you might 'walk' the polymerase from there to position +43 to do these experiments.

B. By what factor, if any, is the correct nucleotide preferred over the incorrect nucleotide? Does the presence of an incorrect nucleotide influence the rate of addition of the next nucleotide? Explain your answer.

6–5 Match the following list of RNAs with their functions.
A.	mRNA	1.	adaptor for protein synthesis
B.	rRNA	2.	chemical modification of other RNAs
C.	snoRNA	3.	codes for proteins
D.	snRNA	4.	components of ribosome
E.	tRNAs	5.	splicing of RNA transcripts

6–6* Imagine that an RNA polymerase is transcribing a segment of DNA that contains the sequence

5′-GTAACGGATG-3′
3′-CATTGCCTAC-5′

If the polymerase transcribes this sequence from left to right, what will the sequence of the RNA be? What will the RNA sequence be if the polymerase moves right to left?

6–7 (**True/False**) The σ subunit is a permanent component of the RNA polymerase from *E. coli*, allowing it to initiate at appropriate promoters in the bacterial genome. Explain your answer.

6–8* In Figure 6–3 the sequences for 13 promoters recognized by the σ_{70} factor of RNA polymerase have been aligned. Deduce the consensus sequences for the –10 and –35 regions of these promoters.

6–9 The trypanosome, which is the microorganism that causes sleeping sickness, can vary its surface glycoprotein coat and thus evade the immune defenses of its host. The promoter for the variable surface glycoprotein (VSG) gene

(A) TEMPLATE FOR TRANSCRIPTION

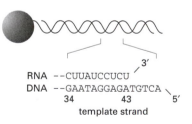

RNA --CUUAUCCUCU
DNA --GAATAGGAGATGTCA
    34   43   5′
        template strand

(B) EXPERIMENTAL DATA

time after adding ATP (seconds)

0   4   6   10   30   60   180

43   44

Figure 6–2 Determination of kinetic parameters for RNA polymerase (Problem 6–4). (A) DNA template attached to an agarose bead. (B) Rate of incorporation of the correct A nucleotide in the nascent RNA chain.

TABLE 6–1 Rate of incorporation of correct and incorrect nucleotides (Problem 6–4).

INCORPORATED NUCLEOTIDE	V_{max} (units/sec)
CUA	0.20
CUG	0.0015
CUAC	0.17
CUGC	0.036

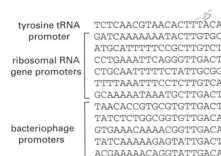

Figure 6–3 Sequences recognized by σ₇₀ factor (Problem 6–8). *Dots represent spaces that have been added to maximize alignment of sequences in the −10 and −35 regions.*

+1

−35   −10

tyrosine tRNA promoter
```
TCTCAACGTAACACTTTACAGCGGCG..CGTCATTTGATATGATGC.GCCCCGCTTCCCGATAAGGG
GATCAAAAAAATACTTGTGCAAAAAA..TTGGGATCCCTATAATGCGCCTCCGTTGAGACGACAACG
```

ribosomal RNA gene promoters
```
ATGCATTTTTCCGCTTGTCTTCCTGA..GCCGACTCCCTATAATGCGCCTCCATCGACACGGCGGAT
CCTGAAATTCAGGGTTGACTCTGAAA..GAGGAAAGCGTAATATAC.GCCACCTCGCGACAGTGAGC
CTGCAATTTTTCTATTGCGGCCTGCG..GAGAACTCCCTATAATGCGCCTCCATCGACACGGCGGAT
TTTTAAATTTCCTCTTGTCAGGCCGG..AATAACTCCCTATAATGCGCCACCACTGACACGGAACAA
GCAAAAATAAATGCTTGACTCTGTAG..CGGGAAGGCGTATTATGC.ACACCCCGCGCCGCTGAGAA
```

bacteriophage promoters
```
TAACACCGTGCGTGTTGACTATTTTA.CCTCTGGCGGTGATAATGG..TTGCATGTACTAAGGAGGT
TATCTCTGGCGGTGTTGACATAAATA.CCACTGGCGGTGATACTGA..GCACATCAGCAGGACGCAC
GTGAAACAAAACGGTTGACAACATGA.AGTAAACACGGTACGATGT.ACCACATGAAACGACAGTGA
TATCAAAAGAGTATTGACTTAAAGT.CTAACCTATAGGATACTTA.CAGCCATCGAGAGGGACACG
ACGAAAAACAGGTATTGACAACATGAAGTAACATGCAGTAAGATAC.AAATCGCTAGGTAACACTAG
GATACAAATCTCCGTTGTACTTTGTT..TCGCGCTTGGTATAATCG.CTGGGGGTCAAAGATGAGTG
```

proved difficult to locate, but was mapped by measuring the sensitivity of the transcript to UV irradiation. Since RNA polymerases cannot transcribe through pyrimidine dimers (the damage produced by UV irradiation), the sensitivity of transcription to UV irradiation is a measure of the distance between the start of transcription (the promoter) and the point where transcription is assayed (VSG gene).

Transcription through rRNA genes was used to calibrate the system. The 5S RNA transcription unit is just over 100 nucleotides long, whereas the 18S, 5.8S, and 28S RNAs are part of a single transcription unit that is about 8 kb long (Figure 6–4A). Trypanosomes were exposed to increasing doses of UV irradiation, their nuclei were then isolated and incubated with ^{32}P-dNTPs. RNA from the nuclei was hybridized to DNA for the 5S RNA gene and various parts of the rRNA gene (Figure 6–4B). Plots of the logarithm of the counts in each spot against the UV dose give straight lines (Figure 6–4C), with slopes that are proportional to the distance from the hybridization probe to the promoter.

When the experiment is done with a probe from the beginning of the VSG gene, transcription is found to be inactivated about seven times faster for the VSG gene than for probe 4 from the ribosomal transcription unit.

A. Why does RNA transcription increase in sensitivity to UV irradiation with increasing distance from the promoter?

B. Roughly how far is the VSG gene from its promoter? What assumption do you have to make in order to estimate this distance?

C. Transcription through another gene, located about 10 kb upstream of the VSG gene, was about 20% less sensitive to UV irradiation than transcription through the VSG gene. Could these two genes be transcribed from the same promoter?

(A) TRANSCRIPTION MAP

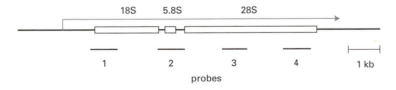

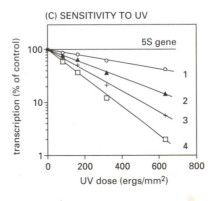

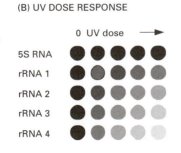

(B) UV DOSE RESPONSE

(C) SENSITIVITY TO UV

Figure 6–4 UV mapping of the promoter in trypanosomes (Problem 6–9). (A) Structure of the ribosomal RNA transcription unit. The positions of the hybridization probes along with a scale marker are shown. The transcription unit begins at the left end of the *arrow*. (B) A dot blot of transcripts from the 5S RNA and rRNA genes. The dot blot is done by placing an excess of the DNA probe in spots on a filter paper and then hybridizing radiolabeled RNA to it. (C) Sensitivities of the transcription units to increasing doses of UV irradiation.

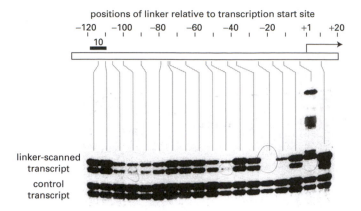

positions of linker relative to transcription start site

−120 −100 −80 −60 −40 −20 +1 +20

10

linker-scanned
transcript

control
transcript

Figure 6–5 Linker-scanning analysis of the *tk* promoter (Problem 6–10). Transcripts from linker-scanned plasmids and control plasmids were analyzed by primer extension, using a radiolabeled primer corresponding to sequences about 80 nucleotides from the 5′ end of the transcript. These primers were extended to the 5′ ends of the transcripts and the products were displayed on a denaturing polyacrylamide gel. Two bands are present for both the control and linker-scanned transcripts because of inefficient extension to the very end of the transcript. The position of each linker is indicated at the center of the segment it replaced. A 10-base pair bar—the length of the replacements—is shown at the top of the figure.

6–10* Deletion analysis of protein-binding sequences in a promoter can be difficult to interpret because altered spacing between elements can critically affect their function. The 'linker-scanning' method eliminates this potential difficulty by replacing 10-nucleotide segments throughout the promoter with oligonucleotide linkers. A classic paper described this method in an analysis of the promoter for the thymidine kinase (*tk*) gene. Plasmids in which 10-nucleotide segments had been replaced with linkers were injected into *Xenopus laevis* oocytes along with control plasmids to serve as a measure of injection efficiency. Results of these injections are shown in Figure 6–5.

A. Estimate from these experiments the locations of sequences that are critical for promoter function, and rank their relative importance.

B. Which, if any, of these elements do you suppose corresponds to the TATA box?

6–11 Purification of transcription factors requires a fast assay, or the factor will be inactive by the time the right fraction is identified. One key technical advance was to use a promoter linked to a DNA sequence that contained no C nucleotides. If GTP is omitted, the only long RNA transcript is made from the synthetic DNA sequence because all other transcripts terminate when a G was required. This set-up allows a rapid assay of specific transcription simply by measuring incorporation of a radioactive nucleotide.

To test this idea originally, two plasmids were constructed that carried the synthetic sequence: one with a promoter from adenovirus (pML1), the other without a promoter (pC1) (Figure 6–6A). Each plasmid was mixed with pure RNA polymerase II, transcription factors, and ^{32}P-CTP. In addition, various combinations of GTP, RNase T1 (which cleaves RNA adjacent to each G nucleotide), and 3′ *O*-methyl GTP (which terminates transcription whenever G is incorporated) were added. The products were measured by gel electrophoresis with the results shown in Figure 6–6B.

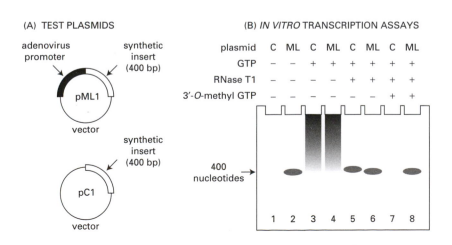

(A) TEST PLASMIDS

adenovirus promoter

synthetic insert (400 bp)

pML1

vector

synthetic insert (400 bp)

pC1

vector

(B) *IN VITRO* TRANSCRIPTION ASSAYS

plasmid	C	ML	C	ML	C	ML	C	ML
GTP	−	−	+	+	+	+	+	+
RNase T1	−	−	−	−	+	+	+	+
3′-*O*-methyl GTP	−	−	−	−	−	−	+	+

400 nucleotides →

1 2 3 4 5 6 7 8

Figure 6–6 Characterization of transcription using a template without C nucleotides (Problem 6–11). (A) Structures of test plasmids. (B) Results of transcription assays under various conditions. All reactions contain RNA polymerase II, transcription factors, and ^{32}P-CTP. Other components are listed *above* each lane.

A. Why is the 400-nucleotide transcript absent in lane 4 but present in lanes 2, 6, and 8?

B. Can you guess the source of the synthesis in lane 3 when the promoterless pC1 plasmid is used?

C. Why is a transcript of about 400 nucleotides present in lane 5 but not in lane 7?

D. The goal in developing this ingenious assay was to aid the purification of transcription factors; however, that process will begin with crude cell extracts, which will contain GTP. How do you suppose you might assay specific transcription in crude extracts?

6–12* Using your rapid assay for specific transcription (see Problem 6–11), you establish that transcripts accumulate linearly for about an hour and then reach a plateau. Your assay conditions use a 25-μL reaction volume containing 16 μg/mL of DNA template (the pML1 plasmid, which is 3.5 kb in length) with all other components in excess. From the specific activity of the ^{32}P-CTP and the total radioactivity in transcripts, you calculate that at the plateau 2.4 pmol of CMP were incorporated. Each transcript is 400 nucleotides long and has an overall composition of C_2AU. (The mass of a nucleotide pair is 660 daltons.)

A. How many transcripts are produced per reaction?

B. How many templates are present in each reaction?

C. How many transcripts are made per template in the reaction?

6–13 What are the roles of general transcription factors in RNA polymerase II-mediated transcription, and why are they referred to as 'general?'

6–14* When RNA polymerases were first being characterized in eucaryotes, three peaks of polymerizing activity (1, 2, and 3) were commonly obtained by fractionating cell extracts on chromatography columns (Figure 6–7A). But it was unclear whether these peaks corresponded to different RNA polymerases or just to different forms of one polymerase. Incubating the three polymerase fractions in the presence of 1 μg/mL or 10 μg/mL α-amanitin (from *Amanita phalloides*, the world's deadliest mushroom, Figure 6–7B) gave the results shown in Figure 6–7A. Do these results argue for different RNA polymerases or different forms of the same RNA polymerase? Explain your answer.

6–15 The large subunit of eucaryotic RNA polymerase II in yeast has a C-terminal domain (CTD) that comprises 27 near-perfect repeats of the sequence YSPTSPS. If the normal RNA polymerase II gene is replaced with one that encodes a CTD with only 11 repeats, the cells are viable at 30°C but are unable to grow at 12°C. This cold sensitivity allows revertants to be selected for growth at 12°C. Some of these revertants proved to be dominant mutations in previously unknown genes such as *SRB2*.

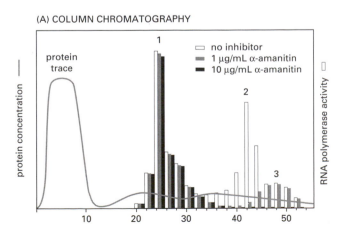

(A) COLUMN CHROMATOGRAPHY

(B) DEADLY MUSHROOMS

Amanita phalloides

Figure 6–7 Characterization of RNA polymerase activities (Problem 6–14). (A) Peaks of RNA polymerase activity after column chromatography. Activities were measured in the presence of various concentrations of α-amanitin. (B) The *Amanita phalloides* mushroom.

Extracts prepared from yeast that are deleted for the *SRB2* gene cannot transcribe added DNA templates, but they can be activated for transcription by addition of Srb2 protein. To test the role of Srb2 in transcription, plasmid DNAs with either a short or a long G-free sequence downstream of a promoter (Figure 6–8A) were incubated separately in the presence or absence of a *limiting* amount of recombinant Srb2 (and in the absence of added NTPs so that transcription could not begin). The reactions were then mixed and transcription was initiated at various times afterward by adding NTPs (Figure 6–8B). After a brief incubation (to prevent reinitiation of transcription) the products were displayed on a gel. Whichever template was preincubated with Srb2 was the one that was transcribed (Figure 6–8C). By contrast, if an *excess* of Srb2 was mixed with one template during the preincubation, transcription was observed from both templates after mixing.

A. Did Srb2 show a preference for either template when the preincubation was carried out with the individual templates or the mixture (Figure 6–8C, lanes 1 to 3)?

B. Do the results indicate that Srb2 acts stoichiometrically or catalytically? How so?

C. Does Srb2 form part of the preinitiation complex or does it act after transcription has begun? How can you tell?

(A) TEMPLATES

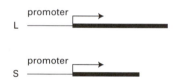

Figure 6–8 Experiments to test the role of Srb2 in transcription (Problem 6–15).

(B) DESIGN OF THE EXPERIMENT

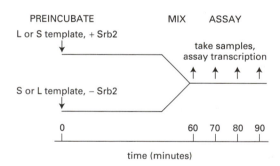

(C) EXPERIMENTAL RESULTS

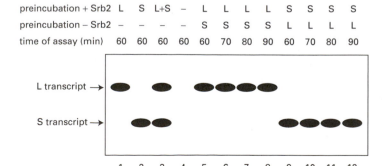

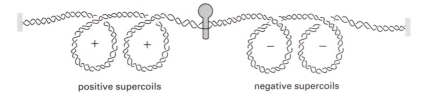

Figure 6–9 Supercoils around a moving RNA polymerase (Problem 6–16).

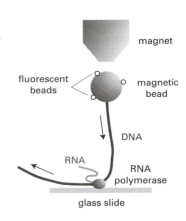

Figure 6–10 System for measuring the rotation of DNA caused by RNA polymerase (Problem 6–17). The magnet holds the bead upright (but doesn't interfere with its rotation) and the attached fluorescent beads allow the direction of motion to be visualized. RNA polymerase is held in place by attachment to the glass slide.

D. What do you think happens during the preincubation that so strongly favors transcription from the template that was included in the preincubation?

E. Do these results indicate that Srb2 binds to the CTD of RNA polymerase II?

6–16* In which direction along the template must the RNA polymerase in Figure 6–9 be moving to have generated the supercoiled structures that are shown?

6–17 Imagine that you have attached an RNA polymerase molecule to a glass slide and allowed it to initiate transcription on a template DNA that is tethered to a magnetic bead as shown in Figure 6–10. If the DNA moves relative to the RNA polymerase as indicated in the figure, in which direction will the bead rotate?

6–18* Why doesn't transcription cause a hopeless tangle? If the RNA polymerase doesn't revolve around the DNA as it moves, then it will cause one DNA supercoil—in front and in back—for every 10 nucleotides it transcribes. If, instead, RNA polymerase revolves around the DNA—avoiding DNA supercoiling—then it will coil the RNA around the DNA duplex, once for every 10 nucleotides it transcribes. Thus, for any reasonable size gene the act of transcription should result in hundreds of coils or supercoils…and that's for every single RNA polymerase! So why doesn't transcription lead to a complete snarl?

Problems 4–63 to 4–67 introduce the principles of supercoiling in DNA.

6–19 Detailed features of the active site of RNA polymerase have been probed using modified RNA nucleotides that can react with nearby targets. In one study an analog of U carrying a reactive chemical group (Figure 6–11A) was incorporated at one of two positions in a radiolabeled RNA and then 'walked' to specific locations relative to the 3' end of the RNA chain using a technique like that described in Problem 6–4 (Figure 6–11B). When the analog was then activated, it reacted almost exclusively with either the adenine it was paired with in the DNA template strand or with the polymerase (Figure 6–11C). Its pattern of reactivity with DNA and protein varied depending on its position relative to the 3' end of the RNA. Why do you suppose that the activated U reacts so strongly with DNA up to a point and then abruptly begins to react more strongly with protein?

Figure 6–11 Probing the active site of RNA polymerase (Problem 6–19). (A) The reactivity of a U analog with adenosine in the template DNA. The *arrow* indicates the site of attack. (B) Locations of the U analog in the RNA chain. *Numbers* refer to the number of nucleotides from the reactive U to the 3' end. (C) The pattern of reaction of the reactive U with DNA and protein.

(A) REACTIVE URIDINE RESIDUE

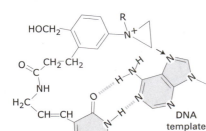

(B) REACTIVE TRANSCRIPTS

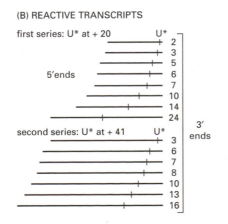

(C) EXPERIMENTAL DATA

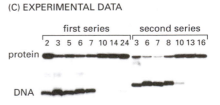

6–20* What are thought to be the various roles of the phosphorylated CTD of RNA polymerase?

6–21 (**True/False**) Eucaryotic mRNA molecules carry 3′ ribosyl OH groups at both their 3′ and 5′ ends. Explain your answer.

6–22* (**True/False**) Since introns are largely genetic 'junk,' they do not have to be removed precisely from the primary transcript during RNA splicing. Explain your answer.

6–23 You are studying a DNA virus that makes a set of abundant proteins late in its infectious cycle. An mRNA for one of these proteins maps to a restriction fragment from the middle of the linear genome. To determine the precise location of this mRNA, you anneal it with the purified restriction fragment under conditions where only DNA–RNA hybrid duplexes are stable and DNA strands do not reanneal. When you examine the reannealed duplexes by electron microscopy, you see structures such as that in Figure 6–12. Why are there single-stranded tails at the ends of the DNA–RNA duplex region?

6–24* The intron–exon structure of eucaryotic genes came as a shock. In the early, skeptical days the most convincing demonstration was visual, as shown, for example, by the electron micrograph in Figure 6–13A, which was obtained by hybridizing ovalbumin mRNA to a long segment of DNA that contained the gene. To those used to looking at single- and double-stranded nucleic acids in the electron microscope, the structure was clear: a set of single-stranded tails and loops emanating from a central duplex segment whose ends corresponded to the ends of the mRNA (Figure 6–13B). To the extent possible, describe the intron–exon structure of this gene.

6–25 Smilin is a (hypothetical) protein that causes people to be happy. It is inactive in many chronically unhappy people. The mRNA isolated from a number of different unhappy persons in the same family was found to lack an internal stretch of 173 nucleotides that are present in the Smilin mRNA isolated from a control group of generally happy people. The DNA sequences of the Smilin genes from the happy and unhappy persons were determined and compared. They differed by just one nucleotide change—and no nucleotides were deleted. Moreover, the change was found in an intron.

　A. Can you hypothesize a molecular mechanism by which a single nucleotide change in a gene could cause the observed *internal* deletion in the mRNA?

　B. What consequences for the Smilin protein would result from removing a 173-nucleotide-long internal stretch from its mRNA? Assume that the 173 nucleotides are deleted from the coding region of the Smilin mRNA.

　C. What can you say about the molecular basis of unhappiness in this family?

6–26* The human α-tropomyosin gene is alternatively spliced to produce several forms of α-tropomyosin mRNA in various cell types (Figure 6–14). For all forms of the mRNA, the encoded protein sequence is the same in the regions

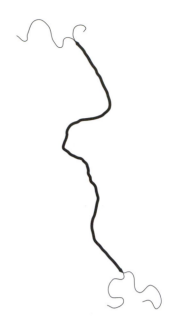

Figure 6–12 DNA–RNA hybrid between an mRNA and a restriction fragment from adenovirus (Problem 6–23).

Figure 6–13 Hybrid between ovalbumin mRNA and a DNA segment containing its gene (Problem 6–24). (A) An electron micrograph. (B) The hybrid with background eliminated. Each of the loops is formed by single-stranded RNA emanating from a central RNA–DNA duplex.

(A) ELECTRON MICROGRAPH

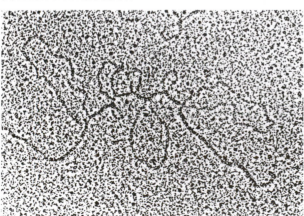

(B) INTERPRETATION

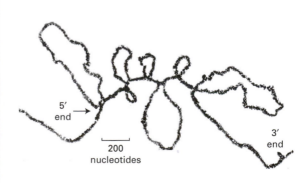

5′ end

200 nucleotides

3′ end

(A) HUMAN α-TROPOMYOSIN GENE

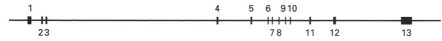

(B) FOUR DIFFERENT SPLICE VARIANTS

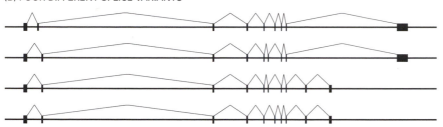

Figure 6–14 Alternatively spliced mRNAs from the human α-tropomyosin gene (Problem 6–26). (A) Exons in the human α-tropomyosin gene. The locations and relative sizes of exons are shown by the *black rectangles*. (B) Splicing patterns for four α-tropomyosin mRNAs. Splicing is indicated by *lines* connecting the exons that are included in the mRNA.

of the mRNA that correspond to exons 1 and 10. Exons 2 and 3 are alternative exons used in different mRNAs, as are exons 7 and 8. Which of the following statements about exons 2 and 3 is the most accurate? Is that statement also the most accurate for exons 7 and 8? Explain your answers.

A. Exons 2 and 3 must have the same number of nucleotides.

B. Exons 2 and 3 must contain an integral number of codons (that is, the number of nucleotides divided by 3 must be an integer).

C. Exons 2 and 3 must contain a number of nucleotides that when divided by 3 leaves the same remainder (that is, 0, 1, or 2).

6–27   You have just printed out a set of DNA sequences around the intron–exon boundaries for genes in the β-globin family, and you take the thick file to the country to study for the weekend. When you look at the printout, however, you discover to your annoyance that there's no indication of where in the gene you are. You know that the sequences in Figure 6–15 come from one of the exon/intron or intron/exon boundaries and that the boundaries lie on the dotted line, but you don't know the order of the intron and exon. You know that introns begin with the dinucleotide sequence GT and end with AG, but you realize that these particular sequences would fit *either* as the start *or* the finish of an intron.

 If you cannot decide which side is the intron, you will have to cut your weekend short and return to the city (or find a neighbor with on-line access). In desperation, you consider the problem from an evolutionary perspective. You know that introns evolve faster (suffering more nucleotide changes) than exons because they are not constrained by function. Does this perspective allow you to identify the intron, or will you have to pack your bags?

6–28*   The interaction of U1 snRNP with the sequences at the 5′ ends of introns is usually shown to involve pairing between the nucleotides in the pre-mRNA

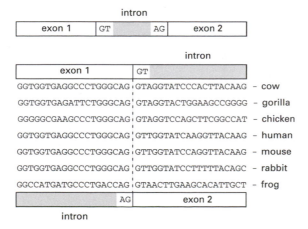

Figure 6–15 Aligned DNA sequences from the β-globin genes in different species (Problem 6–27). As indicated by the gene structures shown *above* and *below*, the DNA sequences could come from the boundary of exon 1 with the intron or from the boundary of the intron with exon 2.

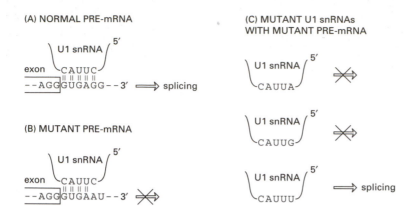

(A) NORMAL PRE-mRNA

(B) MUTANT PRE-mRNA

(C) MUTANT U1 snRNAs
WITH MUTANT PRE-mRNA

Figure 6–16 Characterization of the pairing of U1 snRNP with pre-mRNA (Problem 6–28). (A) Interaction of U1 snRNP with a normal pre-mRNA. (B) Interaction of U1 snRNP with a mutant pre-mRNA. (C) Mutant U1 snRNAs.

and those in the RNA component of the U1 snRNP, as illustrated in Figure 6–16A. But given the relatively small number of interactions, that arrangement is hardly convincing. A series of experiments tested the hypothesis that base-pairing was critical to the function of U1 snRNP in splicing. As shown in Figure 6–16B, a mutant pre-mRNA was generated that could not be spliced. Several mutant U1 RNAs were then tested for their ability to promote splicing of the mutant pre-mRNA, with the results shown in Figure 6–16C. Do these experiments argue that base-pairing is critical to the role of U1 snRNP in splicing? If so, how?

6–29   Many eucaryotic genes contain a large number of exons. Correct splicing of such genes requires that neighboring exons be ligated to one other; if they are not, exons will be left out. Since 5′ splice sites look alike, as do 3′ splice sites, it is remarkable that skipping an exon occurs so rarely. Some mechanism must keep track of neighboring exons and ensure that they are brought together.

One early proposal suggested that the splicing machinery bound to a splice site at one end of an intron and scanned through it to find the splice site at the other end. Such a scanning mechanism would guarantee that an exon was never skipped. This hypothesis was tested with two minigenes: one with a duplicated 5′ splice site and the other with a duplicated 3′ splice site (Figure 6–17). These minigenes were transfected into cells and their RNA products were analyzed to see which 5′ and 3′ splice sites were selected during splicing.

A. Draw a diagram of the products you expect from each minigene if the splicing machinery binds to a 5′ splice site and scans toward a 3′ splice site. Diagram the expected products if the splicing machinery scans in the opposite direction.

B. When the RNA products from the transfected minigenes were analyzed, a mixture of the products diagrammed in part A was generated from each minigene. Are neighboring exons brought together by intron scanning?

6–30*   Nematodes have four genes that encode actin mRNAs: three are clustered on chromosome 5, and one is located on the X chromosome (Figure 6–18). To identify the start site of transcription, you employ S1 mapping and primer extension. For S1 mapping, you anneal a radioactive single-stranded

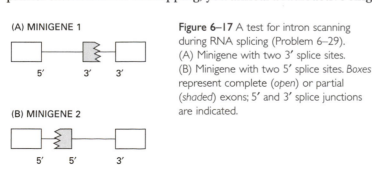

(A) MINIGENE 1

5′   3′   3′

(B) MINIGENE 2

5′   5′   3′

Figure 6–17 A test for intron scanning during RNA splicing (Problem 6–29). (A) Minigene with two 3′ splice sites. (B) Minigene with two 5′ splice sites. Boxes represent complete (open) or partial (shaded) exons; 5′ and 3′ splice junctions are indicated.

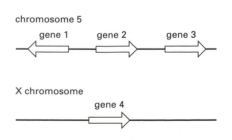

chromosome 5

gene 1   gene 2   gene 3

X chromosome

gene 4

Figure 6–18 Locations of the four actin genes in the nematode genome (Problem 6–30). Genes are indicated by arrows that show their direction of transcription.

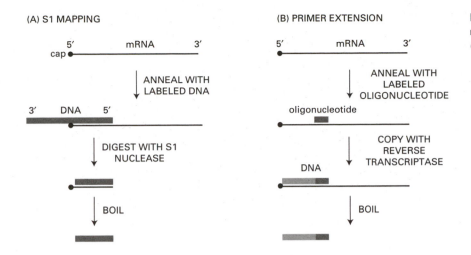

Figure 6–19 Mapping the 5′ ends of actin mRNAs (Problem 6–30). (A) S1 mapping. (B) Primer extension.

segment from the 5′ end of the gene to the mRNAs, digest with S1 nuclease to remove all single strands, and analyze the protected fragments of radioactive DNA on a sequencing gel to determine their length (Figure 6–19A). For primer extension you hybridize specific oligonucleotides to the mRNAs, extend them to the 5′ end, and analyze the resulting DNA segments on sequencing gels (Figure 6–19B).

For gene 4 the two techniques agree, as they usually do. However, for genes 1, 2, and 3, the mRNAs appear to be 20 nucleotides longer when assayed by primer extension. When you compare the mRNA sequences with the sequences for the gene, you find that each has an identical 20-nucleotide segment at its 5′ end that does not match the sequence of the gene (Figure 6–20).

Using an oligonucleotide complementary to this leader RNA segment, you discover that the corresponding DNA is repeated about 100 times in a cluster on chromosome 5, but it's a long way from the actin genes on chromosome 5. This leader gene encodes an RNA about 100 nucleotides long. The 5′ end of the leader RNA is identical to the segment found at the 5′ ends of the actin mRNAs (Figure 6–20).

A. Assuming that the leader RNA and the actin RNAs are joined by splicing according to the usual rules, indicate on Figure 6–20 the most likely point at which the RNAs are joined.

B. The leader gene and actin genes 1, 2, and 3 are all on the same chromosome. Why is it *not* possible that transcription begins at a leader gene and extends through the actin genes to give a precursor RNA that is subsequently spliced to form the actin mRNAs?

C. How do you think the actin mRNAs acquire their common leader segment?

ACTIN GENE 1

DNA: 5′ TATTATCAATTTAATTTTTCAGGTACATTAAAAACTAATCAAA<u>ATG</u>
RNA: 5′ <u>XGUUUAAUUACCCAAGUUUGAGG</u>UACAUUAAAAACUAAUCAAAAUG

ACTIN GENE 2

DNA: 5′ ATAATTCATAATTATTTTGTAGGCTAAGTTCCTCCTAATCTAATAAATC<u>ATG</u>
RNA: 5′ <u>XGUUUAAUUACCCAAGUUUGAGG</u>CUAAGUUCCUCCUAAUCUAAUAAAUCAUG

ACTIN GENE 3

DNA: 5′ TATTATCAATTTAATTTTTCAGGTACATTAAAAACTAATCAAA<u>ATG</u>
RNA: 5′ <u>XGUUUAAUUACCCAAGUUUGAGG</u>UACAUUAAAAACUAAUCAAAAUG

LEADER RNA GENE

DNA: 5′ GGTTTAATTACCCAAGTTTGAGGTAAACATTCAAACTGA
RNA: 5′ <u>XGUUUAAUUACCCAAGUUUGAGG</u>UAAACAUUCAAACUGA

Figure 6–20 RNA and DNA sequences of actin genes and leader RNA gene (Problem 6–30). The start site for translation of the actin genes (ATG) is underlined in the DNA sequence. The leader RNA segments that are present at the 5′ ends of the actin genes and the leader RNA gene are *underlined* in the RNA sequences. The three nucleotides immediately adjacent to the underlined leader RNA segment are identical in all four genes; thus it is unclear whether they were derived from actin RNA or leader gene RNA. The 5′ nucleotide on the RNAs (X) cannot be determined by primer extension.

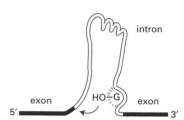

(A) GROUP I INTRON

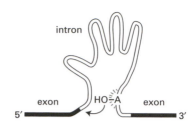

(B) GROUP II INTRON

Figure 6–21 Initial step in self-splicing (Problem 6–31). (A) Reaction catalyzed by a group I self-splicing intron. (B) Reaction catalyzed by a group II self-splicing intron.

6–31   Self-splicing introns use two distinct strategies to accomplish splicing. Group I introns bind a G nucleotide from solution, activating it for attack on the phosphodiester bond that links the intron to the terminal nucleotide of the upstream exon (Figure 6–21A). Group II introns activate a particularly reactive A nucleotide within the intron sequence and use it to attack the phosphodiester bond that links the intron to the terminal nucleotide of the upstream exon (Figure 6–21B). For both types of intron the next step joins the two exons, releasing the 3′ end of the intron. What are the structures of the excised introns in both cases? Which mechanism most closely resembles pre-mRNA splicing catalyzed by the spliceosome?

6–32*   (True/False) The 3′ ends of most RNA polymerase II pre-mRNA transcripts are defined by the termination of transcription, which releases a free 3′ end to which a poly-A tail is quickly added. Explain your answer.

6–33   The AAUAAA sequence just 5′ of the polyadenylation site is a critical signal for polyadenylation, as has been verified in many ways. One elegant confirmation used chemical modification to interfere with specific protein interactions. An RNA containing the signal sequence was treated with diethylpyrocarbonate, which can carboxyethylate A and G nucleotides, rendering them highly sensitive to breakage by aniline treatment. The starting RNA molecules were radiolabeled at one end, treated to contain about one modification per molecule, and then cleaved with aniline. The resulting series of fragments have lengths that correspond to the positions of As and Gs in the RNA (Figure 6–22, lane 1).

To define critical A and G nucleotides, the modified, but still intact RNAs were incubated with an extract capable of cleavage and polyadenylation. The RNAs were then separated into those that had acquired a poly-A tail (poly-A⁺) and those that had not (poly-A⁻). These two fractions were treated with aniline and the fragments were analyzed by gel electrophoresis (Figure 6–22, lanes 2 and 3). In a second reaction EDTA was added to the extract along with the modified RNAs; this does not affect RNA cleavage, but prevents addition of the poly-A tail. These cleaved RNAs were isolated, treated with aniline, and examined by electrophoresis (lane 4).

A. At which end were the starting RNA molecules labeled?
B. Explain why the bands corresponding to the AAUAAA signal (bracket in Figure 6–22) are missing from the poly-A⁺ RNA and the cleaved RNA.
C. Explain why the band at the arrow (the normal nucleotide to which poly A is added) is missing in the poly-A⁺ RNA but is present in the cleaved RNA.
D. Which A and G nucleotides are important for cleavage, and which A and G nucleotides are important for addition of the poly-A tail?

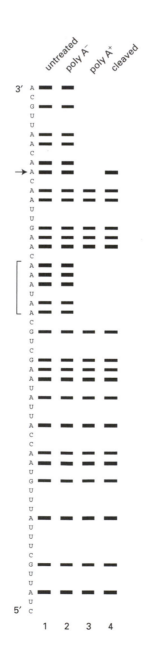

Figure 6–22 Autoradiographic analysis of experiments to define the purines important in RNA cleavage and polyadenylation (Problem 6–33). The sequence of the precursor RNA is shown at the *left* with the 5′ end at the *bottom* and the 3′ end at the *top*. All RNAs were modified by reaction with diethylpyrocarbonate. RNA that was not treated with extract (untreated) is shown in lane 1. RNA that was treated with extract but was not polyadenylated (poly A⁻) is shown in lane 2. RNA that was polyadenylated (poly A⁺) in the extract is shown in lane 3. RNA that was cleaved but not polyadenylated (cleaved) is shown in lane 4. Shorter RNA fragments run faster; that is, they travel farther toward the bottom of the gel during electrophoresis.

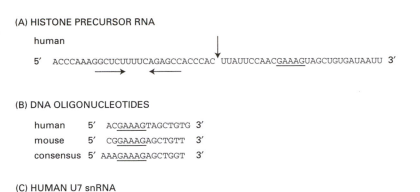

(A) HISTONE PRECURSOR RNA

human

5′ ACCCAAAGGCUCUUUUCAGAGCCACCCAC UUAUUCCAACGAAAGUAGCUGUGAUAAUU 3′

(B) DNA OLIGONUCLEOTIDES

human	5′	ACGAAAGTAGCTGTG 3′
mouse	5′	CGGAAAGAGCTGTT 3′
consensus	5′	AAAGAAAGAGCTGGT 3′

(C) HUMAN U7 snRNA

5′ m₃G–NNGUGUUACAGCUCUUUUAGAAUUUGUCUAGU 3′

Figure 6–23 Processing of histone precursor RNAs (Problem 6–34). (A) Nucleotide sequences of histone precursor RNAs. *Horizontal arrows* indicate the inverted repeat sequences capable of forming a stem-loop structure in the precursor; a *vertical arrow* indicates the site of cleavage, and an *underline* shows the position of the conserved region. (B) DNA oligonucleotides used in the experiments. (C) Sequence of human U7 snRNA. The trimethylated cap, m₃G, is characteristic of 'U' RNAs. N refers to nucleotides whose identity was unknown.

E. What information might be obtained by labeling the RNA molecules at the other end?

6–34* Unlike most mRNAs, histone mRNAs do not have poly-A tails. They are processed from a longer precursor by cleavage just 3′ of a stem-loop structure in a reaction that depends on a conserved sequence near the cleavage site (Figure 6–23A). A classic paper defined the role of U7 snRNP in cleavage of the histone precursor RNA.

When nuclear extracts from human cells were treated with a nuclease to digest RNA, they lost the ability to cleave the histone precursor. Adding back a crude preparation of snRNPs restored the activity. The extract could also be inactivated by treatment with RNase H (which cleaves RNA in an RNA–DNA hybrid) in the presence of DNA oligonucleotides containing the conserved sequence—the suspected site of snRNP interaction in the histone precursor (Figure 6–23B). Somewhat surprisingly, the mouse and mammalian consensus oligonucleotides completely blocked histone processing, but the human oligonucleotide had no effect. The two inhibitory oligonucleotides also caused the disappearance of a 63-nucleotide snRNA, which was then purified and partially sequenced (Figure 6–23C).

A. Explain the design of the oligonucleotide experiment. What were these scientists trying to accomplish by incubating the extract with a DNA oligonucleotide in the presence of RNase H?

B. Since a human extract was used, do you find it surprising that the human oligonucleotide did not inhibit processing, whereas the mouse and consensus oligonucleotides did? Can you offer an explanation for this result?

6–35 What does 'export ready' mRNA mean, and what distinguishes an 'export ready' mRNA from a bit of excised intron that needs to be degraded?

6–36* The nucleolus disappears at each mitosis and then reappears during G₁ of the next cell cycle. How is this reversible process thought to be accomplished?

6–37 In eucaryotes, two distinct classes of snoRNAs, which are characterized by conserved sequence motifs termed boxes, are responsible for 2′-O-methylation and pseudouridylation. Box C/D snoRNAs direct 2′-O-ribose methylation, and box H/ACA snoRNAs direct pseudouridylation of target RNAs. Recently a novel snoRNA, called U85, was shown to contain both box elements (Figure 6–24). You want to know how these box elements participate in the genesis and function of U85 snoRNA. You begin with its mode of synthesis.

In vertebrates most snoRNAs are processed from introns in pre-mRNA. To test whether the processing of U85 snoRNA occurs in this fashion, you

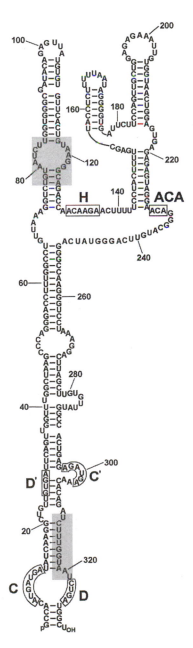

Figure 6–24 Proposed secondary structure of the human U85 snoRNA (Problem 6–37). Box elements are indicated. Regions that are thought to be important for targeting 2′-O-methylation and pseudouridylation are *shaded*.

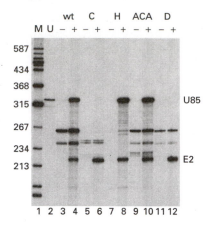

(A) CONSTRUCT

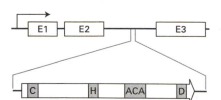

(B) RNase PROTECTION

Figure 6–25 Processing of human U85 snoRNA from the β-globin intron (Problem 6–37). (A) Structure of the G/U85 construct showing location of U85 in the second intron of the β-globin gene. The positions of the box elements, whose sequences were altered in other, analogous constructs, are indicated in the expanded view of the U85 gene shown *below* the β-globin intron. (B) RNase protection by RNA extracted from cells. Radioactively labeled antisense RNA corresponding to U85 RNA and to β-globin exon RNAs were hybridized to RNA extracted from transfected (+) and nontransfected (–) cells, and treated with RNase. Protected RNAs were separated by electrophoresis. Bands corresponding to U85 RNA and exon 2 (E2) of the β-globin gene are indicated. Bands that show up in both transfected and nontransfected cells are artifacts. The lane marked U contains pure U85 RNA. Wild type (wt) refers to the G/U85 construct with no mutational alterations. Marker RNAs (M) are shown on the *left*; their sizes are indicated in base pairs.

deposit the human U85 gene into the second intron of the human β-globin gene to make the plasmid construct G/U85 (Figure 6–25A). You make the additional constructs G/U85-C, G/U85-H, G/U85-ACA, and G/U85-D, in which the indicated box element of U85 was replaced with a stretch of C nucleotides. You transfect these constructs into monkey cells. The accumulated RNAs were analyzed by RNase protection, using U85 and β-globin probes (Figure 6–25B). Are any of the conserved box elements important for the accumulation of U85 snoRNA? Explain your reasoning.

6–38* To identify potential substrate RNAs for the U85 snoRNA, sequences of all known stable cellular RNAs were carefully examined. Sites of 2′-*O*-methylation and pseudouridylation were looked for in sequences that could pair with the putative guide sequences in U85 responsible for such modifications (see Figure 6–24, shaded areas). The search revealed that U85 snoRNA could potentially pair with a region of the U5 spliceosomal snRNA that carries two 2′-*O*-methylated nucleotides, U41 and C45, and two pseudouridines, ψ43 and ψ46 (Figure 6–26A).

To determine whether these U5 snRNA sequences were indeed substrates for modification by U85 snoRNA, the relevant segment of U5 was inserted into a region of the U2 snRNA gene to create a distinctive molecule U2–U5 that could be readily followed. In addition to the normal sequence, a mutant segment of U5 was inserted into U2 to make U2–U5m. A mutant, U85m, was also constructed with compensating changes in the guide sequences adjacent to the box H and box D regions (Figure 6–26B). Expression vectors for U2–U5, U2–U5m, and U2–U5m along with U85m were transfected into cells. All the encoded RNAs were shown to accumulate normally. 2′-*O*-Methylation and pseudouridylation in the critical region in the U2–U5 and U2–U5m molecules were detected by sequencing, as summarized in Table 6–2.

What are the expectations of these experiments if bases in U5 snRNA serve as bona fide targets for U85 snoRNA-dependent modification? Which, if any, of the naturally modified nucleotides in U5—methylated riboses at U41 and C45, and the pseudouridines ψ43 and ψ46—are dependent on U85 snoRNA? Explain your reasoning.

Figure 6–26 Analysis of potential RNA targets for modification by U85 snoRNA (Problem 6–38). (A) Potential pairing between the guide sequences in U85 snoRNA and a segment of U5 snRNA. Pairing with the guide sequence adjacent to box H, which typically directs pseudouridylation, is shown above U5. Pairing with the guide sequence adjacent to box D, which typically directs 2′-*O*-methylation is shown *below* U5. Only one pairing could occur at a time. Pseudouridines are indicated with ψ, sites of 2′-*O*-methylation are indicated with *dots*. (B) Potential pairing between the mutant guide sequences in U85m snoRNA and the mutant segment of U5 in U2–U5m.

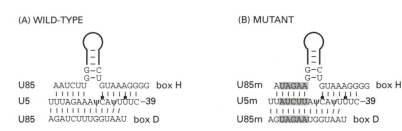

(A) WILD-TYPE

	G–C
	G–U
U85	AAUCUU GUAAAGGGG box H
U5	UUUAGAAAψCAψUUUC–39
U85	AGAUCUUUGGUAAU box D

(B) MUTANT

	G–C
	G–U
U85m	AUAGAA GUAAAGGGG box H
U5m	UUUAUCUUAψCAψUUUC–39
U85m	AGUAGAAUGGUAAU box D

TABLE 6–2 Modification of potential target nucleotides in various transfections (Problem 6–38).

| | TRANSFECTED MOLECULES | | | | MODIFICATIONS[b] | | | |
| | | | | | PSEUDOURIDINE | | 2'-O-METHYL | |
	U85[a]	U85m	U2–U5	U2–U5m	U43	U46	U41	C45
1	+		+		−	ψ	m	m
2	+			+	−	−	m	−
3	+	+		+	−	ψ	m	m

[a]Endogenous U85 is present in all cells.
[b]Nucleotides that have been converted to pseudouridine are indicated by ψ; nucleotides that have been 2'-O-methylated are indicated by an m.

FROM RNA TO PROTEIN

TERMS TO LEARN

aminoacyl-tRNA synthetase	initiator tRNA	ribosome
anticodon	molecular chaperone	ribozyme
codon	prion disease	rRNA (ribosomal RNA)
eucaryotic initiation factor (eIF)	proteasome	translation
genetic code	reading frame	tRNA (transfer RNA)

6–39 For the RNA sequence below indicate the amino acids that are encoded in the three reading frames. If you were told that this segment of RNA was in the middle of an mRNA that encoded a large protein, would you know which reading frame was used? How so? (The genetic code is on the inside back cover.)

AGUCUAGGCACUGA

[handwritten annotations:] UAG   TAA TAG   AU   AUC
  TGA   ACU

6–40* After treating cells with a chemical mutagen, you isolated two mutants. One carries alanine and the other carries methionine at a site in the protein that normally contains valine (Figure 6–27). After treating these two mutants again with the mutagen, you isolate mutants from each that now carry threonine at the site of the original valine (Figure 6–27). Assuming that all mutations involve single nucleotide changes, deduce the codons that are used for valine, methionine, threonine, and alanine at the affected site.

6–41 The genetic code was deciphered in part by experiments in which polynucleotides of repeating sequences were used as mRNAs to direct protein synthesis in cell-free extracts. In the test tube, artificial conditions were used that allowed ribosomes to start protein synthesis anywhere on an RNA molecule, without the need of a translation start codon, as required in a living cell. What types of polypeptides would you expect to be synthesized if the following polynucleotides were used as templates in such a cell-free extract?
A. UUUUUUUUUUUU...
B. AUAUAUAUAUAU...
C. AUCAUCAUCAUC...

6–42* In *B. lichenformis* a few amino acids are removed from the C-terminal end of the β-lactamase enzyme after it is synthesized. The sequence of the original C-terminus can be deduced by comparing it to a mutant in which the reading frame is shifted by insertion or deletion of a nucleotide. The amino acid sequences of the purified wild-type enzyme and the frameshift mutant from amino acid 263 to the C-terminal end are given below.

wild type:	N M N G K
mutant:	N M I W Q I C V M K D

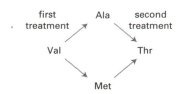

Figure 6–27 Two rounds of mutagenesis and the altered amino acids at a single position in a protein (Problem 6–40).

A. What was the mutational event that gave rise to the frameshift mutant?

B. Deduce the number of amino acids in the synthesized form of the wild-type enzyme and, as far as possible, the actual sequence of the wild-type enzyme.

6–43 Which of the following mutational changes would you predict to be the most deleterious to gene function? Explain your answer.
1. Insertion of a single nucleotide near the end of the coding sequence.
2. Removal of a single nucleotide near the beginning of the coding sequence.
3. Deletion of three consecutive nucleotides in the middle of the coding sequence.
4. Deletion of four consecutive nucleotides in the middle of the coding sequence.
5. Substitution of one nucleotide for another in the middle of the coding sequence.

6–44* Consider the properties of two hypothetical genetic codes constructed with the four common nucleotides: A, G, C, and T.

A. Imagine that the genetic code is constructed so that pairs of nucleotides are used as codons. How many different amino acids could such a code specify?

B. Imagine that the genetic code is a triplet code; that is, it uses three nucleotides to specify each amino acid. In this code, the amino acid specified by each codon depends only on the composition of the codon—not the sequence. Thus, for example, CCA, CAC, and ACC, which all have the composition C_2A, would encode the same amino acid. How many different amino acids could such a code specify?

C. Would you expect to have any difficulty translating these codes, using mechanisms analogous to those used in translating the standard genetic code?

6–45 One remarkable feature of the genetic code is that amino acids with similar chemical properties often have similar codons. Codons with U or C as the second nucleotide, for example, tend to specify hydrophobic amino acids. Can you suggest a possible explanation for this phenomenon in terms of the early evolution of the protein synthesis machinery?

> **Problem 1–5** considers the amazing mutational resistance of the natural genetic code.

6–46* (**True/False**) Wobble pairing occurs between the first position in the codon and the third position in the anticodon. Explain your answer.

6–47 The rules for wobble pairing in bacteria and eucaryotes are shown in Table 6–3. On the left side of the table, the rules are expressed as a wobble codon base and its recognition by possible anticodon bases. Reformulate the rules

TABLE 6–3 Rules for wobble base-pairing between codon and anticodon (Problem 6–47).

	WOBBLE CODON BASE	POSSIBLE ANTICODON BASE		WOBBLE ANTICODON BASE	POSSIBLE CODON BASE
Bacteria	U	A, G, or I	Bacteria	U	
	C	G or I		C	
	A	U or I		A	
	G	C or U		G	
				I	
Eucaryotes	U	G or I	Eucaryotes	U	
	C	G or I		C	
	A	U		A	
	G	C		G	
				I	

as particular anticodon bases and their recognition by possible codon bases, as suggested by the partial information on the right side of the table.

6–48* Given the wobble rules for codon–anticodon pairing in bacteria, the minimum number of different tRNAs that would be required to recognize all 61 codons is 31. What is the minimum number of different tRNAs that is consistent with the wobble rules used in eucaryotes?

6–49 A mutation in a bacterial gene generates a UGA stop codon in the middle of the mRNA coding for the protein product. A second mutation in the cell leads to a single nucleotide change in a tRNA that allows the correct translation of the protein; that is, the second mutation 'suppresses' the defect caused by the first. The altered tRNA translates the UGA as tryptophan. What nucleotide change has probably occurred in the mutant tRNA molecule? What consequences would the presence of such a mutant tRNA have for the translation of the normal genes in this cell?

6–50* In a clever experiment performed in 1962, a cysteine that was already attached to tRNA^{Cys} was chemically converted to an alanine. These alanyl-tRNA^{Cys} molecules were then added to a cell-free translation system from which the normal cysteinyl-tRNA^{Cys} molecules had been removed. When the resulting protein was analyzed, it was found that alanine had been inserted at every point in the protein chain where cysteine was supposed to be. Discuss what this experiment tells you about the role of aminoacyl-tRNA synthetases during the normal translation of the genetic code.

6–51 The charging of a tRNA with an amino acid occurs according to the reaction

$$\text{amino acid} + \text{tRNA} + \text{ATP} \rightarrow \text{aminoacyl-tRNA} + \text{AMP} + \text{PP}_i$$

where PP$_i$ is pyrophosphate, the linked phosphates that were cleaved from ATP to generate AMP. In the aminoacyl-tRNA, the amino acid and tRNA are linked with a high-energy bond. Thus, a large portion of the energy derived from the hydrolysis of ATP is stored in this bond and is available to drive peptide-bond formation at the later stages of protein synthesis. The free-energy change ($\Delta G°$) for the charging reaction shown above is close to zero, so that attachment of the amino acid to tRNA would not be expected to be dramatically favored. Can you suggest a further step that could help drive the reaction to completion?

6–52* Many of the errors in protein synthesis occur because tRNA synthetases have difficulty discriminating between related amino acids. For example, isoleucyl-tRNA synthetase (IleRS) misactivates valine

$$\text{IleRS} + \text{Val} + \text{ATP} \rightarrow \text{IleRS(Val-AMP)} + \text{PP}_i$$

with a frequency about 1/180 that of isoleucine. Protein synthesis is more accurate than that because the synthetase edits out most of its mistakes, in a reaction that depends on the presence of tRNA^{Ile}.

$$\text{IleRS(Val-AMP)} + \text{tRNA}^{Ile} \rightarrow \text{IleRS} + \text{Val} + \text{AMP} + \text{tRNA}^{Ile}$$

The tRNA^{Ile} is, of course, also required for proper aminoacylation (charging) by Ile. Are the parts of tRNA^{Ile} that are required for Ile charging the same as those that are required for Val editing?

One approach to this question is to make changes in the tRNA^{Ile} to see whether the two activities track with one another. Rather than change the sequence nucleotide by nucleotide, blocks of sequence changes were made, using tRNA^{Val} as a donor. tRNA^{Val} by itself does not stimulate Ile charging or Val editing by IleRS. Changing its anticodon from 5'-CAU to 5'-GAU, however, allows it to be charged fairly efficiently by IleRS. A variety of chimeric tRNAs

Problem 3–38 looks at editing and charging from the standpoint of the isoleucyl-tRNA synthetase.

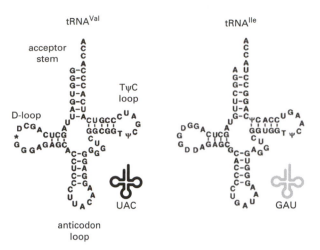

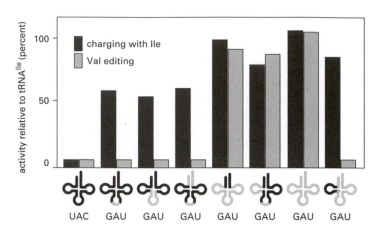

were made by combining bits of tRNA^{Ile} and tRNA^{Val}. Their ability to stimulate Ile charging and Val editing was then tested, as shown in Figure 6–28.

A. What (at a minimum) must be inserted into tRNA^{Val} to permit Ile charging by IleRS?
B. What (at a minimum) must be inserted into tRNA^{Val} to permit Val editing by IleRS?
C. Does IleRS recognize the same features of tRNA^{Ile} when it catalyzes Ile charging that it does when it carries out Val editing? Explain your answer.

Figure 6–28 Portions of tRNA^{Ile} responsible for charging and editing (Problem 6–52). (A) tRNA^{Val} and tRNA^{Ile}. (B) Ile charging and Val editing. Chimeric tRNAs composed of bits from tRNA^{Val} (*black*) and tRNA^{Ile} (*gray*) are shown *below* the results for Ile charging (*black bars*) and Val editing (*gray bars*).

6–53   Consider the following experiment on the coordinated synthesis of the α and β chains of hemoglobin. Rabbit reticulocytes were labeled with ^{3}H-lysine for 10 minutes, which is very long relative to the time required for synthesis of a single globin chain. The ribosomes, with attached nascent chains, were then isolated by centrifugation to give a preparation free of soluble (finished) globin chains. The nascent globin chains were digested with trypsin, which gives peptides ending in C-terminal lysine or arginine. The peptides were then separated by high-performance liquid chromatography (HPLC), and their radioactivity was measured. A plot of the radioactivity in each peptide versus the position of the lysines in the chains (numbered from the N-terminus) is shown in Figure 6–29.

A. Do these data allow you to decide which end of the globin chain (N- or C-terminus) is synthesized first? How so?
B. In what ratio are the two globin chains produced? Can you estimate the relative numbers of α- and β-globin mRNA molecules from these data?
C. How long does a protein chain stay attached to the ribosome once the termination codon has been reached?
D. It was once suggested that heme is added to nascent globin chains during their synthesis and, furthermore, that ribosomes must wait for insertion of heme before they can proceed. The straight lines in Figure 6–29 indicate that

Figure 6–29 Synthesis of α- and β-globin chains (Problem 6–53).

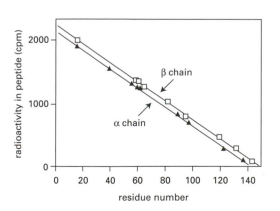

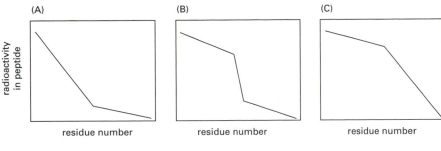

Figure 6–30 Hypothetical curves for globin synthesis with a roadblock to ribosome movement at the midpoint of the mRNA (Problem 6–53). These schematic diagrams are analogous to the graph in Figure 6–29.

ribosomes do not pause significantly, and heme is now thought to be added after synthesis. From among the graphs shown in Figure 6–30, choose the one that would result if there were a significant roadblock to ribosome movement halfway down the mRNA.

6–54* The protein you are studying contains five leucines and consists of a single polypeptide chain. One leucine is C-terminal and another is N-terminal. In a suspension of cells the average time required to synthesize this polypeptide is 8 minutes. At time zero, radioactive leucine is added to five different suspensions of cells that are *already* in the process of synthesizing the protein. You isolate the complete protein from individual suspensions at 2, 4, 6, 8, and 80 minutes. (Any incomplete polypeptide chains are eliminated.) The proteins are then analyzed for N-terminal and total radioactive leucine. With increasing time of exposure of the cells to the radioactive leucine, the ratio of N-terminal radioactivity to total radioactivity in the isolated protein should:

A. Increase to a final value of 0.2.
B. Remain constant at a value of 0.2.
C. Decrease to a final value of 0.2.
D. An answer cannot be determined from this information.

6–55 Rates of peptide chain growth can be estimated from data such as those shown in Figure 6–31. In this experiment, a single TMV mRNA species encoding a 116,000-dalton protein was translated in a rabbit-reticulocyte lysate in the presence of ^{35}S-methionine. Samples were removed at 1-minute intervals and subjected to electrophoresis on SDS-polyacrylamide gels. The separated translation products were visualized by autoradiography. As is apparent in Figure 6–31, the largest detectable proteins get larger with time.

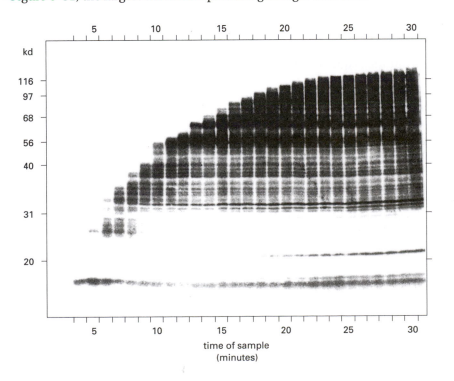

Figure 6–31 Time course of synthesis of a TMV protein in a rabbit-reticulocyte lysate (Problem 6–55). No radioactivity was detected during the first 3 minutes because the short chains ran off the bottom of the gel. SDS denatures proteins so that they run approximately according to their molecular masses. A scale of molecular masses in kilodaltons is shown on the *left*.

A. Is the rate of synthesis linear with time? One simple way to answer this question is to determine the molecular mass of the largest peptide in each sample, as determined by reference to the standards shown on the left in Figure 6–31, and then plot each of these masses against the time at which the relevant sample was taken.

B. What is the rate of protein synthesis (in amino acids/minute) in this experiment? Assume the average molecular mass of an amino acid is 110 daltons.

C. Why does the autoradiograph have so many bands in it rather than just a few bands that get larger as time passes; that is, why does the experiment produce the 'actual' result (Figure 6–23A) rather than the 'theoretical' result (Figure 6–32B)? Can you think of a way to manipulate the experimental conditions to produce the theoretical result?

6–56* The average molecular weight of proteins encoded in the human genome is about 50,000. A few proteins, however, are very much larger that this average. For example, the protein called titin, which is made by muscle cells, has a molecular weight of 3,000,000.

A. Estimate how long it will take a muscle cell to translate an mRNA coding for an average protein and one coding for titin. The average molecular mass of amino acids is about 110 daltons. Assume that the translation rate is two amino acids per second.

B. If the nucleotides in the coding portion of the mRNA constitute 5% of the total that are transcribed, how long will it take a muscle cell to transcribe a gene for an average protein versus the titin gene. Assume that the transcription rate is 20 nucleotides per second.

6–57 Protein synthesis consumes four high-energy phosphate bonds per added amino acid. Transcription consumes two high-energy phosphate bonds per added nucleotide. Calculate how many protein molecules will have been made from an individual mRNA at the point when the energy invested in translation is equal to the energy invested in transcription. Assume that the nucleotides in the coding portion of the mRNA constitute 5% of the total that are transcribed.

6–58* It is commonly reported that 30–50% of a cell's energy budget is spent on protein synthesis. How do you suppose such a measurement might be made?

6–59 One strand of a section of DNA isolated from *E. coli* reads

5'-GTAGCCTACCCATAGG-3'

A. Suppose that an mRNA is transcribed from this DNA using the complementary strand as a template. What will the sequence of the mRNA in this region be?

B. How many different peptides could potentially be made from this sequence of RNA? Would the same peptides be made if the other strand of the DNA served as the template for transcription?

C. What peptide would be made if translation started exactly at the 5' end of the mRNA in part A? When tRNA^Ala leaves the ribosome, what tRNA will be bound next? When the amino group of alanine forms a peptide bond, what bonds, if any, are broken, and what happens to tRNA^Ala?

D. Suppose this stretch of DNA is transcribed as indicated in part A, but you do not know which reading frame is used. Could this DNA segment originate from the beginning of the coding region of a gene? The middle? The end? How can you tell?

6–60* The elongation factor EF-Tu introduces two short delays between codon–anticodon base-pairing and formation of the peptide bond. These delays increase the accuracy of protein synthesis. Describe these delays and explain how they improve the fidelity of translation.

6–61 Polycistronic mRNAs are common in procaryotes, but extremely rare in eucaryotes. Describe the key differences in protein synthesis that underlie this observation.

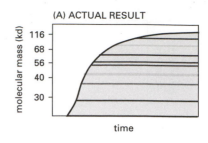

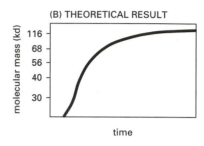

Figure 6–32 Potential outcomes of experiments on rates of protein synthesis (Problem 6–55).

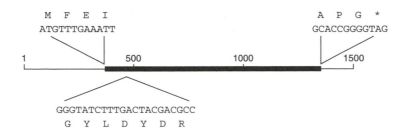

MFEI
ATGTTTGAAATT

APG*
GCACCGGGGTAG

1   500   1000   1500

GGGTATCTTTGACTACGACGCC
G Y L D Y D R

Figure 6–33 Schematic representation of the gene for RF2 (Problem 6–62). The coding sequence is shown as a *thick line*, with sequences at the start and finish shown for reference.

6–62* Termination codons in bacteria are decoded by one of two proteins. Release factor 1 (RF1) recognizes UAG and UAA, whereas RF2 recognizes UGA and UAA. For RF2, a comparison of the nucleotide sequence of the gene with the amino acid sequence of the protein revealed a startling surprise, which is contained within the sequences shown below the gene in Figure 6–33. Sequences of the gene and protein were checked carefully to rule out any artifacts.
   A. What is the surprise?
   B. What hypothesis concerning the regulation of RF2 expression is suggested by this observation?

6–63 The overall accuracy of protein synthesis is difficult to measure because mistakes are very rare. One ingenious approach used flagellin (molecular weight 40,000). Flagellin is the sole protein in bacterial flagella and thus easy to purify. Because flagellin contains no cysteine, it allows for sensitive detection of cysteine that has been misincorporated into the protein.

Problem 3–34 examines the relationship between error rate, length, and fraction of proteins correctly synthesized.

To generate radioactive cysteine, bacteria were grown in the presence of $^{35}SO_4^{2-}$ (specific activity 5.0×10^3 cpm/pmol) for exactly one generation with excess unlabeled methionine in the growth medium (to prevent incorporation of ^{35}S-label into methionine). Flagellin was purified and assayed: 8 μg of flagellin were found to contain 300 cpm of ^{35}S radioactivity.
   A. Of the flagellin molecules that were synthesized during the labeling period, what fraction contained cysteine? Assume that the mass of flagellin doubles during the labeling period and that the specific activity of cysteine in flagellin is equal to the specific activity of the $^{35}SO_4^{2-}$ used to label the cells.
   B. In flagellin, cysteine is misincorporated at the arginine codons CGU and CGC. In terms of anticodon–codon interaction, what mistake is made during misincorporation of cysteine for arginine?
   C. Given that there are 18 arginines in flagellin and assuming that all arginine codons are equally represented, what is the frequency of misreading of each sensitive (CGU and CGC) arginine codon?
   D. Assuming that the error frequency per codon calculated above applies to all amino acid codons equally, calculate the percentage of molecules that is correctly synthesized for proteins 100, 1000, and 10,000 amino acids in length. The probability of synthesizing a correct protein is $P = (1 - E)^n$, where E is the error frequency and n is the number of amino acids added.

6–64* Procaryotes and eucaryotes protect against the dangers of translating broken mRNAs, but they employ different strategies. What are the problems associated with partial mRNAs and how do cells avoid them?

6–65 In procaryotic cells the SmpB protein binds to tmRNA, thereby allowing it to attach a C-terminal tag to proteins whose translation is stalled. You wish to find out the point at which SmpB operates. tmRNA is initially charged with alanine and then it binds to the A-site of a ribosome in which mRNA translation has stalled. The alanine is attached to the end of the protein along with 10 additional amino acids that tmRNA encodes. Finally, the tagged protein is released from the ribosome for destruction by a special protease.

Problem 3–48 characterizes the tight binding between SmpB and tmRNA.

Several experiments were carried out to determine where in this sequence of events SmpB acts. Alanyl-tRNA synthetase was shown to charge tmRNA with alanine equally well in the presence and absence of SmpB. The association of tmRNA with ribosomes, however, was different in the presence (+)

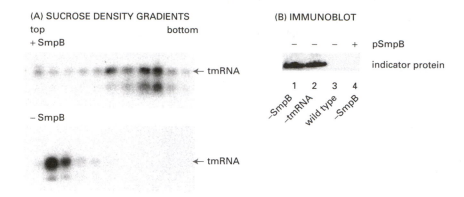

(A) SUCROSE DENSITY GRADIENTS

top   bottom
+ SmpB

← tmRNA

– SmpB

← tmRNA

(B) IMMUNOBLOT

–   –   –   +   pSmpB

indicator protein

1   2   3   4

SmpB –tmRNA wild type –SmpB

Figure 6–34 Analysis of role of SmpB in tmRNA-mediated protein tagging and destruction (Problem 6–65).
(A) Association of tmRNA with ribosomes in normal and SmpB-deficient cells. Cell components were separated by layering an extract on top of a sucrose gradient and then subjecting them to centrifugation. Under the conditions used in these experiments, free tmRNA remains near the top of the gradient and polysomes are driven toward the bottom. The position of tmRNA in the gradient was determined by hybridization to tmRNA-specific oligonucleotides. (B) Destruction of a protein fragment in wild-type cells and in SmpB-deficient (–SmpB) or tmRNA-deficient (–tmRNA) cells. An mRNA that encodes the indicator protein but does not carry a stop codon was included in all lanes. Plasmid pSmpB expresses SmpB; its presence (+) or absence (–) is indicated *above* the lanes.

and absence (–) of SmpB (Figure 6–34A). Destruction of proteins generated by stalled ribosomes, which was detected using a convenient indicator protein, was also different in the presence (+) and absence (–) of SmpB (Figure 6–34B).

To the extent these experiments allow, describe the point at which SmpB is needed to permit tmRNA to carry out its function.

6–66* You are studying protein synthesis in *Tetrahymena*, which is a unicellular ciliate. You have good news and bad news. The good news is that you have the first bit of protein and nucleic acid sequence data for the C-terminus of a *Tetrahymena* protein, as shown below:

```
 I   M   Y   K   Q   V   A   Q   T   Q   L   *
AUU AUG UAU AAG UAG GUC GCA UAA ACA CAA UUA UGA GAC UUA
```

The bad news is that you have been unable to translate purified *Tetrahymena* mRNA in a reticulocyte lysate, which is a standard system for analyzing protein synthesis *in vitro*. The mRNA looks good by all criteria, but the translation products are mostly small polypeptides (Figure 6–35, lane 1).

To figure out what is wrong, you do a number of control experiments using a pure mRNA from tobacco mosaic virus (TMV) that encodes a 116-kd protein. TMV mRNA alone is translated just fine in the *in vitro* system, giving a major band at 116 kd—the expected product—and a very minor band about 50 kd larger (Figure 6–35, lane 2). When *Tetrahymena* RNA is added, there is a decrease in the smaller of the two bands and a significant increase in the larger one (lane 3). When some *Tetrahymena* cytoplasm (minus the ribosomes) is added, the TMV mRNA now gives almost exclusively the higher molecular mass product (lane 4); furthermore, much to your delight, the previously inactive *Tetrahymena* mRNA now appears to be translated (lane 4). You confirm this by leaving out the TMV mRNA (lane 5).

A. What is unusual about the sequence data for the *Tetrahymena* protein?
B. How do you think the minor higher molecular mass band is produced from pure TMV mRNA in the reticulocyte lysate?
C. Explain the basis for the shift in proportions of the major and minor TMV proteins upon addition of *Tetrahymena* RNA alone and in combination with *Tetrahymena* cytoplasm. What *Tetrahymena* components are likely to be required for efficient translation of *Tetrahymena* mRNA?
D. Comment on the evolutionary implications of your results.

6–67 You have isolated an antibiotic, named edeine, from a bacterial culture. Edeine inhibits protein synthesis but has no effect on either DNA synthesis or RNA synthesis. When added to a reticulocyte lysate, edeine stops protein synthesis after a short lag, as shown in Figure 6–36. By contrast, cyclohex-imide stops protein synthesis immediately (Figure 6–36). Analysis of the edeine-inhibited lysate by density-gradient centrifugation showed that no polyribosomes remained by the time protein synthesis had stopped. Instead, all the globin mRNA accumulated in an abnormal 40S peak, which contained equimolar amounts of the small ribosomal subunit and initiator tRNA.

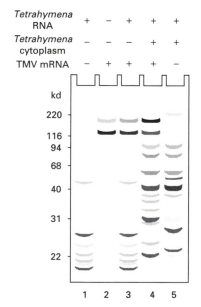

Tetrahymena RNA	+	–	+	+	+
Tetrahymena cytoplasm	–	–	–	+	+
TMV mRNA	–	+	+	+	–

kd

220

116
94

68

40

31

22

1   2   3   4   5

Figure 6–35 Translation of TMV and *Tetrahymena* mRNA in the presence and absence of various components from *Tetrahymena* (Problem 6–66). The molecular masses of marker proteins are indicated in kilodaltons on the *left*.

A. What step in protein synthesis does edeine inhibit?
B. Why is there a lag between addition of edeine and cessation of protein synthesis? What determines the length of the lag?
C. Would you expect the polyribosomes to disappear if you added cycloheximide at the same time as edeine?

6–68* In a reticulocyte lysate the polynucleotide 5′-AUGUUUUUUUUU directs the synthesis of Met–Phe–Phe–Phe. In the presence of farsomycin, a new antibiotic perfected by Fluhardy Pharmaceuticals, this polymer directs synthesis of Met–Phe only. From this information, which of the following deductions could you make about farsomycin?
A. It prevents formation of the 80S initiation complex, which contains the initiator tRNA and both ribosomal subunits.
B. It inhibits binding of aminoacyl-tRNAs to the A-site in the ribosome.
C. It inactivates peptidyl transferase activity of the large ribosomal subunit.
D. It blocks translocation of peptidyl-tRNA from the A-site to the P-site of the ribosome.
E. It interferes with chain termination and release of the peptide.

6–69 Both hsp60-like and hsp70 molecular chaperones share an affinity for exposed hydrophobic patches on proteins, using them as indicators of incomplete folding. Why are hydrophobic patches used as critical signals for the folding status of a protein?

6–70* Hsp70 molecular chaperones are thought to bind to hydrophobic regions of nascent polypeptides on ribosomes. This binding was difficult to demonstrate for DnaK, which is one of the two major hsp70 chaperones in *E. coli*. In one approach nascent proteins were labeled with a 15-second pulse of ^{35}S-methionine, isolated in the absence of ATP, and then incubated with antibodies against DnaK. A collection of labeled proteins was precipitated as shown in Figure 6–37A, lane 1. The proteins were not precipitated if they were treated beforehand with the strong detergent SDS (lane 2), or if they were isolated from a mutant strain missing DnaK (lane 3). If labeled DnaK-mutant cells were mixed with unlabeled wild-type cells before the proteins were isolated, DnaK antibodies did not precipitate labeled proteins (lane 4). Finally, if unlabeled methionine was added in excess after the pulse of ^{35}S-methionine, the labeled proteins disappeared with time (Figure 6–37B).
A. Do the series of control experiments in Figure 6–37A, lanes 2 to 4, argue that DnaK is bound to the labeled proteins in a meaningful way (as opposed to a random aggregation, for example)?
B. When ATP was present during the isolation of the proteins, antibodies against DnaK did not precipitate any proteins. How do you suppose ATP might interfere with precipitation of labeled proteins?
C. Why do you suppose that the labeled proteins disappeared with time in the presence of excess unlabeled methionine?
D. Do these experiments show that DnaK binds to proteins as they are being synthesized on ribosomes? Why or why not?

6–71 DnaK and trigger factor (TF) are the two major hsp70 chaperones in *E. coli*, yet neither gene is essential for growth at 37°C. To investigate this seeming

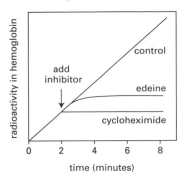

Figure 6–36 Effects of the inhibitors edeine and cycloheximide on protein synthesis in reticulocyte lysates (Problem 6–67).

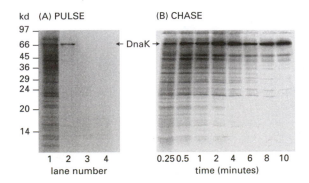

Figure 6–37 Association of DnaK with nascent proteins (Problem 6–70). (A) Pulse-labeled proteins immunoprecipitated by antibodies to DnaK. Wild-type bacteria (lane 1); SDS-treated wild-type bacteria (lane 2); *dnaK*-deletion strain (lane 3); and mixture of labeled *dnaK*-deletion strain and unlabeled wild-type strain (lane 4). Size markers are indicated on the *left* and the position of DnaK is indicated on the *right*. (B) Pulse-chase experiment. Wild-type bacteria labeled for 15 seconds with ^{35}S-methionine were then incubated for varying times in the presence of an excess of unlabeled methionine before immunoprecipitation by antibodies against DnaK.

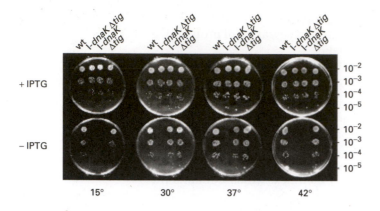

Figure 6–38 Analysis of growth of bacterial strains defective for one or both hsp70 proteins (Problem 6–71). Dilutions (shown on the right) of bacterial cultures (labeled at the top) were spotted on agar plates in the presence and absence of IPTG (indicated on the left) and grown at various temperatures (shown at the bottom). White spots indicate bacterial growth.

paradox, you construct three mutant strains for comparison to the wild-type (wt) parent. One carries a deletion of the TF gene (Δ*tig*); the second carries an altered version of *dnaK* that is expressed from an IPTG-inducible promoter (I-*dnaK*); and the third contains both mutations. You spot various dilutions of the four types of bacteria on plates supplemented with IPTG (which turns on the *dnaK* gene) or lacking IPTG (which turns off the *dnaK* gene) and grow them at a variety of temperatures (Figure 6–38).

A. Compare the growth properties of the strains (Δ*tig* and I-*dnaK*) that express a single hsp70.

B. Contrast the growth properties of the single mutants with those of the double mutant (I-*dnaK* Δ*tig*), which expresses neither heat shock protein. Suggest an explanation for any significant differences in the growth properties of the single and double mutants.

6–72* Hsp60-like molecular chaperones provide a large central cavity in which misfolded proteins can attempt to refold. Two models, which are not mutually exclusive, can be considered for the role of the hsp60-like chaperones in the refolding process. They might act passively to provide an isolation chamber that aids protein folding by preventing aggregation with other proteins. Alternatively, hsp60-like chaperones might actively unfold misfolded proteins to remove stable, but incorrect, intermediate structures that block proper folding. The involvement of ATP binding and hydrolysis and the associated conformational changes of the hsp60-like chaperones could be used in favor of either model.

In bacteria the hsp60-like chaperone GroEL binds to a misfolded protein, then binds ATP and the GroES cap, and after about 15 seconds hydrolyzes the ATP and ejects the protein (Figure 6–39). To distinguish between a passive or active role of GroEL in refolding, you label a protein by denaturing it in tritiated water, ^3H_2O. When the denaturant is removed and the protein is transferred to normal water, ^1H_2O, most of the radioactivity is lost within 10 minutes, but a stable core of 12 tritium atoms exchanges on a much longer timescale—a behavior typical of amide hydrogen atoms involved in stable hydrogen bonds. Disruption of these bonds would allow their exchange within a few milliseconds.

You prepare the radioactive substrate and mix it immediately with a slight molar excess of GroEL, and then wait 10 minutes for the rapidly exchanging

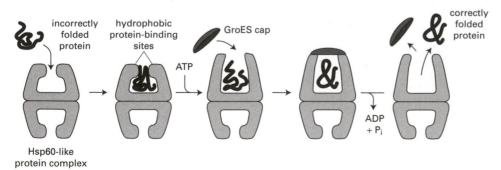

Hsp60-like protein complex

Figure 6–39 Protein refolding by the bacterial GroEL chaperone (Problem 6–72). A misfolded protein is initially captured by hydrophobic interactions along one rim of the barrel. The subsequent binding of ATP plus the GroES cap increases the size of the cavity and confines the protein in the enclosed space, where it has a new opportunity to fold. After about 15 seconds, ATP hydrolysis ejects the protein, whether folded or not, and the cycle repeats.

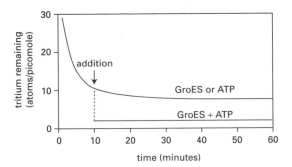

Figure 6–40 Effects of chaperone components on rates of tritium exchange (Problem 6–72).

tritium atoms to be lost. Addition of GroES or ATP alone has no effect on the exchange, as shown by the upper curve in Figure 6–40. Addition of both GroES and ATP, however, causes a rapid loss of tritium (Figure 6–40, lower curve). Addition of GroES plus ADP has no effect, but GroES plus AMPPNP, a nonhydrolyzable analog of ATP, promotes a rapid exchange that is indistinguishable from GroES plus ATP.

A. After addition of components to the complex of tritiated protein and GroEL, it took a minimum of 45 seconds to separate the protein from the freed tritium label. Did the exchange of tritium occur within one cycle of binding and ejection by GroEL, which takes about 15 seconds, or might it have required more than one cycle? Explain your answer.

B. Do the results support a passive isolation-chamber model, or an active-unfolding model, for GroEL action? Explain your reasoning.

6–73 Most proteins require molecular chaperones to assist in their correct folding. How do you suppose the chaperones themselves manage to fold correctly?

6–74* (**True/False**) The only significant difference between simple proteases and the proteasome is that the proteasome targets ubiquitylated proteins. Explain your answer.

6–75 You wish to measure the fraction of newly synthesized proteins that are degraded by proteasomes. Your strategy is to assay newly synthesized proteins in the presence and absence of inhibitors of proteasome function. You pulse-label mouse lymph node cells with ^{35}S-methionine and then chase with excess unlabeled methionine for 30 minutes, all in the presence or absence of proteasome inhibitors. At various times during the chase, you take aliquots of cells and boil them in SDS to denature all proteins. For each sample, you load identical amounts of cellular protein onto SDS-polyacrylamide gels, separate them by electrophoresis, and quantify the radioactivity in each lane. Autoradiographs of the gels and the results of quantification are shown in Figure 6–41.

Problem 15–92 illustrates the role of the proteasome in regulating the activity of β-catenin.

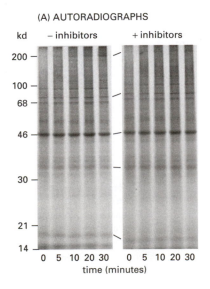
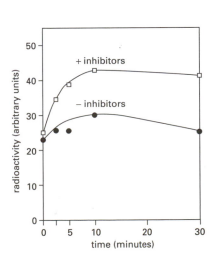

Figure 6–41 Analysis of newly synthesized proteins in the presence and absence of proteasome inhibitors (Problem 6–75). (A) Autoradiographs of SDS-polyacrylamide gel electrophoretic separation of cellular proteins. A combination of inhibitors—carbobenzoxyl-Leu-Leu-leucinal (zLLL), lactacystin, and clasto-lactacystin β-lactone—was used to block proteasome activity. Samples were taken at various times up to 30 minutes after addition of ^{35}S-methionine. Equal amounts of total cellular protein were loaded into each lane. (B) Quantification of newly synthesized proteins in the presence and absence of proteasome inhibitors. The amount of radioactivity in each entire lane in (A) was used as a basis for comparison. Radioactivity was quantified by Phosphorimager analysis.

A. What fraction of newly synthesized proteins is degraded by proteasomes? Explain your reasoning.

B. All proteins appear to be equally affected by proteasome inhibitors; that is, there appears to be a general increase in intensity of radioactivity throughout the gel. How do you suppose that blocking proteasomes might lead to such a generalized increase in intensity of radioactivity?

6–76* Describe the roles of E1, E2, and E3 proteins in conjugating ubiquitin to target proteins. Which components provide the specificity to allow particular proteins or sets of proteins to be targeted for destruction?

6–77 Fission yeast cyclin B, known as Cdc13, must be destroyed at the metaphase–anaphase transition to allow normal progression through the cell cycle. Destruction of Cdc13 requires a nine-amino acid sequence—the so-called destruction box or D-box—near its N-terminus. The D-box mediates an interaction with the anaphase-promoting complex (APC), which is the E3 component of an E2-E3 ubiquitin ligase that adds multi-ubiquitin chains and targets Cdc13 for destruction by the proteasome.

You reason that overexpression of the N-terminal 70 amino acids (N70) of Cdc13, which contains the D-box, might block the ability of APC to trigger destruction of Cdc13 and stop cell proliferation. Sure enough, overexpression of N70 proves to be lethal. As a check on your hypothesis, you assay for accumulation of Cdc13 and another D-box protein (Cut2), which you expect to be protected from destruction, as well as two non-D-box proteins, Rum1 and Cdc18, which you expect to be unaffected because their ubiquitylation is independent of APC. Much to your surprise, all four proteins accumulate when N70 is overexpressed (Figure 6–42). To clarify the situation, you prepare two other versions of N70: one in which all lysines have been replaced by arginines (K0-N70) and one that carries a mutant D-box (dm-N70). Overexpression of these two proteins gives the results shown in Figure 6–42.

Propose an explanation for the differences in the patterns of accumulation of Cdc13, Cut2, Rum1, and Cdc18 after overexpression of N70, K0-N70, and dm-N70.

6–78* The life-spans of proteins are appropriate to their *in vivo* tasks; for example, structural proteins tend to be long-lived, whereas regulatory proteins are usually short-lived. In eucaryotic cells life-spans are strongly influenced by the N-terminal amino acid. The first experiments that revealed these effects used hybrid proteins consisting of ubiquitin fused to β-galactosidase, as shown in Figure 6–43. When plasmids encoding these proteins were introduced into yeast, the hybrid proteins were synthesized, but the ubiquitin was cleaved off exactly at the junction with β-galactosidase, generating proteins with different N-termini (Figure 6–43).

To measure the half-lives of these β-galactosidase molecules, yeast were grown for several generations in the presence of a radioactive amino acid. Protein synthesis was then blocked with the inhibitor cycloheximide (CHX). The rate of degradation of β-galactosidase was determined by removing samples from the cultures at various times, purifying β-galactosidase and measuring the amount of associated radioactivity after SDS-gel electrophoresis. The results at the 5-minute time point are shown in Figure 6–44A, and a graph depicting results for all time points is shown in Figure 6–44B.

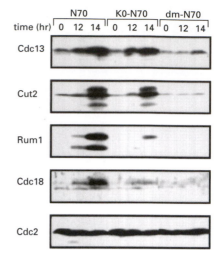

Figure 6–42 Levels of specific proteins after expression of various N-terminal segments of Cdc13 (Problem 6–77). Each version of the Cdc13 N-terminus was expressed in fission yeast from the *nmt1* promoter, which can be induced by adding thiamine to the medium. Induction takes about 12 hours. Proteins were detected by immunoblotting with specific antibodies. Cdc2 is not subject to ubiquitylation and thus serves as a loading control.

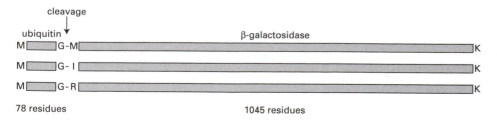

Figure 6–43 Fusion proteins encoded by three yeast plasmids (Problem 6–78). Upon expression in yeast, the fusion proteins are cleaved at the peptide bonds indicated by the *arrow*. The β-galactosidase molecules liberated by cleavage differ only at their N-termini.

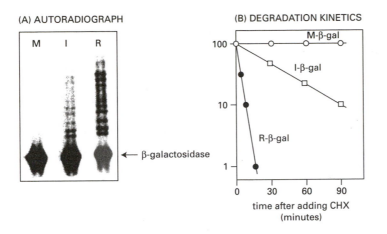

(A) AUTORADIOGRAPH

M   I   R

← β-galactosidase

(B) DEGRADATION KINETICS

M-β-gal

I-β-gal

R-β-gal

time after adding CHX
(minutes)

Figure 6–44 Analysis of half-lives of proteins with different N-termini (Problem 6–78). (A) Electrophoretic separation of radioactive β-galactosidase. Antibodies directed against β-galactosidase were used to precipitate the protein. The N-terminal amino acids are indicated *above* the lanes. (B) Disappearance of β-galactosidases with time after termination of protein synthesis by addition of cycloheximide (CHX). The level of β-galactosidase is expressed as a percentage of that present immediately after protein synthesis was blocked.

A. Using recombinant DNA techniques, it would have been straightforward to generate a series of plasmids in which the first codon in the β-galactosidase gene was changed. Why do you suppose this more direct approach was not tried?

B. Estimate the half-life (time at which half the material has been degraded) of each of the three species of β-galactosidase.

6–79 Two aspects of the work in the previous problem might have made you curious. First, how did the investigators know that ubiquitin was actually removed from the N-terminus? Second, what are the bands above the position of β-galactosidase in Figure 6–44A? Any fragments of β-galactosidase due to degradation should run below β-galactosidase.

To address these questions, the investigators used β-galactosidase antibodies to isolate nonradioactive β-galactosidase from cells transfected with the same three plasmids (see Figure 6–43). These samples were separated by electrophoresis as before, then transferred to a filter paper and reacted with radioactive antibodies specific for ubiquitin. As shown in Figure 6–45, antibodies against ubiquitin did not react with material at the position of β-galactosidase, but did react with a ladder of bands above β-galactosidase, which are at the same positions as those in Figure 6–44A.

A. Do these experiments demonstrate that the ubiquitin at the N-terminus of the hybrid protein is removed?

B. Offer an explanation for the presence of ubiquitin above the position of β-galactosidase in the experiments with isoleucine (I) and arginine (R) at the N-terminus, but not in the experiments with methionine (M) at the N-terminus.

Problem 17–32 considers the consequences of an indestructible cyclin on progress through the cell cycle.

THE RNA WORLD AND THE ORIGINS OF LIFE

TERMS TO LEARN

RNA world

6–80* What is so special about RNA that makes it such an attractive evolutionary precursor to DNA and protein?

6–81 Discuss the following statement: "During the evolution of life on Earth, RNA has been demoted from its glorious position as the first replicating catalyst. Its role now is as a mere messenger in the information flow from DNA to protein."

6–82* Imagine a warm pond on the primordial Earth. Chance processes have just assembled an RNA molecule that can catalyze RNA replication. It can fold into a structure capable of linking nucleotides according to instructions in an RNA template. Given an adequate supply of nucleotides, will this RNA molecule be able to catalyze its own replication? Why or why not?

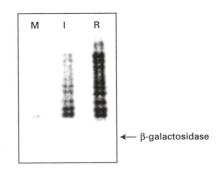

M   I   R

← β-galactosidase

Figure 6–45 Electrophoresis of unlabeled protein expressed from the ubiquitin β-galactosidase fusion gene, followed by reaction with labeled antibodies against ubiquitin (Problem 6–79). The N-terminal amino acids are indicated *above* the lanes. The position at which intact β-galactosidase runs is marked by an *arrow*.

6–83   If an RNA molecule could form a hairpin with a symmetric internal loop, as shown in Figure 6–46, could the complement of this RNA form a similar structure? If so, would there be any regions of the two structures that are identical? Which ones?

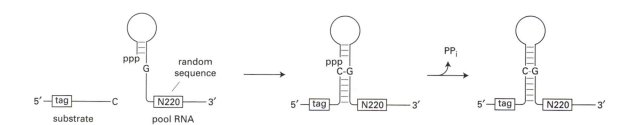

Figure 6–46 An RNA hairpin with a symmetric internal loop (Problem 6–83).

6–84*   An RNA molecule with the ability to catalyze RNA replication—the linkage of RNA nucleotides according to the information in an RNA template—would have been a key ribozyme in the RNA world. Your advisor wants to evolve such an RNA replicase *in vitro* using the techniques of selection and amplification of rare functional molecules from a pool of random RNA sequences. The difficulty has been to devise a strategy that models a reasonable first step and that can be fit into an *in vitro* selection scheme. When you arrived in the lab today, you found the scheme shown in Figures 6–47 and 6–48 on your desk with a note from your advisor that he wants to talk with you about it tomorrow when he returns from an out of town trip. He's even left you some questions to 'focus' the discussion.

A. As shown in Figure 6–47, the reaction being selected for is the attachment of an oligonucleotide 'tag' to the end of the catalytic RNA molecule. How is this reaction an analog of nucleotide addition to the end of a primer on a template RNA?

B. Why is it important in the selection scheme in Figure 6–48 that the 'tag' RNA be linked to the very RNA that catalyzed its attachment?

C. Why does the starting pool of RNA molecules have constant regions at each end and a random segment in the middle? What specific roles do these segments play in the overall scheme?

D. How is a catalytic RNA molecule selected from the pool and specifically amplified?

E. Why do you suppose it is necessary to repeat the cycles of selection and amplification? Why not simply purify the ribozymes at the end of the first cycle?

6–85   Curses! Your advisor has sent you an email with still more questions about his selection scheme (see Figures 6–47 and 6–48).

A. He thinks you will be able to generate about a milligram of RNA to begin the selection. How many molecules will be present in this amount of RNA? (Assume that an RNA nucleotide has a mass of 330 daltons and that the RNA molecules are 300 nucleotides in length.)

B. How many different molecules are possible if the central 220-nucleotide segment is completely random? What fraction of all possible molecules will be present in your 1-mg sample?

C. If the ligation reaction could only be catalyzed by a single, *unique* 50-nucleotide RNA sequence, what do you suppose your chances of success would be? What does the general success of such selection schemes imply about the range of RNA molecules that are capable of catalysis?

6–86*   Once you and your advisor had ironed out the details of the selection scheme, the work went fairly quickly. You have now carried out 10 rounds of selection with the results shown in Table 6–4. You have cloned and sequenced 15 individual RNA molecules from pool 10: no two are the same, although most have very similar sequences. Your advisor is very excited by your results and has asked you to give the next departmental seminar.

Figure 6–47 Ribozyme-catalyzed reaction selected for in this scheme for *in vitro* evolution (Problem 6–84). The random sequence in the pool RNA is 220 nucleotides in length (N220). The 3′ end of the substrate oligonucleotide is complementary to the constant 5′ end of the pool RNA molecules.

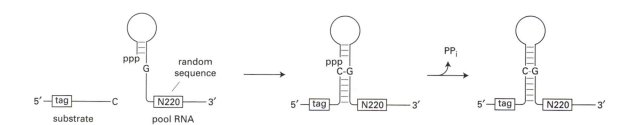

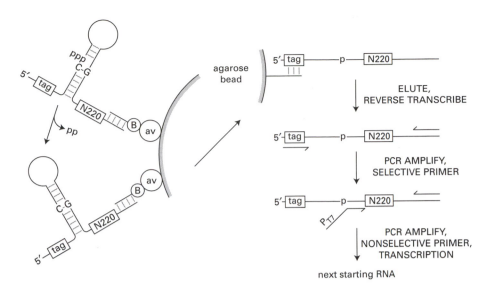

Figure 6–48 One round in the cyclic selection scheme to amplify individual ribozymes from a random pool (Problem 6–84). Each pool RNA is linked at one end to the substrate RNA molecule and at the other end to a complementary DNA oligo linked to an agarose bead for ease of manipulation. In the final PCR amplification with a nonselective primer the noncomplementary portion of the oligo carries the promoter for T7 RNA polymerase (P_{T7}), which allows the final DNA product to be transcribed back into RNA.

You know from your conversations with other students that you will need to prepare careful explanations for the questions listed below.

A. Why did you use error-prone PCR, which can introduce mutations, in some of the rounds of amplification?

B. Why did you reduce the time and Mg^{2+} concentration—both of which increase the difficulty of ligation—in successive rounds of selection?

C. How much of an improvement in ligation rate have you found in your 10 rounds of selection and amplification?

D. Why is there still such diversity among the RNA molecules after 10 rounds of selection and amplification?

6–87   (**True/False**) The main difference between protein enzymes and ribozymes lies in their maximum reaction speed, rather than in the diversity of the reactions they can catalyze. Explain your answer.

6–88*   Why are compartments thought to have been necessary for evolution in the RNA world?

6–89   What is it about DNA that makes it a better material than RNA for storage of genetic information?

TABLE 6–4 Summary of rounds of selection for a ribozyme with RNA ligation activity (Problem 6–86).

		LIGATION CONDITIONS		
ROUND	ERROR-PRONE PCR	MgCl₂ (mM)	TIME (hours)	LIGATION RATE (per hour)
0				0.000003
1	No	60	16	<0.000004
2	No	60	16	0.0008
3	No	60	16	0.0094
4	No	60	16	0.027
5	Yes	60	0.50	0.16
6	Yes	60	0.17	0.40
7	Yes	60	0.02	0.86
8	No	4	0.12	3.2
9	No	2.5	0.17	4.5
10	No	2.5	0.17	8.0

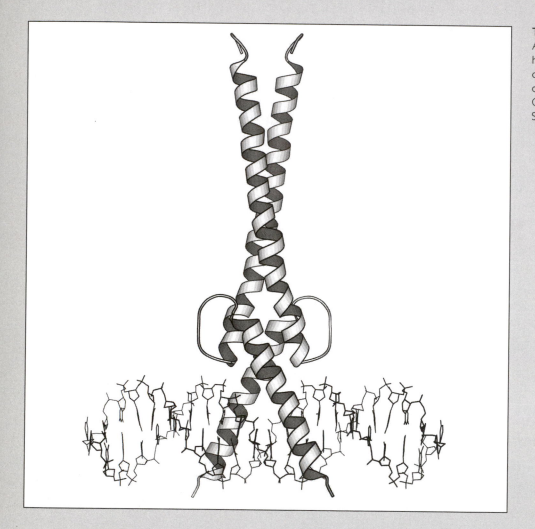

The MAX homodimer bound to DNA.
A perfect illustration of how helix-loop-helix proteins bind to DNA by forming dimers. Coordinates from 1AN2 determined by AR Ferre-D'Amare, GC Prendergast, EB Ziff & SK Burley. *Nature* 363:38–45, 1993.

CONTROL OF GENE EXPRESSION

AN OVERVIEW OF GENE CONTROL

TERMS TO LEARN

mRNA degradation control	RNA processing control	transcriptional control
protein activity control	RNA transport control	translational control
RNA localization control		

7–1 (**True/False**) When the nucleus of a fully differentiated carrot cell is injected into a frog egg whose nucleus has been removed, the injected donor nucleus is capable of programming the recipient egg to produce a normal carrot. Explain your answer.

7–2* In the original cloning of sheep from somatic cells, the success rate was very low. For example, only 1 lamb (Dolly) was born from 277 zygotes that were reconstructed using nuclei derived from breast cells and enucleated, unfertilized eggs. Other experiments using nuclei from embryonic or fetal lamb cells had a higher success rate, albeit still a low one. Given the rarity of successful events, it was critical to eliminate inadvertent mating of either the oocyte donor or the surrogate mother as the source of newborn lambs. To determine whether the cloned animals were derived from the donor nuclei, analysis of DNA microsatellites (short, repetitive DNA sequences) was carried out. Four loci at which many different lengths are present in sheep populations were analyzed in surrogate mothers and donor cells (Figure 7–1).

A. Do the results in Figure 7–1 argue that the lambs were derived from the transplanted nuclei or from an inadvertent mating? Explain your answer.

B. What would the results have looked like for the alternative you did not choose in question A?

---

*Answers to these problems are available to Instructors at www.classwire.com/garlandscience/

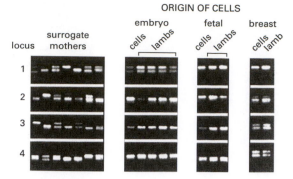

Figure 7–1 Microsatellite analysis of seven surrogate mothers, the three different nuclear donor cell types, and the seven lambs that were born successfully (Problem 7–2). Four polymorphic loci were used in the analysis. The surrogate mothers are arranged *left* to *right* in the same order as the lambs. Nuclear donor cells were derived from embryo, fetus, or breast. At each of the four polymorphic loci, flanking PCR primers were used to amplify DNA that included a particular microsatellite. Microsatellites with different numbers of repeats give rise to different length PCR products.

7–3 Developmentally programmed genome rearrangements occur in mammals during the generation of diversity in the immune system. In B cells, for example, the variable (V) and constant (C) segments of the immunoglobulin gene are juxtaposed by deletion of a long segment of DNA that separates them in other cells. Digestion of unrearranged germline DNA with a restriction nuclease that cuts the DNA flanking the V and C segments generates two bands in a Southern blot when radioactive hybridization probes specific for the V and the C segments are used (Figure 7–2). Would you expect the V- and C-segment probes to hybridize to the same or different DNA fragments after digestion of B cell DNA with the same restriction nuclease? Sketch a possible pattern of hybridization to B-cell DNA that is consistent with your expectations. Explain the basis for your pattern of hybridization. (Without a lot more information you cannot predict the exact pattern, so focus on general features of the patterns.)

7–4* A small portion of a two-dimensional display of proteins from human brain is shown in Figure 7–3. These proteins were separated on the basis of size in one dimension and electrical charge (isoelectric point) in the other dimension. Not all protein spots on such displays are products of different genes; some represent modified forms of the same protein that run at different positions. Pick out a couple of sets of spots that could represent proteins that differ by the number of phosphates they carry. Explain the basis for your selection.

7–5 DNA microarray analysis of the patterns of mRNA abundance in different human cell types shows that the level of expression of almost every active gene is different. The patterns of mRNA abundance are so characteristic of cell type that they can be used to type human cancer cells of unknown tissue origin. By definition, however, cancer cells are different from their noncancerous precursor cells. How do you suppose then that patterns of mRNA expression can be used to determine the tissue source of a human cancer?

7–6* In principle, a eucaryotic cell can regulate gene expression at any step in the pathway from DNA to the active protein (Figure 7–4).

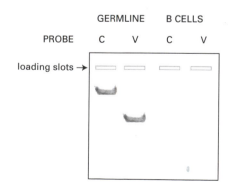

Figure 7–2 Southern blot of DNAs from germline and B cells (Problem 7–3). Germline DNA and B-cell DNA were digested with the same restriction nuclease. Only the hybridization to germline DNA is shown.

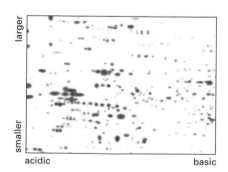

Figure 7–3 Two-dimensional separation of proteins from the human brain (Problem 7–4). The proteins were displayed using two-dimensional gel electrophoresis. Only a small portion of the protein spectrum is shown.

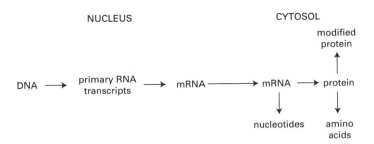

Figure 7–4 Seven steps at which the pathway for eucaryotic gene expression can be controlled (Problem 7–6).

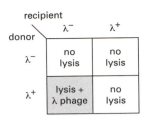

Figure 7–5 Results of matings between bacteria with and without λ prophages (Problem 7–7). λ⁻ indicates the absence of a prophage; λ⁺ indicates its presence.

A. Place the types of control listed below at appropriate places on the diagram in Figure 7–4.
 1. mRNA degradation control
 2. protein activity control
 3. protein stability control
 4. RNA processing control
 5. nuclear export and localization control
 6. transcriptional control
 7. translational control
B. Which of the types of control listed above are unlikely to be used in bacteria?

DNA-BINDING MOTIFS IN GENE REGULATORY PROTEINS

TERMS TO LEARN

chromatin immunoprecipitation	gene regulatory protein	homeodomain
combinatorial control	helix-loop-helix (HLH) motif	leucine zipper motif
DNA affinity chromatography	helix-turn-helix	zinc finger
gel-mobility shift assay		

7–7   When Jacob and Wollman tried to check the genetic linkage between the *gal* gene and a bacteriophage λ prophage (an integrated λ genome), they discovered the surprising phenomenon of 'erotic induction' (which was later called zygotic induction for publication). In a bacterial mating, a portion of the chromosome is transferred via a narrow tube from the donor bacterium to the recipient. Jacob and Wollman found that if the donated chromosome carried a λ prophage, but the recipient cell did not, λ growth was induced in the recipient cell, which then lysed, producing λ phage. If the recipient cell carried the λ prophage, however, no lysis was observed. A summary of results from all their matings is shown in Figure 7–5.
A. Explain how these results are consistent with the notion that a repressor encoded by the prophage normally keeps the bacteriophage's lytic genes turned off.
B. Suppose that the prophage prevented lytic growth by expressing a gene regulatory protein that turned on a gene for an anti-lysis protein. Would the results of the matings have been the same or different? Explain your answer.

7–8*   Figure 7–6 shows a short stretch of a DNA helix displayed as a spacefilling

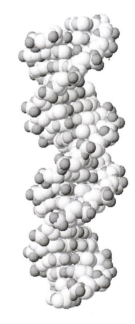

Figure 7–6 A spacefilling model of a DNA duplex (Problem 7–8).

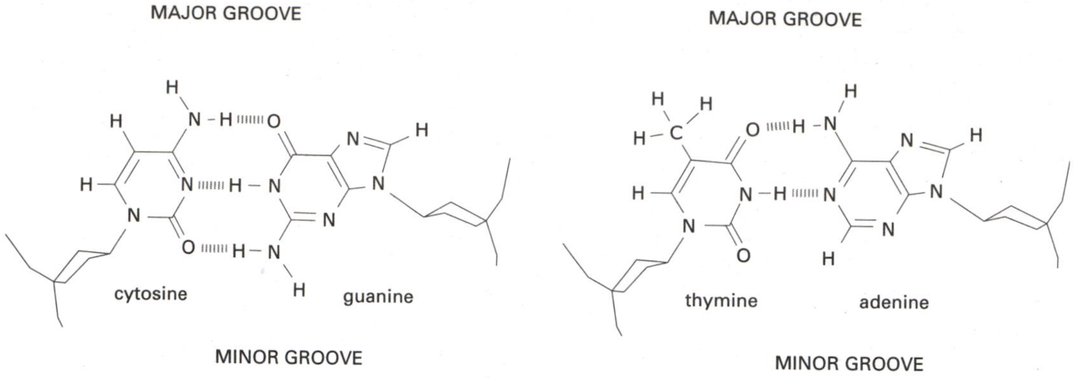

MAJOR GROOVE   MAJOR GROOVE

cytosine   guanine   thymine   adenine

MINOR GROOVE   MINOR GROOVE

Figure 7–7 C–G and T–A base pairs (Problem 7–9).

model. Indicate the major and minor grooves and provide a scale. Is it possible to tell the polarity of each of the strands in this figure?

7–9 Explain how DNA-binding proteins can make sequence-specific contacts to a double-stranded DNA molecule without breaking the hydrogen bonds that hold the bases together. Indicate how, by making such contacts, a protein can distinguish a C–G from a T–A base pair. Use Figure 7–7 to indicate what sorts of noncovalent bonds (hydrogen bonds, ionic bonds, or hydrophobic interactions) could be used to discriminate C–G and T–A. There is no need to specify any particular amino acid on the protein.

7–10* DNA-binding proteins often find their specific sites much faster than would be anticipated by simple three-dimensional diffusion. The lac repressor, for example, associates with the *lac* operator more than 100 times faster than expected. Clearly, the repressor must find the operator by mechanisms that reduce the dimensionality or volume of the search in order to hasten target acquisition.

Several techniques have been used to investigate this problem. One of the most elegant used strongly fluorescent RNA polymerase molecules that could be followed individually. An array of DNA molecules that did not contain promoters was aligned in parallel by an electrophoretic technique, and fluorescent RNA polymerase molecules were flowed across them at an oblique angle (Figure 7–8A). When traces of individual RNA polymerases were followed, about half flowed in the same direction as the bulk and about

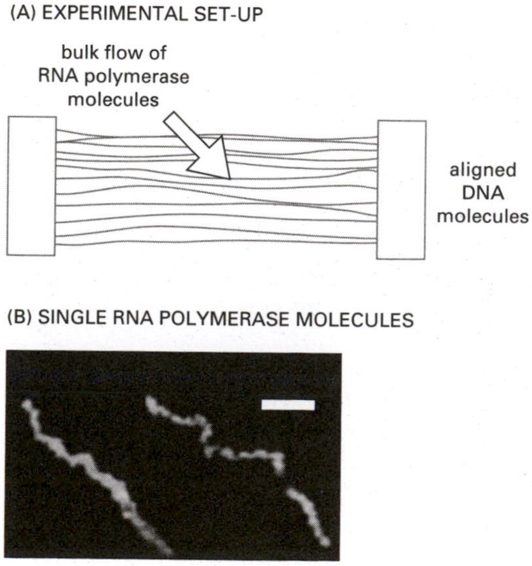

(A) EXPERIMENTAL SET-UP

bulk flow of
RNA polymerase
molecules

aligned
DNA
molecules

(B) SINGLE RNA POLYMERASE MOLECULES

Figure 7–8 Interactions of individual RNA polymerase molecules with DNA (Problem 7–10). (A) Experimental set-up. A population of DNA molecules that do not contain promoters are aligned by dielectrophoresis, and a population of highly fluorescent RNA polymerase molecules is flowed across them. (B) Traces of two individual RNA polymerase molecules. The one on the *left* has traveled with the bulk flow, and the one on the *right* has deviated from it. The scale bar is 10 μm.

half deviated from the bulk flow in a characteristic manner (Figure 7–8B). If the RNA polymerase molecules were first incubated with a short DNA fragment containing a strong promoter, all the traces followed the bulk flow.

A. Offer an explanation for why some RNA polymerase molecules deviated from the bulk flow as shown in Figure 7–8B. Why did incubation with a short DNA fragment containing a strong promoter eliminate traces that deviated from the bulk flow?

B. Do these results suggest an explanation for how site-specific DNA-binding molecules manage to find their sites faster than expected by diffusion?

C. Based on your explanation, would you expect a site-specific DNA-binding molecule to find its target site faster in a population of short DNA molecules or in a population of long DNA molecules? Assume that the concentration of target sites is identical and that there is one target site per DNA molecule.

7–11   The binding of a protein to a DNA sequence can cause the DNA to bend to make appropriate contacts with groups on the surface of the protein. Such protein-induced DNA bending can be readily detected by the way the protein–DNA complexes migrate through polyacrylamide gels. The rate of migration of bent DNA depends on the average distance between its ends as it gyrates in solution: the more bent the DNA, the closer together the ends are on average and the more slowly it migrates. If there are two sites of bending, the end-to-end distance depends on whether the bends are in the same (*cis*) or opposite (*trans*) direction (Figure 7–9A).

You have shown that the catabolite activator protein (CAP) causes DNA to bend by more than 90° when it binds to its regulatory site. You wish to know the details of the bent structure. Specifically, is the DNA at the center of the CAP-binding site bent so that the minor groove of the DNA helix is on the inside, or is it bent so that the major groove is on the inside? To answer this, you prepare two kinds of constructs, as shown in Figure 7–9B. In one, you place two CAP-binding sequences on either side of a central site into which you insert a series of DNA segments that vary from 10 to 20 nucleotides in length. In the other, you flank the insertion site with one CAP-binding

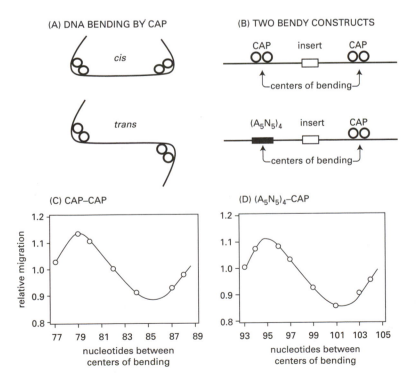

(A) DNA BENDING BY CAP

cis

trans

(B) TWO BENDY CONSTRUCTS

CAP   insert   CAP

centers of bending

$(A_5N_5)_4$ insert   CAP

centers of bending

(C) CAP–CAP

relative migration

nucleotides between
centers of bending

(D) $(A_5N_5)_4$–CAP

nucleotides between
centers of bending

Figure 7–9 Bending of DNA by CAP binding (Problem 7–11). (A) *Cis* and *trans* configurations of a pair of bends. (B) Two constructs used to investigate DNA bending by CAP binding. (C) Relative migration as a function of the number of nucleotides between the centers of bending in the CAP–CAP construct. (D) Relative migration as a function of the number of nucleotides between the centers of bending in the $(A_5N_5)_4$–CAP construct.

sequence and one $(A_5N_5)_4$ sequence, which is known to bend with the major groove on the inside. You now measure the migration of the CAP-bound constructs relative to the corresponding CAP-bound DNA with no insert. You then plot the relative migration versus the number of nucleotides between the centers of bending (Figure 7–9C and D).

A. Assuming that there are 10.6 nucleotides per turn of the DNA helix, estimate the number of turns that separate the centers of bending of the two CAP-binding sites at the point of minimum relative migration. How many helical turns separate the centers of bending at the point of maximum relative migration?

B. Is the relationship between the relative migration and the separation of the centers of bending of the CAP sites what you would expect, assuming that the *cis* configuration migrates slower than the *trans* configuration (Figure 7–9C)? Explain your answer.

C. How many helical turns separate the centers of bending at the point of minimum migration of the construct with one CAP site and one $(A_5N_5)_4$ site (Figure 7–9D)?

D. Which groove of the helix faces the inside of the bend at the center of bending of the CAP site?

7–12* What are the two fundamental components of a genetic switch?

7–13 (**True/False**) Because the individual contacts are weak, the interactions between regulatory proteins and DNA are among the weakest in biology. Explain your answer.

7–14* The nucleus of a eucaryotic cell is much larger than a bacterium, and it contains much more DNA. Therefore, a DNA-binding protein must be able to select its specific binding site from among many more unrelated sequences in a eucaryotic cell than in a bacterium. Does this present any special problems for eucaryotic gene regulation?

Consider the following situation. The eucaryotic nucleus and the bacterial cell each have a single copy of the same DNA-binding site. In addition, the nucleus has both a volume and an amount of DNA 500 times those of the bacterium. Finally, the concentration of the gene regulatory protein that binds the site is the same in the nucleus and in the bacterium. Under these conditions will the regulatory protein find its binding site as well in the eucaryotic nucleus as it does in the bacterium? Explain your answer.

7–15 One type of zinc finger motif consists of an α helix and a β sheet held together by a zinc ion (Figure 7–10). When this motif binds to DNA, the α helix is positioned in the major groove where it makes specific contacts with the bases. Why is this motif thought to enjoy a particular advantage over other DNA-binding motifs when the strength and specificity of the DNA–protein interaction need to be adjusted during evolution?

7–16* Many gene regulatory proteins form dimers of identical or slightly different subunits on the DNA. What are the advantages of dimerization?

7–17 The λ repressor binds as a dimer to critical sites on the λ genome to keep the lytic genes turned off, which stabilizes the prophage (integrated) state. Each molecule of the repressor consists of an N-terminal DNA-binding domain and a C-terminal dimerization domain (Figure 7–11). Upon induction

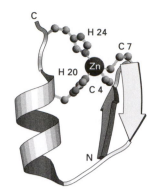

Figure 7–10 One type of zinc finger protein (Problem 7–15). The zinc ion interacts with Cys (C) and His (H) side chains so that the α helix is held tightly to one end of the β sheet.

Problem 3–20 considers how homeodomain proteins with very different sequences can still fold in the same way.

Problem 3–28 examines the symmetry match between the subunits of the Cro repressor and its binding half-sites in DNA.

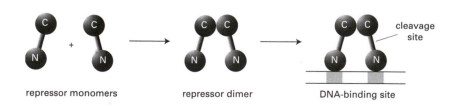

repressor monomers   repressor dimer   DNA-binding site

Figure 7–11 Domains of the λ repressor and the binding of repressor dimers to DNA (Problem 7–17).

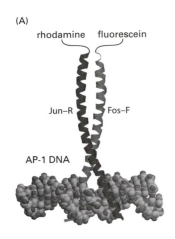

(A)

rhodamine   fluorescein

Jun–R   Fos–F

AP-1 DNA

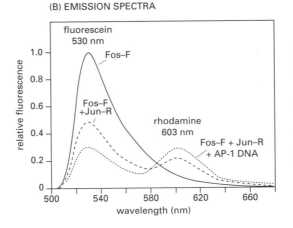

(B) EMISSION SPECTRA

fluorescein
530 nm

Fos–F

Fos–F
+Jun–R

rhodamine
603 nm

Fos–F + Jun–R
+ AP-1 DNA

relative fluorescence

wavelength (nm)

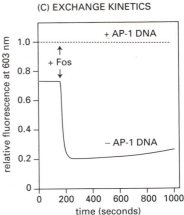

(C) EXCHANGE KINETICS

+ AP-1 DNA

+ Fos

– AP-1 DNA

relative fluorescence at 603 nm

time (seconds)

Figure 7–12 Dynamics of Fos–Jun heterodimerization in the presence and absence of DNA (Problem 7–18). (A) Arrangement of fluorophores in Fos–F and Jun–R. In the dimer, the excited fluorescein on Fos transfers energy to the rhodamine on Jun. (B) Fluorescence spectra of Fos–F alone, Fos–F with Jun–R, and Fos–F with Jun–R and AP-1 DNA. The fluorescence spectra were recorded between 500 and 700 nm after excitation at 490 nm. (C) Analysis of the exchange of Fos–F and Jun–R in the presence and absence of DNA. Rhodamine fluorescence at 603 nm was followed over time after excitation of fluorescein at 490 nm. A 10-fold excess of unlabeled Fos was added at the time indicated by the *arrow*.

(for example, by irradiation with UV light), the genes for lytic growth are expressed, λ progeny are produced, and the bacterial cell lyses. Induction is initiated by cleavage of the λ repressor at a site between the DNA-binding domain and the dimerization domain. In the absence of bound repressor, RNA polymerase binds and initiates lytic growth. Given that the number (concentration) of DNA-binding domains is unchanged by cleavage of the repressor, why do you suppose its cleavage results in its removal from the DNA?

7–18* The *fos* and *jun* oncogenes encode proteins that form a heterodimeric regulator of transcription. Leucine zipper domains in each protein mediate their dimerization through coiled-coil interactions. Dimerization juxtaposes the DNA-binding domains of each protein, positioning them for interaction with regulatory sites in DNA. The dynamics of Fos–Jun interaction in the presence and absence of the AP-1 DNA-binding site were investigated by resonance energy transfer, which is well suited for the rapid measurements that are necessary in such studies.

Fos was tagged with fluorescein (Fos–F) and Jun was tagged with rhodamine (Jun–R), as shown in Figure 7–12A. Fluorescein absorbs light at 490 nm and emits light at 530 nm, whereas rhodamine absorbs light at 530 nm and emits light at 603 nm. When Fos–F and Jun–R are brought into close proximity through heterodimerization, light energy absorbed by fluorescein at 490 nm is efficiently transferred to rhodamine through nonradiative energy transfer and emitted by rhodamine at 603 nm. Dimerization thus decreases fluorescence by fluorescein at 530 nm and increases fluorescence by rhodamine at 603 nm, as shown in Figure 7–12B. In the presence of AP-1 DNA energy transfer is even more efficient (Figure 7–12B), indicating that binding to the DNA brings the two fluorophores into even closer proximity.

Energy transfer was used to measure the ability of Fos–Jun dimers to exchange subunits with monomers in solution in the presence and absence of AP-1 DNA (Figure 7–12C). Fos–F and Jun–R were preincubated in the absence of DNA to allow free heterodimers to form, or in its presence to allow DNA-bound heterodimers to form. A 10-fold excess of Fos (without fluorescein) was then added to both solutions, and rhodamine fluorescence at 603 nm was followed after excitation at 490 nm (Figure 7–12C).

A. Do free heterodimers exchange subunits with added unlabeled Fos? Do heterodimers bound to DNA exchange subunits? How can you tell? Explain any significant differences in the behavior of free and DNA-bound heterodimers.

B. In most cells there are many distinct leucine zipper proteins, several of which can interact to form a variety of heterodimers. If the results in Figure 7–12C were typical of the leucine zipper proteins, what do they imply about the leucine zipper heterodimers in cells?

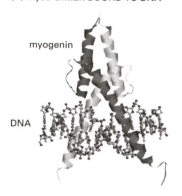

(A) MyoD DIMER BOUND TO DNA

myogenin

DNA

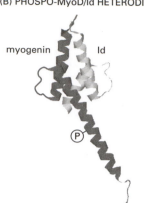

(B) PHOSPO-MyoD/Id HETERODIMER

myogenin   Id

Ⓟ

Figure 7–13 The gene regulatory protein myogenin (Problem 7–19). (A) Myogenin as part of a heterodimer bound to DNA. (B) An inactive form of phosphorylated myogenin bound to Id.

7–19   The differentiation of muscle cells from the somites of the developing embryo is controlled by myogenin, an HLH gene regulatory protein that functions as a heterodimer with another HLH protein (Figure 7–13A). The activity of myogenin must be carefully controlled lest it trigger premature expression of the muscle program of cell differentiation. The *myogenin* gene is turned on in advance of the time when it is needed, but it is prevented from functioning by its tight binding to Id, an HLH protein that lacks a DNA-binding domain, and by phosphorylation in its DNA-binding domain (Figure 7–13B). Explain how dimerization with Id and phosphorylation of the DNA-binding domain might act to keep myogenin nonfunctional.

7–20*   You have cloned four partial cDNAs for a transcription factor to test whether the encoded portions of the factor will bind to the DNA sequence that the complete protein recognizes. The partial cDNA clones extend for different distances toward the 5′ end of the gene (Figure 7–14). These cDNA clones were transcribed and translated *in vitro* and then the translation products were mixed with highly radioactive DNA containing the recognition sequence. When the mixtures were analyzed by polyacrylamide gel electrophoresis, some of the proteins were found to bind to the DNA fragment, retarding its migration (Figure 7–14, lanes 3, 4, and 5). When cDNA clones 3 and 4 were mixed together before transcription and translation, three bands appeared in the gel retardation assay (lane 6).
  A. Why were the retarded bands at different positions on the gel?
  B. Where is the binding domain for the DNA recognition sequence located in the transcription factor?
  C. Why were there three retarded bands when cDNA clones 3 and 4 were mixed together? What does that tell you about the structure of the transcription factor?

7–21   (**True/False**) DNA affinity chromatography allows purification of unlimited amounts of DNA-binding proteins. Explain your answer.

7–22*   You are working on a gene that encodes a transcription factor and you wish to know the DNA sequence to which it binds. Your advisor, who is known for her clever imagination and green thumb when it comes to molecular biology, suggests that you might be able to use PCR to amplify rare DNA molecules that are bound by your protein. Her idea consists of four steps, as outlined in Figure 7–15. (1) Synthesize a population of random-sequence oligonucleotides 26 nucleotides long, flanked by defined 25-nucleotide sequences that can serve as primer sites for PCR amplification (Figure 7–15A). (2) Add these oligonucleotides to a crude cell extract that contains the transcription factor, which can bind to oligonucleotides that contain its binding site. (3) Isolate the oligonucleotides that are bound to the transcription factor using antibodies against the factor. (4) PCR amplify the selected oligonucleotides for sequence analysis.

(A) MAP OF THE CLONES

1
2
3
4
5′   3′

(B) GEL RETARDATION ASSAY

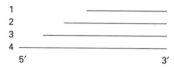

cDNA clone						
1	–	+	–	–	–	–
2	–	–	+	–	–	–
3	–	–	–	+	–	+
4	–	–	–	–	+	+

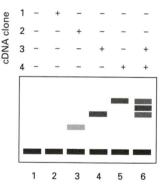

1   2   3   4   5   6

Figure 7–14 Analysis of transcription factor binding to DNA (Problem 7–20). (A) Structure of partial cDNAs encoding portions of the transcription factor. The 5′ end of the gene corresponds to the N-terminal end of the protein. (B) Gel-retardation assays of transcription factor binding to DNA. The grid shows which clones supplied the translation products employed in each of the lanes of the gel. The DNA was radioactively labeled and the bands were visualized by autoradiography. The band at the *bottom* of all lanes is the labeled DNA fragment that contains the binding site for the transcription factor.

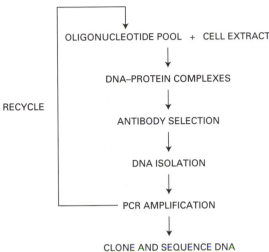

(A) ORIGINAL RANDOM-SEQUENCE OLIGONUCLEOTIDE

            EcoRI   BamHI

5′ GCTGCAGTTGCACT<u>GAATTC</u>GCCTC (N)$_{26}$ CGACA<u>GGATCC</u>GCTGAACTGACCTG 3′

(B) SELECTION AND AMPLIFICATION PROTOCOL

OLIGONUCLEOTIDE POOL + CELL EXTRACT

↓

DNA–PROTEIN COMPLEXES

↓

ANTIBODY SELECTION

↓

DNA ISOLATION

↓

PCR AMPLIFICATION

↓

CLONE AND SEQUENCE DNA

RECYCLE

Figure 7–15 Selection and amplification of specific DNA sequences from a random pool (Problem 7–22). (A) Original random-sequence oligonucleotide used for selection. N stands for any nucleotide. The EcoRI and BamHI sites facilitate cloning of the selected DNA for sequence analysis. (B) Scheme for selecting and amplifying oligonucleotides that bind to a sequence-specific DNA-binding protein.

You begin this procedure with 0.2 ng of single-stranded random-sequence oligo, which you convert to double-stranded DNA using one of the PCR primers. After four rounds of selection and amplification as outlined in Figure 7–15B, you digest the isolated DNA with BamHI and EcoRI, clone the fragments into a plasmid, and sequence 10 individual clones (Table 7–1).

A. What is the consensus sequence to which your transcription factor binds?

B. Does the consensus binding sequence show any signs of symmetry; that is, is any portion of it palindromic? Does this help you decide whether the transcription factor binds to DNA as a monomer or as a dimer?

C. How many single-stranded oligonucleotides were present in the 0.2-ng sample with which you started the experiment? (The average mass of a nucleotide is 330 daltons.)

D. Assuming that the starting population of oligonucleotides was truly random in the 26 central nucleotides, is it likely that all possible 14-base-pair-long sequences were represented in the starting sample?

TABLE 7–1 Nucleotide sequences of selected and amplified DNAs (Problem 7–22).

<u>GAATTC</u>GCCTCGAGCACATCATTGCCCATATATGGCA<u>CGACAGGATCC</u>
<u>GAATTC</u>GCCTCTTCTAATGCCCATATATGGACTTGCT<u>CGACAGGATCC</u>
<u>GGATCC</u>TGTCGGTCCTTTATGCCCATATATGGTCATT<u>GAGGCGAATTC</u>
<u>GAATTC</u>GCCTCATGCCCATATATGGCAATAGGTGTTT<u>CGACAGGATCC</u>
<u>GAATTC</u>GCCTCTATGCCCATATAAGGCGCCACTACCC<u>CGACAGGATCC</u>
<u>GAATTC</u>GCCTCGTTCCCAGTATGCCCATATATGGACA<u>CGACAGGATCC</u>
<u>GGATCC</u>TGTCGACACCATGCCCATATTTGGTATGCTC<u>GAGGCGAATTC</u>
<u>GAATTC</u>GCCTCATTTATGAACATGCCCTTATAAGGAC<u>CGACAGGATCC</u>
<u>GAATTC</u>GCCTCTAATACTGCAATGCCCAAATAAGGAG<u>CGACAGGATCC</u>
<u>GAATTC</u>GCCTCATGCCCAAATATGGTCATCACCTAC<u>CGACAGGATCC</u>

Underlined sequences correspond to PCR primer sites. Two sequences are shown starting at the BamHI end so that the binding sites are oriented the same way in all sequences.

HOW GENETIC SWITCHES WORK

TERMS TO LEARN

coactivator	gene control region	operon
corepressor	gene repressor protein	positive control
enhancer	insulator element	promoter
gene	negative control	regulatory sequence
gene activator protein	operator	tryptophan repressor

7–23 Define negative control and positive control in terms of how the active forms of the gene regulatory proteins function. Explain how an 'inducing' ligand can turn on a gene that is negatively controlled, and how it can turn on a gene that is positively controlled. Explain how an 'inhibitory' ligand can turn off a negatively controlled gene and a positively controlled gene.

7–24* The enzymes for arginine biosynthesis are located at several positions around the genome of *E. coli*, and they are regulated coordinately by a gene regulatory protein encoded by the *argR* gene. The activity of ArgR is modulated by arginine. Upon binding arginine, ArgR alters its conformation, dramatically changing its affinity for the regulatory sequences in the promoters of the genes for the arginine biosynthetic enzymes. Given that ArgR is a gene repressor, would you expect that ArgR would bind more tightly or less tightly to the regulatory sequences when arginine is abundant? If ArgR functioned instead as a gene activator, would you expect the binding of arginine to increase or to decrease its affinity for its regulatory sequences? Explain your answers.

7–25 Bacterial cells can take up the amino acid tryptophan from their surroundings, or, if the external supply is insufficient, they can synthesize tryptophan from small molecules in the cell. The tryptophan repressor inhibits transcription of the genes in the tryptophan operon, which codes for the enzymes required for tryptophan synthesis. Upon binding tryptophan, the tryptophan repressor binds to a site in the promoter of the operon.

A. Why is tryptophan-dependent binding to the operon a useful property for the tryptophan repressor?

B. What would you expect to happen to the regulation of the tryptophan biosynthetic enzymes in cells that express a mutant form of the tryptophan repressor that (i) cannot bind to DNA or (ii) binds to DNA even when no tryptophan is bound to it?

C. What would happen in scenarios (i) and (ii) if the cell produced normal tryptophan repressor from a second unmutated copy of the gene?

7–26* In the absence of glucose, *E. coli* can proliferate on the pentose sugar arabinose, using an inducible set of genes that are arrayed in three groups on the chromosome (Figure 7–16). The *araA, araB,* and *araD* genes encode enzymes for the metabolism of arabinose. The *araC* gene encodes a gene

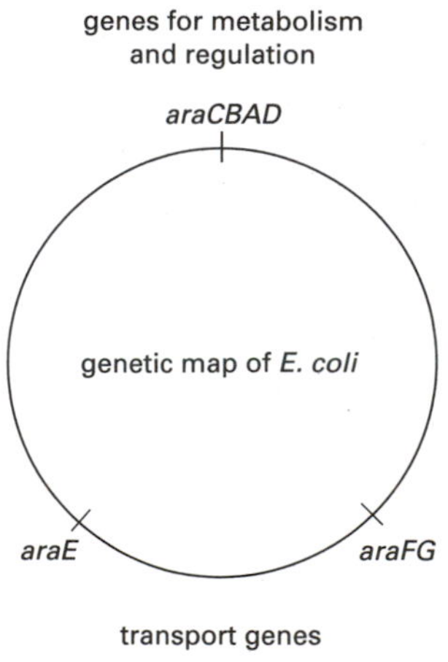

Figure 7–16 Chromosomal locations of the genes involved in arabinose metabolism (Problem 7–26).

TABLE 7–2 Responses of normal and mutant bacteria to the presence and absence of arabinose (Problem 7–26).

GENOTYPE	*araA* GENE PRODUCT	
	− ARABINOSE	+ ARABINOSE
araC⁺	1	1000
araC⁻	1	1

regulatory protein that binds adjacent to arabinose promoters and coordinates the expression of the genes involved in arabinose metabolism. (The other two groups of genes encode proteins involved in arabinose transport.)

To understand the regulatory properties of the AraC protein, you isolate a mutant bacterium with a deletion of the *araC* gene. As shown in Table 7–2, the mutant strain does not induce expression of the *araA* gene when arabinose is added to the medium.

A. Do the results in Table 7–2 indicate that the AraC protein regulates arabinose metabolism by negative control or by positive control? Explain your answer.

B. What would the data in Table 7–2 have looked like if the AraC protein regulated expression of the enzymes of arabinose metabolism by the type of control that you did not choose in part A?

7–27   *E. coli* proliferates faster on the monosaccharide glucose than it does on the disaccharide lactose for two reasons: (1) lactose is taken up more slowly than glucose and (2) lactose must be hydrolyzed to glucose and galactose (by β-galactosidase) before it can be further metabolized.

When *E. coli* is grown on a medium containing a mixture of glucose and lactose, it proliferates with complex kinetics (Figure 7–17, squares). The bacteria proliferate faster at the beginning than at the end, and there is a lag between these two phases when they virtually stop dividing. Assays of the concentrations of the two sugars in the medium show that glucose falls to very low levels after a few cell doublings (Figure 7–17, circles), but lactose remains high until near the end of the experimental time course (not shown). Although the concentration of lactose is high throughout most of the experiment, β-galactosidase, which is regulated as part of the *lac* operon, is not induced until more than 100 minutes have passed (Figure 7–17, triangles).

A. Explain the kinetics of bacterial proliferation during the experiment. Account for the rapid initial rate, the slower final rate, and the delay in the middle of the experiment.

B. Explain why the *lac* operon is not induced by lactose during the rapid initial phase of bacterial proliferation.

> **Problem 3–50** analyzes regulation of the *lac* operon in terms of the quantitative changes in affinity of the lac repressor upon binding the inducing ligand.

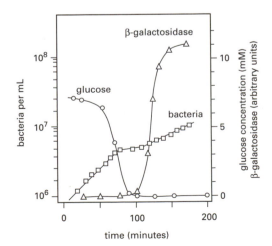

Figure 7–17 Proliferation of *E. coli* on a mixture of glucose and lactose (Problem 7–27).

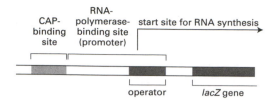

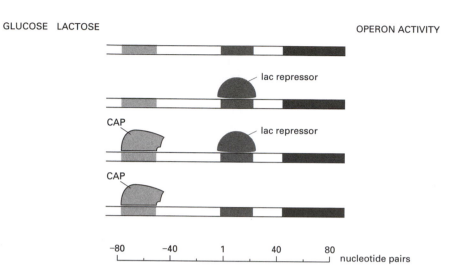

Figure 7–18 Arrangement of binding sites and the four possible combinations of gene regulatory proteins on the promoter for the *lac* operon (Problem 7–28).

GLUCOSE LACTOSE   OPERON ACTIVITY

7–28* In Figure 7–18, CAP and the lac repressor have been placed in the four possible combinations on their binding sites in the promoter for the *lac* operon. Each combination of gene regulatory proteins corresponds to a particular mixture of glucose and lactose. For each of the four combinations, indicate on the left-hand side of the figure which sugars must be present and, on the right-hand side, whether the operon is expected to be on or off.

7–29 Imagine that you have created a fusion between the *trp* operon, which encodes the enzymes for tryptophan biosynthesis, and the *lac* operon, which encodes the enzymes necessary for lactose utilization (Figure 7–19). Under which of conditions (A–F below) will β-galactosidase be expressed in the strain that carries the fused operons?

A. Only when lactose and glucose are absent.
B. Only when lactose and glucose are present.
C. Only when lactose is absent and glucose is present.
D. Only when lactose is present and glucose is absent.
E. Only when tryptophan is absent.
F. Only when tryptophan is present.

7–30* Transcription of the bacterial gene encoding the enzyme glutamine synthetase is regulated by the availability of nitrogen in the cell. The key transcriptional regulator is the NtrC protein, which stimulates transcription only

Figure 7–19 Separated (normal) and fused *trp* and *lac* operons (Problem 7–29).

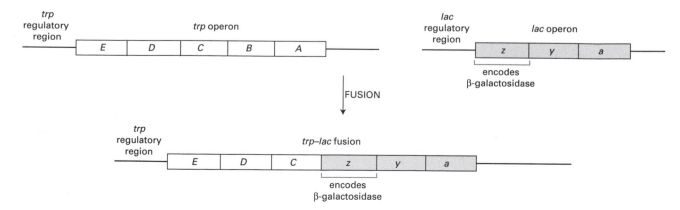

when it is phosphorylated. Phosphorylation of NtrC is controlled by the NtrB protein, which is both a protein kinase and a protein phosphatase. The balance between its kinase and phosphatase activities—hence the level of phosphorylation of NtrC and transcription of glutamine synthetase—is determined by other proteins that respond to the ratio of α-ketoglutarate and glutamine. (This ratio is a sensitive indicator of nitrogen availability because two nitrogens—as ammonia—must be added to α-ketoglutarate to make glutamine.)

Transcription of the gene for glutamine synthetase can be achieved *in vitro* by adding RNA polymerase, a special sigma factor, and phosphorylated NtrC to a linear DNA template containing the gene and its upstream regulatory region. NtrC binds to five sites upstream of the promoter. Although binding of NtrC is only slightly increased by phosphorylation, transcription is absolutely dependent on phosphorylation. RNA polymerase, however, binds strongly to the promoter even in the absence of NtrC.

Activation of transcription by NtrC was further characterized using three different templates: the normal gene with intact regulatory sequences, a gene with all of the NtrC-binding sites deleted, and a gene with three NtrC-binding sites at its 3′ end (Figure 7–20). In the absence of phosphorylated NtrC, no transcription occurred from any of the templates. In the presence of 100 nM phosphorylated NtrC, all three templates supported maximal transcription. However, the three templates differed significantly in the concentration of NtrC required for half-maximal rates of transcription: the normal gene (Figure 7–20A) required 5 nM NtrC, the gene with 3′-binding sites (Figure 7–20C) required 10 nM NtrC, and the gene without NtrC-binding sites (Figure 7–20B) required 50 nM NtrC.

A. If RNA polymerase can bind to the promoter of the glutamine synthetase gene in the absence of NtrC, why do you suppose NtrC is needed to activate transcription?

B. If NtrC can bind to its binding sites regardless of its state of phosphorylation, why is phosphorylation necessary for transcription?

C. If NtrC can activate transcription even when its binding sites are absent, what role do the binding sites play?

7–31   Regulation of arabinose metabolism in *E. coli* is fairly complex. Not only are the genes scattered around the chromosome (see Figure 7–16), but the regulatory protein, AraC, acts as both a positive and a negative regulator. For example, in the regulation of the *araBAD* cluster of genes (Figure 7–21A),

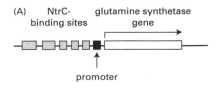

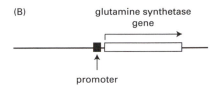

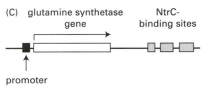

Figure 7–20 Three templates for studying the role of NtrC-binding sites in transcription of the glutamine synthetase gene (Problem 7–30). (A) The normal gene with intact upstream regulatory sequences. (B) A gene with all of the NtrC-binding sites deleted. (C) A gene with three NtrC-binding sites at the 3′ end of the gene.

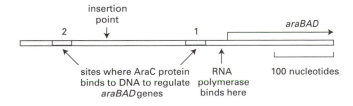

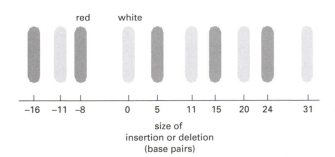

Figure 7–21 Analysis of regulation of the *araBAD* genes (Problem 7–31). (A) Arrangement of AraC-binding sites in the *araBAD* operon. (B) Results of altering the spacing between araC binding sites 1 and 2. For this streak assay the *araBAD* genes were replaced by the *galK* gene. Various numbers of nucleotides were inserted or deleted between the AraC-binding sites at the point indicated by the *arrow* in A. Colonies were streaked out against a scale that showed how many nucleotides had been inserted or deleted. Colonies that do not express galactokinase are *white*; those that do are *red*.

AraC binding at site 1 in the presence of arabinose (and the absence of glucose) increases transcription roughly 100-fold over the level measured in the absence of AraC protein. Binding of AraC at site 2 in the absence of arabinose represses transcription of the *araBAD* genes about 10-fold below the level measured in the absence of AraC protein. The combined effects of negative regulation at site 2 (in the absence of arabinose) and positive regulation at site 1 (in the presence of arabinose) means that addition of arabinose causes a 1000-fold increase in transcription of the *araBAD* genes.

Positive regulation by binding at site 1 seems straightforward since that site lies adjacent to the promoter and presumably facilitates RNA polymerase binding or stimulates open complex formation. You are more puzzled, however, by the results of binding at site 2. Site 2 lies 270 nucleotides upstream from the start site of transcription. Regulatory effects over such distances seem more reminiscent of enhancers in eucaryotic cells. To study the mechanism of repression at site 2 more easily, you move the regulatory region in front of the *galK* gene, whose encoded enzyme, galactokinase, is simple to assay.

To determine the importance of the spacing between the two AraC-binding sites, you insert or delete nucleotides at the insertion point indicated in Figure 7–21A. The activity of the promoter in the absence of arabinose is then assayed by growing the bacteria on special indicator plates. If galacto-kinase is not made, the bacterial colonies are white; if it is produced, the colonies are red. You streak out bacteria containing the altered spacings against a scale that shows how many nucleotides were deleted or inserted (Figure 7–21B). Much to your surprise, red and white streaks are interspersed.

When you show your results to your advisor, she is very pleased and tells you that these experiments distinguish among three potential mechanisms of repression from a distance. (1) An alteration in the structure of the DNA could propagate from the repression site to the transcription site, making the promoter an unfavorable site for RNA polymerase binding. (2) The protein could bind cooperatively (oligomerize) at the repression site in such a way that additional subunits continue to be added until the growing chain covers the promoter, blocking transcription. (3) The DNA could form a loop so that the protein bound at the distant repression site could interact with proteins (or DNA) at the transcription start site.

Which of these general mechanisms do your data support? How do you account for the patterns of red and white streaks?

7–32* (**True/False**) Many gene regulatory proteins in eucaryotes can act even when they are bound to DNA thousands of nucleotide pairs away from the promoter that they influence. Explain your answer.

7–33 When enhancers were initially found to influence activity at remote promoters, two principal models were invoked to explain this action at a distance. In the 'DNA looping' model, direct interactions between proteins bound at enhancers and promoters were proposed to stimulate RNA polymerase. In the 'scanning' or 'entry-site' model, RNA polymerase (or a transcription factor) was proposed to bind at the enhancer and then scan along the DNA until it reached the promoter. These two models were distinguished using an enhancer on one piece of DNA and a β-globin gene and promoter on a separate piece of DNA (Figure 7–22). The β-globin gene was not expressed from the mixture of pieces. When the two segments of DNA were joined via a protein linker, the β-globin gene was expressed.

How does this experiment distinguish between the DNA looping model and the scanning model? Explain your answer.

7–34* Some gene regulatory proteins bind to DNA and cause the double helix to bend at a sharp angle. Such 'bending proteins' can affect the initiation of transcription without contacting the RNA polymerase, any of the general transcription factors, or any other gene regulatory proteins. Can you devise

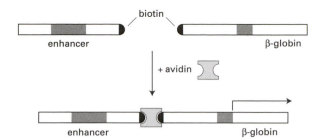

Figure 7–22 Stimulation of β-globin gene expression by an enhancer linked via a protein bridge (Problem 7–33). Each DNA molecule carries biotin attached to one end, as shown. In the presence of the protein avidin, the two molecules are linked together and transcription occurs as shown by the *arrow above* the β-globin gene.

a plausible explanation for how such proteins might work to modulate transcription? Draw a diagram that illustrates your explanation.

7–35 You have developed an *in vitro* transcription system using a defined segment of DNA that is transcribed under the control of a viral promoter. Transcription of this DNA occurs when you add purified RNA polymerase II, TFIID (the TATA-binding factor), and TFIIB and TFIIE (which bind to RNA polymerase). Transcription in the *in vitro* system with purified components occurs at low efficiency relative to a crude nuclear lysate. This suggests that there may be an additional regulatory sequence that binds a transcription factor not present among the purified components. To search for the DNA sequence to which this putative regulatory factor binds, you make a series of deletions upstream of the start site for transcription (Figure 7–23A) and compare their transcriptional activity in the purified system and in crude extracts. As an internal control, you mix each of the deletions with a non-deleted template that encodes a slightly longer transcript (Figure 7–23B). The results of these assays are shown in Figure 7–23C. Roughly ten times less material from the reactions in the nuclear extracts was run on the gel so that the intensities of the bands would be approximately equal. Deletions up to –61 have no effect on transcription and the –11 deletion inactivates transcription in both the purified system and the crude extract. Surprisingly, the –50 deletion is transcribed as efficiently as the control template in the purified system, but not in the nuclear extract.

You purify the protein that is responsible for this effect and show that it stimulates transcription approximately 10-fold from the control template and from the –61 deletion template, but does not stimulate transcription from the –50 deletion template. Footprinting analysis shows that the factor binds to a specific, short sequence upstream of the TATA site. Furthermore,

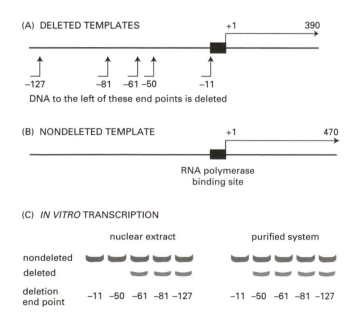

Figure 7–23 Analysis of transcription from a viral promoter (Problem 7–35). (A) Deletion templates. These templates are all truncated at their 3′ ends so that they give rise to slightly shorter transcripts than the longer, control template in B. The deletions remove DNA to the *left* of the indicated end points. Nucleotides are numbered from the start site of transcription (+1); *negative numbers* indicate nucleotides in front of the start site. (B) Nondeleted control template. This template serves as an internal control; it gives rise to a transcript that is 80 nucleotides longer than the experimental templates in A. (C) Results of transcription of a mixture of control and deleted templates in a crude nuclear extract (*left*) and using purified components (*right*). *Negative numbers* identify the particular deletion template that was included in each mixture.

although the factor binds very transiently to its site in the absence of TFIID (a 20-second half-life), in the presence of TFIID it binds stably (with a half-life greater than 5 hours).

A. Where is the binding site for the stimulatory factor located?

B. How is it that a *stimulatory* factor, when added to the other purified transcription components, causes transcripts from the –50 deletion template to be absent from the gel?

C. Why do you think there is such a marked difference in stability of binding of the stimulatory factor in the presence and absence of TFIID?

7–36* Hormone receptors for glucocorticoids alter their conformation upon hormone binding to become DNA-binding proteins that activate a specific set of responsive genes. Genetic and molecular studies indicate that the DNA- and hormone-binding sites occupy distinct regions of the C-terminal half of the glucocorticoid receptor. Hormone binding could generate a functional DNA-binding protein in either of two ways: by altering receptor conformation to create a DNA-binding domain or by altering the conformation to uncover a preexisting DNA-binding domain.

These possibilities have been investigated by comparing the activities of a series of C-terminal deletions (Figure 7–24). Fragments of the cDNA, which correspond to the N-terminal portion of the receptor, were inserted into a vector so they would be expressed upon transfection into appropriate cells. The capacity of the receptor fragments to activate responsive genes was tested by co-transfections with a reporter plasmid carrying a glucocorticoid response element linked to the chloramphenicol acetyltransferase (CAT) gene. As shown in Figure 7–24, cells co-transfected with a cDNA for the full-length receptor (that is, with its C-terminus at position 0) responded as expected. In the absence of glucocorticoid no CAT activity was detected; in the presence of the glucocorticoid dexamethasone CAT activity was readily detected. Six mutant receptors, lacking 27, 101, 123, 180, 287, and 331 C-terminal amino acids, failed to activate CAT expression in the presence or absence of dexamethasone. In contrast, four mutant receptors, lacking 190, 200, 239, and 270 C-terminal amino acids, activated CAT expression in the presence and absence of dexamethasone. Separate experiments indicated that the mutant receptors were synthesized equally.

How do these experiments distinguish between the proposed models for hormone-dependent conversion of the normal receptor to a DNA-binding form? Does hormone binding create a DNA-binding site or does it uncover a preexisting one?

7–37 The yeast Gal4 gene activator protein comprises two domains: a DNA-binding domain and an activation domain. The DNA-binding domain allows it to bind to appropriate sites in the DNA near genes that are required

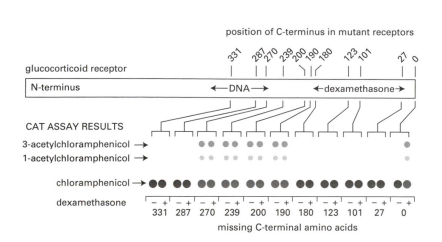

Figure 7–24 Effect of C-terminal deletions on the activity of the glucocorticoid receptor (Problem 7–36). The schematic diagram at the *top* illustrates the position of the DNA-binding site (DNA) and the glucocorticoid-binding site (dexamethasone) in the receptor, as well as the positions of the C-terminal deletions. The *lower* diagram shows the results of a standard CAT assay obtained by mixing cell extracts with ^{14}C-chloramphenicol. The *lowest* spot is unreacted chloramphenicol; the *upper* spots show the attachment of acetyl groups to one or the other of two positions on chloramphenicol. The presence (+) or absence (–) of dexamethasone is indicated *below* appropriate lanes.

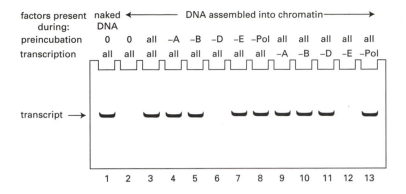

Figure 7–25 Effects of chromatin assembly on transcription (Problem 7–38). Transcription templates were preincubated with none, all, or all minus one of the transcription components (for example, –A means that TFIIA was left out, and –Pol means that RNA polymerase II was left out). The transcription assay was carried out in the presence of all or all minus one of the transcription components.

for metabolism of the sugar galactose. The activation domain binds to a component of the RNA polymerase II holoenzyme complex, attracting it to the promoter so the regulated genes can be turned on. In the absence of Gal4 the galactose genes cannot be turned on. When Gal4 is expressed normally, the genes can be maximally activated. When Gal4 is overexpressed, however, the galactose genes are turned off. Why does too much Gal4 squelch expression of the galactose genes?

7–38*   An early attempt to understand how the packing of DNA into chromatin affects transcription in eucaryotes used a defined template, purified RNA polymerase II, and four transcription factors—TFIIA, TFIIB, TFIID, and TFIIE. If the template was first assembled into nucleosomes, no transcription was detected (Figure 7–25, lane 2). However, if the template was first incubated with the transcription components and then assembled into chromatin, transcription proceeded just as well as it does on the naked DNA template (compare lanes 1 and 3). This suggested that the transcription components bound to the template to keep the promoter accessible.

This phenomenon was investigated in more detail in two additional kinds of experiment. In one, individual transcription components were omitted during the preincubation (Figure 7–25, lanes 4 to 8). In the second individual transcription components were left out during the transcription assay (lanes 9 to 13).

A. Which of the transcription components must be present during the preincubation to keep the template active during chromatin assembly?

B. Which of the transcription components can form a complex with the template that is stable to chromatin formation and subsequent purification?

C. Which of the transcription components must be added during the assay in order to produce a transcript?

7–39   How are histone acetylases and chromatin remodeling complexes recruited to unmodified chromatin, and how are they thought to aid in the activation of transcription from previously silent genes?

Problem 4–47 measures the capacity of chromatin remodeling complexes to move nucleosomes out of the way.

7–40*   How is it that protein–protein interactions that are too weak to cause proteins to assemble in solution can nevertheless allow the same proteins to assemble into complexes on DNA?

7–41   Coactivators do not themselves bind to DNA but instead serve as bridging molecules that link gene-specific activators to general transcription factors at the promoter. The SAGA complex in yeast is a large multiprotein complex that is required for transcription of many genes. SAGA contains a variety of transcription factors, including a histone acetyltransferase and a subset of TATA-binding protein associated factors (TAFs). If SAGA functions as a co-activator, it should be physically present at the promoters it regulates and its recruitment should be governed by an activation signal.

To test this idea, you focus on the regulation of the galactose genes by the Gal4 activator, which binds to DNA sequences (UAS$_G$) adjacent to the

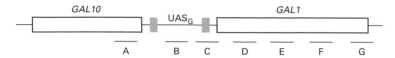

(A) MAP OF THE *GAL1-10* LOCUS

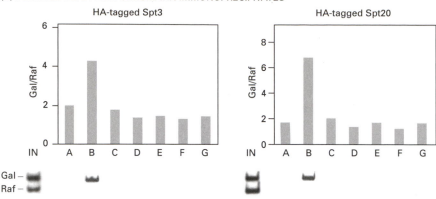

(B) PCR PRODUCTS FROM CHROMATIN IMMUNOPRECIPITATES

Figure 7–26 Analysis of SAGA association with the Gal promoter (Problem 7–41). (A) Organization of the *GAL10* and *GAL1* genes. A common UAS serves both promoters, which are indicated by *shaded regions*. The positions of the PCR fragments that were used for quantification are shown. (B) Results of chromatin immunoprecipitation in strains expressing HA–Spt3 or HA–Spt20. The two strains were grown in medium containing raffinose (Raf) or galactose (Gal), their chromatin was fragmented and immunoprecipitated with HA antibodies, and quantitative PCR was carried out on the isolated DNA. The PCR products from strains grown on raffinose were loaded onto the gels and run for a few minutes before the PCR products from strains grown on galactose, so that the products could be compared within the same lanes. Finally, the amounts of PCR products were measured, normalized to the loading control (IN), and the ratio (Gal/Raf) was calculated. IN refers to the input chromatin before immunoprecipitation; PCR products in those lanes were from an unaffected part of the genome and were used as controls for loading differences. About 50- to 100-fold less input chromatin (IN) was loaded per lane than immunoprecipitated chromatin, reflecting the inefficiency of chromatin immunoprecipitation.

promoters (shaded regions) of galactose genes (Figure 7–26A) and is known to require components of the SAGA complex for activation. When cells are grown on galactose, the activation domain of Gal4 is exposed, allowing the transcription machinery to form at the adjacent galactose promoters. If the cells are grown instead on the sugar raffinose, Gal4 remains bound to the UAS but its activation domain is occluded and transcription does not occur.

To determine whether the SAGA complex associates with Gal4 on the chromosome, you use the method of chromatin immunoprecipitation. You prepare two strains of yeast that each expresses a derivative of a protein in the SAGA complex—either Spt3 or Spt20—that contains the hemaglutinin (HA) epitope tag. The HA tags do not interfere with the function of these proteins and they allow very efficient immunoprecipitation using commercial antibodies against HA. You prepare small chromatin fragments from the two strains after growth on raffinose or galactose, and then precipitate them with HA antibodies. You strip the precipitated chromatin of protein and analyze the DNA by quantitative PCR to measure the amounts of specific DNA segments in the precipitates. The ratio of immunoprecipitated chromatin from galactose-grown cells to that from raffinose-grown cells (Gal/Raf) is shown for several segments around the galactose promoter (Figure 7–26B).

A. Which of the tested fragments would you have expected to be enriched if SAGA behaved as a coactivator? Explain your answer.

B. Does SAGA meet the criteria for a coactivator? Is it physically present at the promoter and does its recruitment depend on an activation signal?

7–42* Consider the following argument: "If the expression of every gene depends on a set of gene regulatory proteins, then the expression of these gene regulatory proteins must also depend on the expression of other gene regulatory proteins, and their expression must depend on the expression of still other gene regulatory proteins, and so on. Cells would therefore need an infinite number of genes, most of which would code for gene regulatory proteins." How does the cell get by without having to achieve the impossible?

7–43 The protein encoded by the *even-skipped (eve)* gene of *Drosophila* is a transcriptional regulator required for proper segmentation in the middle of the body. It first appears about 2 hours after fertilization at a uniform level in all the embryonic nuclei. Not long after that it forms a pattern of seven stripes across the embryo. Each stripe is under the control of a separate module in the promoter, which provides binding sites for both repressors

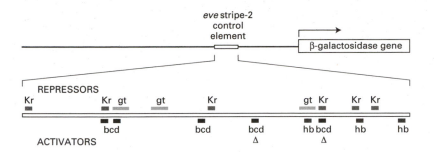

Figure 7–27 Binding sites for repressors and activators of eve stripe 2 (Problem 7–43).

and activators of *eve* transcription. In the case of stripe 2, there are multiple, overlapping binding sites for activators and repressors (Figure 7–27). Two activators, hunchback (hb) and bicoid (bcd), and two repressors, giant (gt) and Krüppel (Kr), are required to give the normal pattern. The binding sites for these proteins have been mapped onto the 670-nucleotide segment shown in Figure 7–27; deletion of this upstream segment abolishes *eve* expression in stripe 2. The patterns of expression of hunchback, bicoid, giant, and Krüppel in the embryo are shown in Figure 7–28. It seems that *eve* expression in stripe 2 occurs only in the region of the embryo that expresses both activators, but neither repressor: a simple enough rule.

To check if this rule is correct, you construct a β-galactosidase reporter gene driven by a 5-kb upstream segment from the *eve* promoter. (This segment also includes the controlling elements for stripes 3 and 7.) In addition to the normal upstream element, you make three mutant versions in which several of the binding sites in the *eve* stripe-2 control segment have been deleted. (Note, however, that because many of the binding sites overlap it is not possible to delete all of one kind of site without affecting some of the other sites.)

Construct 1.   Deletion of all the Krüppel-binding sites
Construct 2.   Deletion of all the giant-binding sites
Construct 3.   Deletion of two bicoid-binding sites (indicated by Δs in Figure 7–27)

You make flies containing these novel genetic constructs integrated into their chromosomes and determine the patterns of β-galactosidase expression in their embryos, which are shown in Figure 7–29.

A. Match the mutant embryos to the mutant constructs.
B. You began these experiments to test the simple rule that *eve* expression in stripe 2 occurs in the embryo where the two activators are present and the two repressors are absent. Do the results with the mutant embryos confirm this rule? Explain your answer.
C. Offer a plausible explanation for why there is no expression of β-galactosidase at the anterior pole of mutant embryo D in Figure 7–29.
D. In the *eve* stripe-2 control segment the binding sites for the two activators do not overlap, nor do the binding sites for the two repressors; however, it is often the case that binding sites for activators overlap binding sites for

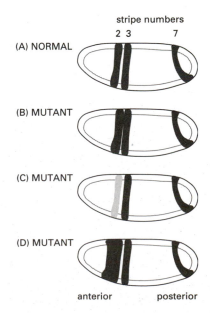

Figure 7–29 Embryonic expression of β-galactosidase from constructs with normal or mutated eve stripe-2 control elements (Problem 7–43).

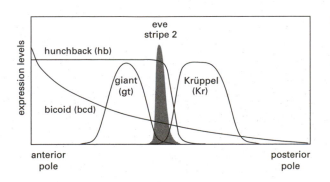

Figure 7–28 Expression of repressors and activators of *eve* stripe 2 in the *Drosophila* embryo (Problem 7–43).

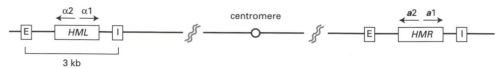

(A) YEAST SILENT MATING TYPE LOCI

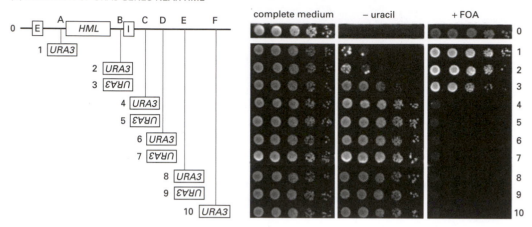

(B) INSERTION OF *URA3* GENES NEAR *HML*

repressors. What does this overlap suggest about the mode of genetic control of *eve* in stripe 2, and what might be the consequences for stripe morphology?

7–44* A locus control region (LCR) that lies far upstream of the cluster of β-like globin genes regulates the entire gene cluster. In skin cells, for example, which do not express the globin genes, the whole gene cluster is tightly packaged into chromatin. By contrast, in erythroid cells the gene cluster is decondensed and individual genes are transcribed in a characteristic developmental sequence. How do you suppose that the LCR might regulate the status of the chromatin over such a large region?

7–45 Genes present at the expressed *MAT* locus on chromosome III determine the mating type of a haploid yeast cell. Identical mating type genes also exist at the *HML* and *HMR* loci on the same chromosome (Figure 7–30A). However, at these loci, the mating type genes are not expressed, even though all the signals for expression are present. Each locus is bracketed by two short DNA sequences, designated E and I, which bind a variety of proteins and serve to establish and maintain repression of genes within each locus.

To investigate the function of these elements, you insert the *URA3* gene at different locations between E and I and outside the *HML* locus, as shown in Figure 7–30B. You then test the growth of these strains in complete medium, in the absence of uracil (–uracil), and in the presence of 5-fluoroorotic acid (+FOA) (Figure 7–30B). These growth conditions test for the activity of the *URA3* gene as follows: in complete medium, expression of *URA3* is irrelevant; in the absence of uracil, expression of *URA3* is required for growth; and in the presence of FOA, expression of *URA3* is lethal to the cell.

Do these control elements specifically repress the mating type genes, or do they act like insulators that control the expression of any gene placed between them?

7–46* (**True/False**) It seems likely that the close-packed arrangement of bacterial genes and genetic switches developed from more extended forms of switches in response to the evolutionary pressure to maintain a small genome. Explain your answer.

Figure 7–30 Effects of E and I elements at *HML* on expression of *URA3* genes inserted nearby (Problem 7–45). (A) Arrangement of E and I elements around *HML* and *HMR*. The mating type genes are shown as α1, α2, *a*1, and *a*2. (B) Locations of *URA3* genes inserted around the *HML* locus. Orientations of the *URA3* genes are indicated by the direction of *lettering*. Growth phenotypes of each strain are indicated on the *right*. Strain 0 is the parental strain with no *URA3* gene; strains 1 to 10 have a *URA3* gene inserted as indicated. Serial dilutions of each strain were spotted on agar plates containing complete medium, complete medium minus uracil (–uracil), and complete medium plus FOA (+FOA). *White areas* indicate growth.

THE MOLECULAR GENETIC MECHANISMS THAT CREATE SPECIALIZED CELL TYPES

TERMS TO LEARN

CG island	lambda repressor protein	phase variation
Cro protein	mating	X-inactivation
DNA methylation	mating-type (MAT) locus	X-inactivation center (XIC)
genomic imprinting		

7–47 (**True/False**) Reversible genetic rearrangements are a common way of regulating gene expression in procaryotes and mammalian cells. Explain your answer.

7–48* It is relatively common for pathogenic organisms to change their coats periodically in order to evade the immune surveillance of their host. *Salmonella* (a bacterium that can cause food poisoning) can exist in two antigenically distinguishable forms, or phases as their discoverer called them in 1922. Bacteria in the two different phases synthesize different kinds of flagellin, which is the protein that makes up the flagellum. Phase 1 bacteria switch to phase 2 and vice versa about once per thousand cell divisions. Two kinds of explanation were originally considered for the switch mechanism: a DNA rearrangement, such as insertion or inversion, and a DNA modification, such as methylation.

The unlinked genes *H1* and *H2* encode the two flagellins responsible for phase variation. The genetic element that enables the bacteria to switch phases is very closely linked to the *H2* gene. To distinguish between the proposed mechanisms of switching, a segment of DNA containing the *H2* gene was cloned. When this segment was introduced into *E. coli*, which has no flagella, most bacteria that picked up the plasmid acquired the ability to swim, indicating that they were synthesizing the flagellin encoded by the *H2* gene. A few colonies, however, were nonmotile even though they carried the plasmid. When DNA was prepared from cultures grown from these nonmotile colonies and introduced into a fresh culture of *E. coli*, some of the transformed bacteria were able to swim, indicating that H2 flagellin synthesis had been switched on.

Plasmid DNA was prepared from these switching cultures, digested with a restriction nuclease, heated to separate the DNA strands, and then slowly cooled to allow DNA strands to reanneal. The DNA molecules were then examined by electron microscopy. About 5% of the molecules contained a bubble, formed by two equal-length single-stranded DNA segments, at a unique position near one end, as shown in Figure 7–31.

A. All *Salmonella* can swim no matter which type of flagellin they are synthesizing. In contrast, the *E. coli* switched between a form that was able to swim and one that was immotile. Why?

B. Explain how these results distinguish between mechanisms of switching that involve DNA rearrangement and ones that involve DNA modification.

C. How do these results distinguish among site-specific DNA rearrangements that involve deletion of DNA, addition of DNA, or inversion of DNA?

7–49 One of the key regulatory proteins produced by the yeast mating-type locus is a repressor protein known as α2. In haploid cells of the α mating type, α2 is essential for turning off a set of genes that are specific for the **a** mating type. In **a**/α diploid cells the α2 repressor collaborates with the product of the **a**1 gene to turn off a set of haploid-specific genes in addition to the **a**-specific genes. Two distinct but related types of conserved DNA sequences are found upstream of these two sets of regulated genes: one in front of the **a**-specific genes and the other in front of the haploid-specific genes. Given the relatedness of these upstream sequences, it is most likely that α2 binds to both; however, its binding properties must be modified in some way by the **a**1 protein before it can recognize the haploid-specific sequence. You wish to understand the nature of this modification. Does **a**1 catalyze

Figure 7–31 Reannealed DNA fragments from cultures of switching *E. coli* (Problem 7–48). *Arrows* indicate single-stranded bubbles.

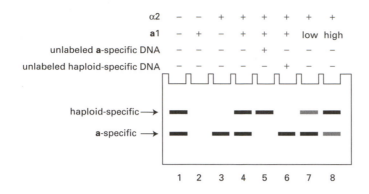

Figure 7–32 Binding of regulatory proteins to fragments of DNA containing the a-specific or haploid specific regulatory sequences (Problem 7–49). Various combinations of regulatory proteins were incubated with a mixture of a-specific and haploid-specific radioactive DNA fragments (shown in lane 1). At the end of the incubation, the samples were precipitated with antibody against the proteins, and the DNA fragments in the precipitate were run on the gel. The gel was then placed against x-ray film to expose the positions of the radioactive DNA fragments.

covalent modification of α2, or does it modify α2 by binding to it stoichiometrically?

To study these questions, you perform three types of experiment. In the first, you measure the binding of a1 and α2, alone and together, to the two kinds of upstream regulatory DNA sites. As shown in Figure 7–32, a1 alone does not bind DNA fragments that contain either regulatory site (Figure 7–32, lane 2), whereas α2 binds to a-specific fragments but not to haploid-specific fragments (lane 3). The mixture of a1 and α2 binds to a-specific *and* haploid-specific fragments (lane 4).

In the second series of experiments you add a vast excess of unlabeled DNA containing the a-specific sequence along with the mixture of a1 and α2 proteins. Under these conditions the haploid-specific fragment is still bound (Figure 7–32, lane 5). Similarly, if you add an excess of unlabeled haploid-specific DNA, the a-specific fragment is still bound (lane 6).

In the third set of experiments you vary the ratio of a1 relative to α2. When α2 is in excess, binding to the haploid-specific fragment is decreased (Figure 7–32, lane 7); when a1 is in excess, binding to the a-specific fragment is decreased (lane 8).

A. In the presence of a1, is α2 present in two forms with different binding specificities or in one form that can bind to both regulatory sequences? How do your experiments distinguish between these alternatives?

B. An α2 repressor with a small deletion in its DNA-binding domain does not bind to DNA fragments containing the haploid-specific sequence. If this mutant protein is expressed in a diploid cell along with normal α2 and a1 proteins, however, the haploid-specific genes are turned on. (These genes are normally off in a diploid.) In the light of this result and your other experiments, do you consider it more likely that a1 catalyzes a covalent modification of α2, or that a1 modifies α2 by binding to it stoichiometrically to form an a1–α2 complex?

7–50*   You have discovered a new strain of yeast with a novel mating system. There are two haploid mating types, called M and F. Cells of opposite mating type can mate to form M/F diploid cells. These diploids can undergo meiosis and sporulate, but they cannot mate with each other or with either haploid mating type.

Your genetic analysis of the strains shows there are four genes that control mating type. When one pair of genes, *M1* and *M2*, are at the mating-type locus, the cells are mating-type M; when a second pair of genes, *F1* and *F2*, are at the mating-type locus, the cells are mating-type F. You have also identified three sets of regulated genes: one that is expressed specifically in M-type haploids (Msg), one in F-type haploids (Fsg), and one in sporulating cells (Ssg). You obtain viable mutants (*M1⁻, M2⁻, F1⁻, F2⁻*) that are defective in each of the mating-type genes, and study their effects—individually and in combination—on the mating phenotype and on expression of their regulated genes. Your results with haploid and diploid cells containing different combinations of mutants are shown in Table 7–3.

TABLE 7–3 Phenotypes of mutants that affect mating in a new strain of yeast (Problem 7–50).

	MUTANT	MATING PHENOTYPE	GENES EXPRESSED
Haploid cells	wild-type M	M	Msg
	M1⁻	nonmating	Msg, Fsg
	M2⁻	M	Msg
	M1⁻M2⁻	nonmating	Msg, Fsg
	wild-type F	F	Fsg
	F1⁻	F	Fsg
	F2⁻	nonmating	Msg, Fsg
	F1⁻F2⁻	nonmating	Msg, Fsg
Diploid cells	wild-type M/F	nonmating	Ssg
	M1⁻/F1⁻	F	Fsg
	M1⁻/F2⁻	nonmating	Msg, Fsg, Ssg
	M2⁻/F1⁻	nonmating	none
	M2⁻/F2⁻	M	Msg
Msg = M-specific genes; Fsg = F-specific genes; Ssg = sporulation-specific genes			

Figure 7–33 Regulation of bacteriophage λ replication by cl and cro (Problem 7–51). (A) The prophage state. (B) The lytic state.

Suggest a regulatory scheme to explain how the *M1, M2, F1,* and *F2* gene products control the expression of the M-specific, F-specific, and sporulation-specific sets of genes. Indicate which gene products are activators and which are repressors of transcription, and decide whether the gene products act alone and/or in combination.

7–51 Bacteriophage λ can replicate as a prophage or lytically. In the prophage state, the viral DNA is integrated into the bacterial chromosome and is copied once per cell division. In the lytic state the viral DNA is released from the bacterial chromosome and replicates many times in the cell. This viral DNA then produces viral coat proteins that enclose replicated viral genomes to form many new virus particles, which are released when the bacterial cell bursts.

These two states are controlled by the gene regulatory proteins cI and cro, which are encoded by the virus. In the prophage state, cI is expressed; in the lytic state, cro is expressed. In addition to regulating the expression of other genes, cI is a repressor of transcription of the gene that encodes cro, and cro is a repressor of the gene that encodes cI (Figure 7–33). When bacteria containing a λ phage in the prophage state are briefly irradiated with UV light, cI protein is degraded.
A. What will happen next?
B. Will the change in (A) be reversed when the UV light is switched off?
C. How is this mechanism beneficial to the virus?

7–52* Imagine the two situations shown in Figure 7–34. In cell I, a transient signal induces the synthesis of protein A, which is a gene activator that turns on many genes including its own. In cell II, a transient signal induces the

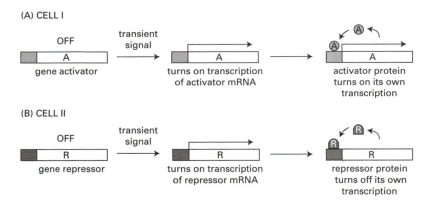

(A) CELL I

OFF → transient signal → turns on transcription of activator mRNA → activator protein turns on its own transcription

(B) CELL II

OFF → transient signal → turns on transcription of repressor mRNA → repressor protein turns off its own transcription

Figure 7–34 Gene regulatory circuits and cell memory (Problem 7–52). (A) Induction of synthesis of gene activator A by a transient signal. (B) Induction of synthesis of gene repressor R by a transient signal.

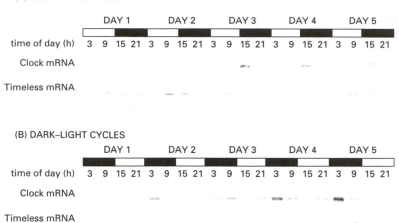

Figure 7–35 Influence of cycles of light and dark on mRNAs for two circadian rhythm proteins in cultured zebrafish hearts (Problem 7–54). (A) Light–dark cycles. (B) Dark–light cycles. Hearts were cultured in light–dark or dark–light cycles for 5 days. RNase protection assays were used to measure the levels of mRNAs for Clock and Timeless.

synthesis of protein R, which is a gene repressor that turns off many genes including its own. In which, if either, of these situations will the descendants of the original cell 'remember' that the progenitor cell had experienced the transient signal? Explain your reasoning.

7–53   Most totally blind people have circadian rhythms that are 'free-running;' that is, they are not synchronized to environmental time cues and they oscillate on a cycle of about 24.5 hours. Why do you suppose the circadian clocks of blind people are not entrained to the same 24-hour clock as the majority of the population? Can you guess what symptoms might be associated with a free-running circadian clock? Do you suppose that totally blind people have trouble sleeping?

7–54*   Expression of the *Clock* gene in zebrafish shows a strong circadian oscillation in many tissues *in vivo* and in culture, showing that endogenous circadian oscillators exist in peripheral tissues. A defining feature of circadian clocks is that they can be set or entrained to local time, usually by the environmental light–dark cycle. An important question is whether peripheral oscillators are entrained to local time by signals from central pacemakers in the brain or are themselves directly light-responsive.

To address this question, you isolate hearts from zebrafish and culture them over a period of 5 days. You expose one group to a light–dark cycle (14 hours light–10 hours dark) and the other to a dark–light cycle (10 hours dark–14 hours light). Periodically, you measure the mRNA levels of two circadian rhythm proteins: Clock, whose mRNA is known to oscillate, and Timeless, whose mRNA does not. Your results are shown in Figure 7–35.

A. Do cycles of light and dark entrain the circadian clock in isolated hearts? Explain your answer.

B. How do you suppose light influences isolated cells and organs?

7–55   (**True/False**) The fibroblasts and other cell types that are converted to muscle cells by expression of myogenic proteins probably have already accumulated a number of gene regulatory proteins that can cooperate with the myogenic proteins to switch on muscle-specific genes. Explain your answer.

7–56*   Figure 7–36 shows a simple scheme by which three gene regulatory proteins might be used during development to create eight different cell types. How many cell types could you create, using the same rules, with four different gene regulatory proteins? MyoD is a regulatory protein that induces muscle-specific gene expression in fibroblasts. How might this observation fit with the scheme shown in Figure 7–36?

7–57   An elegant early study examined whether a transcriptional complex could remain bound to the DNA during DNA replication to serve, perhaps, as a basis for inheritance of the parental pattern of gene expression. An active

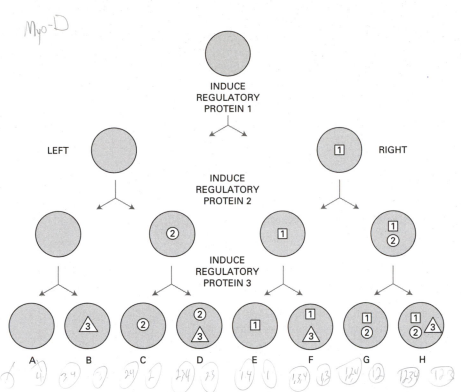

Myo-D

LEFT

INDUCE
REGULATORY
PROTEIN 1

RIGHT

INDUCE
REGULATORY
PROTEIN 2

INDUCE
REGULATORY
PROTEIN 3

A   B   C   D   E   F   G   H

Figure 7–36 An illustration of combinatorial gene control for development (Problem 7–56). In this simple, idealized scheme a 'decision' to make a new gene regulatory protein (*numbered symbols*) is made after each cell division. In this scheme, the daughter cell on the *right* is induced to make the new regulatory protein. Each gene regulatory protein is assumed to stay on after it is induced, thereby allowing different combinations of regulatory proteins to be built up. In this example, eight different cell types have been created.

transcription complex was assembled on the *Xenopus* 5S RNA gene on a plasmid, which was then replicated *in vitro* and tested for transcription.

Replicated and unreplicated templates were distinguished by the restriction nucleases DpnI, MboI, and Sau3A, which are sensitive to the methylation state of their recognition sequence GATC (Figure 7–37A). This sequence is present once at the beginning of the 5S RNA gene and when it is cleaved no transcription occurs. When the plasmid is grown in wild-type *E. coli*, GATC is methylated at the A on both strands by the bacterial *dam* methylase. Replication of fully methylated DNA *in vitro* generates daughter duplexes that are methylated on only one strand (hemimethylated) in the first round and are unmethylated after subsequent rounds. By starting with fully methylated DNA and inducing its replication *in vitro*, transcription could be assayed specifically from the replicated DNA by treating the DNA with DpnI, which selectively cuts unreplicated (methylated) DNA, thus preventing its transcription.

(A) CUTTING PATTERNS

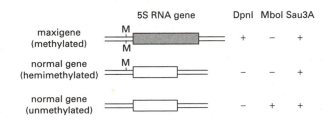

	5S RNA gene	DpnI	MboI	Sau3A
maxigene (methylated)	M / M	+	−	+
normal gene (hemimethylated)	M	−	−	+
normal gene (unmethylated)		−	+	+

(B) TRANSCRIPTIONAL ACTIVITY

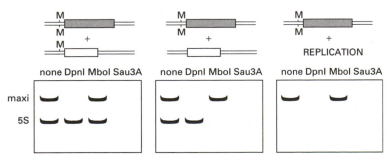

REPLICATION

none DpnI MboI Sau3A   none DpnI MboI Sau3A   none DpnI MboI Sau3A

maxi

5S

Figure 7–37 Analysis of stability of transcription complexes during DNA replication (Problem 7–57). (A) Sensitivity of 5S RNA genes in different methylation states to digestion with restriction enzymes. The 5S RNA maxigene is *shaded* and the normal gene is *white*. M indicates that a strand is methylated. Sensitivity to cleavage by a restriction nuclease is indicated by (+); resistance is indicated by (−). (B) Transcriptional activity in the presence and absence of replication.

Two versions of the 5S RNA gene—a slightly longer 'maxigene' and the normal gene—were constructed so that their transcripts could be distinguished (Figure 7–37B). Mixtures of the fully methylated maxigene with either the hemimethylated or unmethylated normal gene served as controls for the specificity of the restriction digestions (Figure 7–37B). To test the effect of replication on transcription, transcription complexes were assembled on the fully methylated maxigene, replication was induced, and transcriptional activity was assayed before and after cleavage with restriction nucleases (Figure 7–37B).

A. Does the methylation status of the 5S RNA gene affect its level of transcriptional activity? How can you tell?

B. Does the pattern of transcription after cleavage of the mixtures of maxigenes and normal genes with the various restriction nucleases match your expectations? Explain your answer.

C. In your experiment about half of the DNA molecules were replicated. Does the pattern of transcription after replication and cleavage indicate that the transcription complex remains bound to the 5S RNA gene during replication?

D. In these experiments the authors were careful to show that greater than 90% of the molecules were assembled into active transcription complexes and that 50% of the molecules were replicated. How would the pattern have changed if only 10% of the molecules were assembled into active transcription complexes, and only 10% were replicated? Would the conclusions have changed?

7–58*   Explain how cooperatively binding proteins might make a segment of the chromosome distinctive, with the result that a particular chromatin configuration—and pattern of gene expression—could be passed on faithfully from one cell to its daughters.

7–59   In female mammals one X chromosome in each cell is chosen at random for inactivation early in development. X-inactivation, which involves more than a thousand genes in humans, is crucial for equalizing expression of X-chromosome genes in males and females. A critical clue to the mechanism of X-inactivation came from the isolation of a large number of cDNAs for genes on the human X chromosome. Their expression patterns were characterized in cells from normal males and females, in cells from individuals with abnormal numbers of X chromosomes, and in rodent:human hybrid cell lines that retained either one inactive human X chromosome (X_i) or one active human X chromosome (X_a). Among all these cDNAs, there were three patterns of expression, as illustrated in Figure 7–38 for cDNAs A, B, and C.

A. For each pattern of expression decide whether the gene is expressed from the active X, the inactive X, or both. Which pattern do you expect to be the most common? Which pattern is the most surprising?

B. From the results with cells from abnormal individuals, formulate a rule as to how many chromosomes are inactivated and how many remain active during X inactivation.

7–60*   Sexually reproducing organisms use a variety of strategies to compensate for the gene-dosage differences arising from the different numbers of X chromosomes in the two sexes.

A. From the list below, select the methods of dosage compensation that are used in mice, *Drosophila*, and nematodes, and indicate the sex in which it occurs.
1. Two-fold up-regulation of gene expression from one X chromosome.
2. Two-fold down-regulation of gene expression from both X chromosomes.
3. Elimination of gene expression from one X chromosome.

B. Which of these strategies equalize expression of the X chromosome genes in the two sexes?

7–61   One research group directly tested the hypothesis that the *Xist* gene in mice is required for X-inactivation by introducing a targeted deletion of *Xist* into one X chromosome. Using female embryonic stem (ES) cells in which genes

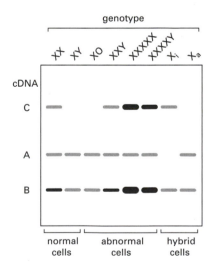

Figure 7–38 Northern analysis of gene expression from cells with different numbers and types of X chromosomes (Problem 7–59). RNA from cells was run out on gels, blotted onto nitrocellulose, probed with a mixture of radioactive cDNAs A, B, and C, and visualized by autoradiography. The positions of the RNA bands that correspond to genes A, B, and C were determined in a separate experiment.

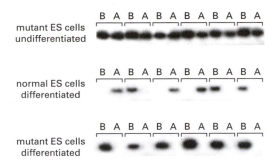

Figure 7–39 Analysis of expression of X chromosome-specific alleles in undifferentiated and differentiated ES cells (Problem 7–61). Analysis of the A and B alleles from individual cells are paired as indicated by the *brackets*.

on the two X chromosomes could be distinguished due to polymorphisms, they followed X-inactivation during differentiation of ES cells *in vitro*. ES cells normally maintain both X chromosomes in the active state; however, when they are induced to differentiate, they randomly inactivate one.

The scientists considered three hypotheses. (1) ES cells mutant for one *Xist* gene would fail to register the presence of two X chromosome and thus fail to undergo X-inactivation. (2) The *Xist* knockout would prevent X-inactivation of the targeted X chromosome, thus predisposing the normal X chromosome to preferential X-inactivation. (3) The mutation would have no effect on X-inactivation at all.

Using allele-specific (X chromosome-specific) oligonucleotide probes, they analyzed the expressed alleles of the *Pgk-1* gene in individual cells. As shown in Figure 7–39, they examined cells from mutant ES cells that were undifferentiated, and from mutant and nonmutant ES cells that had undergone differentiation. Only a few cells are shown in Figure 7–39, but analysis of many more cells confirmed the patterns shown. The A allele marks the X chromosome whose *Xist* gene is intact in the mutant ES cells; the B allele marks the X chromosome from which the *Xist* gene was deleted.

Which of the three hypotheses do these results support? Explain your reasoning.

7–62* Using strand-specific probes of transcription in the X-inactivation center, a second gene was found to be transcribed in the opposite direction to *Xist* and was named *Tsix* to indicate its antisense orientation. The *Tsix* transcript, like the *Xist* transcript, has no significant open reading frame and is thought to function as an RNA. As shown in Figure 7–40, the *Tsix* transcript extends all the way across the *Xist* gene. To determine whether *Tsix* plays a role in counting X chromosomes, choosing which one to inactivate, or in silencing the inactive X, female ES cells were generated in which the promoter for *Tsix* had been deleted from one X chromosome. When these ES cells were allowed to differentiate, it was found that the X chromosome with the *Tsix* deletion was always inactivated.

A. Is *Tsix* important for the counting, choice, or silencing of X chromosomes? Explain your answer.

B. Before X-inactivation, *Tsix* is expressed from both alleles, as is *Xist*. At the onset of X-inactivation *Tsix* expression becomes confined to the future active X, whereas *Xist* expression is restricted to the future inactive X. Can you suggest some possible ways that *Tsix* might regulate *Xist*?

7–63 Maintenance methyltransferase, *de novo* methyltransferases, and demethylating enzymes play crucial roles in the changes in methylation patterns during development. Starting with the unfertilized egg, describe in a general way how these enzymes bring about the observed changes in genomic DNA methylation.

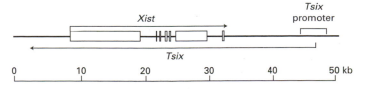

Figure 7–40 Arrangement of *Xist* and *Tsix* transcripts in the X-inactivation center (Problem 7–62). *Boxes* indicate the exons of *Xist*; exons are undefined for *Tsix*. The promoter deletion in the *Tsix* mutant is indicated.

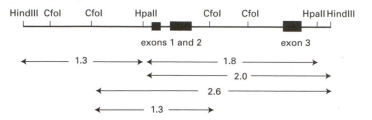

(A) RESTRICTION MAP

exons 1 and 2

exon 3

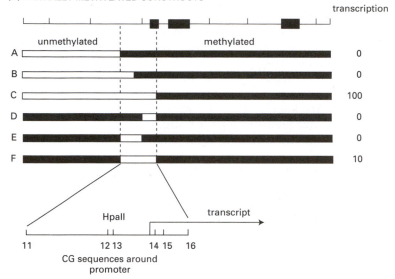

(B) PARTIALLY METHYLATED CONSTRUCTS

transcription

unmethylated   methylated

A   0
B   0
C   100
D   0
E   0
F   10

Hpall

transcript

11   12 13   14 15   16

CG sequences around
promoter

Figure 7–41 Effects of methylation on transcription of the γ-globin gene (Problem 7–64). (A) HindIII fragment containing the γ-globin gene. Sites of cleavage by the methylation-sensitive restriction nucleases, CfoI and HpaII, are indicated along with the sizes of the larger fragments that are observed on the gel in Figure 7–42. (B) Methylated constructs of the γ-globin gene. The methylated segments of the gene are shown in *black*. The six CG sites around the promoter are shown in more detail *below* the constructs. The level of expression of γ-globin RNA from each construct is shown on the *right* as a percentage of the expression from a fully unmethylated construct.

7–64* You are studying the role of DNA methylation in the control of gene expression, using the human γ-globin gene as a test system (Figure 7–41A). Globin mRNA can be detected when this gene is incorporated into the genome of mouse fibroblasts, even though it is expressed at much lower levels than it is in red cells. If the gene is first methylated at all 27 CG sites, however, its expression is blocked completely. You are using this system to decide whether a single critical methylation site is sufficient to determine globin expression.

You use a combination of site-directed mutagenesis and primed synthesis in the presence of 5-methyl dCTP to create several different γ-globin constructs that are unmethylated in various regions of the gene. These constructs are illustrated in Figure 7–41B, with the methylated regions shown in black. The arrangement of six methylation sites around the promoter is shown below the constructs. Sites 11, 12, and 13 are unmethylated in construct E, sites 14, 15, and 16 are unmethylated in construct D, and all six sites are unmethylated in construct F. You incorporate these constructs into mouse fibroblasts, grow cell lines containing individual constructs, and measure γ-globin RNA synthesis relative to cell lines containing the fully unmethylated construct (Figure 7–41B).

To check whether the methylation patterns were correctly inherited, you isolate DNA samples from cell lines containing constructs B, C, or F and digest them with HindIII plus CfoI or HpaII. CfoI and HpaII do not cleave if their recognition sites are methylated. The cleavage sites for these nucleases are shown in Figure 7–41A along with the sizes of relevant restriction fragments larger than 1 kb. You separate the cleaved DNA samples on a gel and visualize them by hybridization to the radiolabeled HindIII fragment (Figure 7–42).

A. To create some of the methylated DNA substrates you used a single-stranded version of the gene as a template and primed synthesis of the second strand using 5-methyl dCTP instead of dCTP in the reaction. This method creates a DNA molecule containing 5-methyl C in place of C in one strand.

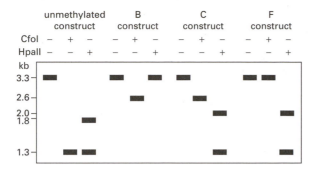

After isolation of colonies containing such constructs, you found that they have 5-methyl C on both strands but only at CG sequences. Explain the retention of 5-methyl C in CG sequences and their loss elsewhere.

B. Do the restriction patterns of the constructs in the isolated cell lines (Figure 7–42) indicate that the CG sequences that were methylated during creation of the constructs (Figure 7–41) were maintained in the cell?

C. Does the γ-globin RNA synthesis associated with the cell lines containing the various constructs (Figure 7–41) indicate that a single critical site of methylation determines whether the gene is expressed?

7–65   The embryonic mouse fibroblast cell line, 10T½ (10 T and a half), is a stable line with cells that look and behave like fibroblasts. If these cells are exposed to medium containing 5-azacytidine (5-aza C) for 24 hours, they will then differentiate into cartilage, fat, or muscle cells when they grow to a high cell density. (Treatment with 5-aza C reduces the general level of DNA methylation, allowing some previously inactive genes to become active.) If the cells are grown at low cell density after the treatment, they retain their original fibroblastlike shape and behavior, but even after many generations they still differentiate when they reach a high cell density. 10T½ cells that have not been exposed to 5-aza C do not differentiate, no matter what the cell density is.

When the treated cells differentiate, about 25% turn into muscle cells (myoblasts). The high frequency of myoblast formation leads your advisor to hypothesize that a single master regulatory gene, which is normally repressed by methylation, may trigger the entire transformation. Accordingly, he persuades you to devise a strategy to clone the gene. You assume the gene is off before treatment with 5-aza C and on in the induced myoblasts. If this assumption is valid, you should be able to find sequences corresponding to the gene among the cDNA copies of mRNAs that are synthesized after 5-aza C treatment. You decide to screen an existing cDNA library from normal myoblasts (which according to your assumption will also express the gene) using a set of radioactive probes to identify likely cDNA clones. You prepare three radioactive probes.

Probe 1. You isolate RNA from 5-aza-C-induced myoblasts and prepare radioactive cDNA copies.

Probe 2. You hybridize the radioactive cDNA from the 5-aza-C-induced myoblasts with RNA from untreated 10T½ cells and discard all the RNA:DNA hybrids.

Probe 3. You isolate RNA from normal myoblasts and prepare radioactive cDNA copies, which you then hybridize to RNA from untreated 10T½ cells; you discard the RNA:DNA hybrids.

The first probe hybridizes to a large number of clones from the cDNA library, but only about 1% of those clones hybridize to probes 2 and 3. Overall, you find four distinct patterns of hybridization (Table 7–4).

A. What is the purpose of hybridizing the radioactive cDNA from the two kinds of myoblasts to the RNA from untreated 10T½ cells? In other words, why are probes 2 and 3 useful?

B. What general kinds of genes might you expect to find in each of the four

TABLE 7–4 Patterns of myoblast cDNA hybridization with radioactive probes (Problem 7–65).

CLASS	PROBE 1	PROBE 2	PROBE 3
A	+	–	–
B	+	–	+
C	+	+	–
D	+	+	+

(A)

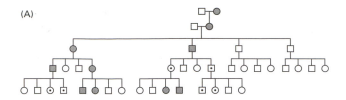

(B)

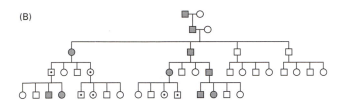

Figure 7–43 Pedigrees reflecting maternal and paternal imprinting (Problem 7–66). In one pedigree the gene is paternally imprinted; in the other it is maternally imprinted. In generations 3 and 4 only one of the two parents in the indicated matings is shown; the other parent is a normal individual from outside this pedigree. Affected individuals are represented by *shaded circles* for females and *shaded squares* for males. *Dotted symbols* indicate individuals that carry the deletion but do not display the phenotype.

classes of cDNA clones (A, B, C, and D in Table 7–4)? Which class of cDNA clone is most likely to include sequences corresponding to the putative muscle regulatory gene you are seeking?

7–66* Examine the two pedigrees shown in Figure 7–43. One results from a deletion of a maternally imprinted, autosomal gene. The other pedigree results from a deletion of a paternally imprinted autosomal gene. In both pedigrees, affected individuals (shaded symbols) are heterozygous for the deletion. These individuals are affected because one copy of the chromosome carries an imprinted, inactive gene, while the other carries a deletion of the gene. Dotted symbols indicate individuals that carry the deleted locus but do not display the mutant phenotype. Decide which of the pedigrees involves paternal imprinting and which involves maternal imprinting. Explain your answer.

7–67 Imprinting only occurs in mammals, and why it should exist at all is a mystery. One idea is that it represents an evolutionary endpoint in a tug of war between the sexes. In most mammalian species a female can mate with multiple males, generating multiple embryos with different fathers. If one father could cause more rapid growth of his embryo, it would prosper at the expense of the other embryos. While this would be good for the father's genes (in an evolutionary sense), it would drain the resources of the mother, potentially putting her life at risk (not good for her genes). Thus, it is in the mother's interest to counter these paternal effects with maternal changes that limit the growth of the embryo.

Based on this scenario, decide whether the gene for insulin-like growth factor-2 (IGF-2), which is required for prenatal growth, is more likely to be imprinted in the male or in the female.

7–68* (**True/False**) CG islands are thought to have arisen during evolution because they were associated with portions of the genome that remained active, hence unmethylated, in the germline. Explain your answer.

POSTTRANSCRIPTIONAL CONTROLS

TERMS TO LEARN

alternative RNA splicing	negative translational control	regulated nuclear transport
exosome	nonsense-mediated	RNA editing
internal ribosome entry site (IRES)	mRNA decay	RNA interference (RNAi)
	posttranscriptional control	transcription attenuation

7–69 RNA polymerase II commonly terminates transcription of the HIV (the human AIDS virus) genome some 50 nucleotides after it begins, unless helped along by a virus-encoded protein called Tat. In the absence of Tat the

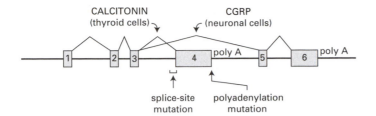

Figure 7–44 Structure and tissue-specific splicing of the gene encoding calcitonin and CGRP (Problem 7–71). *Gray boxes* indicate exons. The splicing/polyadenylation choices used to produce calcitonin and CGRP are shown *above* the gene. *Arrows* mark the positions of the splice-site and polyadenylation mutations.

CTD (C-terminal domain) of RNA polymerase is hypophosphorylated, but in the presence of Tat it becomes hyperphosphorylated and is converted to a more processive form. Tat apparently aids this transition by binding to a specific hairpin structure in the nascent viral RNA. Tat then recruits a collection of proteins that enhance the ability of RNA polymerase to continue transcribing. Among the Tat-recruited proteins is the positive transcription elongation factor b, P-TEFb, which is composed of cyclin T1 and the cyclin-dependent protein kinase Cdk9.

In a random drug screen you have just isolated flavopiridol, the most potent inhibitor of Cdk9 yet discovered. Flavopiridol binds at the site on Cdk9 where ATP would normally bind and blocks Cdk9-mediated phosphorylation of proteins. Would you expect flavopiridol to interfere with HIV transcription? Why or why not?

7–70*   A single *Drosophila* gene—the *DSCAM* gene—has the potential to produce more than 38,000 different proteins by alternative splicing of multiple copies of optional exons. Does that mean that this one gene is as complex as the whole human genome?

7–71   The calcitonin/CGRP gene contains six exons (Figure 7–44). In thyroid cells an mRNA containing exons 1, 2, 3, and 4 encodes calcitonin. In neuronal cells an mRNA containing exons 1, 2, 3, 5, and 6 encodes calcitonin gene-related peptide (CGRP). In both cell types transcription begins in the same place and extends beyond exon 6.

Because the two processing pathways use different poly-A sites and different splice sites, either tissue-specific polyadenylation or splicing factors could regulate calcitonin and CGRP expression. For example, thyroid cells might produce calcitonin because they contain a specific factor that recognizes the poly-A site in exon 4 and causes cleavage of the precursor RNA before splicing of exon 3 to exon 5 can occur. Neuronal cells then would lack this factor with the result that splicing of exon 3 to exon 5 predominates, leading to CGRP mRNA production. Alternatively, thyroid cells might contain a protein that enhances splicing of exon 3 to exon 4, while neuronal cells might contain a protein that enhances splicing of exon 3 to exon 5.

To distinguish these hypotheses—poly-A-site selection versus splice-site selection—the splicing and polyadenylation signals at the ends of exon 4 were altered by mutation (Figure 7–44). The altered genes were transfected into a lymphocyte cell line, which produces only calcitonin from the wild-type gene. The mutant lacking the exon-4 polyadenylation site produced no mRNA at all; the mutant lacking the exon-4 splice site produced only CGRP mRNA.

A. Does the lymphocyte cell line contain the splicing and polyadenylation factors necessary to produce both calcitonin and CGRP mRNAs?

B. If differential processing resulted from polyadenylation-site selection, which mutant would you have expected to produce CGRP mRNA when transfected into the lymphocyte cell line?

C. If differential processing resulted from splice-site selection, which mutant would you have expected to produce CGRP mRNA when transfected into the lymphocyte cell line?

D. Which model for differential processing best explains the ability of the lymphocyte cell line to produce calcitonin mRNA but not CGRP mRNA?

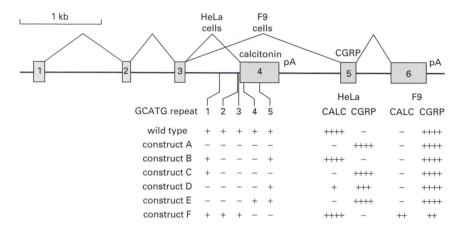

Figure 7–45 Effects of mutations in GCATG repeats on alternative splicing of calcitonin/CGRP pre-mRNA (Problem 7–72). In HeLa cells exon 3 is spliced to exon 4 to make calcitonin mRNA; in F9 cells exon 3 is spliced to exon 5 to make CGRP mRNA. The positions of the GCATG repeats around exon 4 are indicated. In the various constructs the presence of a GCATG repeat is indicated by a *plus*, and its absence by a *minus*. Production of calcitonin- (CALC) and CGRP-specific spliced RNA is indicated relative to wild type as follows: ++++ = 80–100%, +++ = 60–80%, ++ = 40–60%, + = 20–40%, and – = 0–20%.

7–72* A repeated hexanucleotide element, TGCATG, has been shown to regulate the tissue-specific splicing of the fibronectin alternative exon EIIIB. When you examine the calcitonin/CGRP gene you find five copies of a related repeat, GCATG, within 500 nucleotides of the exon-4 splice site (Figure 7–45). To analyze its potential role, you make several calcitonin/CGRP constructs in which the repeats have been altered singly or in combination. You transfect the constructs into HeLa cells, which normally give the calcitonin splicing pattern, and F9 cells, which normally give the CGRP splicing pattern. You find that no single mutation alters the pattern seen with wild type (not shown); however, various combinations of the mutations have dramatic effects (Figure 7–45).

  A. Why do you suppose the most dramatic changes were seen in HeLa cells rather than in F9 cells?

  B. Is there a particular combination of GCATG repeats that is critical for proper calcitonin mRNA production in HeLa cells? If so, what is it?

7–73 (**True/False**) If the site of transcript cleavage and polyadenylation of a particular RNA in one cell is downstream of the site used for cleavage and polyadenylation of the same RNA in a second cell, the protein produced from the longer polyadenylated RNA in the first cell will necessarily contain additional amino acids at it C-terminus. Explain your answer.

7–74* In humans, two closely related forms of apolipoprotein B (apo-B) are found in blood as constituents of the plasma lipoproteins. Apo-B48 is synthesized by the intestine and is a key component of chylomicrons, the large lipoprotein particles responsible for delivery of dietary triglycerides for storage, primarily in adipose. By contrast, apo-B100 is synthesized in the liver for formation of the much smaller, very low-density lipoprotein (VLDL) particles used in the distribution of triglycerides to meet energy needs. A classic set of studies defined the surprising relationship between these two proteins.

  Sequences of cloned cDNA copies of the mRNAs from these two tissues revealed a single difference: cDNAs from intestinal cells had a T, as part of a stop codon, at a point where the cDNAs from liver cells had a C, as part of a glutamine codon (Figure 7–46). To verify the differences in the mRNAs and to search for corresponding differences in the genome, RNA and DNA were isolated from intestinal and liver cells and then subjected to PCR amplification using oligonucleotides that flanked the region of interest. The amplified DNA segments from the four samples were tested for the alteration by hybridization to oligonucleotides containing either the liver cDNA sequence (oligo-Q) or the intestinal cDNA sequence (oligo-STOP). The results are shown in Table 7–5.

  Are the two forms of apo-B produced by transcriptional control from two different genes, by a processing control of the RNA transcript from a single

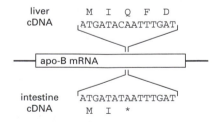

Figure 7–46 Location of the sequence differences in cDNA clones from apo-B RNA isolated from liver and intestine (Problem 7–74). The encoded amino acid sequences, in the one-letter code, are shown aligned with the cDNA sequences.

TABLE 7–5 Hybridization of specific oligonucleotides to the amplified segments from liver and intestine RNA and DNA (Problem 7–74).

	RNA		DNA	
	LIVER	INTESTINE	LIVER	INTESTINE
Oligo-Q	+	–	+	+
Oligo-STOP	–	+	–	–

The oligonucleotide complementary to the sequence derived from liver cDNA is oligo-Q; the oligonucleotide complementary to the intestinal cDNA is oligo-STOP. Hybridizations were carried out under very stringent conditions so that a single nucleotide mismatch was sufficient to prevent hybridization. Hybridization is indicated by +; absence of hybridization is indicated by –.

gene, or by differential cleavage of the protein product from a single gene? Explain your reasoning.

7–75   Transcripts from the mitochondrial DNA of trypanosomes are heavily edited by insertion and occasional deletion of U nucleotides at numerous sites. The information for the editing process is provided by a large number of small guide RNAs that pair at specific places in the transcript. Typically, the guide RNAs pair perfectly at their 5′ ends (as shown by the boxed nucleotides in Figure 7–47), but imperfectly at their 3′ ends. Examine the duplex structure formed by the pairing of the small guide RNAs with the pre-edited transcript (shown in the upper half of Figure 7–47), and then compare the sequences of the pre-edited and edited transcripts. Formulate a set of rules for RNA editing of mitochondrial transcripts in trypanosomes.

Problems 6–37 and 6–38 examine a related type of RNA editing in which modifications are introduced into U5 snRNA by U85 snoRNA.

7–76*   Several distinct mechanisms for mRNA localization have been discovered. They all require specific signals in the mRNA itself, usually in the 3′ UTR. Briefly outline three mechanisms by which different cellular mRNAs might become localized in the cell.

7–77   Ferritin stores iron in many tissues and its synthesis increases up to 2-fold in the presence of iron. You wish to define the molecular mechanism for this induction of ferritin synthesis. You carry out a series of experiments in which you give rats an injection of a ferric salt solution or an injection of saline, each with or without simultaneous administration of actinomycin D, which inhibits RNA synthesis. Three hours later you kill the rats, homogenize their livers, and prepare polysome and supernatant fractions. You extract RNA from each fraction and translate it in a cell-free system in the presence of radioactive amino acids. You measure total protein synthesis by incorporation

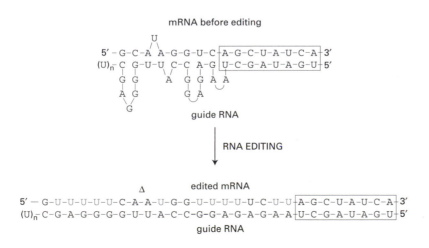

Figure 7–47 Pairing of pre-edited mitochondrial transcripts with guide RNAs in trypanosomes (Problem 7–75). Only a small portion of the pre-edited RNA is shown. The *boxed* nucleotides indicate the portion of the initial structure that is perfectly paired. The 3′ end of the guide RNA consists of a string of U nucleotides (U_n).

TABLE 7–6 Synthesis of ferritin in the rat after various treatments (Problem 7–77).

INJECTION	ACTINOMYCIN D	FRACTION	TOTAL PROTEIN SYNTHESIS	FERRITIN SYNTHESIS
Saline	absent	polysomes	750,000	700
		supernatant	255,000	1400
Iron	absent	polysomes	500,000	900
		supernatant	400,000	500
Saline	present	polysomes	800,000	800
		supernatant	600,000	3000
Iron	present	polysomes	780,000	1380
		supernatant	550,000	700

Numbers show radioactivity (cpm) incorporated into total proteins or into ferritin.

of label, and the synthesis of ferritin by precipitation with ferritin-specific antibodies (Table 7–6).

A. For each sample calculate the percentage of total protein synthesis that is due to synthesis of ferritin. Given that 85% of the bulk mRNA is bound to ribosomes in the polysome fraction and 15% is free (in the supernatant fraction), determine the percentage of total ferritin mRNA in the polysome and supernatant fractions under each condition of treatment. (Assume that the percentages of polysome and supernatant mRNA are not changed by treatment with actinomycin D.)

B. Do your results distinguish between transcriptional and posttranscriptional control of ferritin synthesis by iron? If so, how?

C. How would you account for the 2-fold increase in ferritin synthesis in the presence of iron?

7–78* Regulation of ferritin translation is controlled by the interaction between a hairpin structure in the mRNA, termed an iron-response element (IRE), and an iron-response protein (IRP) that binds to it. When the IRE is bound by the IRP, translation is inhibited. The location of the IRE in the mRNA is critical for its function. To work properly, it must be positioned near the 5′ end of the mRNA. If it is moved more than 60 nucleotides downstream of the cap structure, its ability to regulate translation is dramatically reduced. Why do you suppose there is such a critical position dependence for regulation of translation by IRE and IRP?

7–79 Although essential, iron is also potentially toxic. It is maintained at optimal levels in mammalian cells by the actions of three proteins. Transferrin binds to extracellular iron ions and delivers them to cells; the transferrin receptor binds iron-loaded transferrin and brings it into the cell; and ferritin binds iron (up to 4500 atoms in the internal cavity in each complex) to provide an intracellular storage site.

The regulation of the transferrin receptor, like that of ferritin, is accomplished by IREs and IRPs, and in both cases, iron binding to IRPs prevents their binding to IREs. Nevertheless, the mechanism of transferrin receptor regulation is quite different from that of ferritin. The transferrin receptor mRNA contains five IREs that are all located 3′ of the coding sequence (instead of at the 5′ end as in ferritin). In addition, binding of IRPs to these IREs increases the translation of transferrin receptor (as opposed to a decrease for ferritin).

A. Does this opposite regulation of ferritin and transferrin receptor in response to iron levels make biological sense? Consider the consequences of high and low iron levels.

B. In the presence of iron the transferrin receptor mRNA is rapidly degraded. Can you suggest a mechanism for how iron levels might be linked to the stability of transferrin receptor mRNA?

7–80* The level of tubulin gene expression is established in cells by an unusual

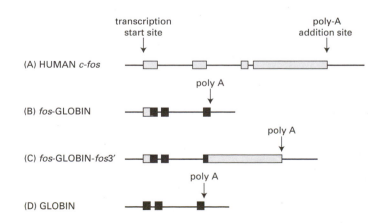

```
     M   R   E   I      regulation
  --ATGAGGGAAATC--          +
  -----T----------          -
  -----C----------          +
  ------C---------          -
  -------A--------          +
  --------T-------          -
  ---------G------          -
  ----------G-----          +
  -----TA---------          -
  -----C-C--------          +
  -----G--A-------          -
  -----G---T------          -
  -----C----G-----          +
```

Figure 7–48 Effects of mutations on the regulation of tubulin mRNA stability (Problem 7–80). The wild-type sequence for the first 12 nucleotides of the coding portion of the gene is shown at the *top*, and the first four amino acids beginning with methionine (M) are indicated *above* the codons. The nucleotide changes in the 12 mutants are shown *below*; only the altered nucleotides are indicated. Regulation of mRNA stability is shown on the *right*: + indicates wild-type response to changes in intracellular tubulin concentration and – indicates no response to changes. *Vertical dotted lines* mark the position of the first nucleotide in each codon.

regulatory pathway in which the intracellular concentration of free tubulin dimers (composed of one α-tubulin and one β-tubulin subunit) regulates the rate of new tubulin synthesis. The initial evidence for such an autoregulatory pathway came from studies with drugs that cause assembly or disassembly of all cellular tubulin. For example, when cells are treated with colchicine, which causes microtubule depolymerization into tubulin dimers, there is a 10-fold repression of tubulin synthesis. Autoregulation of tubulin synthesis by tubulin dimers is accomplished at the level of tubulin mRNA stability. The first 12 nucleotides of the coding portion of the mRNA were found to contain the site responsible for this autoregulatory control.

Since the critical segment of the mRNA involves a coding region, it is not clear whether the regulation of mRNA stability results from the interaction of tubulin dimers with the RNA or with the nascent protein. Either interaction might plausibly trigger a nuclease that would destroy the mRNA.

These two possibilities were tested by mutagenizing the regulatory region on a cloned version of the gene. The mutant genes were then transfected into cells and the stability of their mRNAs was assayed when the intracellular concentration of free tubulin dimers was increased. The results from a dozen mutants that affect a short region of the mRNA are shown in Figure 7–48.

Does the regulation of tubulin mRNA stability result from an interaction with the RNA or from an interaction with the encoded protein? Explain your reasoning.

7–81 The *c-fos* gene is the cellular homolog of the oncogene carried by the FBJ murine osteosarcoma virus. It encodes a gene regulatory protein. Activation of *c-fos* is one of the earliest transcriptional responses to serum growth factors. The *c-fos* transcription rate in mouse cells increases markedly within 5 minutes, reaches a maximum by 10 to 15 minutes, and abruptly decreases thereafter.

To study the mechanism of transient induction of the human *c-fos* gene, you clone a DNA fragment that contains the complete transcription unit starting 750 nucleotides upstream of the transcription start site and ending 1.5 kb downstream of the poly-A addition site (Figure 7–49A). When you

Figure 7–49 Structures of the human *c-fos* gene and the hybrid genes of *c-fos* with the human β-globin gene (Problem 7–81). (A) The cloned *c-fos* gene. (B) A hybrid gene, *fos–globin*, with *fos* sequences at the 5′ end. (C) A hybrid gene, *fos–globin–fos3′*, with *fos* sequences at the 5′ and 3′ ends. (D) The normal β-globin gene. *Open boxes* denote *c-fos* exons; *black boxes* are β-globin exons. The junctions between *fos* and globin sequences in the hybrid genes are located in exons.

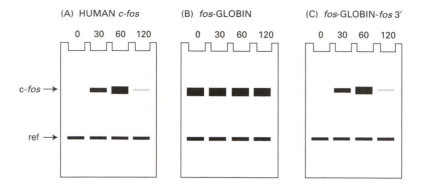

(A) HUMAN c-fos   (B) fos-GLOBIN   (C) fos-GLOBIN-fos 3'

0 30 60 120   0 30 60 120   0 30 60 120

c-fos →

ref →

Figure 7–50 Responses of c-fos and hybrid genes to addition of serum (Problem 7–81). (A) The cloned human c-fos gene. (B) The c-fos–globin hybrid gene with c-fos sequences at the 5' end. (C) The c-fos–globin hybrid gene with c-fos sequences at the 5' and 3' ends. The *upper* band in each gel is the transcript containing c-fos sequences. The *lower* band is a reference transcript (to control for recoveries of RNA) from a gene that does not respond to serum addition. Sampling times (in minutes) are indicated *above* each lane.

transfer this cloned segment into mouse cells (so you can distinguish the human c-fos mRNA from the endogenous mRNA) and stimulate cell growth one day later by adding serum (a rich source of growth factors), you observe the same sort of transient induction (Figure 7–50A), although the timescale is slightly longer than for the mouse gene.

By analyzing a series of deletion mutants, you show that an enhancerlike element (SRE, serum response element) 300 nucleotides upstream of the transcription start site is necessary for increased transcription in response to serum. You also investigate the stability of the c-fos mRNA by creating two hybrid genes containing portions of the human β-globin gene, which encodes a stable mRNA. The structures of the globin gene and the hybrid genes are shown in Figure 7–49B, C, and D. When the hybrid genes are transfected into mouse cells in low serum, they respond to serum added 24 hours later as indicated in Figure 7–50B and C.

A. Which portion of the c-fos gene confers instability on c-fos mRNA?

B. The fos–globin hybrid gene carries all the normal c-fos regulatory elements, including the SRE, and yet the mRNA is present at time zero (24 hours after transfection but before serum addition) and does not increase appreciably after serum addition. Can you account for this behavior in terms of mRNA stability?

7–82*   In nematodes the choice between spermatogenesis and oogenesis in the hermaphrodite germline depends on translational regulation of the tra-2 and fem-3 genes, as shown in Figure 7–51. When they are expressed, tra-2 promotes oogenesis and fem-3 promotes spermatogenesis. Translation of each gene's mRNA is regulated by the binding of proteins to elements within their 3' untranslated regions. In each case, the protein-bound mRNA, although stable, is not efficiently translated and has a short poly-A tail. In each case, the mRNA in its unbound form is translationally active and has a long poly-A tail. How do you suppose that the lengths of the poly-A tails might affect the efficiency of translation of these mRNAs?

7–83   To investigate the molecular mechanism by which the tra-2 regulatory elements (TGEs) control translation, you are using injection into *Xenopus* oocytes. You have shown that an oocyte protein binds to the TGEs and inhibits translation of injected tra-2 mRNA. Now you wish to determine how mRNAs with bound proteins come to have short poly-A tails. To investigate

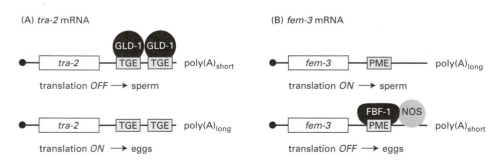

(A) tra-2 mRNA

tra-2 — TGE—TGE — poly(A)short
translation OFF → sperm

tra-2 — TGE—TGE — poly(A)long
translation ON → eggs

(B) fem-3 mRNA

fem-3 — PME — poly(A)long
translation ON → sperm

fem-3 — PME — poly(A)short
translation OFF → eggs

Figure 7–51 Translational control of the choice between spermatogenesis and oogenesis (Problem 7–82). (A) tra-2 mRNA. (B) fem-3 mRNA. Control elements within the 3' untranslated regions of the two genes are shown, along with the specific proteins that bind to those elements.

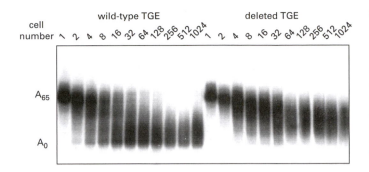

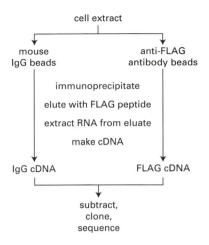

Figure 7–52 Injection of radiolabeled mRNAs with and without TGEs into *Xenopus* one-cell embryos (Problem 7–83). The lengths of the poly-A tails are indicated on the *left*. Cell number refers to the number of cells in the embryos when they were harvested.

Figure 7–53 Scheme for isolating mRNA targets of GLD-1 (Problem 7–84).

this question, you prepare two kinds of radiolabeled *tra-2* mRNA, one with the normal pair of TGE elements and one with those elements deleted (Figure 7–52). Each RNA has a tail of 65 A nucleotides. You inject these mRNAs into one-cell *Xenopus* embryos and then reisolate the mRNAs at various cell stages thereafter. The reisolated RNA was analyzed by gel electrophoresis and autoradiography (Figure 7–52).

A. Does the presence of the TGEs destabilize the mRNAs; that is, are they destroyed? How can you tell?

B. Does the presence of the TGEs influence the length of the poly-A tails?

7–84* Nematode mutants with defects in the germ line-specific protein GLD-1, which binds to the TGE sequences in *tra-2* mRNA (see Figure 7–51A), have phenotypes that suggest that GLD-1 may bind more target mRNAs than just the one from *tra-2*. To try to locate additional mRNA targets of GLD-1, a GLD-1 gene that carries the FLAG epitope (which can be precipitated readily by a commercially available antibody) was constructed and introduced stably into a nematode carrying a nonfunctional GLD-1 gene. The strategy for isolating new GLD-1 targets is shown in Figure 7–53. RNAs were isolated using FLAG antibodies or nonspecific mouse antibodies and then converted to cDNAs. Potential GLD-1 targets were enriched using subtractive hybridization to remove nonspecific cDNAs. From one subtractive hybridization 211 clones were sequenced and found to correspond to 94 different genes. From an independent subtractive hybridization 198 clones were found to correspond to 89 different genes. Comparison of the two data sets indicates that 17 genes were found in common.

A. Why did the investigators focus on the cDNAs that were common to the two independent subtractions? Shouldn't all the subtracted cDNAs be targets for GLD-1?

B. For one specific gene, *rme-2*, the investigators (1) showed that the RME-2 protein was expressed specifically in the germline; (2) demonstrated that the locations of GLD-1 and RME-2 in the germline were nonoverlapping; and (3) identified GLD-1-binding sites in *rme-2* mRNA. How do these observations support their conclusion that *rme-2* mRNA is a bona fide target of GLD-1?

7–85 A very active cell-free protein synthesis system can be prepared from reticulocytes. These immature red blood cells have already lost their nucleus but still contain ribosomes. A reticulocyte lysate can translate each globin mRNA many times provided heme is added (Figure 7–54). Heme serves two functions: it is required for assembly of globin chains into hemoglobin, and, surprisingly, it is required to maintain a high rate of protein synthesis. If heme is omitted, protein synthesis stops after a brief lag (Figure 7–54).

The first clue to the molecular basis for the effect of heme on globin synthesis came from a simple experiment. A reticulocyte lysate was incubated for several hours in the absence of heme. When 5 mL of this preincubated lysate were then added to 100 mL of a fresh lysate in the presence of heme, protein synthesis in this fresh lysate was rapidly inhibited (Figure 7–54). When further characterized, the inhibitor (termed heme-controlled repressor, or HCR) was shown to be a protein with a molecular weight of 180,000. Pure preparations of HCR at a concentration of 1 µg/mL completely inhibit protein synthesis in a fresh, heme-supplemented lysate.

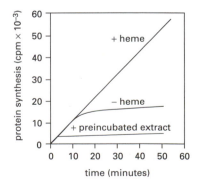

Figure 7–54 Protein synthesis in a reticulocyte lysate (Problem 7–85). The lysate to which preincubated extract was added also contained added heme.

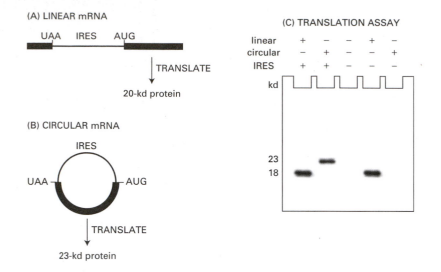

(A) LINEAR mRNA

UAA IRES AUG

↓ TRANSLATE

20-kd protein

(B) CIRCULAR mRNA

IRES

UAA — AUG

↓ TRANSLATE

23-kd protein

(C) TRANSLATION ASSAY

linear	+	–	–	+	–
circular	–	+	–	–	+
IRES	+	+	–	–	–

kd

23
18

Figure 7–55 Analysis of effects of IRESs on translation (Problem 7–87). (A) Linear mRNA that contains an IRES. The structure of the mRNA without the IRES was the same. (B) Circular mRNA that contains an IRES. The structure of the mRNA without the IRES was the same. Circular RNAs were prepared by ligating the ends together; they were purified from the linear starting molecules by gel electrophoresis. (C) Display of translation products from various species of mRNA.

A. Calculate the ratio of HCR molecules to ribosomes and globin mRNA at the concentration of HCR that inhibits protein synthesis. Reticulocyte lysates contain 1 mg/mL ribosomes (molecular weight, 4 million), and the average polysome contains four ribosomes per globin mRNA.

B. Do the results of this calculation favor a catalytic or a stoichiometric mechanism for HCR inhibition of protein synthesis in a reticulocyte lysate?

7–86* Vg1 mRNA encodes a member of the TGF-β family of growth factors, which is important for mesoderm induction. In *Xenopus,* Vg1 mRNA is localized to the vegetal pole of the oocyte. Translation of Vg1 does not occur until a late stage, after localization is complete. Analysis of the 3′ UTR identified two elements: a localization element and a UA-rich translation control element. Translational regulation was shown to be independent of the status of the poly-A tail. It remains unchanged in length throughout oogenesis, and translational regulation could be demonstrated on injected mRNAs that had no poly-A tails. Translational repression was abolished, however, if an IRES sequence was inserted into the mRNA upstream of the coding sequence. How do you suppose translation of Vg1 mRNA is controlled?

7–87 You are skeptical that IRESs really allow direct binding of the eucaryotic translation machinery to the interior of an mRNA. As a critical test of this notion, you prepare a set of linear and circular RNA molecules, with and without IRESs (Figure 7–55A and B). You translate these various RNAs in rabbit reticulocyte lysates and display the translation products by SDS gel electrophoresis (Figure 7–55C). Do these results support or refute the idea that IRESs allow ribosomes to initiate translation of mRNAs in a cap-independent fashion? Explain your answer.

Problem 6–62 considers the translational regulation of *E. coli* release factor 2, a protein required for termination of translation.

HOW GENOMES EVOLVE

TERMS TO LEARN

allele frequency	pseudogene	single-nucleotide polymorphism (SNP)
homologous	purifying selection	synteny
ploidy		

The material dealt with in this section has been covered in bits and pieces in other chapters in the book. Relevant problems are cross-referenced below.

Chapter 1, Problems 1–34 to 1–38; Chapter 2, Problem 2–10; Chapter 4, Problems 4–24, 4–26, 4–30, and 4–43; Chapter 5, Problems 5–53, 5–83, and 5–84; Chapter 6, Problems 6–80 to 6–88.

MANIPULATING PROTEINS, DNA, AND RNA

ISOLATING CELLS AND GROWING THEM IN CULTURE

TERMS TO LEARN

cell line	hybridoma	monoclonal antibody
heterocaryon		

8–1 A common step in the isolation of cells from a sample of animal tissue is to treat it with trypsin, collagenase, and EDTA. Why is such a treatment necessary and what does each component accomplish? And why doesn't this treatment kill the cells?

8–2* You want to isolate rare cells that are present in a population at a frequency of 1 in 10^5 cells, and you need 10 of those cells to do an experiment. If your fluorescence-activated cell sorter can sort cells at the rate of 1000 per second, how long would it take to collect enough rare cells for your experiment?

8–3 Isolation of cells from tissues, fluorescence-activated cell sorting, and laser capture microdissection are just a few of the ways for generating homogeneous cell populations. Why do you suppose it is important to have a homogeneous cell population for many experiments?

8–4* Distinguish among the terms 'primary culture,' 'secondary culture,' and 'cell line.'

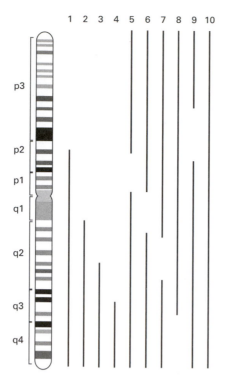

Figure 8–1 Mapping the gene for the FeLV-C receptor using human–rodent hybrid cell lines carrying portions of human chromosome 1 (Problem 8–6). The hybrid cell lines that could be infected by FeLV-C are shown, with the portions of human chromosome 1 retained in the individual hybrid cell lines indicated as *vertical lines*. The p and q designations refer to a standard convention (the Paris nomenclature) for describing chromosome positions. Short arm locations are labeled p (*petit*) and long arms q (*queue*). Each chromosome arm is divided into regions labeled p1, p2, p3, q1, q2, q3, etc., counting outwards from the centromere. Regions are delimited by specific landmarks, which are distinct morphological features, including the centromere and certain prominent bands. Regions are divided into bands labeled p11 (one-one, not eleven), p12, etc. Sub-bands are designated p11.1, p11.2, etc., and sub-sub bands are designated p11.11, p11.12, etc. In all cases the numbers increase from the centromere toward the telomere.

8–5 Why are human embryonic stem cell lines thought to be especially promising from a medical perspective?

8–6* Panels of human–rodent cell hybrids that retain one or a few human chromosomes, parts of human chromosomes, or radiation-induced fragments of human chromosomes have proven enormously useful in mapping genes to defined locations. Now that the human genome has been sequenced, it is a trivial matter to know a gene's location if you have a bit of sequence from the gene, for example, from a cDNA clone. Nevertheless, there are many instances in which such panels of cells are still invaluable.

You wish to map the location of the receptor for feline leukemia virus type C (FeLV-C), which will infect human cells but not rodent cells. Using a panel of hybrid cells carrying whole chromosomes, you have shown that FeLV-C infects only those hybrids carrying human chromosome 1. Using a second panel carrying portions of chromosome 1, you show that FeLV-C infects several of the hybrid cell lines. The segments of chromosome 1 that are present in the infectable hybrid cell lines are shown in Figure 8–1. Where on chromosome 1 is the gene for the FeLV-C receptor located?

8–7 Consider the following two statements. "The most important advantage of the hybridoma technique is that monoclonal antibodies can be made against molecules that constitute only a minor component of a complex mixture." "The most important advantage of the hybridoma technique is that antibodies that may be present as only minor components in conventional antiserum can be produced in quantity in pure form as monoclonal antibodies." Are these two statements equivalent? Why or why not?

8–8* (**True/False**) Because a monoclonal antibody recognizes a specific antigenic site (epitope), it binds to a specific protein. Explain your answer.

8–9 Do you suppose it would be possible to raise an antibody against another antibody? Explain your answer.

Problem 7–59 uses hybrid cell lines carrying either an active or an inactive X chromosome to investigate the mechanism of X-inactivation.

Problems 12–39, 12–71, and 17–26, respectively, exploit heterocaryons to define aspects of nuclear import, to assign mutants in peroxisome biogenesis to complementation groups, and to investigation regulation of the cell cycle.

Problem 3–51 examines the use of monoclonal antibodies for affinity purification of proteins.

Problem 12–73 uses monoclonal antibodies to elucidate the mechanism of protein import into peroxisomes.

Problems 15–12 and 15–13 investigate integrin binding and signaling using monoclonal antibodies with special properties.

FRACTIONATION OF CELLS

8–10* Describe how you would use preparative centrifugation to purify mitochondria from a cell homogenate.

8–11 Distinguish between velocity sedimentation and equilibrium sedimentation. For what general purpose is each technique used? Which do you suppose might be best suited for separating two proteins of different size?

8–12* (**True/False**) It is possible to pellet hemoglobin by centrifugation at sufficiently high speed. Explain your answer.

8–13 Tropomyosin, at 93 kd, sediments at 2.6 S, whereas the 65-kd protein, hemoglobin, sediments at 4.3 S. (The sedimentation coefficient S is a linear measure of the rate of sedimentation.) These two proteins are shown as α-carbon backbone models in Figure 8–2. How is it that the bigger protein sediments slower than the smaller one? Can you think of an analogy from everyday experience that might help you with this problem?

> Problems 5–24, 6–65, 6–67, and 15–34 give examples of the use of velocity sedimentation.

8–14* In the classic paper that demonstrated the semi-conservative replication of DNA, Meselson and Stahl began by showing that DNA itself will form a band when subjected to equilibrium sedimentation. They mixed randomly fragmented *E. coli* DNA with a solution of CsCl so that the final solution had a density of 1.71 g/mL. As shown in Figure 8–3, with increasing length of centrifugation at 70,000 times gravity, the DNA, which was initially dispersed throughout the centrifuge tube, became concentrated over time into a discrete band in the middle.

> Problems 5–14, 5–42, 10–18, and 14–75 give examples of the use of equilibrium sedimentation.

A. Describe what is happening with time and explain why the DNA forms a discrete band.
B. What is the buoyant density of the DNA? (The density of the solution at which DNA 'floats' at equilibrium defines the 'buoyant density' of the DNA.)
C. Even if the DNA were centrifuged for twice as long—or even longer—the width of the band remains about what is shown at the bottom of Figure 8–3. Why doesn't the band become even more compressed? Suggest some possible reasons to explain the thickness of the DNA band at equilibrium.

8–15 Distinguish ion-exchange chromatography, hydrophobic chromatography, gel-filtration chromatography, and affinity chromatography in terms of the column material and the basis for separation of a mixture of proteins.

8–16* The purification of a protein usually requires multiple steps and often involves several types of column chromatography. A key component of any purification is an assay for the desired protein. The assay can be a band on a gel, a structure in the electron microscope, the ability to bind to another molecule, or an enzymatic activity. The purification of an enzyme is particularly instructive because the assay allows one to quantify the extent of purification at each step. Consider the purification of the enzyme shown in

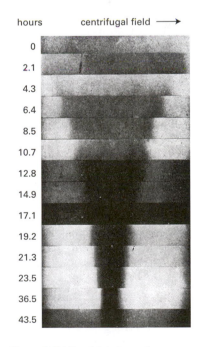

hours   centrifugal field ⟶

0
2.1
4.3
6.4
8.5
10.7
12.8
14.9
17.1
19.2
21.3
23.5
36.5
43.5

Figure 8–3 Ultraviolet absorption photographs showing successive stages in the banding of *E. coli* DNA (Problem 8–14). DNA, which absorbs UV light, shows up as *dark regions* in the photographs. The bottom of the centrifuge tube is on the *right*.

Figure 8–2 Backbone models of tropomyosin and hemoglobin (Problem 8–13).

hemoglobin   tropomyosin

TABLE 8–1 Purification of an enzyme (Problem 8–16).

PROCEDURE	TOTAL VOLUME (mL)	TOTAL PROTEIN (mg)	TOTAL ACTIVITY (units)	SPECIFIC ACTIVITY (units/mg)
1. Crude extract	2,000	15,000	150,000	
2. Ammonium sulfate precipitation	320	4,000	140,000	
3. Ion-exchange chromatography	100	550	125,000	
4. Gel-filtration chromatography	85	120	105,000	
5. Affinity chromatography	8	5	75,000	

Table 8–1. The total volume, total protein, and total enzyme activity are shown at each step.

A. For each step in the purification procedure, calculate the specific activity of the enzyme (units of activity per mg of protein). How can you tell that purification has occurred at each step?

B. Which of the purification steps was most effective? Which was least effective?

C. If you were to carry the purification through additional steps, how would the specific activity change? How could you tell from specific activity measurements that the enzyme was pure? How might you check on that conclusion?

D. If the enzyme is pure at the end of the purification scheme in Table 8–1, what proportion of the protein in the starting cell does it represent?

8–17 The result of gel-filtration chromatography of six, roughly spherical proteins is shown in Figure 8–4. The identities of the proteins, their molecular masses, and their elution volumes are indicated in Table 8–2.

A. Why do the smaller proteins come off the column later than the larger proteins?

B. Plot molecular mass versus elution volume. Now plot the $\log_{10}$ of the molecular mass versus the elution volume. Which plot gives a straight line? What do you suppose is the basis for that result?

8–18* (**True/False**) If the beads used in gel-filtration chromatography had pores of a uniform size, proteins would either be excluded from or included in the pores, but would not be further fractionated. Explain your answer.

8–19 In preliminary studies you've determined that your partially purified protein is stable (retains activity) between pH 5.0 and 7.5. On either side of that pH range the protein is no longer active. Your advisor now wants you to do a quick experiment to determine conditions for ion-exchange chromatography. He has left instructions for you. First, you're supposed to mix a bit of the crude preparation with a small amount of the ion-exchange resin DEAE-Sepharose in a series of buffer solutions that have a pH between 5.0 and 7.5. Next, you are to pellet the resin and assay the supernatant for the presence of your protein. Finally, he tells you to use this information to pick the proper pH to do the ion-exchange chromatography. You have completed the first two steps and have obtained the results shown in Figure 8–5. But you are a

Problems 6–14 and 6–53 give examples of column chromatography and HPLC, respectively.

Problems 3–51 and 15–92 show examples of affinity chromatography for protein purification.

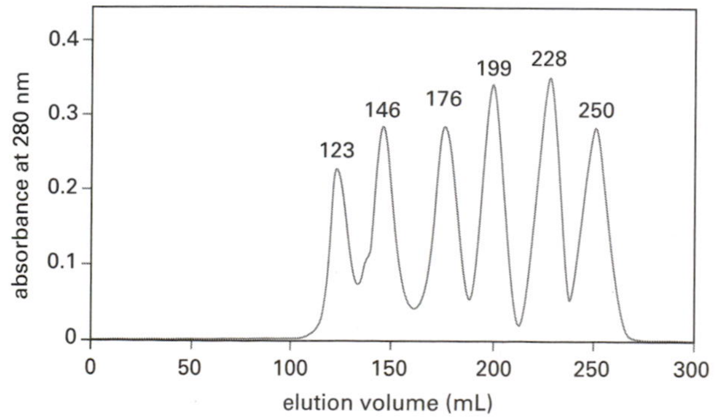

Figure 8–4 Elution profile for proteins fractionated by gel-filtration chromatography (Problem 8–17). The absorbance at 280 nm is a measure of protein concentration. Each of the peaks is identified by its elution volume.

TABLE 8–2 Proteins separated by gel-filtration chromatography (Problem 8–17).

PROTEIN	MOLECULAR MASS (kd)	MOLECULAR MASS (log₁₀)	ELUTION VOLUME (mL)
Ribonuclease A	13	4.14	250
Chymotrypsinogen	25	4.40	228
Ovalbumin	43	4.63	199
Bovine serum albumin	67	4.83	176
Aldolase	158	5.20	146
Catalase	232	5.37	123

little uncertain as to how to use the information to pick the pH for the chromatography.

A. At which end of the pH range is the charge on your protein more positive and at which end is it more negative? (Over this pH range the positively charged amine groups on the DEAE-Sepharose beads are unaffected.)

B. For the chromatography, should you pick a pH at which the protein binds to the beads (pH 6.5 to 7.5) or a pH where it does not bind (pH 5.0 to 6.0)? Explain your choice.

C. Should you pick a pH close to the boundary (that is, pH 6.0 or 6.5) or far away from the boundary (that is, pH 5.0 or pH 7.5)? Explain your reasoning.

D. How will you carry out ion-exchange chromatography of your protein? What are the various steps you will use to accomplish the separation of your protein from others via ion-exchange chromatography?

8–20* How is it that smaller molecules move through a gel-filtration column more slowly than larger molecules, whereas in SDS polyacrylamide-gel electrophoresis (SDS-PAGE) the opposite is true: larger molecules move more slowly?

8–21 You are all set to run your first SDS polyacrylamide-gel electrophoresis. You have boiled your samples of protein in SDS in the presence of mercaptoethanol and loaded them into the wells of a polyacrylamide gel. You are now ready to attach the electrodes. Uh oh, does the positive electrode (the anode) go at the top of the gel, where you loaded your proteins, or at the bottom of the gel?

8–22* You hate the smell of mercaptoethanol. And since there are no disulfide bonds in intracellular proteins (see Problem 3–33), you have convinced yourself that it is not necessary to treat a cytoplasmic homogenate with mercaptoethanol prior to SDS-PAGE. You heat a sample of your homogenate in SDS and subject it to electrophoresis. Much to your surprise, your gel looks horrible; it is an ugly smear! You show your result to a fellow student with a background in chemistry, and she suggests that you treat your sample with N-ethyl maleimide (NEM), which reacts with free sulfhydryls. You run another sample of your homogenate after treating it with NEM and SDS. Now the gel looks perfect!

If intracellular proteins don't have disulfide bonds—and they don't—why didn't your original scheme work? And how does treatment with NEM correct the problem?

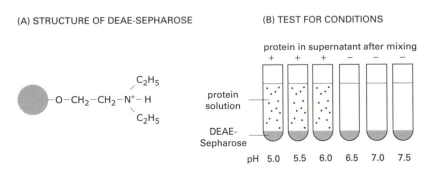

(A) STRUCTURE OF DEAE-SEPHAROSE

(B) TEST FOR CONDITIONS

protein in supernatant after mixing

protein solution

DEAE-Sepharose

pH 5.0 5.5 6.0 6.5 7.0 7.5

Figure 8–5 Preliminary test to determine conditions for ion-exchange chromatography (Problem 8–19). (A) Structure of the charged amine groups attached to Sepharose beads. (B) Results of mixing your protein with DEAE-Sepharose beads. Samples of the protein were mixed with DEAE-Sepharose beads in buffers at a range of pH values, and then the mixtures were centrifuged to pellet the beads. The presence of the protein in the supernatant is indicated by a +; its absence is indicated by a –.

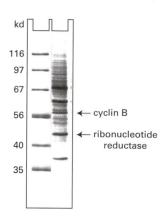

Figure 8–6 Autoradiograph of radiolabeled proteins separated by SDS-PAGE (Problem 8–23). A set of radiolabeled marker proteins with known molecular masses is shown in the *left hand* lane, along with their molecular masses in kd. Radiolabeled proteins from a sea urchin egg extract are shown in the *right hand* lane. *Arrows* mark the bands that correspond to cyclin B and the small subunit of ribonucleotide reductase.

8–23 Figure 8–6 shows an autoradiograph of an SDS-PAGE separation of radiolabeled proteins in a cell-free extract of sea urchin eggs. Alongside are shown a set of radiolabeled marker proteins of defined molecular mass. Two bands that contain known proteins—the small subunit of ribonucleotide reductase and cyclin B—are indicated.

A. Do the standard set of proteins migrate at a rate that is inversely proportional to their molecular masses? That is to say, would you expect a protein of 35 kd, for example, to migrate twice as far down the gel as a protein of 70 kd?

B. How would you use the standard set of proteins to estimate the molecular masses of ribonucleotide reductase and cyclin B? What would you estimate the molecular masses of these two proteins to be?

C. The sequences of the genes for these two proteins give molecular masses of 44 kd for ribonucleotide reductase and 46 kd for cyclin B. Can you offer some possible reasons why the SDS-PAGE estimate for the molecular mass of cyclin B was so far off?

8–24* How many copies of a protein need to be present in a cell in order for it to be visible as a band on a gel? Assume that you can load 100 µg of cell extract onto a gel and that you can detect 10 ng in a single band by silver staining. The concentration of protein in cells is about 200 mg/mL (see Problem 2–53), and a mammalian cell has a volume of about 1000 µm^3 and a bacterium a volume of about 1 µm^3. Given these parameters, calculate the number of copies of a 120-kd protein that would need to be present in a mammalian cell and in a bacterium in order to give a detectable band on a gel. You might try an order-of-magnitude guess before you make the calculations.

8–25 For separation of proteins by two-dimensional polyacrylamide-gel electrophoresis, what are the two types of electrophoresis that are used in each dimension? Do you suppose it makes any difference which electrophoretic method is applied first? Why or why not?

8–26* You want to know the sensitivity for detection of immunoblotting (Western blotting), using an enzyme-linked second antibody to detect the first antibody directed against your protein (Figure 8–7A). You are using the mouse monoclonal antibody 4G10, which is specific for phosphotyrosine residues, to detect phosphorylated proteins. You first phosphorylate the myelin basic protein *in vitro* using a tyrosine protein kinase that adds one phosphate per molecule. You then prepare a dilution series of the phosphorylated protein

Problems 7–4, 16–30, and 16–86 use two-dimensional gel electrophoresis to separate proteins.

Problem 5–44 illustrates the use of two-dimensional gel electrophoresis of DNA in the analysis of origins of DNA replication.

Problems 3–45, 6–77, 12–59, 12–73, and 15–79 give examples of various uses of immunoblotting in cell biological experiments.

(A) SCHEMATIC DIAGRAM

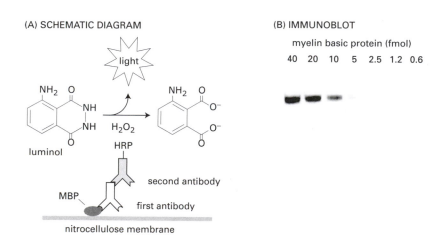

(B) IMMUNOBLOT

myelin basic protein (fmol)

40 20 10 5 2.5 1.2 0.6

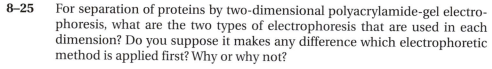

Figure 8–7 Sensitivity of detection of immunoblotting (Problem 8–26). (A) Schematic diagram of experiment. MBP stands for myelin basic protein. In the presence of hydrogen peroxide, HRP converts luminol to a chemiluminescent molecule that emits light, which is detected by exposure of an x-ray film. (B) Exposed film of an immunoblot. The number of femtomoles of myelin basic protein in each band is indicated.

and subject the samples to SDS-PAGE. The protein is then transferred (blotted) onto a nitrocellulose filter, incubated with the 4G10 antibody, and washed to remove unbound antibody. The blot is then incubated with a second goat anti-mouse antibody that carries horseradish peroxidase (HRP) conjugated to it, and any excess unbound antibody is again washed away. You place the blot in a thin plastic bag, add reagents that chemiluminesce when they react with HRP (Figure 8–7A), and place the bag against a sheet of x-ray film. When the film is developed you see the picture shown in Figure 8–7B.

A. Given the amounts of phosphorylated myelin basic protein indicated in each lane in Figure 8–7B, calculate the detection limit of this method in terms of molecules of protein per band.

B. Assuming that you were using monoclonal antibodies to detect proteins, would you expect that the detection limit would depend on the molecular mass of the protein? Why or why not?

8–27 How many molecules of your labeled protein are required for detection by autoradiography? You added 1 μL containing 10 μCi of γ-^{32}P-ATP (a negligible amount of ATP) to 9 μL of cell extract that had an ATP concentration of 1 mM. You incubated the mixture to allow transfer of phosphate to proteins in the extract. You then subjected 1 μL of the mixture to SDS-PAGE, dried the gel, and placed it against a sheet of x-ray film. After an overnight exposure you saw a detectable band in the location of your protein. You know from previous experience that a protein labeled at 1 count per minute per band (1 cpm equals 1 disintegration per minute, dpm, for ^{32}P) will form such a band after an overnight exposure. How many molecules of labeled protein are in the band, if you assume 1 phosphate per molecule? (Some useful conversion factors for radioactivity are shown on the inside of the front cover.)

8–28* You just developed an autoradiograph after a two-week exposure. You had incubated your protein with a cell-cycle kinase in the presence of ^{32}P-ATP in hopes of demonstrating that it was indeed a substrate for the kinase. You see the hint of a band on the gel at the right position, but it is just too faint to be convincing. You show your result to your advisor and tell him that you've put the blot against a fresh sheet of film, which you plan to expose for a longer period of time. He gives you a sideways look and tells you to do the experiment over again and use more radioactivity. What's wrong with your plan to reexpose the blot for a longer time?

8–29 (**True/False**) Given the inexorable progress of technology, it seems inevitable that the sensitivity of detection of molecules will ultimately be pushed beyond the yoctomole level (10^{-24} mole). Explain your answer.

Problems 6–17, 14–17, and 16–68 describe the analysis of single molecules of RNA polymerase, ATP synthase, and a kinesin motor, respectively.

8–30* You have isolated the proteins from two adjacent spots after two-dimensional polyacrylamide-gel electrophoresis and digested them with trypsin. When the masses of the peptides were measured by MALDI-TOF mass spectrometry, the peptides from the two proteins were found to be identical except for one (Figure 8–8). For this peptide, the *m/z* values differed by 79.97, a value that does not correspond to a difference in amino acid sequence. (For example, glutamic acid instead of valine at one position would give an *m/z* difference of around 30.) Can you suggest a possible difference between the two peptides that might account for the observed *m/z* difference?

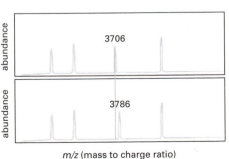

Figure 8–8 Masses of peptides measured by MALDI-TOF mass spectrometry (Problem 8–30).

ISOLATING, CLONING, AND SEQUENCING DNA

TERMS TO LEARN

bacterial artificial chromosome (BAC)	expression vector	plasmid vector
cDNA clone	genomic DNA clone	polymerase chain reaction (PCR)
cDNA library	genomic DNA library	probe
dideoxy sequencing method	hybridization	recombinant DNA
DNA cloning	*in situ* hybridization	restriction nuclease
	Northern blotting	Southern blotting

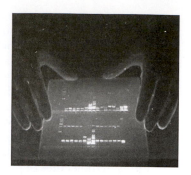

Figure 8–9 DNA fragments separated by gel electrophoresis and stained with ethidium bromide (Problem 8–32). Each of the *bright bands* on the gel represents a site where fragments of DNA have migrated during electrophoresis. Ethidium intercalates between base-pairs in double-stranded DNA. Removal of ethidium from the aqueous environment and fixing its orientation in the nonpolar environment of DNA enhance its fluorescence dramatically. When irradiated with long-wavelength UV light, it fluoresces a bright orange.

8–31 (**True/False**) Bacteria that make a specific restriction nuclease for defense against viruses have evolved in such a way that their own genome does not contain the recognition sequence for that nuclease. Explain your answer.

8–32* Figure 8–9 shows a picture of DNA fragments that have been separated by gel electrophoresis and then stained by ethidium bromide, a molecule that fluoresces intensely under long wavelength UV light when it is bound to DNA. Such gels are a standard way of detecting the products of cleavage by restriction nucleases. For the DNA fragment shown in Figure 8–10, decide whether it will be cut by the restriction nucleases EcoRI (5′-GAATTC), AluI (5′-AGCT), and PstI (5′-CTGCAG). For those that cut the DNA, how many products will be produced?

8–33 The restriction nucleases BamHI and PstI cut their recognition sequences as shown in Figure 8–11.
A. Indicate the 5′ and 3′ ends of the cut DNA molecules.
B. How would the ends be modified if you incubated the cut molecules with DNA polymerase in the presence of all four dNTPs?
C. After the reaction in part B, could you still join the BamHI ends together by incubation with T4 DNA ligase? Could you still join the PstI ends together? (T4 DNA ligase will join blunt ends together as well as cohesive ends.)
D. Will joining of the ends in part C regenerate the BamHI site? Will it regenerate the PstI site?

8–34* The restriction nuclease Sau3A recognizes the sequence 5′-GATC and cleaves on the 5′ side (to the left) of the G. (Since the top and bottom strands of most restriction sites read the same in the 5′-to-3′ direction, only one strand of the site need be shown.) The single-stranded ends produced by Sau3A cleavage are identical to those produced by BamHI cleavage (see Figure 8–11), allowing the two types of ends to be joined together by incubation with DNA ligase. (You may find it helpful to draw out the product of this ligation to convince yourself that it is true.)
A. What fraction of BamHI sites can be cut with Sau3A? What fraction of Sau3A sites can be cut with BamHI?
B. If two BamHI ends are ligated together, the resulting site can be cleaved again by BamHI. The same is true for two Sau3A ends. However, suppose you ligate a Sau3A end to a BamHI end. Can the hybrid site be cut with Sau3A? Can it be cut with BamHI?
C. What is the average size of DNA fragments produced by digestion of chromosomal DNA with Sau3A? What's the average size with BamHI?

8–35 You have purified two DNA fragments that were generated by BamHI digestion of recombinant DNA plasmids. One fragment is 400 nucleotide pairs, and the other is 900 nucleotide pairs. You want to join them together as shown in Figure 8–12 to create a hybrid gene, which, if your speculations are right, will have amazing new properties.

```
5′-AAGAATTGCGGAATTCGAGCTTAAGGGCCGCGCCGAAGCTTTAAA-3′
3′-TTCTTAACGCCTTAAGCTCGAATTCCCGGCGCGGCTTCGAAATTT-5′
```

Figure 8–10 A segment of double-stranded DNA (Problem 8–32).

(A) BamHI CLEAVAGE

```
5′            ↓              3′
-- G G A T C C --
-- C C T A G G --
3′                ↑          5′
```

↓ BamHI

```
-- G              G A T C C --
-- C C T A G              G --
```

(B) PstI CLEAVAGE

```
5′                  ↓        3′
-- C T G C A G --
-- G A C G T C --
3′        ↑                  5′
```

↓ PstI

```
-- C T G C A              G --
-- G              A C G T C --
```

Figure 8–11 Restriction nuclease cleavage of DNA (Problem 8–33). (A) BamHI cleavage. (B) PstI cleavage. Only the nucleotides that form the recognition sites are shown.

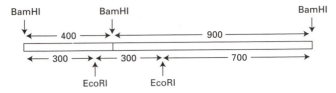

Figure 8–12 Final structure of the desired hybrid gene (Problem 8–35).

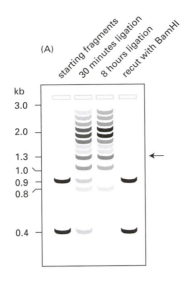

You mix the two fragments together and incubate them in the presence of DNA ligase. After 30 minutes and again after 8 hours, you remove samples and analyze them by gel electrophoresis. You are surprised to find a complex pattern of fragments instead of the 1.3-kb recombinant molecule of interest (Figure 8–13A). You notice that with longer incubation the smaller fragments diminish in intensity and the larger ones become more intense. If you cut the ligated mixture with BamHI, you regenerate the starting fragments (Figure 8–13A).

Puzzled, but undaunted, you purify the 1.3-kb fragment from the gel (arrow in Figure 8–13A) and check its structure by digesting a sample of it with BamHI. As expected, the original two bands are regenerated (Figure 8–13B). Just to be sure it is the structure you want, however, you digest another sample with EcoRI. You expected this digestion to generate two fragments of 300 nucleotides and one fragment of 700 nucleotides. Once again you are surprised by the complexity of the gel pattern (Figure 8–13B).

A. Why are there so many bands in the original ligation mixture?

B. Why are so many fragments produced by EcoRI digestion of the pure 1.3-kb fragment?

8–36* The restriction nuclease EcoRI recognizes the sequence 5′-GAATTC and cleaves between the G and A to leave 5′ protruding single strands (like BamHI, see Figure 8–11). PstI, on the other hand, recognizes the sequence 5′ -CTGCAG and cleaves between the A and G to leave 3′ protruding single strands (see Figure 8–11). These two recognition sites are displayed on the helical representations of DNA in Figure 8–14.

A. For each restriction site indicate the position of cleavage on each strand of the DNA.

B. From the positions of the cleavage sites decide for each restriction nuclease whether you expect it to approach the recognition site from the major-groove side or from the minor-groove side.

8–37 Which, if any, of the restriction nucleases listed in Table 8–3 will *definitely* cleave a segment of cDNA that encodes the peptide KIGPACF?

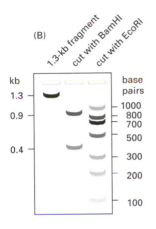

Figure 8–13 Construction of the hybrid gene (Problem 8–35). (A) Ligation of pure DNA 0.9-kb and 0.4-kb fragments. The position of the 1.3-kb fragment is marked by an *arrow*. (B) Diagnostic digestion of the purified 1.3-kb fragment.

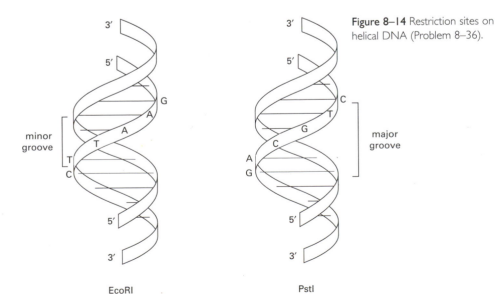

Figure 8–14 Restriction sites on helical DNA (Problem 8–36).

TABLE 8–3 A set of restriction nucleases and their recognition sequences (Problem 8–37).

RESTRICTION NUCLEASE	RECOGNITION SEQUENCE
AluI	AGCT
Sau96I	GGNCC
HindIII	AAGCTT
N stands for any nucleotide.	

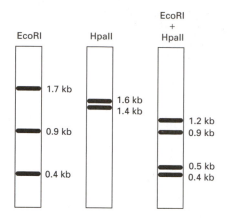

Figure 8–15 Sizes of DNA bands produced by digestion of a 3.0-kb fragment by EcoRI, Hpall, and a mixture of the two (Problem 8–38). Sizes of the fragments are shown in kilobases.

8–38* You wish to make a restriction map of a 3.0-kb BamHI restriction fragment. You digest three samples of the fragment with EcoRI, HpaII, and a mixture of EcoRI and HpaII. You then separate the fragments by gel electrophoresis and visualize the DNA bands by staining with ethidium bromide (Figure 8–15). From these results, prepare a restriction map that shows the relative positions of the EcoRI and HpaII recognition sites and the distances in kilobases (kb) between them.

8–39 *Tetrahymena* is a ciliated protozoan with two nuclei. The smaller nucleus (the micronucleus) maintains a master copy of the cell's chromosomes. The micronucleus participates in sexual conjugation, but not in day-to-day gene expression. The larger nucleus (the macronucleus) maintains a 'working' copy of the cell's genome in the form of a large number of gene-sized double-stranded DNA fragments (minichromosomes), which are actively transcribed. The minichromosome that contains the ribosomal RNA genes is present in many copies. It can be separated from the other minichromosomes by centrifugation and studied in detail.

When examined by electron microscopy, each ribosomal minichromosome is a linear structure 21 kb in length. Ribosomal minichromosomes also migrate at 21 kb when subjected to gel electrophoresis (Figure 8–16, lane 1). However, if the minichromosome is cut with the restriction nuclease BglII, the two fragments that are generated (13.4 kb and 3.8 kb) do not sum to 21 kb (lane 2). When the DNA is cut with other restriction enzymes, the sizes of the fragments always sum to less than 21 kb; moreover, the fragments in each digest add up to different overall lengths.

If uncut minichromosomes are first denatured and reannealed before they are run on a gel, the 21-kb fragments are replaced by double-stranded fragments exactly half their length, 10.5 kb (Figure 8–16, lane 3). Similarly, if the BglII-cut minichromosomes are denatured and reannealed, the 13.4-kb fragments are replaced by double-stranded fragments half their length, 6.7 kb (lane 4).

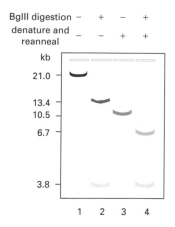

Figure 8–16 Restriction analysis of the *Tetrahymena* ribosomal minichromosome (Problem 8–39). Numbers at the *left* indicate the sizes of the bands in kilobases.

Explain why the restriction fragments do not appear to add up to 21 kb. Why does the electrophoretic pattern change when the DNA is denatured and reannealed? What do you think might be the overall organization of sequences in the ribosomal minichromosome?

8–40* You have cloned a 4-kb segment of a gene into a plasmid vector (Figure 8–17) and now wish to prepare a restriction map of the gene in preparation for other DNA manipulations. Your advisor left instructions on how to do it, but she is now on vacation, so you are on your own. You follow her instructions, as outlined below.
1. Cut the plasmid with EcoRI.
2. Add a radioactive label to the EcoRI ends.
3. Cut the labeled DNA with BamHI.
4. Purify the insert away from the vector.
5. Digest the labeled insert briefly with a restriction nuclease so that on average each labeled molecule is cut about one time.
6. Repeat step 5 for several different restriction nucleases.
7. Run the partially digested samples side by side on an agarose gel.
8. Place the gel against x-ray film so that fragments with a radioactive end can expose the film to produce an autoradiograph.
9. Draw the restriction map.
   Your biggest problem thus far has been step 5; however, by decreasing the amounts of nuclease and lowering the temperature, you were able to find conditions for partial digestion. You have now completed step 8, and your autoradiograph is shown in Figure 8–18.
   Unfortunately, your advisor was not explicit about how to construct a map from the data in the autoradiograph. She is due back tomorrow. Will you figure it out in time?

8–41 (True/False) Pulsed-field gel electrophoresis uses a strong electric field to separate very long DNA molecules, stretching them out so that they travel end-first through the gel at a rate that depends on their length. Explain your answer.

8–42* If you add DNA to wells at the top of a gel, should you place the positive electrode (anode) at the top or at the bottom of the gel? Explain your choice.

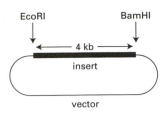

Figure 8–17 Recombinant plasmid containing a cloned DNA segment (Problem 8–40).

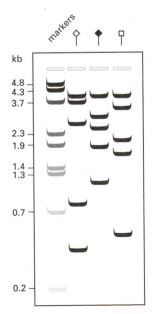

Figure 8–18 Autoradiograph showing the electrophoretic separation of the labeled fragments after partial digestion with the three restriction nucleases represented by the *symbols* (Problem 8–40). Numbers at the *left* indicate the sizes of a set of marker fragments in kilobases.

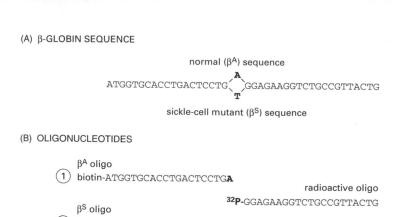

(A) β-GLOBIN SEQUENCE

normal (β^A) sequence

A
ATGGTGCACCTGACTCCTG⟨ ⟩GGAGAAGGTCTGCCGTTACTG
T

sickle-cell mutant (β^S) sequence

(B) OLIGONUCLEOTIDES

β^A oligo
① biotin-ATGGTGCACCTGACTCCTG**A**

radioactive oligo
③ ^32**P**-GGAGAAGGTCTGCCGTTACTG

β^S oligo
② biotin-ATGGTGCACCTGACTCCTG**T**

(C) ASSAY

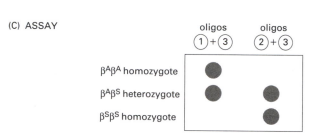

oligos ①+③   oligos ②+③

β^Aβ^A homozygote

β^Aβ^S heterozygote

β^Sβ^S homozygote

Figure 8–19 Oligonucleotide-ligation assay (Problem 8–40). (A) Sequence of the β-globin gene around the site of the sickle-cell (β^S) mutation. (B) Specific oligonucleotides for ligation assay. (C) Assays to detect the single-nucleotide difference between the β^A and β^S sequences. After hybridization to patient DNA, biotinylated oligonucleotides were collected in a spot on a sheet of filter paper and exposed to x-ray film to detect radioactivity, which turns the film black.

8–43 Many mutations that cause human genetic diseases involve the substitution of one nucleotide for another, as is the case for sickle-cell anemia (Figure 8–19A). An assay based on ligation of oligonucleotides provides a rapid way to detect such specific single-nucleotide differences. This assay uses pairs of oligonucleotides: for each pair, one oligonucleotide is labeled with biotin and the other with a radioactive (or fluorescent) tag. In the assay shown in Figure 8–19B for the detection of the mutation responsible for sickle-cell anemia, two pairs of oligonucleotides are hybridized to DNA from an individual and incubated in the presence of DNA ligase. Biotinylated oligonucleotides are then bound to streptavidin on a solid support and any associated radioactivity is visualized by autoradiography, as shown in Figure 8–19C.

A. Do you expect the β^A and β^S oligonucleotides to hybridize to both β^A and β^S DNA?

B. How does this assay distinguish between β^A and β^S DNA?

8–44* The DNA of certain animal viruses can integrate into a cell's DNA as shown schematically in Figure 8–20. You want to know the structure of the viral genome as it exists in the integrated state. You digest samples of viral DNA and DNA from cells that contain the integrated virus with restriction nucleases that cut the viral DNA at known sites (Figure 8–21A). Subsequently you separate the fragments by electrophoresis on agarose gels and visualize the bands that contain viral DNA by Southern blotting, using radioactive viral DNA as a hybridization probe. You obtain the patterns shown in Figure 8–21B.

From this information, decide in which of the five segments of the viral genome (labeled *a* to *e* in Figure 8–21A) the integration event occurred.

8–45 (**True/False**) Imagine that RNA or DNA molecules in a crude mixture are separated by electrophoresis and then hybridized to a probe. If molecules of only one size become labeled, one can be fairly certain that the hybridization was specific. Explain your answer.

(A) MAP OF VIRAL DNA

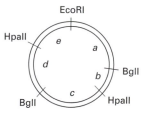

EcoRI

Hpall

e   *a*

d

b   Bgll

c

Bgll   Hpall

(B) RESTRICTION DIGESTS

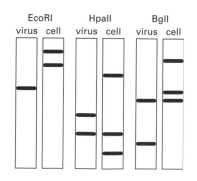

EcoRI   Hpall   Bgll
virus cell virus cell virus cell

Figure 8–21 Southern blots of viral and cell DNA digested with various restriction nucleases and incubated with a viral DNA probe (Problem 8–44). (A) Restriction sites on the viral genome. The DNA segments defined by these sites are indicated by the letters *a* to *e*. (B) Restriction digests of viral DNA and cellular DNA. Agarose gels separate DNA fragments on the basis of size—the smaller the fragment, the farther it moves toward the *bottom* of the gel.

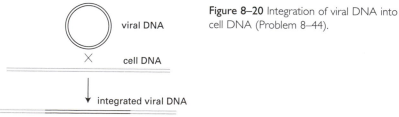

viral DNA

cell DNA

integrated viral DNA

Figure 8–20 Integration of viral DNA into cell DNA (Problem 8–44).

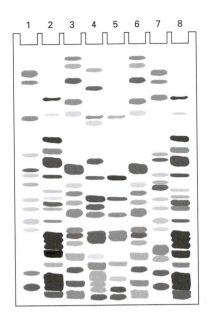

Figure 8–22 DNA fingerprint analysis of shuffled twins (Problem 8–46).

8–46* There has been a colossal snafu in the maternity ward at your local hospital. Four sets of male twins, born within an hour of each other, were inadvertently shuffled in the excitement occasioned by that unlikely event. You have been called in to set things right. As a first step, you want to get the twins matched up. To that end you analyze a small blood sample from each infant using a hybridization probe that detects variable number tandem repeat (VNTR) polymorphisms located in widely scattered regions of the genome. The results are shown in Figure 8–22.

A. Which infants are brothers?

B. How will you match brothers to the correct parents?

8–47 Almost all the cells in an individual animal contain identical genomes. In an experiment, a tissue composed of multiple cell types is fixed and subjected to *in situ* hybridization with a DNA probe to a particular gene. To your surprise, the hybridization signal is much stronger in some cells that in others. Explain this result.

8–48* You want to clone a DNA fragment that has KpnI ends into a vector that has BamHI ends. The problem is that BamHI and KpnI ends are not compatible: BamHI leaves a 5′ overhang and KpnI leaves a 3′ overhang (Figure 8–23). A friend suggests that you try to link them with an oligonucleotide 'splint' as shown in Figure 8–23. It is not immediately clear to you that such a scheme will work because ligation requires an adjacent 5′ phosphate and 3′ hydroxyl. Although molecules that are cleaved with restriction nucleases have appropriate ends, oligonucleotides are synthesized with hydroxyl groups at both ends. Also, although the junction in Figure 8–23 is BamHI–KpnI, the other junction is KpnI–BamHI, and you are skeptical that the same oligonucleotide could splint both junctions.

A. Draw a picture of the KpnI–BamHI junction and the oligo splint that would be needed. Is this oligonucleotide the same or different from the one shown in Figure 8–23?

B. Draw a picture of the molecule after treatment with DNA ligase. Indicate which if any of the nicks will be ligated.

C. Will your friend's scheme work?

8–49 You are constructing a cDNA library in a high-efficiency cloning vector called λYES (Yeast–*E. coli* Shuttle). This vector links a bacteriophage lambda genome to a plasmid that can replicate in *E. coli* and yeast. It combines the advantages of viral and plasmid cloning vectors. cDNAs can be inserted into the plasmid portion of the vector, which can then be packaged into a virus coat *in vitro*. The packaged vector DNA infects *E. coli* much more efficiently than plasmid DNA on its own. Once inside *E. coli*, the plasmid sequence in λYES can be induced to recombine out of the lambda genome and replicate

Problems 4–60, 5–45, 7–3, and 17–38 provide additional examples of Southern blotting.

Problems 7–59 and 12–66 give examples of Northern blotting.

Problem 15–98 employs *in situ* hybridization in *Drosophila* embryos to investigate cell signaling in development.

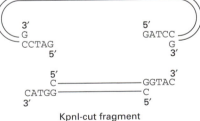

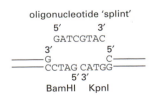

Figure 8–23 Scheme to use an oligonucleotide 'splint' to link incompatible restriction ends (Problem 8–48).

on its own. This allows cDNAs to be isolated on plasmids, which are ideal for subsequent manipulations.

To maximize the efficiency of cloning, both the vector and the cDNAs are prepared in special ways. A preparation of double-stranded cDNAs with blunt ends is ligated to the blunt end of a double-stranded oligonucleotide adaptor composed of paired 5′-CGAGATTTACC and 5′-GGTAAATC oligonucleotides, each of which carries a phosphate at its 5′ end. The vector DNA is cut at its unique XhoI site (5′-C*TCGAG) and then incubated with DNA polymerase in the presence of dTTP only. The vector DNA and cDNAs are then mixed and ligated together. This procedure turns out to be very efficient. Starting with 2 μg of vector and 0.1 μg of cDNA, you make a library consisting of 4×10^7 recombinant molecules.

A. Given that the vector is 43 kb long and the average size of the cDNAs is about 1 kb, estimate the ratio of vector molecules to cDNA molecules in the ligation mixture.

B. What is the efficiency with which vector molecules are converted to recombinant molecules in this procedure? (The average mass of a nucleotide pair is 660 daltons.)

C. Explain how the treatments of the vector molecules and the cDNAs allow them to be ligated together. Can the treated vector molecules be ligated together? Can the cDNAs be ligated together?

D. How does the treatment of the vector and cDNAs improve the efficiency of generating recombinant DNA molecules?

E. Can a cDNA be cut out of a recombinant plasmid with XhoI?

8–50* To prepare a genomic DNA library, it is necessary to fragment the genome so that it can be cloned in a vector. A common method is to use a restriction nuclease.

A. How many different DNA fragments would you expect to obtain if you cleaved human genomic DNA with Sau3A (5′-GATC)? (Recall that there are 3.2×10^9 base pairs in the haploid human genome.) How many would you expect to get with EcoRI (5′-GAATTC)?

B. Human genomic libraries are often made from fragments obtained by cleaving human DNA with Sau3A in such a way that the DNA is only partially digested; that is, so that not all the Sau3A sites have been cleaved. What is a possible reason to do this?

8–51 A degenerate set of oligonucleotide probes for the Factor VIII gene for blood clotting is shown in Figure 8–24. Each of these probes is only 15 nucleotides long. On average how many exact matches to any single 15-nucleotide sequence would you expect to find in the human genome (3.2×10^9 base pairs)? How many matches to the collection of sequences in the degenerate oligonucleotide probe would you expect to find? How might you determine that a match corresponds to the Factor VIII gene?

8–52* It's midnight. Your friend has awakened you with yet another grandiose scheme. He has spent the last two years purifying a potent modulator of the immune response. Tonight he got the first 30 amino acids of the sequence (Figure 8–25). He wants your help in cloning the gene so it can be expressed at high levels in bacteria. He argues that this protein, by stimulating the immune system, could be the ultimate cure for the common cold. He's already picked out a trade name—Immustim.

Even though he gets carried away at times, he is your friend, and you are intrigued by this idea. You promise to call him back in 15 minutes as soon as you have checked out the protein sequence. What two sets of degenerate 20-nucleotide-long oligonucleotide probes will you recommend to your friend as the best hybridization probes for screening a genomic DNA library?

8–53 (**True/False**) By far the most important advantage of cDNA clones over genomic clones is that they can contain the complete coding sequence of a gene. Explain your answer.

8–54* You wish to know whether the cDNA you have isolated and sequenced is the

protein sequence

M Q K F N

ATGCA^A_GA A^A_GA TT^T_CAA^T_C

degenerate oligonucleotide

Figure 8–24 A degenerate oligonucleotide probe for the Factor VIII gene based on a stretch of amino acids from the protein (Problem 8–51). Because more than one DNA triplet can encode each amino acid, a number of different nucleotide sequences are possible for each amino acid sequence. Although only one of these sequences will actually code for the protein in the genomic DNA, it is impossible to tell in advance which one it is. Therefore, a mixture of the possible sequences—called a degenerate oligonucleotide probe—is used to search a genomic library for the gene.

Problems 7–65, 11–64, 15–3, 15–71, 17–9, and 17–22 provide examples of cloning from cDNA libraries.

```
        10         20         30
MFYWMIGRST EDWMPLYMKD FWAKHSLICE
```

Figure 8–25 The first 30 amino acids in your friend's protein (Problem 8–52).

Figure 8–26 A dideoxy sequencing gel of a cloned segment of DNA (Problem 8–55). The lanes are labeled G, A, T, and C to indicate which ddNTP was included in the reaction.

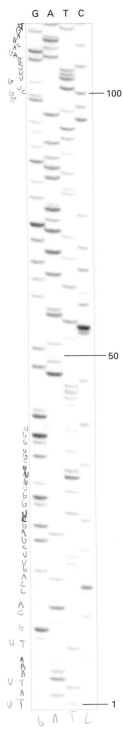

product of a unique gene or is made by a gene that is a member of a family of related genes. To address this question, you digest cell DNA with a restriction nuclease that cleaves the genomic DNA but not the cDNA, separate the fragments by gel electrophoresis, and visualize bands using radioactive cDNA as a probe. The Southern blot shows two bands, one of which hybridizes more strongly to the probe than the other.

You interpret the stronger hybridizing band as the gene that encodes your cDNA and the weaker band as a related gene. When you explain your result to your advisor, however, she cautions that you have not proven that there are two genes. She suggests that you repeat the Southern blot in duplicate, probing one with a radioactive segment from the 5′ end of the cDNA and the other with a radioactive segment from the 3′ end of the cDNA.

A. How might you get two hybridizing bands if the cDNA was the product of a unique gene?

B. What results would you expect from the experiment your advisor proposed if there is a single unique gene? If there are two genes?

8–55 A clear example of a dideoxy sequencing gel is shown in Figure 8–26. Try reading it. As read from the bottom of the gel to the top, the sequence corresponds to the mRNA for a protein. Can you find the open reading frame in this sequence? What protein does it code for?

8–56* How would a DNA sequencing reaction be affected if the ratio of dideoxynucleoside triphosphates (ddNTPs) to deoxynucleoside triphosphates (dNTPs) were increased? What would the consequences be if the ratio were decreased?

8–57 Discuss the following statement: "From the nucleotide sequence of a cDNA clone, the complete amino acid sequence of a protein can be deduced by applying the genetic code. Thus, protein biochemistry has become superfluous because there is nothing more that can be learned by studying the protein."

8–58* DNA sequencing of your own two β-globin genes (one from each of your two copies of chromosome 11) reveals a mutation in one of the genes. Given this information alone, how much should you worry about being a carrier of an inherited disease that could be passed on to your children? What other information would you like to have to assess your risk?

8–59 You want to amplify the DNA between the two stretches of sequence shown in Figure 8–27. Of the listed primers choose the pair that will allow you to amplify the DNA by PCR.

8–60* Duchenne's muscular dystrophy (DMD) is among the most common human genetic diseases, affecting approximately 1 in 3500 male births. One-third of all new cases arise via new mutations. The DMD gene, which is located on the X chromosome, is greater than 2 million base pairs in length and contains at least 70 exons. Large deletions account for about 60% of all cases of

DNA to be amplified

```
5′-GACCTGTGGAAGC ——————————— CATACGGGATTGA-3′
3′-CTGGACACCTTCG ——————————— GTATGCCCTAACT-5′
```

primers

(1) 5′-GACCTGTCCAAGC-3′
(2) 5′-CTGGACACCTTCG-3′
(3) 5′-CGAAGGTGTCCAG-3′
(4) 5′-GCTTCCACAGGTC-3′

(5) 5′-CATACGGGATTGA-3′
(6) 5′-GTATGCCCTAACT-3′
(7) 5′-TGTTAGGGCATAC-3′
(8) 5′-TCAATCCCGTATG-3′

Figure 8–27 DNA to be amplified and potential PCR primers (Problem 8–59).

ISOLATING, CLONING, AND SEQUENCING DNA

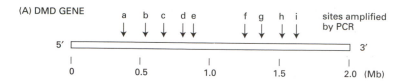

(A) DMD GENE
a b c d e f g h i sites amplified by PCR

5′   3′
| | | | |
0   0.5   1.0   1.5   2.0 (Mb)

(B) MULTIPLEX PCR ANALYSIS

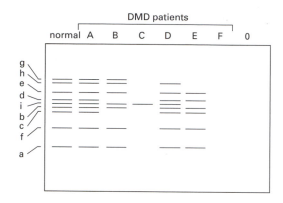

Figure 8–28 Multiplex PCR analysis of six DMD patients (Problem 8–60). (A) The DMD gene with the nine sites amplified by PCR indicated by *arrows*. The sizes of the PCR products are so small on this scale that their location is simply indicated. (B) Agarose gel display of amplified PCR products. 'Normal' indicates a normal male. The lane marked '0' shows a negative control with no added DNA.

the disease, and they tend to be concentrated around two regions of the gene.

The very large size of the DMD gene complicates the analysis of mutations. One rapid approach, which can detect about 80% of all deletions, is termed multiplex PCR. It uses multiple pairs of PCR primers to amplify nine different segments of the gene in the two most common regions for deletions (Figure 8–28A). By arranging the PCR primers so that each pair gives a different size product, it is possible to amplify and analyze all nine segments in one PCR reaction. An example of multiplex PCR analysis of six unrelated DMD males is shown in Figure 8–28B.

A. Describe the deletions, if any, in each of the six DMD patients.

B. What additional control might you suggest to confirm your analysis of patient F?

8–61 (**True/False**) If each cycle of PCR doubles the amount of DNA synthesized in the previous cycle, then 10 cycles will give a 10^3-fold amplification, 20 cycles will give a 10^6-fold amplification, and 30 cycles will give a 10^9-fold amplification. Explain your answer.

8–62* In the very first round of PCR using genomic DNA, the DNA primers prime synthesis that terminates only when the cycle ends (or if a random end of DNA is encountered). Yet by the end of 20 to 30 cycles—a typical amplification—the only visible product is defined precisely by the ends of the DNA primers (Figure 8–29). In what cycle is a double-stranded fragment of the correct size first generated?

8–63 After decades of work, an old classmate of yours has isolated a small amount of attractase, an enzyme producing a powerful human pheromone, from

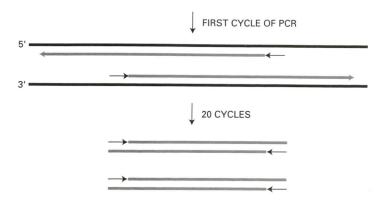

FIRST CYCLE OF PCR

5′

3′

20 CYCLES

Figure 8–29 Products of PCR after 1 and 20 cycles (Problem 8–62).

hair samples of Hollywood celebrities. To produce attractase for his personal use, he obtained a complete genomic clone of the attractase gene, connected it to a strong bacterial promoter on an expression plasmid, and introduced the plasmid into *E. coli* cells. He was devastated to find that no attractase was produced in the cells. What is the likely explanation for his failure?

8–64*   You want to clone a cDNA into an expression vector so you can make large amounts of the encoded protein in *E. coli*. The cDNA is flanked by BamHI sites and you plan to insert it at the BamHI site in the vector. This is your first experience with cloning, so you carefully follow the procedures in the cloning manual.

The manual recommends that you cleave the vector DNA and then treat it with alkaline phosphatase to remove the 5′ phosphates. The next step is to mix the treated vector with the BamHI-cut cDNA fragment and incubate with DNA ligase. After ligation the DNA is mixed with bacterial cells that have been treated to make them competent to take up DNA. Finally, the mixture is spread onto culture dishes filled with a solid growth medium that contains an antibiotic that kills all cells that have not taken up the vector. The vector allows cells to survive because it carries a gene for resistance to the antibiotic.

The cloning manual also suggests four controls.

Control 1.   Plate bacterial cells that have not been exposed to any vector onto the culture dishes.

Control 2.   Plate cells that have been transfected with vector that has not been cut.

Control 3.   Plate cells that have been transfected with vector that has been cut (but not treated with alkaline phosphatase) and then incubated with DNA ligase (in the absence of the cDNA fragment).

Control 4.   Plate cells that have been transfected with vector that has been cut and treated with alkaline phosphatase and then incubated with DNA ligase (in the absence of the cDNA fragment).

For your first attempt at the experiment and all its controls, you borrow a fellow student's competent cells, but all the plates have too many colonies to count (Table 8–4). For your second attempt, you use cells you have prepared yourself, but this time you get no colonies on any plate (Table 8–4). Ever the optimist, you try again, and this time you are rewarded with more encouraging results (Table 8–4). You pick 12 colonies from the experimental sample, prepare plasmid DNA from them, and digest the DNA with BamHI. Nine colonies yield a single band the same size as the linearized vector, but three colonies have, in addition, a fragment the size of the cDNA you wanted to clone. Success is sweet!

A.   What do you think happened in the first experiment? What is the point of control 1?

B.   What do you think happened in the second experiment? What is the point of control 2?

C.   What is the point of doing controls 3 and 4?

D.   Why does the cloning manual recommend treating the vector with alkaline phosphatase?

TABLE 8–4 Results of your cDNA cloning endeavor (Problem 8–64).

PREPARATION OF SAMPLE	RESULTS OF EXPERIMENTS		
	1	2	3
Control 1: cells alone	TMTC	0	0
Control 2: uncut vector	TMTC	0	>1000
Control 3: omit phosphatase, omit cDNA	TMTC	0	435
Control 4: omit cDNA	TMTC	0	25
Experimental sample	TMTC	0	34
TMTC means too many to count.			

ANALYZING PROTEIN STRUCTURE AND FUNCTION

TERMS TO LEARN

co-immunoprecipitation	nuclear magnetic resonance (NMR) spectroscopy	surface plasmon resonance (SPR)
DNA footprinting	phage display	two-hybrid system
fluorescence resonance energy transfer (FRET)	protein affinity chromatography	x-ray crystallography
green fluorescent protein (GFP)		

8–65 Discuss the following statement: "With the ever expanding databases of protein sequences and structures, it will soon be possible to input an amino acid sequence of an unknown protein and, by analogy to known proteins, determine its structure and function. Thus, it will not be long before biochemists are put out of work."

8–66* Hybridoma technology allows one to generate monoclonal antibodies to virtually any protein. Why is it then that epitope tagging is such a commonly used technique, especially since an epitope tag has the potential of interfering with the function of the protein?

8–67 You want to express a rare human protein in bacteria so that you can make large quantities of it. To aid in its purification you decide to add a stretch of six histidines to the N-terminus or the C-terminus of the protein. Such histidine-tagged proteins bind tightly to Ni^{2+} columns, but can be readily eluted with a solution of EDTA or imidazole. This procedure allows an enormous purification in one step.

The nucleotide sequence that encodes your protein is shown in Figure 8–30. Design a pair of PCR primers, each with 18 nucleotides of homology to the gene, that will amplify the coding sequence and add an initiation codon followed by six histidine codons to the N-terminus. Design a pair of primers that will add six histidine codons followed by a stop codon to the C-terminus.

Problems 3–87, 7–41, 12–73, 13–16, 15–71, and 15–92 illustrate experimental uses for a variety of protein tags.

8–68* You have now cloned in an expression vector both versions of the histidine-tagged protein you created in the previous problem. Neither construct expresses particularly strongly in bacteria, but the product is soluble. You pass the crude extract over a Ni^{2+} affinity column, which binds histidine-tagged proteins specifically. After washing the column extensively, you elute your protein from the column using a solution containing imidazole (Figure 8–31), which releases your protein.

When you subject the eluted protein to electrophoresis and stain the gel for protein, you are pleased to find bands in the eluate that are not present

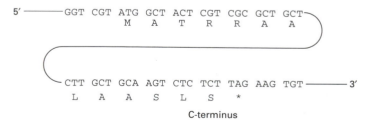

Figure 8–30 Nucleotide sequence around the N- and C-termini of the protein you want to modify (Problem 8–67). The encoded amino acid sequence is indicated *below* each codon using the one-letter code. The *asterisk* (*) indicates the stop codon. Only the top strand of the double-stranded DNA is shown.

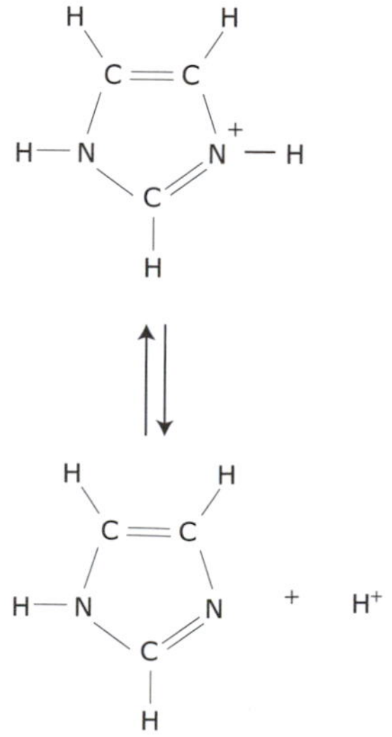

Figure 8–31 Structure of imidazole (Problem 8–68).

when control bacteria are treated similarly. But you are puzzled to see that the construct tagged at the N-terminus gives a ladder of shorter proteins below the full-length protein, whereas the C-terminally tagged construct yields exclusively the full-length protein. The amount of full-length protein is about the same for each construct.

A. Why does a solution of imidazole release a histidine-tagged protein from the Ni^{2+} column?

B. Offer an explanation for the difference in the products generated by the two constructs.

8–69　An adaptation of standard PCR, called recombinant PCR, allows virtually any two nucleotide sequences to be joined any way you want. Imagine, for example, that you want to combine the DNA-binding domain of protein A to the regulatory domain of protein B in order to test your conjectures about how these proteins work. The target domains in the cDNAs and the arrangement of PCR primers needed to join the domains are shown in Figure 8–32.

Recombinant PCR is usually carried out in two steps. In the first step PCR primers 1 and 2 are used to amplify the target segment of gene A, and in a separate reaction, primers 3 and 4 are used to amplify the target sequence in gene B. In the second step the individually amplified products, separated from their primers, are mixed together and amplified using primers 1 and 4. *Voila!* The desired hybrid gene is the major product.

A. Explain how recombinant PCR manages to link the two gene segments together.

B. Illustrate schematically the structure and arrangement of primers you would use to put the regulatory domain of gene B at the N-terminus of the hybrid protein.

8–70*　Consider a fluorescent detector designed to report the cellular location of active protein tyrosine kinases. A blue (cyan) fluorescent protein (CFP) and

Problems 12–25 and 12–88 present two of the earliest uses of synthetic gene fusions to investigate biological questions.

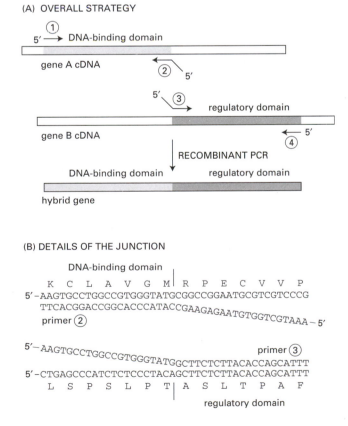

(A) OVERALL STRATEGY

gene A cDNA
DNA-binding domain

gene B cDNA
regulatory domain

RECOMBINANT PCR

DNA-binding domain　regulatory domain
hybrid gene

(B) DETAILS OF THE JUNCTION

DNA-binding domain
```
     K   C   L   A   V   G   M  R   P   E   C   V   V   P
5'–AAGTGCCTGGCCGTGGGTATGCGGCCGGAATGCGTCGTCCCG
   TTCACGGACCGGCACCCATACCGAAGAGAATGTGGTCGTAAA – 5'
   primer ②
```

```
5'–AAGTGCCTGGCCGTGGGTATGGCTTCTCTTACACCAGCATTT       primer ③
5'–CTGAGCCCATCTCTCCCTACAGCTTCTCTTACACCAGCATTT
     L   S   P   S   L   P   T  A   S   L   T   P   A   F
                              regulatory domain
```

Figure 8–32 Recombinant PCR (Problem 8–69). (A) Overall strategy. (B) Details of the junction. Note that primers 2 and 3 are complementary to one another over their entire length. Only one strand of the cDNA is shown; it is the strand with the same sequence as the mRNA. Primer 2 will pair with the cDNA strand that is shown. Primer 3 will pair with the complement of that strand.

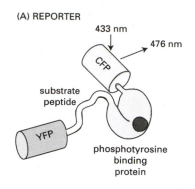

(A) REPORTER

433 nm → 476 nm

CFP

substrate peptide

YFP

phosphotyrosine binding protein

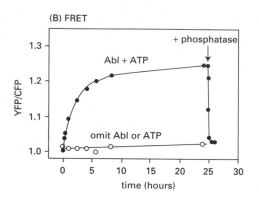

(B) FRET

+ phosphatase

Abl + ATP

omit Abl or ATP

YFP/CFP

time (hours)

Figure 8–33 Fluorescent reporter protein to detect tyrosine phosphorylation (Problem 8–70). (A) Domain structure of reporter protein. Four domains are indicated: CFP, YFP, tyrosine kinase substrate peptide, and a phosphotyrosine-binding domain. (B) FRET assay. YFP/CFP is normalized to 1.0 at time zero. The reporter was incubated in the presence (or absence) of Abl and ATP for the indicated times. *Arrow* indicates time of addition of tyrosine phosphatase.

a yellow fluorescent protein (YFP) were fused to either end of a hybrid protein domain. The hybrid protein segment consisted of a substrate peptide recognized by the Abl protein tyrosine kinase and a phosphotyrosine binding domain (Figure 8–33A). When they are separated, stimulation of the CFP domain does not cause emission by the YFP domain. When the CFP and YFP domains are brought close together, however, fluorescence resonance energy transfer (FRET) allows excitation of CFP to stimulate emission by YFP. FRET shows up experimentally as an increase in the ratio of emission at 526 nm versus 476 nm (YFP/CFP) when CFP is excited by 434-nm light.

Incubation of the reporter protein with Abl protein tyrosine kinase in the presence of ATP gave an increase in YFP/CFP emission (Figure 8–33B). In the absence of ATP or the Abl protein, no FRET occurred. And FRET was eliminated by addition of a tyrosine phosphatase (Figure 8–33B). Describe as best you can how the reporter protein detects active Abl protein tyrosine kinase.

Problems 7–18 and 10–19 provide additional examples of the use of FRET to investigate biological questions.

8–71 Cells activate Abl protein tyrosine kinase in response to platelet-derived growth factor (PDGF). PDGF binds to the PDGF receptor, which activates Src, which then activates Abl. It is unclear where in the cell Abl is active, but one of the consequences of PDGF stimulation is the appearance of membrane ruffles. To investigate this question, the reporter construct described in Problem 8–70 was transfected into cells, which were then stimulated by addition of PDGF. Using fluorescence microscopy, YFP emission in response to CFP excitation (FRET) was followed in different parts of the cell, with the results shown in Figure 8–34. What can you infer about the cellular distribution of active Abl protein tyrosine kinase in response to PDGF?

Problems 12–35, 12–37, 12–39, 16–28, 16–29, 17–33, 17–43, 17–58 and 17–59 show how GFP can be used to track proteins and monitor processes in cells.

8–72* You have raised four different monoclonal antibodies to *Xenopus* Orc1, which is a component of the DNA replication origin recognition complex (ORC) found in eucaryotes. You want to use the antibodies to immunopurify other members of ORC. To decide which of your monoclonal antibodies—TK1, TK15, TK37, or TK47—is best suited for this purpose, you covalently attach them to beads, incubate them with a *Xenopus* egg extract, spin the

Problems 3–41, 6–70, 7–41 and 10–46 demonstrate the use of antibodies to purify associated proteins by co-immunoprecipitation. **Problems 3–87, 13–16 and 15–71** accomplish the same objective using GST 'pull downs.'

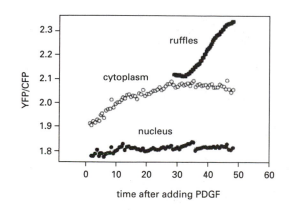

ruffles

cytoplasm

nucleus

YFP/CFP

time after adding PDGF

Figure 8–34 Time course of FRET in various parts of the cell after addition of PDGF (Problem 8–71). FRET was measured as the increase in the ratio of emission at 526 nm/476 nm (YFP/CFP) when CFP was excited by 434-nm light.

Figure 8–35 Immunoaffinity purification of *Xenopus* ORC (Problem 8–72). The monoclonal antibody mAb423 is specific for an antigen not found in *Xenopus* extracts and thus serves as a control. The positions of marker proteins are shown at the *left* with their masses indicated in kilodaltons.

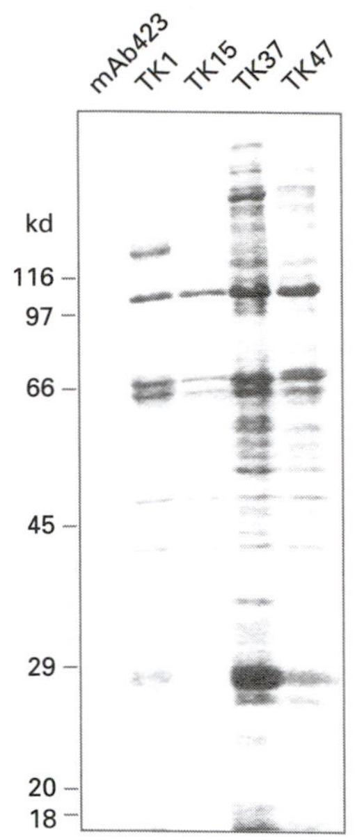

beads down and wash them carefully, and solubilize the bound proteins with SDS. You use SDS-PAGE to separate the solubilized proteins and stain them, as shown in Figure 8–35.

A. From these results which bands do you think arise from proteins that are present in ORC?

B. Why do you suppose the various monoclonal antibodies give such different results?

C. Which antibody do you think is the best one to use in future studies of this kind? Why?

D. How might you determine which band on this gel is Orc1?

8–73 The yeast two-hybrid system was used to find proteins with which Ras interacts. This approach depends on the modular nature of many transcription factors, which have one domain that binds to DNA and another domain that activates transcription. Domains can be interchanged by recombinant DNA methods, allowing hybrid transcription factors to be constructed. Thus, the DNA-binding domain of the *E. coli* LexA repressor can be combined with the powerful VP16 activation domain from herpesvirus to activate transcription of genes downstream of a LexA DNA-binding site (Figure 8–36).

If the two domains of the transcription factor can be brought into proximity by protein–protein interactions, they will activate transcription. This is the key feature of the two-hybrid system. Thus, if one member of an interacting pair of proteins is fused to the DNA-binding domain of LexA (to form the 'bait') and the other is fused to the VP16 activation domain (to form the 'prey'), transcription will be activated when the two hybrid proteins interact inside a yeast cell. It is possible to design powerful screens for protein–protein interactions, if the gene whose transcription is turned on is essential for growth or can give rise to a colored product.

To check out the system, hybrid genes were constructed that contained the LexA DNA-binding domain, one fused to Ras (LexA–Ras) and the other fused to nuclear lamin (LexA–lamin). A second pair of constructs contained the VP16 activation domain alone (VP16) or fused to the adenylyl cyclase gene (VP16–CYR). Adenylyl cyclase is known to interact with Ras and serves as a positive control; nuclear lamins do not interact with Ras and serve as a negative control. These plasmid constructs were introduced into a strain of yeast containing copies of the *HIS3* gene and the *lacZ* gene, both with LexA-binding sites positioned immediately upstream. Individual transformed colonies were tested for growth on a plate lacking histidine, which requires expression of the *HIS3* gene. In addition, they were tested for ability to form blue colonies (as compared to the normal white colonies) when grown in the presence of an appropriate substrate (XGAL) for β-galactosidase. The setup for the experiment is outlined in Table 8–5.

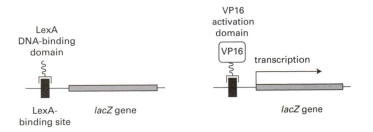

Figure 8–36 Activation of transcription by a hybrid transcription factor (Problem 8–73).

TABLE 8–5 Experiments to test the two-hybrid system (Problem 8–73).

| PLASMID CONSTRUCTS | | GROWTH ON PLATES LACKING HISTIDINE | COLOR ON PLATES WITH XGAL |
BAIT	PREY		
LexA–Ras			
LexA–lamin			
	VP16		
	VP16-CYR		
LexA–Ras	VP16		
LexA–Ras	VP16-CYR		
LexA–lamin	VP16		
LexA–lamin	VP16-CYR		

A. Fill in Table 8–5 with your expectations. Use a plus sign to indicate growth on plates lacking histidine and a minus sign to indicate no growth. Write 'blue' or 'white' to indicate the color of colonies grown in the presence of XGAL.

B. For any entries in the table that you expect to grow in the absence of histidine and form blue colonies with XGAL, sketch the structure of the active transcription factor on the *lacZ* gene.

C. If you want two proteins to be expressed in a single polypeptide chain, what must you be careful to do when you fuse the two genes together?

8–74* To use the two-hybrid system to screen for proteins that interact with Ras, a cDNA library was made with the cDNA inserts positioned at the C-terminus of the *VP16* gene segment. The library was transfected into yeast that already contained the LexA–Ras plasmid described in the preceding problem. The transformed cells were then grown on plates in the presence of XGAL and in the absence of histidine, and blue colonies were isolated for further testing.

The LexA–Ras plasmid was eliminated from such cells by genetic selection, and the 'cured' cells, which contained only a VP16–cDNA plasmid, were checked for growth in the absence of histidine and for color when grown in the presence of XGAL. Cured cells that did not grow in the absence of histidine and formed white colonies in the presence of XGAL were then transformed with the LexA–lamin plasmid (see Problem 8–73) and tested again for growth in the absence of histidine and colony color in the presence of XGAL. Only cells that did not grow in the absence of histidine and formed white colonies in the presence of XGAL are analyzed further.

Since its introduction in 1989, the two-hybrid system has been modified considerably to deal with the problem of false positives; that is, to eliminate cDNA clones that do not really encode a protein that interacts with the protein of interest. Several of these modifications were built into the selection scheme used here.

A. When the cDNA library was initially transfected into yeast containing the LexA–Ras plasmid, many white colonies (in addition to blue colonies) a were observed when the cells were grown in the presence of XGAL and in the absence of histidine. What type of VP16–cDNA might give rise to such white colonies?

B. Among the colonies that were cured of the LexA–Ras plasmid—and thus contained only a VP16–cDNA plasmid—some grew in the absence of histidine and turned blue when grown in the presence of XGAL. What type of VP16–cDNA might give this result in the absence of the LexA–Ras plasmid?

C. When the LexA-lamin plasmid was introduced into cells that contained a VP16–cDNA plasmid (ones that alone did not grow in the absence of histidine and did not turn blue in the presence of XGAL), some formed blue colonies in the absence of histidine and in the presence of XGAL. What type of VP16–cDNA might give these results in the presence of the LexA–lamin plasmid?

8–75 Out of 1.4×10^6 original transformants only 19 clones met the stringent criteria set up in the previous problem. The cDNA inserts downstream of the *VP16* gene segment were sequenced in each. Nine clones had cDNA inserts that corresponded to the N-terminal domain of known Raf serine/threonine protein kinases. This was an exciting finding because other work had shown that immunoprecipitates of Raf could phosphorylate and activate MAP-kinase-kinase, suggesting that Raf was the missing link between Ras and the MAP-kinase cascade.

The scientists that identified Raf by the two-hybrid system did not stop there. They made fusion proteins between the maltose-binding protein and Raf (MBP–Raf) and between glutathione-*S*-transferase and Ras (GST–Ras). They then produced the two fusion proteins in bacteria and tested their ability to bind one another by passing the mixture over an amylose affinity column, which will bind MBP–Raf. When the bound proteins were eluted with maltose, a significant fraction of the input GST–Ras protein was found to elute with the MBP–Raf protein. This result provided a biochemical confirmation of the genetic results with the two-hybrid system.

Why do you think these scientists felt it necessary to demonstrate a direct interaction of Ras and Raf by biochemical studies?

8–76* (**True/False**) In a reverse two-hybrid system, the reporter gene is replaced by a gene whose product can convert a precursor molecule to a toxic product that kills cells that express the gene. When the precursor molecule is added to the medium, mutations or chemicals that disrupt the interaction between the bait and prey proteins can be selected for. Explain your answer.

8–77 The surfaces of different varieties of cyclin A proteins all contain a hydrophobic cleft that is known to bind the sequence PSACRNLFG. This sequence is found close to the N-terminus of the cyclin-dependent kinase inhibitor p27. Looking for potential anti-cancer drugs, you decide to use phage display to hunt for peptides with very high affinity for cyclin A. You attach cyclin A to the bottoms of plastic dishes and 'pan' for phage that will bind (Figure 8–37). You use three different phage M13 libraries that bear randomized 7, 12, or 20 amino acid sequences on one of their coat proteins. You isolate phages that bind to the immobilized cyclin A and sequence the segment of their coat protein gene that encodes the peptide. Sequences for 14 clones are shown in Table 8–6. By comparing all 15 sequences, decide which amino acid residues are likely to be the most critical for binding to cyclin A.

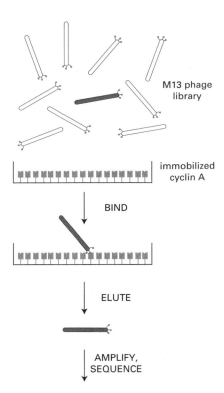

Figure 8–37 Panning a phage-display library for phages that bind to immobilized cyclin A (Problem 8–77). The M13 phage library contains a randomized segment of the coat protein gene so that each phage displays one of a large number of possible amino acid sequences on its surface. The phage library is incubated with the immobilized protein in the plate, unbound phage are washed away, and bound phage are eluted and amplified by growth in *E. coli*. This cycle is repeated 2–3 more times and then individual phages are isolated, amplified, and analyzed by sequencing.

TABLE 8–6 Peptides encoded by phages that were selected by panning of a phage-display library against immobilized cyclin A (Problem 8–77).

CLONE	LIBRARY	PEPTIDE SEQUENCE
1	7-mer	LEPRMLF
2	7-mer	TLPRQLF
3	7-mer	LKPTKLF
4	7-mer	LIPKNLF
5	7-mer	FLPRALF
1	12-mer	NVRVELFPPTKV
2	12-mer	KSSVVRSLFVPT
3	12-mer	ERPSAQRSLVFW
4	12-mer	NLFYPRNLFPEF
5	12-mer	YPSPARNLLPMF
6	12-mer	ATIRELFPPTLP
1	20-mer	HQPESVKRSLFKPAHSALEP
2	20-mer	EVARRELFADHSLVHVGHVR
3	20-mer	EHKALPGKAVTGPKRELVFQ
p27 cyclin A-binding sequence		**PSACRNLFGP**

8–78* In preparing for a phage-display experiment you notice that the supplier created the phage-display libraries by inserting random oligonucleotides of appropriate length into phage genomes. The initial libraries contained 3×10^9 different phages. Is this a sufficient number of phages so that a library could contain all possible peptides seven amino acids in length? Could it contain all possible peptides 12 amino acids in length? Could it contain all the possible peptides 20 amino acids in length?

8–79 In the supplier's information package on phage display, you find that they use a 'reduced' genetic code with only 32 codons (Figure 8–38). They synthesize mixtures of oligonucleotides with all four nucleotides in the first two positions of a codon, but they include only T and G in the third position. In addition, they grow their phage on a strain of bacteria that inserts glutamine (Q) at a TAG codon, which is normally a stop codon.

  A. Why do you suppose they use a reduced genetic code? Can you see any advantages to including just T and G at the third position of codons?

  B. If all four nucleotides were used at each position, all possible codons would be produced equally (assuming no biases). Amino acids would be encoded by such a mixture in the same proportions as their codons are represented in the standard genetic code (see inside back cover). That is to say, cysteine, which has two codons, would be encoded half as often as alanine, which has four codons, and one third as often as leucine, which has six codons. Are the relative frequencies of the various amino acids the same for the reduced genetic code as for the standard genetic code? If not, how do they differ?

8–80* Discuss the following statement: "Surface plasmon resonance (SPR) measures association (k_{on}) and dissociation (k_{off}) rates between molecules in real time using small amounts of unlabeled molecules. If SPR could just measure the binding constant (K), it would be even more useful."

8–81 You want to use SPR to measure the rate of dissociation of your protein from its ligand. You immobilize the ligand on the biosensor surface and allow a solution of your pure protein to flow across it until the binding reaches equilibrium, as indicated by the plateau value for the resonance angle (Figure 8–39). At this point you replace the protein solution with buffer. Now the resonance angle decreases as the protein dissociates from the ligand. When you put these data through the computer algorithm supplied with the machine, you are surprised to find that the dissociation curve has two components. The initial part of the curve is characterized by a high off rate, whereas the final part has a low off rate (Figure 8–39).

  A. Why are you surprised that a protein–ligand complex should have two distinct dissociation rates?

  B. Can you suggest some possible explanations for your results? What kinds of explanations might lead to two off rates, where only one is expected?

8–82* You have determined the DNA-binding site of a protein by DNA footprinting after labeling one strand. To check your results, you repeat the experiment

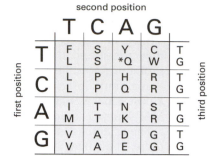

Figure 8–38 Reduced genetic code (Problem 8–79).

Problem 5–47 demonstrates the use of DNA footprinting to identify protein binding sites.

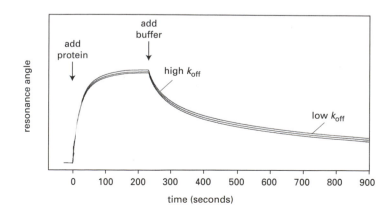

Figure 8–39 SPR measurement of protein association with and dissociation from a ligand immobilized on a biosensor (Problem 8–81).

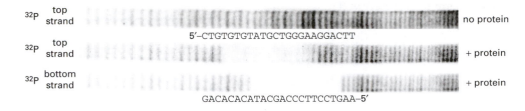

³²P top strand — no protein
5'-CTGTGTGTATGCTGGGAAGGACTT
³²P top strand — + protein
³²P bottom strand — + protein
GACACACATACGACCCTTCCTGAA-5'

Figure 8–40 DNA footprints (Problem 8–82). The sequences of the DNA from the top and bottom strands around the footprint are shown. The 5' ends of the top or bottom strands were labeled with ³²P. Chemical cleavage was used to introduce DNA breaks, which are fairly randomly distributed, as shown by the roughly equal intensities of the individual bands. Each band corresponds to a fragment of DNA that differs from those on either side by one nucleotide.

after labeling the other strand of the duplex. You find that the footprints are slightly offset from one another relative to the sequence of the DNA (Figure 8–40). If the protein binds to the same duplex in both cases, how can the footprints on the two strands be different?

STUDYING GENE EXPRESSION AND FUNCTION

TERMS TO LEARN

Cre/lox	genetics	site-directed mutagenesis
DNA microarrays	genotype	totipotent
embryonic stem cells	phenotype	transgenic organism
genetic screen	reverse genetics	

8–83 Distinguish between the following genetic terms:
A. Locus and allele.
B. Homozygous and heterozygous
C. Genotype and phenotype.
D. Dominant and recessive.

8–84* Explain the difference between a gain-of-function mutation and a dominant-negative mutation. Why are these mutations usually dominant?

8–85 (**True/False**) In an organism whose genome has been sequenced, identifying the mutant gene responsible for an interesting phenotype is as easy for mutations induced by chemical mutagenesis as it is for those generated by insertional mutagenesis. Explain your answer.

8–86* Discuss the following statement: "We would have no idea today of the importance of insulin as a regulatory hormone if its absence were not associated with the devastating human disease diabetes. It is the dramatic consequences of its absence that focused early efforts on the identification of insulin and the study of its normal role in physiology."

Problems 12–25, 12–26, 12–51, 12–69, and 12–84 describe various genetic screens and genetic selections that have been used to find mutants defective in particular cellular processes.

8–87 Early genetic studies in *Drosophila* laid the foundation for our current understanding of genes. *Drosophila* geneticists were able to generate mutant flies with a variety of easily observable phenotypic changes. Alterations from the fly's normal brick-red eye color have a venerable history because the very first mutant found by Thomas Hunt Morgan was a white-eyed fly (Figure 8–41). Since that time a large number of mutant flies with intermediate eye colors have been isolated and given names that challenge your color sense: garnet, ruby, vermilion, cherry, coral, apricot, buff, and carnation. The mutations responsible for these eye-color phenotypes are recessive. To determine whether the mutations affected the same or different

Figure 8–41 *Drosophila* with different color eyes (Problem 8–87). Wild-type flies with brick-red eyes are shown on the *left* and white-eyed flies are shown on the *right*. Flies with eye colors between red and white are shown in between.

brick-red

white

TABLE 8–7 Complementation analysis of *Drosophila* eye-color mutations (Problem 8–87).

MUTATION	WHITE	GARNET	RUBY	VERMILION	CHERRY	CORAL	APRICOT	BUFF	CARNATION
White	−	+	+	+	−	−	−	−	+
Garnet		−	+	+	+	+	+	+	+
Ruby			−	+	+	+	+	+	+
Vermilion				−	+	+	+	+	+
Cherry					−	−	−	−	+
Coral						−	−	−	+
Apricot							−	−	+
Buff								−	+
Carnation									−

genes, homozygous flies for each mutation were bred to one another in pairs and the eye colors of their progeny were noted. In Table 8–7, brick-red wild type eyes are shown as (+) and other colors are indicated as (–).

A. How is it that flies with two different eye colors—ruby and white, for example—can give rise to progeny that all have brick-red eyes?
B. Which mutations affect different genes and which mutations are alleles of the same gene?
C. How can alleles of the same gene give different eye colors? Why don't all the mutations in the same gene give the same phenotype?

8–88* What are single-nucleotide polymorphisms (SNPs), and how can they be used to locate a mutant gene by linkage analysis?

8–89 Fatty acid synthase in mammalian cells is encoded by a single gene. This remarkable protein carries out seven distinct biochemical reactions. The mammalian fatty acid synthase gene is homologous to seven different *E. coli* genes, each of which encodes one of the functions of the mammalian protein. Do you think it is likely that the proteins in *E. coli* function together as a complex? Why or why not?

Problems 7–31, 7–36, 7–43, 12–35, 12–37, 15–22, 16–29, 17–33, 17–43, 17–58, 17–59, and 18–25 give examples of the use of reporter genes to provide information on the timing and location of gene expression.

8–90* You have designed and constructed a DNA microarray that carries 20,000 allele-specific oligonucleotides (ASOs). These ASOs correspond to the wild-type and mutant alleles associated with 1000 human diseases. You have designed the microarray so the ASO that hybridizes to the mutant allele is located right below the ASO that hybridizes to the same site in the wild-type sequence. This arrangement is illustrated in Figure 8–42 for ASOs that are specific for the sickle-cell allele (β^S) and the corresponding site in the wild-type allele (β^A). $ASO^{\beta S}$ hybridizes to the sickle-cell mutation, and $ASO^{\beta A}$ hybridizes to the corresponding position in the wild-type allele. As a test of your microarray, you carry out hybridizations of DNA isolated from individuals who are homozygous for the wild-type allele, homozygous for the sickle-cell allele, or heterozygous for the wild-type and sickle-cell alleles. For each DNA sample draw the expected patterns of hybridization to the globin ASOs on your microarray.

8–91 Now that news of your disease-specific DNA microarray has gotten around, you are being inundated with requests to analyze various samples. Just

Figure 8–42 DNA microarray for detection of disease alleles (Problem 8–90). (A) The β-globin alleles. The wild-type β-globin gene (β^A) and the sickle-cell allele (β^S) are shown. The position of the sickle-cell mutation is shown by a *vertical line*. The ASOs are shown as *short horizontal lines* arranged above the sites in the gene to which they hybridize. Because the ASOs are located at corresponding positions, each is specific for its allele: the wild-type ASO will not hybridize to the globin allele, nor will the sickle-cell allele hybridize to the wild-type allele. (B) DNA microarray. A tiny section of the microarray is enlarged to illustrate the locations of the wild-type and sickle-cell ASOs.

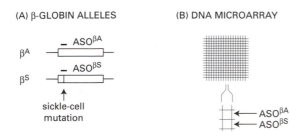

(A) β-GLOBIN ALLELES

β^A — $ASO^{\beta A}$
β^S — $ASO^{\beta S}$
sickle-cell mutation

(B) DNA MICROARRAY

← $ASO^{\beta A}$
← $ASO^{\beta S}$

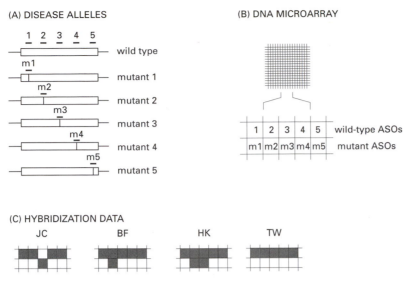

(A) DISEASE ALLELES

1 2 3 4 5

wild type

m1

mutant 1

m2

mutant 2

m3

mutant 3

m4

mutant 4

m5

mutant 5

(B) DNA MICROARRAY

1	2	3	4	5	wild-type ASOs
m1	m2	m3	m4	m5	mutant ASOs

(C) HYBRIDIZATION DATA

JC   BF   HK   TW

Figure 8–43 DNA microarray analysis of alleles present in prenatal samples (Problem 8–91). (A) Wild-type and disease alleles. *Vertical lines* indicate the sites of the mutations in the disease-causing alleles. The ASOs specific for the disease mutations are shown as m1, m2, etc. The corresponding ASOs that hybridize to the wild-type gene at sites that correspond to the position of the mutations are labeled 1, 2, etc. (B) DNA microarray. The arrangement of wild-type and mutant ASOs is indicated. (C) Hybridization data. Samples of fetal DNA were hybridized to DNA arrays. *Dark spots* indicate sites where hybridization occurred. *Letters* identify the patients.

today you received requests from four physicians for help in the prenatal diagnosis of the same disease. Each of the pregnant mothers has a family history of this disease. You included on your array the five alleles known to cause this disease (Figure 8–43). Each of these alleles is recessive. You agree to help. You prepare samples of fetal DNA gotten by amniocentesis and hybridize them to your microarrays. Your data are shown in Figure 8–43C. Assuming that the five disease alleles shown in Figure 8–43A are the only ones in the human population, decide for each sample of DNA whether the individual will have the disease or not. Explain your reasoning.

8–92* How does reverse genetics differ from standard genetics?

8–93 One of the first organisms that was genetically engineered using modern DNA technology was a bacterium that normally lives on the surface of strawberry plants. This bacterium makes a protein, called ice-protein, which causes the efficient formation of ice crystals around it when the temperature drops to just below freezing. Thus, strawberries harboring this bacterium are particularly susceptible to frost damage because the ice crystals destroy their cells. Strawberry farmers have a strong financial interest in preventing such damage.

A genetically engineered version of this bacterium was constructed with the ice-protein gene knocked out. The mutant bacteria were then introduced in large numbers into strawberry fields, where they displaced the normal bacteria by competition for their ecological niche. This approach has been successful: strawberries bearing the mutant bacteria show a much reduced susceptibility to frost damage.

Nevertheless, at the time they were first carried out, the initial open-field trials triggered an intense debate because they represented the first release into the environment of an organism that had been genetically engineered using recombinant DNA techniques. All preliminary experiments were carried out with extreme caution and in strict containment.

Discuss some of the issues that arise from such applications of DNA technology. Do you think that bacteria lacking the ice-protein could be isolated without the use of modern DNA technology? Is it likely that such mutations have already occurred in nature? Would the use of a mutant bacterial strain isolated from nature be of lesser concern? Should we be concerned about the risks posed by the application of genetic engineering techniques in agriculture, medicine, and technology? Explain your answers.

8–94* What are the similarities and differences between antisense RNA and RNA interference (RNAi)?

8–95 The *raf-1* gene encodes a serine/threonine protein kinase that may be important in vertebrate development. In *Drosophila*, when the homolog of

Problems 15–47 and 17–54 illustrate the uses of dominant–negative mutations in analyzing cell signaling pathways.

TABLE 8–8 Injection of *raf-1* and NAF RNA into two-cell embryos (Problem 8–95).

		PHENOTYPE OF TADPOLE		
INJECTION	TOTAL SURVIVORS	NORMAL	TRUNCATED TAIL	OTHER ABNORMALITIES
raf-1 RNA	94	75%	0%	25%
NAF RNA	80	40%	36%	24%
raf-1 + NAF RNA	93	73%	5%	22%
water	101	92%	0%	8%
uninjected	80	99%	0%	1%

this gene is defective, embryos develop through the blastula stage normally, but from then on development is abnormal, leading to a truncated embryo that is lacking the posterior four segments. In vertebrates, fibroblast growth factor (FGF) stimulates mesoderm induction and posterior development, and it is thought that Raf-1 protein participates in the signaling pathway triggered by FGF. The Raf-1 protein consists of a serine/threonine kinase domain that is the C-terminal half of the molecule and an N-terminal regulatory domain that inhibits the kinase activity until a proper signal is received.

To test the role of Raf-1, you construct what you hope will be a dominant-negative *raf-1* cDNA by changing a key residue in the ATP-binding domain. This mutated cDNA encodes a kinase-defective protein, which you term NAF (not a functional Raf). You prepare RNA from the *raf-1* cDNA and from the NAF cDNA and inject them individually or as a mixture into both cells of the two-cell frog embryo to determine their effects on frog development. These experiments are summarized in Table 8–8. The truncated-tail phenotype produced in some injections is very striking (Figure 8–44).

A. Do your experiments provide evidence in support of the hypothesis that Raf-1 participates in posterior development in vertebrates? If so, how?
B. Given the description of the Raf-1 protein, suggest a way that NAF might act as a dominant-negative protein.
C. Why does the co-injection of *raf-1* RNA with NAF RNA result in a low frequency of the truncated-tail phenotype?

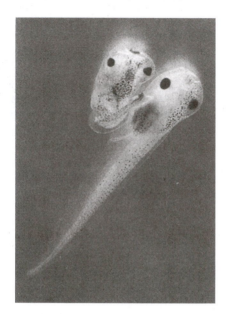

Figure 8–44 A normal tadpole and one with a truncated tail (Problem 8–95).

8–96* From previous work, you suspect that the glutamine (Q) in the protein segment in Figure 8–45 plays an important role at the active site. Your advisor wants you to alter the protein in three ways: change the glutamine to lysine (K), change the glutamine to glycine (G), and delete the glutamine from the protein. You plan to accomplish these mutational alterations on a version of your gene that is cloned into M13 viral DNA. You want to hybridize an appropriate oligonucleotide to the M13 viral DNA, so that when DNA polymerase extends the oligonucleotide around the single-stranded M13 circle, it will complete a strand that encodes the complement of the desired mutant protein. Design three 20-nucleotide-long oligonucleotides that could be hybridized to the cloned gene on single-stranded M13 viral DNA as the first step in effecting the mutational changes.

8–97 A mutation engineered *in vitro* as shown in Figure 8–46 introduces a mismatch into the DNA. When the mismatched DNA is introduced into cells, its replication would be expected to generate an equal mixture of parental and mutant genomes. Would you expect that the mismatched DNA would be recognized and repaired by DNA mismatch repair enzymes? Would you expect mismatch repair to increase, reduce, or not affect the frequency of mutants? Explain your answer.

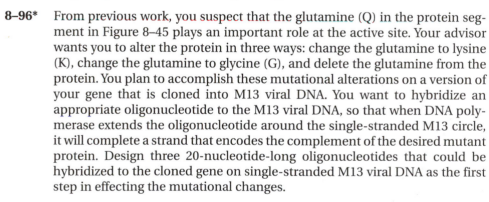

```
      L  R  D  P  Q  G  G  V  I
5' – CTTAGAGACCCGCAGGGCGGCGTCATC – 3'
```

Figure 8–45 Sequence of DNA and encoded protein (Problem 8–96).

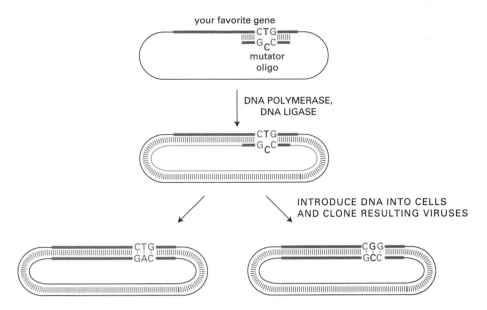

Figure 8–46 Site-directed mutagenesis (Problem 8–97).

8–98* You've heard about this cool technique for targeted recombination into embryonic stem cells that allows you to create a null allele or a conditional allele more or less at the same time. As your friend explained it to you, you first carry out standard gene targeting into ES cells by homologous recombination, as shown in Figure 8–47. The *neo* gene, which codes for resistance to the antibiotic neomycin, allows selection for ES cells that have incorporated the vector. These cells can then be screened by Southern blotting for those that have undergone a targeted event. The really cool part is to flank the *neo* gene and an adjacent exon or two with lox sites. This technique is commonly referred to as 'floxing.' Once the modified ES cells have been identified, they can be exposed to the Cre recombinase, which promotes site-specific recombination between pairs of lox sites. One advantage is that it allows you to get rid of the *neo* gene and any bacterial DNA segments, which can sometimes influence the phenotype.

A. What possible products might you get from expression of Cre in modified ES cells that carry three lox sites, as indicated in Figure 8–47?

B. Which product(s) would be a null allele?

C. Which product(s) would have a pair of lox sites but be an otherwise normal allele?

D. If you had one mouse that expressed Cre under the control of a tissue-specific promoter, can you use the allele in part C (after you've put it into the germline of a mouse) as a conditional allele; that is, one whose defect is expressed only in a particular tissue?

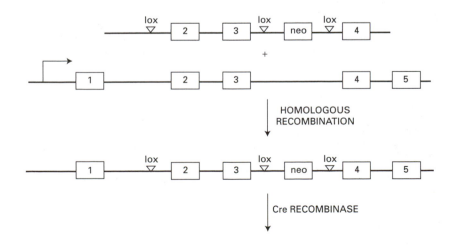

Figure 8–47 Targeted modification of a gene in mouse ES cells (Problem 8–98). Exons are shown as *boxes*. The targeting vector is fully homologous to the gene, except for the presence of the three lox sites and the *neo* gene, all of which are in introns. Cre recombinase, which can be expressed by transfecting in an expression vector, catalyzes a site-specific recombination event between a pair of lox sites in about 20% of transfected cells.

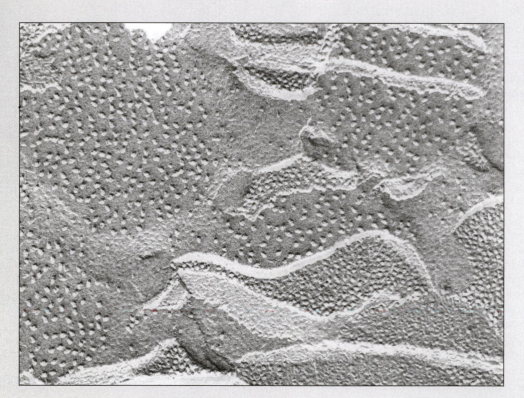

Freeze-fracture electron micrograph of the thylakoid membranes from a chloroplast. What do you suppose all the little lumps are? Which side of the membrane is which? Courtesy of Andrew Staehelin.

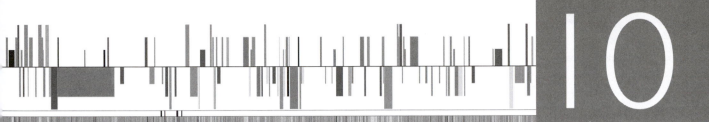

MEMBRANE STRUCTURE

THE LIPID BILAYER

TERMS TO LEARN

amphipathic	glycolipid	lipid raft
black membrane	hydrophilic	liposome
cholesterol	hydrophobic	phospholipid
ganglioside	lipid bilayer	plasma membrane

10–1 Hydrophobic molecules are said to "force adjacent water molecules to reorganize into icelike cages" (Figure 10–1). This statement seems paradoxical because water molecules do not interact with hydrophobic molecules. How could water molecules 'know' about the presence of a hydrophobic molecule and change their behavior to interact differently with one another? Discuss this seeming paradox and develop a clear concept of what is meant by an 'icelike' cage. How does it compare to ice? Why would such a cagelike structure be energetically unfavorable relative to pure water?

10–2* When a lipid bilayer is torn, why doesn't it seal itself by forming a 'hemi-micelle' cap at the edges, as shown in Figure 10–2?

10–3 Five students in your class always sit together in the front row. This could be because (1) they really like each other or (2) nobody else in your class wants to sit next to them. Which explanation holds for the assembly of a lipid bilayer? Explain your answer. If the lipid bilayer assembled for the opposite reason, how would its properties differ?

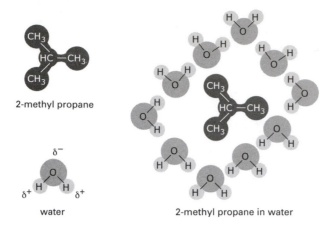

2-methyl propane

water

2-methyl propane in water

10–4* The properties of a lipid bilayer are determined by the structures of its lipid molecules. Predict the properties of the lipid bilayers that would result if the following were true:

A. Phospholipids had only one hydrocarbon chain instead of two.

B. The hydrocarbon chains were shorter than normal, say about 10 carbon atoms long.

C. All of the hydrocarbon chains were saturated.

D. All of the hydrocarbon chains were unsaturated.

E. The bilayer contained a mixture of two kinds of lipid molecule, one with two saturated hydrocarbon tails and the other with two unsaturated hydrocarbon tails.

F. Each lipid molecule were covalently linked through the end carbon atom of one of its hydrocarbon chains to a lipid molecule in the opposite monolayer.

10–5 While crossing the Limpopo River on safari in Africa, a friend of yours was bitten by a poisonous water snake and nearly died from extensive hemolysis. A true biologist at heart, he captured the snake before he passed out and has asked you to analyze the venom to discover the basis of its hemolytic activity. You find that the venom contains a protease, a neuraminidase (which removes sialic acid residues from gangliosides), and a phospholipase (which cleaves bonds in phospholipids). Treatment of isolated red blood cells with these purified enzymes gave the results shown in Table 10–1. Analysis of the products of hemolysis produced by phospholipase treatment showed an enormous increase in free phosphorylcholine (choline with a phosphate group attached) and diacylglycerol (glycerol with two fatty acid chains attached).

A. What is the substrate for the phospholipase, and where is it cleaved?

B. In light of what you know of the structure of the plasma membrane, explain why the phospholipase causes lysis of the red blood cells, but the protease and neuraminidase do not.

TABLE 10–1 Results of treatment of red blood cells with enzymes isolated from snake venom (Problem 10–5).

PURIFIED ENZYME	HEMOLYSIS
Protease	no
Neuraminidase	no
Phospholipase	yes

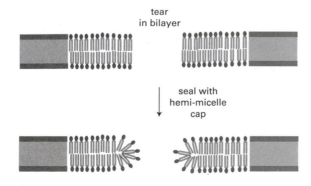

tear in bilayer

seal with hemi-micelle cap

Figure 10–2 A torn lipid bilayer sealed with a hypothetical 'hemi-micelle' cap (Problem 10–2).

10–6* (**True/False**) Although lipid molecules are free to diffuse in the plane of the bilayer, they cannot flip-flop across the bilayer unless enzyme catalysts called phospholipid translocators are present in the membrane. Explain your answer.

10–7 What is meant by the term 'two-dimensional fluid'?

10–8* A classic paper studied the behavior of lipids in the two monolayers of a membrane by labeling individual molecules with nitroxide groups, which are stable free radicals (Figure 10–3). Such spin-labeled lipids can be detected by electron spin-resonance (ESR) spectroscopy, a technique that does not disturb living cells. Spin-labeled lipids are introduced into cell membranes by incorporating them into small lipid vesicles, which are then fused with cells, thereby transferring the labeled lipids into the plasma membrane.

The two spin-labeled phospholipids shown in Figure 10–3 were incorporated into intact human red cell membranes in this way. To determine whether they were introduced equally into the two monolayers of the bilayer, ascorbic acid (vitamin C), which is a water-soluble reducing agent that does not cross membranes, was added to the medium to destroy any nitroxide radicals exposed on the outside of the cell. The ESR signal was followed as a function of time in the presence and absence of ascorbic acid as indicated in Figure 10–4.

A. Ignoring for the moment the difference in extent of loss of ESR signal, offer an explanation for why phospholipid 1 (Figure 10–4A) reacts faster with ascorbate than does phospholipid 2 (Figure 10–4B). Note that phospholipid 1 reaches a plateau in about 15 minutes, whereas phospholipid 2 takes almost an hour.

B. To investigate the difference in extent of loss of ESR signal with the two phospholipids, the experiments were repeated using red cell ghosts that had been resealed to make them impermeable to ascorbate (Figure 10–4C and D). In these experiments the loss of ESR signal for both phospholipids was negligible in the absence of ascorbate and reached a plateau at 50% in the presence of ascorbate. Offer an explanation for the difference in extent of

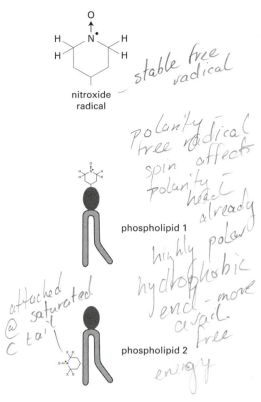

[handwritten annotations: stable free radical; polarity / free radical / spin affects / polarity / head / already; highly polar; hydrophobic / end – more / avail. / free energy; attached @ saturated C tail]

Figure 10–3 Structures of two nitroxide-labeled lipids (Problem 10–8). The nitroxide radical is shown at the *top*, and its position of attachment to the phospholipids is shown *below*.

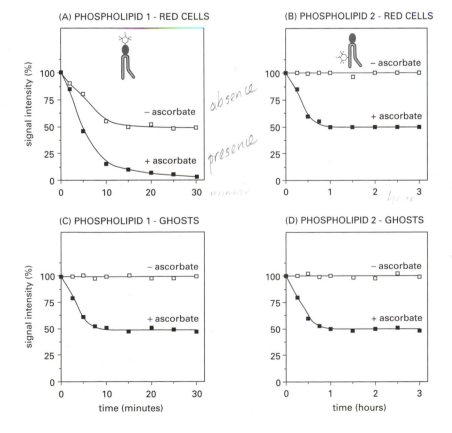

Figure 10–4 Decrease in ESR signal intensity as a function of time in intact red cells and red cell ghosts in the presence and absence of ascorbate (Problem 10–8). (A and B) Phospholipid 1 and phospholipid 2 in intact red cells. (C and D) Phospholipid 1 and phospholipid 2 in red cell ghosts.

[handwritten: signal = spin labled lipids detected by electron spin resonance microscopy ERS.]

loss of ESR signal in experiments with red cell ghosts (Figure 10–4C and D) versus those with intact red cells (Figure 10–4A and B).
C. Were the spin-labeled phospholipids introduced equally into the two monolayers of the red cell membrane?

10–9 Margarine is made from vegetable oil by a chemical process. Does this process convert saturated fatty acids to unsaturated ones or vice versa? Explain your answer.

10–10* Which one of the phospholipids listed below is present in very small quantities in the plasma membranes of mammalian cells, despite its crucial role in cell signaling?
A. Phosphatidylcholine
B. Phosphatidylethanolamine
C. Phosphatidylinositol
D. Phosphatidylserine
E. Sphingomyelin

10–11 You wish to determine the distribution of the phospholipids in the plasma membrane of the human red blood cell. Phospholipids make up 60% of the lipids in the red cell bilayer, with cholesterol (23%) and glycolipids (3%) accounting for most of the rest. To measure the distribution of individual phospholipids, you treat intact red cells and red cell ghosts (1) with two different phospholipases and (2) with a membrane-impermeant fluorescent reagent, abbreviated SITS, which specifically labels primary amine groups.

As summarized in Table 10–2, treatment with sphingomyelinase degrades up to 85% of the sphingomyelin in intact red cells (without causing lysis) and slightly more in red cell ghosts. The phospholipases in sea snake venom release only phosphatidylcholine breakdown products from intact cells (without causing lysis), but in red cell ghosts they degrade phosphatidylserine and phosphatidylethanolamine as well. SITS labels phosphatidylethanolamine and phosphatidylserine almost to completion in red cell ghosts but reacts less than 1% as well with intact red blood cells.
A. From these results deduce the distribution of the four principal phospholipids in red cell membranes. Which of the phospholipids, if any, are located in both monolayers of the membrane?
B. Why did you use red blood cells and not other animal cells for these experiments?

10–12* Predict which one of the following organisms will have the highest percentage of unsaturated fatty acid chains in their membranes. Explain your answer.
A. Antarctic fish
B. Desert iguana
C. Human being
D. Polar bear
E. Thermophilic bacterium

10–13 Within a monolayer, lipid molecules exchange places with their neighbors every 10^{-7} seconds. It takes about 1 second for a lipid molecule to diffuse from one end of a bacterium to the other, a distance of 2 μm.

TABLE 10–2 Sensitivity of phospholipids in human red cells and red cell ghosts to phospholipases and a membrane-impermeant label (Problem 10–11).

PHOSPHOLIPID	SPHINGOMYELINASE		SEA SNAKE VENOM		SITS FLUORESCENCE	
	RED CELLS	GHOSTS	RED CELLS	GHOSTS	RED CELLS	GHOSTS
Phosphatidylcholine	–	–	+	+	–	–
Phosphatidylethanolamine	–	–	–	+	–	+
Phosphatidylserine	–	–	–	+	–	+
Sphingomyelin	+	+	–	–	–	–

A. Are these numbers in agreement? Assume that the diameter of a lipid head group is 0.5 nm. Explain why or why not.

B. To gain an appreciation for the great speed of molecular motions, assume that a lipid molecule is the size of a ping-pong ball (4-cm diameter) and that the floor of your living room (6 m × 6 m) is covered wall to wall in a monolayer of balls. If two neighboring balls exchanged positions once every 10^{-7} seconds, how fast would they be moving in kilometers per hour? How long would it take for a ball to move from one end of the room to the other?

10–14* If a lipid raft is typically 70 nm in diameter and each lipid molecule has a diameter of 0.5 nm, about how many lipid molecules would there be in a lipid raft composed entirely of lipid? At a ratio of 50 lipid molecules per protein molecule (50% protein by mass) how many proteins would be in a typical raft? (Neglect the loss of lipid from the raft that would be required to accommodate the protein.)

10–15 If lipid rafts form because sphingolipid and cholesterol molecules preferentially associate, why don't they all aggregate into a single large raft instead of into multiple tiny rafts?

10–16* Why are lipid rafts thicker than other parts of the bilayer?

10–17 Lipid rafts are rich in both sphingolipids and cholesterol, and it is thought that cholesterol plays a central role in raft formation since lipid rafts apparently do not form in its absence. Why do you suppose that cholesterol is essential for formation of lipid rafts? (Hint: sphingolipids other than sphingomyelin have large head groups composed of several linked sugar molecules.)

10–18* GPI-anchored proteins are insoluble in ice-cold solutions of the detergent, Triton X-100. This is surprising since the hydrophobic portions of the GPI anchors might be expected to partition efficiently in detergent micelles, as indeed they do at higher temperatures. Because GPI anchors are added after protein synthesis during transport to the plasma membrane, you carry out a pulse-labeling experiment to determine when the proteins become insoluble. Cells that express a typical GPI-anchored protein called PLAP were labeled for 5 minutes in ^{35}S-methionine and quickly transferred to nonradioactive medium. At intervals, the cells were treated with ice-cold Triton X-100 and centrifuged. The supernatant (soluble) and pellet (insoluble) fractions were analyzed by immunoprecipitation with an antibody specific for PLAP and electrophoresis in an SDS polyacrylamide gel. The time course for recovery of PLAP in the supernatant (S) and pellet (P) fractions is shown in Figure 10–5A.

(A)

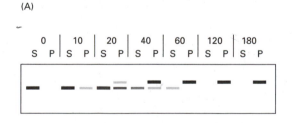

(B)

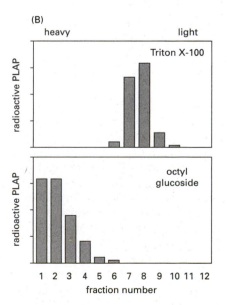

Figure 10–5 Analysis of a GPI-anchored protein (Problem 10–18). (A) Time course of appearance of PLAP in the supernatant and pellet fractions. Numbers indicate minutes after transfer to a nonradioactive medium. (B) Density of PLAP in Triton X-100 and octyl glucoside. Fractions are removed from the bottom of the tube; thus, the *bottom* of the gradient is at the *left* (heavy) and the *top* of the gradient is at the *right* (light).

A. When does PLAP become insoluble in Triton X-100? Why is there a shift in its mobility by SDS-gel electrophoresis? Why do you suppose that there is a time delay before PLAP becomes insoluble?

B. When the composition of the insoluble fraction was analyzed by centrifugation to equilibrium in a sucrose density gradient, PLAP was found to be located near the top of the gradient (low density) (Figure 10–5B). When solubilized by a different detergent, octyl glucoside, PLAP was found near the bottom of the gradient (high density) (Figure 10–5B). Why do you suppose PLAP is located at the top of the sucrose gradient in the Triton X-100-insoluble fraction, but at the bottom of the gradient when extracted with octyl glucoside?

C. Analysis of the lipids in the visibly milky band corresponding to fractions 7 and 8 from the Triton X-100-insoluble PLAP (Figure 10–5B) showed that they were enriched for sphingomyelin, other sphingolipids, and cholesterol, but depleted for phosphatidylethanolamine, phosphatidylcholine, and phosphatidylserine. Do these results support the idea of lipid rafts in the membrane? Why or why not?

D. When fractions 7 and 8 from the Triton X-100-insoluble PLAP were analyzed by electron microscopy, they were found to contain vesicles ranging from 100 to 1000 nm in diameter with PLAP embedded in their membranes. What is the surface area of a 100-nm sphere (Area = $4\pi r^2$)? How does this compare with the surface area of a 70-nm lipid raft (Area = πr^2)? Does this microscopic analysis support the idea of lipid rafts in membranes? Why or why not?

10–19 Sphingolipids, cholesterol, and some proteins form detergent-insoluble complexes in ice-cold Triton X-100, which constituted the first evidence in favor of lipid rafts in membranes. By itself, such evidence is unsatisfying as insoluble complexes could be artifacts of the extraction method. A sensitive form of fluorescence resonance energy transfer (FRET) has been used to investigate the existence of lipid rafts in living cells in a way that avoids this potential artifact.

When a fluorophore on a membrane protein is illuminated with plane-polarized light, its fluorescent emission will have the same polarization as the incident light because the protein in the membrane rotates slowly relative to the time for emission. However, if the absorbed energy is first transferred to a nearby fluorophore (by FRET) the subsequently emitted fluorescent light will be oriented differently than the incident light because the nearby protein will nearly always have a different orientation in the membrane than the original molecule. Thus, the loss of polarization is a sensitive measure of FRET between fluorescently tagged membrane proteins.

To use this technique to investigate lipid rafts, two cell lines were obtained that expressed different forms of the monomeric folate receptor in their plasma membranes: a GPI-anchored form and a transmembrane-anchored form. The folate receptors in each cell line were made fluorescent by addition of a fluorescent folate analog. Cells tagged in this way showed variation in fluorescence intensity over their surface because of chance variations in the distributions of the labeled receptors.

Polarization of the emitted light from individual pixels (1 μm^2, tiny areas of the cell surface, but much larger than individual rafts) was measured with two filters: one parallel to the incident light to capture direct emission; one perpendicular to the incident light to capture emission after FRET. The density (concentration) of receptors in the pixel was measured by the total fluorescence intensity. Polarization of the fluorescent light was measured as intensity detected via the parallel filter (I_{par}) minus intensity detected via the perpendicular filter (I_{perp}) divided by the total intensity.

A. The expectations for dispersed receptors versus receptors in lipid rafts are shown in Figure 10–6. Explain these expectations.

B. The actual experiments showed that transmembrane-anchored folate receptors followed the expectations shown in Figure 10–6A, whereas the

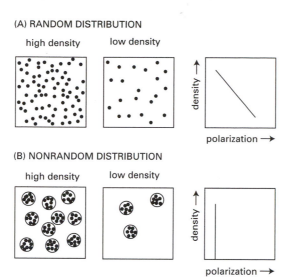

(A) RANDOM DISTRIBUTION

high density   low density

(B) NONRANDOM DISTRIBUTION

high density   low density

Figure 10–6 Expectations for polarization of fluorescence at different densities of receptors (Problem 10–19). (A) Randomly distributed receptors. The *boxed area* represents a pixel and the *solid dots* represent fluorescent receptors. (B) Receptors clustered in microdomains. *Circles* represent microdomains such as lipid rafts.

GPI-anchored folate receptors followed those in Figure 10–6B. Do these experiments provide evidence for the existence of lipid rafts in the plasma membrane? Why or why not?

C. When these same cells were grown in the presence of compactin, which inhibits cholesterol synthesis, both types of folate receptor gave results like those in Figure 10–6A. What bearing does this observation have on the existence of lipid rafts? Explain your reasoning.

10–20* The lipid bilayers found in cells are fluid, yet asymmetrical. Is this a paradox? Explain your answer.

10–21 (**True/False**) All of the common phospholipids—phosphatidylcholine, phosphatidylethanolamine, phosphatidylserine, and sphingomyelin—carry a positively charged moiety on their head group, but none carry a net positive charge. Explain your answer.

10–22* The asymmetric distribution of phospholipids in the two monolayers of the plasma membrane implies that very little spontaneous flip-flop occurs or, alternatively, that any spontaneous flip-flop is rapidly corrected by phospholipid translocators that return phospholipids to their appropriate monolayer. The rate of phospholipid flip-flop in the plasma membrane of intact red blood cells has been measured to decide between these alternatives.

One experimental measurement used the same two spin-labeled phospholipids described in Problem 10–8 (see Figure 10–3). To measure the rate of flip-flop from the cytoplasmic to the outer monolayer, red cells with spin-labeled phospholipids exclusively in the cytoplasmic monolayer were incubated for various times in the presence of ascorbate and the loss of ESR signal was followed. To measure the rate of flip-flop from the outer to the cytoplasmic monolayer, red cells with spin-labeled phospholipids exclusively in the outer monolayer were incubated for various times in the absence of ascorbate and the loss of ESR signal was followed. The results of these experiments are illustrated in Figure 10–7.

A. From the results in Figure 10–7, estimate the rate of flip-flop from the cytoplasmic to the outer monolayer and from the outer to the cytoplasmic monolayer. A convenient way to express such rates is as the half-time of flip-flop—that is, the time it takes for half the phospholipids to flip-flop from one monolayer to the other.

B. From what you learned about the behavior of the two spin-labeled phospholipids in Problem 10–8, deduce which one was used to label the cytoplasmic monolayer of the intact red blood cells, and which one was used to label the outer monolayer.

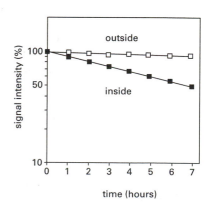

Figure 10–7 Decrease in ESR signal intensity of red cells containing spin-labeled phospholipids in the outer monolayer (outside) and cytoplasmic monolayer (inside) of the plasma membrane (Problem 10–22).

C. Using the information in this problem, propose a method to generate intact red cells that contain spin-labeled phospholipids exclusively in the cytoplasmic monolayer, and a method to generate cells spin labeled exclusively in the outer monolayer.

10–23   Phosphatidylserine, which is normally confined to the cytoplasmic monolayer of the plasma membrane lipid bilayer, is redistributed to the outer monolayer during apoptosis. How is this redistribution accomplished?

10–24*   (**True/False**) Glycolipids are never found on the cytoplasmic face of membranes in living cells. Explain your answer.

MEMBRANE PROTEINS

TERMS TO LEARN

ankyrin	electron crystallography	lectin
bacteriorhodopsin	freeze-fracture electron microscopy	peripheral membrane protein
band 3 protein	glycocalyx	porin
cell coat	GPI anchor	spectrin
detergent	integral membrane protein	transmembrane protein

10–25   (**True/False**) The basic structure of biological membranes is determined by the lipid bilayer, but their specific functions are carried out largely by proteins. Explain your answer.

10–26*   Which of the arrangements of membrane-associated proteins indicated in Figure 10–8 have been found in biological membranes?

10–27   Compare the hydrophobic forces that hold a membrane protein in the lipid bilayer to those that help proteins fold into a unique three-dimensional structure.

10–28*   Which one of the following statements correctly describes the mass ratio of lipids to proteins in membranes?
A. The mass of lipids greatly exceeds the mass of proteins.
B. The mass of proteins greatly exceeds the mass of lipids.
C. The mass of lipids and proteins is about equal.
D. The mass ratio of lipids to proteins varies widely in different membranes.

10–29   Name the three types of lipid anchors that are used to attach proteins to membranes.

10–30*   (**True/False**) Hydropathy plots are useful for identifying hydrophobic polypeptide segments that are long enough to span a membrane as an α helix or as a β sheet. Explain your answer.

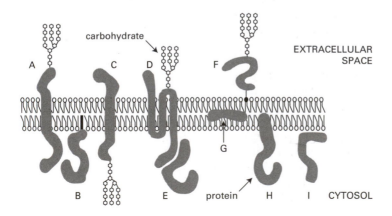

Figure 10–8 A variety of possible associations of proteins with a membrane (Problem 10–26).

10–31 Proteins that span a membrane as an α helix have a characteristic structure in the region of the bilayer. Which of the three 20-amino acid sequences listed below is the most likely candidate for such a transmembrane segment? Explain the reasons for your choice. (See inside back cover for one-letter amino acid code.)

   A. I T L I Y F G V M A G V I G T I L L I S
   B. I T P I Y F G P M A G V I G T P L L I S
   C. I T E I Y F G R M A G V I G T D L L I S

10–32* Consider a transmembrane protein complex that forms a hydrophilic pore across the plasma membrane of a eucaryotic cell. The pore is made of five similar protein subunits, each of which contributes a membrane-spanning α helix to form the pore. Each α helix has hydrophilic amino acid side chains on one side of the helix and hydrophobic amino acid side chains on the opposite side. Propose a possible arrangement of these five α helices in the membrane.

10–33 Proteins that form a β-barrel pore in the membrane have several β strands that span the membrane. The pore-facing side of each strand carries hydrophilic amino acid side chains, whereas the bilayer-facing sides carry hydrophobic amino acid side chains. Which of the three 10-amino acid sequences listed below is the most likely candidate for a transmembrane β strand in a β-barrel pore? Explain the reasons for your choice. (See inside back cover for one-letter amino acid code.)

   A. A D F K L S V E L T
   B. A F L V L D K S E T
   C. A F D K L V S E L T

10–34* Why is it that membrane-spanning protein segments are almost always α helices or β barrels, but never disordered chains?

10–35 Why are α helices more common than β barrels in transmembrane proteins?

10–36* (**True/False**) Whereas all the carbohydrate in the plasma membrane faces outward on the external surface of the cell, all the carbohydrate on internal membranes faces toward the cytosol. Explain your answer.

10–37 You are studying the binding of proteins to the cytoplasmic face of cultured neuroblastoma cells and have found a method that gives a good yield of inside-out vesicles from the plasma membrane. Unfortunately, your preparations are contaminated with variable amounts of right-side-out vesicles. Nothing you have tried avoids this variable contamination. A friend suggests that you pass your vesicles over an affinity column made of lectin coupled to solid beads. What is the point of your friend's suggestion?

10–38* Why is it that intrachain (and interchain) disulfide (S–S) bonds form readily between cysteine side chains outside the cell, but not in the cytosol?

10–39 Detergents are small amphipathic molecules—one end hydrophobic and the other hydrophilic—that tend to form micelles in water. Examine the structures of SDS and Triton X-100 in Figure 10–9 and explain why the black portions are hydrophilic and the gray sections are hydrophobic.

10–40* (**True/False**) Human red blood cells contain no internal membranes other than the nuclear membrane. Explain your answer.

10–41 Why does a red blood cell membrane need proteins?

10–42* Pythagoras forbade his followers to eat fava beans. Beyond the political implications (Greeks voted with beans), there turns out to be a rational basis for this proscription. In the Middle East defective forms of the gene encoding glucose 6-phosphate dehydrogenase (G6PD) are common. These mutant forms of the gene typically reduce G6PD activity to about 10% of normal. They have been selected for in the Middle East, and in other areas

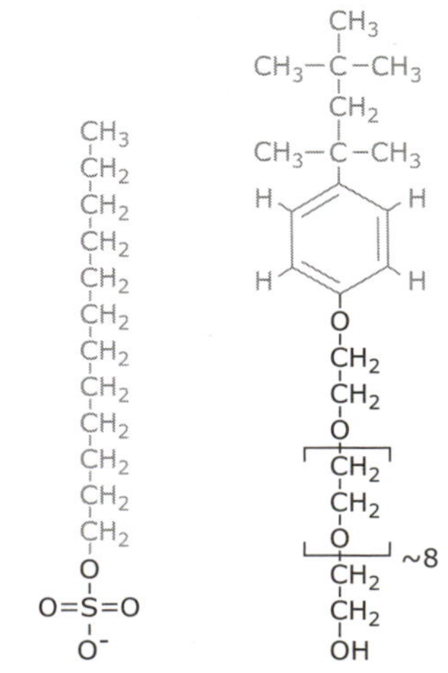

sodium dodecyl sulfate (SDS)   Triton X-100

Figure 10–9 The structures of SDS and Triton X-100 (Problem 10–39).

of the world where malaria is common, because they afford protection against the malarial parasite. G6PD controls the first step in the pathway for NADPH production. A lower-than-normal level of NADPH in red cells creates an environment unfavorable for growth of the protist *Plasmodium falciparum*, which causes malaria.

Although somewhat protected against malaria, G6PD-deficient individuals occasionally have other problems. NADPH is the principal agent required to keep the red cell cytosol in a properly reduced state, constantly converting transient disulfide bonds (–S–S–) back to sulfhydryls (–SH HS–). When a G6PD-deficient individual eats raw or undercooked fava beans, an oxidizing substance in the beans overwhelms the reducing capacity of the red cells, leading to a severe—sometimes life-threatening—hemolytic anemia. How do you suppose eating fava beans leads to anemia?

10–43 In the membrane of a human red blood cell the ratio of the mass of protein (average molecular weight 50,000) to phospholipid (average molecular weight 800) to cholesterol (molecular weight 386) is about 2:1:1. How many lipid molecules are there for every protein molecule?

10–44* Enzymatic digestion of sealed right-side-out red cell ghosts was originally used to determine the sidedness of the major membrane-associated proteins: spectrin, band 3, and glycophorin. These experiments made use of sialidase, which removes sialic acid from protein, and pronase, which cleaves peptide bonds. The proteins from normal ghosts and enzyme-treated ghosts were separated by SDS polyacrylamide-gel electrophoresis and then stained for protein and carbohydrate (Figure 10–10).

A. How does the information in Figure 10–10 allow you to decide whether the carbohydrate of glycophorin is on the cytoplasmic or external surface, and how does it allow you to decide which of the red cell proteins are exposed on the external side of the cell?

B. When you show your deductions to a colleague, she challenges your conclusion that some proteins are not exposed on the external surface and suggests instead that these proteins may be resistant to pronase digestion. What control experiment can you propose to test this possibility?

C. How would you modify this enzymatic approach in order to determine which red cell proteins span the plasma membrane?

10–45 Estimates of the number of membrane-associated proteins per cell and the fraction of the plasma membrane occupied by such proteins provide a useful quantitative basis for understanding the structure of the plasma membrane. These calculations are straightforward for proteins in the plasma membrane of a red blood cell because red cells are readily isolated from blood and they contain no internal membranes to confuse the issue. Plasma membranes are prepared, the membrane-associated proteins are separated by SDS polyacrylamide-gel electrophoresis, and then they are stained with a dye (Coomassie Blue). Because the intensity of color is roughly proportional to the mass of protein present in a band, quantitative estimates can be made as shown in Table 10–3.

A. From the information in Table 10–3, calculate the number of molecules of spectrin, band 3, and glycophorin in an individual red blood cell. Assume that 1 mL of red cell ghosts contains 10^{10} cells and 5 mg of total membrane protein.

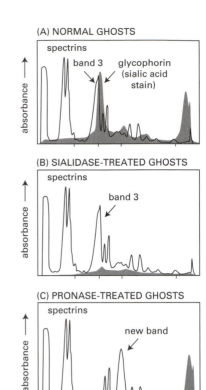

Figure 10–10 Analysis of proteins in the plasma membrane of red blood cells (Problem 10–44). (A) Untreated red cell ghosts. (B) Sialidase-treated red cell ghosts. (C) Pronase-treated red cell ghosts. Membrane-associated proteins were separated by SDS polyacrylamide-gel electrophoresis and then stained for protein and for carbohydrate. *Lines* indicate the distribution of proteins, and *shaded regions* indicate the distribution of carbohydrate.

TABLE 10–3 Proportion of stain associated with three membrane-associated proteins (Problem 10–45).

PROTEIN	MOLECULAR WEIGHT	PERCENT OF STAIN
Spectrin	250,000	25
Band 3	100,000	30
Glycophorin	30,000	2.3

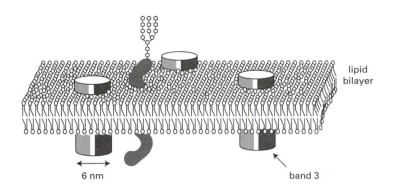

Figure 10–11 Schematic diagram of band 3, represented as a cylinder, in the plasma membrane (Problem 10–45).

B. Calculate the fraction of the plasma membrane that is occupied by band 3. Assume that band 3 is a cylinder 3 nm in radius and 10 nm in height and is oriented in the membrane as shown in Figure 10–11. The total surface area of a red cell is 10^8 nm^2.

10–46* One difficult problem in molecular biology is to define the associations between different proteins in complex assemblies. The associations involving spectrin, ankyrin, band 3, and actin, which generate the filamentous meshwork on the cytoplasmic surface of the red cell plasma membrane, have been investigated in several ways. One general method is to use antibodies that are specific for individual proteins. A mixture of two proteins is incubated together, and then an antibody specific for one of them is added. The resulting antibody–protein complexes are then precipitated and analyzed. This technique, when applied to pairwise mixtures of spectrin, ankyrin, band 3, and actin, yields the results summarized in Table 10–4. From the information in the table, deduce the associations between these proteins.

10–47 Glycophorin in the plasma membrane of the red blood cell normally exists as a homodimer that is held together entirely by interactions between its transmembrane domains. Since transmembrane domains are hydrophobic, how is it that they can associate with one another so specifically?

10–48* (**True/False**) Each molecule of bacteriorhodopsin contains a single chromophore called retinal, which, when activated by a photon of light, causes a series of small conformational change in the protein that results in the transfer of protons from the inside to the outside of the cell. Explain your answer.

10–49 Look carefully at the transmembrane band-3 proteins in Figure 10–12A. Imagine that you could mark these proteins specifically with a fluorescent group and measure their mobility by fluorescence recovery after photobleaching (FRAP). Sketch the recovery curve you would expect to see with time after photobleaching a small spot in the membrane (Figure 10–12B).

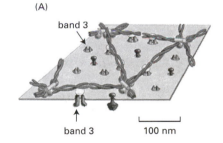

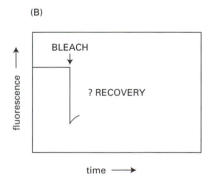

Figure 10–12 Mobility of membrane proteins (Problem 10–49). (A) Model of red cell membrane. (B) Recovery of fluorescence after photobleaching a red cell plasma membrane containing band-3 protein tagged with a fluorescent group.

TABLE 10–4 Precipitation of red cell plasma membrane proteins by antibodies specific for individual proteins (Problem 10–46).

PROTEIN MIXTURE	ANTIBODY SPECIFICITY	PROTEINS IN PELLET
1. Band 3 + actin	actin	actin
2. Band 3 + spectrin	spectrin	spectrin
3. Band 3 + ankyrin	ankyrin	band 3 + ankyrin
4. Actin + spectrin	spectrin	actin + spectrin
5. Actin + ankyrin	ankyrin	ankyrin
6. Spectrin + ankyrin	spectrin	spectrin + ankyrin

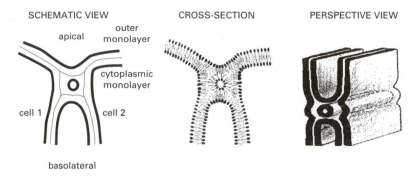

(A) LIPID MODEL

SCHEMATIC VIEW   CROSS-SECTION   PERSPECTIVE VIEW

apical   outer monolayer

cytoplasmic monolayer

cell 1   cell 2

basolateral

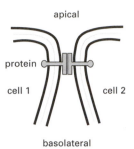

(B) PROTEIN MODEL

apical

protein

cell 1   cell 2

basolateral

Figure 10–13 Models for tight-junction structure (Problem 10–52). (A) Three views of the cylinder of lipids that is proposed to form a tight junction according to the lipid model. (B) Schematic representation of the protein model for tight-junction structure.

10–50* (**True/False**) Although membrane domains with different protein compositions are well known, there are at present no examples of membrane domains that differ in lipid composition. Explain your answer.

10–51 Describe the different methods that cells use to restrict proteins to specific regions of the plasma membrane. Is a membrane with many anchored proteins still fluid?

10–52* In the epithelial cells that line the tubules of the kidney, tight junctions confine certain plasma membrane proteins to the apical surface and others to the basolateral surface. Two structures for tight junctions have been proposed. In one ingenious model, which is based on observations from freeze-fracture electron microscopy, each sealing strand in a tight junction results from a membrane fusion that forms cylinders of lipid at the points of fusion (Figure 10–13A). The other model proposes that each strand of a tight junction is formed by a chain of transmembrane proteins whose extracellular domains bind to one another to seal the epithelial sheet (Figure 10–13B).

As you compare these lipid and protein models for tight junctions, you realize that they might be distinguishable on the basis of lipid diffusion between the apical and basolateral surfaces. Both models predict that lipids in the cytoplasmic monolayer will be able to diffuse freely between the apical and the basolateral surfaces. However, the models suggest different fates for lipids in the outer monolayer. In the lipid model, lipids in the outer monolayer will be confined to either the apical surface or the basolateral surface, since the cylinder of lipids interrupts the outer monolayer, preventing diffusion through it. By contrast, in the protein model the apical and basolateral surfaces appear to be connected by a continuous outer monolayer, suggesting that lipids in the outer monolayer should be able to diffuse freely between the two surfaces.

You have exactly the experimental tools to resolve this issue! You have been working with a line of dog kidney cells that forms an exceptionally tight epithelium with well-defined apical and basolateral surfaces. In addition, after infection with influenza virus, the cells express a fusogenic protein only on their apical surface. This feature allows you to fuse liposomes specifically to the apical surface of infected cells very efficiently by brief exposure to low pH, which activates the fusogenic protein. Thus, you can add fluorescently labeled lipids to the apical surface and detect their migration to the basolateral surface using fluorescence microscopy.

For the experiment you prepare two sets of labeled liposomes: one with a fluorescent lipid only in the outer monolayer, the other with the fluorescent lipid equally distributed between the cytoplasmic and outer monolayers. You fuse these two sets of liposomes to epithelia in which about half the cells were infected with virus. By adjusting the focal plane of the microscope, you examine the apical and basolateral surfaces for fluorescence. As a control,

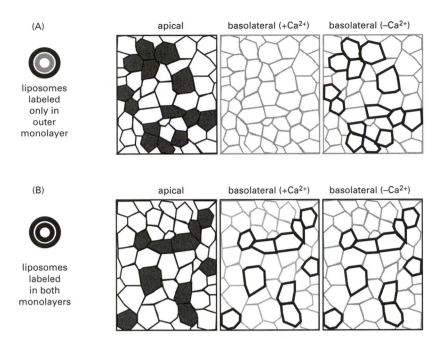

(A)

liposomes labeled only in outer monolayer

apical basolateral (+Ca²⁺) basolateral (–Ca²⁺)

(B)

liposomes labeled in both monolayers

apical basolateral (+Ca²⁺) basolateral (–Ca²⁺)

Figure 10–14 Experimental test of the lipid and protein models for tight-junction structure (Problem 10–52). (A) Liposomes labeled in outer monolayer. (B) Liposomes labeled in both monolayers. Only about half the cells in the epithelium in each experiment were infected with virus. Only infected cells are competent to fuse with the labeled liposomes under the conditions of the experimental protocol.

you remove Ca^{2+} from the medium—a treatment that disrupts tight junctions—and reexamine the basolateral surface. The results are shown in Figure 10–14.

You are delighted! These results show clearly that the lipids in the outer monolayer are confined to the apical surface, whereas lipids in the cytoplasmic monolayer diffuse freely between the apical and basolateral surfaces. Triumphantly, you show these results to your advisor as proof that the lipid model for tight-junction structure is correct. He examines your results carefully, shakes his head knowingly, gives you that penetrating look of his, and tells you that, although the experiments are exquisitely well done, you have drawn exactly the wrong conclusion. These results prove that the lipid model is incorrect.

What has your advisor seen in the data that you have overlooked? How do your results disprove the lipid model? If the protein model is correct, why do you think it is that the fluorescent lipids are confined to the apical surface?

10–53 Many cells interact with the extracellular matrix. Why is there any ambiguity in 'where the plasma membrane ends and the extracellular matrix begins'?

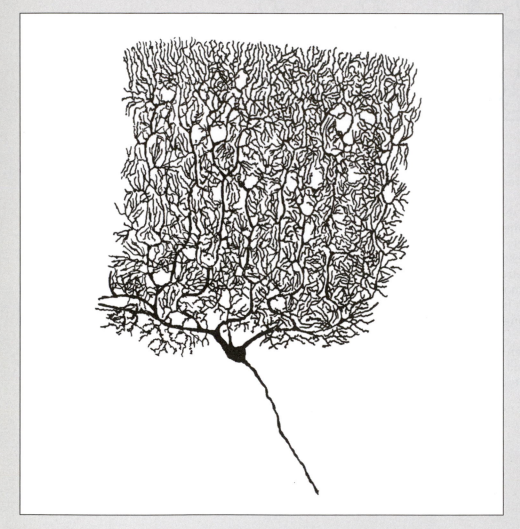

A Purkinje cell of the cerebellum,
"…because of a simple combinatorial trick, each of the 15 million Purkinje cells in the cerebellum is capable of learning over 200 different patterns and discriminating them from unlearned patterns" David Marr, *Vision*. An important point is that this tree is almost completely two-dimensional. The Purkinje cell is essentially flat. Based on drawings by S. Ramón y Cajal, Histologie du Système nerveux de l'Homme et des Vertébrés. Paris: Maline, 1909–1911.

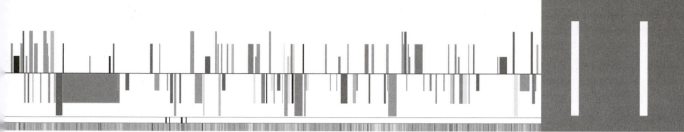

MEMBRANE TRANSPORT OF SMALL MOLECULES AND THE ELECTRICAL PROPERTIES OF MEMBRANES

PRINCIPLES OF MEMBRANE TRANSPORT

TERMS TO LEARN

active transport	channel protein	membrane transport protein
aquaporin	electrochemical gradient	mobile ion carrier
carrier protein	facilitated diffusion	passive transport
channel former	ionophore	valinomycin

11–1   (**True/False**) The plasma membrane is highly impermeable to all charged molecules. Explain your answer.

11–2*   Order the molecules on the following list according to their ability to diffuse through a lipid bilayer, from most to least permeant. Explain your order.
1. Ca^{2+}
2. CO_2
3. Ethanol
4. Glucose
5. RNA
6. H_2O

11–3   If a frog egg and a red blood cell are placed in pure water, the red blood cell will swell and burst, but the frog egg will remain intact. Although a frog egg is about one million times larger than a red cell, they both have nearly identical internal concentrations of ions so that the same osmotic forces are at work in each. Why do you suppose red blood cells burst in water, while frog eggs do not?

11–4*   Carrier proteins were once imagined to operate like revolving doors, transporting their ligands while maintaining a sealed lipid bilayer. Why do you suppose this is no longer considered a likely way for carrier proteins to transport their ligands?

11–5 Why are the maximum rates of transport by carrier proteins and channel proteins thought to be so different?

11–6* Discuss the following analogy: "The differences between transporting a ligand by a channel or a carrier protein are like the differences between crossing a river by a bridge or a ferry."

11–7 A simple enzyme reaction can be represented by the equation

$$E + S \rightleftharpoons ES \rightarrow E + P$$

where E is the enzyme, S is the substrate, P is the product, and ES is the enzyme–substrate complex.

A. Write a corresponding equation describing a carrier protein (CP) that mediates transport of a solute (S) down its concentration gradient.

B. The Michaelis–Menten equation for the simple enzyme reaction above is

$$\text{rate} = V_{\max} \frac{[S]}{[S] + K_M}$$

Where 'rate' is the initial rate of the reaction, $V_{\max}$ is the maximum rate of the enzyme-catalyzed reaction, and K_M is the Michaelis constant. Write the corresponding Michaelis–Menten equation for the process of solute transport by a carrier protein. What do rate, $V_{\max}$, and K_M mean in the equation for transport. (You can review these equations for enzyme kinetics in Problems 3–56 to 3–61.)

C. Would these equations provide an appropriate description for channel proteins? Why or why not?

11–8* Some bacterial cells can grow on either ethanol (CH_3CH_2OH) or acetate (CH_3COO^-) as their only carbon source. Initial rates of entry of these two molecules into bacterial cells as a function of external concentration are shown in Table 11–1, but the identity of the molecules is not shown.

A. Plot the data from the table as initial rate versus concentration.

B. What can you say about the transport of the two molecules from your plot of the data? From the graphs deduce which molecule corresponds with which curve. Rationalize your choice.

C. Estimate the $V_{\max}$ and K_M values for these two molecules.

11–9 Cytochalasin B, which is often used as an inhibitor of actin-based motility systems, is also a very potent competitive inhibitor of D-glucose uptake into mammalian cells. When red blood cell ghosts are incubated with ^{3}H-cytochalasin B and then irradiated with ultraviolet light, the cytochalasin becomes cross-linked to the glucose transporter, GLUT1. Cytochalasin is not cross-linked to the transporter if an excess of D-glucose is present during the labeling reaction; however, an excess of L-glucose (which is not transported) does not interfere with labeling.

If membrane proteins from labeled ghosts are separated by SDS polyacrylamide-gel electrophoresis, GLUT1 appears as a fuzzy radioactive band extending from 45,000 to 70,000 daltons. If labeled ghosts are treated with an

TABLE 11–1 Initial rates of transport of ethanol and acetate (Problem 11–8).

	INITIAL RATE OF TRANSPORT (μmol/min)	
CARBON SOURCE (mM)	MOLECULE A	MOLECULE B
0.1	2.0	18
0.3	6.0	46
1.0	20	100
3.0	60	150
10.0	200	182

enzyme that removes attached sugars before electrophoresis, the fuzzy band disappears and a much sharper band at 46,000 daltons takes its place.

A. Why does an excess D-glucose, but not L-glucose, prevent cross-linking of cytochalasin to GLUT1?

B. Why does GLUT1 appear as a fuzzy band on SDS polyacrylamide gels?

11–10* Insulin is a small protein hormone that binds to a receptor in the plasma membrane of fat cells. This binding dramatically increases the rate of uptake of glucose into the cells. The increase occurs within minutes and is not blocked by inhibitors of protein synthesis or glycosylation. Therefore, insulin must increase the activity of the glucose transporter, GLUT4, in the plasma membrane without increasing the total number of GLUT4 molecules in the cell.

The two experiments described below suggest a possible mechanism for the insulin effect. In the first experiment, the initial rate of glucose uptake in control and insulin-treated cells was measured, with the results shown in Figure 11–1. In the second experiment, the concentration of GLUT4 in fractionated membranes from control and insulin-treated cells was measured, using the binding of radioactive cytochalasin B as the assay (see Problem 11–9), as shown in Table 11–2.

A. Deduce the mechanism by which glucose transport by GLUT4 is increased in insulin-treated cells.

B. Does insulin stimulation alter either the K_M or the V_{max} of GLUT4? How can you tell from these data?

11–11 Brain cells, which depend on glucose for energy, use the glucose transporter GLUT3, which has a K_M of 1.5 mM. Liver cells, which store glucose (as glycogen) after a meal and release glucose between meals, use the glucose transporter GLUT2, which has a K_M of 15 mM.

A. Calculate the rate (as a percentage of V_{max}) of glucose uptake in brain cells and in liver cells at circulating glucose concentrations of 3 mM (starvation conditions), 5 mM (normal levels), and 7 mM (after a carbohydrate-rich meal). Rearranging the Michaelis–Menten equation gives

$$\frac{\text{rate}}{V_{max}} = \frac{[S]}{[S] + K_M}$$

B. Although the concentration of glucose in the general circulation normally doesn't rise much above 7 mM, the liver is exposed to much higher concentrations after a meal. The intestine delivers glucose into the portal circulation, which goes directly to the liver. In the portal circulation the concentration of glucose can be as high as 15 mM. Under these conditions, at what fraction of the maximum rate (V_{max}) do liver cells import glucose?

C. Do these calculations fit with the physiological functions of brain and liver cells? Why or why not?

11–12* Cells use transporters to move nearly all metabolites across membranes. But how much faster is a transporter than simple diffusion? There is sufficient information available for glucose transporters to make a comparison. The normal circulating concentration of glucose in humans is 5 mM, while the intracellular concentration is usually very low.

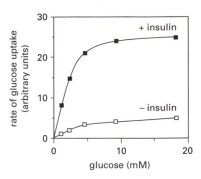

Figure 11–1 Rate of glucose uptake into cells in the presence and absence of insulin (Problem 11–10).

TABLE 11–2 Amount of GLUT4 associated with the plasma membrane and internal membranes in the presence and absence of insulin (Problem 11–10).

MEMBRANE FRACTION	BOUND ³H-CYTOCHALASIN B (cpm/mg vesicle protein)	
	UNTREATED CELLS (– INSULIN)	TREATED CELLS (+ INSULIN)
Plasma membrane	890	4480
Internal membranes	4070	80

(A)

side view

face view

(B)

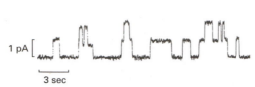

1 pA

3 sec

current trace from patch clamp

Figure 11–2 Channels induced in a membrane by gramicidin A (Problem 11–15). (A) Space-filling model of gramicidin A in side and face views. (B) Currents across a lipid bilayer exposed to a minute amount of gramicidin A. The timescale of the trace is shown in seconds (sec) and the current is shown in picoamperes (pA). A 1-pA current corresponds to about 6×10^6 ions per second.

A. At what rate (molecules/sec) would glucose diffuse into a cell if there were no transporter? The permeability coefficient for glucose is 3×10^{-8} cm/sec. Assume a cell is a sphere with a diameter of 20 μm. The rate of diffusion equals the concentration difference multiplied by the permeability coefficient and the total surface area of the cell (surface area = $4\pi r^2$). (Remember to convert everything to compatible units so that the rate is molecules/sec.)

B. If in the same cell there are 10^5 GLUT3 molecules ($K_M = 1.5$ mM) in the plasma membrane, each of which can transport glucose at a maximum rate of 10^4 molecules per second, at what rate (molecules/sec) will glucose enter the cell? How much faster is transporter-mediated uptake of glucose than entry by simple diffusion?

11–13 How is it possible for some molecules to be at equilibrium across a biological membrane and yet not be at the same concentration on both sides?

11–14* (**True/False**) Transport by carrier proteins can be either active or passive, whereas transport by channel proteins is always passive. Explain your answer.

11–15 Gramicidin A is a channel-forming ionophore. Two gramicidin molecules come together end to end across the bilayer to form a pore that is selective for cations (Figure 11–2A). If a minute amount of gramicidin A is added to an artificial bilayer, movement of cations can be detected as changes in current across the bilayer, as shown in Figure 11–2B. What do the step-wise changes in current tell you about the gramicidin channel? Why are some peaks twice as high as others?

11–16* How would you expect the activities of a channel-forming ionophore and mobile ion carrier to change as you lowered the temperature of a lipid bilayer, increasing its viscosity?

CARRIER PROTEINS AND ACTIVE MEMBRANE TRANSPORT

TERMS TO LEARN

ABC transporter superfamily	Na⁺-K⁺ ATPase	symporter
antiporter	Na⁺-K⁺ pump (Na⁺ pump)	transcellular tranporter
Ca²⁺ pump (Ca²⁺ ATPase)	osmolarity	uniporter
lactose permease	P-type Ca²⁺ ATPase	V-type ATPase
multidrug resistance (MDR) protein		

11–17 What are the three main ways cells carry out active transport? Briefly describe each.

11–18* Which of the ions listed in Table 11–3 could be used to drive an electrically neutral coupled transport of a solute across the plasma membrane? Indicate the direction of movement of the listed ions (inward or outward) and indicate what sort of ion would be co-transported to preserve electrical neutrality.

Incidentally, there is a glaring intracellular deficiency of anions relative to cations in Table 11–3, yet cells are electrically neutral. What anions do you suppose are missing from this table?

TABLE 11–3 A comparison of ion concentrations inside and outside a typical mammalian cell (Problem 11–18).

COMPONENT	INTRACELLULAR CONCENTRATION (mM)	EXTRACELLULAR CONCENTRATION (mM)
Cations		
Na^+	5–15	145
K^+	140	5
Mg^{2+}	0.5	1–2
Ca^{2+}	10^{-4}	1–2
H^+	7×10^{-5} ($10^{-7.2}$ M or pH 7.2)	4×10^{-5} ($10^{-7.4}$ M or pH 7.4)
Anions		
Cl^-	5–15	110

11–19 (**True/False**) A symporter would function as an antiporter if its orientation in the membrane were reversed (that is, if the portion of the protein normally exposed to the cytosol faced the outside of the cell instead). Explain your answer.

11–20* A transmembrane protein has the following properties: it has two binding sites, one for solute A and one for solute B. The protein can undergo a conformational change to switch between two states: either both binding sites are exposed exclusively on one side of the membrane or both binding sites are exposed exclusively on the other side of the membrane. The protein can switch between the two conformational states only if both binding sites are occupied or if both binding sites are empty, but cannot switch if only one binding site is occupied.

   A. What kind of a transporter do these properties define?
   B. Do you need to specify any additional properties to turn this protein into a symporter that couples the movement of solute A up its concentration gradient to the movement of solute B down its electrochemical gradient?
   C. Write a set of rules that defines the properties of an antiporter.

11–21 A model for a uniporter that could mediate passive transport of glucose down its concentration gradient is shown in Figure 11–3. How would you need to change the diagram to convert the carrier protein into a pump that transports glucose up its concentration gradient by hydrolyzing ATP? Explain the need for each of the steps in your new illustration.

11–22* Ion transporters are 'linked' together—not physically, but as a consequence of their actions. For example, cells can raise their intracellular pH, when it becomes too acidic, by exchanging external Na^+ for internal H^+, using a Na^+-H^+ antiporter. The change in internal Na^+ is then redressed using the Na^+-K^+ pump.

   A. Can these two transporters, operating together, normalize both the H^+ and the Na^+ concentrations inside the cell?

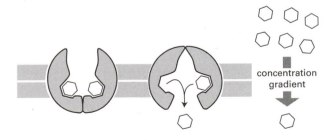

concentration gradient

Figure 11–3 Hypothetical model showing how a conformational change in a carrier protein could mediate passive transport of glucose (Problem 11–21). The transition between the two conformational states is proposed to occur randomly and to be completely reversible, regardless of binding site occupancy.

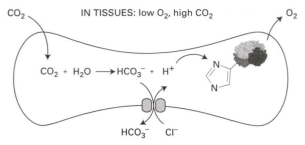 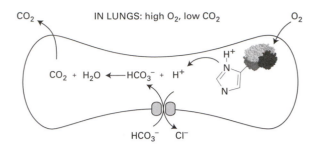

Figure 11–4 Red blood cell-mediated transport of CO_2 from the tissues to the lungs (Problem 11–25). Low and high O_2 and CO_2 refer to their concentrations, or partial pressures.

B. Why does the linked action of these two pumps cause imbalances in both the K^+ concentration and the membrane potential? Explain how the K^+ leak channel corrects these imbalances?

11–23 Why does export of HCO_3^- out of a cell via the Cl^--HCO_3^- exchanger lower the intracellular pH?

11–24* Cells continually generate CO_2 as a consequence of the oxidation of glucose and fatty acids required to meet their energy needs.
A. Unopposed by any transport process, what effect would this process have on intracellular pH?
B. How does the Na^+-driven Cl^--HCO_3^- exchanger counteract the changes in intracellular pH?
C. Does the combination of metabolic oxidation and the action of the Na^+-driven Cl^--HCO_3^- exchanger increase the total concentration of $CO_2 + HCO_3^-$ in the cell, decrease it, or keep it constant?
D. How do you suppose CO_2 exits the cell?

11–25 CO_2 is removed from the body through the lungs in a process that is mediated by red blood cells, as summarized in Figure 11–4. Transport of CO_2 is coupled to the transport of O_2 through hemoglobin. Upon release of O_2, hemoglobin undergoes a conformational change that raises the pK of a histidine side chain, allowing it to bind an H^+.
A. To what extent does the intracellular pH of the red blood cell vary during its movement from the tissues to the lungs and back, and why is this so?
B. In what form and where, is the CO_2 during its movement from the tissues to the lungs?
C. How is it that the Cl^--HCO_3^- exchanger operates in one direction in the tissues and in the opposite direction in the lungs?

11–26* If you have ever used the standard probe on a pH meter, you may well wonder how pH could possibly be measured in the tiny volumes inside cellular compartments. The recent development of pH-sensitive fluorophores has simplified this difficult task immensely. One such fluorescent indicator is a hydrophobic ester of SNARF-1, which can enter cells by passive diffusion and then is trapped inside after intracellular enzymes hydrolyze the ester bonds to liberate SNARF-1 (Figure 11–5). SNARF-1 absorbs light at 488 nm and emits fluorescent light with peaks at 580 nm and 640 nm. Emission spectra for SNARF-1 at pH 6.0 and pH 9.0 are shown in Figure 11–6. The pK of SNARF-1 is 7.5.
A. Explain why the ester of SNARF-1 diffuses through membranes, whereas the cleaved form stays inside cells.

Figure 11–5 Structure of the ester and free forms of SNARF-1 (Problem 11–26). The blocking ester groups are shown as R. The acid (HA) and salt (A⁻) forms of SNARF-1 are indicated. The two different resonance structures for the salt form of SNARF-1 are shown.

B. Why are there two peaks of fluorescence (at 580 nm and at 640 nm) that change so dramatically in intensity with a change in pH (see Figure 11–6)?

C. What forms of SNARF-1 are present at pH 6.0 and what are their relative proportions? At pH 9.0? You will recall that the Henderson–Hasselbalch equation is pH = pK + $\log_{10}$ ([salt]/[acid]).

D. Sketch an approximate curve for the SNARF-1 emission spectrum inside a cell at pH 7.2. (All such curves pass through the point where the two curves in Figure 11–6 cross.)

E. Why do you suppose emission spectra with two peaks such as that for SNARF-1 are preferred to those that have a single peak?

11–27   A principal function of the plasma membrane is to control the entry of nutrients into the cell. This function is especially important for the epithelial cells that line the gut because they are responsible for absorbing virtually all the nutrients that enter the body. As befits this important role, their plasma membranes are specialized so that the surface facing the gut is folded into numerous fingerlike projections, termed microvilli. Microvilli increase the surface area of the intestinal cells, providing more efficient absorption of nutrients. Microvilli are shown in profile and cross-section in Figure 11–7. From the dimensions given in the figure, estimate the increase in surface area that microvilli provide (for the portion of the plasma membrane in contact with the lumen of the gut) relative to the corresponding surface of a cell with a 'flat' plasma membrane.

11–28*   (**True/False**) Without a continual input of energy, cells will burst. Explain your answer.

11–29   A rise in the intracellular Ca^{2+} concentration causes muscle cells to contract. In addition to an ATP-driven Ca^{2+} pump, heart muscle cells, which contract quickly and regularly, have an antiporter that exchanges Ca^{2+} for extracellular Na^+ across the plasma membrane. This antiporter rapidly pumps most of the entering Ca^{2+} ions back out of the cell, allowing the cell to relax. Ouabain and digitalis, drugs that are used in the treatment of patients with heart disease, make the heart contract more strongly. Both drugs function by partially inhibiting the Na^+-K^+ pump in the membrane of the heart muscle cell. Can you propose an explanation for the effects of these drugs in patients? What will happen if too much of either drug is taken?

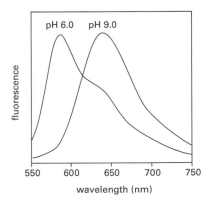

Figure 11–6 Emission spectra of SNARF-1 at pH 6.0 and pH 9.0 (Problem 11–26). SNARF-1 in solution at pH 6.0 or pH 9.0 was excited by light at 488 nm and the intensity of fluorescence was determined at wavelengths from 550 nm to 750 nm.

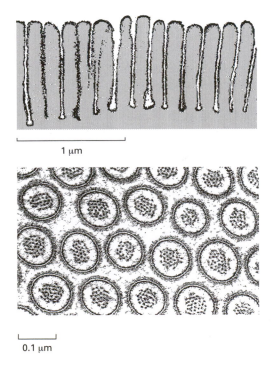

Figure 11–7 Microvilli of intestinal epithelial cells in profile and cross-section (Problem 11–27).

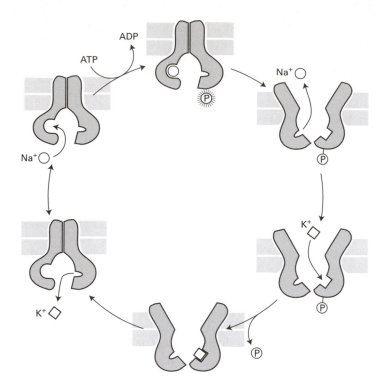

Figure 11–8 The Na⁺-K⁺ pump (Problem 11–30).

11–30* You have prepared lipid vesicles (spherical lipid bilayers) that contain Na⁺-K⁺ pumps as the sole membrane protein. Assume for the sake of simplicity that each pump transports one Na⁺ one way and one K⁺ the other way in each pumping cycle, as illustrated in Figure 11–8. All of the Na⁺-K⁺ pumps are oriented so that the portion of the molecule that normally faces the cytosol faces the outside of the vesicle. Predict what would happen under each of the following conditions.

A. The solution inside and outside the vesicles contains both Na⁺ and K⁺ ions but no ATP.

B. The solution inside the vesicles contains both Na⁺ and K⁺ ions; the solution outside contains both ions as well as ATP.

C. The solution inside contains Na⁺; the solution outside contains Na⁺ and ATP.

D. The solution is as in B, but the Na⁺-K⁺ pump molecules are randomly oriented, some facing one direction, some the other.

11–31 How much energy does it take to pump substances across membranes? Or, to put it another way, since active transport is usually driven directly or indirectly by ATP, how steep a gradient can ATP hydrolysis maintain for a particular solute? For transport into the cell the free-energy change (ΔG_{in}) per mole of solute moved across the plasma membrane is

$$\Delta G_{in} = -2.3RT \log_{10} \frac{C_o}{C_i} + zFV$$

Where R = the gas constant, 1.98×10^{-3} kcal/°K mole
T = the absolute temperature in °K (37°C = 310°K)
C_o = solute concentration outside the cell
C_i = solute concentration inside the cell
z = the valence (charge) on the solute
F = Faraday's constant, 23 kcal/V mole
V = the membrane potential in volts (V).

Since $\Delta G_{in} = -\Delta G_{out}$, the free-energy change for transport out of the cell is

$$\Delta G_{out} = 2.3RT \log_{10} \frac{C_o}{C_i} - zFV$$

At equilibrium, where $\Delta G = 0$, the equations can be rearranged to the more familiar form known as the Nernst equation.

$$V = 2.3 \, \frac{RT}{zF} \, \log_{10} \frac{C_o}{C_i}$$

For the questions below, assume that hydrolysis of ATP to ADP and P_i proceeds with a ΔG of −12 kcal/mole; that is, ATP hydrolysis can drive active transport with a ΔG of +12 kcal/mole. Assume that V is −60 mV.

A. What is the maximum concentration gradient that can be achieved by the ATP-driven active transport into the cell of an uncharged molecule such as glucose, assuming that 1 ATP is hydrolyzed for each solute molecule that is transported?

B. What is the maximum concentration gradient that can be achieved by active transport of Ca^{2+} from the inside to the outside of the cell? How does this maximum compare with the actual concentration gradient observed in mammalian cells (see Table 11–3)?

C. Calculate how much energy it takes to drive the Na^+-K^+ pump. This remarkable molecular device transports five ions for every molecule of ATP that is hydrolyzed: 3 Na^+ out of the cell and 2 K^+ into the cell. The pump typically maintains internal Na^+ at 10 mM, external Na^+ at 145 mM, internal K^+ at 140 mM, and external K^+ at 5 mM. As shown in Figure 11–9, Na^+ is transported against the membrane potential, whereas K^+ is transported with it. (The ΔG for the overall reaction is equal to the sum of the ΔG values for transport of the individual ions.)

D. How efficient is the Na^+-K^+ pump? That is, what fraction of the energy available from ATP hydrolysis is used to drive transport?

11–32* (True/False) The light-activated proton pump of *Halobacterium* synthesizes ATP from ADP and P_i when H^+ ions are pumped out of the cell across the membrane. Explain your answer.

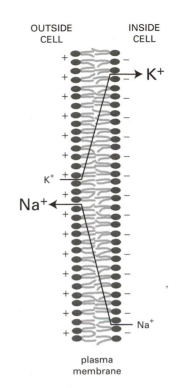

OUTSIDE CELL   INSIDE CELL

K^+

K^+

Na^+

Na^+

plasma membrane

Figure 11–9 Na^+ and K^+ gradients and direction of pumping across the plasma membrane (Problem 11–31). *Large letters* symbolize high concentrations and *small letters* symbolize low concentrations. Both Na^+ and K^+ are pumped against chemical concentration gradients; however, Na^+ is pumped up the electrical gradient, whereas K^+ runs down the electrical gradient.

ION CHANNELS AND THE ELECTRICAL PROPERTIES OF MEMBRANES

TERMS TO LEARN

acetylcholine receptor	long-term potentiation (LTP)	patch-clamp recording
action potential	membrane potential	resting membrane potential
adaptation	myelin sheath	Schwann cell
axon	Nernst equation	selectivity filter
dendrite	neuron	spatial summation
excitatory neurotransmitter	neuromuscular junction	synapse
glial cell	neurotransmitter	temporal summation
inhibitory neurotransmitter	NMDA receptor	transmitter-gated ion channel
ion channel	oligodendrocyte	voltage-gated channel
K^+ leak channel		

11–33 According to Newton's laws of motion, an ion exposed to an electric field in a vacuum would experience a constant acceleration from the electric driving force, just as a falling body in a vacuum constantly accelerates due to gravity. In water, however, an ion moves at constant velocity in an electric field. Why do you suppose that is?

11–34* What two properties distinguish an ion channel from a simple aqueous pore?

11–35 Name the three ways in which an ion channel can be gated.

11–36* (True/False) Carrier proteins saturate at high concentrations of the transported molecule when all their binding sites are occupied; channel proteins,

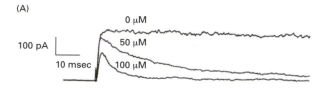
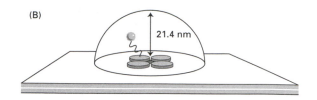

on the other hand, do not bind the ions they transport and thus the flux of ions through a channel does not saturate. Explain your answer.

11–37 K^+ channels typically transport K^+ ions more than a 1000-times better than Na^+ ions, yet these two ions carry the same charge and are about the same size (0.133 nm for K^+ and 0.095 nm for Na^+). Briefly describe how a K^+ channel manages to discriminate against Na^+ ions.

11–38*   The 'ball-and-chain' model for the rapid inactivation of voltage-gated K^+ channels has been elegantly confirmed for the *shaker* K^+ channel from *Drosophila melanogaster*. (The *shaker* K^+ channel in *Drosophila* is so named because of a mutant form, which causes generally excitable behavior—even anesthetized flies keep twitching.) Deletion of the N-terminal amino acids from the normal *shaker* channel gives rise to a channel that opens in response to membrane depolarization, but stays open instead of rapidly closing as the normal channel does (Figure 11–10A, 0 µM). A peptide (MAAVAGLYGLGEDRQHRKKQ) that corresponds to the deleted N-terminus can nearly completely correct this deficiency—inactivate the open channel—when applied at 100 µM to the cytoplasmic surface of the membrane in a patch clamp experiment (Figure 11–10A).

Is the concentration of free peptide (100 µM) required to inactivate the defective K^+ channel anywhere near the normal local concentration of the tethered ball on a normal channel? Assume that the tethered ball can explore a hemispherical volume (Volume = $2/3\ \pi r^3$) with a radius of 21.4 nm, the length of the polypeptide 'chain' (Figure 11–10B). Calculate the concentration for one ball in this hemisphere. How does that value compare to the concentration of free peptide needed to inactivate the channel?

11–39   The *shaker* K^+ channel in *Drosophila* opens in response to membrane depolarization and then rapidly inactivates via a ball-and-chain mechanism. The *shaker* K^+ channel assembles as a tetramer composed of four subunits, each with its own ball and chain. Do multiple balls in the tetrameric channel act together to inactivate the channel, or is one ball sufficient?

This question has been answered by mixing subunits from two different forms of the K^+ channel: the normal subunits with balls and scorpion toxin-resistant subunits without balls (Figure 11–11A). Scorpion toxin prevents the opening of normal (toxin-sensitive) K^+ channels, but not channels composed entirely of toxin-resistant subunits. Moreover, hybrid channels containing even a single toxin-sensitive subunit fail to open in the presence of toxin.

Mixed K^+ channels were created by injecting a mixture of mRNAs for two types of subunit into frog oocytes. After expression, patches of membrane

(A) MIXTURE OF K+ CHANNELS

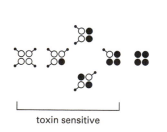

toxin sensitive

(B) PATCH-CLAMP RECORDINGS

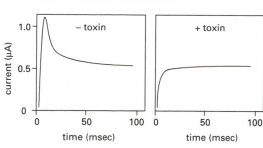

Figure 11–10 Inactivation of voltage-gated K^+ channels (Problem 11–38). (A) Patch-clamp recording of a defective *shaker* K^+ channel in the absence and presence of inactivating peptide. (B) A 'ball' tethered by a 'chain' to a normal channel.

Figure 11–11 Analysis of the inactivation of the *shaker* K^+ channel (Problem 11–39). (A) Mixture of normal subunits with balls (*white*) and toxin-resistant subunits without balls (*black*). If the two subunits were expressed in equal amounts, the different forms of the K^+ channel would be present in the ratio of 1:4:6:4:1; however, higher levels of toxin-resistant subunit mRNA were used, significantly skewing the ratios in favor of channels with toxin-resistant subunits. (B) Patch-clamp recordings in the presence and absence of scorpion toxin.

containing several hundred channels were studied by patch-clamp recording in the absence or presence of scorpion toxin (Figure 11–11B).

A. Sketch the expected patch-clamp recordings, in the presence and absence of toxin, for a pure population of K+ channels composed entirely of toxin-resistant subunits without balls.

B. Sketch the expected patch-clamp recordings, in the presence and absence of toxin, for a pure population of K+ channels composed entirely of normal (toxin-sensitive) subunits with balls.

C. Sketch the expected patch-clamp recordings, in the presence and absence of toxin, for a mixed population of K+ channels, 50% composed entirely of normal (toxin-sensitive) subunits with balls and 50% composed entirely of toxin-resistant subunits without balls.

D. The key observation in Figure 11–11B is that the final plateau values of the currents in the absence and presence of toxin are the same. Does this observation argue that a single ball can close a channel or that multiple balls must act in concert? (Think about what the curves would look like if one, two, three, or four balls were required to close a channel.)

11–40* (**True/False**) The resting membrane potential of a typical animal cell arises predominantly through the action of the Na+-K+ pump, which in each cycle transfers 3 Na+ ions out of the cell and 2 K+ into the cell, leaving an excess of negative charges inside the cell. Explain your answer.

11–41 You have prepared lipid vesicles that contain molecules of the K+-leak channel, all oriented so that their cytosolic surface faces the outside of the vesicle. Predict how K+ ions will move under the following conditions and what sort of membrane potential will develop.

A. Equal concentrations of K+ ion are present inside and outside the vesicle.

B. K+ ions are present only inside the vesicle.

C. K+ ions are present only outside the vesicle.

11–42* If the resting membrane potential of a cell is –70 mV and the thickness of the lipid bilayer is 4.5 nm, what is the strength of the electric field across the membrane in V/cm? What do you suppose would happen if you applied this voltage to two metal electrodes separated by a 1-cm air gap?

11–43 The squid giant axon occupies a unique position in the history of our understanding of cell membrane potentials and action potentials. Its large size (0.2–1.0 mm in diameter and 5–10 cm in length) allowed electrodes, large by modern standards, to be inserted so that intracellular voltages could be measured. When an electrode is stuck into an intact giant axon, a membrane potential of –70 mV is registered. When the axon, suspended in a bath of seawater, is stimulated to conduct a nerve impulse, the membrane potential changes transiently from –70 mV to +40 mV.

The Nernst equation relates equilibrium ionic concentrations to the membrane potential.

$$V = 2.3 \, \frac{RT}{zF} \, \log_{10} \frac{C_o}{C_i}$$

For univalent ions and 20°C (293°K),

$$V = 58 \text{ mV} \times \log_{10} \frac{C_o}{C_i}$$

A. Using this equation, calculate the potential across the resting membrane (1) assuming that it is due solely to K+ and (2) assuming that it is due solely to Na+. (The Na+ and K+ concentrations in axon cytoplasm and in seawater are given in Table 11–4.) Which calculation is closer to the measured resting potential? Which calculation is closer to the measured action potential? Explain why these assumptions approximate the measured resting and action potentials.

TABLE 11–4 Ionic composition of seawater and of cytoplasm from the squid giant axon (Problem 11–43).

ION	CYTOPLASM	SEAWATER
Na+	65 mM	430 mM
K+	344 mM	9 mM

B. If the solution bathing the squid giant axon is changed from seawater to artificial seawater in which NaCl is replaced with choline chloride, there is no effect on the resting potential, but the nerve no longer generates an action potential upon stimulation. What would you predict would happen to the magnitude of the action potential if the concentration of Na^+ in the external medium were reduced to a half or a quarter of its normal value, using choline chloride to maintain osmotic balance?

11–44* Intracellular changes in ion concentration often trigger dramatic cellular events. For example, when a clam sperm contacts a clam egg, it triggers ionic changes that result in the breakdown of the egg nuclear envelope, condensation of chromosomes, and initiation of meiosis. Two observations confirm that ionic changes initiate these cellular events: (1) suspending clam eggs in seawater containing 60 mM KCl triggers the same intracellular changes as do sperm; (2) suspending eggs in artificial seawater lacking calcium prevents activation by 60 mM KCl.

A. How does 60 mM KCl affect the resting potential of eggs? The intracellular K^+ concentration is 344 mM and that of normal seawater is 9 mM. Remember from Problem 11–43 that

$$V = 58 \text{ mV} \times \log_{10} \frac{C_o}{C_i}$$

B. What does the lack of activation by 60 mM KCl in calcium-free seawater suggest about the mechanism of KCl activation?

C. What would you expect to happen if the calcium ionophore, A23187, was added to a suspension of eggs (in the absence of sperm) in (1) regular seawater and (2) calcium-free seawater?

11–45 (True/False) The membrane potential arises from movements of charge that leave ion concentrations practically unaffected and result in only a very slight discrepancy in the number of positive and negative ions on the two sides of the membrane. Explain your answer.

11–46* The number of Na^+ ions entering the squid giant axon during an action potential can be calculated from theory. Because the cell membrane separates positive and negative charges, it behaves like a capacitor. From the known capacitance of biological membranes, the number of ions that enter during an action potential can be calculated. Starting from a resting potential of –70 mV, it can be shown that 1.1×10^{-12} moles of Na^+ must enter the cell per cm^2 of membranes during an action potential.

To determine experimentally the number of entering Na^+ during an action potential, a squid giant axon (1 mm in diameter and 5 cm in length) was suspended in a solution containing radioactive Na^+ (specific activity = 2×10^{14} cpm/mole) and a single action potential was propagated down its length. When the cytoplasm was analyzed for radioactivity, a total of 340 cpm were found to have entered the axon.

A. How well does the experimental measurement match the theoretical calculation?

B. How many moles of K^+ must cross the membrane of the axon, and in which direction, to reestablish the resting potential after the action potential is over?

C. Given that the concentration of Na^+ inside the axon is 65 mM, calculate the fractional increase in internal Na^+ concentration that results from the passage of a single action potential down the axon.

D. At the other end of the spectrum of nerve sizes are small dendrites about 0.1 μm in diameter. Assuming the same length (5 cm), the same internal Na^+ concentration (65 mM), and the same resting and action potentials as for the squid giant axon, calculate the fractional increase in internal Na^+ concentration that would result from the passage of a single action potential down a dendrite.

E. Is the Na^+-K^+ pump more important for the continuing performance of a giant axon or a dendrite?

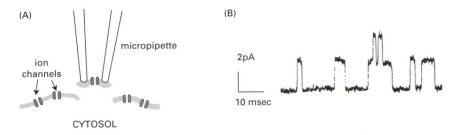

Figure 11–12 Analysis of acetylcholine-gated ion channels (Problem 11–51). (A) A micropipette with a patch of membrane. (B) Patch-clamp recording of current through acetylcholine-gated ion channels.

11–47 (**True/False**) Upon stimulation of a nerve cell, two processes limit the entry of Na⁺ ions: (1) the membrane potential reaches the Na⁺ equilibrium potential, which stops further net entry of Na⁺, and (2) the Na⁺ channels are inactivated and cannot reopen until the original resting potential has been restored. Explain your answer.

11–48* Explain in 100 words or less how an action potential is passed along an axon.

11–49 The myelin sheath that insulates many vertebrate axons changes the way in which an action potential is conducted, allowing it to jump from node to node in a process called saltatory conduction. What are the two main advantages of saltatory conduction?

11–50* (**True/False**) The aggregate current crossing the membrane of an entire cell indicates the degree to which individual channels are open. Explain your answer.

11–51 A recording from a patch of membrane (Figure 11–12A) measured current through acetylcholine-gated ion channels, which open when acetylcholine is bound (Figure 11–12B). To obtain the recording, acetylcholine was added to the solution in the micropipette (Figure 11–12A). Describe what you can learn about the channels from this recording. How would the recording differ if acetylcholine were (a) omitted or (b) added only to the solution outside the micropipette?

11–52* Cytosolic Ca²⁺ concentrations typically rise about 50-fold when Ca²⁺ channels open in the plasma membrane. Assume that 1000 Ca²⁺ channels open in a cell with a volume of 1000 μm^3 and an internal Ca²⁺ concentration of 100 nM. Each Ca²⁺ channel passes 10^6 Ca²⁺ ions per second. For how long would the channels need to stay open in order to raise the cytosolic Ca²⁺ concentration 50-fold to 5 μM?

11–53 (**True/False**) Transmitter-gated ion channels open in response to specific neurotransmitters in their environment but are insensitive to the membrane potential; therefore, they cannot by themselves (in the absence of ligand) generate an action potential. Explain your answer.

11–54* The neurotransmitter acetylcholine is made in the cytosol and then transported into synaptic vesicles, where its concentration is more than 100-fold higher than in the cytosol. Synaptic vesicles isolated from neurons can take up additional acetylcholine if it is added to the solution in which they are suspended, but only in the presence of ATP. Na⁺ ions are not required for acetylcholine uptake, but, curiously, raising the pH of the solution in which the synaptic vesicles are suspended increases acetylcholine uptake. Furthermore, transport is inhibited in the presence of drugs that make the membrane permeable to H⁺ ions. Suggest a mechanism that is consistent with all these observations.

11–55 Excitatory neurotransmitters open Na⁺ channels while inhibitory neurotransmitters open either Cl⁻ or K⁺ channels. Rationalize this observation in terms of the effects of these ions on the firing of an action potential.

11–56* One important parameter for understanding any particular membrane transport process is to know the number of copies of the specific transport protein present in the cell membrane. To measure the number of voltage-gated Na⁺

channels in the rabbit vagus nerve, you use a potent toxin, saxitoxin, a shellfish poison that specifically inactivates voltage-gated Na^+ channels in these nerve cells. Assuming that each channel binds one toxin molecule, the number of Na^+ channels in a segment of vagus nerve will be equal to the maximum number of bound toxin molecules.

You incubate identical segments of nerve for 8 hours with increasing amounts of ^{125}I-labeled toxin. You then wash the segments to remove unbound toxin and measure the radioactivity associated with the nerve segments to determine the toxin-binding curve (Figure 11–13). You are puzzled because you expected to see binding reach a maximum (saturate) at high concentrations of toxin; however, no distinct end point was reached, even at higher concentrations of toxin than those shown in the figure. After careful thought, you design a control experiment in which the binding of labeled toxin is measured in the presence of a large molar excess of unlabeled toxin. The results of this experiment, which are shown in the lower curve in Figure 11–13, make everything clear and allow you to calculate the number of Na^+ channels in the membrane of the vagus nerve axon.

A. Why does binding of the labeled toxin not saturate? What is the point of the control experiment, and how does it work?

B. Given that 1 gram of vagus nerve has an axonal membrane area of 6000 cm^2 and assuming that the Na^+ channel is a cylinder with a diameter of 6 nm, calculate the number of Na^+ channels per square micrometer of axonal membrane and the fraction of the cell surface occupied by the channel. (Use 100 pmol as the amount of toxin specifically bound to the receptor.)

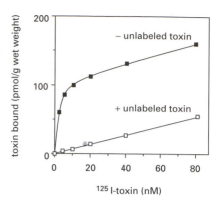

Figure 11–13 Toxin-binding curves in the presence and absence of unlabeled saxitoxin (Problem 11–56).

11–57 Acetylcholine-gated cation channels do not discriminate among Na^+, K^+, and Ca^{2+} ions, allowing all to pass through freely. How is it then that when acetylcholine receptors in muscle cells open there is a large net influx principally of Na^+?

11–58* In the disease myasthenia gravis, the human body makes—by mistake—antibodies to its own acetylcholine receptors molecules. These antibodies bind to and inactivate acetylcholine receptors on the plasma membrane of muscle cells. The disease leads to a progressive weakening of the patient's motor responses. Early on, they may have difficulty opening their eyelids, for example. As the disease progresses, most muscles weaken, and patients have difficulty speaking and swallowing. Eventually, impaired breathing can cause death. Explain which step of muscle function is affected.

11–59 To make antibodies against the acetylcholine receptor from the electric organ of electric eels, you inject the purified receptor into mice. You note an interesting correlation: mice with high levels of antibodies against the receptor appear weak and sluggish; those with low levels are lively. You suspect that the antibodies against the eel acetylcholine receptors are reacting with the mouse acetylcholine receptors, causing many of the receptors to be destroyed. Since a reduction in the number of acetylcholine receptors is also the basis for the human autoimmune disease myasthenia gravis, you wonder whether an injection of the drug neostigmine might give a temporary restoration of strength, as it does for myasthenic patients. Neostigmine inhibits acetylcholinesterase, the enzyme responsible for hydrolysis of acetylcholine in the synaptic cleft. Sure enough, when you inject your mice with neostigmine, they immediately stand up and become very lively. Propose an explanation for how neostigmine restores temporary function to a neuromuscular synapse with a reduced number of acetylcholine receptors.

11–60* Acetylcholine-gated cation channels at the neuromuscular junction open in response to acetylcholine released by the nerve terminal and allow Na^+ ions to enter the muscle cell, which causes membrane depolarization and ultimately leads to muscle contraction. Patch-clamp measurements in young rat muscle show that there are two classes of channel—a 4-pA channel and a 6-pA channel—that respond to acetylcholine (Figure 11–14).

Calculate the number of ions that enter through each type of channel in one millisecond. (One Amp is a current of one Coulomb per second; one pA

2 pA
40 msec

Figure 11–14 Patch-clamp measurements of acetylcholine-gated cation channels in young rat muscle (Problem 11–60).

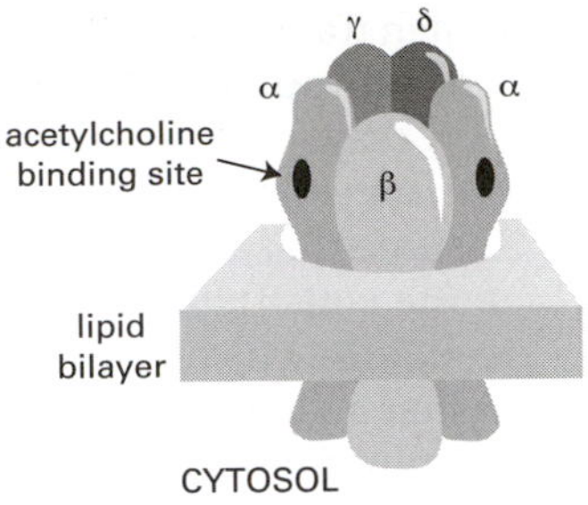

equals 10^{-12} Amp. An ion with a single charge such as Na^+ carries a charge of 1.6×10^{-19} Coulomb.)

11–61   The acetylcholine-gated cation channel is a pentamer of homologous proteins containing two copies of the α subunit and one copy each of the β, γ, and δ subunits (Figure 11–15). The two distinct acetylcholine-gated channels in young rat muscle (see Problem 11–60) differ in their γ subunits. The 6-pA channel contains the γ_1 subunit, whereas the 4-pA channel contains the γ_2 subunit.

   To pinpoint the portions of the γ_1 and γ_2 subunits responsible for the conductance differences of the two channels, you prepare a series of chimeric cDNAs in which different portions of the γ_1 and γ_2 cDNAs have been swapped (Figure 11–16). Since it seems likely that the transmembrane segments determine channel conductance, you have designed the chimeric cDNAs to test different combinations of the four transmembrane segments (M1 to M4 in Figure 11–16A). You mix mRNA made from each of the chimeric cDNAs with mRNA from the α, β, and δ subunits, inject the mRNAs into *Xenopus* oocytes, and measure the current through individual channels by patch clamping (Figure 11–16B).

Figure 11–15 A model for the structure of the acetylcholine receptor (Problem 11–61).

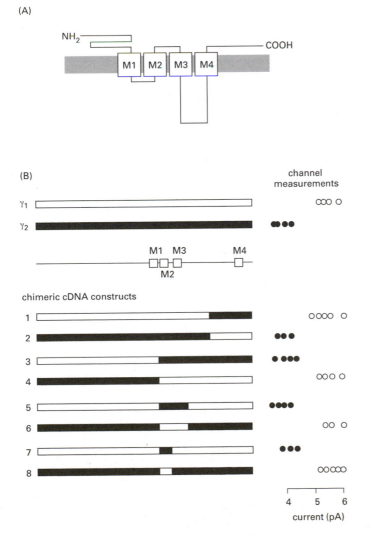

Figure 11–16 Analysis of the structural basis for conductance differences in two acetylcholine-gated channels (Problem 11–61). (A) Arrangement of a γ subunit in the membrane. (B) Current through channels made with the γ_1 subunit, the γ_2 subunit, or chimeric subunits. The current through the different channels is indicated on the *right*. *Open circles* indicate channels with conductance like those made from the γ_1 subunit; *filled circles* indicate channels with conductance like those made from the γ_2 subunit. M1, M2, M3, and M4 indicate the four transmembrane segments in the γ subunits.

A. Which transmembrane segment is responsible for the differences in conductance through the two types of acetylcholine-gated cation channels?

B. Why do you suppose that the difference in channel conductance is all due to one transmembrane segment?

C. Assume that a threonine in the critical transmembrane segment contributes to the constriction that forms the ion-selectivity filter for the channel. How would the current through the channel change if a glycine were substituted at that position? If leucine were substituted at that position? How might you account for the different currents through channels constructed with γ_1 and γ_2 subunits?

11–62* The ion channels for neurotransmitters such as acetylcholine, serotonin, GABA, and glycine have a similar overall structure. Each class comprises an extremely diverse set of channel subtypes, with different ligand affinities, different channel conductances, and different rates of opening and closing. Why is such extreme diversity a good thing from the standpoint of the pharmaceutical industry?

11–63 (**True/False**) When an action potential depolarizes the muscle cell membrane, the Ca^{2+} pump is responsible for pumping Ca^{2+} from the sarcoplasmic reticulum into the cytosol to initiate muscle contraction. Explain your answer.

11–64* Glutamate is a major excitatory neurotransmitter in the central nervous system. Glutamate binds to a variety of receptors, some of which control intracellular signaling via G proteins and others that control cation channels. One of the latter, the NMDA receptor, opens a Ca^{2+} channel in response to N-methyl-D-aspartate (NMDA), which is an analog of glutamate.

The NMDA receptor was cloned from mRNA isolated from rat forebrain. The mRNA was first size fractionated on a sucrose gradient and individual fractions were then injected into *Xenopus* oocytes. A 3–5 kb fraction conferred NMDA responsiveness, as assessed by electrophysiological measurements, on the injected oocytes, which normally do not respond to NMDA. A cDNA library was prepared from the active mRNA fraction and then subdivided into 10 aliquots containing 1000 clones each. RNA was made from each aliquot by *in vitro* transcription and then injected into oocytes. Aliquots whose RNA conferred NMDA responsiveness were further divided into aliquots containing fewer cDNA clones and retested for the ability of the transcribed RNA to confer NMDA responsiveness to injected oocytes. Repeated subdivision of NMDA-responsive cDNA aliquots allowed the cDNA for the NMDA receptor to be cloned as outlined in Figure 11–17.

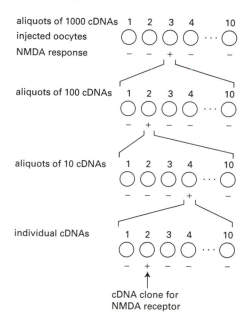

Figure 11–17 Pooling strategy for isolating cDNA for NMDA receptor (Problem 11–64).

TABLE 11–5 Pharmacology of the cloned NMDA receptor (Problem 11–64).

AGONIST		ANTAGONIST		
FOR NMDA RECEPTOR	FOR OTHER GLUTAMATE RECEPTORS	FOR NMDA RECEPTOR	FOR OTHER GLUTAMATE RECEPTORS	RESPONSE
NMDA (100 µM)				100%
Ibotenate (100 µM)				71%
L-HCA (100 µM)				85%
	Kainate (500 µM)			<5%
	AMPA (50 µM)			<5%
NMDA (100 µM)		D-APV (10 µM)		27%
NMDA (100 µM)		7-Cl-KYNA (50 µM)		2%
NMDA (100 µM)			GAMS (100 µM)	97%
NMDA (100 µM)			JSTX (100 µM)	83%

Oocytes injected with the putative cDNA for the cloned NMDA receptor were tested with a variety of pharmacological agents whose receptor specificities and effective concentrations were already defined from previous work.

L-HCA = homocysteate; AMPA = α-amino-3-hydroxy-5-methyl-4-isoxazolepropionate; D-APV = 2-amino-5-phosphonovalerate; 7-Cl-KYNA = 7-chlorokynurenate; GAMS =γ-D-glutamylaminomethyl sulphonate; JSTX = Joro spider toxin.

The identity of the NMDA receptor was confirmed using a variety of pharmacological agents that are known agonists (activators) or antagonists (inhibitors) of the NMDA receptor or other glutamate receptors (Table 11–5).

A. The pooling strategy outlined in Figure 11–17 is an efficient way to find one particular cDNA in a complex mixture. How many oocyte injections and analyses for NMDA responsiveness would have been required if the original cDNA library were tested one cDNA at a time? How many were required using the pooling strategy?

B. How do the data in Table 11–5 confirm that the cDNA for the NMDA receptor was cloned?

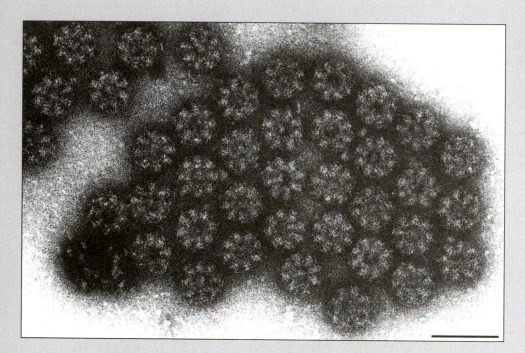

Nuclear pore complexes: face-on electron micrographs of negatively stained, detergent extracted nuclear membranes. Scale bar is 0.1 μm. Courtesy of Ron Milligan.

INTRACELLULAR COMPARTMENTS AND PROTEIN SORTING

THE COMPARTMENTALIZATION OF CELLS

TERMS TO LEARN

cytoplasm	organelle	sorting signal
cytosol	signal patch	transmembrane transport
gated transport	signal peptidase	vesicular transport
lumen	signal sequence	

12–1   (**True/False**) The membranes that partition the cell into functionally distinct compartments are impermeable. Explain your answer.

12–2*   A typical animal cell is said to contain some 10 billion protein molecules that need to be sorted into their proper compartments. That's a lot of proteins. Can 10 billion protein molecules even fit into a cell? An average protein encoded by the human genome is 450 amino acids in length, and the average mass of an amino acid in a protein is 110 daltons. Given that the average density of a protein is 1.4 g/cm^3, what fraction of the volume of a cell would 10 billion protein molecules occupy? Consider a liver cell, which has a volume of about 5000 μm^3, and a pancreatic exocrine cell, which has a volume of about 1000 μm^3.

12–3   Discuss the following statement: "The plasma membrane is only a minor membrane in most eucaryotic cells."

12–4*   In a typical animal cell, which of the following compartments is the largest? Which compartment is present in the greatest number?
  A. Cytosol
  B. Endoplasmic reticulum
  C. Endosome
  D. Golgi apparatus
  E. Lysosome
  F. Mitochondria
  G. Nucleus
  H. Peroxisome

*Answers to these problems are available to Instructors at www.classwire.com/garlandscience/

12–5   Is it really true that *all* human cells contain the same basic set of membrane-enclosed organelles? Do you know of any examples of human cells that do not have a complete set of organelles?

12–6*   The lipid bilayer, which is 5 nm thick, occupies about 60% of the volume of typical cell membranes. (Lipids and proteins contribute equally on a mass basis, but lipids are less dense and therefore account for more of the volume.) For liver cells and pancreatic exocrine cells, the total area of all cell membranes is estimated at about 110,000 μm^2 and 13,000 μm^2, respectively. What fraction of the total volumes of these cells is accounted for by lipid bilayers? Recall that the volumes of liver cells and pancreatic exocrine cells are about 5000 μm^3 and 1000 μm^3, respectively.

12–7   When cells are treated with drugs that depolymerize microtubules, the Golgi apparatus is fragmented into small vesicles and dispersed throughout each cell. When such drugs are removed, cells typically recover and grow normally. Does this mean that the Golgi apparatus has formed again from the vesicles? If so, how do you suppose it might happen?

12–8*   (**True/False**) Like the lumen of the ER, the interior of the nucleus is topologically equivalent to the outside of the cell. Explain your answer.

12–9   Why do eucaryotic cells require a nucleus as a separate compartment when procaryotic cells manage perfectly well without?

12–10*   A decrease in surface-to-volume ratio relative to bacteria is one rationale for why the much larger eucaryotic cells evolved to possess such a profusion of internal membranes. But how much different are the surface-to-volume ratios for bacteria and eucaryotic cells? A liver cell has about 110,000 μm^2 of membrane, 2% of which is plasma membrane, and a volume of about 5000 μm^3. A pancreatic exocrine cell has about 13,000 μm^2 of membrane, 5% of which is plasma membrane, and a volume of about 1000 μm^3. The rod-shaped bacterium, *E. coli*—essentially a cylinder 2 μm long and 1 μm in diameter—has a surface area of 7.85 μm^2 and a volume of 1.57 μm^3.

What are the surface-to-volume ratios for these cells? How much smaller are they in the eucaryotic cells versus *E. coli*? What would the surface-to-volume ratio for *E. coli* be if its diameter and length were increased by a factor of 10, so that its volume was increased 1000-fold to 1570 μm^3—about the same size as the eucaryotic cells? (For a cylinder, volume is $\pi r^2 h$ and surface area is $2\pi r^2 + 2\pi rh$.)

12–11   What is the fate of a protein with no sorting signal?

12–12*   (**True/False**) Membrane-bound and free ribosomes, which are structurally and functionally identical, differ only in the proteins they happen to be making at a particular time. Explain your answer.

12–13   Protein synthesis in a liver cell occurs nearly exclusively on free ribosomes in the cytosol and on ribosomes that are bound to the ER membrane. (A small fraction of total protein synthesis is directed by the mitochondrial genome and occurs on ribosomes in the mitochondrial matrix.) Which type of protein synthesis—in the cytosol or on the ER—do you think is responsible for the majority of protein synthesis in a liver cell? Assume that the average density and lifetimes of proteins are about the same in all compartments. Explain the basis for your answer. Would your answer change if you took into account that some proteins are secreted from liver cells?

12–14*   List the organelles in an animal cell that obtain their proteins via gated transport, via transmembrane transport, or via vesicular transport.

12–15   Although the vast majority of transmembrane proteins are inserted into membranes with the help of dedicated protein-translocation machines, a few proteins can insert into membranes on their own. Such proteins may provide a window into how proteins entered membranes in the days before complex translocators had evolved.

You are studying a protein that inserts itself into the bacterial membrane independent of the normal translocation machinery. This protein has an N-terminal, 18-amino acid hydrophilic segment that is located on the outside of the membrane, a 19-amino acid hydrophobic transmembrane segment flanked by negatively and positively charged amino acids, and a C-terminal domain that resides inside the cell (Figure 12–1A). If the protein is properly inserted in the membrane, the N-terminal segment is exposed externally where it can be clipped off by a protease, allowing you to quantify insertion.

To examine the roles of the hydrophobic segment and its flanking charges, you construct a set of modified genes that express mutant proteins with altered charges in the N-terminal segment, altered lengths of the hydrophobic segment, and combinations of the two (Figure 12–1B). For each gene you measure the fraction of the total protein that is cleaved by the protease, which is the fraction that was inserted correctly (Figure 12–1B). To assess the contribution of the normal membrane potential (positive outside, negative inside), you repeat the measurements in the presence CCCP, an ionophore that eliminates the charge on the membrane (Figure 12–1B).

A. Which of the two N-terminal negative charges is the more important for insertion of the protein in the presence of the normal membrane potential (minus CCCP)? Explain your reasoning.

B. In the presence of the membrane potential (minus CCCP) is the hydrophobic segment or the N-terminal charge more important for insertion of the protein into the membrane? Explain your reasoning.

C. In the absence of the membrane potential (plus CCCP) is the hydrophobic segment or the N-terminal charge more important for insertion of the protein into the membrane? Explain your reasoning.

12–16* **(True/False)** Depending on the individual protein, a signal peptide or a signal patch may direct the protein to the ER, mitochondria, chloroplasts, or peroxisomes. Explain your answer.

12–17 Imagine that you have engineered a set of genes, each encoding a protein with a pair of conflicting signal sequences that specify different compartments. If the genes were expressed in a cell, predict which signal would win out for the following combinations. Explain your reasoning.

A. Signals for import into nucleus and import into ER.

B. Signals for import into peroxisomes and import into ER.

C. Signals for import into mitochondria and retention in ER.

D. Signals for import into nucleus and export from nucleus.

12–18* The rough ER is the site of synthesis of many different classes of membrane proteins. Some of these proteins remain in the ER, whereas others are sorted to compartments such as the Golgi apparatus, lysosomes, and the plasma membrane. One measure of the difficulty of the sorting problem is the degree of 'purification' that must be achieved during transport from the ER to the other compartments. For example, if membrane proteins bound for the plasma membrane represented 90% of all proteins in the ER, then only a small degree of purification would be needed (and the sorting problem would appear relatively easy). On the other hand, if plasma membrane proteins represented only 0.01% of the proteins in the ER, a very large degree of purification would be required (and the sorting problem would appear correspondingly more difficult).

What is the magnitude of the sorting problem? What fraction of the membrane proteins in the ER are destined for other compartments? A few simple considerations allow one to estimate the answers to these questions. Assume that all proteins on their way to other compartments remain in the ER 30 minutes on average before exiting, and that the ratio of proteins to lipids in the membranes of all compartments is the same.

A. In a typical growing cell that is dividing once every 24 hours the equivalent of one new plasma membrane must transit the ER every day. If the ER membrane is 20 times the area of a plasma membrane, what is the ratio of plasma membrane proteins to other membrane proteins in the ER?

(A)

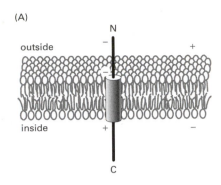

(B)

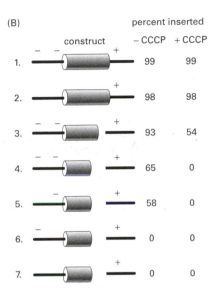

Figure 12–1 Insertion of a small protein into the bacterial membrane (Problem 12–15). (A) Normal orientation of the protein in the membrane. (B) Mutant proteins used to investigate the contributions of the N-terminal negative charges and length of the hydrophobic segment to membrane insertion. The presence of negative charges is indicated by –; *cylinders* indicate the length of α-helical, transmembrane hydrophobic segments; deletions of the hydrophobic segments are shown as *gaps*. Percent inserted refers to the proportion of the protein whose N-termini are sensitive to protease digestion.

B. If in the same cell the Golgi membrane is three times the area of the plasma membrane, what is the ratio of Golgi membrane proteins to other membrane proteins in the ER?

C. If the membranes of all other compartments (lysosomes, endosomes, inner nuclear membrane, and secretory vesicles) that receive membrane proteins from the ER are equal in total area to the area of the plasma membrane, what fraction of the membrane proteins in the ER of this cell are permanent residents of the ER membrane?

12–19   If you think of the protein as a traveler, what kind of vehicle would best describe the sorting receptor: a private car, a taxi, or a bus? Explain your choice.

12–20*   If membrane proteins are integrated into the ER membrane by means of protein translocators, which are themselves membrane proteins, how do the first protein translocators become incorporated into the ER membrane?

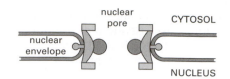

Figure 12–2 Cross-section through a nuclear pore complex, showing continuity of inner and outer nuclear membranes (Problem 12–22).

THE TRANSPORT OF MOLECULES BETWEEN THE NUCLEUS AND THE CYTOSOL

TERMS TO LEARN

inner nuclear membrane	nuclear import receptor	nuclear transport receptor
nuclear envelope	nuclear lamin	nucleoporin
nuclear export receptor	nuclear localization signal (NLS)	outer nuclear membrane
nuclear export signal	nuclear pore complex	Ran

12–21   (**True/False**) The nuclear membrane is freely permeable to ions and other small molecules under 5000 daltons. Explain your answer.

12–22*   As shown in Figure 12–2, the inner and outer nuclear membranes form a continuous sheet, connecting through the nuclear pores. Continuity implies that membrane proteins can move freely between the two nuclear membranes by diffusing through the bilayer at the nuclear pores. Yet the inner and outer nuclear membranes have different protein compositions, as befits their different functions. How do you suppose this apparent paradox is reconciled?

12–23   Before nuclear pore complexes were well understood, it was unclear whether nuclear proteins diffused passively into the nucleus and accumulated there by binding to residents of the nucleus such as chromosomes, or were actively transported and accumulated regardless of their affinity for nuclear components.

A classic experiment that addressed this problem used several forms of radioactive nucleoplasmin, which is a large pentameric protein involved in chromatin assembly. In this experiment, either the intact protein or the nucleoplasmin heads, tails, or heads with a single tail were injected into the cytoplasm or into the nucleus of frog oocytes (Figure 12–3). All forms of nucleoplasmin, except heads, accumulated in the nucleus when injected in the cytoplasm, and all forms were retained in the nucleus when injected there.

A. What portion of the nucleoplasmin molecule is responsible for localization in the nucleus?

B. How do these experiments distinguish between active transport, in which a nuclear localization signal triggers transport by the nuclear pore complex, and passive diffusion, in which a binding site for a nuclear component allows accumulation in the nucleus?

12–24*   Proteins have to be unfolded during their transport into most organelles.

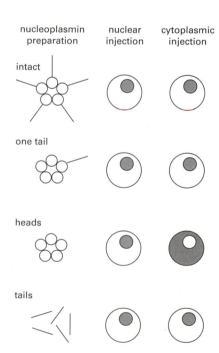

Figure 12–3 Cellular location of injected nucleoplasmin and nucleoplasmin components (Problem 12–23). Schematic diagrams of autoradiographs show the cytoplasm and nucleus with the location of nucleoplasmin indicated by the *shaded* areas.

TABLE 12–1 Results of proliferation experiments with yeast carrying plasmids pNL⁺ or pNL⁻ (Problem 12–25).

PLASMID	GLUCOSE MEDIUM	GALACTOSE MEDIUM
pNL⁺	proliferation	death
pNL⁻	proliferation	proliferation

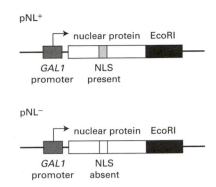

Figure 12–4 Two plasmids for investigating nuclear localization in yeast (Problem 12–25). Plasmids are shown as linear molecules for clarity. *Arrows* indicate direction of transcription from the *GAL1* promoter.

What is the key observation indicating that nuclear pores can dilate to accommodate fully folded proteins?

12–25 You have just joined a laboratory that is engaged in defining the nuclear transport machinery in yeast. Your advisor, who is known for her extraordinarily clever ideas, has given you a project with enormous potential. In principle, it would allow a genetic selection for conditional-lethal mutants in the nuclear transport apparatus.

She gave you two plasmids. Each plasmid consists of a hybrid gene under the control of a regulatable promoter (Figure 12–4). The hybrid gene is a fusion between a gene whose product is normally imported into the nucleus and the gene for the restriction nuclease EcoRI. The plasmid pNL⁺ contains a functional nuclear localization signal (NLS); the plasmid pNL⁻ does not have an NLS. The promoter, which is from the yeast *GAL1* gene, allows transcription of the hybrid gene only when the sugar galactose is present in the growth medium.

Following her instructions, you introduce the plasmids into yeast (in the absence of galactose) and then assay the transformed yeast in medium containing glucose and in medium containing galactose. Your results are shown in Table 12–1. You don't remember what your advisor told you to expect, but you know you will be expected to explain these results at the weekly lab meeting. Why do yeasts with the pNL⁺ plasmid proliferate in the presence of glucose but die in the presence of galactose?

12–26* Now that you understand why plasmid pNL⁺ (see Figure 12–4) kills cells in galactose-containing medium, you begin to understand how your advisor intends to exploit its properties to select mutants in the nuclear transport machinery. You also understand why she emphasized that the desired mutants would have to be *conditionally* lethal. Since nuclear import is essential to the cell, a fully defective mutant could never be grown and thus would not be available for study. By contrast, conditional-lethal mutants can be grown perfectly well under one set of conditions (permissive conditions). Under a different set of conditions (restrictive conditions), however, the cells exhibit the defect, which can then be studied.

With this overall strategy in mind, you design a selection scheme for temperature-sensitive (ts) mutants in the nuclear translocation machinery. You want to find mutants that proliferate well at low temperature (the permissive condition) but are defective at high temperature (the restrictive condition). You plan to mutagenize cells containing the pNL⁺ plasmid at low temperature and then shift them to high temperature in the presence of galactose. You reason that at the restrictive temperature the nuclear transport mutants will not take up the killer protein encoded by pNL⁺ and, therefore, will not be killed. Normal cells, however, will transport the killer protein into the nucleus and die. After one or two hours at the high temperature to allow selection against normal cells, you intend to lower the temperature and put the surviving cells on nutrient agar plates containing glucose. You anticipate that the nuclear translocation mutants will recover at the low temperature and form colonies.

When you show your advisor your scheme, she is very pleased at your initiative. However, she sees a critical flaw in your experimental design that would prevent isolation of nuclear translocation mutants, but she won't tell you what it is—she hints that it has to do with the killer protein made during the high-temperature phase of the experiment.

A. What is the critical flaw in your original experimental protocol?
B. How might you modify your protocol to correct this flaw?
C. Assuming for the moment that your original protocol would work, can you think of any other types of mutants (not involved in nuclear transport) that would survive your selection scheme?

12–27 Assuming that 32 million histone octamers are required to package the human genome, how many histone molecules must be transported per second per pore complex in cells whose nuclei contain 3000 nuclear pores and is dividing once per day?

12–28* How is it that a single nuclear pore complex can efficiently transport proteins that possess quite different kinds of nuclear localization signal?

12–29 (**True/False**) To avoid the inevitable congestion that would occur if two-way traffic through a single pore were allowed, nuclear pore complexes appear to be specialized so that some mediate import while others mediate export. Explain your answer.

12–30* Nuclear import and export are typically studied in cells treated with mild detergents under carefully controlled conditions, so that the plasma membrane is freely permeable to proteins, but the nuclear envelope remains intact. Because the cytoplasm leaks out, nuclei take up proteins only when the cell remnants are incubated with fresh cytoplasm as a source of critical soluble proteins. One common substrate for uptake studies is serum albumin that is cross-linked to a nuclear localization signal peptide and tagged with a fluorescent label such as fluorescein, so that its uptake into nuclei can be observed by microscopy.

The small GTPase, Ran, was shown to be required for nuclear uptake using this assay, and not surprisingly, it requires GTP to function. Since neither Ran-GDP nor Ran-GTP binds to nuclear localization signals, other factors must be responsible for identifying proteins to be imported into the nucleus.

Using this assay system, you purify a protein, which you name importin, that in the presence of Ran and GTP promotes uptake of the labeled substrate into nuclei to about the same extent as crude cytosol (Figure 12–5A). No uptake occurs in the absence of GTP, and Ran alone is unable to promote nuclear uptake. Importin by itself causes a GTP-independent accumulation of substrate at the nuclear periphery, but does not promote nuclear uptake (Figure 12–5A).

To define the steps in the uptake pathway, you first incubate nuclei with substrate in the presence of importin. You then wash away free importin and substrate and incubate a second time with Ran and GTP (Figure 12–5B).

A. Why do you think the substrate accumulates at the nuclear periphery, as is seen in the absence of GTP or with importin alone?
B. From these data decide whether importin is a nuclear import receptor, a Ran-GAP, or a Ran-GEF. Explain your reasoning.
C. To the extent these data allow, define the order of events that leads to uptake of substrate into the nucleus.

12–31 The structures of Ran-GDP and Ran-GTP (actually Ran-GppNp) are strikingly different, as shown in Figure 12–6. Not surprisingly, Ran-GDP binds to a different set of proteins than does Ran-GTP.

To look at the uptake of Ran itself into the nuclei of permeabilized cells, you attach a red fluorescent tag to a cysteine side chain in Ran to make it visible. This modified Ran supports normal nuclear uptake. Fluorescent Ran-GDP is taken up by nuclei only if cytoplasm is added, whereas a mutant form, RanQ69L-GTP, which is unable to hydrolyze GTP, is not taken up in the presence or absence of cytoplasm. To identify the cytoplasmic protein that is crucial for Ran-GDP uptake, you construct affinity columns with either

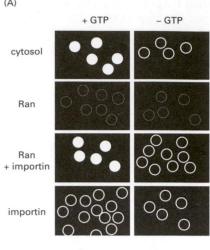

(A)

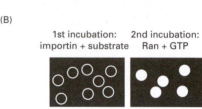

(B)

Figure 12–5 Effects of Ran, importin, and GTP on nuclear uptake of flourescein-labeled substrate (Problem 12–30). (A) Comparison of various combinations of Ran and importin relative to cytosol in the presence and absence of GTP. (B) Two-stage incubation of cell remnants with importin and substrate and then with Ran and GTP. *Circles* are the nuclei; *very light circles* are nuclei without bound substrate.

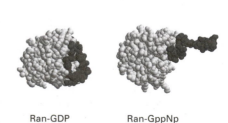

Figure 12–6 Structures of Ran-GDP and Ran-GppNp (Problem 12–31). The *shaded portion* of the structures shows the segment of Ran that differs most dramatically when GDP or the GTP analog is bound.

Ran-GDP or RanQ69L-GTP and pass cytoplasm through them. Cytoplasm passed over a Ran-GDP column no longer supports nuclear uptake of Ran-GDP, whereas cytoplasm passed over a column of RanQ69L-GTP retains this activity. You elute the bound proteins from each column and analyze them on an SDS polyacrylamide gel, looking for differences that might identify the factor that is required for nuclear uptake of Ran (Figure 12–7).

A. Why did you use RanQ69L-GTP instead of Ran-GTP in these experiments?

B. Which of the proteins eluted from the two different affinity columns is a likely candidate for the factor that promotes nuclear import of Ran-GDP?

C. What other protein or proteins would you predict the Ran-GDP import factor would be likely to bind, in order to carry out its function?

D. How might you confirm that the factor you have identified is necessary for promoting the nuclear uptake of Ran?

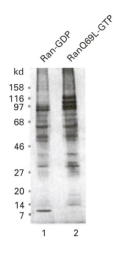

Figure 12–7 Proteins eluted from Ran-GDP (lane 1) and RanQ69L-GTP (lane 2) affinity columns (Problem 12–31). The molecular masses of marker proteins are shown on the *left*.

12–32* The original experiments on nuclear uptake uniformly—but incorrectly—concluded that GTP hydrolysis was essential for import. This conclusion was largely based on results with nonhydrolyzable analogs of GTP such as GppNp, which efficiently blocked nuclear import. It is now known that a Ran-GEF, which converts Ran-GDP to Ran-GTP, exists in the nucleus and a Ran-GAP, which converts Ran-GTP to Ran-GDP, exists in the cytosol, often associated with the pore complexes. These two proteins ensure that Ran-GTP is present primarily in the nucleus and that Ran-GDP is present primarily in the cytoplasm.

The experiments that cleared up this situation—and helped define our present view of nuclear import—showed that some Ran-GEF (a protein known as RCC1) leaks out of nuclei during preparation of cell remnants. How might the presence of Ran-GEF outside the nucleus lead to a spurious requirement for GTP hydrolysis in nuclear import? Why does a nonhydrolyzable analog of GTP such as GppNp block import under these conditions while GTP does not?

12–33 The Rev protein encoded by the HIV virus is required for nuclear export of viral mRNAs for certain structural proteins of the virus. Rev normally shuttles back and forth from nucleus to cytoplasm to accomplish mRNA export. Under some conditions Rev is found primarily in the cytoplasm (Figure 12–8A). In the presence of the broad-spectrum antibiotic leptomycin B, however, Rev is found exclusively in the nucleus (Figure 12–8B). Suggest some possible ways by which leptomycin B might alter the distribution of Rev.

12–34* How do you suppose that proteins with a nuclear export signal get into the nucleus?

12–35 How does leptomycin B work? In the yeast *S. pombe* resistance to leptomycin B can arise by mutations in the *CRM1* gene, which encodes a nuclear export receptor for proteins with leucine-rich nuclear export signals. To look at nuclear export directly, you modify the green fluorescent protein (GFP) by adding a nuclear export signal (NES). In both wild-type and mutant cells that are resistant to leptomycin B the modified NES-GFP is found exclusively in the cytoplasm in the absence of leptomycin B (Figure 12–9). In the presence of leptomycin B, however, NES-GFP is present in the nuclei of wild-type cells, whereas it is located in the cytoplasm of mutant cells (Figure 12–9). Is this result what you would expect if leptomycin B blocks nuclear export? Why or why not?

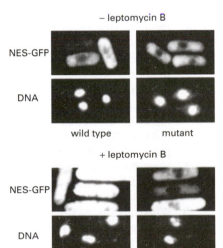

Figure 12–9 Distribution of NES-GFP in *S. pombe* in the presence and absence of leptomycin B (Problem 12–35). *Light areas* in the NES-GFP panels show the position of GFP. *Light areas* in the DNA panels result from a stain that binds to DNA and marks the position of the nuclei in the cells in the NES-GFP panels.

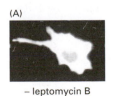
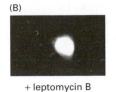

Figure 12–8 The cellular distribution of Rev (Problem 12–33).
(A) Absence of leptomycin B.
(B) Presence of leptomycin B.

12–36* The δ-lactone ring of leptomycin B (Figure 12–10A) is highly reactive, suggesting that it could form a covalent adduct with suitably reactive cysteines in proteins. One of the most resistant forms of Crm1 has serine in place of cysteine at position 529, suggesting the possibility that leptomycin B blocks Crm1 activity by covalent attachment. To test this possibility, you add a biotin moiety to the nonreactive end of leptomycin B (the right-hand end of the molecule as shown in Figure 12–10A). You add this modified leptomycin B (biotin-LMB) to a culture of human cells, wait 2 hours, pass the cell extracts over a strepavidin column, which binds biotin tenaciously, and then elute the bound proteins and display them on an SDS polyacrylamide gel (Figure 12–10B, lane 1). As controls, you incubate in the absence of biotin-LMB (lane 4), in the presence of a 10-fold molar excess of leptomycin B added 1 hour before the biotin-LMB (lane 2), and in the presence of an excess of an inactive analog of leptomycin B, LMB* (lane 3).

A. Figure 12–10, lanes 1 and 3 show a strong new band of 102 kd that is retained on the affinity column when biotin-LMB is added to the cell culture medium. How might you show that this protein is Crm1?

B. Why was the 102-kd protein not seen in Figure 12–10, lanes 2 and 4? What was the point of performing these particular control experiments?

C. Several proteins that bind to the streptavidin column are present in all four lanes in Figure 12–10B. Can you suggest what kinds of proteins they might be? (Think about why they might have been retained on the streptavidin affinity column.)

12–37 You have modified the green fluorescent protein (GFP) to carry a nuclear localization signal at its N-terminus and a nuclear export signal at its C-terminus. When this protein is expressed in cells, you find that about 60% of it is located in the nucleus and the rest in the cytoplasm. You assume that the protein is shuttling between the nucleus and the cytoplasm and that its rates of import and export determine the observed distribution. Nevertheless, you are concerned that the modified GFP might not fold properly even though it retains its fluorescent properties. Specifically, you worry that the 40% in the cytoplasm is not being transported at all because neither signal is exposed, and the 60% in the nucleus has only the nuclear import signal available. How might addition of leptomycin B to such cells resolve this issue?

12–38* Frog oocytes are a useful experimental system for studying nuclear export because they are large cells with large nuclei. It is easy (with practice) to inject oocytes with labeled RNA and to separate the nucleus and cytoplasm to follow the fate of the injected label. You inject a mixture of various ^{32}P-labeled RNA molecules into the nucleus in the presence and absence of leptomycin B to study its effect on nuclear export of RNA. Immediately after injection and three hours later, you analyze total (T), cytoplasmic (C) and nuclear (N) contents by polyacrylamide-gel electrophoresis and autoradiography (Figure 12–11).

A. How good was your injection technique? Did you actually inject into the nucleus? Did you rip apart the nuclear envelope when you injected the RNAs? How do you know?

B. Which, if any, of the RNAs are normally exported from the nucleus?

C. Is the export of any of the RNAs inhibited by leptomycin B? What does your answer imply about export of this collection of RNAs.

12–39 Your advisor is explaining his latest results in your weekly lab meeting. By fusing his protein of interest to GFP, he has shown that it is located entirely in the nucleus. But he wonders if it is a true nuclear protein or a shuttling protein that just spends most of its time in the nucleus. He is unsure how to resolve this issue. He knows that leptomycin B will not be helpful in this instance and asks for input. Having just read an article about how a similar problem was answered, you suggest that he make a heterocaryon by fusing cells that are expressing his tagged protein with an excess of cells that are not expressing it. You tell him that in the presence of a protein synthesis inhibitor to block new synthesis of the tagged protein he can resolve the

(A)

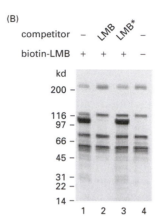

(B)

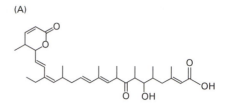

Figure 12–10 Target for reaction with leptomycin B (Problem 12–36). (A) Structure of leptomycin B. (B) Proteins eluted from a streptavidin affinity column after various cell treatments. Molecular masses of marker proteins are shown at the *left* in kilodaltons.

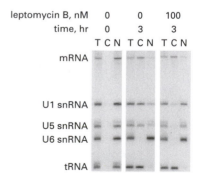

Figure 12–11 Effects of leptomycin B on export of various RNAs from frog oocytes (Problem 12–38). Total (T), cytoplasmic (C), and nuclear (N) fractions are indicated.

issue by examining fused cells with two nuclei. He gives you a puzzled look and asks how does that help. You tell him what he has so often told you "Think about it."

A. Why wouldn't addition of leptomycin B answer the question?
B. How would examining the two nuclei in a heterocaryon answer the question? What results would you expect if the protein were a true nuclear protein? What would your expect if it were a shuttling protein?
C. Why did you suggest that a protein synthesis inhibitor would be needed in this experiment?

12–40*   To test the hypothesis that the directionality of transport across the nuclear membrane is determined primarily by the gradient of the Ran-GDP outside the nucleus and Ran-GTP inside the nucleus, you decide to reverse the gradient to see if you can force the import of a protein that is normally exported from the nucleus. You add a well-defined nuclear export substrate, fluorescent BSA coupled with a nuclear export signal (NES-BSA), to the standard permeabilized cell assay. Sure enough, it is excluded from the nuclei (Figure 12–12A). Now you add Crm1, the nuclear export receptor that recognizes the export signal, and RanQ69L-GTP, the mutant form of Ran that cannot hydrolyze GTP. With these additions, the tagged BSA now enters the nuclei (Figure 12–12B). Unlike conventional nuclear import (see Figure 12–5, Problem 12–30), the maximum nuclear concentration of NES-BSA is no higher than in the cytoplasm surrounding the nucleus.

A. Why doesn't NES-BSA accumulate to a higher concentration in the nucleus than in the cytoplasm in these experiments?
B. In a standard nuclear import assay with added cytoplasm and GTP, proteins with a nuclear localization signal accumulate to essentially 100% in the nucleus (see Figure 12–5, Problem 12–30). How is it that the standard assay allows 100% accumulation in the nucleus?

12–41   The bright ring of fluorescent NES-BSA around the nuclei in Figure 12–12B suggests that there is a substantial accumulation of transport substrate at the nuclear periphery. How many molecules of NES-BSA would need to be bound per nuclear pore complex to give such a bright ring? As a basis for estimating the answer, calculate the concentration of NES-BSA in the nuclear membrane, assuming that one NES-BSA is bound per pore complex, that there are 3000 pores per nucleus, and that the volume of the nuclear membrane is 16 fL (16×10^{-15} L). How does this calculated concentration compare to the concentration of NES-BSA used in these experiments (0.3 µM)? If the ring of fluorescence in Figure 12–12B was ten times brighter than the background, how many NES-BSA molecules would you calculate are bound per nuclear pore complex in that experiment? How does this estimate fit with the structure of the pore complex?

12–42*   By following the increase in nuclear fluorescence over time in the forced nuclear import experiments shown in Figure 12–12B, the authors were able to show that nuclear fluorescence reached half its maximal value in 60 seconds. Since the added concentration of NES-BSA was 0.3 µM, its concentration in the nucleus after 60 seconds was 0.15 µM. Given that the volume of the nucleus is 500 fL and that each nucleus contains 3000 nuclear pores, calculate the rate of import of NES-BSA per pore in this experiment. (For this calculation, neglect export of NES-BSA.) Is your answer physiologically reasonable? Why or why not?

12–43   (True/False) Some proteins are kept out of the nucleus, until needed, by inactivating their nuclear localization signals by phosphorylation. Explain your answer.

12–44*   The gene for the κ-light chain of immunoglobulins contains a DNA sequence element (an enhancer) in one of its introns that regulates its expression. A protein called NF-κB binds to this enhancer and can be found in nuclear extracts of B cells, which actively synthesize immunoglobulins. By contrast, NF-κB activity is absent from precursor B cells (pre-B cells), which

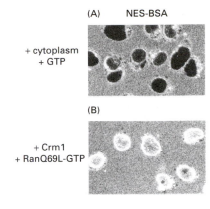

(A)   NES-BSA

+ cytoplasm
+ GTP

(B)

+ Crm1
+ RanQ69L-GTP

Figure 12–12 Directionality of nuclear transport (Problem 12–40). (A) Exclusion of NES-BSA from the nucleus in a standard import assay. (B) Import of NES-BSA in the presence of Crm1 and RanQ69L-GTP.

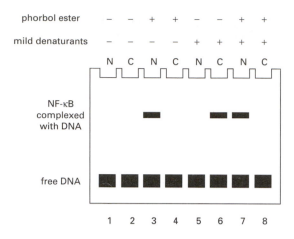

phorbol ester   −   −   +   +   −   −   +   +

mild denaturants   −   −   −   −   +   +   +   +

N C N C N C N C

NF-κB complexed with DNA

free DNA

1 2 3 4 5 6 7 8

Figure 12–13 Gel retardation assay of NF-κB activity under various conditions (Problem 12–44). N and C refer to nuclear and cytoplasmic fractions, respectively. The mild denaturants are a combination of formamide (27%) and sodium deoxycholate (0.2%). When NF-κB binds to the DNA, it forms a complex that migrates more slowly on the gel.

do not express immunoglobulins. You are interested in the strategy by which NF-κB regulates expression of κ-light-chain genes.

If pre-B cells are treated with phorbol ester (which activates protein kinase C), NF-κB activity appears within a few minutes in parallel with the transcriptional activation of the κ-light-chain gene. This activation of NF-κB is not blocked by cycloheximide, which is an inhibitor of protein synthesis.

You find that cytoplasmic extracts of unstimulated pre-B cells contain an inactive form of NF-κB that can be activated by treatment with mild denaturants. (Denaturants disrupt noncovalent bonds but do not break covalent bonds.) To understand the relationship between NF-κB activation and transcriptional regulation of κ-light-chain genes, you isolate nuclear (N) and cytoplasmic (C) fractions from pre-B cells before and after stimulation by phorbol ester. You then measure NF-κB activity, before and after treatment with mild denaturants, by its ability to form a complex with labeled DNA fragments that carry the enhancer. The results are shown in Figure 12–13.

A. How does the subcellular localization of NF-κB change in pre-B cells in response to phorbol ester treatment?

B. Do you think that the activation of NF-κB in response to phorbol ester occurs because NF-κB is phosphorylated?

C. Outline a molecular mechanism to account for activation of NF-κB by treatment with phorbol ester.

12–45   Nuclear localization signals are not cleaved off after transport into the nucleus, whereas the signal sequences for import into other organelles are often removed after import. Why do you suppose it is critical that nuclear localization signals remain attached to their proteins?

12–46*   (**True/False**) All cytosolic proteins have nuclear export signals that allow them to be removed from the nucleus when it reassembles after mitosis. Explain your answer.

THE TRANSPORT OF PROTEINS INTO MITOCHONDRIA AND CHLOROPLASTS

TERMS TO LEARN

chloroplast	mitochondrial hsp70	stroma
inner membrane	mitochondrial precursor protein	thylakoid
intermembrane space	outer membrane	TIM complex
matrix space	OXA complex	TOM complex
mitochondria	protein translocator	

12–47   To aid your studies of protein import into mitochondria, you treat yeast cells with cycloheximide, which blocks ribosome movement along mRNA. When you examine these cells in the electron microscope, you are surprised to find cytosolic ribosomes attached to the outside of the mitochondria. By contrast, you have never seen attached ribosomes in the absence of cycloheximide. To investigate this phenomenon further, you prepare mitochondria from cells that have been treated with cycloheximide and then extract the mRNA that is bound to the ribosomes associated with the mitochondria. You translate this mRNA *in vitro* and compare the protein products with similarly translated mRNA from the cytosol. The results are clear-cut: the mitochondria-associated ribosomes are translating mRNAs that encode mitochondrial proteins.

You are astounded! Here, clearly visible in the electron micrographs, seems to be proof that protein import into mitochondria occurs during translation. How might you reconcile this result with the prevailing view that mitochondrial proteins are imported after they have been synthesized and released from ribosomes?

12–48*　The vast majority of mitochondrial proteins are imported through the outer membrane by the multisubunit protein translocators known as TOM complexes. Since the number of TOM complexes in yeast mitochondria is known (10 pmole/mg of mitochondrial protein), it is possible to calculate whether a co-translational mechanism could account for the bulk of mitochondrial protein import. If mitochondrial proteins were all imported co-translationally, then 10 pmol of TOM complexes would need to import 1 mg of protein each generation. Given that mitochondria double every 3 hours and that the rate of protein synthesis is 3 amino acids per second (which is therefore the maximum rate of import through a single TOM complex), how many mg of mitochondrial protein could 10 pmol of TOM complexes import in one generation? (On average an amino acid has a mass of 110 daltons.)

12–49　Glutamine synthetase is an enzyme involved in ammonia detoxification. In chickens the enzyme is targeted to mitochondria, whereas in humans it is cytosolic. Amphipathic α helices with positively charged amino acids on one side and hydrophobic amino acids on the other are the key features of signal peptides used for protein import into mitochondria. Using the helix-wheel projection in Figure 12–14A, decide which of the N-terminal sequences in Figure 12–14B is from the chicken enzyme and which is from the human enzyme.

Problem 3–10 describes the use of a helix-wheel projection.

12–50*　You have made a peptide that contains a functional mitochondrial import signal. Would you expect the addition of an excess of this peptide to affect the import of mitochondrial proteins? Why or why not?

12–51　Components of the TIM complex, the multisubunit protein translocator in the mitochondrial inner membrane, are much less abundant than those of the TOM complex. They were initially identified using a genetic trick. The yeast *URA3* gene, whose product is normally located in the cytosol where it is essential for synthesis of uracil, was modified to carry an import signal for the mitochondrial matrix. A population of cells carrying the modified *URA3* gene was then grown in the absence of uracil. Most cells died, but the rare cells that grew in the absence of uracil were shown to be defective for mitochondrial import. Explain how this selection identifies cells with defects in components required for mitochondrial import. Why don't normal cells with the modified *URA3* gene grow in the absence of uracil? Why do cells that are defective for mitochondrial import grow in the absence of uracil?

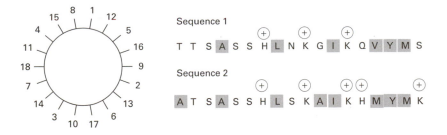

(A) HELIX WHEEL

(B) N-TERMINAL SEQUENCES

Sequence 1

T T S A S S H L N K G I K Q V Y M S

Sequence 2

A T S A S S H L S K A I K H M Y M K

Figure 12–14 Analysis of the N-terminal sequences for human and chicken glutamine synthetase (Problem 12–49). (A) Helix-wheel projection of an α helix. *Numbers* show the positions of the first 18 amino acids of an α helix; amino acid 19 would occupy the same position as amino acid 1. (B) N-terminal amino acid sequences for glutamine synthetases from humans and chickens (unidentified). The N-termini are shown at the *left*; hydrophobic amino acids are *shaded*; positively charged amino acids are indicated by +.

TABLE 12–2 Mitochondrial compartments and translocator complexes (Problem 12–54).

COMPARTMENT	TOM	TIM22	TIM23	OXA
Outer membrane				
Intermembrane space				
Inner membrane				
Matrix				

12–52* Mitochondria normally provide cells with most of the ATP they require to meet their energy needs. However, mitochondria that cannot import proteins—like those in Problem 12–51—are defective for ATP synthesis. How is it that cells with import-defective mitochondria can survive at all? How do they get the ATP they need to function?

12–53 (**True/False**) The TOM complex is required for the import of all nucleus-encoded mitochondrial proteins. Explain your answer.

12–54* Four protein translocator complexes direct proteins to various compartments in mitochondria. Indicate on Table 12–2 the mitochondrial compartment(s) into which each translocator inserts its transport substrates.

12–55 Describe in a general way how you might use radiolabeled proteins and proteases to study import processes in isolated, intact mitochondria. What sorts of experimental controls might you include to ensure that the results you obtain mean what you think they do?

12–56* If the enzyme dihydrofolate reductase (DHFR), which is normally located in the cytosol, is engineered to carry a mitochondrial targeting sequence at its N-terminus, it is efficiently imported into mitochondria. If the modified DHFR is first incubated with methotrexate, which binds tightly to the active site, the enzyme remains in the cytosol. How do you suppose that the binding of methotrexate interferes with mitochondrial import?

12–57 Barnase is a 110-amino acid bacterial ribonuclease that is much used as a model for studies of protein folding and unfolding. It forms a compact folded structure that has a high energy of activation for unfolding (about 20 kcal/mole). Can such a protein be imported into mitochondria? To the N-terminus of barnase, you add 35, 65, or 95 amino acids from the N-terminus of pre-cytochrome *b2*, all of which include the cytochrome's mitochondrial import signal. N35-barnase is not imported, N65-barnase is imported at a low rate, and N95-barnase is imported very efficiently into isolated mitochondria. None of these N-terminal extensions have any measurable effect on the stability of the barnase domain. If these proteins are denatured before testing for import, they are all imported at the same high rate. How do you suppose that longer N-terminal extensions facilitate the import of barnase?

12–58* Are proteins imported into mitochondria as completely unfolded polypeptide chains, or can the translocation apparatus accommodate fully or partially folded structures? That is, is the protein sucked up like a noodle, or is it swallowed whole, as a python devours its prey? It is possible to engineer cysteine amino acids into barnase and then cross-link them to make disulfide bonds either between C5 and C78 or between C43 and C80 (Figure 12–15). Import of N95-barnase (see Problem 12–57) was tested in the presence and absence of disulfide cross-links at these two positions. Its import was unaffected by either cross-link. By contrast, import of N65-barnase was blocked by the C5-C78 cross-link but unaffected by the C43-C80 cross-link. Do these results allow you to distinguish between import of extended polypeptide chains or of folded structures? Why or why not?

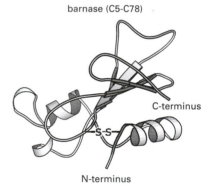

barnase (C5-C78)

C-terminus

N-terminus

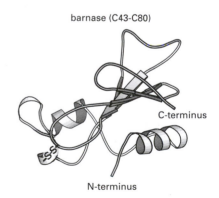

barnase (C43-C80)

C-terminus

N-terminus

Figure 12–15 Structures of barnase molecules with disulfide bonds between cysteines at positions 5 and 78 (C5-C78) or between positions 43 and 80 (C43-C80) (Problem 12–58).

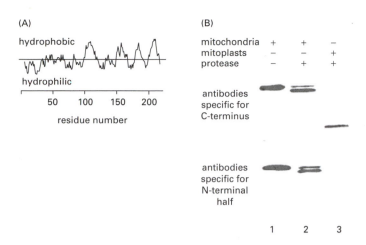

(A)

hydrophobic

hydrophilic

50   100   150   200

residue number

(B)

mitochondria	+	+	−
mitoplasts	−	−	+
protease	−	+	+

antibodies specific for C-terminus

antibodies specific for N-terminal half

1   2   3

Figure 12–16 Arrangement of Tim23 in mitochondrial membranes (Problem 12–59). (A) Hydropathy plot for Tim23. (B) Sensitivity of Tim23 in mitochondria and mitoplasts to digestion with a protease.

12–59 Tim23 is a key component of the mitochondrial TIM23 protein translocator complex. The N-terminal half of Tim23 is hydrophilic, while the C-terminus is hydrophobic and probably spans the membrane four times, as suggested by the hydropathy plot in Figure 12–16A. To determine the arrangement of Tim23 in mitochondrial membranes, a protease was added to intact mitochondria or to mitoplasts, which are mitochondria from which the outer membranes have been removed. The mobility of Tim23 was detected on SDS polyacrylamide gels by immunoblotting with antibodies specific for the N-terminal half or for the extreme C-terminus of Tim23 (Figure 12–16B). Normal-sized Tim23 was present in both mitochondria and mitoplasts (as shown for mitochondria in Figure 12–16B), but Tim23 was partially digested when mitochondria and mitoplasts were treated with a protease (Figure 12–16B).

A. Is Tim23 an integral component of the inner or outer mitochondrial membrane? Explain your reasoning.

B. To the extent the information in this problem allows, diagram the arrangement of Tim23 in mitochondrial membranes.

12–60* In principle, could a protein be imported into the mitochondrial matrix in the absence of mitochondrial ATP hydrolysis according to the thermal-ratchet model for import? Could it be imported in the absence of mitochondrial ATP hydrolysis according to the cross-bridge-ratchet model?

12–61 (**True/False**) The two signal sequences required for transport of nucleus-encoded proteins into the mitochondrial inner membrane via the TIM23 complex are cleaved off the protein in different mitochondrial compartments. Explain your answer.

12–62* Why do mitochondria need a fancy translocator to import proteins across the outer membrane? Their outer membranes already have large pores formed by porins.

12–63 (**True/False**) Import of proteins into mitochondria and chloroplasts is very similar, even the individual components of their transport machinery are homologous, as befits their common evolutionary origin. Explain your answer.

12–64* Chloroplasts contain six compartments—outer membrane, intermembrane space, inner membrane, stroma, thylakoid membrane, and thylakoid lumen (Figure 12–17)—each of which is populated by specific sets of proteins. To investigate the import of nucleus-encoded proteins into chloroplasts, you have chosen to study ferredoxin (FD), which is located in the stroma, and plastocyanin (PC), which is located in the thylakoid lumen, as well as two hybrid genes: ferredoxin with the plastocyanin signal peptide (pcFD) and

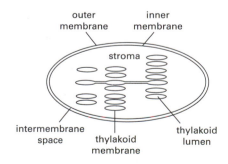

Figure 12–17 The six compartments of a chloroplast (Problem 12–64).

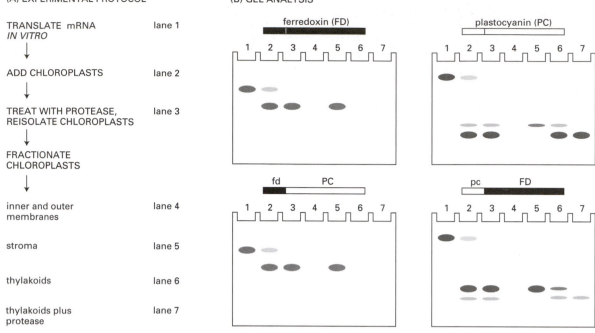

(A) EXPERIMENTAL PROTOCOL

TRANSLATE mRNA *IN VITRO* lane 1

↓

ADD CHLOROPLASTS lane 2

↓

TREAT WITH PROTEASE, REISOLATE CHLOROPLASTS lane 3

↓

FRACTIONATE CHLOROPLASTS

↓

inner and outer membranes lane 4

stroma lane 5

thylakoids lane 6

thylakoids plus protease lane 7

(B) GEL ANALYSIS

plastocyanin with the ferredoxin signal peptide (fdPC). You translate mRNAs from these four genes *in vitro*, mix the translation products with isolated chloroplasts for a few minutes, reisolate the chloroplasts after protease treatment, and fractionate them to find which compartments the proteins have entered (Figure 12–18A). The status of the normal and hybrid proteins at each stage of the experiment are shown in Figure 12–18B: each lane in the gels corresponds to a stage of the experiment as indicated alongside the experimental protocol in Figure 12–18A.

A. How efficient is chloroplast uptake of ferredoxin and plastocyanin in your *in vitro* system? How can you tell?

B. Are ferredoxin and plastocyanin localized to their appropriate chloroplast compartments in these experiments? How can you tell?

C. Are the hybrid proteins imported as you would expect if the N-terminal signal peptides determined their final location? Comment on any significant differences.

D. Why are there three bands in experiments with plastocyanin and pcFD but only two bands in experiments with ferredoxin and fdPC? To the extent you can, identify the molecular species in the bands and their relationships to one another.

E. Based on your experiments, propose a model for the import of proteins into the stroma and thylakoid lumen.

Figure 12–18 Import of ferredoxin and plastocyanin into chloroplast compartments (Problem 12–64). (A) Experimental protocol. (B) Gel analysis. Samples from each stage in the experimental protocol were analysed by gel electrophoresis. Each lane in (B) corresponds to a particular experimental treatment in (A). Ferredoxin gene segments are *black*; plastocyanin gene segments are *white*. The signal peptides in all genes are located at the *left* (N-terminal) end. The ferredoxin signal peptide in the hybrid gene is shown as fd; the plastocyanin signal peptide in the hybrid gene is shown as pc.

PEROXISOMES

TERMS TO LEARN

peroxin   peroxisome

12–65 (**True/False**) Peroxisomes are found in only a few types of eucaryotic cells. Explain your answer.

12–66* Trypanosomes are single-celled parasites that cause sleeping sickness when they infect humans. Trypanosomes from humans carry the enzymes for a portion of the glycolytic pathway in a peroxisomelike organelle, termed the glycosome. By contrast, trypanosomes from the tsetse fly—the intermediate host—carry out glycolysis entirely in the cytosol. This intriguing difference has alerted the interest of the pharmaceutical company that employs you. Your company wishes to exploit this difference to control the disease.

You decide to study the enzyme phosphoglycerate kinase (PGK) because it is in the affected portion of the glycolytic pathway. Trypanosomes from the tsetse fly express PGK entirely in the cytosol, whereas trypanosomes from humans express 90% of the total PGK activity in glycosomes and only 10% in the cytosol. When you clone PGK genes from trypanosomes, you find three forms that differ slightly from one another. Exploiting these small differences, you design three oligonucleotides that hybridize specifically to the mRNAs from each gene. Using these oligonucleotides as probes, you determine which genes are expressed by trypanosomes from humans and from tsetse flies. The results are shown in Figure 12–19.

A. Which PGK genes are expressed in trypanosomes from humans? Which are expressed in trypanosomes from tsetse flies?
B. Which PGK gene probably encodes the glycosomal form of PGK?
C. Do you think that the minor cytosolic PGK activity in trypanosomes from humans is due to inaccurate sorting into glycosomes? Explain your answer.

12–67   Primary hyperoxaluria type 1 (PH1) is a lethal autosomal recessive disease caused by a deficiency of the liver-specific peroxisomal enzyme alanine: glyoxylate aminotransferase (AGT). About one-third of PH1 patients possess significant levels of AGT protein and enzyme activity. Analysis of these patients shows that their AGT contains two critical single amino acid changes: one that interferes with peroxisomal targeting and a second that allows the N-terminus to form an amphipathic α helix with a positively charged side. Where do you suppose this mutant AGT is found in cells from these patients? Explain your reasoning.

12–68*   Catalase, an enzyme normally found in peroxisomes, is present in normal amounts in cells that do not have visible peroxisomes. It is possible to determine the location of catalase in such cells, using immunofluorescence microscopy with antibodies specific for catalase. Fluorescence micrographs of normal cells and two peroxisome-deficient cell lines are shown in Figure 12–20. Where is catalase located in cells without peroxisomes (Figure 12–20B and C)? Why does catalase show up as small dots of fluorescence in normal cells (Figure 12–20A)?

12–69   You have developed an assay for peroxisomal function. Mutagenized cell colonies are incubated with a soluble radioactive precursor that is converted into a readily detectable insoluble product by a peroxisomal enzyme. A laborious screen of 25,000 colonies is finally rewarded by the discovery of two colonies that do not have the insoluble radioactive product. Sure enough, these mutant cells lack typical peroxisomes, as judged by electron microscopy.

When you test these cells for peroxisomal enzymes, you find that catalase activity is virtually the same as in normal cells. By contrast, acyl CoA oxidase activity is absent in both mutant cell lines. To investigate the acyl CoA oxidase deficiency, you perform a pulse-chase experiment: you grow cells for 1 hour in medium containing ^{35}S-methionine, then transfer them to unlabeled medium and immunoprecipitate acyl CoA oxidase at various

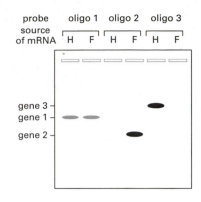

Figure 12–19 Hybridization of specific oligonucleotide probes to mRNA isolated from trypanosomes from humans (H) and tsetse flies (F) (Problem 12–66). The intensity of the bands on the autoradiograph reflects the concentrations of the mRNAs.

(A)   (B)   (C)

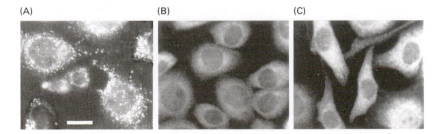

Figure 12–20 Location of catalase in cells as determined by immunofluorescence microscopy (Problem 12–68). (A) Normal cells. (B and C) Peroxisome-deficient cell lines. Cells were reacted with antibodies specific for catalase, washed, and then stained with a fluorescein-labeled second antibody that is specific for the catalase-specific antibody. All three panels are the same size; a 20-μm scale bar is shown in (A).

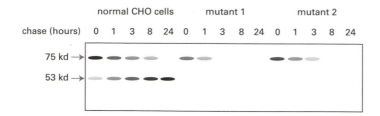

Figure 12–21 Pulse-chase experiments with normal and mutant cells (Problem 12–69).

times after transfer (Figure 12–21). You observe two forms of oxidase in normal cells, but only one in the mutant cell lines. To clarify the relationship between the 75-kd and 53-kd forms of the oxidase, you isolate mRNA from wild-type and the mutant cell lines, translate it *in vitro*, and immunoprecipitate acyl CoA oxidase. All three sources of mRNA give similar levels of the 75-kd form, but none of the 53-kd form.

A. How do you think the two forms of acyl CoA oxidase in normal cells are related? Which one, if either, do you suppose is the active enzyme?

B. Why do the mutant cells have only the 75-kd form of acyl CoA oxidase, and why do you think it disappears during the pulse-chase experiment? If you had done a similar experiment with catalase, do you suppose it would have behaved the same way?

12–70* Cells with functional peroxisomes incorporate 9-(1′-pyrene)nonanol (P9OH) into membrane lipids. Exposure of such cells to UV light causes cell death because excitation of the pyrene moiety generates reactive oxygen species, which are toxic to cells. Cells that do not make peroxisomes lack missing a critical enzyme responsible for incorporating P9OH into membrane lipids. How might you make use of P9OH to select for cells that are missing peroxisomes?

12–71 Through cleverness and hard work, you have isolated 14 cell lines that are missing peroxisomes, as judged by electron microscopy. You wish to know how many different genes are represented in this collection of mutant cell lines. You carry out two types of complementation analysis—fusion of cells to form heterocaryons and transfection with cDNAs for defined peroxisomal genes—and test for the appearance of peroxisomes. Your results are shown in Table 12–3, where the presence of peroxisomes is indicated by + and their absence by –.

As is apparent from the blank spaces in Table 12–3, you haven't done all the possible heterocaryon analyses (nor all the possible transfection analyses, for that matter). Is your present data set adequate to allow you to group all 14 mutant cell lines unambiguously into complementation groups? To the

TABLE 12–3 Complementation analysis of peroxisome-deficient cell lines (Problem 12–71).

MUTANTS	HETEROCARYON ANALYSIS								cDNA TRANSFECTION		
	Z65	ZP105	ZP92	Z24	ZP109	ZP110	ZP114	ZP165	PEX2	PEX5	PEX6
ZP109				+					–	–	–
ZP110				+	+				–	–	–
ZP114				+	+	+			–	–	–
ZP116	–	+	+		+				+	–	–
ZP119	+	+	+	+	+	+	+	–	–	–	–
ZP160	–								+	–	–
ZP161						–			–	–	–
ZP162		–							–	+	–
ZP164			–						–	–	+
ZP165				+				–	–	–	–

extent the data allow, sort mutant cell lines into complementation groups and indicate which groups are mutant in the *PEX2*, *PEX5*, and *PEX6* genes.

12–72* If peroxisomes come from preexisting peroxisomes by growth and fission, how is it that mutant yeast cells with no visible peroxisomes, peroxisome ghosts, or peroxisome remnants, can suddenly form fully functional peroxisomes when the missing gene is introduced?

12–73 Proteins that are imported into the peroxisome matrix using a C-terminal tripeptide signal are recognized by the cytosolic receptor Pex5, which docks at a complex of proteins in the peroxisomal membrane. Delivery by a cytosolic receptor distinguishes peroxisomal import from import into mitochondria, chloroplasts, and the endoplasmic reticulum. Moreover, unlike import into those organelles, peroxisomes can import fully folded proteins and protein oligomers. You wish to distinguish three potential modes of action for Pex5: (1) it delivers its cargo to the peroxisomal membrane but remains cytosolic, (2) it enters the peroxisome along with its cargo; or (3) it cycles between the cytosol and the peroxisomal matrix.

To define the mechanism for Pex5-mediated import, you modify the *PEX5* gene to encode an N-terminal peptide segment that includes a cleavage site for a protease localized exclusively to the matrix of the peroxisome (Figure 12–22A). Immediately adjacent to the cleavage site is a sequence of amino acids (the so-called FLAG tag) that can be recognized by commercial antibodies. One antibody, mAb2, binds the FLAG tag in any context, whereas another, mAb1, binds the FLAG tag only when it is at the N-terminus and thus will detect only cleaved Pex5. By preparing a whole-cell extract (WCE) and fractionating it into a pellet (P), which contains the peroxisomes, and supernatant (S), which contains the cytosol, you can distinguish the three possible mechanisms of Pex5-mediated import—cytosolic, imported, cycling—by using the mAb1 and mAb2 antibodies, as shown in Figure 12–22B.

When you express the modified *PEX5* gene in cells, prepare cell fractions, separate the proteins by electrophoresis, and react them with mAb1 and mAb2 antibodies, you obtain the results shown in Figure 12–22C.

A. Explain the theoretical results (Figure 12–22B) expected for each of the three possible mechanisms of Pex5-mediated import.

B. Based on the results in Figure 12–22C, how does Pex5 mediate the import of proteins into the matrix of the peroxisome? Explain your reasoning.

C. Pex5-mediated import into peroxisomes resembles most closely import into what other cellular organelle?

THE ENDOPLASMIC RETICULUM

TERMS TO LEARN

BiP	free ribosome	Sec61 complex
calnexin	glycoprotein	signal hypothesis
calreticulin	GPI anchor	signal-recognition particle (SRP)
co-translational	membrane-bound ribosome	smooth ER
dolichol	microsome	SRP receptor
endoplasmic reticulum (ER)	phospholipid exchange protein	start-transfer signal
ER lumen	polyribosome	stop-transfer signal
ER resident protein	posttranslational	transmembrane protein
ER retention signal	protein glycosylation	unfolded protein response
ER signal sequence	rough ER	

12–74* Explain how an mRNA molecule can remain attached to the ER membrane while the individual ribosomes translating it are released and rejoin the cytosolic pool of ribosomes after each round of translation.

(A) CONSTRUCT

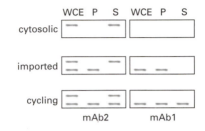

(B) THEORETICAL

(C) EXPERIMENTAL

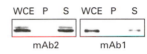

Figure 12–22 Mechanism of Pex5-mediated peroxisomal import (Problem 12–73). (A) Modified Pex5. (B) Expectations for different mechanisms of Pex5-mediated peroxisomal import. (C) Analysis of transfection of the modified *PEX5* gene into cells. 'WCE' stands for whole-cell extract, 'P' for pellet, and 'S' for supernatant. Proteins were detected by immunoblotting using mAb1 or mAb2, followed by reaction with a second antibody that binds mAb1 and mAb2 and also carries bound horseradish peroxidase, which then converts an added precursor into a light-emitting molecule that can be detected on a photographic film. *Black bands* correspond to sites where the film has been exposed by the light-emitting molecule. In (B) uncleaved Pex5 (*upper band*) and cleaved Pex5 (*lower band*) are separated by a larger distance than they are in the experiment (C) for clarity.

12–75   Why are cytosolic hsp70 chaperone proteins required for import of proteins into mitochondria and chloroplasts, but not for co-translational import into the ER?

12–76*   Where would you expect to see microsomes in an electron micrograph of a liver cell?

12–77   If smooth ER and rough ER are continuous membranes, how is it that rough microsomes (derived from rough ER) can have more than 20 proteins that are not present in smooth microsomes (derived from smooth ER)?

12–78*   Translocation of proteins across rough microsomal membranes can be judged by several experimental criteria: (1) the newly synthesized proteins are protected from added proteases, unless detergents are present to solubilize the protecting lipid bilayer; (2) the newly synthesized proteins are glycosylated by oligosaccharide transferases, which are localized exclusively to the lumen of the ER; (3) the signal peptides are cleaved by signal peptidase, which is active only on the lumenal side of the ER membrane.

   Use these criteria to decide whether a protein is translocated across rough microsomal membranes. The mRNA is translated into protein in a cell-free system in the absence or presence of microsomes. Samples from these translation reactions are treated in four different ways: (1) no treatment, (2) addition of a protease, (3) addition of a protease and detergent, and (4) disruption of microsomes and addition of endoglycosidase H (endo H), which removes *N*-linked sugars that are added in the ER. An electrophoretic analysis of these samples is shown in Figure 12–23.

   A. Explain the experimental results that are seen in the absence of microsomes (Figure 12–23, lanes 1 to 4).
   B. Using the three criteria outlined in the problem, decide whether the experimental results in the presence of microsomes (Figure 12–23, lanes 5 to 8) indicate that the protein is translocated across microsomal membranes. How would you account for the migration of the proteins in Figure 12–23, lanes 5, 6, and 8.
   C. Is the protein anchored in the membrane, or is it translocated all the way through the membrane?

12–79   (**True/False**) The signal peptide, when it emerges from the ribosome, binds to a hydrophobic site on the ribosome causing a pause in protein synthesis, which resumes when SRP binds to the signal peptide. Explain your answer.

12–80*   An elegant early study of co-translational import of protein into the ER focused on the source of energy for translocation: does the ribosome 'push' the protein across or does the translocation machinery 'pull' it? The authors reasoned that if the ribosome pushed the protein across they should not be able to uncouple translation from translocation, whereas if the translocation machinery pulled the protein across, they might be able to.

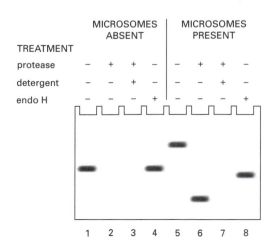

Figure 12–23 Results of translation of a pure mRNA in the presence and absence of microsomal membranes (Problem 12–78). Treatments of the products of translation before electrophoresis are indicated at the *top* of each lane. Electrophoresis was on an SDS polyacrylamide gel, which separates proteins on the basis of size, with lower molecular weight proteins migrating farther down the gel.

They cloned a gene for an ER protein onto a plasmid so they could transcribe it *in vitro* by adding RNA polymerase (Figure 12–24). By cutting the plasmid at different positions within the gene, they could also generate two mRNAs that were shorter than normal (Figure 12–24). These three mRNAs were translated *in vitro*. In one experiment rough microsomes were added before translation began. In a second experiment microsomes were added after translation was complete (along with cycloheximide to inhibit any additional protein synthesis). Translocation was assessed by comparing samples that were untreated, treated directly with protease, or treated with endoglycosidase H (endo H) after disruption of the microsomes. The products were then displayed by SDS-gel electrophoresis (Figure 12–25).

A. Were the proteins from each of the mRNAs translocated when the microsomes were present during translation? How can you tell?

B. Was translocation uncoupled from translation in any of the experiments? Explain your answer.

C. From these results the authors concluded that the translocation machinery pulls proteins across the membrane. What was their reasoning?

D. It is now well accepted that ribosomes push their proteins across the ER membrane during co-translational insertion. How do you suppose the authors might reinterpret their results in hindsight, with the benefit of nearly two decades of additional experimentation?

12–81 You are studying posttranslational translocation of a protein into purified microsomes, but you find that import is very inefficient. Which one of the following might be expected to increase the efficiency of import if added to the mixture of protein and microsomes? Explain your answer.

A. BiP
B. Cytosolic hsp70
C. Free ribosomes
D. Sec61 complex
E. SRP

12–82* Compare and contrast protein import into the ER and into the nucleus. List at least two major differences in the mechanisms and speculate on why the nuclear mechanism might not work for ER import and vice versa.

12–83 (**True/False**) Nascent polypeptide chains are transferred across the ER membrane through a pore in the Sec61 protein translocator complex. Explain your answer.

12–84* Things are not going well. You've just had a brief but tense meeting with your research advisor that you will never forget, and it is clear that your future in his lab is in doubt. He is not fond of your habit of working late and sleeping late; he thinks it is at the root of your lack of productivity (as he perceives it). A few days after the meeting, you roll in at the bright and early hour of noon, to be informed by your student colleagues that your advisor is "looking all over for you." You are certain the axe is going to fall. But it turns out he is excited, not upset. He's just heard a seminar that reminded him of the note

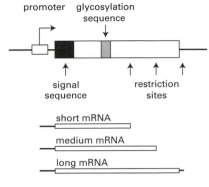

Figure 12–24 A cloned gene for testing the coupling of translation and translocation (Problem 12–80). The plasmid is shown as a linear sequence for simplicity. The protein coding segment is shown by the *large rectangle*; promoter sequences are indicated by the *small rectangle*. Cleavage sites for restriction nucleases used to truncate the transcription template are indicated. The three different mRNA products of transcription from the truncated templates are indicated *below* the map of the gene.

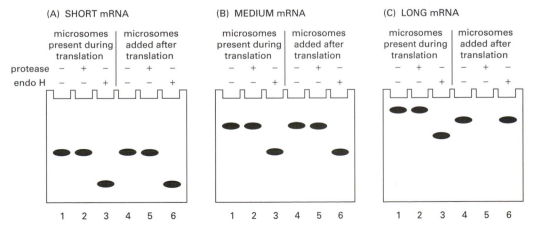

Figure 12–25 Results of experiments to test the coupling of translation and translocation (Problem 12–80). (A) Short mRNA. (B) Medium mRNA. (C) Long mRNA. Treatments of samples before electrophoresis are indicated above the gels. Endo H treatment was applied after the microsomes were disrupted. Endo H removes sugars of the type added in the ER.

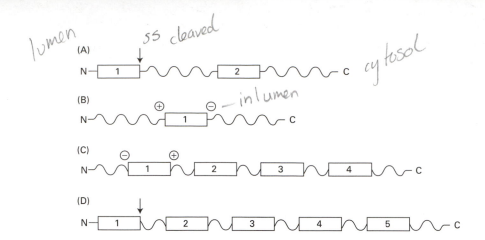

Figure 12–26 Distribution of the membrane-spanning segments in proteins to be inserted into the ER membrane (Problem 12–86). *Boxes* represent membrane-spanning segments and *arrows* indicate sites at which signal sequences are cleaved. The *pluses* and *minuses* indicate the charges at the ends of some transmembrane segments.

you'd left on his desk the month before, describing a selection scheme you had crafted (late one night, of course) for isolating mutants in the ER translocation machinery. You are completely flabbergasted.

You quickly settle on the details for the selection, which involves fusing an ER import signal to the N-terminus of the *HIS4* gene product. His4 is a cytosolic enzyme that converts histidinol to histidine. Yeast strains that are defective for an early step in the histidine biosynthetic pathway can grow on added histidinol, if His4 is present. You decide to look for temperature sensitive (ts) mutants, which are normal at 24°C but defective at 37°C. Using a strain that expresses His4 with an ER import signal, you select for cells that grow on histidinol at 30°C and screen them for ones that die at 37°C. The first mutant you isolate is in the *SEC61* gene (later shown to encode a principal component of the translocator through which ribosomes insert nascent proteins across the ER membrane.) You are back in your advisor's good graces!

A. Why is it that normal cells with the modified His4 cannot grow on histidinol, whereas cells with a defective ER-import apparatus can?

B. Why did the selection scheme set out to find ts mutants? Why was selection applied at the intermediate temperature of 30°C, rather than at 24°C or 37°C?

12–85 **(True/False)** In multipass transmembrane proteins the odd-numbered transmembrane segments (counting from the N-terminus) act as start-transfer signals and the even-numbered segments act as stop-transfer signals. Explain your answer.

12–86* Four membrane proteins are represented schematically in Figure 12–26. The boxes represent membrane-spanning segments and the arrows represent sites for cleavage of the signal sequence. Predict how each of the mature proteins will be arranged across the membrane of the ER. Indicate clearly the N-and C-termini relative to the cytosol and the lumen of the ER, and label each box as a start-transfer or stop-transfer signal.

12–87 Examine the multipass transmembrane protein shown in Figure 12–27. What would you predict would be the effect of converting the first

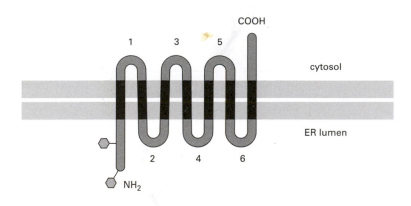

Figure 12–27 Arrangement of a multipass transmembrane protein in the ER membrane (Problem 12–87).

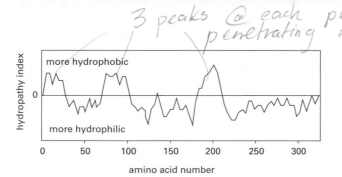

3 peaks @ each peak. start stop, sequence
penetrating memb. in hydrophobic

Figure 12–28 Hydropathy plot of a membrane protein (Problem 12–88). The three hydrophobic peaks indicate the positions of three potential membrane-spanning segments.

hydrophobic transmembrane segment to a hydrophilic segment? Sketch the arrangement of the modified protein in the ER membrane.

12–88* A classic paper describes a genetic method for determining the organization of a bacterial protein in the membrane of *E. coli*. The hydropathy plot of the protein in Figure 12–28 indicated three potential membrane-spanning segments. Hybrid fusion proteins of different lengths, some with internal deletions, were made with the membrane protein at the N-terminus and alkaline phosphatase at the C-terminus (Figure 12–29). Alkaline phosphatase is easy to assay in whole cells and has no significant hydrophobic stretches. Moreover, when it is on the cytoplasmic side of the membrane, its activity is low, and when it is on the external side of the membrane (in the periplasmic space), its activity is high. The assayed levels of alkaline phosphatase activity are indicated (HIGH or LOW) in Figure 12–29.

A. How is the protein organized in the membrane? Explain how the results with the fusion proteins indicate this arrangement.

B. How is the organization of the membrane protein altered by the deletion? Are your measurements of alkaline phosphatase activity in the internally deleted plasmids consistent with the altered arrangement?

C. Are the N-terminus and the C-terminus of the mature membrane protein (the normal, nonhybrid protein) on the same side of the membrane?

12–89 (**True/False**) The ER lumen contains a mixture of thiol-containing reducing agents that prevent the formation of S—S linkages (disulfide bonds) by maintaining the cysteine side chains of lumenal proteins in reduced (–SH) form. Explain your answer.

12–90* Why might it be advantageous to add a preassembled block of 14 sugars to a protein in the ER, rather than building the sugar chains step-by-step on the surface of the protein by the sequential addition of sugars by individual enzymes?

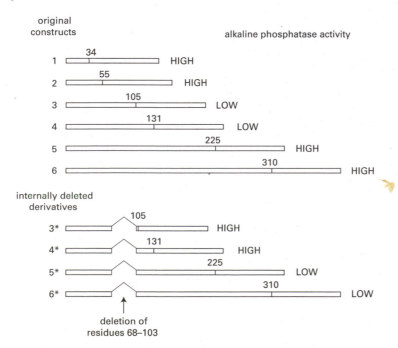

Figure 12–29 Structures of hybrid proteins used to determine the organization of a membrane protein (Problem 12–88). The membrane protein (*unshaded segment*) is at the N-terminus and alkaline phosphatase (*shaded segment*) is at the C-terminus of the protein. The *inverted V* indicates the site from which amino acids were deleted from modified hybrid proteins. The most C-terminal amino acid of the membrane protein is *numbered* in each hybrid protein. The activity of alkaline phosphatase in each hybrid protein is shown on the *right*.

12–91 Discuss the following statement: "The reason why the sugar tree on dolichol phosphate is synthesized in such a baroque fashion, with extensive flipping of dolichol-phosphate-sugar intermediates across the ER membrane, is because there are no nucleoside triphosphates in the ER lumen. The activated forms of the sugars are all generated by reaction with UTP or GTP in the cytosol."

12–92* Outline the steps by which misfolded proteins in the ER trigger synthesis of additional ER chaperone proteins. How does this response benefit the cell?

12–93 (**True/False**) Some membrane proteins are attached to the cytoplasmic surface of the plasma membrane through a C-terminal linkage to a glyco-sylphosphatidylinositol (GPI) anchor. Explain your answer.

12–94* All new phospholipids are added to the cytoplasmic leaflet of the ER membrane, yet the ER membrane has a symmetrical distribution of different phospholipids in its two leaflets. By contrast, the plasma membrane, which receives all its membrane components ultimately from the ER, has a very asymmetrical distribution of phospholipids in the two leaflets of its lipid bilayer. How is symmetry generated in the ER membrane, and how is asymmetry generated in the plasma membrane?

12–95 Mitochondria and peroxisomes, as opposed to most other cellular membranes, acquire new phospholipids in soluble form from phospholipid exchange proteins. One such protein, PC exchange protein, specifically transfers phosphatidylcholine (PC) between membranes. Its activity can be measured by mixing red blood cell ghosts (intact plasma membranes with cytoplasm removed) with synthetic phospholipid vesicles containing radioactively labeled PC in both monolayers of the vesicle bilayer. After incubation at 37°C, the mixture is centrifuged briefly so that ghosts form a pellet, whereas the vesicles stay in the supernatant. The amount of exchange is determined by measuring the radioactivity in the pellet.

Figure 12–30 shows the result of an experiment along these lines, using labeled (donor) vesicles with an outer radius of 10.5 nm and a bilayer 4.0 nm in thickness. No transfer occurred in the absence of the exchange protein, but in its presence up to 70% of the labeled PC in the vesicles could be transferred to the red cell membranes.

Several control experiments were performed to explore the reason why only 70% of the label in donor vesicles was transferred.

1. Five times as many membranes from red cell ghosts were included in the incubation: the transfer still stopped at the same point.
2. Fresh exchange protein was added after 1 hour: it caused no further transfer.
3. The labeled lipids remaining in donor vesicles at the end of the reaction were extracted and made into fresh vesicles: 70% of the label in these vesicles was exchangeable.

When the red cell ghosts that were labeled in this experiment were used as donor membranes in the reverse experiment (that is, transfer of PC from red cell membranes to synthetic vesicles), 96% of the label could be transferred to the acceptor vesicles.

A. What possible explanations for the 70% limit do each of the three control experiments eliminate?
B. What do you think is the explanation for the 70% limit? (HINT: the area of the outer surface of these small donor vesicles is about 2.6 times larger than the area of the inner surface.)
C. Why do you think that almost 100% of the label in the red cell membrane can be transferred back to the vesicle?

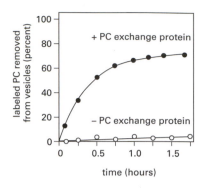

Figure 12–30 Transfer of labeled PC from donor vesicles to red cell membranes by PC exchange protein (Problem 12–95).

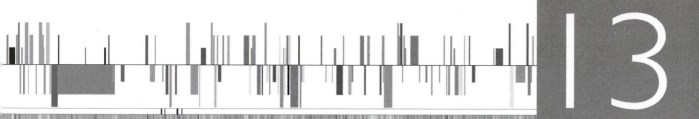

INTRACELLULAR VESICULAR TRAFFIC

THE MOLECULAR MECHANISMS OF MEMBRANE TRANSPORT AND THE MAINTENANCE OF COMPARTMENT DIVERSITY

TERMS TO LEARN

adaptin	COPI-coated vesicle	SarI protein
ARF protein	COPII-coated vesicle	SNARE
clathrin	dynamin	t-SNARE
clathrin-coated vesicle	NSF	v-SNARE
coated vesicle	Rab effector	

13–1   In a nondividing cell such as a liver cell, why must the flow of membrane between compartments be balanced, with retrieval pathways matching the outward flow? Would you expect the same balanced flow in a gut epithelial cell, which is actively dividing?

13–2*   At least three different coats form around transport vesicles. What two principal functions do these different coats have in common?

13–3   (**True/False**) In all events involving fusion of a vesicle to a target membrane, the cytosolic leaflets of the vesicle and target bilayers always fuse together, as do the leaflets that are not in contact with the cytosol. Explain your answer.

*Answers to these problems are available to Instructors at www.classwire.com/garlandscience/

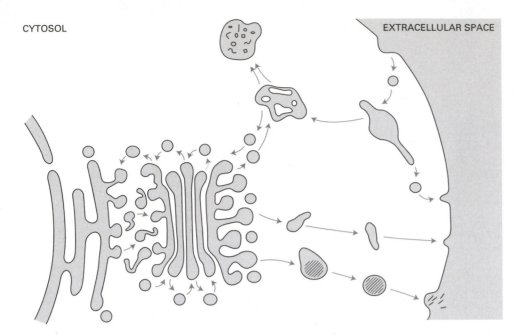

EXTRACELLULAR SPACE

Figure 13–1 The intracellular compartments in the biosynthetic-secretory, endocytic, and retrieval pathways, with flow between compartments indicated (Problem 13–4).

13–4* The diagram in Figure 13–1 shows the various intracellular compartments involved in the biosynthetic-secretory, endocytic, and retrieval pathways.

A. Label various compartments in the diagram.

B. Indicate on the arrows whether they are part of the biosynthetic-secretory pathway, the endocytic pathway, or a retrieval pathway.

13–5 Discuss the following analogy: "Receptors competing to be transported by the coated pit system can be compared to skiers joining a cable-car network. Entry is permitted to ticket holders only, but there is no guarantee of who is found with whom in a particular cable car, although all travelers, hopefully, will reach the next station."

13–6* The clathrin coat on a vesicle is made up of numerous triskelions that form a cage 60–200 nm in diameter, composed of both pentagonal and hexagonal faces, just like C60 fullerene (Figure 13–2). Sketch the location of an individual triskelion in the clathrin-coated vesicle in Figure 13–2B. At what point in its structure does a triskelion have to be most flexible to accommodate changes in size of the vesicle? At what point does it have to be most flexible to fit into both the pentagonal and hexagonal faces?

13–7 Yeast, and many other organisms, make a single type of clathrin heavy chain and a single type of clathrin light chain; thus, they make a single kind of clathrin coat. How is it then that a single clathrin coat can be used for three different transport pathways (Golgi to late endosomes, plasma membrane to early endosomes, and immature secretory vesicles to Golgi) that each involves different specialized cargo proteins?

13–8* Clathrin-coated vesicles bud from eucaryotic plasma membrane fragments

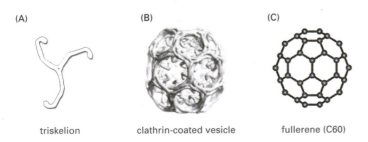

(A) triskelion

(B) clathrin-coated vesicle

(C) fullerene (C60)

Figure 13–2 Structure of a clathrin coat (Problem 13–6). (A) A triskelion subunit. (B) A clathrin-coated vesicle. (C) A C60 fullerene.

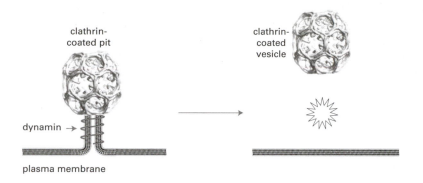

Figure 13–3 Dynamin-mediated step in clathrin-coated vesicle formation (Problem 13–9).

clathrin-coated pit

clathrin-coated vesicle

dynamin →

plasma membrane

when adaptins, clathrin, and dynamin-GTP are added. What would you expect to observe if the following modifications were made to the experiment? Explain your answers.

A. Adaptins were omitted.

B. Clathrin was omitted.

C. Dynamin was omitted.

D. Procaryotic membrane fragments were used.

13–9 The molecular details of how dynamin mediates the final membrane-fusion step in clathrin-coated vesicle formation (Figure 13–3) are controversial. Although dynamin has a GTPase domain, it is not clear how GTP is used to accomplish membrane fusion. One view is that dynamin uses the energy of GTP hydrolysis to pinch off the neck of the vesicle; that is, that dynamin itself is a mechanochemical 'pinchase' powered by GTP hydrolysis. An alternative view is that the GTPase domain of dynamin behaves more like a conventional small GTPase, regulating the activities of other proteins, which are the true pinchases. In this view the binding of GTP to dynamin serves as the 'ON' signal to recruit the proteins that pinch off the vesicle.

One attempt to distinguish these alternatives used a mutant form of dynamin in which a threonine at position 65, which is in the GTPase domain, was changed to alanine. This T65A mutant dynamin cannot hydrolyze GTP, although it is normal by all other criteria. It was then tested for its ability to support formation of clathrin-coated vesicles in an *in vitro* system. The mutant dynamin blocked formation of clathrin-coated vesicles. Which model for dynamin function does this result support? What result would have been expected according to the alternative hypothesis? Explain your answers.

13–10* Correct the following description, as necessary. "Sar1 protein is a COPII-recruitment GTPase that facilitates the unidirectional transfer of COPII vesicles from the ER membrane to the Golgi membrane. A unique directionality is imposed on the transfer by the locations of a guanine-nucleotide exchange factor (GEF) and a GTPase activating protein (GAP). The GEF, which is located in the ER membrane, stimulates vesicle formation by mediating attachment of Sar1-GTP to the ER membrane where it recruits COPII subunits. The GAP, which is located in the Golgi membrane, stimulates vesicle docking by stimulating hydrolysis of Sar1-bound GTP. Sar1-GDP causes disassembly of the COPII coat, which prepares the vesicle for fusion with the Golgi membrane."

13–11 When the fungal metabolite brefeldin A is added to cells, the Golgi apparatus largely disappears and the Golgi proteins intermix with those in the ER. Brefeldin-A treatment also causes the rapid dissociation of some Golgi-associated peripheral membrane proteins, including subunits of the COPI coat. This implies that brefeldin A prevents transport involving COPI-coated vesicles by blocking the assembly of coats and thus the budding of transport vesicles. In principle, brefeldin A could block formation of COPI-coated

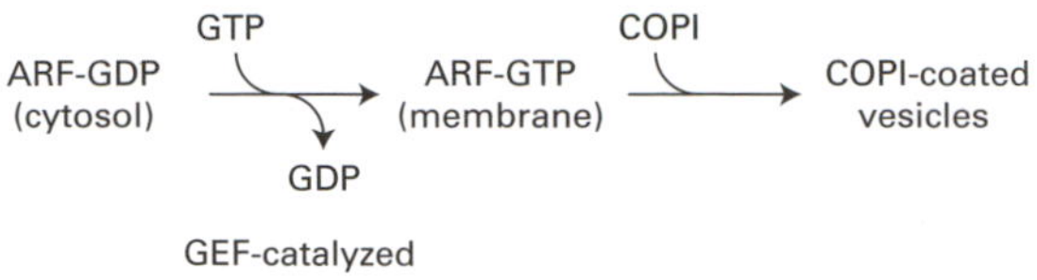

Figure 13–4 Normal pathway for formation of COPI-coated vesicles (Problem 13–11). The small GTPase ARF carries a bound GDP in its cytosolic form. In response to a GEF, ARF releases GDP and picks up GTP. Binding of GTP causes a conformational change that exposes a fatty acid tail on ARF, which promotes binding of ARF–GTP to the membrane. COPI subunits bind to ARF–GTP to form COPI-coated vesicles.

vesicles at any point in the normal scheme for assembly, which is shown in Figure 13–4. The following observations identify the point of action of brefeldin A.

1. ARF with bound GTPγS (a nonhydrolyzable analog of GTP) causes COPI-coated vesicles to form when added to Golgi membranes. Formation of vesicles in this way is not affected by brefeldin A.

2. ARF with bound GDP exchanges GDP for GTP when added to Golgi membranes. Trypsin-treated Golgi membranes do not stimulate GTP-for-GDP exchange. The exchange reaction with normal Golgi membranes does not occur in the presence of brefeldin A.

3. ARF with bound GTP can be made to exchange GDP for GTP in the absence of Golgi membranes by treatment with phosphatidylcholine and cholate. This artificial exchange is not affected by brefeldin A.

Given these experimental observations, how do you think brefeldin A blocks formation of COPI-coated vesicles?

13–12* Small GTPases are generally active in the GTP-bound state and inactive when the GTP is hydrolyzed to GDP. In the absence of a GTPase activating protein (GAP), small GTPases typically hydrolyze GTP very slowly. The mechanism by which GAP stimulates GTP hydrolysis is known for the small GTPase Ras. When Ras–GAP binds to Ras, it alters the conformation of Ras and provides a critical, catalytic arginine 'finger' that stabilizes the transition state for GTP hydrolysis, thereby stimulating hydrolysis by several orders of magnitude.

During assembly of COPI-coated vesicles, ARF1—a small GTPase—binds to ARF1–GAP, which locks ARF1 into its active catalytic conformation but does not supply the catalytic arginine. Since COPI subunits also bind to ARF1, you wonder if they might affect GTP hydrolysis. To test this possibility, you mix ARF1, ARF1–GAP, and COPI subunits in various combinations and measure GTP hydrolysis (Table 13–1).

How would you interpret these results? How might you further test your conclusions?

13–13 Imagine that ARF1 protein was mutated so that it could not hydrolyze GTP, regardless of its binding partners. Would you expect COPI-coated vesicles to form normally? How would you expect transport mediated by COPI-coated vesicles to be affected? If this were the only form of ARF1 in a cell, would you expect it to be lethal? Explain your answers.

13–14* SNAREs exist as complementary partners that carry out membrane fusions between appropriate vesicles and their target membranes. In this way, a vesicle with a particular variety of v-SNARE will fuse only with a membrane

TABLE 13–1 Rates of GTP hydrolysis by various combinations of ARF1, ARF1–GAP, and COPI subunits (Problem 13–12).

COMPONENTS ADDED	RATE OF GTP HYDROLYSIS
ARF1	0
ARF1 + ARF1–GAP	1
ARF1 + COPI subunits	0
ARF1 + ARF1–GAP + COPI subunits	1000

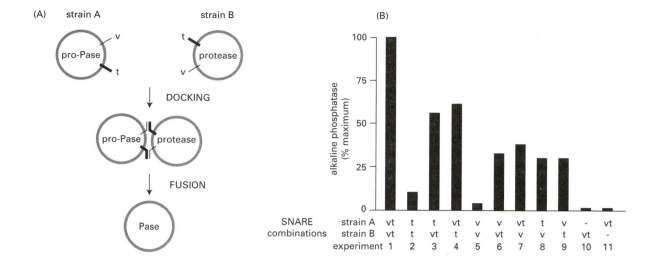

Figure 13–5 SNARE requirements for vesicle fusion (Problem 13–14). (A) Scheme for measuring fusion of vacuolar vesicles. (B) Results of fusions of vesicles with different combinations of v-SNAREs and t-SNAREs. The SNAREs present on the vesicles of the two strains are indicated as v (v-SNARE) and t (t-SNARE).

that carries the complementary t-SNARE. In some instances, however, fusions of identical membranes (homotypic fusions) are known to occur. For example, when a yeast cell forms a bud, vesicles derived from the mother cell's vacuole move into the bud where they fuse together to form a new vacuole. These vesicles carry both v-SNAREs and t-SNAREs. Are both types of SNARE essential for this homotypic fusion event?

To test this point, you have developed an ingenious assay for fusion of vacuolar vesicles. You prepare vesicles from two different mutant strains of yeast: strain A has a defective gene for vacuolar alkaline phosphatase (Pase); strain B is defective for the protease that converts the precursor of phosphatase (pro-Pase) into its active form (Pase). Neither strain has active alkaline phosphatase, but when extracts of the strains are mixed, fusion of vesicles generates active alkaline phosphatase, which can be easily and sensitively measured (Figure 13–5A).

Now you delete the genes for the vacuolar v-SNARE, t-SNARE, or both in each of the two yeast strains. You prepare vacuolar vesicles from each and test them for their ability to fuse, using the alkaline phosphatase assay (Figure 13–5B).

What do these data say about the requirements for v-SNAREs and t-SNAREs in fusion of vacuolar vesicles? Does it matter which kind of SNARE is on which vesicle?

13–15 How can it possibly be true that complementary pairs of specific SNAREs uniquely mark vesicles and their target membranes? Once vesicles fuse with their target, the target membrane will contain a mixture of t-SNAREs and v-SNAREs. Initially, these SNAREs are tightly bound to one another, but NSF can pry them apart, reactivating them. Why don't all target membranes accumulate a population of v-SNAREs equal to or greater than their population of t-SNAREs?

13–16* A complex of syntaxin (t-SNARE), Snap25 (t-SNARE), and synaptobrevin (v-SNARE) are responsible for docking synaptic vesicles at the plasma membrane of a nerve terminal. But what is the next step in fusion?

You have discovered a protein (synaphin) that interacts with syntaxin and wonder if it also binds to the complex. By making a fusion of synaphin to glutathione S-transferase (GST), you can rapidly test this possibility. You incubate GST–synaphin with purified Snap25, synaptobrevin, or both in the presence of increasing concentrations of syntaxin. You then add beads with attached glutathione to the mixture, and allow GST–synaphin, along with any associated proteins, to bind via GST to the beads. You wash the beads to eliminate any unbound proteins, and then incubate the beads with free glutathione to release GST–synaphin and any associated proteins. (This technique of using a GST fusion protein to detect interactions of a fused protein with other proteins is commonly known as a GST 'pull down.') You measure the amount of syntaxin that was bound under the various conditions using syntaxin-specific antibodies, as shown in Figure 13–6.

A. In this experiment can you tell whether synaphin binds to Snap25 or to synaptobrevin in the absence of syntaxin? Why or why not?

B. How does the binding of synaphin to syntaxin compare with binding of synaphin to syntaxin complexed with other molecules?

C. What do you suppose is the target for synaphin binding at a synaptic terminal?

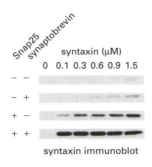

Figure 13–6 Detection of syntaxin in a GST–synaphin pull down (Problem 13–16). After the GST pull down, samples were boiled in SDS to disrupt protein–protein interactions and then separated by electrophoresis on an SDS-containing gel. Syntaxin was detected by reaction with an antibody specific for it.

13–17 To learn more about the role of synaphin in synaptic vesicle fusion, you incubate the complete SNARE complex (consisting of the two t-SNAREs, syntaxin and Snap25, and the v-SNARE, synaptobrevin) with increasing concentrations of synaphin. You then separate the products by electrophoresis under conditions where the SNARE complex remains intact, and probe a blot of the gel with antibodies specific for syntaxin (Figure 13–7). In the absence of synaphin, the SNARE complex runs predominantly at about 60 kd, as expected for the intact complex. In the presence of high concentrations of synaphin, however, the 60-kd band disappears and syntaxin is detected in slower-migrating (larger) complexes. If a 15-amino acid peptide, corresponding to the portion of synaphin that binds to syntaxin, is added to the incubation mixture at high concentration, only the 60-kd complex is detected (this result is not shown in Figure 13–7).

A. What do you suppose the larger complexes are, and how do you think synaphin promotes their formation?

B. How do you suppose the peptide interferes with formation of the larger complexes?

C. Injection of the peptide into a squid giant axon causes a complete block of synaptic transmission. What do you suppose this means in terms of the normal role of synaphin in synaptic vesicle fusion?

D. Can you formulate a model of how synaphin might promote synaptic vesicle fusion?

13–18* Consider the v-SNAREs that direct transport vesicles from the *trans* Golgi network to the plasma membrane. They, like all other v-SNAREs, are membrane proteins that are integrated into the membrane of the ER during their biosynthesis and are then transported by vesicles to their destination. Thus, transport vesicles budding from the ER contain at least two kinds of v-SNAREs—those that target the vesicles to the *cis* Golgi network and those that are in transit to the *trans* Golgi network to be packaged into different transport vesicles destined for the plasma membrane.

A. Why might this be a problem?

B. How do you suppose the cell might solve this problem?

13–19 You wish to identify the target proteins that are bound by NSF and its adaptors. You incubate purified NSF and adaptors with a crude detergent extract of synaptic membranes, and then add an NSF-specific antibody attached to beads. By centrifuging the mixture, you can readily separate the beads and any attached proteins from the rest of the crude extract. The proteins attached to the beads can be analyzed by SDS gel electrophoresis. If the incubation is carried out in the presence or absence of ATP, you find that NSF alone is present on the beads. However, if you incubate in the presence of ATPγS, a nonhydrolyzable analog of ATP, the beads bring down NSF, its adaptors, syntaxin, Snap25, and synaptobrevin. What is the substrate for NSF and its adaptors? Why does the experiment work when ATPγS is present, but not in the presence or absence of ATP?

13–20* (**True/False**) Complementary Rab proteins on transport vesicles and target membranes bind to one another to allow transport vesicles to dock selectively at their appropriate target membranes. Explain your answer.

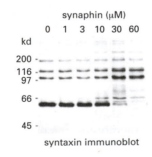

Figure 13–7 Results of incubating increasing concentrations of synaphin with the SNARE complex from nerve terminals (Problem 13–17). Gels were run under conditions that keep the SNARE complex intact. Syntaxin was detected using antibodies specific for it.

13–21   The *SEC4* gene of budding yeast encodes a small GTPase that plays an essential role in the secretion pathway that forms the daughter bud. Normally, about 80% of the Sec4 protein is found on the cytosolic surface of transport vesicles and 20% is free in the cytosol. When temperature-sensitive *SEC4* mutants of yeast (*SEC4*^ts) are incubated at high temperature, growth ceases and small vesicles accumulate in the daughter bud.

To define the role of Sec4 in secretion, you engineer two specific *SEC4* mutants based on the way other small GTPases work. One mutant, *SEC4-ccΔ*, lacks two cysteines at its C-terminus, which you anticipate will prevent attachment of the fatty acid required for membrane binding. The second mutant, *SEC4N133I*, encodes an isoleucine in place of the normal asparagine at position 133; you anticipate that this protein will be locked into its active state, even though it should not be able to bind GTP or GDP.

You find that Sec4-ccΔ binds GTP but remains entirely cytosolic with none bound to vesicles. When expressed at high levels in yeast with a normal *SEC4* gene, it does not inhibit their growth. In contrast, Sec4N133I is located almost entirely on vesicles, and when it is expressed at high levels in normal yeast, it completely inhibits growth and the yeast are packed with small vesicles.

A. Do you think Sec4 is required for formation of vesicles, for vesicle fusion with target membranes, or for both? Based on its function, would you guess it was analogous to mammalian ARF, Sar1, or Rab proteins?

B. Outline how you think normal Sec4 functions in vesicle formation and fusion. Why is some Sec4 free in the cytosol of wild-type cells? How does removal of the C-terminal cysteines prevent Sec4-ccΔ from carrying out its function?

C. Why do you think expression of Sec4N133I in the presence of wild-type Sec4 inhibits growth of the yeast?

13–22*   For fusion of a vesicle with its target membrane to occur, the membranes have to be brought to within 1.5 nm so that the two bilayers can join (Figure 13–8). Assuming that the relevant portions of the two membranes at the fusion site are circular regions 1.5 nm in diameter, calculate the number of water molecules that would remain between the membranes. (Water is 55.5 M and the volume of a cylinder is $\pi r^2 h$.) Given that an average phospholipid occupies a membrane surface area of 0.2 nm^2, how many phospholipids would be present in each of the opposing monolayers at the fusion site? Are there sufficient water molecules to bind to the hydrophilic head groups of this number of phospholipids? (It is estimated that 10–12 water molecules are normally associated with each phospholipid head group at the exposed surface of a membrane.)

13–23   Viruses are the ultimate scavengers—a necessary consequence of their small genomes. Wherever possible they make use of the cell's machinery to accomplish the steps involved in their own reproduction. Many different viruses have membrane coverings. These so-called enveloped viruses gain access to the cytosol by fusing with a cell membrane. Why do you suppose that each of these viruses encodes its own special fusion protein, rather than making use of a cell's SNAREs?

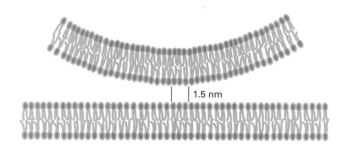

Figure 13–8 Close approach of a vesicle and its target membrane in preparation for fusion (Problem 13–22).

1.5 nm

TRANSPORT FROM THE ER THROUGH THE GOLGI APPARATUS

TERMS TO LEARN

cis-face	Golgi complex	proteoglycan
cis Golgi network (CGN)	high-mannose oligosaccharide	*trans*-face
cisternal maturation model	*N*-linked oligosaccharide	*trans* Golgi network (TGN)
complex oligosaccharide	*O*-linked glycosylation	vesicular transport model
Golgi apparatus		

13–24* How is it that soluble proteins in the ER can be selectively recruited into vesicles destined for the Golgi?

13–25 (**True/False**) There is one strict requirement for the exit of a protein from the ER: it must be correctly folded. Explain your answer.

13–26* Isn't quality control always a good thing? How can quality control in the ER be detrimental to cystic fibrosis patients?

13–27 In the assay for the homotypic fusion of yeast vacuolar vesicles described in Problem 13–14, it was necessary to incubate the vesicles from the two strains with NSF and ATP before they were mixed together. If they were not pretreated in this way, the vesicles would not fuse. Why do you suppose that step was necessary? Would you expect that such a treatment would be required for vesicles that each carried just one kind of SNARE (a t-SNARE or a v-SNARE)?

13–28* The C-terminal 40 amino acids of three ER-resident proteins—calnexin, calreticulin, and HMG CoA reductase—are shown in Figure 13–9. Decide for each protein whether it is likely to be transmembrane or soluble. Explain your answer.

13–29 If you were to remove the ER retrieval signal from protein disulfide isomerase (PDI), which is normally a soluble resident of the ER lumen, where would you expect the modified PDI to be located?

13–30* The KDEL receptor must cycle back and forth between the ER and the Golgi apparatus in order to accomplish its task of ensuring that soluble ER proteins are retained in the ER lumen. In which compartment does the KDEL receptor bind its ligands more tightly? In which compartment does it bind its ligands more weakly? What is thought to be the basis for its different binding affinities in the two compartments? If you were designing the system, in which compartment would you have the highest concentration of KDEL receptor? Would you predict that the KDEL receptor, which is a transmembrane protein, would itself possess an ER retrieval signal?

13–31 If the KDEL retrieval signal is added to rat growth hormone or human chorionic gonadotropin, two proteins that are normally secreted, the proteins are still secreted, but about six times more slowly. If the C-terminal L is changed to V, the proteins are once again secreted at their normal rate. By contrast, bona fide ER resident proteins rarely, if ever, are secreted from the cell; they are usually captured and returned very efficiently. How is it, do you suppose, that normal resident proteins with a KDEL signal are efficiently retained in the ER, whereas secreted proteins to which a KDEL signal has been added

Calnexin C-terminus

...KDKGDEEEEGEEKLEEKQKSDAEEDGGTVSQEEEDRKPKAEEDEILNRSPRNRKPRRE

Calreticulin

...KQDEEQRLKEEEEDKKRKEEEEAEDKEDDEDKDEDEEDEEDKEEDEEEDVPGQAKDEL

HMG CoA reductase

...PGENARQLARIVCGTVMAGELSLMAALAAGHLVKSHMIHNRSKINLQDLQGACTKKTA

Figure 13–9 C-terminal amino acids of proteins that are residents of the ER (Problem 13–28).

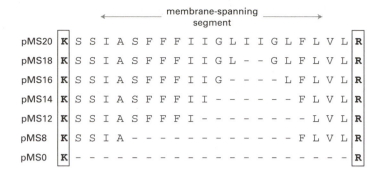

Figure 13–10 The membrane-spanning domains of normal and mutant VSV G proteins (Problem 13–32). Plasmid numbers indicate the number of amino acids in the membrane-spanning segment; for example, pMS20 contains the wild type, 20-amino acid segment. *Dashed lines* indicate amino acids that are missing in the other plasmids. *Boxed letters* indicate the basic amino acids that flank the membrane-spanning segment.

are not efficiently retained? Is this what you would expect if the KDEL signal and the KDEL receptor accounted entirely for retention of soluble proteins in the ER?

13–32* The vesicular stomatitis virus (VSV) G protein is a typical membrane glycoprotein. In addition to its signal peptide, which is removed after import into the ER, the G protein contains a single membrane-spanning segment that anchors the protein in the plasma membrane. The membrane-spanning segment consists of 20 uncharged and mostly hydrophobic amino acids that are flanked by basic amino acids (Figure 13–10). Twenty amino acids arranged in an α helix is just sufficient to span the 3-nm thickness of the lipid bilayer of the membrane.

To test the length requirements for membrane-spanning segments, you modify a cloned version of the G protein to generate a series of mutants in which the membrane-spanning segment is shorter, as indicated in Figure 13–10. When you introduce the modified plasmids into cultured cells, roughly the same amount of G protein is synthesized from each mutant as from wild-type cells. You analyze the cellular distribution of the altered G proteins in several ways.

1. You examine the cellular location of the modified G proteins by immuno-fluorescence microscopy, using G-specific antibodies tagged with fluorescein.
2. You characterize the attached oligosaccharide chains by digestion with endoglycosidase H (endo H), which cleaves off *N*-linked oligosaccharides up to the point at which the first mannose is removed in the medial portion of the Golgi apparatus (see Figure 13–11, Problem 13–34).
3. You determine whether the altered G proteins retain the small C-terminal cytoplasmic domain (which characterizes the normal G protein) by treating isolated microsomes with a protease. In the normal G protein this domain is sensitive to protease treatment and removed.

The results of these experiments are summarized in Table 13–2.

A. To the extent these data allow, deduce the intracellular location of each altered G protein that fails to reach the plasma membrane.

TABLE 13–2 Results of experiments characterizing the cellular distribution of G proteins from normal and mutant cells (Problem 13–32).

PLASMID	CELLULAR LOCATION	ENDO H TREATMENT	PROTEASE TREATMENT
pMS20	plasma membrane	resistant	sensitive
pMS18	plasma membrane	resistant	sensitive
pMS16	plasma membrane	resistant	sensitive
pMS14	plasma membrane	resistant	sensitive
pMS12	intracellular	+/– resistant	sensitive
pMS8	intracellular	sensitive	sensitive
pMS0	intracellular	sensitive	resistant

TABLE 13–3 Analysis of the sugars present in the N-linked oligosaccharides from wild-type and mutant cell lines defective in oligosaccharide processing (Problem 13–34).

CELL LINE	Man	GlcNAc	Gal	NANA	Glc
Wild type	3	5	3	3	0
Mutant A	3	5	0	0	0
Mutant B	5	3	0	0	0
Mutant C	9	2	0	0	3
Mutant D	9	2	0	0	0
Mutant E	5	2	0	0	0
Mutant F	3	3	0	0	0
Mutant G	8	2	0	0	0
Mutant H	9	2	0	0	2
Mutant I	3	5	3	0	0

Abbreviations: Man = mannose; GlcNAc = N-acetylglucosamine; Gal = galactose; NANA = N-acetylneuraminic acid, or sialic acid; Glc = glucose. Numbers indicate the number of sugar monomers in the oligosaccharide.

B. For the VSV G protein, what is the minimum length of the membrane-spanning segment that is sufficient to anchor the protein in the membrane?

C. What is the minimum length of the membrane-spanning segment that is consistent with proper sorting of the G protein? How is it that transmembrane segments shorter than this can make it to the Golgi apparatus, but then not be able to exit?

13–33 (**True/False**) All of the glycoproteins and glycolipids in intracellular membranes have their oligosaccharide chains facing the lumenal side, whereas those in the plasma membrane have their oligosaccharide chains facing the outside of the cell. Explain your answer.

13–34* You have isolated several mutant cell lines that are defective in their ability to add carbohydrate to exported proteins. Using an easily purified protein that carries only N-linked complex oligosaccharides, you have analyzed the sugar monomers that are added in the different mutant cells. Each mutant is unique in the kinds and numbers of different sugars contained in its N-linked oligosaccharides (Table 13–3).

A. Arrange the mutants in the order that corresponds to the steps in the pathway for processing N-linked oligosaccharides (Figure 13–11). (Assume that each mutant cell line is defective for a single enzyme required to construct the N-linked oligosaccharide.)

B. Which of these mutants are defective in processing events that occur in the ER? Which mutants are defective in processing steps that occur in the Golgi?

C. Which of the mutants are likely to be defective in a processing enzyme that is directly responsible for modifying N-linked oligosaccharides? Which mutants might not be defective in a processing enzyme, but rather in another enzyme that affects oligosaccharide processing indirectly?

13–35 Processing of N-linked oligosaccharides is not uniform among species. Most mammals, with the exception of humans and Old World primates, occasionally add galactose, in place of an N-acetylneuraminic acid, to a galactose, forming a terminal Gal(α1–3)Gal disaccharide on some branches of an N-linked oligosaccharide. How does this explain the preferred use of Old World primates for production of recombinant proteins for therapeutic use in humans?

13–36* (**True/False**) The Golgi apparatus confers the heaviest glycosylation of all on proteoglycan core proteins, which are converted into proteoglycans by the addition of one or more O-linked glycosaminoglycan chains. Explain your answer.

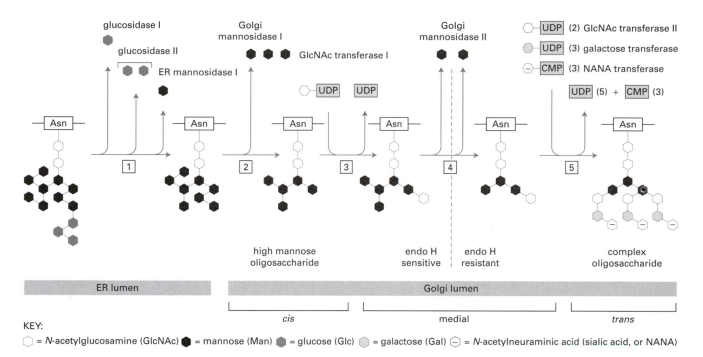

Figure 13–11 Oligosaccharide processing in the ER and the Golgi apparatus (Problem 13–34). The reactions listed in step 5 occur in two different compartments; addition of GlcNAc occurs in the medial compartment, whereas addition of Gal and NANA occur in the *trans* compartment.

KEY:
⬡ = *N*-acetylglucosamine (GlcNAc) ⬤ = mannose (Man) ⬤ = glucose (Glc) ⬡ = galactose (Gal) ⊖ = *N*-acetylneuraminic acid (sialic acid, or NANA)

13–37 Cells have evolved a set of complicated pathways for addition of carbohydrates to proteins, implying that carbohydrates serve some important function, yet for the most part these functions are not known. List three functions that carbohydrates on proteins are known to serve.

13–38* Two extreme models—vesicular transport and cisternal maturation—have been proposed to account for the movement of molecules across the polarized structure of the Golgi apparatus. In the vesicular transport model the individual Golgi cisternae remain in place as proteins move through them (Figure 13–12A). By contrast, in the cisternal maturation model the individual Golgi cisternae move across the stack, carrying the proteins with them (Figure 13–12B). Transport vesicles serve critical functions in both models,

Figure 13–12 Two models for the movement of molecules through the Golgi apparatus (Problem 13–38). (A) The vesicular transport model. (B) The cisternal maturation model. In (B) the individual cisternae have been separated for illustration purposes.

(A) VESICULAR TRANSPORT MODEL

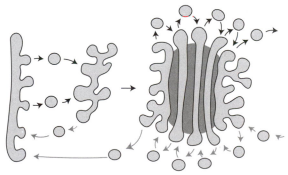

(B) CISTERNAL MATURATION MODEL

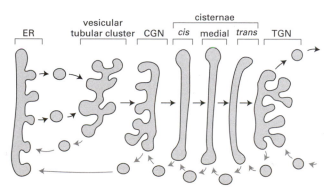

TABLE 13–4 Addition of galactose to VSV G protein after fusion of VSV-infected cells with uninfected cells (Problem 13–39).

INFECTED CELLS	UNINFECTED CELLS	PRECIPITATE	SUPERNATANT
Mutant cells	wild-type cells	45%	55%
Mutant cells	mutant cells	5%	95%
Wild-type cells	wild-type cells	85%	15%

but their roles are distinctly different. Describe the roles of the transport vesicles in each of the two models. Comment specifically on the roles of vesicles in the forward movement of proteins across the Golgi stack, in the retention of Golgi resident proteins in individual cisternae, and on the return of escaped ER proteins to the ER.

13–39   One early test of the vesicular transport and cisternal maturation models (see Figure 13–12) looked for the movement of a protein between Golgi cisternae. This study made use of mutant cells that cannot add galactose to proteins, which normally occurs in the *trans* compartment of the Golgi (see Figure 13–11). The mutant cells were infected with vesicular stomatitis virus (VSV) to provide a convenient marker protein, the viral G protein. At an appropriate point in the infection an inhibitor of protein synthesis was added to stop further synthesis of G protein. The infected cells were then incubated briefly with a radioactive precursor of GlcNAc, which is added only in the medial cisterna of the Golgi (see Figure 13–11). Next, the infected mutant cells were fused with uninfected wild-type cells to form a common cytoplasm containing both wild-type and mutant Golgi stacks. After a few minutes the cells were dissolved with detergent and all the G protein was precipitated using G-specific antibodies. After separation from the antibodies the G proteins carrying galactose were precipitated with a lectin that binds galactose. The radioactivity in the precipitate and in the supernatant was measured. The results of this experiment along with control experiments (which used mutant cells only or wild-type cells only) are shown in Table 13–4.

A.   Between which two compartments of the Golgi apparatus is the movement of proteins being tested in this experiment? Explain your answer.

B.   If proteins moved through the Golgi apparatus by cisternal maturation, what would you predict for the results of this experiment? If proteins moved through the Golgi via vesicular transport, what would you predict?

C.   Which model is supported by the results in Table 13–4?

13–40*   A second test of the vesicular transport and cisternal maturation models used electron microscopy to follow movement of procollagen type I (PC) across the Golgi apparatus. PC is a long (300 nm) rodlike protein that is composed of three chains folded into an uninterrupted triple helix. During its assembly, proline amino acids within the individual chains are hydroxylated in the ER. This allows them to form the helix, which is the signal that the PC is ready to exit the ER. The assembled PC, however, is too large to fit into transport vesicles, which are typically 70–90 nm in diameter.

   PC can be detected by electron microscopy using antibodies specific for the triple helix, which detect only the assembled form, or antibodies specific for its globular head groups, which detect both the individual chains and the fully assembled form. Its movement through the Golgi apparatus was followed after two synchronization procedures, both of which yielded the same result. When assembly in the ER was blocked by a reversible inhibitor of proline hydroxylase, PC that was already in the Golgi apparatus was seen to disappear first from the *cis* cisterna, then the medial cisterna, and finally from the *trans* cisterna. If the block was maintained until the Golgi apparatus was empty of PC and then the block was removed, PC appeared first in the *cis* cisterna, then the medial cisterna, and finally the *trans* cisterna. Both antibodies yielded the same result and no PC was observed except within Golgi cisternae. Three-dimensional reconstructions from serial sections show that PC was always located in Golgi cisternae, never in vesicles or in budding vesicles.

A. Why was it important in these experiments to use two antibodies, one specific for triple helical PC and the other that could also detect individual chains? What possibility were the authors concerned about?

B. Which model for movement of proteins through the Golgi apparatus—vesicular transport or cisternal maturation—do these results support? Explain your reasoning.

13–41 Most exported proteins move across the Golgi apparatus in 5–10 minutes, but large proteins, like procollagen type I (PC), take about an hour. If the movement of PC marks the rate of cisternal maturation, how is it that other proteins can move more quickly?

TRANSPORT FROM THE *TRANS* GOLGI NETWORK TO LYSOSOMES

TERMS TO LEARN

acid hydrolase	lysosomal storage disease	M6P receptor protein
autophagy	lysosome	vacuole
late endosome		

13–42* (**True/False**) Lysosomal membranes contain a proton pump that utilizes the energy of ATP hydrolysis to pump protons out of the lysosome, thereby maintaining the lumen at a low pH. Explain your answer.

13–43 How is it that the low pH of lysosomes protects the rest of the cell from lysosomal enzymes in case the lysosome breaks?

13–44* Imagine that an autophagosome is formed by engulfment of a mitochondrion by the ER membrane. How many layers of membrane separate the matrix of the mitochondrion from the cytosol outside the autophagosome? Identify the source of each membrane and the spaces between the membranes.

13–45 (**True/False**) Late endosomes are converted to mature lysosomes by the loss of distinct endosomal membrane proteins and a further decrease in their internal pH. Explain your answer.

13–46* The principal pathway for transport of lysosomal hydrolases from the *trans* Golgi network (pH 6.6) to the late endosomes (pH 6) and the recycling of M6P receptors back to the Golgi depends on the pH difference between those two compartments. From what you know about M6P receptor binding and recycling and the pathways for delivery of material to lysosomes, describe the consequences of changing the pH in those two compartments.

A. What do you think would happen if the pH in late endosomes were raised to pH 6.6?

B. What do you think would happen if the pH in the *trans* Golgi network were lowered to pH 6?

13–47 Most lysosomal hydrolases are tagged with several oligosaccharide chains that can acquire multiple M6P groups. This multiplicity of M6P groups substantially increases the affinity of these hydrolases for M6P receptors and markedly improves the efficiency of sorting to lysosomes. The reasons for the increase in affinity are 2-fold: one relatively straightforward and the other more subtle. Both can be appreciated at a conceptual level by considering a hydrolase with one M6P group or one with four equivalent M6P groups.

Problem 3–53 looks at multivalent interactions in the context of the attachment of phage T4 to *E. coli*.

The binding of a hydrolase (H) to the M6P receptor (R) to form a hydrolase–receptor complex (HR) can be represented as

$$H + R \underset{k_{off}}{\overset{k_{on}}{\rightleftharpoons}} HR$$

At equilibrium the rate of association of hydrolase with the receptor ($k_{on}[H][R]$) equals the rate of dissociation ($k_{off}[HR]$)

$$k_{on}[H][R] = k_{off}[HR]$$

$$\frac{k_{on}}{k_{off}} = \frac{[HR]}{[H][R]} = K$$

where K is the equilibrium constant. Because the equilibrium constant is a measure of the strength of binding between two molecules, it is sometimes called the affinity constant: the larger the value of the affinity constant, the stronger the binding.

A. Consider first the hypothetical situation in which the hydrolase and the M6P receptor are both soluble—that is, the receptor is not in a membrane. Assume that the M6P receptor has a single binding site for M6P groups, and think about the interaction between a single receptor and a hydrolase molecule. How do you think the rate of association will change if the hydrolase has one M6P group or four equivalent M6P groups? How will the rate at which the hydrolase dissociates from a single receptor change if the hydrolase has one M6P group or four? Given the effect on association and dissociation, how will the affinity constants differ for a hydrolase with one M6P group versus a hydrolase with four M6P groups?

B. Consider the situation in which a hydrolase with four equivalent M6P groups has bound to one receptor already. Assuming that the first receptor is locked in place, how will the binding to the first receptor influence the affinity constant for binding to a second receptor? (For simplicity, assume that the binding to the first receptor does not interfere with the ability of other M6P groups on the hydrolase to bind to a second receptor.)

C. In the real situation, as it occurs during sorting, the M6P receptors are in the Golgi membrane, whereas the hydrolases are initially free in the lumen of the Golgi. Consider the situation in which a hydrolase with four M6P groups has bound to one receptor already. In this case how do you think the binding of the hydrolase to the first receptor will influence the affinity constant for binding to a second receptor? (Think about this question from the point of view of how the binding changes the distribution of the hydrolase with respect to the lumen and the membrane.)

Problem 3–75 examines the effect of attachment of v-Src to the membrane on its affinity for a membrane-bound target.

13–48* (True/False) If cells were treated with a weak base such as ammonia or chloroquine, which raises the pH of organelles toward neutrality, M6P receptors would be expected to accumulate in the Golgi because they could not bind to the lysosomal enzymes. Explain your answer.

13–49 Patients with Hunter's syndrome or with Hurler's syndrome rarely live beyond their teens. These patients accumulate glycosaminoglycans in lysosomes due to the lack of specific lysosomal enzymes necessary for their degradation. When cells from patients with the two syndromes are fused, glycosaminoglycans are degraded properly, indicating that the cells are missing different degradative enzymes. Even if the cells are just cultured together, they still correct each other's defects. Most surprising of all, the medium from a culture of Hurler's cells corrects the defect in Hunter's cells (and vice versa). The corrective factors in the media are inactivated by treatment with proteases, by treatment with periodate, which destroys carbohydrate, and by treatment with alkaline phosphatase, which removes phosphates.

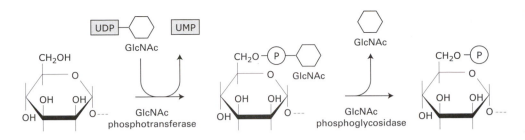

α-D-mannose

mannose 6-phosphate

A. What do you think the corrective factors are, and what is the route by which they correct the lysosomal defects?

B. Why do you think the treatments with protease, periodate, and alkaline phosphatase inactivate the corrective factors?

C. Would you expect a similar sort of correction scheme to work for mutant or missing cytosolic enzymes?

13–50* Children with I-cell disease synthesize perfectly good lysosomal enzymes but secrete them outside the cell instead of sorting them to lysosomes. The mistake occurs because the cells lack GlcNAc phosphotransferase, which is required to create the M6P marker that is essential for proper sorting (Figure 13–13). In principle, I-cell disease could also be caused by deficiencies in GlcNAc phosphoglycosidase, which removes GlcNAc to expose M6P (Figure 13–13), or in the M6P receptor itself. Thus, there are three potential kinds of I-cell disease, which could be distinguished by the ability of various culture supernatants to correct defects in mutant cells. Imagine that you have three cell lines (A, B, and C), each of which derives from a patient with one of the three hypothetical I-cell diseases. Experiments with supernatants from these cell lines give the results below.

1. The supernatant from normal cells corrects the defects in B and C but not the defect in A.

2. The supernatant from A corrects the defect in Hurler's cells, which are missing a specific lysosomal enzyme, but the supernatants from B and C do not.

3. If the supernatants from the mutant cells are first treated with phosphoglycosidase to remove GlcNAc, then the supernatants from A and C correct the defect in Hurler's cells, but the supernatant from B does not.

From these results deduce the nature of the defect in each of the mutant cell lines.

13–51 Patients with I-cell disease are missing the enzyme GlcNAc phosphotransferase, which controls the first step in addition of phosphate to mannose to create the M6P marker (see Figure 13–13). How is it then that the lysosomes in some types of their cells—liver cells, for example—can contain a normal complement of lysosomal enzymes?

13–52* Melanosomes are specialized lysosomes that store pigments for eventual release by exocytosis. Various cells such as skin and hair cells then take up the pigment, which accounts for their characteristic pigmentation. Mouse mutants that have defective melanosomes often have pale or unusual coat colors. One such mouse, the *mocha* mouse, has a defect in the gene for one of the subunits of the adaptin AP-3. *Mocha* mice have a light coat color (Figure 13–14). How might the loss of this specific adaptin cause a defect in melanosomes?

13–53 More than 50 different genes are known to affect coat color in mice. Three of them—*dilute*, *leaden*, and *ashen*—are grouped together because of their highly similar phenotypes. Although these mice have normal melanosomes in their melanocytes, the pigment in the melanosomes is not delivered correctly to hair cells, giving rise to pale coats, as shown for *ashen* mice in

normal mouse • *mocha* mouse

Figure 13–14 A normal mouse and the *mocha* mouse (Problem 13–52). In addition to its light coat color, the *mocha* mouse has a poor sense of balance.

(A)

normal and *ashen* mice

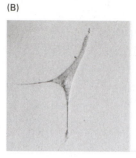

(B)

normal melanocyte

(C)

ashen melanocytes

Figure 13–15 Pigmentation defects in *ashen* mice (Problem 13–53). (A) Normally pigmented mice and pale *ashen* mice. (B) A melanocyte from a normal mouse. (C) Melanocytes from an *ashen* mouse.

Figure 13–15A. *Dilute* mice lack an unconventional myosin heavy chain, MyoVa, which interacts with a microtubule-based transport motor, whereas *ashen* mice carry a mutation in the gene for the Rab protein, Rab27a. The molecular defect in *leaden* mice is not yet known.

Melanocytes from normal mice have a characteristic branch morphology (Figure 13–15B) and normally discharge their melanosomes near the tips of the branches. As shown in Figure 13–15C, melanocytes from *ashen* mice have a normal morphology but their melanosomes surround the nucleus. Melanocytes from *dilute* mice have the same appearance as those from *ashen* mice. Try to put these observations together to formulate a hypothesis to account for the defects in melanosome function in these mice.

TRANSPORT INTO THE CELL FROM THE PLASMA MEMBRANE: ENDOCYTOSIS

TERMS TO LEARN

caveolae	fluid-phase endocytosis	phagosome
caveolin	low-density lipoprotein (LDL)	pinocytosis
clathrin-coated pit	macrophage	receptor-mediated endocytosis
dendritic cell	multivesicular body	recycling endosome
early endosome	neutrophil	transcytosis
endocytic-exocytic cycle	phagocytosis	transferrin receptor
endocytosis		

13–54* **(True/False)** Any particle that is bound to the surface of a phagocyte will be ingested by phagocytosis. Explain your answer.

13–55 A macrophage ingests the equivalent of 100% of its plasma membrane each half hour by endocytosis. What is the rate at which membrane is returned by exocytosis?

13–56* Cells take up extracellular molecules by receptor-mediated endocytosis and by fluid-phase endocytosis. A classic paper compared the efficiencies of these two pathways by incubating human epithelial carcinoma cells for various periods of time in a range of concentrations of either ^{125}I-labeled epidermal growth factor (EGF) to measure receptor-mediated endocytosis or horseradish peroxidase (HRP) to measure fluid-phase endocytosis. Both EGF and HRP were present in small vesicles with an internal radius of 20 nm. The uptake of HRP was linear (Figure 13–16A), while that of EGF was initially linear but reached a plateau at higher concentrations (Figure 13–16B).

A. Explain why the shapes of the curves in Figure 13–16 are different for HRP and EGF.

B. From the curves in Figure 13–16, estimate the difference in the uptake rates for HRP and EGF when both are present at 40 nM. What would the difference be if both were present at 40 μM?

C. Calculate the average number of HRP molecules that get taken up by each endocytic vesicle (radius 20 nm) when the medium contains 40 μM HRP. (The volume of a sphere is $4/3\pi r^3$.)

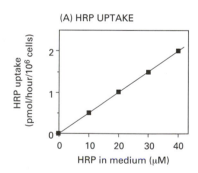

(A) HRP UPTAKE

HRP uptake (pmol/hour/10^6 cells)

HRP in medium (μM)

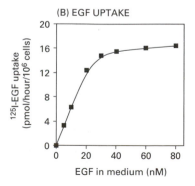

(B) EGF UPTAKE

^{125}I-EGF uptake (pmol/hour/10^6 cells)

EGF in medium (nM)

Figure 13–16 Uptake of HRP and EGF as a function of their concentration in the medium (Problem 13–56).

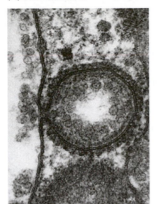
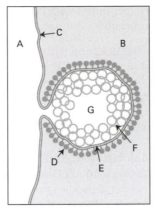

(A) MICROGRAPH (B) DRAWING

⊢—⊣
100 nm

Figure 13–17 A coated pit about to bud from the membrane (Problem 13–57). (A) An electron micrograph. (B) A schematic drawing.

D. The scientists who did these experiments said at the time, "These calculations clearly illustrate how cells can internalize EGF by endocytosis while excluding all but insignificant quantities of extracellular fluid." Explain what they meant.

13–57 The electron micrograph in Figure 13–17A is illustrated schematically by the drawing in Figure 13–17B. Name the structures that are labeled in the drawing.

13–58* Caveolae are thought to form from lipid rafts, which are patches of the plasma membrane that are especially rich in cholesterol and glycosphingolipids. Caveolae may collect cargo proteins by virtue of the lipid composition of their membrane, rather than by assembly of a cytosolic protein coat. What might you predict would be a characteristic of the structure of transmembrane proteins that collect in caveolae?

13–59 Cholesterol is an essential component of the plasma membrane, but people who have very high levels of cholesterol in their blood (hypercholesterolemia) tend to have heart attacks. Blood cholesterol is carried in the form of cholesterol esters in low-density lipoprotein (LDL) particles. LDL binds to a high-affinity receptor on the cell surface, enters the cell via a coated pit, and ends up in lysosomes. There its protein coat is degraded, and cholesterol esters are released and hydrolyzed to cholesterol. The released cholesterol enters the cytosol and inhibits the enzyme HMG CoA reductase, which controls the first unique step in cholesterol biosynthesis. Patients with severe hypercholesterolemia cannot remove LDL from the blood. As a result, their cells do not turn off normal cholesterol synthesis, which makes the problem worse.

LDL metabolism can be conveniently divided into three stages experimentally: binding of LDL to the cell surface, internalization of LDL, and regulation of cholesterol synthesis by LDL. Skin cells from a normal person and two patients suffering from severe familial hypercholesterolemia were grown in culture and tested for LDL binding, LDL internalization, and LDL regulation of cholesterol synthesis. The results are shown in Figure 13–18.

A. In Figure 13–18A the surface binding of LDL by normal cells is compared with LDL binding by cells from patients FH and JD. Why does binding by normal cells and by JD's cells reach a plateau? What explanation can you suggest for the lack of LDL binding by FH's cells?

B. In Figure 13–18B internalization of LDL by normal cells increases as the external LDL concentration is increased, reaching a plateau 5-fold higher than the amount of externally bound LDL. Why does LDL not enter cells from patients FH or JD at a significant rate?

C. In Figure 13–18C the regulation of cholesterol synthesis by LDL in normal cells is compared with that in cells from FH and JD. Why does increasing the external LDL concentration inhibit cholesterol synthesis in normal cells but not significantly affect it in cells from FH or JD?

Problems 10–14 to 10–19 examine various properties of lipid rafts in membranes.

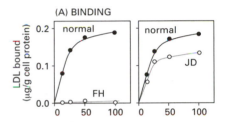

(A) BINDING

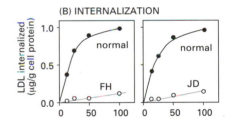

(B) INTERNALIZATION

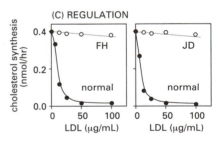

(C) REGULATION

Figure 13–18 LDL metabolism in normal cells and in cells from patients with severe familial hypercholesterolemia (Problem 13–59). (A) Surface binding of LDL. Assays at 4°C allow binding but not internalization. (B) Internalization of LDL. After binding at 4°C, the cells are warmed to 37°C. Binding and uptake of LDL can be followed by labeling LDL either with ferritin particles, which can be seen by electron microscopy, or with radioactive iodine, which can be measured in a gamma counter. (C) Regulation of cholesterol synthesis by LDL.

D. How would you expect the rate of cholesterol synthesis to be affected if normal cells and cells from FH or JD were incubated with cholesterol itself? (Free cholesterol crosses the plasma membrane by diffusion.)

13–60* Referring to the data in Problem 13–59, what is wrong with JD's metabolism of LDL? JD's cells bind LDL with the same affinity as normal cells and in almost the same amounts, but the binding does not lead to internalization of LDL. Two classes of explanation could account for JD's problem:
1. JD's LDL receptors are defective in a way that prevents internalization, even though the LDL-binding domains on the cell surface are un-affected.
2. JD's LDL receptors are entirely normal, but there is a mutation in the cellular internalization machinery such that loaded LDL receptors cannot be brought in.

To distinguish between these explanations, JD's parents were studied. Since the gene encoding the LDL receptor is autosomal, each parent must have donated one defective gene to JD. JD's mother suffered from mildly elevated blood cholesterol. Her cells bound only half as much LDL as normal cells, but the bound LDL was internalized at the same rate as in normal cells. JD's father also had mild hypercholesterolemia, but his cells bound even more LDL than normal cells. Of the bound LDL, less than half the label could be internalized; the rest remained on the cell surface.

The association of this family's LDL receptors with coated pits was studied by electron microscopy, using LDL that was labeled with ferritin. The results are shown in Table 13–5.

A. Why does JD's mother have mild hypercholesterolemia? Based on the LDL-binding and internalization studies and on the EM observations, decide what kind of defective LDL-receptor gene she passed to JD?
B. Why does JD's father have mild hypercholesterolemia? Based on the LDL-binding and internalization studies and on the EM observations, decide what kind of defective LDL-receptor gene he passed to JD?
C. Can you account for JD's hypercholesterolemia from the behavior of the LDL receptors in his parents? In particular, how is it that JD binds nearly a normal amount of LDL, but has severe hypercholesterolemia.
D. At the beginning of this problem two possible explanations—defective receptor or defective internalization machinery—were proposed to account for the lack of internalization by JD's LDL receptors in the face of nearly normal LDL binding. Do these studies allow you to decide between these alternative explanations?

13–61 (**True/False**) Like the LDL receptor, most of the more than 25 different receptors known to participate in receptor-mediated endocytosis enter coated pits only after they have bound their specific ligands. Explain your answer.

13–62* A ligand for receptor-mediated endocytosis circulates at a concentration of 1 nM (10^{-9} M). It is taken up in coated vesicles with a volume of 1.66×10^{-18} L (about 150 nm in diameter). On average there are 10 of its receptors in each coated vesicle. If all the receptors were bound to the ligand, how much more concentrated would the ligand be in the vesicle than it was in the extracellular fluid? What would the dissociation constant (K_d) for the receptor–ligand binding need to be in order to concentrate the ligand 1000-fold in the vesicle? (You may wish to review the discussion of K_d in Problem 3–49.)

13–63 (**True/False**) All the molecules that enter early endosomes ultimately reach late endosomes where they become mixed with newly synthesized acid hydrolases and end up in lysosomes. Explain your answer.

13–64* Influenza viruses are surrounded by a membrane that contains a fusion protein, which is activated by acidic pH. Upon activation, the protein causes the viral membrane to fuse with cell membranes. An old folk remedy against flu recommends that one should spend a night in a horse stable. Odd as it

TABLE 13–5 Distribution of LDL receptors on the surface of cells from JD and his parents as compared with normal individuals (Problem 13–60).

INDIVIDUAL	NUMBER OF LDL RECEPTORS	
	IN PITS	OUTSIDE PITS
Normal male	186	195
Normal female	186	165
JD	10	342
JD's father	112	444
JD's mother	91	87

may sound, there is a rational explanation for this advice. Air in stables contains ammonia (NH_3) generated by bacteria from the horses' urine. Sketch a diagram showing the pathway (in detail) by which flu virus enters cells and speculate how NH_3 may protect cells from virus infection. (Hint: NH_3 can neutralize acidic solutions by the reaction $NH_3 + H^+ \rightarrow NH_4^+$.)

13–65   Iron (Fe) is an essential trace metal that is needed by all cells. It is required, for example, for the synthesis of the heme groups that are part of cytochromes and hemoglobin. Iron is taken into cells via a two-component system. The soluble protein transferrin circulates in the bloodstream, and the transferrin receptor is a membrane protein that is continually endocytosed and recycled to the plasma membrane. Fe ions bind to transferrin at neutral pH but not at acidic pH. Transferrin binds to the transferrin receptor at neutral pH only when it has bound an Fe ion, but it binds to the receptor at acidic pH even in the absence of bound iron. From these properties, describe how iron is taken up and discuss the advantages of this elaborate scheme.

13–66*   The recycling of transferrin receptors has been studied by labeling the receptors on the cell surface and following their fate at 0°C and 37°C. A sample of intact cells at 0°C was reacted with radioactive iodine under conditions that label cell-surface proteins. If these cells were kept on ice and treated with trypsin, which destroys the receptors without damaging the integrity of the cell, the radioactive transferrin receptors were completely degraded. If the cells were first warmed to 37°C for 1 hour and then treated with trypsin on ice, about 70% of the initial radioactivity was resistant to trypsin. At both temperatures, however, most of the receptors were not labeled and most remained intact, as apparent from a protein stain.

A second sample of cells that had been surface-labeled at 0°C and incubated at 37°C for 1 hour was analyzed with transferrin-specific antibodies, which identify transferrin receptors via their linkage to Fe–transferrin complexes. If intact cells were reacted with antibody, 0.54% of the labeled proteins were bound by antibody. If the cells were first dissolved in detergent, 1.76% of the labeled proteins were bound by antibody.

A. When the cells were kept on ice, why did trypsin treatment destroy the labeled transferrin receptors, but not the majority of receptors? Why did most of the labeled receptors become resistant to trypsin when the cells were incubated at 37°C?

B. What fraction of all the transferrin receptors is on the cell surface after a 1-hour incubation at 37°C? Do the two experimental approaches agree?

13–67   The average time for transferrin receptors to cycle between the cell surface and endosomes has been determined by labeling cell-surface receptors with radioactive iodine at 0°C and then following their fate at 37°C. At various times after shifting labeled cells to 37°C, samples were diluted into ice-cold medium that contained trypsin. The amount of radioactivity in trypsin-resistant transferrin receptors was measured after separation from other membrane components by two-dimensional polyacrylamide-gel electrophoresis. The results are shown in Figure 13–19.

A. The initial rate of internalization of labeled transferrin receptors is indicated by the dashed line. Why does the rate of internalization decline with time?

B. Using the initial rate of internalization, estimate the fraction of surface receptors that were internalized each minute.

C. What fraction of the *total* receptor population was internalized each minute?

D. At the rate determined in part C, how many minutes would it take for the equivalent of the entire population of receptors to be internalized? Explain why this time equals the average time for a receptor to cycle from the cell surface through the endosomal compartment and back to the cell surface.

E. On average, how long does each transferrin receptor spend on the cell surface?

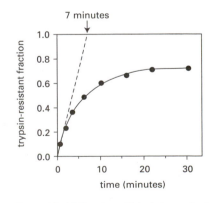

Figure 13–19 Fraction of labeled transferrin receptor that was trypsin resistant as a function of time after labeling (Problem 13–67). The *dashed line* indicates the initial rate of internalization.

13–68* **(True/False)** During transcytosis, vesicles that form from coated pits on the apical surface fuse with the plasma membrane on the basolateral surface and in that way transport molecules across the epithelium. Explain your answer.

TRANSPORT FROM THE *TRANS* GOLGI NETWORK TO THE CELL EXTERIOR: EXOCYTOSIS

TERMS TO LEARN

constitutive secretory pathway	immature secretory vesicle	regulated secretory pathway
default pathway	lipid raft	secretory vesicle
exocytosis	mast cell	synaptic vesicle

13–69 In a cell capable of regulated secretion, what are the three main classes of proteins that must be separated before they leave the *trans* Golgi network?

13–70* You are interested in exocytosis and endocytosis in a line of cultured liver cells that secrete albumin and take up transferrin. To distinguish between these events, you tag transferrin with colloidal gold and prepare ferritin-labeled antibodies that are specific for albumin. You add the tagged transferrin to the medium, and then after a few minutes you fix the cells, prepare thin sections, and react them with ferritin-labeled antibodies against albumin. Colloidal gold and ferritin are both electron dense and therefore readily visible when viewed by electron microscopy; moreover, they can be easily distinguished from one another on the basis of size.

A. Will this experiment allow you to identify vesicles in the exocytic and endocytic pathways? How?

B. Not all the gold-labeled vesicles are clathrin coated. Why?

13–71 Liver cells secrete a broad spectrum of proteins into the blood via the constitutive pathway. You are interested in how long it takes for different proteins to be secreted. Accordingly, you add ^{35}S-methionine to cultured liver cells to label proteins as they are synthesized. You then sample the medium at various times to measure the appearance of individual labeled proteins. As shown in Figure 13–20, albumin appears after 20 minutes, transferrin appears after 50 minutes, and retinol-binding protein appears after 90 minutes. You are surprised at the variability in secretion rates, which bear no obvious relationship to the size, function, or quantity of the individual proteins.

   Why do transferrin and the retinol-binding protein take so much longer than albumin to be secreted? You suspect that the slow step in secretion occurs either in the ER or in the Golgi apparatus. To determine which, you label cells for 4 hours, which is long enough for the labeled proteins to reach the same steady-state distribution as unlabeled proteins. (At steady state the influx into a pathway exactly equals efflux from the pathway.) You then homogenize the cells to break the ER and Golgi into vesicles and separate the vesicles by density on a sucrose gradient. You measure the amount of labeled albumin and transferrin that are associated with the two types of vesicles (Figure 13–21).

   Does the slow step in the constitutive secretion of transferrin occur in the ER or in the Golgi? Where does the slow step in the constitutive secretion of albumin occur? How do these experiments allow you to decide?

13–72* Proteins without special signals are transported between cisternae in Golgi stacks and onward to the plasma membrane via the nonselective constitutive secretory pathway, or the default pathway as it is commonly known. This transport is also sometimes referred to as bulk transport because the Golgi substituents do not become concentrated in the vesicles. Given that transport in clathrin-coated vesicles is so highly concentrating, you are skeptical that no concentration occurs in the default pathway for secretion.

   To determine whether vesicles in the default pathway concentrate their

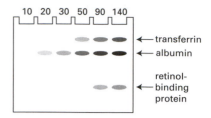

minutes after addition of label

10 20 30 50 90 140

← transferrin
← albumin
retinol-binding protein

Figure 13–20 Time of appearance of secreted proteins in the medium (Problem 13–71). At various times after labeling, proteins were immunoprecipitated with specific antibodies, separated by gel electrophoresis, and subjected to autoradiography.

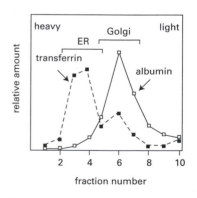

Figure 13–21 Distribution of albumin and transferrin in vesicles derived from the ER and Golgi (Problem 13–71). Labeled albumin and transferrin were assayed by immunoprecipitation, electrophoresis, and autoradiography.

TABLE 13–6 Relative densities of G protein in Golgi and vesicle lumens and membranes (Problem 13–72).

SOURCE OF GOLGI	SITE MEASURED	PARAMETER MEASURED	MEAN DENSITY
Uninfected cells	whole Golgi	surface density	$5/\mu m^2$
Infected cells	whole Golgi	surface density	$271/\mu m^2$
Infected cells	Golgi buds and vesicles	surface density	$233/\mu m^2$
Infected cells	Golgi cisternal membranes	linear density	$6/\mu m$
Infected cells	Golgi buds and vesicles	linear density	$4/\mu m$

contents, you infect cells with vesicular stomatitis virus (VSV) and follow the viral G protein. Your idea, an ambitious one, is to compare the concentration of G protein in the lumen of the Golgi stacks with that in the associated transport vesicles. You intend to measure G-protein concentration by preparing thin sections of VSV-infected cells and incubating them with G-specific antibodies tagged with gold particles. Since the gold particles are visible in electron micrographs as small black dots, it is relatively straight-forward to count dots in the lumens of transport vesicles (fully formed and just budding) and of the Golgi apparatus. You make two estimates of G-protein concentration: (1) the number of gold particles per cross-sectional area and (2) the number of gold particles per linear length of membrane. Your results are shown in Table 13–6.

Do the vesicles involved in bulk transport concentrate their contents or not? Explain your reasoning.

13–73 (**True/False**) When a gene encoding a secretory protein is transferred to a secretory cell that normally does not make the protein, the foreign protein is not packaged into secretory vesicles. Explain your answer.

13–74* What would you expect to happen in cells that secrete large amounts of protein through the regulated secretory pathway if the ionic conditions in the ER lumen could be changed to resemble those of the *trans* Golgi network?

13–75 Insulin is synthesized as a pre-pro-protein in the β-cells of the pancreas. Its pre-peptide is cleaved off after it enters the ER lumen. To define the cellular location at which its pro-peptide is removed, you have prepared two anti-bodies: one that is specific for pro-insulin and one that is specific for insulin. You've tagged the anti-pro-insulin antibody with a red fluorophore and the anti-insulin antibody with green fluorophore, so that you can follow them independently in the same cell. When you incubate a pancreatic β-cell with a mixture of your two antibodies, you obtain the results shown in Table 13–7. In what cellular compartment is the pro-peptide removed from pro-insulin?

13–76* Antitrypsin, which inhibits certain proteases, is normally secreted into the bloodstream by liver cells. Antitrypsin is absent from the bloodstream of patients who carry a mutation that results in a single amino acid change in the protein. Antitrypsin deficiency causes a variety of severe problems, partic-ularly in lung tissue, because of uncontrolled protease activity. Surprisingly,

TABLE 13–7 Fluorescence associated with various compartments of β-cells after reaction with fluorescent antibodies directed against pro-insulin and insulin (Problem 13–75).

COMPARTMENT	FLUORESCENCE
Endoplasmic reticulum	red
cis Golgi network	red
Golgi cisternae	red
trans Golgi network	red
Immature secretory vesicles	yellow
Mature secretory vesicles	green

when the mutant antitrypsin is synthesized in the laboratory, it is as active as the normal antitrypsin at inhibiting proteases. Why then does the mutation cause the disease? Think of more than one possibility and suggest ways in which you could distinguish among them.

13–77 (**True/False**) Once a secretory vesicle is properly positioned beneath the plasma membrane, it will fuse with the membrane and release its contents to the cell exterior. Explain your answer.

13–78* Polarized epithelial cells must make an extra sorting decision since their plasma membranes are divided into apical and basolateral domains, which are populated by distinctive sets of proteins. Proteins destined for the apical or basolateral domain seem to travel there directly from the *trans* Golgi network. One way to sort proteins to these domains would be to use a specific sorting signal for one class of proteins, which would then be actively recognized and directed to one domain, and to allow the other class to travel via a default pathway to the other domain.

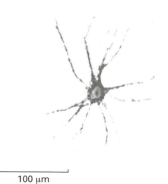

100 µm

Figure 13–22 A hippocampal neuron (Problem 13–79).

Consider the following experiment to identify the default pathway. The cloned genes for several foreign proteins were engineered by recombinant DNA techniques so that they could be expressed in the polarized epithelial cell line MDCK. These proteins are secreted in other types of cells but are not normally expressed in MDCK cells. The cloned genes were introduced into the polarized MDCK cells, and their sites of secretion were assayed. Although the cells remained polarized, the foreign proteins were delivered in roughly equal amounts to the apical and basolateral domains.

A. What is the expected result of this experiment, based on the hypothesis that targeting to one domain of the plasma membrane is actively signaled and targeting to the other domain is via a default pathway?

B. Do these results support the concept of a default pathway as outlined above?

13–79 Neurons are difficult to study because of their excessively branched structure and long thin dendrites, as shown in Figure 13–22. Fluorescently tagged antibodies are powerful tools for investigating certain aspects of neuron structure. Synaptic vesicles, for example, were shown to be concentrated in the presynaptic cells at nerve synapses in this way. A culture of neurons was first exposed for 1 hour to a fluorescently tagged antibody specific for the lumenal domain of synaptotagmin, a transmembrane protein that resides exclusively in the membranes of synaptic vesicles. The culture was then washed thoroughly to remove all synaptotagmin antibodies. Next the cells were fixed with formaldehyde, treated with detergent to make them permeable to antibodies, and exposed to a fluorescent antibody specific for a microtubule-associated protein found only in dendrites—the portions of the neuron with which the presynaptic cells form synapses. The microtubule-specific antibody beautifully outlines the dendrites, and dots of color from the synaptotagmin-specific antibody mark the positions of the synaptic vesicles in the nerve terminals. The rest of the presynaptic cell remains invisible in this procedure.

If antibodies do not cross intact membranes, how do the synaptic vesicles get labeled? When the procedure was repeated using an antibody specific for the cytoplasmic domain of synaptotagmin, the nerve terminals did not become labeled. Explain the results with the two different antibodies for synaptotagmin.

13–80* Dynamin was first identified as a microtubule-binding protein, and its sequence correctly predicted that it was a GTPase. The key to its function came from neurobiological studies in *Drosophila*. *Shibire* mutant flies, which carry a mutation in the dynamin gene, are rapidly paralyzed when the temperature is elevated. They recover quickly once the temperature is lowered. The complete paralysis at the elevated temperature suggested that synaptic transmission between nerve and muscle cells was blocked. Electron micrographs of synapses of the paralyzed flies showed a loss of synaptic

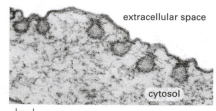

extracellular space

cytosol

100 nm

Figure 13–23 Electron micrograph of a nerve terminal from a *shibire* mutant fly at elevated temperature (Problem 13–80).

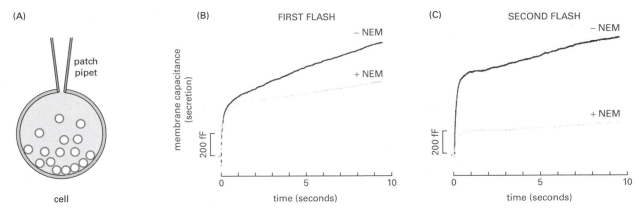

(A) patch pipet

cell

(B) FIRST FLASH

membrane capacitance (secretion)

200 fF

– NEM

+ NEM

0   5   10
time (seconds)

(C) SECOND FLASH

200 fF

– NEM

+ NEM

0   5   10
time (seconds)

Figure 13–24 Analysis of the role of NSF in vesicle fusion (Problem 13–81). (A) Whole-cell patch-clamp. (B) Responses to the first flash-mediated release of Ca^{2+}. (C) Reponses to the second flash-mediated release of Ca^{2+}. Secretion was measured as an increase in membrane capacitance (in femtoFarads, fF) in the absence (–NEM) or presence (+NEM) of an inhibitor of NSF.

vesicles and a tremendously increased number of coated pits relative to normal synapses (Figure 13–23).

Suggest an explanation for the paralysis shown by the *shibire* mutant flies, and indicate why signal transmission at a synapse might require dynamin.

13–81 The original version of the SNARE hypothesis suggested that the ATP-dependent disassembly of SNAREs by NSF provided the energy necessary for membrane fusion, and thus that NSF should act at the last step in secretion. More recent evidence suggests that NSF acts at an earlier step to prime the vesicle for secretion, and that SNAREs alone are sufficient to catalyze membrane fusion at the last step in secretion. It is important to know what really happens, and you have the means at hand to answer the question.

Using a whole-cell patch-clamp protocol (Figure 13–24A), you can diffuse cytosolic components into the cell through the pipet and assay exocytosis by changes in capacitance, which is a measure of the increase in the area of the plasma membrane. To control precisely the timing of vesicle fusion, you enclose Ca^{2+} in a photosensitive chemical 'cage,' from which it can be released with a flash of light. In response to Ca^{2+} release, there is a rapid burst of vesicle fusion (indicated by a rapid rise in capacitance), followed by a longer, slower fusion process (Figure 13–24B, –NEM). The initial burst represents the fusion of vesicles that were just waiting for the Ca^{2+} trigger. Because Ca^{2+} is rapidly removed from the cell, the procedure can be repeated with a second flash of light 2 minutes later; it yields the same rapid and slow components (Figure 13–24C, –NEM).

To test the role of NSF in vesicle fusion, you first diffuse *N*-ethylmaleimide (NEM) into cells to inhibit NSF. (The name 'NSF' stands for NEM-sensitive factor.) You then repeat the flash protocols as before. In response to the first flash, the rapid component is unaffected, but the slow component is decreased (Figure 13–24B, +NEM). In response to the second flash, both components are inhibited (Figure 13–24C, +NEM).

A. What do you suppose the slow component of the fusion process represents?
B. Why does inhibition of NSF affect the slow component after both flashes, but inhibit the rapid component only after the second flash?
C. Which of the alternatives for the role of NSF in vesicle fusion—acting at the last step or an early step—do these experiments support? Explain your reasoning.
D. Propose a model for the molecular role of NSF in fusion of secretory vesicles.

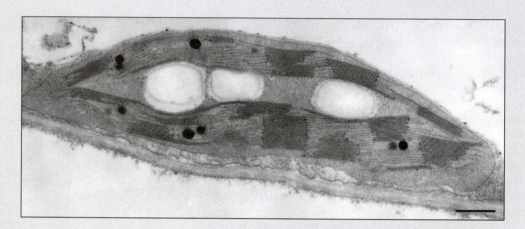

Thin section of a chloroplast. You should be able to identify the cell wall, chloroplast envelope, thylakoid membranes, starch granules and lipid droplets. Scale bar is 0.5 μm. Courtesy of K. Plaskitt.

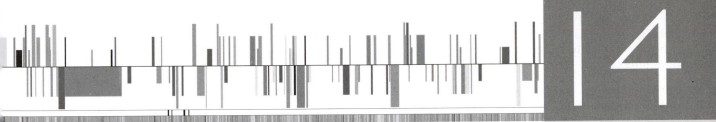

ENERGY CONVERSION: MITOCHONDRIA AND CHLOROPLASTS

THE MITOCHONDRION

TERMS TO LEARN

ATP synthase	inner membrane	outer membrane
chemiosmotic coupling	intermembrane space	oxidative phosphorylation
citric acid cycle	matrix	proton-motive force
cristae	mitochondria	respiratory chain
electrochemical proton gradient		

14–1 Mitochondria in liver cells appear to move freely in the cytosol, whereas those in cardiac muscle are immobilized at positions between adjacent myofibrils. Do you suppose these differences are a trivial consequence of cell architecture or do they reflect some underlying functional advantage? Explain your answer.

14–2* Electron micrographs show that mitochondria in heart muscle have a much higher density of cristae than mitochondria in skin cells. Propose an explanation for this observation.

14–3 (**True/False**) Due to the many specialized transport proteins in the mitochondrial outer membrane, the intermembrane space is chemically equivalent to the cytosol with respect to small molecules. Explain your answer.

14–4* There are substantial differences in the mitochondrial enzymes in different cell types. Yet mitochondria from virtually all cells have the enzymes for the citric acid cycle and for oxidation of fatty acids. Enzymes for what sorts of processes might you suppose would be present in mitochondria in some types of cells, but not others?

14–5   (**True/False**) The most important contribution of the citric acid cycle to energy metabolism is the extraction of high-energy electrons during the oxidation of acetyl CoA to CO_2. Explain your answer.

14–6*   In the 1860s Louis Pasteur noticed that when he added O_2 to a culture of bacteria growing anaerobically on glucose, the rate of glucose consumption declined dramatically. Explain the basis for this result, which is known as the Pasteur effect.

14–7   During a marathon, a runner's leg muscles are well supplied with oxygen and thus abstract all the energy available from the oxidation of glucose to CO_2. By contrast, in a sprint, which requires even more ATP hydrolysis per second, the extreme contraction of the leg muscles restricts blood flow and severely reduces the oxygen available for oxidative phosphorylation. As a result, nearly all of the sprinter's ATP comes from glycolysis alone. Glycolysis releases only about 1/15 of the total energy present in a glucose molecule. With such a reduced energy yield from glucose, how is it possible for a sprinter to sprint?

14–8*   The citric acid cycle generates NADH and $FADH_2$, which are then used in the process of oxidative phosphorylation to make ATP. If the citric acid cycle, which does not use oxygen, and oxidative phosphorylation are separate processes, as they are, then why is it that the citric acid cycle stops almost immediately upon removal of O_2?

14–9   (**True/False**) Each respiratory enzyme complex in the electron-transport chain has a greater affinity for electrons than its predecessors, so that electrons pass sequentially from one complex to another until they are finally transferred to oxygen, which has the greatest electron affinity of all. Explain your answer.

14–10*   The respiratory chain is relatively inaccessible to experimental manipulation in intact mitochondria. After disrupting mitochondria with ultrasound, however, it is possible to isolate functional submitochondrial particles, which consist of broken cristae that have resealed inside out into small closed vesicles. In these vesicles the components that originally faced the matrix are now exposed to the surrounding medium. How might such an arrangement aid in the study of electron transport and ATP synthesis?

14–11   As electrons move down the respiratory chain, protons are pumped across the inner membrane. Are those protons confined to the intermembrane space? Why or why not?

14–12*   In actively respiring liver mitochondria, the pH inside the matrix is about one pH unit higher than that in the cytosol. Assuming that the cytosol is at pH 7 and the matrix is a sphere with a diameter of 1 μm ($V = 4/3 \, \pi r^3$), calculate the total number of protons in the matrix of a respiring liver mitochondrion. If the matrix began with a pH equal to that in the cytosol, how many protons would have to be pumped out to establish a pH difference of one pH unit?

14–13   You have reconstituted into the same membrane vesicles purified bacteriorhodopsin, which is a light-driven H^+ pump from a photosynthetic bacterium, and purified ATP synthase from ox heart mitochondria. Assume that all molecules of bacteriorhodopsin and ATP synthase are oriented as shown in Figure 14–1, so that protons are pumped into the vesicle and ATP synthesis occurs on the outer surface.

A. If you add ADP and phosphate to the external medium and shine light into the suspension of vesicles, would you expect ATP to be synthesized? Why or why not?

B. If you prepared the vesicles without being careful to remove all the detergent, which makes the bilayer leaky to protons, would you expect ATP to be synthesized?

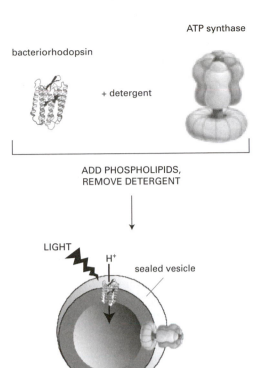

ATP synthase

bacteriorhodopsin

+ detergent

ADD PHOSPHOLIPIDS,
REMOVE DETERGENT

LIGHT

H+

sealed vesicle

Figure 14–1 Reconstitution of bacteriorhodopsin and ATP synthase into lipid vesicles (Problem 14–13).

C. If the ATP synthase molecules were randomly oriented so that about half faced the outside of the vesicle and half faced the inside, would you expect ATP to be synthesized? If the bacteriorhodopsin molecules were randomly oriented, would you expect ATP to be synthesized? Explain your answers.

D. You tell a friend over dinner about your new experiments. He questions the validity of an approach that utilizes components from so widely divergent, unrelated organisms. As he so succinctly puts it, "Why would anybody want to mix vanilla pudding with brake fluid?" Defend your approach against his criticism.

14–14* (**True/False**) If the flow of protons through ATP synthase is blocked by an inhibitor, addition of a small amount of oxygen into an anaerobic preparation of submitochondrial particles (which have their matrix surface exposed to the surrounding medium) will result in a burst of respiration that will cause the medium to become more basic. Explain your answer.

14–15 When dinitrophenol (DNP) is added to mitochondria, the inner membrane becomes permeable to protons. When the drug nigericin is added to mitochondria, the inner membrane becomes permeable to K^+.

A. How will the electrochemical proton gradient change in response to DNP?
B. How will it change in response to nigericin?

14–16* It was difficult to define the molecular mechanism by which the electrochemical proton gradient is coupled to ATP synthesis. A fundamental question at the outset was whether ATP was synthesized directly from ADP and inorganic phosphate or by transfer of a phosphate from an intermediate source such as a phosphoenzyme or some other phosphorylated molecule.

One elegant approach to this question analyzed the stereochemistry of the reaction mechanism. As is often done, the investigators studied the reverse reaction (ATP hydrolysis into ADP and phosphate) to gain an understanding of the forward reaction. (A basic principle of enzyme catalysis is that the forward and reverse reactions are precisely the opposite of one another.) All enzyme-catalyzed phosphate transfers occur with inversion of configuration about the phosphate atom; thus one-step mechanisms, in which the

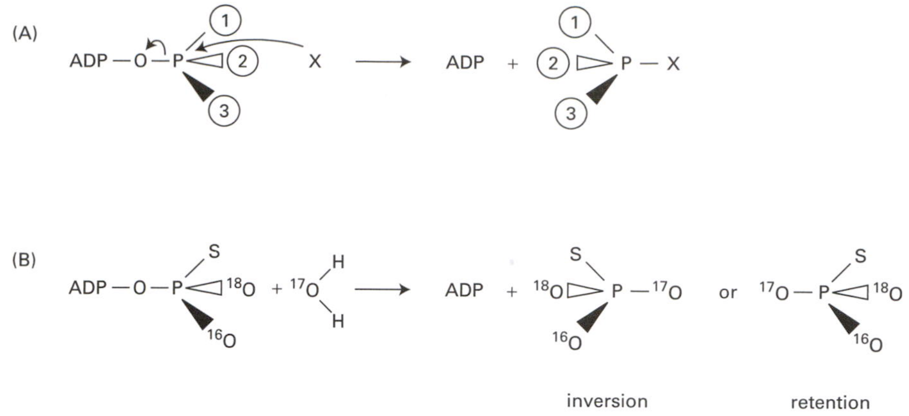

(A)

ADP — O — P (1, 2, 3) X ⟶ ADP + (2) ▷ P — X (1, 3)

(B)

ADP — O — P ◁ (S, ^{18}O, ^{16}O) + ^{17}O (H, H) ⟶ ADP + ^{18}O ▷ P — ^{17}O (S, ^{16}O) or ^{17}O — P ◁ ^{18}O (S, ^{16}O)

inversion   retention

Figure 14–2 Stereochemistry of phosphate transfer reactions (Problem 14–16). (A) Inversion of configuration by a one-step phosphate transfer reaction. (B) Experimental setup for assaying stereochemistry of ATP synthesis by ATP synthase. *Thin* bonds are in the same plane as the page; *thick white* bonds point behind the page; and *thick black* bonds project out of the page. Oxygen atoms are indicated by their atomic number.

phosphate is transferred directly between substrates, result in inversion of the final product (Figure 14–2A).

To analyze the stereochemistry of ATP hydrolysis, the investigators first generated a version of ATP with three distinct atoms (S, ^{16}O, and ^{18}O) attached stereospecifically to the terminal phosphorus atom (Figure 14–2B). They then hydrolyzed this compound to ADP and inorganic phosphate using purified ATP synthase in the presence of H_2O that was enriched for ^{17}O. Using NMR to analyze the resulting inorganic phosphate, they could determine whether the configuration about the phosphorus atom had been inverted or retained (Figure 14–2B).

A. How does this experiment distinguish between synthesis of ATP directly from ADP and inorganic phosphate and synthesis of ATP through a phosphorylated intermediate?

B. Their analysis showed that the configuration had been inverted. Does this result support direct synthesis of ATP or synthesis of ATP through a phosphorylated intermediate?

14–17   ATP synthase is the world's smallest rotary motor. Passage of H^+ ions through the membrane-embedded portion of ATP synthase (the F_0 component) causes rotation of the single, central axle-like γ subunit inside the head group. The tripartite head is composed of the three αβ dimers, the β subunit of which is responsible for synthesis of ATP. The rotation of the γ subunit induces conformational changes in the αβ dimers that allow ADP and P_i to be converted into ATP. A variety of indirect evidence had suggested rotary catalysis by ATP synthase, but seeing is believing.

To demonstrate rotary motion, a modified form of the $α_3β_3γ$ complex was used. The β subunits were modified (on their inner membrane-distal ends) so they could be firmly anchored to a solid support. The γ subunit was modified (on the end that normally inserts into the F_0 component in the inner membrane) so that a fluorescently tagged, readily visible filament of actin could be attached (Figure 14–3A). This arrangement allows rotations of the γ subunit to be visualized as revolutions of the long actin filament. In these experiments ATP synthase was studied in the reverse of its normal mechanism by allowing it to hydrolyze ATP. At low ATP concentrations the actin filament was observed to revolve in steps of 120° and then pause for variable lengths of time, as shown in Figure 14–3B.

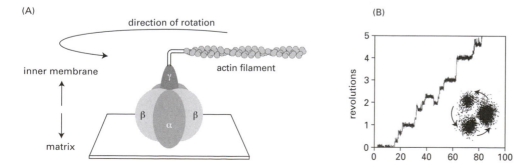

(A)
direction of rotation
inner membrane
actin filament
γ
β   β
α
matrix

(B)
revolutions
5
4
3
2
1
0
0   20   40   60   80   100

Figure 14–3 Experimental set-up for observing rotation of the γ subunit of ATP synthase (Problem 14–17). (A) The immobilized $α_3β_3γ$ complex. The β subunits are anchored and a fluorescent actin filament is attached to the γ subunit. (B) Stepwise revolution of the actin filament. The inset shows the positions in the revolution at which the actin filament pauses.

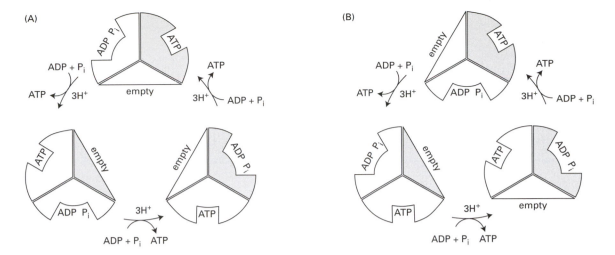

(A)   (B)

A. Why does the actin filament revolve in steps with pauses in between? What does this rotation correspond to in terms of the structure of the $\alpha_3\beta_3\gamma$ complex?

B. In its normal mode of operation inside the cell, how many ATP molecules do you suppose would be synthesized for each complete 360° rotation of the γ subunit? Explain your answer.

14–18* The three αβ dimers in each ATP synthase normally exist in three different conformations: one empty, one with ADP and P_i bound, and one with ATP bound. The conformational changes are driven sequentially by rotation of the γ subunit, which in turn is driven by the flow of protons through the ATP synthase. As viewed from the inner membrane, looking at the underside of the $\alpha_3\beta_3$ complex, these sites could be arranged in either of two ways around the central γ subunit (Figure 14–4). The sequential, linked conformational changes driven by proton flow are also shown for the two arrangements in the figure. In Problem 14–17 the revolutions of the attached actin filaments during ATP hydrolysis were shown to be counterclockwise when viewed from the same perspective. Which of the two arrangements of conformations of αβ dimers shown in Figure 14–4 is correct? Explain your answer.

14–19 A manuscript has been submitted to a prestigious scientific journal. In it the authors describe an experiment using an immobilized $\alpha_3\beta_3\gamma$ complex with an attached actin filament like that shown in Figure 14–3A. The authors show that they can mechanically rotate the γ subunit by applying force to the actin filament. Moreover, in the presence of ADP and phosphate each 120° clockwise rotation of the γ subunit is accompanied by the synthesis of one molecule of ATP. Is this result at all reasonable? What would such an observation imply about the mechanism of ATP synthase? Should this manuscript be considered for publication in one of the best journals?

14–20* What is the role in ATP synthesis of the stator that links the $\alpha_3\beta_3$ complex to the membrane-embedded base of the ATP synthase. If this connection between the base and the $\alpha_3\beta_3$ complex were missing, would ATP be synthesized in response to the proton flow? Why or why not?

14–21 A variety of coupled transport processes that occur across the inner mitochondrial membrane are illustrated in Figure 14–5. For each one decide whether transport is with the electrochemical proton gradient, against it, or unaffected by it. For those transport processes that are affected by the gradient, identify which component of the gradient (membrane potential or ΔpH) affects transport.

14–22* The ADP-ATP antiporter in the mitochondrial inner membrane can exchange ATP for ATP, ADP for ADP, and ATP for ADP. Even though mitochondria can transport both ADP and ATP, there is a strong bias in favor of

Figure 14–4 The two possible arrangements of conformations of the three αβ dimers in ATP synthase, along with the linked conformational changes driven by proton flow (Problem 14–18). One αβ dimer is *shaded* to emphasize that its position remains fixed as it changes conformation in response to proton flow. The perspective illustrated is from the *inner* membrane, looking at the *underside* of the tripartite $\alpha_3\beta_3$ complex.

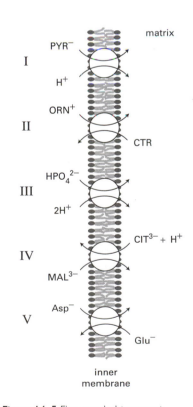

Figure 14–5 Five coupled transport processes that occur across the inner mitochondrial membrane (Problem 14–21). PYR is pyruvate; ORN is ornithine; CTR is citrulline; CIT is citrate; MAL is malate; Asp is aspartic acid; and Glu is glutamic acid.

TABLE 14-1 Entry of ADP and ATP into isolated mitochondria (Problem 14-22).

EXPERIMENT	SUBSTRATE	INHIBITOR	RELATIVE RATES OF ENTRY
1	absent	none	ADP = ATP
2	present	none	ADP > ATP
3	present	dinitrophenol	ADP = ATP
4	present	oligomycin	ADP > ATP

In all cases the initial rates of entry of ATP and ADP were measured.

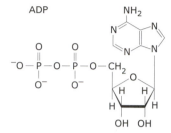

Figure 14-6 Structures of ATP and ADP (Problem 14-22).

exchange of external ADP for internal ATP in actively respiring mitochondria. You suspect that this bias is due to the conversion of ADP into ATP inside the mitochondrion. ATP synthesis would continually reduce the internal concentration of ADP and thereby create a favorable concentration gradient for import of ADP. The same process would increase the internal concentration of ATP, thereby creating favorable conditions for export of ATP.

To test your hypothesis, you conduct experiments on isolated mitochondria. In the absence of substrate (when the mitochondria are not respiring and the membrane is uncharged), you find that ADP and ATP are taken up at the same rate. When you add substrate, the mitochondria begin to respire, and ADP enters mitochondria at a much faster rate than ATP. As you expected, when you add an uncoupler (dinitrophenol, which collapses the pH gradient) along with the substrate, ADP and ATP again enter at the same rate. However, when you add an inhibitor of ATP synthase (oligomycin) along with the substrate, ADP is taken up much faster than ATP. Your results are summarized in Table 14-1. You are puzzled by the results with oligomycin, since your hypothesis predicted that the rates of uptake would be equal.

When you show the results to your advisor, she compliments you on your fine experiments and agrees that they disprove the hypothesis. She suggests that you examine the structures of ATP and ADP (Figure 14-6) if you wish to understand the behavior of the antiporter. What is the correct explanation for the biased exchange by the ADP-ATP antiporter under some of the experimental conditions and an unbiased exchange under others?

14-23 How many molecules of ATP are formed from ADP + P_i when a pair of electrons from NADH is passed down the electron-transport chain to oxygen? Is the number an integer or not? These deceptively simple questions are difficult to answer from purely theoretical considerations, but they can be measured directly with an oxygen electrode, as illustrated in Figure 14-7. At the indicated time, a suspension of mitochondria was added to a phosphate-buffered solution containing β-hydroxybutyrate, which can be oxidized by mitochondria to generate NADH + H^+. After an initial rapid burst, oxygen consumption slowed to a background rate. When the rate of oxygen consumption stabilized, 500 nmol of ADP were added, causing a rapid increase in the rate of consumption until all the ADP had been converted into ATP, at which point the rate of oxygen consumption again slowed to the background rate.

A. Why did the rate of oxygen consumption increase dramatically over the background rate when ADP was added? Why did oxygen consumption return to the background rate when all the ADP had been converted to ATP?

B. Why do you think it is that mitochondria consume oxygen at a slow background rate in the absence of added ADP?

C. How many ATP molecules were synthesized per pair of electrons transferred down the electron-transport chain to oxygen (P/2 e^- ratio)? How many ATP molecules were synthesized per oxygen atom consumed (P/O ratio)? (Remember that ½ O_2 + 2 e^- → H_2O.)

D. In experiments like this one, what processes in addition to ATP production are driven by the electrochemical proton gradient?

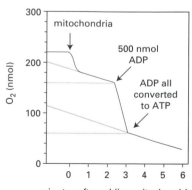

Figure 14-7 Consumption of oxygen by mitochondria under various experimental conditions (Problem 14-23).

14–24* *Thiobacillus ferrooxidans* is a bacterium that lives on slag heaps at pH 2. It is an important organism in the mining industry, as we will cover in Problems 14–32 and 14–67. Is it possible for *T. ferrooxidans* to make ATP for free using the natural pH gradient supplied by the environment? The ΔG available from the transport of protons from the outside (pH 2) to the inside (pH 6.5) of *T. ferrooxidans* is given by the Nernst equation

$$\Delta G = 2.3 \, RT \log_{10} \frac{[\text{H}^+]_{in}}{[\text{H}^+]_{out}} + nFV$$

where $R = 1.98 \times 10^{-3}$ kcal/°K mole, T = temperature in °K, n = the number of electrons transferred, $F = 23$ kcal/V mole, and V = the membrane potential.

In *T. ferrooxidans* grown at pH 2, the membrane potential is zero. Assuming that $T = 310$°K and that $\Delta G = 11$ kcal/mole for ATP synthesis under the prevailing intracellular conditions, how many protons (to the nearest integer) would have to enter the cell through the ATP synthase to drive ATP synthesis? In order for proton transport to be coupled to ATP synthesis, could the protons pass through the ATP synthase one at a time, or would they all have to pass through at the same time?

14–25 Heart muscle gets most of the ATP needed to power its continual contractions through oxidative phosphorylation. When oxidizing glucose to CO_2, heart muscle consumes O_2 at a rate of 10 μmol/min g of tissue, in order to replace the ATP used in contraction and give a steady-state ATP concentration of 5 μmol/g of tissue. At this rate, how many seconds would it take the heart to consume an amount of ATP equal to its steady-state levels? (Complete oxidation of one molecule of glucose to CO_2 yields 30 ATP, 26 of which are derived by oxidative phosphorylation using the 12 pairs of electrons captured in the electron carriers NADH and $FADH_2$.)

14–26* The relationship of free-energy change (ΔG) to the concentrations of reactants and products is important because it predicts the direction of spontaneous chemical reactions. Familiarity with this relationship is essential for understanding energy conversions in cells. Consider, for example, the hydrolysis of ATP to ADP and inorganic phosphate (P_i).

$$\text{ATP} + \text{H}_2\text{O} \rightarrow \text{ADP} + \text{P}_i$$

The free-energy change due to ATP hydrolysis is

$$\Delta G = \Delta G° + RT \ln \frac{[\text{ADP}][\text{P}_i]}{[\text{ATP}]}$$
$$= \Delta G° + 2.3 \, RT \log_{10} \frac{[\text{ADP}][\text{P}_i]}{[\text{ATP}]}$$

where the concentrations are expressed as molarities (by convention, the concentration of water is not included in the expression). R is the gas constant (1.98×10^{-3} kcal/°K mole), T is temperature (assume 37°C, which is 310°K), and $\Delta G°$ is the standard free-energy change (–7.3 kcal/mole for ATP hydrolysis to ADP and P_i).

A. Calculate ΔG for ATP hydrolysis when the concentrations of ATP, ADP, and P_i are all equal to 1 M. What is the ΔG when the concentrations of ATP, ADP, and P_i are all equal to 1 mM?

B. In a resting muscle, the concentrations of ATP, ADP, and P_i are approximately 5 mM, 1 mM, and 10 mM, respectively. What is ΔG for ATP hydrolysis in resting muscle?

C. What will ΔG equal when the hydrolysis reaction reaches equilibrium? At $[P_i] = 10$ mM, what will be the ratio of [ATP] to [ADP] at equilibrium?

D. Show that, at constant $[P_i]$, ΔG decreases by 1.4 kcal/mole for every 10-fold increase in the ratio of [ATP] to [ADP], regardless of the value of $\Delta G°$. (For example, ΔG decreases by 2.8 kcal/mole for a 100-fold increase, by 4.2 kcal/mole for a 1000-fold increase, and so on.)

14–27 Suspensions of isolated mitochondria will synthesize ATP until the [ATP]/[ADP][P_i] ratio is about 10^4. If each proton that moves into the matrix down an electrochemical proton gradient of 200 mV liberates 4.6 kcal/mole of free energy ($\Delta G = -4.6$ kcal/mole), what is the minimum number of protons that will be required to synthesize ATP ($\Delta G° = 7.3$ kcal/mole) at 37°C?

ELECTRON-TRANSPORT CHAINS AND THEIR PROTON PUMPS

TERMS TO LEARN

cytochrome	iron–sulfur center	redox potential
cytochrome b-c_1 complex	NADH dehydrogenase complex	redox reaction
cytochrome oxidase complex	quinone (Q)	respiratory control
electron-transport chain	redox pairs	respiratory enzyme complex

14–28* Both H^+ and Ca^{2+} are ions that move through the cytosol. Why is the movement of H^+ ions so much faster than that of Ca^{2+} ions? How would the speed of these two ions be affected by freezing the solution? Would you expect them to move faster or slower? Explain your answer.

14–29 Distinguish between a proton, a hydrogen atom, a hydride ion, and a hydrogen molecule.

14–30* One of the problems in understanding redox reactions is coming to grips with the language. Consider the reduction of pyruvate by NADH

$$\text{pyruvate} + \text{NADH} + H^+ \rightleftharpoons \text{lactate} + \text{NAD}^+$$

In redox reactions, oxidation and reduction necessarily occur together; however, it is convenient to list the two halves of a redox reaction separately. By convention, each half reaction is written as a reduction: oxidant + $e^- \rightarrow$ reductant. For the reduction of pyruvate by NADH the half reactions are

$$\text{pyruvate} + 2\,H^+ + 2\,e^- \rightarrow \text{lactate} \qquad E_0' = -0.19\,\text{V}$$
$$\text{NAD}^+ + H^+ + 2\,e^- \rightarrow \text{NADH} \qquad E_0' = -0.32\,\text{V}$$

Where E_0' is the standard redox potential and refers to a reaction occurring under standard conditions (25°C or 298°K, all concentrations at 1 M, and pH 7). To obtain the overall equation for reduction of pyruvate by NADH, it is necessary to reverse the NAD^+/NADH half reaction and change the sign of E_0'.

$$\text{pyruvate} + 2\,H^+ + 2\,e^- \rightarrow \text{lactate} \qquad E_0' = -0.19\,\text{V}$$
$$\text{NADH} \rightarrow \text{NAD}^+ + H^+ + 2\,e^- \qquad E_0' = +0.32\,\text{V}$$

Summing these two half reactions and their E_0' values gives the overall equation and its $\Delta E_0'$ value.

$$\text{pyruvate} + \text{NADH} + H^+ \rightleftharpoons \text{lactate} + \text{NAD}^+ \qquad \Delta E_0' = +0.13\,\text{V}$$

When a redox reaction takes place under nonstandard conditions, the tendency to donate electrons (ΔE) is equal to $\Delta E_0'$ modified by a concentration term

$$\Delta E = \Delta E_0' - \frac{2.3\,RT}{nF} \log_{10} \frac{[\text{lactate}][\text{NAD}^+]}{[\text{pyruvate}][\text{NADH}]}$$

where $R = 1.98 \times 10^{-3}$ kcal/°K mole, T = temperature in °K, n = the number of electrons transferred, and F = 23 kcal/V mole.

ΔG is related to ΔE by the equation

$$\Delta G = -nF\,\Delta E$$

Since the signs of ΔG and ΔE are opposite, a favorable redox reaction has a positive ΔE and a negative ΔG.

A. Calculate ΔG for reduction of pyruvate to lactate at 37°C with all reactants and products at a concentration of 1 M.
B. Calculate ΔG for the reaction at 37°C under conditions where the concentrations of pyruvate and lactate are equal and the concentrations of NAD^+ and NADH are equal.
C. What would the concentration term need to be for this reaction to have a ΔG of zero at 37°C?
D. Under normal conditions in vascular smooth muscle (at 37°C), the concentration ratio of NAD^+ to NADH is 1000, the concentration of lactate is 0.77 μmol/g, and the concentration of pyruvate is 0.15 μmol/g. What is ΔG for reduction of pyruvate to lactate under these conditions?

14–31 All the redox reactions in the citric acid cycle use NAD^+ to accept electrons except for the oxidation of succinate to fumarate, where FAD is used.

$$\text{succinate} + \text{FAD} \rightleftharpoons \text{fumarate} + \text{FADH}_2$$

Is FAD a more appropriate electron acceptor for this reaction than NAD^+? The relevant half reactions are given below. The half reaction for $FAD/FADH_2$ is for the cofactor as it exists covalently attached to the enzyme that catalyzes the reaction.

$$\text{fumarate} + 2\,H^+ + 2\,e^- \rightarrow \text{succinate} \qquad E_0' = 0.03\ \text{V}$$
$$\text{FAD} + 2\,H^+ + 2\,e^- \rightarrow \text{FADH}_2 \qquad E_0' = 0.05\ \text{V}$$
$$\text{NAD}^+ + H^+ + 2\,e^- \rightarrow \text{NADH} \qquad E_0' = -0.32\ \text{V}$$

A. From these E_0' values decide which, if either, cofactor (FAD or NAD^+) would yield a negative $\Delta G°$ value for the reaction?
B. What is the concentration ratio of products to reactants using FAD that would give a ΔG value of zero for the reaction at 37°C? What concentration ratio with NAD^+ would give a ΔG value of zero for the reaction?
C. Is FAD a more appropriate electron acceptor for this reaction than NAD^+? Explain your answer.

14–32* *Thiobacillus ferrooxidans*, the bacterium that lives on slag heaps at pH 2, is used by the mining industry to recover copper and uranium from low-grade ore by an acid leaching process. The bacteria oxidize Fe^{2+} to produce Fe^{3+}, which in turn oxidizes (and solubilizes) these minor components of the ore. It is remarkable that the bacterium can live in such an environment. It does so by exploiting the pH difference between the environment and its cytoplasm (pH 6.5) to drive synthesis of ATP (see Problem 14–24) and NADPH, which it can then use to fix CO_2 and nitrogen. In order to keep its cytoplasmic pH constant, *T. ferrooxidans* uses electrons from Fe^{2+} to reduce O_2 to water, thereby removing the protons.

$$4\,Fe^{2+} + O_2 + 4\,H^+ \rightarrow 4\,Fe^{3+} + 2\,H_2O$$

What are the energetics of these processes? Is the flow of electrons from Fe^{2+} to O_2 energetically favorable? How difficult is it to reduce $NADP^+$ using electrons from Fe^{2+}? These are key questions for understanding how *T. ferrooxidans* manages to thrive in such an unlikely niche.

A. What are ΔE and ΔG for the reduction of O_2 by Fe^{2+}, assuming that the reaction occurs under standard conditions? The half reactions are

$$Fe^{3+} + e^- \rightarrow Fe^{2+} \qquad E_0' = 0.77\ \text{V}$$
$$O_2 + 4\,H^+ + 4\,e^- \rightarrow 2\,H_2O \qquad E_0' = 0.82\ \text{V}$$

B. Write a balanced equation for the reduction of $NADP^+ + H^+$ by Fe^{2+}. What is

TABLE 14–2 Standard redox potentials for electron carriers in the respiratory chain (Problem 14–34).

HALF REACTION	E_0' (V)
ubiquinone + 2 H$^+$ + 2 e^- → ubiquinol	0.045
cytochrome b (Fe^{3+}) + e^- → cytochrome b (Fe^{2+})	0.077
cytochrome c_1 (Fe^{3+}) + e^- → cytochrome c_1 (Fe^{2+})	0.22
cytochrome c (Fe^{3+}) + e^- → cytochrome c (Fe^{2+})	0.25
cytochrome a (Fe^{3+}) + e^- → cytochrome a (Fe^{2+})	0.29
cytochrome a_3 (Fe^{3+}) + e^- → cytochrome a_3 (Fe^{2+})	0.55

ΔG for this reaction under standard conditions? The half reaction for NADP$^+$ is

$$NADP^+ + H^+ + 2\ e^- \rightarrow NADPH \qquad E_0' = -0.32\ V$$

What is ΔG for the reduction of NADP$^+$ + H$^+$ by Fe^{2+}, if the concentrations of Fe^{3+} and Fe^{2+} are equal, the concentration of NADPH is10-fold greater than that of NADP$^+$, and the temperature is 310°K? (Note: adjusting the number of atoms and electrons in order to balance the chemical equation does not affect E_0' or $\Delta E_0'$ values. It does, however, affect ΔE by its influence on the concentration term: each concentration term must be raised to an exponent equal to the number of atoms or molecules used in the balanced equation.)

14–33 (**True/False**) Most cytochromes have a higher redox potential (higher affinity for electrons) than iron-sulfur centers, which is why the cytochromes tend to serve as electron carriers near the O$_2$ end of the respiratory chain. Explain your answer.

14–34* The half reactions for some of the carriers in the respiratory chain are given in Table 14–2. From their E_0' values what would you guess is their order in the chain? What would you need to know before you were more certain of their order?

14–35 In 1925 David Keilin used a simple spectroscope to observe the characteristic absorption bands of the cytochromes that make up the electron-transport chain in mitochondria. A spectroscope passes a very bright light through the sample of interest and then through a prism to display the spectrum from red to blue. If molecules in the sample absorb light of particular wavelengths, dark bands will interrupt the colors of the rainbow. Keilin found that tissues from a wide variety of animals all showed the pattern in Figure 14–8. (This pattern had actually been observed several decades before by an Irish physician named MacMunn, but he thought all the bands were due to a single pigment. His work was all but forgotten by the 1920s.)

The different heat stabilities of the individual absorption bands and their different intensities in different tissues led Keilin to conclude that the absorption pattern was due to three components, which he labeled cytochromes a, b, and c (Figure 14–8). His key discovery was that the absorption bands disappeared when oxygen was introduced (Figure 14–9A) and then reappeared when the samples became anoxic (Figure 14–9B). He later confessed, "This visual perception of an intracellular respiratory process

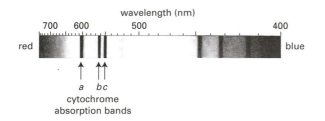

cytochrome
absorption bands

Figure 14–8 Cytochrome absorption bands (Problem 14–35). *Numbers* are the wavelengths of light in nanometers.

was one of the most impressive spectacles I have witnessed in the course of my work."

Keilin subsequently discovered that cyanide prevented the bands from disappearing when oxygen was introduced (Figure 14–9C). When urethane (a no longer used inhibitor of electron transport) was added, bands *a* and *c* disappeared in the presence of oxygen, but band *b* remained (Figure 14–9D). Finally, using cytochrome *c* extracted from dried yeast, he showed that the band due to cytochrome *c* remained when oxygen was present (Figure 14–9E).

A. Is it the reduced (electron-rich) or the oxidized (electron-poor) forms of the cytochromes that give rise to the bands Keilin observed?

B. From Keilin's observations, deduce the order in which the three cytochromes carry electrons from intracellular substrates to oxygen.

C. One of Keilin's early observations was that the presence of excess glucose prevented the disappearance of the absorption bands when oxygen was added. How do you think that rapid glucose oxidation to CO_2 might explain this observation?

14–36* If isolated mitochondria are incubated with a source of electrons such as succinate, but without oxygen, electrons enter the respiratory chain, reducing each of the electron carriers almost completely. When oxygen is then introduced, the carriers oxidize at different rates (Figure 14–10). How does this result allow you to order the electron carriers in the respiratory chain? What is their order?

14–37 You mix components of the respiratory chain in a solution as indicated below. Assuming that the electrons must follow the standard path through the electron-transport chain, in which experiments would you expect a net transfer of electrons to cytochrome *c*? Discuss why no electron transfer occurs in the other experiments.

A. Reduced ubiquinone and oxidized cytochrome *c*
B. Oxidized ubiquinone and oxidized cytochrome *c*
C. Reduced ubiquinone and reduced cytochrome *c*
D. Oxidized ubiquinone and reduced cytochrome *c*
E. Reduced ubiquinone, oxidized cytochrome *c*, and cytochrome *b*-c_1 complex
F. Oxidized ubiquinone, oxidized cytochrome *c*, and cytochrome *b*-c_1 complex
G. Reduced ubiquinone, reduced cytochrome *c*, and cytochrome *b*-c_1 complex
H. Oxidized ubiquinone, reduced cytochrome *c*, and cytochrome *b*-c_1 complex.

14–38* Inhibitors have provided extremely useful tools for analyzing mitochondrial function. Figure 14–11 shows three distinct patterns of oxygen electrode traces obtained using a variety of inhibitors. In all experiments mitochondria were added to a phosphate-buffered solution containing succinate as the sole source of electrons for the respiratory chain. After a short interval ADP was added followed by an inhibitor, as indicated in Figure 14–11. The rates of oxygen consumption at various times during the experiment are shown by downward sloping lines, with faster rates shown by steeper lines (see Figure 14–7).

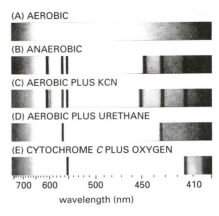

Figure 14–9 Cytochrome absorption bands under a variety of experimental conditions (Problem 14–35).

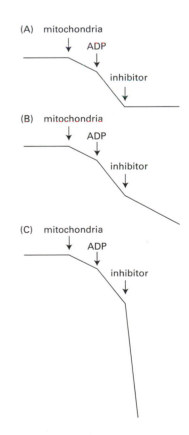

Figure 14–11 Oxygen traces showing three patterns of inhibitor effects on oxygen consumption by mitochondria (Problem 14–38).

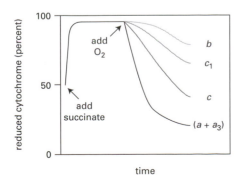

Figure 14–10 Rapid spectrophotometric analysis of the rates of oxidation of electron carriers in the respiratory chain (Problem 14–36). Cytochromes *a* and a_3 cannot be distinguished and thus are listed as cytochrome ($a + a_3$).

TABLE 14–3 Effects of a variety of inhibitors of mitochondrial function (Problem 14–38).

INHIBITOR	FUNCTION
1. FCCP	makes membranes permeable to protons
2. Malonate	prevents oxidation of succinate
3. Cyanide	inhibits cytochrome oxidase
4. Atractyloside	inhibits the ADP-ATP antiporter
5. Oligomycin	inhibits ATP synthase
6. Butylmalonate	blocks mitochondrial uptake of succinate

A. Based on the descriptions of the inhibitors in Table 14–3, assign each inhibitor to one of the oxygen traces in Figure 14–11. All these inhibitors stop ATP synthesis.

B. Using the same experimental protocol indicated in Figure 14–11, sketch the oxygen traces that you would expect for the sequential addition of the pairs of inhibitors in the list below.
   1. FCCP followed by cyanide
   2. FCCP followed by oligomycin
   3. Oligomycin followed by FCCP.

14–39   The cytochrome oxidase complex is strongly inhibited by cyanide, which binds to the Fe^{3+} form of cytochrome a_3. Carbon monoxide (CO) also inhibits the cytochrome oxidase complex by binding to cytochrome a_3, but it binds to the Fe^{2+} form. Cyanide kills at very low concentration because of its effects on cytochrome oxidase. By contrast, much larger amounts of CO must be inhaled to cause death, and its toxicity is due to its binding to the heme group of hemoglobin, which also carries an Fe^{2+}. Thus, in a sense hemoglobin sops up CO to prevent its killing effects on cytochrome oxidase. One treatment for cyanide poisoning—if administered quickly enough—is to give sodium nitrite, which oxidizes Fe^{2+} to Fe^{3+}. How do you suppose sodium nitrite protects against the effects of cyanide?

14–40*   (**True/False**) The three respiratory enzyme complexes in the mitochondrial inner membrane exist in structurally ordered arrays that facilitate the correct transfer of electrons between appropriate complexes. Explain your answer.

14–41   The two different diffusible electron carriers, ubiquinone and cytochrome c, shuttle electrons between the three protein complexes of the electron-transport chain. In principle, could the same diffusible carrier be used for both steps? If not, why not? If it could, what characteristics would it need to possess and what would be the disadvantages of such a situation?

14–42*   During operation of the respiratory chain, cytochrome c accepts an electron from the cytochrome b-c_1 complex and transfers it to the cytochrome oxidase complex. What is the relationship of cytochrome c to the two complexes it connects? Does cytochrome c simultaneously connect to both complexes, serving like a wire to allow electrons to flow through it from one complex to the other, or does cytochrome c move between the complexes like a ferry, picking up an electron from one and handing it over to the next? An answer to this question is provided by the experiments described below.

   Individual lysines on cytochrome c were modified to replace the positively charged amino group with a neutral group or with a negatively charged group. The effects of these modifications on electron transfer from the cytochrome b-c_1 complex to cytochrome c and from cytochrome c to the cytochrome oxidase complex were then measured. Some modifications had no effect (Figure 14–12, open circles), whereas others inhibited electron transfer from the cytochrome b-c_1 complex to cytochrome c (Figure 14–12A, shaded circles) or from cytochrome c to the cytochrome oxidase complex (Figure 14–12B, shaded circles). In an independent series of experiments, several lysines that inhibited transfer of electrons to and from cytochrome c were shown to be protected from acetylation when cytochrome c was bound either to the cytochrome b-c_1 complex or to the cytochrome oxidase complex.

(A) FROM CYTOCHROME b-c_1 COMPLEX

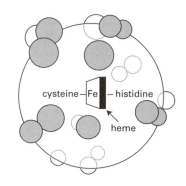

(B) TO CYTOCHROME OXIDASE COMPLEX

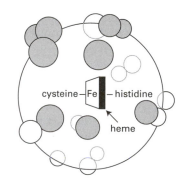

Figure 14–12 Analysis of electron transfer by cytochrome c (Problem 14–42). (A) Lysines that inhibit the transfer of electrons from the cytochrome b-c_1 complex to cytochrome c. (B) Lysines that inhibit the transfer of electrons from cytochrome c to the cytochrome oxidase complex. One edge of the heme group protrudes from the front surface of cytochrome c toward the reader. *Circles* show the positions of lysines. *Solid circles* are on the front surface of the molecule; *dashed circles* are on the back surface. The *larger* the circle, the *closer* it is to the reader. *Shaded circles* indicate lysines that inhibit electron transfer; *open circles* indicate lysines that do not.

How do these results distinguish the possibilities that cytochrome *c* connects the two complexes by binding to them simultaneously or by shuttling between them?

14–43 What is the standard free energy change ($\Delta G°$) associated with transfer of electrons from NADH to O_2, according to the balanced equation below?

$$O_2 + 2\ NADH + 2\ H^+ \rightleftharpoons 2\ H_2O + 2\ NAD^+$$

The half reactions are

$$O_2 + 4\ H^+ + 4\ e^- \rightarrow 2\ H_2O \qquad\qquad E_0' = 0.82\ V$$
$$NAD^+ + H^+ + 2\ e^- \rightarrow NADH \qquad\qquad E_0' = -0.32\ V$$

A common way of writing this equation is

$$\tfrac{1}{2}\ O_2 + NADH + H^+ \rightleftharpoons H_2O + NAD^+$$

What is $\Delta G°$ for this equation? Do the two calculations give the same answer? Explain why they do or why they don't.

14–44* The uncoupler dinitrophenol was once prescribed as a diet drug to aid in weight loss. How would an uncoupler of oxidative phosphorylation promote weight loss? Why do you suppose that it is no longer prescribed?

14–45 (**True/False**) Lipophilic weak acids short-circuit the normal flow of protons across the inner membrane, thereby eliminating the proton-motive force, stopping ATP synthesis, and blocking the flow of electrons. Explain your answer.

14–46* If you were to impose an artificially large electrochemical proton gradient across the mitochondrial inner membrane, would you expect electrons to flow up the respiratory chain, in the reverse of their normal direction? Why or why not?

14–47 Methanogenic bacteria produce methane as the end-product of electron transport. *Methanosarcina barkeri*, for example, when grown under hydrogen in the presence of methanol, transfers electrons from hydrogen to methanol, producing methane and water.

$$CH_3CH + H_2 \rightarrow CH_4 + H_2O$$

This reaction is analogous to those used by aerobic bacteria and mitochondria, which transfer electrons from carbon compounds to oxygen, producing water and carbon dioxide. Given the peculiar biochemistry involved in methanogenesis, however, it was initially unclear whether methanogenic bacteria synthesized ATP by electron transport-driven phosphorylation.

In the experiments depicted in Figure 14–13, methanol was added to cultures of methanogenic bacteria grown under hydrogen. The production of CH_4, the magnitude of the electrochemical proton gradient, and the intracellular concentration of ATP were then assayed in the presence of the inhibitors TCS and DCCD. Based on the effects of these inhibitors, and by analogy to mitochondria, decide for each inhibitor if it blocks electron transport, inhibits ATP synthase, or uncouples electron transport from ATP synthesis. Explain your reasoning.

14–48* The electrochemical proton gradient is responsible not only for ATP production in bacteria, mitochondria, and chloroplasts, but also for powering bacterial flagella. The flagellar motor is thought to be driven directly by the flux of protons through it. To test this idea, you analyze a motile strain of *Streptococcus*. These bacteria swim when glucose is available for oxidation, but they do not swim when glucose is absent (and no other substrate is available for oxidation). Using a series of ionophores that alter the pH gradient or the membrane potential (the two components of the electrochemical proton gradient), you make several observations.

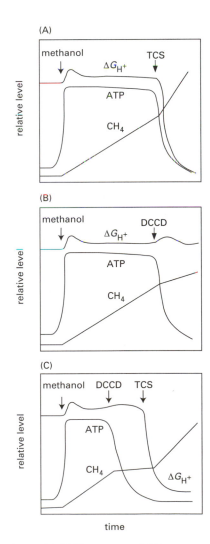

Figure 14–13 Time course of CH_4 production after addition of methanol to a culture of methanogenic bacteria growing under H_2 (Problem 14–47). The magnitude of the electrochemical proton gradient (ΔG_{H^+}), the intracellular ATP concentration, and the production of CH_4 are expressed in arbitrary units.

TABLE 14–4 Effects of ionophores on the swimming of normal bacteria (Problem 14–48).

OBSERVATION	GLUCOSE	IONOPHORE	ION IN MEDIUM	EFFECT ON BACTERIA
1	present	FCCP (H^+)	—	stop swimming
2	present	valinomycin (K^+)	K^+	keep swimming
3	absent	valinomycin (K^+)	K^+	remain motionless
4	absent	valinomycin (K^+)	Na^+	swim briefly

The specificity of each ionophore is shown in parentheses.

1. Bacteria that are swimming in the presence of glucose stop swimming upon addition of the proton ionophore FCCP.
2. Bacteria that are swimming in the presence of glucose in a medium containing K^+ are unaffected by addition of the K^+ ionophore valinomycin.
3. Bacteria that are motionless in the absence of glucose in a medium containing K^+ remain motionless upon addition of valinomycin.
4. Bacteria that are motionless in the absence of glucose in a medium containing Na^+ swim briefly upon addition of valinomycin and then stop. These observations are summarized in Table 14–4.

A. Explain how each of these observations is consistent with the idea that the flagellar motor is driven by a flux of protons. (The concentration of K^+ inside these bacteria is lower than the concentration of K^+ used in the medium.)

B. Wild-type bacteria can swim in the presence or absence of oxygen. Mutant bacteria that are missing ATP synthase, which couples proton flow to ATP production, can swim only in the presence of oxygen. How are normal bacteria able to swim in the absence of oxygen when there is no electron flow? How do you think the loss of the ATP synthase prevents swimming in mutant bacteria?

14–49 Some bacteria have become specialized to live in an alkaline environment at pH 10. They maintain their internal environment at pH 7. Why is it that they cannot exploit the pH difference across their membrane to get ATP for free using a standard ATP synthase? Can you suggest an engineering modification to ATP synthase that would allow it to generate ATP from proton flow in such an environment?

CHLOROPLASTS AND PHOTOSYNTHESIS

TERMS TO LEARN

antenna complex	cyclic photophosphorylation	photosystem
carbon fixation	noncyclic photophosphorylation	plastid
carbon-fixation reactions	photochemical reaction center	starch
carbon-fixation cycle (Calvin cycle)	photorespiration	stroma
chlorophyll	photosynthetic electron-transfer	sucrose
chloroplast		

14–50* (**True/False**) In a general way, one might view the chloroplast as a greatly enlarged mitochondrion in which the cristae have been pinched off to form a series of interconnected submitochondrial particles in the matrix space. Explain your answer.

14–51 Both mitochondria and chloroplasts use electron transport to pump protons, creating an electrochemical proton gradient, which drives ATP synthesis. Are protons pumped across the same (analogous) membranes in the two organelles? Is ATP synthesized in the analogous compartments? Explain your answers.

14–52* How much energy is available in visible light? How much energy does sunlight deliver to the earth? How efficient are plants at converting light energy into chemical energy? The answers to these questions provide an important backdrop to the subject of photosynthesis.

Each quantum or photon of light has energy $h\nu$, where h is Planck's constant

$(1.58 \times 10^{-37}$ kcal sec/photon) and ν is the frequency in sec^{-1}. The frequency of light is equal to c/λ, where c is the speed of light $(3.0 \times 10^{17}$ nm/sec) and λ is the wavelength in nm. Thus, the energy (E) of a photon is

$$E = h\nu = hc/\lambda$$

A. Calculate the energy of a mole of photons $(6 \times 10^{23}$ photons/mole) at 400 nm (violet light), at 680 nm (red light), and at 800 nm (near infrared light).

B. Bright sunlight strikes Earth at the rate of about 0.3 kcal/sec per square meter. Assuming for the sake of calculation that sunlight consists of monochromatic light of wavelength 680 nm, how many seconds does it take for a mole of photons to strike a square meter?

C. Assuming that it takes eight photons to fix one molecule of CO_2 as carbohydrate under optimal conditions (8–10 photons is the currently accepted value), calculate how long it would take a tomato plant with a leaf area of 1 square meter to make a mole of glucose from CO_2. Assume that photons strike the leaf at the rate calculated above and, furthermore, that all the photons are absorbed and used to fix CO_2.

D. If it takes 112 kcal/mole to fix a mole of CO_2 into carbohydrate, what is the efficiency of conversion of light energy into chemical energy after photon capture? Assume again that eight photons of red light (680 nm) are required to fix one molecule of CO_2.

14–53 Deriving a balanced equation from the numerous individual reactions of the carbon-fixation cycle (Calvin cycle) should be straightforward, but evidently is not. A selection of prominent biochemistry texts lists a variety of equations for the balanced reaction (Table 14–5). All agree on the amounts of CO_2, NADPH, and ATP required to produce glucose ($C_6H_{12}O_6$), but disagree on water and protons. Part of the difficulty is that explicit chemical structures are not shown for ATP, ADP, and P_i. P_i (HPO_4^{2-}) incorporates an additional O and H that are not present in ATP, but must be accounted for in a balanced equation. Which equation in Table 14–5 is the balanced equation for the Calvin cycle?

> **Problem 1–17** presents a classic experiment from the 1930s that reasoned correctly that photosynthetic fixation of CO_2 released O_2 by photolysis of H_2O, not CO_2 as was widely believed.

14–54* Recently, your boss expanded the lab's interest in photosynthesis from algae to higher plants. You have decided to study photosynthetic carbon fixation in the cactus, but so far you have had no success: cactus plants do not seem to fix $^{14}CO_2$, even in direct sunlight. Under the same conditions your colleagues studying dandelions get excellent incorporation of $^{14}CO_2$ within seconds of adding it and are busily charting new biochemical pathways.

Depressed, you leave the lab one day without dismantling the labeling chamber. The following morning you remove the cactus and, much to your surprise, find it has incorporated a great deal of $^{14}CO_2$. Evidently, the cactus fixed carbon during the night. When you repeat your experiments at night in complete darkness, you find that cactus plants incorporate label splendidly.

Although you are forced to shift your work habits, at least now you can make some progress. A brief exposure to $^{14}CO_2$ labels one compound—malate—almost exclusively. During the night, labeled malate builds up to very high levels in specialized vacuoles inside chloroplast-containing cells. In addition, the starch in these same cells disappears. During the day the malate disappears, and labeled starch is formed in a process that requires light. Furthermore, you find that $^{14}CO_2$ reappears in these cells during the day.

TABLE 14–5 A selection of 'balanced' equations for the reactions of the Calvin cycle (Problem 14–53).

EQUATIONS FOR THE CALVIN CYCLE
A. $6\,CO_2 + 12\,NADPH + 18\,ATP + 12\,H_2O \rightarrow C_6H_{12}O_6 + 12\,NADP^+ + 18\,ADP + 18\,P_i + 6\,H^+$
B. $6\,CO_2 + 12\,NADPH + 18\,ATP + 12\,H^+ + 12\,H_2O \rightarrow C_6H_{12}O_6 + 12\,NADP^+ + 18\,ADP + 18\,P_i$
C. $6\,CO_2 + 12\,NADPH + 18\,ATP + 12\,H^+ + 12\,H_2O \rightarrow C_6H_{12}O_6 + 12\,NADP^+ + 18\,ADP + 18\,P_i + 6\,O_2$
D. $6\,CO_2 + 12\,NADPH + 18\,ATP + 12\,H_2O \rightarrow C_6H_{12}O_6 + 12\,NADP^+ + 18\,ADP + 18\,P_i$
E. $6\,CO_2 + 12\,NADPH + 18\,ATP + 12H^+ \rightarrow C_6H_{12}O_6 + 12\,NADP^+ + 18\,ADP + 18\,P_i + 6\,H_2O$

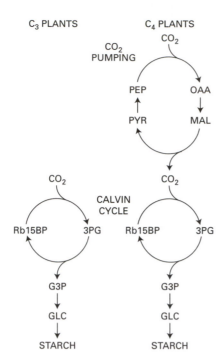

C_3 PLANTS C_4 PLANTS

Figure 14–14 Fixation of CO_2 in C_3 and C_4 plants (Problem 14–54). In these simplified diagrams the usage of ATP and NADPH in various reactions is not shown. PEP is phosphoenolpyruvate; OAA is oxaloacetate; MAL is malate; PYR is pyruvate; Rb15BP is ribulose 1,5-bisphosphate; 3PG is 3-phosphoglycerate; G3P is glyceraldehyde 3-phosphate; and GLC is glucose.

These results remind you in some ways of CO_2 pumping in C_4 plants (Figure 14–14), but they are quite distinct in other ways.

A. Why is light required for starch formation in the cactus? Is it required for starch formation in C_4 plants? In C_3 plants?

B. Using the reactions of the CO_2 pump in C_4 plants as a starting point (Figure 14–14), sketch a brief outline of CO_2 fixation in the cactus. Which reactions occur during the day and which at night?

C. A cactus depleted of starch could not fix CO_2, but a C_4 plant could. Why is starch required for CO_2 fixation in the cactus but not in C_4 plants?

D. Can you offer an explanation for why this method of CO_2 fixation is advantageous for cactus plants?

14–55   A suspension of the cyanobacterium *Chlamydomonas* is actively carrying out photosynthesis in the presence of light and CO_2. If you turned off the light, how would you expect the amounts of ribulose 1,5-bisphosphate and 3-phosphoglycerate to change over the next minute? How about if you left the light on but removed the CO_2?

14–56*   Recalling Joseph Priestley's famous experiment in which a sprig of mint saved the life of a mouse in a sealed chamber, you decide to do an analogous experiment to see how a C_3 and a C_4 plant do when confined together in a sealed environment. You place a corn plant (C_4) and a geranium (C_3) in a sealed plastic chamber with normal air (300 parts per million CO_2) on a windowsill in your laboratory. What will happen to the two plants? Will they compete or collaborate? If they compete, which one wins and why?

14–57   (**True/False**) When a molecule of chlorophyll in an antenna complex absorbs a photon, the excited electron is rapidly transferred from one chlorophyll molecule to another until it reaches the photochemical reaction center. Explain your answer.

14–58*   What fraction of the free energy of light at 700 nm is captured when a chlorophyll molecule (P700) at the photochemical reaction center in photosystem I absorbs a photon? The equation for calculating the free energy available in one photon of light is given in Problem 14–52. If one assumes standard conditions, the captured free energy ($\Delta G = -nF\Delta E_0'$) can be calculated from the standard redox potential for P700* (excited) → P700 (ground state), which can be gotten from the half reactions.

$$P700^+ + e^- \rightarrow P700 \qquad\qquad E_0' = 0.4 \text{ V}$$
$$P700^+ + e^- \rightarrow P700^* \qquad\qquad E_0' = -1.2 \text{ V}$$

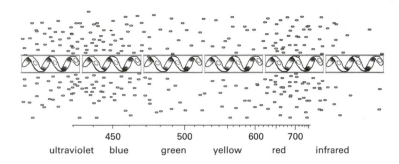

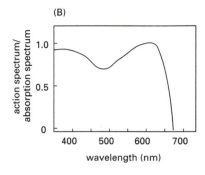

Figure 14–15 Experiment to measure the action spectrum of a filamentous green alga (Problem 14–59). Bacteria, which are indicated by the *tiny rectangles*, were distributed evenly throughout the test tube at the beginning of the experiment.

Figure 14–16 Analysis of photosynthesis in the alga *Chlorella* (Problem 14–61). (A) Absorption and action spectra. The action spectrum shows the evolution of O_2 at different wavelengths. (B) The ratio of the action spectrum to the absorption spectrum. The ratio is shown on an arbitrary scale.

14–59 Careful experiments comparing absorption and action spectra of plants ultimately led to the notion of two cooperating photosystems in chloroplasts. The absorption spectrum is the amount of light captured by photosynthetic pigments at different wavelengths. The action spectrum is the rate of photosynthesis (for example, O_2 evolution or CO_2 fixation) resulting from the capture of photons.

T.W. Englemann, who used simple equipment and an ingenious experimental design, probably made the first measurement of an action spectrum in 1882. He placed a filamentous green alga into a test tube along with a suspension of oxygen-seeking bacteria. He allowed the bacteria to use up the available oxygen and then illuminated the alga with light that had been passed through a prism to form a spectrum. After a short time he observed the results shown in Figure 14–15. Sketch the action spectrum for this alga and explain how this experiment works.

14–60* Why are plants green?

14–61 If all pigments captured energy and delivered it to the photosystem with equal efficiency, then the absorption spectrum and the action spectrum would have the same shape; however, the two spectra differ slightly (Figure 14–16A). When a ratio of the two spectra is displayed (Figure 14–16B), the most dramatic difference is the so-called 'red drop' at long wavelengths. In 1957 Emerson found that if shorter wavelength light (650 nm) was mixed with the less effective longer wavelength light (700 nm), the rate of O_2 evolution was much enhanced over either wavelength given alone. This result, along with others, suggested that two photosystems (now called photosystem I and photosystem II) were cooperating with one another, and it led to the familiar Z scheme for photosynthesis.

One clue to the order in which the two photosystems are linked came from experiments in which illumination was switched between 650 nm and 700 nm. As shown in Figure 14–17, a shift from 700 nm to 650 nm was accompanied by a transient burst of O_2 evolution, whereas a shift from 650 nm to 700 nm was accompanied by a transient depression in O_2 evolution.

Using your knowledge of the Z scheme of photosynthesis, explain why these so-called chromatic transients occur, and deduce whether photosystem II,

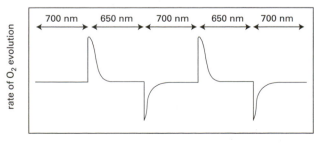

Figure 14–17 Chromatic transients observed upon switching between 650-nm light and 700-nm light (Problem 14–61). The intensities of light at the two wavelengths were adjusted beforehand so that alone each gave the same rate of O_2 evolution.

which accepts electrons from H_2O, is more responsive to 650-nm light or to 700-nm light.

14–62* Treatment of chloroplasts with the herbicide DCMU stops O_2 evolution and photophosphorylation. If an artificial electron acceptor is added that accepts electrons from plastoquinone (Q), oxygen evolution is restored but not photophosphorylation. Propose a site at which DCMU acts in the flow of electrons through photosystems I and II (Figure 14–18). Explain your reasoning. Why is DCMU an herbicide?

14–63 The most compelling early evidence for the Z scheme of photosynthesis came from measuring the oxidation states of the cytochromes in algae under different regimes of illumination (Figure 14–19). Illumination with light at 680 nm caused oxidation of cytochromes (indicated by the upward trace in Figure 14–19A). Additional illumination with light at 562 nm caused reduction of the cytochromes (indicated by the downward trace in Figure 14–19A). If the lights are then turned off, both effects are reversed (Figure 14–19A). In the presence of the herbicide DCMU (see Problem 14–62), no reduction with 562-nm light occurred (Figure 14–19B).

A. In these algae, which wavelength stimulates photosystem I and which stimulates photosystem II?

B. How do these results support the Z scheme for photosynthesis; that is, how do they support the idea that there are two photosystems that are linked by cytochromes?

C. On which side of the cytochromes does DCMU block electron transport—on the side nearer photosystem I, or the side nearer photosystem II?

14–64* Photosystem II accepts electrons from water, generating O_2, and donates them via the electron-transport chain to photosystem I. Each photon absorbed by photosystem II transfers only a single electron, and yet four electrons must be removed from water to generate a molecule of O_2. Thus, four photons are required to evolve a molecule of O_2.

$$2 H_2O + 4 \, h\nu \rightarrow 4 \, e^- + 4 \, H^+ + O_2$$

How do four photons cooperate in the production of O_2? Is it necessary that four photons arrive at a single reaction center simultaneously? Can four activated reaction centers cooperate to evolve a molecule of O_2? Or is there some sort of 'gear wheel' that collects the four electrons from H_2O and transfers them one at a time to a reaction center?

To investigate this problem, you expose dark-adapted spinach chloroplasts to a series of brief saturating flashes of light (2 μsec) separated by short periods of darkness (0.3 sec) and measure the evolution of O_2 that results

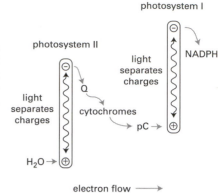

Figure 14–18 Flow of electrons through photosystems I and II during photosynthesis in chloroplasts (Problem 14–62). Electrons from photosystem II flow to plastoquinone (Q), then to the cytochrome b_6-f complex (cytochromes), and then to plastocyanin (pC), after which they enter photosystem I. The protons pumped by the cytochrome b_6-f complex generate an electrochemical gradient, which is used to drive ATP synthesis.

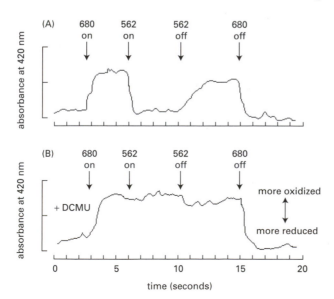

Figure 14–19 Oxidation state of cytochromes after illumination of algae with different wavelength light (Problem 14–63). (A) In the absence of DCMU. (B) In the presence of DCMU. An *upward* trace indicates oxidation of the cytochromes; a *downward* trace indicates reduction of the cytochromes.

from each flash. Under this lighting regime most photosystems capture a photon during each flash. As shown in Figure 14–20, O_2 is evolved with a distinct periodicity: the first burst of O_2 occurs on the third flash, and subsequent peaks occur every fourth flash thereafter. If you first inhibit 97% of the photosystem II reaction centers with DCMU and then repeat the experiment, you observe the same periodicity of O_2 production, but the peaks are only 3% of the uninhibited values.

A. How do these results distinguish among the three possibilities posed at the outset (simultaneous action, cooperation among reaction centers, and a gear wheel)?

B. Why do you think it is that the first burst of O_2 occurs after the third flash, whereas additional peaks occur at four-flash intervals? (Consider what this observation implies about the dark-adapted state of the chloroplasts.)

C. Can you suggest a reason why the periodicity in O_2 evolution becomes less pronounced with increasing flash number?

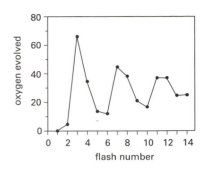

Figure 14–20 Oxygen evolution by spinach chloroplasts in response to saturating flashes of light (Problem 14–64). The chloroplasts were placed in the dark for 40 minutes prior to the experiment to allow them to come to the same 'ground' state. Oxygen production is expressed in arbitrary units.

14–65 The balanced equation for production of NADPH by noncyclic photophosphorylation is

$$2\ H_2O + 2\ NADP^+ \rightarrow 2\ NADPH + 2\ H^+ + O_2$$

How many photons must be absorbed to generate two NADPH and a molecule of O_2? (Assume one photon excites one electron.)

14–66* What is the standard free energy ($\Delta G°$) for the reduction of $NADP^+$ by H_2O? The half reactions are

$$
\begin{array}{ll}
NADP^+ + H^+ + 2\ e^- \rightarrow NADPH & E_0' = -0.32\ V \\
O_2 + 4\ H^+ + 4\ e^- \rightarrow 2\ H_2O & E_0' = 0.82\ V
\end{array}
$$

14–67 *T. ferrooxidans*, the slag heap bacterium that lives at pH 2, fixes CO_2 like photosynthetic organisms but uses the abundant Fe^{2+} in its environment as a source of electrons instead of H_2O. *T. ferrooxidans* oxidizes Fe^{2+} to Fe^{3+} to reduce $NADP^+$ to NADPH, a very unfavorable reaction with a ΔE of about –1.1 V. It does so by coupling production of NADPH to the energy of the natural proton gradient across its membrane, which has a free energy (ΔG) of –6.4 kcal/mole H^+. What is the smallest number of protons to the nearest integer that would be required to drive the reduction of $NADP^+$ by Fe^{2+}? How do you suppose proton flow is mechanistically coupled to the reduction of $NADP^+$?

14–68* Conversion of 6 CO_2 to one molecule of glucose via the Calvin cycle requires 18 ATP and 12 NADPH. Assuming that both noncyclic and cyclic photophosphorylation generate 1 ATP per pair of electrons, what is the ratio of electrons that travel the noncyclic path to those that travel the cyclic path to meet the needs of the Calvin cycle? What is the ratio of electrons processed in photosystem I to those processed in photosystem II to meet the needs of the Calvin cycle?

14–69 What is the free energy change (ΔG) for ATP synthesis under typical conditions in which ATP, ADP, and P_i are at 3 mM, 0.1 mM, and 10 mM, respectively, at 37°C? The standard free energy change for ATP synthesis is 7.3 kcal/mole.

14–70* In chloroplasts, protons are pumped out of the stroma across the thylakoid membrane, whereas in mitochondria, they are pumped out of the matrix across the inner membrane. Explain how this arrangement allows chloroplasts to generate a larger proton gradient across the thylakoid membrane than mitochondria can generate across the inner membrane.

14–71 In an insightful experiment performed in the 1960s, chloroplasts were first soaked in an acidic solution at pH 4, so that the stroma and thylakoid space

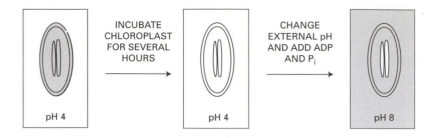

Figure 14–21 Soaking chloroplasts in acidic and basic solutions (Problem 14–71). *White areas are at pH 4.*

became acidified (Figure 14–21). They were then transferred to a basic solution (pH 8). This rapidly increased the pH of the stroma to 8, while the thylakoid space temporarily remained at pH 4. A burst of ATP synthesis was observed, and the pH difference between the thylakoid space and the stroma quickly disappeared.

A. Explain why these conditions lead to ATP synthesis.
B. Is light needed for the experiment to work? Why or why not?
C. What would happen if the solutions were switched so that the first incubation was in the pH 8 solution and the second one was in the pH 4 solution? Explain your answer.
D. Does the experiment support or question the chemiosmotic model?

14–72* In one of the earliest and most convincing tests of the chemiosmotic model, formation of ATP was assayed in a suspension of thylakoid vesicles (stromal surface facing outward, that is, right-side-out). These vesicles were first acidified to pH 4 and then made alkaline in the presence of ADP and $^{32}PO_4$. As shown in Figure 14–22A, the yield of ATP was greater when the vesicles were acidified with succinic acid than when they were acidified with HCl. Furthermore, the yield of ATP increased with increasing concentration of succinic acid (even though pH 4 solutions were used in all experiments). As shown in Figure 14–22B, the yield of ATP also depended on the pH of the alkaline stage of the experiment, with the yield of ATP increasing up to pH 8.5.

A. Why does acidification with succinic acid yield more ATP than acidification with HCl? And why does the yield of ATP increase with increasing concentrations of succinic acid? (Succinic acid has two carboxylic acid groups with pKs of 4.2 and 5.5.)
B. Why is the yield of ATP so critically dependent on the pH of the alkaline stage of the experiment?
C. Predict the effects of treatments with FCCP, an uncoupler of electron transport and ATP synthesis, and DCMU, which blocks electron transport, during the alkaline incubation. Explain the basis for your predictions.

14–73 Unlike mitochondria, chloroplasts do not have a transporter that allows them to export ATP to the cytosol. How then does the rest of the cell get the ATP it needs to survive?

THE GENETIC SYSTEMS OF MITOCHONDRIA AND PLASTIDS

TERMS TO LEARN

cytoplasmic inheritance	maternal inheritance	non-Mendelian inheritance
endosymbiont hypothesis	mitotic segregation	

14–74* (**True/False**) Mitochondria and chloroplasts replicate their DNA in synchrony with the nuclear DNA and divide when the cell divides, thereby ensuring a constant number of organellar genomes. Explain your answer.

14–75 Replication of an individual mitochondrial genome takes about an hour in

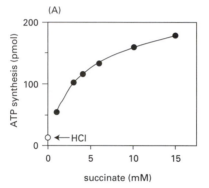

(A)

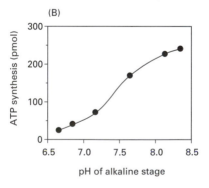

(B)

Figure 14–22 Test of the chemiosmotic model in thylakoid vesicles (Problem 14–72). (A) Yield of ATP with HCl and with increasing concentrations of succinic acid. The yield of ATP with HCl is shown on the y axis, where the concentration of succinic acid is zero. In all cases the thylakoid suspension was treated with acid for 60 seconds, then treated with alkali for 15 seconds in the presence of ADP and $^{32}PO_4$, at which point the reaction was stopped and the yield of radioactive ATP was measured. (B) Yield of ATP at different pHs during the alkaline stage of the experiment.

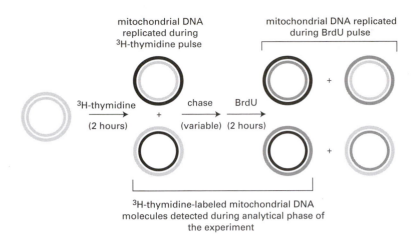

mitochondrial DNA
replicated during
³H-thymidine pulse

mitochondrial DNA replicated
during BrdU pulse

³H-thymidine → + → chase → BrdU →
(2 hours)   (variable) (2 hours)

³H-thymidine-labeled mitochondrial DNA
molecules detected during analytical phase of
the experiment

Figure 14–23 Experimental design to assess the timing of mitochondrial DNA replication (Problem 14–75). Unlabeled mitochondrial DNA is indicated with *light gray lines*; DNA labeled with ³H-thymidine is indicated with *black lines*; DNA labeled with BrdU is indicated with *dark gray lines*.

mouse cells, which is about 5% of the 20-hour cell-generation time. Since the average number of mitochondrial genomes per cell is constant, they must replicate, on average, once per cell cycle. Do mitochondrial genomes replicate at random times throughout the cell cycle or is their replication confined to one particular stage, for example, S phase?

Your elegant approach to this question is to label mouse cell mitochondrial DNA briefly (2 hours) with ³H-thymidine, to chase with nonradioactive thymidine for various times, and finally to label with 5-bromodeoxyuridine (BrdU) for 2 hours (Figure 14–23). Any DNA that was replicated during both exposures will be radioactive (due to the ³H-thymidine) and denser than normal mitochondrial DNA (due to the BrdU). Using density-gradient centrifugation, you separate heavy (BrdU-labeled) mitochondrial DNA from light (non-BrdU-labeled) mitochondrial DNA and measure the radioactivity associated with each (Figure 14–24). Your results show that a relatively constant fraction of ³H-labeled mitochondrial DNA was shifted to a higher density, regardless of the time between the labeling periods (the chase times).

A. Do these results fit better with your expectations for mitochondrial replication during a specific part of the cell cycle or at random times? Why?

B. One of your colleagues criticizes the design of these experiments. He suggests that you cannot distinguish between random or timed replication because the cells were growing asynchronously (that is, within the cell population all different stages of the cell cycle were represented). How does this concern affect your interpretation?

C. What results would you expect for a similar analysis of nuclear DNA? Assume that the DNA synthesis phase—S phase—of the 20-hour cell cycle is 5 hours long. Explain your answer.

D. What results would you expect if mitochondrial DNA were replicated at all times during the cell cycle, but once an individual molecule was replicated, it had to wait exactly one cell cycle before it was replicated again? Would they be the same as the results in Figure 14–24, the same as you expect for nuclear DNA, or would they give you a new pattern entirely? Why?

14–76* How many genomes are there, on average, in a single mitochondrion in a human liver cell? Assume that the weight of mitochondrial DNA is 1% the

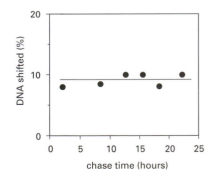

chase time (hours)

Figure 14–24 The fraction of mitochondrial DNA labeled with ³H-thymidine that is shifted to heavy density after various lengths of chase (Problem 14–75). The fraction of DNA that is density shifted is the radioactivity at the heavy density divided by the total radioactivity. The length of chase is the time from the end of the ³H-thymidine labeling to the beginning of the BrdU labeling (see Figure 14–23).

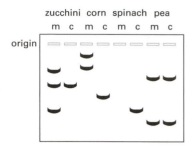

zucchini corn spinach pea
m c m c m c m c

origin

Figure 14–25 Patterns of hybridization of a probe from spinach chloroplast DNA to mitochondrial and chloroplast DNAs from zucchini, corn, spinach, and pea (Problem 14–78). Lanes labeled *m* contain mitochondrial DNA; lanes labeled *c* contain chloroplast DNA. Restriction fragments to which the probe hybridized are shown as *dark bands*.

weight of the nuclear DNA and that there are 1000 mitochondria per liver cell. The human mitochondrial genome is 16,569 bp and the diploid human genome is 6.4×10^9 bp.

14–77 You have discovered a remarkable, one-celled protozoan that lives in an anaerobic environment. It has caught your attention because it has absolutely no mitochondria, an exceedingly rare condition among eucaryotes. If you could show that this organism derives from an ancient lineage that split off from the rest of eucaryotes before mitochondria were acquired, it would truly be a momentous discovery. You sequence the organism's genome so you can make detailed comparisons. It is clear from sequence comparisons that your organism does indeed derive from an ancient lineage. But here and there, scattered around the genome are bits of DNA that in aggregate resemble the bacterial genome from which mitochondria evolved. Propose a plausible evolutionary history for your organism.

Problem 1–30 examines the evolutionary history of *Giardia*, a parasitic eucaryote that is missing mitochondria, ER, and the Golgi apparatus.

14–78* It is well accepted that transfer of DNA from organellar genomes to nuclear genomes is common during evolution. Do transfers between organellar genomes also occur? One experiment to search for genetic transfers between organellar genomes used a defined restriction fragment from spinach chloroplasts, which carried information for the gene for the large subunit of ribulosebisphosphate carboxylase. This gene has no known mitochondrial counterpart. Mitochondrial and chloroplast DNAs were prepared from zucchini, corn, spinach, and pea. All these DNAs were digested with the same restriction nuclease, and the resulting fragments were separated by electrophoresis. The fragments were then transferred to a filter and hybridized to a radioactive preparation of the spinach fragment. A schematic representation of the autoradiograph is shown in Figure 14-25.

Problems 1–32 and 1–33 present evidence for transfer of DNA from mitochondrial to nuclear genomes and reveal some of the mechanisms involved.

A. It is very difficult to prepare mitochondrial DNA that is not contaminated to some extent with chloroplast DNA. How do these experiments control for contamination of the mitochondrial DNA preparation by chloroplast DNA?

B. Which of these plant mitochondrial DNAs appear to have acquired chloroplast DNA?

14–79 At the cellular level evolutionary theories are particularly difficult to test since fossil evidence is lacking. The possible evolutionary origins of mitochondria and chloroplasts must be sought in living organisms. Fortunately, living forms resembling the ancestral types required by the endosymbiotic theory for the origin of mitochondria and chloroplasts can be found today. For example, the plasma membrane of the free-living aerobic bacterium *Paracoccus denitrificans* contains a respiratory chain that is nearly identical to the respiratory chain of mammalian mitochondria—both in the types of respiratory components present and in its sensitivity to respiratory inhibitors such as antimycin and rotenone. Indeed, no significant feature of the mammalian respiratory chain is absent from *Paracoccus*. *Paracoccus* effectively assembles in a single organism all those features of the mitochondrial inner membrane that are otherwise distributed at random among other aerobic bacteria.

Imagine that you are a protoeucaryotic cell looking out for your evolutionary future. You have been observing proto-*Paracoccus* and are amazed at

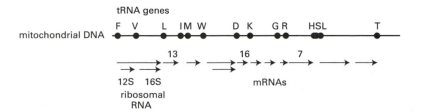

Figure 14–26 Transcription map of human mitochondrial DNA (Problem 14–83). Individual tRNA genes are indicated by *black circles*; the amino acids they carry are shown in the one-letter code. The three mRNAs whose detailed sequences are shown in Figure 14–27 are indicated by *number*.

its incredibly efficient use of oxygen in generating ATP. With such a source of energy your horizons would be unlimited. You plot to hijack a proto-*Paracoccus* and make it work for you and your descendants. You plan to take it into your cytoplasm, feed it any nutrients it needs, and harvest the ATP. Accordingly, one dark night you trap a philandering proto-*Paracoccus*, surround it with your plasma membrane, and imprison it in a new cytoplasmic compartment. To your relief, the proto-*Paracoccus* seems to enjoy its new environment. After a day of waiting, however, you feel as sluggish as ever. What has gone wrong with your scheme?

14–80* (**True/False**) The mitochondrial genetic code differs slightly from the nuclear code, but it is identical in mitochondria from all species that have been examined. Explain your answer.

14–81 How many codons specify STOP in human mitochondria?

14–82* (**True/False**) The presence of introns in organellar genes is not surprising since similar introns have been found in related genes from bacteria whose ancestors are thought to have given rise to mitochondria and chloroplasts. Explain your answer.

14–83 The majority of mRNAs, tRNAs, and rRNAs in human mitochondria are transcribed from one strand of the genome. These RNAs are all present initially on one very long transcript, which is 93% the length of the DNA strand. During mitochondrial protein synthesis, however, these RNAs function as separate, independent species of RNA. The relationship of the individual RNAs to the primary transcript and many of the special features of the mitochondrial genetic system have been revealed by comparing the sequences of the RNAs with the nucleotide sequence of the genome. An overview of the transcription map is shown in Figure 14–26.

Three segments of the nucleotide sequence of the human mitochondrial genome are shown in Figure 14–27 along with the three mRNAs that are generated from those regions. The nucleotides that encode tRNA species are underlined; the amino acids encoded by the mRNAs are indicated below the center base of the codon.

A. In terms of codon usage and mRNA structure, in what two ways does initiation of protein synthesis in mitochondria differ from initiation in the cytoplasm?

```
tRNA^L                                         tRNA^I
TTCTTAACAACATACCCAT........CTCAAACCTAAGAAATATG      DNA
       ACAUACCCAU........CUCAAACCUAAAAAAAAAA          mRNA 13
       M  P                   E  T  *                protein


tRNA^D                                         tRNA^K
TATATCTTAATGGCACATG........CTCTAGAGCCCACTGTAAA       DNA
        AUGGCACAUG........CUCUAGAGCCAAAAAAAAA         mRNA 16
        M  A  H              S  *                    protein


tRNA^R                                         tRNA^H
ATTTACCAAATGCCCCTCA........TTTTCCTCTTGTAAATATA       DNA
        AUGCCCCUCA........UUUUCCUCUUAAAAAAAAA         mRNA 7
        M  P  L             F  S  S  *               protein
```

Figure 14–27 Arrangements of tRNA and mRNA sequences at three places on the human mitochondrial genome (Problem 14–83). *Underlined* sequences indicate tRNA genes. The sequences of the mRNAs are shown *below* the corresponding genes. The middle portions of the mRNAs and their genes are indicated by *dots*. The 5′ ends of the sequences are shown at the *left*. The 5′ ends of the mRNAs are unmodified and the 3′ ends have poly-A tails. The encoded protein sequences are indicated below the mRNAs, with the letter for the amino acid immediately under the center nucleotide of the codon. An asterisk (*) indicates a termination codon.

Figure 14–28 A variegated leaf of *Aucuba japonica* with green (*dark*) and yellow (*light*) patches (Problem 14–85).

B. In what two ways are the termination codons for protein synthesis in mitochondria unusual? (The termination codons are shown in Figure 14–27 as asterisks.)

C. Does the arrangement of tRNA and mRNA sequences in the genome suggest a possible mechanism for processing the primary transcript into individual RNA species?

14–84* (**True/False**) Mutations that are inherited according to Mendelian rules affect nuclear genes; mutations whose inheritance violates Mendelian rules are likely to affect organelle genes. Explain your answer.

14–85 Examine the variegated leaf shown in Figure 14–28. The dark areas are green and the light areas are white. White areas surrounded by green are common, but there are no green areas surrounded by white. Propose an explanation for this phenomenon.

14–86* A friend of yours has been studying a pair of mutants in the fungus *Neurospora*, which she has whimsically named *poky* and *puny*. Both mutants grow at about the same rate, but much more slowly than wild type. Your friend has been unable to find any supplement that improves their growth rates. Her biochemical analysis shows that each mutant displays a different abnormal pattern of cytochrome absorption. To characterize the mutants genetically, she crossed them to wild type and to each other and tested the growth rates of the progeny. She has come to you because she is puzzled by the results.

She explains that haploid nuclei from the two parents fuse during a *Neurospora* mating and then divide meiotically to produce four haploid spores, which can be readily tested for their growth rates. The parents contribute unequally to the diploid: one parent (the protoperithecial parent) donates a nucleus and the cytoplasm; the other (the fertilizing parent) contributes little more than a nucleus—much like egg and sperm in higher organisms. As shown in Table 14–5, the 'order' of the crosses sometimes makes a difference: a result she has not seen before.

Can you help your friend understand these results?

TABLE 14–6 Genetic analysis of *Neurospora* mutants (Problem 14–86).

| | | | SPORE COUNTS | |
CROSS	PROTOPERITHECIAL PARENT	FERTILIZING PARENT	FAST GROWTH	SLOW GROWTH
1	*poky*	wild	0	1749
2	wild	*poky*	1334	0
3	*puny*	wild	850	799
4	wild	*puny*	793	801
5	*poky*	*puny*	0	1831
6	*puny*	*poky*	754	710
7	wild	wild	1515	0
8	*poky*	*poky*	0	1389
9	*puny*	*puny*	0	1588

14–87 The pedigrees in Figure 14–29 show one example each of the following types of mutation: mitochondrial mutation, autosomal recessive mutation, autosomal dominant mutation, and X-linked recessive mutation. In each family the parents have had nine children. Assign each pedigree to one type of mutation. Explain the basis for your assignments.

14–88* (True/False) Mitochondria, which divide by fission, can apparently grow and divide indefinitely in the cytoplasm of proliferating eucaryotic cells even in the complete absence of a mitochondrial genome. Explain your answer.

14–89 Mutants of yeast that are defective in mitochondrial function grow on fermentable substrates such as glucose, but they fail to grow on nonfermentable substrates such as glycerol. Genetic crosses can distinguish nuclear and mitochondrial mutations. Nearly 200 different nuclear genes have been defined by complementation analysis of nuclear petite (*pet*) mutants. Surprisingly, mutations in about a quarter of these genes affect the expression of single, or restricted sets of, mitochondrial gene products without blocking overall gene expression.

For example, *pet494* mutants contain normal levels of all the known mitochondrial gene products except subunit III of cytochrome oxidase (coxIII). However, *pet494* mutants contain normal levels of coxIII mRNA, which appears normal in size. Thus, some posttranscriptional step in expression of the gene for coxIII appears to be regulated by the normal *PET494* gene product. It could be required to promote translation of coxIII mRNA or to stabilize coxIII during assembly of the cytochrome oxidase complex. To distinguish between these possibilities, mitochondrial mutations that suppress the respiratory defect of *pet494* mutations were selected and analyzed. All these mutations were deletions that fused the coxIII coding region to the 5′ untranslated region of another gene, as illustrated in Figure 14–30.

A. How do these results distinguish between a requirement for translation of coxIII mRNA and a requirement for stabilization of the coxIII protein?

B. Each of the mitochondrial deletions eliminates one or more genes that are essential for mitochondrial function, yet these yeast strains contain all the normal mitochondrial mRNAs and make normal colonies when grown on nonfermentable substrates. How can this be?

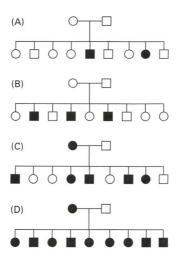

Figure 14–29 Hypothetical pedigrees representing four patterns of inheritance (Problem 14–87). Males are shown as *squares*; females as *circles*. Affected individual are shown as *filled symbols*; unaffected individual are shown as *empty symbols*.

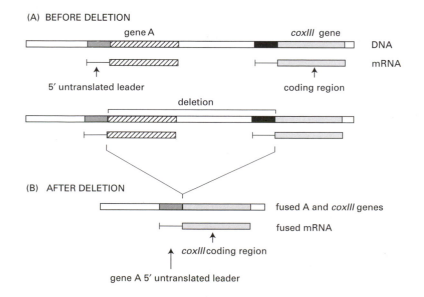

Figure 14–30 Schematic representation of the mitochondrial deletion mutations that suppress the nuclear mutation *pet494* (Problem 14–89).

The semaphore alphabet as seen by the recipient. You may like to consider the similarities and differences between this human mode of communication and the signaling networks used by cells.

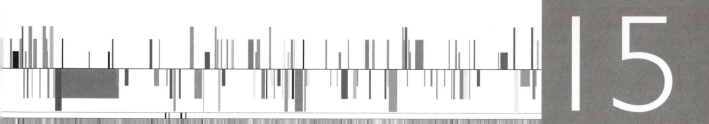

CELL COMMUNICATION

GENERAL PRINCIPLES OF CELL COMMUNICATION

TERMS TO LEARN

adaptation	intracellular signaling protein	phosphorylation cascade
autocrine signaling	ion-channel-linked receptor	receptor
binding domain	lipid raft	scaffold protein
contact-dependent signaling	local mediator	second messenger
desensitization	molecular switch	signal molecule
endocrine cell	neurotransmitter	signaling cascade
gap junction	nitric oxide (NO)	small intracellular mediator
GTP-binding protein	nuclear receptor superfamily	synaptic signaling
hormone	paracrine signaling	

15–1 (**True/False**) There is no fundamental distinction between signaling molecules that bind to cell-surface receptors and those that bind to intracellular receptors. Explain your answer.

15–2* Suppose that the circulating concentration of hormone is 10^{-10} M and the K_d for binding to its receptor is 10^{-8} M. What fraction of the receptors will have hormone bound? If a meaningful physiological response occurs when 50% of the receptors have bound a hormone molecule, how much will the concentration of hormone have to rise to elicit a response? Recall that the fraction of receptors (R) bound to hormone (H) to form a receptor–hormone complex (R–H) is $[R–H]/([R] + [R–H]) = [R–H]/[R]_{TOT} = [H]/([H] + K_d)$ (see Problem 3–49).

15–3 There are roughly 10 billion protein molecules per cell, but cell-surface receptors are typically present at only 1000 to 100,000 copies per cell. Their low abundance makes them extremely difficult to purify by normal biochemical methods. One common approach is to clone the gene for the receptor and then express the gene to make the receptor. Cloning is usually done using a pooling strategy. A cDNA library is made from a cell that expresses the receptor. The library is divided into pools of 1000 plasmid clones and each pool is transfected into mammalian cells that normally do not express the receptor. Cells that have received plasmids survive and form colonies on a petri dish. A solution containing a radioactive ligand specific for the receptor is then added to the dishes, incubated for a time and then washed off. Cells are fixed in ice-cold methanol and the dishes are then

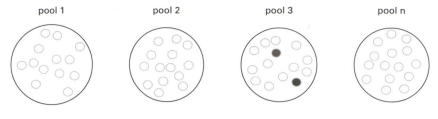

pool 1　　pool 2　　pool 3　　pool n

Figure 15–1 Autoradiograph of colonies transfected with pools from a cDNA library (Problem 15–3). In such an autoradiograph unlabeled colonies would not be visible; they have been included here for completeness.

subjected to autoradiography to identify colonies that have bound the ligand (Figure 15–1).

A. Which pool contains a cDNA for the receptor?

B. Why don't all the colonies on the petri dish transfected with the positive pool have bound radioactive ligand?

C. How would you complete the cloning of the receptor gene so that you have a pure plasmid containing the receptor cDNA?

15–4*　Surgeons use succinylcholine, which is an acetylcholine analog, as a muscle relaxant. Care must be taken because some individuals recover abnormally slowly from this paralysis, with life-threatening consequences. Such individuals are deficient in an enzyme called pseudocholinesterase, which is normally present in the blood.

　　　　If succinylcholine is an analog of acetylcholine, why do you think it causes muscles to relax and not contract as acetylcholine does?

15–5　The cellular slime mold *Dictyostelium discoideum* is a eucaryote that lives on the forest floor as independent motile cells called amoebae, which feed on bacteria and yeast. When their food supply is exhausted, the amoebae stop dividing and gather together to form tiny, multicellular, wormlike structures, which crawl about as glistening slugs and leave trails of slime behind them. How do individual amoebae know when to stop dividing and how to find their way into a common aggregate? A set of classic experiments investigated this phenomenon more than half a century ago.

　　　　It was shown that amoebae aggregate if placed on a glass coverslip under water, provided that simple salts are present. The center of the aggregation pattern can be removed with a pipette and placed in a field of fresh amoebae, which immediately start streaming toward it. Thus, the center is emitting some sort of attractive signal. Four experiments were designed to determine the nature of the signal by using an existing center of aggregation as the source of the signal and previously unexposed amoebae as the target cells. The arrangements of aggregation centers and test amoebae at the beginning and end of the experiments are shown in Figure 15–2.

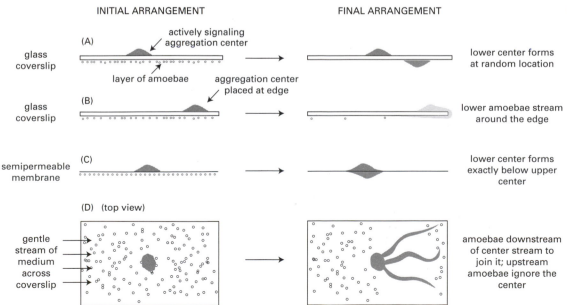

Figure 15–2 Four experiments to study the nature of the attractive signal generated by aggregation centers (Problem 15–5).

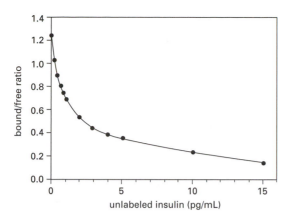

Figure 15–3 Calibration curve for radioimmunoassay of insulin (Problem 15–6).

Do these results show that *Dictyostelium discoideum* aggregates through the action of a secreted chemical signal? Explain your reasoning.

15–6* Radioimmunoassay (RIA) is a powerful tool for quantifying virtually any substance of biological interest because it is sensitive, accurate, and fast. RIA technology arose from studies on adult onset diabetes. Some patients had antibodies with high affinity for insulin and RIA was developed as a method to distinguish free insulin from antibody-bound insulin.

How can high-affinity antibodies be exploited to measure low concentrations of insulin? When a small amount of insulin-specific antiserum is mixed with an equally small amount of very highly radioactive insulin, some binds and some remains free according to the equilibrium.

$$\text{Insulin (I)} + \text{Antibody (A)} \rightleftharpoons \text{Insulin–Antibody (I–A) Complex}$$

$$K = \frac{[\text{I–A}]}{[\text{I}][\text{A}]}$$

When increasing amounts of unlabeled insulin are added to a fixed amount of labeled insulin and anti-insulin antibody, the ratio of bound to free radioactive insulin decreases as expected from the equilibrium expression. If the concentration of the unlabeled insulin is known, then the resulting curve serves as a calibration against which other unknown samples can be compared (Figure 15–3).

You have three samples of insulin whose concentrations are unknown. When mixed with the same amount of radioactive insulin and anti-insulin antibody used in Figure 15–3, the three samples gave the following ratios of bound to free insulin:

Sample 1	0.67
Sample 2	0.31
Sample 3	0.46

A. What is the concentration of insulin in each of these unknown samples?
B. What portion of the standard curve is the most accurate, and why?
C. If the antibodies were raised against pig insulin, which is similar but not identical to human insulin, would the assay still be valid for measuring human insulin concentrations?

15–7 The optimal sensitivity of radioimmunoassay (RIA) occurs when the concentrations of the unknown and the radioactive tracer are equal. Thus the sensitivity of RIA is limited by the specific activity of the radioactive ligand. The more highly radioactive the ligand, the smaller the amount needed per assay and, therefore, the smaller amount of unknown that can be detected.

You wish to measure an unknown sample of insulin (molecular weight, 11,466) using an insulin tracer labeled with radioactive iodine, which has a half-life of 7 days. Assume that you can attach one atom of radioactive iodine per molecule of insulin, that iodine does not interfere with antibody binding, that each radioactive disintegration has a 50% probability of being detected as a 'count,' and that the limit of detection in the RIA requires a total of at least 1000 counts per minute (cpm) distributed between the bound and free fractions.

A. Given these parameters, calculate how many picograms of radioactive insulin will be required for an assay.

B. At optimal sensitivity, how many picograms of unlabeled insulin will you be able to detect?

15–8* To remain a local response, paracrine signal molecules must be prevented from straying too far from their point of origin. Suggest different ways by which this could be accomplished.

15–9 (**True/False**) The receptors involved in paracrine, synaptic, and endocrine signaling all have very high affinity for their respective signaling molecules. Explain your answer.

15–10* Compare and contrast signaling by neurons to signaling by endocrine cells. What are the relative advantages of these two mechanisms for cellular communication?

15–11 Cells communicate in ways that are analogous to human communication. Decide which of the following forms of human communication are analogous to autocrine, paracrine, endocrine, and synaptic signaling by cells.

A. A telephone conversation
B. Talking to people at a cocktail party
C. A radio announcement
D. Talking to yourself

15–12* Integrins are important transmembrane receptors that bind and respond to the extracellular matrix. K562 cells are a line of mouse erythroleukemia cells that are capable of differentiating into red blood cells in response to certain stimuli. These cells express a single type of integrin, $\alpha_5\beta_1$, on their surface, which endows the cells with the ability to bind the RGD motifs of the extracellular matrix protein fibronectin.

Most high-affinity monoclonal antibodies against the separated α_5 and β_1 chains reduce the amount of fibronectin that binds to the cells; however, one rare antibody specific for the β_1 chain gives a 20-fold increase in binding (Table 15–1). Experiments using mixtures of anti-integrin antibodies, anti-fibronectin antibodies, and peptides with and without the RGD motif confirmed that the observed binding was due to interactions of fibronectin with the integrin (Table 15–1). In the presence of the anti-β_1 antibody the integrin has an increased affinity for fibronectin (Figure 15–4). Any intracellular influences on the stimulated binding in the presence of the anti-β_1 antibody were ruled out by showing that the same results were obtained in the presence of various metabolic poisons such as azide.

TABLE 15–1 Effects of antibodies on the binding of fibronectin to cells (Problem 15–12).

ADDITIONS	FIBRONECTIN BOUND (cpm)
None	2,000
Anti-α_5 antibody	0
Anti-α_4 antibody	2,000
Anti-β_1 antibody	40,000
Anti-β_1 antibody + anti-α_5 antibody	0
Anti-β_1 antibody + anti-α_4 antibody	40,000
Anti-β_1 antibody + anti-fibronectin antibody	500
Anti-β_1 antibody + GRGDSP peptide	3,000
Anti-β_1 antibody + GRGESP peptide	40,000

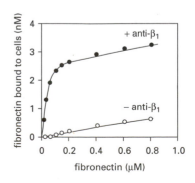

Figure 15–4 Detailed studies of fibronectin binding to K562 cells (Problem 15–12).

TABLE 15–2 Fibrinogen-dependent aggregation of CHO cells expressing various wild-type and mutant α_{IIb} and β_3 subunits (Problem 15–13).

α_{IIb} CHAIN	β_3 CHAIN	AGGREGATION	
		NO ANTIBODY	WITH MAb 62
Normal	normal	–	+++
Truncated	normal	+++	+++
Normal	truncated	–	+++
Truncated	truncated	+++	+++

A. How do the experiments in Table 15–1 confirm that anti-β_1 stimulates fibronectin binding due to its interaction with $\alpha_5\beta_1$ integrin?

B. Which experiments rule out the possibility that the apparent increase in binding affinity in Figure 15–4 is really due to increased numbers of integrin molecules on the cell surface: either newly synthesized or newly transferred from some internal compartment?

C. How do you suppose this particular anti-β_1 antibody increases the affinity of the $\alpha_5\beta_1$ integrin for fibronectin?

15–13 The ability of a cell to control integrin–ligand interactions from within is termed inside-out signaling. The major surface protein of blood platelets, $\alpha_{IIb}\beta_3$ integrin, binds to fibrinogen when platelets are stimulated with clotting factors such as thrombin. By binding to a receptor on the cell surface, thrombin triggers an intracellular signaling pathway that activates $\alpha_{IIb}\beta_3$ integrin, allowing platelets to aggregate to form blood clots. Platelets do not bind fibrinogen or aggregate until stimulated, although $\alpha_{IIb}\beta_3$ integrin is always present on their surface. What regulates the activity of this all-important integrin?

If the genes for the subunits of $\alpha_{IIb}\beta_3$ are expressed in Chinese hamster ovary (CHO) cells, the cells fail to aggregate when incubated with fibrinogen in the presence or absence of thrombin. When the cells are first incubated with MAb 62 antibodies, which bind to $\alpha_{IIb}\beta_3$ integrin and activate it (analogous to the anti-β_1 antibody described in Problem 15–12), the cells aggregate within minutes of adding fibrinogen. CHO cells without $\alpha_{IIb}\beta_3$ do not aggregate when treated this way.

By deleting the short cytoplasmic domains of α_{IIb} and β_3, various combinations of truncated and wild-type α_{IIb} and β_3 can be tested in CHO cells. All combinations of the α_{IIb} and β_3 chains allow cells to aggregate in the presence of fibrinogen and MAb 62; however, the truncated α_{IIb} chain allows aggregation even in the absence of MAb 62 (Table 15–2).

A. Why do you suppose that truncating the cytoplasmic domain of the α_{IIb} subunit increases the affinity of the integrin for fibrinogen and allows the cells to aggregate?

B. The $\alpha_{IIb}\beta_3$ integrin is accessible on the surface of the CHO cells, as revealed by the various aggregation studies. Why then does thrombin not stimulate the cells to aggregate?

C. There are two genes for α_{IIb} in diploid human cells. If one of the two genes suffered a mutation of the kind described in this problem, do you think the individual would show any blood-clotting problems?

15–14* Why do signaling responses that require changes in proteins already present in the cell occur in milliseconds to seconds, whereas responses that involve changes in gene expression usually require hours?

15–15 Cells in a developing embryo make and break gap-junction connections in specific and interesting patterns, suggesting that gap junctions play an important role in the signaling processes that occur between these cells. At the eight-cell stage mouse embryos undergo a process known as compaction, as illustrated in Figure 15–5. Although the mechanism is not clear,

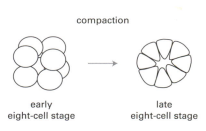

compaction

early
eight-cell stage

late
eight-cell stage

Figure 15–5 Compaction of the eight-cell mouse embryo (Problem 15–15).

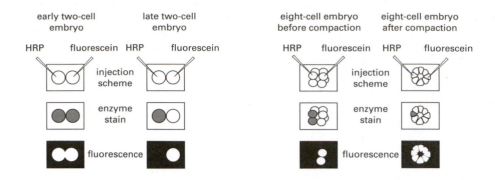

early two-cell embryo | late two-cell embryo

eight-cell embryo before compaction | eight-cell embryo after compaction

Figure 15–6 Microinjection of HRP and fluorescein in two-cell and eight-cell mouse embryos (Problem 15–15).

the cells appear to adhere to one another more strongly, changing from a clump of loosely associated cells to a tightly sealed ball. You wish to know whether gap junctions are present before or after this change in adhesion.

Using very fine glass micropipettes, you can measure electrical events and at the same time microinject the enzyme horseradish peroxidase (HRP), 40,000 daltons, or the fluorescent dye fluorescein, 330 daltons. Fluorescein glows bright green under UV illumination, and HRP can be detected by fixing the cells and incubating them with appropriate substrates.

You inject embryos at various stages of development. At both the two-cell and eight-cell stages, different results are obtained, depending on whether the injections are made immediately after cell division or later (Figure 15–6). Immediately after cell division cytoplasmic bridges linger for a while before cytokinesis is completed.

A. Why do both HRP and fluorescein enter neighboring cells early, but not late, at the two-cell stage?

B. At what stage of embryo development do gap junctions form? Explain your reasoning.

C. In which of the four stages of development diagrammed in Figure 15–6 would you detect electrical coupling if you injected current from the HRP injection electrode and recorded voltage changes in the fluorescein electrode?

15–16* It seems counterintuitive that a cell with an abundant supply of nutrients would commit suicide if not constantly stimulated by signals from other cells. What do you suppose might be the purpose of such regulation?

15–17 How is it that different cells can respond in different ways to exactly the same signaling molecule even when they have identical receptors?

15–18* Two intracellular molecules, A and B, are normally synthesized at a constant rate of 1000 molecules per second per cell. Each molecule of A survives an average of 100 seconds, while each molecule of B survives an average of 10 seconds.

A. How many molecules of A and B will a cell contain?

B. If the rates of synthesis of both A and B were suddenly increased 10-fold to 10,000 molecules per second—without any change in their average lifetime—how many molecules of A and B would there be after one second?

C. Which molecule would be preferred for rapid signaling? Explain your answer.

15–19 A rise in the cyclic GMP levels in smooth muscle cells causes relaxation of the blood vessels in the penis, resulting in an erection. Explain how the natural signal molecule NO and the drug Viagra produce an increase in cyclic GMP.

15–20* Nuclear receptors carry a binding site for a signal molecule and for a DNA sequence. How is it that identical nuclear receptors in different cells can activate different genes when they bind the same signal molecule?

15–21 The steroid hormones, cortisol, estradiol, and testosterone, are all derived from cholesterol by modifications that introduce polar groups such as –OH and =O (Figure 15–7). If cholesterol itself were not normally found in cell

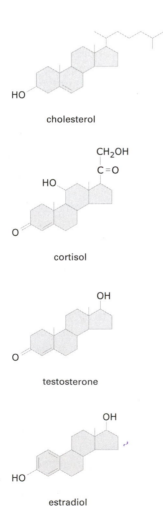

cholesterol

cortisol

testosterone

estradiol

Figure 15–7 Steroid hormones and their parent molecule, cholesterol (Problem 15–21).

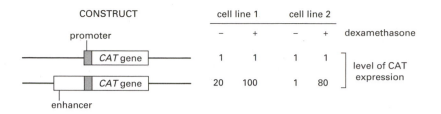

Figure 15–8 Transfection of constructs into two different cell lines (Problem 15–22). Presence (+) and absence (–) of the hormone (dexamethasone) are indicated. *Numbers* indicate expression of the *CAT* gene product; in all cases, the values are expressed relative to a construct without the viral enhancer.

membranes, do you suppose it could be used effectively as a hormone, provided that an appropriate intracellular receptor were available?

15–22* A segment of the Moloney murine sarcoma virus carries an enhancer of gene expression, which confers glucocorticoid responsiveness on genes that are linked to it. Figure 15–8 shows the activity of a reporter gene (chloramphenicol acetyl transferase, *CAT*) after transfection into two cell lines in the presence and absence of dexamethasone, a glucocorticoid. You are puzzled by the results with cell line 1 because the viral enhancer increased CAT expression 20-fold in the absence of dexamethasone.

A. Do both cell lines contain glucocorticoid receptors? How can you tell?

B. Propose an explanation for the difference in the CAT activity in cell lines 1 and 2 after transfection with the construct containing the viral enhancer in the absence of dexamethasone.

C. Based on your explanation, predict the outcome of an experiment in which a variety of shorter pieces of the viral enhancer are placed in front of the *CAT* gene and tested for CAT activity in these two cell lines.

15–23 Glucocorticoids induce transcription of mouse mammary tumor virus (MMTV) genes. Using hybrid constructs containing the MMTV long terminal repeat (LTR) linked to an easily assayable gene, you have shown that the LTR contains regulatory elements that respond to glucocorticoids. To map the glucocorticoid response elements within the LTR, you delete different portions of the LTR (Figure 15–9) and measure the binding of glucocorticoid receptors to the remaining DNA segments. To measure binding, you cut each of the mutant DNAs into fragments, purify the fragments containing LTR sequences, label the ends with ^{32}P, and then incubate a mixture of the labeled fragments with the purified glucocorticoid receptor. You assess binding by passing the incubation mixture through a nitrocellulose filter, which binds protein (and any attached DNA) but not free DNA. You display the DNA fragments that were bound to the receptor by agarose gel electrophoresis. The electrophoretic patterns of the starting mixture of fragments and the bound fragments are shown in Figure 15–9.

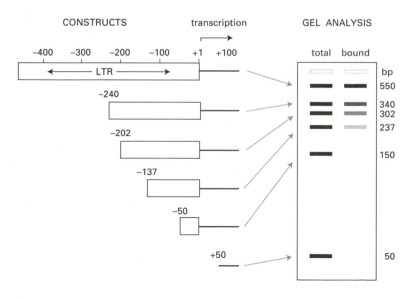

Figure 15–9 Diagram of the LTR deletions and the electrophoretic patterns of the starting and bound mixtures of DNA fragments (Problem 15–23). Nucleotides are numbered relative to the start site for transcription, which is at +1. Negative numbers are in front of the start site, and positive numbers are after it. The electrophoretic pattern is an autoradiograph of the agarose gel.

```
1.    CCAAGGAGGGGACAGTGGCTGGACTAATAG
2.    GGACTAATAGAACATTATTCTCCAAAAACT
3.    TCGTTTTAAGAACAGTTTGTAACCAAAAAC
4.    AGGATGTGAGACAAGTGGTTTCCTGACTTG
5.    AGGAAAATAGAACACTCAGAGCTCAGATCA
6.    CAGAGCTCAGATCAGAACCTTTGATACCAA
7.    CATGATTCAGCACAAAAAGAGCGTGTGCCA
8.    CTGTTATTAGGACATCGTCCTTTCCAGGAC
9.    CCTAGTGTAGATCAGTCAGATCAGATTAAA
10.   GATCAGTCAGATCAGATTAAAAGCAAAAAG
11.   TTCCAAATAGATCCTTTTTGCTTTTAATCT
```

Figure 15–10 DNA sequences that are bound by the glucocorticoid receptor (Problem 15–23). The DNA sequences have been aligned by their consensus binding sequences.

A. Where in the LTR are the glucocorticoid response elements located?

B. A list of 11 MMTV sequences (several from the LTR) that are bound by the glucocorticoid receptor are shown in Figure 15–10. For ease of analysis the sequences have already been aligned via the consensus sequence to which the receptor binds. Identify the consensus sequence for glucocorticoid receptor binding. (A consensus sequence is the average or most typical form of a sequence that is reproduced with minor variations. It shows the nucleotide most often found at each position.)

15–24* Studies with the fruit fly *Drosophila* provided initial clues to the complex changes in patterns of gene expression that a simple hormone can trigger. *Drosophila* larvae molt in response to an increase in the concentration of the steroid hormone ecdysone. The polytene chromosomes of the *Drosophila* salivary glands are an excellent experimental system in which to study the pattern of gene activity initiated by the hormone because active genes enlarge into puffs that are visible in the light microscope. Furthermore, the size of a puff is proportional to the rate at which it is transcribed. Prior to addition of ecdysone, a few puffs—termed intermolt puffs—are already active. Upon exposure of dissected salivary glands to ecdysone, these intermolt puffs regress, and two additional sets of puffs appear. The early puffs arise within a few minutes after addition of ecdysone; the late puffs arise within 4–10 hours. The concentration of ecdysone does not change during this time period. The pattern of puff appearance and disappearance is illustrated for a typical puff in each category in Figure 15–11A.

Two critical experiments helped to define the relationships between the different classes of puff. In the first, cycloheximide, which blocks protein synthesis, was added at the same time as ecdysone. As illustrated in Figure 15–11B, under these conditions the early puffs did not regress and the late puffs were not induced. In the second experiment, ecdysone was washed out after a 2-hour exposure. As illustrated in Figure 15–11C, this treatment caused an immediate regression of the early puffs and a *premature induction* of the late puffs.

A. Why do you think the early puffs didn't regress and the late puffs weren't induced in the presence of cycloheximide? Why do you think the intermolt puffs were unaffected?

B. Why do you think the early puffs regressed immediately when ecdysone was removed? Why do you think the late puffs arose prematurely under these conditions?

C. Outline a model for ecdysone-mediated regulation of the puffing pattern.

15–25 The signaling mechanisms used by a steroid-hormone receptor and by an ion-channel-linked receptor have very few components. Can either mechanism lead to an amplification of the initial signal? If so, how?

15–26* (**True/False**) All small intracellular mediators (second messengers) are water soluble and diffuse freely through the cytosol. Explain your answer.

15–27 Ordering the action of the individual components is an essential step in defining a signaling pathway. Imagine that two protein kinases, PK1 and PK2, act sequentially in an intracellular signaling pathway. If either kinase

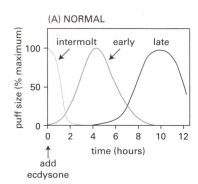

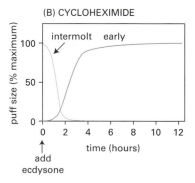

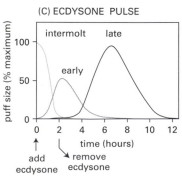

Figure 15–11 Puffing patterns in salivary gland giant chromosomes (Problem 15–24). (A) Normal puffing pattern. (B) Puffing pattern in the presence of cycloheximide. (C) Puffing pattern after removal of ecdysone.

contains a mutation that permanently inactivates its function, no response is seen in cells when an extracellular signal is received. A different mutation in PK1 makes it permanently active, so that in cells containing that mutation a response is observed even in the absence of an extracellular signal. In a doubly mutant cell that contains inactivated PK2 and permanently active PK1, you observe the response in the absence of a signal.

In the normal signaling pathway, does PK1 activate PK2 or does PK2 activate PK1? What would you have predicted as the outcome in a doubly mutant cell line with an activating mutation in PK2 and an inactivating mutation in PK1? Explain your reasoning.

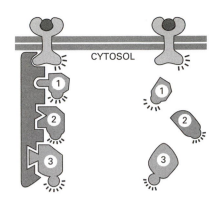

Figure 15–12 A protein kinase cascade organized by a scaffolding protein or composed of freely diffusing components (Problem 15–31).

15–28* Proteins play a variety of roles in signaling pathways. Match the following list of signaling proteins with their functions.

A. Adaptor Proteins	1. Bind multiple signaling proteins together in a functional complex
B. Amplifier Proteins	2. Carry the signal from one part of the cell to another
C. Anchoring Proteins	3. Combine signals from two or more pathways before passing it onward
D. Bifurcation Proteins	4. Convert the signal to a different form
E. Integrator Proteins	5. Greatly increase the signal they receive
F. Latent Gene Regulatory Proteins	6. Link one signaling protein to another, without themselves conveying a signal
G. Messenger Proteins	7. Maintain specific signaling proteins at a precise location in the cell
H. Modulator Proteins	8. Migrate to the nucleus and stimulate transcription after activation in the cytosol
I. Relay Proteins	9. Modify the activity of signaling proteins to regulate signal strength
J. Scaffold Proteins	10. Pass the message to the next signaling component in the pathway
K. Transducer Proteins	11. Spread the signal from one signaling pathway to another

15–29 Why do you suppose that phosphorylation/dephosphorylation, as opposed to allosteric binding of small molecules, for example, plays such a prominent role in switching proteins on and off in signaling pathways?

15–30* The two main classes of molecular switches involve changes in phosphorylation state or changes in guanine nucleotide binding. Comment on the following statement. "In the regulation of molecular switches, protein kinases and guanine nucleotide exchange factors (GEFs) turn proteins on and protein phosphatases and GTPase activating proteins (GAPs) turn proteins off."

Problem 3–84 asks how molecular on/off switches can give rise to intermediate responses

15–31 Consider a signaling pathway that proceeds through three protein kinases that are sequentially activated by phosphorylation. In one case the kinases are held in a signaling complex by a scaffolding protein; in the other, the kinases are freely diffusing (Figure 15–12). Discuss the properties of these two types of organization in terms of signal amplification, speed, and potential for cross talk between signaling pathways.

15–32* Proteins in signaling pathways use a variety of binding domains to assemble into signaling complexes. Match the following domains with their binding targets. (A binding target can be used more than once.)

Problem 3–41 analyzes the binding of an SH2 domain to its target protein

A. PH domain   1. phosphorylated tyrosines
B. PTB domain   2. proline-rich sequences
C. SH2 domain   3. phosphorylated inositol phospholipids
D. SH3 domain

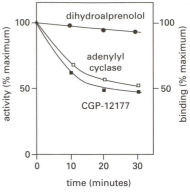

Figure 15–13 Adenylyl cyclase activity, ³H-CGP-12177 binding, and ³H-dihydroalprenolol binding at various times after treatment with isoproterenol (Problem 15–34). All activities are expressed as a percentage of the values at time zero.

15–33 Briefly describe three ways in which a gradual increase in an extracellular signal can be sharpened to produce an abrupt or nearly all-or-none cellular response.

15–34* After prolonged exposure to isoproterenol, which binds to β-adrenergic receptors and activates adenylyl cyclase, cells become refractory and cease responding. To investigate the basis for this desensitization phenomenon, you treat cells for various times with isoproterenol, wash it out, and then assay for β-adrenergic receptors by measuring the binding of dihydro-alprenolol, which is hydrophobic, and CGP-12177, which is hydrophilic. You find that CGP-12177 binding decreases in parallel to adenylyl cyclase activity, whereas dihydroalprenolol binding remains high (Figure 15–13). When you lyse isoproterenol-treated (desensitized) cells and fractionate membrane-enclosed vesicles by centrifugation through sucrose-density gradients, you find that CGP-12177 binds to vesicles derived from the plasma membrane (as indicated by the presence of the marker enzyme, 5′-nucleotidase), whereas dihydroalprenolol binds to an additional population of vesicles (Figure 15–14).

A. Give an explanation for the differences in binding by dihydroalprenolol and CGP-12177.

B. What do you think might be the basis for isoproterenol-induced desensitization in these cells?

15–35 The nicotinic acetylcholine receptor is a neurotransmitter-dependent ion channel, which is composed of four types of subunit. Phosphorylation of the receptor by protein kinase A attaches one phosphate to the γ subunit and one phosphate to the δ subunit. Fully phosphorylated receptors desensitize much more rapidly than unmodified receptors. To study this process in detail, you phosphorylate two preparations of receptor to different extents (0.8 mole phosphate/mole receptor and 1.2 mole phosphate/mole receptor) and measure desensitization over several seconds (Figure 15–15). Both preparations behave as if they contain a mixture of receptors; one form that is rapidly desensitized (the initial steep portion of the curves) and another form that is desensitized at the same rate as the untreated receptor.

A. Assuming that the γ and δ subunits are independently phosphorylated at equal rates, calculate the percentage of receptors that carry zero, one, and two phosphates per receptor at the two extents of phosphorylation.

B. Do these data suggest that desensitization requires one phosphate or two phosphates per receptor? If you decide that desensitization requires only one phosphate, indicate whether the phosphate has to be on one specific subunit or can be on either of the subunits.

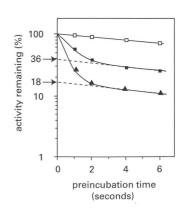

Figure 15–15 Desensitization rates of untreated acetylcholine receptor and two preparations of phosphorylated receptor (Problem 15–35). *Open squares* represent untreated receptors; *filled squares* represent receptors with 0.8 mole phosphate/mole receptor; and *filled triangles* represent receptors with 1.2 mole phosphate/mole receptor. *Arrows* indicate the fractions of the phosphorylated preparations that behaved like the untreated receptor.

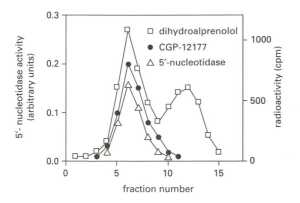

Figure 15–14 Ligand binding in sucrose-gradient fractions from isoproterenol-treated cells (Problem 15–34).

SIGNALING THROUGH G-PROTEIN-LINKED CELL-SURFACE RECEPTORS

TERMS TO LEARN

adenylyl cyclase	cyclic GMP phosphodiesterase	protein kinase A (PKA)
arrestin	diacylglycerol	protein kinase C (PKC)
calmodulin	G_q	protein phosphatase
CaM-kinase	G-protein-linked receptor	RGS protein
CaM-kinase II	inhibitory G protein (G_i)	rhodopsin
CRE-binding (CREB) protein	IP_3	rod photoreceptor (rod)
cyclic AMP (cAMP)	olfactory receptors	stimulatory G protein (G_s)
cyclic AMP phosphodiesterase	phospholipase C-β	trimeric G protein
cyclic GMP	PI(4,5)P$_2$	

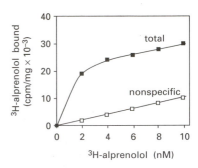

Figure 15–16 Binding of ^{3}H-alprenolol to frog erythrocyte membranes (Problem 15–36).

15–36* In a classic paper the number of β-adrenergic receptors on the membranes of frog erythrocytes was determined using a competitive inhibitor of adrenaline, alprenolol, which binds to the receptors 500 times more tightly than adrenaline. These receptors normally bind adrenaline and stimulate adenylyl cyclase activity. Labeled alprenolol was mixed with erythrocyte membranes, left for 10 minutes at 37°C, and then the membranes were pelleted by centrifugation and the radioactivity in the pellet was measured. The experiment was done in two ways. The binding of increasing amounts of ^{3}H-alprenolol to a fixed amount of erythrocyte membranes was measured to determine total binding. The experiment was repeated in the presence of a vast excess of unlabeled alprenolol to measure nonspecific binding. The results are shown in Figure 15–16.

A. On Figure 15–16 sketch in the curve for specific binding of alprenolol to β-adrenergic receptors. Has alprenolol binding to the receptors reached saturation?

B. Assuming that one molecule of alprenolol binds per receptor, calculate the number of β-adrenergic receptors on the membrane of a frog erythrocyte. The specific activity of the labeled alprenolol is 1×10^{13} cpm/mmol, and there are 8×10^8 frog erythrocytes per milligram of membrane protein.

15–37 G-protein-linked receptors activate G proteins by reducing the strength of GDP binding, allowing GDP to dissociate and GTP, which is present at much higher concentrations, to bind. How do you suppose the activity of a G protein would be affected by a mutation that caused its affinity for GDP to be reduced without significantly changing its affinity for GTP?

15–38* When adrenaline binds to adrenergic receptors on the surface of a muscle cell, it activates a G protein, initiating a signaling pathway that results in breakdown of muscle glycogen. How would you expect glycogen breakdown to be affected if muscle cells were injected with a nonhydrolyzable analog of GTP, which can't be converted to GDP? Consider what would happen in the absence of adrenaline and after a brief exposure to it.

15–39 Should RGS (regulator of G protein signaling) proteins be classified as GEFs (guanine nucleotide exchange factors) or GAPs (GTPase activating proteins)? Explain what role this activity plays in modulating G-protein-mediated responses in animals and yeasts.

15–40* The mating behavior of yeast depends on signaling peptides termed pheromones that bind to G-protein-linked pheromone receptors (Figure 15–17). When the α-factor pheromone binds to a wild-type yeast cell, it blocks cell-cycle progression, arresting proliferation until a mating partner is found. Yeast mutants with defects in one or more of the components of the G protein have characteristic phenotypes in the absence and in the presence of the α-factor pheromone (Table 15–3). Strains with defects in any of these genes cannot undergo the mating response and are therefore termed sterile.

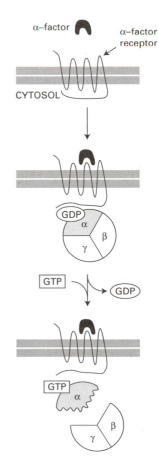

Figure 15–17 G-protein-linked α-factor pheromone receptor (Problem 15–40).

TABLE 15-3 Mating phenotypes of various mutant and nonmutant strains of yeast (Problem 15–40).

| | PHENOTYPE | |
MUTATION	MINUS α FACTOR	PLUS α FACTOR
None (wild type)	normal proliferation	arrested proliferation, mating response
α subunit deleted	normal proliferation	arrested proliferation, sterile
β subunit deleted	normal proliferation	normal proliferation, sterile
γ subunit deleted	normal proliferation	normal proliferation, sterile
α and β deleted	normal proliferation	normal proliferation, sterile
α and γ deleted	normal proliferation	normal proliferation, sterile
β and γ deleted	normal proliferation	normal proliferation, sterile

A. Based on genetic analysis of the yeast mutants, decide which component of the G protein normally transmits the mating signal to the downstream effector molecules.

B. Predict the proliferation and mating phenotypes in the absence and presence of the α-factor pheromone of strains with the following mutant α subunits:
1. An α subunit that can bind GTP but cannot hydrolyze it.
2. An α subunit with an altered N-terminus to which the fatty acid myristoylate cannot be added, thereby preventing its localization to the plasma membrane.
3. An α subunit that cannot bind to the activated pheromone receptor.

15–41 What is 'cyclic' about cyclic AMP?

15–42* Explain why cyclic AMP must be broken down rapidly in a cell to allow rapid signaling.

15–43 You are trying to purify adenylyl cyclase from brain. The assay is based on the conversion of α-^{32}P-ATP to cAMP. You can easily detect activity in crude brain homogenates stimulated by isoproteronol, which binds to β-adrenergic receptors, but the enzyme loses activity when low molecular weight cofactors are removed by dialysis. What single molecule do you think you could add back to the dialyzed homogenate to restore activity?

15–44* You are baffled. You are studying the control of cyclic AMP levels in brain slices. You have confirmed that signal molecules such as isoproteronol that act through β-adrenergic receptors cause a modest increase in cyclic AMP, as expected from G-protein-mediated coupling between the receptor and adenylyl cyclase. You find a puzzling synergy, however, between isoproteronol and a number of pharmacological agents that by themselves have no effect on cyclic AMP levels. What is the basis for this paradoxical augmentation of cyclic AMP levels?

A biochemist friend of yours has suggested a possible explanation: she has found in *in vitro* experiments that βγ subunits from inhibitory trimeric G proteins stimulate type II adenylyl cyclase, which is expressed in brain. To test this idea in cells, you plan to express the cDNAs encoding the component proteins in human kidney cells, which lack the receptors found in brain. In this way you hope to reconstruct the effects you observed in brain slices, but in a much simpler background.

You transfect the kidney cells with various combinations of cDNAs encoding type II adenylyl cyclase, the dopamine receptor (which interacts with an inhibitory G protein), and a mutated (constitutively active) α_s* subunit. You measure the levels of cyclic AMP in the resulting cell lines in the absence or presence of quinpirole, which activates the dopamine receptor (Figure 15–18). You also measure the effects of pertussis toxin, which blocks the signal from G_i-coupled receptors by modifying the α_i subunit in such a way that it can no longer bind GTP or dissociate from its βγ subunit (Figure 15–18).

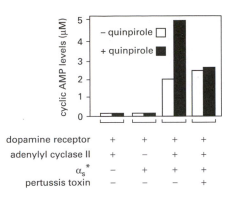

Figure 15–18 Measurements of cyclic AMP levels in cells transfected with cDNAs for the dopamine receptor, type II adenylyl cyclase, and α_s* (Problem 15–44).

A. Did you succeed in reproducing the paradoxical result you observed in brain slices in the transfected kidney cells? How so?

B. Explain the effects of pertussis toxin in your experiments.

C. What do your experiments indicate is required for maximal activation of type II adenylyl cyclase? Propose a molecular explanation for the augmented activation of type II adenylyl cyclase.

D. Predict the results of expressing the cDNA for the α subunit of transducin, which does not bind to adenylyl cyclase but binds tightly to free $\beta\gamma$ subunits.

15–45 (**True/False**) Protein kinase A itself is different in different cell types, which explains why the effects of cyclic AMP vary depending on the target cell. Explain your answer.

15–46* Propose specific types of mutation in the gene for the regulatory subunit of PKA that could lead to either a permanently active PKA or a permanently inactive PKA.

15–47 Does cyclic AMP work exclusively through PKA or do cyclic AMP-binding proteins have other critical roles, for example, as DNA-binding proteins as in bacteria? You are working with a hamster cell line that makes genetic analysis of this question possible. Because of chromosomal rearrangements, this cell line possesses only one functional copy of many genes, making isolation of recessive mutations much easier than in a fully diploid line. In addition, high intracellular levels of cyclic AMP stop their proliferation. By stimulating adenylyl cyclase with cholera toxin and by inhibiting cyclic AMP phosphodiesterase with theophylline, cyclic AMP can be artificially elevated. Under these conditions only cells that are resistant to the effects of cyclic AMP can grow.

In this way you isolate several resistant colonies that grow under the selective conditions and assay them for PKA activity: they are all defective. About 10% of the resistant lines are completely missing PKA activity. The remainder possess PKA activity, but a very high level of cyclic AMP is required for activation. To characterize the resistant lines further, you fuse them with the parental cells and test the hybrids for resistance to cholera toxin. Hybrids between parental cells and PKA-negative cells are sensitive to cholera toxin, which indicates that these mutations are recessive. By contrast, hybrids between parental cells and resistant cells with altered PKA responsiveness are resistant to cholera toxin, which indicates that these mutations are dominant.

A. Is PKA an essential enzyme in these hamster cells?

B. PKA is a tetramer consisting of two catalytic (protein kinase) and two regulatory (cyclic AMP-binding) subunits. Propose an explanation for why the mutations in some cyclic AMP-resistant cell lines are recessive, while others are dominant.

C. Do these experiments support the notion that all cyclic AMP effects in hamster cells are mediated by PKA?

15–48* (**True/False**) The activity of any protein regulated by phosphorylation depends on the balance at any instant between the activities of the kinases that phosphorylate it and the phosphatases that dephosphorylate it. Explain your answer.

15–49 During a marathon, runners draw heavily on their internal reserves of glycogen and triglycerides to fuel muscle contraction. Initially, energy is derived mostly from carbohydrates with increasing amounts of fat being used as the race progresses. If runners use up their muscle glycogen reserves before they finish the race, they hit what is known as 'the wall,' a point of diminished performance that arises because fatty acids from triglyceride breakdown cannot be delivered to the muscles quickly enough to sustain maximum effort. One trick that marathon runners use to avoid the wall is to drink a cup of strong black coffee an hour or so before the race begins. Coffee contains caffeine, which is an inhibitor of cyclic AMP phosphodiesterase. How do you suppose inhibition of this enzyme helps them avoid the wall?

15–50* A particularly graphic illustration of the subtle, yet important, role of cyclic AMP in the whole organism comes from studies of the fruit fly *Drosophila melanogaster*. In search of the gene for cyclic AMP phosphodiesterase, one laboratory measured enzyme levels in flies with chromosomal duplications or deletions and found consistent alterations in flies with mutations involving bands 3D3 and 3D4 on the X chromosome. Duplications in this region have about 1.5 times the normal activity of the enzyme; deletions have about half the normal activity.

An independent laboratory in the same institution was led to the same chromosomal region through work on behavioral mutants of fruit flies. The researchers had developed a learning test in which flies were presented with two metallic grids, one of which was electrified. If the electrified grid was painted with a strong-smelling chemical, normal flies quickly learned to avoid it, even when it was no longer electrified. The mutant flies, on the other hand, never learned to avoid the smelly grid; they were aptly called *dunce* mutants. The *dunce* mutation was mapped genetically to bands 3D3 and 3D4.

Is the learning defect really due to lack of cyclic AMP phosphodiesterase or are the responsible genes simply closely linked? Further experiments showed that the level of cyclic AMP in *dunce* flies was 1.6 times higher than in normal flies. Furthermore, sucrose gradient analysis of homogenates of *dunce* and normal flies revealed two cyclic AMP phosphodiesterase activities, one of which was missing in *dunce* flies (Figure 15–19).

A. Why do *dunce* flies have higher levels of cyclic AMP than normal flies?
B. Explain why homozygous (both chromosomes affected) duplications of the nonmutant *dunce* gene cause cyclic AMP phosphodiesterase levels to be elevated 1.5-fold and why homozygous deletions of the gene reduce enzyme activity to half the normal value?
C. What would you predict would be the effect of caffeine, a phosphodiesterase inhibitor, on the learning performance of normal flies?

15–51 In a cell line derived from normal rat thyroid, stimulation of the α_1-adrenergic receptor increases both inositol trisphosphate (IP_3) formation and release of arachidonic acid. IP_3 elevates intracellular Ca^{2+}, which mediates thyroxine efflux, whereas arachidonic acid serves as a source of prostaglandin E_2, which stimulates DNA synthesis. How is arachidonic acid release connected to the adrenergic receptor? Arachidonic acid could arise by cleavage from the diacylglycerol that accompanies IP_3 production. Alternatively, it could arise through an independent effect of the receptor on phospholipase A_2, which can directly release arachidonic acid from intact phosphoglycerides. Consider the following experimental observations:
1. Addition of noradrenaline to cell cultures stimulates production of both IP_3 and arachidonic acid.
2. If the α_1-adrenergic receptors are made unresponsive to noradrenaline by treatment with phorbol esters (which act through protein kinase C to cause phosphorylation, hence inactivation, of the receptor), addition of noradrenaline causes no increase in either IP_3 or arachidonic acid.

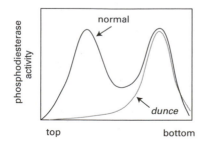

Figure 15–19 Sucrose gradient analysis of cyclic AMP phosphodiesterase activity in homogenates of normal and *dunce* flies (Problem 15–50).

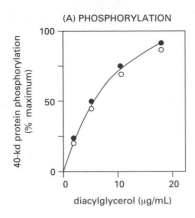

(A) PHOSPHORYLATION

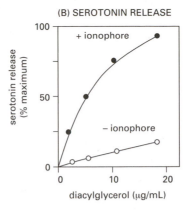

(B) SEROTONIN RELEASE

Figure 15–20 Treatments of platelets with calcium ionophore and diacylglycerol (Problem 15–52). (A) Effects on phosphorylation of the 40-kd protein. (B) Effects on serotonin release. *Filled circles* indicate the presence of calcium ionophore and *open circles* indicate its absence.

3. When cells are made permeable and treated with GTPγS (a nonhydrolyzable analog of GTP), production of both IP_3 and arachidonic acid is increased.

4. If cells are treated with neomycin (which blocks the action of phospholipase C), subsequent treatment of permeabilized cells with GTPγS stimulates arachidonic acid production but causes no increase in IP_3.

5. If cells are treated with pertussis toxin, subsequent treatment of permeabilized cells with GTPγS stimulates production of IP_3 but causes no increase in arachidonic acid.

A. Which of the two proposed mechanisms for arachidonic acid production in these cells do the observations support?

B. Describe a molecular pathway for activation of arachidonic acid production that is consistent with the experimental results.

15–52* The primary role of platelets is to control blood clotting. When they encounter the exposed basement membrane (collagen fibers) of a damaged blood vessel or a newly forming fibrin clot, they change their shape from round to spiky and stick to the damaged area. At the same time, they begin to secrete serotonin and ATP, which accelerate similar changes in newly arriving platelets, leading to the rapid formation of a clot. The platelet response is regulated by protein phosphorylation. Significantly, platelets contain high levels of two protein kinases: PKC, which initiates serotonin release, and myosin light-chain kinase, which mediates the change in shape.

When platelets are stimulated with thrombin, the light chain of myosin and an unknown protein of 40,000 daltons are phosphorylated. When platelets are treated with a calcium ionophore, only the myosin light chain is phosphorylated; when they are treated with diacylglycerol, only the 40-kd protein is phosphorylated. Experiments using a range of concentrations of diacylglycerol in the presence or absence of calcium ionophore show that the extent of phosphorylation of the 40-kd protein depends only on the concentration of diacylglycerol (Figure 15–20A). Serotonin release, however, depends on diacylglycerol and the calcium ionophore (Figure 15–20B).

A. Based on these experimental observations, describe the normal sequence of molecular events that leads to phosphorylation of the myosin light chain and the 40-kd protein. Indicate how the calcium ionophore and diacylglycerol treatments interact with the normal sequence of events.

B. Why do you think serotonin release requires both calcium ionophore and diacylglycerol?

15–53 How is an IP_3-triggered Ca^{2+} response terminated?

15–54* Why do you suppose cells use Ca^{2+} (intracellular concentration 10^{-7} M) for signaling rather than the more abundant Na^+ (intracellular concentration 10^{-3} M)?

15–55 EGTA chelates Ca^{2+} with high affinity and specificity. How would microinjection of EGTA affect glucagon-triggered breakdown of glycogen in liver? How would it affect vasopressin-triggered breakdown of glycogen in liver?

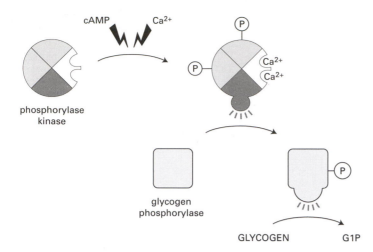

Figure 15–21 Integration of cyclic AMP-dependent and Ca²⁺-dependent signaling pathways by phosphorylase kinase in liver and muscle cells (Problem 15–56). G1P is glucose 1-phosphate, the product of the cleavage of a glucose from glycogen using a phosphate.

15–56* Phosphorylase kinase integrates signals from cyclic AMP-dependent and Ca²⁺-dependent signaling pathways that control glycogen breakdown in liver and muscle cells (Figure 15–21). Phosphorylase kinase is composed of four subunits. One is the protein kinase that catalyzes addition of phosphate to glycogen phosphorylase to activate it for glycogen breakdown. The other three subunits are regulatory proteins that control the activity of the catalytic subunit. Two contain sites for phosphorylation by PKA, which is activated by cyclic AMP. The remaining subunit is calmodulin, which binds Ca²⁺ when its cytosolic concentration rises. The regulatory subunits control the equilibrium between the active and inactive conformations of the catalytic subunit. How does this arrangement allow phosphorylase kinase to serve its role as an integrator protein for the multiple pathways that stimulate glycogen breakdown?

15–57 Unlike myosin from skeletal muscle, smooth muscle myosin interacts with actin only when its light chains are phosphorylated. Phosphorylation is controlled by variations in the intracellular concentration of Ca²⁺, which is mediated through calmodulin. If myosin light-chain kinase is purified from smooth muscle in the absence of protease inhibitors, its kinase activity is the same in the presence or absence of Ca²⁺/calmodulin. In the presence of protease inhibitors, however, the purified kinase is inactive unless Ca²⁺/calmodulin is present. As shown in Table 15–4, the purified kinase now behaves as expected; it shows an absolute dependence on Ca²⁺/calmodulin.

   A. Why do you suppose the original purified enzyme was active independent of Ca²⁺/calmodulin?

   B. Propose a sequence of molecular events that leads to contraction of smooth muscles. Begin with the entry of Ca²⁺ into the cytosol.

TABLE 15–4 Activities of myosin light-chain kinase purified in the presence and absence of protease inhibitors (Problem 15–57).

PURIFICATION SCHEME	ADDITIONS TO ASSAY MIX	RELATIVE ACTIVITY
Minus inhibitors	none	50
Minus inhibitors	Ca²⁺	50
Minus inhibitors	calmodulin	50
Minus inhibitors	Ca²⁺/calmodulin	50
Plus inhibitors	none	1
Plus inhibitors	Ca²⁺	1
Plus inhibitors	calmodulin	1
Plus inhibitors	Ca²⁺/calmodulin	100

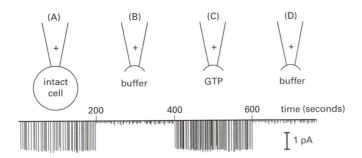

Figure 15–22 Experimental setup and typical results of patch-clamp analysis of K+ channel activation by acetylcholine (Problem 15–59). The buffer is a salts solution that does not contain nucleotides or Ca²⁺. In all these experiments, acetylcholine is present inside the pipet, as indicated by the *plus* sign. The current through the membrane is measured in picoamps (pA). In (C) the GTP is added to the buffer.

15–58* CaM kinase II is a remarkable molecular memory device. How does CaM kinase II 'remember' its exposure to Ca²⁺/calmodulin and why does it eventually 'forget?'

15–59 Acetylcholine acts on G-protein-linked muscarinic receptors in the heart to open K+ channels, thereby slowing the heart rate. This process can be directly studied using the inside-out membrane patch-clamp technique. The external surface of the membrane is in contact with the solution in the bore of the pipet, and the cytoplasmic surface faces outward and can be exposed readily to a variety of solutions (Figure 15–22). Receptors, G proteins, and K+ channels remain associated with the membrane patch.

When acetylcholine is added to a pipet with a whole cell attached, K+ channels open as indicated by the flow of current (Figure 15–22A). Under similar circumstances with a patch of membrane inserted into a buffered salts solution, no current flows (Figure 15–22B). When GTP is added to the buffer, however, current resumes (Figure 15–22C). Subsequent removal of GTP stops the current (Figure 15–22D). The results of several similar experiments to test the effects of different combinations of components are summarized in Table 15–5.

A. Why do you think it is that $G_{\beta\gamma}$ activated the channel when the complete G protein did not? Is the active component of the G protein in this system the same as the one that activates adenylyl cyclase in other cells?

B. Addition of GppNp (a nonhydrolyzable analog of GTP) causes the K+ channel to open in the absence of acetylcholine (Table 15–5, line 4). The flow of current, however, rose very slowly and reached its maximum only after a minute (compare with the immediate rise in Figure 15–22A and C). How do you suppose GppNp causes the channels to open slowly in the absence of acetylcholine?

C. To the extent these experiments allow, draw a scheme for the activation of K+ channels in heart cells in response to acetylcholine.

15–60* In visual transduction one activated rhodopsin molecule leads to hydrolysis of 5×10^5 cyclic GMP molecules per second. One stage in this enormous signal amplification is achieved by cyclic GMP phosphodiesterase, which

TABLE 15–5 Responses of K+ channel to various experimental manipulations (Problem 15–59).

	ADDITIONS			
	ACETYLCHOLINE	SMALL MOLECULES	G-PROTEIN COMPONENTS	K+ CHANNEL
1	+	none	none	closed
2	+	GTP	none	open
3	–	GTP	none	closed
4	–	GppNp	none	open
5	–	none	G protein	closed
6	–	none	G_α	closed
7	–	none	$G_{\beta\gamma}$	open
8	–	none	boiled G protein	closed

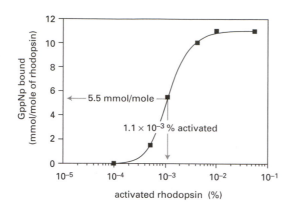

Figure 15–23 Binding of GppNp to rod cell membranes as a function of the fraction of activated rhodopsin (Problem 15–60). Background binding of GppNp to rod cell membranes in the dark has been subtracted from the values shown.

hydrolyzes 1000 molecules of cyclic GMP per second. The additional factor of 500 could arise because one activated rhodopsin activates 500 transducin (G_t), or because one activated transducin activates 500 cyclic GMP phosphodiesterases, or through a combination of both effects. One experiment to address this question measured the amount of GppNp (a nonhydrolyzable analog of GTP) that is bound by transducin in the presence of different amounts of activated rhodopsin. As indicated in Figure 15–23, 5.5 mmol of GppNp were bound per mole of total rhodopsin when 0.0011% of the rhodopsin was activated.

A. Assuming each transducin molecule binds one molecule of GppNp, calculate the number of transducin molecules that are activated by each activated rhodopsin molecule. Which mechanism of amplification does this measurement support?

B. Binding studies have shown that transducin-GDP has a high affinity for activated rhodopsin and that transducin-GTP has a low affinity; conversely, transducin-GTP has a high affinity and transducin-GDP has a low affinity for cyclic GMP phosphodiesterase. Are these affinities consistent with the mechanism of amplification you deduced from the above experiment? Explain your reasoning.

15–61   The outer segments of rod photoreceptor cells can be isolated and used to study the effects of small molecules on visual transduction because the ends of the segments remain unsealed. How would you expect the visual response to be affected by the following additions?

A. An inhibitor of cyclic GMP phosphodiesterase.

B. A nonhydrolyzable analog of GTP.

C. An inhibitor of rhodopsin-specific kinase.

15–62* Patients with Oguchi's disease have an inherited form of nightblindness. After a flash of bright light these individuals recover their night vision (become dark adapted) very slowly. Night vision depends almost entirely on the visual responses of rod photoreceptor cells. What aspect of the visual response in these patients' rod cells do you suppose is defective? What genes, when defective, might give rise to Oguchi's disease?

15–63   (True/False) In contrast to the more direct signaling pathways used by nuclear receptors, catalytic cascades of intracellular mediators provide numerous opportunities for amplifying the responses to extracellular signals. Explain your answer.

15–64* In muscle cells adrenaline binds to the β-adrenergic receptor to initiate a signaling cascade that leads to the breakdown of glycogen (Figure 15–24). At what points in this pathway is the signal amplified?

15–65   A critical feature of all signaling cascades is that they must be turned off rapidly when the extracellular signal is removed. Examine the signaling cascade in Figure 15–24. Describe how each component of this signaling pathway is returned to its inactive state when adrenaline is removed.

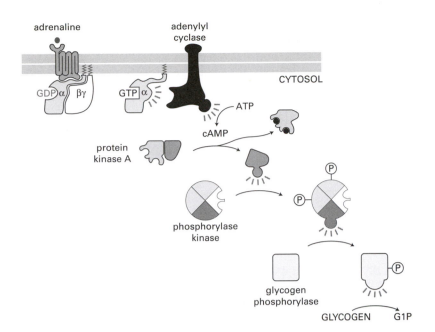

Figure 15–24 Signaling cascade for activation of glycogen breakdown by adrenaline in muscle cells (Problem 15–64). G1P is glucose 1-phosphate; cAMP bound to the regulatory subunits of PKA is shown as *black balls*.

SIGNALING THROUGH ENZYME-LINKED CELL-SURFACE RECEPTORS

TERMS TO LEARN

BTK	guanylyl cyclase	Ras
CD45 protein	histidine kinase	receptor tyrosine kinase
chemotaxis receptor	Jak-STAT signaling pathway	S6 kinase
cytokine receptor	Janus kinase (Jak)	SH2 domain
cytoplasmic tyrosine kinase	MAP-kinase	SHP-1
enzyme-linked receptor	PDK1	SHP-2
Eph receptor	PH domain	Smad
ephrin	PI 3-kinase	Sos
focal adhesion kinase (FAK)	PLC-γ	Src
Grb-2 protein	protein kinase B (PKB)	STATs
GTPase-activating	protein tyrosine	TGF-β
protein (GAP)	phosphatase (PTP)	two-component signaling pathway
guanine nucleotide	Raf	tyrosine phosphatase
exchange factor (GEF)		

15–66* Antibodies are Y-shaped molecules that carry two identical binding sites. Imagine that you have obtained an antibody that is specific for the extracellular domain of a receptor tyrosine kinase. If cells were exposed to the antibody, would you expect the receptor tyrosine kinase to be activated, inactivated, or unaffected? Explain your reasoning.

15–67 What is bidirectional signaling and how do ephrins and ephrin receptors mediate it?

15–68* Genes encoding mutant forms of a receptor tyrosine kinase can be introduced into cells that also express the normal receptor from their own genes. If the mutant genes are expressed at considerably higher levels than the normal genes, what will be the consequences for receptor-mediated signaling of introducing genes for the following mutant receptors?
A. A mutant receptor tyrosine kinase that lacks its extracellular domain.
B. A mutant receptor tyrosine kinase that lacks its intracellular domain.

15–69 (**True/False**) It is thought that extracellular ligand binding to a receptor tyrosine kinase activates the intracellular catalytic domain by propagating a conformational change across the lipid bilayer through the single transmembrane α helix. Explain your answer.

15–70* What does autophosphorylation mean? When a receptor tyrosine kinase binds its ligand and forms a dimer, do the individual receptor molecules phosphorylate themselves or does one receptor cross-phosphorylate the other, and vice versa? To investigate this question, you've constructed genes for three forms of a receptor tyrosine kinase: the normal form with an active kinase domain and three sites of phosphorylation; a large form that carries an inactivating point mutation in the kinase domain but retains the three phosphorylation sites; and a short version that has an active kinase domain but is lacking the sites of phosphorylation (Figure 15–25A). You express the genes singly and in combination in a cell line that lacks this receptor tyrosine kinase, and then break open the cells and add the ligand for the receptor in the presence of radioactive ATP. You immunoprecipitate the receptors and analyze them for expression levels by staining for protein (Figure 15–25B) and for phosphorylation by autoradiography (Figure 15–25C).

A. What results would you expect on the autoradiograph if individual receptors phosphorylated themselves?
B. What would you expect if receptors cross-phosphorylated each other?
C. Which model for autophosphorylation do your data support?

15–71 The SH3 domain, which comprises about 60 amino acids, recognizes and binds to structural motifs in other proteins. The motif recognized by SH3 domains was found by constructing a fusion protein between an SH3 domain and glutathione-S-transferase (GST). GST fusions allow for easy purification using a glutathione-affinity column, which binds GST specifically. After tagging the purified GST-SH3 protein with biotin to make it easy to detect, it was used to screen filters containing *E. coli* colonies expressing a cDNA library. Two different clones were identified that bound to SH3 domains: in both cases binding was shown to occur at short proline-rich sequences.

A. Could you use this method to find cDNAs for proteins that bind to SH2 domains? Why or why not?
B. Many proteins bind to short strings of amino acids in other proteins. How do you think these kinds of interactions differ from the kinds of interactions found between the protein subunits of multisubunit enzymes?

15–72* When activated, the platelet-derived growth factor (PDGF) receptor phosphorylates itself on multiple tyrosines. These phosphorylated tyrosines serve as assembly sites for several SH2-domain-containing proteins that include phospholipase C-γ (PLC-γ), a Ras-specific GTPase activating protein (GAP), a subunit of phosphatidylinositol 3-kinase (PI3K), and a phospho-tyrosine phosphatase (PTP) (Figure 15–26). PDGF binding stimulates several changes in the target cell, one of which is an increase in DNA synthesis, as measured by incorporation of radioactive thymidine or bromodeoxyuridine into DNA.

To determine which of the bound proteins is responsible for activation of DNA synthesis, you construct several mutant genes for the PDGF receptor

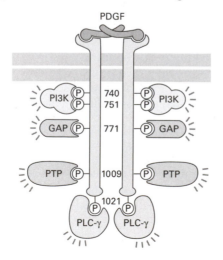

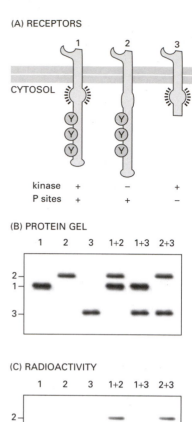

(A) RECEPTORS

	kinase	P sites
1	+	+
2	–	+
3	+	–

(B) PROTEIN GEL

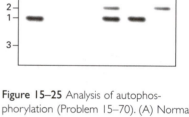

(C) RADIOACTIVITY

Figure 15–25 Analysis of autophosphorylation (Problem 15–70). (A) Normal and mutant receptor tyrosine kinases. P sites refers to the sites of phosphorylation. (B) Expression of receptor tyrosine kinases. (C) Phosphorylation of receptor tyrosine kinases.

Figure 15–26 The signaling complex assembled on the PDGF receptor (Problem 15–72).

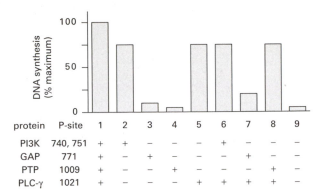

protein	P-site	1	2	3	4	5	6	7	8	9
PI3K	740, 751	+	+	−	−	−	+	−	−	−
GAP	771	+	−	+	−	−	−	+	−	−
PTP	1009	+	−	−	+	−	−	−	+	−
PLC-γ	1021	+	−	−	−	+	+	+	+	−

Figure 15–27 Stimulation of DNA synthesis by the normal PDGF receptor and by receptors missing some phosphorylation sites (Problem 15-72). Stimulation by the normal receptor is set arbitrarily at 100%.

that retain individual or combinations of tyrosine phosphorylation sites. When expressed in cells that do not make a PDGF receptor of their own, each of the receptors is expressed normally and phosphorylated at its tyrosines upon binding of PDGF. As shown in Figure 15–27, DNA synthesis is stimulated to different extents in cells expressing the mutant receptors.

What roles do PI3K, GAP, PTP, and PLC-γ play in the stimulation of DNA synthesis by PDGF?

15–73 The Ras protein functions as a molecular switch that is set to its on state by a guanine-nucleotide exchange factor (GEF) that causes it to bind GTP. A GTPase-activating protein (GAP) resets the switch to the off state by inducing Ras to hydrolyze its bound GTP to GDP much more rapidly than in the absence of the GAP. Thus Ras works like a light switch that one person turns on and another turns off. In a cell line that lacks the Ras-specific GAP, what abnormalities would you expect to find in the way Ras activity responds to extracellular signals?

15–74* What are the similarities and differences between the reactions that lead to the activation of G proteins and the reactions that lead to the activation of Ras?

15–75 In principle, the activated, GTP-bound form of Ras could be increased by activating a guanine-nucleotide exchange factor (GEF) or by inactivating a GTPase activating protein (GAP). Why do you suppose that Ras-mediated signaling pathways always increase Ras–GTP by activating a GEF, rather than inactivating a GAP?

15–76* A single amino acid change in Ras eliminates its ability to hydrolyze GTP, even in the presence of a GTPase activating protein (GAP). Roughly 30% of human cancers have this change in Ras. You have just identified a small molecule that prevents dimerization of a receptor tyrosine kinase that signals via Ras. Would you expect this molecule to be effective in the treatment of cancers with mutant Ras proteins? Why or why not?

15–77 MAP-kinase-kinase-kinase (MAPKKK) activates MAP-kinase-kinase (MAPKK) by phosphorylation of two serine side chains. Doubly phosphorylated (active) MAPKK, in turn, activates MAP-kinase (MAPK) by phosphorylation of a threonine and a tyrosine. The doubly phosphorylated MAPK then phosphorylates a variety of target proteins to bring about complex changes in cell behavior. It is possible to write down all of the rate equations for the individual steps in this activation cascade, as well as for the removal of the phosphates (inactivation) by protein phosphatase, and to solve them by making reasonable assumptions about concentrations of the proteins. The resulting plot of activation of the kinases versus input stimulus is shown in Figure 15–28. Why is the very steep response curve for MAPK a good thing for this signaling pathway?

15–78* An explicit assumption in the analysis in Problem 15–77 is that the components of the MAP-kinase cascade operate independently of one another, so that the dual phosphorylation events that activate MAPKK and MAPK occur

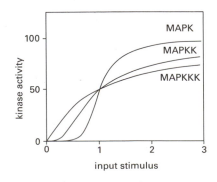

Figure 15–28 Stimulus–response curves for the components of the MAPK cascade (Problem 15–77). For ease of comparison the curves have been normalized so that an input stimulus of 1 gives 50% activation of the kinases.

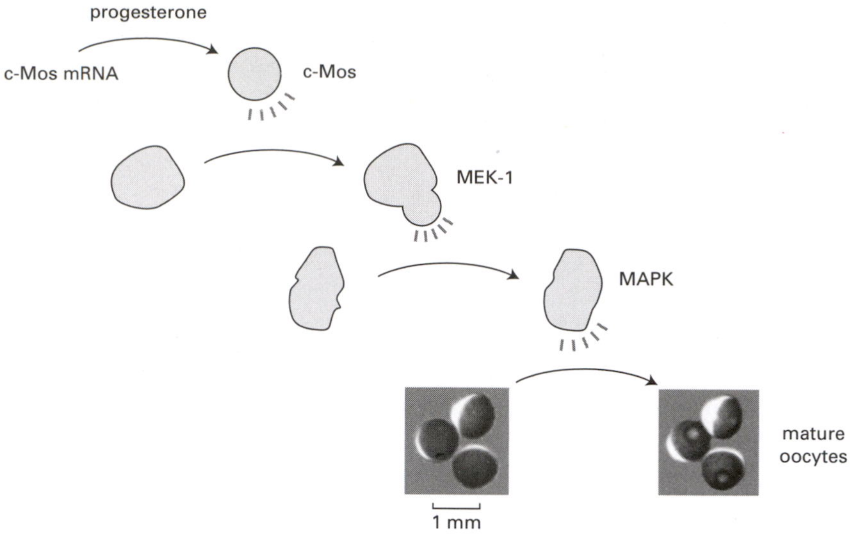

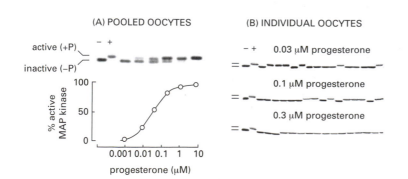

progesterone

c-Mos mRNA → c-Mos

MEK-1

MAPK

mature oocytes

1 mm

Figure 15–29 Progesterone-induced MAP-kinase cascade, leading to oocyte maturation (Problem 15–79).

one at a time as molecules collide in solution. How do you suppose the curves in Figure 15–28 would change if a scaffold protein held the kinases of the MAP-kinase cascade together? Most MAP-kinase cascades are scaffolded. What is the advantage of linking these kinases onto scaffold proteins?

15–79 Activation ('maturation') of frog oocytes is signaled through the MAP-kinase cascade. An increase in the hormone progesterone triggers the cascade by stimulating the translation of c-Mos mRNA, which is the frog's MAP-kinase-kinase-kinase (Figure 15–29). Maturation is easy to score visually by the presence of a white spot in the middle of the brown surface of the oocyte (Figure 15–29). To determine the dose–response curve for progesterone-induced activation of MAP-kinase, you place 16 oocytes in each of six plastic dishes and add various concentrations of progesterone. After an overnight incubation, you crush the oocytes, prepare an extract, and determine the state of MAP-kinase phosphorylation (hence, activation) by SDS-polyacrylamide gel electrophoresis (Figure 15–30A). This analysis shows a graded response of MAPK to increasing concentrations of progesterone.

Before you crushed the oocytes you noticed that not all oocytes in individual dishes had white spots. Have some oocytes undergone partial activation and not yet reached the white-spot stage? To answer this question, you repeat the experiment but this time you analyze MAP-kinase activation in individual oocytes. You are surprised to find that each oocyte has either a fully activated or a completely inactive MAP-kinase (Figure 15–30B). How can an all-or-none response in individual oocytes give rise to a graded response in the population?

15–80* Phosphorylation of MAP-kinase at a threonine (T) and a tyrosine (Y) residue causes a visible conformational change in the enzyme (Figure 15–31). Can you guess where the active enzyme binds ATP and its target proteins?

15–81 (**True/False**) PI3-kinase phosphorylates the inositol head groups of

(A) POOLED OOCYTES

active (+P)
inactive (−P)

% active MAP kinase

100

50

0

0.001 0.01 0.1 1 10

progesterone (μM)

(B) INDIVIDUAL OOCYTES

− + 0.03 μM progesterone

0.1 μM progesterone

0.3 μM progesterone

Figure 15–30 Activation of frog oocytes (Problem 15–79). (A) Phosphorylation of MAP-kinase in pooled oocytes. (B) Phosphorylation of MAP-kinase in individual oocytes. MAP-kinase was detected by immunoblotting using a MAP-kinase-specific antibody. The first two lanes in each gel contain nonphosphorylated, inactive MAP-kinase (−) and phosphorylated, active MAP-kinase (+).

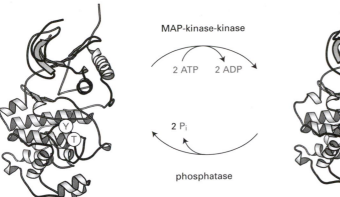

MAP-kinase-kinase

2 ATP → 2 ADP

2 Pi

phosphatase

Figure 15–31 Conformational change of MAP-kinase upon phosphorylation (Problem 15–80).

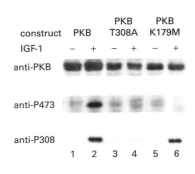

Figure 15–32 Expression levels of various forms of PKB and their degree of phosphorylation in the presence and absence of IGF-1 (Problem 15–82). Anti-PKB recognizes all three forms of PKB regardless of their phosphorylation state; anti-P473 specifically recognizes the phosphorylated serine at position 473; anti-P308 specifically recognizes the phosphorylated threonine at position 308.

phospholipids at the 3 position of the ring so that they can be cleaved by phospholipase C to produce IP$_3$. Explain your answer.

15–82* Protein kinase B (PKB) is a key enzyme in the signaling pathway that leads to cell growth. PKB is activated by a phosphatidylinositol-dependent protein kinase (PDK1), which phosphorylates threonine 308. At the same time, serine 473 is phosphorylated. Your advisor has been unsuccessful in purifying the protein kinase responsible for phosphorylation of serine 473, but you think you know what is going on. You construct genes encoding two mutant forms of PKB: one that carries a point mutation in the kinase domain, PKB-K179M, which renders it kinase-dead, and the other that carries a point mutation in the domain required to bind to PDK1 (PKB-T308A), which cannot be activated by PDK1. You transfect each of these constructs, and a construct for wild-type PKB, into cells that do not express their own PKB. You treat a portion of the cells with an insulin-like growth factor (IGF-1), which activates PDK1, and analyze the phosphorylation state of the various forms of PKB using antibodies specific for PKB or for particular phosphorylated amino acids (Figure 15–32).

What is the identity of the enzyme that phosphorylates serine 473 on PKB?

15–83 Interferon-γ (IFN-γ) is a cytokine produced by activated T lymphocytes. It binds to surface receptors on macrophages and stimulates their efficient scavenging of invading viruses and bacteria via a Jak-STAT signaling pathway. A number of genes are activated in response to IFN-γ binding, all of which contain a DNA sequence element with partial dyad symmetry (TTCCXGTAA) that is required for the IFN-γ response.

You have cloned the gene for the STAT transcription factor that is activated in response to INF-γ binding. The sequence of the gene indicates that the protein contains several heptad repeat sequences near its N-terminus—a common dimerization domain in many transcription factors—and SH2 and SH3 domains adjacent to a site for tyrosine phosphorylation near the C-terminus (Figure 15–33). By making antibodies to the protein, you show that it is normally located in the cytosol; however, after 15 minutes exposure to IFN-γ, the protein becomes phosphorylated on a tyrosine and moves to the nucleus.

Suspecting that tyrosine phosphorylation is the key to the regulation of this transcription factor, you assay its ability to bind the DNA sequence element in the presence of high concentrations of free phosphotyrosine or when mixed with anti-phosphotyrosine antibodies. Both treatments inhibit binding of the protein to DNA, as does treatment with a protein phosphatase. Finally, you measure the molecular weight of the cytosolic and

heptad repeats

SH3 SH2 (P)

Figure 15–33 Sequence elements in the transcription factor that responds to IFN-γ (Problem 15–83).

nuclear forms of the protein, which suggest that the cytosolic form is a monomer and the nuclear form is a dimer.

A. Do you think that phosphorylation of the transcription factor is necessary for the factor to bind to DNA, or do you think phosphorylation is required to create an acidic activation domain to promote transcription?

B. Bearing in mind that SH2 domains bind phosphotyrosine, how do you think free phosphotyrosine interferes with the activity of the transcription factor?

C. How might tyrosine phosphorylation of the protein promote its dimerization? How do you think dimerization enhances its binding to DNA?

15–84* (True/False) Protein tyrosine phosphatases are much like serine/threonine protein phosphatases; they each display a broad specificity and remove phosphates from a wide range of phosphorylated cellular proteins. Explain your answer.

Problem 3–87 illustrates a strategy used to define target substrates for a protein phosphatase

15–85 Receptorlike tyrosine phosphatases are single-pass transmembrane proteins whose ligands and functions are largely unknown. Explain in a general way how the signaling pathway triggered by these putative receptors can be studied by replacing their extracellular domains with the extracellular domain from a growth factor receptor such as the EGF receptor.

15–86* What is meant by the statement that inhibitory Smads act as decoys? What is being decoyed and what is the functional consequence?

15–87 (True/False) Atrial natriuretic peptides bind to a receptor that activates a G protein, which in turn activates guanylyl cyclase to produce cyclic GMP, which then activates a cyclic GMP-dependent protein kinase (PKG). Explain your answer.

15–88* Four types of chemotaxis receptors have been identified in *E. coli*. These receptors mediate chemotactic responses to two different amino acids, to sugars, and to dipeptides. As part of a practical demonstration in bacterial chemotaxis, your instructor has given you a wild-type strain with all four receptors intact and four mutant strains with one or more of the receptors missing (Table 15–6). Your assignment is to identify the receptor that mediates the response to each attractant. The experimental assay is very simple. You fill a capillary pipet with a solution of the attractant, dip it into a buffered solution containing bacteria, remove it after 5 minutes, and count the number of bacteria in the capillary. Your results are shown in Table 15-6. Identify each attractant and its appropriate receptor.

15–89 To clarify the relationship between the structure of a chemotaxis receptor and the functions of stimulus recognition, signal transduction, and adaptation in *Salmonella typhimurium*, you have cloned the normal gene for the aspartate receptor and a mutant that is missing 35 C-terminal amino acids. By introducing the wild-type and truncated forms of the gene back into *Salmonella* that are missing the normal gene, you can test for functional differences between the two cloned genes. Both the wild-type and the truncated receptor are overexpressed about 15-fold above normal—not uncommon for genes expressed from plasmids—but both bind aspartate normally.

TABLE 15–6 Chemotaxis in wild-type and mutant strains of *E. coli* (Problem 15–88).

STRAIN	INTACT RECEPTORS	NUMBER OF CELLS (1000s) IN CAPILLARY				
		SERINE	ASPARTATE	RIBOSE	PRO-GLY	NONE
1	Tap, Tar, Trg, Tsr	59	105	95	6.6	0.5
2	Tap, Tar, Trg	0.7	84	77	13	0.8
3	Trg, Tsr	34	0.7	59	0.6	0.6
4	Tap, Trg, Tsr	55	0.6	65	4.1	0.5
5	Tar, Trg, Tsr	70	59	85	0.9	0.8

The only major difference is that the truncated receptor cannot be methylated.

Wild-type bacteria in the absence of an attractant change their direction of rotation (tumble) every few seconds. Upon exposure to an attractant, however, the changes in direction of rotation are suppressed, which leads to a period of smooth swimming. If the concentration of attractant remains constant (even if high), wild-type bacteria quickly adapt and begin again to tumble every few seconds. To observe these behavioral changes experimentally, you tether bacteria by their flagella to coverslips so that you can observe their direction of rotation. You then expose them to aspartate and count the number of bacteria that do not reverse their direction of rotation in 1-minute intervals. The results for wild-type cells and for cells containing the cloned aspartate receptors are shown in Figure 15–34.

A. Is signal transduction by the two cloned receptors normal?

B. Are the adaptive properties of the cloned receptors normal?

C. Suggest molecular explanations for why the cloned normal receptor and the cloned truncated receptor, when introduced into bacteria, respond differently from the normal receptor in wild-type cells (Figure 15–34).

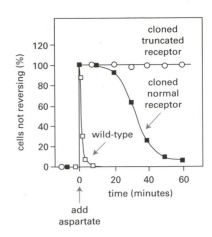

Figure 15–34 Behavior of wild-type cells and cells containing the cloned receptors (Problem 15–89).

SIGNALING PATHWAYS THAT DEPEND ON REGULATED PROTEOLYSIS

TERMS TO LEARN

adenomatous polyposis coli (APC)	glycogen synthase kinase-3β (GSK-3β)	NF-κB
β-catenin	Hedgehog	Notch
Cubitus interruptus (Ci)	IκB	Patched
Delta	IκB kinase (IKK)	Smoothened
Frizzled		Wnt

15–90* One of the most difficult aspects of signaling through pathways that involve proteolysis is keeping track of the names of the components and their functions. Sort the following list of proteins into the Notch, Wnt, and Hedgehog signaling pathways, list them in the order that they function, and select one or more appropriate descriptors that identify their role.

PROTEIN

A. Adenomatous polyposis coli (APC)
B. Axin
C β-Catenin
D. CSL
E. Cubitus interruptus (Ci)
F. Delta
G. Dishevelled
H. Frizzled
I. Glycogen synthase kinase-3β
J. Hedgehog
K. Notch
L. Patched
M. Presenilin-1
N. Smoothened
O. Wnt

FUNCTION

1. Adaptor protein
2. Amplifier protein
3. Anchoring protein
4. Bifurcation protein
5. Extracellular signal protein
6. Gene regulatory protein
7. Integrator protein
8. Latent gene regulatory protein
9. Messenger protein
10. Modulator protein
11. Protease
12. Receptor protein
13. Relay protein
14. Scaffold protein
15. Transducer protein

15–91 Like Notch, the β-amyloid precursor protein (APP) is cleaved near its transmembrane segment to release an extracellular and an intracellular component. Explain how the fragments of APP relate to the amyloid plaques that are characteristic of Alzheimer's disease.

15–92* β-Catenin is a target for phosphorylation by glycogen synthase kinase-3β and also a substrate for degradation in proteasomes. Treatment of mouse fibroblasts with a proteasome inhibitor, ALLN, increases the stability of

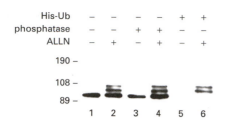

His-Ub   –   –   –   –   +   +
phosphatase   –   –   +   +   –   –
ALLN   –   +   –   +   –   +

190 –

108 –

89 –

   1   2   3   4   5   6

Figure 15–35 Electrophoretic analysis of β-catenin (Problem 15–92). β-Catenin was detected using β-catenin-specific antibodies. Size markers are shown on the left.

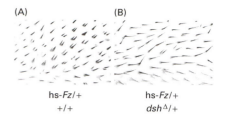

(A)   (B)

hs-*Fz*/+   hs-*Fz*/+
+/+   *dsh*Δ/+

Figure 15–36 Pattern of hair growth on wing cells in genetically different *Drosophila* (Problem 15–94).

β-catenin and causes the appearance of new, slower migrating forms of the protein on SDS-polyacrylamide gels (Figure 15–35, lanes 1 and 2). Do these new bands represent phosphorylated proteins, which often run more slowly on such gels (see Figure 15–30), or do they arise by addition of ubiquitin, which would increase their size. To test for phosphorylation, you treat samples with a protein phosphatase that efficiently removes phosphates from other proteins in the same sample (not shown) and run them on gels (lanes 3 and 4). To test for ubiquitylation, you express His-tagged ubiquitin in cells, treat with ALLN (or not), and then purify His-ubiquitylated proteins by Ni^{2+}-column chromatography before running them on gels (lanes 5 and 6). In all cases you detect β-catenin specifically, using antibodies directed against it.

Are the slower migrating forms of β-catenin due to phosphorylation or ubiquitylation? Explain your answer.

15–93   β-Catenin can be phosphorylated by glycogen synthase kinase-3β (GSK-3β) and it can be degraded in proteasomes. β-Catenin could be sensitized for degradation by phosphorylation, it could be protected from degradation by phosphorylation, or its phosphorylation status could be irrelevant for degradation. To distinguish among these alternatives, you generate cell lines that express either a mutant GSK-3β that cannot carry out phosphorylation or a mutant β-catenin that is missing its site of phosphorylation. In the presence and absence of the proteasome inhibitor, ALLN, both cell lines yield β-catenin that migrates as a single band, with no slower migrating bands visible, in contrast to the situation with nonmutant β-catenin and GSK-3β (which is shown in Figure 15–35, lanes 1 and 2). What is the relationship between β-catenin phosphorylation and its degradation in proteasomes? Explain your answer.

15–94*   The Wnt signaling pathway normally ensures that each wing cell in *Drosophila* has a single hair. Overexpression of the *frizzled* gene from a heat-shock promoter (hs-*Fz*) causes multiple hairs to grow from many cells (Figure 15–36A). This phenotype is suppressed if hs-*Fz* is combined with a heterozygous deletion (*dsh*Δ) of the *disheveled* gene (Figure 15–36B). Do these results allow you to order the action of Frizzled and Dishevelled? If so, what is the order? Explain your reasoning.

15–95   There are two mutational routes to the uncontrolled cell proliferation and invasiveness that characterize cancer cells. The first is to make a stimulatory gene (a proto-oncogene) hyperactive: this type of mutation has a dominant effect so that only one of the cell's two gene copies needs to undergo change. The second is to make an inhibitory gene (a tumor suppressor gene) inactive: this type of mutation usually has a recessive effect so that both the cell's gene copies must be inactivated.

Mutations of the *APC* (adenomatous polyposis coli) gene occur in 80% of human colon cancers. Normal APC increases the affinity of the degradation complex for β-catenin, which in excess can enter the nucleus and promote transcription of key target genes for cell proliferation. Given this information, which category—oncogene or tumor suppressor—would you expect the *APC* gene to belong to? Why?

15–96*   The *hedgehog* gene encodes a protein 471 amino acids long. The hedgehog precursor protein (Figure 15–37A) is normally cleaved between glycine 257 (G257) and cysteine 258 (C258) to generate a fragment that is active in local

(A) HEDGEHOG PRECURSOR

G257   C258

1   471
N—————————————C

(B) TIME COURSE

0   0.5 1 2 4 8 16
hours

(C) CONCENTRATION DEPENDENCE

0.05   0.2   0.8   3.2   12.8
μM

Figure 15–37 Mechanism of cleavage of the hedgehog precursor protein (Problem 15–96). (A) Site of cleavage in the hedgehog precursor protein. (B) Time course of cleavage of the fragment of the precursor protein. (C) Dependence of cleavage on concentration of the precursor protein fragment.

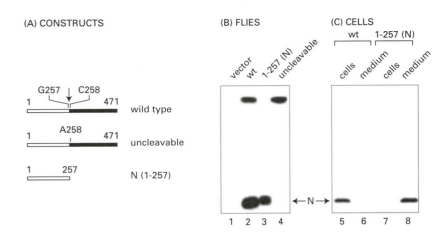

Figure 15–38 Fate of the fragments of hedgehog after cleavage (Problem 15–97). (A) Constructs encoding different forms of the hedgehog precursor protein. (B) Results of expression in *Drosophila* embryos. (C) Results of expression in insect cells. Hedgehog fragments were detected using antibodies specific for the N-terminal segment.

and long-range signaling. Cleavage is essential for signaling. After expression in *E. coli*, you purify a segment of the *hedgehog* gene encoding a portion of the protein that includes the cleavage site and the entire C-terminus. When you incubate this purified protein in buffer, you observe cleavage over the course of several hours, as shown in Figure 15–37B. If you vary its concentration over a 256-fold range and assay cleavage after 4 hours of incubation, you observe the results shown in Figure 15–37C.

A. Explain how these data support the idea that the hedgehog precursor protein cleaves itself. How do they rule out the possibility that the purified protein is contaminated with a bacterial protease, for example?

B. Does a molecule of precursor protein cleave itself, or does it cleave another molecule of the precursor; that is, is the reaction intramolecular or intermolecular?

15–97   To find out what happens to the fragments of hedgehog after cleavage, you express three versions: wild-type hedgehog precursor, an uncleavable form, and the N-terminal cleavage product (Figure 15–38A). In fly embryos the constructs behave as expected: wild-type hedgehog is cleaved, the uncleavable version is not, and the N-terminal segment is expressed (Figure 15–38B). When wild-type hedgehog and the N-terminal segment are expressed in insect cells, however, the N-terminal segment from wild-type hedgehog remains associated with the cells, while the synthesized N-terminal segment is secreted into the medium (Figure 15–38C). Can you suggest possible explanations for the difference in localization of the N-terminal segment?

15–98*   If you overexpress various hedgehog contructs (see Figure 15–38A) in flies and examine the pattern of Wnt expression (a well-characterized target of hedgehog signaling), you observe a striped pattern of expression in all cases, but some constructs lead to thicker stripes than normal (Figure 15–39).

A. Which part of the hedgehog molecule is responsible for signaling?

B. All the cells in the embryo are overexpressing the various hedgehog constructs. Why is it, do you suppose, that you observe the same basic striped pattern of Wnt expression in all of them?

C. Why do you see stripes of Wnt expression even in the absence of hedgehog overexpression?

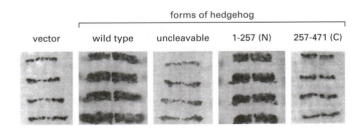

Figure 15–39 Patterns of Wnt expression in *Drosophila* embryos that are overexpressing various hedgehog constructs (Problem 15–98). Wnt expression was detected by *in situ* hybridization.

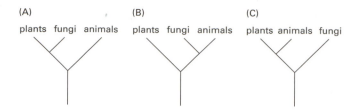

15–99   Latent gene regulatory proteins are prevented from entering the nucleus until the cell receives an appropriate signal. List four ways by which cells keep gene regulatory proteins out of the nucleus, and give an example of a latent gene regulatory protein that is controlled by each mechanism.

Problem 12–44 examines the spatial control of the latent gene regulatory protein, NF-κB

SIGNALING IN PLANTS

TERMS TO LEARN

CLAVATA 1 (CLV1)	growth regulator	phototropin
cryptochrome	leucine-rich repeat (LRR) protein	phytochrome
ethylene		

15–100* The last common ancestor to plants and animals was a unicellular eucaryote. Thus, it is thought that multicellularity and the attendant demands for cell communication arose independently in these two lineages. This evolutionary viewpoint accounts nicely for the vastly different mechanisms that plants and animals use for cell communication. Fungi use signaling mechanisms and components that are very similar to those used in animals. Which of the phylogenetic trees shown in Figure 15–40 does this observation support?

15–101   If signaling arose as a solution to the demands of multicellularity, how then do you account for the very similar mechanisms of signaling that are used in animals and the unicellular fungus *Saccharomyces cerevisiae*?

15–102* In plant cells the cortical array of microtubules determines the orientation of cellulose microfibrils, which in turn fixes the direction of cell expansion. Cells elongate perpendicular to the orientation of the cellulose microfibrils. The plant growth factors ethylene and gibberellic acid have opposite effects on the orientation of microtubule arrays in epidermal cells of young pea shoots. Gibberellic acid promotes an orientation of the cortical microtubule array that is perpendicular to the long axis of the cell, whereas ethylene treatment causes the microtubule arrays to orient parallel to the long axis of the cell (Figure 15–41).

Which treatment do you think would produce short, fat shoots and which treatment would produce long, thin shoots?

15–103   The ripening of fruit is a complicated process of development, differentiation, and death (except for the seeds, of course). The process is triggered by minute amounts of ethylene gas. (This was discovered by accident many years ago; the paraffin stoves used to heat greenhouses in the olden days gave off enough ethylene to initiate the process.) The ethylene is normally produced by the fruits themselves in a biochemical pathway, the rate-limiting step of which is controlled by ACC synthase, which converts *S*-adenosylmethionine to a cyclopropane compound that is the immediate precursor of ethylene. Ethylene initiates a program of sequential gene expression that includes production of several new enzymes, including polygalacturonase, which probably contributes to softening the cell wall.

Your company, Agribucks, is trying to make mutant tomatoes that cannot synthesize their own ethylene. Such fruit could be allowed to stay longer on the vine, developing their flavor while remaining green and firm. They could

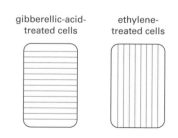

gibberellic-acid-treated cells   ethylene-treated cells

Figure 15–41 Effects of gibberellic acid and ethylene on the orientation of cortical arrays of microtubules (Problem 15–102).

be shipped in this robust unripe state and exposed to ethylene just before arrival at market. This should allow them to be sold at the peak of perfection, and the procedure involves no artificial additives of any kind.

You decide to use an antisense approach, which works especially well in plants. You place an ACC synthase cDNA into a plant expression vector so that the gene will be transcribed backward, introduce it into tomato cells, and regenerate whole tomato plants. Sure enough, ethylene production is inhibited by 99.5% in these transgenic tomato plants, and their fruit fails to ripen. But when placed in air containing a small amount of ethylene, they turn into beautiful, tasty ripe red fruit in about 2 weeks.

A. How do you imagine that transcribing the ACC synthase gene backward blocks the production of ethylene?
B. Will you be a millionaire before you are 30?

15–104* How is it that plant growth regulators can be present throughout a plant and yet have specific effects on particular cells and tissues?

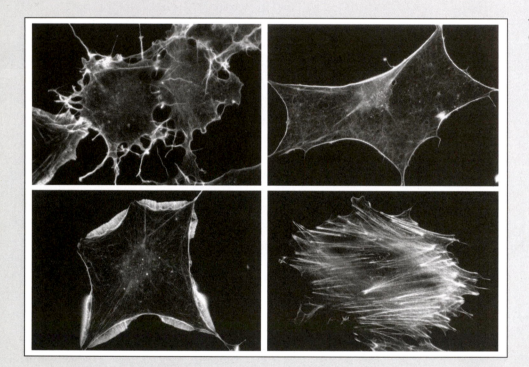

The actin cytoskeleton is highly plastic.
Courtesy of Kate Nobes.

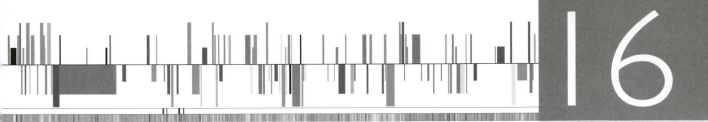

THE CYTOSKELETON

THE SELF-ASSEMBLY AND DYNAMIC STRUCTURE OF CYTOSKELETAL FILAMENTS

TERMS TO LEARN

actin	keratin	protofilament
cytoskeleton	minus end	tubulin
dynamic instability	neurofilament	treadmilling
intermediate filament	plus end	

16–1   In general terms what are the cellular functions of intermediate filaments, microtubules, and actin filaments?

16–2*   If each type of cytoskeletal filament is made up of subunits that are held together by weak noncovalent bonds, how is it possible for a human being to lift heavy objects?

16–3   A typical time course of polymerization of actin filaments from actin subunits is shown in Figure 16–1.
   A. Explain the properties of actin polymerization that account for each of the three phases of the polymerization curve.
   B. How would the curve change if you doubled the concentration of actin? Would the concentration of free actin be higher or lower than in the original experiment, or would it be the same in both?

---

*Answers to these problems are available to Instructors at www.classwire.com/garlandscience/

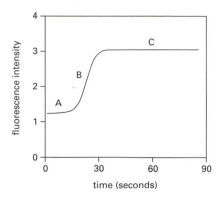

Figure 16–1 Formation of actin filaments with time, starting with purified actin monomers that are labeled with a fluorescent probe (Problem 16–3). Upon polymerization the probe fluorescence increases, which allows polymerization to be measured. The intensity of fluorescence at zero seconds is due to the background fluorescence of the actin monomers. The three phases of polymerization are indicated as A, B, and C.

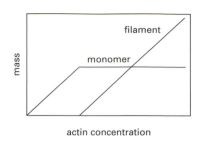

Figure 16–2 Mass of actin monomers and filaments as a function of actin concentration (Problem 16–4).

16–4*   Figure 16–2 shows the mass of actin in free subunits (monomers) and filaments at equilibrium as a function of actin concentration. Indicate the critical concentration of actin on this diagram.

16–5   Why do you suppose it is much easier to add tubulin to existing microtubules than to start a new microtubule from scratch?

16–6*   (True/False) The structural polarity of all microtubules is such that α-tubulin is exposed at one end and β-tubulin is exposed at the opposite end. Explain your answer.

16–7   In a 13-filament microtubule the majority of lateral interactions are between like subunits, with α-tubulin binding to α-tubulin and β-tubulin binding to β-tubulin. Between the first and thirteenth protofilaments, however, there is a seam at which α-tubulin interacts with β-tubulin (Figure 16–3). Are these heterotypic interactions (α with β) likely to be stronger than, weaker than, or the same strength as homotypic interactions (α with α or β with β)? Explain your reasoning.

16–8*   If you add short actin filaments marked by bound myosin heads (myosin-decorated filaments) to a solution with an excess of actin monomers, wait for a few minutes and then examine the filaments by electron microscopy, you see the picture shown in Figure 16–4.
   A. Which is the plus end of the myosin-decorated filament and which is the minus end? Which is the 'barbed' end and which is the 'pointed' end? How can you tell?
   B. If you now diluted the mixture so that the actin concentration was below the critical concentration, which end would depolymerize more rapidly?
   C. When the actin filament depolymerizes, why are subunits removed exclusively from the ends and not from the middle of the filament?

16–9   The orientation of the αβ-tubulin dimer in a microtubule was determined in several ways. For example, GTP-coated fluorescent beads were found to bind exclusively at the plus ends of microtubules, whereas gold beads coated with antibodies specific for a peptide of α-tubulin bound exclusively at the minus end. How do these observations define the orientation of the αβ-tubulin dimer in the microtubule? Which tubulin subunit is at which end? Explain your reasoning.

16–10*   At 1.4 mg/mL pure tubulin, microtubules grow at a rate of about 2 μm/min. At this growth rate how many αβ-tubulin dimers (8 nm in length) are added to the ends of a microtubule each second?

16–11   Imagine that the polymer in Figure 16–5A can add subunits at either end, just like actin filaments and microtubules. Imagine also three hypothetical types of free subunit, as shown in Figure 16–5B. Each type of subunit can add to the polymer and, once added, it adopts the conformation of the other subunits in the polymer (Figure 16–5A). For each of these subunits, decide

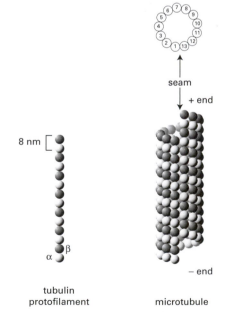

Figure 16–3 Structure of a 13-protofilament microtubule, showing the seam between the first and thirteenth protofilaments (Problem 16–7).

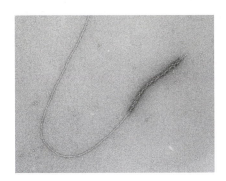

Figure 16–4 Myosin-decorated actin filament after a few minutes in a solution with excess actin monomers (Problem 16–8).

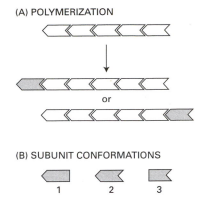

(B) SUBUNIT CONFORMATIONS

Figure 16–5 Polymerization of a polymer (Problem 16–11). (A) Addition of a subunit to a polymer. (B) Conformations of three hypothetical subunits.

which end of the polymer, if either, will grow the faster when the concentration of that subunit is higher than the critical concentration required for polymerization. Explain your reasoning. For any of the subunits, will there be a concentration at which one end will preferentially grow while the other shrinks? Why or why not?

16–12* The complex kinetics of microtubule assembly make it hard to predict the behavior of individual microtubules. Some microtubules in a population can grow, even as the majority shrink to nothing. One simple hypothesis to explain this behavior is that growing ends are protected from disassembly by a GTP cap and that faster growing ends have a longer GTP cap. Real-time video observations of changes in length with time are shown for two individual microtubules in Figure 16–6. Measurements of their rates of growth and shrinkage show that the plus end of each microtubule grows three times faster and shrinks at half the rate of the minus end.

A. Are changes in length at the two ends of a microtubule dependent or independent of one another? How can you tell?

B. What does the simple GTP-cap hypothesis predict about the rate of switching between growing and shrinking states at the fast-growing end relative to the slow-growing end? Does the outcome of this experiment support the simple GTP-cap hypothesis?

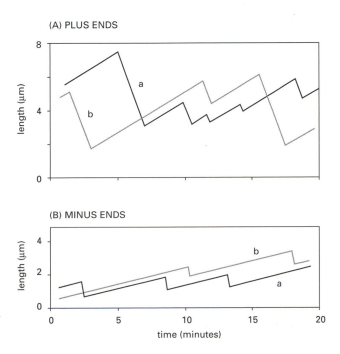

Figure 16–6 Analysis of growth kinetics of individual microtubules (Problem 16–12). (A) Changes in length at the plus ends. (B) Changes in length at the minus ends. Results from two individual microtubules are identified with a and b.

THE SELF-ASSEMBLY AND DYNAMIC STRUCTURE OF CYTOSKELETAL FILAMENTS

16–13 The growth rates at the plus and minus ends of actin filaments as a function of actin concentration are shown in Figure 16–7A and, on an expanded scale, in Figure 16–7B.

  A. The data in Figure 16–7A were gathered by measuring initial growth rates at each actin concentration. Similar data gathered for any Michaelis–Menten enzyme would generate a hyperbolic plot, instead of the linear plots shown here. Why does the growth rate of actin filaments continue to increase linearly with increasing actin concentration, while an enzyme-catalyzed reaction reaches a plateau with increasing substrate concentration?

  B. Figure 16–7B shows the filament growth rates at low actin concentration on an expanded scale. Imagine that you could add actin filaments of defined length to a solution of actin subunits at the concentrations indicated as A, B, C, D, and E. For each of these concentrations decide whether the added actin filament would grow or shrink at its plus and minus ends. What is the critical concentration for the plus end? What is the critical concentration for the minus end? Would treadmilling occur at any of these concentrations?

16–14* The microtubules in Figure 16–8A were obtained from a population that was growing rapidly, whereas the one in Figure 16–8B came from microtubules undergoing catastrophic shrinkage. Comment on any differences between the two images and suggest likely explanations for those you observe.

16–15 Dynamic instability causes microtubules either to grow or shrink rapidly. Consider an individual microtubule that is in its shrinking phase.

  A. What must happen at the end of the microtubule in order for it to stop shrinking and start growing?

  B. How would a change in the tubulin concentration affect this switch?

  C. What would happen if GDP, but no GTP, were present in the solution?

  D. What would happen if the solution contained an analog of GTP that cannot be hydrolyzed?

16–16* Comparisons of microtubule behavior between species point to differences that raise questions about the biological importance of dynamic instability. Notothenioid fishes, for example, which live in the Southern Ocean at a constant temperature of –1.8°C, have remarkably stable microtubules compared with warm-blooded vertebrates such as the cow. This is an essential modification for notothenioid fish because normal microtubules disassemble completely into αβ-tubulin dimers at 0°C. Measurements on individual microtubules in solutions of pure tubulin show that fish microtubules grow at a much slower rate, shrink at a much slower rate, and only rarely switch from growth to shrinkage (catastrophe) or from shrinkage to growth (rescue) (Table 16–1).

  A. The amino acid sequences of the α- and β-tubulin subunits from notothenioid fish differ from those of the cow at positions and in ways that might reasonably be expected to stabilize the microtubule, in accord with the data in Table 16–1. Would you expect these changes to strengthen the interactions between the α- and β-tubulin subunits in the αβ-dimer, between adjacent dimers in the protofilament, or between tubulin subunits in adjacent protofilaments? Explain your reasoning.

  B. Dynamic instability is thought to play a fundamental role in the rapid microtubule rearrangements that occur in cells. How do you suppose cells in these notothenioid fishes manage to alter their microtubule architecture quickly enough to accomplish essential cell functions? Or do you suppose that these cells exist with a stable microtubule cytoskeleton that only slowly rearranges itself?

16–17 A standard purification scheme for tubulin is to prepare a cell extract, chill it to 0°C, spin it at high speed and save the supernatant. The supernatant is then warmed to 37°C and incubated in the presence of GTP. The mixture is then spun at high speed and the pellet is saved and redissolved. Then the cycle is repeated: chill the dissolved pellet, spin at high speed, save the supernatant, incubate with GTP at 37°C, spin at high speed, save the pellet.

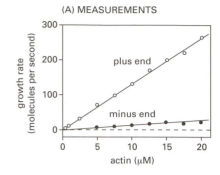

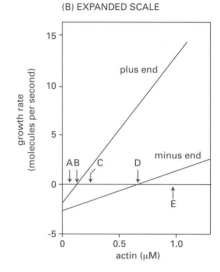

Figure 16–7 Growth rates at the plus and minus ends of actin filaments as a function of actin concentration (Problem 16–13). (A) Measurements of growth rates over a broad range of actin concentrations. (B) Growth rates at low actin concentrations, shown on an expanded scale.

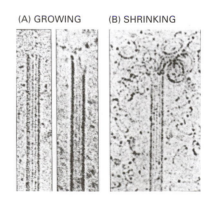

Figure 16–8 Electron microscopic analysis of microtubule dynamics (Problem 16–14). (A) Rapidly growing microtubules. (B) Catastrophically shrinking microtubules.

TABLE 16–1 Properties of individual microtubules in notothenioid fish and the domestic cow (Problem 16–16).

MICROTUBULES	GROWTH RATE (µm/min)	SHRINKAGE RATE (µm/min)	CATASTROPHE FREQUENCY (min⁻¹)	RESCUE FREQUENCY (min⁻¹)
Notothenioid fish	0.27	−0.8	0.008	<0.0004
Domestic cow	2.18	−61.2	0.52	3.1

Multiple individual microtubules were observed by videomicroscopy near the body temperature for each species: 5°C for fish and 37°C for cow. Average growth rates were calculated for growing microtubules, and average shrinkage rates were calculated for shrinking microtubules. Changes from growth to shrinkage (catastrophe) and from shrinkage to growth (rescue) were averaged over the observation period and expressed as frequency of events per minute.

After a few cycles one obtains a pure preparation of tubulin. Explain how this procedure yields pure tubulin.

16–18* The β-tubulin subunit of an αβ-tubulin dimer retains its bound GTP for a short time after it is added to a microtubule, yielding a GTP cap whose size depends on the relative rates of polymerization and GTP hydrolysis. A simple notion about microtubule growth dynamics is that the ends with GTP caps grow, whereas ends without GTP caps shrink. To test this idea, you allow microtubules to form under conditions where you can watch individual microtubules. You then sever one microtubule in the middle using a laser beam. Would you expect the newly exposed plus and minus ends to grow or to shrink? Explain your answer.

16–19 Your ultimate goal is to understand human consciousness, but first, your advisor wants you to understand some basic facts about actin assembly. He tells you that ATP binds to actin monomers and is required for assembly, but that ATP hydrolysis is not necessary for polymerization, since ADP can, under certain circumstances, substitute for the ATP requirement. ADP filaments, however, are much less stable than ATP filaments, supporting your secret suspicion that the free energy of ATP hydrolysis really is used to drive actin assembly.

Your advisor suggests that you make careful measurements of the quantitative relationship between the number of ATP molecules hydrolyzed and the number of actin monomers linked into polymer. The experiments are straightforward. To measure ATP hydrolysis, you add ATPγ³²P to a solution of polymerizing actin, take samples at short intervals, and determine how much radioactive phosphate has been produced. To assay polymerization, you measure the increase in light scattering that is caused by formation of the actin filaments. Your results are shown in Figure 16–9. Your light-scattering measurements indicate that 20 µmoles of actin monomers were polymerized. Since the number of polymerized actin monomers matches exactly the number of ATP molecules hydrolyzed, you conclude that one ATP is hydrolyzed as each new monomer is added to an actin filament.

When you show your advisor the data and tell him your conclusions, he smiles and very gently tells you to look more closely at the graph. He says your data prove that actin can polymerize without ATP hydrolysis.

A. What does your advisor see in the data that you have overlooked?
B. What do your data imply about the distribution of ATP and ADP in polymerizing actin filaments?

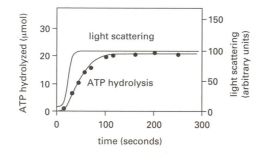

Figure 16–9 The kinetics of actin polymerization and ATP hydrolysis (Problem 16–19).

16–20* (**True/False**) The role of ATP hydrolysis in actin polymerization is similar to the role of GTP hydrolysis in tubulin polymerization: both serve to weaken the bonds in the polymer and thereby promote depolymerization. Explain your answer.

16–21 The average time taken for particles to diffuse a distance of x cm is given by the formula

$$t = x^2/2D$$

where t is the time in seconds and D is the diffusion coefficient, which is a constant that depends on the size and shape of the particle.
A. Using this formula, calculate the average time it would take for a small molecule, a protein molecule, and a membrane-enclosed vesicle to diffuse across a cell 10 μm in diameter. A typical diffusion coefficient for a small molecule is 5×10^{-6} cm^2/sec, for a protein molecule 5×10^{-7} cm^2/sec, and for a membrane vesicle 5×10^{-8} cm^2/sec.
B. Why do you suppose a cell relies on the strategy of polymerizing and depolymerizing cytoskeletal filaments rather than on diffusion of the filaments themselves to accomplish its cytoskeletal rearrangements?

16–22* Tubulin is distantly related to the GTPase family that includes Ras. The β-tubulin in an αβ-tubulin dimer hydrolyzes GTP at a slow intrinsic rate. When it is incorporated into a growing microtubule, however, its rate of GTP hydrolysis increases dramatically: like Ras when it interacts with a GTPase activating protein (GAP). What is the GAP for β-tubulin?

16–23 List differences between bacteria and animal cells that could have depended on the appearance during evolution of some or all of the components of the present eucaryotic cytoskeleton. Explain why a cytoskeleton might have been crucial for these differences to evolve.

16–24* The amino acid sequences of actins and tubulins from all eucaryotes are remarkably well conserved, yet the large numbers of proteins that interact with these filaments are no more conserved than most other proteins in different species. How can it be that the filament proteins themselves are highly conserved while the proteins that interact with these filaments are not?

16–25 (**True/False**) Like actin filaments and microtubules, cytoplasmic intermediate filaments are found in all eucaryotes. Explain your answer.

16–26* Why is it that intermediate filaments have identical ends and lack polarity, whereas actin filaments and microtubules have two distinct ends with a defined polarity?

Problem 3–31 examines the sequence of an intermediate filament protein for the characteristic coiled-coil heptad repeat motif.

16–27 Which of the following types of cell would you expect to contain a high density of cytoplasmic intermediate filaments? Explain your answers.
A. *Amoeba proteus* (a free-living amoeba).
B. Human skin epithelial cell.
C. Smooth muscle cell in the digestive tract of a vertebrate.
D. Nerve cell in the spinal cord of a mouse.
E. Human sperm cell.
F. Plant cell.

16–28* A particular antibody specific for intermediate filaments causes fragmentation of vimentin filaments formed *in vitro*. A few hours after this antibody was injected into cells, the intermediate filaments, which were initially distributed throughout the cytosol, were observed at the nuclear periphery. One explanation for this observation is that the network of intermediate filaments disassembled after antibody injection and reassembled in the perinuclear region. An alternative hypothesis is that the antibody disrupted the interactions between intermediate filaments and the rest of the cytoskeleton, and the network of intermediate filaments collapsed to the nuclear periphery. Assuming that you could construct a vimentin–GFP (green

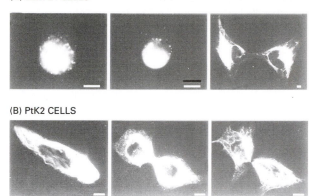

(A) BHK-21 CELLS

(B) PtK2 CELLS

Figure 16–10 Vimentin networks during cell division (Problem 16–29). (A) BHK–21 cells. The three images from *left* to *right* correspond to prometaphase and anaphase of mitosis and to the daughter cells. (B) PtK2 cells. The images from *left* to *right* correspond to prometaphase, telophase, and late cytokinesis. Note that in late cytokinesis the cells are still connected by a bridge of cytoplasm. The scale bar in all pictures is 5 μm.

fluorescent protein) fusion that could be expressed in cells and become incorporated into vimentin filaments, how might you use real-time video recording of antibody-treated cells to distinguish between these alternative hypotheses?

16–29   The intermediate filament networks in cells must be dealt with in some way when a cell divides. The vimentin networks (tagged with vimentin–GFP) in two different cell types—BHK-21 (baby hamster kidney) cells and PtK2 (*Potorous tridactylis*—rat kangaroo—kidney) cells—are shown undergoing division in Figure 16–10. By examining these photographs, decide how each of these cell types handles its vimentin network.

16–30*   Although the mechanism of disassembly of most intermediate filaments is unclear, it is well defined for their ancestors, the nuclear lamins. The nuclear envelope is strengthened by a fibrous meshwork of lamins (the nuclear lamina), which supports the membrane on the nuclear side. When cells enter mitosis, the nuclear envelope breaks down and the nuclear lamina disassembles. Assembly and disassembly of the nuclear lamina may be controlled by reversible phosphorylation of lamins A, B, and C, since the lamins from cells that are in mitosis carry significantly more phosphate than do the lamins from cells that are in interphase.

   To investigate the role of phosphorylation, you label cells with ^{35}S-methionine and then purify lamins A, B, and C from mitotic cells and from interphase cells. You then analyze each of the purified lamins and a mixture of the lamins from mitotic and interphase cells by two-dimensional gel electrophoresis (Figure 16–11A). You also treat identical samples with alkaline phosphatase, which removes phosphates from proteins, and analyze them in the same way (Figure 16–11B).

A. Why does treatment with alkaline phosphatase reduce the number of lamin spots to three, regardless of the number seen in the absence of phosphatase treatment?

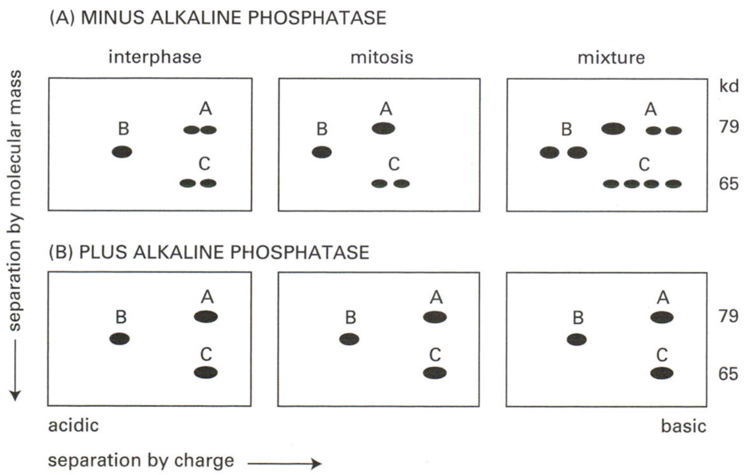

(A) MINUS ALKALINE PHOSPHATASE

separation by molecular mass

interphase   mitosis   mixture

(B) PLUS ALKALINE PHOSPHATASE

acidic   basic

separation by charge

Figure 16–11 Two-dimensional separation of nuclear lamins from cells in interphase and mitosis (Problem 16–30). (A) No treatment with alkaline phosphatase. (B) Treatment with alkaline phosphatase. *Letters* identify the positions of lamins A, B, and C. The purified lamins from interphase and mitotic cells were added together to create the mixture. Acidic proteins are more negatively charged; basic proteins are more positively charged.

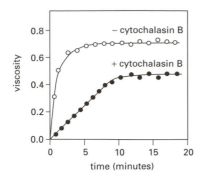

Figure 16–12 Increase in the viscosity of actin solutions in the presence and absence of cytochalasin B (Problem 16–33).

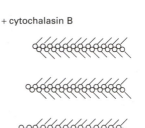

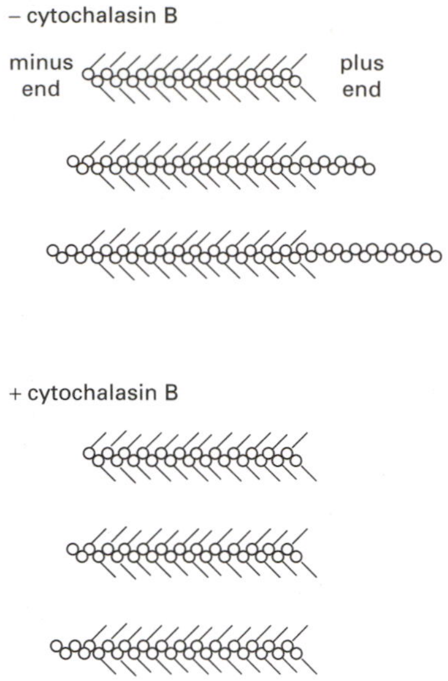

Figure 16–13 Appearance of typical actin filaments formed in the presence and absence of cytochalasin B (Problem 16–33). The decorated actin filaments present before addition of actin monomers are shown at the *top*. Filaments present after increasing times of incubation in the presence of actin monomers are shown *below*.

B. How many phosphate groups are attached to lamins A, B, and C during interphase? How many are attached during mitosis? How can you tell?

C. Why was ^{35}S-methionine rather than ^{32}P-phosphate used to label lamins in experiments designed to measure phosphorylation differences? How would the autoradiograms have differed if ^{32}P-phosphate had been used instead?

D. Do you think that these results prove that lamin disassembly during mitosis is caused by their reversible phosphorylation? Why or why not?

16–31   Disulfide bonds do not form in the cytosol of eucaryotic cells (see Problem 3–33). Yet keratin intermediate filaments in the skin are cross-linked by disulfide bonds. How can that be?

16–32*   Although knockouts of genes for some intermediate filaments have detectable phenotypes in mice, knockouts of genes for vimentin and glial fibrillary acid protein (GFAP) appear normal. Mice with a combined deficiency of vimentin and GFAP, however, exhibit impaired function of their astrocytes, which are accessory cells in the central nervous system. Why do you suppose that individual knockouts for vimentin and GFAP are normal, whereas the combined knockout has a demonstrable deficiency?

16–33   Cytochalasin B strongly inhibits certain forms of cell motility, such as cytokinesis and the ruffling of growth cones, and it dramatically decreases the viscosity of gels formed with mixtures of actin and a wide variety of actin-binding proteins. These observations suggest that cytochalasin B interferes with the assembly of actin filaments. In the classic experiment that defined its mechanism, short lengths of actin filaments were decorated with myosin heads and then mixed with actin subunits in the presence or absence of cytochalasin B. Assembly of actin filaments was measured by assaying the viscosity of the solution (Figure 16–12) and by examining samples by electron microscopy (Figure 16–13).

A. Suggest a plausible mechanism to explain how cytochalasin B inhibits actin filament assembly. Account for the appearance of the filaments in the electron micrographs and the viscosity measurements (both the altered rate and extent).

B. The normal growth characteristics of an actin filament and the actin-binding properties of cytochalasin B argue that actin monomers undergo a conformational change upon addition to an actin filament. How so?

16–34*   Phalloidin, which is a toxic peptide from the mushroom *Amanita phalloides*, binds to actin filaments. Phalloidin tagged with a fluorescent probe is commonly used to stain actin filament assemblies in cells (Figure 16–14A).

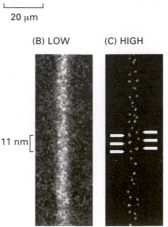

(A) CELLS

20 μm

(B) LOW   (C) HIGH

11 nm

Figure 16–14 Binding of phalloidin to actin filaments (Problem 16–34). (A) The actin cytoskeleton stained with fluorescent phalloidin. (B) An actin filament bound by gold-tagged phalloidin at low contrast. (C) The same actin filament as in (B), but at high contrast. *Bright bands* mark the positions of six gold particles.

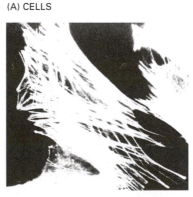

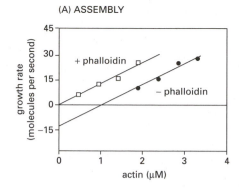

(A) ASSEMBLY

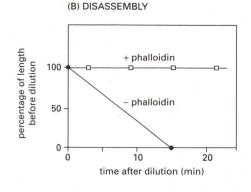

(B) DISASSEMBLY

Figure 16–15 Effects of phalloidin on actin filaments (Problem 16–35). (A) Growth rates at the minus ends of acrosomal bundles in the presence and absence of phalloidin. (B) Disassembly of actin filaments upon dilution in the presence and absence of phalloidin.

If phalloidin is attached to a gold particle instead, its binding to actin filaments can be examined at high resolution by scanning transmission electron microscopy. Figure 16–14B shows a micrograph of an actin filament with bound phalloidin and Figure 16–14C shows the same picture with the contrast adjusted so that only the points of highest intensity (the gold particles) are visible. Does phalloidin bind to every actin subunit? How can you tell?

16–35   Isolated bundles of actin filaments from the acrosomal processes of *Limulus polyphemus* (horseshoe crab) sperm have readily distinguishable plus ends (tapered) and minus ends (blunt). Assembly at the ends of such bundles was used to determine the mechanism of action of phalloidin, which has a marked effect on actin assembly. When phalloidin is mixed with actin in a molar ratio of at least 1:1, the growth rate increases at both ends, as shown for minus ends in Figure 16–15A. Because growth rate = $k_{on}[\text{actin}]_{initial} - k_{off}$, these plots have the form $y = mx + b$, so that the slope of the line equals k_{on} and the y intercept equals $-k_{off}$.

A. How does phalloidin increase the growth rate of actin filaments? Explain your reasoning.

B. In Figure 16–15B actin filaments grown in the presence or absence of phalloidin were diluted in the absence of actin monomers and their disassembly was assayed. Do these results confirm or contradict your conclusions from part A? Explain your answer.

C. What is the critical concentration for actin assembly at the minus end in the absence of phalloidin? What is the critical concentration for actin assembly at the minus end in the presence of phalloidin?

D. Propose a molecular mechanism for the effects of phalloidin on actin assembly.

16–36* Swinholide A is a member of a class of lipophilic compounds termed macrolides, which include a number of useful antibiotics such as erythromycin, that are derived naturally from Actinomycetes. Swinholide A is a 'twin' molecule, composed of two identical halves (Figure 16–16A). When

Figure 16–16 Effects of swinholide A on actin filaments (Problem 16–36). (A) Structure of swinholide A. The identical halves of swinholide are arranged head to tail, so that if the molecule were rotated 180° about the indicated axis (circle with an X in it), it would superimpose on itself. For this reason it is said to have a twofold axis of symmetry. (B) Time course of actin filament depolymerization in the presence and absence of swinholide A. *Numbers* indicate the concentration of swinholide A (nM) used in each depolymerization assay. Depolymerization was measured by stopped-flow kinetics, which allows extremely rapid measurements. (C) Initial rates of depolymerization as a function of swinholide A concentration.

(A) SWINHOLIDE STRUCTURE

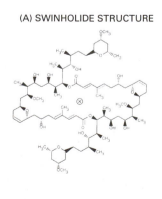

(B) DISASSEMBLY ASSAY

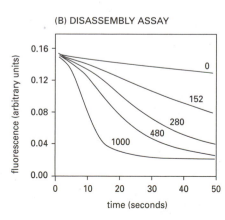

(C) SWINHOLIDE TITRATION

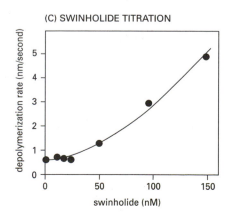

added to cells growing in culture, swinholide A disrupts the actin cyto-skeleton. Your advisor has shown conclusively that swinholide A binds a pair of actin monomers. He suspects that swinholide A causes actin filaments to depolymerize by sequestering actin subunits in a nonfunctional dimeric form and thus accelerating depolymerization through mass action effects. He wants you to test this hypothesis.

You prepare actin filaments tagged with a probe that fluoresces intensely in the filament, but much less so on the free subunits (or swinholide-bound subunits). This allows you to follow depolymerization readily and rapidly as a loss of fluorescence (see Problem 16–3). Just as your advisor predicted, depolymerization increases in the presence of increasing concentrations of swinholide A (Figure 16–16B). But you notice two features of these curves that suggest to you that swinholide A may actually sever actin filaments. One of these features is illustrated in Figure 16–16C, which shows a nonlinear dependence of the initial rate of depolymerization on the concentration of swinholide A. A simple mass-action effect—the sequestering of actin monomers by binding to swinholide A—predicts a linear dependence; however, increasing increments in swinholide A concentration have a progressively greater effect on depolymerization.

A. In Figure 16–16B, why does fluorescence reach a plateau value (at about 0.03) instead of decreasing to zero?

B. The other odd feature you noticed about depolymerization in the presence of swinholide A (Figure 16–16B) is that the lines have a 'hump' in them in the first few seconds (before they reach the plateau at later times). Why does this hump suggest that swinholide A severs actin filaments?

C. Assuming that swinholide A does sever actin filaments, is one molecule enough, or are multiple molecules needed? How do you know?

16–37   The drug taxol, extracted from the bark of yew trees, has the opposite effect of the drug colchicine, an alkaloid from autumn crocus. Taxol binds tightly to microtubules and stabilizes them. When added to cells, it causes much of the free tubulin to assemble into microtubules. In contrast, colchicine prevents microtubule formation. Taxol is just as toxic to dividing cells as colchicine, and both are used as anticancer drugs. Based on your knowledge of microtubule dynamics, suggest why both drugs are toxic to dividing cells despite their opposite modes of action.

16–38*   The common laboratory reagent acrylamide, used as a precursor in making polyacrylamide gels, is a potent neurotoxin. One hypothesis for its toxic effects is that it destroys neurofilament bundles by binding to the subunits of these intermediate filaments, causing their depolymerization. To test this possibility, you compare acrylamide toxicity in normal mice and knockout mice that are lacking neurofilaments. Surprisingly, although these knockout mice have neurons with smaller diameters, they have no obvious mutant phenotype. You find that acrylamide is an equally potent neurotoxin in normal mice and the knockout mice. Is acrylamide toxicity mediated through its effects on neurofilaments?

HOW CELLS REGULATE THEIR CYTOSKELETAL FILAMENTS

TERMS TO LEARN

ARP complex	focal contact	Rho protein family
cell cortex	γ-tubulin ring complex (γ-TuRC)	WASP protein
centriole	microtubule-associated protein (MAP)	
centrosome	microtubule-organizing center (MTOC)	

16–39   (**True/False**) All microtubule-organizing centers contain centrioles that help nucleate microtubule polymerization. Explain your answer.

16–40* The function of microtubules depends on their specific spatial organization within the cell. How are specific arrangements created, and what determines the formation and disappearance of individual microtubules?

To address these questions, investigators have studied the *in vitro* assembly of αβ-tubulin dimers into microtubules. Below 15 µM αβ-tubulin no microtubules are formed, but above 15 µM, microtubules form readily (Figure 16–17A). If centrosomes are added to the solution of tubulin, microtubules begin to form at less than 5 µM (Figure 16–17B). (Different assays were used in the two experiments—total weight of microtubules in Figure 16–17A and the average number of microtubules per centrosome in Figure 16–17B—but the lowering of the critical concentration for microtubule assembly in the presence of centrosomes is independent of the method of assay.)

A. Why do you think that the concentration at which microtubules begin to form (the critical concentration) is different in the two experiments?

B. Why do you think that the plot in Figure 16–17A increases linearly with increasing tubulin concentration above 15 µM, whereas the plot in Figure 16–17B reaches a plateau at about 25 µM?

C. The concentration of αβ-tubulin dimers (the subunits for assembly) in a typical cell is 1 mg/mL and the molecular weight of a tubulin dimer is 11,000. What is the molar concentration of tubulin dimers in cells? How does the cellular concentration compare with the critical concentrations in the two experiments in Figure 16–17? What are the implications for assembly of microtubules in cells?

16–41 In addition to centrosomes, flagellar axonemes and kinetochores also can serve as nucleation sites for microtubule assembly. The following experiment was designed to determine whether these two structures nucleate microtubule growth by binding to the plus end or to the minus end of the nascent microtubule. Flagellar axonemes were included as a control since their plus and minus ends can be distinguished. Centrosomes and kinetochores (and flagellar axonemes) were incubated briefly in unlabeled tubulin to nucleate microtubule growth. A high concentration of biotin-labeled tubulin was then added and the incubation was continued for 10 minutes. At that point the preparations were fixed and the biotin-labeled segments were visualized by adding fluorescein-labeled antibodies specific for biotin. The lengths of the biotin-labeled segments were measured and plotted as shown in Figure 16–18.

A. Which end of a newly assembled microtubule is attached to the plus end of the flagellar axoneme?

B. Which end of a microtubule assembled on a flagellar axoneme grows faster?

C. Which end of an assembled microtubule is attached to a centrosome? To a kinetochore? Explain your reasoning.

16–42* The γ-tubulin ring complex (γ-TuRC), which nucleates microtubule assembly in cells, derives from a smaller complex by condensation and association with other proteins. To get at the mechanism of nucleation, you have prepared monomeric γ-tubulin by *in vitro* translation and purification. You measure

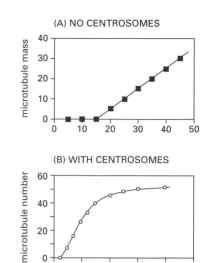

Figure 16–17 Analysis of microtubule assembly (Problem 16–40). (A) Mass of microtubules assembled in the absence of centrosomes as a function of tubulin concentration. (B) Average number of microtubules per centrosome as a function of tubulin concentration. Concentrations refer to αβ-tubulin dimers, which are the subunit of assembly.

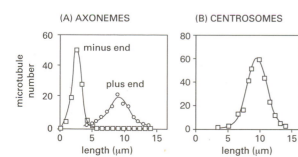
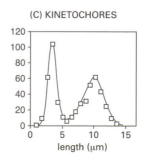

Figure 16–18 Length distributions of microtubules (Problem 16–41). (A) Nucleated by axonemes. (B) Nucleated by centrosomes. (C) Nucleated by kinetochores.

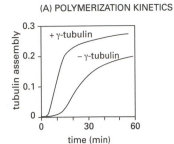

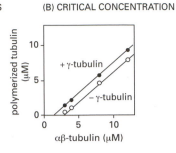

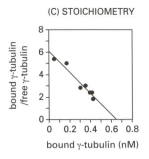

(A) POLYMERIZATION KINETICS

(B) CRITICAL CONCENTRATION

(C) STOICHIOMETRY

Figure 16–19 Effects of γ-tubulin on microtubule polymerization (Problem 16–42). (A) Kinetics of polymerization in the presence and absence of γ-tubulin. (B) The critical concentration for microtubule assembly in the presence and absence of γ-tubulin. (C) Stoichiometry of γ-tubulin binding as determined by a Scatchard plot.

the effect of adding monomeric γ-tubulin to a solution of αβ-tubulin dimers, as shown in Figure 16–19.

A. In the presence of monomeric γ-tubulin, the lag time for assembly of microtubules is decreased and assembly occurs more rapidly (Figure 16–19A). How would you account for these two effects of γ-tubulin?

B. The critical concentration of αβ-tubulin needed for assembly of microtubules is reduced from about 3.2 μM in the absence of γ-tubulin to about 1.7 μM in the presence of 0.6 nM γ-tubulin (Figure 16–19B). How do you suppose γ-tubulin lowers the critical concentration? How does this account for the greater extent of polymerization in Figure 16–19A? Which end (plus or minus) is polymerization occurring at and how can you tell?

C. A Scatchard plot (see Problem 3–47) of the bound over free γ-tubulin versus bound γ-tubulin in the presence of microtubules at 0.62 nM is shown in Figure 16–19C. What is the K_d for γ-tubulin binding to microtubules? (The slope of the line in a Scatchard plot is $-1/K_d$.) How many γ-tubulin monomers are bound per microtubule in these experiments? (The x-intercept in a Scatchard plot is equal to the total number of binding sites.) How many γ-tubulin monomers do you think it takes to nucleate one microtubule?

16–43 A solution of pure αβ-tubulin dimers is thought to nucleate microtubules by forming a linear protofilament about 7 dimers in length. At that point, the probabilities that the next αβ-dimer will bind laterally or to the end of the protofilament are about equal. The critical event for microtubule formation is thought to be the first lateral association (Figure 16–20). How does this event promote the subsequent rapid formation of a microtubule?

16–44* In the presence of monomeric γ-tubulin, αβ-tubulin dimers nucleate microtubules by forming protofilament about 3 dimers in length, as opposed to 7 dimers in the absence of γ-tubulin (see Problem 16–43). By separating α-tubulin from β-tubulin on polyacrylamide gels, transferring the proteins to nitrocellulose paper, and blotting with ^{35}S-γ-tubulin, you show that γ-tubulin binds very tightly to β-tubulin (Figure 16–21). Based on this information, propose a model for nucleation of microtubules by monomeric

LINEAR GROWTH

LATERAL ASSOCIATION

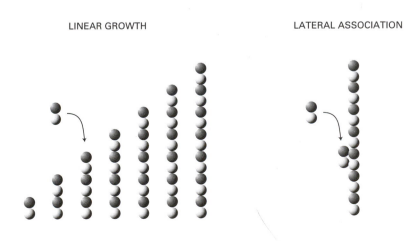

Figure 16–20 Model for microtubule nucleation by pure αβ-tubulin dimers (Problem 16–43).

γ-tubulin. Incorporate into your model whatever is needed to ensure that γ-tubulin caps the nucleated end (that is, permits growth in only one direction)?

16–45 In the paper that defined the γ-tubulin ring complex (γ-TuRC), the authors purified the complex from *Xenopus* oocytes and showed that it dramatically stimulated nucleation of microtubules. To determine whether nucleation occurred at plus ends or at minus ends, they polymerized microtubules in two steps. In the first step, microtubules nucleated with or without γ-TuRC were allowed to form in the presence of a small amount of rhodamine-labeled αβ-tubulin, which makes the microtubules fluoresce dimly. In the second step these microtubules were extended in the presence of a higher concentration of rhodamine-tagged tubulin to label the ends brightly. The longer bright segment identifies the plus end; the shorter segment, the minus end (Figure 16–22A). Measurements of the lengths of a large number of bright segments in individual microtubules yielded the data in Figure 16–22B and C. At which end of the microtubule does γ-TuRC nucleate growth? Explain your reasoning.

16–46* How does a centrosome 'know' when it has found the center of the cell?

16–47 Some actin-binding proteins significantly increase the rate at which formation of actin filaments is initiated in the cytosol. How might these proteins do this? What must they *not* do when binding the actin monomers?

16–48* How are γ-TuRC and the ARP complex similar, and how are they different?

16–49 Accessory proteins that regulate nucleation of actin filaments promote binding of the ARP complex to actin filaments so that most new filaments form as branches from existing ones. These proteins could stimulate ARP binding to the sides of existing filaments or to the plus end of a growing filament in a way that doesn't interfere with growth. Both possibilities would yield the final characteristic branched network of filaments. To distinguish between these alternatives, you mix the regulatory proteins with the ARP complex and actin subunits in the presence of actin filaments that are capped at their plus ends. After a short incubation you examine the resulting structures by electron microscopy. How will this experiment distinguish between these alternatives? What structures would you expect to see according to each model for nucleation by the ARP complex?

Figure 16–21 Blot of separated α- and β-tubulin by ³⁵S-γ-tubulin (Problem 16–44). (A) Protein stain showing positions of α- and β-tubulin. (B) Autoradiograph of blot with ³⁵S-γ-tubulin.

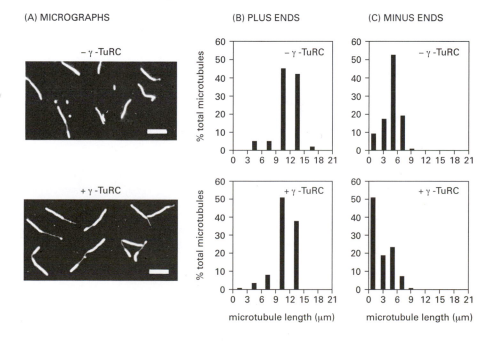

Figure 16–22 Effects of γ-TuRC on microtubule assembly (Problem 16–45). (A) Example of microtubules grown in the presence and absence of γ-TuRC. Scale bar is 10 μm. (B) Distribution of the lengths of bright segments at microtubule plus ends in the presence and absence of γ-TuRC. (C) Distributions of the lengths of bright segments at microtubule minus ends in the presence and absence of γ-TuRC. In (B) and (C) only microtubules with a defined dim segment and one or two bright terminal segments were counted.

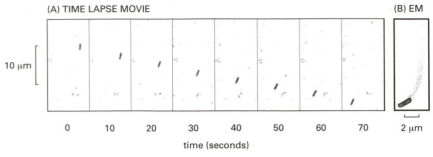

(A) TIME LAPSE MOVIE

10 μm

0   10   20   30   40   50   60   70

time (seconds)

(B) EM

2 μm

Figure 16–23 Movement of a bacterium through the cytosol on a comet tail of actin filaments (Problem 16–50). (A) Time lapse movie. (B) Electron micrograph. The bacterium is 2 μm in length.

16–50* The intracellular pathogenic bacterium *Listeria monocytogenes* propels itself through the cytosol on a comet tail of actin filaments (Figure 16–23). Remarkably, only a single bacterial protein, the transmembrane protein ActA, is required for this motility. ActA is distributed unequally on the surface of the bacterium, with maximum concentrations at the pole in contact with the actin tail. The effects of ActA and the ARP complex on actin polymerization are shown in Figure 16–24A, and the first few seconds of the reactions are shown on an expanded scale in Figure 16–24B. Polymerization of actin was followed using pyrene-actin, which exhibits much higher fluorescence intensity when actin is polymerized.

A. What are the effects of ActA and the ARP complex, separately and together, on the rate of nucleation of actin filaments? Explain your answer.

B. Explain how the effects of ActA and the ARP complex can propel the bacterium across the cell.

16–51 The concentration of actin in cells is 50–100 times greater than the critical concentration observed for pure actin in a test tube. How is this possible? Why don't the actin subunits in cells polymerize into filaments? Why is it advantageous to the cell to maintain such a large pool of actin subunits?

16–52* The observations on individual microtubules in Figure 16–6 (see Problem 16–12) were made with microtubules polymerized in a solution of pure αβ-tubulin. How do you think the observations would change if centrosomes were used to nucleate microtubule growth? What would happen if you added microtubule-associated proteins (MAPs) to the solution of pure tubulin?

16–53 Cofilin binds to actin filaments, as well as to actin monomers. How then does it promote depolymerization from the minus end and how does it distinguish old filaments from new ones?

16–54* You have two proteins that you suspect cap the ends of actin filaments. To determine whether they do and, if so, which protein caps which end, you measure filament formation as a function of actin concentration in the absence of either protein, in the presence of protein 1, and in the presence of protein 2 (Figure 16–25). Which protein caps the plus end and which caps the minus end? How can you tell? Give examples of proteins in the cell that you would expect to behave like protein 1 and protein 2.

16–55 (**True/False**) All proteins that bind to the ends of microtubules or actin filaments cap the ends to prevent further polymerization. Explain your answer.

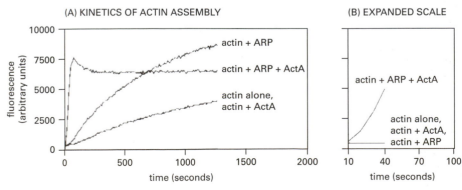

(A) KINETICS OF ACTIN ASSEMBLY

10000
7500
5000
2500
0

fluorescence (arbitrary units)

0   500   1000   1500   2000

time (seconds)

actin + ARP
actin + ARP + ActA
actin alone, actin + ActA

(B) EXPANDED SCALE

actin + ARP + ActA

actin alone, actin + ActA, actin + ARP

10   40   70   100

time (seconds)

Figure 16–24 Effects of ActA and the ARP complex on actin polymerization (Problem 16–50). (A) Kinetics of polymerization of actin in the presence of ActA and the ARP complex. (B) Kinetics of polymerization on an expanded scale. In all cases actin was present at 2 μM, and ActA and the ARP complex were present at 30 nM.

(A) NO ADDED PROTEIN   (B) PROTEIN 1   (C) PROTEIN 2

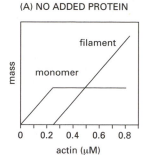
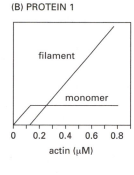
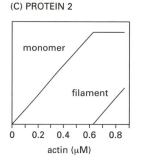

Figure 16–25 Effects of two proteins on actin polymerization (Problem 16–54). (A) Polymerization of pure actin. (B) Actin polymerization in the presence of protein 1. (C) Actin polymerization in the presence of protein 2. The mass of actin, as polymer or subunits, was determined at equilibrium.

16–56* Filamin cross-links actin filaments at roughly right angles to produce a viscous gel that is required for cells to extend thin sheetlike lamellipodia. Why is the loss of filamin in melanoma cells bad news for the melanoma cells, but good news for the patient?

16–57 The haploid slime mold, *Dictyostelium discoideum*, lives on the forest floor as independently motile cells called amoebae, which feed on bacteria and yeast. When the food supply is exhausted, the amoebae come together to form multicellular wormlike slugs that crawl around, eventually differentiating into a plantlike structure with a fruiting body full of spores. When growth conditions become favorable once again, the spores germinate to form amoebae, completing the developmental life cycle.

 Dictyostelium has two major actin filament cross-linking proteins: α-actinin and gelation factor. You have isolated two mutant strains of *Dictyostelium*, one that fails to express α-actinin, and the other that is missing gelation factor. Surprisingly, neither of these single mutants shows any defect in cell growth, motility, or development. The double mutants, however, are clearly defective. Although they move well as amoebae, as slugs they stay put wherever they first form, instead of crawling around as they normally would.

A. Your studies with the single mutants defective in either α-actinin or gelation factor suggest that neither gene is essential. Why then do you think that the loss of both genes causes such dramatic defects in the cell movements associated with development?

B. Why do you suppose that cells from the double mutant move all right as amoebae, but not as slugs?

16–58* When cells enter mitosis, their existing array of cytoplasmic microtubules has to be rapidly broken down and replaced with the mitotic spindle, which pulls the chromosomes into the daughter cells. The enzyme katanin, named after Japanese samurai swords, is activated during the onset of mitosis and chops microtubules into short pieces. What do you suppose is the fate of the microtubule fragments created by katanin?

16–59 A resting platelet is discoid in shape, and it contains short actin filaments capped by CapZ, surrounded by a large pool of actin monomers bound to profilin. When the platelet is activated by physical contact with a damaged blood vessel, a rapid signal cascade triggers a series of events that activates the platelet and allows it to spread out lamellipodia and filopodia and to extend itself across the clot and pull the edges together. Order the following molecules and discuss their roles in the sequence of events that allows resting platelets to accomplish this feat.

A. α-Actinin
B. Ca^{2+}
C. CapZ
D. Filamin
E. Fimbrin
F. Gelsolin
G. Myosin II
H. PiP_2
I. Profilin

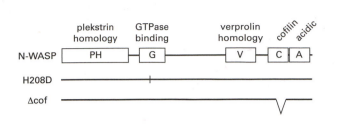

(A) DOMAINS OF N-WASP

N-WASP | plekstrin homology [PH] | GTPase binding [G] | verprolin homology [V] | cofilin [C] | acidic [A]

H208D

Δcof

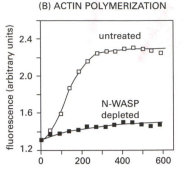

(B) ACTIN POLYMERIZATION

untreated

N-WASP depleted

fluorescence (arbitrary units)

time (seconds)

Figure 16–26 The role of N-WASP in Cdc42-triggered actin polymerization and bundling (Problem 16–61). (A) Domain structure of N-WASP and two mutants defective in individual domains. (B) Actin polymerization in untreated and N-WASP-depleted egg extracts. In the polymerization assays, the extracts were supplemented with pyrene-actin, which fluoresces much more highly in the filament than it does on the subunit.

16–60* How is it possible for actin filaments, which are located entirely inside the cell, to pull on the extracellular matrix?

16–61 Activation of Cdc42, a monomeric GTPase, triggers actin polymerization and bundling to form either filopodia or shorter cell protrusions called microspikes. These effects of Cdc42 could be mediated by N-WASP, which is a multifunctional protein. As shown in Figure 16–26A, N-WASP contains a plekstrin homology (PH) domain, which binds to PiP$_2$, a Cdc42-binding domain (G), a verprolin homology domain (V), which binds to actin, a cofilin-homology domain (C), which can bind to actin filaments, and a C-terminal acidic domain (A), which binds the ARP complex.

In *Xenopus* egg extracts, a convenient source of components, addition of Cdc42 charged with GTPγS, a nonhydrolyzable analog of GTP, stimulates actin polymerization (Figure 16–26B). If the extract is depleted of N-WASP using N-WASP-specific antibodies, no actin polymerization is observed when Cdc42-GTPγS is added (Figure 16–26B). Actin polymerization can be restored by addition of purified N-WASP, but not by addition of either of two mutant forms of N-WASP: one (H208D) that cannot bind to Cdc42, and a second (Δcof) that eliminates the function of the cofilin domain (Figure 16–26A).

Do these experiments support a role for N-WASP in rearrangement of actin filaments in response to Cdc42 activation? Explain your reasoning. Include a discussion of why the two mutant forms of N-WASP do not restore actin polymerization.

16–62* To determine the mechanism by which N-WASP mediates activation by Cdc42, polymerization was measured in the presence of purified components. In the presence of the ARP complex, N-WASP stimulates actin polymerization substantially over the ARP complex or N-WASP alone, but not nearly so dramatically as the C-terminal segment of N-WASP that contains just the verprolin (V), cofilin (C), and acidic (A) domains (Figure 16–27A). To account for the difference between N-WASP and its C-terminal VCA segment,

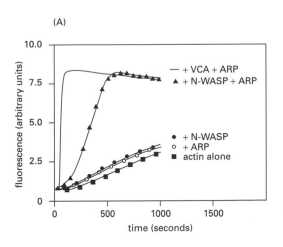

(A)

fluorescence (arbitrary units)

time (seconds)

— + VCA + ARP
▲ + N-WASP + ARP
● + N-WASP
○ + ARP
■ actin alone

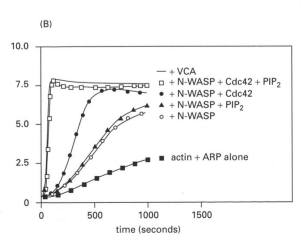

(B)

fluorescence (arbitrary units)

time (seconds)

— + VCA
□ + N-WASP + Cdc42 + PIP$_2$
● + N-WASP + Cdc42
▲ + N-WASP + PIP$_2$
○ + N-WASP
■ actin + ARP alone

Figure 16–27 Polymerization of actin in the presence of various purified components (Problem 16–62). (A) Mixtures of actin (2.5 μM), N-WASP (400 nM), the ARP complex (250 nM), and the VCA C-terminal segment of N-WASP (200 nM). (B) Mixtures of actin (2.5 μM), N-WASP (200 nM), the ARP complex (60 nM), Cdc42-GTPγS (500 nM), and PiP$_2$-containing vesicles (100 μM). Vesicles without PiP$_2$ do not stimulate in any combination with the other components.

0 sec   24 sec   48 sec   72 sec

Figure 16–28 Movement of microtubules on a bed of microtubule motor molecules (Problem 16–65). *Black arrows* mark the movement of a microtubule with a gold bead attached via antibodies to the minus end of a microtubule; *white arrows* mark the movement of a microtubule without an attached bead. Pictures were taken using video-enhanced differential interference contrast (VE-DIC) microscopy.

N-WASP and the ARP complex were mixed with combinations of Cdc42-GTPγS and vesicles containing PiP$_2$, as shown in Figure 16–27B.

A. What is required for N-WASP to stimulate polymerization as efficiently as its C-terminal VCA segment? Explain your reasoning.

B. Based on these results propose a model for the activation of N-WASP and its stimulation of actin polymerization.

MOLECULAR MOTORS

TERMS TO LEARN

axoneme	flagella	motor protein
cilia	kinesin	myofibril
dynein	kinesin-related protein (KRP)	myosin

16–63 There are no known motor proteins that move on intermediate filaments. Suggest an explanation for this observation.

16–64* (**True/False**) Myosin II molecules have two motor domains and a rodlike tail that allows them to assemble into bipolar filaments, which are crucial for efficient sliding of oppositely oriented actin filaments past each other. Explain your answer.

16–65 A useful technique for studying a microtubule motor is to attach the motor proteins by their tails to a glass coverslip (the tails stick avidly to a clean glass surface) and then to allow microtubules to settle onto them. In the light microscope, the microtubules can be seen to move over the surface of the coverslip as the heads of the motors propel them (Figure 16–28).

A. Since the motor proteins attach in random orientations to the coverslip, how can they generate coordinated movement of individual microtubules, rather than engaging in a tug-of-war?

B. In which direction will microtubules crawl on a bed of dynein motor molecules (that is, will they move plus end first or minus end first)?

C. In the experiment shown in Figure 16–28, some of the microtubules were marked by gold beads that were bound by minus end-specific antibodies. Is the motor protein on the coverslip a plus end or minus end-directed motor? How can you tell?

16–66* In Problem 16–9 the orientation of the αβ-tubulin dimer in the microtubule was determined by showing that α-tubulin antibody-coated gold beads bound to the minus end. The electron micrographs, however, just showed microtubules with beads at one end (Figure 16–29). How do you suppose the investigators knew which end was which? Design an experiment to determine the orientation of the microtubules shown in Figure 16–29.

Figure 16–29 Microtubules with α-tubulin antibody-coated gold beads attached to one end (Problem 16–66). At the *vertical line* a section of each microtubule has been removed so that the two ends can be displayed side by side.

16–67 Kinesin carries vesicles for long distances along microtubule tracks in the cell. Are the two motor domains of a kinesin molecule essential to accomplish this task, or could a one-headed motor protein function just as well? Using recombinant DNA techniques, a version of kinesin was prepared that was identical to normal kinesin except that one motor domain was absent. Wild-type kinesin with two motor domains and recombinant kinesin with one were attached to coverslips at a variety of densities and the rate at which microtubules were bound and moved (collectively, the landing rate) was measured (Figure 16–30).

A. Why do you suppose that the curves at low motor densities are so different?
B. What do these experiments say about the design of the kinesin motor: are two heads required for vesicle transport or is only one? Explain your reasoning.

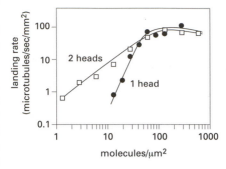

Figure 16–30 Landing rates—binding and moving—of microtubules as a function of motor protein density (Problem 16–67). Results with wild-type kinesin are shown as *open squares*, while those with recombinant kinesin are shown as *solid circles*.

16–68* The movements of single motor-protein molecules can be analyzed directly. Using polarized laser light, it is possible to create interference patterns that exert a centrally directed force, ranging from zero at the center to a few piconewtons at the periphery (about 200 nm from the center). Individual molecules that enter the interference pattern are rapidly pushed to the center, allowing them to be captured and moved at the experimenter's discretion.

Using such 'optical tweezers,' single kinesin molecules can be positioned on a microtubule that is fixed to a coverslip. Although a single kinesin molecule cannot be seen optically, it can be tagged with a silica bead and tracked by following the bead (Figure 16–31A). In the absence of ATP the kinesin molecule remains at the center of the interference pattern, but with ATP it moves toward the plus end of the microtubule. As kinesin moves along the microtubule, it encounters the force of the interference pattern, which simulates the load kinesin carries during its actual function in the cell. Moreover, the pressure against the silica bead counters the effects of Brownian (thermal) motion so that the position of the bead more accurately reflects the position of the kinesin molecule on the microtubule.

Traces of the movements of two kinesin molecules along a microtubule are shown in Figure 16–31B.

A. As shown in Figure 16–31B, all movement of kinesin is in one direction (toward the plus end of the microtubule). What supplies the free energy needed to ensure a unidirectional movement along the microtubule?
B. What is the average rate of movement of each kinesin along the microtubule?
C. What is the length of each step a kinesin takes as it moves along a microtubule?
D. From other studies it is known that kinesin has two globular domains that each can bind to β-tubulin and that kinesin moves along a single protofilament in a microtubule. In each protofilament the β-tubulin subunit repeats at 8-nm intervals. Given the step length and the interval between β-tubulin subunits, how do you suppose a kinesin molecule moves along a microtubule?
E. Is there anything in the data in Figure 16–31B that tells you how many ATP molecules are hydrolyzed per step?

16–69 Living systems continually transform chemical free energy into motion. Muscle contraction, ciliary movement, cytoplasmic streaming, cell division, and active transport are examples of the ability of cells to transduce chemical free energy into mechanical work. In all these instances a protein motor harnesses the free energy released in a chemical reaction to drive an

Figure 16–31 Movement of kinesin along a microtubule (Problem 16–68). (A) Experimental setup with kinesin, which is linked to a silica bead, moving along a microtubule. (B) Position of kinesin (as visualized by position of silica bead) relative to center of interference pattern as a function of time of movement along the microtubule. The jagged nature of the trace results from Brownian motion of the bead. The movements of two different kinesin molecules are shown.

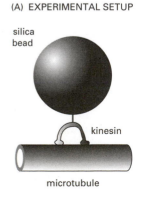

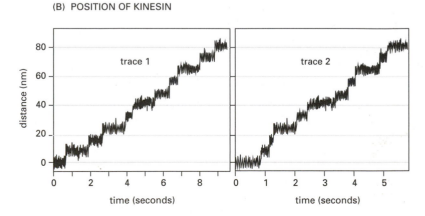

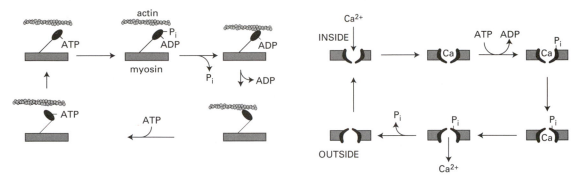

(A) SLIDING FILAMENT

actin

ATP

P_i
ADP

myosin

ADP

P_i

ADP

ATP

ATP

(B) ACTIVE TRANSPORT

Ca²⁺

INSIDE

Ca

ATP ADP

Ca

P_i

P_i

P_i

P_i

Ca

OUTSIDE

Ca²⁺

attached molecule (the ligand) in a particular direction. Analysis of free-energy transduction in favorable biological systems suggests that a set of general principles governs the process in cells.

1. A cycle of reactions is used to convert chemical free energy into mechanical work.
2. At some point in the cycle a ligand binds very tightly to the protein motor.
3. At some point in the cycle the motor undergoes a major conformational change that alters the physical position of the ligand.
4. At some point in the cycle the binding constant of the ligand markedly decreases, allowing the ligand to detach from the motor.

These principles are illustrated by the two cycles for free-energy transduction shown in Figure 16–32: (1) the sliding of actin and myosin filaments against each other and (2) the active transport of Ca²⁺ from inside the cell, where its concentration is low, to the cell exterior, where its concentration is high. An examination of these cycles underscores the principles of free-energy transduction.

A. What is the source of chemical free energy that powers these cycles, and what is the mechanical work that each cycle accomplishes?
B. What is the ligand that is bound tightly and then released in each of the cycles? Indicate the points in each cycle where the ligand is bound tightly.
C. Identify the conformational changes in the protein motor that constitute the 'power stroke' and 'return stroke' of each cycle.

16–70* Kinesin motors are highly processive, moving long distances on microtubule tracks without dissociating. By contrast, myosin II motors in skeletal muscle cannot move processively and take only one or a few steps before letting go. How are these different degrees of processivity adapted to the biological functions of kinesin and myosin II?

16–71 (**True/False**) Minus end-directed microtubule motors deliver their cargo to the periphery of the cell, whereas plus end-directed microtubule motors deliver their cargo to the interior of the cell. Explain your answer.

16–72* Many lower vertebrates such as fish and amphibians control their color by regulating specialized pigment cells called melanophores. These cells contain small, pigmented organelles, termed melanosomes, that can be dispersed throughout the cell, making the cell darker, or aggregated in the center to make the cell lighter (Figure 16–33A and B). Melanosomes can be dragged along microtubules in either direction in a matter of minutes in response to hormonal stimulation.

Aggregated and dispersed melanosomes have two attached motors—dynein and kinesin II—whose activities are regulated in response to

Figure 16–32 Transduction of chemical free energy into mechanical work (Problem 16–69). (A) Sliding of actin filaments relative to myosin filaments. (B) Active transport of Ca²⁺ from the inside to the outside of the cell. In both cycles *arrows* are drawn in only one direction to emphasize their normal operation. The phosphorylation and dephosphorylation steps in the active transport cycle are catalyzed by enzymes that are not shown in the diagram.

(A) DISPERSED

(B) AGGREGATED

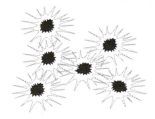

(C) MICROTUBULE BINDING

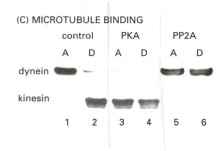

	control		PKA		PP2A	
	A	D	A	D	A	D
dynein						
kinesin						
	1	2	3	4	5	6

Figure 16–33 Control of color by regulation of melanophores (Problem 16–72). (A) Dispersed melanosomes. (B) Aggregated melanosomes. (C) Microtubule-binding activities of dynein and kinesin II. 'A' stands for aggregated and 'D' for dispersed.

hormones. If aggregated or dispersed melanosomes are isolated, the activities of the motor proteins can be assessed by their ability to bind to microtubules (Figure 16–33C, lanes 1 and 2). The activities of the motors can be altered by treatment with protein kinase A (PKA, lanes 3 and 4) or protein phosphatase 2A (PP2A, lanes 5 and 6). From these data decide how dynein and kinesin II regulate the behavior of the melanosomes, and how phosphorylation and dephosphorylation regulate dynein and kinesin II.

16–73 Compare the structure of intermediate filaments to that of the myosin II filaments in skeletal muscle cells. What are the major similarities? What are the major differences? How do the differences in structure relate to their function?

16–74* Which one of the following changes takes place when a skeletal muscle contracts?

A. Z discs move farther apart.
B. Actin filaments contract.
C. Myosin filaments contract.
D. Sarcomeres become shorter.

16–75 Two electron micrographs of striated muscle in longitudinal section are shown in Figure 16–34. The sarcomeres in these micrographs are in two different stages of contraction.

A. Using the micrograph in Figure 16–34A, identify the location of the following:

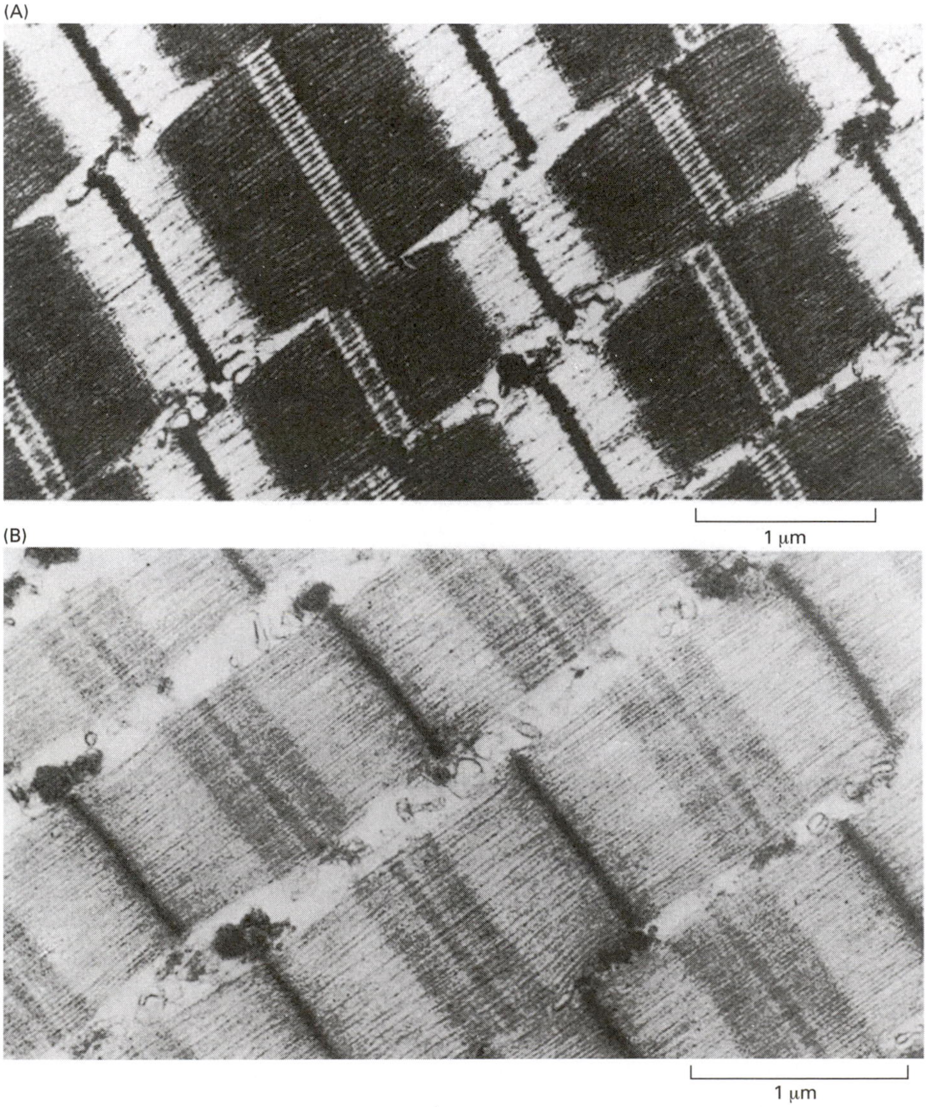

Figure 16–34 Two electron micrographs of striated muscle in longitudinal section (Problem 16–75). The micrographs have been photographed at different exposures.

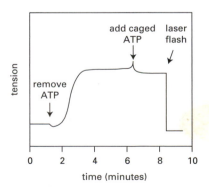

Figure 16–35 Caged ATP (Problem 16–76).

CAGED ATP

ATP

1. Dark band
2. Light band
3. Z disc
4. Myosin II filaments
5. Actin filaments (show plus and minus ends)
6. α-Actinin
7. Nebulin
8. Titin

B. Locate the same features on the micrograph in Figure 16–34B. Be careful!

Problem 3–32 describes the basis for the springlike behavior of titin, which keeps the myosin thick filament centered in the sarcomere.

16–76* As a laboratory exercise, you and your classmates are carrying out experiments on isolated muscle fibers using a new compound, called 'caged' ATP (Figure 16–35). Since caged ATP does not bind to muscle components, it can be added to a muscle fiber without stimulating activity. Then, at some later time it can be split by a flash of laser light to release ATP instantly throughout the muscle fiber.

To begin the experiment, you treat an isolated, striated muscle fiber with glycerol to make it permeable to nucleotides. You then suspend it in a buffer containing ATP in an apparatus that allows you to measure any tension generated by fiber contraction. As illustrated in Figure 16–36, you measure the tension generated after several experimental manipulations: removal of ATP by dilution, addition of caged ATP, and activation of caged ATP by laser light. You are somewhat embarrassed because your results are very different from everyone else's. In checking over your experimental protocol, you realize that you forgot to add Ca^{2+} to your buffers. The teaching assistant in charge of your section tells you that your experiment is actually a good control for the class but you will have to answer the following questions to get full credit.

A. Why did the ATP in the suspension buffer not cause the muscle fiber to contract?

B. Why did the subsequent removal of ATP generate tension? Why did tension develop so gradually? (If our muscles normally took a full minute to contract, we would move very slowly.)

C. Why did laser illumination of a fiber containing caged ATP lead to relaxation?

16–77 The change in sarcomere length during muscle contraction was one of the key observations that suggested a sliding filament model. The degree of tension generated at different sarcomere lengths also changes in a way that is consistent with the model. Detailed measurements of sarcomere length and tension during isometric contraction in a striated muscle are shown

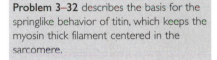

Figure 16–36 Tension in a striated muscle fiber as a result of various experimental manipulations (Problem 16–76).

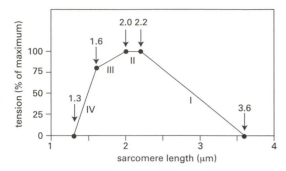

Figure 16–37 Tension as a function of sarcomere length during isometric contraction (Problem 16–77).

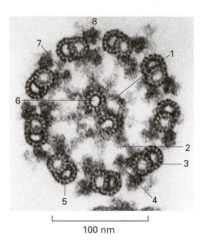

in Figure 16–37. In this muscle the length of the myosin filament is 1.6 μm and the lengths of the actin thin filaments that project from the Z discs are 1.0 μm.

Based on your understanding of the sliding filament model and the structure of a sarcomere, present a molecular explanation for the relationship of tension to sarcomere length in the portions of the graph marked I, II, III, and IV in Figure 16–37.

16–78* **(True/False)** A signal from the motor nerve triggers an action potential that opens voltage-sensitive Ca^{2+} channels in the muscle cell plasma membrane, allowing Ca^{2+} to flow into the cytosol from outside the cell. Explain your answer.

16–79 **(True/False)** When the cytosolic level of Ca^{2+} rises, troponin C causes troponin I to release its hold on actin, thereby allowing the tropomyosin molecules to shift their positions slightly so that the myosin heads can bind to the actin filaments. Explain your answer.

16–80* Troponin molecules are evenly spaced along an actin filament with one troponin bound to every seventh actin molecule. How do you suppose troponin molecules can be positioned this regularly?

16–81 What two major roles does ATP hydrolysis play in muscle contraction?

16–82* An electron micrograph of a cross-section through a flagellum is shown in Figure 16–38.
A. Assign the following components to the indicated positions on the figure.
 A tubule
 B tubule
 Outer dynein arm
 Inner dynein arm
 Inner sheath
 Nexin
 Radial spoke
 Singlet microtubule
B. Which of the above structures are composed of tubulin?

16–83 On ciliated cells the beating of individual cilia is usually coordinated so that the cilia move in the same direction, thereby imparting unidirectional motion to the cell (or to the surrounding fluid). In principle, the coordinated, unidirectional beating of adjacent cilia could be determined by some feature of their structure or, alternatively, by some cellular control mechanism that is independent of ciliary structure.

An electron micrograph of a cross-section through the cortex of the ciliated protozoan *Tetrahymena* is shown in Figure 16–39. The plane of sections

Figure 16–38 Electron micrograph of a cross-section through a flagellum of *Chlamydomonas reinhardtii* (Problem 16–82).

Figure 16–39 Electron micrograph of a cross-section through the cortex of a ciliated protozoan *Tetrahymena* (Problem 16–83).

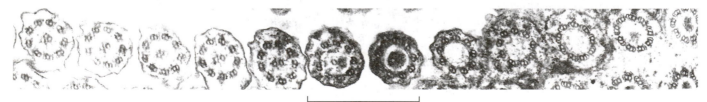

500 nm

(A) WILD TYPE

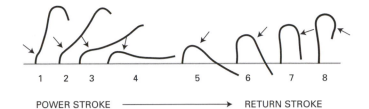

1 2 3   4   5   6   7   8

POWER STROKE ⟶ RETURN STROKE

(B) DOUBLE MUTANT

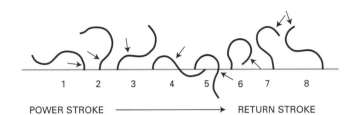

1   2   3   4   5   6   7   8

POWER STROKE ⟶ RETURN STROKE

Figure 16–40 Analysis of flagellar beat cycles in *Chlamydomonas* (Problem 16–84). (A) Wild-type flagella. (B) Double-mutant flagella. The beating flagella shown on the *left* were photographed with stroboscopic illumination under a microscope. Individual frames from these pictures represent flagella at successive stages in the beat cycle. *Arrows* point to the tip-proximal end of a bent segment of flagellum as it moves from the bottom of the flagellum to the tip during the beat cycle.

(A)

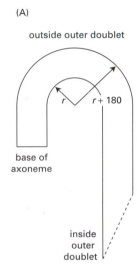

grazes the surface of the cell, showing in successive sections how the '9 + 2' arrangement of microtubules in the axoneme leads into the nine triplet microtubules of the basal body. Are there any clues in the micrograph that allow you to decide whether axoneme structure or cellular control is the basis for the unidirectional beating of adjacent cilia? Explain your reasoning.

16–84* When analyzed in detail, the rhythmic beating of a cilium is revealed as a series of precisely repeated movements. In *Chlamydomonas* the flagellar beat cycle is straightforward (Figure 16–40A). The beat cycle begins with a power stroke, which is initiated by a bend near the base of the flagellum (arrow at base of flagellum 1 in Figure 16–40A). The power stroke ends when the bent segment of flagellum extends roughly through half the circumference of a circle (flagellum 5 in Figure 16–40A). The return stroke is formed by the movement of the semicircular segment of the flagellum outward toward the tip, which is accomplished by further bending at the leading edge of the semicircle and relaxation at the trailing edge (flagella 6 to 8 in Figure 16–40A).

A. How much sliding of microtubule doublets against one another is required to account for the observed bending of the flagellum into a semicircle? Calculate how much farther the doublet on the inside of the semicircle protrudes beyond the doublet on the outside of the semicircle at the tip of the flagellum (Figure 16–41A). The width of a flagellum is 180 nm.

B. The elastic nexin molecules that link adjacent outer doublets must stretch to accommodate the bending of a flagellum into a semicircle. If the length of an unstretched nexin molecule at the base of a flagellum is 30 nm, what is the length of a stretched nexin molecule at the tip of a flagellum (Figure 16–41B)? Adjacent doublets are 30 nm apart.

C. *Chlamydomonas* mutants that are missing radial spokes have paralyzed flagella. The paralysis can be overcome by mutations in a second gene (called sup_{pf} for suppressor of paralyzed flagella), which encodes a component of the outer dynein arm. Although the flagella now move, their beat pattern is aberrant (Figure 16–40B). At the gross level, how does the beat stroke of this double mutant differ from that of the wild type? What does this gross difference suggest for the function mediated by the radial spokes in *Chlamydomonas*?

16–85 The sliding microtubule mechanism for ciliary bending is undoubtedly correct. The consequences of sliding are straightforward when a pair of outer doublets is considered in isolation. The dynein arms are arranged so that, when activated, they push their neighboring outer doublet outward toward the tip of the cilium (see Figure 16–41B). It is confusing, however, to think about sliding in the circular array of outer doublets in the axoneme.

(B)

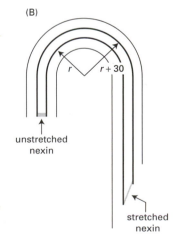

Figure 16–41 Flagella bent into half circles (Problem 16–84). (A) Representation showing the 'inside' and 'outside' doublets, which are 180 nm apart. (B) Representation showing adjacent doublets, which are 30-nm apart, and the nexin molecules that link them.

If all the dynein arms in a circular array were equally active, there could be no significant relative motion. (The situation is equivalent to a circle of strongmen, each trying to lift his neighbor off the ground; if they all succeeded, the group would levitate.)

Devise a pattern of dynein activity (consistent with axoneme structure and the directional pushing of dynein) that can account for bending of the axoneme in one direction. How would this pattern change for bending in the opposite direction?

16–86* The structure of the ciliary axoneme, which is composed of more than 200 different proteins, is exceedingly complex. *Chlamydomonas reinhardtii*, which bears two flagella, is an extremely useful organism for analyzing axoneme structure. Physiological and microscopic observation are straightforward, but even more important is the ease of genetic and biochemical analysis. It is a simple matter to isolate mutants with paralyzed flagella because they cannot move. These mutants can then be assigned to specific genes by genetic crosses, which are routine with *Chlamydomonas*. Finally, wild-type and mutant flagella can be detached readily (for example, by pH shock) and recovered for easy biochemical analysis.

Many of the mutants with paralyzed flagella are missing one or another of the major substructures of the axoneme, such as the radial spokes, the outer dynein arms, the inner dynein arms, or one or both central microtubules. In most cases loss of the axonemal substructure is caused by mutation of a single gene, and yet biochemical analysis shows that the defective axoneme is missing multiple proteins. Consider, for example, the mutant *pf*14 (paralyzed flagella); electron micrographs of its flagella show a complete absence of radial spokes (Figure 16–42A), and two-dimensional electrophoretic analysis shows that it lacks 17 different proteins (Figure 16–42B).   *Motor proteins*

A single-gene defect that results in multiple protein deficiencies could have two underlying explanations: (1) the defect is in a regulatory gene that controls the synthesis of the missing proteins or (2) the defect is in a gene whose product must be present in the structure before the other proteins can be added. Two approaches have been used to evaluate these possibilities.

A. The first method takes advantage of a feature of the regular mating cycle of *Chlamydomonas*. In the mating reaction, biflagellate gametes fuse efficiently to give a population of temporary dikaryons with four flagella. In a mating of

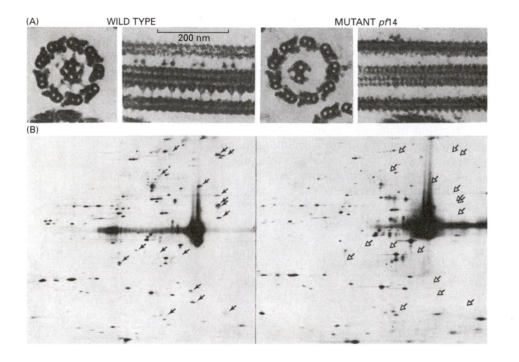

(A)   WILD TYPE   MUTANT *pf*14
200 nm

(B)

Figure 16–42 Comparison of flagella from wild-type *Chlamydomonas* with those from the mutant, *pf*14 (Problem 16–86). (A) Electron micrographs showing flagella in transverse and longitudinal sections. Notice the absence of radial spokes in *pf*14. (B) Analysis of flagella by two-dimensional gel electrophoresis. The first dimension (*horizontal*) is separation by isoelectric focusing with the more acidic proteins on the *right*; the second dimension is separation by molecular mass using SDS-gel electrophoresis. *Arrows* indicate the positions of proteins that are present in wild type but missing in *pf*14. The highly exposed areas correspond to α- and β-tubulin, which far outnumber the other axonemal proteins.

*pf*14 with wild-type gametes, the paralyzed flagella recover function after fusion, indicating that the defective structures can be repaired without completely rebuilding them. (In normal cells there is a pool of flagellar components sufficient to rebuild an entire flagellum in the absence of protein synthesis.) To distinguish between the possible explanations for flagellar defects, investigators labeled mutant cells by growth in $^{35}SO_4$ and then fused the labeled gametes to nonradioactive wild-type gametes in the presence of an inhibitor of protein synthesis. After recovery of function, the flagella were isolated and the radioactive proteins were analyzed by two-dimensional gel electrophoresis followed by autoradiography.

Predict the expected electrophoretic pattern of *radioactive* proteins from the dikaryon if the affected gene controlled the synthesis of the missing proteins. How would it differ from the electrophoretic pattern that would be expected if the affected gene product participated in assembly?

B. The second approach was to expose *pf*14 to a mutagen to generate revertants that regained flagellar function not because the original defect has been corrected, but because a second alteration within the gene compensates for the first one. (Depending on the gene, such intragenic revertants can be very common.) The proteins from several such revertants were compared to those in the wild type by two-dimensional gel electrophoresis.

How might this method distinguish between the two possible explanations for the defect in *pf*14?

THE CYTOSKELETON AND CELL BEHAVIOR

TERMS TO LEARN

axon	filopodium	neurite
dendrite	lamellipodium	pseudopodium

16–87 Distinguish between the three processes—protrusion, attachment, and traction—that make up the crawling movements of cells.

16–88* The locomotion of fibroblasts in culture is immediately halted by the drug cytochalasin B, whereas colchicine causes fibroblasts to cease to move directionally and to begin extending lamellipodia in seemingly random directions. Injection of fibroblasts with antibodies to vimentin has no discernible effect on their migration. What do these observations suggest to you about the involvement of the three different cytoskeletal filaments in fibroblast locomotion?

16–89 Actin filaments are said to 'push' on the cell membrane to cause it to form a protrusion. There are problems, however, at both ends of the filaments. When a plus end reaches the membrane and abuts it, how are new subunits added to extend the filament? And how is the minus end of the filament anchored so that the filament isn't simply pushed back into the cell's interior? What do you suppose might be the answers to these questions?

16–90* One of the most striking examples of a purely actin-based cellular movement is the extension of the acrosomal process of a sea cucumber sperm. The sperm contains a store of unpolymerized actin in its head. When a sperm makes contact with a sea cucumber egg, the actin polymerizes rapidly to form a long spearlike extension. The tip of the acrosomal process penetrates the egg, and it is probably used to pull the sperm inside.

Are actin monomers added to the base or to the tip of the acrosomal bundle of actin filaments during extension of the process? If the supply of monomers to the site of assembly depends on diffusion, it should be possible to distinguish between these alternatives by measuring the length of the acrosomal process with increasing time. If actin monomers are added to the base of the process, which is inside the head, the rate of growth should be linear because the distance between the site of assembly and the pool of monomers does not change with time. On the other hand, if the subunits are

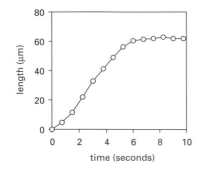
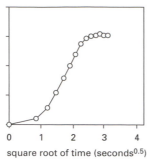

Figure 16–43 Plots of the length of the acrosome versus time and the square root of time (Problem 16–90).

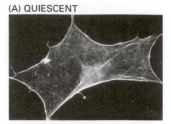

(A) QUIESCENT

(B) STRESS FIBERS

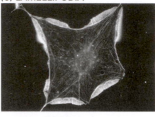

(C) LAMELLIPODIA

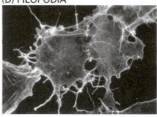

(D) FILOPODIA

added to the tip, the rate of growth should decline progressively as the acrosomal process gets longer because the monomers must diffuse all the way down the shaft of the process. In this case the rate of extension should be proportional to the square root of time. Plots of the length of the acrosomal process versus time and the square root of time are shown in Figure 16–43.

A. Are the ascending portions of the plots in Figure 16–43 more consistent with addition of actin monomers to the base or to the tip of the acrosomal process?

B. Why do you suppose the process grows so slowly at the beginning and at the end of the acrosomal reaction?

16–91 The characteristic actin staining in a quiescent cell is shown in Figure 16–44A. If such cells are injected with a constitutively activated forms of Rac, Rho, or Cdc42 monomeric GTPases, they dramatically alter their actin cytoskeletons. Which GTPase is associated with formation of stress fibers (Figure 16–44B), lamellipodia (Figure 16–44C), and filopodia (Figure 16–44D)?

16–92* How is the unidirectional motion of a lamellipodium maintained?

16–93 (**True/False**) Neutrophils move toward a source of bacterial infection by chemotaxis using receptors on their surface to respond to a gradient of *N*-formylated peptides derived from bacterial proteins. Explain your answer.

16–94* In addition to conducting impulses in both directions, nerve axons carry vesicles to and from the cell body along microtubule tracks. Do outbound vesicles move along microtubules that are oriented in one direction and incoming vesicles move along oppositely oriented microtubules? Or are microtubules all oriented in the same direction with different motor proteins providing the directionality?

To distinguish between these possibilities, you prepare a cross-section through a nerve axon and decorate the microtubules with tubulin, which binds to the tubulin subunits of the microtubule to form hooks. The decorated microtubules are illustrated in Figure 16–45. Do all the microtubules run in the same direction or not? How can you tell?

16–95 Using the equation for diffusion given in Problem 16–21, calculate the average time it would take for a vesicle to diffuse to the end of an axon 10 cm in length.

16–96* A mitochondrion 1 μm long can travel the 1 meter length of the axon from the spinal cord to the big toe in a day. The Olympic men's freestyle swimming record for 200 meter is 1.77 minutes. In terms of body lengths per day, who is moving faster: the mitochondrion or the swimmer? (Assume the swimmer is 2 meters tall.)

16–97 Mice that are homozygous for a knockout of the gene for the kinesin motor protein KIF1B die at birth. Heterozygous knockouts survive, but suffer from a progressive muscle weakness similar to human neuropathies. Humans

Figure 16–44 Actin cytoskeleton in different cells (Problem 16–91). (A) Quiescent cells. (B) Cells with prominent stress fibers. (C) Cells with multiple lamellipodia. (D) Cells with many long filopodia. Cells in B, C, and D were injected with an activated form of a monomeric GTPase.

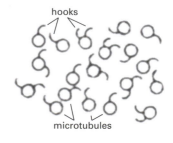

hooks

microtubules

Figure 16–45 Tubulin-decorated microtubules in a cross-section through a nerve axon (Problem 16–94). The hooks represent the tubulin decoration.

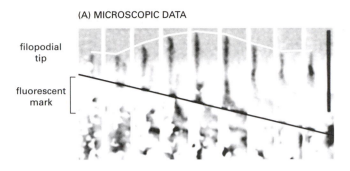

(A) MICROSCOPIC DATA

filopodial tip

fluorescent mark

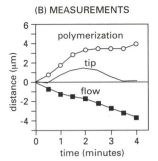

(B) MEASUREMENTS

polymerization

tip

flow

distance (μm)

time (minutes)

Figure 16–46 Regulation of filopodial extension and retraction (Problem 16–99). (A) Time-lapse observations of a single filopodium. The *white line* identifies the position of the tip of the filopodium; the *black line* marks the position of the fluorescent segment. The scale bar at the right is 5 μm. (B) Summary of the data. For the data summary the positions of the tip and the fluorescent segment were arbitrarily set at zero at zero minutes, as was the difference between the tip and the fluorescent mark. The position of the filopodium is labeled 'tip;' the position of the fluorescent mark is labeled 'flow' for retrograde flow; and the difference is labeled 'polymerization' for actin polymerization.

with Charcot-Marie-Tooth disease type 2A have a mutation in one copy of the gene for KIF1B that prevents the protein from binding to ATP. The surviving mice and the human patients have very similar progressive neuropathies. How do you suppose that loss of one copy of a gene for a kinesin motor can have such profound effects on nerve function?

16–98* Axonal transport is divided into two overall categories based on relative speed: membranous organelles move by fast axonal transport; cytoskeletal and associated proteins move by slow axonal transport. Kinesin motors transport oligomers of neurofilament proteins down axonal microtubules to sites where they are used in construction or repair of neurofilaments. Classic studies that followed pulse-labeled neurofilament proteins during axonal transport agree that the peak of radioactivity broadens markedly during transport. More recent studies demonstrated that unphosphorylated neurofilament proteins bind strongly to kinesin motors and weakly to existing neurofilaments. By contrast, the phosphorylated forms bind weakly to kinesin motors and strongly to neurofilaments.

A. How might the phosphorylation dependence of oligomers binding to kinesin motors and neurofilaments account for the broadening of the transport wave?

B. If you could track the movement of single oligomers down an axon, how would you expect them to move? Consider an oligomer at the leading edge of the transport wave and one at the trailing edge.

16–99 Nerve growth cones navigate along stereotyped pathways during development by continually extending and retracting slender filopodia to sense directional cues in the environment. The bundled actin cytoskeleton in such filopodia grows at the tip by actin polymerization, and is pulled back into the cell over time, a phenomenon known as retrograde flow. In principle, extension and retraction of filopodia could be controlled by regulating the rate of actin polymerization or the rate of retrograde flow. The experiments below were carried out to determine which rate—polymerization or retrograde flow—is regulated.

Actin monomers tagged with caged rhodamine were injected into cells and allowed to incorporate into actin filaments. The rhodamine in a narrow segment of the actin bundles near the tip of a single filopodium was uncaged by brief irradiation, yielding a fluorescent mark that allowed retrograde flow to be observed directly over time (Figure 16–46A). Extension and retraction of the tip of the filopodium was followed microscopically (Figure 16–46A). Actin polymerization was taken as the distance between the mark and the tip of the filopodium (Figure 16–46A). A summary of the data for this single filopodium is shown in Figure 16–46B. Are extension and contraction of this filopodium regulated by the rate of actin polymerization or by the rate of retrograde flow? Explain your reasoning.

16–100* Yeast cells choose bud sites on their surface in two distinct spatial patterns: axial for **a** and α haploid cells and bipolar for **a**/α diploid cells (Figure 16–47A). The selection of a new bud site establishes a cell polarity that involves the cytoskeleton and determines the site of new cell growth

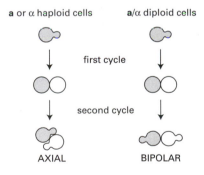

(A) BUDDING PATTERNS

a or α haploid cells

a/α diploid cells

first cycle

second cycle

AXIAL

BIPOLAR

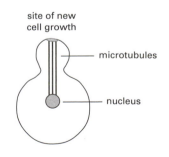

(B) CELL POLARIZATION

site of new cell growth

microtubules

nucleus

Figure 16–47 Budding in yeast (Problem 16–100). (A) Natural patterns of budding. (B) Cell polarization during bud formation.

(Figure 16–47B). To find the genes responsible for bud-site selection, mutagenized α cells were examined visually to identify mutant cells with altered budding patterns. Five genes, *BUD1–5*, were identified this way: all are nonessential genes that have no effect on cell growth or morphology. Further genetic tests indicated that these genes are involved in a single pathway for bud-site selection. In the absence of *BUD1, BUD2,* or *BUD5* the budding pattern is random; in the absence of *BUD2* or *BUD3* the budding pattern is bipolar. All five genes are required for axial budding.

RANDOM   ⟶   BIPOLAR   ⟶   AXIAL
               BUD 1   *BUD 3*
               BUD 2   *BUD 4*
               BUD 5

Sequence analysis of the genes and biochemical characterization of the proteins indicate that *BUD1* encodes a monomeric GTPase, *BUD2* encodes a GTPase-activating protein (GAP) that stimulates GTP hydrolysis by the Bud1 protein, and *BUD5* encodes a guanine-nucleotide-exchange factor (GEF) that catalyzes exchange of GTP for the GDP bound to the Bud1 monomeric GTPase.

A. By analogy to the use of monomeric GTPases, GAPs, and GEFs in directing vesicular transport between membrane-bounded compartments, design a plausible scheme by which the BUD1, BUD2, and BUD5 proteins might be used to deliver critical cytosolic proteins to a bud site.

B. Does your scheme account for the initial selection of the site for bud formation?

THE CELL CYCLE AND PROGRAMMED CELL DEATH

AN OVERVIEW OF THE CELL CYCLE

TERMS TO LEARN

budding yeast	G_1 phase	restriction point
cell cycle	G_2 phase	S phase
cell-division-cycle gene (cdc gene)	M phase	Start
fission yeast		

17–1   There are about 10^{13} cells in an adult human and about 10^{10} cells die and are replaced each day. Does this mean we become new people every three years?

17–2*   If the most basic function of the cell cycle is to duplicate accurately the DNA in the chromosomes and then distribute the copies precisely to the daughter cells, why are there gaps between S phase and M phase?

17–3   (**True/False**) Although the lengths of all phases of the cycle are variable to some extent, by far the greatest variation occurs in the duration of G_1. Explain your answer.

17–4*   Why is it remarkable that many cell-cycle genes from human cells function perfectly well when expressed in yeast cells? After all, many human genes encoding enzymes for metabolic reactions also function in yeast and no one considers that remarkable.

17–5   What determines the length of S phase? One possibility is that it depends on how much DNA the nucleus contains. As a test, you measure the length of S phase in dividing cells of a lizard, a frog, and a newt, each one of which has a different amount of DNA. As shown in Table 17–1, the length of S phase does increase with increasing DNA content.

Even though these organisms are similar, they are different species. You recall that it is possible to obtain haploid embryos of frogs and repeat your measurements with haploid and diploid frog cells. Haploid frog cells have the same length S phase as diploid frog cells. Further research in the literature shows that in plants, tetraploid strains of beans and oats have the same length S phase as their diploid cousins.

TABLE 17–1 Correlation between length of S phase and DNA content (Problem 17–5).

ORGANISM	DNA CONTENT OF NUCLEUS (pg)	LENGTH OF S PHASE (hr)
Lizard	3.2	15
Frog	15	26
Newt	45	41

Propose an explanation to reconcile these apparently contradictory results. Why is it that the length of S phase increases with increasing DNA content in different species but remains constant with increasing DNA content in the same species?

17–6* The budding yeast *Saccharomyces cerevisiae* and the fission yeast *Schizosaccharomyces pombe* provide facile genetic systems for studying a wide range of eucaryotic cell biological processes. If cell-cycle progression is essential for cell viability, however, as it is in these yeasts, how is it possible to isolate cells that are defective in cell-cycle genes?

17–7 A common first step in characterizing cell-division-cycle (*cdc*) mutants is to define the phase of the cell cycle at which the mutational block stops the cell's progress. Temperature-sensitive *cdc* mutants are particularly useful because they grow and divide normally at one temperature (the permissive temperature) but express a mutant phenotype at a higher temperature (the restrictive temperature). One method for characterizing temperature-sensitive *cdc* mutants uses the drug hydroxyurea, which blocks DNA synthesis by inhibiting ribonucleotide reductase (which provides deoxyribonucleotide precursors). Simply changing the incubation medium can reverse hydroxyurea blockade of DNA synthesis. Consider the following results with the hypothetical mutants *cdc*101 and *cdc*102.

You incubate a culture of a yeast *cdc*101 mutant at its restrictive temperature (37°C) for 2 hours (the approximate length of the cell cycle in yeasts) so that its mutant phenotype is expressed. Then you transfer it to medium containing hydroxyurea at the permissive temperature (20°C). None of the cells divide.

You now reverse the order of treatment. You incubate *cdc*101 at 20°C for 2 hours in medium containing hydroxyurea and then transfer it to medium without hydroxyurea at 37°C. The cells undergo one round of division.

You repeat these two experiments with the *cdc*102 mutant. The cells do not divide in either case.
A. In what phase of the cell cycle is *cdc*101 blocked at the restrictive temperature? Explain the results of the reciprocal temperature-shift experiments.
B. In what phase of the cell cycle is *cdc*102 blocked at the restrictive temperature? Explain the results of the reciprocal temperature-shift experiments.

17–8* You have isolated a temperature-sensitive mutant of budding yeast. It proliferates well at 25°C, but at 35°C all the cells develop a large bud and then halt their progression through the cell cycle. The characteristic morphology of the cells at the time they stop cycling is known as the landmark morphology.

It is very difficult to obtain synchronous cultures of this yeast, but you would like to know exactly where in the cell cycle the temperature-sensitive gene product must function—its execution point, in the terminology of the field—in order for the cell to complete the cycle. A clever friend, who has a good microscope with a heated stage and a video camera, suggests that you take movies of a field of cells as they experience the temperature increase and follow the behavior of individual cells as they stop cycling. Since the cells do not move much, it is relatively simple to study individual cell behavior. To make sense of what you see, you arrange a circle of pictures of cells at the start of the experiment in order of the size of their daughter buds. You then find the corresponding pictures of those same cells 6 hours later, when growth and division has completely stopped. The results with your mutant are shown in Figure 17–1.

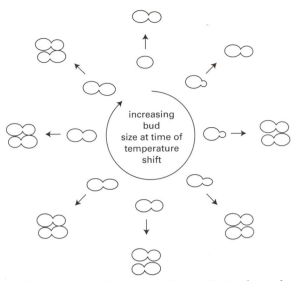

Figure 17–1 Time-lapse photography of a temperature-sensitive mutant of yeast (Problem 17–8). Cells on the *inner ring* are arranged in order of their bud sizes, which corresponds to their position in the cell cycle. After 6 hours at 37°C, they have given rise to the cells shown on the *outer ring*. No further growth or division occurs.

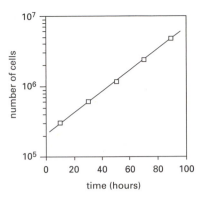

Figure 17–2 Increase in the number of mouse L cells with time (Problem 17–12).

A. Indicate on the diagram in Figure 17–1 where the execution point for your mutant lies.

B. Does the execution point correspond to the time at which the cell cycle is arrested in your mutant? How can you tell?

17–9 You have isolated a new *cdc* mutant of budding yeast that forms colonies at 25°C but not at 37°C. You would now like to isolate the wild-type gene that corresponds to the defective gene in your *cdc* mutant. How might you isolate the wild-type gene using a plasmid-based cDNA library prepared from wild-type yeast cells?

17–10* Fertilized eggs from the frog *Xenopus*, which contain 100,000 times more cytoplasm than a typical mammalian cell, are a favorite choice for studying the biochemistry of the cell cycle. Why isn't it just as easy to study these biochemical questions by simply growing large numbers of mammalian cells, which is a straightforward process?

17–11 The frequency of cells in a population that are undergoing mitosis (the mitotic index) is a convenient way to estimate the length of the cell cycle. You have decided to measure the cell cycle in the liver of the adult mouse by measuring the mitotic index. Accordingly, you have prepared liver slices and stained them to make cells in mitosis easy to recognize. After 3 days of counting, you have found only 3 mitoses in 25,000 cells. Assuming that M phase lasts 30 minutes, calculate the length of the cell cycle in the liver of an adult mouse.

17–12* The overall length of the cell cycle can be measured from the doubling time for a population of exponentially proliferating cells. The doubling time of a population of mouse L cells was determined by counting the number of cells in samples of culture fluid at various times (Figure 17–2). What is the overall length of the cell cycle in mouse L cells?

17–13 Cells that grow and divide in medium containing radioactive thymidine covalently incorporate the thymidine into their DNA during S phase. Consider a simple experiment in which cells are labeled by a brief (30 minute) exposure to radioactive thymidine. The medium is then replaced with one containing unlabeled thymidine, and the cells grow and divide for some additional time. At different time points after replacement of the medium, cells are examined in a microscope. Cells in mitosis are easy to recognize by their condensed chromosomes, and the fraction of mitotic cells that have radioactive DNA can be determined by autoradiography and plotted as a function of time after the thymidine labeling (Figure 17–3).

A. Would all cells (including cells at all phases of the cell cycle) contain radioactive DNA after the labeling procedure?

B. Note that initially there are no mitotic cells that contain radioactive DNA. Why is this?

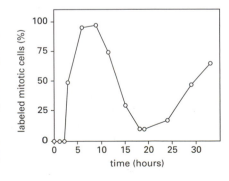

Figure 17–3 Percentage of mitotic cells that are labeled as a function of time after brief incubation with radioactive thymidine (Problem 17–13).

C. Explain the rise and fall of the curve.

D. Given that mitosis lasts 30 minutes, estimate the lengths of the G_1, S, and G_2 phases from these data. (Hint: Use the points where the curves correspond to 50% labeled mitoses to estimate the lengths of phases in the cell cycle.)

17–14* The phases of the cell cycle can also be determined using a continuous labeling protocol if the overall length of the cell cycle is known (for example, as determined in Problem 17–12). ^{3}H-thymidine is added to an asynchronous culture of cells (randomly distributed throughout the cell cycle); at various times thereafter cells are stained and prepared for autoradiography. Cells that incorporated ^{3}H-thymidine expose the photographic emulsion and are covered by silver grains. In Figure 17–4A the fraction of mitotic cells that are labeled is plotted as a function of time after addition of ^{3}H-thymidine. In Figure 17–4B the average number of silver grains above mitotic cells is plotted as a function of time after addition of ^{3}H-thymidine. Assuming the M phase lasts 30 minutes and that the overall length of the cell cycle was measured at 20 hours, deduce the duration of the G_1, S, and G_2 phases in these cells. Explain your reasoning.

17–15 For many experiments it is desirable to have a population of cells that are traversing the cell cycle synchronously. One of the first, and still often used, methods for synchronizing cells is the so-called double thymidine block. If high concentrations of thymidine are added to the culture fluid, cells in S phase stop DNA synthesis, though other cells are not affected. The excess thymidine blocks the enzyme ribonucleotide reductase, which is responsible for converting ribonucleotides into deoxyribonucleotides. When this enzyme is inhibited, the supply of deoxyribonucleotides falls and DNA synthesis stops. When the excess thymidine is removed by changing the medium, the supply of deoxyribonucleotides rises and DNA synthesis resumes normally.

For a cell line with a 22-hour cell cycle divided so that M phase = 0.5 hour, G_1 phase = 10.5 hours, S phase = 7 hours, and G_2 phase = 4 hours, a typical protocol for synchronization by a double thymidine block would be as follows:

1. At 0 hours ($t = 0$ hours) add excess thymidine.
2. After 18 hours ($t = 18$ hours) remove excess thymidine.
3. After an additional 10 hours ($t = 28$ hours) add excess thymidine.
4. After an additional 16 hours ($t = 44$ hours) remove excess thymidine.

A. At what point in the cell cycle is the cell population when the second thymidine block is removed?

B. Explain how the times of addition and removal of excess thymidine synchronize the cell population.

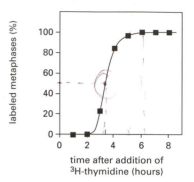

(A) LABELED MITOSES

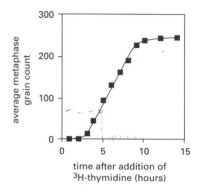

(B) NUMBER OF SILVER GRAINS

Figure 17–4 Labeled mitotic cells as a function of time after addition of ^{3}H-thymidine (Problem 17–14).
(A) Fraction of labeled mitotic cells.
(B) Average number of silver grains above labeled mitotic cells.

COMPONENTS OF THE CELL-CYCLE CONTROL SYSTEM

TERMS TO LEARN

Cdc25	cyclin–Cdk complex	M-cyclin (M-Cdk)
Cdk-activating kinase (CAK)	cyclin-dependent kinase (Cdk)	S-cyclin (S-Cdk)
Cdk inhibitor protein (CKI)	G_1-cyclin (G_1-Cdk)	ubiquitin ligase
checkpoint	G_1/S-cyclin (G_1/S-Cdk)	Wee1
cyclin		

17–16* Hoechst 33342 is a membrane-permeant dye that fluoresces when it binds to DNA. If a population of cells is incubated briefly with Hoechst dye and then sorted in a flow cytometer, the cells display various levels of fluorescence as shown in Figure 17–5.

A. Which cells in Figure 17–5 are in the G_1, S, G_2, and M phases of the cell cycle. Explain the basis for your answer.

B. Sketch the sorting distributions you would expect for cells that were treated with inhibitors that block the cell cycle in the G_1, S, or M phase. Explain your reasoning.

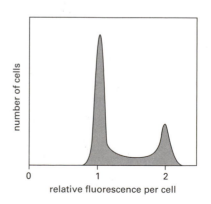

Figure 17–5 Analysis of Hoechst fluorescence in a population of cells sorted in a flow cytometer (Problem 17–16).

17–17 There is a point in G_1 of the cell cycle where a cell checks that it is of sufficient size before committing to the next step. A possible clue regarding the mechanism by which this checkpoint operates comes from cells that have an abnormal number of copies of their genome. For a given cell type, it seems to be a general rule that cell size is roughly proportional to the number of chromosome sets in the cell. With this observation in mind, suggest a mechanism by which a cell measures its own size in G_1.

17–18* As director of the Royal Ballet, you are intrigued by the similarity between your nightly show and the cell cycle. Each night the theater fills, the lights go out, the curtain rises, the dancers perform, the curtain falls, the lights go on, and the theater empties—a defined sequence of events, just like the cell cycle. You wonder whether checkpoint mechanisms might enhance theater operations, as they do in the cell cycle. As a first step, you set out to design a system so that when all the seats are filled, the curtains will automatically rise. You have the engineers place sensors in each seat to detect a seated person. But you haven't yet decided how to connect the sensors to the curtains. Do you want the sensors to send out a positive signal that in aggregate will be enough to raise the curtain, or do you want each sensor to send out a negative signal that will stop the curtain going up until every seat is filled? Which do you think would be the more reliable system? How would a cell do it?

17–19 Frog oocytes mature into eggs when incubated with progesterone. This maturation is characterized by disappearance of the nucleus (termed germinal vesicle breakdown) and formation of a meiotic spindle. The requirement for progesterone can be bypassed by microinjecting 50 nL of egg cytoplasm directly into a fresh oocyte (1000 nL), which then matures normally (Figure 17–6). The control experiment of microinjecting cytoplasm from untreated oocytes into other oocytes causes no maturation, as expected. Maturation-promoting factor (MPF) activity—later called mitosis-promoting factor and shown to be M-Cdk—in the egg cytoplasm is responsible for maturation.

Progesterone-induced maturation requires protein synthesis, as indicated by its sensitivity to cycloheximide; however, MPF-induced maturation does not. By placing progesterone-stimulated oocytes into cycloheximide at different times after stimulation, it can be shown that maturation becomes cycloheximide independent (no longer inhibited by cycloheximide) a few hours before the oocytes become eggs. In addition, the time at which the oocytes become cycloheximide independent corresponds to the appearance of MPF activity.

Is synthesis of MPF itself the cycloheximide-sensitive event? To test this possibility, you transfer MPF serially from egg to oocyte to test whether its activity diminishes with dilution. You first microinject 50 nL of cytoplasm from an activated egg into an immature oocyte as shown in Figure 17–6; when the oocyte matures into an egg, you transfer 50 nL of its cytoplasm into another immature oocyte; and so on. Surprisingly, you find that you can continue this process for at least 10 transfers, even if the recipient oocytes are bathed in cycloheximide! Moreover, the apparent MPF activity in the last egg is equal to that in the first egg.

Problem 3–86 describes the activation of Cdk2 by phosphorylation.

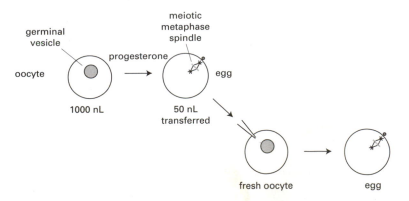

Figure 17–6 Progesterone- and MPF-induced maturation of oocytes (Problem 17–19).

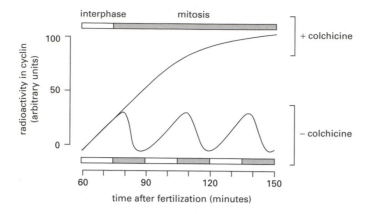

Figure 17–7 Effects of colchicine treatment on cyclin B destruction and cell-cycle progression (Problem 17–20). *Boxes indicate position in the cell cycle; white segments represent interphase and gray segments represent mitosis.*

A. What dilution factor is achieved by 10 serial transfers of 50 nL into 1000 nL? Do you consider it likely that a molecule might have an undiminished biological effect over this concentration range?

B. How can MPF activity, which is due to a protein, be absent from immature oocytes yet appear in activated eggs, even when protein synthesis has been blocked by cycloheximide?

C. Propose a means by which MPF might maintain its activity through repeated serial transfers.

D. Propose a role for the cycloheximide-sensitive factor that is required for the appearance of MPF activity in a progesterone-stimulated oocyte.

17–20* You are studying the synthesis and destruction of cyclins in dividing clam embryos. You suspect that colchicine, a drug that binds to tubulin and arrests cells in mitosis, might work by inhibiting the normal destruction of cyclin B at the metaphase–anaphase transition. To test this idea, you add ^{35}S-methionine to a suspension of fertilized clam eggs, divide the suspension in two and add colchicine to one. You take duplicate samples at 5-minute intervals, one for analysis of cyclin B by gel electrophoresis and the other for analysis of mitotic chromosomes by fixing and staining the cells. Untreated cells alternated between interphase and mitosis every 30 minutes, whereas colchicine-treated cells entered mitosis normally but remained there for hours (Figure 17–7). Moreover, colchicine treatment abolished the normal disappearance of cyclin B that precedes the metaphase–anaphase transition.

To get a clearer picture of how colchicine inhibits cyclin B destruction, you repeat the experiment, but add the protein synthesis inhibitor emetine just before cells enter mitosis. Emetine-treated cells in the absence of colchicine entered and exited mitosis normally, and divided into two daughter cells, which then remained indefinitely in interphase (Figure 17–8). In the presence of colchicine, the emetine-treated cells stayed in mitosis for about 2 hours and then decondensed their chromosomes and re-formed nuclei without dividing. In the presence and absence of colchicine, the exit from mitosis coincided with the disappearance of cyclin B.

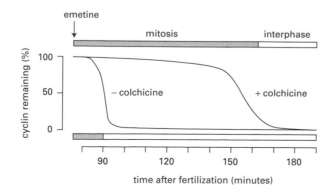

Figure 17–8 Effects of emetine treatment on cyclin B destruction and cell-cycle progression in normal cells and colchicine-treated cells (Problem 17–20).

A. What effect does colchicine have on cyclin B synthesis and destruction? Can these effects explain how colchicine causes metaphase arrest?

B. How does inhibition of protein synthesis eventually reverse the metaphase arrest produced by colchicine?

17–21 Vertebrate cells use several different Cdks to manage various transitions in the cell cycle, yet budding yeast is able to get by with a single Cdk. How do budding yeast cells manage that neat trick?

17–22* You have isolated a temperature-sensitive mutant of the *CDC28* gene, which encodes Cdk1 in budding yeast. At the restrictive temperature the mutant Cdk1 binds the G_1 cyclins Cln1 and Cln2 so weakly that colonies do not form. When a cDNA library on a high copy-number plasmid was expressed in the mutant strain, three different cDNAs were found that allowed colony formation at the restrictive temperature. One was the cDNA from the wild-type *CDC28* gene; the other two were cDNAs that encoded cyclins Cln1 or Cln2. If the cDNAs for the cyclin genes were transferred to a plasmid that was maintained at one copy per cell, neither allowed the *CDC28* mutant to grow at high temperature. Suggest a mechanism whereby high-level expression of Cln1 or Cln2 would allow survival of mutant *CDC28* cells at high temperature, whereas low-level expression would not.

17–23 Imagine that you've placed the *CLN3* gene, which encodes the cyclin component of G_1-Cdk, on a plasmid under the control of a regulatable promoter and introduced the plasmid into wild-type yeast cells. With the promoter turned off, the cells progressed through the cell cycle in the normal way as determined by flow cytometry (Figure 17–9A). When the promoter was turned on, however, the distribution of cells was significantly altered (Figure 17–9B).

A. How was the distribution of cells in the phases of the cell cycle altered by overexpression of the cyclin Cln3? How do you suppose that increased expression of the cyclin Cln3 causes these changes?

B. When Cln3 was overexpressed, would you have expected the cells to be smaller than normal, the same size, or larger than normal? Explain your reasoning.

17–24* (**True/False**) The regulation of cyclin–Cdk complexes depends entirely on phosphorylation and dephosphorylation. Explain your answer.

17–25 You have isolated two temperature-sensitive strains of yeast (called '*giant*' and '*tiny*') that show very different responses to elevated temperature. *Giant* cells grow until they become enormous but no longer divide. *Tiny* cells have very short cell cycles and divide when they are very much smaller than usual. You are amazed to discover that these strains arose by different mutations in the very same gene. Propose a model to explain these observations, and suggest what might be the function of the normal protein encoded by this gene.

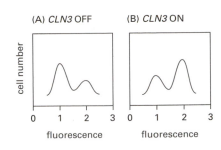

Figure 17–9 Flow cytometry of wild type yeast cells (Problem 17–23). (A) Cln3 not expressed from plasmid. (B) Cln3 overexpressed from plasmid.

INTRACELLULAR CONTROL OF CELL-CYCLE EVENTS

TERMS TO LEARN

anaphase-promoting complex (APC)	E2F	Polo kinase
Cdc6	Mcm protein	pre-replicative complex (pre-RC)
Cdc20	Mdm2	retinoblastoma protein (Rb)
Cln3	origin recognition complex (ORC)	securin
cohesins	p21	separase
condensins	p53	sister chromatid
DNA damage checkpoint		spindle-attachment checkpoint
DNA replication checkpoint		

17–26* Early clues about the regulation of S phase came from studies in which human cells at various cell-cycle stages were fused to form single cells with

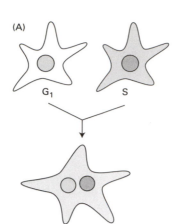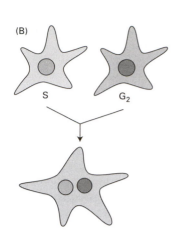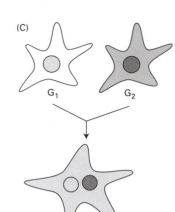

(A)

G₁   S

G₁-phase nucleus
immediately enters S phase;
S-phase nucleus continues
DNA replication

(B)

S   G₂

G₂-phase nucleus
stays in G₂ phase;
S-phase nucleus continues
DNA replication

(C)

G₁   G₂

G₂-phase nucleus
stays in G₂ phase;
G₁-phase nucleus enters
S-phase according
to its own timetable

Figure 17–10 Results of cell-fusion experiments using mammalian cells at different stages of the cell cycle (Problem 17–26). (A) Fusion of S and G₁ cells. (B) Fusion of S and G₂ cells. (C) Fusion of G₁ and G₂ cells.

two nuclei. Figure 17–10 shows the outcome of pairwise fusions between G₁, S, and G₂ cells. Given what we now know about the roles of cyclin–Cdk complexes in progression of the cell cycle, how would you interpret the outcomes of each of these experiments? Do these experiments suggest that there may be a block to re-replication in the cell cycle?

Problems 5–40 and 5–42 Analyze the block to re-replication in eucaryotic cells.

17–27 In budding yeast a pre-replicative complex, consisting of ORC, Cdc6, and Mcm proteins, is established at origins of replication during the G₁ phase. S-Cdk then triggers origin firing and helps to prevent re-replication? But not all yeast origins begin replication at the same time: some fire early in S phase while others fire late. How is it possible for S-Cdk to trigger origin firing at a variety of times and also prevent re-replication? The details of this process are largely unknown. Propose a scheme that could account for this behavior of S-Cdk.

17–28* It is remarkable that the concentration of cyclin B in the cleaving clam egg rises very slowly and steadily throughout the cell cycle, whereas M-Cdk activity increases suddenly at mitosis (Figure 17–11). How is the activity of M-Cdk so sharply regulated in the presence of a gradual increase in cyclin B?

17–29 The activities of Wee1 tyrosine kinase and Cdc25 tyrosine phosphatase determine the state of phosphorylation of tyrosine 15 in the Cdk1 component of M-Cdk. When tyrosine 15 is phosphorylated, M-Cdk is inactive; when tyrosine 15 is not phosphorylated, M-Cdk is active (Figure 17–12). Just as the activity of M-Cdk itself is controlled by phosphorylation, so too are the activities of Wee1 kinase and Cdc25 phosphatase.

The regulation of these various activities can be studied in extracts of frog

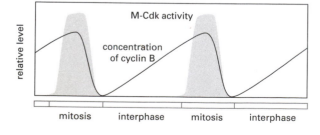

Figure 17–11 The rise and fall of M-Cdk activity and cyclin B concentration during the cell cycle in a cleaving clam egg (Problem 17–28).

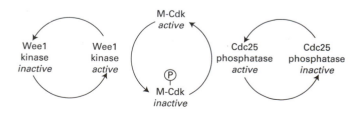

Figure 17–12 Control of M-Cdk activity by Wee1 tyrosine kinase and Cdc25 tyrosine phosphatase (Problem 17–29).

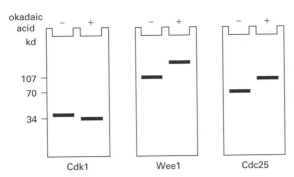

Figure 17–13 Effects of okadaic acid on the phosphorylation states of Cdk1, Wee1, and Cdc25 (Problem 17–20). Molecular mass markers are shown in kilodaltons on the *left*.

oocytes. In such extracts Wee1 tyrosine kinase is active and Cdc25 tyrosine phosphatase is inactive. As a result M-Cdk is inactive because its Cdk1 component is phosphorylated on tyrosine 15. M-Cdk in these extracts can be rapidly activated by addition of okadaic acid, which is a potent inhibitor of serine/threonine protein phosphatases. Using antibodies specific for each component, it is possible to examine their phosphorylation states by changes in mobility upon gel electrophoresis (Figure 17–13). (Phosphorylated proteins generally run slower than their nonphosphorylated counterparts.)

A. Based on the results with okadaic acid decide whether the active forms of Wee1 tyrosine kinase and Cdc25 tyrosine phosphatase are phosphorylated or nonphosphorylated. In Figure 17–12 indicate the phosphorylated forms of Wee1 and Cdc25 and label the arrows connecting their active and inactive forms to show which transitions are controlled by protein kinases and which are controlled by protein phosphatases.

B. Are the protein kinases and phosphatases that control Wee1 and Cdc25 specific for serine/threonine side chains or for tyrosine side chains? How do you know?

C. How does addition of okadaic acid cause an increase in phosphorylation of Wee1 and Cdc25, but a decrease in phosphorylation of Cdk1?

D. If you assume that Cdc25 and Wee1 are targets for phosphorylation by active M-Cdk, can you explain how the appearance of a small amount of active M-Cdk would lead to its rapid and complete activation?

17–30* When activated, the DNA replication checkpoint blocks activation of M-Cdk and thereby prevents entry into mitosis. The target of this checkpoint mechanism is the inhibitory phosphate on M-Cdk, which is added by Wee1 tyrosine kinase and removed by Cdc25 tyrosine phosphatase. The level of phosphorylation on M-Cdk is a balance between the activities of Wee1 kinase and Cdc25 phosphatase. The checkpoint mechanism inhibits Cdc25 phosphatase. Why do you suppose it inhibits the phosphatase instead of activating Wee1 kinase?

17–31 If high doses of caffeine (Figure 17–14) interfere with the DNA replication checkpoint mechanism in mammalian cells, why do you suppose the Surgeon General hasn't yet issued an appropriate warning to heavy coffee and cola drinkers? A typical cup of coffee (150 mL) contains 100 mg of caffeine (196 g/mole). How many cup(s) of coffee would you have to drink to reach the dose (10 mM) required to interfere with the DNA replication checkpoint mechanism? (A typical adult contains about 40 liters of water.)

17–32* At the transition from metaphase to anaphase M-Cdk is inactivated and chromosomes begin to separate into sister chromatids. M-Cdk is inactivated by the anaphase-promoting complex (APC), which destroys the cyclin-B component of M-Cdk, eliminating its kinase activity. You want to know how the separation of sister chromatids is related to M-Cdk inactivation. To answer this question, you make cell-free extracts from unfertilized frog eggs. When nuclei are added to the extract, they spontaneously form spindles with condensed chromosomes aligned on the metaphase plate. Anaphase and the separation of sister chromatids can be triggered by addition of Ca^{2+}, which activates APC and turns off M-Cdk.

To investigate the control of sister chromatid separation, you make use of

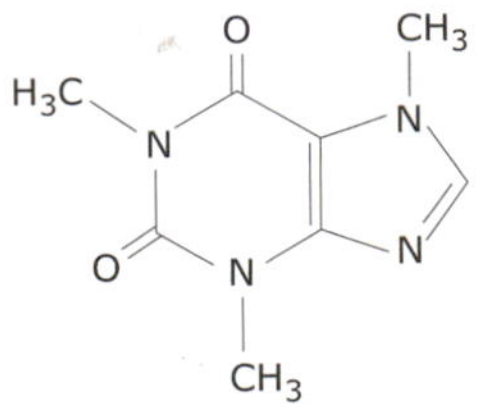

Figure 17–14 Structure of caffeine (Problem 17–31).

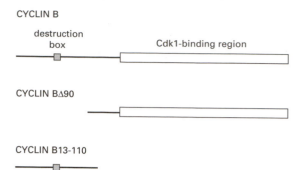

CYCLIN B

destruction
box　　　　　　　　　Cdk1-binding region

CYCLIN BΔ90

CYCLIN B13-110

Figure 17–15 Cyclin B and two mutants (Problem 17–32).

two mutant forms of cyclin B (Figure 17–15). Cyclin BΔ90 is missing the destruction box, a sequence of amino acids required for inactivation by APC, but it retains its ability to bind to Cdk1 and make functional M-Cdk. Cyclin B13-110 retains the destruction box but cannot bind to Cdk1. When either protein is added in excess to the extract, M-Cdk activity remains high after addition of Ca^{2+}. The two proteins differ, however, in their effects on chromatid separation. In the presence of cyclin BΔ90, sister chromatids separate normally; in the presence of cyclin B13-110, sister chromatids remain linked.

A. Why does M-Cdk remain active in the presence of Ca^{2+} when cyclin BΔ90 is added to the extract?

B. Why does M-Cdk remain active when cyclin B13-110 is added to the extract?

C. How is the separation of sister chromatids related to M-Cdk inactivation? Do sister chromatids separate because a linker protein must be in its phosphorylated state in order to hold the chromatids together? Or do chromatids separate because APC degrades the linker protein?

17–33　Using a clever genetic screen, you have identified a temperature-sensitive mutant in a yeast gene (*SCC1*) that appears to be required for sister chromatid cohesion. To assay directly for sister chromatid cohesion, you insert a tandem array of 336 short DNA sequences, to which a bacterial protein can bind tightly, adjacent to the centromere of chromosome V. You then express a fusion of the bacterial protein with GFP in the same cells. When the GFP fusion protein binds to its recognition sequences, it creates a bright dot of green fluorescence on the chromosome. To test for the effects of mutant Scc1 on sister chromatid cohesion, you isolate unbudded cells from wild-type and *scc1*^{ts} cells that were grown at 25°C and grow them at 37°C for various times. Representative examples of small-budded cells in S phase and large-budded cells that have passed the metaphase–anaphase transition are shown for both strains in Figure 17–16.

A. Do sister chromatids in wild-type cells adhere to each other normally during S phase and separate normally during mitosis? How can you tell?

B. Do sister chromatids in *scc1*^{ts} cells adhere normally in S phase and separate normally during mitosis? How can you tell?

C. In the large-budded cells from the *scc1*^{ts} strain, why do both sister chromatids remain in one cell?

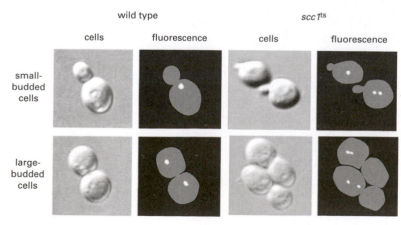

Figure 17–16 Small- and large-budded cells from wild type and *scc1*^{ts} grown at 37°C (Problem 17–33). For each strain a matched set of pictures shows the appearance of the cells and the corresponding sites of fluorescence.

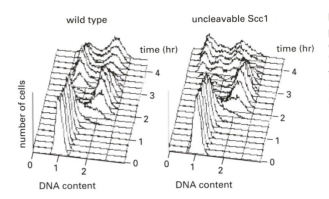

wild type   uncleavable Scc1

Figure 17–17 Analysis of cell-cycle progression in wild-type cells and cells expressing an uncleavable form of Scc1 (Problem 17–34). Samples were taken at 15-minute intervals after release from a G₁ block. Successive samples have been 'stacked' for clarity.

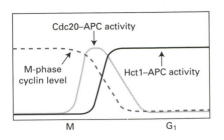

Figure 17–18 Levels of M-phase cyclins, Cdc20–APC, and Hct1–APC at the transition between M and G₁ (Problem 17–37).

17–34* Scc1 protein is cleaved by the protease separase. By altering the sites in yeast Scc1 that are recognized and cleaved by separase, you have created an uncleavable form of Scc1. To analyze the cellular consequences of uncleavable Scc1, you release cells from a G₁ cell-cycle block and at the same time turn on expression of uncleavable Scc1 from a regulatable promoter. As expected, sister chromatids do not separate (as analyzed by the technique described in Problem 17–33). You are puzzled, however, by their progress through the cell cycle, as analyzed by flow cytometry (Figure 17–17). Up to about 3 hours cells with unclearable Scc1 are indistinguishable from wild-type cells, but after that they are very different.

A. Did you expect them to be the same as wild-type cells up the 3-hour time point? Why or why not?

B. How do cells that express uncleavable Scc1 differ from wild-type cells after the 3-hour time point? Can you suggest an explanation for their behavior?

17–35 Nocodazole reversibly inhibits microtubule polymerization, which is essential for formation of the mitotic spindle. By treating a population of mammalian cells with nocodazole for a time and then washing it out of the medium, it is possible to synchronize the cell population. In the presence of nocodazole, where in the cell cycle do the cells accumulate? What mechanism do you suppose is responsible for stopping cell-cycle progression in the presence of nocodazole?

17–36* Budding yeast cells that are deficient for Mad2, a component of the spindle-attachment checkpoint, are killed by treatment with benomyl, which causes microtubules to depolymerize. In the absence of benomyl, however, the cells are perfectly viable. Explain why this is so.

17–37 Toward the end of M phase the concentration of M-phase cyclins drops precipitously and remains low until the next cell cycle (Figure 17–18). Explain how M–Cdk activity coordinates the activities of Cdc20–APC and Hct1–APC to bring about this sharp drop in cyclin levels.

17–38* Retinoblastoma is an extremely rare cancer of the nerve cells in the eye. The disease mainly affects children up to the age of five years because it can only occur while the nerve cells are still dividing. In some cases tumors occur in only one eye, but in other cases tumors develop in both eyes. The bilateral cases all show a familial history of the disease; most of the cases affecting only one eye arise in families with no previous history of the disease.

   An informative difference between unilateral and bilateral cases becomes apparent when the fraction of still undiagnosed cases is plotted against the age at which diagnosis is made (Figure 17–19). The regular decrease with time shown by the bilateral cases suggests that a single chance event is sufficient to trigger onset of bilateral retinoblastoma. By contrast, the presence of a 'shoulder' on the unilateral curve suggests that multiple events in one neuron are required to trigger unilateral retinoblastoma. (A shoulder arises because the events accumulate over time. For example, if two events are required, most affected cells at early times will have suffered only a

Problems 18–1, 18–2, and 18–4 explore the roles of various proteins in sister-chromatid cohesion.

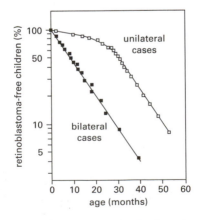

Figure 17–19 Time of onset of unilateral and bilateral cases of retinoblastoma (Problem 17–38). A population of children who ultimately developed retinoblastoma is represented in this graph. The fraction of the population that is still tumor free is plotted against the time after birth.

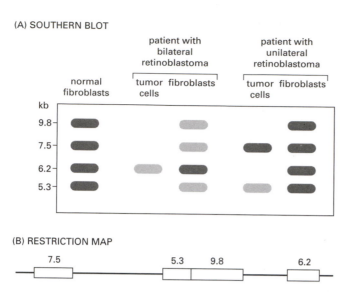

(A) SOUTHERN BLOT

patient with bilateral retinoblastoma

patient with unilateral retinoblastoma

normal fibroblasts

tumor cells   fibroblasts

tumor cells   fibroblasts

kb
9.8
7.5
6.2
5.3

(B) RESTRICTION MAP

7.5   5.3   9.8   6.2

Figure 17–20 Patterns of blot hybridization of restriction fragments from the retinoblastoma gene (Problem 17–38). (A) Southern blot for normal individuals and for patients with unilateral and bilateral retinoblastoma. *Lighter shading* of some bands indicates half the normal number of copies. (B) The order of the restriction fragments in the *Rb* gene. Fragments that contain exons (*rectangles*) hybridize to the cDNA clone that was used as a probe in these experiments.

single event and will not generate a tumor. With time the probability increases that a second event will occur in an already affected cell and, therefore, cause a tumor.)

One possible explanation for these observations is that tumors develop when both copies of the critical gene (the retinoblastoma, *Rb*, gene) are lost or mutated. In the inherited (bilateral) form of the disease, a child receives a defective *Rb* gene from one parent: tumors develop in an eye when the other copy of the gene in any nerve cell in the eye is lost through somatic mutation. In fact, the loss of a copy of the gene is frequent enough that tumors usually occur in both eyes. If a person starts with two good copies of the *Rb* gene, tumors arise in an eye only if both copies are lost *in the same cell*. Since such double loss is very rare, it is usually confined to one eye.

To test this hypothesis, you use a cDNA clone of the *Rb* gene to probe the structure of the gene in cells from normal individuals and from patients with unilateral or bilateral retinoblastoma. As illustrated in Figure 17–20, normal individuals have four restriction fragments that hybridize to the cDNA probe (which means each of these restriction fragments contains at least one exon). Fibroblasts (nontumor cells) from the two patients also show the same four fragments, although three of the fragments from the child with bilateral retinoblastoma are present in only half the normal amount. Tumor cells from the two patients are missing some of the restriction fragments.

A. Explain why fibroblasts and tumor cells from the same patient show different band patterns.

B. What are the structures of the *Rb* genes in the fibroblasts from the two patients? What are their structures in the tumor cells from the two patients?

C. Are these results consistent with the hypothesis that retinoblastoma is due to the loss of the *Rb* gene?

D. Suggest a plausible explanation for how the loss of the Rb protein might cause retinoblastoma.

17–39   *Rb* is one example of a category of antiproliferative genes in humans. Typically, when both copies of such genes are lost, cancers develop. Do you suppose that cancer could be eradicated if tumor-suppressor genes such as *Rb* could be expressed at abnormally high concentrations in all human cells? Explain your answer.

17–40* (**True/False**) In order for proliferating cells to maintain a relatively constant size, the length of the cell cycle must match the time it takes for the cell to double in size. Explain your answer.

17–41   One model for coordination of growth with the cell cycle in budding yeast is that the G_1 cyclin Cln3 steadily increases with cell size until it has bound to

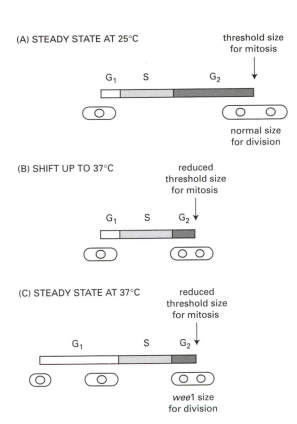

(A) STEADY STATE AT 25°C

threshold size for mitosis

G_1 S G_2

normal size for division

(B) SHIFT UP TO 37°C

reduced threshold size for mitosis

G_1 S G_2

(C) STEADY STATE AT 37°C

reduced threshold size for mitosis

G_1 S G_2

*wee*1 size for division

Figure 17–21 Lengths of the cell cycle and sizes of *Wee1*ts cells grown at 25°C and at 37°C (Problem 17–42).

all available Cln3-specific DNA-binding sites. The excess Cln3 would then bind to Cdk1 to activate G_1-Cdk and trigger a round of cell division. According to this model, if you were to increase the number of DNA-binding sites for Cln3, would you expect that the cells would be smaller than normal, the same size, or larger than normal? Explain your reasoning.

17–42* In the cell cycle of the fission yeast *S. pombe*, G_1 is very short and G_2 is very long (just the opposite of the situation in budding yeast). When a *Wee1*ts mutant is grown at 25°C, it has a normal cell cycle and a normal size (Figure 17–21A). When shifted to 37°C, the mutant cells undergo a shortened first cell cycle because inactivation of Wee1 kinase reduces the threshold size for mitosis, generating smaller than normal cells (Figure 17–21B). Surprisingly, in subsequent cell cycles at 37°C the cell cycle is of normal length, but now with a long G_1 and a short G_2; nevertheless, small cells are still generated. What do you suppose would happen to *Wee1*ts cells if the cell cycle did not increase in length? Propose a molecular explanation for how G_1 might be lengthened in *Wee1*ts cells grown at high temperature.

17–43 Unlike mice and humans, *Drosophila* has a single gene for cyclin D and for its binding partner Cdk4, which greatly simplifies analysis and interpretation. To investigate the role of the cyclin D–Cdk4 complex in the cell cycle, you overexpress both genes using a genetic trick. You deposit the genes into the genome using a transposon and then activate them at various times in development by site-specific recombination (see Problem 5–85). You have arranged it so that activation of these two genes is accompanied by activation of GFP, so that cells expressing cyclin D and Cdk4 also fluoresce. When you examine the pattern of fluorescent cells in developing wings in a large number of flies you find that the clones of marked cells average 16 cells in size, whereas clones of cells in control flies (expressing GFP alone) average just 12 cells in size. The cells that are overexpressing cyclin D and Cdk4 are exactly the same size as normal cells. (Parallel experiments in which one or the other gene was overexpressed gave results that matched the controls.)

(A) NORMAL EYE

(B) EYE WITH OVERPRODUCED CYCLIN D AND Cdk4

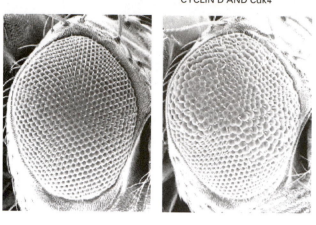

Figure 17–22 Effects of overexpression of cyclin D and Cdk4 (Problem 17–44). (A) A normal eye. (B) An eye with a patch of cells overexpressing cyclin D and Cdk4.

From these results, would you conclude that overexpression of cyclin D and Cdk4 altered the growth rate of the cells, the duration of their cell cycles, or both? Explain your reasoning.

17–44*   To try to distinguish whether overexpression of cyclin D and Cdk4 in flies results primarily from an effect on growth rate or, alternatively, primarily from an effect on cell cycle, you examine terminally differentiated cells that are no longer dividing. You find that in differentiating *Drosophila* eyes, over-expression of cyclin D and Cdk4 causes cell enlargement in post-mitotic (nondividing) cells (Figure 17–22).

Do these results support a primary role of cyclin D and Cdk4 on growth, or do they support a primary effect on cell cycle? Explain your reasoning.

17–45   One important biological effect of a large dose of ionizing radiation is to halt cell division.
   A. How does a large dose of ionizing radiation stop cell division?
   B. What happens if a cell has a mutation that prevents it from halting cell division after being irradiated?
   C. What might be the effects of such a mutation if the cell was not irradiated?
   D. An adult human who has reached maturity will die within a few days of receiving a radiation dose large enough to stop cell division. What does this tell you (other than that one should avoid large doses of radiation)?

17–46*   (True/False) Budding yeast and mammalian cells respond to DNA damage in the same way: they transiently arrest their cell cycles to repair the damage and if repair cannot be completed, they resume their cycles despite the damage. Explain your answer.

17–47   You have found a new way to study radiation-sensitive yeast mutants. By culturing cells on a thin layer of agar on a microscope slide you can record the fate of individual cells. When you irradiate wild-type cells with x-rays, which cause chromosome breaks, and follow their growth for the next 10 hours, you find that most of the cells arrest temporarily at the large-bud (dumbbell) stage, but about half the cells eventually recover and form small viable colonies after 10 hours (Figure 17–23). The fraction that is still arrested at the dumbbell stage after 10 hours is equal to the fraction of nonviable cells (Table 17–2).

WILD-TYPE CELLS

time = 0   time = 10 hr

*rad*9 CELLS

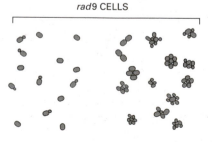

time = 0   time = 10 hr

Figure 17–23 Time-lapse pictures of wild-type and *rad9* cells at zero and 10 hours after x-irradiation (Problem 17–47).

TABLE 17–2 Fractions of arrested and nonviable cells after x-ray treatment (Problem 17–47).

STRAIN	ARRESTED AT 10 HOURS (%)	NONVIABLE (%)
Wild type	50	50
rad52	90	95
rad9	20	70

TABLE 17–3 Mutant strains that affect the mitotic entry checkpoint (Problem 17–48).

MUTANT STRAINS	MITOTIC DELAY IN RESPONSE TO	
	DAMAGED DNA	UNREPLICATED DNA
rad24	No	Yes
cdc2-3w	Yes	No
hus1	No	No
hus2	No	No
rad1	No	No
cdc2-F15	Yes	No

You repeat these experiments with seven different radiation-sensitive (*rad*) mutants. For six of the mutants you observe a similar equality of 10-hour arrested cells and nonviable cells, as shown in Table 17–2 for the *rad52* mutant, although relative to wild-type cells a higher fraction of *rad* cells are still arrested in the dumbbell stage.

One of the seven *rad* mutants, however, has a strikingly different phenotype. Many fewer *rad9* cells arrest even temporarily at the dumbbell stage, and after 10 hours only 20% are still arrested (Table 17–2). Although many of the cells divide once or twice, they mostly form nonviable microcolonies (Figure 17–23, Table 17–2).

A. For the wild-type cells decide which cells in the population appear most likely to remain arrested at the dumbbell stage after 10 hours (Figure 17–23). Given that the cells used in these experiments are haploid and that x-ray-induced breaks are repaired by homologous recombination, decide which stage of the cell cycle the sensitive cells are in.

B. By staining with DNA-binding reagents and tubulin-specific antibodies, you show that cells in the dumbbell stage have a single nucleus stuck in the neck and no visible spindle. Given what you know about cell-cycle checkpoints, in what stage of the cell cycle do you think the cells are arrested?

C. Why do half of the wild-type cells temporarily arrest at the dumbbell stage but then go on to form viable colonies after 10 hours?

D. Why do you think that so many more *rad52* mutant cells (relative to wild-type cells) are arrested at the dumbbell stage after 10 hours?

E. Why do you think that so few *rad9* mutant cells arrest even temporarily at the dumbbell stage? Why are so many of the *rad9* cells nonviable?

F. Would you expect the fraction of nonviable *rad9* cells to increase, decrease, or stay the same if the cells were artificially delayed for a couple of hours using a microtubule inhibitor that reversibly prevents spindle formation? Explain your reasoning?

17–48* Fission yeast respond to damaged DNA and unreplicated DNA by delaying entry into mitosis. You want to know how the signals from such DNA interact with the mitotic entry checkpoint. You screen a large number of mutant yeast strains and find six that do not delay mitosis in response to DNA damage, unreplicated DNA, or both, as shown in Table 17–3. Which of the signaling pathways shown in Figure 17–24 is supported by your data? On the pathway you choose, indicate where each of the mutant genes acts.

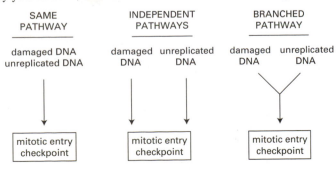

Figure 17–24 Possible pathways by which signals from damaged DNA and unreplicated DNA interact with the mitotic entry checkpoint (Problem 17–48).

17–49 What do you suppose happens in mutant cells with the following defects?
  A. Cannot degrade M-phase cyclins.
  B. Always express high levels of p21.
  C. Cannot phosphorylate Rb.

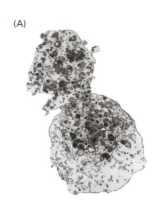

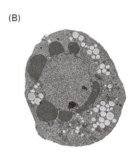

PROGRAMMED CELL DEATH (APOPTOSIS)

TERMS TO LEARN

adaptor protein	Bcl-2 family	Fas ligand
Apaf-I	caspases	IAP (inhibitor of apoptosis) family
apoptosis	Fas	programmed cell death

17–50* (**True/False**) In adult tissues cell death exactly balances cell division. Explain your answer.

17–51 Why do you think programmed cell death occurs by a different mechanism from the cell death that occurs in necrosis? In apoptosis the cell destroys itself from within and avoids leakage of the cell contents into the extracellular space. What might be the consequences if programmed cell death were not achieved in so neat and orderly a fashion?

17–52* Compare the rules of cell proliferation and apoptosis in an animal to the rules that govern human behavior in society. What would happen to an animal if its cells behaved like people normally behave in our society? Could the rules that govern cell proliferation be applied to human societies?

17–53 Look carefully at the electron micrographs in Figure 17–25. Describe the differences between the cell that died by necrosis and the one that died by apoptosis. How do the pictures confirm the differences between the two processes? Explain your answer.

Figure 17–25 Cell death (Problem 17–53). (A) By necrosis. (B) By apoptosis.

17–54* Fas ligand is a trimeric, extracellular protein that binds to its receptor, Fas, which is composed of three identical transmembrane subunits (Figure 17–26). The binding of Fas ligand alters the conformation of Fas so that it binds an adaptor protein, which then recruits and activates procaspase-8, triggering a caspase cascade that leads to cell death. In humans the autoimmune lymphoproliferative syndrome (ALPS) is associated with dominant mutations in Fas that include point mutations and C-terminal truncations. In individuals that are heterozygous for such mutations, lymphocytes do not die at their normal rate and accumulate in abnormally large numbers,

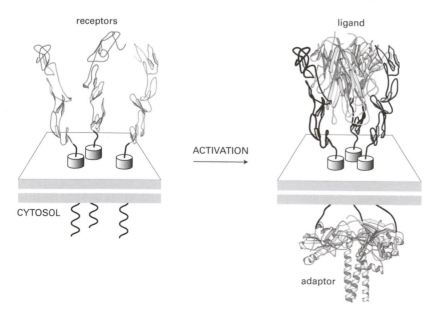

receptors

ligand

ACTIVATION

CYTOSOL

adaptor

Figure 17–26 The binding of trimeric Fas ligand to Fas (Problem 17–54).

causing a variety of clinical problems. In contrast to these patients, individuals that are heterozygous for mutations that eliminate Fas expression entirely have no clinical symptoms.

    A. Assuming that normal and mutant Fas are expressed to the same level and bind Fas ligand equally, what fraction of Fas–Fas ligand complexes on a lymphocyte from a heterozygous ALPS patient would be expected to be composed entirely of normal Fas subunits?

    B. In an individual heterozygous for a mutation that eliminates Fas expression, what fraction of Fas–Fas ligand complexes would be expected to be composed entirely of normal Fas subunits?

    C. Why are the Fas mutations that are associated with ALPS dominant, while those that eliminate expression of Fas are recessive?

17–55   One important role of Fas and Fas ligand is to mediate elimination of tumor cells by killer lymphocytes. In a study of 35 primary lung and colon tumors, half the tumors were found to have amplified and overexpressed a gene for a secreted protein that binds to Fas ligand. How do you suppose that overexpression of this protein might contribute to the survival of these tumor cells? Explain your reasoning.

17–56*   An early argument against the involvement of mitochondria in apoptosis was based on the observation that mammalian cells selected to lack mitochondrial DNA undergo apoptosis normally and can be protected from apoptosis by overexpression of Bcl-2. These cells contain mitochondria that lack all proteins encoded by the mitochondrial genome. As a result they cannot make ATP by oxidative phosphorylation and survive by relying exclusively on glycolysis to meet their energy needs. Yet these cells carry out cytochrome *c*-mediated apoptosis. How do you suppose that they do it?

17–57   Development of the nematode *Caenorhabditis elegans* generates exactly 959 somatic cells; it also produces an additional 131 cells that are later eliminated by programmed cell death. The facile genetic manipulation that is possible in *C. elegans* allowed isolation of the first mutations that affected programmed cell death. Of the many mutant genes affecting apoptosis in the nematode, however, none have ever been found in the gene for cytochrome *c*. Why do you suppose that such a central effector molecule in apoptosis was not found in the many genetic screens for 'death' genes that have been carried out in *C. elegans*?

17–58*   When HeLa cells are exposed to UV light at 90 mJ/cm^2, most of the cells undergo apoptosis within 24 hours. Release of cytochrome *c* from mitochondria can be detected as early as 6 hours after exposure of a population of cells to UV light and it continues to increase for more than 10 hours thereafter. Does this mean that individual cells slowly release their cytochrome *c* over this time period, or do individual cells release their cytochrome *c* more rapidly but with different cells being triggered over the longer time period?

    To answer this fundamental question, you have fused the gene for green fluorescent protein (GFP) to the gene for cytochrome *c*, so that you can observe the behavior of individual cells by confocal fluorescence microscopy. In cells that are expressing the cytochrome *c*–GFP fusion, fluorescence shows the punctate pattern typical of mitochondrial proteins. You then irradiate these cells with UV light and observe individual cells for changes in the punctate pattern of fluorescence. Two such cells (outlined in white) are shown in Figure 17–27A and B. Release of cytochrome *c*–GFP is detected as a change from a punctate to a diffuse pattern of fluorescence. Times after UV exposure are indicated as hours:minutes below the individual panels.

    Which model for cytochrome *c* release do these observations support? Explain your reasoning.

17–59   One common cell strategy for generating a rapid response is to use a positive-feedback loop. Positive feedback could explain the rapid release of

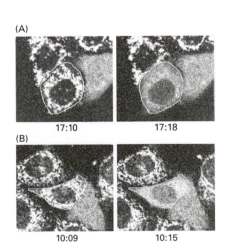

Figure 17–27 Time-lapse analysis of cytochrome *c*–GFP release from mitochondria of individual cells (Problem 17–58). (A) Apoptosis of a cell after 17 hours. (B) Apoptosis of a cell after 10 hours. One cell in (A) and one in (B), each *outlined in white*, have undergone apoptosis during the time frame of the observation, which is shown as hours:minutes below each panel.

cytochrome *c* from mitochondria, for example, if the caspases activated by cytochrome *c* acted on mitochondria to release additional cytochrome *c*. To test this possibility, cells expressing the cytochrome *c*–GFP fusion were incubated in the presence or absence of the broad-spectrum caspase inhibitor zVAD prior to exposure to a variety of apoptosis-inducing agents. The average time for release of cytochrome *c*–GFP from all the mitochondria in individual cells was then compared, as shown in Figure 17–28.

If a caspase-mediated, positive-feedback loop was involved in the rapid release of cytochrome *c*–GFP from mitochondria, what results would you have expected for cells that were incubated in the presence of the caspase inhibitor? Do your expectations match the observed results? Is a caspase-mediated, positive-feedback loop involved in cytochrome *c* release?

17–60* Activation of Fas activates caspase-8, which cleaves the protein Bid (in addition to other targets) to produce an active fragment, tBid, that binds to the mitochondrial membrane. tBid promotes oligomerization of Bax and of Bak, which stimulates the release of cytochrome *c* into the cytosol to trigger events leading to apoptosis (Figure 17–29). To study this pathway in more detail, you've generated mouse embryo fibroblasts (MEFs) that are *Bax*⁻/⁻, or *Bak*⁻/⁻, or knockouts for both genes, *Bax*⁻/⁻*Bak*⁻/⁻. You also construct a vector that expresses tBid so that you can study the process independent of Fas.

In untreated MEFs and in MEFs treated with the empty vector, very few cells are observed to undergo apoptosis (Figure 17–30). By contrast, when MEFs are treated with the vector that expresses tBid, the wild-type cells and the cells individually defective for Bax or Bak show a dramatic increase in apoptotic cells. MEFs that are defective for both Bax and Bak, however, are resistant to tBid expression, showing no increase in the number of apoptotic cells (Figure 17–30). Among the various mutant cells, cytochrome *c* is retained in the mitochondria only in the double knockout MEFs treated with tBid. What do these results tell you about the requirements for Bax and Bak in cytochrome *c*-induced apoptosis? Explain your reasoning.

17–61 Imagine that you could microinject cytochrome *c* into the cytosol of wild-type cells and of cells that were doubly defective for Bax and Bak. Would you expect one, both, or neither of the cell lines to undergo apoptosis? Explain your reasoning.

17–62* (**True/False**) Cells that do not have cytochrome *c* should be resistant to apoptosis induced by UV light but not by apoptosis induced by the binding of Fas ligand. Explain your answer.

17–63 A variety of treatments can cause cells to undergo apoptosis. You wish to know which of these signals are processed through Bid, Bax, and Bak.

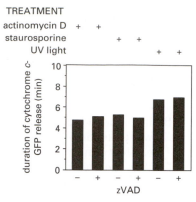

Figure 17–28 Duration of cytochrome *c*–GFP release from mitochondria in cells treated with various apoptosis-inducing agents in the presence and absence of a caspase inhibitor (Problem 17–59). *Bars represent the average of measurements on multiple individual cells.*

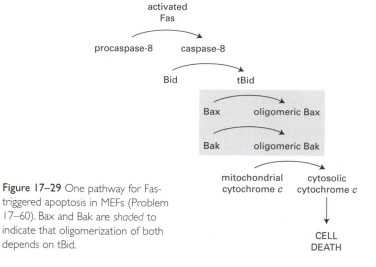

Figure 17–29 One pathway for Fas-triggered apoptosis in MEFs (Problem 17–60). Bax and Bak are *shaded* to indicate that oligomerization of both depends on tBid.

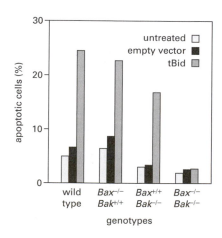

Figure 17–30 Apoptosis induced by tBid expression in wild-type, Bax-deficient, Bak-deficient, and doubly deficient MEFs (Problem 17–60).

TABLE 17–4 Results in *Bid*$^{-/-}$ and *Bax*$^{-/-}$*Bak*$^{-/-}$ MEFs of various treatments that cause apoptosis in wild-type cells (Problem 17–63).

		APOPTOSIS	
TREATMENT	EFFECT	*Bid*$^{-/-}$	*Bax*$^{-/-}$*Bak*$^{-/-}$
Fas ligand	activates Fas	Yes	Yes
Staurosporine	inhibits protein kinases	Yes	No
UV light	damages DNA	Yes	No
Etoposide	inhibits topoisomerase II	Yes	No
Tunicamycin	blocks *N*-linked glycosylation	Yes	No
Thapsigargin	inhibits Ca^{2+} pump in ER	Yes	No

You generate *Bid*$^{-/-}$ and *Bax*$^{-/-}$*Bak*$^{-/-}$ mouse embryo fibroblasts (MEFs) and test them for apoptosis in response to several treatments with the results shown in Table 17–4.

A. Based on these results, indicate where each of the signals enters the apoptotic pathway relative to Bid, Bax, and Bak.

B. How do you suppose that Fas ligand, which binds to Fas, manages to cause apoptosis in these Bid-deficient and Bax- and Bak-deficient cells?

17–64* The JNK subfamily of MAP kinases is essential for neuronal apoptosis in response to certain stimuli. To test whether JNK family members are required for apoptosis in mouse embryo fibroblasts (MEFs), you knock out the genes for the two JNK proteins—JNK1 and JNK2—that are expressed in MEFs. As shown in Figure 17–31A, *Jnk1*$^{-/-}$ and *Jnk2*$^{-/-}$ cells are about as sensitive to UV light as wild-type cells, whereas *Jnk1*$^{-/-}$*Jnk2*$^{-/-}$ cells are totally resistant. The UV-induced cell death occurs by apoptosis since death is prevented by the generalized caspase inhibitor zVAD. Apoptosis is accompanied by release of cytochrome *c* into the cytosol, which does not occur in *Jnk1*$^{-/-}$*Jnk2*$^{-/-}$ cells. As measured by DNA fragmentation, *Jnk1*$^{-/-}$*Jnk2*$^{-/-}$ cells are protected not only against apoptosis triggered by treatment with UV light, but also against apoptosis induced by treatment with anisomycin (which blocks protein synthesis) and methyl methanesulfonate (MMS, which methylates DNA); however, they are not protected from apoptosis induced by activation of Fas (Figure 17–31B).

A. Based on these results, and those in the preceding Problem, decide where the JNK proteins fit in the intracellular pathway of apoptosis.

B. Assuming that JNK1 and JNK2 produce their effects by phosphorylating one or more target proteins, propose how addition of a phosphate to specific targets might bring about apoptosis.

17–65 Mice that are defective for Apaf-1 (*Apaf1*$^{-/-}$) or for caspase-9 (*Casp9*$^{-/-}$) die around the time of birth and exhibit a characteristic set of abnormalities, including brain overgrowth and cranial protrusions. Why do you suppose such abnormalities arise in these deficient mice?

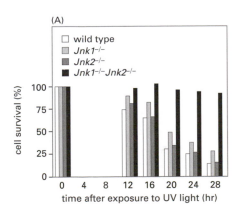

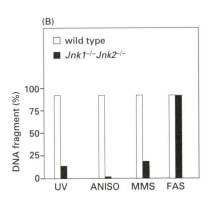

Figure 17–31 Roles of JNK1 and JNK2 in apoptosis in MEFs (Problem 17–64). (A) Cell survival with time after UV irradiation. (B) DNA fragmentation in response to various apoptosis-inducing treatments. ANISO stands for anisomycin.

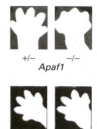

Figure 17–32 Appearance of paws in *Apaf1*$^{-/-}$ and *Casp9*$^{-/-}$ new born mice relative to normal mice (Problem 17–66).

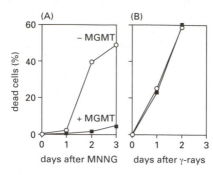

Figure 17–33 Apoptosis induced in MGMT-deficient and MGMT-overexpressing cells (Problem 17–67). (A) By MNNG. (B) By γ-irradiation.

17–66* In contrast to their similar brain abnormalities, mice deficient in Apaf-1 or caspase-9 have distinctive abnormalities in their paws. Apaf-1-deficient mice fail to eliminate the webs between their developing digits, whereas caspase-9-deficient mice have normally formed digits (Figure 17–32). If Apaf-1 and caspase-9 function in the same apoptotic pathway, how is it possible for these deficient mice to differ in web-cell apoptosis?

17–67 Alkylating agents are commonly used for cancer chemotherapy because they are highly cytotoxic, inducing death by apoptosis. Agents such as *N*-methyl-*N*'-nitro-*N*-nitrosoguanidine (MNNG) alkylate a variety of cellular targets including DNA, RNA, proteins, and lipids. Which of these targets generates the signal for apoptosis? The most common mutagenic lesion to DNA, O^6-methylguanine, can be removed by the enzyme O^6-methylguanine methyltransferase (MGMT). To test the possibility that alkylation of DNA is responsible for the apoptotic signal, you compare MNNG-induced apoptosis in cells that are deficient for MGMT with apoptosis in cells that overexpress MGMT (Figure 17–33A). As a control, you compare apoptosis mediated by γ-irradiation (Figure 17–33B). Do these results support the idea that alkylation of DNA leads to the apoptotic signal? Why or why not?

17–68* O^6-methylguanine in DNA creates an abnormal base pair that can be recognized by the mismatch repair machinery in mammalian cells in addition to O^6-methylguanine methyltransferase (MGMT) (see Problem 17–67). Since MGMT protects against alkylation-induced apoptosis, you reason that mismatch repair may also protect against alkylation-induced apoptosis. To test your hypothesis, you compare alkylation-induced apoptosis in a pair of MGMT-deficient cell lines, one of which is proficient for mismatch repair (MMR$^+$) and one of which is deficient (MMR$^-$). As a measure of apoptosis, you assay for alkylation-induced DNA fragmentation, which appears as a smear of fragments less than 50 kb in length (Figure 17–34). Much to your

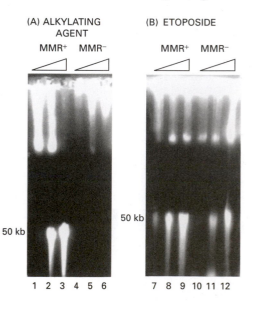

Figure 17–34 Apoptosis-induced DNA fragmentation in MMR$^+$ and MMR$^-$ cells (Problem 17–68). (A) In response to an alkylating agent. (B) In response to etoposide. *Triangles* above the lanes indicate increasing concentration of the agents. DNA fragmentation was assayed 72 hours after exposure to the agents.

surprise, mismatch-repair proficient cells undergo apoptosis in response to alkylation, whereas mismatch-repair deficient cells are resistant (Figure 17–34A). Both cell lines undergo DNA fragmentation when the nonalkylating agent etoposide is used to induce apoptosis (Figure 17–34B).

Propose an explanation for these results. What do these results suggest about the nature of the apoptosis-inducing lesion?

EXTRACELLULAR CONTROL OF CELL DIVISION, CELL GROWTH, AND APOPTOSIS

TERMS TO LEARN

G_0	$p19^{ARF}$	S6-kinase
growth factor	PI 3-kinase	survival factor
mitogen	platelet-derived growth factor (PDGF)	telomerase
Myc	Ras	telomere
nerve growth factor	replicative cell senescence	

17–69 How do mitogens, growth factors, and survival factors differ from one another?

17–70* In his highly classified research laboratory Dr M. is charged with the task of developing a strain of dog-sized rats to be deployed behind enemy lines. In your opinion, which of the following strategies should Dr M. pursue to increase the size of rats?
A. Block programmed cell death.
B. Delay cell senescence.
C. Overproduce mitogens.
D. Overproduce growth factors.
E. Overproduce survival factors.
F. Obtain a taxi driver's license and switch careers.

17–71 For each of the following decide whether such cells exist in humans and give examples.
A. Cells that do not grow and do not divide.
B. Cells that grow, but do not divide.
C. Cells that divide, but do not grow.
D. Cells that grow and divide.

17–72* Vertebrate cells pause in the G_1 phase of the cell cycle until conditions are appropriate for their entry into S phase with subsequent cell division. Some of the requirements for the passage of fibroblasts through G_1 have been defined using mouse 3T3 cells, which are a fibroblastlike cell line. In the absence of serum these cells do not enter S phase. If serum is added to a culture of such arrested cells, they progress through G_1 and begin to enter S phase 12 hours later. The serum requirement can be met by supplying three extracellular factors: PDGF, EGF, and IGF-1. When these factors are mixed with appropriate nutrients and added to quiescent cells, the cells begin to enter S phase 12 hours later. If any one of the factors is left out, the cells do not enter S phase.

Do all three factors have to be present at the same time? Is their stimulation of cells independent of one another? Or do they stimulate cells in an ordered sequence? To address these questions, you pretreat cells with the factors in a defined order and then add complete medium (containing serum and nutrients) in the presence of ^{3}H-thymidine. At various times thereafter you fix cells and subject them to autoradiography. You define the time of appearance of the first labeled nuclei as the time of entry into S phase. The results of these experiments are given in Table 17–5.

Do the cells require these factors simultaneously, independently, or in an ordered sequence? Explain your answer.

Problem 15–72 examines the cell signaling pathways that are triggered by the binding of PDGF to the PDGF receptor.

	ORDER OF ADDITION			
EXPERIMENT	1	2	3	S PHASE ENTRY
1	EGF	PDGF	IGF-1	12 hours
2	EGF	IGF-1	PDGF	12 hours
3	PDGF	EGF	IGF-1	1 hour
4	PDGF	IGF-1	EGF	6 hours
5	IGF-1	EGF	PDGF	12 hours
6	IGF-1	PDGF	EGF	6 hours

Cells were treated for 6 hours with the indicated factors in the order listed. They were thoroughly washed to remove one factor before the next one was added. After the regimen of factor pretreatment, complete medium with ^{3}H-thymidine was added and the time before labeled nuclei appeared was determined.

17–73   A classic paper that first characterized the EGF receptor made use of the mouse fibroblast A-431 cell line, which fortuitously carries enormously increased numbers of EGF receptors. The following experimental observations were made.

1. A plasma membrane preparation contained many proteins as shown on the SDS gel in Figure 17–35A. When ^{125}I-EGF was added in the presence of a protein cross-linking agent, two proteins were labeled (Figure 17–35B, lane 1). When excess unlabeled EGF was included in the incubation mixture, the labeled band at 170 kd disappeared (lane 2).

2. If the membrane preparation was incubated with γ-^{32}P-ATP, several proteins, including the 170-kd protein, became phosphorylated. Inclusion of EGF in the incubation mixture significantly stimulated phosphorylation.

3. When antibodies specific for the 170-kd protein were used to precipitate the protein, and the incubation with γ-^{32}P-ATP was repeated with the precipitate, the 170-kd protein was phosphorylated in an EGF-stimulated reaction (Figure 17–35C, lanes 3 and 4).

4. If the antibody-precipitated protein was first run on the SDS gel and then renatured in the gel, subsequent incubation with γ-^{32}P-ATP in the presence and absence of EGF yielded the same pattern shown in Figure 17–35C.

A. Which of these experiments demonstrated most clearly that the 170-kd protein is the EGF receptor?

B. Is the EGF receptor a substrate for an EGF-stimulated protein kinase? How do you know?

C. Which experiments showed that the EGF receptor is a protein kinase?

D. Is it clear that the EGF receptor is a substrate for its own protein kinase activity?

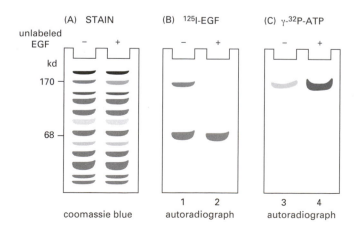

Figure 17–35 Analysis of the EGF receptor (Problem 17–73). (A) SDS gel of a membrane preparation from A-431 cells. (B) Autoradiograph of an SDS gel of a membrane preparation from A-431 cells incubated with ^{125}I-EGF and a protein cross-linking agent in the presence and absence of excess unlabeled EGF. (C) Autoradiograph of an SDS gel of antibody-precipitated EGF-receptor incubated with γ-^{32}P-ATP in the presence and absence of EGF.

17–74* Why do you suppose cells have evolved a special G_0 state to exit the cell cycle, rather than just stopping in G_1 at a G_1 checkpoint?

17–75 (**True/False**) Serum deprivation causes proliferating cells to stop wherever they are in the cell cycle and enter G_0. Explain your answer.

17–76* When cells in G_0 are exposed to mitogenic growth factors, they enter S phase about 20 hours after stimulation, as can be detected by incorporation of the nucleoside analog BrdU. If antibodies to cyclin D are microinjected into cells up to 12 hours after adding mitogenic growth factors, very few incorporate BrdU. Cyclin D-specific antibodies have little effect on BrdU incorporation, however, when injected more than 14 hours after exposure to factors (Figure 17–36). What critical event in G_1 do antibodies against cyclin D block? Why do antibodies against cyclin D have no effect after 14 hours?

17–77 Overexpression of Myc is a common feature of many types of cancer cells, contributing to their excessive cell growth and proliferation. By contrast, when Myc is hyperactivated in most normal cells, the result is not excessive proliferation but cell-cycle arrest or apoptosis. How is it that overexpression of Myc can have such different outcomes in normal cells and cancer cells?

17–78* PDGF is encoded by a gene that can cause cancer when expressed inappropriately. Why then do cancers not arise at wounds when PDGF is released from platelets?

17–79 The formation of tumors is a multistep process that may involve the successive activation of several oncogenes. This notion is supported by the discovery that certain pairs of oncogenes, of which *Ras* and *Myc* are among the best-studied examples, transform cultured cells more efficiently than either alone. Similar experiments with pairs of oncogenes have been carried out in transgenic mice. In one experiment the *Ras* oncogene was placed under the control of the MMTV (mouse mammary tumor virus) promoter and incorporated into the germlines of several transgenic mice. In a second experiment the *Myc* oncogene under control of the same MMTV promoter was incorporated into the germlines of several transgenic animals. In a third experiment mice containing the individual oncogenes were mated to produce mice with both oncogenes.

All three kinds of mice developed tumors at a higher frequency than normal animals. Female mice were most rapidly affected because the MMTV promoter, which is responsive to steroid hormones, turns on the transferred oncogenes in response to the hormonal changes at puberty. In Figure 17–37 the rate of appearance of tumors is plotted as the percent of tumor-free females as a function of time after puberty.

A. Assume that the lines drawn through the data points are an accurate representation of the data. How many events in addition to expression of the oncogenes are required to generate a tumor in each of the three kinds of mice? (You may wish to reread Problem 17–38.)
B. Is activation of the cellular *Ras* gene the event required to trigger tumor formation in mice that are already expressing the MMTV-regulated *Myc* gene (or vice versa)?
C. Why do you think the rate of tumor production is so high in the mice containing both oncogenes?

17–80* It has been suggested that normal cells are limited to about 50 cell divisions in order to restrict the maximum size of tumors, thus affording some protection against cancer. Assuming that 10^8 cells have a mass of 1 gram, calculate the mass of a tumor that originated from 50 doublings of a single cancerous cell.

17–81 Replicative cell senescence occurs at a characteristic number of population doublings, typically about 40 for cells taken from normal human tissue. This observation suggests that in some way individual cells can 'count' the

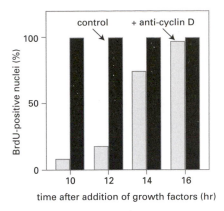

Figure 17–36 Percentage of control cells and cells injected cyclin D antibodies that have incorporated BrdU by 26 hours after addition of mitogenic growth factors (Problem 17–76). Cells were injected with cyclin D antibodies (or not) at the times indicated. *Black bars* indicate results with control cells that were not injected with cyclin D antibodies. *Gray bars* indicate results with cells that were injected with cyclin D antibodies.

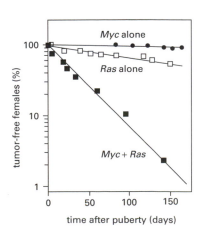

Figure 17–37 Fraction of tumor-free female mice as a function of time after puberty (Problem 17–79).

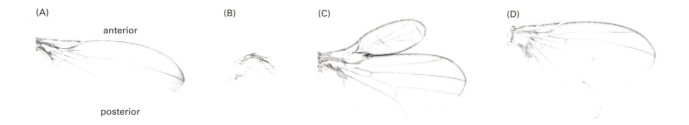

(A)

anterior

posterior

(B)

(C)

(D)

Figure 17–38 Effects of DPP expression on wing development in *Drosophila* (Problem 17–84). (A) Normal DPP expression. (B) Absence of DPP expression. (C) Additional anterior DPP expression. (D) Additional posterior DPP expression.

number of times they have divided. How does the structure of telomeres figure into a cell's calculations?

17–82* Active telomerase comprises two components—a protein and an RNA—that are encoded by distinct genes. Most human somatic cells do not have telomerase activity. If the gene for the protein component of telomerase, hTRT, is expressed in such cells, however, they regain telomerase activity. How is it that expression of just one component can give telomerase activity?

17–83 (**True/False**) If we could turn on telomerase activity in all our cells, we could prevent aging. Explain your answer.

17–84* The extracellular protein factor Decapentaplegic (DPP) is critical for proper wing development in *Drosophila* (Figure 17–38A). It is normally expressed in a narrow stripe in the middle of the wing, along the anterior-posterior boundary. Flies that are defective for DPP form stunted 'wings' (Figure 17–38B). If an additional copy of the gene is placed under control of a promoter that is active in the anterior part of the wing, or in the posterior part of the wing, a large mass of wing tissue composed of normal-looking cells is produced at the site of DPP expression (Figure 17–38C and D). Does DPP stimulate cell division, cell growth, or both? How can you tell? Give some examples of proteins that you might expect to be activated in response to DPP stimulation.

17–85 Liver cells proliferate both in patients with alcoholism and in patients with liver tumors. What are the differences in the mechanisms by which cell proliferation is induced in these diseases?

17–86* You have been led by a bizarre accident to a productive line of experimentation. You lost one of your contact lenses while transferring a line of tissue culture cells; a few days later you found the lens on the bottom of a petri dish. Interestingly, the cells attached to the lens were rounded up and very sparsely distributed, whereas those on the rest of the dish were flat and had grown to near confluency with most cells touching their neighbors. Ah ha! you thought, perhaps this observation can be used to investigate the relationship between cell shape and growth control. As someone once said, "Chance favors the prepared mind."

The manufacturer graciously sends you a supply of the plastic—poly(HEMA)—from which your soft contact lenses were made. When an alcoholic mixture of poly(HEMA) is pipetted into a plastic culture dish, a thin, hard, sterile film of optically clear polymer remains bound to the plastic surface after the alcohol evaporates. Serial dilutions of the alcohol-polymer solution, introduced into each dish at a constant volume, result in decreasing thicknesses of the polymer film. The thinner the film, the more strongly cells adhere to the dish. Moreover, there is a gradual change in cell shape from round to flat with decreasing thickness of the film. Using the height of the cells as an indicator of cell shape (no mean technical feat), you demonstrate that there is a smooth relationship between cell shape and growth potential: the flatter the cells, the better they incorporate ^{3}H-thymidine.

Now for the big question: Is density-dependent inhibition of cell growth mediated by changes in cell shape? You grow cells to different densities

TABLE 17–6 Incorporation of ^{3}H-thymidine by cells grown on normal dishes and on poly(HEMA)-treated dishes (Problem 17–86).

TYPE OF DISH	CELL DENSITY (cells/dish)	CONFLUENCY	CELL HEIGHT (μm)	^{3}H INCORPORATION (cpm/dish)
Normal	60,000	subconfluent	6	15,200
Normal	200,000	confluent	15	11,000
Normal	500,000	confluent	22	3,500
Poly(HEMA)	30,000	sparse	6	7,500
Poly(HEMA)	30,000	sparse	15	1,500
Poly(HEMA)	30,000	sparse	22	210

(different degrees of confluency) on normal plastic dishes and measure the height of the cells and their ability to incorporate ^{3}H-thymidine. As shown in Table 17–6, the more confluent the cells, the greater their height and the lower their incorporation of ^{3}H-thymidine. Is the decrease in growth due to the increase in cell crowding or to the change in cell shape? To answer this question, you distribute cells at a low density on plates with poly(HEMA) films of different thickness, such that the height of the cells in the sparse cultures matches the height of cells in the various confluent cultures. Your measurements of ^{3}H-thymidine incorporation in these sparse cultures are shown in Table 17–6.

Based on the results in Table 17–6, would you conclude that density-dependent inhibition of cell growth correlates completely, partially, or not at all with changes in cell shape? Explain your reasoning.

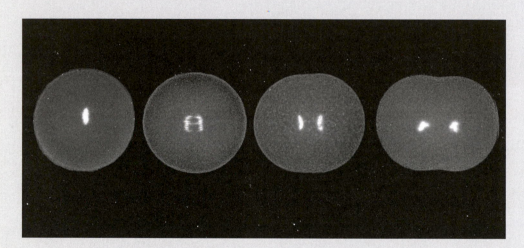

Metaphase to anaphase and the onset of cleavage at first division in a fertilized sea urchin egg. The chromosomes were stained with Hoechst 33342, a DNA-binding dye that can penetrate living cells.

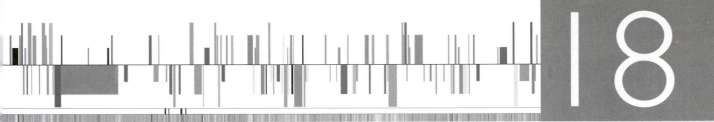

THE MECHANICS OF CELL DIVISION

AN OVERVIEW OF M PHASE

TERMS TO LEARN

aster	cohesin	M phase
centrosome	condensin	mitosis
centrosome cycle	γ-tubulin ring complex	sister chromatid
chromosome condensation	interphase	

18–1 The yeast cohesin Scc1 can be artificially regulated to be expressed at any point in the cell cycle. If expression is turned on by the beginning of S phase, all the cells survive. By contrast, if Scc1 expression is turned on only after S phase is completed, all the cells die, even though Scc1 accumulates in the nucleus and binds efficiently to chromosomes. Why do you suppose that cohesin must be present during S phase in order to keep sister chromatids together?

18–2* Scc1, Scc3, Smc1, and Smc3 are subunits of the yeast cohesin complex that is required to hold sister chromatids together until their separation at mitosis. A fifth protein, Eco1, is also required for sister-chromatid cohesion, but it is not a part of the cohesion complex. To investigate the role of Eco1, you compare the viability of temperature-sensitive mutants of *Scc1* and *Eco1*. You synchronize cultures of *Scc1*^{ts} and *Eco1*^{ts} in G$_1$ and divide them into three aliquots. You incubate one aliquot at 25°C (the permissive temperature), one at 37°C (the restrictive temperature), and one you incubate at 25°C until the

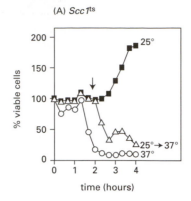

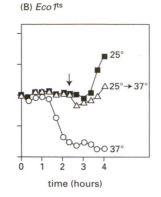

(A) Scc1ts   (B) Eco1ts

Figure 18–1 Analysis of the requirement for proteins involved in sister-chromatid cohesion during the cell cycle (Problem 18–2). (A) Viable cells in cultures of Scc1 ts mutants grown at various temperatures. (B) Viable cells in cultures of Eco1 ts mutants grown at various temperatures. Arrows indicate the time at which an aliquot was shifted from 25°C to 37°C. Viable cells were assayed by their ability to form colonies after plating at 25°C. Number of colonies at time zero was defined as 100%.

end of S phase and then shift it to 37°C. At different times you measure the number of viable cells in each aliquot by counting colonies that grow at 25°C (Figure 18–1).

A. At 37°C replication begins at 1.5 hours in Scc1ts and Eco1ts mutants. Are Scc1 and Eco1 required during replication? How can you tell from these experiments?

B. At 25°C S phase ends at about 2 hours. Are Scc1 and Eco1 required after S phase? How can you tell?

C. Propose an explanation for the behavior of Scc1ts mutants in these experiments.

D. Propose an explanation for the behavior of Eco1ts mutants in these experiments.

Problems 17–33 and 17–34 analyze the role of Scc1 in the cohesion and separation of sister chromatids.

18–3   If cohesins join sister chromatids all along their length, how is it possible for condensins to generate mitotic chromosomes such as that shown in Figure 18–2, which clearly shows the two sister chromatids as separate domains?

18–4*   Cohesins and condensins are very similar in structure yet carry out quite different biochemical tasks: cohesion of sister chromatids and condensation of chromosomes, respectively. You are skeptical that such similar molecules can perform such distinct functions, and set out to determine if they are truly different using purified components. You incubate pure cohesin or condensin with nicked circular DNA and ATP. You then add topoisomerase II to link duplexes that have been juxtaposed. (Topoisomerase II binds to one duplex and breaks both strands, attaching itself covalently to the ends and holding them together. The topoisomerase II complex can gate the passage of a second duplex through the break and then reseal the original duplex. By linking—or unlinking—duplexes, topoisomerase II can alter their topology in informative ways.)

As shown in Figure 18–3, condensin and cohesin yield very different results upon incubation with topoisomerase II and analysis by gel electrophoresis. Incubation with condensin followed by topoisomerase generates a particular kind of trefoil knot (Figure 18–3A), whereas incubation with cohesin followed by topoisomerase generates a series of catenanes (circles joined like links of a chain, Figure 18–3B).

A. Do these results support the proposed roles of cohesin and condensin? How so?

B. Suggest a plausible mechanism by which binding of cohesin might allow topoisomerase II to link molecules into catenanes.

C. Knots are much more difficult to think about, but often are very revealing of mechanistic details. See if you can figure out a way to use the binding of condensin molecules, coupled with one duplex crossing event catalyzed by topoisomerase II, to tie a circular molecule into any kind of a knot.

Problem 4–67 probes the function of condensin by analyzing its effects on DNA supercoiling.

1 μm

Figure 18–2 A scanning electron micrograph of a fully condensed mitotic chromosome from vertebrate cells (Problem 18–3).

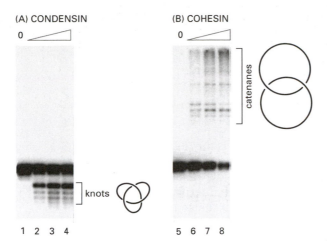

(A) CONDENSIN

0

knots

(B) COHESIN

0

catenanes

1 2 3 4   5 6 7 8

Figure 18–3 Topological analysis of the functions of condensin and cohesin (Problem 18–4). (A) Electrophoretic analysis of migration of circular DNA incubated with condensin and topoisomerase II. The knots formed by incubation are all like the one illustrated. (B) Electrophoretic analysis of migration of circular DNA incubated with cohesin and topoisomerase II. The products of incubation are catenanes; a dimeric catenane is shown. In both gels the heavy band of material corresponds to the nicked circles that were added to the incubation mixture.

18–5   What are the two distinct cytoskeletal machines that are assembled to carry out the mechanical processes of mitosis and cytokinesis in animal cells?

18–6*   It would be catastrophic if the two major events of M phase—mitosis and cytokinesis—occurred out of sequence. What two mechanisms does the cell employ to ensure that these two events occur in the correct order?

18–7   Examine the schematic representation of centrosome duplication in Figure 18–4. By analogy with DNA replication, would you classify centrosome duplication as conservative or semi-conservative? Explain your reasoning.

18–8*   If the centrosome cycle is disrupted so that centrosomes are not duplicated, continued progression of the cell cycle will ultimately lead to formation of a monopolar spindle rather than a bipolar one. The pole of a monopolar spindle could, in principle, contain a single centriole, a centrosome with the usual pair of centrioles, or multiple centrosomes, depending on where the centrosome cycle was interrupted (see Figure 18–4).
A. When defective, the Zyg-1 protein kinase of *C. elegans* leads to monopolar spindles that have a single centriole at the pole. At what point is the centrosome cycle interrupted in the absence of functional Zyg-1? Explain your reasoning.
B. When a mutant, nonphosphorylatable form of nucleophosmin, a component of centrosomes and a target of G_1/S-Cdk, is expressed in cells, monopolar spindles form with a single centrosome (with two centrioles) at the pole. At what point does the mutant form of nucleophosmin interrupt the centrosome cycle? Explain your reasoning.

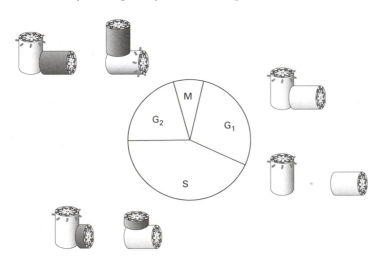

Figure 18–4 The centrosome duplication cycle (Problem 18–7). The individual centrioles that make up the centrosome are shown as *cylinders*.

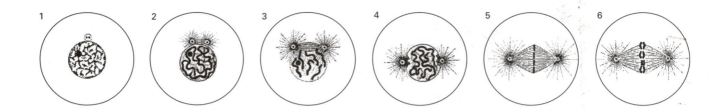

18–9 One of the least well-understood aspects of the cell cycle is the duplication of the spindle poles. As illustrated in Figure 18–5, the centrosome normally splits at the beginning of mitosis to form the two spindle poles, which orchestrate chromosome segregation. During the next interphase, the centriole pair within the centrosome is duplicated so that the centrosome can split at the next mitosis. The cycles of centrosome duplication and splitting normally keep step with cell division so that all cells have the capacity to produce bipolar spindles. It is possible, however, to throw the two cycles out of phase as indicated by experiments first performed in the late 1950s.

If a fertilized sea urchin egg at the metaphase stage of the first mitotic division is exposed to mercaptoethanol (MSH), the mitotic spindle disassembles (Figure 18–6A). (It is not known how mercaptoethanol causes this, but the effect is reversible.) While the eggs are kept in mercaptoethanol, the nucleus does not re-form, no DNA synthesis occurs, and the chromosomes stay condensed. When mercaptoethanol is washed out, the spindle re-forms and cell division takes place. Although some eggs re-form a bipolar spindle and divide normally, the majority of eggs form a tetrapolar spindle and divide into four daughter cells (Figure 18–6A). No matter how long the eggs are arrested in mercaptoethanol, they never divide into more than four cells.

The daughter cells from a four-way division re-form the nucleus and traverse the next cell cycle; however, at mitosis they form a monopolar spindle. In a majority of cases these cells stay in mitosis a little longer than usual and then decondense their chromosomes, disassemble the spindle, and re-form a nucleus (Figure 18–6B). At the next mitosis these cells form a bipolar spindle and divide normally (Figure 18–6B). More rarely, the cells with monopolar spindles stay in mitosis much longer, the monopole splits to form a bipolar spindle, and the cell divides normally (Figure 18–6B). The daughter cells from such a division once again form a monopolar spindle at the next mitosis (Figure 18–6B).

Describe patterns of centriole duplication and splitting that can account for the observations shown in Figure 18–6.

Figure 18–5 Normal process of centrosome splitting during mitosis to form a bipolar spindle (Problem 18–9).

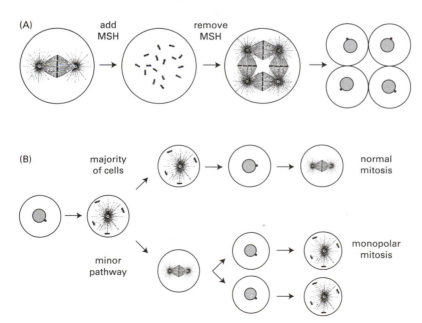

Figure 18–6 Abnormal spindles and cell divisions induced by mercaptoethanol (MSH) treatment (Problem 18–9). (A) Formation of tetrapolar spindles upon treatment of fertilized sea urchin eggs with mercaptoethanol followed by four-way division. With the breakdown of the spindle, chromosomes disperse while remaining condensed. (B) Major and minor pathways for spindle formation and cell division among the daughter cells from a four-way division. Note that the condensed chromosomes are located at the periphery of a monopolar spindle.

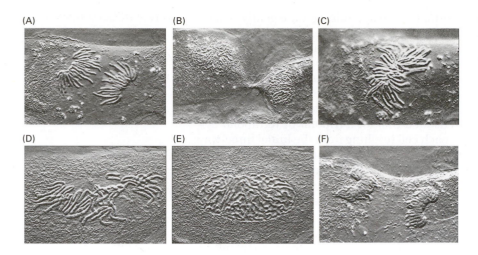

Figure 18–7 Light micrographs of a single cell at different stages of M phase (Problem 18–11).

18–10* **(True/False)** The six stages of M phase—prophase, prometaphase, metaphase, anaphase, telophase, and cytokinesis—occur in strict sequential order. Explain your answer.

18–11 A living cell from the lung epithelium of a newt is shown at different stages in M phase in Figure 18–7. Order these light micrographs into the correct sequence and identify the stage in M phase that each represents.

18–12* Imagine that you could color one sister chromatid and follow it through mitosis and cytokinesis. What event commits this chromatid to a particular daughter cell? Once initially committed, can its fate be reversed? What is the main influence on its initial commitment?

18–13 Order the following events in animal cell division.
A. Alignment of chromosomes at the spindle equator.
B. Attachment of microtubules to chromosomes.
C. Breakdown of nuclear envelope.
D. Condensation of chromosomes.
E. Decondensation of chromosomes.
F. Duplication of centrosome.
G. Elongation of the spindle.
H. Pinching of cell in two.
I. Re-formation of nuclear envelope.
J. Separation of centrosomes.
K. Separation of sister chromatids.

18–14* By the turn of the century it was clear that chromosomes were the carriers of hereditary information. But a fundamental question remained: Does each chromosome carry the total hereditary information or does each one carry a different portion of the hereditary information? According to the first view, multiple chromosomes were required to raise the total quantity of hereditary material above the threshold value needed for proper development. According to the second view, multiple chromosomes were needed so that all portions of the hereditary information would be represented. This question was answered definitively in a classic series of experiments carried out by Theodor Boveri from 1901 to 1905.

 As an experimental system, Boveri followed the development of sea urchin eggs that had been fertilized by two sperm, a frequent occurance during artificial fertilization when large quantities of sperm are added. In contrast to the normal bipolar mitotic spindle and division into two cells, eggs fertilized by two sperm form a tetrapolar mitotic spindle and then divide into four cells. The three sets of chromosomes—one from the egg and two from the sperm—are distributed randomly among the four spindles as shown for four chromosomes in Figure 18–8A. If the dispermic eggs are gently shaken immediately after fertilization, one of the spindle poles often fails to form, resulting in a tripolar mitotic spindle followed by division into three cells (Figure 18–8B).

The species of sea urchin that Boveri studied, *Echinus microtuberculatus*, has a diploid chromosome number of 18, but will develop normally to a pluteus larva—a free-swimming stage in sea urchin development—with a haploid number of 9. Boveri reasoned that for a tripolar or tetrapolar egg to develop to a normal pluteus, each cell resulting from the initial three-way or four-way division would need to have either 9 total chromosomes or 9 different chromosomes—depending on which view of chromosomes was correct. Boveri followed the development of 695 tripolar eggs and found that 58 developed into a normal pluteus. Among 1170 tetrapolar eggs, none formed a normal pluteus.

A. To set up the expectations for this experiment, it is instructive to consider first a hypothetical case in which the egg and two sperm each contribute a single chromosome. For tripolar spindles there are 10 different arrangements of three chromosomes on three spindles. Sketch these 10 arrangements. Upon separation of sister chromatids and cell division, how many of these arrangements would be expected to produce three cells that each carry at least one chromosome? One arrangement and its division into three cells is shown in Figure 18–8C. (If you want to try your hand at tetrapolar spindles, there are 20 arrangements.)

B. If the total number of chromosomes is the critical factor, the number of tripolar eggs in which each cell gets the minimum number of chromosomes will be the same as that calculated in part A regardless of the number of chromosomes. By contrast, if the distribution of chromosomes is the critical factor, the number of tripolar eggs that generate three cells, each with at least one copy of each different chromosome, will decrease with increasing numbers of chromosomes. The number of plutei should decrease according to the fraction calculated in part A raised to the power of 9 (the number of different chromosomes). Which hypothesis—total number of chromosomes or distribution of chromosomes—do Boveri's observations support?

18–15 In a classic experiment, Barbara McClintock studied the genetics of color variegation in corn. She generated speckled kernels by crossing a strain that carries an x-ray-induced rearrangement of chromosome 9 (Figure 18–9). This chromosome carries a color marker (*C*, colored; recessive form *c*, colorless), which allowed her to follow its inheritance in individual kernels. When strains carrying the rearranged chromosome 9 bearing the dominant *C* allele were crossed with wild-type corn bearing the recessive *c* allele, a small number of kernels in the progeny ears of corn had a speckled appearance.

This color variegation arises by an interesting mechanism. In meiosis, recombination within the rearranged segment generates a chromosome with two centromeres as shown in Figure 18–10. In a fraction of these meioses the recombined chromosome with two centromeres gets strung out between the two poles at the first meiotic anaphase, forming a bridge between the two meiotic poles. Some time in anaphase to telophase the

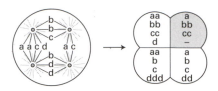

(A) TETRAPOLAR MITOSIS

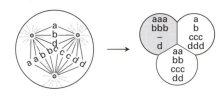

(B) TRIPOLAR MITOSIS: 4 CHROMOSOMES

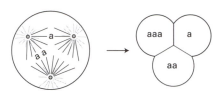

(C) TRIPOLAR MITOSIS: 1 CHROMOSOME

Figure 18–8 Distributions of chromosomes among multiple spindles (Problem 18–14). (A) Example of a random arrangement of three sets of four chromosomes among the four mitotic spindles in a tetrapolar egg. Note that in this example, all four cells have at least four chromosomes. If total number were the critical aspect of chromosomes in heredity, these cells should develop into a pluteus. On the other hand, if each of the four cells have to have at least one copy of each different chromosome, these cells would not be expected to form a pluteus because one cell (*shaded*) is missing a chromosome. (B) Example of a random arrangement of three sets of four chromosomes among the three mitotic spindles in a tripolar egg. Once again, if total number of chromosomes were critical, these cells should develop into a pluteus; but if chromosome type is critical, they will not since the *shaded* cell is missing one chromosome. (C) Example of one arrangement—out of 10 possible—of three chromosomes on a tripolar spindle.

Figure 18–9 Normal and rearranged chromosomes 9 in corn (Problem 18–15). *Arrows* above the normal chromosome indicate sites of breakage that gave rise to the rearranged chromosome.

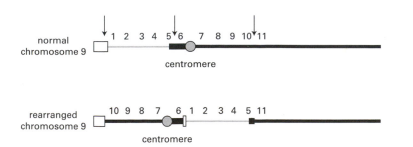

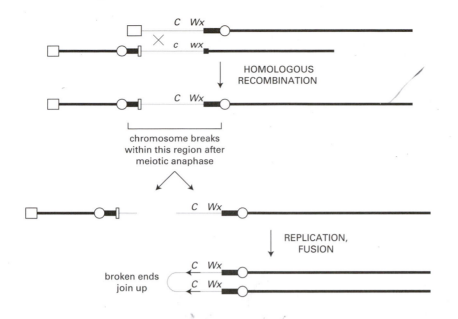

Figure 18–10 Recombination between a normal and a rearranged chromosome 9 to give a dicentric chromosome (Problem 18–15). Homologous recombination occurs at the X. Only the product with two centromeres is shown; the other product, which has no centromere, will be lost because it cannot attach to the spindle. Breakage of the original dicentric chromosome followed by replication and fusion of the ends gives rise to a second dicentric chromosome.

strained chromosome breaks at a random position between the duplicated centromeres. The broken ends of the chromosome tend to fuse together after the next S phase, which generates a new dicentric chromosome whose structure depends on where the previous break occurred (Figure 18–10). The forces acting during the subsequent mitosis in turn will break this chromosome, and the fusion–bridge–breakage cycle will repeat itself in the next cell cycle, unless a repair mechanism adds a telomere to the broken end.

Figure 18–10 also shows the chromosomal location of another genetic marker on chromosome 9, which can cause a 'waxy' alteration to the starch deposited in the kernels. The waxy allele can be detected by staining with iodine. Waxy (*wx*) is recessive to the normal, nonwaxy (*Wx*) allele. By following the inheritance of the *C* and *Wx* markers in the kernels, McClintock gained an understanding of the behavior of broken chromosomes. She observed three types of patches within the otherwise colored, nonwaxy (*C-Wx*) kernels: colorless, nonwaxy (*c-Wx*) patches; colorless, nonwaxy (*c-Wx*) patches containing one or more colorless, waxy (*c-wx*) spots; and intensely colored, nonwaxy (*?-Wx*) patches (Figure 18–11).

A. Patches arise because cells with a different genetic constitution divide to give identical neighbors that remain together in a cluster. Starting with the dicentric chromosome shown at the bottom of Figure 18–10, show how fusion–bridge–breakage cycles might account for the three types of patches shown in Figure 18–11. What is the genetic constitution of the intensely colored patches? (In these crosses, the dominant alleles—*C* and *Wx*—are carried on the rearranged chromosome, and the recessive alleles—*c* and *wx*—are carried on the normal chromosome.)

B. Would you ever expect to see colored spots within colorless patches? Why or why not?

C. Would you ever expect to see colorless spots within the intensely colored patches? Why or why not?

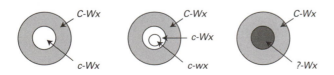

Figure 18–11 Three types of patches observed in speckled kernels (Problem 18–15).

MITOSIS

18–16* Describe the three main classes of spindle microtubules in animal cells and their functions during mitosis.

18–17 The lifetime of a microtubule in a mammalian cell, from its formation by polymerization to its spontaneous disappearance by depolymerization, varies with the stage of the cell cycle. For an actively proliferating cell, the average lifetime is 5 minutes in interphase and 15 seconds in mitosis. If the average length of a microtubule in interphase is 20 μm, what will it be during mitosis, assuming that the rates of microtubule elongation due to the addition of tubulin subunits in the two phases are the same? If a typical centrosome in an interphase cell has 100 nucleation sites for microtubules, how many sites would you expect to find in centrosomes in a mitotic cell, assuming that the total length of microtubules is the same in both phases?

18–18* The lengths of microtubules in various stages of mitosis depend on the balance between the activities of catastrophins, which destabilize microtubules, and microtubule-associated proteins (MAPs), which stabilize them. If you overexpressed catastrophins, would you expect the length of the mitotic spindle to be longer, shorter, or unchanged relative to the corresponding stage of mitosis in wild-type cells? What would you expect if you overexpressed MAPs? Explain your reasoning.

18–19 The balance between plus-end-directed and minus-end-directed motors determines spindle length. Where in the spindle do these motors operate to control spindle length?

18–20* If multimeric minus-end-directed motors are incubated with microtubules, an astral array will be generated. A similar astral array will be generated if multimeric plus-end-directed motors are incubated with microtubules. How do the two astral arrays differ, if at all?

18–21 (**True/False**) After the nuclear envelope breaks down, microtubules gain access to the chromosomes and, every so often, a randomly probing microtubule connects with a kinetochore and captures the chromosome. Explain your answer.

18–22* A classic paper clearly distinguished the properties of astral microtubules from those of kinetochore microtubules. Centrosomes were used to initiate microtubule growth, and then chromosomes were added. The chromosomes bound to the free ends of the microtubules, as illustrated in Figure 18–12. The complexes were then diluted to very low tubulin concentration (well below the critical concentration for assembly) and examined again (Figure 18–12). As is evident, only the kinetochore microtubules were stable to dilution.

   A. Why do you think kinetochore microtubules are stable?

   B. Explain the disappearance of the astral microtubules after dilution. Do they detach from the centrosome, depolymerize from an end, or disintegrate along their length at random?

   C. How would a time course after dilution help to distinguish among these possible mechanisms for disappearance of the astral microtubules?

18–23 In higher eucaryotes, rare chromosomes containing two centromeres at different locations are highly unstable: they are literally torn apart at

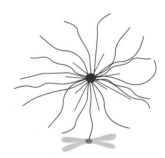

before dilution

after dilution

Figure 18–12 Arrangements of centrosomes, chromosomes, and microtubules before and after dilution to low tubulin concentration (Problem 18–22).

anaphase when the chromosomes separate (see Problem 18–15). You wonder whether the same phenomenon occurs in yeast, whose chromosomes are too small to analyze microscopically.

You construct a plasmid with two centromeres, as shown in Figure 18–13. Growth of this plasmid in bacteria requires the bacterial origin of replication (*ori*) and a selectable marker (*amp^R*); its growth in yeast requires the yeast origin of replication (*ARS1*) and a selectable marker (*TRP1*). You prepare a plasmid stock by growth in *E. coli*. This dicentric plasmid transforms yeast with about the same efficiency as a plasmid that contains a single centromere. Individual colonies, however, contained plasmids with a single centromere or no centromere, but never a plasmid with two centromeres. By contrast, colonies that arose after transformation with a monocentric plasmid invariably contained intact plasmids.

A. Considering their extreme instability in yeasts, why are dicentric plasmids stable in bacteria?

B. Why do you think that the dicentric plasmid is unstable in yeasts?

C. Suggest a mechanism for deletion of one of the centromeres from a dicentric plasmid grown in yeast. Can this mechanism account for loss of both centromeric sequences from some of the plasmids?

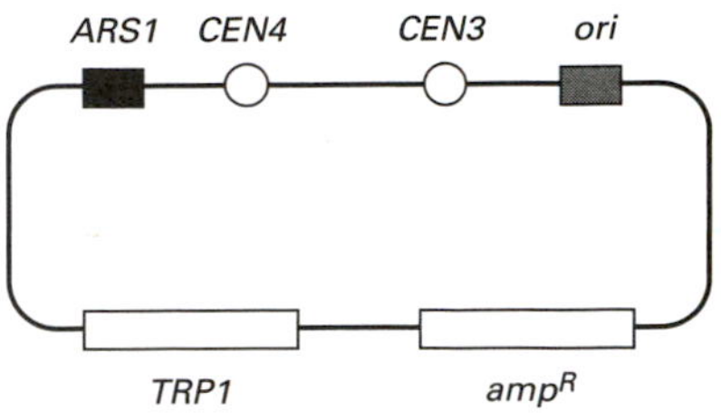

18–24* Circular yeast plasmids that lack a centromere are distributed among individual cells in a peculiar way. Only 5% to 25% of the cells harbor the plasmids, yet the plasmid-bearing cells contain 20 to 50 copies of the plasmid. To investigate the apparent paradox of a high copy number in a small fraction of cells, you perform a pedigree analysis to determine the pattern of plasmid segregation during mitosis. You use a yeast strain that requires histidine for growth and a plasmid that provides the missing histidine gene. The strain carrying the plasmid grows well under selective conditions; that is, when histidine is absent from the medium. By micromanipulation you separate mother and daughter cells for five divisions under selective conditions, and then score for those cells that can form a colony in the absence of histidine. In Figure 18–14 cells that formed colonies are indicated with heavy lines and cells that failed to form colonies are shown with light lines.

A. From the pedigree analysis, it is apparent that cells lacking the plasmid can grow for several divisions in selective medium. How can this be?

B. Does this plasmid segregate equally to mother and daughter cells?

C. Assuming that plasmids in yeast cells replicate only once per cell cycle as the chromosomes do, how can there be 20 to 50 molecules of the plasmid per plasmid-bearing cell?

D. When grown under selective conditions, cells containing plasmids with one centromere (1–2 plasmids per cell) form large colonies, whereas cells containing plasmids with no centromere (20–50 plasmids per cell) form small colonies. Does the pedigree analysis help to explain the difference?

18–25 A strong promoter that directs transcription across a centromere can inactivate the kinetochore. Likewise, kinetochores can provide a strong block to transcription. Evidently, centromere function and transcription are mutually interfering processes. Based on this mutual interference, your advisor has devised a clever scheme that exploits the transcriptional block to measure the strength of DNA–protein interactions in the kinetochore.

To carry out this scheme, you construct a test system consisting of the yeast actin gene fused to the *E. coli* β-galactosidase gene, with the hybrid

Figure 18–13 Structure of a dicentric plasmid (Problem 18–23). *CEN3* and *CEN4* refer to centromeres from yeast chromosomes 3 and 4, respectively.

Figure 18–14 A pedigree analysis showing the inheritance of a plasmid that contains an origin of replication and a selectable histidine marker (Problem 18–24). The *heavy lines* show cells containing the plasmid, and the *light lines* show cells lacking the plasmid. At each division mother cells are shown to the *left* and daughter cells are shown to the *right*.

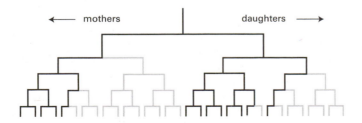

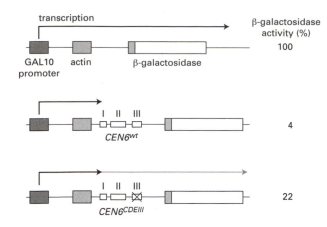

Figure 18–15 Constructs for testing centromere function (Problem 18–25). β-galactosidase activity observed when these constructs are introduced into cells is shown on the *right*.

gene under control of the strong galactose-inducible *GAL10* promoter (Figure 18–15). Into the intron in the actin gene you insert a functional yeast centromere (*CEN6^{wt}*) or a nonfunctional version with an inactivating mutation (*CEN6^{CDEIII}*). When you introduce these plasmids into cells growing on galactose and measure β-galactosidase activity, you are encouraged to find that the wild-type centromere blocks transcription and that the mutant centromere partially relieves the transcription block (Figure 18–15).

You next test a collection of chromosome-transmission-fidelity (*ctf*) mutants that show elevated rates of chromosome loss. You expect that *ctf* mutants defective in kinetochore assembly or function will show enhanced expression of β-galactosidase like the mutation that inactivates the centromere. As shown in Table 18–1, some *ctf* mutants give substantially elevated levels of β-galactosidase.

A. The test plasmid used in these experiments already contained its own centromere. Why does the introduction of a second centromere in the intron of the hybrid gene not lead to breakage of the plasmid and extreme plasmid instability as described in Problem 18–23?

B. How do you think a functional kinetochore might interfere with transcription?

C. Some of the *ctf* mutants (*ctf9*, for example) do not show enhanced β-galactosidase activity, yet they do show elevated rates of chromosome loss. What other mechanisms, apart from failure to assemble a functional kinetochore, might lead to the *ctf* phenotype?

18–26* The results obtained from the transcription assay described in the previous problem suggest that *ctf13* may encode a kinetochore protein. As a second and independent check on kinetochore function, you set up a functional assay for kinetochores using dicentric minichromosomes that are not essential for cell survival. One of the two centromeres on the minichromosome is perfectly normal. The other is conditionally active: it is inactive when

TABLE 18–1 Transcription through centromeres in *ctf* mutants (Problem 18–25).

HOST STRAIN	TRANSCRIPTION BLOCK	β-GALACTOSIDASE ACTIVITY (nmol/min/mg protein)
Wild type	*CEN6^{wt}*	25
Wild type	*CEN6^{CDEIII}*	135
ctf7	*CEN6^{wt}*	86
ctf8	*CEN6^{wt}*	50
ctf9	*CEN6^{wt}*	19
ctf13	*CEN6^{wt}*	110
ctf17	*CEN6^{wt}*	160

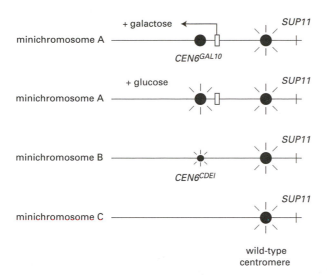

Figure 18–16 Minichromosomes used to test kinetochore function (Problem 18–26).

transcription is directed across it by the galactose-inducible *GAL10* promoter, but fully active when transcription is turned off by addition of glucose (Figure 18–16). As controls, you make two other minichromosomes: one with a weak second centromere due to a small deletion in the centromere sequence (minichromosome B) and the other with a single centromere (minichromosome C) (Figure 18–16).

If both centromeres are functional, they will cause the minichromosomes to break. The loss of an intact minichromosome can be followed visually due to a suppressor tRNA gene (*SUP11*) located next to the normal centromere (Figure 18–16). When the minichromosome breaks, the daughter cell that does not inherit the suppressor tRNA gene will turn red, owing to a suppressible mutation (that is no longer suppressed) in a gene required for adenine biosynthesis. Minichromosome loss gives rise to white colonies containing red sectors. Examining the number and size of such sectors is commonly referred to as a sectoring assay.

To test the effects of the *ctf13* mutation on the stability of dicentric minichromosomes, you introduce the three minichromosomes into yeast carrying the *ctf13* mutation and into yeast with the normal wild-type gene. In all cases you grow the cells in the presence of galactose and then test for sectoring on agar plates in the presence of glucose (Figure 18–17).

A. Explain why you placed the second centromere in minichromosome A under control of the *GAL10* promoter, and why you grew the yeasts in galactose before switching to glucose for the sectoring assay.

B. Why do you think minichromosome B, which has a weak second centromere, gives less sectoring in the wild-type strain than does minichromosome A, which has two fully functional centromeres?

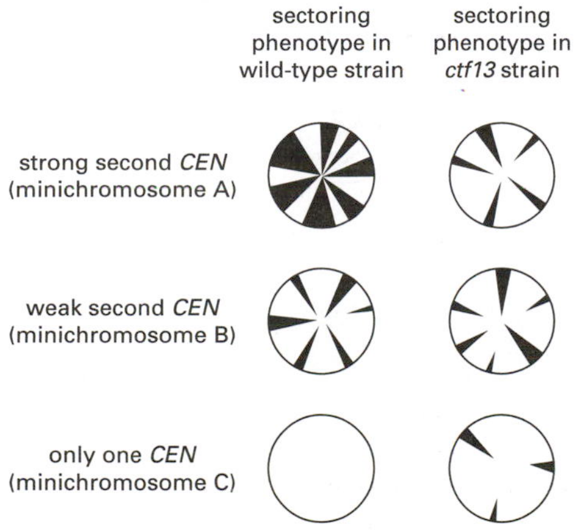

	sectoring phenotype in wild-type strain	sectoring phenotype in *ctf13* strain
strong second *CEN* (minichromosome A)		
weak second *CEN* (minichromosome B)		
only one *CEN* (minichromosome C)		

Figure 18–17 Sectoring patterns in colonies of wild-type and *ctf13* yeast carrying monocentric and dicentric minichromosomes (Problem 18–26).

TABLE 18–2 DNA content, haploid number of chromosomes, and microtubules per chromosome in a variety of organisms (Problem 18–28).

TYPE OF ORGANISM	SPECIES	DNA CONTENT (bp)	NUMBER OF CHROMOSOMES	MICROTUBULES/ CHROMOSOME
Yeast	*S. cerevisiae*	1.4×10^7	16	1
Yeast	*S. pombe*	1.4×10^7	3	3
Protozoan	*Chlamydomonas*	1.1×10^8	19	1
Fly	*Drosophila*	1.7×10^8	4	10
Human	*Homo sapiens*	3.2×10^9	23	25
Plant	*Haemanthus*	1.1×10^{11}	18	120

   C. Do your results support the notion that the *ctf13* gene encodes a kinetochore protein? In your answer account for the difference in sectoring observed when minichromosome A is introduced into the *ctf13* strain versus the wild-type strain.

   D. The *ctf13* mutant strain was isolated on the basis of an elevated rate of chromosome loss. Yet in your experiments it apparently lowers the rate of loss of minichromosome A. How can it be that a mutation that enhances the loss of normal chromosomes actually stabilizes dicentric chromosomes?

18–27   How many kinetochores are there in a human cell at mitosis?

18–28*   How much DNA does a single microtubule carry in mitosis? From the information in Table 18–2, calculate the average length of chromosomes in each organism in base pairs and in millimeters (1 bp = 0.34 nm), and then calculate how much DNA (in base pairs) each microtubule carries on average in mitosis. Do microtubules carry about the same amount of DNA or does it vary widely in different organisms?

18–29   Rarely, both sister chromatids of a chromosome end up in one daughter cell. Suggest some possible causes for such an event. What could be the consequences of this event occurring in mitosis?

18–30*   How can there be a constant poleward flux of tubulin subunits in the absence of any visible change in the appearance of the spindle?

18–31   (**True/False**) Chromosomes are positioned on the metaphase plate by equal and opposite forces that pull them toward the two poles of the spindle. Explain your answer.

18–32*   Among the variety of microtubule-dependent motors associated with mitotic spindles are ones that bind to chromosomes arms. The role of one such motor protein, Xkid, during spindle assembly was investigated by removing the protein (by immunodepletion) from frog egg extracts, which will form spindles under defined conditions. Extracts that have Xkid and immunodepleted extracts, which are missing Xkid, both assemble normal looking spindles, as assessed by tubulin staining (Figure 18–18). In the presence of Xkid the chromosomes are aligned on the metaphase plate (Figure 18–18A), whereas in its absence the chromosomes are dispersed throughout the spindle (Figure 18–18B).

   A. Suggest a mechanism by which Xkid might function to align chromosomes

(A) + Xkid

tubulin   DNA

(B) – Xkid

tubulin   DNA

Figure 18–18 Assembly of mitotic spindles in extracts (Problem 18–32). (A) In the presence of Xkid. (B) In the absence of Xkid. The spindle microtubules were made visible with fluorescent tubulin and the DNA, with a fluorescent stain.

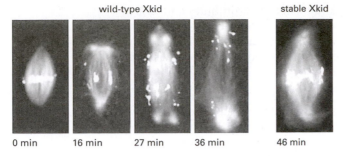

wild-type Xkid | stable Xkid

0 min   16 min   27 min   36 min   46 min

Figure 18–19 Anaphase in the presence of wild-type and stable Xkid (Problem 18–33). Metaphase spindles were assembled and then anaphase was initiated by addition of Ca^{2+} at time zero. *Faint fluorescence* marks the position of microtubules and *bright spots* mark the position of the chromosomes.

on the metaphase plate. Include in your description whether you think Xkid is a plus-end- or a minus-end-directed motor, and why.

B. Is Xkid a plausible candidate for the mediator of the astral ejection force that pushes chromosomes away from the poles? Explain your reasoning.

18–33   At the transition from metaphase to anaphase Xkid is normally degraded. Is degradation critical for anaphase to proceed? To answer this question, you assemble metaphase spindles in the presence of wild-type Xkid or a functional, but nondegradable (stable) form of Xkid. You then initiate anaphase and examine the spindles at various times thereafter. Spindles assembled in the presence of wild-type Xkid progress through anaphase normally with the chromosomes reaching the poles within 36 minutes (Figure 18–19). By contrast, the chromosomes on spindles assembled in the presence of stable Xkid remain at the equator even after 46 minutes (Figure 18–19). Examination of the chromosomes shows that in all cases the sister chromatids have separated. Why do you suppose anaphase does not proceed in the absence of Xkid degradation?

18–34*   Bipolar spindles assemble in the absence of centrosomes in *Sciara* when development occurs parthenogenetically. (Normally, the sperm delivers a centrosome to the egg along with a haploid genome.) These spindles look normal except that they lack astral microtubules (Figure 18–20). Moreover, they can support the rapid, synchronous series of early nuclear divisions that occur in a common cytoplasm in *Sciara* (similar to the early nuclear divisions in *Drosophila*). The products of these early mitotic events, however, are clearly different in normal embryos and parthenogenetic ones. The nuclei in normal embryos are well distributed in the common cytoplasm (Figure 18–20A), but those in parthenogenetic embryos are clustered together (Figure 18–20B). Can you suggest a way in which astral microtubules might function to keep nuclei well distributed in the common cytoplasm?

(A) NORMAL

spindle with asters

embryo after division 1

after division 5

(B) PARTHENOGENETIC

anastral spindle

embryo after division 1

after division 5

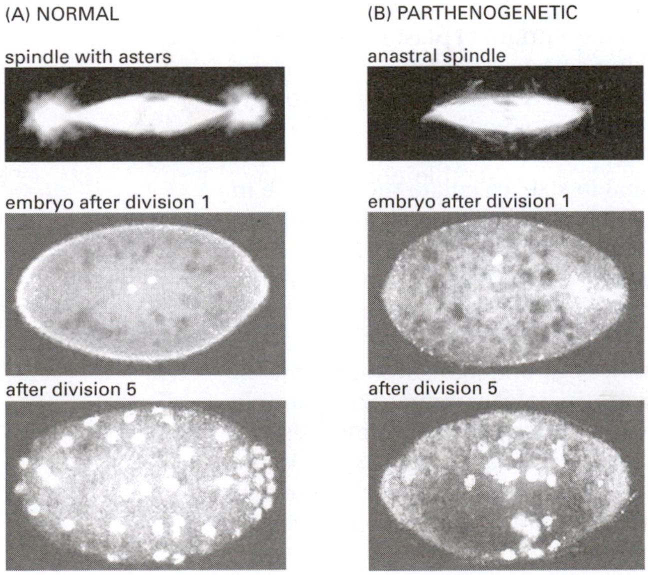

Figure 18–20 Bipolar spindles and nuclear divisions in *Sciara* (Problem 18–34). (A) In normal embryos. (B) In parthenogenetic embryos.

18–35 It is estimated that as many as 25% of kinetochore microtubules and 75% of overlap microtubules are not anchored to the centrosome. In spite of that, all the microtubules are focused tightly at the spindle pole. Why do you suppose the microtubule ends that weren't nucleated by centrosomes don't splay out away from the poles?

18–36* If a fine glass needle is used to manipulate a chromosome inside a living cell during early M phase, it is possible to trick the kinetochores on the two sister chromatids into attaching to the same spindle pole. This arrangement is normally unstable and is rapidly converted to the standard arrangement with sister chromatids attached to opposite poles. The abnormal attachment can be stabilized, however, if the needle is used to gently pull the chromosome so that the microtubules that attach it to the same pole are under tension. What does this suggest to you about the mechanism by which kinetochores normally become attached and stay attached to microtubules from opposite spindle poles during M phase? Explain your answer.

18–37 Compare and contrast the movements of chromosomes and spindle poles, and their underlying mechanisms, during anaphase A and anaphase B.

18–38* Discuss the following analogy: "Chromosomes are pulled to the spindle pole like fish on a line."

18–39 (**True/False**) Once formed, kinetochore microtubules depolymerize at the plus ends (the ends attached to the kinetochores) throughout mitosis. Explain your answer.

18–40* Consider the events that lead to formation of the new nucleus at telophase. How do nuclear and cytosolic proteins become properly sorted so that the new nucleus contains nuclear proteins but not cytosolic proteins?

CYTOKINESIS

TERMS TO LEARN

binary fission	cytokinesis	phragmoplast
cell plate	FtsZ protein	polo-like kinase
cellularization	midbody	preprophase band
contractile ring		

18–41 (**True/False**) Cytokinesis follows mitosis as inevitably as night follows day. Explain your answer.

18–42* When cells divide after mitosis, their surface area increases—a natural consequence of dividing a constant volume into two compartments. The increase in surface requires an increase in the amount of plasma membrane. One can estimate this increase by making certain assumptions about the geometry of cell division. Assuming that the parent cell and the two progeny cells are spherical, one can apply the familiar equations for the volume $(4/3\pi r^3)$ and surface area $(4\pi r^2)$ of a sphere.
   A. Assuming that the progeny cells are equal in size, calculate the increase in plasma membrane that accompanies cell division. (Although this problem can be solved algebraically, you may find it easier to substitute real numbers. For example, let the volume of the parent cell equal 1.) Do you think that the magnitude of this increase is likely to cause a problem for the cell? Explain your answer.
   B. During early development many fertilized eggs undergo several rounds of cell division without any overall increase in total volume. For example *Xenopus* eggs undergo 12 rounds of division before growth commences and the total cell volume increases. Assuming once again that all cells are spherical and equal in size, calculate the increase in plasma membrane that accompanies development of the early embryo, going from one large cell (the egg) to 4096 small cells (12 divisions).

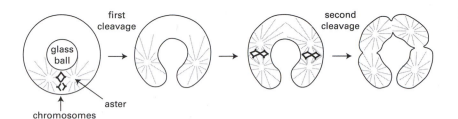

Figure 18–21 First and second divisions in a torus-shaped sand dollar egg (Problem 18–43).

18–43 Cytokinesis—the actual process of cell division—has attracted theorists for well over 100 years. It has been said that all possible explanations of cytokinesis have been proposed; the problem is to decide which one is correct. Consider the following three hypotheses:

1. *Chromosome signaling:* When chromosomes split at anaphase, they emit a signal to the nearby cell surface to initiate furrowing.

2. *Polar relaxation:* Asters relax the tension in the region of the cell surface nearest the spindle poles (the polar region), allowing the region of the membrane farthest from the poles (the equatorial plane) to contract and initiate furrowing.

3. *Aster stimulation:* The asters stimulate contraction in the region of the cell surface where oppositely oriented spindle fibers overlap (that is, the equatorial region), thereby initiating furrowing.

These hypotheses have been tested in a number of ways. One particularly informative experiment involved pressing a glass ball onto the center of a dividing sand dollar egg so as to deform it into a torus (donut shape). As illustrated in Figure 18–21, at the first division the egg divided into a single sausage-shaped cell; at the second division it divided into four cells.

A. What does the chromosome-signaling hypothesis predict for the result of this experiment? Do the predictions match the experimental observations?

B. What does the polar-relaxation hypothesis predict for the experimental outcome? Do the predictions match the experimental observations?

C. What does the aster-stimulation hypothesis predict? Do the predictions match the experimental observations?

18–44* (**True/False**) Whether cells divide symmetrically or asymmetrically, the mitotic spindle positions itself centrally in the cytoplasm.

18–45 You have obtained an antibody to myosin that prevents the movement of myosin molecules along actin filaments. If this antibody were injected into cells, would you expect the movement of chromosomes at anaphase to be affected? How would you expect antibody injection to affect cytokinesis? Explain your answers.

18–46* What observation does not support a simple 'purse-string' mechanism for constriction of the contractile ring?

18–47 If a cell just entering mitosis is treated with nocodazole, which destabilizes microtubules, the nuclear envelope breaks down and chromosomes condense, but no spindle forms and mitosis arrests. In contrast, if such a cell is treated with cytochalasin D, which destabilizes actin filaments, mitosis proceeds normally but a binucleate cell is generated and proceeds into G_1. Explain the basis for the different outcomes of these treatments with cytoskeleton inhibitors. What do these results tell you about cell-cycle checkpoints in M phase?

18–48* At telophase, the cleavage furrow narrows to form the midbody, which contains the remains of the central spindle. This fine intercellular bridge persists for a time before the connection is broken and the membranes reseal to complete the separation of the daughter cells. Little is known about this final phase of cytokinesis in animal cells beyond a requirement for microtubules and possibly centrosomes. Curiously, if exocytosis is blocked by treatment with brefeldin A, cleavage furrows form normally and seem complete, but

0 min   17 min   20 min   24 min   32 min   39 min   41 min

then regress to give a binucleate cell (Figure 18–22). Propose a mechanism that links vesicle secretion and microtubules to the final step in cytokinesis.

Figure 18–22 Video recording of a two-cell *C. elegans* embryo treated with brefeldin A (Problem 18–48). *Triangles* mark spindle poles, the *vertical line* indicates spindle orientation, *arrows* show cleavage furrow, forming and regressing. The two nuclei in the final binucleate cell are *circled*.

18–49 Globoid cell leukodystrophy (GLD, also known as Krabbe's disease) is a hereditary metabolic disorder, characterized morphologically by distinctive multinucleated globoid cells in the white matter of the brain. Deficiency of an enzyme of sphingolipid catabolism leads to accumulation of psychosine in the brain (Figure 18–23A). Psychosine binds to a G protein-coupled receptor that is expressed in only a few cell types. To test whether there might be a relationship between psychosine, its receptor, and multinucleate cells, you express the psychosine receptor in cells that normally lack it and measure the effects of psychosine treatment by FACS (fluorescence-activated cell sorting) analysis (Figure 18–23B).

Do these results support the idea that psychosine acts through its receptor to inhibit cytokinesis? Explain your reasoning.

18–50* Megakaryocytes, which are the precursor cells of blood platelets, undergo a unique differentiation program, becoming polyploid through repeated cycles of DNA synthesis without concomitant cell division. Such cells contain some 4 to 128 times the normal DNA content in a single large nucleus. Ultimately, mature megakaryocytes begin to bud off platelets as shown in

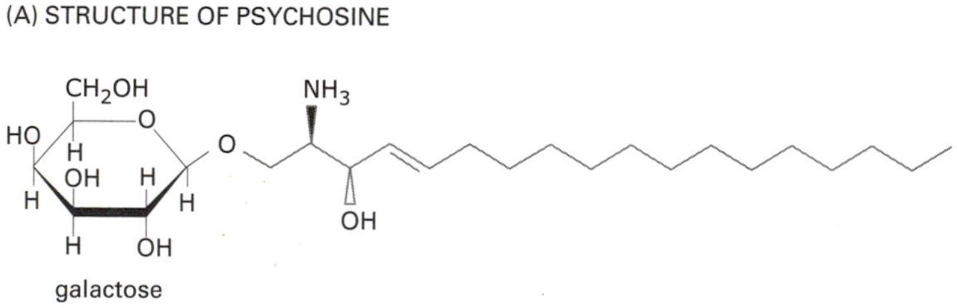

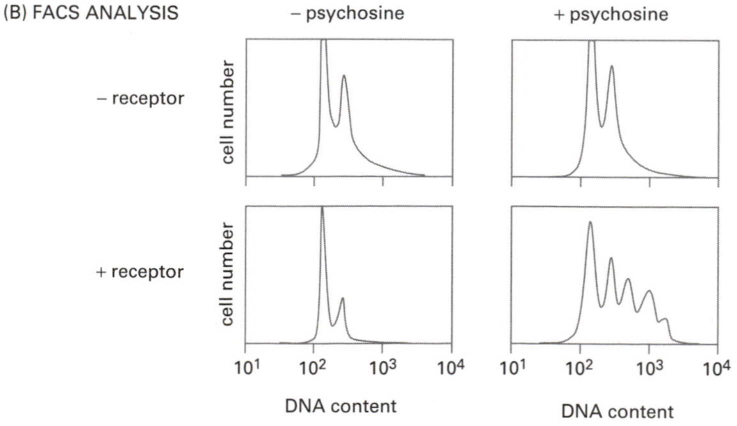

(A) STRUCTURE OF PSYCHOSINE

galactose

Figure 18–23 Analysis of role of psychosine in generation of multinucleate cells (Problem 18–49). (A) Structure of psychosine. (B) Effects of psychosine on cells that do or do not express the psychosine receptor. FACS analysis measures the DNA content of individual cells using a fluorescent DNA dye.

(B) FACS ANALYSIS

– psychosine   + psychosine

– receptor

+ receptor

cell number

10^1 10^2 10^3 10^4

DNA content

10^1 10^2 10^3 10^4

DNA content

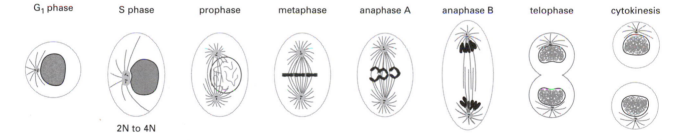

Figure 18–24 A megakaryocyte budding off platelets (Problem 18–50).

Figure 18–24. Careful observations of individual precursor cells that were stimulated to undergo polyploidization show the sequence of events in Figure 18–25. How do these events differ from the normal sequence in cell division? At what stage in M phase do these cells deviate from normal cells? What sorts of molecular differences might you expect to find among the components involved in M phase in these cells versus normal cells?

18–51 Sketch the formation of the new cell wall that separates the two daughter cells when a plant cell divides. In particular, show where the membrane proteins of the Golgi-derived vesicles end up, indicating what happens to the part of a protein in the Golgi vesicle membrane that is exposed to the interior of the Golgi vesicle.

Figure 18–25 Schematic diagrams of cell cycles (Problem 18–50). (A) Cell cycle in normal cells. (B) Two cell cycles in a megakaryocyte precursor cell.

(A) NORMAL CELL CYCLE

G$_1$ phase	S phase	prophase	metaphase	anaphase A	anaphase B	telophase	cytokinesis

2N to 4N

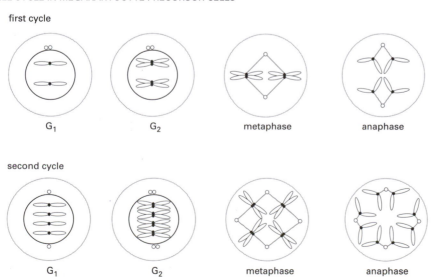

(B) CELL CYCLE IN MEGAKARYOCYTE PRECURSOR CELLS

first cycle

G$_1$ G$_2$ metaphase   anaphase

second cycle

G$_1$ G$_2$ metaphase   anaphase

Answers

Four heroes of molecular cell biology: Louis Pasteur, Frederick Gowland-Hopkins, Barbara McClintock and Fred Sanger. This book contains problems based directly on the work of all except one of these. Louis Pasteur courtesy of the Pasteur Institute, Frederick Gowland Hopkins, Barbara McClintok and Frederick Sanger © The Nobel Foundation.

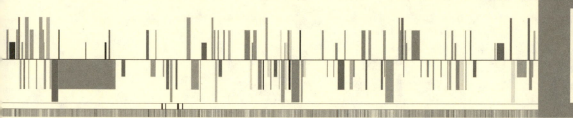

CELLS AND GENOMES

THE UNIVERSAL FEATURES OF CELLS ON EARTH

1–1 Trying to define life in terms of properties is an elusive business, as suggested by this scoring exercise (Table 1–3). Cars are highly organized objects, take energy from the environment and transform gasoline into motion, responding to stimuli from the driver as they do so. On the other hand, they cannot reproduce themselves, or grow and develop—but then neither can old animals. Cacti are not particularly responsive to stimuli, and so on. It is curious that standard definitions of life usually do not mention that living organisms on Earth are largely made of organic molecules, that life is carbon based. The first few pages of MBOC emphasize this point and discuss the properties of living cells mainly in terms of their 'informational macromolecules'—DNA, RNA, and protein.

Reference: Pace NR (2001) The universal nature of biochemistry. *Proc. Natl. Acad. Sci. USA* 98, 805–808.

1–2*

1–3

A. The number (n) of generations of cell divisions required to produce 10^{13} cells is

$$2^n = 10^{13}$$

It is useful to remember that $2^{10} \cong 10^3$ (2^n produces the series: 2, 4, 8, 16, 32, 64, 128, 256, 512, 1024; thus, $2^{10} = 1024 \cong 10^3$). If 10^3 cells result from ten generations of dividing, 10^{12} cells will result from $4 \times 10 = 40$ generations. Thus, you can estimate quickly that it will take a little over 40 generations to

TABLE 1–3 Plausible 'life' scores for car, cactus, and humans (Answer 1–1).

CHARACTERISTIC	CAR	CACTUS	HUMAN
1. Organization	Yes	Yes	Yes
2. Homeostasis	Yes	Yes	Yes
3. Reproduction	No	Yes	Yes
4. Development	No	Yes	Yes
5. Energy	Yes	Yes	Yes
6. Responsiveness	Yes	No	Yes
7. Adaptation	No	Yes	Yes

reach 10^{13} cells. You can get a more accurate answer, 43.2, by plugging different values of n into your calculator. Alternatively, you can solve the equation for n, which tests your familiarity with logarithms. Remember that

$$2 = 10^{\log 2} \text{ and } 2^n = 10^{n\log 2}$$

Substituting,

$$10^{n\log 2} = 10^{13}$$

Taking the log of both sides,

$$n\log 2 = 13$$
$$n = 13/\log 2 = 13/0.301$$
$$n = 43.2$$

B. If cells divided once per day and all cells continued to divide, it would take 43.2 days to generate the number of cells in an adult human.

C. Obviously we don't become adults in 43 days. The simple answer is that all cells don't continue to divide once per day and some cells are programmed to die. As cells differentiate, they generally slow their rate of division, ultimately in the adult dividing just often enough to replace cells that are lost or die. Of course, the real answer is much more complex, involving time for cell movements, for local environments to be established, for extracellular matrices to be laid down, for cells to differentiate, for global patterns to develop, and so on.

1–4* Figure 1–7.

1–5 On the surface, the extraordinary mutation resistance of the genetic code argues that it was subjected to the forces of natural selection. An underlying assumption, which seems reasonable, is that resistance to mutation is a valuable feature of a genetic code, one that would allow organisms to maintain sufficient information to specify complex phenotypes. This reasoning suggests that it would have been a lucky accident indeed—roughly a one-in-a-million chance—to stumble on a code as error proof as our own.

But all is not so simple. If resistance to mutation is an essential feature of any code that can support the complexity of organisms such as humans, then the only codes we *could* observe are ones that are error resistant. A less favorable frozen accident, giving rise to a more error-prone code, might limit the complexity of life to organisms that would never be able to contemplate their genetic code. This is akin to the anthropic principle of cosmology: many universes may be possible, but few are compatible with life that can ponder the nature of the universe.

Beyond these considerations, there is ample evidence that the code is not static, and thus could respond to the forces of natural selection. Deviant versions of the standard genetic code have been identified in the mitochondrial and nuclear genomes of several organisms. In each case one or a few codons have taken on a new meaning.

Reference: Freeland FJ & Hurst LD (1998) The genetic code is one in a million. *J. Mol. Evol.* 47, 238–248.

1–6*

1–7 In double-stranded DNA, which is the form of the genomes in all cellular life, G pairs with C, and A pairs with T. It is this requirement for base pairing that necessitates that the number of Gs will equal the number of Cs, and that the numbers of As and Ts will be the same. In bulk samples of DNA this translates into equivalent mole percents of G and C and of A and T.

The virus φX174 does not obey the 'rules' because its genome is single-stranded DNA. In the absence of a requirement for systematic base pairing there is no constraint on the relative amounts of G and C or of A and T.

1–8*

1–9 True. Even in eucaryotes where the coding regions of a gene are often interrupted by noncoding segments, the *order* of codons in the DNA is still the same as the order of amino acids in the protein.

1–10*

1–11 The surface-to-volume ratio for a sphere is $4\pi r^2/(4/3\pi r^3) = 3/r$; thus, the ratio is inversely proportional to radius. Consequently, relative to a human cell a bacterium has 10 times more surface per volume of cytoplasm to allow the passage of nutrients in and waste products out. The bacterium, however, grows 72 times faster than the human cell, which suggests that something besides the available surface limits the rate of growth.

1–12*

1–13
   A. During replication, parental DNA serves as a template for synthesis of new DNA.
   B. During transcription, DNA serves as a template for synthesis of RNA.
   F. During translation, RNA (mRNA) serves as the template for synthesis of protein.

   Two other processes, D. RNA → DNA, called reverse transcription, and E. RNA → RNA, called RNA replication, occur in the life cycles of RNA viruses such as HIV and poliovirus.

THE DIVERSITY OF GENOMES AND THE TREE OF LIFE

1–14*

1–15 The hemoglobin of the giant tube worms binds O_2 and H_2S and transports them to the symbiotic bacteria, which use the H_2S as an electron donor and the O_2 as an electron acceptor to generate ATP and reducing power to meet their energy needs. The resulting growth of the bacteria benefits the worms by providing increased waste products and dead bodies to live on. Moreover, in the process the toxic H_2S is rendered harmless by oxidation to elemental sulfur, thereby preventing it from poisoning the worms.

1–16*

1–17 The balanced equation for oxygenic photosynthesis, derived from experiments using water with isotopically labeled oxygen, is

$$6 \; CO_2 + 12 \; H_2\mathbf{O} + \text{light} \rightarrow C_6H_{12}O_6 + 6 \; H_2O + 6 \; \mathbf{O}_2$$

In this form of the equation it is apparent that the O_2 derives from H_2O, and that all the oxygen in glucose derives from CO_2.

1–18* Figure 1–8.

1–19 It is unlikely that any gene came into existence perfectly optimized for its function. It is thought that highly conserved genes such as ribosomal RNA genes were optimized by more rapid evolutionary change during the evolution of the common ancestor to the archaea, eubacteria, and eucaryotes. Since ribosomal RNAs (and the products of most highly conserved genes) participate in fundamental processes that were optimized early, there has been no evolutionary pressure (and little leeway) for change. By contrast, less conserved—more rapidly evolving—genes have been continually presented with opportunities to fill new functional niches. Consider, for example, the evolution of distinct globin genes that are optimized for oxygen delivery to embryos, fetuses, and adult tissues in placental mammals.

1–20*

1–21
B. It is not thought that formation of genes de novo from the vast amount of unused, noncoding DNA typical of eucaryotic genomes is a significant process in evolution. Mutation to generate a coding sequence complete with regulatory elements is too slow a process to account for the observed rates of evolutionary change.

1–22*

1–23
A. Since it appears that genes involved in informational processes are less subject to horizontal transfer, evolutionary trees derived from such genes should provide a more reliable estimate of evolutionary relationships. Thus, archaea most likely separated from eucaryotes after the archaea–eucaryote lineage separated from eubacteria.
B. Complexity is a logical explanation for the difference in rates of horizontal gene transfer (and it may even be right, although there are other possibilities). Successful transfer of an 'informational' gene would require that the new gene product fit into a preexisting, functional complex, perhaps supplanting the original related protein. In order for a new protein to fit into a complex with other proteins, it would need to have binding surfaces that would allow it to interact with the right proteins in the appropriate geometry. If a new protein had one good binding surface, but not others, it would most likely disrupt the complex and put the recipient at a selective disadvantage. By contrast, a gene product that carries out a metabolic reaction on its own would be able to function in any organism. So long as the metabolic reaction conferred some advantage on the recipient (or at least no disadvantage) the gene transfer could be accommodated.

Reference: Jain R, Rivera MC & Lake JA (1999) Horizontal gene transfer among genomes: The complexity hypothesis. *Proc. Natl. Acad. Sci. USA* 96, 3801–3806.

1–24*

1–25 It takes only 20 hours—less than a day—before the mutant cells become more abundant in the culture. From the equation provided in the question, the number of the original ('wild-type') bacterial cells at time t minutes after the mutation occurred is $10^6 \times 2^{t/20}$. The number of mutant cells at time t is $1 \times 2^{t/15}$. At the time the mutant cells 'overtake' the wild-type cells, these two numbers are equal.

$$10^6 \times 2^{t/20} = 2^{t/15}$$

Converting to base 10 (see Answer 1–3),

$$10^6 \times 10^{(t/20)\,\log 2} = 10^{(t/15)\,\log 2}$$

Taking the log of both sides and substituting for log2 (0.301),

$$6 + (t/20)(0.301) = (t/15)(0.301)$$

Multiplying both sides by 60 and solving for t,

$$360 + 0.9t = 1.2t$$
$$0.3t = 360$$
$$t = 1200 \text{ minutes, or 20 hours}$$

Note that it is also possible to solve this problem quickly using the useful relationship $2^{10} \cong 10^3$ by realizing that after one hour the mutant cells have doubled one more time than the wild-type cells. Thus, the mutant cells double relative to the wild-type cells once per hour. After 10 hours (2^{10}) the mutant cells would have gained a factor of a thousand (10^3), and after 20 hours (2^{20}), a factor of a million (10^6), at which time they would be equal in number to the wild-type cells.

Incidentally, when the two populations of cells are equal, the culture would contain 2×10^{24} cells [$(10^6 \times 2^{60}) + (1 \times 2^{80}) = (10^6 \times 10^{18}) + 10^{24} = 2 \times 10^{24}$], which at 10^{-12} g per cell, would weigh 2×10^{12} g, or two million tons! This can only have been a thought experiment.

1–26*

GENETIC INFORMATION IN EUCARYOTES

1–27   False. Plant cells contain both mitochondria and chloroplasts.

1–28*

1–29   True. Bacterial genomes seem to be pared down to the essentials: most of the DNA sequences encode proteins, a small amount of DNA is devoted to regulating gene expression, and there are very few extraneous, nonfunctional sequences. By contrast, only about 1.5% of the DNA sequences in the human genome is thought to code for proteins; even allowing for large amounts of regulatory DNA, much of the human genome is composed of DNA with no apparent function.

1–30*   **Reference:** Roger AJ, Svard SG, Tovar J, Clark CG, Smith MW, Gillin FD & Sogin ML (1998) A mitochondrial-like chaperonin 60 gene in *Giardia lamblia*: Evidence that diplomonads once harbored an endosymbiont related to the progenitor of mitochondria. *Proc. Natl. Acad. Sci. USA* 95, 229–234.

1–31   False. In addition to transfers from the mitochondrial genome, there are many examples of transfers of viral genomes; for example, some 1% of the mouse genome arose from copies of a sequence that originated as the genome of the mouse mammary tumor virus. What is rare is the transfer of genes from other species.

1–32*   Figure 1–9.

   Reference: Adams KL, Song K, Roessler PG, Nugent JM, Doyle JL, Doyle JJ & Palmer JD (1999) Intracellular gene transfer in action: Dual transcription and multiple silencings of nuclear and mitochondrial *cox2* genes in legumes. *Proc. Natl. Acad. Sci. USA* 96, 13863–13868.

1–33   If the intermediary in transfer were DNA, you would expect that the nuclear copy of the gene would have Cs at the sites of RNA editing. If the intermediary were RNA, you would expect Ts at the sites of RNA editing.

When sequences of nuclear *Cox2* genes were examined, they were found to resemble more closely the edited RNA transcript. This observation suggests that RNA was an intermediary in the transfer process. At some point the RNA was presumably copied back into DNA by reverse transcription. Whether this is a general feature of transfer is unclear.

Reference: Nugent JM & Palmer JD (1991) RNA-mediated transfer of the gene *coxII* from the mitochondrion to the nucleus during flowering plant evolution. *Cell* 66, 473–481.

1–34* Figure 1–10.

1–35 For five species there are 10 equations, which is more than enough to solve for the 7 line segments that make up the tree.

1.	Human/Whale	$= a + b$	$= 8$
2.	Human/Chicken	$= a + c + d$	$= 11$
3.	Human/Frog	$= a + c + e + f$	$= 18$
4.	Human/Fish	$= a + c + e + g$	$= 17$
5.	Whale/Chicken	$= b + c + d$	$= 12$
6.	Whale/Frog	$= b + c + e + f$	$= 17$
7.	Whale/Fish	$= b + c + e + g$	$= 17$
8.	Chicken/Frog	$= d + e + f$	$= 17$
9.	Chicken/Fish	$= d + e + g$	$= 20$
10.	Frog/Fish	$= f + g$	$= 20$

B. Solutions to the two three-at-a-time equations are shown below. In each case the three equations are summed in such a way as to eliminate all but one variable. For Human/Whale/Chicken this involves subtracting equation 5 from the sum of equations 1 and 2.

Human/Whale/Chicken

1.	Human/Whale	$= a + b$	$= 8$
2.	Human/Chicken	$= a + c + d$	$= 11$
5.	Whale/Chicken	$= \underline{b + c + d}$	$= \underline{12}$
	$(1. + 2. – 5.)$	$= 2a$	$= 7$
			$a = 3.5, b = 4.5$

Human/Whale/Frog

1.	Human/Whale	$= a + b$	$= 8$
3.	Human/Frog	$= a + c + e + f$	$= 18$
6.	Whale/Frog	$= \underline{b + c + e + f}$	$= \underline{17}$
	$(1. + 3. – 6.)$	$= 2a$	$= 9$
			$a = 4.5, b = 3.5$

Note that the values for *a* and *b* are not the same. Because there are more equations than unknowns, multiple values for the unknowns are obtained. These are averaged to get the distances represented in the phylogenetic trees. Three common representations of the phylogenetic tree obtained from the data in this problem are shown in Figure 1–11.

1–36* **Reference:** Li WH (1997) Molecular Evolution. Sinauer Associates, Inc.: Sunderland MA.

1–37

A. The data in the phylogenetic tree (see Figure 1–6) refutes the hypothesis that plant hemoglobin genes arose by horizontal transfer. Looking at the more familiar parts of the tree, we see that the vertebrates (fish to human) cluster together as a closely related set of species. Moreover, the relationships in the unrooted tree shown in Figure 1–6 are compatible with the order of branching we know from the evolutionary relationships among these species: fish split off before amphibians, reptiles before birds, and mammals last of all in

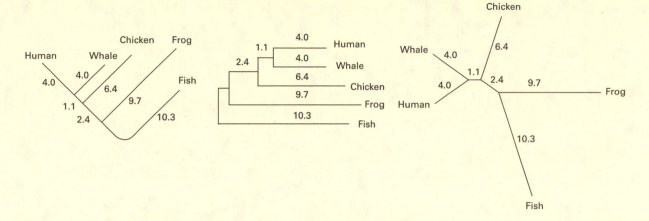

a tightly knit group. Plants also form a distinct group that displays accepted evolutionary relationships, with barley, a monocot, diverging before bean, alfalfa, and lotus, which are all dicots (and legumes). The sequences of the plant hemoglobins appear to have diverged long ago in evolution, at or before the time that mollusks, insects, and nematodes arose. The relationships in the tree indicate that the hemoglobin genes arose by descent from some common ancestor.

B. Had the plant hemoglobin genes arisen by horizontal transfer from a parasitic nematode, then the plant sequences would have clustered with the nematode sequences in the phylogenetic tree in Figure 1–6.

1–38* **Reference:** Li WH (1997) Molecular Evolution, pp 228–230. Sinauer Associates, Inc.: Sunderland MA.

Figure 1–11 Three representations of the phylogenetic tree derived from the data in this problem (Answer 1–35). These trees are termed unrooted trees because the node where this cluster of species diverged from more primitive species is not shown, nor can it be derived from the data in the problem.

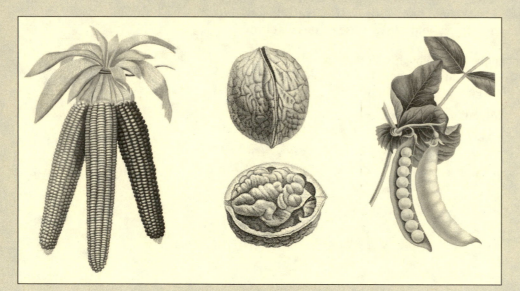

Corn, nuts, and peas… the starch and protein in these foods is formed from air, water, and earth by the action of light. Humans eat the food and transform it into their own flesh and blood, but in the end, the elements return to the air and the earth. Biochemistry is what happens in between. Courtesy of the John Innes Foundation.

CELL CHEMISTRY AND BIOSYNTHESIS

THE CHEMICAL COMPONENTS OF A CELL

2–1 Organic chemistry in laboratories—even the very best—is rarely carried out in a water environment because of low solubility of some components and because water is reactive and usually competes with the intended reaction. The most dramatic difference, however, is the complexity. It is critical in laboratory organic chemistry to use pure components and very few of them to ensure a high yield of the intended product. By contrast, living cells carry out thousands of different reactions simultaneously with good yield and virtually no interference between reactions. The key, of course, is that cells use enzyme catalysts, which bind substrate molecules in an active site, where they are isolated from the rest of the environment and the reactivity of individual atoms can be manipulated to encourage the correct reaction. It's the ability of enzymes to provide such special environments—miniature reaction chambers—that allows the cell to carry out an enormous number of reactions simultaneously without cross talk between them.

2–2*

2–3

    A. The atomic number of carbon, which is the number of protons, is six. The atomic weight, which is the number of protons plus neutrons, is 12.

    B. The number of electrons, which equals the number of protons, is six.

    C. The first shell can accommodate two electrons and the second shell, eight. Carbon therefore needs four additional electrons (or would have to give up four electrons) to obtain a full outermost shell. Carbon is most stable when it shares four additional electrons with other atoms (including other carbon atoms) by forming four covalent bonds.

    D. Carbon 14 has two additional neutrons in its nucleus. Because its electrons determine the chemical properties of an atom, carbon 14 is chemically identical to carbon 12.

2–4*

TABLE 2–7 Decay constants for radioactive isotopes (Answer 2–9).

RADIOACTIVE ISOTOPE	HALF-LIFE	DECAY CONSTANTS (λ)	
^{14}C	5730 years	1.21×10^{-4}/year	2.30×10^{-10}/min
^{3}H	12.3 years	5.63×10^{-2}/year	1.07×10^{-7}/min
^{35}S	87.4 days	7.93×10^{-3}/day	5.51×10^{-6}/min
^{32}P	14.3 days	4.85×10^{-2}/day	3.37×10^{-5}/min

2–5   Glucose ($C_6H_{12}O_6$) has a molecular weight of 180 [(6 × 12) + (12 × 1) + (6 × 16)] and therefore a mass of 180 g/mole. A concentration of 90 mg/dL corresponds to 5×10^{-3} M or 5 mM.

$$[\text{glucose}] = \frac{90 \text{ mg}}{\text{dL}} \times \frac{\text{mole}}{180 \text{ g}} \times \frac{10 \text{ dL}}{\text{L}} \times \frac{\text{g}}{10^3 \text{ mg}}$$

$$= 5 \times 10^{-3} \text{ mole/L, which is } 5 \times 10^{-3} \text{ M, or 5mM}$$

2–6*   Figure 2–38.

2–7   You would expect the DNA backbone to break. The chemistry of an atom is determined by its number of protons, which determines the number and reactivity of electrons in the outer electron shell. A phosphorus atom is comfortable, chemically speaking, bonded to the four oxygen atoms in the arrangement that makes up the phosphodiester bond in DNA, whereas a sulfur atom is not.

2–8*

2–9   The values of λ for each of the isotopes are shown in Table 2–7. At $t_{1/2}$

$$2.303 \log_{10}(0.5) = -\lambda t_{1/2}$$

Rearranging,

$$\lambda = \frac{-2.303 \log_{10}(0.5)}{t_{1/2}} = \frac{-2.303(-0.301)}{t_{1/2}}$$

$$\lambda = \frac{0.693}{t_{1/2}}$$

For ^{14}C,

$$\lambda = \frac{0.693}{5730 \text{ years}}$$

$$\lambda = 1.21 \times 10^{-4}/\text{year}$$

Since there are 5.26×10^5 min/year, λ equals 2.30×10^{-10}/min.

2–10*

2–11   True. With each half-life, half the remaining radioactivity decays. After 10 half-lives $(1/2)^{10}$, about 1/1000 (exactly 1/1024) will remain. (It is useful to remember that 2^{10} is about 1000.)

2–12*

2–13   The specific activity of glycine would be 0.12 Ci/mmol if all carbons were ^{14}C. The number of radioactive atoms in a mmol of glycine is 12×10^{20} [(6 × 10^{20} molecules/mmol) × (2 ^{14}C/molecule)]. With a decay constant, λ, of 2.30×10^{-10}/min, this number of ^{14}C atoms corresponds to 2.76×10^{11} dpm [(12×10^{20}) × (2.30×10^{-10})], which is 0.124 Ci [(2.76×10^{11} dpm)/(2.22×10^{12} dpm/Ci)].

At a specific activity of 200 μCi/mmol about 1/300 molecules of glycine would carry a ^{14}C. 200 μCi/mmol is 1/600 of the maximum specific activity $[(200 \times 10^{-6} \text{ Ci/mmol})/(0.12 \text{ Ci/mmol})]$, and since there are 2 carbons per glycine, about 1/300 molcules will have a ^{14}C.

2–14*

2–15

A. From the innermost to the outermost, the first three electron shells can carry 2, 8, and 8 electrons.

B. H can gain one electron to fill the first shell, or it can lose one electron to leave a completely empty first shell. C can gain or lose four electrons to generate a filled outer shell. N and P will gain three electrons to fill their outer shells. O and S will gain two electrons to fill their outer shells. In general, it is energetically most favorable for an atom to lose or gain the fewest number of electrons required to generate a filled outer shell.

C. The valences for these atoms equal the number electrons that must be gained or lost to complete the outer shell. Thus, the valences are the same as the numbers in part B.

2–16*

2–17 The statement is correct. Both ionic and covalent bonds are based on the same principles: electrons can be shared equally between two interacting atoms, forming a nonpolar covalent bond; electrons can be shared unequally between two interacting atoms, forming a polar covalent bond; or electrons can be completely lost from one atom and gained by the other, forming an ionic bond. There are bonds of every conceivable intermediate state, and for borderline cases it becomes arbitrary whether a bond is described as a very polar covalent bond or as an ionic bond.

2–18*

2–19 Permanent dipoles are critical in biology because they allow molecules to interact through electrical forces. Any large molecule with many polar groups will have a pattern of partial positive and negative charges on its surface. When such a molecule encounters a second molecule with a complementary set of charges, the two molecules will be attracted to one another by interactions between their permanent dipoles. Such interactions resemble ionic bonds but are weaker.

2–20*

2–21 Because of its larger size, the outermost electrons in a sulfur atom are not as strongly attracted to the nucleus as they are in an oxygen atom. Consequently, the hydrogen–sulfur bond is much less polar than the hydrogen–oxygen bond. Because of the reduced polarity, the sulfur in H_2S is not strongly attracted to hydrogen atoms in adjacent H_2S molecules, and hydrogen bonds do not form. It is the lack of hydrogen bonds in H_2S that allows it to be a gas, and the presence of strong hydrogen bonds in water that makes it a liquid.

2–22*

2–23 Although the symbol 'p' in common usage denotes the 'negative logarithm of,' what it stands for is unclear. In the original 1909 paper in which the concept of pH was developed, the author—Danish chemist Soren P.L. Sorensen—was not explicit. In textbooks where it is commented on at all, it is most commonly reputed to stand for the French or German words for power or potential. Close examination of the original paper reveals that the 'p' in pH is likely a consequence of the author's arbitrary choice to call two solutions by the letters 'p' and 'q'. The q solution had the known H^+ concentration of 1, the p solution had the unknown H^+ concentration. If the solutions had been switched, do you think qH would ever have caught on?

Reference: Norby JG (2000) The origin and the meaning of the little p in pH. *Trends Biochem. Sci.* 25, 36–37.

2–24*

2–25

A. A solution is said to be neutral when the concentrations of H^+ and OH^- are exactly equal. This occurs when the concentration of each ion is 10^{-7} M, so that their product is 10^{-14} M^2.

B. In a 1 mM solution of NaOH, the concentration of OH^- is 10^{-3} M. Thus, the concentration of H^+ is 10^{-11} M, which is pH 11.

$$[H^+] = \frac{K_w}{[OH^-]}$$
$$= \frac{10^{-14}\,M^2}{10^{-3}\,M} = 10^{-11}\,M$$

C. A pH of 5.0 corresponds to an H^+ concentration of 10^{-5} M. Thus, the OH^- concentration is 10^{-9} M (10^{-14} $M^2/10^{-5}$ M).

2–26*

2–27 A solution of sodium chloride will be neutral. Neither ion binds H^+ or OH^- ions and thus do not influence the dissociation of water.

A solution of potassium acetate (the salt of a weak acid) will be basic because the acetate ion will steal sufficient numbers of protons from water to satisfy equilibrium

$$CH_3COO^- + H_2O \rightleftharpoons CH_3COOH + OH^-$$

The increase in hydroxyl ions will cause the number of protons to decrease, satisfying the equilibrium for water ionization ($[OH^-][H^+] = 10^{-14}$) and making the solution basic.

A solution of ammonium chloride (the salt of a weak base) will be acidic because the ammonium ion will dissociate sufficiently to satisfy the equilibrium

$$NH_4^+ + H_2O \rightleftharpoons NH_3 + H_3O^+$$

The increase in hydronium ions lowers the pH and makes the solution more acidic.

2–28* Figure 2–39.

2–29

A. The values for $\log_{10} [A^-]/[HA]$, $[A^-]/[HA]$, and the percentage of the acid that has dissociated are shown in Table 2–8. Included in the table are a set of

TABLE 2–8 Dissociation of a weak acid at pH values above and below the pK (Answer 2–29).

pH	$\log_{10}\frac{[A^-]}{[HA]}$	$\frac{[A^-]}{[HA]}$	% DISSOCIATION	'RULE-OF-THUMB' % DISSOCIATION
pK+4	4	10^4	99.99	99.99
pK+3	3	10^3	99.9	99.9
pK+2	2	10^2	99	99
pK+1	1	10^1	91	90
pK	0	10^0	50	50
pK–1	–1	10^{-1}	9.1	10
pK–2	–2	10^{-2}	0.99	1
pK–3	–3	10^{-3}	0.099	0.1
pK–4	–4	10^{-4}	0.0099	0.01

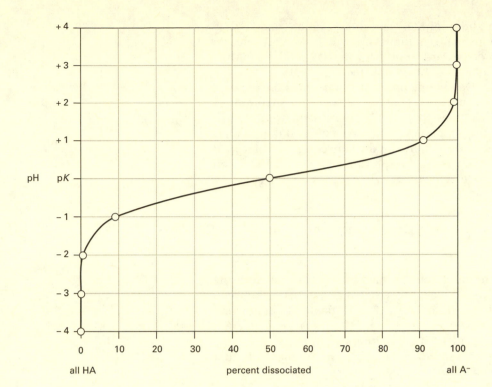

'rule-of-thumb' values that may be easier to remember, and are handy to have mentally available for estimating answers.

B. A plot of pH versus percentage dissociation of the weak acid, HA, is shown in Figure 2–40. All weak acids, regardless of pK, yield titration curves that are identical to this one. The curves for different weak acids are shifted along the pH scale depending on their pK values.

This titration curve is fundamentally similar to protein–ligand binding curves and to enzyme activity curves. As pointed out in Problem 3–49, all three phenomena—titration of weak acids, protein–ligand binding, and enzyme activity—generate identical curves. Most importantly, the 'rule-of-thumb' values pertain to each, allowing rapid estimates in all three situations.

2–30*

2–31   You could start with any of the individual solutions of phosphate and make a 1 mM solution by adding 10 mL to a liter of water. (You would want to start with a little less than a liter of water so that you could adjust the pH by addition of HCl or KOH, as needed. When the pH was adjusted to 6.9, you would then bring the solution to a final volume of 1 liter by addition of water.) If you start with H_3PO_4, you would need to add 1.5 mL of 1 M KOH (1.5 equivalents of OH^-) to bring the solution to pH 6.9. If you start with KH_2PO_4, you would need to add 0.5 mL KOH (0.5 equivalents of OH^-) to bring the solution to pH 6.9. If you start with K_2HPO_4, you would need to add 0.5 mL of 1 M HCl (0.5 equivalents of H^+) to bring the pH to 6.9. If you start with K_3PO_4, you would need to add 1.5 mL HCl (1.5 equivalents of H^+) to bring the pH to 6.9.

In all these cases you are moving phosphate along its titration curve to reach 6.9 (the pK for $H_2PO_4^- \rightleftharpoons H^+ + HPO_4^{2-}$) at which point $[H_2PO_4^-]$ equals $[HPO_4^{2-}]$. You could reach that same point by mixing equal amounts of the species directly. Thus, you would get a 1 mM solution at pH 6.9 by mixing 5 mL aliquots of the KH_2PO_4 and K_2HPO_4 solutions in 1 liter of water. In practice you would still measure the pH to make sure it was 6.9 in case one of your original solutions was not exactly 'as advertised.' Similarly, you could also achieve the same end by mixing 5 mL aliquots of the H_3PO_4 and K_3PO_4 solutions.

The solutions will not all be the same. While they will all be 1 mM phosphate, they can differ in the amount of K^+ and Cl^- ions. Solutions that are brought up to pH 6.9 by addition of KOH and those that are obtained by direct mixing of different phosphate solutions will be 1.5 mM K$^+$. A solution of K$_2$HPO$_4$ that is adjusted downward to pH 6.9 with HCl will be 2 mM K$^+$ and 0.5 mM Cl$^-$. A solution of K$_3$PO$_4$ that is adjusted to pH 6.9 with HCl will be 3 mM K$^+$ and 1.5 mM Cl$^-$. Depending on the uses for which the buffer is intended, these differences could be crucial. It is especially important to note that these calculations assume that the pH adjustments with KOH or HCl were precise. If one is sloppy, and overshoots the pH and then undershoots it several times before hitting it exactly, the concentrations of K$^+$ and Cl$^-$ can be arbitrarily high—even though the phosphate is exactly 1 mM and the pH is exactly 6.9. This can lead to very puzzling results...and has!

2–32* Figure 2–41.

2–33 The structures of these three forms of glycine are shown in Figure 2–42.

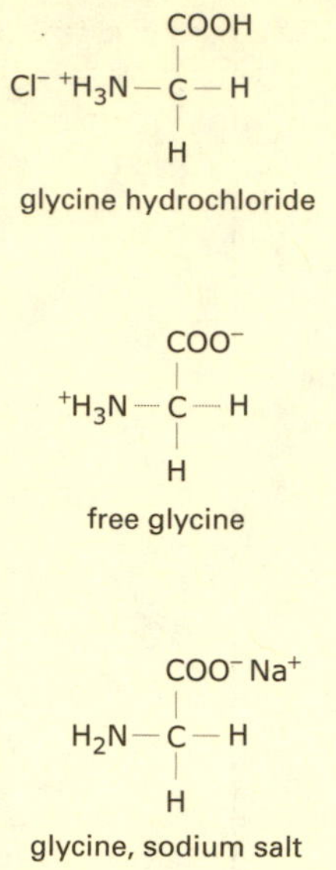

2–34*

2–35 The effects on pK values are due to electrostatic interactions between the carboxyl and amino groups. In alanine a large electrostatic attraction between the –NH$_3^+$ and the –COO$^-$ is present at pH 7. This favorable interaction makes it more difficult to remove a proton from –NH$_3^+$, raising its pK, and more difficult to add a proton to –COO$^-$, lowering its pK. The electrostatic attraction decreases as the amino and carboxyl groups are moved farther and farther away from one another in the oligomers of alanine, virtually disappearing by Ala$_4$, as reflected in the changes in pK values.

Reference: Cantor CR & Schimmel PR (1980) Biophysical Chemistry, Part 1: The Conformation of Biological Macromolecules, pp 42–46. San Francisco: WH Freeman and Company.

2–36*

2–37 Assuming that the change in enzyme activity is due to the change in protonation state of histidine, the enzyme must require histidine in the protonated, charged state. The enzyme is active only below the pK of histidine (which is typically around 6.5 to 7.0 in proteins), where the histidine is expected to be protonated.

2–38*

2–39

A. The ratio of [HCO$_3^-$] to [CO$_2$(dis)] at pH 7.4 is 20.

$$pH = pK' + \log_{10}\frac{[HCO_3^-]}{[CO_2(dis)]}$$
$$7.4 = 6.1 + \log_{10}\frac{[HCO_3^-]}{[CO_2(dis)]}$$
$$\log_{10}\frac{[HCO_3^-]}{[CO_2(dis)]} = 1.3 \text{ and } \frac{[HCO_3^-]}{[CO_2(dis)]} = 20$$

Since the total carbonate is 25 mM, HCO$_3^-$ is 23.8 mM [(20/21) × 25 mM] and CO$_2$(dis) is 1.2 mM [(1/21) × 25 mM].

B. Addition of 5 mM of H$^+$ would drive 5 mM of HCO$_3^-$ to CO$_2$(dis), thereby maintaining the equilibrium for hydration and dissociation of CO$_2$. Thus, addition of 5 mM H$^+$ would reduce [HCO$_3^-$] to 18.8 mM, and increase [CO$_2$(dis)] to 6.2 mM. At these concentrations the pH would be 6.6.

$$pH = pK' + \log_{10}\frac{[HCO_3^-]}{[CO_2(dis)]}$$
$$= 6.1 + \log_{10}\frac{18.8}{6.2}$$
$$pH = 6.1 + 0.48 = 6.6$$

Figure 2–42 Three forms of glycine (Answer 2–33).

Thus, in a closed system bicarbonate/CO_2 would provide a very weak buffering system with a very small buffering capacity.

C. In an open system, addition of 5 mM H^+ would cause the same changes as above except that the excess CO_2 would be removed by exhalation, maintaining its concentration at 1.2 mM. Under these conditions the pH would be 7.3.

$$pH = pK' + \log_{10}\frac{[HCO_3^-]}{[CO_2(dis)]}$$
$$= 6.1 + \log_{10}\frac{18.8}{1.2}$$
$$pH = 6.1 + 1.19 = 7.3$$

Thus, in an open system the pH decreases by only about 0.1 pH unit. The beauty of this buffering system is that HCO_3^- is constantly being added back to the system through metabolism, which generates CO_2 that is then hydrated to HCO_3^-. Moreover, the two components of the system are independently regulated: CO_2 exhalation by the lungs can be controlled by the rate of breathing, and HCO_3^- can be excreted or retained by the kidneys.

2–40*

2–41 False. Strong acids bind protons weakly and give them up readily in a water environment.

2–42*

2–43 The statement is correct. The hydrogen–oxygen bond in water molecules is polar; thus, the oxygen atom carries a partial negative charge and the hydrogen atoms carry partial positive charges. The partial negative charges on the oxygen atoms are attracted to the positive charges on the sodium ions, but are repelled by the negative charges on the chloride ions.

2–44*

2–45 The functional groups on the three molecules are indicated and named in Figure 2–43.

2–46*

2–47 Both amylose and cellulose are polymers of glucose. Amylose is a polymer of α-D-glucose, linked together by α1→4 glycosidic bonds. Cellulose is a polymer of β-D-glucose linked together by β1→4 glycosidic bonds. It is more difficult to discern this for cellulose, as every other monomer is flipped 180° about an axis that passes through carbons 1 and 4. The structures of α-D-glucose and β-D-glucose and the linked dimers in amylose and cellulose, respectively, are shown in Figure 2–44.

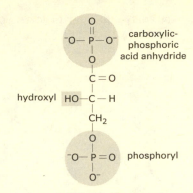

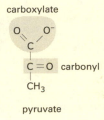

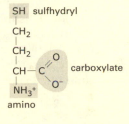

Figure 2–43 The functional groups in 1,3-bisphosphoglycerate, pyruvate, and cysteine (Answer 2–45).

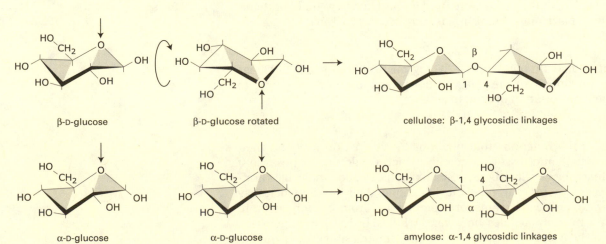

Figure 2–44 The structures of α-D-glucose and β-D-glucose and their linkage into dimers (Answer 2–47). *Arrows* point to ring oxygens to show alignment of monomers that are joined into dimers. The *curved arrow* indicates a 180° rotation of the sugar.

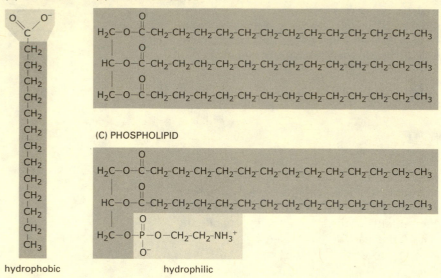

(A) FATTY ACID   (B) TRIACYLGLYCEROL

(C) PHOSPHOLIPID

hydrophobic   hydrophilic

Figure 2–46 A fatty acid, a triacylglycerol, and a phospholipid (Answer 2–49). The hydrophilic and hydrophobic portions of the molecules are indicated by *light* and *dark shading*, respectively.

2–48* Figure 2–45.

2–49 A molecule is amphipathic if its hydrophobic and hydrophilic portions are segregated into two distinct regions of the molecule. As indicated in Figure 2–46, fatty acids and phospholipids are amphipathic because one end of those molecules is hydrophilic and the other is hydrophobic. By contrast, triacylglycerols are relatively hydrophobic throughout. Because fatty acids and phospholipids are amphipathic, collections of these molecules can form distinctive kinds of structures, including lipid monolayers (at an air interface), lipid bilayers (the essence of membrane structure), and micelles (aggregates with a hydrophilic surface and a hydrophobic center. Triacylglycerols, which are hydrophobic throughout, separate from a water solution, forming a lipid droplet (the storage form of fat in adipose cells).

2–50* Figure 2–47.

2–51 The synthesis of a macromolecule with a unique structure requires that only one stereoisomer is used in each position. Changing one amino acid from its L- to its D-form would result in a different protein. Thus, if a random mixture of the D- and L-forms were used to build a protein, its amino acid sequence would not specify a single structure, but rather many different structures (2^N different structures would be formed, where N is the number of amino acids in the protein).

Why L-amino acids were selected in evolution as the exclusive building blocks of proteins is a mystery; we could easily imagine a cell in which certain (or even all) amino acids were used in the D-forms to build proteins, as long as these particular stereoisomers were used exclusively.

2–52* Figure 2–48.

2–53 The concentration of protein is about 200 mg/mL (0.18×1.1 g/mL = 198 mg/mL). Given the density of the cell, you don't need to know its volume to calculate the concentration of protein.

2–54*

2–55 False. Many of the functions that macromolecules perform rely on their ability to associate and dissociate readily, which would not be possible if they were linked by covalent bonds. By linking their macromolecules non-covalently, cells can, for example, quickly remodel their interior when they move or divide, and easily transport components from one organelle to another. Some macromolecules, however, are linked by covalent bonds. This occurs primarily in situations where extreme structural stability is required, such as in the cell walls of many bacteria, fungi, and plants, and in the extra-cellular matrix that provides the structural support for most animal cells.

CATALYSIS AND THE USE OF ENERGY BY CELLS

2–56*

2–57   The second law of thermodynamics applies to closed systems—for example, a chamber in a scientist's laboratory or the entire universe. Closed systems do not exchange matter or energy with their surroundings. Living organisms such as cells and human beings are not closed systems; they continually exchange matter and energy with their surroundings. It is perfectly permissible for a portion of a closed system—a human being in the universe—to increase its order, provided that the rest of the system (the rest of the universe) becomes disordered to a greater extent. This is what living organisms do; they take in food and use the energy to increase their order. But to do so they release waste products that are less complex (less ordered) than the food they took in, and much of the energy in the food is released in its most disordered form—as heat. Whatever order is created within a cell or an organism is more than paid for by the disorder introduced into its environment.

2–58*

2–59   True. The difference between plants and animals is in how they obtain their food molecules. Plants make their own from sunlight, CO_2, and H_2O, whereas animals must forage for theirs.

2–60*

2–61   If the reaction is rewritten as its two half reactions, it is then clear that Na is oxidized and Cl is reduced.

$$2\,Na \rightarrow 2\,Na^+ + 2\,e^-$$
$$\underline{2\,e^- + Cl_2 \rightarrow 2\,Cl^-}$$
$$\text{NET:}\quad 2\,Na + Cl_2 \rightarrow 2\,Na^+ + 2\,Cl^-$$

Electrons are removed from sodium; therefore it is oxidized. Electrons are added to chlorine; therefore it is reduced.

2–62*   Figure 2–49.

2–63
    A. The first and third reactions in the series (succinate → fumarate and malate → oxaloacetate) are redox reactions. In the first, the two central carbons of succinate are oxidized by the introduction of a double bond—they have each lost one electron (their oxidation states have increased from –2 to –1). In the third reaction, the carbon attached to the hydroxyl group has been oxidized by conversion to a carbonyl group—it has lost a pair of electrons (its oxidation state has increased from 0 to +2). The second reaction is not a redox reaction, since the oxidation states of the two central carbons have changed in compensating ways: the upper one from –1 to 0 and the lower one from –1 to –2.
    B. Since the molecules in the pathway have been oxidized, the unseen electron carriers must have been reduced; that is, they must have accepted a pair of electrons from pathway intermediates. It is instructive to examine the reactions a little closer. In addition to losing a pair of electrons, both succinate and malate also lose a pair of hydrogens. One of the hydrogens and both electrons travel together as a hydride ion (H^-) to the electron carrier; the other hydrogen is released as a proton. The electron carrier in the oxidation of succinate is FAD, which picks up both the hydride ion and the proton to become $FADH_2$. The electron carrier in the oxidation of malate is NAD^+, which picks up the hydride ion to become NADH, but leaves the proton in solution.

2–64*

2–65   The instantaneous velocities are $H_2O = 3.8 \times 10^4$ cm/sec, glucose = 1.2 ×

10^4 cm/sec, and globin = 1.3×10^3 cm/sec. The calculation for a water molecule, which has a mass of 3×10^{-23} g [(18 g/mole) × (mole/6 × 10^{23} molecules)], is shown below.

$$v = (kT/m)^{1/2}$$
$$v = \left(\frac{1.38 \times 10^{-16}\, \text{g cm}^2}{\text{°K sec}^2} \times 310\, \text{°K} \times \frac{1}{3 \times 10^{-23}\, \text{g}}\right)^{1/2}$$
$$v = 3.78 \times 10^4\, \text{cm/sec}$$

When these numbers are converted to km/hr the results are fairly astounding. Water moves at 1,360 km/hr, glucose at 428 km/hr, and globin at 47 km/hr. Thus, even the largest (slowest) of these molecules is moving faster than the swiftest human sprinter! Unlike a human sprinter, however, these molecules make forward progress only slowly because they are constantly colliding with other molecules in solution.

Reference: Berg HC (1993) Random Walks in Biology, Expanded Edition, pp 5–6. Princeton, NJ: Princeton University Press.

2–66* **Reference:** Berg HC (1993) Random Walks in Biology, Expanded Edition, pp 5–6. Princeton, NJ: Princeton University Press.

2–67

A. In order for polymerization to be favored at high temperature and depolymerization to be favored at low temperature, ΔH and ΔS must both be positive. At high temperature, where polymerization is favored, the $-T\Delta S$ term becomes large enough to overcome the positive ΔH term, yielding a negative, favorable ΔG for polymerization. At low temperature, where depolymerization is favored, the $-T\Delta S$ becomes small enough that it is outweighed by the positive ΔH term, giving rise to a positive, unfavorable ΔG for polymerization.

B. It seems counterintuitive that polymerization of free tubulin subunits into highly ordered microtubules should occur with an overall increase in entropy (decrease in order). But it is counterintuitive only if one considers the subunits in isolation. Remember that thermodynamics refers to the whole system, which includes the water molecules. The increase in entropy is due largely to the effects of polymerization on water molecules. The surfaces of the tubulin subunits that bind together to form microtubules are fairly hydrophobic, and constrain (order) the water molecules in their immediate vicinity. Upon polymerization, these constrained water molecules are freed up to interact with other water molecules. Their newfound disorder much exceeds the increased order of the protein subunits, and thus the net increase in entropy (disorder) favors polymerization.

2–68* Figure 2–50.

Reference: Howard J (2001) Mechanics of Motor Proteins and the Cytoskeleton, pp 151–163. Sunderland, MA: Sinauer Associates, Inc.

2–69

A. Reaction rates could be limited by any combination of the following factors: (1) the frequency of collision with the active site of the enzyme; (2) the proportion of molecules that are energetic enough to undergo reaction; or (3) the rate of release of the products to free the active site.

B. The 10^7–fold rate enhancement corresponds to the ratio of the areas under the curve to the right of the thresholds in Figure 2–26. The number of molecules with sufficient energy to undergo a catalyzed reaction (the area to the right of threshold A) divided by the number of molecules with sufficient energy to undergo an uncatalyzed reaction (the area to the right of threshold B) is equal to 10^7.

2–70*

2–71   The statement is correct. A reaction with a negative $\Delta G°$, for example, would not proceed spontaneously under conditions where there is already an excess of products, that is, an excess over those that would be present at equilibrium. Conversely, a reaction with a positive $\Delta G°$ would proceed spontaneously under conditions where there is an excess of substrates (over those present at equilibrium).

2–72*

2–73

    A. At equilibrium ΔG is zero (for any reaction); there is no tendency for the reaction to proceed in one direction over the other direction. Substituting K for [F6P]/[G6P] gives

$$0 = \Delta G° + 2.3\ RT \log_{10} K$$
$$\Delta G° = -2.3\ RT \log_{10} K,\ \text{or}\ -1.41\ \text{kcal/mole}\ \log_{10} K$$

    B. At equilibrium ΔG is zero and $\Delta G°$ is 0.42 kcal/mole

$$\Delta G° = -2.3\ RT \log_{10} \frac{[\text{F6P}]}{[\text{G6P}]}$$

Substituting 1.41 kcal/mole for 2.3 RT,

$$= \frac{-1.41\ \text{kcal}}{\text{mole}} \log_{10} (0.5)$$
$$= 0.42\ \text{kcal/mole}$$

    C. Since $\Delta G°$ relates to the equilibrium, it is unchanged; that is, $\Delta G° = 0.42$ kcal/mole. At $\Delta G = -0.6$ kcal/mole, the ratio of [F6P] to [G6P] is

$$\Delta G = \Delta G° + 2.3\ RT \log_{10} \frac{[\text{F6P}]}{[\text{G6P}]}$$
$$\frac{-0.6\ \text{kcal}}{\text{mole}} = \frac{0.42\ \text{kcal}}{\text{mole}} + \frac{1.41\ \text{kcal}}{\text{mole}} \log_{10} \frac{[\text{F6P}]}{[\text{G6P}]}$$
$$\log_{10} \frac{[\text{F6P}]}{[\text{G6P}]} = \frac{-1.02\ \text{kcal/mole}}{1.41\ \text{kcal/mole}} = -0.72$$
$$\frac{[\text{F6P}]}{[\text{G6P}]} = 0.19$$

2–74*

2–75

    A. The equilibrium constant $K = 4.6 \times 10^{-3}$ M^{-1}.

$$\Delta G° = -2.3\ RT \log_{10} K$$
$$3.3\ \text{kcal/mole} = -1.41\ \text{kcal/mole} \log_{10} K$$
$$\log_{10} K = \frac{3.3\ \text{kcal/mole}}{-1.41\ \text{kcal/mole}} = -2.34$$
$$K = 4.57 \times 10^{-3} = 4.57 \times 10^{-3}\ M^{-1}$$

    (You will note that this expression gives the numerical value of K because it enters the equation as a log value, where units are meaningless. Because of this and because the units derived from the equilibrium expression can vary so widely (for example, M^2, M^1, no units, M^{-1}, M^{-2}, etc., depending on the reaction), some sources treat the equilibrium constant as a unitless number by convention. Throughout this book we have added back the appropriate units so that they can be used as a guide in other calculations such as the one below.)

    B. The equilibrium concentration of G6P would be 1.1×10^{-7} M, or 0.11 µM, if [GLC] and [P$_i$] are at 5×10^{-3} M (5 mM). Remember that in the expression for equilibrium all concentrations are M.

$$K = \frac{[\text{G6P}]}{[\text{GLC}][\text{P}_i]}$$

$$[\text{G6P}] = (4.57 \times 10^{-3}\,\text{M}^{-1}) \times (5 \times 10^{-3}\,\text{M}) \times (5 \times 10^{-3}\,\text{M})$$

$$[\text{G6P}] = 1.14 \times 10^{-7}\,\text{M} = 0.11\,\mu\text{M}$$

This would be a low concentration for an intermediate in glucose metabolism, most of which are in the range of 10 to 100 μM or so. However, given that one intermediate in the glycolytic pathway, 1,3-bisphosphoglycerate, is present at about 1 μM, perhaps this concentration for G6P is not out of the question. Nevertheless, cells have devised a different way to carry out this reaction.

C. The value of $\Delta G°$ for the net reaction is obtained by adding together the $\Delta G°$ values for the individual reactions; thus, $\Delta G°$ for the net reaction is –4.0 kcal/mole (3.3 kcal/mole – 7.3 kcal/mole). The equilibrium constant K is 6.9×10^2.

$$\Delta G° = -2.3\,RT \log_{10} K$$

$$-4.0\,\text{kcal/mole} = -1.41\,\text{kcal/mole}\,\log_{10} K$$

$$\log_{10} K = \frac{-4.0\,\text{kcal/mole}}{-1.41\,\text{kcal/mole}} = 2.84$$

$$K = 6.87 \times 10^2$$

D. The equilibrium [G6P] would be 10.3 M, which is an improbably high concentration for a cell. It far exceeds the amount of available phosphate and would be more viscous than maple syrup.

$$K = \frac{[\text{G6P}][\text{ADP}]}{[\text{GLC}][\text{ATP}]}$$

$$[\text{G6P}] = \frac{(6.87 \times 10^2) \times (5 \times 10^{-3}\,\text{M}) \times (3 \times 10^{-3}\,\text{M})}{(10^{-3}\,\text{M})}$$

$$[\text{G6P}] = 10.3\,\text{M}$$

E. The cellular [G6P] of 200 μM is about 50,000-fold lower than the equilibrium concentration (10.3 M) calculated in part D. Obviously, in cells the phosphorylation of glucose to glucose 6-phosphate is not at equilibrium. In cells ΔG for the reaction is –6.6 kcal/mole.

$$\Delta G = \Delta G° + 2.3\,RT \log_{10} \frac{[\text{G6P}][\text{ADP}]}{[\text{GLC}][\text{ATP}]}$$

$$= -4.0\,\text{kcal/mole} + 1.41\,\text{kcal/mole}\,\log_{10} \frac{(200 \times 10^{-6})(10^{-3})}{(5 \times 10^{-3})(3 \times 10^{-3})}$$

$$= -4.0\,\text{kcal/mole} + (1.41\,\text{kcal/mole})(-1.88)$$

$$\Delta G = -4.0\,\text{kcal/mole} - 2.64\,\text{kcal/mole} = -6.64\,\text{kcal/mole}$$

2–76*

2–77 A positive ΔG for D $\rightarrow$ E and a negative ΔG for E $\rightarrow$ F is an unstable situation. A positive ΔG for D $\rightarrow$ E means that E will be converted to D. This will reduce the concentration of E and increase that of D. Meanwhile more D will be added from the upstream reaction, and E will be removed by the downstream reaction. These effects, which increase D and decrease E, continue until the concentration ratio [E]/[D] is sufficient to drive the reaction in the forward (D $\rightarrow$ E) direction, at which point the ΔG becomes negative.

2–78* **References**: Berg JM, Tymoczko JL & Stryer L (2002) Biochemistry, Fifth Edition, pp 436–437. New York: WH Freeman and Co.

Minakami S, Suzuki C, Saito T & Yoshikawa H (1965) Studies on erythrocyte glycolysis. I. Determination of the glycolytic intermediates in human erythrocytes. *J. Biochem. (Tokyo)* 58, 543–550.

2–79 The free energy ΔG (–11 to –13 kcal/mole) derived from ATP hydrolysis depends on both $\Delta G°$ (–7.3 kcal/mole) and the concentrations of the substrates and products.

$$\Delta G = \Delta G° + 1.41 \text{ kcal/mole} \log_{10} \frac{[ADP][P_i]}{[ATP]}$$

Given that $\Delta G°$ is –7.3 kcal, the ratio of $[ADP][P_i]/[ATP]$ in cells must range from a little more than 10^{-3} ($\Delta G = -11.5$) to a little less than 10^{-4} ($\Delta G = -12.9$ kcal/mole), as they do under different cellular conditions.

2–80*

2–81 Enzyme A is beneficial. It allows the interconversion of two energy carrier molecules, both of which are required in the triphosphate form for many metabolic reactions. Any ADP that is formed is quickly converted to ATP by oxidative phosphorylation, and thus the cell maintains a high [ATP]/[ADP] ratio. Because of enzyme A, called nucleotide phosphokinase, some of the ATP is used to keep the [GTP]/[GDP] ratio similarly high.

Enzyme B would be highly detrimental to the cell. Cells use NAD^+ as an electron acceptor in catabolic reactions and must maintain a high $[NAD^+]/$ [NADH] ratio to support the breakdown of glucose and fats to make ATP. By contrast, NADPH is used as an electron donor in biosynthetic reactions; cells thus maintain a high $[NADPH]/[NADP^+]$ ratio so as to allow the synthesis of various biomolecules. Since enzyme B would bring both ratios to 1, it would reduce the rates of both catabolic *and* anabolic reactions.

2–82*

2–83 Reactions B, D, and E all require coupling to other, energetically favorable reactions. In each case, molecules are made that have higher-energy bonds. In A and C, however, simpler molecules (A) or lower-energy bonds (C) are made.

HOW CELLS OBTAIN ENERGY FROM FOOD

2–84*

2–85 At this rate of ATP regeneration, the cell will consume oxygen at 6.7×10^{-15} L/min [$(0.9 \times 10^9 \text{ ATP/min}) \times (1 O_2/5 \text{ ATP}) \times (22.4 \text{ L}/6 \times 10^{23} O_2) = 6.72 \times 10^{-15}$]. The volume of the cell is 10^{-12} L [$(1000 \text{ μm}^3) \times (\text{cm}/10^4 \text{ μm})^3 \times (\text{mL/cm}^3) \times$ (L/1000 mL)]. Dividing the cell volume by the rate of consumption of O_2 [$(10^{-12} \text{ L})/(6.7 \times 10^{-15} \text{ L/min})$] indicates that the cell will consume its own volume of oxygen in 149 minutes, or about 2.5 hours. (Since air contains only 20% oxygen, a cell would consume its own volume of air in about 30 minutes.)

2–86*

2–87 You would need to expend 496 kcal in climbing from Zermatt to the top of the Matterhorn, a vertical distance of 2818 m. Substituting into the equation for work

$$\text{work} = 75 \text{ kg} \times \frac{9.8 \text{ m}}{\text{sec}^2} \times 2818 \text{ m} \times \frac{\text{J}}{\text{kg m}^2/\text{sec}^2} \times \frac{\text{kcal}}{4.18 \times 10^3}$$
$$= 495.5 \text{ kcal}$$

This is equal to about 1.5 Snickers™ (496 kcal/325 kcal), so you would be well advised to plan a stop at Hörnli Hut to eat another one.

In reality the human body does not convert chemical energy into external work at 100% efficiency, as assumed in this answer, but rather at an efficiency of around 25%. Thus you would need 6 Snickers™ to make it all the way.

Reference: Frayn KN (1996) Metabolic Regulation: A Human Perspective, p 179. London: Portland Press.

2–88*

2–89 The extreme conservation of glycolysis is one form of evidence that all present cells derive from a single founder cell. In this view the elegant reactions of glycolysis would have evolved only once, and then they would have been inherited as cells evolved. The later invention of oxidative phosphorylation allowed follow-up reactions to capture 15 times more energy than is possible by glycolysis alone. This remarkable efficiency is close to the theoretical limit and hence virtually eliminates the opportunity for further improvements. The generation of alternative pathways would result in no obvious growth advantage that could have been selected in evolution.

2–90*

2–91

A. The $\Delta G°$ for conversion of 3PG to pyruvate (PYR) and phosphate is the sum of $\Delta G°$ values for the individual steps in the reaction.

$$\Delta G°_{3PG \to PYR} = \Delta G°_{3PG \to PEP} + \Delta G°_{PEP \to PYR}$$
$$= 0.42 \text{ kcal/mole} - 14.8 \text{ kcal/mole}$$
$$\Delta G°_{3PG \to PYR} = -14.4 \text{ kcal/mole}$$

B. The $\Delta G°$ for conversion of 3PG to pyruvate and phosphate is independent of the pathway for the conversion. Thus, the $\Delta G°$ is –14.4 kcal/mole.

   The $\Delta G°$ value for conversion of glycerate to pyruvate is obtained by subtracting $\Delta G°$ for 3PG to glycerate (GLY) from the overall $\Delta G°$.

$$\Delta G°_{GLY \to PYR} = \Delta G°_{3PG \to PYR} - \Delta G°_{3PG \to GLY}$$
$$= -14.4 \text{ kcal/mole} + 3.3 \text{ kcal/mole}$$
$$\Delta G°_{GLY \to PYR} = -11.1 \text{ kcal/mole}$$

C. The analysis above indicates that a very large standard free-energy change occurs between glycerate and pyruvate. Removal of water ($\Delta G° = -0.5$ kcal/mole) does not account for very much of this free-energy change. Thus it appears that the conversion of enolpyruvate to pyruvate is accompanied by a large standard free-energy change of around –10.6 kcal/mole. This reasoning suggests that the majority of the standard free-energy change associated with conversion of PEP to pyruvate (–10.6 kcal/mole out of –14.8 kcal/mole) comes from the conversion of enolpyruvate to pyruvate and not from the hydrolysis of the phosphate bond.

   In fact, the standard free-energy change for PEP to pyruvate (–14.8 kcal/mole) is close to the sum of the enolpyruvate to pyruvate step (about –11 kcal/mole) and a normal standard free-energy change for hydrolysis of a simple phosphate ester bond (about –3.0 kcal/mole). Thus, the phosphate bond in PEP is a high-energy bond because its hydrolysis is linked to the very favorable conversion of enolpyruvate to pyruvate.

D. The $\Delta G°$ for the linked conversion of PEP to pyruvate and of ADP to ATP is –7.5 kcal/mole. The $\Delta G°$ for the linked reaction can be obtained by adding together the $\Delta G°$ values for the individual reactions.

PEP → PYR + P$_i$	$\Delta G° = -14.8$ kcal/mole
ADP + P$_i$ → ATP	$\Delta G° = 7.3$ kcal/mole
NET: PEP + ADP → PYR + ATP	$\Delta G° = -7.5$ kcal/mole

Reference: Lipmann F (1941) Metabolic generation and utilization of phosphate bond energy. *Adv. Enzymol.* 1, 99–162.

2–92*

2–93   Under anaerobic conditions, cells are unable to make use of pyruvate—the end product of the glycolytic pathway—and NADH. The electrons carried in NADH are normally delivered to the electron transport chain for oxidative phosphorylation, but in the absence of oxygen the carried electrons are a waste product, just like pyruvate. Thus, in the absence of oxygen, pyruvate and NADH accumulate. Fermentation combines these waste products into a single molecule, either lactate or ethanol, which is shipped out of the cell.

The flow of material through the glycolytic pathway could not continue in the absence of oxygen in cells that cannot carry out fermentation. Because NAD$^+$ + NADH is present in cells in limited quantities, anaerobic glycolysis in the absence of fermentation would quickly convert the pool largely to NADH. The change in ratio of NAD$^+$/NADH would stop glycolysis at the step in which glyceraldehyde 3-phosphate (G3P) is converted to 1,3-bisphospho-glycerate (1,3BPG), a step with only a small negative ΔG normally (see Table 2–4). The purpose of fermentation is to regenerate NAD$^+$ by transferring the pair of carried electrons in NADH to pyruvate and excreting the product. Thus, fermentation allows glycolysis to continue.

2–94*

2–95   In the presence of arsenate, 1-arseno-3-phosphoglycerate is formed instead of 1,3-bisphosphoglycerate (Figure 2–51). Because it is sensitive to hydrolysis in water, the arsenate high-energy bond is destroyed before the molecule that contains it can diffuse to reach the next enzyme. The product of the hydrolysis, 3-phosphoglycerate, is the same product normally formed, but because it is formed nonenzymatically, the reaction is not coupled to ATP formation. Arsenate wastes metabolic energy by uncoupling many phosphotransfer reactions by the same mechanism, and that is why it is so poisonous.

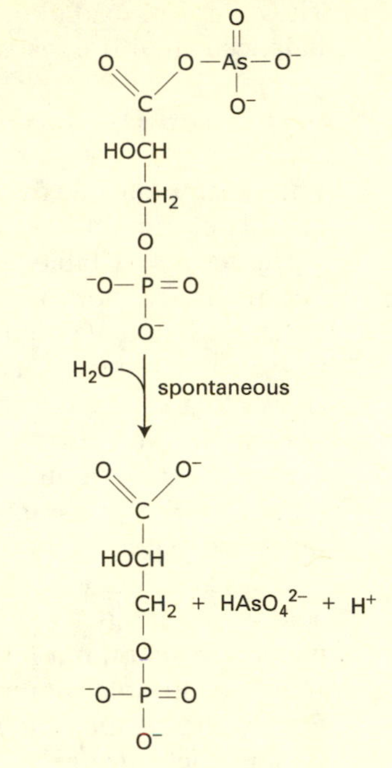

2–96*

2–97   The reverse of the forward reaction is not a feasible step for export of glucose from glycogen. It's not that it would be inefficient, it is simply not a possibility under physiological conditions. Recall from Problems 2–77 and 2–78 that flow of material through a pathway requires that the ΔG values for *every* step must be negative. Thus, for a flow from liver glycogen to serum glucose the step from glucose 6-phosphate to glucose must have a negative ΔG. To simply reverse the forward reaction (that is, G6P + ADP → GLC + ATP, $\Delta G° = 4.0$ kcal/mole) would require that the concentration ratio of [GLC][ATP]/[G6P][ADP] be less than $10^{-2.84}$ (0.0015) in order to bring the reaction to equilibrium ($\Delta G = 0$).

$$\Delta G = \Delta G° + 1.41\ \text{kcal/mole}\ \log_{10}\frac{[\text{GLC}][\text{ATP}]}{[\text{G6P}][\text{ADP}]}$$

$$= 4.0\ \text{kcal/mole} + 1.41\ \text{kcal/mole}\ \log_{10}\frac{[\text{GLC}][\text{ATP}]}{[\text{G6P}][\text{ADP}]}$$

$$= 4.0\ \text{kcal/mole} + 1.41\ \text{kcal/mole}\ \log_{10} 0.00145$$

$$= 4.0\ \text{kcal/mole} + 4.0\ \text{kcal/mole}$$

$$\Delta G = 0\ \text{kcal/mole}$$

Inside a functioning cell, such as a liver cell exporting glucose, the concentration of ATP always exceeds that of ADP, but just for illustrative purposes let's assume the ratio is 1. Under these conditions the ratio of [GLC]/[G6P] must be 0.00145; that is the concentration of G6P must be nearly 700 times higher than that of GLC. Since the circulating concentration of glucose is maintained at between 4 and 5 mM, this corresponds to about 3 M G6P, an impossible concentration given that the total concentration of cellular phosphate is less than about 25 mM.

2–98*   **Reference:** Knoop F (1905) Der Abbau aromatischer Fettsäuren im Tierkörper. *Beitr. Chem. Physiol.* 6, 150–162.

1-arseno-3-phosphoglycerate

Figure 2–51 Hydrolysis of 1-arseno-3-phosphoglycerate (Problem 2–95).

A. The complete oxidation of citrate to CO_2 and H_2O occurs according to the balanced chemical reaction.

$$C_6H_8O_7 + 4.5\ O_2 \rightarrow 6\ CO_2 + 4\ H_2O$$

Thus each molecule of citrate would require 4.5 molecules of oxygen for its complete oxidation.

The results in Table 2–5 were surprising to Krebs and others at the time because much more oxygen is consumed (40 mmol) than could be accounted for by oxidation of citrate itself. Only 13.5 mmol of oxygen would be required to oxidize 3 mmol of citrate completely (3 mmol × 4.5). This calculation shows that citrate is acting catalytically in the oxidation of carbohydrates (which in these experiments were endogenous in the minced pigeon breasts). Although others were aware of the catalytic nature of other intermediates, Krebs was the first person to complete the circle of chemical reactions that constitute the citric acid cycle.

Krebs's experimental rationale is clearly laid out in the paper: "Since citric acid reacts catalytically in the tissue, it is probable that it is removed by a primary reaction but regenerated by a subsequent reaction. In the balance sheet no citrate disappears and no intermediate products accumulate. The first object of the study of intermediates is therefore to find conditions under which citrate disappears in the balance sheet."

B. The consumption of oxygen is low in the presence of the metabolic poisons because citrate is prevented from acting catalytically. The balanced equations for the conversion of citrate to α-ketoglutarate and succinate show that the amount of oxygen consumed is approximately what is expected. For citrate conversion to α-ketoglutarate, half a molecule of oxygen is consumed.

$$C_6H_8O_7 + 0.5\ O_2 \rightarrow C_5H_6O_5 + CO_2 + H_2O$$

For citrate conversion to succinate, one molecule of oxygen is consumed.

$$C_6H_8O_7 + O_2 \rightarrow C_4H_6O_4 + 2\ CO_2 + H_2O$$

Thus the observed stoichiometry of oxygen consumption matches the expectations.

C. The absence of oxygen is crucial for demonstrating an accumulation of citrate from an intermediate in the cycle. In the presence of oxygen, citrate acts catalytically—is consumed and then regenerated—so that it does not accumulate no matter what intermediate is added. In the absence of oxygen, however, the conversion of citrate to α-ketoglutarate is blocked, since that conversion requires oxygen, as described below. Under these conditions citrate will accumulate if an appropriate intermediate is present. Of all the intermediates, only conversion of oxaloacetate to citrate does not require oxygen. The immediate precursor of oxaloacetate is malate. Since the conversion of malate to citrate requires oxygen, all other intermediates also must require oxygen to be converted to citrate.

The requirement for oxygen is indirect and is mediated through the cofactors NAD^+ and FAD; they accept electrons from the substrates and transfer them to the electron-transport chain and ultimately to oxygen. In the absence of oxygen, all of the NAD will quickly be converted to NADH, and all of the FAD will quickly be converted to $FADH_2$. In the absence of NAD^+ and FAD, the reactions of the cycle cannot proceed.

D. *E. coli* and yeast do indeed use the citric acid cycle. Krebs got this point wrong because he did not realize (nor did anyone for a long time) that citrate cannot get into these cells. Therefore, when he added citrate to intact *E. coli* and yeast, he found no stimulation of oxygen consumption. Passage of citrate across a membrane requires a transport system, which is present in mitochondria but is absent from yeast and *E. coli* plasma membranes.

References: Krebs HA & Johnson WA (1973) The role of citric acid in intermediate metabolism in animal tissues. *Enzymologia* 4, 148–156.

Stare FJ & Baumann CA (1936) The effect of fumarate on respiration. *Proc. Roy. Soc. Lon. Ser.* B121, 338–357.

Szent-Györgyi AV (1924) Über den mechanismus de Succin- und Paraphenylendiaminoxydation. Ein Betrag der Zellatmung. *Biochem Z.* 150, 141–149.

See also Albert Szent-Györgyi's Nobel Lecture (1937) at www.nobel.se/medicine/laureates/1937/szent-gyorgyi-lecture.pdf

2–100*

2–101

A. The mitochondrion obtains a flow through this reaction by maintaining high concentrations of substrates and low concentrations of products. The concentration ratio of products to substrates, $[OAA][NADH]/[MAL][NAD^+]$, must be sufficiently small to overcome a positive $\Delta G°$ value, as calculated in part B.

B. The minimum ratio of $[MAL]/[OAA]$ is the ratio that will make ΔG zero, which is 1.1×10^4.

$$\Delta G = \Delta G° + 1.41 \text{ kcal/mole} \log_{10} \frac{[OAA][NADH]}{[MAL][NAD^+]}$$

$$= 7.1 \text{ kcal/mole} + 1.41 \text{ kcal/mole} \log_{10} \frac{[OAA](1)}{[MAL](10)}$$

$$= 7.1 \text{ kcal/mole} + 1.41 \text{ kcal/mole} \log_{10} 0.1 + 1.41 \text{ kcal/mole} \log_{10} \frac{[OAA]}{[MAL]}$$

if $\Delta G = 0$,

$$\log_{10} \frac{[OAA]}{[MAL]} = \frac{-7.1 \text{ kcal/mole} - 1.41 \text{ kcal/mole} \log_{10} 0.1}{1.41 \text{ kcal/mole}}$$

$$\log_{10} \frac{[OAA]}{[MAL]} = \frac{-5.69 \text{ kcal/mole}}{1.41 \text{ kcal/mole}} = -4.04$$

$$\frac{[OAA]}{[MAL]} = 9.2 \times 10^{-5}, \text{ and thus } [MAL]/[OAA] = 1.1 \times 10^4$$

2–102*

2–103 The carbon atoms in a sugar molecule are already partially oxidized, in contrast to all but the very first carbon in a fatty acid. Thus, more electrons and more energy can be abstracted per carbon from a fatty acid than from a sugar. One consequence of this can be seen just by looking at what fraction of a fatty acid versus a sugar enters the citric acid cycle as acetyl CoA. Two carbon atoms are lost from glucose in its conversion to acetyl CoA; thus, only two-thirds of its carbons enter the citric acid cycle. By contrast, all of the carbons of fatty acids enter the citric acid cycle as acetyl CoA.

2–104*

2–105

A. The cross-feeding experiments indicate that the three steps controlled by the products of the *trpB*, *trpD*, and *trpE* genes are arranged in the order

$$X \xrightarrow{trpE} Y \xrightarrow{trpD} Z \xrightarrow{trpB} \text{tryptophan}$$

where X, Y, and Z are undefined intermediates in the pathway.

The ability of the *trpE⁻* strain to be cross-fed by the other two strains indicates that the *trpD⁻* and *trpB⁻* strains accumulate intermediates that are farther along the pathway than the step controlled by the *trpE* gene. The ability of the *trpD⁻* strain to be cross-fed by the *trpB⁻* strain but not the *trpE⁻* strain places it in the middle. The inability of the *trpB⁻* strain to be cross-fed by either of the other strains is consistent with its controlling the step closest to tryptophan.

B. The patterns of growth on minimal medium supplemented with known intermediates in the tryptophan biosynthetic pathway are consistent with the order deduced from cross-feeding

$$\text{chorismate} \xrightarrow{\textit{trpE}} \text{anthranilate} \xrightarrow{\textit{trpD}} \text{indole} \xrightarrow{\textit{trpB}} \text{tryptophan}$$

In reality, of course, the intermediates for the pathway were unknown (or not fully known) at the time the cross-feeding experiments were done. The intermediates were worked out by a combination of educated guesses at the likely intermediates, which could then be tested on mutant strains, and of analysis of the compounds that accumulated in the mutants.

Reference: Yanofsky C (2001) Advancing our knowledge in biochemistry, genetics, and microbiology through studies on tryptophan metabolism. *Annu. Rev. Biochem.* 70, 1–37.

2–106* Figure 2–52.

2–107 The citric acid cycle continues because intermediates are replenished as necessary by reactions leading to the citric acid cycle (instead of away from it). One of the most important reactions of this kind is the conversion of pyruvate to oxaloacetate by the enzyme pyruvate carboxylase (Figure 2–53).

$$\text{pyruvate} + CO_2 + ATP + H_2O \rightarrow \text{oxaloacetate} + ADP + P_i + 2\ H^+$$

2–108*

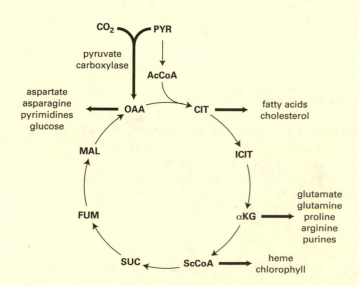

Figure 2–53 An important reaction for replenishing citric acid cycle intermediates that are removed for biosynthesis (Answer 2–107).

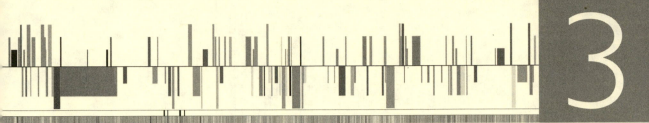

PROTEINS

THE SHAPE AND STRUCTURE OF PROTEINS

3–1   Free amino acids have an amino group and a carboxylate group, both of which are charged at neutral pH. In proteins these groups are involved in peptide bonds, which are uncharged. Thus, the hydrophobicity/hydrophilicity of a free amino acid is not the same as that of its side chain in a protein.

    To measure the hydrophobicity/hydrophilicity of the side chains, it is common to measure the properties of side-chain analogs. Thus, for alanine one would use methane, for threonine, ethanol, for aspartic acid, acetic acid, and so on. There are several ways to measure hydrophilicity and hydrophobicity. One approach for assessing hydrophilicity is to measure the solubility of the side chains in water. In general, this is done by measuring how the side-chain analog partitions between a vapor of the analog and water. Hydrophobicity can be measured by assessing the way a side-chain analog partitions between water and a nonpolar solvent such as cyclohexane. One might imagine that the rank order for hydrophilicity would be the reverse of that for hydrophobicity. From the rank order lists in Figure 3–35, that is mostly true, but there are differences, most notably tyrosine (Y) and tryptophan (W).

Reference: Creighton TE (1993) Proteins, 2nd edn, pp 153–155. New York: WH Freeman.

hydrophobicity (dark bars)		hydrophilicity (light bars)	
amino acid	kcal/mole	amino acid	kcal/mole
L	–3.98	G	0.00
I	–3.98	L	–0.11
V	–3.10	I	–0.24
F	–2.04	V	–0.40
M	–1.41	A	–0.45
W	–1.39	F	–3.15
A	–0.87	C	–3.63
C	–0.34	M	–3.87
G	0.00	T	–7.27
Y	1.08	S	–7.45
T	3.51	W	–8.27
S	4.34	Y	–8.50
H	5.60	Q	–11.77
Q	6.48	K	–11.91
K	6.49	N	–12.07
N	7.58	E	–12.63
E	7.75	H	–12.66
D	9.66	D	–13.34
R	15.86	R	–22.31

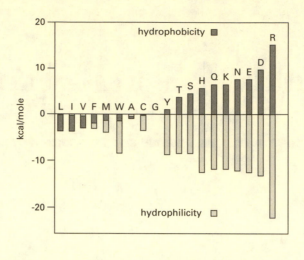

Figure 3–35 Measurements of hydrophilicity and hydrophobicity of side-chain analogs (Answer 3–1). Partition coefficients are expressed as kcal/mole. The rank order for hydrophobicity is listed in descending order, while that for hydrophilicity is listed in ascending order. If there were a perfect negative correspondence between hydrophobicity and hydrophilicity, the amino acids would appear in the same order in the two lists. The data are plotted with the amino acids arranged in order of decreasing hydrophobicity (*dark bars*) from left to right. Hydrophilicity is indicated by the overlaid *gray bars*.

3–2*

3–3 At equilibrium there would be 1 unfolded protein for every 10^7 folded proteins. This ratio comes from substituting values for $\Delta G°$ (9.9 kcal/mole), R, and T into the equation and solving for $\log K$.

$$\log K = (9.9)/[(-2.3) \times (1.98 \times 10^{-3}) \times (310)] = 10^{-7}$$

Since $K = [U]/[F]$,

$$\log K = \log ([U]/[F]) = -7$$

Taking the log of both sides,

$$[U]/[F] = 10^{-7}, \text{ or } [U] = 10^{-7} [F]$$

3–4* **Reference:** Creighton TE (1993) Proteins, 2nd edn, pp 297–300. New York: WH Freeman.

3–5

A. Heating egg-white proteins denatures them, allowing them to interact with one another in ways that were not possible at the lower temperature of the hen's oviduct. This process forms a tangled meshwork of polypeptide chains. In addition to these interactions, interchain disulfide bonds also form, so that hard-boiled egg white becomes one giant macromolecule.

B. Dissolving hard-boiled egg white requires a strong detergent to overcome the noncovalent interchain interactions and mercaptoethanol to break the covalent disulfide bonds. Together, but not separately, the two reagents eliminate the bonds that hold the tangled protein chains in place. Try it for yourself!

3–6* **Reference:** Creighton TE (1993) Proteins, 2nd edn, pp 288–289. New York: WH Freeman.

3–7 True. In both states—stretched like a string and properly folded—a protein has a highly ordered arrangement of it atoms. A folded protein is stable at a near entropy minimum because the entropic cost is more than balanced by the contributions of weak bonds. A stretched out protein, however, is not stable at this entropy minimum and will assume a more disordered state; that is, maximize its entropy.

3–8* Figure 3–36.

3–9 The ends of α helices, like polar amino acids, are almost always found at the surface of a protein where they can interact with polar water molecules. In addition to their partial charge, the backbones of the four amino acids at either end of the helix carry hydrogen-bonding groups that are unsatisfied by hydrogen bonding within the helix (Figure 3–37). These groups also add to the polarity of the termini of α helices.

3–10* Figure 3–38.

3–11 As illustrated in Figure 3–39, the first three strands of the sheet are antiparallel to their neighbors, whereas the fourth strand is parallel to the third.

3–12*

3–13

    D. Because the side chains of the amino acids alternately project above and below the sheet, a sequence that could form a strand in an amphipathic β sheet should have alternating hydrophobic and hydrophilic amino acids. Only choice D satisfies this condition.

3–14*

3–15 Antiparallel strands are commonly formed by a polypeptide chain that folds back on itself. Thus, only a few amino acids are required to allow the polypeptide chain to make the turn. By contrast, parallel strands must be connected by a polypeptide chain that is at least as long as the strands. For a long peptide, a common solution for satisfying backbone hydrogen-bonding requirements is an α helix (Figure 3–40).

3–16*

3–17 True. Protruding loops allow a large number of chemical groups to surround a molecule so the protein can link to it with many weak bonds.

3–18* **Reference:** Fersht A (1999) Structure and Mechanism in Protein Science, pp 575–600. New York: WH Freeman.

3–19 Since there are 20 possible amino acids at each position in a 300 amino acid long protein, there are 20^{300} or 10^{390} possible proteins. The mass of one copy of each possible protein would be

$$\text{mass} = \frac{110\ \text{d}}{\text{aa}} \times \frac{300\ \text{aa}}{\text{protein}} \times 10^{390}\ \text{protein} \times \frac{\text{g}}{6 \times 10^{23}\ \text{d}}$$

$$\text{mass} = 5.5 \times 10^{370}\ \text{g}$$

Thus, the mass of protein would exceed the mass of the observable universe (10^{80} g) by a factor of about 10^{290}!!

Figure 3–37 Representation of an α helix showing dipole and unsatisfied hydrogen bonding groups at its ends (Answer 3–9). Non-hydrogen bonded Os and Ns are labeled.

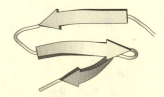

from lysozyme

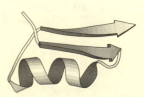

from alcohol dehydrogenase

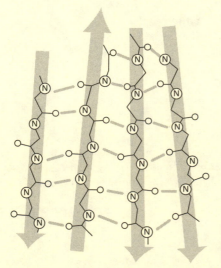

Figure 3–39 A segment of β sheet showing the polarity (N to C) of the individual strands (Answer 3–11).

Figure 3–40 Connections between parallel and antiparallel strands of a β sheet (Answer 3–15).

3–21   The statement is correct. Evolutionary separation is twice the time from the last common ancestor, that is, counting backward from one kind of organism to the common ancestor and then forward to the other kind of organism.

3–22*   Figure 3–41.

3–23   Perhaps this is so. Nevertheless, it seems likely that new, and useful, protein folds have been invented during evolution by the chance fusion of genes. A distribution of protein folds within the tree of life would be informative. Protein folds that are distributed in all divisions of the tree—archaea, eubacteria, and eucaryotes—were undoubtedly present in the last common ancestor. More recently invented protein folds would likely be confined to a single division, or a few branches. Even if all protein folds were present in the last common ancestor, it seems unlikely that there would be a one-to-one correspondence between folds and genes. Surely, the evolution that led up to the last common ancestor would already have exploited some of the benefits of gene duplication and refinement of function that lead to families of related genes.

  The limited number of protein folds raises a more fundamental question about the total number of protein folds that are possible. It may be that evolutionary processes have already exploited most of the stable folds that are possible. Alternatively, it may be that the number of possible stable folds greatly exceeds those currently used on Earth. The simple cataloguing of natural protein folds will not address this more basic question.

3–24*

3–25   False. Although the technique of threading has proven useful in the special case of an amino acid sequence that is homologous to a family of proteins with a known fold, it is not yet robust enough to deal with the more general problem of testing an amino acid sequence against all possible folds. It is thought that in time this will be so, but at present it is limited in two ways: the computational methods for threading are not adequate and not all the natural protein folds are known.

3–26*

3–27   As shown in Figure 3–42, the three protein monomers have distinctly different assembly properties because of the three-dimensional arrangement of their complementary binding surfaces. Monomer A would assemble into a sheet; monomer B would assemble into a long chain; monomer C would assemble into a ring composed of four subunits.

3–28*

3–29   'Head-to-tail' dimers have unsatisfied binding sites at each end, which would lead to formation of chains (see Figure 3–42B).

3–30*   Figure 3–43.

3–31   The coil 1A segment of nuclear lamin C matches the heptad repeat at 9 of 11 positions (Figure 3–44), which is very good. The match need not be perfect to allow formation of a coiled-coil. The matches to the heptad repeat in the

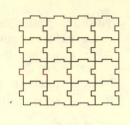

(A)

(B)

(C)

Figure 3–42 Assembly of protein monomers (Answer 3–27).

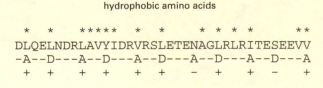

hydrophobic amino acids

```
 *    *    *****  *    *     *  *  *    **
DLQELNDRLAVYIDRVRSLETENAGLRLRITESEEVV
-A--D---A--D---A--D---A--D---A--D---A
 +  +  +  +    +  +    -  +   +  -    +
```

match with heptad repeat

Figure 3–44 Heptad repeat motif in the coil 1A region of nuclear lamin C (Answer 3–31). Hydrophobic amino acids are marked with an asterisk (*). When a hydrophobic amino acid occurs at an A or D in the heptad repeat, it is assigned a +. The start of the heptad repeat was positioned to maximize matches.

other two marked segments (coil 1B and coil 2, see Figure 3–10) are not as good, but they are still acceptable for formation of a coiled-coil.

Reference: McKeon FD, Kirschner MW & Caput D (1986) Homologies in both primary and secondary structure between nuclear envelope and intermediate filament proteins. *Nature* 31, 463–468.

3–32* Figure 3–45.

Reference: Rief M, Gutel M, Oesterhelt F, Fernandez JM & Gaub HE (1997) Reversible folding of individual titin immunoglobulin domains by AFM. *Science* 276, 1109–1112.

3–33

A. None are detected in this experiment. Treating first with radiolabeled NEM shows that many cytosolic proteins have cysteines that are not linked by disulfide bonds. Treating first with unlabeled NEM to block these sites, followed by DTT to break disulfide bonds, should expose any sulfhydryls that were linked by disulfide bonds. These newly exposed sulfhydryls should be labeled by subsequent treatment with radiolabeled NEM. The absence of labeling indicates that no cysteines were involved in disulfide bonds.

B. BSA and insulin are labeled extensively only after their disulfide bonds have been broken by treatment with DTT. In the absence of DTT treatment BSA is weakly labeled. Since BSA has an odd number of cysteines, at least one cannot be involved in disulfide bonds, and structural analysis confirms that only one of its 37 cysteines is not involved in a disulfide bond.

C. Because the ER is the site where disulfide bond formation is catalyzed in preparation for export of proteins, it is expected that lysates from cells that have internal membranes would have many proteins with disulfide bonds.

3–34*

3–35

A. As calculated in the previous problem, synthesis of one 10,000 amino acid protein would be expected to occur correctly 37% of the time. By contrast, synthesis of each 200 amino acid subunit would be expected to occur correctly 98% of the time $[P_C = (f_C)^n = (0.9999)^{200} = 0.98]$. Assembly of the mixture of subunits into correct ribosomes follows the same equation $[P_C = (f_C)^n = (0.98)^{50} = 0.37]$. Thus, making a ribosome from subunits gives the same final fraction of correct ribosomes, 37%, as making them from one long protein.

B. The assumption in part A that correct and incorrect subunits are assembled into ribosomes with equal likelihood is not true. Any mistake that interferes with the correct folding of a subunit, or that interferes with the ability of the subunit to bind to other subunits, would eliminate the subunit from assembly into a ribosome. As a result, the fraction of correctly assembled ribosomes would be higher than calculated in part A. Thus, the value of subunit synthesis lies not in more accurate synthesis, but rather in permitting quality control mechanisms to reject incorrect subunits efficiently.

3–36* Figure 3–46.

PROTEIN FUNCTION

3–37 Antifreeze proteins function by binding to tiny ice crystals and arresting their growth, thereby preventing the fish from freezing. Ice crystals that form in the presence of antifreeze proteins are abnormal in that their surfaces are curved instead of straight. The various forms of the antifreeze proteins in these fishes are all composed of repeats of a simple glycotripeptide (Thr-Ala/Pro-Ala) with a disaccharide attached to each threonine. The genes for these antifreeze proteins were apparently derived by repeated duplication of a small segment of a protease gene.

References: Cheng CHC & Chen L (1999) Evolution of an antifreeze protein. *Nature* 401, 443–444.

Jia Z & Davies PL (2002) Antifreeze proteins: an unusual receptor–ligand interaction. *Trends Biochem. Sci.* 27, 101–106.

3-38* Figure 3–47.

3–39 False. The pKs of specific side-chain groups depend critically on their environment. On the surface of a protein, in the absence of surrounding charged groups, the pK of a carboxyl group is usually close to that of the free amino acid. In the neighborhood of negatively charged groups, the pK of a carboxyl group is usually higher; that is, the proton dissociates less readily since the increase in local density of negative charge is not favored. The opposite is true in a positively charged environment. In a hydrophobic environment the dissociation of a proton can be substantially suppressed, since the presence of a naked charge in such an environment is highly disfavored. It is this ability to alter the reactivities of individual groups that allows proteins to fine-tune their biological functions.

3–40*

3–41
A. Your results support the idea that the PI3-kinase interacts with the activated PDGF receptor through its SH2 domains. The interaction is blocked specifically by the phosphorylated pentapeptides 708 and 719. In their nonphosphorylated forms these same pentapeptides do not block the association.
B. The common features of the peptides that can bind to PI3-kinase are a phosphotyrosine and a methionine (M) located three positions away in the C-terminal direction. (Although not shown explicitly here, there seems to be no requirement for specific amino acids on the N-terminal side of the phosphotyrosine.)
C. Recognition of a couple of amino acids in a short sequence is characteristic of a string–surface interaction. Indeed, recognition of sequences by SH2 domains is often cited as a prime example of such an interaction.

3–42* Figure 3–48.

3–43
A. The equilibrium constant, K, equals 10^6 M^{-1}.

$$K = \frac{[AB]}{[A][B]} = \frac{10^{-6}\,M}{(10^{-6}\,M) \times (10^{-6}\,M)}$$

$$K = 10^6\,M^{-1}$$

B. The same calculation as above, when each component is present at 10^{-9} M, gives an equilibrium constant of 10^9 M^{-1}.
C. This example illustrates that interacting cellular proteins present at low concentrations need to bind to each other with high affinities if a high proportion of the molecules are to be bound together. A three-order of magnitude decrease in the equilibrium constant corresponds to a free-energy difference of about –4.2 kcal/mole. Thus, effective binding at the lower concentration would require the equivalent of 4–5 extra hydrogen bonds. The free-energy difference between the two equilibrium constants can be calculated. For an equilibrium constant of 10^6 M^{-1}

$$\Delta G° = -2.3\,RT \log_{10} K$$

Substituting,

$$\Delta G° = -1.41\,\text{kcal/mole} \times (6)$$

$$\Delta G^\circ = -8.46 \text{ kcal/mole}$$

For an equilibrium constant of 10^9 M^{-1}

$$\Delta G^\circ = -1.41 \text{ kcal/mole} \times (9)$$
$$\Delta G^\circ = -12.69 \text{ kcal/mole}$$

Thus, the higher equilibrium constant corresponds to a free-energy difference that is 4.2 kcal/mole more negative. To supply this amount of binding energy with hydrogen bonds (about 1 kcal/mole) would require about 4–5 extra hydrogen bonds.

3–44*

3–45   The immunoblot shows that antibodies BPA1 and BPA2 react specifically with a 220-kd protein, which is likely to be Brca1. By contrast, although antibody C20 reacts with the same protein, it seems to react even more strongly with a second protein of about 180 kd. Thus, a likely explanation for the contradictory cell-localization experiments is that C20 antibodies were identifying the location of the 180-kd protein, whereas BPA1 and BPA2 were showing the location of Brca1. Brca1 is now thought to function in the nucleus. Additional experiments have identified the epidermal growth factor (EGF) receptor as the protein with which C20 cross-reacts. The EGF receptor has a couple of regions of similarity to the peptide that was used to generate the C20 antibody. Cross-reactivity of antibodies is a not uncommon problem. For this reason, cell-localization studies are usually performed with antibodies raised against more than one region of a protein. Agreement with different antibodies decreases the likelihood that cross-reactivity is a problem.

References: Jensen RA, Thompson ME, Jetton TL, Szabo CI, van der Meer R, Helou B, Tronick SR, Page DL, King MC & Holt JT (1996) BRCA1 is secreted and exhibits properties of a granin. *Nature Genet.* 12, 303–308.

Thomas JE, Smith M, Rubinfeld B, Gutowski M, Beckmann RP & Polakis P (1996) Subcellular localization and analysis of apparent 180-kDa and 220-kDa proteins of the breast cancer susceptibility gene, BRCA1. *J. Biol. Chem.* 271, 28630–28635.

3–46*

3–47
   A. The slopes $(-1/K_d)$ of the lines in Figure 3–18 can be estimated by taking the difference between two points on the y axis divided by the difference between the corresponding points on the x axis. Thus, the slope of line A is

$$-1/K_d = (0.08 - 0.35)/(5 \times 10^{-7} \text{ M} - 1 \times 10^{-7} \text{ M})$$
$$= -6.75 \times 10^5 \text{ M}^{-1}$$
$$K_d = 1.5 \times 10^{-6} \text{ M}$$

For line B,

$$-1/K_d = (0.20 - 0.90)/(5 \times 10^{-7} \text{ M} - 1 \times 10^{-7} \text{ M})$$
$$K_d = 5.7 \times 10^{-7} \text{ M}$$

The precise values are dependent on your estimate of the corresponding values on the x axis.
   B. The lower the K_d, the tighter the binding; thus, the tighter IPTG-binding mutant of the lactose repressor corresponds to line B ($K_d = 5.7 \times 10^{-7}$ M) and the wild-type lactose repressor corresponds to line A ($K_d = 1.5 \times 10^{-6}$ M). That a lower value corresponds to tighter binding is apparent from the definition of K_d in the problem. Tighter binding will give more complex (Pr–L) and fewer free components (Pr + L), thus, the ratio of concentrations, K_d, will be smaller.

References: Gilbert W & Muller-Hill B (1996) Isolation of the Lac Repressor. *Proc. Natl. Acad. Sci. USA* 56, 1891–1898.

Kyte J (1995) Mechanism in Protein Chemistry, pp 175–177. New York: Garland Publishing, 1995.

3–48* **Reference:** Karzai AW, Susskind MM & Sauer RT (1999) SmpB, a unique RNA-binding protein essential for the peptide-tagging activity of SsrA (tmRNA). *EMBO J.* 18, 3793–3799.

3–49 The calculated values of fraction bound versus protein concentration are shown in Table 3–9. Also shown are the rule-of-thumb values that are easier to remember.

TABLE 3–9 Calculated values for fraction bound versus protein concentration (Answer 3–49).

[PROTEIN]	FRACTION BOUND (%)	RULE-OF-THUMB
$10^4 K_d$	99.99	99.99
$10^3 K_d$	99.9	99.9
$10^2 K_d$	99	99
$10^1 K_d$	91	90
K_d	50	50
$10^{-1} K_d$	9.1	10
$10^{-2} K_d$	0.99	1
$10^{-3} K_d$	0.099	0.1
$10^{-4} K_d$	0.0099	0.01

These relationships are useful not only for thinking about K_d, but also for enzyme kinetics, which we cover in later problems. The rate of a reaction expressed as a fraction of the maximum rate is

$$\frac{\text{rate}}{V_{\max}} = \frac{[S]}{[S] + K_M}$$

which has the same form as the equation for fraction bound. Thus, for example, when the concentration of substrate, [S], is 10-fold above the Michaelis constant, K_M, the rate is 90% of the maximum, $V_{\max}$. When [S] is 100-fold below K_M, the rate is 1% $V_{\max}$.

The relationship also works for the fractional dissociation of an acidic group, HA, as a function of pH. When the pH is 2 units above the pK, 99% of the acidic group is ionized. When the pH is 1 unit less than pK, 10% is ionized.

3–50*

3–51 The problem is that the off rate for the antibody–enzyme complex is too slow. In order for the peptide to displace the enzyme from the column, the enzyme must first dissociate from the antibody. The antibody binding sites would then be quickly bound by the peptide, which in high concentration would prevent the enzyme from reattaching to the antibody (any newly exposed antibody binding site would be bound by peptide rather than enzyme). In principle, you could soak the column with peptide for several days (for several dissociation half-times, see Problem 3–46), but this usually has adverse consequences for the quality and activity of the enzyme preparation. In general, high-affinity antibodies have slow off rates and are unsuitable for affinity chromatography.

Special procedures have been devised for preparing or identifying antibodies that work in such experiments. Usually, lower-affinity antibodies are used, or chromatography is carried out under special conditions that reduce the affinity of the antibody.

Reference: Thompson NE & Burgess RR (1996) Immunoaffinity purification of RNA polymerase II and transcription factors using polyol-responsive monoclonal antibodies. *Methods Enzymol.* 274, 513–526.

3–52* **References:** Pierschbacher MD & Ruoslahti E (1984) Cell attachment activity of fibronectin can be duplicated by small synthetic fragments of the molecule. *Nature* 309, 30–33.

Pytela R, Pierschbacher MD & Ruoslahti E (1985) Identification and isolation of a 140 kd cell surface glycoprotein with properties expected of a fibronectin receptor. *Cell* 40, 191–198.

Ruoslahti E & Pierschbacher MD (1986) Arg-Gly-Asp: a versatile cell recognition signal. *Cell* 44, 517–518.

3–53

A. The fraction of the phage population that will be attached by at least one tail fiber at any one instant is equal to one minus the fraction not attached by any tail fibers, which is $(0.5)^{12} = 0.00024$ for wild-type bacteria and $(0.5)^6 = 0.016$ for $ompC^-$ bacteria. Thus, at any instant 99.98% of the phage population will be attached to wild-type bacteria and 98.4% will be attached to $ompC^-$ bacteria.

B. The very small difference in the fraction of the phage population attached to wild-type and $ompC^-$ bacteria at first seems too little to account for the 1000-fold difference in infectivity. However, since T4 must wander around the surface of a bacterium to find an appropriate place to attach its baseplate, the instantaneous calculation is misleading. If, for example, T4 must stay bound to the bacterial surface for 500 'instants' during its wandering, then $(0.9998)^{500} = 90\%$ will remain attached to wild-type bacteria, but only $(0.984)^{500} = 0.03\%$ will remain attached to $ompC^-$ bacteria. This difference would be more than enough to account for the 1000-fold difference in infectivity.

By associating with the bacterial surface through multiple weak interactions, bacteriophage T4 can wander around the surface without falling off. This allows a search for relatively rare injection sites, which are at points of connection between the inner and outer membranes.

Reference: Goldberg E (1983) Recognition, attachment, injection. In Bacteriophage T4 (CK Mathews, EM Kutter, G Mosig, PB Berget eds.), pp 32–39. Washington, DC: American Society for Microbiology.

3–54*

3–55 The reaction rate for the altered enzyme would be substantially slower than for the normal enzyme. The reaction rate is related to the free-energy difference (ΔG) between the trough labeled ES in Figure 3–22 and the transition state: the larger the free-energy difference the slower the rate. If the altered enzyme bound the substrate with higher affinity (a lower ES trough), then the free-energy difference between ES and the transition state would be increased; hence, the reaction would occur at a slower rate.

3–56*

3–57

A. It is important that only a small quantity of product is made, because otherwise the rate of the reaction would decrease as the substrate was depleted and product accumulated. Thus, the measured rates would be lower than they should be, and the kinetic parameters would be incorrect.

B. The Michaelis–Menten plot, shown in Figure 3–49 is a rectangular hyperbola, as expected if this enzyme obeys Michaelis–Menten kinetics. To determine values for K_M and V_{max} from this plot by visual inspection, you must estimate the rate at infinite substrate concentration. From the curve of the line in the figure you might reasonably estimate the V_{max} as anywhere from 1.8 to 2.0 µmol/min. (As developed in the answer to Problem 3–49, a useful rule of thumb is that at a concentration of substrate 10-fold above K_M, the rate is about 90% of V_{max}.) If you chose 2.0 µmol/min, then 1/2 V_{max} (1.0 µmol/min) corresponds to a substrate concentration of 1.0 µM, which is the value of K_M. It's the visual uncertainty in this plot that led early researchers to transform the equation into a straight-line form so a line could be fitted to the data and the kinetic parameters could be more accurately determined.

C. As indicated in the Lineweaver–Burke plot in Figure 3–49, the y intercept is 0.5 ($1/V_{max}$) and the x intercept is –1.0 ($-1/K_M$). Thus, V_{max} equals 2.0 µmol/min and K_M equals 1.0 µM. Although this form of a straight-line plot is commonly discussed in textbooks, it is rarely used in practice because the data points that are most reliable are tightly grouped at one end of the line. Consequently, the slope of the line is unduly influenced by the low (and usually less accurately determined) rates at low substrate concentration. Other straight-line transformations of the Michaelis–Menten equation such as the

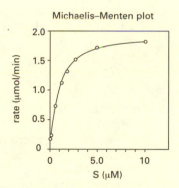

Michaelis–Menten plot

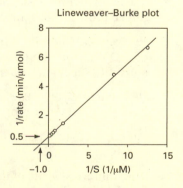

Lineweaver–Burke plot

Figure 3–49 Michaelis–Menten and Lineweaver–Burke plots of the data in Table 3–6 (Answer 3–57). The x and y intercepts are indicated on the Lineweaver–Burke plot.

Eadie–Hofstee plot, which is analogous in form to the Scatchard plot shown in Problem 3–47, are generally preferred. In this era of computers, however, the data can be fit perfectly well to the nonlinear Michaelis–Menten equation, although it is still common to present such data in a linear form.

3–58*

3–59

D. Because an enzyme has a fixed number of active sites, the rate of the reaction cannot be further increased once the substrate concentration is sufficient to bind to all the sites. It is the saturation of binding sites that leads to an enzyme's saturation behavior. The other statements are all true, but none is relevant to the question of saturation.

3–60*

3–61

A. Since k_1 corresponds to the on rate and k_2 corresponds to the off rate

$$K_d = [E][S]/[ES] = k_{off}/k_{on} = k_2/k_1$$

B. K_M is approximately equal to K_d when k_3 is much less than k_2; that is to say, when the ES complex dissociates much more rapidly than substrate is converted to product. This is true for many enzymes, but not all.

C. Because k_3 is in the numerator of the expression, K_M will always be somewhat larger that K_d. Since lower values of K_d indicate higher binding affinity, K_M will always underestimate the binding affinity. When k_3 is much less than k_2, the underestimate will be slight and K_M will be essentially equal to K_d.

3–62*

3–63 False. The turnover number is constant since it is V_{max} divided by enzyme concentration. For example, a 2-fold increase in enzyme concentration would give a 2-fold higher V_{max}, but it would give the same turnover number: $2 V_{max}/2 [E] = k_3$.

3–64*

3–65

A. An enzyme composed entirely of mirror-image amino acids would be expected to fold stably into a mirror-image conformation; that is, it would look like the normal enzyme when viewed in a mirror.

B. A mirror-image enzyme would be expected to recognize the mirror image of its normal substrate. Thus, 'D' hexokinase would be expected to add a phosphate to L-glucose and ignore D-glucose.

This experiment has actually been done for HIV protease. The mirror-image protease recognizes and cleaves a mirror-image substrate.

Reference: Milton RC, Milton SC & Kent SB (1992) Total chemical synthesis of a D-enzyme: the enantiomers of HIV-1 protease show reciprocal chiral substrate specificity. *Science* 256, 1445–1448.

3–66* **References:** Kyte J (1995) Mechanism in Protein Chemistry, pp 207–208. New York: Garland Publishing.

Pauling L (1948) Chemical achievement and hope for the future. *Am. Sci.* 36, 50–58.

3–67 The numerical value of the product of the K_ds for the substrates (3.0×10^{-7}) is 9-fold less than that of the K_d for PALA (2.7×10^{-8}), suggesting that PALA might be a transition-state analog. PALA, however, is composed not of aspartate plus carbamoyl phosphate, but of succinate plus a close analog of carbamoyl phosphate. If one substitutes the K_d for succinate, 0.9 mM, into the calculation, the product of the K_ds is 2.4×10^{-8}, which is very close to the K_d for PALA. Thus, PALA is likely to be a bisubstrate analog rather than a transition-state analog.

You may have wondered whether it is valid to compare K_ds in this way. Recall that $\Delta G^\circ = -2.3RT\log_{10}K$. Using this equation, we could have converted K_ds to ΔG°s and compared the sum of ΔG°s for aspartate and carbamoyl phosphate with the ΔG° for PALA. However, since ΔG° is proportional to $\log_{10}K$, this is equivalent to comparing the products of the K_ds for aspartate and carbamoyl phosphate with the K_d for PALA. Do the calculation for ΔG°s and convince yourself that this is true.

Reference: Fersht A (1999) Structure and Mechanism in Protein Science, pp 360–361. New York: W.H. Freeman.

3–68* **Reference:** Wahl GM, Padgett RA & Stark GR (1979) Gene amplification causes overproduction of the first three enzymes of UMP synthesis in *N*-(phosphonacetyl)-L-aspartate (PALA)–resistant hamster cells. *J. Biol. Chem.* 254, 8679–8689.

3–69

A. Amino acid side chains in proteins often have quite different pKs than they do in solution. Glu 35 is uncharged because its local environment is nonpolar, which makes ionization less favorable (raises its pK). The local environment of Asp 52 is more polar, permitting ionization near its solution pK.

B. As the pH drops below 5, Asp 52 picks up a proton and become nonionized, interfering with the mechanism. As the pH rises above 5, Glu 35 begins to release its proton, also interfering with the mechanism.

3–70*

3–71 Water from rusty pipes provides iron, which is essential for all forms of life. Egg white contains a special protein, ovotransferrin, which binds iron very tightly, analogous to the binding of biotin by avidin. Washing eggs in rusty water allows iron to enter in sufficient quantities to exceed the binding capacity of ovotransferrin, thereby making free iron available to the microorganisms.

3–72*

3–73 Although the *rate* of diffusion cannot be altered by changes to the enzymes, the average *distance* over which a molecule must diffuse can be altered. Linking the two enzymes together decreases the average distance for diffusion of the first product to the second enzyme. A decrease in the distance of diffusion decreases the *time* for diffusion and, thus, increases the overall rate of the reactions catalyzed by the pair of enzymes.

3–74*

3–75

A. The relative concentrations of the normal and mutant v-Src proteins are inversely proportional to the volumes in which they are distributed. The mutant v-Src is distributed throughout the volume of the cell, which is

$$V_{cell} = 4/3\,\pi r^3 = 4/3\,\pi\,(10 \times 10^{-6}\text{ m})^3 = 4.1888 \times 10^{-15}\text{ m}^3$$

Normal v-Src is confined to the 4-nm layer beneath the membrane, which has a volume equal to the volume of the cell minus the volume of a sphere with a radius 4 nm less than that of the cell.

$$\begin{aligned}
V_{layer} &= V_{cell} - 4/3\,\pi\,(r - 4\text{ nm})^3 \\
&= V_{cell} - 4/3\,\pi\,[(10 \times 10^{-6}\text{ m}) - (4 \times 10^{-9}\text{ m})]^3 \\
&= (4.1888 \times 10^{-15}\text{ m}^3) - (4.1838 \times 10^{-15}\text{ m}^3) \\
V_{layer} &= 0.0050 \times 10^{-15}\text{ m}^3
\end{aligned}$$

Thus, the volume of the cell is 838 times greater than the volume of a 4-nm-thick layer beneath the membrane (4.1888×10^{-15} m^3/0.0050×10^{-15} m^3).

Even allowing for the interior regions of the cell from which it would be excluded (nucleus and organelles), the mutant v-Src would still be a couple of orders of magnitude less concentrated in the neighborhood of the membrane than the normal v-Src.

B. Its lower concentration in the region of its target X at the membrane is the reason why mutant v-Src does not cause cell proliferation. This notion can be quantified by a consideration of the binding equilibrium for v-Src and X.

$$v\text{-Src} + X \rightarrow v\text{-Src–X}$$
$$K = \frac{[v\text{-Src–X}]}{[v\text{-Src}]\,[X]}$$

The lower concentration of the mutant v-Src in the region of the membrane will shift the equilibrium toward the free components, reducing the amount of complex. If the concentration is on the order of 100-fold lower, the amount of complex will be reduced up to 100-fold. Such a large in decrease in complex formation could readily account for the lack of effect of the mutant v-Src on cell proliferation.

3–76*

3–77 One reasonable proposal would be for excess AMP to feedback inhibit the enzyme for converting *E* to *F*, and excess GMP to feedback inhibit the step from *E* to *H*. Intermediate *E*, which would then accumulate, would be proposed to feedback inhibit the step from R5P to *A*. Some branched pathways are regulated in just this way. Purine nucleotide synthesis is regulated somewhat differently, however (Figure 3–50). AMP and GMP regulate the steps from *E* to *F* and from *E* to *H*, as above, but they also regulate the step from R5P to *A*. Regulation by AMP and GMP at this step might seem problematical since it suggests that a rise in AMP, for example, could shut off the entire pathway even in the absence of GMP. The cell uses a very clever trick to avoid this problem. Individually, excess AMP or GMP can inhibit the enzyme to about 50% of its normal activity; together, however, they can completely inhibit it.

3–78*

3–79 True. The term cooperativity embodies the idea that changes in the conformation of one subunit are communicated to the other subunits so that all of the subunits in any given molecule are in the same conformation.

3–80*

3–81 The first MWC postulate, which states that the subunits are arranged symmetrically, rules out all arrangements except those shown in the left-most and right-most columns of the diagram. If ligand binds much more tightly to circles, then the allowed arrangements are those shown in Figure 3–51. If the ligand binds equally to both subunit conformations, then all the arrangements in the left-most and right-most columns would be allowed, consistent with MWC postulate 1.

Detailed studies on a few cooperative enzymes such as aspartate transcarbamoylase have found no evidence for intermediate, nonsymmetrical conformations.

3–82* **References:** Cantor CR & Schimmel PR (1980) Biophysical Chemistry, pp 944–945. New York: WH Freeman.

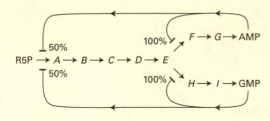

Figure 3–50 Pattern of inhibition in the metabolic pathway for purine nucleotide synthesis (Answer 3–77).

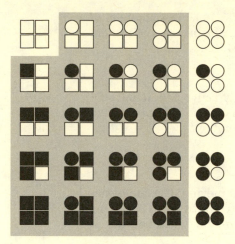

Figure 3–51 Arrangements of subunit conformations that are consistent with the MWC postulates (Answer 3–81). *Shaded* area indicates those arrangements that are excluded by the MWC postulates for a ligand with affinity to one conformation of subunit (*circle*).

Gerhart JH & Pardee AB (1963) The effect of the feedback inhibitor, CTP, on subunit interactions in aspartate transcarbamylase. *Cold Spring Harbor Symp. Quant. Biol.* 28, 491–496.

3–83 These changes are exactly what you would expect for an allosteric enzyme such as ATCase. Because binding at one active site (one of six per ATCase molecule) is sufficient to shift the conformation of a molecule of ATCase, the change in global conformation (change in sedimentation) is expected to 'lead' the change in occupancy of binding sites (change in spectral measurement).

References: Cantor CR & Schimmel PR (1980) Biophysical Chemistry, pp 954–956. New York: W.H. Freeman.

Kirschner MW & Schachman HK (1973) Local and gross conformational changes in aspartate transcarbamylase. *Biochemistry* 12, 2997–3004.

3–84*

3–85 True. Each cycle of phosphorylation–dephosphorylation hydrolyzes one molecule of ATP; however, it is not wasteful in the sense of having no benefit. Constant cycling allows the regulated protein to switch quickly from one state to another in response to stimuli that require rapid adjustments of cellular metabolism or function. This is the essence of effective regulation.

3–86* **Reference:** Brown NR, Noble MEM, Lawrie AM, Morris MC, Tunnah P, Divita G, Johnson LN & Endicott JA (1999) Effects of phosphorylation of threonine 160 on cyclin-dependent kinase 2 structure and activity. *J. Biol. Chem.* 274, 8746–8756.

3–87
A. Mutant 2 is Asp-181 → Ala (D181A). It is the best candidate because it has a K_M close to that of the wild-type enzyme, but a very low turnover number (k_{cat}). With these kinetic parameters it might be expected to bind normally to its target substrates but not remove phosphate from tyrosine. The next most likely candidate would be Arg-221→ Lys, which has a slightly lower K_M than wild-type PTP1B and turns over slowly, although about 20 times faster than D181A. (Further studies identified the band at 180 kd as the epidermal growth factor—EGF—receptor.)
B. C215S showed no activity as expected since the sulfhydryl of cysteine is required for catalysis. Since it gave no activity, however, it was not possible to determine its K_M, which might have been similar to the wild-type enzyme. (Measurements of K_d were not made.) Thus, C215S was a reasonable candidate to test. Lack of success with C215S suggests that it binds phosphotyrosine-containing proteins very poorly.

Reference: Flint AJ, Tiganis T, Barford D & Tonks NK (1997) Development of substrate-trapping mutants to identify physiological substrates of protein tyrosine phosphatases. *Proc. Natl. Acad. Sci. USA* 94, 1680–1685.

3–88*

3–89   A nonfunctional GAP (choice A) or a permanently active GEF (choice D) would allow Ras to remain in the active state (with GTP bound) longer than normal, and thus might cause excessive cell proliferation.

3–90*

3–91

A. In the absence of ATP a motor protein would stop moving. The conformational shifts that are required for movement are triggered by ATP binding and hydrolysis. In the absence of ATP the motor protein would be stuck in its lowest-energy conformation.

B. If the free-energy change for the hydrolysis of ATP by the motor protein was zero—conditions under which ATP is as easily made as hydrolyzed—the motor protein would wander back and forth. With zero free-energy change there would be no barrier between conformations.

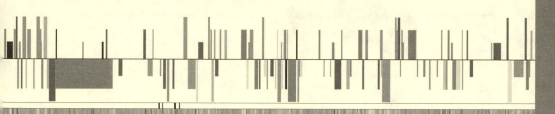

DNA AND CHROMOSOMES

THE STRUCTURE AND FUNCTION OF DNA

4–1 The complementary strand for this DNA is 5′-GTGCACCAT-3′. By convention, all DNA strands are written 5′ to 3′, so that the complement of 5′-ATGGTGCAC is 5′-GTGCACCAT. Keeping polarities in mind is a key to following DNA structure, DNA replication, DNA transcription, DNA repair, and recombination—virtually all aspects of DNA metabolism.

4–2*

4–3 The base pairs are TA (see Figure 4–1A) and CG (see Figure 4–1B). The components of the base pairs, along with their stick representations, are shown in Figure 4–31. Thymine and cytosine are pyrimidines; adenine and guanine are purines.

4–4* Figure 4–32.

4–5 Because C always pairs with G in duplex DNA, their mole percents must be equal. Thus, the mole percent of G, like C, is 20%. The mole percents of A and T account for the remaining 60%. Since A and T always pair, their mole percents are equal to half this value: 30%.

4–6*

4–7 The segment of DNA in Figure 4–3 reads from top to bottom 5′-ACT-3′. The carbons in the ribose sugar are numbered clockwise around the ring, starting with C1′, the carbon to which the base is attached, and ending with C5′, the carbon that lies outside the ribose ring.

---

*Answers to these problems are available to Instructors at www.classwire.com/garlandscience/

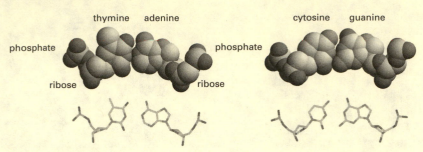

(A) TA BASE PAIR

thymine   adenine

phosphate

ribose   ribose

(B) CG BASE PAIR

cytosine   guanine

phosphate

Figure 4–31 Space-filling and stick representations of two base pairs (Answer 4–3). (A) TA base pair. (B) CG base pair. Individual bases, ribose, and phosphates are labeled. The stick representations *below* each base pair are shown in the same orientations as the base pairs (carbon and phosphorus atoms are *light gray*, nitrogen atoms are *intermediate*, and oxygen atoms are *dark gray*).

4–8*   **Reference:** Hershey AD & Chase M (1952) Independent functions of viral protein and nucleic acid in growth of bacteriophage. *J. Gen. Physiol.* 36, 39–56.

4–9

A. The DNA in a human cell is about 2.2 meters in length.

$$\text{length} = 6.4 \times 10^9 \text{ bp} \times \frac{0.34 \text{ nm}}{\text{bp}} \times \frac{\text{m}}{10^9 \text{ nm}}$$

$$\text{length} = 2.18 \text{ m}$$

B. The DNA occupies about 9% of the volume of the nucleus. The volume of the nucleus is

$$V = 4/3 \times 3.14 \times (3 \times 10^3 \text{ nm})^3$$
$$V = 1.13 \times 10^{11} \text{ nm}^3$$

The volume of DNA is

$$V = 3.14 \times (1.2 \text{ nm})^2 (2.2 \times 10^9 \text{ nm})$$
$$V = 9.95 \times 10^9 \text{ nm}^3$$

The ratio of DNA volume to nuclear volume is about 0.09; thus, the DNA occupies about 9% of the nuclear volume.

4–10*

4–11   Helix A is right handed. Helix C is left handed. Helix B has one right-handed strand and one left-handed strand. There are several ways to tell the handedness of a helix. For a vertically oriented helix, like the ones in Figure 4–4, if the strands in front point up to the right, the helix is right handed; if they point up to the left, the helix is left handed. Once you are comfortable identifying the handedness of a helix, you will be amused to note that nearly 50% of the 'DNA' helices in advertisements are left handed, as are a surprisingly high number of the ones in books. Amazingly, a version of Helix B was used in advertisements for a prominent international conference, celebrating the 30-year anniversary of the discovery of the DNA helix.

4–12*

CHROMOSOMAL DNA AND ITS PACKAGING IN THE CHROMATIN FIBER

4–13   The DNA molecules in chromosomes are long and exceedingly thin, and therefore very fragile. The techniques in use in the 1950s, which were gentle enough for the isolation of proteins, were much too harsh for DNA. For example, the shearing force exerted by pipetting DNA—sucking it through a small aperture—was sufficient to break it into the observed small pieces. It was a major technical achievement to demonstrate that chromosomes contain a single long DNA molecule.

4–14*

4–15 True. The human karyotype comprises 22 autosomes and the two sex chromosomes, X and Y. Females have 22 autosomes and two X chromosomes for a total of 23 different chromosomes. Males also have 22 autosomes, but have an X and a Y chromosome for a total of 24 different chromosomes.

4–16*

4–17 The total length of DNA in chromosome 1 is 9.5×10^7 nm [$(2.8 \times 10^8$ bp) $\times (0.34$ nm/bp)], which is 9.5×10^4 µm. In mitosis the chromosome measures 10 µm. Therefore the DNA molecule in chromosome 1 is compacted 9500-fold (9.5×10^4 µm/10 µm) at mitosis.

4–18*

4–19
A. The digestion patterns with the various restriction nuclease indicate that the *U2* genes are tandemly repeated in the human genome, like the two copies that are present in the bacteriophage lambda clone. Restriction nucleases such as HindIII, HincII, and KpnI, which cut once within the 6-kb repeat, cut the entire cluster into 6-kb fragments. (These enzymes also produce a junction fragment at one end of the cluster; however, it is too faint to show up under these hybridization conditions.) Restriction nucleases such as EcoRI, BglII, and XbaI, which do not cut within the 6-kb repeat, leave the entire cluster on one fragment. Partial digestion with HindIII creates a ladder of bands that are multiples of 6 kb.
B. Two bands are visible in the bacteriophage lambda digest because KpnI cleaves the cloned DNA twice, generating a 6-kb fragment containing one *U2* gene and a larger fragment from the right side of the clone (see Figure 4–7), which also carries one copy of the *U2* gene. Cleavage of any tandem repeat will generate so-called junction fragments from the ends of the repeated region. Only one of the junction fragments shows up in these experiments because the *U2* gene is used as a probe. The junction fragment is not visible in the genomic digest because it is too faint—there is only one junction fragment, but there are many internal fragments. A longer exposure of the autoradiograph would have allowed the junction fragment to be seen.
C. Since 2 ng of cloned DNA gives the same intensity 6-kb band as 10 µg of genomic DNA, a simple proportionality can be set up: 2 ng of cloned DNA, which contains 1 hybridizing *U2* gene per 43 kb, is equal to 10 µg of genomic DNA, which carries an unknown number (n) of *U2* genes per 3.2×10^6 kb. (Although there are two *U2* genes on the cloned DNA, only one contributes to the hybridization of the 6-kb band.)

$$\frac{n \, U2 \text{ genes}}{3.2 \times 10^6 \text{ kb}} \times 10^4 \text{ ng} = \frac{1 \, U2 \text{ gene}}{43 \text{ kb}} \times 2 \text{ ng}$$

$$n = \frac{3.2 \times 10^6 \text{ kb}}{43 \text{ kb}} \times \frac{2 \text{ ng}}{10^4 \text{ ng}}$$

$$n = 14.9$$

Taking into account the one *U2* gene associated with the invisible junction fragment, there are about 16 *U2* genes in the human genome.
D. The basis for the different estimates of numbers of *U2* genes comes down to the definition of 'gene.' For the hybridization studies a *U2* gene was defined as a stretch of DNA that would hybridize to a U2 probe. The sequence analysis used a more stringent definition: a stretch of DNA was identified as a *U2* gene only if it encoded U2 snRNA. By this criterion, there are only three true *U2* genes in the haploid human genome. Most of the copies identified by hybridization have suffered mutations and no longer serve as a source of U2 snRNA. Such nonfunctional copies are referred to as pseudogenes.

References: van Arsdell SW & Weiner AM (1984) Human genes for U2 small nuclear RNA are tandemly repeated. *Mol. Cell. Biol.* 4, 492–499.

Lander ES *et al.* (2001) Initial sequencing and analysis of the human genome. *Nature* 409, 860–921.

4–20*   Figure 4–33.

References: Nathans J, Thomas D & Hogness DS (1986) Molecular genetics of human color vision: the genes encoding blue, green, and red pigments. *Science* 232, 193–202.

Nathans J, Piantanida TP, Eddy RL, Shows TB & Hogness DS (1986) Molecular genetics of inherited variation in human color vision. *Science* 232, 203–210.

Vollrath D, Nathans J & Davis RW (1988) Tandem array of human visual pigment genes at Xq28. *Science* 240, 1669–1671.

Lander ES *et al.* (2001) Initial sequencing and analysis of the human genome. *Nature* 409, 860–921.

4–21   Extrapolating from the number of genes on chromosome 22 to the whole genome gives an estimate of about 47,000 genes (700/0.015 = 46,667), which is well in excess of the accepted value of about 30,000 genes. The error in the calculation is the assumption that the *density* of genes on chromosome 22 is the same as that for the whole genome. Genes are present at the genome-wide average density of about 11.1 genes per Mb, which is equal to a total of 35,500 genes (11.1 genes/Mb × 3200 Mb = 35,520). (The true value is thought to be somewhat less because there are about 200 Mb of heterochromatin and short-arm sequences, which are gene poor.) The gene density on chromosome 22 is higher than this average (700 genes/48 Mb = 14.6 genes/Mb). Gene density varies considerably among different chromosomes, with chromosome 19 the highest at 26.8 genes per Mb and chromosome Y the lowest at 6.4 per Mb.

References: Lander ES *et al.* (2001) Initial sequencing and analysis of the human genome. *Nature* 409, 860–921.

Dunham I *et al.* (1999) The DNA sequence of human chromosome 22. *Nature* 402, 489–495.

4–22*   **Reference:** Dunham I *et al.* (1999) The DNA sequence of human chromosome 22. *Nature* 402, 489–495.

4–23   True. Overall only a couple of percent of the human genome is present in mRNA and only about one third of the genome is transcribed into RNA. Even allowing for regulatory regions and other critical sequences, it still appears that more than half the genome has no function—is unimportant junk.

4–24*   **Reference:** Lander ES *et al.* (2001) Initial sequencing and analysis of the human genome. *Nature* 409, 860–921.

4–25   True. Humans and mice diverged from a common ancestor long enough ago for roughly 2 out of 3 nucleotides to have been changed by random mutation. The only regions that have been conserved are those with important functions—exons and regulatory sequences. Mutations in such regions were eliminated by natural selection. Noncoding regions—regions without a critical function—have not been conserved because natural selection cannot operate to eliminate changes in nonfunctional DNA.

4–26*   Figure 4–34.

Reference: Konkel DA, Maizel JV & Leder P (1979) The evolution and sequence comparison of two recently diverged mouse chromosomal β-globin genes. *Cell* 18, 865–873.

4–27   True. Currently, it is very difficult to identify a human gene by simply searching the sequence of the DNA. In large part this problem derives from the low frequency of coding information (2–3%) in the human genome and the small size of exons relative to introns. Current predictive methods identify about 70% of individual exons in the human genome and about 20% of complete genes. By far the most reliable method is to compare sequences of human mRNA (as determined from cloned cDNAs) with the genome sequence, a method that identifies exons directly. Using mRNA sequences from organisms such as mice and searching for conserved homologies commonly supplement this method. Direct comparisons of the mouse and human genomes will identify highly conserved regions, which will include both genes and their regulatory sequences.

4–28* **Reference:** Lander ES *et al.* (2001) Initial sequencing and analysis of the human genome. *Nature* 409, 860–921.

4–29 In a random sequence of DNA, each of the 64 different codons will be generated with equal frequency. Since 3 of the 64 are STOP codons, they will be expected to occur every 21 codons (64/3 = 21.3) on average.

4–30* Figure 4–35.

4–31 A gene is any DNA sequence that is transcribed as a single unit and produces a functional RNA or encodes one or a set of closely related polypeptide chains (protein isoforms). Note that the explicit requirement for transcription makes the control region a part of the gene as well.

4–32*

4–33

A. The restriction analysis of the plasmid indicates that it is a linear molecule. If the plasmid were a circle, digestion with single-cut restriction nucleases would have generated only one band at 12 kb in each case—which was not observed. On the other hand, a linear molecule generates two bands when cut once, and they should sum to 12 kb. Since two bands that summed to 12 kb were generated with each of the three single-cut restriction nucleases, the plasmid must be a linear molecule.

B. Digestions with BamHI or BglII yield DNA fragments with identical 5′ extensions (5′-GATC). Thus, a mixture of these fragments can join in all possible combinations. Neither enzyme, however, can cut the hybrid sites created by joining a telomere fragment to a plasmid end (a BamHI end to a BglII end, 5′-AGATCC and 5′-GGATCT). By contrast, the appropriate enzyme can cut the sites created by joining two BamHI ends or two BglII ends. Therefore, in the presence of DNA ligase, BamHI, and BglII, joints that create a BamHI site or a BglII site are quickly recut, whereas the hybrid sites, which cannot be cut, accumulate. This strategy neatly selects for formation of hybrid joints—in this case, the telomere fragment joined to the plasmid.

Reference: Szostak JW & Blackburn EH (1982) Cloning yeast telomeres on linear plasmid vectors. *Cell* 19, 245–255.

4–34*

4–35 The essence of your proposed cure is that telomerase is essential for cancer cells and that its inhibition would ultimately stop their growth. The troubling aspect of the rival company's observations is that telomerase-knockout mice still get cancers, indicating that there are nontelomerase-dependent routes to tumor formation. Thus, there is no assurance that if you inhibited the growth of cancer cells with a telomerase inhibitor, the cancer cells wouldn't find another, telomerase-independent way around the block.

Reference: Artandi SE, Chang S, Lee S-L, Alson S, Gottlieb GJ, Chin L & DePinho RA (2000) Telomere dysfunction promotes non-reciprocal translocations and epithelial cancers in mice. *Nature* 406, 641–645.

4–36*

4–37   The packing ratio within a nucleosome core is 4.5 [(146 bp × 0.34 nm/bp)/ (11 nm) = 4.5]. If there are an additional 54 bp of linker DNA, then the packing ratio for 'beads-on-a-string' DNA is 2.3 [(200 bp × 0.34 nm/bp)/ (11 nm + {54 bp × 0.34 nm/bp}) = 2.3]. This first level of packing represents only 0.023% (2.3/10,000) of the total condensation that occurs at mitosis.

4–38*   Figure 4–36.

Reference: Prunell A, Kornberg RD, Lutter L, Klug A, Levitt M & Crick FH (1979) Periodicity of deoxyribonuclease I digestion of chromatin. *Science* 204, 855–858.

4–39   Histone octamers occupy about 9% of the volume of the nucleus. The volume of the nucleus is

$$V = 4/3 \times 3.14 \times (3 \times 10^3 \text{ nm})^3$$
$$V = 1.13 \times 10^{11} \text{ nm}^3$$

The volume of the histone octamers is

$$V = 3.14 \times (4.5 \text{ nm})^2 \times (5 \text{ nm}) \times (32 \times 10^6)$$
$$V = 1.02 \times 10^{10} \text{ nm}^3$$

The ratio of the volume of histone octamers to the nuclear volume is 0.09; thus, histone octamers occupy about 9% of the nuclear volume. Since the DNA also occupies about 9% of the nuclear volume, together they occupy about 18% of the volume of the nucleus.

4–40*

4–41   The presence of a nuclease-resistant fraction in chromatin—but not in naked DNA—suggests that the Martian DNA is associated with a nucleosomelike structure that protects it from micrococcal nuclease. Since extensive digestion produced a limit product of about 300 nucleotides, the nucleosomelike structure must protect about this length of DNA. The smear of digestion products indicates that the nucleosomelike structures are not regularly spaced along the DNA as Earthly nucleosomes are. If they were regularly spaced, they would have given a ladder of bands analogous to those seen in rat liver.

4–42*

4–43   In contrast to most proteins, which accumulate amino acid changes over evolutionary time, the functions of histone proteins must involve nearly all of their amino acids, so that a change in any position is deleterious to the cell. Histone proteins are exquisitely refined for their function.

4–44*   **Reference:** Wang YH & Griffith J (1995) Expanded CTG triplet repeat blocks from the myotonic dystrophy gene create the strongest known natural nucleosome positioning elements. *Genomics* 25, 570–573.

4–45   With these assumptions, the DNA is compacted 27-fold in 30-nm fibers relative to the extended DNA. The total length of duplex DNA in 50 nm of the fiber is 1360 nm [(20 nucleosome) × (200 bp/nucleosome) × (0.34 nm/bp) = 1360 nm]. 1360 nm of duplex DNA reduced to 50 nm of chromatin fiber represents a 27-fold condensation [(1360 nm/50 nm) = 27.2]. This level of packing represents only 0.27% (27/10,000) of the total condensation that occurs at mitosis, still a long way off what is needed.

4–46*

4–47   These results argue strongly that the SWI/SNF complex slides nucleosomes along the DNA in an ATP-dependent manner. Two key observations support this model. First, incubation with SWI/SNF causes the nucleosome to disappear from the small fragment released by NheI (see Figure 4–17B, lane 6), but the nucleosome remains associated with the large fragment released by EcoRI. Second, cleavage with NheI before incubation prevents loss of the nucleosome from the small fragment (see lane 4), suggesting that the nucleosome can be moved only if there is contiguous DNA.

If the mechanism of action of the SWI/SNF complex were to release the nucleosome from the DNA, then the nucleosome should have been released regardless of the restriction enzyme used or the order of incubation. Retention of the nucleosome on the large EcoRI fragment and on the pre-cleaved NheI fragment argues against this mechanism.

If the SWI/SNF complex transferred the nucleosome from one duplex to another, then the nucleosome should have been released from the precleaved NheI fragment and transferred to the other duplex (the rest of the DNA in the substrate). This mechanism cannot strictly be ruled out by these experiments because of concentration effects. When the NheI fragment is attached to the rest of the DNA, the local concentration of the receptor duplex (the rest of the DNA) is very high; however, when it is detached, the concentration of receptor duplex is low. The authors of this study ruled out this possibility by placing a barrier to sliding (but presumably not to transfer) adjacent to the NheI site. Under these conditions the nucleosome remained associated with the NheI fragment when the substrate was incubated with the SWI/SNF complex before cleavage with NheI (in contrast to the results in the absence of a barrier, as shown in Figure 4–17B, lane 6).

Reference: Whitehouse I, Flaus A, Cairns BR, White MF, Workman JL & Owen-Hughes T (1999) Nucleosome mobilization catalysed by the yeast SWI/SNF complex. *Nature* 400, 784–787.

4–48*   Figure 4–37.

4–49   True. Deacetylation increases the positive charge on the histone tails by unmasking the positive charges on lysines. The increased charge tends to stabilize chromatin structure, perhaps by allowing the tails to interact more strongly with the DNA.

4–50*   **Reference:** Weintraub H & Groudine M (1976) Chromosomal subunits in active genes have an altered conformation. *Science* 193, 848–856.

THE GLOBAL STRUCTURE OF CHROMOSOMES

4–51   False. The loops are actively transcribed, but contain only a minority of the DNA. Most of the DNA in lampbrush chromosomes is highly condensed in the chromomeres, and transcriptionally inactive.

4–52*   Figure 4–38.

References: Callan HG (1963) The nature of lampbrush chromosomes. *Int. Rev. Cytol.* 15, 1–34.

4–53   In amphibian oocytes the loops in lampbrush chromosomes are actively transcribed, suggesting that loops may correspond to active genes. Based on this analysis of lampbrush chromosomes, it is thought that interphase chromosomes in most animals may adopt such a looped structure, but the structure is too fragile to be observed. If a human chromosome were to form a visibly looped, lampbrushlike structure in amphibian oocytes, it should be possible by hybridization, for example, to map the sequences that are present in the loops and determine directly whether loops correspond with known transcription units in the human genome.

4–54*   **Reference:** Gall JG & Murphy C (1998) Assembly of lampbrush chromosomes from sperm chromatin. *Mol. Biol. Cell* 9, 733–747.

4–55 False. Classical genetic studies suggested that each band might correspond to a gene; however, analysis of the *Drosophila* genome indicates that there are three times as many genes as bands. Thus, the original 'one-band, one-gene' hypothesis has been disproved.

4–56* **Reference:** Spierer A & Spierer P (1984) Similar levels of polyteny in bands and interbands of *Drosophila* giant chromosomes. *Nature* 307, 176–178.

4–57 These statements are not necessarily contradictory. It could be that two-thirds of the genes in *Drosophila* are in interbands. However, that conclusion would be highly surprising, since 95% of the DNA in polytene chromosomes is in bands and only 5% is in interbands. Although some genes are present in interbands, it is also true that loops of chromatin often include more than a single gene. The true relationship of genes to bands and interbands is not clear.

4–58*

4–59

A. It is apparent by visual inspection of the chromosomes in Figure 4–23A that there is a higher density of genes with increased expression (black bars) near telomeres than elsewhere in the chromosomes after depletion of histone H4. This impression is confirmed by a more rigorous analysis of the data, as shown in Figure 4–39. This analysis indicates that the fraction of telomere-proximal genes that have increased expression is more than 3-fold higher than the genome-wide average of 15%.

B. It is more difficult to be certain by visual inspection alone that deletion of the *SIR3* gene preferentially increases expression of genes near telomeres. Statistical analysis of the data, as described above, makes it clearer (Figure 4–39). The fraction of telomere-proximal genes that have increased expression is nearly 10-fold higher than the genome average of 1.5%.

C. The loss of Sir3 would inactivate the Sir protein complex, which would dramatically inhibit deacylation of histones in the region of the telomere. The normally deacylated histones near telomeres allow the nucleosomes to pack together into tighter arrays, which are associated with lower levels of expression. In the absence of deacylation, nucleosomes would be expected to pack less tightly near telomeres and gene expression should be increased, as it is. From the analysis shown in Figure 4–39, the effect of Sir3 is most apparent within 6–10 kb of the chromosome ends.

Depletion of histone H4 is thought to cause a general, genome-wide reduction in the number of nucleosomes. A specific effect on gene expression of telomere-proximal genes was not expected. The effect extends out to 15 kb or so away from the telomere, which is farther than the effect of loss of Sir3. This result suggests that some genes near telomeres are normally repressed by a nucleosome-specific mechanism that may be independent of the effects of the Sir protein complex. The mechanism of this effect is not yet defined. These observations suggest that a special form of chromatin may extend beyond the region to which the Sir protein complex is bound.

Reference: Wyrick JJ, Holstege FC, Jennings EG, Causton HC, Shore D, Grunstein M, Lander ES & Young RA (1999) Chromosomal landscape of nucleosome-dependent gene expression and silencing in yeast. *Nature* 402, 418–421.

4–60* Figure 4–40 and Figure 4–41.

Reference: Bloom KS & Carbon J (1982) Yeast centromere DNA is in a unique and highly ordered structure in chromosomes and small circular minichromosomes. *Cell* 29, 305–317.

4–61 A dicentric chromosome is unstable because two kinetochores have the potential to interfere with one another. Normally, microtubules from the two poles of the spindle apparatus attach to opposite faces of a single kinetochore

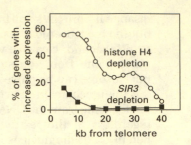

Figure 4–39 Extent of telomeric gene activation after depletion of histone H4 or deletion of the *SIR3* gene (Answer 4–59). For this analysis all the chromosomes were aligned by their telomeres and the results were summed. Windows 50-genes wide were moved one gene at a time across the chromosomes starting at the telomere. At each position the fraction of genes in the window with increased expression was determined and plotted as kb from the midpoint of the window to the telomere.

in order to separate the individual chromatids at mitosis. If a chromosome contains two centromeres, half of the time the microtubules from one of the poles will attach to the two kinetochores associated with one chromatid, while the microtubules from the other pole will attach to the two kinetochores associated with the other chromatid. Division can then occur satisfactorily. However, the other half of the time, the microtubules from each pole will attach to kinetochores that are associated with different chromatids. When that happens, an *individual* chromatid will be pulled to opposite spindle poles with enough force to snap it in two. Thus, two centromeres are bad for a chromosome, causing chromosome breaks—rendering it unstable.

4–62*

4–63

A. A few of the untreated plasmid molecules are relaxed because they contain one or more single-strand breaks in the DNA. DNA that is carefully handled during the isolation procedure is mostly supercoiled, as shown in Figure 4–27, lane 1. Harsh treatment during isolation will yield a DNA preparation that is almost entirely relaxed.

B. The bands that run at intermediate positions are a collection of topoisomers that differ only in their linking number; that is, the number of times one strand is wrapped around the other. Adjacent bands on the gel contain topoisomers that differ by a linking number of one. The rate at which DNA molecules migrate through a gel depends on how compact they are, with more compact molecules moving faster. Relaxed circular molecules are the least compact and therefore move the slowest, whereas highly supercoiled molecules are the most compact and run the fastest. Treatment with topoisomerase removes supercoils one at a time, making the molecule progressively less compact and slower migrating.

C. The number of supercoils in the original plasmid can be estimated by counting the number of bands between the highly supercoiled and relaxed positions of the gel. About 8 intermediate bands can be counted in the treated samples; thus, there must be at least 9 supercoils in the original plasmid. This number is likely to be an underestimate because of the limited resolving power of such gels. Once a molecule reaches a certain degree of compactness, a gel cannot resolve molecules with further increases in supercoiling.

D. Since the supercoils are removed by *E. coli* topoisomerase I, the original supercoils must have been negative. (*E. coli* topoisomerase I will not relax positive supercoils.)

4–64*

4–65 In order for the circular DNA molecules to retain a net supercoiling of zero, plectonemic supercoils and solenoidal supercoils must have the opposite sign. Thus, a negative solenoidal supercoil can be compensated by a positive plectonemic supercoil, and vice versa. Arrangements B and C are the only structures with zero net supercoiling, as shown in Figure 4–42.

4–66* Figure 4–43.

4–67

A. Since treatment with both *E. coli* and calf thymus resulted in supercoils in the final molecules (see Figure 4–30, lanes 5 and 9), the compensating plectonemic supercoils in the condensin-treated DNA must have been negative, which both topoisomerases can relax. When those were removed by topoisomerase treatment, only the positive supercoils would have remained; hence, the plectonemic supercoils in Figure 4–30, lanes 5 and 9, are expected to be positive. Since the compensating plectonemic supercoils were negative, the solenoidal supercoils formed by condensin were positive (right handed).

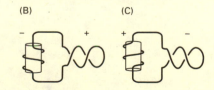

Figure 4–42 The two compensating arrangements of solenoidal and plectonemic supercoils (Answer 4–65). The signs of the solenoidal and plectonemic supercoils are indicated by '+' and '–'.

B. The results in Figure 4–30 are not consistent with condensin acting as an ATP-driven topoisomerase. If condensin introduced a net negative plectonemic supercoiling, those supercoils would have been removed by both topoisomerases to leave relaxed molecules, contrary to what was observed (see Figure 4–30, lanes 5 and 9). If condensin introduced a net positive plectonemic supercoiling, the supercoils would have been removed by calf thymus topoisomerase, but not by *E. coli* topoisomerase, contrary to what was observed (see lanes 5 and 9).

Reference: Kimura K & Hirano T (1997) ATP-dependent positive supercoiling of DNA by 13S condensin: A biochemical implication for chromosome condensation. *Cell* 90, 625–634.

4–68*

4–69 Although chromosomes in individual cells occupy small discrete areas, the particular location is unlikely to be important. The key observation is that the locations of chromosomes and the nearest neighbors differ in otherwise identical cells in a tissue.

DNA REPLICATION, REPAIR, AND RECOMBINATION

THE MAINTENANCE OF DNA SEQUENCES

5–1 In the initial population there are 10^6 copies of your 1000-bp gene, or a total of 10^9 bp to be copied when the population doubles. At a mutation rate of 1 mutation per 10^9 bp per generation, you might expect 1 copy of your 1000-bp gene to carry a mutation (one mutant cell) in the population of 2×10^6 cells, which is a frequency of 5×10^{-7} [1 mutant cell/(2×10^6 total cells)].

After the first doubling there will be 2×10^6 copies of your 1000-bp gene, for a total of 2×10^9 bp to be copied at the next population doubling. At the same rate of mutation, you would now expect 2 mutant copies of the gene to be generated, which is a frequency of new mutants of 5×10^{-7} [2 mutant cells/(4×10^6 total cells)]. The frequency of mutations in the population, however, is greater because the mutant cell that was generated in the first doubling will divide to produce two mutant cells in the second generation, for an overall frequency of mutants equal to 10^{-6} [4/(4×10^6)].

After the second doubling there will be 4×10^6 copies of your 1000-bp gene for a total of 4×10^9 bp to be copied. You would expect 4 mutant copies of the gene to be generated, which is a frequency of new mutants of 5×10^{-7} [4/(8×10^6)] in the third generation. The four mutant cells present after the second doubling would also double to generate 8 mutant cells; thus, the overall frequency of mutants would be 1.5×10^{-6} [12/(8×10^6)].

This exercise illustrates a key difference between *rates* of mutation and *frequencies* of mutation. Rates are constant whereas frequencies increase with increasing growth of the cell population.

5–2* **Reference:** Luria SE & Delbrück M (1943) Mutations of bacteria from virus sensitivity to virus resistance. *Genetics* 28, 491–511.

5–3 For either hypothesis you might expect to see about 10 surviving colonies per plate. If the bacteriophages induce resistance, the surviving colonies would appear in random positions on each of the replica plates. If the mutations preexist, however, the resistant colonies would appear at the same locations on each of the three replica plates. In actual experiments of this kind, the surviving colonies appear at the same locations, indicating that the mutations preexist in the population.

Reference: Hartwell LH, Hood L, Goldberg ML, Reynolds AE, Silver LM & Veres RC (2000) Genetics: From Genes to Genomes, pp 217–218. New York: McGraw Hill.

5–4*

5–5   For most proteins, some mutations will alter the amino acid sequence in such a way that the protein no longer functions. Such mutations will tend to be lost through natural selection; that is, through the preferential death of organisms that carry the mutations. This preferential loss leads to an underestimate of mutation rate because some mutations are not counted. Fibrinopeptides, however, are much less sensitive to such effects because their function does not depend on their amino acid sequence. Thus, they can tolerate almost any amino acid change and, as a result, estimates of mutation rates are more accurate. Even so, corrections still must be made for mutations that occur at the site of a previous mutation, changing it to the normal sequence or to a new mutant sequence. In the first case two mutations would be counted as none; in the second, two mutations would be counted as one. The corrections become increasingly important with greater evolutionary separation.

5–6*

5–7   The incidence of cancer increases dramatically with age because it takes mutations in several critical genes to disable a cell's normal mechanisms for controlling its growth. Since growing cells are continually accumulating mutations, which they pass on to their progeny cells, the chance that a cell will accumulate a critical set of mutations increases with age. The steep rise in cancer incidence in older women seen in Figure 5–1 reveals that colon cancer increases as the sixth power of age, suggesting that it arises only after mutations have occurred in six or so genes that regulate cell growth in the colon.

5–8*

DNA REPLICATION MECHANISMS

5–9   The complementary strand is 5′-TGATTGTGGACAAAAATCC-3′. Recall that the two strands of a DNA double helix run in opposite directions. By convention, sequences of single strands are written in the 5′-to-3′ direction.

5–10*

5–11   False. The sequence of nucleotides in a newly synthesized strand is very different from that in the parental strand; it is complementary to the 3′-to-5′ sequence of the parental strand.

5–12*   Figure 5–54 and Figure 5–55.

  Reference: Inman RB & Schnos M (1971) Structure of branch points in replicating DNA: presence of single-stranded connections in lambda DNA branch points. *J. Mol. Biol.* 56, 319–325.

5–13   True. At each replication fork the leading strand is synthesized continuously and the lagging strand is synthesized as Okazaki fragments. Since half the DNA at each replication fork is stitched together from Okazaki fragments, half the genome must be made this way.

5–14*   Figure 5–56.

5–15

  A. Dideoxycytidine triphosphate (ddCTP) is identical to dCTP except that it lacks the 3′ hydroxyl group on the sugar ring. ddCTP is recognized by DNA polymerase as dCTP and becomes incorporated into DNA; however, because it lacks the crucial 3′-hydroxyl group, its addition to a growing DNA strand creates a dead end to which no further nucleotides can be added. Thus, if ddCTP is added in large excess, strands will be synthesized until the first G is encountered in the template strand. ddCTP will then be incorporated instead of C, and the extension of this strand will be terminated.

B. If ddCTP is added at 10% the concentration of dCTP, there is a 1 in 10 chance of its being incorporated whenever a G is encountered on the template strand. Thus, a population of DNA fragments will be synthesized, and from their lengths the location of the G nucleotides in the template strand can be deduced. These results form the basis of methods used to determine the sequence of nucleotides in a stretch of DNA, as covered elsewhere in this book.

C. Dideoxycytidine monophosphate (ddCMP) lacks the 5′-triphosphate group as well as the 3′-hydroxyl group of the sugar ring. The absence of the triphosphate means that ddCMP cannot provide the free energy that drives the polymerization of nucleotides into DNA. It is not a substrate for DNA polymerase and will not be incorporated into the replicating DNA. The molecule, at either concentration, is therefore not expected to affect DNA replication.

5–16*

5–17

A. The different labels used for the T and C nucleotides make it easy to measure their respective losses from the polymer. The radioactive disintegrations from the energetic ^{32}P and from the rather weak ^{3}H can be distinguished using a liquid scintillation counter.

B. Because the nuclease activity of DNA polymerase I is an exonuclease (that is, it removes nucleotides from the ends of strands), the T nucleotides cannot be released until all the C nucleotides have been removed, hence the lag.

C. When dTTP is added to the reaction, polymerization will begin just as soon as a proper AT nucleotide pair is uncovered at the end of the primer. Polymerization does not occur efficiently from a mismatched AC pair. Since the rate of polymerization exceeds the rate of exonuclease digestion by two or three orders of magnitude, the labeled Ts will be buried quickly and will be unavailable to the exonuclease.

D. The results will not be affected by the presence of dCTP. Since the template is poly(dA), C nucleotides cannot be incorporated. If they are incorporated by mistake, the mismatched C will not serve as a primer for polymerization.

Reference: Brutlag D & Kornberg A (1972) Enzymatic synthesis of deoxyribonucleic acid. 36. A proofreading function for the 3′ to 5′ exonuclease activity in deoxyribonucleic acid polymerases. *J. Biol. Chem.* 247, 241–248.

5–18*

5–19   While the process may seem wasteful, it is not possible to proofread during primer formation. To start a new primer on a piece of single-stranded DNA, one nucleotide needs to be put in place and then linked to a second and then to a third and so on. Even if these first nucleotides were perfectly matched to the template strand, such short oligonucleotides bind with very low affinity and it would consequently be difficult to distinguish the correct from incorrect bases by a hypothetical proofreading activity. The task of the primase is to 'just get anything down that binds reasonably well and not worry about accuracy.' Later, these sequences are removed and replaced by DNA polymerase, which uses the accurate, correctly proofread newly synthesized DNA of the adjacent Okazaki fragment as its primer. The latter enzyme has the advantage—which primase does not have—of putting the new nucleotides onto the end of an already existing strand. The newly added nucleotide is held firmly in place, and the accuracy of its base pairing to the next nucleotide on the template strand can be checked. Therefore, as DNA polymerase fills the gap, it can proofread the new DNA strand that it makes. What appears at first glance as energetically wasteful is really just a necessary price to be paid for accuracy.

5–20*   Figure 5–57.

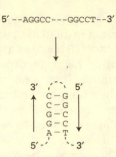

5′ --AGGCC---GGCCT--3′

```
           3′  ⌐⁻⁻⌐  5′
          ↑   C – G   ↓
              C – G
              G – C
              G – C
              A – T
           5′ ⌐⁻    ⌐⁻ 3′
```

Figure 5–58 An example of a sequence of single-stranded DNA that could form a hairpin helix (Answer 5–23).

Reference: Cha TA & Alberts BM (1986) Studies of the DNA helicase–RNA primase unit from bacteriophage T4. A trinucleotide sequence on the DNA template starts RNA primer synthesis. *J. Biol. Chem.* 261, 7001–7010.

5–21   True. If the replication fork moves forward at 500 nucleotide pairs per second, the DNA ahead of it must rotate at 48 revolutions per second (500 nucleotides per second/10.5 nucleotides per helical turn) or 2880 revolutions per minute.

5–22*   **Reference:** LeBowitz JH & McMacken R (1986) The *Escherichia coli* DnaB replication protein is a DNA helicase. *J. Biol. Chem.* 261, 4738–4748.

5–23   Sequences in single-stranded DNA that can form hairpin helices are self-complementary, which means that they can base-pair and that the resulting duplex will have strands running in opposite directions. An example of such a sequence is shown in Figure 5–58.

5–24*   **Reference:** Alberts BM & Frey L (1970) T4 bacteriophage gene 32: a structural protein in the replication and recombination of DNA. *Nature* 227, 1313–1318.

5–25

A. In general, proteins that are required for movement of the replication fork will display a quick-stop phenotype because the fork will be unable to progress in the absence of their function. Thus, temperature-sensitive mutants of DNA topoisomerase I (inability to relieve winding tension ahead of the replication fork), SSB protein (inability to stabilize the single-stranded DNA at the fork), DNA helicase (inability to melt the DNA ahead of the replication fork), and DNA primase will display the quick-stop phenotype. Of these, only the phenotype of DNA primase is difficult to predict. DNA primase directly affects synthesis of the lagging strand, but it is not required for synthesis of the leading strand. Its quick-stop phenotype may result from either of two indirect effects: (1) exposure of sufficient single-stranded DNA to use up all the SSB protein or (2) interference with the DNA helicase, which is linked to DNA primase as part of the primosome.

   Proteins that are not involved in the movement of the replication fork will display a slow-stop phenotype. Thus, a temperature-sensitive initiator protein would show the slow-stop pattern of replication, since DNA molecules that had passed the initiation step when the temperature was increased would continue to replicate around the chromosome until the next initiation step was reached. Similarly, a temperature-sensitive DNA ligase would show a slow-stop phenotype, since the progress of the replication fork would not be stopped. Replication would cease only during the next cycle when the nicks were 'uncovered' on the template strand.

B. The mixed extracts should be fully competent for DNA replication at 42°C; that is, the mixture should exhibit a nonmutant phenotype. The defective DNA helicase extract would provide normal DNA ligase, and the defective DNA ligase extract would provide normal DNA helicase. Thus, the entire complement of normal proteins would be present in the mixed extract. This mutual correction by extracts with different deficiencies is called complementation. (Because of the extreme complexity of DNA replication and the large number of proteins involved, cell-free extracts are not capable of maintaining DNA replication indefinitely. In practice, the behaviors of extracts from slow-stop mutants and from nonmutant cells are often difficult to distinguish.)

5–26*

5–27   If the old strand were 'repaired' using the new strand that contains a replication error as the template, then the error would become a permanent mutation in the genome. The old 'correct' information would be erased in the process. Therefore, if repair enzymes did not distinguish between the two strands, there would be only a 50% chance that any given replication error would be corrected.

Overall, such indiscriminate repair would be the same as no repair at all. In the absence of repair a mismatch would persist until the next replication. When the replication fork passed the mismatch, and the strands were separated, properly paired nucleotides would be inserted opposite each of the nucleotides involved in the mismatch. A normal, nonmutant duplex would be made from the strand containing the original information; a mutant duplex would be made from the strand that carried the misincorporated nucleotide. Thus, the original misincorporation event would lead to 50% mutants and 50% nonmutants in the progeny. This is equivalent to the outcome of indiscriminate repair: averaged over all misincorporation events, indiscriminate repair would yield 50% mutants and 50% nonmutants among the progeny.

5–28* **Reference:** Johnson KA (1993) Conformational coupling in DNA polymerase fidelity. *Annu. Rev. Biochem.* 62, 685–713.

5–29

A. There is a consistent ~2-fold increase in reversions of the *lacZ* allele in the R orientation in mismatch-repair deficient (1.9-fold) and proofreading deficient (2.5-fold) strains of *E. coli*. For the *rif* gene, however, there was no difference in mismatch-repair deficient (1.1-fold) and proofreading deficient (1.0-fold) strains, as expected. Since reversion results from misincorporation of G opposite T, and this occurs on the leading strand in orientation R (see Figure 5–12), leading-strand DNA synthesis appears to be slightly less accurate than lagging-strand synthesis.

B. The reason for the apparent difference in fidelity of DNA synthesis on the leading and lagging strands is not clear. (Four different alleles of *lacZ* were tested in this study; all showed the same 2- to 5-fold lower fidelity of synthesis on the leading strand.) The difference is unlikely to be due to the intrinsic properties of the DNA polymerase since the same polymerase is used to make both strands. If you thought about transcription of the *lacZ* gene, good for you. But transcription occurs at a very low level under the conditions used here and direction of transcription did not correlate with mutation frequency for the four alleles that were studied.

  The authors suggested the following explanation. Because the polymerase on the lagging strand must dissociate and rebind each time it comes to the end of an Okazaki fragment, it might dissociate with greater ease than the polymerase on the leading strand. If the polymerase on the lagging strand dissociated from mismatches more readily, as well, the mismatched primer would be exposed more often on the lagging strand than on the leading strand. They argue that such an exposed mismatch might be subject to repair by other 3′-to-5′ exonucleases in the cell. In effect this would mean that the lagging strand has two ways to repair a mismatched primer, while the leading strand has only one. Such an explanation might account for the difference in fidelity, but it remains to be proven.

Reference: Fijalkowska IJ, Jonczyk P, Tkaczyk MM, Bialoskorska M & Schaaper RM (1998) Unequal fidelity of leading strand and lagging strand DNA replication on the *Escherichia coli* chromosome. *Proc. Natl. Acad. Sci. USA* 95, 10020–10025.

5–30* Figure 5–59.

5–31 True. When topoisomerase I cleaves DNA, it stores the energy of the phosphodiester backbone in a phosphotyrosine bond to the enzyme, which it then uses to remake the phosphodiester bond in DNA.

5–32*

5–33 The enzyme topoisomerase II is responsible for unlinking SV40 daughter duplexes. Topoisomerase II introduces a transient double-strand break into one circle, allowing the second duplex to pass through the first before the break is resealed.

THE INITIATION AND COMPLETION OF DNA
REPLICATION IN CHROMOSOMES

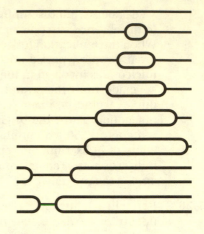

5–34*

5–35 As always, you come through with flying colors. Although you were initially bewildered by the variety of structures, you quickly realized that H forms were just like the bubbles except that cleavage occurred within the bubble instead of outside it. Next you realized that by reordering the molecules according to the increasing size of the bubble (and flipping some structures end-for-end), you could present a convincing visual case for bidirectional replication away from a unique origin of replication (Figure 5–60). The case for bidirectional replication is clear since unidirectional replication would give a set of bubbles with one end in common. Replication from a unique origin is likely, but not certain, because you cannot rule out the possibility that there are two origins on either side of and equidistant from the restriction site. Repeating the experiment using a different restriction nuclease will resolve this issue and define the exact position of the origin(s) on the viral DNA. Your advisor is pleased.

Figure 5–60 Bidirectional replication from a unique origin (Answer 5–35).

5–36*

5–37
   A. The regions of the tracks that are dense with silver grains correspond to those segments of DNA that were replicated when the concentration of the label was high. The less dense regions mark segments of DNA that were replicated when the concentration of label was low.
   B. The difference in the arrangements of the dark and light sections of the tracks derives from the difference in the labeling schemes in the two experiments. In the first experiment (see Figure 5–15A) ^{3}H-thymidine was added immediately after release of the synchronizing block. Thus, replication initiated at origins in the presence of label, giving a continuous dark section on both sides of the origin. When the concentration of label was lowered, replication proceeded in both directions away from the origin, leaving light sections at both ends of the dark sections. In the second experiment (see Figure 5–15B) replication began at origins in the absence of ^{3}H-thymidine so that the origin was unlabeled. Addition of a high concentration of label followed by a low concentration gave rise to a dark section with a light section at one end. Adjacent dark sections are part of the same replicating DNA molecule; they are linked by the unlabeled (therefore invisible) segment that contains the replication origin.
   C. The approximate rate of fork movement can be estimated from the labeling times and the lengths of the labeled sections. In the first experiment, segments roughly 100 μm in length were labeled during the 45-minute labeling period. Because two replication forks were involved in synthesizing each labeled segment, each replication fork synthesized about 50 μm of DNA in 45 minutes. Therefore, the rate of fork movement is about 1.1 μm/min (50 μm/45 min = 1.1 μm/min). In the second experiment segments roughly 50 μm in length were labeled; however, each was synthesized by only one replication fork. Thus, the rate of fork movement was also about 1.1 μm/min.

   This information is not sufficient to estimate the time required to replicate the entire genome. The missing information is the number of active origins of replication and their distribution.

   Reference: Huberman JA & Riggs AD (1968) On the mechanism of DNA replication in mammalian chromosomes. *J. Mol. Biol.* 32, 327–341.

5–38* Figure 5–61.

5–39 If there were no time constraints on replication, one origin would be required for each chromosome; thus, a minimum of 46 origins, equal to the number of chromosomes in a human cell, would be needed.

In 8 hours (28,800 seconds) the two replication forks from one origin would synthesize 2.88×10^6 nucleotides [(2 forks) × (50 nucleotides/second) × (2.88×10^4 seconds)]. To replicate the entire genome would require about 2200 equally spaced origins [(6.4×10^9 nucleotides)/(2.88×10^6 nucleotides/origin) = 2222 origins]. It is estimated that the human genome has about 10,000 origins of replication, more than enough to finish replication within the time allotted in the cell cycle.

5–40*   Figure 5–62 and Figure 5–63.

Reference: Dodson M, Dean FB, Bullock P, Echols H & Hurwitz J (1987) Unwinding of duplex DNA from the SV40 origin of replication by T-antigen. *Science* 238, 964–967.

5–41

F.   Each newly synthesized strand in a daughter duplex was synthesized by a mixture of continuous and discontinuous DNA synthesis from multiple origins. Consider a single replication origin. The fork moving in one direction synthesizes a daughter strand continuously as part of leading-strand synthesis. The fork moving in the opposite direction synthesizes a portion of the same daughter strand discontinuously as part of lagging-strand synthesis.

5–42*   Figure 5–64.

Reference: Harland RM & Laskey RA (1980) Regulated replication of DNA microinjected into eggs of *Xenopus laevis. Cell* 21, 761–771.

5–43   False. Expressed genes tend to replicate early in S phase, whereas genes that are not expressed tend to replicate late in S phase. Thus, genes will replicate early in S phase in those cell types in which they are expressed, but will replicate later in S phase in those cell types in which they are not expressed.

5–44*   Figure 5–65.

Reference: Brewer BJ & Fangman WL (1987) The localization of replication origins on ARS plasmids in *S. cerevisiae. Cell* 51, 463–471.

5–45

A.   The DNA from an amplified cluster is an 'onion-skin' structure as illustrated in Figure 5–66. When examined by electron microscopy, DNA from late-stage follicle cells shows multiple, nested replication forks, just as expected for this mechanism of amplification.
B.   If every origin is activated, each round of replication would double the number of chorion genes. Therefore, it would take six rounds of replication to achieve a 60-fold amplification ($2^6 = 64$).
C.   Given the overreplication of the chorion gene cluster, the 510-nucleotide amplification-control element is probably an origin of replication. It cannot, however, be a standard origin; it must also contain a sequence that allows it to escape the block to re-replication in follicle cells at specific stages. It is known that ORC binds throughout the nucleus until a specific stage of development at which it is cleared from all origins except those that are to be amplified. The clearing of ORC from most origins and its continued binding at amplification sites are dependent on the activities of other proteins, but the details of these processes are not yet defined.

References: Orr-Weaver TL & Spradling AD (1986) *Drosophila* chorion gene amplification requires an upstream region regulating s18 transcription. *Mol. Cell. Biol.* 6, 4624–4633.

Royzman I, Austin RJ, Bosco G, Bell SP & Orr-Weaver TL (1999) ORC localization in *Drosophila* follicle cells and the effects of mutations in *dE2F* and *dDP. Genes Dev.* 13, 827–840.

5–46*

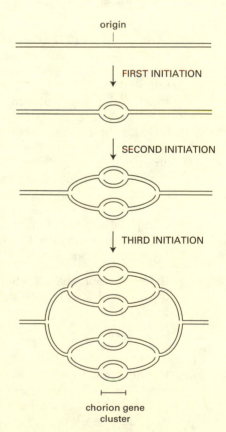

Figure 5–66 Onion-skin structure of an amplified cluster of chorion genes (Answer 5–45). Three initiation events are illustrated; six would be required to amplify the chorion gene cluster 60-fold.

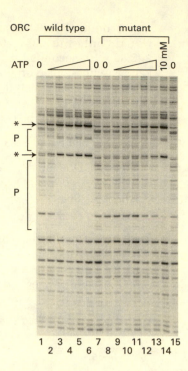

Figure 5–67 ORC binding sites on origin DNA as revealed by DNase footprinting (Answer 5–47). 'P' indicates protected areas; * indicates enhanced cleavage.

5–47

A. ORC binding protects two neighboring locations on the origin DNA, as indicated by blank regions where bands that were visible in the absence of ORC are missing in its presence (Figure 5–67, regions marked with P). Note that a couple of bands (marked by an *) are more intense when ORC is bound, indicating that they are *more* accessible to DNase I in the complex than in native DNA.

B. ATP is required for binding by both the wild-type and the mutant ORCs. For the wild-type ORC, about 100 nM ATP gives full binding (Figure 5–67, lane 3); for the mutant ORC a whopping 10 mM (10^5 more than for wild-type ORC) is required (lane 14).

C. Because exactly the same results were obtained with ATP and the nonhydrolyzable analog ATPγS, ATP hydrolysis cannot be required for ORC binding. The authors of this study suggest that ATP hydrolysis is important for a subsequent step in the complex process that enables an origin for replication.

D. The Walker motif in Orc1 is important to the function of ORC, as evidenced by the dramatically different binding results when the motif is mutated. The binding of mutant ORC at very high ATP concentrations suggests that the mutation lowers the affinity of Orc1 for ATP, but probably does not compromise other functions of the protein.

Reference: Klemm RD, Austin RJ & Bell SP (1997) Coordinate binding of ATP and origin DNA regulates the ATPase activity of the origin recognition complex. *Cell* 88, 493–502.

5–48*   **Reference:** Kobayashi T, Rein T & DePamphilis ML (1998) Identification of primary initiation sites for DNA replication in the hamster dihydrofolate reductase gene initiation zone. *Mol. Cell. Biol.* 18, 3266–3277.

5–49

A. Most of the SV40 DNA is not replicated. This is most clear in the CAF-1-treated sample. CAF-1 treatment moves most of the replicated (labeled) DNA to the supercoiled position, but does not significantly alter the distribution of bulk DNA, as indicated by the stained gel. The small increase in stained DNA at the supercoiled position in the CAF-1-treated sample shows how little of the total DNA has been replicated.

B. CAF-1 assembles nucleosomes only on replicated DNA, as indicated by the dramatic change in the migration of replicated (labeled) DNA, in the absence of any significant effect on the bulk (stained) DNA.

C. Because CAF-1 specifically targets replicated DNA for assembly into nucleosomes, the replicated DNA must bear some 'mark' of replication. Further experiments by the authors of this study identified the 'mark' as the sliding clamp (PCNA) that tethers DNA polymerase to the duplex. The interaction between the clamp and CAF-1 is clearly useful, allowing nucleosome assembly to occur immediately in the wake of the DNA polymerase.

Reference: Shibahara K & Stillman B (1999) Replication-dependent marking of DNA by PCNA facilitates CAF-1-coupled inheritance of chromatin. *Cell* 96, 575–585.

5–50*   Figure 5–68.

5–51

A. Hybridization to telomeres in the unaffected spores (1 and 3) extends from less than 200 nucleotides to just over 300 nucleotides, averaging about 250 nucleotides. Since the cleavage site is 35 nucleotides from the beginning of the telomere repeats, the average length of telomere repeat in fission yeast is just over 200 nucleotides.

B. The descendants of spores 2 and 4 show telomere shortening with time, while the descendants of spores 1 and 3 remain the same size. Thus, spores 2 and 4 appear to lack telomerase, and it looks as though your identification of the fission telomerase gene was correct. It is worth noting, however, that a number of genes in yeast cause a similar phenotype of telomere shortening, but only one of them encodes the catalytic subunit of telomerase. Such genes are known as *EST* genes, for ever shorter telomeres.

C. Although it is somewhat difficult to estimate precisely, it looks as though telomeres lose about 60 nucleotides every 3 days. At four generations per day [(24 hours/day)/(6 hours/generation)] the yeast go through about 12 generations in 3 days. Thus, they lose about 5 nucleotides per generation (60 nucleotides/12 generations).

D. Though you don't have a basis for prediction, the majority of fission yeast that lose their telomeres stop dividing but continue to grow in size, forming abnormally long cells.

Reference: Nakamura TM, Morin GB, Chapman KB, Weinrich SL, Andrews WH, Lingner J, Harley CB & Cech TR (1997) Telomerase catalytic subunit homologs from fission yeast and human. *Science* 277, 955–959.

DNA REPAIR

5–52*

5–53

A. The majority of sequences isolated from the Neanderthal bone differ by seven substitutions and a single base insertion compared to the human reference sequence. Although several clones show additional individual differences, these differences, because they are represented infrequently, are most likely to be due to misincorporation of nucleotides during the PCR reaction. They are probably caused by damage to the original template and so can be ignored. The variability among contemporary human lineages, as shown in Figure 5–28, is too little to encompass the sequence obtained from Neanderthal bone. Thus, it is very likely that you have indeed determined the sequence of Neanderthal DNA.

B. Eight of the 44 sequences show an identical match, or differ only by a single base from the human sequence. These sequences almost certainly arose by amplification of contaminating human DNA. In the paper on which this problem is based, the authors estimated that they were amplifying from a starting population of about 50 DNA molecules in 5 μL of bone extract. A single contaminating human cell would have added 50–1000 additional contaminating sequences, so their precautions worked very well.

C. The main reason for using the mitochondrial DNA for these studies is its abundance compared with nuclear DNA. Cells typically contain 500–1000 copies of mitochondrial DNA molecules, compared with 2 copies of nuclear DNA molecules. With the accumulation of DNA damage over the 30,000 years or so since this individual died, abundance is critical for success. The second reason is that mitochondrial DNA is much more variable than nuclear DNA. The ability to detect multiple differences is also critical for showing that this one Neanderthal sample is different from human DNA.

D. The most important way to verify these results is to isolate DNA from a second Neanderthal sample. Three years after this initial report, the same segment of mitochondrial DNA from a second sufficiently preserved specimen was sequenced. Although these two individuals differed at several positions in the sequence, as expected, they shared 19 substitutions relative to the reference human sample. By sequence analysis these two Neanderthals form a group that is distinct from modern humans. It is estimated that the modern human and Neanderthal lineages diverged between 360,000 and 850,000 years ago.

References: Krings M, Stone A, Schmitz RW, Krainitzki H, Stoneking M & Paabo S (1997) Neanderthal DNA sequences and the origin of modern humans. *Cell* 90, 19–30.

Ovchinnikov IV, Gotherstrom A, Romanova GP, Kharitonov VM, Liden K & Goodwin W (2000) Molecular analysis of Neanderthal DNA from the northern Caucasus. *Nature* 404, 490–493.

5–54*

5–55   The statement is incorrect. DNA defects introduced by deamination and depurination reactions occur spontaneously. They are not the result of replication and are therefore equally likely to occur on either strand. If DNA repair enzymes recognized such defects only on newly synthesized DNA strands, half of the defects would go uncorrected. Also there is no fundamental reason to link such repair events to replication. The bases produced by deamination and depurination are distinct from the normal bases and can be recognized correctly in any sequence context. By contrast, misincorporation during replication adds *normal* bases that are mispaired. The only way to identify them correctly is to search the newly synthesized strand.

5–56*

5–57   True. Both spontaneous depurination and removal of deaminated C by uracil DNA glycosylase leave a sugar that is missing its base, which is the substrate recognized by AP endonuclease.

5–58*

5–59

A. Excision repair of UV damage, which is initiated by the UvrABC endonuclease, is very accurate. It is the only pathway that functions in the *recA* strain, thereby accounting for the low frequency of mutations. RecA participates in two pathways: a recombinational-repair pathway, which is very accurate, and the multifaceted SOS response, one component of which is error prone. It is this error-prone component that yields the high frequency of mutations in the *uvrA* strain.

   Since the wild-type strain yields only 1% of the frequency of mutations that the *uvrA* strain does, the *uvr* pathway of repair must predominate in wild-type *E. coli.*

B. Incorporation of adenine nucleotides opposite pyrimidine dimers should yield only one-third of the number of mutations generated by random incorporation. Incorporation of a random nucleotide at each ambiguous site would be correct only one out of four times (25% correct). By contrast, all-A incorporation would be correct 75% of the time because Ts account for 75% of the pyrimidines in pyrimidine dimers. All the As incorporated opposite CC (10%) would be incorrect, and half the As incorporated opposite TC and CT (15%) would be incorrect. Thus, A incorporation is a good strategy for dealing with UV-induced DNA damage.

5–60*

5–61

A. The fairly even distribution of frameshift mutations indicates that UV damage is distributed throughout the gene. A frameshift mutation anywhere in the gene would be detected by the gene-fusion assay, which is independent of repressor function. If UV damage is evenly distributed, the nonrandom distribution of missense mutations in the *lacI* gene must reflect the functional importance of the ends of the LacI protein. Most mutations in the ends yield a nonfunctional protein, allowing their detection as mutants. The middle of the gene, however, being less critical for function, can accommodate some alterations and still produce a functional protein. These 'silent' mutations would not be detected in an assay that depends on loss of function.

B. The common deletion of one nucleotide in response to UV damage is thought to occur as a mistake during DNA synthesis opposite a pyrimidine dimer. Presumably, in response to the abnormal spacing of bases caused by the pyrimidine dimer, DNA polymerase inserts a single nucleotide rather than two. The frameshift hot spots are hot because they contain runs of Ts and therefore multiple possibilities for dimer formation.

Reference: Miller JH (1985) Mutagenic specificity of ultraviolet light. *J. Mol. Biol.* 182, 45–65.

5–62* **Reference:** Masutani C, Araki M, Yamada A, Kusumoto R, Nogimori T, Maekawa T, Iwai S & Hanaoka F (1999) Xeroderma pigmentosum variant (XP-V) correcting protein from HeLa cells has a thymine dimer bypass DNA polymerase activity. *EMBO J.* 18, 3491–3501.

5–63

A. The adaptive response to low levels of MNNG must require synthesis of new proteins, since the response is blocked by chloramphenicol (see Figure 5–33). If activation of a preexisting protein were all that was required, chloramphenicol would not be expected to block adaptation.

B. The adaptive response might be short-lived for several reasons. Presumably, once the signal for adaptation (ultimately MNNG) is removed, the induced synthesis of new proteins would come to a stop. The resistance to MNNG mutagenesis and killing would then depend on the stability of the induced proteins. If they were relatively unstable, the resistant state would decay rapidly as the proteins became inactive. Even if the proteins were stable, the resistant state of the population of bacteria would decay fairly rapidly due to their growth and the resulting dilution of the protein.

Reference: Teo I, Sedgwick B, Kilpatrick MW, McCarthy TV & Lindahl T (1986) The intracellular signal for induction of resistance to alkylating agents in *E. coli. Cell* 45, 315–324.

5–64* **Reference:** Lindahl T, Demple B & Robins P (1982) Suicide inactivation of the *E. coli* O^6-methylguanine-DNA methyltransferase. *EMBO J.* 1, 1359–1363.

5–65 High-level expression of MGMT confers resistance to killing by the alkylating agent MNNG, but not to killing by γ-irradiation, as shown in Figure 5–36. Because MGMT is specific for removal of O^6-methylguanine and its overexpression confers resistance, O^6-methylguanine must be responsible for cell killing.

Reference: Meikrantz W, Bergom MA, Memisoglu A & Samson L (1998) O^6-Alkylguanine DNA lesions trigger apoptosis. *Carcinogenesis* 19, 369-372.

5–66*

5–67

A. As evident in the micrograph in Figure 5–37A, hRad52 binds to the ends of DNA. Also, it apparently binds more effectively to ends with single-strand tails than it does to blunt-ended molecules, since DNA molecules with hRad52 bound to the ends are commonly observed only when the DNA has single-stranded tails.

B. By binding to the ends of the linear DNA, hRad52 prevents access by the exonuclease, which requires a free end, but does not interfere with digestion by the endonuclease, which can cleave in the interior of the DNA.

C. The preferential binding of hRad52 to ends with single-strand tails, as opposed to blunt ends, suggests that broken ends are processed first by an exonuclease to create single strands to which hRad52 can bind. The binding of hRad52 then protects against further exonuclease action. At this point it is thought that hRad52 loads hRad51, a RecAlike recombinase, onto the single strands, leading to the subsequent step of strand invasion on a homologous duplex, as shown in Figure 5–69.

Reference: Van Dyck E, Stasiak AJ, Stasiak A & West SC (1999) Binding of double-strand breaks in DNA by human Rad52 protein. *Nature* 398, 728–731.

5–68*

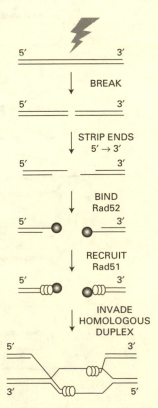

Figure 5–69 The initial steps in hRad52-promoted repair of DNA breaks by homologous recombination (Answer 5–67).

5–69   The variable in these experiments is light: the brighter the light, the less the observed killing. Thus, visible light can reverse the effects of UV irradiation. Direct reversal of UV damage is common in microorganisms and is called enzymatic photoreactivation. The enzyme from *E. coli* has two chromophores that cooperate in capturing photons from sunlight and using their energy to unlink pyrimidine dimers.

The account here is not much different from the original discovery of photoreactivation by Albert Kelner in the 1940s. While investigating the effects of postirradiation temperature on UV survival, Kelner was plagued by another variable. In his own words:

"Careful consideration was made of variable factors which might have accounted for such tremendous variation. We were using a glass-fronted water bath placed on a table near a window, in which were suspended transparent bottles containing the irradiated spores. The fact that some of the bottles were more directly exposed to light than others suggested that light might be a factor... Experiments showed that exposure of UV-irradiated suspensions to light resulted in an increase in survival rate or a recovery of 100,000- to 400,000-fold. Controls kept in the dark... showed no recovery at all."

Reference: Friedberg EC, Walker GC & Siede W (1995) DNA Repair and Mutagenesis, pp 92–103. New York: WH Freeman.

GENERAL RECOMBINATION

5–70*

5–71   The recombination substrates and products are shown in Figure 5–70. The

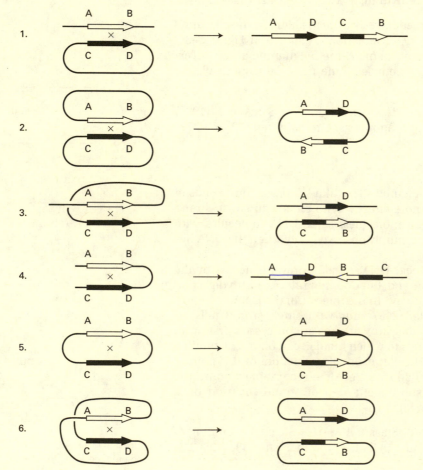

Figure 5–70 Alignment and crossovers in various recombination substrates (Answer 5–71).

first rule for deducing the recombination products is to align the homologous segments, that is, to draw the arrows one above the other so that they are pointing in the same direction. Alignment requires a twisting of substrates 3, 4, and 6. Alignment is necessary in order to form a Holliday junction, as would be more apparent if real sequences were used instead of arrows.

Substrates 3 and 4 illustrate a useful rule. Recombination between direct repeats in a chromosome (as in substrate 3) deletes one copy of the repeat and the intervening DNA. Recombination between inverted repeats in a chromosome (as in substrate 4) simply inverts the DNA between the repeats.

5–72* **Reference:** Potter H & Dressler D (1976) On the mechanism of genetic recombinaton: electron microscopic observation of recombination intermediates. *Proc. Natl. Acad. Sci. USA* 73, 3000–3004.

5–73

A. RecBCD must cut the DNA at (or near) the Chi site, since a 400-nucleotide fragment was generated from a substrate labeled at the left end (see Figure 5–42, lane 2) and a 100-nucleotide fragment was generated from a substrate labeled at the right end (lane 5). These lengths mark the position of the Chi site in the original DNA fragment (see Figure 5–41). Since shorter labeled fragments were generated only when the top strand (5′L and 3′R) was labeled, RecBCD must cut only the top strand. This result means that RecBCD recognizes the orientation of a Chi site—if the Chi site were flipped in this experiment, the bottom strand would have been cleaved.

B. The ability of RecBCD to separate DNA strands is shown by the unboiled control, which was indistinguishable from the boiled sample (see Figure 5–42, compare lanes 5 and 6). If RecBCD simply bound to the Chi site and introduced a nick, the 100-nucleotide single strand would have remained attached to the original fragment.

C. These results suggest that RecBCD stimulates homologous recombination in the region of the Chi site by nicking one strand and melting off a single-strand whisker. This single strand in the presence of RecA can be used to search for homology on another duplex and thereby initiate formation of a cross-strand exchange.

From these and other experiments it is known that RecBCD binds to the end of a linear, double-stranded DNA (a broken duplex in bacterial cells, since the chromosome is normally circular) and proceeds along the DNA molecule, unwinding the strands of the duplex. When it reaches a properly oriented Chi site, it cuts one strand. Subsequent unwinding past the Chi site produces a single-stranded tail, which then initiates homologous recombination in the presence of RecA by invading a homologous duplex. The overall result is a repair of the original double-strand break.

Reference: Ponticelli AS, Schultz DW, Taylor AF & Smith GR (1985) Chi-dependent DNA strand cleavage by RecBCD enzyme. *Cell* 41, 145–151.

5–74* Figure 5–71.

Reference: Cox M & Lehman IR (1981) The polarity of the recA protein-mediated branch migration. *Proc. Natl. Acad. Sci. USA* 78, 6023–6027.

5–75 This statement is incorrect. Crossing and noncrossing pairs of strands can be interconverted by rotational movements that do not require strand breakage.

5–76* Figure 5–72.

5–77

A. RuvC can cleave only 4 of the 256 (4^4) possible 4-nucleotide sequences, which is 1/64 of all possible sequences.

B. No. One of the four 4-nucleotide sequences would be expected on average every 64 nucleotides. Thus, only a small amount of branch migration will juxtapose a Holliday junction with an appropriate cleavage sequence. In cells RuvC operates in conjunction with RuvAB, which is a helicase that drives branch migration of Holliday junctions.

C. Evidently, the two subunits of RuvC coordinate their cleavages. Only when both have encountered an appropriate cleavage sequence does either site get cleaved. This conclusion is apparent in the results with the hybrid junction in Figure 5–46B. When one duplex carries a resolution sequence but the other does not, the sequence is not cleaved. This indicates that the two subunits do not operate independently of one another. Additional experiments in the reference below showed that in Holliday junctions with two cleavable but nonidentical sequences, both sequences were cleaved.

D. The duplexes generated by cleavage of the indicated strands of the Holliday junction in Figure 5–46A would have the same sequences shown in the figure, except that segment *a* would be connected to segment *d*, and segment *c* would be connected to segment *b*. Thus, a crossover would be generated. In the absence of any proteins but RuvC, the two product duplexes would each carry a nick at the site of RuvC cleavage.

Reference: Shah R, Cosstick R & West SC (1997) The RuvC protein dimer resolves Holliday junctions by a dual incision mechanism that involves base-specific contacts. *EMBO J.* 16, 1464–1472.

5–78* Figure 5–73.

5–79 True. Conversion is the change in frequency of markers (nucleotide differences) during recombination. In the starting duplexes each marker is present equally: once in each duplex (or twice in each duplex if the individual strands are counted). In the products of a recombination event associated with conversion, the frequency of markers is altered, so that they are no longer equal. Instead of 1:1 (or 2:2) as in the input duplexes, the frequency becomes 2:0 (or 4:0 or 3:1) in the output duplexes. The two common mechanisms for generating this change—mismatch repair and DNA synthesis—both involve some amount of DNA synthesis.

5–80*

SITE-SPECIFIC RECOMBINATION

5–81 Transposable elements integrate nearly randomly and genes often are destroyed or altered by the integration event. While it is true that some of these events are lethal to the cell and to the transposable element, most events are not. Spreading throughout the genome, even at the cost of a few cells (and transposons), ensures that the transposable element will survive with the species.

5–82* **Reference:** Bender J & Kleckner N (1986) Genetic evidence that Tn10 transposes by a nonreplicative mechanism. *Cell* 45, 801–815.

5–83

A. Transposition of the Ty element depends on reverse transcription of an RNA intermediate. Normally, reverse transcriptase is expressed at a very low level. Your modified plasmid, however, places the gene under control of the galactose control elements. In the presence of glucose (absence of galactose) the galactose control elements turn the gene off and, as a result, the expression of reverse transcriptase is very low. In the presence of galactose the reverse transcriptase gene is expressed at very high levels. Thus, the frequency of transposition increases substantially.

B. The frequency of Ty-induced *HIS⁺* colonies is low because a very specific kind of transposition event is required to activate the defective histidine gene: the Ty element must transpose to a site near the 5′ end of the gene.

Thus, even though nearly all cells show evidence for transposition, insertion near the defective histidine gene is still relatively rare.

C. The data in Figure 5–50 indicate that nearly every cell harboring the Ty-bearing plasmid suffers one or more transposition events when grown on galactose. Each Ty transposition has the potential for altering the function or expression of genes near the site of integration. If the element integrates into the coding portion of a gene, it can eliminate the encoded function; if it integrates in the noncoding region near a gene, it may alter the gene's expression. In organisms such as yeasts, which have been finely tuned to their environmental niche by evolutionary pressure, it is unlikely that random insertion of a Ty element will improve growth characteristics. Thus it is not unreasonable that a high rate of transposition should cause cells to grow poorly.

These data do not prove that the cells grow more slowly because of the high rate of transposition, even though that explanation is very likely to be correct. As the authors point out, the high level of expression of reverse transcriptase might interfere directly with RNA metabolism. For example, mRNA molecules could be inactivated by reverse transcription. Alternatively, the reverse transcripts of the cellular mRNAs could be mutagenic to the nuclear genes. Errors introduced during reverse transcription into DNA could be incorporated into the nuclear genes by recombination.

Reference: Boeke JD, Garfinkel DJ, Styles CA & Fink GR (1985) Ty elements transpose through an RNA intermediate. *Cell* 40, 491–500.

5–84* Figure 5–74.

Reference: Ruffner DE, Sprung CN, Minghetti PP, Gibbs PEM & Dugaiczyk A (1987) Invasion of the human albumin-α-fetoprotein gene family by *Alu, Kpn,* and two novel repetitive DNA elements. *Mol. Biol. Evol.* 4, 1–9.

5–85 Cre-mediated recombination between oppositely oriented LoxP sites inverts the sequences between the sites, whereas recombination between loxP sites in the same orientation deletes the sequences between the sites (Figure 5–75). This result should remind you of the similar results obtained for homologous recombination between direct repeats and inverted repeats in Problem 5–71 (see Figure 5–39, substrates 3 and 4). As in that problem, the easiest way to work out the products is to align the LoxP sites and then follow the crossover between them.

5–86*

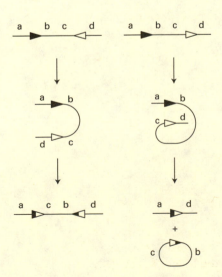

Figure 5–75 Products of Cre-mediated recombination between oppositely oriented and directly repeated LoxP sites (Answer 5–85).

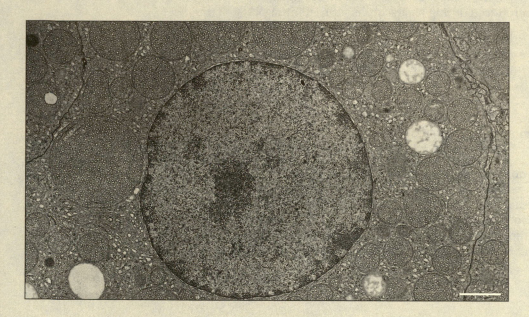

The cell nucleus. Scale bar is 1 μm.
Courtesy of Daniel S. Friend.

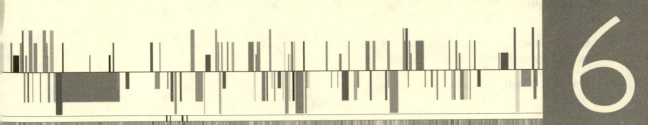

HOW CELLS READ THE GENOME: FROM DNA TO PROTEIN

FROM DNA TO RNA

6–1   The answer is best given by Francis Crick himself, who coined the terms 'the sequence hypothesis,' which proposes that genetic information is encoded in the sequence of the DNA bases, and 'the central dogma,' which states that DNA makes RNA makes protein, in 1957.

"I called this idea the central dogma, for two reasons, I suspect. I had already used the obvious word hypothesis in the sequence hypothesis, and in addition I wanted to suggest that this new assumption was more central and more powerful. I did remark that their speculative nature was emphasized by their names.

As it turned out, the use of the word dogma caused almost more trouble than it was worth. Many years later Jacques Monod pointed out to me that I did not appear to understand the correct use of the word dogma, which is a belief *that cannot be doubted*. I did apprehend this in a vague sort of way but since I thought that *all* religious beliefs were without serious foundation, I used the word in the way I myself thought about it, not as most of the rest of the world does, and simply applied it to a grand hypothesis that, however plausible, had little direct experimental support."

Reference: Crick F (1988) What Mad Pursuit: A Personal View of Scientific Discovery, p 109. New York: Basic Books, Inc.

6–2*

6–3   Actually, the RNA polymerases are not moving at all because they have been fixed and coated with metal to prepare the sample for viewing in the electron microscope. Before they were fixed, however, they were moving from left to right, as indicated by the gradual lengthening of the RNA transcripts. The RNA transcripts are shorter than the DNA that encodes them because they begin to fold up (acquire a three-dimensional structure) as they are synthesized, whereas the DNA is an extended double helix.

6–4* **Reference:** Thomas MJ, Platas AA & Hawley DK (1998) Transcriptional fidelity and proofreading by RNA polymerase II. *Cell* 93, 627–637.

6–5 A with 3, B with 4, C with 2, D with 5, and E with 1.

6–6*

6–7 False. The σ subunit associates with the bacterial RNA polymerase only during the initiation phase of RNA synthesis. The σ subunit helps the RNA polymerase bind to the promoter and stays associated with the polymerase until RNA synthesis has been properly initiated.

6–8* Figure 6–49.

6–9

A. Since RNA polymerase is blocked by pyrimidine dimers, the sensitivity of transcription of a gene will depend on the distance between the promoter and the probe. It is a simple matter of the target size for UV damage. If the polymerase must travel twice as far to make a transcript, the chances of its encountering a block to transcription are twice as great.

B. Transcription through the VSG gene is seven times more sensitive to UV irradiation than transcription through the ribosomal transcription unit at the site of rRNA probe 4, which is about 7 kb from its promoter. Thus, the beginning of the VSG gene is located about 50 kb (7 × 7 kb) away from its promoter. This calculation assumes that the DNA between the VSG promoter and the VSG gene has about the same sensitivity to UV light as the DNA in the ribosomal RNA transcription unit. It also assumes that multiple UV-induced pyrimidine dimers are not common enough to skew the linear relationship between UV dose and distance.

C. If the nearby gene is 20% less sensitive to UV irradiation than the VSG gene, it is inactivated at 80% the rate of the VSG gene. Therefore, its promoter is 40 kb away (0.80 × 50 kb). Given that the nearby gene is 10 kb in front of the VSG gene, its promoter must map very near the promoter for the VSG gene. Thus, it is likely that the two genes are transcribed from the same promoter.

Reference: Johnson PJ, Kooter JM & Borst P (1987) Inactivation of transcription by UV irradiation of *T. brucei* provides evidence for a multicistronic transcription unit including a VSG gene. *Cell* 51, 273–281.

6–10* **Reference:** McKnight SL & Kingsbury R (1982) Transcriptional control signals of a eukaryotic protein-coding gene. *Science* 217, 316–324.

6–11

A. The 400-nucleotide transcript is absent from lane 4, Figure 6–6B, because GTP was included in the reaction mixture. GTP allows transcription to proceed beyond the C-minus sequence (the synthetic sequence lacking C nucleotides), thereby generating transcripts longer than 400 nucleotides. In the absence of GTP (see lane 2) transcription cannot proceed beyond the C-minus sequence. In the presence of GTP and RNase T1 (see lanes 6 and 8) the longer transcripts are cleaved at the first G to yield the 400-nucleotide transcript.

B. One of the difficulties in assaying promoter function *in vitro* is the high background of nonspecific initiation of transcription. It is this background that is so evident in Figure 6–6B, lane 3. Its source is not altogether clear, but transcription may start at sequences in the rest of the plasmid that weakly resemble true RNA polymerase II promoters.

C. A transcript of about 400 nucleotides is present in Figure 6–6B, lane 5, because cleavage with RNase T1 liberates it from any randomly initiated transcript that has traversed the C-minus sequence. It is actually a few nucleotides longer than the specifically initiated transcript since its 5′ end is defined by the first G that precedes the C-minus sequence.

The 400-nucleotide transcript is absent from Figure 6–6B, lane 7, because 3′ O-methyl GTP will terminate most transcripts that are initiated in front of the C-minus sequence. The combination of 3′ O-methyl GTP and RNase T1 eliminates virtually all the background synthesis from the control plasmid.

D. As shown in Figure 6–6B, lanes 7 and 8, specific transcription can be assayed in the presence of G nucleotides if 3′ O-methyl GTP and RNase T1 are included (to inhibit background transcription and to cleave any random transcripts to small pieces).

Reference: Sawadogo M & Roeder RG (1985) Factors involved in specific transcription by human RNA polymerase II: analysis by a rapid and quantitative *in vitro* assay. *Proc. Natl. Acad. Sci. USA* 82, 4394–4398.

6–12* **Reference:** Sawadogo M & Roeder RG (1985) Factors involved in specific transcription by human RNA polymerase II: analysis by a rapid and quantitative *in vitro* assay. *Proc. Natl. Acad. Sci. USA* 82, 4394–4398.

6–13 General transcription factors play several roles in promoting transcription by RNA polymerase II. They help position the RNA polymerase correctly at the promoter, they aid in pulling apart the two strands of DNA to allow transcription to begin, and they release RNA polymerase from the promoter once transcription has begun. They are called 'general' because they assemble on all promoters used by RNA polymerase II; they are identified by names beginning with TFII (transcription factor for RNA polymerase II). That label also serves to distinguish them from more specialized factors that enhance transcription at selected promoters in certain cell types.

6–14* **Reference:** Roeder RG (1974) Multiple forms of deoxyribonucleic acid-dependent ribonucleic acid polymerase in *Xenopus laevis*. Isolation and partial characterization. *J. Biol. Chem.* 249, 241–248.

6–15

A. Since equal amounts of transcription from each template were observed when the preincubation was carried out with the individual templates or a mixture (see Figure 6–8C, lanes 1 to 3), Srb2 protein does not show a preference for either template.

B. These results indicate that Srb2 acts stoichiometrically. If Srb2 acted catalytically, it should have been able to modify the second template after the two were mixed. Catalytic activity would have produced transcripts from both templates regardless of which one was originally included in the preincubation with Srb2. When *excess* Srb2 was added at the beginning of the preincubation with one template, transcription was observed from both templates after mixing. This is consistent with a stoichiometric requirement for Srb2 and rules out the possibility that Srb2 was inactivated during the preincubation—and for that reason unable to act (catalytically) on the second template after mixing.

C. The production of transcripts solely from the template that was preincubated with Srb2 indicates that Srb2 is part of the preinitiation complex. If Srb2 were able to act after transcription had begun, transcripts would have been produced from both templates regardless of which one was included in the preincubation.

D. During preincubation of the template with the extract and Srb2, a number of proteins including Srb2 bind to the promoter to form a preinitiation complex. Evidently, the preinitiation complex, once formed, is stable and does not readily exchange proteins with other templates that are added later.

E. Although the *SRB2* gene was originally identified as a suppressor of the cold-sensitive phenotype of yeast carrying an RNA polymerase II gene with a short CTD, neither the genetic results nor these transcription assays provide

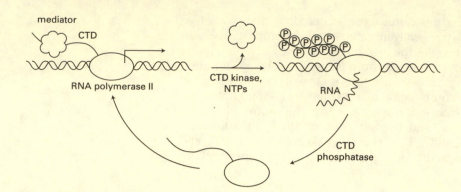

Figure 6–50 A summary of the proposed role of the mediator complex in initiation of transcription (Answer 6–15).

evidence that Srb2 binds to the CTD. (Nor do they argue against direct interaction; they simply do not speak to the issue.) Additional experiments have shown that Srb2 is part of a complex of proteins known as the mediator. The mediator binds to the dephosphorylated CTD, entering and leaving initiation complexes at every round of transcription in a process that may be coupled to C-terminal domain phosphorylation and the release of RNA polymerase at initiation of transcription (Figure 6–50).

References: Koleske AJ, Buratowski S, Nonet M & Young RA (1992) A novel transcription factor reveals a functional link between the RNA polymerase II CTD and TFIID. *Cell* 69, 883–894.

Thompson CM, Koleske AJ, Chao DM & Young RA (1993) A multisubunit complex associated with the RNA polymerase II CTD and TATA-binding protein in yeast. *Cell* 73, 1361–1375.

Svejstrup JQ, Li Yang, Fellows J, Gnatt A, Bjorklund S & Kornberg RD (1997) Evidence for a mediator cycle at the initiation of transcription. *Proc. Natl. Acad. Sci. USA* 94, 6075–6078.

6–16*

6–17 The bead would rotate clockwise from the perspective of the magnet, as shown in Figure 6–51A. The effect of polymerase relative motion on the helix is shown in Figure 6–51B.

Reference: Harada Y, Ohara O, Takatsuki A, Itoh H, Shimamoto N & Kinosita K (2001) Direct observation of DNA rotation during transcription by *Escherichia coli* RNA polymerase. *Nature* 409, 113–115.

6–18*

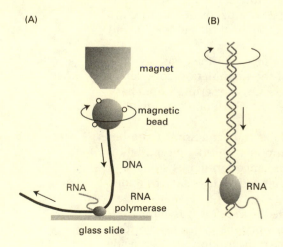

(A) (B)

Figure 6–51 Rotation of duplex due to movement of RNA polymerase (Answer 6–17). (A) Direction of rotation of the magnetic bead. (B) Direction of rotation of the DNA duplex.

6–19   The pattern of reaction with DNA and protein is strikingly clear. When the U analog was closer than 10 nucleotides to the 3′ end of the RNA, it reacted predominantly with its pairing partner in the template strand. By contrast, when it was 10 nucleotides or farther from the 3′ end, it no longer reacted with DNA at all, and reacted more strongly with protein. These patterns of reactivity indicate that the newly synthesized RNA remains paired with the DNA template over a stretch of 8–9 nucleotides from the 3′ end, and then separates from the template strand. It also seems that the U analog must be closely associated with the RNA polymerase even when it is up to 24 nucleotides from the 3′ end.

Reference: Nudler E, Mustaev A, Lukhtanov E & Goldfarb A (1997) The RNA–DNA hybrid maintains the register of transcription by preventing backtracking of RNA polymerase. *Cell* 89, 33–41.

6–20*

6–21   True. At its 3′ end the mRNA has a string of adenine nucleotides, the last of which has a terminal ribose with a free 3′-OH group. At its 5′ end the mRNA carries a 7-methylguanosine that is linked 5′ to 5′ with the first nucleotide in the mRNA. This linkage leaves a free 3′-OH group on the ribose of the capping nucleotide.

6–22*

6–23   The tails arise because the ends of the mRNA are not complementary to the ends of the restriction fragment. Thus, one of the single-strand tails at each end is DNA from the restriction fragment. A single-strand tail at one end corresponds to the 5′ end of the mRNA, which must come from an upstream exon that is not present in the restriction fragment. A single-strand tail at the other end corresponds to the 3′ end of the mRNA, which may come from a downstream exon not present in the restriction fragment or simply be the poly-A tail itself. Without additional information you cannot identify which single strand comes from which source.

Reference: Berget SM, Berk AJ, Harrison T & Sharp PA (1977) Spliced segments at the 5′ termini of adenovirus-2 late mRNA: a role for heterogeneous nuclear RNA in mammalian cells. *Cold Spring Harbor Symp. Quant. Biol.* 42, 523–529.

6–24*   Figure 6–52.

References: Dugaiczyk A, Woo SL, Lai EC, Mace Jr ML, McReynolds L & O'Malley BW (1978) The natural ovalbumin gene contains seven intervening sequences. *Nature* 274, 328–333.

Garapin AC, Cami B, Roskam W, Kourilsky P, Le Pennec JP, Perrin F, Gerlinger P, Cochet M & Chambon P (1978) Electron microscopy and restriction enzyme mapping reveal additional intervening sequences in the chicken ovalbumin split gene. *Cell* 14, 629–639.

6–25
A. A single nucleotide change in a gene could cause an internal deletion in the mRNA if it altered splicing so that an exon that was usually incorporated was skipped instead.
B. Removal of 173 nucleotides from the protein-coding portion of the mRNA would cause a shift in the reading frame for translation into amino acids. Because a codon for an amino acid is three nucleotides, a loss of 173 nucleotides does not correspond to an integral number of codons. Thus, the Smilin encoded by the deleted mRNA would be fine up to the missing exon, but would encode an unrelated sequence of amino acids thereafter.
C. The simplest explanation is that the Smilin gene contains a 173-nucleotide-long exon (exon 2 in Figure 6–53A) that is lost during the processing of the

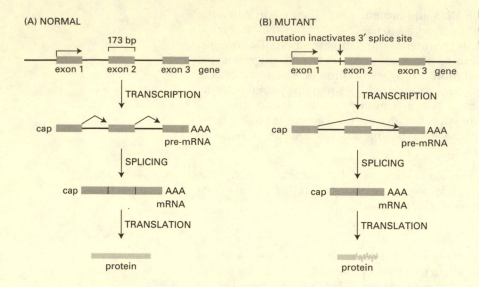

Figure 6–53 Splicing of the Smilin transcript (Answer 6–25). (A) Normal transcript. (B) Mutant transcript.

mutant precursor mRNA. This could occur, for example, if the mutation changed the 3′ splice site in the preceding intron so that it was no longer recognized by the splicing machinery (a change in the conserved AG at the intron–exon boundary could do this). Use of the next available 3′ splice site—adjacent to exon 3—would cause loss of exon 2 from the mutant mRNA (Figure 6–53B). During protein synthesis, the absence of exon 2 (173 nucleotides) would throw the ribosomes out of the correct reading frame as they moved from exon 1 to exon 3. At that junction the ribosomes would begin synthesizing a protein sequence unrelated to that normally encoded by exon 3.

6–26*

6–27   Since introns evolve faster than exons, the introns of the different species will be more variable than the exons. It is difficult to scan these sequences by eye and decide, with confidence, which side is the more conserved. One way to quantify the differences is to pick one sequence, for example, the cow, and count up how often the other sequences differ at each position, as shown in Figure 6–54. Summing the differences on either side of the junction makes it clear that sequences on the left are much more similar to one another than are the sequences on the right. (A similar difference exists no matter which sequence is chosen for comparison.) Thus, the more conserved sequences, which are on the left in Figure 6–15, correspond to exons, and the less conserved sequences, which are on the right, correspond to introns.

   This simple exercise emphasizes how essential computers are for evaluating relatedness of different sequences.

6–28*   **Reference:** Zhuang Y & Weiner AM (1986) A compensatory base change in U1 snRNA suppresses a 5′ splice site mutation. *Cell* 46, 827–835.

6–29

A. If the splicing machinery binds to one splice site and scans across the intron to find its complementary splice site, it must use the first appropriate splice site it encounters. (Not using the first splice site is equivalent to skipping an exon.) The expected products from intron scanning in the two minigenes are

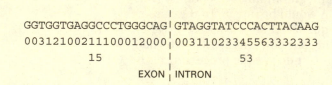

```
GGTGGTGAGGCCCTGGGCAG GTAGGTATCCCACTTACAAG
00312100211100012000 00311023345563332333
         15                    53
         EXON | INTRON
```

Figure 6–54 Sum of differences from the cow β-globin sequence (Answer 6–27). The β-globin sequence from the cow was compared nucleotide by nucleotide with the β-globin sequences from the other six species. The number of differences at each position is summed *below* each nucleotide of the β-globin sequence. The total number of differences on each side of the exon–intron boundary (*dashed line*) is shown at the *bottom* (15 and 53).

shown in Figure 6–55. If the splicing machinery binds to a 5′ splice site and scans toward a 3′ splice site, minigene 1 should generate one product (Figure 6–55A) and minigene 2 should generate two products (Figure 6–55B). By contrast, if the splicing machinery binds to a 3′ splice site and scans toward a 5′ splice site, minigene 1 should generate two products (Figure 6–55A) and minigene 2 should generate one product (Figure 6–55B).

B. The results of this experiment do not match the expectations for either direction of intron scanning. Therefore, selection of splice sites by a mechanism involving unidirectional scanning through introns in either a 5′-to-3′ or a 3′-to-5′ direction seems unlikely. The ordering mechanism by which cells avoid exon skipping probably depends on two factors: first, that assembly of the spliceosome occurs as the pre-mRNA emerges from the RNA polymerase; and second, that exons may be defined as an independent step prior to removal of introns.

Reference: Kuhne T, Wieringa B, Reiser J & Weissmann C (1983) Evidence against a scanning model for RNA splicing. *EMBO J.* 2, 727–733.

6–30* Figure 6–56.

Reference: Krause M & Hirsh D (1987) A trans-spliced leader sequence on actin mRNA in *C. elegans. Cell* 49, 753–761.

6–31 Group I excised introns are linear, and they carry the activated G in covalent linkage at their 5′ ends. Group II excised introns are lariats, with the activated A having reacted with the 5′-most nucleotide of the intron. The mechanism of pre-mRNA splicing catalyzed by the spliceosome is most similar to the mechanism used by the Group II self-splicing introns.

6–32*

6–33

A. The 5′ ends of the RNA molecules were labeled. Only labeled fragments show up in the autoradiograph (see Figure 6–22). Thus, if the shortest fragments (those that were at the bottom of the gel) are from the 5′ end, the 5′ end must have been labeled.

B. The bands corresponding to the As in the AAUAAA signal sequence are missing from the ladder of bands in polyadenylated and cleaved RNA (see Figure 6–22, lanes 3 and 4) because modification of any one of those As interfered with cleavage and with polyadenylation. Thus, RNA molecules that carry a single modification in the signal sequence are not recognized by the components of the extract and, as a result, do not show up in the population of molecules that carry poly-A tails (see lane 3) or in the population of molecules that are cleaved (see lane 4).

C. The band at the arrow in Figure 6–22 is absent in the polyadenylated RNA but present in the cleaved RNA because modification of this A does not prevent cleavage, but it does prevent polyadenylation. Thus, RNA molecules with this A modified are present in the cleaved molecules (see Figure 6–22, lane 4) but are not present in the polyadenylated molecules (see lane 3).

D. The analysis of the missing bands in parts B and C indicates that the AAUAAA signal sequence is important for the cleavage of precursor RNAs and that the AAUAAA sequence and the single A are required for polyadenylation.

E. If the other end—the 3′ end—of the RNA molecules were labeled, it would have been possible to determine whether any of the As or Gs on the 3′ side of the cleavage site were important for polyadenylation. These experiments have been done; they show that no single modification 3′ of the polyadenylation site prevents polyadenylation. The sequence requirements on the 3′ side of the cleavage site (GU- or U-rich) are not so specific as those on the 5′ side and would not be expected to be inactivated by single changes.

Reference: Conway L & Wickens M (1987) Analysis of mRNA 3′-end formation by modification interference: the only modifications which prevent processing lie in AAUAAA and the poly (A) site. *EMBO J.* 6, 4177–4184.

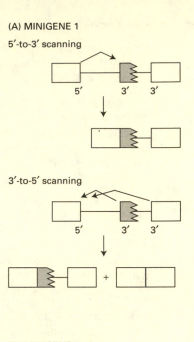

(A) MINIGENE 1

5′-to-3′ scanning

3′-to-5′ scanning

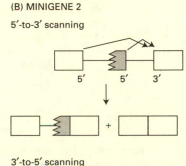

(B) MINIGENE 2

5′-to-3′ scanning

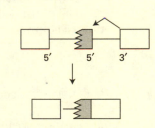

3′-to-5′ scanning

Figure 6–55 Expected products in a test of intron scanning (Answer 6–29). (A) Expected products for 5′-to-3′ scanning and 3′-to-5′ scanning of intron in minigene 1. (B) Expected products for 5′-to-3′ scanning and 3′-to-5′ scanning of intron in minigene 2. *Open boxes* indicate complete exons; *shaded boxes* represent partial exons.

6–34* Figure 6–57.

Reference: Mowry KL & Steitz JA (1987) Identification of human U7 snRNP as one of several factors involved in the 3′-end maturation of histone pre-messenger RNAs. *Science* 238, 1682–1687.

6–35 To be 'export ready,' it seems that an mRNA must be bound by the appropriate set of proteins. Proteins such as the cap-binding complex must be present, while proteins such as spliceosome components must be absent. RNA fragments from excised introns do not acquire the necessary set of proteins and are thus doomed to degradation.

6–36*

6–37 These results indicate that box elements C and D are important for the accumulation of U85 RNA. The presence of E2 RNA in Figure 6–25B, lanes 6 and 12, shows that the transfections were successful. Thus the absence of U85 RNA from those lanes is meaningful. It is unclear from these studies whether the altered box elements prevent processing from the intronic RNA or render the processed RNA unstable.

Reference: Jady B & Kiss T (2001) A small nucleolar guide RNA functions both in 2′-*O*-ribose methylation and pseudouridylation of the U5 spliceosomal RNA. *EMBO J.* 20, 541–551.

6–38* **Reference:** Jady B & Kiss T (2001) A small nucleolar guide RNA functions both in 2′-*O*-ribose methylation and pseudouridylation of the U5 spliceosomal RNA. *EMBO J.* 20, 541–551.

FROM RNA TO PROTEIN

6–39 The amino acids encoded in each of the three reading frames are shown in Figure 6–58. If this segment of RNA encoded part of a larger protein, it would have to be translated in reading frame 1, which is the only one that does not contain a stop codon.

6–40*

6–41
   A. UUUUUUUUUUU… codes for FFFF…, that is, a polymer of phenylalanine.
   B. AUAUAUAUAUAU…codes for IYIY…, that is, a polymer of alternating isoleucines and tyrosines. Because the start point of the ribosome on the RNA is random, however, the ribosomes will generate a mixture of polymers, some of which start with isoleucine and some with tyrosine.
   C. AUCAUCAUCAUC… codes for a mixture of three different polymers. Ribosomes randomly start translation in each of the three reading frames: AUC–AUC–AUC–AUC-…codes for IIII…, a polymer of isoleucine; UCA–UCA–UCA–UCA-…codes for SSSS…, a polymer of serine; and CAU–CAU–CAU–CAU-…, codes for HHHH…, a polymer of histidine.

6–42*

6–43 Mutations of the type described in 2 and 4 are often the most harmful. In both cases, the reading frame would be changed. Because these frameshifts occur early in the coding sequence, the encoded protein will contain a nonsensical and usually truncated sequence of amino acids. In contrast, a reading-frameshift that occurs toward the end of the coding sequence, as described in 1, will result in a largely correct protein that may be functional.

Deletion of three consecutive nucleotides, as in scenario 3, leads to the deletion of one amino acid, but does not alter the reading frame. The deleted amino acid may or may not be important for the folding or activity of the protein. In many cases such mutations are silent; that is, they have insignificant consequences for the organism.

FRAME 1

```
AGU CUA GGC ACU GA-3′
 S   L   G   T
```

FRAME 2

```
A GUC UAG GCA CUG A-3′
   V   *   A   L
```

FRAME 3

```
AG UCU AGG CAC UGA-3′
    S   R   H   *
```

3-FRAME TRANSLATION

```
  AGUCUAGGCACUGA-3′
1  S  L  G  T
2   V  *  A  L
3    S  R  H  *
```

Figure 6–58 Amino acids encoded in the three reading frames of an RNA (Answer 6–39). The amino acids encoded in each reading frame are shown separately and all together as they are usually represented.

Substitution of one nucleotide for another, as in scenario 5, is often completely harmless, because it does not change the encoded amino acid. In other cases it may change an amino acid, which may be deleterious or benign, depending on the location and functional significance of that amino acid. Often, the most deleterious kind of single-nucleotide change creates a new stop codon, which gives rise to a truncated protein.

6–44*

6–45   In present-day cells, there is some wobble in the matching of codons to anti-codons. In a number of cases, the same tRNA can pair with several codons that specify the same amino acid but differ slightly in their nucleotide sequence. It seems likely that in the early world, without such highly evolved ribosomes to help in the pairing process, the converse may also have been true: several different tRNAs, with slightly different anticodons, may have been able to bind to the same codon. This would have played havoc with the translation of the genetic message into protein, unless the amino acids carried by these tRNAs were chemically similar. In this way natural selection may have ensured that tRNAs with similar anticodons carried chemically similar amino acids.

Perhaps in the early world, before modern aminoacyl-tRNA synthetases had evolved, there was also some 'wobble' in the matching of tRNAs with appropriate amino acids. The same tRNA might have become coupled to any of a number of amino acids that were chemically similar. One can imagine the evolution of the genetic code by refinement of a matching process that was originally imprecise and gave only a blurred relationship between sets of roughly similar codons and sets of roughly similar amino acids.

6–46*

6–47   The rules for wobble pairing between the anticodon and the codon, expressed both ways, are shown in Table 6–5. It appears strange that A in the wobble position in eucaryotes does not have a pairing partner. It is thought that A is not used in the wobble position in the tRNAs in eucaryotes. When-ever A is encoded in the wobble position, it is changed to inosine (I) after transcription, giving rise to a mature tRNA that recognizes U or C in the wobble position of the codon.

Reference: Lander ES et al (2001) Initial sequencing and analysis of the human genome. *Nature* 409, 860–921.

6–48*

6–49   The single codon for tryptophan is 5′-UGG. The anticodon of the normal tryptophan tRNA is 5′-CCA, which pairs specifically with this codon.

TABLE 6–5 Rules for wobble base-pairing between codon and anticodon (Answer 6–47).

	WOBBLE CODON BASE	POSSIBLE ANTICODON BASE		WOBBLE ANTICODON BASE	POSSIBLE CODON BASE
Bacteria	U	A, G, or I	Bacteria	U	A or G
	C	G or I		C	G
	A	U or I		A	U
	G	C or U		G	U or C
				I	U, C, or A
Eucaryotes	U	G or I	Eucaryotes	U	A
	C	G or I		C	G
	A	U		A	
	G	C		G	U or C
				I	U or C

A mutation that changes the anticodon to 5'-*UCA* would allow the tRNA to recognize the 5'-UGA. That would lead to insertion of tryptophan at the UGA stop codon, preventing termination of translation. Because of wobble, the mutant anticodon would also recognize the normal 5'-UGG codon, so that, in principle, its ability to insert tryptophan at the normal UGG codons would not be compromised.

Many protein-encoding sequences, however, contain UGA codons as their natural stop sites. These stop codons would also be affected by the mutant tRNA. In reality there is a competition between the mutant tRNA and the termination factors. Whenever the tRNA wins the race, however, the affected proteins would be made with additional amino acids at their C-terminal ends. The additional lengths would depend on the number of codons before the ribosomes encounter a non-UGA stop codon in the mRNA in the reading frame in which the protein is translated. Of course, there is also the possibility that the ribosomes might encounter another UGA stop codon, that termination factors might win the race this time, and that translation would then stop. The potential chaos that such mutations might cause is mitigated by two factors: the efficiency of translation of stop codons by such mutant tRNAs is usually low, and many bacterial genes are 'protected' by double stop codons at their ends. In truth, such suppressors have been invaluable in bacterial genetics.

6–50* **Reference:** Chapeville F, Lipmann F, von Ehrenstein G, Weisblum B, Ray WJ & Benzer S (1962) On the role of soluble ribonucleic acid in coding for amino acids. *Proc. Natl. Acad. Sci. USA* 48, 1086–1092.

6–51 One effective way of driving a reaction to completion is to remove one of the products. This has the effect of increasing the flow of substrates to products to reestablish the equilibrium ratio—the principle of mass action. All three of the products of this reaction are removed. AMP is constantly reconverted to ADP and then to ATP by other reactions in the cell; thus the concentration of AMP is maintained at a low level. Similarly, the aminoacyl-tRNAs are used in protein synthesis, constantly decreasing their concentrations. But by far the most dramatic influence is the removal of PP_i by hydrolysis to two phosphates. That reaction yields as much free energy as the hydrolysis of ATP to ADP, which means that essentially all of the PP_i will be converted to free phosphates. As a result, the linked reactions for charging a tRNA and hydrolyzing PP_i—the reactions as they occur in cells—have a $\Delta G°$ of –6.9 kcal/mole.

6–52* Figure 6–59.

Reference: Hale SP, Auld DS, Schmidt E & Schimmel P (1997) Discrete determinants in transfer RNA for editing and aminoacylation. *Science* 276, 1250–1252.

6–53

A. The data in Figure 6–29 indicate that the N-terminus of the protein is synthesized first. The steadily decreasing level of radioactivity from the N-terminus to the C-terminus is exactly what you would expect if synthesis begins at the N-terminus. As illustrated in Figure 6–60A and B, all the ribosomes carry a labeled lysine at position 8 in their nascent chains, but the ribosome at the 5' end of the mRNA has not yet reached the lysine at position 16. Thus, when digested with trypsin, all of these nascent chains will yield a labeled N-terminal peptide, but a smaller fraction will yield the second peptide. Fewer still will contain the third peptide, and so on. Almost none of the ribosomes will carry a nascent chain with the labeled lysine nearest the C-terminus.

B. The lines for the α and β chains are very similar, with intercepts on both axes very close together, which indicates that roughly equal *numbers* of each chain are being synthesized. However, there is not enough information to decide whether the numbers of α- and β-globin mRNA molecules are equal. You would need to know how many ribosomes there were on each

(A) RIBOSOMES WITH ATTACHED PEPTIDES

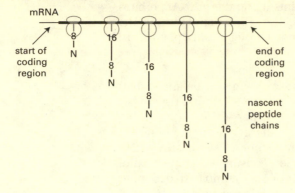

mRNA

start of
coding
region

end of
coding
region

nascent
peptide
chains

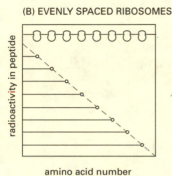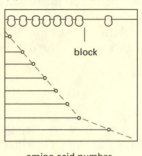

(B) EVENLY SPACED RIBOSOMES

(C) BLOCKED RIBOSOMES

radioactivity in peptide

amino acid number

block

amino acid number

Figure 6–60 Relationship of ribosome position to peptide length and labeling pattern (Answer 6–53). (A) Lengths of peptides associated with ribosomes at various positions along β-globin mRNA. Numbers refer to positions of the first two lysines. (B) Pattern of peptide labeling for evenly spaced ribosomes. (C) Pattern of peptide labeling for ribosomes whose movement is inhibited at a point midway down the mRNA. Peptides associated with each ribosome are shown on the graph as *lines*. *Small circles* correspond to the C-termini of the polypeptides and are aligned immediately *below* the ribosome on which the peptides are synthesized. *Dashed lines* through the small circles show the expected patterns of peptide labeling.

mRNA—the average polyribosome size for α- and β-globin mRNAs—to deduce their relative abundance from these kinds of data. Actually, there is about twice as much α-globin mRNA as β-globin mRNA, but the α-globin mRNA is less efficiently translated; that is, fewer ribosomes initiate synthesis on α-globin mRNA per unit time than on β-globin mRNA. These factors cancel out to give a fairly balanced production of the two chains.

C. The graph in Figure 6–29 hits zero right at the end of the coding region, which indicates that chains are released from ribosomes as soon as they encounter the stop codon—or at least they do so without a measurable pause on this time scale.

D. If there were a significant roadblock to ribosome movement, the data would resemble that in Figure 6–30A. A roadblock would result in more densely packed ribosomes in front of the block and less densely packed ribosomes beyond the block. The consequences of inhibited ribosome movement are illustrated schematically in Figure 6–60C.

6–54*

6–55

A. As shown in Figure 6–61, the rate of synthesis is linear with time. The curvature so apparent in the autoradiograph in Figure 6–31 results from the nonlinear migration of proteins in SDS polyacrylamide gels.

B. The rate of protein synthesis can be determined from the slope of the line in Figure 6–61. This system is synthesizing roughly 52,000 daltons of protein per 10 minutes, or 5200 daltons per minute, which corresponds to about 47 amino acids per minute [(5200 daltons/minute)/(110 daltons/amino acid)]. This rate is less than the rate in *E. coli*, which is about 10 times greater. The rate is also about three times less than that of globin synthesis in the same reticulocyte lysate. As discussed in part C, part of the reason for the low rates may be that the mix of tRNAs in the reticulocyte lysate is not optimal for this plant virus protein.

C. The autoradiograph contains many bands, rather than just a few, because ribosomes keep loading onto the mRNA throughout the course of the experiment. You could obtain the theoretical result in Figure 6–32B by adding an inhibitor of initiation after 5 minutes or, alternatively, by adding unlabeled methionine in vast excess after 5 minutes.

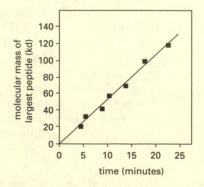

Figure 6–61 Rate of synthesis of a TMV protein (Answer 6–55).

The presence of discrete bands rather than a continuous background fuzz suggests that there are specific hang-up points along the mRNA, perhaps where ribosomes must wait for rare tRNAs. The tRNA population in the reticulocyte is specialized for making globin, not a protein from a plant virus!

6–56*

6–57   The energy invested in translation and transcription will balance when 30 protein molecules have been made from one mRNA. Protein synthesis requires four high-energy phosphate bonds per codon (per three nucleotides). Transcription consumes six high-energy phosphate bonds to make a codon, but also consumes 19 times more energy synthesizing RNA that will be discarded (95%). Thus, transcription consumes a grand total of 120 high-energy phosphate bonds per codon (6 + 114), compared to four per codon for translation. The ratio of energies invested per codon (120/4 = 30) defines the number of proteins that will have been made when the energy invested in translation matches that invested in transcription. Because most mRNAs are used to make hundreds to thousands of proteins, translation consumes a much higher fraction of the cell's energy than does transcription.

6–58*

6–59
   A. 5'-GUAGCCUACCCAUAGG-3'
   B. This short mRNA encodes three different peptides because there are three different reading frames. In the second reading frame, the first codon is the stop codon UAG; however, the subsequent codons can be translated.

>   5'-GUAGCCUACCCAUAGG-3'
>   Frame 1   V   A   Y   P   *
>   Frame 2   *   P   T   H   R
>   Frame 3   S   L   P   I

The other possible mRNA from this DNA would read

>   5'-CCUAUGGGUAGGCUAC-3'
>   Frame 1   P   M   G   R   L
>   Frame 2   L   W   V   G   Y
>   Frame 3   Y   G   *   A

Thus, the sequence of the peptides would be completely different. Be careful to keep the polarity of the strands correct; don't fall into the trap of thinking that the complementary sequence of the first mRNA is 5'-CAUCGGAUGGGUAUCC-3', which is incorrect because the strands of DNA run in opposite directions.
   C. If translation begins at the 5' end of the RNA, the synthesized protein would be valine-alanine-tyrosine-proline (VAYP). Only after a peptide bond has been formed between alanine and tyrosine will tRNA^{Ala} leave the ribosome. Thus, the next tRNA that will bind to the ribosome after tRNA^{Ala} has left is tRNA^{Pro}. When the amino group of alanine forms a peptide bond, the ester bond between valine and tRNA^{Val} is broken, tRNA^{Val} moves from the P-site to the E-site (exit site), and tRNA^{Ala} moves from the A-site to the P-site.
   D. Since there is no AUG (or even GUG, which is sometimes used as a start signal for translation in *E. coli*), this sequence cannot come from the beginning of the coding region of a gene. It could be from the end of a gene, if the first or second reading frames were used, or from the middle of a gene, if the third reading frame were used. More information is required to distinguish between these possibilities.

6–60*

6–61   In eucaryotic cells protein synthesis is normally initiated by scanning from the 5' end of the mRNA until the first AUG codon is found. This mechanism of initiation ensures that ribosomes will all start translating near the 5' end of the mRNA. When the ribosomes complete synthesis of the protein, they fall off the mRNA and must reinitiate by scanning from the 5′ end. By contrast, in procaryotic cells protein synthesis is initiated by base-pairing between mRNA sequences adjacent to an initiation AUG codon and sequences in the 16S rRNA of the small ribosomal subunit. This key difference in the mechanism by which procaryotic cells identify a start codon underlies their ability to make several proteins from a single polycistronic mRNA.

6–62* **References:** Craigen WJ, Cook RG, Tate WP & Caskey CT (1985) Bacterial peptide chain release factors: conserved primary structure and possible frameshift regulation of release factor 2. *Proc. Natl. Acad. Sci. USA* 82, 3616–3620.

Jacks T & Varmus H (1985) Expression of the Rous sarcoma virus *pol* gene by ribosomal frameshifting. *Science* 230, 1237–1242.

6–63

A. Since the bacteria were labeled for one generation, which represents a doubling in mass, 4 μg of the 8 μg of flagellin isolated from the gel were synthesized in the presence of ^{35}S cysteine. The amount of radioactivity in the sample indicates that about 1 out of every 1670 flagellin (flgn) molecules contains a cysteine.

$$\frac{Cys}{flgn} = \frac{300 \text{ cpm Cys}}{4 \text{ μg flgn}} \times \frac{\text{pmol Cys}}{5 \times 10^3 \text{ cpm}} \times \frac{4 \times 10^4 \text{ μg flgn}}{\text{μmol flgn}} \times \frac{\text{μmol}}{10^6 \text{ pmol}}$$

$$= \frac{6 \times 10^{-2} \text{ pmol Cys}}{100 \text{ pmol flgn}}$$

$$\frac{Cys}{flgn} = 6 \times 10^{-4}$$

which is equal to 1 cysteine per 1670 flagellin molecules $[1/(6 \times 10^{-4})]$.

B. The normal codons for cysteine are UGU and UGC. Thus, the error in anti-codon–codon interaction is a mistake at the first position of the codon (third position of the anticodon). The experiment described, and other experiments too, suggest that ribosomes tend to mistake U for C and C for U in the first two positions of the codon, and C and U for A in the first position.

C. Assuming that all six arginine codons are equally likely, there should be six sensitive (CGC and CGU) arginine codons $[(2/6) \times 18]$ in a flagellin molecule. Therefore, the actual error frequency per codon-at-risk is

$$\text{error frequency} = \frac{1 \text{ cysteine}}{1670 \text{ flagellin molecules}} \times \frac{\text{flagellin molecule}}{6 \text{ sensitive codons}}$$

$$\text{error frequency} = 10^{-4}$$

D. If the probability of making a mistake at each codon is 10^{-4}, the probability of not making a mistake at each codon is $(1 - 10^{-4})$. The probability of not making a mistake at n codons is then $(1 - 10^{-4})^n$. Thus, the percentage of correctly synthesized molecules 100 amino acids in length is $(1 - 10^{-4})^{100}$, or 99%. For a protein 1000 amino acids long, 90% are correct. For a molecule 10,000 amino acids long, only 37% are correct. Given these sorts of estimates, it is perhaps not surprising that proteins more than 3000 amino acids long are rare. Referring back to Problem 6–56, what fraction of titin molecules would you expect to be made correctly?

Reference: Edelman P & Gallant J (1977) Mistranslation in *E. coli. Cell* 10, 131–137.

6–64*

6–65 SmpB evidently plays no role in the charging of tmRNA with alanine, since that reaction is unaffected by the presence or absence of SmpB. SmpB is, however, critical for the association of tmRNA with ribosomes (see Figure 6–34A), which presumably explains why protein fragments are not tagged and degraded in SmpB-deficient cells (see Figure 6–34B). Although these studies identify roughly where in the process SmpB acts, they do not define its precise function.

Reference: Karzai AW, Susskind MM & Sauer RT (1999) SmpB, a unique RNA-binding protein essential for the peptide-tagging activity of SsrA (tmRNA). *EMBO J.* 18, 3793–3799.

6–66* **References:** Horowitz S & Gorovsky MA (1985) An unusual genetic code in nuclear genes of *Tetrahymena. Proc. Natl. Acad. Sci. USA* 82, 2452–2455.

Andreasen PH, Dreisig H & Kristiansen K (1987) Unusual ciliate-specific codons in *Tetrahymena* mRNAs are translated correctly in a rabbit reticulocyte lysate supplemented with a subcellular fraction from *Tetrahymena. Biochem. J.* 244, 331–335.

6–67

A. Edeine specifically inhibits initiation of protein synthesis by preventing the joining of the 60S ribosomal subunit to the 40S subunit/mRNA/initiator tRNA complex. Since elongation is not blocked, ribosomes that have already begun synthesis complete their individual chains and fall off the mRNA, leaving attached only the small subunit and the initiator tRNA. Edeine is an antibiotic produced by certain strains of *Bacillus brevis.*

B. A lag occurs before protein synthesis shuts off because edeine inhibits initiation but has no effect on elongation. Thus, a ribosome that has just started making a new polypeptide is free to complete it. Incorporation of label continues for just the length of time it takes to complete the protein (in this case, the globin chains of hemoglobin), which takes about a minute.

C. If cycloheximide (or any other elongation inhibitor) is added at the same time as an initiation inhibitor, the polyribosomes are 'frozen.' Polyribosome breakdown by initiation inhibitors requires ribosome movement, which is blocked by elongation inhibitors.

Reference: Safer B, Kemper W & Jagus R (1978) Identification of a 48S preinitiation complex in reticulocyte lysate. *J. Biol. Chem.* 253, 3384–3386.

6–68*

6–69 Exposed hydrophobic patches indicate that a protein is abnormal in some way. In a well-folded protein the majority of hydrophobic amino acids will be sequestered in the interior away from water. Some proteins initially fold with exposed hydrophobic patches that are used in binding to other proteins, ultimately burying those hydrophobic amino acids as well. Thus, hydrophobic amino acids are usually not exposed on the surface of a protein, and any significant patch is a good indicator that something has gone awry. The protein may have failed to fold properly after leaving the ribosome, suffered an accident that partly unfolded it at a later time, or failed to find its normal partner subunit in a larger protein complex.

6–70* **References:** Teter SA, Houry WA, Ang D, Tradler T, Rockabrand D, Fischer G, Blum P, Georgopoulos C & Hartl FU (1999) Polypeptide flux through bacterial Hsp70: DnaK cooperates with trigger factor in chaperoning nascent chains. *Cell* 97, 755–765.

Deuerling E, Schulze-Specking A, Tomoyasu T, Mogk A & Bukau B (1999) Trigger factor and DnaK cooperate in folding of newly synthesized proteins. *Nature* 400, 693–696.

6–71

A. When IPTG is present, the strain that does not express TF (Δ*tig*) grows as well as the strain that does not express DnaK (I-*dnaK*), and also as well as the wild-type strain, at all temperatures. When IPTG is absent, the Δ*tig* strain continues to grow whereas, in contrast, the I-*dnaK* strain does not do so at either 15°C or 42°C, although it grows fine at the intermediate temperatures. Thus, DnaK appears to be the more critical chaperone under the stressful conditions of high and low temperature.

Figure 16–28 Movement of microtubules on a bed of microtubule motor molecules (Problem 16–65).

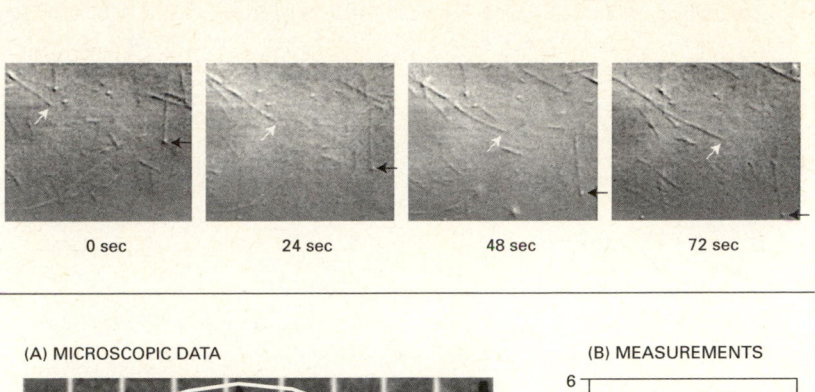

0 sec   24 sec   48 sec   72 sec

Figure 16–46 Regulation of filopodial extension and retraction (Problem 16–99).

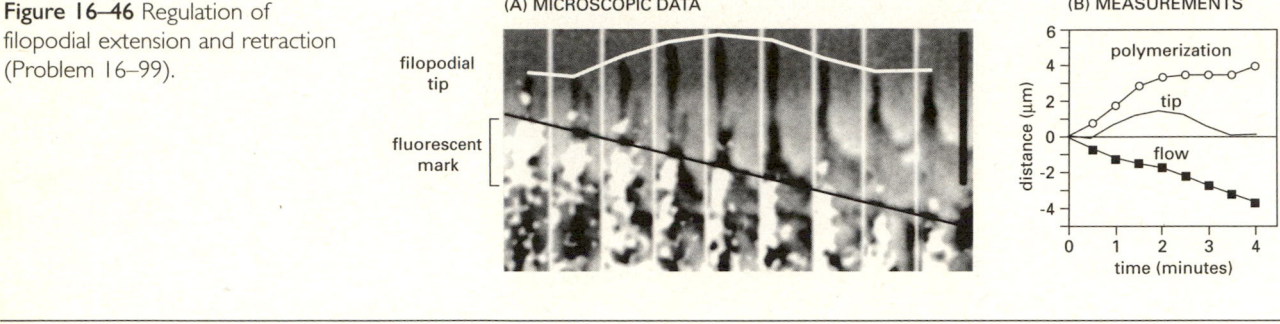

(A) MICROSCOPIC DATA

filopodial tip

fluorescent mark

(B) MEASUREMENTS

polymerization

tip

flow

distance (μm)

time (minutes)

Figure 17–38 Effects of DPP expression on wing development in *Drosophila* (Problem 17–84)

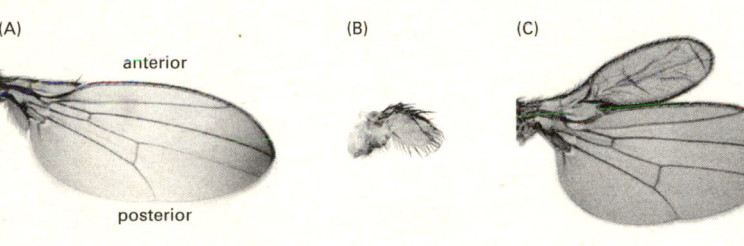

(A)

anterior

posterior

(B)

(C)

(D)

Figure 18–20 Bipolar spindles and nuclear divisions in *Sciara* (Problem 18–34).

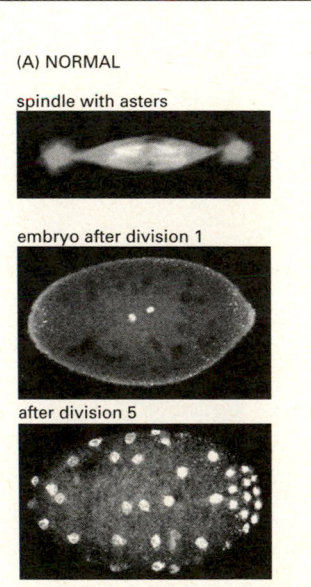

(A) NORMAL

spindle with asters

embryo after division 1

after division 5

(B) PARTHENOGENETIC

anastral spindle

embryo after division 1

after division 5

Figure 12–8 The cellular distribution of Rev (Problem 12-33).

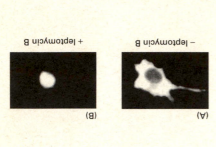

(A) – leptomycin B

(B) + leptomycin B

Figure 8–35 Immunoaffinity purification of *Xenopus* ORC (Problem 8-72).

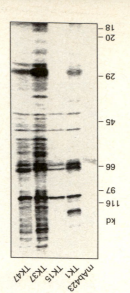

kd — 116, 97, 66, 45, 29, 20, 18

Lanes: mAb423, TK1, TK37, TK15, TK47

Figure 7–35 Influence of cycles of light and dark on mRNAs for two circadian rhythm proteins in cultured zebrafish hearts (Problem 7-54).

(A) LIGHT–DARK CYCLES

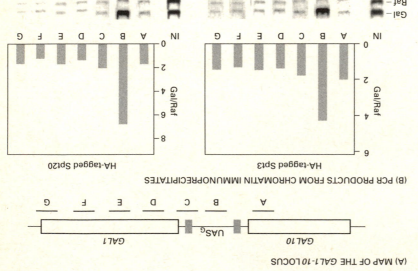

DAY 1 DAY 2 DAY 3 DAY 4 DAY 5
time of day (h) 3 9 15 21 3 9 15 21 3 9 15 21 3 9 15 21 3 9 15 21
Clock mRNA
Timeless mRNA

(B) DARK–LIGHT CYCLES

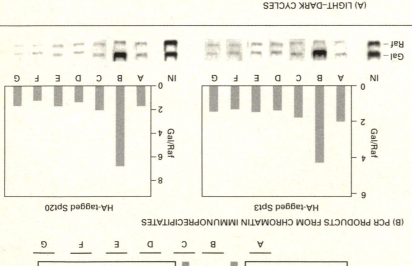

DAY 1 DAY 2 DAY 3 DAY 4 DAY 5
time of day (h) 3 9 15 21 3 9 15 21 3 9 15 21 3 9 15 21 3 9 15 21
Clock mRNA
Timeless mRNA

Figure 7–26 Analysis of SAGA association with the Gal promoter (Problem 7-41).

(A) MAP OF THE *GAL1-10* LOCUS

GAL10   UAS$_G$ GAL1
A   B   C   D   E   F   G

(B) PCR PRODUCTS FROM CHROMATIN IMMUNOPRECIPITATES

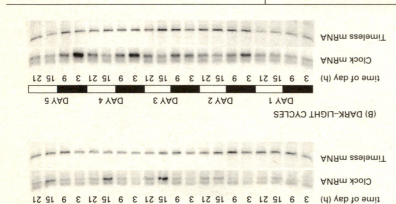
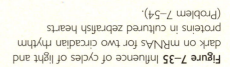

HA-tagged Spt3   —   HA-tagged Spt20
Gal/Raf
IN A B C D E F G
Gal–
Raf–

Molecular Biology of the Cell, Fourth Edition: A Problems Approach. © John Wilson and Tim Hunt. ISBN 0-8153-3577-6. © 2002 Garland Science.

Figures 7–26, 7–35, 8–35, 12–8, 16–28, 16–46, 17–38 and 18–20 may not have been printed clearly in your copy of this book. We provide the figures on this sheet for clarification. The table of radioactive isotopes inside the back cover contains a printing error. The isotopes should be ^{14}C, ^{3}H, ^{35}S, ^{125}I, ^{32}P, and ^{131}I.

B. Unlike the single mutants, the double mutant does not grow at any temperature in the absence of both TF and DnaK (that is, when there is no IPTG in the plate). The lethality of the double mutant at 30°C and 37°C, where both of the single mutants grow perfectly well, is termed synthetic lethality. A synthetic lethality indicates that the two gene products cooperate in some way, which need not be direct. However, because both of these genes encode hsp70 chaperones, it seems likely that they collaborate or cover for each other in some way in protein folding. DnaK can fully compensate for the loss of TF throughout the temperature ranges tested in these experiments, and TF can compensate for the loss of DnaK at intermediate, but not extreme temperatures. When neither chaperone is functional, misfolding of proteins even at intermediate temperatures is lethal to the cells. Additional experiments demonstrated that in the absence of both chaperones cytosolic proteins undergo massive aggregation.

References: Deuerling E, Schulze-Specking A, Tomoyasu T, Mogk A & Bukau B (1999) Trigger factor and DnaK cooperate in folding of newly synthesized proteins. *Nature* 400, 693–696.

Teter SA, Houry WA, Ang D, Tradler T, Rockabrand D, Fischer G, Blum P, Georgopoulos C & Hartl FU (1999) Polypeptide flux through bacterial Hsp70: DnaK cooperates with trigger factor in chaperoning nascent chains. *Cell* 97, 755–765.

6–72* **References:** Shtilerman M, Lorimer GH & Englander SW (1999) Chaperonin function: Folding by forced unfolding. *Science* 284, 822–825.

Brinker A, Pfeifer G, Kerner MJ, Naylor DJ, Hartl FU & Hayer-Hartl M (2001) Dual function of protein confinement in chaperonin-assisted protein folding. *Cell* 107, 223–233.

6–73 Molecular chaperones fold like any other protein. Molecules in the act of synthesis on ribosomes are bound by hsp70 chaperones. And incorrectly folded molecules are helped by hsp60-like chaperones. That they function as chaperones when they have folded correctly makes no difference to the way they are treated before they reach their final, functional conformation.

6–74*

6–75
A. In this experiment the plateau value of the radioactivity in the absence of proteasome inhibitors was about 30% lower than that in their presence, suggesting that 30% of newly synthesized proteins are degraded in proteasomes in lymph node cells. Similar results in other cell types suggest that substantial degradation of newly synthesized proteins may be common to most cells.
B. The absence of differential affects on some proteins is surprising. One possibility is that a constant fraction (about 30% of all proteins) misfolds and is degraded by proteasomes; however, given the variety of proteins it seems unlikely that they would all misfold to the same degree. Another possibility is that newly synthesized proteins are sampled randomly for degradation to serve some other biological function. (One such function is to provide an array of peptides for display on the cell surface to inform the immune system of the proteins the cell is currently making—which is thought to give it a head start in identifying infected cells.) Alternatively, it may be that ribosomes make a high fraction of mistakes (30%) in the form of peptide fragments and misincorporations that cannot fold properly and are normally removed. These errors would contribute to a background of radioactivity throughout the gel, accounting for an increased overall intensity in the absence of proteasomes.

Reference: Schubert U, Antón LC, Gibbs J, Norbury CC, Yewdell JW & Bennink JR (2000) Rapid degradation of a large fraction of newly synthesized proteins by proteasomes. *Nature* 404, 770–774.

6–76*

6–77   Overexpression of dm-N70 does not cause accumulation of any of the proteasome substrates. That is the expected result if the D-box is critical for its interaction with APC. Because it doesn't bind to APC, it does not influence ubiquitylation.

Overexpression of K0-N70 gives the result that was initially anticipated: D-box proteins accumulate, but non-D-box proteins do not. This result suggests that K0-N70 specifically interferes with destruction of D-box proteins, probably by competing for binding to APC.

The most difficult result to understand is the original one: overexpression of N70 causes accumulation of both D-box and non-D-box proteins. Because removal of its lysines eliminates this effect, it seems likely that the effect is caused by ubiquitylation of N70. Overexpression of N70 and its ubiquitylation is thought to sequester a large fraction of the cellular supply of ubiquitin; thus, interfering with ubiquitylation of other proteins by decreasing the availability of ubiquitin.

Reference: Yamano H, Tsurumi C, Gannon J & Hunt T (1998) The role of the destruction box and its neighbouring lysine residues in cyclin B for anaphase ubiquitin-dependent proteolysis in fission yeast: defining the D-box receptor. *EMBO J.* 17, 5670–5678.

6–78* **Reference:** Bachmair A, Finley D & Varshavsky A (1986) *In vivo* half-life of a protein is a function of its amino-terminal residue. *Science* 234, 179–186.

6–79

A. The absence of radioactivity at the position of β-galactosidase in Figure 6–45 indicates that the labeled antibodies against ubiquitin do not react with the protein at that position. Thus, the band that is marked β-galactosidase in Figure 6–44A does not carry any attached ubiquitin, indicating that ubiquitin was removed from the N-terminus of the fusion protein.

B. The ubiquitin above the position of β-galactosidase in Figure 6–45 must be attached to β-galactosidase, since the enzyme was purified by binding to antibodies specific for β-galactosidase. Ubiquitin attached to the two unstable enzymes—but not to the stable enzyme—suggests that ubiquitin marks the protein for degradation. The ladderlike appearance of the bands suggests that a variable number of copies of ubiquitin are attached to the protein (at lysines) before the enzyme is degraded.

Reference: Bachmair A, Finley D & Varshavsky A (1986) *In vivo* half-life of a protein is a function of its amino-terminal residue. *Science* 234, 179–186.

THE RNA WORLD AND THE ORIGINS OF LIFE

6–80*

6–81   Although RNA is thought to have played an important role in the evolution of life on Earth, possibly as a replicating catalyst, it is unclear that it was the *first* replicating catalyst. Other less efficient replicating catalysts may have preceded RNA. It is clear, however, that RNA now plays a greater role than that of mere messenger in information flow; RNA plays critical roles in replication, splicing, translation, peptide-bond formation, membrane transport of proteins, and telomere maintenance.

6–82* **Reference:** Bartel DP & Szostak JW (1993) Isolation of new ribozymes from a large pool of random sequences. *Science* 261, 1411–1418.

6–83   The complement of a hairpin RNA could also form a similar hairpin, as shown in Figure 6–62. The two structures would be identical in the double-stranded regions that involved canonical GC and AU base pairs. They would differ in the sequence of the single-stranded regions. Because GU base pairs

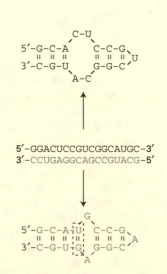

Figure 6–62 Hairpins formed by an RNA strand and its complement (Answer 6–83). An RNA and its complement are shown as double-stranded RNA in the middle. The structures formed by each strand are shown *above* and *below* the duplex.

are stable in RNA, whereas CA base pairs are not, there can also be regions that are unpaired in one strand but base-paired in the other, as shown by the boxed GU base pair.

6–84* Figure 6–63.

Reference: Bartel DP & Szostak JW (1993) Isolation of new ribozymes from a large pool of random sequences. *Science* 261, 1411–1418.

6–85

A. There are 6×10^{15} molecules, 300 nucleotides (nt) in length, in 1 mg of RNA.

$$\text{number} = \frac{1 \text{ RNA molecule}}{300 \text{ nt}} \times \frac{\text{nt}}{330 \text{ d}} \times \frac{6 \times 10^{20} \text{ d}}{1 \text{ mg}}$$
$$= 6 \times 10^{15} \text{ RNA molecules}$$

B. If the 220 nucleotide segment is completely random, there will be four choices of nucleotide at each of 220 positions, which is 4^{220} or about 3×10^{132} possible different RNA molecules. Thus, in a 1-mg sample, there will be 2×10^{-117} [$(6 \times 10^{15})/(3 \times 10^{132})$] of all possible sequences represented…a trivial fraction of the whole. (A sample large enough to have one copy of each possible RNA would outweigh the known universe by more than 30 orders of magnitude.)

C. If a single 50-nucleotide RNA were required to catalyze the ligation, your chances of success would be close to nil. There are about 10^{18} different 50-nucleotide sequences represented in a 1-mg sample of RNA. Considering just the random 220 nucleotides, there would be about 170 different 50-mers in each of 6×10^{15} molecules (imagine sliding a 50-nucleotide window across the 220 nucleotides one nucleotide at a time) for a total of $170 \times 6 \times 10^{15}$, or 10^{18} different molecules. Since there are 4^{50} or about 10^{30} different 50-mers, your chances would be roughly 1 in a trillion (10^{-12}) of having the unique catalytic RNA in your sample. Are you feelin' lucky?

That so many ribozymes have been successfully isolated from such pools argues that a very large number of different sequences must be able to catalyze any given reaction, or that the catalytic RNAs must be very small. Since the identified ribozymes are not particularly small, it must be that many different sequences are capable of catalysis.

Reference: Bartel DP & Szostak JW (1993) Isolation of new ribozymes from a large pool of random sequences. *Science* 261, 1411–1418.

6–86* **Reference:** Bartel DP & Szostak JW (1993) Isolation of new ribozymes from a large pool of random sequences. *Science* 261, 1411–1418.

6–87 True. Although only a few types of reaction are represented among the ribozymes in present-day cells, ribozymes that have been selected in the laboratory can catalyze a wide variety of biochemical reactions.

6–88* **Reference:** Szostak JW, Bartel DP & Luisi L (2001) Synthesizing life. *Nature* 409, 387–390.

6–89 The deoxyribose sugar of DNA makes the molecule much less susceptible to breakage. The hydroxyl group on carbon 2 of the ribose sugar is an agent for catalysis of the adjacent 3'-5' phosphodiester bond that links nucleotides together in RNA. Its absence from DNA eliminates that mechanism of chain breakage. In addition, the double helical structure of DNA provides two complementary strands, which allows damage in one strand to be repaired accurately by reference to the sequence of the second strand. Finally, the use of T in DNA instead of U, as in RNA, builds in a protection against the effects of deamination—a common form of damage. Deamination of T produces an aberrant base (methyl C), whereas deamination of U generates C, a normal base. The presence of an abnormal base eases the cell's job of recognizing the damaged strand.

Dolly and her first lamb, Bonnie. Courtesy of Roddy Field, Photographer Roslin Institute.

CONTROL OF GENE EXPRESSION

AN OVERVIEW OF GENE CONTROL

7–1   False. Carrots can be grown from single carrot cells and tadpoles can be gotten by injecting differentiated frog nuclei into frog eggs. But carrots cannot be gotten from frog eggs no matter what.

7–2*   **Reference:** Wilmut I, Schnieke AE, McWhir J, Kind AJ & Campbell KHS (1997) Viable offspring derived from fetal and adult mammalian cells. *Nature* 385, 810–813.

7–3   The V and C segments in the germline are separated by a very long segment of DNA, which contains sites for the restriction nuclease used in the digestion. As a result, the V- and C-segment probes each hybridize to a band of a different size (Figure 7–56). After the two segments are brought into proximity by the deletion of the intervening DNA in B cells, the probes might be expected to hybridize to new bands. If the chosen restriction nuclease does not cleave between the two segments, the situation shown in Figure 7–56, both probes will hybridize to the same band. For the data in Figure 7–56, a restriction nuclease was found that didn't cut between the rearranged segments, because that gave the clearest demonstration that a rearrangement had indeed occurred.

There are three additional possible patterns, depending on the sites at which the restriction nuclease cleaves relative to the endpoints of the rearrangement. If the deleted DNA removed neither of the two restriction sites that lie between the V and C segments, then the hybridization pattern in B cells would be the same as in the germline. If the interior site near the

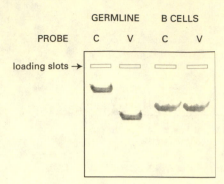

GERMLINE   B CELLS

PROBE   C   V   C   V

loading slots →

Figure 7–56 Pattern of hybridization to V- and C-segment probes in germline and B cells (Answer 7–3).

V segment was deleted, the C segment would be on the same fragment as in the germline, but the V segment would be on a fragment of a different size. If the interior site near the C segment was deleted, the V segment would be on the same fragment as in the germline, but the C segment would be on a fragment of a new size.

Reference: Tonegawa, S. (1983) Somatic generation of antibody diversity. *Nature* 302, 575–581.

7–4*   Figure 7–57.

7–5   Although it is true that cancer cells differ from their normal precursors, they typically differ in their expression of a relatively few genes (oncogenes and tumor suppressor genes). When the abundances of hundreds to thousands of mRNAs are compared, as they are in DNA microarray analysis, the patterns of mRNAs from the unaffected genes (the vast majority) allow a tumor to be definitively assigned to a particular tissue type.

7–6*   Figure 7–58.

DNA-BINDING MOTIFS IN GENE REGULATORY PROTEINS

7–7

A. The results of the matings in Figure 7–5 show that bacteriophage proliferation and cell lysis occur only when the donor carries a λ prophage and the recipient does not. These results are consistent with the idea that a repressor keeps the lytic genes of the prophage turned off. In bacteria that harbor a prophage, the presence of the repressor ensures that the prophage remains quiescent. When the prophage is transferred into a bacterium that does not carry a prophage, it finds itself in a repressor-free environment and its lytic program is induced. If the prophage is instead transferred into a bacterium that harbors a prophage of its own, the presence of the repressor keeps the prophage from being induced.

B. Exactly the same results would be expected if lysis were controlled by a gene regulatory protein that activated expression of an anti-lysis protein. If the prophage enters a prophage-negative recipient, the activator would be absent, preventing expression of the anti-lysis protein and allowing prophage induction and lysis to occur. It was only with additional genetic and biochemical experiments that the anti-lysis factor was identified as a repressor.

Reference: Echols H (2001) Operators and Promoters: The Story of Molecular Biology and Its Creators, pp 46–47. Berkeley: University of California Press.

7–8*   Figure 7–59.

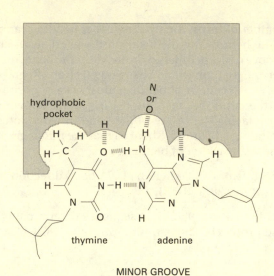

Figure 7–60 Arrangements of noncovalent bonds that allow proteins to recognize C–G and T–A base pairs (Answer 7–9).

cytosine   guanine

MINOR GROOVE

hydrophobic pocket

thymine   adenine

MINOR GROOVE

7–9   Contacts can form between the protein and the edges of the base pairs that are exposed in the major groove of the DNA. The types of contacts that can form are shown in Figure 7–60. Hydrogen bonds provide sequence-specific contacts. In addition, a hydrophobic interaction can sense the methyl group on the pyrimidine ring of T. The arrangement of hydrogen-bond donors and acceptors of a T–A base pair is different from that of a C–G base pair. Similarly, the arrangement of hydrogen-bond donors and acceptors of A–T and G–C base pairs are different from one another and from the two base pairs shown in Figure 7–60. In addition to the specific contacts shown in the figure, ionic interactions between positively charged amino acid side chains and the negatively charged phosphate groups in the DNA backbone usually stabilize DNA–protein interactions.

7–10*   **References:** Kabata H, Kurosawa O, Arai I, Washizu M, Margarson SA, Glass RE & Shimamoto N (1993) Visualization of single molecules of RNA polymerase sliding along DNA. *Science* 262, 1561–1563.

Shimamoto N (1999) One-dimensional diffusion of proteins along DNA. *J. Biol. Chem.* 274, 15293–15296.

7–11

A. At the point of minimum relative migration, the CAP sites are separated by 85 nucleotides (see Figure 7–9C). At 10.6 nucleotides per turn, this number of nucleotides corresponds to 8 helical turns (85/10.6 = 8). At the point of maximum relative migration, the CAP sites are separated by 79 nucleotides, which equals 7.5 helical turns.

B. Yes. The two CAP sites must be bent exactly the same way since they are identical. Therefore, they will both have the same groove of the helix facing the inside of the bend at the center of bending. In order for the DNA to bend into the *cis* configuration, the two centers of bending must be on the same side of the helix. The major grooves (or minor grooves) are on the same side of the helix at integral numbers of helical turns. Thus, it is expected that the point of minimum relative migration (the *cis* configuration) will occur after an integral number of helical turns. Similarly, the point of maximum relative migration (the *trans* configuration) will occur when the centers of bending are on opposite sides of the helix, that is, at half integral numbers of helical turns.

C. At the point of minimum migration of the construct with one CAP site and one $(A_5N_5)_4$ site, the centers of bending are separated by 101 nucleotides (see Figure 7–9D). At 10.6 nucleotides per helical turn, the centers of bending are separated by 9.5 helical turns (101/10.6 = 9.5).

D. Since the point of minimum relative migration (the *cis* configuration) occurs at a half-integral number of turns, the two centers of bending cannot have the same groove of the helix facing the inside of the bend. As discussed in part B, if the same groove of the helix faced the inside of the bend, the centers of bending in the *cis* configuration would be separated by an integral number of turns. Therefore, the two centers of bending must have opposite grooves facing the inside of the bend. Because the $(A_5N_5)_4$ site is known to bend with the major groove facing the inside of the bend (as was stated in the problem), the CAP-binding site must be bent so that the minor groove faces the inside of the bend at the center of bending.

References: Zinkel SS & Crothers DM (1987) DNA bend direction by phase sensitive detection. *Nature* 328, 178–181.

Gartenberg MR & Crothers DM (1988) DNA sequence determinants of CAP-induced bending and protein binding affinity. *Nature* 333, 824–829.

7–12*

7–13   False. Although the individual contacts are weak, the 20 or so contacts that are typically formed at a protein–DNA interface add together to ensure that the interaction is both highly specific and very strong. In fact, DNA–protein interactions include some of the tightest and most specific molecular interactions known in biology.

7–14*   **Reference:** Ptashne M (1986) A Genetic Switch: Gene Control and Phage λ, p 114. Oxford, UK: Blackwell Scientific Press.

7–15   The strength and specificity of the DNA–protein interaction can be adjusted with this motif by changing the number of zinc finger repeats. The other types of DNA-binding motifs, which function primarily as head-to-head dimers, cannot be so readily formed into repeating chains.

7–16*

7–17   The affinity of the dimeric λ repressor for its binding site is the sum of all the interactions made by each DNA-binding domain. An individual DNA-binding domain will make just half the contacts and provide just half the binding energy as the dimer. Thus, although the concentration of binding domains is unchanged, their binding as monomers is sufficiently weak that they do not compete with the binding of RNA polymerase. As a result, the genes for lytic growth are turned on.

7–18*   **Reference:** Patel LR, Curran T & Kerppola TK (1994) Energy transfer analysis of Fos–Jun dimerization and DNA binding. *Proc. Natl. Acad. Sci. USA* 91, 7360–7364.

7–19   The phosphate interferes with the function of the DNA-binding domain by adding a negative charge and by creating steric problems. Typically, DNA-binding domains are positively charged, which helps them bind to the negatively charged DNA. Addition of a negative charge would increase charge repulsion between the DNA and the protein, interfering with its function. Moreover, if the phosphate is added to the binding surface itself, it would likely interfere directly with the interaction of myogenin with the DNA.

A heterodimer formed between myogenin and a truncated HLH protein lacking a DNA-binding domain would be unable to bind to DNA tightly because it would make only half of the necessary contacts.

References: Benezra R, Davis RL, Lockshon D, Turner DL & Weintraub H (1990) The protein Id: a negative regulator of helix–loop–helix DNA binding proteins. *Cell* 61, 49–59.

Li L, Zhou J, James G, Heller-Harrison R, Czech MP & Olsen EN (1992) FGF inactivates myogenic helix–loop–helix proteins through phosphorylation of a conserved protein kinase C site in their DNA-binding domains. *Cell* 71, 1181–1194.

7–20* **References:** Treisman R (1987) Identification and purification of a polypeptide that binds to the c-fos serum response element. *EMBO J.* 6, 2711–2717.

Norman C, Runswick M, Pollock R & Treisman R (1988) Isolation and properties of cDNA clones encoding SRF, a transcription factor that binds to the *c-fos* serum response element. *Cell* 55, 989–1003.

7–21 False. Although DNA affinity chromatography can achieve purifications as great as 10,000-fold with relative ease, the amounts of pure protein are usually small. They are adequate, however, to obtain amino acid sequence information by mass spectrometry or other means. Armed with a bit of the amino acid sequence, it is usually possible to clone the gene. Once the gene is cloned, the protein can be produced in virtually unlimited amounts.

7–22* Table 7–7.

Reference: Pollock R & Treisman R (1990) A sensitive method for the determination of protein–DNA binding specificities. *Nucleic Acids Res.* 18, 6197–6204.

HOW GENETIC SWITCHES WORK

7–23 If the active, DNA-binding form of a gene regulatory protein (a gene repressor protein) serves to turn genes off, the mode of gene regulation is called negative control. If the active, DNA-binding form of a gene regulatory protein (a gene activator protein) serves to turn genes on, the mode of regulation is called positive control.

An inducing ligand can turn on a negatively controlled gene by binding to the gene repressor protein, causing an alteration in its affinity for the DNA that allows it to fall off. In the absence of the repressor the gene is turned on. An inducing ligand can turn on a positively controlled gene by binding to the gene activator protein, increasing its affinity for the DNA and allowing it to bind, which would turn the gene on.

An inhibitory ligand can turn off a negatively controlled gene by binding to the gene repressor protein and increasing its affinity for DNA, allowing it to bind and turn the gene off. An inhibitory ligand can turn off a positively controlled gene by binding to the gene activator protein and decreasing its affinity for DNA, allowing it to dissociate from the DNA and turning the gene off.

7–24* **Reference:** Glandsdorff N (1987) Biosynthesis of arginine and polyamines. In *Eschericia coli* and *Salmonella typhimurium*: Cellular and Molecular Biology (FC Neidhardt ed), pp 321–344. Washington DC: American Society for Microbiology.

7–25

A. If sufficient tryptophan is present in the cells, the tryptophan repressor will block the synthesis of enzymes that would make more tryptophan. Likewise, if cells are starved for tryptophan, the unoccupied repressor would not bind to the DNA and the enzymes that synthesize tryptophan would be induced. This simple and elegant form of feedback inhibition allows cells to adjust the rate of tryptophan synthesis to their needs.

B. In both scenarios, transcription of the genes encoding the tryptophan biosynthetic enzymes would no longer be regulated by the absence or presence of tryptophan. The enzymes would be permanently on in scenario (i) because the repressor could not bind to the DNA. The enzymes would be permanently off in scenario (ii) because the repressor would always be bound to the DNA.

C. In scenario (i), the normal tryptophan repressor molecules would completely restore the regulation of the tryptophan biosynthetic enzymes. Because the mutant repressor does not bind to the DNA, it would not affect the function of the normal repressor. By contrast, expression of the normal tryptophan

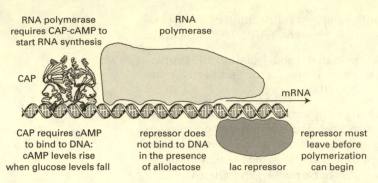

RNA polymerase requires CAP-cAMP to start RNA synthesis

RNA polymerase

CAP

mRNA

CAP requires cAMP to bind to DNA: cAMP levels rise when glucose levels fall

repressor does not bind to DNA in the presence of allolactose

lac repressor

repressor must leave before polymerization can begin

repressor would have no effect in scenario (ii) because the binding sites on the DNA would remain permanently occupied by the mutant repressor.

7–26* **Reference:** Schleif RF (1986) Genetics and Molecular Biology, Chap. 13. Reading, MA: Addison Wesley.

7–27

A. The rapid bacterial growth at the beginning of the experiment results from the metabolism of glucose. The slower growth at the end results from metabolism of lactose. The bacteria stopped growing in the middle of the experiment because they ran out of glucose but did not yet possess the enzymes necessary for lactose metabolism. Before they could utilize the lactose in the medium, they had to induce the *lac* operon. The delay in growth represents the time required for the induction.

B. Induction of the *lac* operon requires that two conditions be met: lactose must be present and glucose must be absent. During the first part of the experiment, glucose and lactose are both present; therefore, the conditions for induction are not met. Only when glucose is exhausted are the requirements for induction satisfied.

CAP and the lactose repressor mediate induction (Figure 7–61). For the operon to be on, CAP must be bound and the lactose repressor must not be bound. The presence of lactose in the medium increases the intracellular concentration of allolactose, which binds to the lactose repressor, thereby lowering its affinity for its binding site and causing its release from the DNA. Removal of the lactose repressor satisfies one condition for induction. The second condition is tied to the concentration of glucose. When the concentration of glucose falls, the intracellular level of cAMP rises. cAMP binds to CAP and alters its conformation so that it can bind to its binding site. When CAP is in place (and the lactose repressor is absent), RNA polymerase can bind to the promoter and initiate transcription.

Reference: Monod J (1947) The phenomenon of enzymatic adaptation. Growth Symposium XI:223–289. [Reprinted in Selected Papers in Molecular Biology by Jacques Monod (A Lwoff, A Ullmann, eds), pp 68–134. New York: Academic Press, 1947.]

7–28* Figure 7–62.

7–29

E. In the fused operons the genes in the *lac* operon have come under control of the regulatory region of the *trp* operon. Thus, expression of β-galactosidase, the product of the *lacZ* gene, will be regulated by the tryptophan repressor. Because the tryptophan repressor requires tryptophan in order to bind to its regulatory region and shut off the operon, expression of β-galactosidase (and the other genes in the fused operons) will occur only when tryptophan is absent from the medium.

7–30* **Reference:** Ninfa AJ, Reitzer LJ & Magasanik B (1987) Initiation of transcription at the bacterial *glnAp2* promoter by purified *E. coli* components is facilitated by enhancers. *Cell* 50, 1039–1046.

7–31 Neither the propagation of an altered DNA structure nor the oligomerization of a protein from the repression site would be expected to be sensitive to small changes in the spacing between the repression site and the start site of transcription. Nor is there any obvious reason in those mechanisms why some insertions and deletions would prevent repression, while other interspersed insertions and deletions would maintain repression.

The third mechanism—formation of a loop in the DNA—is consistent with the observations. Repression of the *galK* gene occurs when integral numbers of helical turns (multiples of 10.6 nucleotides, which is the number of nucleotide pairs per helical turn) are added to or deleted from the DNA. By contrast, when half turns are involved, repression is prevented. This is exactly the behavior expected if DNA must bend into a tight loop to allow AraC at site 2 to interact with another protein near the transcription start site. As shown in Figure 7–63, nonintegral numbers of turns would place the AraC protein on the wrong face of the DNA, which would require that the DNA twist half a turn to allow proper positioning of the proteins. Although twisting a DNA helix by half a turn may not seem difficult, it actually requires about 4 kcal/mole for a DNA 200 nucleotides in length. The binding energy available from typical protein–DNA interactions is about 10–15 kcal/mole. Since a substantial fraction of the binding energy would be required to twist the DNA, it is not unreasonable to expect that twisting the DNA could alter a delicately balanced interaction required for repression.

Other experiments suggest that AraC at site 2 may interact with AraC at site 1 to form the DNA loop. In the absence of arabinose, the *araBAD* genes are fully repressed, presumably by interference of the DNA loop with the binding of RNA polymerase. If AraC-binding site 2 is deleted (or if the AraC protein is absent because of mutation), the DNA loop cannot form and the binding of RNA polymerase is not prevented, thereby leading to a 10-fold elevation in transcription of the *araBAD* cluster of genes. When arabinose is present (and glucose is absent), the AraC protein undergoes a conformational change that prevents formation of the DNA loop and also facilitates the binding of RNA polymerase or aids its conversion to an open complex, thereby increasing transcription 1000-fold.

Reference: Dunn TM, Hahn S, Ogden S & Schleif RF (1984) An operator at −280 base pairs that is required for repression of *araBAD* operon promoter: addition of DNA helical turns between the operator and promoter cyclically hinders repression. *Proc. Natl. Acad. Sci. USA* 81, 5017–5020.

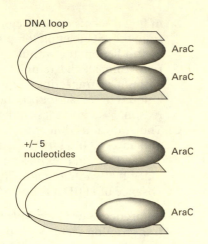

Figure 7–63 Diagram showing consequences of introducing a half turn of 5 nucleotides into a DNA loop formed by interaction between two proteins (Answer 7–31).

7–32*

7–33   The results of this experiment favor the DNA looping model, which would not be affected by the protein bridge (so long as it allowed the DNA to bend, which it does). The scanning or entry site model, however, is likely to be affected by the nature of the linkage between the enhancer and the promoter. If the proteins enter at the enhancer and scan to the promoter, they would have to traverse the protein bridge. If such proteins are geared to scan on DNA, they are likely to have difficulty scanning across a protein.

Reference: Mueller-Storm HP, Sogo JM & Schaffner W (1989) An enhancer stimulates transcription in trans when attached to the promoter via a protein bridge. *Cell* 58, 767–777.

7–34*   Figure 7–64.

7–35
A. The ability of the transcription factor to stimulate transcription from the −61 deletion, but not from the −50 deletion, indicates that a critical portion of the binding site for the factor resides in the interval between −61 and −50 nucleotides upstream from the start site of transcription. Deletion studies alone do not define the end of the binding site closest to the start site since removal of a part of the binding site would prevent binding of the factor. DNA footprinting experiments show that the factor protects a 12-nucleotide-long sequence from −63 to −52.

B. At first it seems contradictory that a *stimulatory* factor causes the transcript from the –50 deletion template to disappear from the gel. In fact, there is no contradiction: the stimulatory factor causes a 10-fold increase in transcription from all the templates that retain the binding sequence. Thus the amount of transcript from the –50 deletion template falls *relative* to the amount of transcript produced from the stimulated templates (and for that reason, is so much less intense on the gel that it seems to be absent).

C. The enormous difference in the stability of factor binding in the presence and absence of TFIID suggests that it binds not only to the DNA but also to TFIID. Thus, in the presence of TFIID the stimulatory factor is anchored in place by two interactions (with the DNA and with TFIID), whereas in the absence of TFIID it is bound by a single interaction (with the DNA).

Reference: Sawadogo M & Roeder RG (1985) Interaction of a gene-specific transcription factor with the adenovirus major late promoter upstream of the TATA box region. *Cell* 43, 165–175.

7–36* **Reference:** Godowski PJ, Rusconi S, Miesfeld R & Yamamoto KR (1987) Glucocorticoid receptor mutants that are constitutive activators of transcriptional enhancement. *Nature* 325, 365–368.

7–37 In order for Gal4 to work properly, the DNA-bound Gal4 must recruit the RNA polymerase holoenzyme to the promoter. If there is too much Gal4 in the cell, there will be a competition between free and DNA-bound Gal4 for binding to a limiting quantity of the RNA polymerase II holoenzyme complex. In the presence of excess Gal4, the holoenzyme complex is tied up in unproductive complexes with free Gal4, thereby preventing its recruitment to the promoter.

Reference: Ptashne M & Gann A (2002) Genes & Signals, pp 75–76. Cold Spring Harbor, New York: Cold Spring Harbor Laboratory Press.

7–38* **Reference:** Workman JL & Roeder RG (1987) Binding of transcription factor TFIID to the major late promoter during *in vitro* nucleosome assembly potentiates subsequent initiation by RNA polymerase II. *Cell* 51, 613–622.

7–39 Histone acetylases and chromatin remodeling complexes are recruited to specific regions of chromatin by gene activator proteins that can bind to DNA in unmodified chromatin. Once bound, histone acetylases add acetyl groups to histone tails, altering their packing properties and providing binding sites for some specific transcription factors. Similarly, chromatin remodeling complexes, once recruited, alter the local chromatin structure. This facilitates the binding of additional activator proteins that cannot bind to unmodified chromatin. Together, these changes in histone acetylation and chromatin structure allow the transcription machinery to be assembled at specific promoters and to initiate transcription.

7–40*

7–41
A. If the SAGA complex serves as a coactivator that bridges the Gal4 activator and the general transcription factors at the promoter, you might expect that fragment B, which carries the UAS to which Gal4 binds, would be enriched by chromatin immunoprecipitation of SAGA components. A several-fold enrichment of this fragment was observed in cells grown on galactose (see Figure 7–26B). You might also have expected fragment C to be enriched, since it contains the promoter region and carries the binding site for the TATA-binding protein where the transcription machinery assembles. In experiments not discussed in this problem, the authors showed that SAGA binds to the TATA-binding protein, but that interaction does not survive the steps of chromatin immunoprecipitation.
B. SAGA meets the criteria for a coactivator. It is demonstrably present at the UAS in cells grown on galactose, but is absent when the cells are grown on raffinose.

Reference: Larschan E & Winston F (2001) The *S. cerevisiae* SAGA complex functions *in vivo* as a coactivator for transcriptional activation by Gal4. *Genes Dev.* 15, 1946–1956.

7–42*

7–43

A. Construct 1 corresponds to mutant embryo B, construct 2 corresponds to mutant embryo D, and construct 3 corresponds to mutant embryo C.

B. With the exception of one aspect of embryo D, as discussed below, the results with the various constructs validate the simple rule for *eve* expression in stripe 2. When the Krüppel-binding sites are removed (construct 1), thereby eliminating the effects of the Krüppel repressor, the stripe-2 expression of β-galactosidase expands slightly in the posterior direction (see Figure 7–29B). This is as expected according to the rule because the activators hunchback and bicoid are both present slightly beyond the posterior end of stripe 2.

When two of the bicoid-binding sites are removed (construct 3), making the construct less sensitive to the effects of the bicoid activator, the stripe-2 expression of β-galactosidase is lessened but appears at its normal position (see Figure 7–29C).

When the giant-binding sites are removed (construct 2), eliminating the effects of the giant repressor, the stripe-2 expression of β-galactosidase expands substantially in the anterior direction (see Figure 7–29D). The simple rule for *eve* stripe-2 expression, however, along with the expression patterns for the hunchback and bicoid activators, predicts that β-galactosidase might be expressed in the entire anterior end of the mutant embryo. The activators bicoid and hunchback are both present in this region, in the absence of an effective repressor. This result suggests that something is missing from the formulation of the simple rule.

C. The reason why β-galactosidase is not expressed in the anterior end of the embryo is not clear. There are several plausible mechanisms that fall into two general categories. It is possible that another protein whose expression is confined to the anterior end modifies one or both activators to render them nonfunctional for binding to the *eve* stripe-2 control element. The modification could take the form of a phosphorylation, for example, or occur by direct binding to the activators. The second general mechanism would be the expression of another, as yet undefined repressor that binds to the *eve* stripe-2 control element and prevents expression.

D. The overlap of repressor-binding sites and activator-binding sites in the *eve* stripe-2 control element is striking. It suggests that the repressors function primarily by preventing activator binding. The extensive competition for binding between activators and repressors presumably sharpens the boundary between *eve* expression and nonexpression, giving rise to very well defined stripes in the embryo.

Reference: Stanojevic D, Small S & Levine M (1991) Regulation of a segmentation stripe by overlapping activators and repressors in the *Drosophila* embryo. *Science* 254, 1385–1387.

7–44*

7–45   These results show that E and I inhibit expression of *URA3* genes placed between them (see Figure 7–30, sites A and B), but do not affect expression of *URA3* genes placed outside (sites C through F). Whereas all strains grow in complete medium, strains with *URA3* between E and I do not grow in the absence of uracil but do grow in the absence of FOA, indicating that expression of these *URA3* genes is inhibited. By contrast, all the strains with *URA3* inserted outside of E and I grow in the absence of uracil and do not grow in the presence of FOA, demonstrating expression of these *URA3* genes. Thus, E and I can apparently silence genes placed between them, insulating them from any effects of enhancers outside the region.

Reference: Bi X, Braunstein M, Shei G-J & Broach J (1999) The yeast *HML* I silencer defines a heterochromatin domain boundary by directional establishment of silencing. *Proc. Natl. Acad. Sci. USA* 96, 11934–11939.

7–46*

THE MOLECULAR GENETIC MECHANISMS THAT CREATE SPECIALIZED CELL TYPES

7–47   False. Although there are numerous examples of reversible gene rearrangements as gene regulatory mechanisms in procaryotes, there are no known examples of regulation by reversible rearrangements in mammalian cells.

7–48*   Figure 7–65.

Reference: Zieg J, Silverman M, Hilmen M & Simon M (1977) Recombinational switch for gene expression. *Science* 196, 170–172.

7–49

A. In the presence of **a**1, α2 exists in two forms with different binding specificities. Your second set of experiments (see Figure 7–32, lanes 5 and 6) supports this interpretation and rules out the possibility that α2 is present in a single form that can bind to both **a**-specific and haploid-specific regulatory sequences. Addition of excess unlabeled **a**-specific DNA eliminates binding to the radioactive **a**-specific fragments, but not to the radioactive haploid-specific fragments. Similarly, addition of excess unlabeled haploid-specific DNA eliminates binding to the radioactive haploid-specific fragments. If a single form of α2 were able to bind to both regulatory sequences, either unlabeled site in excess would have eliminated binding to both kinds of radioactive fragments.

   Your third experiment (see Figure 7–32, lanes 7 and 8) shows that the ratio of binding activities for **a**-specific and haploid-specific sequences varies depending on the amount of **a**1 protein. If the α2 repressor in a diploid cell were shifted entirely into a form that could bind both sites, the ratio should be independent of the amount of **a**1 protein.

B. Your experiments are most easily explained if **a**1 acts stoichiometrically to alter the binding specificity of the α2 repressor, presumably by binding to it. In the fragment binding experiments shown in lanes 7 and 8 of Figure 7–32, when **a**1 is low, the binding to haploid-specific fragments is low; when **a**1 is high, the binding to haploid-specific fragments is high. The simplest explanation for the effects of the defective α2 repressor is that it binds to **a**1 protein, thereby reducing the availability of the **a**1 protein for binding to the normal α2 repressor.

   Note that neither of these experiments rules out the possibility that **a**1 protein acts catalytically on the α2 repressor. However, a catalytic mechanism for **a**1 protein would require special assumptions to account for the experimental observations. For these reasons it is considered most likely that the form of the repressor that binds to haploid-specific sequences will be found to contain both the **a**1 protein and the α2 repressor.

Reference: Goutte C & Johnson AD (1988) **a**1 protein alters the DNA-binding specificity of α2 repressor. *Cell* 52, 875–882.

7–50*   Figure 7–66.

7–51

A. UV light throws the switch from the prophage to the lytic state. When cI is destroyed, cro is made and turns off the production of new cI. The virus starts to produce coat proteins, and new virus particles are released.

B. When the UV light is switched off, the virus remains in the lytic state. Thus, cI and cro form a gene regulatory switch that, once thrown, is not reversible.

C. This switch makes sense in the viral life cycle. UV light is likely to damage the bacterial DNA, thereby rendering the bacterium an unreliable host for the virus. A prophage will switch to the lytic state and leave an irradiated cell in search of new host cells to infect.

7–52*

7–53 A specialized group of cells in the hypothalamus—the cells of the suprachiasmatic nucleus (SCN)—regulates our circadian rhythm. These cells receive neural cues from the retina, entraining them to the daily cycle of light and dark. In totally blind people, information about the light and dark cycle does not reach the SCN cells. As a consequence, they operate on their own inherent rhythm, which is slightly longer than 24 hours. Blind people typically report recurrent periods of insomnia and daytime sleepiness, as their circadian rhythms drift in and out of phase with the normal 24-hour cycle.

Reference: Sack RL, Brandes RW, Kendall AR & Lewy AJ (2000) Entrainment of free-running circadian rhythms by melatonin in blind people. *N. Engl. J. Med.* 343, 1114–1116.

7–54* **References:** Whitmore D, Foulkes NS & Sassone-Corsi P (2000) Light acts directly on organs and cells in culture to set the vertebrate circadian clock. *Nature* 404, 87–91.

Schibler U (2000) Heartfelt enlightenment. *Nature* 404, 25–28.

7–55 True. This is thought to be the likely way that myogenic proteins turn on the program of muscle development. All genes whose regulation is understood depend on a group of gene regulatory proteins that act collectively to determine whether a gene will be transcribed. It is thought that most, if not all, the required regulatory proteins are in place and that the myogenic proteins complete the signal, turning some genes on and others off.

7–56*

7–57
A. The methylation status of the 5S RNA gene does not affect its transcriptional activity, as indicated by the equal intensities of the maxigene RNA bands and the 5S RNA bands in the starting mixtures of templates (the lanes marked 'none' in Figure 7–37B).
B. The patterns of transcription after cleavage (see Figure 7–37B) are exactly as expected. As shown in Figure 7–37B, transcription from the fully methylated 5S RNA maxigene is specifically abolished after cleavage with DpnI, which cleaves only fully methylated restriction sites. Cleavage with MboI specifically abolishes transcription from unmethylated 5S RNA genes. And cleavage with Sau3A, which is insensitive to the methylation status of the restriction site, abolishes transcription from all 5S RNA genes.
C. The pattern of transcription after replication and cleavage indicates that transcription complexes do not remain associated with the 5S RNA maxigene during replication. If the replicated molecules retained an active transcription complex, transcriptional activity would have been evident after DpnI digestion, which does not cleave the replicated molecules. The activity that remains after MboI cleavage derives from fully methylated DNA molecules, which were not replicated. If transcriptional complexes were not erased by replication, there would have been activity after DpnI cleavage.
D. If only 10% of the molecules had been assembled into active transcription complexes, the pattern would have been unchanged (except for a reduction in the intensity of the bands). You could no longer have concluded, however, that replication eliminated the transcription complexes. If only 10% of the molecules were assembled into complexes and only 10% were replicated, then a skeptic (that is, a good scientist) would raise the possibility that only the unassembled molecules were replicated. Thus, the experiment would fail to test what it was designed to test. For the conclusion to be valid, it is

essential to prove that molecules carrying transcriptional complexes were replicated. Although there might be ways to test specifically for replication of transcriptional complexes, the easiest way is to show that the fraction of molecules that are replicated is significantly greater than the fraction of molecules that are not assembled into transcription complexes.

Reference: Wolffe AP & Brown DD (1986) DNA replication *in vitro* erases a *Xenopus* 5S RNA gene transcription complex. *Cell* 47, 217–227.

7–58*

7–59

A. Gene A is expressed only from active X chromosomes, as indicated by equal expression in males and females and by the pattern of expression in the hybrid cells. Gene B is expressed from both active and inactive X chromosomes, as indicated by the results with the hybrid cell lines and by the levels of expression in the other cells, which correlate with the total number of X chromosomes. Gene C is expressed only from inactive X chromosomes, as indicated by its expression in females but not males and by the pattern of expression in the hybrid cell lines.

The most common pattern of expression is like that of gene A; the vast majority of genes on the inactive X chromosome are turned off. A few genes like gene B are expressed from both active and inactive X chromosomes. The pattern for gene C was very surprising. The gene that encodes this RNA is called *XIST* (for X_i-specific transcripts); it is expressed exclusively from the inactive X chromosome. It is expressed from the X-inactivation center, and plays a direct role in X-inactivation.

B. The rule for X-inactivation is that only one X chromosome remains active— all the rest are inactivated. This is apparent in the Northern analysis in Figure 7–38. Gene A, which is expressed from the active X, is expressed at uniform levels regardless of the number of X chromosomes, indicating that only one X remains active. By contrast, gene C, which is expressed only from the inactive X, is expressed in all cells that have more than one X chromosome, at levels that depend on the number of X chromosomes. (Analysis of gene B, which is expressed from both active and inactive X chromosomes, provides no information about X-inactivation.)

These rules for X-inactivation were worked out from cytological observations long before the advent of molecular biology techniques. The inactive X chromosome is highly condensed and easily visualized as a distinct entity in the nucleus—the so-called Barr body. It was noted early on that female cells have one Barr body and male cells have none. Abnormal individuals with extra X chromosomes have a number of Barr bodies equal to the number of X chromosomes minus one.

Reference: Brown CJ, Ballabio A, Rupert JL, Lafreniere RG, Grompe M, Tonlorenzi R & Willard HF (1993) A gene from the region of the human X inactivation center is expressed exclusively from the inactive X chromosome. *Nature* 349, 38–44.

7–60*

7–61 The results of single-cell analysis support the second hypothesis; namely, that deletion of *Xist* prevents inactivation of the X chromosome from which it is deleted. Thus, in mutant ES cells that have undergone differentiation it is always allele A—the allele on the nonmutant X chromosome—that is inactivated.

If the first hypothesis—no X-inactivation—had been correct, the differentiated mutant ES cells would have looked the same as the undifferentiated cells; that is, both alleles would have been expressed. If the third hypothesis— no effect on X-inactivation—had been correct, the differentiated mutant ES cells would have resembled the nonmutant ES cells; that is, the A and B alleles would have been expressed in different individual cells.

Reference: Penny GD, Kay GF, Sheardown SA, Rastan S & Brockdorff N (1996) Requirement for *Xist* in X chromosome inactivation. *Nature* 379, 131–137.

7–62* **References:** Lee JT & Lu N (1999) Targeted mutagenesis of *Tsix* leads to non-random X inactivation. *Cell* 99, 47–57.

Mlynarczyk SK & Panning B (2000) X inactivation: *Tsix* and *Xist* as yin and yang. *Curr. Biol.* 10, R899–R903.

7–63 The DNA in the unfertilized egg is methylated in a characteristic way. Shortly after fertilization there is a genome-wide wave of demethylation, when the vast majority of methyl groups are lost from the DNA. This demethylation may occur passively as a consequence of inhibition of the maintenance methyltransferase, resulting in the loss of methyl groups during each round of DNA replication. Alternatively, demethylation could occur by activation of a specific demethylating enzyme. Later in development, at the time of implantation in the uterine wall, new methylation patterns are established by several *de novo* methyltransferases that methylate specific CG dinucleo-tides. Once the new patterns of methylation are established, they are pre-served by the maintenance methyltransferase, which adds methyl groups to an unmethylated CG opposite a methylated CG.

7–64* **Reference:** Murray EJ & Grosveld F (1987) Site-specific demethylation in the promoter of human γ-globin gene does not alleviate methylation mediated suppression. *EMBO J.* 6, 2329–2335.

7–65

A. Probes 2 and 3 are useful because they are specific for mRNAs that are not present in untreated 10T½ cells. You are searching for an mRNA that is present in both kinds of myoblast, but is not present in 10T½ cells. The reason for hybridizing the radioactive myoblast cDNA probes with RNA from 10T½ cells is to remove from the probe all the sequences that correspond to mRNAs that are common between the 10T½ cells and the myoblasts. This subtractive-hybridization procedure makes the probe more specific; it eliminates from analysis a large class of cDNA clones that you do not think will contain the gene of interest.

B. The A class of clones includes cDNAs corresponding to RNAs that are com-mon to 10T½ cells and to induced and normal myoblasts. In this class are all the normal housekeeping genes present in all cells. This is the class of clones that subtractive hybridization was meant to eliminate. Since probes 2 and 3 hybridized to only 1% of the cDNA clones identified by probe 1, these house-keeping RNAs must represent the vast majority of the mRNA species in cells.

The B class of clones includes cDNAs corresponding to RNAs that are induced by 5-aza C treatment, but are not present in normal myoblasts or in the 10T½ cells. Since 5-aza C causes widespread demethylation, it is not surprising that it activates some genes that are not normally expressed in myoblasts.

The C class of clones includes cDNAs corresponding to RNAs that are present in normal myoblasts but are not present in 5-aza-C-induced myoblasts. If the induced myoblasts were identical to the normal myoblasts, this class of clones should not exist. It would be natural to suspect some deficiency in the induced myoblasts; however, the problem seems to lie with the normal myoblasts. The normal myoblasts contain a small fraction of fully differentiated myotubes, whereas the induced myoblasts do not. Thus the RNA isolated from the normal myoblast cell population includes RNA from more differentiated cell types. When analyzed, the C class of clones are found to encode muscle-specific gene products, such as troponin I and myosin heavy and light chains.

The D class of clones includes cDNAs that correspond to myoblast-specific RNAs. It is among these clones that you expect to find the regulatory gene that controls myoblast differentiation.

The prototype for the myogenic genes that control myoblast differentiation

(*MyoD*) was found among the D class of clones. It was selected out of this class after further tests that were based on four assumptions. (1) The regulatory gene should not be expressed at all in 10T½ cells. (2) Its expression should reach a maximum in myoblasts. (3) Its expression should decline in myotubes. (4) It should not be expressed in variants of 10T½ cells that do not differentiate after treatment with 5-aza C. The final test was that the cloned cDNA, when introduced into 10T½ cells in an expressed form, should induce the cells to differentiate into myoblasts—which it does!

Reference: Davis RL, Weintraub H & Lassar AB (1987) Expression of a single transfected cDNA converts fibroblasts to myoblasts. *Cell* 51, 987–1000.

7–66* **Reference:** Hartwell LH, Hood L, Goldberg ML, Reynolds AE, Silver LM & Veres RC (2000) Genetics: From Genes to Genomes, pp 408–410. Boston: McGraw-Hill Companies.

7–67 IGF-2 is an imprinted gene. It is expressed from the paternal chromosome, but the gene on the female chromosome is imprinted and turned off. This pattern of expression is just what would be expected based on the tug-of-war hypothesis. The gene is expressed from the male, which would tend to make the embryo larger; however, it is not expressed (because of imprinting) from the female chromosome, which would tend to limit growth of the embryo. This expected pattern of parental imprinting of growth-promoting and growth-retarding genes is common, but there are exceptions that call the general hypothesis into question. Nevertheless, it provides a provocative perspective on the phenomenon of imprinting.

Reference: Moore T & Haig D (1991) Genomic imprinting in mammalian development: a parental tug of war. *Trends Genet.* 7, 45–49.

7–68*

POSTTRANSCRIPTIONAL CONTROLS

7–69 Flavopiridol blocks the ability of Cdk9 to phosphorylate its target proteins. If the CTD is a target for Cdk9, then you might expect flavopiridol to interfere with the conversion of RNA polymerase to the processive form required for productive HIV transcription. Although this seems likely from the brief summary of the effects of Tat on HIV transcription given in the problem, there are other protein kinases in the transcription complex that might carry out CTD phosphorylation. The only way to know for sure is to test the effects of flavopiridol directly. Such experiments show that flavopiridol effectively blocks Tat-activated transcription and interferes with HIV replication.

References: Jones KA (1997) Taking a new TAK on Tat transactivation. *Genes Dev.* 11, 2593–2599.

Chao S-H, Fujinaga K, Marion JE, Taube R, Sausville EA, Senderowicz AM, Peterlin BM & Price DH (2000) Flavopiridol inhibits P-TENb and blocks HIV-1 replication. *J. Biol. Chem.* 275, 28345–28348.

7–70*

7–71
A. Because calcitonin mRNA is produced when the cells are transfected with the wild-type gene, and CGRP mRNA is produced when they are transfected with the exon-4 splice-site mutant, the lymphocytes must contain all the processing factors necessary to generate both mRNAs.
B. If selection of a polyadenylation site was the critical choice in the expression of calcitonin mRNA in the lymphocyte cell line, then the mutant that was missing the exon-4 polyadenylation site would be expected to produce CGRP mRNA. If the splicing of exon 3 to exon 5 (to produce CGRP mRNA)

was precluded by use of the polyadenylation site in exon 4, then removal of the site should permit CGRP mRNA production. By contrast, the mutant lacking the exon-4 splice site might still be expected to be preferentially polyadenylated at exon 4, which would prevent production of CGRP mRNA.

C. If selection of the exon-4 splice site was the critical choice in the expression of calcitonin mRNA in the lymphocyte cell line, then the mutant that was missing the exon-4 splice site would be expected to produce CGRP mRNA. If the splicing of exon 3 to exon 4 is favored in lymphocytes, then removal of the exon-4 splice site should permit the exon-5 splice site to be used, thus generating CGRP mRNA. By contrast, the mutant lacking the exon-4 polyadenylation site might still be expected to splice exon 3 to exon 4 preferentially, which might be expected to lead to an aberrant RNA containing the fourth intron along with exons 5 and 6.

D. The predictions of the splice-site-selection model best match the results from the two mutants. As explained in parts B and C, the splice-site-selection model correctly predicted that the mutant lacking the exon-4 splice site would make CGRP mRNA. The polyadenylation-site-selection model, by contrast, predicts incorrectly that the mutant that is missing the exon-4 polyadenylation site will make CGRP mRNA. Thus, these results favor splice-site selection as the critical choice that enables the lymphocyte cell line to produce calcitonin mRNA instead of CGRP mRNA.

Reference: Leff SE, Evans RM & Rosenfeld MG (1987) Splice commitment dictates neuron-specific alternative RNA processing in calcitonin/CGRP gene expression. *Cell* 48, 517–524.

7–72*   **Reference:** Hedjran F, Yeakley JM, Huh GS, Hynes RO & Rosenfeld MG (1997) Control of alternative pre-mRNA splicing by distributed repeats. *Proc. Natl. Acad. Sci. USA* 94, 12343–12347.

7–73   False. In order for there to be a change in the amino acid sequence at the end of the protein, a new splicing event would need to accompany the change in polyadenylation site. Splicing would be required to eliminate one stop codon and bring a new one into play. In eucaryotic cells, the transcription of many genes yields a set of mRNAs that differ only in the position of their poly-A tails. For most such genes ribosomes translate the mRNA until they reach the stop codon and ignore any differences in lengths of the 3′ untranslated regions.

7–74*   **References:** Powell LM, Wallis SC, Pease RJ, Edwards YH, Knott TJ & Scott J (1987) A novel form of tissue-specific RNA processing produces apolipoprotein-B48 in intestine. *Cell* 50, 831–840.

Chen S-H, Habib G, Yang CY, Gu ZW, Lee BR, Weng S-A, Silbermann SR, Cai S-J, Deslypere JP, Rosseneu M, Gotto AM, Li W-H & Chan L (1987) Apolipoprotein B-48 is the product of a messenger RNA with an organ-specific in-frame stop codon. *Science* 238, 363–366.

7–75   The guide RNA is positioned by antiparallel pairing between anchoring sequences at its 5′ end and the pre-edited RNA. Unpaired bases in the pre-edited RNA are removed during editing, resulting in deletions in the edited RNA. Unpaired bases in the guide RNA direct the incorporation of U nucleotides into the edited RNA. The U nucleotides can pair with the looped out A and G nucleotides in the guide RNA. Several proteins are required for RNA editing. An endonuclease cleaves the pre-edited RNA at the mismatch, terminal uridylyl transferase adds U nucleotides, and RNA ligase rejoins the two halves of the mRNA. RNA editing proceeds from 3′ to 5′ along the mRNA until all mismatches are resolved.

Reference: Madison-Antenucci S, Grams J & Hajduk SL (2002) Editing machines: the complexities of *Trypanosome* RNA editing. *Cell* 108, 435–438.

7–76*

TABLE 7–8 Synthesis of ferritin in the rat liver after various treatments (Answer 7–77).

INJECTION	ACTINOMYCIN	FRACTION	TOTAL SYNTHESIS	FERRITIN SYNTHESIS	PERCENT FERRITIN	ADJUSTED FERRITIN	DISTRIBUTION OF FERRITIN
Saline	absent	polysomes	750,000	700	0.093	0.079	49%
		supernatant	255,000	1400	0.549	0.082	51%
						0.161	
Iron	absent	polysomes	500,000	900	0.180	0.153	89%
		supernatant	400,000	500	0.125	0.019	11%
						0.172	
Saline	present	polysomes	800,000	800	0.100	0.085	53%
		supernatant	600,000	3000	0.500	0.075	47%
						0.160	
Iron	present	polysomes	780,000	1380	0.177	0.150	89%
		supernatant	550,000	700	0.127	0.019	11%
						0.169	

Numbers show radioactivity (cpm) incorporated into total proteins or into ferritin.

7–77

A. For each sample the percentage of ferritin synthesis is equal to ferritin synthesis divided by the total protein synthesis. For the first sample, this value is 0.093% (700/750,000). All other values are listed under percent ferritin in Table 7–8. Percent ferritin is then adjusted for the distribution of bulk mRNA in the polysome (0.85) and supernatant (0.15) fractions. For the first sample this would give an 'adjusted ferritin' of 0.079 (0.093 × 0.85). These values are listed in Table 7–8 under the column labeled 'adjusted ferritin.' Finally, the percentage of total ferritin mRNA in the polysome fraction from a particular treatment is equal to the 'adjusted ferritin' divided by the total 'adjusted ferritin' (the sum of the polysome and supernatant values in Table 7–8). For the first sample this value is 49% (0.079/0.161). The distribution values are listed in the last column in Table 7–8.

B. Two features of the data show clearly that iron does not control ferritin synthesis by regulating the rate of transcription. First, injection of iron does not increase the amount of ferritin mRNA. As shown under the 'adjusted ferritin' column, the total amount of ferritin mRNA present after saline injection or iron injection is the same (as assayed by translation). Second, the RNA synthesis inhibitor, actinomycin D, has no effect on the amount of total ferritin mRNA. If the increased synthesis of ferritin induced by iron were due to an increase in the amount of ferritin mRNA, the increase in ferritin mRNA should have been detectable and it would have been blocked by actinomycin D.

C. The major effect of iron is to alter the distribution of ferritin mRNA between the polysome and supernatant fractions. In the absence of iron only 50% of the ferritin mRNA is bound to polysomes. But when iron is present, 90% of the ferritin mRNA is in the polysome fraction. Thus, the fraction of ferritin mRNA on polysomes increases by nearly a factor of two in the presence of iron. This shift from free mRNA to polysomal mRNA accounts nicely for the 2-fold increase in ferritin synthesis in the presence of iron. Only mRNAs that are in polysomes are translated into protein.

References: Zahringer J, Baliga BS & Munro HN (1976) Novel mechanism for translational control in regulation of ferritin synthesis by iron. *Proc. Natl. Acad Sci. USA* 73, 857–861.

Liebold EA & Munro HN (1988) Cytoplasmic protein binds *in vitro* to a highly conserved sequence in the 5′ untranslated region of ferritin heavy- and light-subunit mRNAs. *Proc. Natl. Acad. Sci. USA* 85, 2171–2175.

7–78* **Reference:** Hentze MW & Kühn LC (1996) Molecular control of vertebrate iron metabolism: mRNA-based regulatory circuits operated by iron, nitric oxide, and oxidative stress. *Proc. Natl. Acad. Sci. USA* 93, 8175–8182.

7–79

A. The opposite regulation of ferritin and the transferrin receptor make perfect biological sense. When iron levels are high, cells prevent toxicity in two ways. They decrease the amount of iron they take in by reducing their synthesis of transferrin receptor, and they increase the amount of iron that is safely sequestered inside ferritin by increasing ferritin synthesis. When iron levels are low, cells avoid iron deficiency by reversing these two effects. They increase the amount of transferrin receptor, allowing them to bring more iron into the cells. And they decrease ferritin synthesis, thereby reducing their levels of sequestered iron.

B. At low levels of iron, when the IRPs are bound to the IREs in the mRNA, the mRNA is stable. At high levels of iron, when IRPs are bound to the iron and are released from the mRNA, the mRNA is rapidly degraded. The observation that unbound mRNA is subject to rapid degradation led to the suggestion that a sensitive cleavage site was being masked by IRPs, leading to message stability in the absence of iron, and to rapid degradation when the cleavage site was exposed in the presence of iron. Specific cleavage sites have been mapped in the mRNA, but the precise mechanism of RNA degradation remains elusive.

Reference: Hentze MW & Kühn LC (1996) Molecular control of vertebrate iron metabolism: mRNA-based regulatory circuits operated by iron, nitric oxide, and oxidative stress. *Proc. Natl. Acad. Sci. USA* 93, 8175–8182.

7–80* **References:** Gay DA, Yen TJ, Lau JTY & Cleveland DW (1987) Sequences that confer β-tubulin autoregulation through modulated mRNA stability reside within exon 1 of a β-tubulin mRNA. *Cell* 50, 671–679.

Yen TJ, Machlin PS & Cleveland DW (1988) Autoregulated instability of β-tubulin mRNAs by recognition of the nascent amino terminus of β-tubulin. *Nature* 334, 580–585.

Bachurski CJ, Theodorakis NG, Coulson RM & Cleveland DW (1994) An amino-terminal tetrapeptide specifies cotranslational degradation of β-tubulin but not α-tubulin mRNAs. *Mol. Cell Biol.* 14, 4076–4086.

7–81

A. The mRNA from the *fos–globin–fos* 3′ hybrid gene, which includes the 3′ end of the *c-fos* gene, has the same stability characteristics as the mRNA from the normal *c-fos* gene (see Figure 7–50A and C). By contrast, the *fos–globin* hybrid gene, which includes the same 5′ *fos* sequences as the *fos–globin–fos* 3′ gene (but is missing the 3′ *fos* sequences) is very stable (see Figure 7–50B). Thus, the 3′ end of the human *c-fos* gene confers instability on the *c-fos* mRNA. The instability element must be included in the mRNA to make it unstable. Therefore, the element is presumably located in the 3′ exon of the *c-fos* gene, rather than in the 3′ flanking sequences.

B. Although not obvious at first, the behavior of the mRNA from the *fos–globin* hybrid gene can be accounted for in terms of mRNA stability. If the *fos–globin* mRNA is very stable—like normal globin mRNA—then a low rate of transcription in the absence of serum is sufficient to allow the mRNA to accumulate to high levels in the 24-hour period before serum was added and the first measurement was made. If the *fos–globin* mRNA is already present at high levels, then the transient burst of transcription that follows serum addition will not appreciably increase the total amount of *fos–globin* mRNA. Thus, the enhanced stability of the mRNA from the *fos–globin* hybrid gene can account for both the high initial levels and the lack of induction observed with the *fos–globin* hybrid gene.

These results emphasize the need for instability if a system must respond rapidly to change. This requirement has many familiar analogs in everyday life. For example, if images on a TV screen persisted for more than a fraction of a second, moving objects would be trailed by their ghosts. In a similar way, echoing acoustics blur the perception of both speech and music. Biological signaling pathways have built-in mechanisms to return the system to the starting state: old signals are continually 'erased' so that they do not blur the perception of new signals.

References: Treisman R (1985) Transient accumulation of *c-fos* RNA following serum stimulation requires a conserved 5′ element and *c-fos* 3′ sequences. *Cell* 42, 889–902.

Wilson T & Treisman R (1988) Removal of poly(A) and consequent degradation of c-*fos* mRNA facilitated by 3′ AU-rich sequences. *Nature* 336, 396–399.

7–82* **Reference:** Wickens M, Goodwin EB, Kimble J, Strickland S & Hentze M (2000) Translational control of developmental decisions. In Translational Control of Gene Expression (N Sonenberg, JWB Hershey, MB Mathews eds), pp 295–370. Cold Spring Harbor, NY: Cold Spring Harbor Laboratory Press.

7–83

A. The mRNAs in the presence or absence of the TGEs seem to be perfectly stable. There appears to be about as much label present at the 1024-cell stage as there was at the one-cell stage, indicating that little, if any, of the input RNA has been lost.

B. The presence of the TGEs has a dramatic effect on the lengths of the poly-A tails. In the absence of the TGEs, the poly-A tails are slowly shortened, but retain 15–40 A nucleotides even at the 1024-cell stage. By contrast, in the presence of the TGEs, the poly-A tails are rapidly shortened and RNAs without tails begin to accumulate as early as the four-cell stage. Thus, it appears that the binding of proteins to the TGEs promotes deadenylation.

Reference: Thompson SR, Goodwin EB & Wickens M (2000) Rapid deadenylation and poly(A)-dependent translational repression mediated by the *Caenorhabditis elegans tra-2* 3′ untranslated region in *Xenopus* embryos. *Mol. Cell. Biol.* 20, 2129–2137.

7–84* **Reference:** Lee M-H & Schedl T (2001) Identification of in vivo mRNA targets of GLD-1, a maxi-KH motif containing protein required for *C. elegans* germ cell development. *Genes Dev.* 15, 2408–2420.

7–85

A. The concentration of ribosomes in a reticulocyte lysate is 2.5×10^{-7} M, and the concentration of HCR that completely blocks protein synthesis is 5.6×10^{-9} M. Thus, when protein synthesis is completely blocked, the ratio of HCR molecules to ribosomes is 1 to 45 ($2.5 \times 10^{-7}/5.6 \times 10^{-9} = 45$). Since there are an average of four ribosomes per globin mRNA, the ratio of HCR molecules to globin mRNA molecules is about 1 to 10.

B. The ratios of HCR molecules to ribosomes and to globin mRNA indicate that HCR is unlikely to inhibit protein synthesis by a stoichiometric interaction with either of these components. These ratios, however, do not rule out the possibility that HCR stoichiometrically inactivates some other factor that is essential for protein synthesis, and which is present at a concentration 10-fold below that of globin mRNA.

It is now known that HCR is a protein kinase that inactivates protein synthesis by a catalytic mechanism. HCR phosphorylates the initiation factor eIF-2.

Reference: Farrell PJ, Balkow K, Hunt T, Jackson RJ & Trachsel H (1977) Phosphorylation of initiation factor eIF-2 and the control of reticulocyte protein synthesis. *Cell* 11, 187–200.

7–86* **Reference:** Otero LJ, Devaux A & Standart N (2001) A 250-nucleotide UA-rich element in the 3′ untranslated region of *Xenopus laevis* Vg1 mRNA represses translation both *in vivo* and *in vitro*. *RNA* 7, 1753–1767.

7–87   These experiments provide a very convincing demonstration that IRESs allow ribosomes to initiate translation in the absence of a cap (or an end of any kind). The linear and circular mRNAs were arranged so that they would give different translation products: a 20-kd protein from the linear mRNA and a 23-kd protein from the circular mRNA. The gels confirm that a circular mRNA with an IRES is translated into the 23-kd product, whereas a circular mRNA without an IRES gives no product at all. Arranging the linear and circular mRNAs so that they would give different sized fragments was a key aspect of the experimental design. Had they been constructed to give the same size fragment, it would have been very difficult to rule out contamination of the circular mRNA by the linear molecule.

Reference: Chen C-Y & Sarnow P (1995) Initiation of protein synthesis by the eucaryotic translation apparatus on circular RNAs. *Science* 268, 415–417.

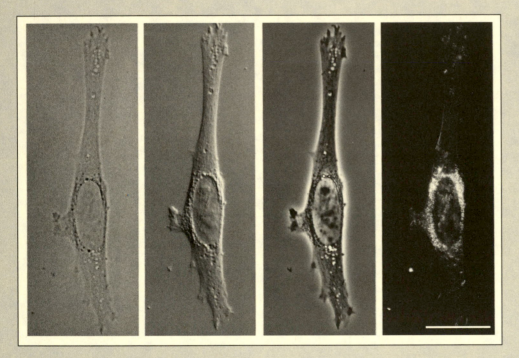

Four ways of looking at a cell. Can you tell what kind of microscopy was used for each image? Scale bar is 50 μm. Courtesy of Keith Roberts.

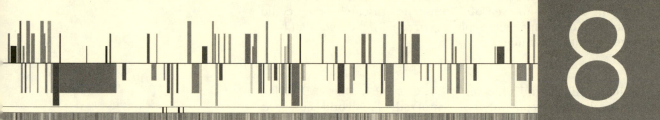

MANIPULATING PROTEINS, DNA, AND RNA

ISOLATING CELLS AND GROWING THEM IN CULTURE

8–1 Cells in a tissue are bound together by protein-mediated attachments to one another and to an extracellular matrix containing collagen. Treatment with trypsin, collagenase, and EDTA disrupts these attachments. Trypsin is a protease that will cleave most proteins, but generally only those portions of a native protein that are unstructured. The triple helical structure of collagen is a poor substrate for trypsin. Collagenase is specific for collagen. EDTA chelates Ca^{2+}, which is required for the cell-surface proteins, known as cadherins, to bind to one another to link cells together. Removal of Ca^{2+} prevents this binding and thereby loosens cellular attachments.

The treatment doesn't kill the cells because all the damage occurs to extracellular components, which can be replaced. So long as the plasma membrane isn't breached, the cells will survive.

8–2*

8–3 It is the general goal of cell biology research to discover how individual cells work, but it is very difficult to study most processes in single cells. If one has a population of identical cells, however, then its analysis can yield valid conclusions about the workings of the individual cells. By contrast, if the population is a mixture of different cell types, its analysis will give properties of the mixture, which may or may not accurately describe the individual cells. Consider an analogy. We know from looking at individual human eyes that they are various shades of brown or blue or green. Yet if we could tell eye color only by looking at 1000 at a time, and if we started with a random population, we might conclude that eyes were a bluish brown—a color that doesn't apply to any single individual.

8–4*

8–5   Human embryonic stem cell lines, which are derived from the inner cell mass of the early embryo, can proliferate indefinitely and retain the ability to give rise to any part of the body. If their differentiation can be guided appropriately in culture—an area of very intense research—they could provide a source of cells capable of replacing or repairing tissues that have been damaged by disease or injury.

8–6*   Figure 8–48.

Reference: Rasko JEJ, Battini J-L, Kruglyak L, Cox DR & Miller AD (2000) Precise gene localization by phenotypic assay of radiation hybrid cells. *Proc. Natl. Acad. Sci. USA* 97, 7388–7392.

8–7   The two statements are not exactly the same. Both emphasize the essential feature of the hybridoma technique: its ability to generate antibodies of a single type in pure form. The two statements differ in exactly why this might be useful. The first statement stresses the possibility of obtaining a monoclonal antibody directed against a minor component of the mixture of molecules used to elicit an immune response. Thus, if a protein of interest were present as a minor component in an impure mixture, the hybridoma technique would allow specific monoclonal antibodies to be generated against it.

The second statement highlights another advantage of the hybridoma technique. Even when a pure protein is used to elicit an immune response, a variety of different antibodies are generated in different proportions. These antibodies are typically directed at different antigenic sites (epitopes) on the protein. In some cases a minor type of antibody directed against a particular epitope can have very useful properties; the hybridoma technique allows such an antibody to be generated in a pure form as a monoclonal antibody.

8–8*

8–9   It is common to raise antibodies against other antibodies. Usually, this is done by introducing antibodies from one species into a second species, for example, by injecting mouse antibodies into goats. In this example, the mouse antibodies are recognized as foreign proteins in the goat, which mounts an immune response and generates goat antibodies that bind to the mouse antibodies. It is also possible to raise antibodies against antibodies from the same species. In the same species most parts of the injected antibody molecules will be indistinguishable from the host antibodies and thus will be treated as 'self.' But the portion of the antibody molecule that binds to an antigen (the so-called idiotype) can be recognized as foreign and elicit an immune response, generating antibodies directed against the antigen-combining site of the injected antibody.

FRACTIONATION OF CELLS

8–10*

8–11   Velocity sedimentation is used to separate components that differ in size and/or shape. It is carried out by layering a solution containing the components to be separated on top of a shallow density gradient formed by increasing concentrations—from top to bottom—of a small molecule such as sucrose. Upon centrifugation individual components will move through the gradient according to their size and shape. Because identical components have the same properties, they move as a defined band, which can be collected.

Equilibrium sedimentation is used to separate components that differ in their buoyant density. Components to be separated are most often layered on top of a steep sucrose gradient and centrifuged until the components move to their equilibrium density. (The components can also be mixed into the gradient to start with, but for sucrose density gradients it is more common to establish the gradient and then layer the components on top.) Molecules with the same density will form defined bands, which can be collected.

Because most proteins have about the same density, velocity sedimentation would be preferred over equilibrium sedimentation for the separation of two proteins of different size.

8–12*

8–13   The rate of sedimentation of a protein is based on size *and* shape. The nearly spherical hemoglobin will sediment faster than the more rod-shaped tropomyosin, even though tropomyosin is the larger protein. Shape comes into play because molecules that are driven through a solution by centrifugal force experience the equivalent of frictional drag. A spherical protein, with its smaller surface-to-volume ratio, will experience less drag than a rod, and therefore sediments faster. You can demonstrate this difference using two sheets of paper. Crumple one into a sphere and roll the other into a tube. Now drop them. The ball will hit the ground faster than the tube. In this demonstration, the centrifugal force is replaced by gravity and the friction with molecules in solution is replaced by friction with air. The underlying principles are the same.

8–14*   Figure 8–49.

References: Meselson M & Stahl FW (1958) The replication of DNA in *Escherichia coli. Proc. Natl. Acad. Sci. USA* 44, 671–682.

Cantor CR & Schimmel PR (1980) Biophysical Chemistry, pp 632–634. New York: WH Freeman and Company.

8–15   All these methods employ small beads that are packed into columns to which a solution of proteins is applied. Ion-exchange chromatography uses beads that carry positive charges (anion exchangers) or negative charges (cation exchangers). Proteins spend more or less time associated with the beads, depending on the arrangement of charged groups on their surfaces. Weaker binding proteins elute from the column earlier and tighter binding proteins elute later. Because the strength of association varies with pH and ionic strength of the solution that is passing down the column, the association between proteins and the beads can be varied systematically to find the best conditions for purification of a particular protein.

Hydrophobic chromatography uses beads that have hydrophobic groups protruding from their surfaces. These hydrophobic groups can interact with hydrophobic regions on the surfaces of proteins and delay their progress through the column. Once again, the stronger the interaction with the beads the longer the protein remains on the column.

Gel-filtration chromatography (also known as size-exclusion chromatography) uses beads with pores. Proteins that are too large to fit into the beads pass unretarded through the column, whereas proteins that can enter the beads are retarded by the time they spend inside the bead. For proteins that can enter the beads, larger proteins come off the column earlier than smaller proteins. Beads with a variety of pore sizes are available so that a gel-filtration column can be tailored to purification of a particular protein.

Affinity chromatography uses beads to which specific molecules, small or large, have been attached. The choice of molecule depends on the particular protein whose purification is desired. One common application, for example, uses glutathione Sepharose (the small molecule glutathione attached to Sepharose beads) to capture proteins fused to GST (glutathione-*S*-transferase). Passing glutathione through the column, which displaces the GST-tagged protein, elutes the bound protein. Another common example attaches antibodies that are specific for a particular protein to the beads, allowing a specific protein to be bound to the column; these interactions can be disrupted with high salt or changes in pH to allow the protein to be

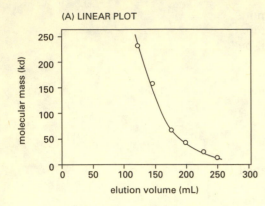

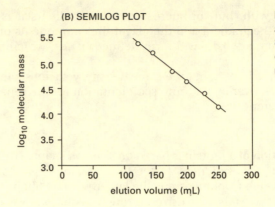

Figure 8–50 Plots of molecular mass against elution volumes (Answer 8–17). (A) Molecular mass versus elution volume. (B) Log₁₀ molecular mass versus elution volume.

eluted. Proteins that are known to associate with other proteins can also be used for affinity chromatography.

8–16* Table 8–9.

8–17

A. The smaller proteins come off the column later than the larger proteins because they experience a larger fraction of the internal volume of the column; that is, that portion of the column that is inside the beads. Smaller proteins can fit into more of the pores in the beads than can the larger proteins; hence their migration through the column will be retarded relative to larger proteins.

B. Plots of molecular mass and $\log_{10}$ of molecular mass versus elution volumes are shown in Figure 8–50A and B, respectively. It is clear that the plot of $\log_{10}$ of molecular mass gives a much closer approximation to a straight line than does the plot of molecular mass. The reason is that the time a protein spends on the column is related to its radius, which determines its ability to enter the pores. Since these proteins are roughly spherical, their radii will be related approximately to the cube roots of their volumes, which is equivalent to the cube roots of their molecular masses. A more rigorous treatment of this question shows that the Stokes radius, which is the effective hydrated radius of a protein and takes into account the shape of the protein, is the best predictor of elution volume.

Reference: Cantor CR & Schimmel PR (1980) Biophysical Chemistry, pp 674–675. New York: WH Freeman and Company.

8–18* **Reference:** Cantor CR & Schimmel PR (1980) Biophysical Chemistry, pp 670–675. New York: WH Freeman and Company.

8–19

A. The protein will become more negative at higher pHs, as more ionizable groups give up their protons. Over the indicated range, the major effect will be on histidine residues, which have pK values around 6.5. These groups will be positively charged at pH 5.0 and uncharged at pH 7.5. This loss of positive charge means the protein will become more negative as the pH is raised.

B. You want to pick a pH at which your protein binds to the DEAE-Sepharose beads. If it does not bind to the beads, the protein will pass through the column with all the other proteins that don't bind. Ion-exchange chromatography is carried out under conditions in which the protein interacts with the beads. The nature of the interaction determines how the protein is fractionated relative to other proteins.

C. In general, you want to pick a pH at which the protein of interest binds to the column, but doesn't bind too strongly. Thus, pH 6.5 would be the best choice for your initial studies. If you pick too high a pH, the protein may bind to the column too strongly, requiring harsh conditions to remove it.

D. Your preliminary experiments have given you the proper pH to bind your protein to the column. Your first step will be to equilibrate the column with a buffer at that pH. Next you will bring your protein sample to the same pH in the same buffer, and allow it to flow into the column. The column is then washed with one or two column volumes of buffer, which washes out all of the protein that does not bind to the column. Finally, you elute your protein using a gradient of increasing ionic strength (salt concentration). At a characteristic ionic strength the salt in the buffer will compete effectively with your protein for the charged groups on the column. The released protein will then flow out of the column in a relatively tight band.

8–20*

8–21 The detergent SDS carries a negative charge so that when it binds, the proteins become highly negatively charged. Because they are negatively charged, the proteins will move toward the positive electrode (the anode). Thus, it is critical that you place the positive electrode at the bottom of the gel. (Nearly everyone who has ever run a gel has attached the electrodes incorrectly, but usually once is enough.)

8–22*

8–23
A. No, the standard proteins do not migrate as an inverse function of their molecular masses. It is clear in this set of proteins that the 67-kd protein has not migrated half as far as the 35-kd protein.
B. Because there is not a linear relationship between molecular mass and distance of migration, it is difficult to estimate masses with confidence directly from the gel in Figure 8–6. It is best to plot migration of the standard proteins as a function of their molecular masses to give a line that can then be used to estimate the molecular masses of the unknown proteins. One such curve is shown in Figure 8–51, which plots the $\log_{10}$ of the molecular mass against the distance of migration. A plot using $\log_{10}$ values gives a much straighter line than a plot using molecular mass directly; this makes for more accurate estimates of the molecular masses. As shown in Figure 8–51, the estimated molecular mass of ribonucleotide reductase is 44 kd and that of cyclin B is 54 kd.
C. The molecular mass of the small subunit of ribonucleotide reductase, as estimated from the SDS gel, matches the calculated molecular mass very well. The estimate for cyclin B is much higher than expected from the calculated molecular mass. Possible reasons for this large difference are posttranslational

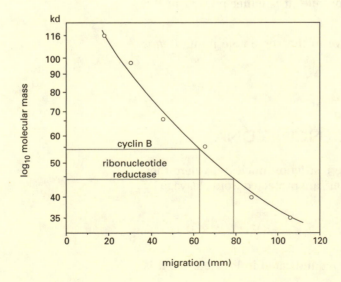

Figure 8–51 A plot of $\log_{10}$ molecular mass versus distance of migration (Answer 8–23).

modifications that are not accounted for by the calculated molecular mass, or peculiarities of cyclin B sequence that make it migrate anomalously. Examples of both effects are known (and not uncommon). In the case of cyclin B, the basis for its slow migration is its large number of positive charges, which partially offset the charges added by SDS. As a result, cyclin B carries less overall negative charge than would an average SDS-coated protein of its size. Histones, which are also highly positively charged, likewise run anomalously slowly on SDS gels.

8–24*

8–25 Isoelectric focusing in the first dimension and SDS-PAGE in the second dimension are used to separate proteins by two-dimensional polyacrylamide-gel electrophoresis. Isoelectric focusing must be carried out first. If SDS-PAGE were carried out first, all the proteins would be coated with negatively charged SDS, which would eliminate the effectiveness of isoelectric focusing.

8–26*

8–27 This problem involves a two-part calculation. If you know the specific activity of the labeled ATP (μCi/mmol), then you can use that value to convert 1 cpm per band (1 dpm/band for ^{32}P) into the number of proteins per band. The specific activity of ATP is 1.1×10^6 μCi/mmol. Since ATP is 0.9 mM in the extract after addition of the label,

$$\text{specific activity} = \frac{10\,\mu\text{Ci}}{10\,\mu\text{L}} \times \frac{\text{L}}{0.9\,\text{mmol ATP}} \times \frac{10^6\,\mu\text{L}}{\text{L}}$$
$$= 1.11 \times 10^6\,\mu\text{Ci/mmol ATP}$$

The specific activity of ATP is equal to the specific activity of phosphate on the labeled proteins. Using this specific activity (and various conversion factors shown on the inside of the front cover), you can calculate that 1 dpm/band corresponds to 2.4×10^8 protein molecules per band.

$$\frac{\text{proteins}}{\text{band}} = \frac{1\,\text{dpm}}{\text{band}} \times \frac{\text{mmol}}{1.11 \times 10^6\,\mu\text{Ci}} \times \frac{\mu\text{Ci}}{3.7 \times 10^4\,\text{Bq}} \times \frac{\text{Bq}}{60\,\text{dpm}} \times \frac{6 \times 10^{20}\,\text{molecules}}{\text{mmol}}$$
$$= 2.4 \times 10^8\,\text{molecules/band}$$

8–28*

8–29 False. There are 6×10^{23} molecules per mole; hence, only 0.6 molecules in a yoctomole. The limit of detection is one molecule, or 1.7 yoctomole. No instrument can detect less than one molecule (it is either present in the instrument or it is not).

Reference: Castagnola M (1998) Sensitive to the yoctomole limit. *Trends Biochem. Sci.* 23, 283.

8–30*

ISOLATING, CLONING, AND SEQUENCING DNA

8–31 False. The recognition sequences for the restriction nuclease, where they occur in the genome of the bacterium itself, are protected from cleavage by methylation at an A or a C residue.

8–32*

8–33
A. The 5′ and 3′ ends of the cut molecules are indicated in Figure 8–52. It is

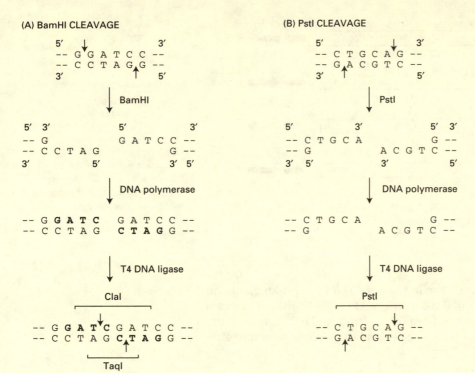

Figure 8–52 Cleavage, modification, and joining of DNA containing a BamHI site or a PstI site (Answer 8–33). *Brackets* indicate recognition sites for restriction nucleases.

standard practice to represent DNA sequences so that the 5′ end of the top strand is on the left.

B. As indicated in Figure 8–52, the BamHI ends can be filled in by DNA polymerase, but the PstI ends cannot. These different fates follow from the requirements of DNA polymerase: a primer with a 3′-OH to which dNTPs can be added and a template strand to specify correct addition. These requirements are met by the BamHI ends but not by the PstI ends, which have recessed 5′ ends that cannot serve as primers and, thus, cannot be filled in.

A standard technique in recombinant DNA technology is to use T4 DNA polymerase to blunt both types of ends. It will blunt BamHI ends by filling them in, as indicated in Figure 8–52A. It will also blunt PstI ends by virtue of an associated 3′-to-5′ exonuclease activity, which removes the 3′ extension, leaving a blunt end.

C. As indicated in Figure 8–52, the blunted BamHI ends and the unmodified PstI ends can both be joined by T4 DNA ligase.

D. Joining of the treated ends regenerates the PstI site but not the BamHI site. Joining of the filled-in BamHI ends generates two new restriction sites, as indicated in Figure 8–52A. Cleavage, filling in the ends, and rejoining often generates new restriction sites that sometimes are useful for further manipulation of the DNA.

8–34*

8–35

A. The complexity of the original ligation pattern arises because any two BamHI ends can join together. Thus, the 0.4-kb fragments can join together to generate a set of fragments with sizes 0.8 kb, 1.2 kb, 1.6 kb, 2.0 kb, and so forth. Similarly, the 0.9-kb fragments will produce a set of fragments with sizes 1.8 kb, 2.7 kb, 3.6 kb, and so on. Finally, combinations of the two fragments generate a third set of fragments with sizes 1.3 kb, 1.7 kb, 2.1 kb, 2.2 kb, etc. (The actual pattern often is more complicated still, since these fragments can circularize by joining their ends.)

B. Since any two BamHI ends can join, even one size of fragment can have

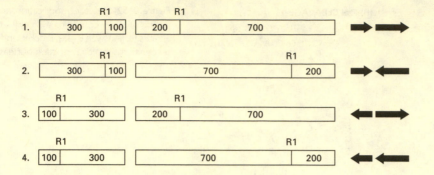

Figure 8–53 Various arrangements of 0.4-kb and 0.9-kb BamHI fragments to form a 1.3-kb fragment (Answer 8–35). *Arrows* on the *right* show the four arrangements of the large and small fragments.

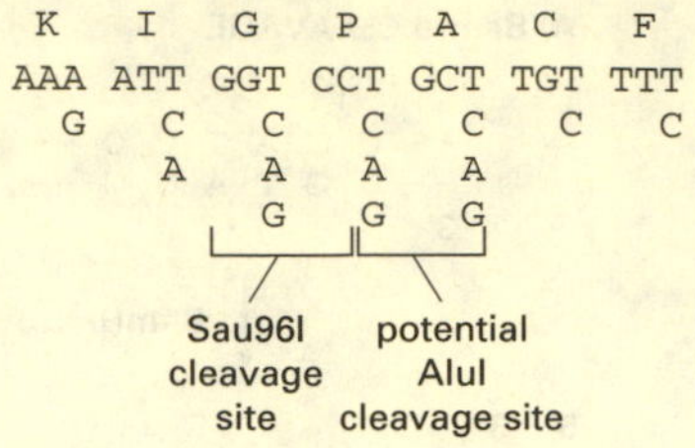

K	I	G	P	A	C	F
AAA	ATT	GGT	CCT	GCT	TGT	TTT
G	C	C	C	C	C	C
	A	A	A	A		
		G		G		

Sau96I cleavage site   potential AluI cleavage site

Figure 8–55 Representation of the nucleotide sequences that could encode the peptide KIGPACF (Answer 8–37). Only the DNA strand that corresponds to the mRNA sequence is shown.

several different structures, as shown in Figure 8–53 for the 1.3-kb fragment. Digestion of the population of 1.3-kb fragments with EcoRI generates a variety of fragments, which range from 0.1 kb (from the left ends of structures 3 and 4, Figure 8–53) to 1.0 kb (the internal fragment in structure 4).

8–36* Figure 8–54.

References: Anderson JE (1993) Restriction endonucleases and modification methylases. *Curr. Opin. Struct. Biol.* 3, 24–30.

Newman M, Lunnen K, Wilson G, Greci J, Schildkraut I & Phillips SEV (1998) Crystal structure of restriction endonuclease BglI bound to its interrupted DNA recognition sequence. *EMBO J.* 17, 5466–5476.

8–37 As indicated in Figure 8–55, the restriction nuclease Sau96I will definitely cleave the cDNA sequence at the segment corresponding to the adjacent codons for glycine and proline, regardless of the particular codon used for glycine. AluI might cut this DNA if the right codons were used for both proline and alanine. HindIII definitely will not cut this sequence.

8–38* Figure 8–56.

8–39 The lengths of the restriction fragments do not sum to 21 kb and the electrophoretic patterns change with denaturation and reannealing because the ribosomal minichromosome is an inverted dimer, or a palindrome (Figure 8–57). That is to say, the sequences on the left half of the minichromosome are repeated in the opposite orientation on the right half. Thus, a restriction nuclease cuts each arm (that is, each repeated segment) at sites that are the same distance from the ends, thereby generating one central fragment and two identical copies of each terminal fragment. For BglII digestion, the central fragment is 13.4 kb and the two terminal fragments are each 3.8 kb. The fragments on the gel did not add up to 21 kb because the terminal fragment should have been counted twice: 2 times 3.8 kb plus 13.4 kb does sum to 21 kb.

Denaturing and reannealing the minichromosome or its digestion products alters the electrophoretic pattern because the left and right halves of the single strands derived from the central fragment are complementary and, therefore, can reanneal internally to form a hairpin. Because the two complementary halves are part of the same molecule, they reanneal internally very much faster than they do with other molecules. Self-reannealing

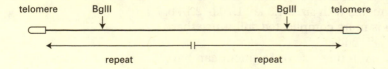

Figure 8–57 Palindromic structure of the ribosomal minichromosome from *Tetrahymena* (Answer 8–39).

reduces the apparent size of the fragment by a factor of 2. Thus, the uncut minichromosome, when denatured and reannealed, is reduced in size from 21 kb to 10.5 kb. Similarly, the 13.4-kb central BglII fragment is reduced to 6.7 kb. Of course, the single strands from the terminal fragments are not self-complementary and so can reanneal only with other complementary single strands to reform their usual double-stranded structure.

Most unexpectedly, the ribosomal genes of *Tetrahymena* exist in a micronuclear chromosome as one-half of the inverted repeat that makes up the macronuclear minichromosome. Since the micronucleus generates a new macronucleus after conjugation, the ribosomal genes must be cleaved out of the chromosome in a way that specifically generates an inverted dimer. You might try to imagine how this could be accomplished before looking up the speculation by scientists involved in some of these studies.

References: Karrer KM & Gall JG (1976) The macronuclear ribosomal DNA of *Tetrahymena pyriformis* is a palindrome. *J. Mol. Biol.* 104, 421–453.

Yao M-C, Zhu S-G & Yao C-H (1985) Gene amplification in *Tetrahymena thermophila*: formation of extrachromosomal palindromic gene coding for rRNA. *Mol. Cell. Biol.* 5, 1260–1267.

8–40* Figure 8–58.

8–41 False. When DNA molecules travel end-first through the gel in a snakelike configuration, their rates of movement are independent of length. In pulsed-field gel electrophoresis the direction of the field is changed periodically, which forces the molecules to reorient before continuing to move snakelike through the gel. This reorientation takes much more time for larger molecules, so that progressively larger molecules move more and more slowly.

8–42*

8–43

A. Both oligonucleotides will hybridize to β^A and β^S DNA. The β^A oligo, for example, is a perfect (20-of-20) match for β^A DNA and a 19-of-20 match for β^S DNA. The situation is equivalent for the β^S oligo. The difference in hybridization between a 19-of-20 match and a 20-of-20 match is difficult to detect when the mismatch is at one end of the oligonucleotide. Under normal hybridization conditions both oligonucleotides hybridize equally well to β^A and β^S DNA.

B. Although hybridization itself is not affected by the mismatch, the reaction catalyzed by DNA ligase is exquisitely sensitive to it. As shown in Figure 8–59 for the β^S oligo, ligation can occur only if the bases on both sides of the nick are properly paired with the complementary bases in the target DNA. Ligation is essential if the radioactive oligo is to be linked to the biotin-labeled oligo and bound to the solid support where it can expose the x-ray film.

Reference: Landegren U, Kaiser R, Sanders J & Hood L (1988) A ligase-mediated gene detection technique. *Science* 241, 1077–1080.

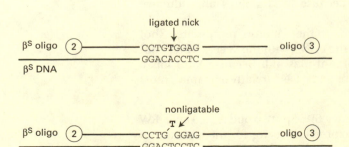

Figure 8–59 Sensitivity of ligation to terminal mismatches (Answer 8–43).

8–44*   Figure 8–60.

8–45   True. If the probe was not specific (usually because it was not made from a pure source), then it would hybridize to multiple bands.

8–46*   **Reference:** Gelehrter TD & Collins FS (1990) Principles of Medical Genetics, p 80. Baltimore: Williams & Wilkins.

8–47   Although several explanations are possible, the simplest is that the DNA probe has hybridized predominantly with its corresponding mRNA. When a gene is expressed, its mRNA is present in many more copies than the gene. The strongly hybridizing cells probably express the gene at high levels and therefore have high levels of the mRNA.

8–48*   Figure 8–61.

8–49

A. The ratio of vector molecules to cDNA molecules is about 1 to 2 in the ligation mixture. The vector is about 40 times longer than the cDNA (43 kb vs. 1 kb), but there is only 20 times as much vector as cDNA by weight (2 µg vs. 0.1 µg). Thus there are half as many vector molecules as cDNA molecules in the ligation mixture.

B. The number of vector molecules in 2 µg is 4.2×10^{10}.

$$\text{molecules} = \frac{1 \text{ molecule}}{43 \text{ kb}} \times \frac{\text{kb}}{1000 \text{ bp}} \times \frac{\text{bp}}{660 \text{ d}} \times \frac{6 \times 10^{17} \text{ d}}{\text{µg}} \times 2 \text{ µg}$$
$$= 4.2 \times 10^{10}$$

Since this number of vector molecules gave 4×10^7 recombinants, the efficiency of generating recombinants is about 0.1% ($4 \times 10^7 / 4.2 \times 10^{10}$).

C. Incubation of the XhoI-cut vector DNA with dTTP and DNA polymerase adds a single T to each end.

```
5'-CT              TCGAG-3'
3'-GAGCT              TC-5'
```

The double-stranded adaptor oligonucleotide has single-strand tails that are complementary to the modified vector ends.

```
    Vector              Adaptor
    5'-CT          CGAGATTTACC-3'
    3'-GAGCT          CTAAATGG-5'
```

This allows the modified vector ends to be ligated to the adaptor-modified cDNAs, after the blunt end of the adaptor has been ligated to the blunt end of the cDNAs.

   After modification the vector ends are not complementary to each other and thus cannot be ligated together. The modified cDNAs are also not self-complementary and cannot be ligated together.

D. After modification of the vector and cDNAs, the vector molecules can ligate only to the cDNAs and vice versa. This greatly improves the efficiency of formation of recombinant molecules because it eliminates unproductive vector-to-vector ligations.

E. The ligation of the modified vector ends to the adaptors regenerates XhoI sites (5'-CTCGAG) at both junctions of the cDNA with the vector. Thus XhoI will cut the cDNA out of the recombinant plasmid. Because XhoI has a six-base-pair recognition site, and thus cuts DNA relatively rarely, most cDNAs will be cut out intact.

Reference: Elledge SJ, Mulligan JT, Ramer SW, Spottswood M & Davis RW (1991) λYES: a multifunctional cDNA expression vector for the isolation of genes by complementation of yeast and *Escherichia coli* mutations. *Proc. Natl. Acad. Sci. USA* 88, 1731–1735.

8–50*

8–51 There are approximately 10^9 different possible sequences formed by 15 nucleotides ($4^{15} = 1.07 \times 10^9$). By chance, therefore, you would expect to find any given 15-nucleotide an average of 3 times ($3.2 \times 10^9/1.07 \times 10^9$) in the human genome. Since the degenerate probe is a mixture of 16 different 15-nucleotide sequences, you would expect to find about 48 matches (16×3) in the genome, of which only one will correspond to the Factor VIII gene.

To determine that a sequence identified by the probe indeed corresponds to the Factor VIII gene, one usually repeats the procedure using a second degenerate probe from elsewhere in the protein sequence. If both degenerate probes hybridize to one genomic DNA fragment, it is a strong indication that the Factor VIII gene has been identified.

8–52* Figure 8–62.

8–53 True. Since eucaryotic coding sequences are usually in pieces in genomic DNA, cDNA clones allow expression of the protein directly and permit one to deduce the amino acid sequence of the protein.

8–54*

8–55 The sequence of the cloned DNA and the amino acids encoded by the open reading frames are shown in Figure 8–63. Stop codons are underlined and labeled 2 or 3 to indicate their reading frame. As you can appreciate from this exercise, the sequencing gel must be read very carefully; omission of a single nucleotide would have disastrous consequences for determining the open reading frame. To minimize this problem, it is best to determine the sequence of both strands of DNA. Since the two strands are complementary, any mistakes will be readily apparent. The sequence encodes the canine SRP receptor.

8–56*

8–57 Protein biochemistry is still critically important, mostly because it provides the link between the amino acid sequence (which can be deduced from DNA sequences) and the function of the protein. We are, for example, still not able to predict the folding of a polypeptide chain from its amino acid sequence. Thus, in many cases information regarding the function of the protein, such as its catalytic activity, cannot be deduced but must instead be determined experimentally by analyzing the properties of proteins biochemically. Furthermore, the structural information that can be deduced from DNA sequences is necessarily incomplete. We cannot obtain information, for example, about modifications of protein side chains (such as phosphorylation), proteolytic processing, the presence of tightly bound cofactors, or the association of the protein with other polypeptide chains.

8–58*

8–59 The appropriate PCR primers are primer 1 (5'-GACCTGTGGAAGC) and primer 8 (5'-TCAATCCCGTATG). The first primer will hybridize to the bottom strand and prime synthesis in the rightward direction. The second primer will hybridize to the top strand and prime synthesis in the leftward direction. (Remember that strands pair antiparallel.)

```
    Y K L D N Q F E L V F V V G F Q K I L T
TATAAACTGGACAACCAGTTCGAGCTGGTGTTCGTGGTCGGTTTTCAGAAGATCCTAAC
   3                                                      2

    L T Y V D K L I D D V H R L F R D K
GCTGACGTACGTAGACAAGTTGATAGATGATGTGCATCGGCTGTTTCGAGACAAGTA
   2        2        2   2   3
```

Figure 8–63 Nucleotide and amino acid sequence of a cloned gene (Answer 8–55). Stop codons are *underlined* and the reading frame they affect is indicated *below*. The amino acids encoded in frame 1 are *centered* over their codons.

The middle two primers in each list (primers 2, 3, 6, and 7) would not hybridize to either strand. The remaining pair of primers (4 and 5) would hybridize, but would prime synthesis in the wrong direction—that is, outward, away from the central segment of DNA. Each of these wrong choices has been made at one time or another in most laboratories that use PCR. In most cases the confusion arises because the conventions for writing nucleotide sequences have been ignored. By convention, nucleotide sequences are written 5′ to 3′ with the 5′ end on the left. For double-stranded DNA the 5′ end of the top strand is on the left.

8–60*   Figure 8–64.

Reference: Chamberlain JS, Gibbs RA, Ranier JE & Caskey CT (1990) Multiplex PCR for the diagnosis of Duchenne muscular dystrophy. In PCR Protocols (MA Innis, DH Gelfand, JJ Sninsky & TJ White eds), pp 272–281. San Diego, CA: Academic Press.

8–61   True. If each cycle doubles the amount of DNA, then 10 cycles equal a 2^{10}-fold amplification (which is 1024), 20 cycles equal a 2^{20}-fold amplification (which is 1.05×10^6), and 30 cycles equal a 2^{30}-amplification (which is 1.07×10^9). (It is useful to remember that 2^{10} is roughly equal to 10^3 or 1000. This simple relationship allows you to estimate the answer to this problem rapidly without resorting to your calculator. It comes in handy in a variety of contexts.)

8–62*   Figure 8–65.

8–63   Like the vast majority of mammalian genes, it is likely that the attractase gene contains introns. Bacteria do not have the splicing machinery required to remove introns, and therefore the correct protein cannot be expressed from the gene. For expression of most mammalian genes in bacterial cells, a cDNA version of the gene must be used.

8–64*

ANALYZING PROTEIN STRUCTURE AND FUNCTION

8–65   It is difficult to overestimate the value of the current (and future) databases of protein structure and function. It seems entirely likely that such databases will yield increasingly detailed information about the basic properties of previously unknown proteins. Yet as valuable as such information is, it is unlikely to provide sufficient detail to permit one to know how the protein functions in its biological context. Knowing that a protein is a protein kinase, for example, does not define its targets of phosphorylation and the range of its biological effects. Far from putting biochemists out of work, such databases will allow them to tackle ever more interesting questions.

8–66*

8–67   The pairs of PCR primers for adding a stretch of six histidines to either the N-terminus or C-terminus of your protein are illustrated in Figure 8–66. In both cases the primer on the left is complementary to the bottom strand of DNA (not shown) and primes synthesis in the rightward direction. The primer on the right is complementary to the top strand of DNA and primes synthesis in the leftward direction.

If you are set up to do database searches, you might try to find out what protein is being modified in this problem.

8–68*

8–69

A. The critical event in recombinant PCR occurs when the two individually amplified products are mixed. In the denaturation–renaturation cycle the strands from the two products can hybridize in the region corresponding to

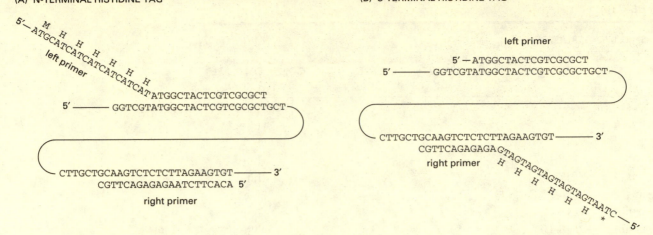

(A) N-TERMINAL HISTIDINE TAG

```
    M   H   H   H   H   H
5'—ATGCATCATCATCATCATCAT
       left primer      ATGGCTACTCGTCGCGCT
              5'——GGTCGTATGGCTACTCGTCGCGCTGCT
```

```
——CTTGCTGCAAGTCTCTCTTAGAAGTGT———3'
  CGTTCAGAGAGAATCTTCACA 5'
         right primer
```

(B) C-TERMINAL HISTIDINE TAG

```
                    left primer
          5'—ATGGCTACTCGTCGCGCT
5'———————GGTCGTATGGCTACTCGTCGCGCTGCT—
```

```
CTTGCTGCAAGTCTCTCTTAGAAGTGT———3'
CGTTCAGAGAGAGTAGTAGTAGTAGTAGTAATC—
      right primer    H   H   H   H   H   H   H
                                              *   5'
```

primers 2 and 3, as shown in Figure 8–67A. The hybridization product in which the 3′ ends overlap can be extended to make the full-length hybrid sequence. From then on, all that is required is amplification with primers 1 and 4 in the standard way.

B. The structure and arrangement of primers needed to put the regulatory domain at the N-terminus of the hybrid gene are shown in Figure 8–67B. When the individually amplified domains are mixed and denatured and reannealed, they will overlap in the regions corresponding to primers 1 and 4. The hybrid gene can then be amplified using primers 2 and 3.

Figure 8–66 PCR primers to add hexa-histidine tags to your protein (Answer 8–67). (A) Primers to add histidines to the N-terminus. (B) Primers to add histidines to the C-terminus. In these primers only one codon for histidine is used, but the other histidine codon, or a mixture could have been used. For the C-terminal histidine tag the complement of the histidine codon is used.

Figure 8–67 Recombinant PCR (Answer 8–69). (A) Critical steps in recombinant PCR. (B) Structure and arrangement of primers to generate hybrid gene with the regulatory domain at the N-terminus.

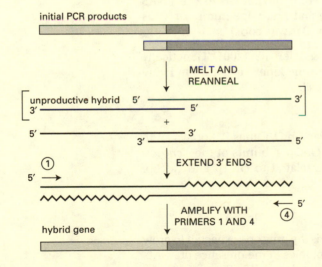

(A) THE CRITICAL STEPS IN RECOMBINANT PCR

initial PCR products

MELT AND REANNEAL

unproductive hybrid

EXTEND 3′ ENDS

AMPLIFY WITH PRIMERS 1 AND 4

hybrid gene

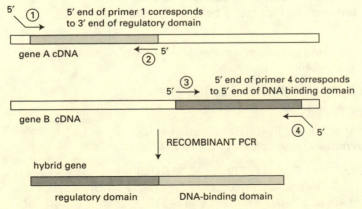

(B) REVERSING THE ORDER

5′ end of primer 1 corresponds to 3′ end of regulatory domain

gene A cDNA

5′ end of primer 4 corresponds to 5′ end of DNA binding domain

gene B cDNA

RECOMBINANT PCR

hybrid gene

regulatory domain   DNA-binding domain

ANSWERS

TABLE 8–10 Expected results of experiments to test the two-hybrid system (Answer 8–73).

PLASMID CONSTRUCTS		GROWTH ON PLATES LACKING HISTIDINE	COLOR ON PLATES WITH XGAL
BAIT	PREY		
LexA–Ras		–	white
LexA–lamin		–	white
	VP16	–	white
	VP16–CYR	–	white
LexA–Ras	VP16	–	white
LexA–Ras	VP16–CYR	+	blue
LexA–lamin	VP16	–	white
LexA–lamin	VP16–CYR	–	white

Reference: Higuchi R (1990) Recombinant PCR. In PCR Protocols (MA Innis, DH Gelfand, JJ Sninsky & TJ White eds), pp 177–183. San Diego, CA: Academic Press.

8–70* Figure 8–68.

8–71 The data in Figure 8–34 indicate the location of the reporter protein in the cell, and by inference the location of active Abl. The reporter is not phosphorylated (activated) in the nucleus. It is activated in the cytoplasm, but is most highly activated in membrane ruffles. These results are most simply consistent with the idea that active Abl is most prevalent in the membrane ruffles. (You can watch this process in a color movie. Go to the PNAS website, www.pnas.org, type in the volume and first page number of this article, and select 'supporting movies.' Movie 2 is way cool.)

Reference: Ting AY, Kain KH, Klemke RL & Tsien RY (2001) Genetically encoded fluorescent reporters of protein tyrosine kinase activities in living cells. *Proc. Natl. Acad. Sci. USA* 98, 15003–15008.

8–72* Figure 8–69.

Reference: Tugal T, Zou-Yang XH, Gavin K, Pappin D, Canas B, Kobayashi R, Hunt T & Stillman B (1998) The Orc4p and Orc5p subunits of the *Xenopus* and human origin recognition complex are related to Orc1p and Cdc6p. *J. Biol. Chem.* 273, 32421–32429.

8–73

A. The expected results are indicated in Table 8–10.
B. The combination of LexA–Ras and VP16–CYR is the only one expected to grow in the absence of histidine and to give blue colonies in the presence of XGAL. The structure of the active transcription factor is sketched in Figure 8–70.
C. In order for the two proteins of a fusion gene to be expressed as a single polypeptide chain, it is essential that there be no translational stop codons between the two coding segments and that the two genes be fused in the same translational reading frame.

References: Fields S & Song O (1989) A novel genetic system to detect protein–protein interactions. *Nature* 340, 245–246.

Vojtek AB, Hollenberg SM & Cooper JA (1993) Mammalian Ras interacts directly with the serine/threonine kinase Raf. *Cell* 74, 205–214.

8–74* **References:** Vojtek AB, Hollenberg SM & Cooper JA (1993) Mammalian Ras interacts directly with the serine/threonine kinase Raf. *Cell* 74, 205–214.

Zhang X-F, Settleman J, Kyriakis JM, Takenchi-Suzuki E, Elledge SJ, Marshall MS, Bruder JT, Rapp UR & Avruch J (1993) Normal and oncogenic p21ras proteins bind to the amino-terminal domain of c-Raf-1. *Nature* 364, 308–313.

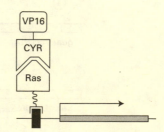

Figure 8–70 Structure of active transcription factor in yeast transformed with the LexA–Ras and VP16-CYR fusion genes (Answer 8–73).

8–75 Results from the two-hybrid system must be confirmed in some way. Even with all the layers of selection and screening built into it, the two-hybrid system can still give false positives. The 9 out of 19 clones with homology to Raf may seem very convincing given what was known about the biology of cell signaling. But it was still possible that the interaction detected by the two-hybrid system was an indirect interaction; for example, it could have been that there was a natural yeast protein that could bind to both Ras and Raf. Such a protein could bring the LexA–Ras and VP16–Raf hybrid proteins together to activate transcription in the absence of any direct interaction between Ras and Raf. The biochemical studies, which were carried out using proteins grown in *E. coli*, eliminated the possibility of this kind of indirect interaction.

References: Vojtek AB, Hollenberg SM & Cooper JA (1993) Mammalian Ras interacts directly with the serine/threonine kinase Raf. *Cell* 74, 205–214.

Zhang X-F, Settleman J, Kyriakis JM, Takenchi-Suzuki E, Elledge SJ, Marshall MS, Bruder JT, Rapp UR & Avruch J (1993) Normal and oncogenic p21^{ras} proteins bind to the amino-terminal domain of c-Raf-1. *Nature* 364, 308–313.

8–76* **Reference:** Vidal M & Legrain P (1999) Yeast forward and reverse "n"-hybrid systems. *Nucleic Acids Res.* 27, 919–929.

8–77 As shown in Table 8–11, 8 of the 14 phages have a peptide sequence that contains an R-LF match to the cyclin A-binding sequence in p27. Thus, this sequence is likely to be the critical portion of proteins that bind to the hydrophobic cleft in cyclin A.

8–78*

8–79

A. Using a reduced genetic code avoids two stop codons (TAA and TGA) of the three in the standard genetic code. The third stop codon is avoided by growing the phage in bacteria that insert glutamine in response to the TAG codon. This strategy ensures that all inserted nucleotides will encode peptide sequences.

B. Remarkably, the reduced genetic code encodes most amino acids in the same relative frequencies as the standard genetic code. Cysteine (C), for example, which has one codon in the reduced genetic code, is present half as often as alanine (A), which has two codons, and one third as often as

TABLE 8–11 Peptide sequences in bound phages optimally aligned with the cyclin A-binding sequence in p27 (Answer 8–77).

CLONE	LIBRARY	PEPTIDE SEQUENCE
1	7-mer	LEP**RM**L**F**
2	7-mer	TLP**RQ**L**F**
3	7-mer	LKPTK**LF**
4	7-mer	LIPKN**LF**
5	7-mer	FLP**RA**L**F**
1	12-mer	NVRVE**LF**PPTKV
2	12-mer	KSSVV**RS**L**F**VPT
3	12-mer	ERPSAQ**RS**L**V**FW
4	12-mer	NLFYP**RN**L**F**PEF
5	12-mer	YPSPA**RN**L**L**PMF
6	12-mer	ATI**RE**L**F**PPTLP
1	20-mer	HQPESVK**RS**L**F**KPAHSALEP
2	20-mer	EVA**RRE**L**F**ADHSLVHVGHVR
3	20-mer	EHKALPGKAVTGPK**RE**L**V**FQ
p27 cyclin A-binding sequence		PSAC**RN**L**F**GP

leucine (L), which has three codons. Both tryptophan (W) and methionine (M) are overrepresented in the reduced genetic code; each has one codon out of 64 in the standard code, but one out of 32 in the reduced code. Glutamine (Q) is also overrepresented because the TAG codon has essentially been turned into a glutamine codon.

8–80*

8–81

A. A uniform protein–ligand complex is expected to have a single dissociation rate, which is independent of concentration of the complex. So long as the buffer and temperature remain the same, the off rate should be constant. Two distinct off rates suggests an inhomogeneity in the system somewhere.

B. There are many possible explanations that fall into two general classes: problems with the protein and problems with the apparatus. It may be that your pure protein contains two components that differ slightly in their folding and for that reason bind to the ligand with different affinities. It may be that a proportion of your protein carries a posttranslational modification such as phosphorylation that affects its binding affinity. In terms of the apparatus, it could be that the ligand is attached in two different ways, one of which can be bound tightly and the other only weakly. It may be that your protein (perhaps as a dimer) has two binding sites for the ligand; proteins bound to one ligand on the biosensor surface would release more readily than proteins attached to two ligands.

Reference: Nagata K & Handa H (2000) Real-Time Analysis of Biomolecular Interactions: Applications of BIACORE. Tokyo, Japan: Springer.

8–82* Figure 8–71.

STUDYING GENE EXPRESSION AND FUNCTION

8–83

A. The locus is the site of the gene in the genome, more particularly the location of the gene in a chromosome. Each gene has one defined locus in the genome. If the locus is on an autosome, there are two loci in the genome; if it is on a sex chromosome such the X chromosome, there will be two loci in the female and one in the male. An allele is an alternative form of a gene at a locus. Within the population, there are often several 'normal' alleles, whose functions are indistinguishable. In addition, there may be many rare alleles that are defective to varying degrees. An individual, however, normally has a maximum of two alleles of a gene.

B. An individual is said to be homozygous if the two alleles at a locus are the same. An individual is said to be heterozygous if the two alleles at a locus are different.

C. Genotype is the specific set of alleles forming the genome of an individual; it is an enumeration of all the particular forms of each gene in the genome. In practice, for organisms studied in a laboratory, the genotype is usually specified as a list of the known differences between the individual and the wild type, which is the standard, naturally occurring type. Phenotype is a description of the visible characteristics of the individual. In practice, phenotype is usually a list of the differences in visible characteristics between the individual and the wild type.

D. An allele is dominant (relative to second allele) if the phenotype is the same when the allele is homozygous and when it is heterozygous. In that case the second allele, whose presence makes no difference to the phenotype, is said to be recessive (to the first allele). If the phenotype of the heterozygous individual differs from the phenotypes of individuals that are homozygous for either allele, the alleles are said to be co-dominant.

8–84*

8–85   False. Even with so many sequenced genomes, it is still a laborious process to identify a gene that has been mutated by chemical mutagenesis. Such a mutation must be mapped to a chromosomal location (a time consuming process) and then candidate genes in that region (many of which are known from genome sequencing efforts) can be screened for the presence of the mutation. Insertional mutagenesis, by contrast, places a known sequence—often a transposable element—into the mutated gene. It is a simple process to obtain sequence information adjacent to a segment of known sequence. In a sequenced genome, a bit of sequence is all that is needed to identify the location of the inserted DNA and the mutated gene.

8–86*

8–87

A. As outlined in Figure 8–72, if flies are defective in different genes their progeny will have one normal gene at each locus. In the case of a mating between a ruby-eyed fly and a white-eyed fly, every progeny fly will inherit one functional copy of the white gene from one parent and one functional copy of the ruby gene from the other parent. Since each of the mutant alleles is recessive to the corresponding wild-type allele, the progeny will have the wild-type phenotype—brick-red eyes.

B. Garnet, ruby, vermilion, and carnation complement one another and the various alleles of the white gene; thus they define separate genes. White, cherry, coral, apricot, and buff do not complement each other; thus, they must be alleles of the same gene, which has been named the *white* gene. Thus, these nine different eye-color mutants define five different genes.

C. Different alleles of the same gene, like the five alleles of the *white* gene, often have different phenotypes. Different mutations compromise the function of the gene product to different extents, depending on the location of the mutation. Different null alleles of the same gene, which have no function at all, do have the same phenotype.

Reference: Hartwell LH, Hood L, Goldberg ML, Reynolds AE, Silver LM & Veres RC (2000) Genetics: From Genes to Genomes, pp 191–195. Boston, Massachusetts: The McGraw Hill Companies, Inc.

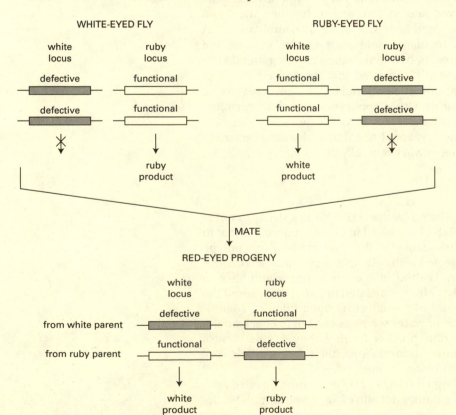

Figure 8–72 Complementation between white and ruby (Answer 8–87).

8–88*

8–89   Knowledge that the seven proteins in *E. coli* correspond to seven domains of the mammalian fatty acid synthase means that they are likely to be functionally associated, and probably work together in a protein complex.

8–90*   Figure 8–73.

8–91   The sample marked JC has two copies of mutant allele 3 and thus will have the disease. The sample marked BF has a single copy of mutant allele 2 and thus is heterozygous; this individual should not have the disease. BF will, however, be a carrier of the mutation and should be apprized of that fact eventually. The sample marked HK has one copy of mutant allele 2 and one copy of mutant allele 3. Since both copies of the gene are defect, this individual will have the disease. (You may have considered the possibility that a recombination event in one of the parental lineages placed both mutations on one chromosome, generating a new allele that carries both mutations. For mutations in the same gene, this is an extremely remote possibility in humans.) The sample marked TW has no mutant alleles; thus will not have the disease and will not be a carrier of the mutation.

8–92*

8–93   Mutant bacteria that do not produce ice-protein have probably arisen many times in nature. However, bacteria that produce ice-protein have a slight growth advantage over bacteria that do not, so it would be difficult to find such mutants in the wild (and it would probably be difficult for the ice-protein-free mutant bacteria to survive in the long run facing the competition of their natural counterparts). Genetic engineering, using genes deliberately mutated *in vitro*, simply makes these mutants much easier to obtain. The consequences, both advantageous and disadvantageous, of using a genetically engineered organism for a practical application, therefore, are nearly indistinguishable from those that would follow the use of a natural mutant. Indeed, bacterial and yeast strains have been selected for centuries for desirable genetic traits that make them suitable for industrial-scale applications such as cheese and wine production. The possibilities of genetic engineering are endless; however, and as with any technology, there is a finite risk that unforeseen consequences will arise. Recombinant DNA experimentation, therefore, is regulated, and review panels assess the potential risks of individual projects before permissions are granted. The state of our knowledge is sufficiently advanced that the consequences of some changes, such as the disruption of a bacterial ice-protein gene, can be predicted with reasonable certainty. Other applications, such as germline gene therapy to correct human disease, may have far more complex outcomes, and it will take many more years of research and ethical debate to determine whether such treatments will eventually be used.

8–94*

8–95

A. The truncated-tail phenotype induced by the NAF RNA is a striking confirmation of your hypothesis that Raf-1 is involved in posterior development in vertebrates. The reduction in frequency of the truncated-tail phenotype when normal *raf-1* RNA is co-injected with NAF RNA is an important control that strengthens the conclusion. Without this control, you would have to worry whether NAF protein had an effect, unrelated to Raf-1, that caused the truncated-tail phenotype. Reversal of the effect by bona fide Raf-1 eliminates that concern. The injection of water serves as control to check on the deleterious effects of the injection process itself. Although this control shows that injection causes some abnormalities, their frequency is well below that seen when the various RNAs are injected.

B. In response to the intracellular signal triggered by FGF, some positive activator presumably binds to the regulatory domain of Raf-1 and stimulates the protein kinase activity. In the presence of NAF, which is in excess in these experiments, most of the positive activator is tied up in unproductive interactions with NAF. By soaking up the positive activator in this way, NAF interferes with the normal activation of the Raf-1 protein in the embryos.

C. When *raf-1* RNA is injected in equal amounts with NAF RNA, the positive activator is distributed between Raf-1 and NAF, which are present in more nearly equal amounts. Thus, under these conditions some Raf-1 is activated, propagating the signal that leads to proper posterior development.

Reference: MacNicol A, Muslin AJ & Williams LT (1993) Raf-1 kinase is essential for early *Xenopus* development and mediates the induction of mesoderm by FGF. *Cell* 73, 571–583.

8–96*   Figure 8–74.

8–97   The mismatch would be expected to be recognized by the cell's mismatch repair machinery. Whether it would be repaired depends on a race between the repair machinery and the replication machinery. If replication wins, the mismatch is eliminated and an equal mixture of the two possible genomes would be generated (see Figure 8–46).

The outcome would be essentially the same if mismatch repair won the race. Because two normal bases are involved in the mismatch, the repair enzymes cannot distinguish which strand of the DNA contains the mutation and which one contains the normal nucleotide. Therefore, in half the cells that have been transformed with the mismatched plasmid, a normal gene is restored. In the other half, however, the normal strand is converted to match the mutated strand and the mutation is thus propagated. Within the population, roughly equal numbers of mutant and wild-type genomes would be present.

Cells containing a plasmid with the desired mutation can be identified, for example, by hybridization with a single-stranded oligonucleotide probe that allows one to distinguish between the normal and mutant genes.

8–98*   Figure 8–75.

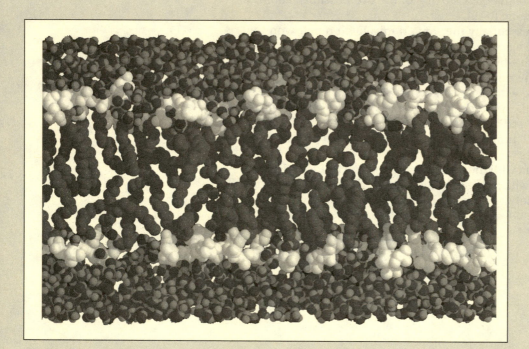

A computer-generated model of a lipid bilayer. Adapted from RM Venable, Y Zhang, BJ Hardy & RW Pastor. *Science*, 262:223–228, 1993.

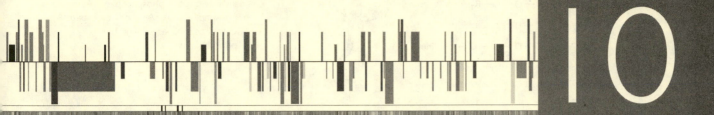

MEMBRANE STRUCTURE

THE LIPID BILAYER

10–1   Water is a liquid, and thus hydrogen bonds between water molecules are not static; they are continually formed and broken again by thermal motion. When a water molecule happens to be next to a hydrophobic molecule, it is more restricted in motion and has fewer neighbors with which it can interact because it cannot form any hydrogen bonds in the direction of the hydrophobic molecule. It will therefore form hydrogen bonds to the more limited number of water molecules in its proximity. Bonding to fewer partners results in a more ordered water structure, which constitutes the icelike cage in Figure 10–1. The true cage of water molecules is substantially different from that represented in the figure; the real cage exists in three-dimensions, forming a pentagonal dodecahedron (like a soccer ball) or clusters of them that enclose the hydrophobic solute. The structure is similar to ice, although it is a more transient, less organized, and less extensive network than even a tiny ice crystal. The formation of any ordered structure decreases the entropy of the system, which is energetically unfavorable.

10–2*

10–3   Lipid bilayers assemble because the surrounding water molecules exclude the component lipids; thus, analogy (2) is the correct one. If bilayers formed because of attractive forces among the lipids—analogy (1)—the properties of the bilayer would likely be quite different. Molecules 'attract' one another by forming specific bonds that hold them together. Such bonding among lipids would make the bilayer less fluid, perhaps even rigid, depending on the strength of the interaction.

10–4*

---

*Answers to these problems are available to Instructors at www.classwire.com/garlandscience/

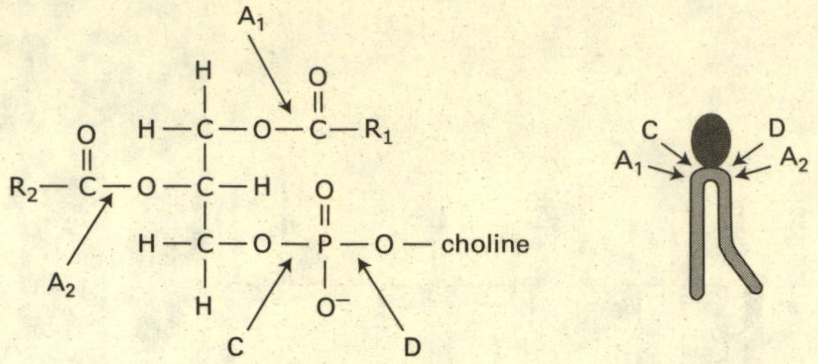

Figure 10–15 Cleavage specificity of several phospholipases (Answer 10–5). Susceptible bonds are indicated by *arrows*. Letters indicate specific phospholipases: A_1 is phospholipase A_1; A_2 is phospholipase A_2; C is phospholipase C; and D is phospholipase D.

10–5

A. The phospholipase cleaves phosphatidylcholine into phosphorylcholine and diacylglycerol by breaking the ester bond that links the phosphate group to glycerol. This phospholipase is like phospholipase C in its specificity (Figure 10–15). Several other types of phospholipase are indicated in the figure: phospholipase A_1 and A_2 remove fatty acid chains at specific positions, and phospholipase D removes the head group but leaves the phosphate group still attached to glycerol. These enzymes are extremely useful tools for lipid analysis. Several come from the venoms of snakes.

B. It may strike you as odd that removing the hydrophilic head group from phosphatidylcholine causes complete breakdown of the red cell membrane, whereas removing the exterior sialic acid groups or the external protein domains does not. After all, about half of the membrane is protein, and less than half of the outer lipid monolayer is phosphatidylcholine. Removing the choline head groups destabilizes the lipid bilayer by converting a large number of the constituent phospholipids, which are amphipathic, to diacylglycerols, which are almost entirely nonpolar. This result emphasizes that the lipid bilayer is crucial to the integrity of the membrane and further stresses that a hydrophilic head group is essential if a lipid is to form a stable bilayer. (Not all phospholipase C enzymes cause hemolysis, by the way. Some cannot act on the phospholipids in intact membranes, presumably because they cannot reach the sensitive bonds.)

10–6*

10–7   In a two-dimensional fluid the molecules are free to move only in one plane; the molecules in a normal fluid can move in three dimensions.

10–8*   **Reference:** Rousselet A, Guthmann C, Matricon J, Bienvenue A & Devaux PF (1976) Study of the transverse diffusion of spin labeled phospholipids in biological membranes: 1. Human red blood cells. *Biochim. Biophys. Acta* 426, 357–371.

10–9   Vegetable oil is converted to margarine by reduction of double bonds (by hydrogenation), which converts unsaturated fatty acids to saturated ones. This change allows the fatty acids chains in the lipid molecules to pack more tightly against one another, increasing the viscosity: turning oil into margarine.

10–10*

10–11

A. Only phosphatidylserine and phosphatidylethanolamine have primary amino groups, which can react with SITS. Since these phospholipids are labeled only when the red cells are made permeable (ghosts), they presumably reside in the cytoplasmic monolayer. This conclusion is supported by the results from experiments with sea snake venom, which degrades phosphatidylserine and phosphatidylethanolamine only in ghosts. These results, taken together, indicate that the phosphatidylserine and phosphatidylethanolamine are localized almost exclusively in the cytoplasmic monolayer of red cell membranes.

Phospholipase degradation of phosphatidylcholine and sphingomyelin in intact cells indicates that they are present in the outer monolayer. This conclusion depends on the red cell remaining intact during the treatment. In the case of sea snake venom, the absence of degradation of phosphatidylserine and phosphatidylethanolamine in intact cells provides an internal control. In the case of sphingomyelinase, there is no internal control, but the absence of lysis indicates that the membrane is intact.

The results in Table 10–2 do not exclude the possibility that phosphatidylcholine and sphingomyelin are also located in the cytoplasmic monolayer. However, the quantification of sphingomyelin degradation by sphingomyelinase (provided in the body of the problem) indicates that sphingomyelin is localized almost entirely in the outer monolayer of the membrane. No such data are provided for phosphatidylcholine; thus, it is incorrect to conclude from the data given that phosphatidylcholine is located exclusively in the outer monolayer. However, other experiments not reported here do suggest that phosphatidylcholine is found almost entirely in the outer monolayer.

B. You chose red cells for these experiments because they contain no internal membranes. If the same experiments were performed on cells with internal membranes, it would have been impossible to measure directly the phospholipid composition of the cytoplasmic monolayer, since phospholipids from the cytoplasmic monolayer would have been hopelessly confused with those from internal membranes.

References: Bretscher M (1972) Asymmetrical lipid bilayer structure for biological membranes. *Nature New Biol.* 236, 11–12.

Deenen LLM & DeGier J (1974) Lipids of the red cell membrane. In The Red Blood Cell (D. MacN. Surgenor, ed), pp 147–211. New York: Academic Press.

10–12*

10–13

A. When lined up, it would take 4000 lipid molecules to reach from one end of a bacterium to the other (2 μm/0.5 nm = 4000). If a lipid molecule at one end moved directionally by exchanging places with its immediate neighbor down the line every 10^{-7} seconds, it would reach the other end in 4×10^{-4} seconds (4000×10^{-7} seconds). This is some 2500 times faster than the measured rate of about 1 second. These numbers do not agree because lipid molecules do not diffuse in a straight line; they move along random paths so that it takes much longer to travel from one end of the cell to the other.

B. If a 4-cm ping-pong ball exchanged places with a neighbor every 10^{-7} seconds, it would travel at a speed of 1,440,000 km/hr [(4 cm/10^{-7} sec) × (km/10^5 cm) × (3600 sec/hr)]. If its movement were only in one direction, it would reach the other wall in 1.5×10^{-5} seconds [6 m × (10^{-7} sec/0.04 m)]; in a random walk it would take considerably longer.

10–14*

10–15 The size of a lipid raft depends on the affinity of the sphingolipids and cholesterol molecules for one another. If they bound one another sufficiently tightly, they would aggregate into a single domain in the membrane. If they bound one another with the same affinity as they bind to other species of lipid molecules, they would remain dispersed. The small size of the lipid rafts indicates that sphingolipids and cholesterol molecules have only a slightly higher affinity for one another than for other lipids. Presumably at this size the aggregated sphingolipids and cholesterol molecules are in equilibrium with their free forms, so that lipids are added to and leave the raft at equal rates.

10–16*

10–17 The large head groups of most sphingolipids prevent the close packing of their fatty acid tails in the membrane. The planar cholesterol molecules are postulated to fill the voids that form underneath the large head groups of the sphingolipids, thereby packing tightly against fatty acid chains that could not otherwise approach closely enough to bind.

Reference: Harder T & Simons K (1997) Caveolae, DIGs, and the dynamics of sphingolipid-cholesterol microdomains. *Curr. Opin. Cell Biol.* 9, 534–542.

10–18* **Reference:** Brown DA & Rose JK (1992) Sorting of GPI-anchored proteins to glycolipid-enriched membrane subdomains during transport to the apical cell surface. *Cell* 68, 533–544.

10–19

A. For randomly dispersed receptors (see Figure 10–6A), the polarization of the fluorescent light will depend critically on the concentration of the receptors in the membrane. At high density (pixels with high intensity fluorescence) there will be efficient FRET (high intensity fluorescence detected by the perpendicular filter), giving rise to a low value for polarization of the fluorescence $[(I_{par} - I_{perp})/(total intensity)]$. By contrast, at low density the overall intensity will be lower, but FRET will be much less efficient since molecules are on average farther away from one another. As a result, what fluorescence there is will be more polarized.

For receptors that are confined to microdomains such as lipid rafts (see Figure 10–6B), the overall fluorescence intensity will decrease with decreasing density of the rafts, which is determined just by chance distribution of rafts relative to the very small window (a pixel) being examined. The polarization of the fluorescence, however, will be independent of concentration. At high density and at low density of rafts, the receptors in microdomains will always be equally close to their neighbors; thus, a constant proportion of the absorbed light energy will be transferred by FRET. As a result, receptors in microdomains will give the same low value for polarization of fluorescence regardless of the total fluorescence intensity in a pixel.

B. The results suggest that transmembrane-anchored folate receptors are randomly dispersed in the membrane, while GPI-anchored receptors are clustered in microdomains. Although such microdomains are likely to be lipid rafts, these experiments do not prove that point.

C. Both types of receptor behave as if randomly dispersed in cells grown in the presence of compactin, which reduces the amount of cholesterol in the membrane by blocking its synthesis. These observations suggest more strongly that the microdomains in the previous experiments were indeed lipid rafts, which are known to require cholesterol for their formation.

Reference: Varma R & Mayor S (1998) GPI-anchored proteins are organized in submicron domains at the cell surface. *Nature* 394, 798–801.

10–20*

10–21 True. The positively charged moieties in all cases are balanced by the negative charge on the phosphate group; thus, none of the common phospholipids carries a net positive charge.

10–22* **Reference:** Rousselet A, Guthmann C, Matricon J, Bienvenue A & Devaux PF (1976) Study of the transverse diffusion of spin labeled phospholipids in biological membranes: 1. Human red blood cells. *Biochim. Biophys. Acta* 426, 357–371.

10–23 The redistribution of phosphatidylserine from the cytoplasmic to the outer monolayer of the plasma membrane bilayer occurs by two mechanisms: (1) the phospholipid translocators that normally transport this lipid from the noncytoplasmic monolayer to the cytoplasmic monolayer is inactivated in apoptotic cells; and (2) a 'scramblase' that transfers phospholipid nonspecifically in both directions between the two monolayers is activated.

10–24*

MEMBRANE PROTEINS

10–25 True. The lipid bilayer defines the structure of the membrane and provides a permeability barrier that separates the inside from the outside of the cell. Specific membrane proteins allow particular solutes to enter and leave the cell, bind signaling molecules, and mediate attachment to the extracellular matrix.

10–26*

10–27 The principles are the same for both. The exposure of hydrophobic amino acid side chains to water is energetically unfavorable. There are two ways that such side chains can be sequestered from water to achieve an energetically more favorable state. One, they can form transmembrane segments that span a lipid bilayer, which requires that about 20 hydrophobic amino acid side chains be located sequentially in a polypeptide chain. Two, the hydrophobic amino acid side chains can be sequestered in the interior of the folded polypeptide chain, which is one of the major forces that lock a polypeptide chain into a unique three-dimensional structure.

10–28*

10–29 Fatty acid chains, prenyl groups, and glycosylphosphatidylinositol (GPI) anchors are the three common lipid anchors for membrane proteins.

10–30*

10–31
A. Sequence A is the actual membrane-spanning α-helical segment of glycophorin, a transmembrane protein from red blood cells. It is predominantly nonpolar, although it does contain the uncharged polar amino acids threonine (T) and serine (S), which are not uncommon in membrane-spanning α helices.

    Sequence B is unlikely to be a membrane-spanning segment because it contains three prolines (P), which would disrupt an α helix and thereby expose hydrogen-bonding moieties to the nonpolar environment of the lipid bilayer.

    Sequence C is also unlikely to be a transmembrane segment because it contains three charged amino acids, glutamic acid (E), arginine (R), and aspartic acid (D), whose presence in the nonpolar lipid bilayer would be energetically unfavorable.

10–32* Figure 10–16.

10–33
A. In a β strand, adjacent amino acid side chains protrude from opposite sides of the strand; thus, every other amino acid side chain will face the same side of the strand. If a β strand is part of a β-barrel pore, its amino acid side chains will alternate between hydrophobic and hydrophilic, so that one side of the strand will be hydrophobic and the other side will be hydrophilic. Only choice A has alternating hydrophobic and hydrophilic amino acids.

10–34*

10–35 It is thought that transmembrane α helices are more common than transmembrane β barrels because they provide a more flexible arrangement of transmembrane segments. Because α helices can slide against one another, they allow the protein to undergo conformational changes that can be exploited to gate ion channels, transport solutes, or transmit information. By contrast, the β strands in a β barrel are rigidly fixed to their neighbors by hydrogen bonds that lock the protein into a single conformation.

10–36*

10–37 Your friend's suggestion is based on an important difference between inside-out and right-side-out vesicles. The contaminating right-side-out vesicles will carry carbohydrate on their exposed surface and, therefore, should be retained on a lectin affinity column. Inside-out vesicles, by contrast, will lack carbohydrate on their exposed surface and, therefore, should pass through the column.

10–38*

10–39 The sulfate group in SDS is charged and therefore hydrophilic. The OH group and the C–O–C groups in Triton X-100 are polar; they can form hydrogen bonds with water and are therefore hydrophilic. The gray portions of these detergents are either hydrocarbon chains or aromatic rings, neither of which have polar groups that can hydrogen bond to water molecules; they are therefore hydrophobic.

10–40*

10–41 Membrane proteins anchor the lipid bilayer to the cytoskeleton, which strengthens the plasma membrane so that it can withstand the forces on it when the red blood cell is pumped through small blood vessels. Membrane proteins also transport nutrients and ions across the plasma membrane.

10–42*

10–43 Using these average molecular weights, there are 84 lipid molecules (phospholipid + cholesterol) for every protein molecule [$2 \times 50,000/(800 + 386) = 84$]. A similar lipid to protein ratio is present in many cell membranes.

10–44* **References:** Bennett V & Stenbuck PJ (1979) The membrane attachment protein for spectrin is associated with band 3 in human erythrocyte membranes. *Nature* 280, 468–473.

Bennett V & Stenbuck PJ (1980) Association between ankyrin and the cytoplasmic domain of band 3 isolated from the human erythrocyte membrane. *J. Biol. Chem.* 255, 6424–6432.

10–45

A. The calculation for the number of spectrin molecules per red blood cell is shown in detail below. In essence, one first calculates the fraction of total protein that is spectrin and then converts that number into the number of spectrin molecules using the molecular weight of spectrin and Avogadro's number

$$\frac{\text{spectrin}}{\text{cell}} = \frac{5 \text{ mg protein}}{10^{10} \text{ cells}} \times \frac{0.25 \text{ spectrin}}{\text{total protein}} \times \frac{\text{mmol spectrin}}{250,000 \text{ mg}} \times \frac{6 \times 10^{20} \text{ molecules}}{\text{mmol spectrin}}$$

$$= 3 \times 10^5 \text{ molecules}$$

The analogous calculation gives values of 9×10^5 molecules of band 3 and 2.3×10^5 molecules of glycophorin per red cell. The calculated number of glycophorin molecules per cell is too low by a factor of 2.5 because about 60% of the molecular weight of glycophorin is carbohydrate, which is not stained by Coomassie Blue.

B. The fraction of the plasma membrane that is occupied by band 3 is the area of the face of a single band 3 molecule (πr^2) times the total number of band 3 molecules per cell (9×10^5) divided by the total area of the red blood cell (10^8 nm^2). Note that the height of the molecule is irrelevant to the calculation.

$$\frac{\text{band 3}}{\text{plasma membrane}} = \frac{3.14 \times (3 \text{ nm})^2}{\text{molecule}} \times \frac{9 \times 10^5 \text{ molecules}}{\text{cell}} \times \frac{1 \text{ cell}}{10^8 \text{ nm}^2}$$

$$= 0.25$$

Thus band 3 occupies about 25% of the surface area of a red blood cell. This somewhat surprising result is consistent with freeze-fracture electron micrographs of red blood cells, which show a high density of intramembranous particles that are thought to be dimers of band 3.

10–46* Figure 10–17.

10–47 Transmembrane domains that are composed entirely of hydrophobic amino acid side chains obviously cannot interact with one another via hydrogen bonds or ionic interactions, two of the more important ways to link proteins together noncovalently. They can, however, interact specifically via van der Waals bonds. If their surfaces are complementary, they can fit together well enough to make a large number of van der Waals contacts, which can hold them together. It should be noted, however, that the transmembrane segment of glycophorin does contain a few polar amino acids that may participate in dimerization (see Problem 10–46).

10–48*

10–49 There are two populations of band-3 proteins in the red cell membrane: one population is immobilized by attachment to the spectrin-based cytoskeleton; the other is freely mobile. Only the freely mobile population will be able to diffuse into the bleached spot and contribute to recovery of fluorescence. Thus, the curve for recovery of fluorescence will reach a plateau below the original level of fluorescence (Figure 10–18). The extent of recovery will correspond to the proportion of band-3 proteins that are freely mobile.

10–50*

10–51 Proteins can be restricted to specific regions of the plasma membrane in several ways: by attachment to extracellular or intracellular proteins, by attachment to proteins in other cells, and by molecular fences that in some way corral proteins in specific membrane domains. The fluidity of the lipid bilayer is not significantly affected by the anchoring of membrane proteins; the lipid molecules flow around anchored proteins like water around rocks in a stream.

10–52* **Reference:** Van Meer G & Simons K (1986) The function of tight junctions in maintaining differences in lipid composition between the apical and the basolateral cell surface domains of MDCK cells. *EMBO J.* 5, 1455–1464.

10–53 The carbohydrate-rich zone on the cell surface is termed the cell coat, or glycocalyx. Although most of the carbohydrate is attached to intrinsic plasma membrane molecules, the glycocalyx usually also contains glycoproteins and proteoglycans that have been secreted into the extracellular space and then adsorbed onto the cell surface. Many of the adsorbed macromolecules are components of the extracellular matrix. So are these bridging molecules part of the plasma membrane or part of the extracellular matrix? This is the ambiguity that makes 'where the plasma membrane ends and the extracellular matrix begins' largely a matter of semantics.

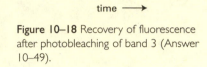

Figure 10–18 Recovery of fluorescence after photobleaching of band 3 (Answer 10–49).

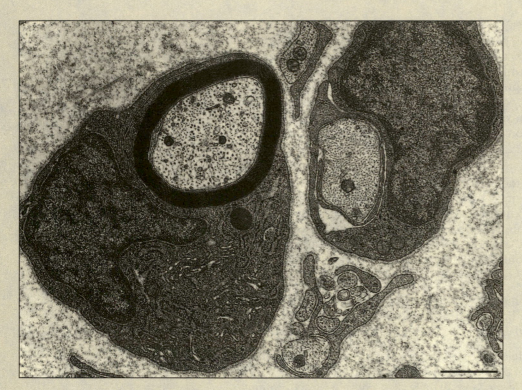

Electron micrograph of a section of a nerve in the leg of a young rat. Scale bar is 1 μm. Courtesy of Cedric S Raine, in Myelin (P Morell, ed). New York: Plenum, 1976.

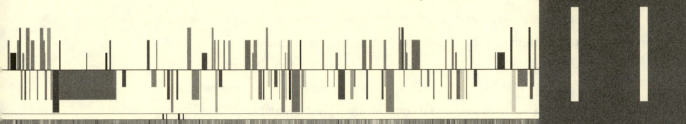

MEMBRANE TRANSPORT OF SMALL MOLECULES AND THE ELECTRICAL PROPERTIES OF MEMBRANES

PRINCIPLES OF MEMBRANE TRANSPORT

11–1 False. Lipid bilayers are impermeable to ions, but the plasma membrane contains specific ion channels and carriers that make it very permeable to particular ions and charged solutes under certain circumstances.

11–2*

11–3 Red blood cells have water channels—aquaporins—that make them about 10-fold more permeable to water than a lipid bilayer. Frog eggs do not express aquaporins and thus their permeability to water is roughly that of a lipid bilayer. However, if there were only a 10-fold difference in permeability to water, wouldn't a frog egg still burst, but just take ten times as long to do so? A part of the answer lies in the enormous volume difference between a red cell and a frog egg, more specifically in the surface-to-volume ratio. Assuming both are spheres, which of course red cells are not, the 10^6-fold difference in volume translates into a 100-fold lower surface-to-volume ratio in the egg. (This is an underestimate of the difference in the surface-to-volume ratio because the red cell biconcave disc has a larger surface-to-volume ratio than does a sphere of the same volume.) Thus, in the absence of aquaporins the egg should take up water at more than a 1000-fold lower rate than a red cell. If aquaporins are engineered to be expressed into frog eggs, they also swell and burst when placed in water.

Reference: Abrami L, Simon M, Rousselet G, Berthonaud V, Buhler JM & Ripoche P (1994) Sequence and functional expression of an amphibian water channel, FA-CHIP: a new member of the MIP family. *Biochim. Biophys. Acta* 1192, 147–151.

11–4*

11–5 Carrier proteins transport their specific solutes through the membrane much more slowly than channel proteins because they must bind the solute and then undergo a series of conformational changes to transfer the bound molecule across the membrane. Transport through channel proteins is much faster because they are ion-specific pores that neither bind the ion nor undergo any conformational changes in order to move it across the membrane.

11–6*

11–7

A. Transport by a carrier protein can be described by a strictly analogous equation:

$$CP + S_{high} \rightleftharpoons CPS \rightarrow CP + S_{low}$$

Where S_{high} is the solute on the high side of the concentration gradient and S_{low} is the solute on the low side. For a transport process the solute remains unchanged—unlike the substrate in an enzyme reaction—but it is located on the other side of the membrane.

B. The Michaelis–Menten equation for carrier-mediated transport is also strictly analogous:

$$\text{rate} = V_{max} \frac{[S_{high}]}{[S_{high}] + K_M}$$

The terms also have analogous meanings. The 'rate' is the initial rate of transport; V_{max} is the maximum rate of transport; and K_M is the concentration of solute at which the rate of transport is half maximal. The accuracy of the analogy allows one to apply classical enzyme 'thinking' to carrier-mediated transport.

C. The equations do not describe the behavior of channels, because solutes passing through channels do not bind to them in the way that a substrate binds to an enzyme (or a solute to a carrier).

11–8*   Figure 11–18.

11–9

A. Cytochalasin B inhibits glucose transport competitively, suggesting that it binds at or near the site of D-glucose binding on GLUT1. If an excess of D-glucose is present, the binding site on GLUT1 will be occupied by D-glucose, preventing cytochalasin from binding and thereby interfering with cross-linking. On the other hand, L-glucose does not interfere with cross-linking because it does not bind to the transporter and protect it from the binding of cytochalasin.

B. Since treatment of the native GLUT1 with an enzyme that removes oligosaccharide side chains sharpens the electrophoretic band, the fuzziness must be due to heterogeneity of the carbohydrate moieties attached to the protein. Whether this heterogeneity represents variable occupancy of potential oligosaccharide addition sites on the protein or is due to actual variability in the length or sequence of the oligosaccharide side chains is not known. This degree of heterogeneity is unusual; most glycoproteins form much sharper bands on SDS polyacrylamide gels. Although there are about 350,000 molecules of GLUT1 in each red cell (about the same as glycophorin and band 3), it went unnoticed for many years. In addition, there was considerable controversy about the molecular identity of the glucose transporter, with estimates of its molecular weight going as high as 200,000. All because it was a fuzzy band.

Reference: Allard WJ & Lienhard GE (1985) Monoclonal antibodies to the glucose transporter from human erythrocytes: identification of the transporter as a Mr = 55,000 protein. *J. Biol. Chem.* 260, 8668–8675.

11–10*   **Reference:** Oka Y & Czech MP (1984) Photoaffinity labeling of insulin-sensitive hexose transporters in intact rat adipocytes: direct evidence that latent transporters become exposed to the extracellular space in response to insulin. *J. Biol. Chem.* 259, 8125–8133.

11–11

A. The rates of glucose uptake in brain and liver cells at various concentrations

TABLE 11–6 Uptake of glucose as a percentage of V_{max} (Answer 11–11).

| GLUCOSE CONCENTRATION (mM) | RATE/V_{max} = [S]/([S] + K_M) | |
	BRAIN CELLS GLUT3 (K_M = 1.5 mM)	LIVER CELLS GLUT2 (K_M = 15 mM)
3	67%	17%
5	77%	25%
7	82%	32%
15	91%	50%

of glucose are shown in Table 11–6.

B. At 15 mM glucose in the portal circulation, the liver transports glucose at 50% of the maximum rate.

C. These calculations fit with the physiological functions of brain and liver. Brain cells depend on glucose as their primary energy source. Only during starvation is a significant fraction of its energy from another source, namely, ketone bodies. Thus, brain cells have to be able to take up glucose efficiently to meet their energy needs, and the low-K_M transporter, GLUT3, seems well matched to the brain's physiological role of glucose consumer.

By contrast, the physiological role of the liver is more complex. The liver serves as the body's storehouse for glucose (in the form of glycogen), which is built up after a meal and then doled out between meals to meet the rest of the body's needs for glucose. The high-K_M transporter, GLUT2, is matched to these needs. It helps to ensure that at low circulating glucose concentration the liver does not compete with the rest of the body for glucose; its role is to supply glucose, not consume it, under these conditions. At higher glucose concentration, however, it can take up glucose to build its glycogen reserves. In liver these physiological functions are enforced by hormone-induced changes to key regulatory enzymes in the pathways of glucose metabolism.

11–12*

11–13 The equilibrium distribution of a molecule across a membrane depends on the chemical gradient (concentration) and on the electrical gradient (membrane potential). An uncharged molecule does not experience the electrical gradient and, thus, will be at equilibrium when it's at the same concentration on both sides of the membrane. A charged molecule, however, responds to both components of the electrochemical gradient and will distribute accordingly. K^+ ions, for example, are nearly at their equilibrium distribution across the plasma membrane even though they are nearly 30-fold more concentrated inside the cell. The difference in concentration is nearly exactly balanced by the membrane potential (negative inside), which opposes the movement of cations to the outside of the cell.

11–14*

11–15 The step-wise changes in current indicate that the gramicidin A channel is transient: the two halves continually pair and fall apart in the membrane. Each time a channel forms, a current of about 1 pA flows across the bilayer; when the two halves of the channel dissociate, the current stops. Peaks that are twice as high arise when a second channel forms in the bilayer before the first has fallen apart. As is apparent in Figure 11–2B, individual channels vary somewhat—from less than 1 second to about 3 seconds—in how long they remain intact. A key to the success of this experiment was to use a minute amount of gramicidin A so that on average only one channel is open at a time.

Reference: Bamberg E & Lauger P (1974) Temperature-dependent properties of gramicidin A channels. *Biochim. Biophys. Acta* 367, 127–133.

11–16*

CARRIER PROTEINS AND ACTIVE MEMBRANE TRANSPORT

11–17 The three main ways in which cells carry out active transport are by (1) coupled transporters, (2) ATP-driven pumps, and (3) light-driven pumps. Coupled transporters link the uphill transport of one solute across the membrane with the downhill transport of another. ATP-driven pumps couple the uphill transport of a solute to hydrolysis of ATP. Light-driven pumps couple the uphill transport of a solute to input of energy from light.

11–18*

11–19 False. A symporter binds two different solutes on the *same side* of the membrane. Turning it around would not change it into an antiporter, which must bind two different solutes on *opposite sides* of the membrane.

11–20*

11–21 One model for incorporating ATP into the cycle of conformational changes necessary to drive glucose transport against its concentration gradient is shown in Figure 11–19. ATP is hydrolyzed and donates a phosphate group to the carrier protein when—and only when—it has glucose bound on the inside face of the membrane. The binding of glucose is shown as inducing a small conformational change in the transporter, which signals to the kinase that the transporter is ready to be phosphorylated (step 1 → 2). The attachment of the phosphate would trigger an immediate conformational change, thereby capturing the glucose and exposing it to the outside (step 2 → 3). The phosphate would be removed from the protein when—and only when—the solute has dissociated, and the now empty, nonphosphorylated carrier protein would switch back to the starting position (step 3 → 4).

11–22*

11–23 Export of HCO_3^- out of the cell affects the equilibrium position of the reaction catalyzed by carbonic anhydrase ($CO_2 + H_2O \rightleftharpoons H^+ + HCO_3^-$). Decreasing the amount of HCO_3^- in the cell drives the equilibrium to the right (by mass action), thereby increasing the concentration of H^+ and lowering the intracellular pH.

11–24*

11–25
A. The intracellular pH of the red blood cell changes very little because the histidine group on hemoglobin buffers it effectively.
B. CO_2 enters the red cell in the tissues but then is immediately converted to HCO_3^-, which is transported out of the cell by the Cl^--HCO_3^- exchanger. Thus, CO_2 is carried from the tissues to the lungs as HCO_3^- in the plasma outside the red cell.
C. In both the tissues and the lungs the Cl^--HCO_3^- exchanger moves HCO_3^- down its concentration gradient. In the tissues the hydration of CO_2 by carbonic anhydrase increases the intracellular concentration, allowing the exchanger to transport it out of the cell down its concentration gradient. In the lungs the removal of CO_2 by exhalation lowers the CO_2 concentration in

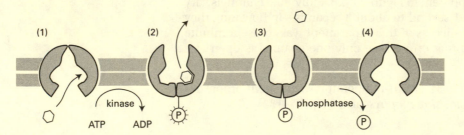

Figure 11–19 Coupling ATP hydrolysis to a hypothetical carrier protein to convert it from passive to active transport (Answer 11–21).

the cell, pulling the carbonic anhydrase reaction ($CO_2 + H_2O \rightleftharpoons H^+ + HCO_3^-$) to the left. The resultant lowering of the HCO_3^- concentration allows the exchanger to transport HCO_3^- into the cell down its concentration gradient.

11–26* Figure 11–20.

Reference: Molecular Probes Handbook (www.probes.com/handbook).

11–27 Each microvillus approximates a cylinder 0.1 μm in diameter and 1.0 μm in height. The ratio of the area of the sides of a cylinder, which represent new membrane (new surface area), to the top of a cylinder (which is equivalent to the plasma membrane that would have been present anyway had the microvillus not been extruded) gives the increase in surface area due to an individual microvillus. The area of the sides of a cylinder ($2\pi rh$, where r is the radius and h is the height) is 0.31 μm^2; the area of the top of the cylinder (πr^2) is 0.0079 μm^2. Thus, the increase in surface area for one microvillus is 0.31 μm^2/0.0079 μm^2 or 40. However, this value overestimates the increase for the entire plasma membrane, since the microvilli occupy only a portion of the surface. An estimate of the fraction of plasma membrane occupied by microvilli can be obtained from the cross-section in Figure 11–7. A conservative estimate is that about half the plasma membrane is covered by microvilli. Thus, microvilli increase the surface area in contact with the lumen of the gut by approximately 40/2 or 20-fold.

Reference: Adapted from Kristic RV (1997) Ultrastructure of the Mammalian Cell, p 207. Berlin, Germany: Springer-Verlag.

11–28*

11–29 If the Na^+-K^+ pump is not working at full capacity because it is partially inhibited by ouabain or digitalis, it generates an electrochemical gradient of Na^+ that is less steep than normal. Consequently, the Ca^{2+}-Na^+ antiporter works less efficiently, and Ca^{2+} is removed from the cell more slowly. When the next cycle of muscle contraction begins, there is still an elevated level of Ca^{2+} left in the cytosol. The entry of the same number of Ca^{2+} ions into the cell leads therefore to a higher Ca^{2+} concentration than in untreated cells, which in turn leads to a stronger and longer lasting contraction. Because the Na^+-K^+ pump fulfills essential functions in all animal cells, both to maintain osmotic balance and to generate the Na^+ gradient used to power many transporters, the drugs are deadly poisons at higher concentrations.

11–30*

11–31
   A. If the entire free-energy change due to ATP hydrolysis ($\Delta G = -12$ kcal/mole) could be used to drive transport, then the maximum concentration gradient that could be achieved by ATP hydrolysis would have a free-energy change of +12 kcal/mole.

$$\Delta G_{in} = -2.3RT \log_{10} \frac{C_o}{C_i} + zFV$$

Rearranging the equation gives

$$\log_{10} \frac{C_o}{C_i} = \frac{-\Delta G_{in} + zFV}{2.3RT}$$

For an uncharged solute, the electrical term (zFV) drops to zero. Thus,

$$\log_{10} \frac{C_o}{C_i} = \frac{-\Delta G_{in}}{2.3RT}$$

Substituting for ΔG_{in}, R, and T gives

$$\log_{10} \frac{C_o}{C_i} = \frac{-12 \text{ kcal/mole}}{2.3 \times (1.98 \times 10^{-3} \text{ kcal/}^\circ\text{K mole}) \times 310^\circ\text{K}}$$

$$\log_{10}\frac{C_o}{C_i} = -8.50$$

$$\log_{10}\frac{C_i}{C_o} = 8.50$$

$$\frac{C_i}{C_o} = 3.2 \times 10^8$$

Thus, for an uncharged solute a transport system that couples hydrolysis of 1 ATP to transport of 1 solute molecule could, in principle, drive a concentration difference across the membrane of more than eight orders of magnitude. Amazing!

B. If the entire free-energy change due to ATP hydrolysis ($\Delta G = -12$ kcal/mole) could be used to drive transport of Ca^{2+} out of the cell, then the maximum concentration gradient would yield a free-energy change of $+12$ kcal/mole.

$$\Delta G_{out} = 2.3RT \log_{10}\frac{C_o}{C_i} - zFV$$

Rearranging the equation gives

$$\log_{10}\frac{C_o}{C_i} = \frac{\Delta G_{out} + zFV}{2.3RT}$$

Since Ca^{2+} is charged, the electrical term must be included. Substituting for ΔG_{out}, R, T, z, F, and V, gives

$$\log_{10}\frac{C_o}{C_i} = \frac{12 \text{ kcal/mole} + (2 \times 23 \text{ kcal/V mole} \times -0.06 \text{ V})}{2.3 \times (1.98 \times 10^{-3} \text{ kcal/}^\circ\text{K mole}) \times 310^\circ\text{K}}$$

$$\log_{10}\frac{C_o}{C_i} = 6.54$$

$$\frac{C_o}{C_i} = 3.5 \times 10^6$$

Thus a transport system that couples hydrolysis of 1 ATP to transport of 1 Ca^{2+} ion to the outside of the cell could, in principle, drive a concentration difference across the membrane of more than six orders of magnitude. Note, by comparison with uncharged solute, that pumping against the membrane potential reduces the theoretical limit by two orders of magnitude. The difference in Ca^{2+} concentration across a typical mammalian plasma membrane is more than four orders of magnitude, but well within the theoretical limit (see Table 11–3).

C. The free-energy change for transporting Na^+ out of the cell is

$$\Delta G_{out} = 2.3RT \log_{10}\frac{C_o}{C_i} - zFV$$

Substituting (with $2.3\ RT = 1.41$ kcal/mole), the free-energy change for transporting K^+ into the cell is

$$\Delta G_{in} = -2.3RT \log_{10}\frac{C_o}{C_i} + zFV$$

Substituting (with $2.3\ RT = 1.41$ kcal/mole), the overall free-energy change for the Na^+-K^+ pump is

$$\Delta G = \Delta G_{out} + \Delta G_{in}$$
$$\Delta G = 9.0 \text{ kcal/3 mole } Na^+ + 1.3 \text{ kcal/2 mole } K^+$$
$$\Delta G = 10.3 \text{ kcal/(3 mole } Na^+ \text{ and 2 mole } K^+)$$

D. Since the hydrolysis of ATP provides 12 kcal/mole and the pump requires 10.3 kcal to transport 3 Na^+ out and 2 K^+ in, the efficiency of the Na^+-K^+ pump is

$$\text{eff} = \frac{10.3}{12.0}$$

$$= 86\%$$

Even with this remarkable efficiency the Na^+-K^+ pump typically accounts for a third of a mammalian cell's energy requirements and thus, presumably, a corresponding fraction of a mammal's total caloric intake.

11–32*

ION CHANNELS AND THE ELECTRICAL PROPERTIES OF MEMBRANES

11–33   Just as a falling body in air reaches a terminal velocity due to friction, an ion in water also reaches a terminal velocity (drift velocity) due to friction with water molecules. An ion in water will accelerate for less than 10 nanoseconds before it reaches drift velocity.

11–34*

11–35   Ion channels can be gated by a change in voltage across a membrane (voltage-gated channels), by a change in mechanical stress (mechanically gated channels), and by the binding of a ligand (ligand-gated channels).

11–36*

11–37   The key feature of a K^+ channel is its selectivity filter. For the bacterial K^+ channel, whose three-dimensional structure is known, the selectivity filter is a short, rigid, narrow pore, which is lined by carbonyl oxygens from the polypeptide backbone. The carbonyl groups are spaced exactly to accommodate a K^+ ion, allowing it to shed its water molecules and interact with the oxygen atoms of the carbonyls. By contrast, the oxygen atoms are spaced slightly too far apart to accommodate a Na^+ ion, providing too little binding energy to make up for the loss of water molecules that must be shed to pass through the filter.

11–38*   **Reference:** Zagotta WN, Hoshi T & Aldrich RW (1990) Restoration of inactivation in mutants of *shaker* potassium channels by a peptide derived from ShB. *Science* 250, 568–570.

11–39

A. K^+ channels composed of toxin-resistant subunits without balls will open and stay open in the presence or absence of toxin (Figure 11–21A).

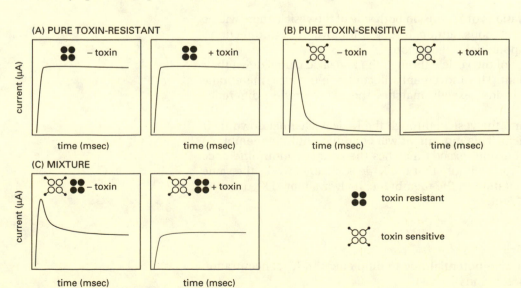

Figure 11–21 Expected patch-clamp recordings for mutant K^+ channels in the absence and presence of scorpion toxin (Answer 11–39). (A) Toxin-resistant K^+ channels without balls. (B) Toxin-sensitive K^+ channels with balls. (C) A 50:50 mix of the K^+ channels in A and B.

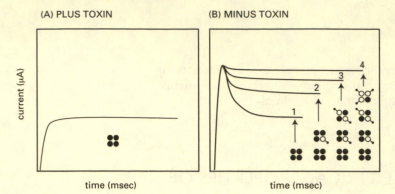

(A) PLUS TOXIN

(B) MINUS TOXIN

current (μA)

time (msec)

time (msec)

Figure 11–22 Results in the presence of toxin compared to those expected in the absence of toxin if 1, 2, 3, or 4 balls were required to close a channel (Answer 11–39). Pictures illustrate the types of channels that would remain open if 1, 2, 3, or 4 balls were required.

B. K⁺ channels composed of toxin-sensitive subunits with balls will open and then inactivate normally in the absence of toxin, but they will not open at all in the presence of toxin (Figure 11–21B).

C. A 50:50 mixture of K⁺ channels composed of toxin-resistant subunits without balls or toxin-sensitive subunits with balls will give a mixture of the curves shown in Figure 11–21A and B. In the absence of toxin all the channels will open; the 50% with balls will close and the 50% without balls will remain open (Figure 11–21C). In the presence of toxin, the 50% composed of toxin-sensitive subunits will not open at all; the 50% composed of toxin-resistant subunits without balls will open and stay open (Figure 11–21C).

D. The patch-clamp recording in Figure 11–11B indicates that a single ball is sufficient to close a channel. The reasoning is as follows. When the mixture of K⁺ channels in the oocyte membrane was subjected to membrane depolarization in the absence of toxin, all the channels opened and then those without a sufficient number of balls remained opened to give a plateau current. Thus, the plateau current in the absence of toxin is due to all forms of the channel that cannot close—but those forms are undefined. In the presence of toxin, only those channels composed of four toxin-resistant subunits can open and those stay open because they have no balls. Since the plateau current in the presence of toxin, which is entirely due to channels without balls, exactly equals the plateau current in the absence of toxin, channels with a single ball must be able to close. If a channel required two, three, or four balls to close, the plateau current in the absence of toxin would be higher than in the presence of toxin (Figure 11–22).

Reference: MacKinnon R, Aldrich RW & Lee AW (1993) Functional stoichiometry of *shaker* potassium channel inactivation. *Science* 262, 757–759.

11–40*

11–41

A. With equal concentrations of K⁺ ions on both side of the vesicle there will be no net movement of K⁺ ions, although they will move in both directions. Also, no membrane potential will develop.

B. K⁺ ions will move out of the vesicle through the K⁺-leak channel down their concentration gradient. This movement will continue until the membrane potential (negative inside) exactly matches the concentration difference (higher K⁺ inside).

C. K⁺ ions will move into the vesicle through the K⁺-leak channel down their concentration gradient. This movement will continue until the membrane potential (negative outside) exactly matches the concentration difference (higher K⁺ outside). You will note that the K⁺-leak channel works the same regardless of its orientation in the membrane; it always moves K⁺ down its electrochemical gradient.

11–42*

11–43

A. The expected membrane potential due to differences in K⁺ concentration across the resting membrane is

$$V = 58 \text{ mV} \times \log_{10} \frac{C_o}{C_i}$$

$$V = 58 \text{ mV} \times \log_{10} \frac{9 \text{ mM}}{344 \text{ mM}}$$

$$V = -92 \text{ mV}$$

For Na^+, the equivalent calculation gives a value of +48 mV.

The assumption that the membrane potential is due solely to K^+ leads to a value near that of the resting potential. The assumption that the membrane potential is due solely to Na^+ leads to a value near that of the action potential.

These assumptions approximate the resting potential and action potential because K^+ *is* primarily responsible for the resting potential and Na^+ *is* responsible for the action potential. A resting membrane is 100-fold more permeable to K^+ than it is to Na^+ because of the presence of K^+ leak channels. The leak channel allows K^+ to leave the cell until the membrane potential rises sufficiently to oppose the K^+ concentration gradient. The theoretical maximum gradient (based on calculations like those above) is lowered somewhat by the entrance of Na^+, which carries positive charge into the cell (compensating for the positive charges on the exiting K^+). Were it not for the Na^+-K^+ pump, which continually removes Na^+, the resting membrane potential would be dissipated completely.

The action potential is due to a different channel, a voltage-gated Na^+ channel. These channels open when the membrane is stimulated, allowing Na^+ ions to enter the cell. The magnitude of the resulting membrane potential is limited by the difference in the Na^+ concentrations across the membrane. The influx of Na^+ reverses the membrane potential locally, which opens adjacent Na^+ channels and ultimately causes an action potential to propagate away from the site of original stimulation.

B. The substitution of choline chloride for sodium chloride eliminates the action potential, as expected, since the action potential is due to specific Na^+ channels. As illustrated in the calculation above, the difference in concentrations of Na^+ across a membrane determines the magnitude of the action potential that results from Na^+ influx. Thus, if the external Na^+ concentration was reduced to half or one quarter the normal value, the calculated membrane potentials would be reduced to 30 mV and 13 mV, respectively. Measurements of the action potential for various mixtures of choline chloride and sodium chloride match these expectations.

Reference: Hille B (1992) Ionic Channels of Excitable Membranes, 2nd edn, pp 23–58. Sunderland, MA: Sinauer.

11–44*

11–45 True. It takes a difference of only a minute number of ions to set up the membrane potential.

11–46* **Reference:** Hille B (1992) Ionic Channels of Excitable Membranes, 2nd edn, pp 1–20. Sunderland, MA: Sinauer.

11–47 True. The Na^+ influx cannot exceed its equilibrium potential since Na^+ flows down its electrochemical gradient. The quick inactivation of the Na^+ channels then allows the nerve to recover its membrane potential. In the absence of this second limitation nerve cells would suffer permanent electrical spasm.

11–48*

11–49 Saltatory conduction increases the speed of conduction so that the action potential travels faster, and it conserves metabolic energy by confining the region of active excitation to the small regions of axonal plasma membranes at nodes of Ranvier. You will recall that pumping out Na^+ requires hydrolysis of ATP.

11–50*

11–51   Each of the rectangular peaks corresponds to the opening of a single channel, which allows a small current to pass. Individual channels in the patch of membrane open and close frequently, remaining open for a very short, somewhat variable time, averaging about 5 milliseconds. When open, the channels allow a small current with a unique amplitude (4 pA) to pass. In one instance, the current doubles, indicating that two channels in the same membrane patch opened simultaneously.

If acetylcholine were omitted or added to the solution outside the micropipette, no peaks of current would be seen: only the baseline would be observed. Acetylcholine must bind to the extracellular portion of the acetylcholine receptor to allow the channels to open. In the membrane patch in Figure 11–12A the binding sites for acetylcholine are exposed only to the solution in the micropipette.

11–52*

11–53   True. In the absence of a specific ligand, such ion channels will remain closed, preventing them from generating an action potential.

11–54*

11–55   Opening Na^+ channels allows an influx of Na^+ ions that depolarizes the membrane toward the threshold potential for firing an action potential. By contrast, opening either Cl^- or K^+ channels opposes membrane depolarization. Both Cl^- and K^+ ions are near their equilibrium distribution across the membrane: the resting membrane potential (negative inside) balances their concentration differences across the membrane (Cl^- high outside and K^+ high inside). As the membrane begins to depolarize (that is, as the membrane potential becomes more positive), both ions will tend to move down their concentration gradients (Cl^- ions into the cell, K^+ ions out of the cell). If a channel for either ion is opened, its movement across the membrane will make the inside of the cell more negative, tending to reestablish the original membrane potential and suppressing the firing of an action potential.

11–56*   Figure 11–23.

Reference: Hille B (1992) Ionic Channels of Excitable Membranes, pp 315–336. Sunderland, MA: Sinauer.

11–57   There is little net movement of K^+ because it is nearly at its equilibrium distribution; the membrane potential opposes movement out of the cell down its concentration gradient. By contrast, Na^+ is not at its equilibrium distribution; both the concentration difference and the membrane potential tend to push it into the cell. The same is true for Ca^{2+}; however, its external concentration is only about 1 mM versus about 145 mM for Na^+. Thus, when an acetylcholine-gated channel opens, the great majority of the cation that enters is Na^+.

11–58*

11–59   When acetylcholine is released from the synaptic vesicles of the neurons, some of the acetylcholine finds target receptors, some diffuses away, but most is rapidly hydrolyzed to acetate and choline, which are taken up by the nerve terminal. When the density of receptors is reduced by reaction with antibodies, the probability diminishes that an acetylcholine molecule will find its receptor before it is hydrolyzed. The suboptimal transmission of the signal is responsible for the muscular weakness of myasthenic patients. One way to overcome their muscular weakness is to increase the concentration of acetylcholine to compensate for the reduced number of receptors. By inhibiting acetylcholinesterase, neostigmine increases the effective concentration of acetylcholine, thereby increasing the efficiency of signal transmission across the synapse.

11–60* **Reference:** Sakmann B (1992) Elementary steps in synaptic transmission revealed by currents through single ion channels. *Science* 256, 503–512.

11–61

A. The transmembrane segment M2 is responsible for the differences in conductance through the two types of acetylcholine-gated channels in young rat muscle. The results with chimeric cDNAs 3 and 4 indicate that the difference lies in transmembrane segments M2, M3, and M4, but the result with chimeric cDNAs 1 and 2 rule out M4. Results with chimeric cDNAs 5 and 6 verify that the difference lies in segments M2 and M3. Finally, results with chimeric cDNAs 7 and 8 pinpoint the difference as transmembrane segment M2.

B. It is likely that the differences in channel conductance are due to the M2 transmembrane segment alone because that segment forms the actual lining of the pore through which the ions flow. The other transmembrane segments presumably serve to hold the M2 segment in the proper orientation. Since the 5 subunits of the acetylcholine-gated cation channel are homologous, it is thought that the M2 transmembrane segments from each subunit line the pore.

C. As illustrated in Figure 11–24, glycine has a smaller side chain than threonine, while leucine has a larger side chain. If the size of the side chain determines the critical constriction in the pore, substitution of a glycine for threonine should increase the size of the pore and permit a freer flow of ions, hence increased current through the channel. By contrast, substitution of leucine for threonine should narrow the pore further and reduce the current.

The M2 transmembrane segments of the γ_1 and γ_2 subunits may differ in analogous ways. At some critical point in the M2 segment the γ_2 subunit may contain a bulkier amino acid side chain than the γ_1 subunit. Such a difference could account for the lower conductance of channels made with the γ_2 subunit relative to those made with the γ_1 subunit.

Reference: Sakmann B (1992) Elementary steps in synaptic transmission revealed by currents through single ion channels. *Science* 256, 503–512.

11–62*

11–63 False. Voltage-gated Ca^{2+} channels and Ca^{2+}-release channels in the sarcoplasmic reticulum control the entry of Ca^{2+} into the cytoplasm when the plasma membrane is depolarized. The Ca^{2+} pump, which requires ATP, is responsible for the reverse process: pumping cytosolic Ca^{2+} back up its concentration gradient into the sarcoplasmic reticulum.

11–64* **Reference:** Moriyoshi K, Masu M, Ishii T, Shigemoto R, Mizuno N & Nakanishi S (1991) Molecular cloning and characterization of the rat NMDA receptor. *Nature* 354, 31–37.

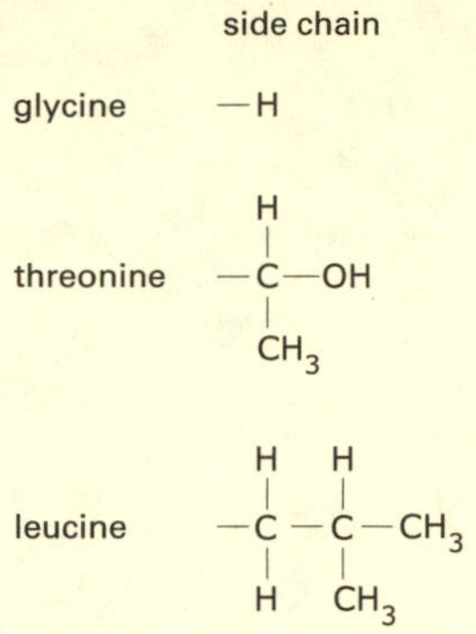

Figure 11–24 Amino acid side chains (Answer 11–61).

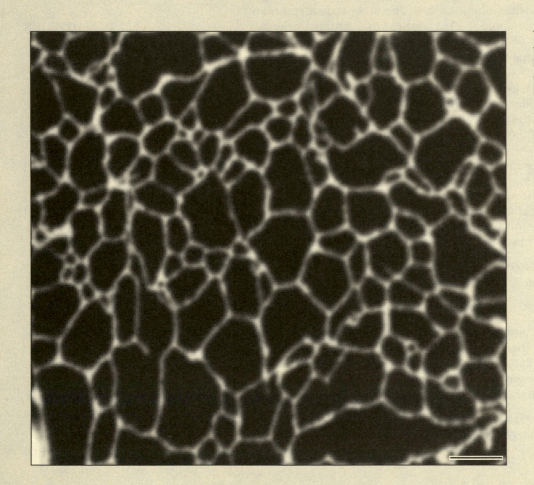

The endoplasmic reticulum (ER) in a tobacco cell made visible by a GFP-tagged ER-resident protein. Scale bar is 5 μm. Courtesy of Petra Boevink and Chris Hawes.

INTRACELLULAR COMPARTMENTS AND PROTEIN SORTING

THE COMPARTMENTALIZATION OF CELLS

12–1   False. Lipid bilayers by themselves are impermeable to hydrophilic molecules, but biological membranes, which contain proteins in addition to the bilayer, are not. Membranes in cells contain various transport proteins that make them selectively permeable, allowing certain small molecules and particular proteins to cross. It is this selective permeability that establishes the unique chemical identity of each compartment.

12–2*

12–3   In terms of its functional importance to a cell, the plasma membrane is anything but minor. It is the boundary that separates the cell from the outside world, it controls selective entry and exit of molecules, and it is the principal site at which intercellular communications are received. In terms of its surface area and mass, however, it is a minor fraction (2–5%) of all the membranes in a cell.

12–4*

12–5   While the vast majority of cells in the human body do have a complete set of membrane-enclosed organelles, certain specialized cells do not. A prime example is the red blood cell. At a late stage in its development the precursor of the red blood cell—the reticulocyte—jettisons all of its internal membrane-enclosed organelles, leaving just the plasma-membrane enclosed cytosol. The cells that make up the lens of the eye are similar. But in a way these are exceptions that prove the rule; these cells are derived from cells that do carry the complete set of membrane-enclosed organelles.

12–6*

12–7   If cells that have recovered from such treatment are examined by electron microscopy, they are found to contain a perfectly normal looking Golgi apparatus. The Golgi apparatus reassembles itself from the dispersed vesicles. Once the microtubule skeleton of the cell has been reestablished, the vesicles apparently use them to track back to their normal location in the cell, where they fuse with one another to re-form the Golgi apparatus. This same process of fragmentation and reassembly occurs in normal cells at each cell division, when the cytoskeleton breaks down in preparation for mitosis.

12–8*

12–9   Eucaryotic gene expression is more complicated than procaryotic gene expression. In particular, procaryotic cells do not have introns that interrupt the coding sequences of their genes, so that an mRNA can be translated immediately after it is transcribed, without further processing. In fact, in procaryotic cells ribosomes start translating most mRNAs before transcription is finished. This would have disastrous consequences in eucaryotic cells, because most RNA transcripts have to be spliced before they can be translated. The nuclear envelope separates the transcription and translation processes in space and time. A primary RNA transcript is held in the nucleus until it is properly processed to form an mRNA, and only then is it allowed to leave the nucleus so that ribosomes can translate it.

12–10*

12–11   In the absence of a sorting signal, a protein will remain in the cytosol.

12–12*

12–13   One way to approach this problem is to compare the relative volumes of the compartments that are served by cytosolic and ER protein synthesis. Assuming that the average density and lifetimes of proteins are about the same in all compartments—a reasonable first approximation—their relative volumes would provide a rough estimate of the amount of protein synthesis. The compartments served by cytosolic protein synthesis, which include the cytosol, nucleus, mitochondria, and peroxisomes, account for more than 80% of the cell volume. The compartments that depend on ER protein synthesis—the ER, Golgi apparatus, endosomes, and lysosomes—account for less than 20% of the cell volume. On this basis, then, one would conclude that cytosolic protein synthesis is responsible for the majority of cellular protein synthesis.

In cells that do not secrete large amounts of protein, the majority of protein synthesis is likely to occur in the cytosol. One of the main functions of liver cells, however, is to export proteins such as albumin, which makes up about half of the total serum protein. The fraction of liver ribosomes engaged in synthesizing albumin is probably less than 10%. At this level a liver cell would still be carrying out the majority of its protein synthesis on cytosolic ribosomes, but in other, specialized secretory cells (like those of the pancreas) ER protein synthesis may exceed cytosolic protein synthesis.

12–14*

12–15

A.   The negative charge nearest to the transmembrane segment is the more important. In the presence of a normal-length hydrophobic segment, neither N-terminal negative charge is essential for membrane insertion, as shown by the results with constructs 1 and 2. Proteins with shortened hydrophobic segments, however, depend on the negative charges for proper insertion, as shown by comparison of constructs 4 through 7. Construct 5 with the single negative charge adjacent to the hydrophobic segment is inserted nearly as efficiently as construct 4, which has both negative charges. By contrast, construct 6 with the single negative charge near the N-terminus is not inserted at all.

B.   In the presence of the membrane potential (minus CCCP), the hydrophobic segment is more important than the N-terminal negative charge. Construct 2, which has neither negative charge, is inserted as efficiently as construct 1, which has both negative charges. In addition, when both negative charges are present, the amount of inserted protein decreases as the length of the hydrophobic segment is reduced (compare constructs 1, 3, and 4).

C.   In the absence of the membrane potential (plus CCCP), the hydrophobic segment is still the most important determinant of insertion efficiency: construct 2 with no negative charges is inserted as efficiently as construct 1. As the hydrophobic helix is made shorter, however, insertion comes to depend much more on the presence of the negative charges. Construct 3 is only about half as efficient at insertion when CCCP is added, and construct 4, with a slightly shorter hydrophobic sequence, is absolutely reliant on the membrane potential for insertion.

Thus, in the absence of a membrane potential—presumably the case in the earliest cells—a sufficiently long hydrophobic segment may have been adequate to accomplish insertion of a protein into a membrane. In the presence of a membrane potential a second feature—the distribution of charges around a transmembrane segment—would have been available for translocator-independent insertion of membrane proteins.

Reference: Delagado-Partin VM & Dalbey RE (1998) The proton motive force, acting on acidic residues, promotes translocation of amino-terminal domains of membrane proteins when the hydrophobicity of the translocation signal is low. *J. Biol. Chem.* 273, 9927–9934.

12–16*

12–17

A.   The protein would enter the ER. The signal for import into the ER is located at the N-terminus of the protein and functions before the protein is fully synthesized, whereas the signal for nuclear import is internal and functions after the protein has been released from the ribosome.

B.   The protein would enter the ER. The signal for import into the ER is located at the N-terminus of the protein and functions before the protein is fully synthesized, whereas the signal for peroxisome import is at the C-terminus and functions after the protein has been released from the ribosome.

C.   The protein would enter the mitochondria. In order to be retained in the ER the protein must first be imported into the ER. Without a signal for ER import, the ER retention signal could not function.

D.   A protein with signals for both nuclear import and nuclear export would shuttle between the cytosol and the nucleus. Unlike the other pairs of signals, these signals are not necessarily in conflict. A number of cellular proteins, whose function requires shuttling in and out of the nucleus, are designed in just this way.

12–18*

12–19   A taxi is the closest analogy. Anyone who has the fare—the sorting signal—is taken on the journey. A private car implies a specific relationship between the traveler (the sorted protein) and the vehicle (the sorting receptor) that doesn't apply to protein sorting; namely, that there would be specific sorting receptors for each different kind of protein. In reality, all the proteins destined for the ER, for example, use the same sorting receptor. A bus would imply that travelers—sorted proteins—were carried in groups. Sorting receptors handle proteins one at a time.

12–20*

THE TRANSPORT OF MOLECULES BETWEEN THE NUCLEUS AND THE CYTOSOL

12–21 True. The nuclear pore complex contains one or more open aqueous channels through which small water-soluble molecules can passively diffuse.

12–22*

12–23

A. The portion of nucleoplasmin responsible for localization in the nucleus must reside in the tail. The nucleoplasmin head does not localize to the nucleus when injected into the cytoplasm, and it is the only injected component that is missing a tail.

B. These experiments suggest that the nucleoplasmin tail carries a nuclear localization signal and that accumulation in the nucleus is not the result of passive diffusion. The observations involving complete nucleoplasmin or fragments that retain the tail do not distinguish between passive diffusion and active transport; they say only that the tail carries the important part of nucleoplasmin—be it a localization signal or a binding site. The key observations that argue against passive diffusion are the results with the nucleoplasmin heads. They do not diffuse into the cytoplasm when they are injected into the nucleus, nor into the nucleus when injected into the cytoplasm, suggesting that the heads are too large to pass through the nuclear pores. Since the more massive forms of nucleoplasmin with tails do pass through the nuclear pores, passive diffusion of nucleoplasmin is ruled out.

Reference: Dingwall C, Sharnick SV & Laskey RA (1982) A polypeptide domain that specifies migration of nucleoplasmin into the nucleus. *Cell* 30, 449–458.

12–24*

12–25 The rationale for the experiment is that the restriction nuclease EcoRI will cleave the cell's DNA into pieces if it gains access to the nucleus, thereby killing the cell. In glucose-containing medium, the hybrid gene is not transcribed and the fusion protein is not made; thus, both types of yeast proliferate perfectly well. However, in the presence of galactose, the hybrid gene is expressed. In yeasts carrying the pNL⁻ plasmid, the hybrid protein, although expressed, cannot enter the nucleus because it lacks an NLS and thus does no harm to the cell. However, in yeasts carrying the pNL⁺ plasmid, the hybrid protein, which has a functional NLS, enters the nucleus and cuts up the cell's DNA; thus, the cells die.

Reference: Barnes G & Rine J (1985) Regulated expression of endonuclease EcoRI in *Saccharomyces cerevisiae*: nuclear entry and biological consequences. *Proc. Natl. Acad. Sci. USA* 82, 1354–1358.

12–26* **Reference:** Barnes G & Rine J (1985) Regulated expression of endonuclease EcoRI in *Saccharomyces cerevisiae*: nuclear entry and biological consequences. *Proc. Natl. Acad. Sci. USA* 82, 1354–1358.

12–27 Each nuclear pore complex must transport about 1 histone molecule per second on average throughout a day.

$$\text{transport} = \frac{32 \times 10^6 \text{ octamers}}{\text{day}} \times \frac{8 \text{ histones}}{\text{octamer}} \times \frac{\text{day}}{8.64 \times 10^4 \text{ sec}} \times \frac{1}{3000 \text{ pores}}$$

$$= 0.99 \text{ histones/second/pore}$$

Because histones are synthesized and imported into nuclei only during S phase, which is typically about 8 hours long, the transport rate is about 3 histones per second during S phase (and none during the rest of the cell cycle).

12–28*

12–29 False. Individual nuclear pores mediate transport in both directions. It is unclear how pores coordinate this two-way traffic so as to avoid head-on collisions and congestion.

12–30* **References:** Görlich D, Prehn S, Laskey RA & Hartmann E (1994) Isolation of a protein that is essential for the first step of nuclear protein import. *Cell* 79, 767–776.

Moore MS & Blobel G (1993) The GTP-binding protein Ran/TC4 is required for protein import into the nucleus. *Nature* 365, 661–663.

Moore MS & Blobel G (1994) Purification of a Ran-interacting protein that is required for protein import into the nucleus. *Proc. Natl. Acad. Sci. USA* 91, 10212–10216.

12–31

A. Ran is a GTPase and will slowly convert GTP to GDP. Thus, if you had prepared Ran-GTP to start with, by the time you did the experiments you would have had an undefined mixture of Ran-GTP and Ran-GDP, which would have confused the results. By using a form of Ran that cannot hydrolyze GTP, you guaranteed that Ran was in its Ran-GTP conformation.

B. Since the Ran-GDP column removed the nuclear import factor, whereas the RanQ69L-GTP column did not, you are looking for a protein that is present in lane 1 but not in lane 2 (see Figure 12–7). One such protein is present, between the 7-kd and 14-kd markers. Note that the RanQ69L-GTP column binds a set of proteins between the 97-kd and 116-kd markers that the Ran-GDP column does not. They are members of the importin family of nuclear import receptors, which are evidently not required for the nuclear uptake of Ran-GDP.

C. The small protein that binds to Ran-GDP is known as NTF2. Besides binding tightly to Ran-GDP, NTF2 also binds to the FG motifs present in the nucleoporins of the nuclear pore complex. It is the progressive movement of the NTF2-Ran-GDP complex along the FG tracks in the nucleoporins that allows Ran-GDP to be delivered to the nucleus. In the nucleus the Ran-GEF converts Ran-GDP to Ran-GTP, causing it to dissociate from NTF2. NTF2 then recycles to the cytoplasm to bring in another Ran-GDP.

D. The information in the problem says only that cytoplasm passed over a Ran-GDP column is depleted in some factor that is essential for nuclear uptake, and the experimental results shown in Figure 12–7 indicate that NTF2 binds to Ran-GDP. The inference is that NTF2 is the critical factor necessary for nuclear uptake of Ran-GDP. To prove that NTF2 is the import factor, you would need to show that purified or recombinant NTF2 can promote uptake of Ran-GDP into nuclei. The authors of this study went even further. They made use of information from the crystal structure of the NTF2-Ran-GDP complex and mutated the glutamate at position 42 in NTF2 to lysine, thereby disrupting a key salt bridge between the two proteins. This NTF2E42K mutant no longer promoted nuclear uptake of Ran-GDP. These additional experiments demonstrate that NTF2 is necessary for nuclear uptake of Ran-GDP.

Reference: Ribbeck K, Lipowsky G, Kent HM, Stewart M & Görlich D (1998) NTF2 mediates nuclear import of Ran. *EMBO J.* 17, 6587–6598.

12–32* **Reference:** Schwoebel ED, Talcott B, Cushman I & Moore MS (1998) Ran-dependent signal-mediated nuclear import does not require GTP hydrolysis by Ran. *J. Biol. Chem.* 273, 35170–35175.

12–33 Leptomycin B in some way must interfere with the normal shuttling of Rev from the cytoplasm to the nucleus, trapping it in the nucleus. There are several possible ways that leptomycin B might accomplish that. (1) Leptomycin B could bind to Rev and mask its nuclear export signal, thereby directly preventing Rev from exiting the nucleus. (2) If Rev's nuclear export signal is normally unmasked by binding to viral mRNA, leptomycin B could prevent its export by binding to Rev in such a way as to prevent binding of Rev to the RNA. (3) Leptomycin B could inactivate the nuclear export receptor that mediates the export of Rev. (4) Leptomycin B could bind to a component of the nuclear pore complex responsible for Rev export. Note that it would have to be a component specifically involved in export since import was not affected. (5) Leptomycin B could bind to Ran-GTP to promote its conversion to Ran-GDP, thereby eliminating the form of Ran required for nuclear export. This is probably not a complete list of possible effects of leptomycin B on nuclear export.

Thus, the simple observation that leptomycin B traps Rev in the nucleus does not go very far toward defining the mechanism of leptomycin B action.

Reference: Wolff B, Sanglier J-J & Wang Y (1997) Leptomycin B is an inhibitor of nuclear export: inhibition of nucleo-cytoplasmic translocation of the human immunodeficiency virus type 1 (HIV-1) Rev protein and Rev-dependent mRNA. *Chem. Biol.* 4, 139–147.

12–34*

12–35 These results are more or less what you would expect if leptomycin B blocks nuclear export. In the absence of leptomycin B, NES-GFP is excluded from the nuclei, as shown by the dark areas that correspond to the positions of the nuclei in the DNA panels. This result indicates that NES-GFP is efficiently exported from nuclei. (Of course, it could also mean that NES-GFP never entered the nuclei in the first place.) The same result is observed in leptomycin B resistant cells in the presence of leptomycin B, as expected if it is without effect in the mutant cells. The presence of NES-GFP in the nuclei of wild-type cells treated with leptomycin B confirms that NES-GFP can enter the nucleus and that leptomycin B prevents its export. The presence of NES-GFP in the cytoplasm, as well, indicates either that NES-GFP doesn't enter the nucleus very well or that leptomycin B doesn't completely block nuclear export.

Reference: Kudo N, Matsumori N, Taoka H, Fujiwara D, Schreiner EP, Wolff B, Yoshida M & Horinouchi S (1999) Leptomycin B inactivates CRM1/exportin 1 by covalent modification at a cysteine residue in the central conserved region. *Proc. Natl. Acad. Sci. USA* 96, 9112–9117.

12–36* **Reference:** Kudo N, Matsumori N, Taoka H, Fujiwara D, Schreiner EP, Wolff B, Yoshida M & Horinouchi S (1999) Leptomycin B inactivates CRM1/exportin 1 by covalent modification at a cysteine residue in the central conserved region. *Proc. Natl. Acad. Sci. USA* 96, 9112–9117.

12–37 If the modified GFP is actively shuttling between the nucleus and the cytoplasm, addition of leptomycin B, which blocks nuclear export, should trap all the protein in the nucleus. If the cellular distribution arises because there are two forms of modified GFP—one with neither functional signal and one with a functional import signal—then leptomycin B should have no effect on the cellular distribution.

12–38* **Reference:** Fornerod M, Ohno M, Yoshida M & Mattaj IW (1997) CRM1 is an export receptor for leucine-rich nuclear export signals. *Cell* 90, 1051–1060.

12–39

A. Addition of leptomycin B to block nuclear export wouldn't be expected to alter the distribution of the GFP-tagged protein, regardless of whether it's a true nuclear protein or a shuttling protein that spends most of its time in the nucleus. In either case it would still be located in the nucleus in the presence of leptomycin B.

B. In a heterocaryon with two nuclei, one expressing the GFP-tagged protein and the other not, it is possible to decide whether the protein is a true nuclear protein or a shuttling protein. If it is a true nuclear protein, the GFP should remain associated with a single nucleus. If it is a shuttling protein, the GFP will redistribute to both nuclei.

C. It is critical in this experiment to block synthesis of new GFP-tagged protein.

If new protein were made, it would enter both nuclei, regardless of whether it was a true nuclear protein or a shuttling protein.

12–40* **Reference:** Nachury MV & Weis K (1999) The direction of transport through the nuclear pore can be inverted. *Proc. Natl. Acad. Sci. USA* 96, 9622–9627.

12–41 If one NES-BSA were bound to each of 3000 nuclear pore complexes, its concentration in the nuclear envelope would be 0.31 μM, which is the same as the concentration of NES-BSA (0.3 μM) used in these experiments.

$$[\text{NES-BSA}] = \frac{3000 \text{ NES-BSA}}{16 \times 10^{-15} \text{ L}} \times \frac{\text{mole}}{6 \times 10^{23} \text{ NES-BSA}} \times \frac{10^6 \text{ μmol}}{\text{mole}}$$

$$[\text{NES-BSA}] = 0.31 \text{ μmol/L} = 0.31 \text{ μM}$$

Thus, at one NES-BSA per pore complex the fluorescence of the nuclear membrane would approximate that of the nucleus and its surroundings. If the ring of fluorescence is 10 times brighter than the background, that would mean that about 10 NES-BSA molecules were bound per pore complex. This is not an unreasonable number since there are eight fibers on each side of the nuclear pore complex, and each fiber is thought to contain multiple binding sites for nuclear import receptors. (The volume of the nuclear envelope, which is about 50 nm in width, was calculated by subtracting the volume of the interior of the nucleus from that of the whole nucleus. The nucleus was assumed to be a sphere with a 10 μm outer diameter and a 9.9 μm inner diameter.)

Reference: Nachury MV & Weis K (1999) The direction of transport through the nuclear pore can be inverted. *Proc. Natl. Acad. Sci. USA* 96, 9622–9627.

12–42* **Reference:** Nachury MV & Weis K (1999) The direction of transport through the nuclear pore can be inverted. *Proc. Natl. Acad. Sci. USA* 96, 9622–9627.

12–43 True. Gene regulatory proteins in particular are subject to this kind of regulation, as a way of preventing gene activation (or repression) until the proper time.

12–44* **Reference:** Baeuerle PA & Baltimore D (1988) Activation of DNA-binding activity in an apparently cytoplasmic precursor of the NF-κB transcription factor. *Cell* 53, 211–217.

12–45 At each mitosis, the contents of the nucleus and the cytosol mix when the nuclear envelope is disassembled. When the nucleus is then reassembled, the nuclear proteins must be selectively reimported. If the nuclear localization signals were removed upon import, the proteins would be trapped in the cytosol after the next mitosis. By contrast, the contents of other organelles never mix with the cytosol. At mitosis organelles such as the Golgi apparatus and the ER break up into vesicles, which retain the lumenal contents of their larger parents. Because of this, their resident proteins have to be imported only once, and their signal sequences are therefore dispensable.

12–46*

THE TRANSPORT OF PROTEINS INTO MITOCHONDRIA AND CHLOROPLASTS

12–47 Import of mitochondrial proteins occurs posttranslationally. Normally, translation is much faster than mitochondrial import, so that proteins completely clear the ribosome before interacting with the mitochondrial membrane. However, by blocking protein synthesis with cycloheximide, you have made the rate of translation artificially slower than the rate of import.

Since the signal peptide for protein import into mitochondria resides at the N-terminus, some of the partially synthesized mitochondrial proteins, which are still attached to ribosomes, will be able to interact with the mitochondrial membrane. The attempted import of even one such protein will attach the ribosome and the mRNA (and all other ribosomes translating the same mRNA molecule) to the mitochondrial membrane.

Reference: Kellems RE, Allison VF & Butow RA (1975) Cytoplasmic type 80S ribosomes associated with yeast mitochondria. IV. Attachment of ribosomes to the outer membrane of isolated mitochondria. *J. Cell Biol.* 65, 1–14.

12–48* **Reference:** Neupert W (1997) Protein import into mitochondria. *Annu. Rev. Biochem.* 66, 863–917.

12–49 As illustrated in Figure 12–31B, sequence 2 forms a well-defined amphipathic helix with a positively charged surface. Thus, it is likely to be imported into mitochondria. Sequence 2 corresponds to the N-terminal segment of glutamine synthetase from chickens, which import this enzyme into mitochondria. By contrast, sequence 1 does not form a particularly well-defined amphipathic helix and it carries two fewer positive charges (Figure 12–31A). Thus, it is not surprising that the glutamine synthetase with this N-terminus, which is from the human enzyme, remains in the cytosol. Four amino acid differences contribute significantly to the mitochondrial targeting of the chicken enzyme: addition of two hydrophobic amino acids at positions 1 and 11, and addition of two positively charged amino acids at positions 14 and 18.

12–50*

12–51 Normal cells that carry the modified *URA3* gene make Ura3 that gets imported into mitochondria and is therefore unavailable to carry out an essential reaction in the metabolic pathway for uracil synthesis. They might as well not have the enzyme at all, and they will grow only when uracil is supplied in the medium. By contrast, in cells that are defective for mitochondrial import, Ura3 is prevented from entering mitochondria and remains in the cytosol where it can function normally in the pathway for uracil synthesis. Thus, cells with defects in mitochondrial import can grow in the absence of added uracil because they can make their own.

Reference: Maarse AC, Blom J, Grivell LA & Meijer M (1992) MPI1, an essential gene encoding a mitochondrial membrane protein, is possibly involved in protein import into yeast mitochondria. *EMBO J.* 11, 3619–3628.

12–52*

12–53 True. Regardless of their final destination in the mitochondrion, all proteins that are synthesized in the cytosol (that is, all nucleus-encoded mitochondrial proteins) must first enter the TOM complex. After the TOM complex, the pathways of import diverge as proteins are sorted to their appropriate mitochondrial compartment.

12–54* Table 12–4.

12–55 Incubate the radiolabeled proteins with isolated mitochondria under conditions you wish to test, allow a sufficient time for import, and then treat the mixture with a protease. Proteins that are not imported will be digested by the protease. Proteins that have been imported will be resistant to the protease. Protease-resistant proteins could be assayed by reisolating the mitochondria and measuring the counts associated with them. Alternatively, they could be assayed by solubilizing the entire mixture and separating the proteins by electrophoresis on a gel. Protease-resistant proteins would run at the same position as untreated proteins.

These analyses assume that proteins are protease resistant because they are sequestered inside mitochondria, meaning they have been imported.

(A) SEQUENCE 1

human glutamine synthetase (cytosolic)

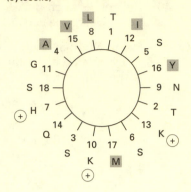

(B) SEQUENCE 2

chicken glutamine synthetase (mitochondrial)

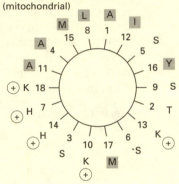

Figure 12–31 Helix-wheel projections of the N-terminal sequences of glutamine synthetases (Answer 12–49). (A) Sequence 1 is human glutamine synthetase. (B) Sequence 2 is chicken glutamine synthetase. Hydrophobic amino acids are *shaded* and positively charged amino acids are indicated.

You would need to include several controls, however, before you could make this conclusion. You would need to know that the protease is working, which could be measured by leaving the mitochondria out of the incubation mixture. You would need to know that the protein is stable in the absence of the protease, which you could assay by leaving the protease out of the incubation mixture. You would need to know that protease-resistant proteins are in the mitochondria, which could be assayed by solubilizing the mitochondria with a detergent to show that protease-resistant proteins now become protease sensitive. Appropriate controls are essential for informative research into any biological problem.

12–56* **Reference:** Eilers M & Schatz G (1986) Binding of a specific ligand inhibits import of a purified precursor protein into mitochondria. *Nature* 322, 228–232.

12–57 Since each modified barnase includes an import signal and the length of the N-terminal extension does not affect the stability of the barnase domain, the dependence of import on the length of the extension presumably reflects some process inside mitochondria. The most likely possibility is that only the longer extensions can span both mitochondrial membranes and project into the matrix. There they can be bound by the mitochondrial hsp70, which can use the hydrolysis of ATP to help drive import. Presumably, the 95-amino acid extension is long enough to be efficiently engaged by hsp70, whereas the 65-amino acid extension must be less efficiently bound. Hsp70 and the energy of ATP hydrolysis are required for import of barnase because of its extremely stable folded structure. If the protein is first denatured, all three N-terminal extensions can facilitate its import at the same high rate because the unfolded protein does not hinder entry into the matrix.

Reference: Huang S, Ratliff KS, Schwartz MP, Spenner JM & Matouschek A (1999) Mitochondria unfold precursor proteins by unraveling them from their N-termini. *Nat. Struct. Biol.* 6, 1132–1138.

12–58* **Reference:** Huang S, Ratliff KS, Schwartz MP, Spenner JM & Matouschek A (1999) Mitochondria unfold precursor proteins by unraveling them from their N-termini. *Nat. Struct. Biol.* 6, 1132–1138.

12–59

A. Tim23 appears to be an integral component of both mitochondrial membranes. In intact mitochondria a small portion of Tim23 is digested by the protease, indicating that a segment of Tim23 is exposed outside mitochondria. This result implies that a portion of Tim23 extends through the outer mitochondrial membrane. This portion of Tim23 must be the N-terminus because it is still recognized by antibodies specific for the C-terminus (see Figure 12–16, lane 2). In mitoplasts a larger N-terminal segment of Tim23 is digested by the protease, but the C-terminal portion is protected indicating that it is in the inner membrane or inside the mitoplasts (see lane 3). In combination these results indicate that Tim23 must extend through both mitochondrial membranes.

B. The pattern of protease sensitivity of Tim23 in mitochondria and mitoplasts combined with the information in the hydropathy plot suggest that Tim23 is arranged as shown in Figure 12–32.

References: Donzeau M, Káldi K, Adam A, Paschen S, Wanner G, Guiard B, Bauer MF, Neupert W & Brunner M (2000) Tim23 links the inner and outer mitochondrial membranes. *Cell* 101, 401–412.

Paschen SA, Rothbauer U, Kaldi K, Bauer MF, Neupert W & Brunner M (2000) The role of the TIM8-13 complex in the import of Tim23 into mitochondria. *EMBO J.* 19, 6392–6400.

12–60*

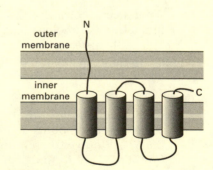

Figure 12–32 Arrangement of Tim23 in the inner and outer mitochondrial membranes (Answer 12–59).

12–61   False. Only one of the two signal sequences is cleaved. The N-terminal signal is cleaved off the imported protein when it reaches the mitochondrial matrix. The second signal—a very hydrophobic sequence at the new N-terminus—directs the protein to the inner membrane through either the TIM23 complex or the OXA complex. The second signal is not cleaved; it anchors the protein in the inner membrane.

12–62*

12–63   False. Although import of proteins is similar, the components of the import machinery in chloroplasts and mitochondria are not related. The functional similarities appear to have arisen by convergent evolution, reflecting the common requirements for translocation across a double membrane system.

12–64*   Figure 12–33.

> **Reference:** Smeekens S, Bauerle C, Hageman J, Keegstra K & Weisbeek P (1986) The role of the transit peptide in the routing of precursors toward different chloroplast compartments. *Cell* 46, 365–375.

PEROXISOMES

12–65   False. All eucaryotic cells contain peroxisomes.

12–66*   **Reference:** Osinga KA, Swinkels BW, Gibson WC, Borst P, Veeneman GH, Van Boom JH, Michels PAM & Opperdoes FR (1985) Topogenesis of microbody enzymes: sequence comparison of the genes for the glycosomal (microbody) and cytosolic phosphoglycerate kinases of *Trypanosoma brucei*. *EMBO J.* 4, 3811–3817.

12–67   The AGT enzyme is found predominantly (95%) in the mitochondria of these patients. A positively charged amphipathic α helix is a signal for mitochondrial import. The amphipathic helix formed by the substitution of a leucine for a proline at position 11 is a weak mitochondrial targeting signal. In fact, this particular mutation exists as a polymorphism in the human population with an allelic frequency of about 10%. By itself, it mistargets about 10% of AGT to mitochondria with the rest being accurately targeted to peroxisomes. Combined with a second mutation that inhibits peroxisomal targeting, however, this weak mitochondrial targeting signal misdirects about 95% of AGT to mitochondria.

> **Reference:** Purdue PE, Allsop J, Isaya G, Rosenberg LE & Danpure CJ (1991) Mistargeting of peroxisomal L-alanine:glyoxylate aminotransferase to mitochondria in primary hyperoxaluria patients depends upon activation of a cryptic mitochondrial targeting sequence caused by a point mutation. *Proc. Natl. Acad. Sci. USA* 88, 10900–10904.

12–68*   **Reference:** Kinoshita N, Ghaedi K, Shimozawa N, Wanders RJA, Matsuzono Y, Imanaka T, Okumoto K, Suzuki Y, Kondo N & Fujiki Y (1998) Newly identified Chinese hamster ovary cell mutants are defective in biogenesis of peroxisomal membrane vesicles (peroxisome ghosts), representing a novel complementation group in mammals. *J. Biol. Chem.* 273, 24122–24130.

12–69

A.   Translation of mRNA from normal cells and each of the mutant cell lines yielded equal amounts of the 75-kd form but none of the 53-kd form. Since the 53-kd form is present only in the normal cells, it is likely that it arises from the 75-kd form during the process of import into peroxisomes. This reasoning suggests that the 53-kd form is the active form of the enzyme. This conclusion is also supported by the observation that the mutant cells, which have no acyl CoA oxidase, have none of the 53-kd form.

B.   The mutant cells have only the 75-kd form of the enzyme because their defective peroxisomes cannot import it and process it to the active 53-kd form. Because the 75-kd form disappears so quickly in the pulse-chase experiments in the mutant cells (without giving rise to the 53-kd form), the 75-kd form must be unstable in the cytosol and rapidly degraded. A similar experiment performed with catalase would be expected to show no degradation since catalase activity is normal in the mutant cells.

Reference: Tsukamoto T, Yokota S & Fujiki Y (1990) Isolation of Chinese hamster ovary cell mutants defective in assembly of peroxisomes. *J. Cell. Biol.* 110, 651–660.

12–70* **Reference:** Morand OH, Allen LA, Zoeller RA & Raetz CR (1990) A rapid selection for animal cell mutants with defective peroxisomes. *Biochem. Biophys. Acta* 1034, 132–141.

12–71 The data set is adequate to allow you to sort the mutant cell lines unambiguously into complementation groups. The mutant cell lines fall into 8 complementation groups, as listed below. Three cell lines (Z65, ZP116, and ZP160) are mutant for the *PEX2* gene; two (ZP105 and ZP162) are mutant for the *PEX5* gene; and two (ZP164 and ZP92) are mutant for the *PEX6* gene.

1. Z65/ZP116/ZP160 = *PEX2*
2. ZP105/ZP162 = *PEX5*
3. ZP164/ZP92 = *PEX6*
4. ZP110/ZP161
5. ZP119/ZP165
6. Z24
7. ZP109
8. ZP114

The presence (+) or absence (–) of peroxisomes in the two types of complementation have different meanings in terms of the genes that are involved. For example, by heterocaryon analysis Z65 does not complement ZP116 or ZP160, as indicated by – in Table 12–3. This result indicates that these three cell lines each have defects in the same gene; hence, are in the same complementation group. In the transfection analysis a + indicates that the cell line is defective in the gene that was transfected. For example, ZP116 and ZP160 developed peroxisomes when they were transfected with *PEX2* cDNA; thus, these two cell lines must have defects in the *PEX2* gene. Since Z65 is in the same complementation group as ZP116 and ZP160, it must also be defective in the *PEX2* gene. This kind of reasoning applied to the entire data set allows all the mutant cell lines to be sorted into distinct complementation groups.

Reference: Kinoshita N, Ghaedi K, Shimozawa N, Wanders RJA, Matsuzono Y, Imanaka T, Okumoto K, Suzuki Y, Kondo N & Fujiki Y (1998) Newly identified Chinese hamster ovary cell mutants are defective in biogenesis of peroxisomal membrane vesicles (peroxisome ghosts), representing a novel complementation group in mammals. *J. Biol. Chem.* 273, 24122–24130.

12–72* **Reference:** Fujiki Y (2000) Peroxisome biogenesis and peroxisome biogenesis disorders. *FEBS Let.* 476, 42–46.

12–73

A. If the modified Pex5 remains in the cytosol, it will not be cleaved by the peroxisomal protease and thus will be recognized only by mAb2, which will detect it only in the cell extract and in the supernatant.

If the modified Pex5 can be imported into peroxisomes but not exported, the cleaved form (recognized by mAb1) will be found in the cell extract and pellet, which contain peroxisomes, but not in the supernatant, which does not contain peroxisomes. Any modified Pex5 that has not entered peroxisomes, and thus has not been cleaved, will be detected by mAb2 in the cell extract and supernatant. The cleaved Pex5 in the cell extract and pellet will also be recognized by mAb2.

If the modified Pex5 can cycle between the peroxisomal matrix and the cytosol, the cleaved form will be detected by mAb1 in cell extract and pellet, as expected for imported Pex5, but will also be detected in the supernatant, indicating that cleaved Pex5 was exported from the peroxisome. Any modified Pex5 that has not entered peroxisomes will be detected by mAb2 in the cell extract and supernatant; mAb2 will also detect cleaved Pex5 in all three fractions.

B. The results in Figure 12–22C match the expectations for the cycling mechanism for Pex5-mediated import of proteins into peroxisomes. The critical observation is that the cleaved form of Pex5, which is bound by mAb1, is found in the cell extract, pellet, and supernatant. Modified Pex5 must have entered the peroxisome to be cleaved, and it must also exit the peroxisome since it is found in the supernatant.

C. Pex5-mediated import into peroxisomes resembles import into the nucleus, which is mediated by nuclear import receptors. In both cases a cytosolic receptor binds to the cargo, accompanies it through a complex of membrane proteins to the interior of the compartment, drops off the cargo and exits from the compartment back to the cytosol. The results in Figure 12–22C do not directly answer the question of whether cleaved Pex5 continues to cycle between the cytosol and matrix, but other experiments in the paper suggest that Pex5 does function by cycling between the cytosol and the peroxisomal matrix.

Reference: Dammai V & Subramani S (2001) The human peroxisomal targeting signal receptor, Pex5p, is translocated into the peroxisomal matrix and recycled to the cytosol. *Cell* 105, 187–196.

THE ENDOPLASMIC RETICULUM

12–74*

12–75   Protein import into mitochondria and chloroplasts occurs as a posttranslational event and cytosolic hsp70 chaperones are required to keep the newly made proteins in an unfolded conformation so they can be imported. By contrast, during co-translational import into the ER the protein is imported as it is being made; thus, there is no possibility that it will fold and no need to involve hsp70 proteins to keep it unfolded.

12–76*

12–77   This observation shows that there must be some mechanism for separating a subset of ER membrane proteins; however, the mechanism of separation is unclear. Some of the proteins specific for rough microsomes help bind ribosomes to the rough ER, while others may be responsible for the distinctive morphology of the rough ER. Other proteins may be specific for the rough ER by virtue of interactions with cytosolic or lumenal proteins that are apposed to the rough ER.

12–78*

12–79   False. When the hydrophobic signal peptide emerges from the ribosome, it is bound by SRP, which causes a pause in protein synthesis. Synthesis resumes when the ribosome with a bound SRP then binds to the SRP receptor on the cytosolic surface of the rough ER.

12–80*   **Reference:** Perara E, Rothman RE & Lingappa VR (1986) Uncoupling translocation from translation: implications for transport of proteins across membranes. *Science* 232, 348–352.

12–81

B. The cytosolic hsp70 chaperone protein would be expected to bind to the import protein and aid its unfolding, which should increase its efficiency of uptake.

Because the protein to be imported has already been synthesized neither free ribosomes (C), nor the Sec61 complex (D), nor SRP (E) would have any effect. The BiP chaperone (A) might help if it were inside the microsomes where it could be loaded onto the incoming protein to initiate the cycle of BiP binding and ATP hydrolysis that drives import. Outside the microsome, however, it is unclear that it would have any effect.

12–82*

12–83   True. The Sec61 complex is composed of several proteins that assemble into a donutlike structure. The central pore in the complex lines up with a tunnel in the large ribosomal subunit through which the growing polypeptide chain exits from the ribosome.

12–84*   **Reference:** Deshaies RJ & Schekman R (1987) A yeast mutant defective at an early stage in import of secretory protein precursors into the endoplasmic reticulum. *J. Cell Biol.* 105, 633–645.

12–85   False. The first (most N-terminal) transmembrane segment that exits the ribosome initiates translocation (acts as a start-transfer signal). Its orientation in the ER membrane fixes the reading frame for insertion of subsequent transmembrane segments. If the first transmembrane segment is oriented with its N-terminus in the cytosol, even-numbered segments will act as stop-transfer signals and odd-numbered segments will act as start-transfer signals. If the first segment is oriented with its N-terminus in the lumen, then the second segment and subsequent even-numbered segments will act as start-transfer signals. Subsequent odd-numbered segments will act as stop-transfer signals.

12–86*   Figure 12–34.

12–87   As shown in Figure 12–35, elimination of the first transmembrane segment (by making it hydrophilic) would be expected to reverse the orientation of the protein in the membrane. What was the second transmembrane segment (a stop-transfer signal), would now be read as a start-transfer signal and would have the opposite orientation in the membrane—as would all the remaining transmembrane segments. Although the N-terminus would still be in the ER lumen, all the rest of the external parts of the protein would swap positions so that what was in the cytosol would now be in the ER lumen, and vice versa.

12–88*   Figure 12–36.

Reference: Manoil C & Beckwith J (1986) A genetic approach to analyzing membrane protein topology. *Science* 233, 1403–1408.

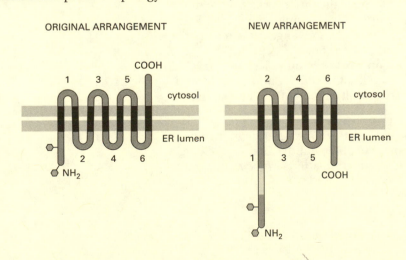

ORIGINAL ARRANGEMENT

NEW ARRANGEMENT

Figure 12–35 Arrangement of the original multipass transmembrane protein and of the new protein after the first hydrophobic segment was converted to a hydrophilic segment (Answer 12–87).

12–89 False. The ER lumen does not contain reducing agents (they are in the cytosol) and, therefore, S—S bonds can form in the ER.

12–90*

12–91 The activated forms of the sugars used to construct the sugar tree on dolichol phosphate are all generated in the cytosol by reaction with cytosolic UTP or GTP. (The same is true for addition of sugars in the Golgi apparatus.) At least in one case, however, it is clear that there are nucleoside triphosphates in the ER lumen. Posttranslational insertion of proteins into the ER, for example, requires BiP-mediated hydrolysis of ATP in the ER lumen. Thus, there seems to be no fundamental reason why nucleoside triphosphates could not be present in organellar lumen. Wherever they are present, however, there needs to be a transporter to allow them to move across the membrane, since NTPs themselves cannot pass through the lipid bilayer.

12–92*

12–93 False. Proteins that are linked to GPI anchors are attached to the external surface of the plasma membrane. The attachment reaction occurs in the lumen of the ER, which is topologically equivalent to the outside of the cell.

12–94*

12–95

A. Experiment 1 tests whether the acceptor membranes (red cell ghosts) are in excess. Since the PC exchange protein catalyzes an *exchange* reaction, there is a simple theoretical limit to how much transfer can occur at equilibrium. If the amount of donor and acceptor membranes were equal, for example, the limit of possible transfer would be 50%. Doubling the amount of acceptor membrane would raise the limit to 67% (a 2 to 1 ratio of donor to acceptor); tripling the acceptor membranes would raise the limit to 75% (a 3 to 1 ratio); and so on. Since adding more acceptor membranes made no difference, the red cell membranes must be in excess. Thus the 70% limit is not an equilibrium point for the exchange.

Experiment 2 rules out the possibility that the enzyme is inactivated during the reaction, since addition of fresh exchange protein causes no further exchange.

Experiment 3 eliminates the possibility that the starting labeled material was impure (that is, untransferable by the PC exchange protein, which is specific for PC) or was somehow altered during the course of the incubation.

B. The apparent explanation for the 70% limit is that the PC exchange protein transfers PC only from the outer monolayer of the vesicle bilayer. The area of the outside face of the donor vesicles is about 2.6 times the area of the inner face. The area of the surface of a sphere is $4\pi r^2$. Thus the ratio of the areas of the outer and inner faces of the donor vesicle is the ratio of squares of their radii, which is $10.5^2/(10.5 - 4.0)^2$ or 2.6. Since the outer surface is 2.6 times the inner surface, 74% (2.6/3.5) of the lipid is in the outer monolayer. Thus, 70% transfer is close to the expected limit if the exchange protein can only exchange PC from the outer leaflet and PC does not flip-flop.

C. If the exchange protein exchanges PC only between outer leaflets, the label in the acceptor red cell membranes will all be in the outer leaflet and, therefore, all available for transfer. This result supports the idea that the PC exchange protein only transfers PC between outer monolayers.

Reference: Rothman JE & Dawidowicz EA (1975) Asymmetric exchange of vesicle phospholipids catalyzed by the phosphatidylcholine exchange protein. Measurement of inside-outside transitions. *Biochemistry* 14, 2809–2816.

INTRACELLULAR VESICULAR TRAFFIC

THE MOLECULAR MECHANISMS OF MEMBRANE TRANSPORT AND THE MAINTENANCE OF COMPARTMENT DIVERSITY

13–1 If the flow of membrane between cellular compartments were not balanced in a nondividing liver cell, some compartments would grow in size and others would shrink (in the absence of new membrane synthesis.) Keeping all the membrane compartments the same relative size is essential for proper functioning of a liver cell. In a growing cell such as a gut epithelial cell the situation is different. Over the course of a single cell cycle, all of the compartments must double. Thus, there will be an imbalance in favor of the outward flow, which will be supported by new membrane synthesis equal to the sum total of all the cell's membrane.

13–2*

13–3 True. The cytosolic leaflets of the two membrane bilayers are the first to come into contact and fuse, followed by the noncytosolic leaflets. It is this pattern of leaflet fusion that maintains the topology of membrane proteins, so that protein domains that face the cytosol always do so, regardless of what compartment they occupy.

13–4* Figure 13–25.

13–5 This is an apt analogy. Receptors can be incorporated into a coated vesicle only if they can bind to adaptins—'have a ticket'—which allows them to enter a coated pit and be incorporated into a vesicle—'the cable car.' Receptors—'like the skiers'—are mixed together without any guaranteed traveling companions, but all get to the next compartment—'station.'

Reference: Pearse BMF, Smith CJ & Owen DJ (2000) Clathrin coat construction in endocytosis. *Curr. Opin. Struc. Biol.* 10, 220–228.

13–6* Figure 13–26.

References: Kirchhausen T (2000) Clathrin. *Annu. Rev. Biochem.* 69, 699–727.

Ybe JA, Brodsky FM, Hofmann K, Lin K, Liu SH, Chen L, Earnest TN, Fletterick RJ & Hwang RK (1999) Clathrin self-assembly is mediated by a tandemly repeated superhelix. *Nature* 399, 371–375.

13–7 The specificity for both the transport pathway and the transported cargo come not from the clathrin coat, but from the adaptins that link the clathrin to the transmembrane receptors for specific cargo proteins. The several varieties of adaptins allow different cargo receptors, hence different cargo proteins, to be specifically transported along different transport pathways.

Incidentally, humans are different from most other organisms in that they have two heavy chain genes. Like other mammals, they have two light chain genes. In addition, in the neurons of mammals the light chain transcripts are differentially spliced. Thus, there exists the potential in humans for additional complexity of clathrin coats; the functional consequences of this potential variability are not clear.

References: Kirchhausen T (2000) Clathrin. *Annu. Rev. Biochem.* 69, 699–727.

Pearse BMF, Smith CJ & Owen DJ (2000) Clathrin coat construction in endocytosis. *Curr. Opin. Struc. Biol.* 10, 220–228.

13–8*

13–9 Assuming that the altered dynamin functioned normally in all ways except for the ability to hydrolyze GTP, this result supports the action of dynamin as a mechanochemical pinchase, perhaps acting as shown in Figure 13–27. In the absence of GTP hydrolysis, the mutant dynamin would not be expected to complete the final step in vesicle formation since that step requires GTP hydrolysis to power it.

According to the alternative hypothesis, the inability of dynamin to hydrolyze GTP would have locked the mutant dynamin into its active 'ON' state, which, if anything, might have been expected to increase vesicle formation. Thus, the observed inhibition of vesicle formation argues against this hypothesis.

Reference: Marks B, Stowell MHB, Vallis Y, Mills IG, Gibson A, Hopkins CR & McMahon HT (2001) GTPase activity of dynamin and resulting conformation changes are essential for endocytosis. *Nature* 410, 231–235.

13–10*

13–11 Brefeldin A appears to block COPI-coated vesicle formation by interfering with the exchange of GTP for GDP, which is essential for ARF to bind to the Golgi membrane and initiate formation of coated vesicles. Brefeldin A probably interferes with the GEF, which catalyzes GTP-for-GDP exchange, rather than with ARF itself. In fact, recent experiments show that brefeldin A binds to the complex of ARF and GEF, locking it into a nonproductive conformation.

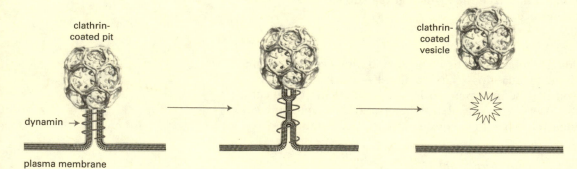

clathrin-coated pit

clathrin-coated vesicle

dynamin →

plasma membrane

Figure 13–27 One model for how dynamin might pinch a vesicle off the membrane (Answer 13–9). In response to GTP hydrolysis dynamin, in this model, uncoils like a spring under tension and pops the vesicle from the membrane.

Observation 1 shows that brefeldin A does not block coated vesicle formation if ARF is first locked into its active form by GTPγS. Thus, the assembly of the COPI coats and formation of vesicles are not affected by brefeldin A if the active membrane-bound form of ARF is present.

Observation 2 shows that a trypsin-sensitive protein in the Golgi membrane catalyzes the exchange of GTP for GDP. This exchange reaction is blocked by brefeldin A. This observation alone does not distinguish whether the effect of brefeldin A is on the GEF or on ARF: binding of brefeldin A to either protein could interfere with the exchange reaction.

Observation 3 shows that ARF, even in the presence of brefeldin A, is capable of exchanging guanine nucleotides in a GEF-independent exchange reaction. This suggests that the most likely effect of brefeldin A is on the GEF rather than on ARF. The artificial nature of the exchange reaction, however, makes this conclusion less certain.

References: Donaldson JG, Finazzi D & Klausner RD (1992) Brefeldin A inhibits Golgi membrane-catalyzed exchange of guanine nucleotide into ARF protein. *Nature* 360, 350–352.

Helms JB & Rothman JE (1992) Inhibition by brefeldin A of a Golgi membrane enzyme that catalyzes exchange of guanine nucleotide bound to ARF. *Nature* 360, 352–354.

Chardin P & McCormick F (1999) Brefeldin A: the advantage of being uncompetitive. *Cell* 97, 153–155.

13–12*   **Reference:** Goldberg J (1999) Structural and functional analysis of the ARF1–ARFGAP complex reveals a role for coatomer in GTP hydrolysis. *Cell* 96, 893–902.

13–13   If ARF1 were mutated so that it could not hydrolyze GTP, ARF1 would exist in a cell as ARF1–GTP. Since ARF1–GTP promotes assembly of COPI-coated vesicles, you would expect such vesicles to form readily. However, they might not form at the right place in the cell. Normally, ARF1 is delivered specifically to the Golgi membrane by a Golgi-bound ARF1–GEF, which converts the cytosolic ARF1–GDP to ARF1–GTP, exposing a fatty acid tail that allows it to bind to the membrane. The mutant ARF1, which would always have GTP bound and its fatty acid tail exposed, might bind inappropriately to other cell membranes, thus promoting COPI-coated vesicle formation at inappropriate places in a cell.

Disassembly of the COPI coat requires hydrolysis of GTP by ARF1. Thus, in the presence of the mutant form of ARF1, the COPI coat would not be able to disassemble. Since the uncoated vesicle is the substrate for the fusion reaction with the target membrane, the mutant ARF1–GTP would be expected to block ARF1-mediated transport.

If the mutant ARF1 were the only form of ARF1 in the cell, it is likely that it would prove lethal. Under these conditions, all ARF1-mediated transport involving COPI-coated vesicles should be blocked. Furthermore there should be an accumulation of COPI-coated vesicles that cannot be uncoated, which might reduce the availability of free COPI subunits necessary for other transport pathways. What is unclear is the extent to which other members of the ARF family of proteins might substitute for ARF1. It would be necessary to do the experiment to know the result for certain.

13–14*   **Reference:** Nichols BJ, Undermann C, Pelham HRB, Wickner WT & Haas A (1997) Homotypic vacuolar fusion mediated by t- and v-SNAREs. *Nature* 387, 199–202.

13–15   There will always be some v-SNAREs in the target membrane. Immediately after fusion, the v-SNAREs will be in inactive complexes with t-SNAREs. Once NSF pries the complexes apart, v-SNAREs may be kept inactive by binding to inhibitory proteins. Accumulation of v-SNAREs in the target membrane beyond some minimal population is prevented by active retrieval pathways that incorporate v-SNAREs into vesicles for redelivery to the original donor membrane.

13–16* **Reference:** Tokumaru H, Umayahara K, Pellegrini LL, Ishizuka T, Saisu H, Betz H, Augustine GJ & Abe T (2001) SNARE complex oligomerization by synaphin/complexin is essential for synaptic vesicle exocytosis. *Cell* 104, 421–432.

13–17

A. The larger complexes (at apparent sizes of 100 kd, 120 kd, and 190 kd) are oligomers of the SNARE complex that are formed by the binding of synaphin, which presumably links the complexes together. The apparent sizes are not exact multiples of the 60-kd SNARE complex because the gels are run under nondenaturing conditions where rate of migration is especially sensitive to shape.

B. The synaphin-derived peptide prevents oligomerization of SNARE complexes by competing with synaphin for binding to syntaxin. If the synaphin binding site on syntaxin is already occupied by the peptide, synaphin will be unable to bind. This is analogous to the use of a competitive inhibitor of an enzyme to block its activity.

C. The ability of the peptide to interfere with synaptic transmission in the squid giant axon means that the binding of synaphin to syntaxin is critical for synaptic vesicle fusion. This is a key experiment because it tells you that synaphin plays an important biological role in synaptic vesicle fusion. In any biochemical experiment such as that depicted in Figure 13–7, it is a concern that the interactions observed in the test tube may not reflect the true situation in the cell. This observation indicates that the biochemical results are truly relevant to the cellular process of synaptic fusion. (The squid giant axon is a favored tool for studies of nerve function because its large size simplifies injection experiments of the type done here.)

D. The biochemical results indicate that synaphin promotes oligomerization of SNARE complexes, and the injection studies indicate that the binding of synaphin to syntaxin is essential for vesicle fusion. Taken together, these results suggest that oligomerization of SNARE complexes may be a critical step in synaptic vesicle fusion. By forming oligomers of SNARE complexes—perhaps even a circle around the site of fusion—synaphin may bring the vesicle membrane and the plasma membrane into close enough proximity that they can fuse spontaneously. These experiments do not rule out the possibility that additional proteins are required for fusion after the synaphin-promoted oligomerization of the SNARE complexes.

Reference: Tokumaru H, Umayahara K, Pellegrini LL, Ishizuka T, Saisu H, Betz H, Augustine GJ & Abe T (2001) SNARE complex oligomerization by synaphin/complexin is essential for synaptic vesicle exocytosis. *Cell* 104, 421–432.

13–18*

13–19 Syntaxin and Snap25 are t-SNAREs and synaptobrevin is a v-SNARE. NSF and its adaptors recognize complexes of t- and v-SNAREs, using the energy of ATP hydrolysis to pry them apart. Binding of NSF and its adaptors to a SNARE complex depends on the presence of ATP. In the absence of ATP the complex does not form; thus, the beads did not bring down anything other than NSF. In the presence of ATP the complex forms, but NSF then hydrolyzes ATP to separate the SNAREs and release them. Thus, in the presence of ATP only NSF is attached to the beads. In the presence of ATPγS, however, the complex can form, but it cannot dissociate because NSF cannot hydrolyze ATPγS. As a result, all members of the complex remain attached to NSF and are brought down by the beads.

Reference: Söllner T, Whiteheart SW, Brunner M, Erdjument-Bromage H, Geromanos S, Tompst P & Rothman JE (1993) SNAP receptors implicated in vesicle targeting and fusion. *Nature* 362, 318–324.

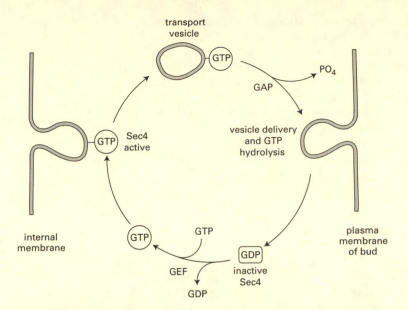

Figure 13–28 Outline of Sec4 function in delivery of transport vesicles from an internal membrane to the bud membrane (Answer 13–21). Additional proteins that are involved are not shown.

13–20*

13–21

A. Since vesicles form and accumulate when the function of Sec4 is impaired (as in *SEC4*ts and *SEC4N133I*), Sec4 cannot be involved in vesicle formation. Accumulation of vesicles in these mutants suggests that the vesicles can no longer deliver their cargo to the growing bud when Sec4 is not working properly. Thus Sec4 seems to be involved in vesicle targeting and fusion.

Functionally, Sec4 resembles mammalian Rab proteins, which are also required for proper delivery of transport vesicles to their target membrane. Indeed, Sec4 was the first identified member of the Rab family of proteins. Sec4 is unlike mammalian Sar1 and ARF proteins, which are required for formation of coated vesicles.

B. From the description of the defects in the presence of the mutant Sec4 proteins and by analogy to Rab proteins, it is possible to outline the way normal Sec4 functions in delivery of vesicles to the bud membrane (Figure 13–28).

The presence of some Sec4 (20% of total) in the cytosol of wild-type cells presumably represents Sec4 that is recycling after delivery of vesicles to the bud membrane. Removal of the C-terminal cysteines presumably prevents attachment of a lipid that is essential for the binding of Sec4 to the forming vesicle. If Sec4 cannot bind to the vesicle, it cannot carry out its function.

C. The inhibitory properties of Sec4N133I are very interesting and not altogether easy to interpret. Since Rab proteins function as monomers, it is unlikely that there is a direct effect of Sec4N133I on normal Sec4. More likely, there is an indirect effect that prevents normal Sec4 from carrying out its function. For example, if a vesicle component such as v-SNARE were present in limiting amounts, the accumulation of vesicles carrying Sec4N133I might deplete the supply and thereby interfere with the proper fusion of vesicles carrying normal Sec4. Alternatively, Sec4N133I may bind too tightly to its Rab-like effector on the target membrane, preventing normal Sec4 from gaining access to the docking machinery.

References: Walworth NC, Goud B, Kabcenell AK & Novick PJ (1989) Mutational analysis of *SEC4* suggests a cyclical mechanism for the regulation of vesicular traffic. *EMBO J.* 8, 1685–1693.

Guo W, Roth D, Walch-Solimena C & Novick P (1999) The exocyst is an effector for Sec4p, targeting secretory vesicles to sites of exocytosis. *EMBO J.* 18, 1071–1080.

13–22* **Reference:** Meuse CW, Krueger S, Majkrzak CF, Dura JA, Fu J, Connor JT & Plant AL (1998) Hybrid bilayer membranes in air and water: infrared spectroscopy and neutron reflectivity studies. *Biophys. J.* 74, 1388–1398.

13–23 The cell's SNAREs are all bound to the cytosolic surface of whatever membrane they are in. They function by juxtaposing the cytosolic surfaces of the two membranes to be fused. By contrast, enveloped viruses must fuse with a cell membrane by bringing together its external surface with an external surface of a cell membrane. Thus, enveloped viruses cannot make use of a cell's SNAREs because they are located on the wrong side of the membrane. It is for this reason that enveloped viruses make their own fusion proteins that are properly situated on their external surface.

TRANSPORT FROM THE ER THROUGH THE GOLGI APPARATUS

13–24*

13–25 True. A misfolded protein is selectively retained in the ER by binding to chaperone proteins such as BiP or calnexin. Only after it has been released from such a chaperone protein—and thus approved as properly folded—does a protein become a substrate for exit from the ER.

13–26*

13–27 Since yeast vacuolar vesicles normally carry a mixture of t-SNAREs and v-SNAREs, you might imagine that they would bind to one another on the same vesicle, rendering them unavailable for vesicle docking and fusion. In that case, NSF and ATP would be required to dissociate the complex so that individual SNAREs would be available to dock the vesicles in preparation for vesicle fusion.

If prying apart t- and v-SNARE complexes were the only role of NSF and ATP, you might not expect them to be required for fusion of vesicles that each had a single type of SNARE. Surprisingly, treatment of vesicles with NSF and ATP was also required for mixtures in which one vesicle carried just a t-SNARE and the other vesicle carried just a v-SNARE. This result suggests that NSF may have additional roles beyond that of untangling complexes of v-SNAREs and t-SNAREs. One possibility is that NSF is also required to remove fusion inhibitors that bind to individual SNAREs to keep them inactive till needed.

Reference: Nichols BJ, Undermann C, Pelham HRB, Wickner WT & Haas A (1997) Homotypic vacuolar fusion mediated by t- and v-SNAREs. *Nature* 387, 199–202.

13–28* **Reference:** Zerangue N, Malan MJ, Fried SR, Dazin PF, Jan YN, Jan LY & Schwappach B (2001) Analysis of endoplasmic reticulum trafficking signals by combinatorial screening in mammalian cells. *Proc. Natl. Acad. Sci. USA* 98, 2431–2436.

13–29 It would be located outside the cell. If PDI were missing the ER retrieval signal, its gradual flow out of the ER to the Golgi apparatus would not be countered by its capture and return to the ER, as normally occurs. Similarly, it would be expected to leave the Golgi apparatus by the default pathway, mixed with the other proteins the cell is secreting. It would not be expected to be retained anywhere else along the secretory pathway because it presumably has no signals to promote such localization.

Reference: Munro S & Pelham HR (1987) A C-terminal signal prevents secretion of luminal ER proteins. *Cell* 48, 899–907.

13–30* **Reference:** Teasdale RD & Jackson MR (1996) Signal-mediated sorting of membrane proteins between the endoplasmic reticulum and the Golgi apparatus. *Annu. Rev. Cell Dev. Biol.* 12, 27–54.

13–31 If the KDEL signal and the KDEL receptor were all that was required to retain a protein in the ER, then addition of KDEL to a secreted protein should result in its retention in the ER. Clearly, addition of KDEL to rat growth hormone or human chorionic gonadotropin did not result in their efficient retention in the ER. Presumably their slower rate of secretion was due to the KDEL system, since changing KDEL to KDEV abolished the effect. A comparable effect is also seen for ER residents that have had their KDEL signals removed; they are secreted but at significantly slower rates than true secreted proteins. One explanation that might account for both these effects is kin recognition, which embodies the idea that residents of the ER might have a general affinity for one another, making it more difficult for them to leave the compartment. According to this idea, ER proteins that are missing their KDEL signal are secreted slowly because the proteins still retain their affinity for other ER residents. Similarly, secreted proteins with an added KDEL signal would not be expected to have a general affinity for ER residents, and thus, would escape the ER at a higher rate than true ER residents.

Reference: Zagouras P & Rose JK (1989) Carboxy-terminal SEKDEL sequences retard but do not retain two secretory proteins in the endoplasmic reticulum. *J. Cell Biol.* 109, 2633–2640.

13–32* **Reference:** Adams GA & Rose JK (1985) Structural requirements of a membrane-spanning domain for protein anchoring and cell surface transport. *Cell* 41, 1007–1015.

13–33 True. The oligosaccharide chains are added in the lumens of the ER and Golgi apparatus, which are topologically equivalent to the outside of the cell. This basic topology is conserved in all membrane budding and fusion events. Thus, oligosaccharide chains are always topologically outside the cell, whether they are in a lumen or on the cell surface.

13–34* Table 13–8.

13–35 If therapeutic proteins with *N*-linked oligosaccharides were produced in nonprimate cells, they would carry occasional oligosaccharides with Gal(α1–3)Gal linkages. Since such linkages are not present on normal human proteins, the protein might be recognized as foreign by the immune system, triggering production of antibodies against the protein, which would lead to clearing of the protein with no therapeutic benefit. Moreover, in reality, the treatment is likely to be even less effective. Humans, who are periodically infected with microorganisms that contain Gal(α1–3)Gal linkages, already have circulating antibodies to this disaccharide.

Reference: Takeuchi Y, Porter CD, Strahan KM, Preece AF, Gustafsson K, Cosset FL, Weiss RA & Collins MK (1996) Sensitization of cells and retroviruses to human serum by (α1-3) galactosyltransferase. *Nature* 379, 85–88.

13–36*

13–37 (1) Attached carbohydrates serve as a marker for protein folding in the ER. (2) Attached carbohydrates serve as a recognition marker for transport from the ER and for protein sorting in the *trans* Golgi network. (3) Oligosaccharides on proteins provide protection against proteases. (4) Oligosaccharides on cell surface proteins can function in cell–cell adhesion.

13–38*

13–39
    A. The radioactive label (GlcNAc) is added in the medial compartment, and the lectin precipitation depends on the presence of galactose, which is added in the *trans* compartment. Therefore, this experiment follows the movement of material between the medial and the *trans* compartments of the Golgi apparatus.

B. If proteins moved through the Golgi apparatus by cisternal maturation, then a protein that entered the Golgi in a mutant cell should remain with that stack and mature as the newly formed cisterna moves through the stack. Thus, the cisternal maturation model predicts that none of the labeled G protein (which was labeled in the medial compartment of the Golgi apparatus in the mutant cell) should have galactose attached to it (which could only have been added in the Golgi apparatus from the wild-type cell). For this model the fusion of the infected mutant cells to uninfected wild-type cells (see Table 13–4, line 1) should be the same as the fusion of infected mutant cells to uninfected mutant cells (line 2).

   By contrast, if material moved through the Golgi apparatus by vesicular transport, there is the possibility that proteins might move between separated Golgi stacks inside transport vesicles. The vesicular transport model predicts that some labeled G protein may acquire galactose in this way. For this model the fusion of infected mutant cells to uninfected wild-type cells (line 1) should yield more radioactive precipitate than fusion of infected mutant cells to uninfected mutant cells (line 2) but less than fusion of infected wild-type cells to uninfected wild-type cells (line 3).

C. The results in Table 13–4 support the vesicular transport model, since nearly half the labeled G protein acquired galactose. The extent of galactose addition is surprising because it suggests that once a vesicle leaves a cisterna, it has roughly an equal chance of fusing with a cisterna in the same or different Golgi stack. A number of other control experiments showed that the morphology of the Golgi stacks was unaltered by the fusion procedure, that the mutant and wild-type Golgi stacks remained distinct from one another, and that G protein did move into the wild-type Golgi stack.

Reference: Rothman JE, Miller RL & Urbani LJ (1984) Intercompartmental transport in the Golgi complex is a dissociative process: facile transfer of membrane protein between two Golgi populations. *J. Cell Biol.* 99, 260–271.

13–40* **Reference:** Bonfanti L, Mironov Jr. AA, Martinez-Menarguez JA, Martella O, Fusella A, Baldassarre M, Buccione R, Geuze HJ, Mironov AA & Luini A (1998) Procollagen traverses the Golgi stack without leaving the lumen of cisternae: evidence for cisternal maturation. *Cell* 95, 993–1003.

13–41 If cisternal maturation were the sole way for proteins to move through the Golgi apparatus, then all proteins, including PC, would move at the same rate. The existence of faster rates for most proteins suggests that they probably travel by another pathway; namely, by vesicular transport. The presence of exported molecules in transport vesicles, as suggested by the experiments in Problem 13–39 (and many other experiments), also argues strongly for the vesicular transport pathway. Thus, both models to explain movement of proteins through the Golgi apparatus are probably correct.

Reference: Pelham HRB & Rothman JE (2000) The debate about transport in the Golgi—two sides of the same coin? *Cell* 102, 713–719.

TRANSPORT FROM THE *TRANS* GOLGI NETWORK TO LYSOSOMES

13–42*

13–43 The lysosomal enzymes are all acid hydrolases, which have optimal activity at the low pH (about 5.0) in the interior of lysosomes. If a lysosome were to break, the acid hydrolases would find themselves at pH 7.2, the pH of the cytosol, and would therefore do little damage to cellular constituents.

13–44* Figure 13–29.

13–45 True. Endosomal membrane proteins are selectively retrieved from late endosomes by transport vesicles that deliver the proteins back to endosomes or to the *trans* Golgi network. The interior of late endosomes is mildly acidic (about pH 6) and as they mature into lysosomes the pH drops to the lysosomal value of pH 5.0.

13–46*

13–47

A. In the hypothetical situation in which both the hydrolase and the M6P receptor are soluble, the rate of association will increase in direct proportion to the number of M6P groups. Each additional M6P group gives the hydrolase one additional way to bind to the receptor. Although the concentration of the hydrolase ([H]) remains the same, the concentration of M6P groups (which is what the receptor binds to) increases by a factor of four when the hydrolase has four M6P groups attached to it instead of one, thereby increasing the rate of association by a factor of four.

   The rate of dissociation of a hydrolase from a receptor is the same whether the hydrolase has one M6P group or four. The rate of dissociation is related to the stability of the interaction between a single M6P group and the M6P receptor. That interaction is unaffected by other, unbound M6P groups on the hydrolase.

   If the rate of association increases by a factor of four while the rate of dissociation remains unchanged, the affinity constant for the binding of a hydrolase with four M6P groups must be four times larger than the affinity constant for binding of a hydrolase with one M6P group.

B. If the first receptor is assumed to be locked in place and does not interfere with binding of a second receptor to other M6P groups on the hydrolase, the affinity constant for binding a second receptor will be three-quarters that of the affinity constant calculated in part A. The presence of one receptor on each hydrolase covers one M6P group, reducing by one-quarter the number available for subsequent binding to a second receptor. This would reduce the rate of association with the second receptor by one-quarter, giving an affinity constant that is also one-quarter less.

C. In the real situation with a soluble hydrolase and a membrane-bound receptor, binding to the first receptor would cause the hydrolase to become localized to a thin layer of the lumen adjacent to the membrane. This would have the effect of substantially increasing the concentration of the hydrolase in the immediate neighborhood of the M6P receptors. The increased local concentration would increase the rate of association with a second receptor correspondingly (but would not affect the rate of dissociation). As a result, the affinity constant for binding to a second receptor would increase substantially. The actual magnitude of the increase would depend on the volume of the lumen versus the volume of the thin layer adjacent to the membrane.

   In framing this problem we have skirted several important issues that are essential to a detailed understanding of the true situation. For example, a hydrolase with four M6P groups can interact with as many as four M6P receptors (provided they do not interfere with one another), and at equilibrium (which the real system may never achieve) there would be hydrolases in the population with zero to four bound receptors, with the various forms interconvertible by appropriate rate constants. We have avoided this complexity to try to bring out two conceptual points. The presence of multiple M6P groups on lysosomal hydrolases increases their affinity for M6P receptors (and improves the efficiency of lysosomal targeting) in two distinct ways. First, multiple M6P groups increase the rate of association of the hydrolases with M6P receptors by providing more opportunities for binding. Second, multiple M6P groups increase the concentration of hydrolases near the membrane, giving rise to a much tighter overall binding than could be achieved if hydrolases had only a single M6P group. The situation is not unlike that of a climber on a sheer rock face: one toe- or finger-hold is good, but four are better.

13–49

A. The corrective factors are the lysosomal enzymes themselves. Hurler's cells supply the enzyme missing from Hunter's cells, and Hunter's cells supply the enzyme missing from Hurler's cells. These enzymes are present in the medium because of inefficiency in the sorting process. Since they carry M6P, which normally should direct them to lysosomes, they presumably escaped capture by the lysosomal pathway and were secreted. They are taken into cells and delivered to lysosomes by receptor-mediated endocytosis, which operates due to a small number of M6P receptors on the cell surface. The degradative enzymes, bound to receptors, are taken up through coated pits into endosomes and are eventually delivered to lysosomes. Since lysosomes are the normal site of action for these degradative enzymes, the defect is thereby corrected.

B. Protease treatment destroys the lysosomal enzymes themselves. Periodate treatment and alkaline phosphatase treatment both remove the M6P signal that is required for binding to the receptor, thus preventing the enzymes (which are still active) from entering the cell.

C. Such a scheme is unlikely to work for defects in cytosolic enzymes. External proteins normally do not cross membranes; thus, even when they are taken into cells, they remain in the lumen of a membrane-bounded compartment. In addition, foreign proteins are usually delivered to lysosomes and degraded.

Reference: Kaplan A, Achord DT & Sly WS (1977) Phosphohexosyl components of a lysosomal enzyme are recognized by pinocytosis receptors on human fibroblasts. *Proc. Natl. Acad. Sci. USA* 74, 2026–2030.

13–50*

13–51 This striking result indicates that there must be a lysosomal pathway that is independent of M6P and the M6P receptor. The nature of the pathway is unknown. It is not even clear whether the M6P-independent pathway operates inside the cell to accomplish sorting—presumably—from the *trans* Golgi to lysosomes, or as some sort of scavenger pathway that picks up lysosomal enzymes from outside the cell and delivers them to lysosomes, where they are perfectly happy. Studies with M6P-receptor deficient mice indicate that both types of pathways may operate. In thymocytes from such mice, lysosomal enzymes appear to be delivered via an intracellular route, whereas liver and skin cells can pick them up via an extracellular route.

Reference: Dittmer F, Ulbrich EJ, Hafner A, Schmahl W, Meister T, Pohlmann R & von Figura K (1999) Alternative mechanisms for trafficking of lysosomal enzymes in mannose 6-phosphate receptor-deficient mice are cell type specific. *J. Cell Sci.* 112, 1591–1597.

13–52* **References:** Kantheti P, Qiao X, Diaz ME, Peden AA, Meyer GE, Carskadon SI, Kapfhamer D, Sufalko D, Robinson MS, Noebels JL & Burmeister M (1998) Mutation in AP-3 delta in the *mocha* mouse links endosomal transport to storage deficiency in platelets, melanosomes, and synaptic vesicles. *Neuron* 21, 111–122.

Zhen L, Jiang S, Feng L, Bright NA, Peden AA, Seymour AB, Novak EK, Elliott R, Gorin MB, Robinson MS & Swank RT (1999) Abnormal expression and subcellular distribution of subunit proteins of the AP-3 adaptor complex lead to platelet storage pool deficiency in the *pearl* mouse. *Blood* 94, 146–155.

13–53 From the position of the melanosomes in *ashen* and *dilute* melanocytes, it appears that both mice have defects in transporting melanosomes to the tips of the branches so that the pigment can be properly released. The defect in *dilute* mice, which lack a myosin heavy chain that interacts with a microtubule-based motor, suggests that melanosomes may move to their proper location in the cell by being dragged along microtubules. Rab proteins, in general, bind to transport vesicles and ensure correct targeting by subsequently binding to a Rab effector protein at the target membrane. Given the phenotype of *ashen* mice, it is likely that Rab27a plays a similar role in correct targeting of melanosomes to the plasma membrane at the tips of the branches. But if that were their only defect, the microtubule-based motor system that is defective in *dilute* mice might still be expected to deliver the melanosomes to the tips of the branches. The central location of melanosomes in *ashen* mice suggests that they, like the melanosomes in *dilute* mice, may not bind to the microtubule-based motor. Thus, Rab27a (or a protein it interacts with) may be required for binding to some component of the microtubule-based motor.

Reference: Wilson SM, Yip R, Swing DA, O'Sullivan TN, Zhang Y, Novak EK, Swank RT, Russell LB, Copeland NG & Jenkins NA (2000) A mutation in *Rab27* causes the vesicle transport defects observed in *ashen* mice. *Proc. Natl. Acad. Sci. USA* 97, 7933–7938.

TRANSPORT INTO THE CELL FROM THE PLASMA MEMBRANE: ENDOCYTOSIS

13–54*

13–55   Since the surface area and volume of a macrophage do not change significantly over this time, the rate of exocytosis must also equal 100% of the plasma membrane.

13–56*   **Reference:** Haigler HT, McKanna JA & Cohen S (1979) Rapid stimulation of pinocytosis in human A-431 carcinoma cells by epidermal growth factor. *J. Cell Biol.* 83, 82–90.

13–57
- A. Extracellular space
- B. Cytosol
- C. Plasma membrane
- D. Clathrin coat
- E. Membrane of deeply invaginated clathrin-coated pit
- F. Captured cargo particles
- G. Lumen of deeply invaginated clathrin-coated pit

13–58*

13–59
- A. Binding of LDL by normal cells and JD's cells reaches a plateau because there are a limited number of LDL receptors per cell and they become saturated at high levels of LDL. The slope of the binding curve gives a measure of the binding affinity and the plateau gives a measure of the total number of binding sites (about 20,000 to 50,000, though you could not calculate this from the data shown here). JD has slightly fewer receptors on his cells, but they have an affinity similar to those in normal cells.

     Cells from patient FH bind essentially no LDL even at saturating external LDL levels. Either these cells completely lack the LDL receptor or the receptor is defective, so that its affinity for LDL is drastically reduced. It could also be that the cells do contain receptors, but for some reason they fail to appear on the surface of the cell.
- B. Cells from neither of the hypercholesterolemic patients take up any LDL. Lack of entry is readily explained for patient FH because no LDL bound to the cells. This result indicates that the receptor is crucial for LDL cholesterol to enter cells. Since LDL is not taken up by JD's cells, his LDL receptors must

also be defective, in a different way from FH's LDL receptors. JD's cells bind LDL with the same affinity as normal and almost to the same level. Although his receptors are normal as far as LDL binding is concerned, the bound LDL does not get in. Thus, mere possession of a receptor on the cell surface is no guarantee of entry.

C. LDL must enter cells in order for the cholesterol esters to be released and hydrolyzed to cholesterol, which causes inhibition of cholesterol synthesis. In the affected patients LDL does not enter the cells and, therefore, does not inhibit cholesterol synthesis.

D. If the defects in the hypercholesterolemic patients are due to defects in their LDL receptors, then free cholesterol should inhibit cholesterol synthesis in their cells as well as in normal cells. Free cholesterol does inhibit cholesterol synthesis in all these cells, strongly supporting the idea that the defects in the patients are due solely to problems with their LDL receptors.

Reference: Brown MS & Goldstein JL (1979) Receptor-mediated endocytosis: insights from the lipoprotein receptor system. *Proc. Natl. Acad. Sci. USA* 76, 3330–3337.

13–60* **Reference:** Brown MS & Goldstein JL (1979) Receptor-mediated endocytosis: insights from the lipoprotein receptor system. *Proc. Natl. Acad. Sci. USA* 76, 3330–3337.

13–61 False. The LDL receptor and many other receptors enter coated pits irrespective of whether they have bound their specific ligands.

13–62*

13–63 False. Many molecules that enter early endosomes are specifically diverted from the journey to late endosomes and lysosomes; they are recycled instead from early endosomes back to the plasma membrane via transport vesicles. Only those molecules that are not retrieved from endosomes are delivered to lysosomes for degradation.

13–64* Figure 13–30.

13–65 In the absence of bound Fe, transferrin does not interact with its receptor and circulates in the bloodstream until it catches an Fe ion. Once iron is bound, the iron–transferrin complex can bind to the transferrin receptor on the surface of a cell and be endocytosed. Under the acidic conditions of the endosome, the transferrin releases its iron, but the transferrin remains bound to the transferrin receptor, which is recycled back to the cell surface. The neutral pH of the blood causes the receptor to release the transferrin into the circulation, where it can pick up another Fe ion to repeat the cycle. The iron released in the endosome moves on to lysosomes, and from there it is transported into the cytosol.

   This system allows cells to take up iron efficiently even though the concentration of iron in the blood is extremely low. The iron bound to transferrin is concentrated at the cell surface by binding to transferrin receptors; it becomes further concentrated in clathrin-coated pits, which collect the transferrin receptors. In this way, transferrin cycles between the blood and endosomes, delivering the iron that cells need to grow.

13–66* **Reference:** Bleil JD & Bretscher MS (1982) Transferrin receptor and its recycling in HeLa cells. *EMBO J.* 1, 351–355.

13–67
A. At 37°C transferrin receptors recycle between the plasma membrane and the endosomal compartment. In the beginning all the plasma membrane receptors are labeled and all the endosomal receptors are unlabeled. Therefore, every receptor that is internalized initially is labeled. Initially, whenever a labeled receptor is internalized, it is replaced on the cell surface by an unlabeled receptor, thereby diluting the label on the surface. For that reason the rate of entry of labeled receptors is rapid at first but then slows down,

even though the flow of receptors into the cell in constant. At later times, labeled receptors begin to reappear on the surface as they are cycled back to the membrane. Eventually, the labeled and unlabeled receptors in the plasma membrane and endosomes become thoroughly mixed. At that point the rates of internalization and reappearance of labeled receptors become equal and a plateau level is reached.

B. Initially, all the cell-surface receptors are labeled. The initial rate of internalization of labeled receptors is 100% in 7 minutes, or 14% of the cell-surface receptors per minute.

C. At the plateau, 30% of the labeled receptors are sensitive to trypsin, indicating that 30% of the receptors are on the cell surface. Since 14% of the cell-surface receptors are internalized per minute, 4.2% of the total receptor population (30% × 14%) is internalized per minute.

D. At 4.2% of the total receptors per minute, an equivalent of the entire population of receptors would be internalized in 24 minutes (100%/4.2% per minute). Since the rates of internalization and reappearance are equal, an equivalent of the entire population of receptors would also be returned to the cell surface in the same interval. Thus, 24 minutes is the average time it takes for transferrin receptors to cycle from the cell surface through the endosomal compartment and back to cell surface.

E. If the mean cycle time is 24 minutes and 30% of the receptors are on the cell surface at any given time, then each receptor spends, on average, about 7 minutes (24 minutes × 30%) on the surface (and about 17 minutes inside the cell).

Reference: Bleil JD & Bretscher MS (1982) Transferrin receptor and its recycling in HeLa cells. *EMBO J.* 1, 351–355.

13–68*

TRANSPORT FROM THE *TRANS* GOLGI NETWORK TO THE CELL EXTERIOR: EXOCYTOSIS

13–69 The three main classes of proteins that must be sorted before they leave the *trans* Golgi network in a cell capable of regulated secretion are (1) those destined for lysosomes, (2) those destined for secretory vesicles, and (3) those destined for immediate delivery to the cell surface.

13–70*

13–71 The slow step in the constitutive secretion of transferrin occurs in the ER. The slow step in the constitutive secretion of albumin occurs in the Golgi. From Figure 13–21 it is clear that most of the transferrin in the cell is in the ER and most of the albumin is in the Golgi. The steady-state distribution of proteins along the constitutive pathway tells you where the proteins spend the majority of their time. As with any pathway, an accumulation occurs at the slow step. Therefore, the location of the majority of material corresponds to the slow step.

The constitutive secretion of transferrin is slow relative to albumin because it is delayed in the ER. This result appears to be general: if the constitutive secretion of a protein is slow, the protein is delayed for some reason in the ER.

References: Fries E, Gustafsson L & Peterson PA (1984) Four secretory proteins synthesized by hepatocytes are transported from the endoplasmic reticulum to Golgi complex at different rates. *EMBO J.* 3, 147–152.

Lodish HF, Kong N, Snider M & Strous GJAM (1983) Hepatoma secretory proteins migrate from rough endoplasmic reticulum to Golgi at characteristic rates. *Nature* 304, 80–83.

13–72* **Reference:** Orci L, Glick BS & Rothman JE (1986) A new type of coated vesicular carrier that appears not to contain clathrin: its possible role in protein transport within the Golgi stack. *Cell* 46, 171–184.

13–73 False. Secretory proteins, even those that are not normally expressed in a given secretory cell, are appropriately packaged into secretory vesicles. For this reason it is thought that the sorting signal, which is as yet undefined, is common to proteins in this class.

13–74*

13–75 The pro-peptide is removed from pro-insulin in immature secretory vesicles. The red fluorescence in compartments from the ER through the *trans* Golgi network indicates that they contain only pro-insulin. The green fluorescence in mature secretory vesicles indicates that they contain insulin. The yellow fluorescence, which arises when both the red and green fluorophores are excited in the same place—the combination of red and green light is yellow—indicates that pro-insulin and insulin are both present in immature secretory vesicles. Thus, immature secretory vesicles must be where the pro-peptide is removed.

13–76*

13–77 False. Once positioned beneath the plasma membrane, a secretory vesicle waits until the cell receives an appropriate signal—often a rise in Ca^{2+} concentration—before fusing with the membrane and releasing its contents.

13–78* **References:** Gottlieb TA, Beaudry G, Rizzolo L, Colman A, Rindler MJ, Adesnik M & Sabatini DD (1986) Secretion of endogenous and exogenous proteins from polarized MDCK monolayers. *Proc. Natl. Acad. Sci. USA* 83, 2100–2104.

Kondor-Koch C, Bravo R, Fuller SD, Cutler D & Garoff H (1985) Protein secretion in the polarized epithelial cell line MDCK. *Cell* 43, 297–306

13–79 Antibodies specific for the cytoplasmic domain of synaptotagmin do not stain the nerve terminals because the cytoplasmic domain is never exposed on the outside of the cell. The lumenal domain, however, is exposed to the outside of the cell when the synaptic vesicle fuses with the plasma membrane to release neurotransmitter molecules into the synaptic cleft. At that time the antibody can bind to the lumenal domain of synaptotagmin. The membrane of the synaptic vesicle is quickly retrieved from the plasma membrane and reused to form new synaptic vesicles that contain bound antibodies within them. If the fusion of synaptic vesicles with the plasma membrane is stopped by lowering the temperature to 0°C, no labeling is observed.

Reference: Matteoli M, Takei K, Perin MS, Südhof TC & DeCamilli P (1992) Exo-endocytotic recycling of synaptic vesicles in developing processes of cultured hippocampal neurons. *J. Cell Biol.* 117, 849–861.

13–80* **Reference:** Koenig JH & Ikeda K (1999) Contribution of active zone subpopulation of vesicles to evoked and spontaneous release. *J. Neurophysiol.* 81, 1495–1505.

13–81
A. The rapid component of the fusion response is due to vesicles that are already docked at the membrane and waiting for the signal to fuse and release their contents. The slow component of the fusion process is due to those vesicles that are not already docked and waiting; they are in various states of preparation for the 'ready-to-go' state.
B. NSF in some way must be required for a step in the preparation for the ready-to-go state. Thus, when NSF is inhibited, the slow component of the fusion process is also inhibited. Interference with a preparation step would

also explain why inhibition of NSF blocks the rapid component only after the second step. Nearly all the ready-to-go vesicles fuse after the first flash. In the absence of NEM (active NSF) the pool of ready-to-go vesicles is repopulated in the 2 minutes between flashes. But when NSF is inhibited, the pool remains depleted, giving rise to a much reduced rapid component in response to the second flash.

C. These results show very clearly that NSF does not control the final step in fusion, otherwise its inhibition would have blocked the initial rapid fusion in response to the first flash (see Figure 13–24B, +NEM). They also point to a role of NSF in a preparatory (early) step by showing that the slow component is inhibited after both flashes, and that the rapid component is inhibited after the second flash.

D. NSF-mediated ATP hydrolysis is required to disentangle v- and t-SNAREs after the fusion event. Vesicles that are docked and waiting already have their SNAREs paired, but are probably kept from fusing by a regulatory protein. In response to Ca^{2+} the regulatory protein is thought to release the brake on the paired SNAREs so they can complete the fusion event. For a vesicle to re-form so it can dock and fuse again, its SNAREs must be pried apart. If they are not, the vesicle cannot dock with the membrane. It is this step in the recycling of SNAREs that requires NSF.

Reference: Xu T, Ashery U, Burgoyne RD & Neher E (1999) Early requirement for α-SNAP and NSF in the secretory cascade in chromaffin cells. *EMBO J.* 18, 3293–3304.

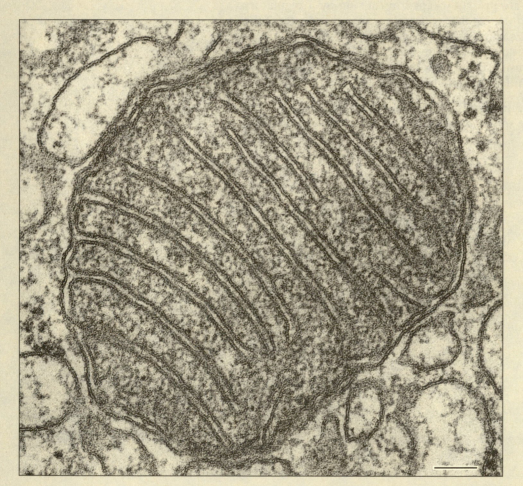

Electron micrograph of a mitochondrion.
Scale bar 100 µm. Courtesy of Daniel
S Friend.

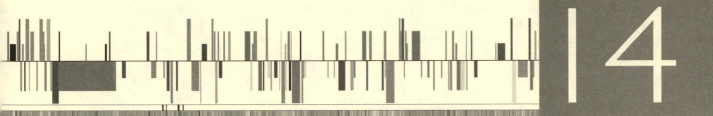
ENERGY CONVERSION: MITOCHONDRIA AND CHLOROPLASTS

THE MITOCHONDRION

14–1 The characteristic organization of mitochondria in liver cells and cardiac muscle cells is likely to reflect functional differences in these cells. Most of the energy needs of liver cells derive from metabolic processes in the cytosol, whereas the energy needs of cardiac muscle cells result primarily from contractions of the myofibrils. Mitochondria, thus, are positioned in these cells at the sites where energy, in the form of ATP, is most required.

14–2*

14–3 False. The intermembrane space is chemically equivalent to the cytosol with respect to small molecules, not because of the presence of any specialized transport proteins in the mitochondrial outer membrane, but because this membrane contains many copies of the channel-forming protein porin, which forms large aqueous channels. These channels convert the outer membrane into a sieve that allows free passage of all molecules less than 5000 daltons, which includes all the ions and metabolites in a cell and some small proteins.

14–4*

14–5 True. When the citric acid cycle is operating as a cycle of reactions, its sole products are CO_2 and reduced forms of the electron carriers NADH and $FADH_2$. The electrons in NADH and $FADH_2$ are passed into the electron transport chain to generate ATP via oxidative phosphorylation.

14–6*

14–7 The sprinter generates sufficient ATP to power muscle contraction during a sprint by the simple expedient of running enormous amounts of glucose through the glycolytic pathway. Thus, a sprinter metabolizes 15 molecules of glucose to harvest the same amount of energy that a marathon runner gets from a single molecule, and even more glucose to meet the increased per-second energy demand of sprinting over marathon running. One measure of this difference is the length of time the muscle's glycogen reserve—the intracellular source of glucose—will last during the two types of running. At maximum power output it would last a sprinter less than 80 seconds, whereas the glycogen reserve (supplemented by circulating glucose and fatty acid oxidation) lasts the marathon runner more than 2 hours.

14–8*

14–9 True. The natural (thermodynamic) tendency of electrons is to move from low-affinity to high-affinity carriers.

14–10*

14–11 As soon as protons enter the intermembrane space, they move rapidly throughout the cytosol. They are not confined to the intermembrane space because of the large aqueous channels that are present in the outer mitochondrial membrane.

14–12*

14–13

A. You would expect ATP to be synthesized, as shown in Figure 14–31. When the vesicles are exposed to light, protons will be pumped inside by bacteriorhodopsin, creating a pH difference across the vesicle membranes. The proton-motive force represented by the pH difference would drive the protons back out of the vesicles through the ATP synthase, causing ATP to be synthesized from ADP and phosphate in the external medium.

B. If the vesicles were leaky to protons, no pH difference would be generated, hence no ATP could be synthesized. The protons pumped into the vesicles by bacteriorhodopsin would immediately leak back out without generating a pH difference.

C. If the ATP synthase molecules were randomly oriented, you would still expect ATP to be synthesized, although at about half the rate. The molecules that were oriented correctly would make ATP; the oppositely oriented ATP synthase molecules would be inert.

　If bacteriorhodopsin was randomly oriented, you would expect much less ATP to be synthesized. In vesicles with equal numbers of oppositely oriented bacteriorhodopsin molecules, no pH difference would be generated upon exposure to light because the proton pumping in both directions would be equal. In vesicles with an excess of outwardly directed proton pumps, the pH difference would be in the wrong direction to be utilized by ATP synthase and, thus, no ATP would be made. In vesicles with an excess of inwardly directed proton pumps, a pH difference of the right orientation would be generated; thus, those vesicles would be capable of synthesizing some ATP.

D. Using components from widely divergent organisms can be a very powerful experimental tool. Because the two proteins come from such different sources, it is very unlikely that they form a direct functional interaction. The experiment therefore strongly suggests that the pumping of protons (normally carried out by the respiratory chain) and the synthesis of ATP are separate events. The notion that electron transport and ATP synthesis are separate events is now clearly established (this experiment was one of the earliest demonstrations); thus, this approach is a valid one.

Reference: Racker E & Stoeckenius W (1974) Reconstitution of purple membrane vesicles catalyzing light-driven proton uptake and adenosine triphosphate formation. *J. Biol. Chem.* 249, 662–663.

14–14*

14–15

A. DNP will collapse the electrochemical proton gradient completely. Protons

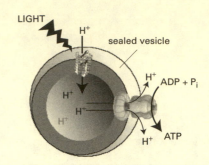

Figure 14–31 ATP synthesis in reconstituted lipid vesicles containing bacteriorhodopsin and ATP synthase (Answer 14–13).

that are pumped across the inner membrane will flow back freely, and therefore no energy can be stored across the membrane.

B. The electrochemical proton gradient is made up of two components: an H^+ concentration difference and an electrical potential. If the membrane is made permeable to K^+ with nigericin, K^+ ions will be driven into the matrix by the electrical potential of the inner membrane (minus inside, positive outside). The influx of positively charged K^+ ions will abolish the membrane's electrical potential. By contrast, the concentration component of the proton gradient (the pH difference) is unaffected by nigericin. As a result, only part of the driving force that makes it energetically favorable for protons to flow back into the matrix will be lost.

14–16* Figure 14–32.

Reference: Webb MR, Grubmeyer C, Penefsky HS & Trentham DR (1980) The stereochemical course of phosphoric residue transfer catalyzed by beef heart mitochondrial ATPase. *J. Biol. Chem.* 255, 11637–11639.

14–17

A. Presumably the hydrolysis of an individual ATP molecule provides the driving force for the 120° rotation of the γ subunit, hence the corresponding revolution of the actin filament. Since a low concentration of ATP was used in these experiments, the pauses represent the variable times it takes for the next molecule of ATP to bind. Rotation of 120° corresponds to one αβ dimer, the unit of ATP hydrolysis (or of synthesis in the normal direction).

B. If three ATP molecules must be hydrolyzed to drive one complete rotation of the γ subunit, then in its normal operation ATP synthase must synthesize three molecules per rotation of the γ subunit.

Reference: Masaike T, Mitome N, Noji H, Muneyuki E, Yasuda R, Kinosita K & Yoshida M (2000) Rotation of F_1-ATPase and the hinge residues of the β subunit. *J. Exp. Biol.* 203, 1–8.

14–18* Figure 14–33.

References: Noji H, Yasuda R, Yoshida M & Kinosita K (1997) Direct observation of the rotation of F_1-ATPase. *Nature* 386, 299–302.

Abrahams JP, Leslie AG, Lutter AGW & Walker JE (1994) Structure at 2.8 Å resolution of F1-ATPase from bovine heart mitochondria. *Nature* 370, 621–628.

14–19 Such a result is entirely reasonable; mechanical force has just been substituted for the proton-motive force to turn the axle-like γ subunit. This experiment would suggest a two-step model for ATP synthase: (1) proton flow causes rotation of the γ subunit and (2) rotation of the γ subunit inside the $\alpha_3\beta_3$ complex causes synthesis of ATP. In this experiment the authors have succeeded in uncoupling these two steps: mechanically rotating the γ subunit inside the $\alpha_3\beta_3$ complex is sufficient to cause synthesis of ATP. This would be a very exciting experiment, indeed, because it would directly demonstrate the relationship between mechanical movement and enzymatic activity. There is no doubt that it should be published and that it would become a 'classic.'

14–20*

14–21 The antiport of the amino acids Asp and Glu (V) is electrically neutral and doesn't involve a proton; thus, it is unaffected by the electrochemical proton gradient. All the rest are affected by the gradient. The symport of pyruvate and H^+ (I) is electrically neutral but because a proton is transported into the matrix it is driven by the ΔpH component of the electrochemical gradient. The antiport of citrulline and ornithine (II), which are components of the urea cycle, is driven by the membrane potential component of the electrochemical gradient. The symport of phosphate with two protons (III) is driven by the ΔpH component of the electrochemical gradient. The antiport of citrate plus a proton with malate (IV) occurs against the ΔpH component of the electrochemical proton gradient.

14–22* **References:** Nicholls DG & Ferguson SJ (1992) Bioenergetics 2, pp 215–218. London: Academic Press.

Tzagoloff A (1982) Mitochondria, pp 212–213. New York: Plenum Press.

14–23

A. The rate of oxygen consumption is determined by the rate of electron transport down the respiratory chain. Electron transport generates an electrochemical proton gradient, which opposes the flow of electrons. In the complete absence of a way to dissipate the gradient, the flow of electrons ultimately would stop when the electron pressure balances the opposing electrochemical proton gradient. In the experiment in Figure 14–7, the electrochemical proton gradient is dissipated at a slow background rate, which accounts for the slow background rate of oxygen consumption. Addition of ADP and its subsequent conversion to ATP allows protons to flow back into the mitochondria, dramatically reducing the electrochemical proton gradient and permitting the rapid transport of electrons to oxygen. The increased rate of electron transport produces an increased rate of oxygen consumption. When all the ADP is converted to ATP, proton flow again slows to the background rate, and the increased electrochemical proton gradient once again reduces the flow of electrons.

B. The slow background rate of oxygen consumption by mitochondria in the absence of added ADP indicates that electrons continue to flow down the electron-transport chain to oxygen in the absence of ATP synthesis. Such a flow can continue only if the electrochemical proton gradient is slowly being dissipated. If the mitochondrial inner membrane was completely impermeable to protons, the rate of oxygen consumption would drop to zero when proton pumping due to electron transport was balanced by the back-pressure of the electrochemical proton gradient. Thus, the protons must be crossing the membrane in the absence of ATP synthesis.

   Several processes other than ATP synthesis from added ADP might account for the slow passage of protons across the membrane and the slow background rate of oxygen consumption. (1) The mitochondrial inner membrane is not completely impermeable to protons, which can slowly cross the membrane even in the absence of ATP synthesis. (2) The internal mitochondrial supply of ATP may be hydrolyzed to ADP and then reconverted to ATP using the proton-motive force. (3) If some mitochondria in the preparation are damaged so that their inner membranes are not intact, they will transport electrons to oxygen continuously because there will be no electrochemical proton gradient to oppose electron flow.

C. Since each pair of electrons that flows down the respiratory chain from NADH to oxygen reduces one oxygen atom, the $P/2e^-$ ratio is equivalent to the P/O ratio. The P/O ratio, as calculated below, is between 2.5 and 2.8 molecules of ATP per O atom. Uncertainty in the P/O ($P/2e^-$) ratio arises from the uncertainty in how much oxygen is consumed during conversion of 500 nmol ADP to ATP. If oxygen consumption is calculated as the difference between the dotted, horizontal lines in Figure 14–7, which is 100 nmol O_2, then the P/O ratio is 2.5 (500 nmol ATP/200 nmol O). On the other hand, if oxygen consumption is calculated as the difference between the dashed, slanted lines in Figure 14–7, which is 90 nmol O_2, then the P/O ratio is 2.8 (500 nmol ATP/180 nmol O). The latter calculation makes the implicit assumption that the background rate of oxygen consumption continues during the conversion of ADP to ATP, which is a perfectly reasonable assumption. It turns out, however, that the natural slow flow of protons across intact inner membranes is quite sensitive to the size of the electrochemical proton gradient. The slight decline in the proton-motive force during ATP synthesis may reduce the leakage to nearly zero, in which case the larger value for oxygen consumption may be the more valid one (giving a P/O ratio of 2.5). A P/O ratio of 2.5 is now accepted as the value for the number of ATP molecules formed by flow of a pair of electrons down the electron transport chain from NADH to O_2.

D. Several processes in these kinds of experiments, in addition to ATP synthesis, are driven by the electrochemical proton gradient. The uptake of substrate (β-hydroxybutyrate) into mitochondria may require symport with protons. The import of phosphate into mitochondria also requires symport with a proton. Finally, the membrane potential—one component of the electrochemical proton gradient—drives the exchange of internal ATP for external ADP. Given that several processes are driven by the electrochemical proton gradient, it is not surprising that the P/O ratio is not an integer. Before the chemiosmotic theory, when chemical coupling hypotheses were fashionable, integral values were expected and values of 2.5 or 2.8 were assumed to 'really' mean 3.

Reference: Nicholls DG & Ferguson SJ (1992) Bioenergetics 2, pp 65–104. London: Academic Press.

14–24* **Reference:** Ingledew JW (1982) *Thiobacillus ferrooxidans*: the bioenergetics of an acidophilic chemolithotroph. *Biochim. Biophys. Acta* 683, 89–117.

14–25 It would take the heart 6 seconds to consume its steady-state levels of ATP. Because each pair of electrons reduces one atom of oxygen, the 12 pairs of electrons generated by oxidation of one glucose molecule would reduce $6\ O_2$. Thus, 30 ATP are generated per $6\ O_2$ consumed. At steady state, the rate of ATP production equals its rate of consumption. The time in seconds required to consume the steady-state level of ATP is

$$\text{time} = \frac{5\ \mu\text{mol ATP}}{\text{g}} \times \frac{6\ O_2}{30\ \text{ATP}} \times \frac{\text{min g}}{10\ \mu\text{mol}\ O_2} \times \frac{60\ \text{sec}}{\text{min}}$$
$$= 6\ \text{sec}$$

14–26*

14–27 At a $[\text{ATP}]/[\text{ADP}][\text{P}_i]$ ratio of 10^4, three protons would be required to drive synthesis of one ATP. The cost of synthesis of ATP under the specified conditions is

$$\Delta G = \Delta G^\circ + 2.3RT \log_{10} \frac{[\text{ATP}]}{[\text{ADP}][\text{P}_i]}$$
$$= \frac{7.3\ \text{kcal}}{\text{mole}} + 2.3 \times \frac{1.98 \times 10^{-3}\ \text{kcal}}{^\circ\text{K mole}} \times 310\ ^\circ\text{K} \times \log_{10}(10^4)$$
$$= 12.9\ \text{kcal/mole}$$

At least three protons (3 × –4.6 kcal/mole = –13.8 kcal/mole) would be required to balance the cost of ATP synthesis under these conditions.

ELECTRON-TRANSPORT CHAINS AND THEIR PROTON PUMPS

14–28*

14–29 A proton is a hydrogen atom that has lost its single electron and thus is positively charged. A hydride ion is a hydrogen atom that has gained an extra electron and thus is negatively charged. A hydrogen atom is a proton plus one electron; it is neutral. A hydrogen molecule is a pair of hydrogen atoms that share their two electrons in a covalent bond; it is neutral.

14–30* **Reference:** Barron JT, Gu L & Parrillo JE (2000) NADH/NAD redox state of cytoplasmic glycolytic compartments in vascular smooth muscle. *Am. J. Physiol. Heart Circ. Physiol.* 279, H2872–H2878.

14–31

A. Reversing the fumarate/succinate half reaction and summing gives $\Delta E_0'$ values of 0.02 V for FAD and –0.35 V for NAD^+. Since a positive $\Delta E_0'$ value corresponds to a negative $\Delta G°$ value, only the reaction using FAD would yield a favorable (negative) $\Delta G°$ value.

B. The concentration ratio needed to yield a ΔG value of zero is 4.4 for FAD and 5.1×10^{-12} for NAD^+. When ΔG is zero, ΔE is also zero. Thus for FAD

$$\Delta E = \Delta E_0' - \frac{2.3\,RT}{nF} \log_{10} \frac{[\text{fumarate}][\text{FADH}_2]}{[\text{succinate}][\text{FAD}]} = 0$$

$$\Delta E_0' = \frac{2.3\,RT}{nF} \log_{10} \frac{[\text{fumarate}][\text{FADH}_2]}{[\text{succinate}][\text{FAD}]}$$

$$0.02\,\text{V} = 0.031\,\text{V} \log_{10} \frac{[\text{fumarate}][\text{FADH}_2]}{[\text{succinate}][\text{FAD}]}$$

$$0.65 = \log_{10} \frac{[\text{fumarate}][\text{FADH}_2]}{[\text{succinate}][\text{FAD}]}$$

$$4.4 = \frac{[\text{fumarate}][\text{FADH}_2]}{[\text{succinate}][\text{FAD}]}$$

With NAD^+ as a cofactor the analogous calculation gives a concentration ratio of 5.1×10^{-12} as the ratio where ΔG is zero.

C. From these calculations it is clear that FAD is a more appropriate cofactor for oxidation of succinate than NAD^+. Using FAD the concentration ratio would have to be maintained at the reasonable value of less than 4.4 to allow the reaction to proceed. By contrast, with NAD^+ the concentration ratio would have to be maintained at less than 5.1×10^{-12}. Such a ratio could be achieved, for example, if the succinate concentration were 4.4×10^5 times the fumarate concentration, and the FAD concentration were 4.4×10^5 times the $FADH_2$ concentration. Such an excess of reactants over products is highly unlikely, if not impossible, for a biological reaction pathway.

14–32* **Reference:** Ingledew JW (1982) *Thiobacillus ferrooxidans*: the bioenergetics of an acidophilic chemolithotroph. *Biochim. Biophys. Acta* 683, 89–117.

14–33 True. The flow of electrons down the respiratory chain moves from carriers with lower affinity for electrons to ones with higher affinity and finally to oxygen, which has the highest affinity of all. Thus, the relative positions of cytochromes and iron-sulfur centers make sense in terms of the natural flow of electrons to oxygen.

14–34*

14–35

A. Oxygen accepts electrons from the electron-transport chain and is reduced to H_2O. Therefore, in the presence of oxygen the cytochromes would be drained of their electrons, that is, oxidized. Since the absorption bands do not show up in the presence of oxygen, the oxidized forms must not absorb light. The reduced forms of the cytochromes absorb light and are responsible for the characteristic absorption patterns. In the absence of oxygen the cytochromes pick up electrons from substrates (become reduced) but cannot get rid of them by transfer to oxygen. In the presence of oxygen, the electrons are transferred efficiently, leaving the cytochromes in their electron-deficient or oxidized state.

B. Keilin's observations indicate that the order of electron flow through the cytochromes is

$$\begin{array}{c}\text{reduced}\\\text{substrates}\end{array} \rightarrow \begin{array}{c}\text{cytochrome}\\b\end{array} \rightarrow \begin{array}{c}\text{cytochrome}\\c\end{array} \rightarrow \begin{array}{c}\text{cytochrome}\\a\end{array} \rightarrow O_2$$

This order can be deduced from Keilin's results. Since the bands become visible in the absence of oxygen, they represent the reduced (electron-rich) forms of the cytochromes. When oxygen is added, they are all converted to the oxidized (electron-poor) form. When cyanide is added, all the cytochromes are reduced, indicating that cyanide blocks the flow of electrons from the cytochromes to oxygen; that is, all the cytochromes are 'upstream' of oxygen (in the sense of electron flow).

When urethane is added, cytochrome b remains reduced but cytochromes a and c become oxidized. Thus, urethane interrupts the flow of electrons from cytochrome b to cytochromes a and c, indicating that cytochrome b is 'upstream' of cytochromes a and c.

These results indicate that either cytochrome a or c transfers electrons to oxygen. The inability of oxygen to oxidize a preparation of cytochrome c suggests, by elimination, that cytochrome a is responsible for transfer of electrons to oxygen. This ordering of cytochromes a and c is weak since it is based on a negative result (which could have other interpretations). Keilin himself confirmed this order by observing subtle spectral shifts in the cytochrome a band in the presence of cyanide under reducing conditions; he named the active component cytochrome a_3. We now know that cytochrome a is a large complex with several redox centers, one of which reacts with molecular oxygen.

C. The rapid oxidation of glucose to CO_2 prevents the disappearance of the absorption bands by providing a source of reduced substrates (ultimately NADH and $FADH_2$) that transfer electrons into the electron-transport chain faster than oxygen can remove them. Under these conditions the cytochromes remain reduced (electron rich) and therefore continue to absorb light.

Reference: Keilin D (1966) The History of Cell Respiration and Cytochrome. Cambridge, U.K.: Cambridge University Press.

14–36*

14–37 Only in mixture E would electron transfer occur, with cytochrome c becoming reduced. In this mixture a portion of the respiratory chain has been reconstituted, so that electrons can flow in the energetically favored direction from reduced ubiquinone to the cytochrome b-c_1 complex to cytochrome c.

Although energetically favorable, the transfer in mixture A cannot occur spontaneously in the absence of the cytochrome b-c_1 complex, which catalyzes this reaction. No electron flow occurs in the other mixtures, whether the cytochrome b-c_1 complex is present or not: in mixtures B and F both ubiquinone and cytochrome c are oxidized; in mixtures C and G both are reduced; and in mixtures D and H electron flow is energetically disfavored because reduced cytochrome c has a lower free energy than oxidized ubiquinone.

14–38* Figure 14–34.

Reference: Nicholls DG & Ferguson SJ (1992) Bioenergetics 2, pp 82–87. London: Academic Press.

14–39 Treatment with sodium nitrite oxidizes the Fe^{2+} in a proportion of hemoglobin molecules to Fe^{3+}. Because there is so much more hemoglobin than cytochrome oxidase, the Fe^{3+} form of hemoglobin competes effectively with cytochrome oxidase for the binding of cyanide. Treatment with sodium nitrite also impairs the oxygen-carrying function of hemoglobin, which will bind oxygen only when the heme group carries an Fe^{2+} ion. The treatment is therapeutically effective because there is a useful middle range where sufficient hemoglobin has been converted to the Fe^{3+} form to bind up the cyanide, but adequate Fe^{2+} hemoglobin remains to carry oxygen.

14–40*

14–41 In principle, the same diffusible carrier could be used for both steps. It would need to have a redox potential and a binding specificity that would allow it to accept electrons from either the NADH dehydrogenase complex or the cytochrome b-c_1 complex and pass them on to the cytochrome oxidase complex. Neither of these requirements is unreasonable. Such an arrangement, however, would be counterproductive because it would allow frequent bypass of the cytochrome b-c_1 complex: electrons could be accepted from the NADH dehydrogenase complex and passed directly to the cytochrome oxidase complex. This short-circuit of the standard flow would waste the energy normally harvested by the cytochrome b-c_1 complex, releasing it as heat instead.

14–42* **References:** Smith HT, Ahmed AJ & Millet F (1981) Electrostatic interaction of cytochrome c with cytochrome c_1 and cytochrome oxidase. *J. Biol. Chem.* 256, 4984–4990.

Capaldi RA, Darley-Usmar V, Fuller S & Millet F (1982) Structural and functional features of the interaction of cytochrome c with complex III and cytochrome c oxidase. *FEBS Lett.* 138, 1–7.

14–43 Reversing the NAD^+/NADH half reaction and summing E_0' values give a $\Delta E_0'$ value of 1.14 V. Noting that the balanced equation is a four-electron reaction, the standard free-energy change is

$$\Delta G° = -nF\Delta E_0'$$
$$= -4 \times \frac{23\ \text{kcal}}{\text{V mole}} \times 1.14\ \text{V}$$
$$= -105\ \text{kcal/mole}$$

The balanced equation that uses $\frac{1}{2}\ O_2$ has the same $\Delta E_0'$ value of 1.14 V, but because it is a two-electron reaction, the standard free-energy change is only –52.4 kcal/mole, just half the value calculated above. When the values from these calculations are expressed per mole of electrons, the values agree: –26.2 kcal/mole of electrons.

14–44*

14–45 False. Lipophilic weak acids act as uncoupling agents that dissipate the proton-motive force and stop ATP synthesis; however, they increase the flow of electrons through the respiratory chain by eliminating the respiratory control imposed by the electrochemical proton gradient. Normally, the electrochemical proton gradient exerts a backpressure that restricts the flow of electrons down the electron-transport chain. In the absence of the gradient electron transport runs unchecked at the maximum rate.

14–46*

14–47 An inhibitor of electron transport (like cyanide in mitochondria) would be expected to block electron flow, which would rapidly lead to collapse of the proton gradient (as it was used up to make ATP) and a drop in ATP levels (as it was consumed in cellular processes). Neither TCS nor DCCD cause all three effects in methanogenic bacteria.

An inhibitor of ATP synthase (like oligomycin in mitochondria) would be expected to stop ATP synthesis, leading to a drop in intracellular ATP levels. Blocking ATP synthase, which would prevent the flow of protons across the membrane, would lead to a backpressure that would restrict the flow of electrons down the respiratory chain. As shown in Figure 14–13B, DCCD behaves like an inhibitor of ATP synthase: ATP levels fall, the proton gradient remains high, and the rate of production of CH_4 (the indicator of electron flow) decreases substantially.

An inhibitor that uncouples electron transport from ATP synthesis (like dinitrophenol in mitochondria) would be expected to collapse the proton gradient, which would prevent ATP synthesis. In addition, collapse of the proton gradient would allow electrons (unopposed by the proton gradient) to flow more readily down the electron transport chain. As shown in Figure 14–13A, TCS behaves like an uncoupler: the proton gradient collapses, ATP levels fall, and CH_4 production accelerates.

The sequential addition of DCCD and TCS (see Figure 14–13C) confirms these assignments. Addition of an uncoupler (TCS) to a system in which ATP synthase is inhibited (by DCCD) would be expected to collapse the proton gradient and restore electron flow (CH_4 production).

Reference: Nicholls DG & Ferguson SJ (1992) Bioenergetics 2, pp 149–154. London: Academic Press.

14–48* **Reference:** Manson MD, Tedesco P, Berg HC, Harold FM & van der Drift C (1977) A protonmotive force drives bacterial flagella. *Proc. Natl. Acad. Sci. USA* 74, 3060–3064.

14–49 For all ATP synthases that use a proton-motive force to generate ATP, the flow of protons is to the same side of the membrane on which ATP is made. In a bacterium that lives in an alkaline environment, the flow of protons would be from the inside of the cell to the outside. Thus, the proton flow is in the wrong direction to allow a normally oriented ATP synthase to make ATP; in fact, outward proton flow causes ATP hydrolysis. Even if the ATP synthase were reversed in the membrane, so that it could use the proton flow, it would then make ATP on the outside of the cell—not a very useful arrangement.

Bacteria of this kind have adapted to use a gradient of another ion such as Na^+ to drive their ATP synthases, suggesting that using another ion probably is the simplest adaptation accessible to evolution. As a purely theoretical engineering problem, however, other solutions can be imagined. Structurally, the ATP synthase consists of a proton-powered motor embedded in the membrane, an ADP to ATP converter, and a drive shaft that connects the two. Normally, an *inward* proton flow through the motor drives a *counterclockwise* rotation of the drive shaft inside the converter to cause the *normal* sequence of conformational changes that converts ADP to ATP. (An *outward* flow drives a *clockwise* rotation of the shaft that causes the *reverse* sequence of conformational changes, converting ATP to ADP.) With this scheme in mind, you might consider re-engineering the motor so that an *outward* flow drives a *counterclockwise* rotation of the drive shaft. Alternatively, you might re-engineer the converter so that a *clockwise* rotation of the drive shaft inside it would drive the *normal* sequence of conformational changes. Either of these changes would allow the re-engineered ATP synthase to make ATP from the proton gradient in an alkaline environment.

CHLOROPLASTS AND PHOTOSYNTHESIS

14–50*

14–51 Protons are pumped across the inner membrane into the intermembrane space in mitochondria. By contrast, they are pumped across the thylakoid membrane into the thylakoid space in chloroplasts. ATP is synthesized in the corresponding compartments in the two organelles: in the matrix in mitochondria and in the stroma in chloroplasts.

14–52*

14–53 Choice A in Table 14–5 gives the true balanced equation for the Calvin cycle. You can decide which is balanced by seeing whether the oxygen and hydrogen atoms on each side of the equation are equal. For equation A there are 12 Hs from NADPH and 24 Hs from H_2O for 36 total on the left, and 12 Hs from glucose, 18 from P_i, and 6 from H^+ for 36 total on the right; thus, H atoms balance. There are 12 Os from CO_2 and 12 from H_2O for 24 total on the left, and 6 Os from glucose and 18 from P_i for 24 total on the right; thus,

O atoms balance. For none of the other equations do both the H atoms and O atoms balance. (A difficulty in deciding on the balanced equation is the form of P_i the authors intend in their equations. The pK for $H_2PO_4^- \rightarrow HPO_4^{2-} + H^+$ is 6.9, which means that at intracellular pH—slightly above 7—most of the P_i will be in the HPO_4^{2-} form: the form you were directed to use in this problem. If $H_2PO_4^-$ is used instead, then equation B in Table 14–5 is the balanced equation. Because cells contain a mixture of HPO_4^{2-} and $H_2PO_4^-$ at intracellular pH, the actual balanced equation is somewhere in between A and B.)

14–54* Figure 14–35.

Reference: Foyer CH (1984) Photosynthesis, pp 176–195. New York: Wiley.

14–55 In the absence of light, but presence of CO_2, ribulose 1,5-bisphosphate will decrease and 3-phosphoglycerate will increase. The presence of CO_2 allows ribulose 1,5-bisphosphate to be converted to 3-phosphoglycerate but in the absence of light (and therefore reduced amounts of NADPH and ATP) 3-phosphoglycerate will accumulate because subsequent reactions require NADPH and ATP.

In the absence of CO_2 ribulose 1,5-bisphosphate will accumulate (and 3-phosphoglycerate will decrease) because its conversion to 3-phosphoglycerate is dependent on CO_2.

14–56*

14–57 False. When an electron in a chlorophyll molecule in the antenna complex is excited, it transfers its energy—not the electron—from one chlorophyll molecule to another by resonance energy transfer.

14–58*

14–59 One measure of photosynthesis is the evolution of O_2. In 1882 none of the sensitive devices now used for measuring O_2 were available. Instead, Englemann made use of bacteria that grow best in the presence of O_2 and actively seek it. When the alga was illuminated with a spectrum of light, only those portions that received light at the blue or red ends of the spectrum were able to carry out photosynthesis and evolve O_2. Bacteria that use O_2 tend to collect around those portions of the alga that give off O_2. Thus the density of bacteria is a crude measure of the rate of O_2 evolution. The action spectrum (the rate of O_2 evolution at different wavelengths) can be approximated by the density of the bacteria at different places in the spectrum (Figure 14–36).

14–60*

14–61 The burst of oxygen production when the illumination is switched to 650 nm suggests that this wavelength stimulates photosystem II, which accepts electrons directly from water and generates oxygen. Similarly, the dip in oxygen production when the illumination is switched to 700 nm suggests that this wavelength stimulates photosystem I, which accepts electrons from the electron-transport chain and, thus, is farther removed from the reactions that generate oxygen. This interpretation is supported by the more detailed analysis of the chromatic transients below.

The chromatic transients result because the two photosystems are out of

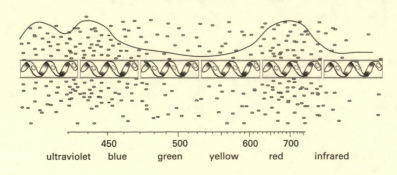

Figure 14–36 Action spectrum of a filamentous green alga (Answer 14–59).

balance with one another. Each separate wavelength preferentially (but not absolutely) stimulates one of the two photosystems. Thus, when photosystem I is stimulated (by 700-nm light), it pulls electrons out of the electron-transport chain that links the two photosystems, leaving them in a relatively oxidized state, primed to accept electrons from photosystem II. When the light is switched to 650 nm (which stimulates photosystem II), there is an initial rush of electrons (from H_2O) into the cytochrome chain that causes a burst of O_2 evolution. The flow of electrons through the cytochromes is quickly limited, however, by the electrons' ability to be transferred to photosystem I, which is suboptimally stimulated, and O_2 evolution slows.

When the light is switched back to 700 nm, the electron pressure from photosystem II (which is now suboptimally stimulated) is insufficient to push electrons into the relatively reduced (electron-rich) cytochromes. As a result, O_2 evolution is depressed transiently while electrons are bled off from the cytochromes. Once the cytochromes have been partially drained of their electrons, they can accept new electrons from photosystem II, thereby reestablishing the normal level of oxygen production.

References: Emerson R (1958) The quantum yield of photosynthesis. *Annu. Rev. Plant Physiol.* 9, 1–24.

Lawlor DW (1987) Photosynthesis: Metabolism, Control, and Physiology. New York: Wiley.

14–62*

14–63
A. Since stimulation by 680-nm light removes electrons from the cytochromes, causing their oxidation, 680-nm light must preferentially stimulate photosystem I, which transports electrons from the cytochromes to $NADP^+$ (Figure 14–37). The subsequent stimulation by 562-nm light causes electrons to flow into the cytochromes at a faster rate than before, thereby causing them to become more reduced. Consequently, 562-nm light must stimulate photosystem II, which transfers electrons from water to the cytochromes (Figure 14–37). Thus, in these algae, as in most plants, the longer wavelength preferentially stimulates photosystem I, and the shorter wavelength preferentially stimulates photosystem II.
B. These results support the Z scheme of photosynthesis in several ways. First, the different effects at the two wavelengths suggest that there are at least two components that differ in their responses to different wavelengths of light. Second, the two wavelengths of light have opposite effects on the redox poise of the cytochromes—680-nm light causing oxidation and 562-nm light causing reduction. Finally, the effects at the two wavelengths could be separated by DCMU, which indicates that the two photosystems communicate through the cytochromes (Figure 14–37).
C. These results indicate that DCMU blocks electron transport through the cytochromes on the upstream side, that is, on the side nearer photosystem II (Figure 14–37). Thus, when photosystem I is stimulated by 680-nm light in the presence of DCMU, it transfers what electrons are available out of the cytochromes, causing their oxidation. In addition, in the presence of DCMU electrons cannot be transferred into the cytochromes by stimulation of photosystem II by 562-nm light (see Figure 14–19B). These two effects indicate that DCMU blocks electron transport very near the beginning of the cytochrome chain.

Reference: Duysens LNM, Amesz J & Kamp BM (1961) Two photochemical systems in photosynthesis. *Nature* 190, 510–511.

14–64* **Reference:** Forbush B, Kok B & McGloin M (1971) Cooperation of charges in photosynthetic oxygen evolution II. Damping of flash yield, oscillation and deactivation. *Photochem. Photobiol.* 14, 307–321.

14–65 Eight photons are required: four by photosystem II and four by photosystem I.

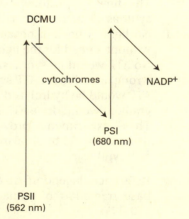

Figure 14–37 Simplified Z scheme of photosynthesis showing the relationship of the two photosystems, the cytochromes, and the point of action of DCMU (Answer 14–63).

Four electrons (ultimately from water, generating O_2) are excited by absorption of four photons in photosystem II and then re-excited by absorption of four photons in photosystem I. The four electrons from photosystem I are used to reduce two molecules of $NADP^+$ to NADPH.

14–66*

14–67 A minimum of eight protons (8×-6.4 kcal/mole = 51.2 kcal/mole) would be required to drive electrons from Fe^{2+} to $NADP^+$, a process with an unfavorable free energy of 50.6 kcal/mole. The free energy for transfer of two electrons from Fe^{2+} to $NADP^+$ ($\Delta E = -1.1$ V) is

$$\Delta G = -nF\Delta E$$
$$= -2 \times \frac{23 \text{ kcal}}{\text{V mole}} \times -1.1 \text{ V}$$
$$= 50.6 \text{ kcal/mole}$$

These thermodynamic considerations say nothing about the mechanism of coupling. Reasoning from what is known in other organisms, it is likely that an electron transport chain links the Fe^{2+}/Fe^{3+} redox pair to the $NADP^+/$ NADPH redox pair. In other organisms the flow of electrons in the favorable direction is coupled to the pumping of protons across a membrane. This is likely to be the arrangement in *T. ferrooxidans* as well; however, the naturally imposed pH difference across the membrane drives an inward proton flow that reverses the proton pumps to cause an upward (otherwise unfavorable) flow of electrons from Fe^{2+} to $NADP^+$ (see Problem 14–46).

14–68*

14–69 The free energy change for ATP synthesis in chloroplasts under these conditions is 12.2 kcal/mole, about the same as it is in mitochondria.

$$\Delta G = \Delta G° + 2.3RT \log_{10} \frac{[\text{ATP}]}{[\text{ADP}][\text{P}_i]}$$
$$= 7.3 \text{ kcal/mole} + 2.3 \times \frac{1.98 \times 10^{-3} \text{ kcal}}{°\text{K mole}} \times 310 °\text{K} \times \log_{10} \frac{3 \times 10^{-3}}{(10^{-2})(10^{-4})}$$
$$= 7.3 \text{ kcal/mole} + 4.9 \text{ kcal/mole} = 12.2 \text{ kcal/mole}$$

14–70*

14–71
A. The switch in solutions creates a pH gradient across the thylakoid membrane. The flow of protons down their electrochemical potential drives ATP synthase, which converts ADP to ATP.
B. No light is needed because the proton gradient is established artificially without a need for the light-driven electron-transport chain.
C. No ATP would be synthesized because the proton gradient would be in the wrong direction for ATP synthase to make ATP. In fact, one might expect that ATP would be hydrolyzed in this experiment because the backwards proton gradient would drive ATP synthase in reverse, causing hydrolysis of ATP.
D. These experiments provided early supporting evidence for the chemiosmotic model by showing that a pH difference alone is sufficient to drive ATP synthesis.

14–72* **Reference:** Jagendorf AT & Uribe E (1966) ATP formation caused by acid-base transition of spinach chloroplasts. *Proc. Natl. Acad. Sci. USA* 55, 170–177.

14–73 Even during daylight hours in chloroplast-containing cells, mitochondria are required to supply the cell with ATP derived by oxidative phosphorylation. Glyceraldehyde 3-phosphate made by photosynthesis in chloroplasts moves to the cytosol and is eventually used as a source of energy to drive ATP production in mitochondria.

THE GENETIC SYSTEMS OF MITOCHONDRIA AND PLASTIDS

14–74*

14–75

A. The results in Figure 14–24 are exactly what you would expect if mitochondrial DNA was replicated at random times throughout the cell cycle. Regardless of the length of the chase, a constant fraction of the labeled DNA is triggered to replicate. Even the fraction of the DNA that is shifted (about 10%) is what you expect, since the labeling time with BrdU (2 hours) is about 10% of the 20-hour cell cycle.

   If replication were confined to a specific part of the cell cycle, then DNA that was labeled with ^{3}H-thymidine in the first pulse would not be replicated a second time until the critical phase of the cell cycle came around again. As a result, very little of the labeled DNA should be shifted in density until that time. The critical time in the cell cycle would show up in this experiment as a high fraction of labeled DNA that was density shifted at a particular chase time.

B. It is true that the cell population is asynchronous, but asynchrony has no bearing on the interpretation of the experiment. If the mitochondrial DNA is replicating at random times, then the synchrony of the cell population is irrelevant. Your colleague's concern is directed at the possibility that an asynchronous cell population might obscure your ability to detect a timed replication of mitochondrial DNA. Your elegant experimental design, however, nicely gets around that potential objection. If mitochondria replicated at a specific time during the cell cycle, the brief pulse of ^{3}H-thymidine would label only those cells in that portion of the cycle. Since in the remainder of the experiment you follow only the radioactive mitochondrial DNA molecules, the brief labeling period has, in essence, synchronized the cell population—you are blind to what happens in any cells that were outside the critical labeling period.

C. A similar analysis of nuclear DNA would be expected to show a peak of density shifting between 15 and 20 hours, with very little shifting of radioactive DNA at shorter chase periods (Figure 14–38). A pulse of ^{3}H-thymidine would label nuclear DNA only in S phase. The labeled DNA would not replicate again until the next S phase. If the DNA were radioactively labeled at the end of one S phase, it would come to the beginning of the next S phase after 15 hours or so—at which point its density would be shifted by exposure to BrdU. If the DNA were labeled at the beginning of S phase, it would not enter S phase again for nearly 20 hours. In actual experiments, the peak of density shifting of labeled nuclear DNA was indeed observed between 15 and 20 hours.

D. If mitochondrial DNA molecules were replicated at all times, but individual molecules had to wait one entire cell cycle between replication events, there would be a peak of density shifting after 18 to 20 hours, when the labeled molecules returned to their critical time in the cell cycle. These results would resemble those for nuclear DNA replication, since nuclear DNA also has to wait one full cycle; however, the timing would be slightly different because mitochondrial DNA replicates in 2 hours whereas nuclear DNA takes 5 hours.

   Reference: Bogenhagen D & Clayton DA (1977) Mouse L cell mitochondrial DNA molecules are selected randomly for replication throughout the cell cycle. *Cell* 11, 719–727.

14–76*

14–77 It is likely that your organism derived from an ancient eucaryote that once possessed an endosymbiont. Transfer of DNA from the endosymbiont to the nuclear genome occurred, giving rise to the precursors of the scattered bits of 'bacterial' DNA you found in your organism's genome. At some later point

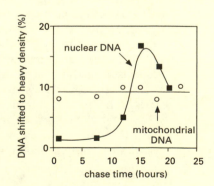

Figure 14–38 Peak of density-shifted nuclear DNA (Answer 14–75).

the endosymbiont-derived organelles (mitochondria) were lost, perhaps in adaptation to the anaerobic niche in which it now lives. An alternative hypothesis, which is difficult to rule out, is that the bits of bacterial DNA were picked up by lateral gene transfer directly from other bacteria; that is, that there was no endosymbiont stage.

This hypothetical organism resembles *Giardia*, one of the rare eucaryotes without mitochondria, which is thought to have been derived from a more typical eucaryotic cell that lost its mitochondria (see Problem 1–30).

14–78* **Reference:** Stern DB & Palmer JD (1984) Extensive and widespread homologies between mitochondrial DNA and chloroplast DNA in plants. *Proc. Natl. Acad. Sci. USA* 81, 1946–1950.

14–79 What you have neglected in your scheme is a mechanism to get the ATP out of the proto-*Paracoccus*. An adenine nucleotide carrier is absent from all free-living bacteria, as expected, since they must retain ATP inside if it is to do them any good. All mitochondria have an ATP-ADP antiporter in their membranes to allow free exchange of ATP and ADP between the cytoplasm and the mitochondrial matrix. Only after the acquisition of this carrier would ATP synthesized by the endosymbiont be available to the host.

14–80*

14–81 Four. Two of the three in the standard code (UAA and UAG) plus two (AGA and AGG) that encode arginine in the standard code.

14–82*

14–83
A. Initiation of protein synthesis in mitochondria differs from that in the cytoplasm in two distinct ways. The first is straightforward: the codon AUA in mitochondria can serve as an initiation codon and encodes methionine (see Figure 14–27, mRNA 13). In cytoplasmic protein synthesis, AUA encodes isoleucine and does not serve as an initiator of protein synthesis. The second difference is more subtle: the encoded protein can begin immediately at the 5′ end of the mRNA (see Figure 14–27, mRNAs 7 and 16). Cytoplasmic (and bacterial) mRNAs typically have a short stretch of untranslated nucleotides at their 5′ ends that are thought to help guide ribosomes onto the mRNA. In bacteria there is a short sequence in front of the start codon to which a ribosomal RNA hybridizes. In the cytoplasm ribosomes bind to a 5′ cap (a modified G attached posttranscriptionally to the end of the mRNA) and thread onto the mRNA for some distance before reaching the first start codon. Neither of these features is present at the 5′ ends of mitochondrial mRNA.
B. The termination codons for protein synthesis in mitochondria are unusual in two ways. First, the termination codon in mRNA 16 is AGA (see Figure 14–27), which in the nucleus encodes arginine. Second, the termination codons in mRNAs 7 and 13 are not completely encoded in the DNA; instead, they are generated by addition of the poly-A tail (see Figure 14–27). In mRNA 7 only the initial U of the UAA stop codon is encoded; in mRNA 13 only the initial UA of the stop codon is encoded.
C. The presence of tRNA genes at the exact boundaries of the mRNA genes suggests that they might be involved in processing the mRNAs out of the single, long primary transcript. The tRNAs are thought to serve as structural signposts for the processing of the primary transcript. The folding of the tRNAs into cloverleaf structures would place distinctive structures at the ends of the mRNAs. It is thought that the tRNA structures are recognized and cleaved at their ends to remove them from the primary transcript. The mRNAs are then the remains of tRNA processing. This scheme is referred to as the tRNA punctuation model of RNA processing.

References: Montoya J, Ojala D & Attardi G (1981) Distinctive features of the 5′-terminal sequences of the human mitochondrial mRNAs. *Nature* 290, 465–470.

Ojala D, Montoya J & Attardi G (1981) tRNA punctuation model of RNA processing in human mitochondria. *Nature* 290, 470–474.

14–84*

14–85 Variegation occurs because the plants have a mixture of normal and defective chloroplasts. These sort out by mitotic segregation to give patches of green and white in leaves. Many of the green patches have cells that still retain defective chloroplasts in addition to the normal ones. As such areas grow, they can segregate additional cells that have only defective chloroplasts, giving rise upon cell division to an island of white cells in a sea of green ones. By contrast, white areas are due to cells that retain only defective chloroplasts. Thus, white cells cannot give rise to green cells; hence, there are no green islands surrounded by white.

14–86* **References:** Mitchell MB & Mitchell HK (1952) A case of "maternal" inheritance in *Neurospora crassa. Proc. Natl. Acad. Sci. USA* 38, 442–449.

Mitchell MB, Mitchell HK & Tissieres A (1953) Mendelian and non-Mendelian factors affecting the cytochrome system in *Neurospora crassa. Proc. Natl. Acad. Sci. USA* 39, 605–613.

14–87 Pedigree A is for an autosomal recessive mutation. Both the mother and the father must be heterozygotes that carry one copy of the chromosome. One quarter of their children, irrespective of gender, are expected to get a defective copy of the autosome from each parent, thereby becoming homozygous (and affected).

Pedigree B is for an X-linked recessive mutation. The mother carries the mutation on one copy of her two X chromosomes, and is thus unaffected. She passes the X chromosome with the mutation randomly to half her offspring. The males who get that X chromosome are affected because it's their only X chromosome. Females who get the mutant X chromosome are unaffected because they have a normal X chromosome from their father. Note that without additional information (a larger pedigree) it would be difficult to be certain that this pedigree truly resulted from an X-linked recessive mutation and not from an autosomal recessive mutation. The distinguishing feature—a small number of affected individuals who are all male—could have resulted by chance in a small pedigree for an autosomal recessive mutation.

Pedigree C is for an autosomal dominant mutation. The mother is heterozygous, but affected because one copy of a pair of autosomal chromosomes carries a dominant mutation. She will pass the affected chromosome to half her children, regardless of gender, all of whom will be affected because the mutation is dominant.

Pedigree D is for a mitochondrial mutation. The mother is affected and passes on her defective mitochondria to all the children because the fertilized egg contains only her mitochondria. In reality, mitochondrial inheritance is rarely as clear-cut as this example would indicate. The mother's mitochondria are rarely all of the defective type; thus, mitotic segregation can give rise to a range of mixtures of mutant and normal mitochondria in the children, who may display phenotypes that range from unaffected to severely affected. Real pedigrees for mitochondrial mutations are thus sometimes difficult to distinguish from pedigrees for autosomal dominant mutations.

14–88*

14–89

A. The deletions that suppress the *pet494* mutation do not affect the coding portion of the mRNA. Therefore, they do not alter the *coxIII* gene product and cannot affect its stability. By contrast, the replacement of the normal 5′ untranslated region with one from any of several other genes is perfectly consistent with an alteration in translation. These results suggest that the normal *PET494* gene product promotes translation of coxIII from the wild-type mitochondrial mRNA.

B. The rearrangements of the 5′ end of the *coxIII* gene result in deletion of essential mitochondrial genes, which normally would produce a cytoplasmic petite strain of yeast. Yet these strains have all the usual mRNAs and grow perfectly well. These observations suggest that the deleted DNA must be present somewhere else in the mitochondria. It turns out that the *pet494* suppressor strains contain both wild-type and deleted mitochondrial genomes. Although it is not uncommon for mitochondria to contain more than one DNA molecule, the mixture of DNAs is quite unusual. This so-called heteroplasmic state is normally unstable, and the individual mitochondrial genomes segregate rapidly. However, by demanding that the cells grow on glycerol, it is possible to maintain the heteroplasmic state indefinitely. In the presence of the nuclear *pet494* mutation, both mitochondrial genomes are required for growth on glycerol: the deleted genome provides translatable coxIII mRNA and the wild-type genome provides all other essential gene products.

Reference: Fox TD (1986) Nuclear gene products required for translation of specific mitochondrially coded mRNAs in yeast. *Trends Genet.* 2, 97–100.

CELL COMMUNICATION

GENERAL PRINCIPLES OF CELL COMMUNICATION

15–1 False. Signaling molecules that bind to cell-surface receptors do not have to cross the plasma membrane; thus, they can be large or small, hydrophilic or hydrophobic. By contrast, signaling molecules that bind to intracellular receptors must be sufficiently small and hydrophobic to diffuse across the plasma membrane.

15–2*

15–3

A. Only pool 3 has colonies that have bound the radioactive ligand, as shown by the black circles. Thus, one or more of the plasmids in this pool contain the cDNA for the receptor.

B. Not all of the colonies on the dish transfected with pool-3 plasmid DNA are radioactive because not all of the transfected cells received the plasmid carrying the receptor cDNA. In a pool of 1000 plasmids, it is expected that only a few of the transfected cells will have received a plasmid carrying the receptor cDNA.

C. This experiment identifies a pool of plasmids, at least one of which carries the cDNA for the receptor. The usual next step is to subdivide this pool into smaller pools containing, for example, 10 plasmids each and then repeat the transfection. Once again a positive pool is identified by autoradiography. Finally, individual plasmids from the small pool are isolated and tested by transfection and autoradiography. At this final stage a positive autoradiograph identifies a plasmid that carries the receptor cDNA.

15–4*

15–5 The results shown in Figure 15–2 all are consistent with chemical signaling: in particular, the limited range of the signal (experiments A and B); the ability to bend around corners (experiment B); the ability to penetrate a semipermeable membrane but not a glass coverslip (experiments A and C); and the profound influence of a gentle stream of liquid (experiment D). These experiments go a long way toward demonstrating a secreted chemical signal, but the final proof is the isolation of the signal molecule, which was accomplished many years afterward and shown to be cyclic AMP.

Reference: Bonner JT & Savage LJ (1947) Evidence for the formation of cell aggregates by chemotaxis in the development of the slime mold *Dictyostelium discoideum. J. Exp. Zool.* 106, 1–26.

15–6* **Reference:** Yalow RS (1978) Radioimmunoassay: a probe for the fine structure of biologic systems. *Science* 200, 1236–1245.

15–7

A. On average, two atoms of radioactive iodine (half-life of 7 days) will give rise to one disintegration in a week. Thus, to determine the required number of picograms of labeled insulin, it is necessary to convert 1000 counts per minute to disintegrations per week, multiply by 2 to get the number of radioactive iodine atoms, and then to calculate the weight of the same number of insulin molecules.

1000 counts per minute is 2.0×10^7 disintegrations (disint) per week.

$$\frac{disint}{week} = \frac{1000\ counts}{min} \times \frac{2\ disint}{count} \times \frac{60\ min}{hr} \times \frac{24\ hr}{day} \times \frac{7\ days}{week}$$
$$= 2.0 \times 10^7\ disintegrations\ per\ week$$

Thus, 1000 cpm corresponds to 4.0×10^7 atoms of radioactive iodine. The weight of this number of insulin molecules is 0.76 picograms.

$$insulin = 4.0 \times 10^7\ molecules \times \frac{moles}{6 \times 10^{23}\ molecules} \times \frac{11{,}446\ g}{mole} \times \frac{10^{12}\ pg}{g}$$
$$= 0.76\ pg$$

B. For optimal sensitivity of RIA, the amounts of the tracer and unknown should be the same. Thus, given the starting assumptions, at optimum sensitivity you will be able to detect 0.76 pg of unlabeled insulin.

15–8*

15–9 False. The concentration of a neurotransmitter in the synaptic cleft is much higher than, for example, the concentration of a circulating hormone in the blood. Correspondingly, neurotransmitter receptors have a relatively low affinity for their ligand, which allows the neurotransmitter to dissociate rapidly from the receptor to terminate a response. As a rule of thumb, the K_d for binding to a receptor is about equal to the concentration of signaling molecule to which the receptor is exposed to produce a physiological response.

15–10*

15–11

A. A telephone conversation is analogous to synaptic signaling in the sense that it is a private communication from one person to another. It differs from synaptic signaling because it is (usually) a two-way exchange, whereas synaptic signaling is a one-way communication.

B. Talking to people at a cocktail party is analogous to paracrine signaling, which occurs between different cells (individuals) and is locally confined.

C. A radio announcement is analogous to an endocrine signal, which is sent out to the whole body (the audience) with only target cells (affected individuals) responding to it.

D. Talking to yourself is analogous to an autocrine signal, which is a signal that is sent and received by the same cell.

15–12* **Reference:** Faull RJ, Kovach NL, Harlan JM & Ginsberg MH (1993) Affinity modulation of integrin $\alpha_5\beta_1$: regulation of the functional response by soluble fibronectin. *J. Cell Biol.* 121, 155–162.

15–13

A. The $\alpha_{IIb}\beta_3$ integrin, like the $\alpha_5\beta_3$ integrin described in Problem 15–12 can exist in two conformations: one with very low affinity for fibrinogen and the other with very high affinity. Evidently, the cytoplasmic domain α_{IIb} controls the affinity of this integrin. In CHO cells the cytoplasmic domain holds the integrin in its low-affinity conformation. Truncation of the cytoplasmic domain allows the integrin to flip into its high-affinity conformation.

B. If the affinity of the $\alpha_{IIb}\beta_3$ integrin is regulated by the cytoplasmic domain of α_{IIb}, then the factors that stimulate clotting presumably initiate a cell-signaling pathway that alters the affinity of the integrin via effects on the cytoplasmic domain of α_{IIb}. Thrombin has no effect on CHO cells that display $\alpha_{IIb}\beta_3$ integrin on their surface because the thrombin receptor or a portion of the thrombin-induced signaling pathway is missing in CHO cells.

C. Individuals who carry one α_{IIb} gene with a mutation like the one described in this problem would be expected to have blood-clotting problems. Although half the $\alpha_{IIb}\beta_3$ integrin in their platelets would be normal, half would be expected to have a high affinity for fibrinogen and cause platelet aggregation and blood clotting in the absence of thrombin stimulation. Thus, such a mutation would be expected to be dominant and cause serious problems even when heterozygous.

Reference: O'Toole TE, Mandelman D, Forsyth J, Shattil SJ, Plow EF & Ginsberg MH (1991) Modulation of the affinity of integrin $\alpha_{IIb}\beta_3$ (GPIIb-IIIa) by the cytoplasmic domain of α_{IIb}. *Science* 254, 845–847.

15–14*

15–15

A. HRP and fluorescein enter both cells at the early two-cell stage (but not at the late two-cell stage) because the cells are still connected by cytoplasmic bridges, which allow passage of large molecules.

B. Gap junctions form at the compaction stage of embryo development. As a result, fluorescein can enter all cells in the compacted eight-cell embryo. Since gap junctions permit passage only of molecules less than 1000 daltons or so, HRP is confined to the cell into which it was initially injected. The different result before and after compaction indicates that formation of gap junctions is associated with compaction.

C. If you inject current from the HRP electrode, you would detect it in the fluorescein electrode only in the early two-cell embryo and the compacted eight-cell embryo. Only at these two stages are the adjacent cells electrically coupled. At the two-cell stage, the cytoplasmic bridge remaining from cell division mediates the coupling; at the eight-cell stage, gap junctions mediate the coupling.

Reference: Lo CW & Gilula NB (1979) Gap junctional communication in the preimplantation mouse embryo. *Cell* 18, 399–409.

15–16*

15–17 Cells with identical receptors can respond differently to the same signal molecule because of differences in the internal machinery to which the receptors are coupled. Even when the entire pathway is the same, cells can respond differently if they express different target proteins for the pathway.

15–18*

15–19 The concentration of cyclic GMP in the smooth muscle cells lining the blood vessels of the penis is controlled by its rate of synthesis by guanylyl cyclase and its rate of degradation by cyclic GMP phosphodiesterase. The natural signal molecule NO binds to guanylyl cyclase and stimulates its activity, thereby increasing the concentration of cyclic GMP by increasing its rate of synthesis. The drug Viagra binds to cyclic GMP phosphodiesterase and inhibits its activity, thereby increasing the concentration of cyclic GMP by decreasing its rate of degradation.

15–20*

15–21 The modifications of cholesterol to make steroid hormones increase the hydrophilicity of the molecules by removing the hydrocarbon tail of cholesterol and by introducing polar groups. These modifications make the molecules sufficiently hydrophilic to diffuse from their carrier molecules in the bloodstream to cells, but not so hydrophilic as to prevent their crossing of the plasma membrane to enter cells. By contrast, cholesterol is so hydrophobic that it normally spends all its time in the membrane. A lipid that is virtually insoluble in water could not move readily as a messenger from one cell to another via the extracellular fluid.

15–22* **Reference:** DeFranco D & Yamamoto KR (1986) Two different factors act separately or together to specify functionally distinct activities at a single transcriptional enhancer. *Mol. Cell. Biol.* 6, 993–1001.

15–23

A. The glucocorticoid response elements are located primarily between nucleotides –240 and –50 on the LTR. The electrophoretic pattern of the starting mixture shows that all the fragments from the mutant LTRs are equally represented initially. In the electrophoretic pattern of the bound mixture, however, the fragments are unequally represented: the fragment from the –240 deletion is nearly equal to the fragment from the complete LTR; the fragments from the –50 and +50 deletions are totally absent; and the fragments from deletions –202 and –137 are intermediate. The apparent gradient of binding suggests that several glucocorticoid response elements are located between nucleotides –240 and –50. This suggestion was confirmed by experiments like the ones summarized in part B.

B. A consensus sequence for glucocorticoid receptor binding can be obtained by listing the frequencies of G, C, A, and T at each position in the bound sequences, as shown in Figure 15–42. Individual nucleotides that are present in more than half the sequences are shown immediately below the frequencies. Two consensus glucocorticoid-binding sequences (AGAA/TCAGT/A and AGAACA) have been reported in the literature. The common nucleotides that are outside the central group are not part of the true binding site, as additional experiments have shown.

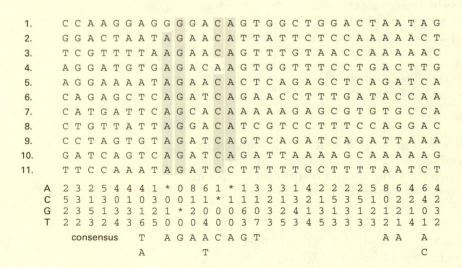

```
 1.   C C A A G G A G G G G A C A G T G G C T G G A C T A A T A G
 2.   G G A C T A A T A G A A C A T T A T T C T C C A A A A A C T
 3.   T C G T T T T A A G A A C A G T T T G T A A C C A A A A A C
 4.   A G G A T G T G A G A C A A G T G G T T T C C T G A C T T G
 5.   A G G A A A A T A G A A C A C T C A G A G C T C A G A T C A
 6.   C A G A G C T C A G A T C A G A A C C T T T G A T A C C A A
 7.   C A T G A T T C A G C A C A A A A A G A G C G T G T G C C A
 8.   C T G T T A T T A G G A C A T C G T C C T T T C C A G G A C
 9.   C C T A G T G T A G A T C A G T C A G A T C A G A T T A A A
10.   G A T C A G T C A G A T C A G A T T A A A A G C A A A A A G
11.   T T C C A A A T A G A T C C T T T T T G C T T T T A A T C T

  A   2 3 2 5 4 4 4 1 * 0 8 6 1 * 1 3 3 3 1 4 2 2 2 2 5 8 6 4 6 4
  C   5 3 1 3 0 1 0 3 0 0 1 1 * 1 1 1 2 1 3 2 1 5 3 5 1 0 2 2 4 2
  G   2 3 5 1 3 3 1 2 1 * 2 0 0 0 6 3 2 4 1 3 1 3 1 2 1 2 1 0 3
  T   2 2 3 2 4 3 6 5 0 0 0 4 0 0 3 7 3 5 3 4 5 3 3 3 3 2 1 4 1 2

consensus   T   A G A A C A G T                 A A   A
            A           T                              C
```

Figure 15–42 Frequencies of G, C, A, and T at each of the positions in the bound sequences (Answer 15–23). An *asterisk* (*) indicates nucleotides that are present in 10 or more of the sequences. The consensus sequence shows those nucleotides that are represented in more than half the sequences in this set of 11; whenever, two nucleotides are listed, the lower one was also prevalent.

References: Payvar F, DeFranco D, Firestone GL, Edgar B, Wrange O, Okret S, Gustafsson J-A & Yamomoto KR (1983) Sequence-specific binding of glucocorticoid receptor to MMTV DNA at sites within and upstream of the transcribed region. *Cell* 35, 381–392.

Scheidereit C, Geisse S, Westphal HM & Beato M (1983) The glucocorticoid receptor binds to defined nucleotide sequences near the promoter of mouse mammary tumor virus. *Nature* 304, 749–752.

15–24* Figure 15–43.

Reference: Ashburner M, Chihara C, Meltzer P & Richards G (1973) Temporal control of puffing activity in polytene chromosomes. *Cold Spring Harbor Symp. Quant. Biol.* 38, 655–662.

15–25 In the case of the steroid receptor, a one-to-one complex of steroid and receptor binds to DNA to activate transcription; thus, there is no amplification between ligand binding and activation of the target gene. By contrast, for ion-channel-linked receptors, a single ion channel will let through thousands of ions in the time it remains open, which serves as an amplification step in this type of signaling pathway.

15–26*

15–27 In the normal signaling pathway PK2 activates PK1. If PK1 is permanently activated, a response is observed independent of the status of PK2. If the order were reversed—that is, PK1 activates PK2—signaling would not occur when PK2 carried an inactivating mutation.

   If the experimental setup were changed so that PK1 was mutationally inactive and PK2 carried an activating mutation, no response would be observed since PK2 would not be able to activate PK1, which carries the inactivating mutation.

15–28*

15–29 Phosphorylation/dephosphorylation offers a simple, universal solution to the problem of controlling protein activity. In a signaling pathway the activities of several proteins must be readily switched from the off state to the on state, or vice versa. Attaching a negatively charged phosphate to a protein is an effective way to alter its conformation and activity. And it is an easy modification to reverse. It is a universal solution in the sense that one activity—that of a protein kinase—can be used to attach a phosphate, and a second activity—a protein phosphatase—can be used to remove it. During evolution the binding of kinases and phosphatases can be altered to make them specific for particular target proteins. About 2% of the genes in the human genome encode protein kinases, which presumably arose by gene duplication and modification to create appropriate specificity. Because serine, threonine, and tyrosine are common amino acids on the surfaces of proteins, target proteins can evolve to have appropriate phosphorylation sites at places that will alter their conformations. Finally, phosphorylation/dephosphorylation provides a flexible response that can be adjusted to give rapid on/off switches or more long lasting changes.

   All of these attributes of phosphorylation/dephosphorylation are missing with allosteric regulators. While it is possible in principle for small molecules to turn proteins on or off, it is not a universal solution. Specific molecules would have to be 'designed' for each target protein, which would require evolution of a metabolic pathway for synthesis and degradation of each regulatory molecule. Even if such a system evolved for one target protein, that specific solution would not help with the evolution of a system for any other target protein. In addition, regulation by binding of small molecules is very sensitive to the concentration of the regulator. For a monomeric target protein the concentration of small molecule would have to change by 100-fold to go from 9% bound to 91% bound—a minimal molecular switch (see Problem 3–49). Few metabolites in cells vary by such large amounts.

15–30*

15–31 The use of a scaffolding protein to hold the three kinases into a signaling complex increases the speed of signal transmission and eliminates cross talk between pathways; however, there is no amplification of the signal from the receptor to the third kinase. Freely diffusing kinases offers the possibility for significant signal amplification since the first kinase can phosphorylate many molecules of the second kinase, which in turn can phosphorylate many molecules of the third kinase. The speed of signal transmission is likely to be slower unless the concentration of kinases (and the potential for amplification) is high enough to compensate for their separateness. Finally, free kinases offer the potential for interaction with other signaling pathways. The organization a cell uses for a particular signaling pathway depends on what the pathway is intended to accomplish.

15–32*

15–33

1. If more than one effector molecule must bind to activate the target molecule, the response will be sharpened in a way that depends on the number of required effector molecules. At low concentrations of the effector most target proteins will have a single effector bound (and therefore be inactive). At increasing concentrations of effector the target proteins with the requisite number of bound effectors will rise sharply, giving a correspondingly sharp increase in the cellular response.
2. If the effector activates one enzyme and inhibits another enzyme that catalyzes the reverse reaction, the forward reaction will respond sharply to a gradual increase in effector concentration. This is a common strategy employed in metabolic pathways involved in energy production and consumption.
3. The above mechanisms give sharp responses, but a true all-or-none response can be generated if the effector molecule triggers a positive feedback loop so that an activated target molecule contributes to its own further activation. If the product of an activated enzyme, for example, binds to the enzyme to activate it, a self-accelerating, all-or-none response will be produced.

15–34* **Reference:** Hertel C, Muller P, Portenier M & Staehelin M (1983) Determination of the desensitization of β-adrenergic receptors by [³H]CGP-12177. *Biochem. J.* 216, 669–674.

15–35

A. If phosphorylation of the two subunits occurs independently and at equal rates, four different types of receptor will exist: nonphosphorylated receptor, receptor phosphorylated on the γ subunit, receptor phosphorylated on the δ subunit, and receptor phosphorylated on both subunits. At 0.8 mole phosphate/mole receptor, each subunit would be 40% phosphorylated and 60% nonphosphorylated. Thus the ratio of the various receptor forms would be 36% with no phosphate (0.6×0.6), 24% with only the γ subunit phosphorylated (0.6×0.4), 24% with only the δ subunit phosphorylated (0.6×0.4), and 16% with both subunits phosphorylated (0.4×0.4). At 1.2 mole phosphate/mole receptor, the ratios would be: 16% with no phosphate, 24% with the γ subunit phosphorylated, 24% with the δ subunit phosphorylated, and 36% with both subunits phosphorylated.
B. These experiments suggest that desensitization requires only one phosphate per receptor and that phosphorylation of either the γ or the δ subunit is sufficient for desensitization. For both preparations, the fraction that behaved like the untreated receptor matched best the fraction calculated to carry no phosphate: 36% versus 36% at 0.8 mole phosphate/mole receptor

and 18% versus 16% at 1.2 mole phosphate/mole receptor. This result suggests that phosphorylation of either subunit is sufficient to promote desensitization. If a specific subunit were required to be phosphorylated, then the expected fractions behaving like the untreated receptor would have been 60% (24% + 36%) at 0.8 mole phosphate/mole receptor and 40% (24% + 16%) at 1.2 mole phosphate/mole receptor.

Reference: Huganir RL, Delcour AH, Greengard P & Hess GP (1986) Phosphorylation of the nicotine acetycholine receptor regulates its rate of desensitization. *Nature* 321, 774–776.

SIGNALING THROUGH G-PROTEIN-LINKED CELL-SURFACE RECEPTORS

15–36* Figure 15–44.

Reference: Lefkowitz RJ, Limbird LE, Mukherjee C & Caron MG (1976) The β-adrenergic receptor and adenylate cyclase. *Biochim. Biophys. Acta* 457, 1–39.

15–37 The mutant G protein would be constantly active. Each time the α subunit hydrolyzed GTP to GDP, the GDP would spontaneously dissociate, allowing GTP to bind and reactivate the α subunit. Normally, GDP is tightly bound by the α subunit, which keeps the G protein in its inactive state until release of GDP is stimulated by interaction with an appropriate G-protein-linked receptor.

15–38*

15–39 RGS proteins are GAPs that have a critical role in shutting off G protein responses in animals and yeasts. They stimulate the GTPase activity of G proteins, converting them to their inactive (GDP-bound form), and thereby limit the duration of a response.

15–40* **Reference:** Kurjan I (1992) Pheromone response in yeast. *Annu. Rev. Biochem.* 61, 1097–1129.

15–41 The 'cyclic' in cyclic AMP refers to the ring of atoms formed by the phosphorus atom, its two oxygen atoms, and the carbons at the 3′, 4′, and 5′ positions of the ribose sugar (Figure 15–45). The ball-and-stick representations above and below the chemical formula give a more accurate representation of the chemical structure. The six-member phosphodiester ring is fused to the five-member ribose ring, forming a fairly planar structure that resembles the adenine ring in size and shape. In the more common representation (center) the phosphodiester ring looks very strained, but in reality it's not.

15–42*

15–43 Since β-adrenergic receptors are coupled to adenylyl cyclase through a G protein, a reasonable guess would be GTP. Activation of the receptor by isoproteronol would stimulate the G protein to exchange bound GDP for free GTP. In the absence of free GTP the G protein would not be activated and thus would not in turn activate adenylyl cyclase. The requirement for GTP in the receptor-mediated activation of adenylyl cyclase was one of the original clues that led to the discovery of G proteins.

15–44* **References:** Federman AD, Conklin BR, Schrader KA, Reed RR & Bourne HR (1992) Hormonal stimulation of adenylyl cyclase through G_i protein βγ subunits. *Nature* 356, 159–161.

Tang W-J & Gilman AG (1991) Type-specific regulation of adenylyl cyclase by G protein βγ subunits. *Science* 254, 1500–1503.

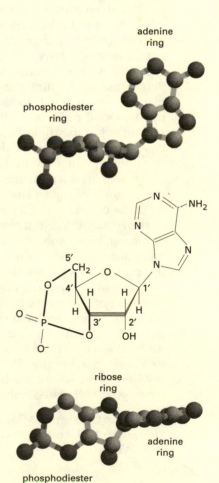

Figure 15–45 Chemical formula and ball-and-stick representations of cyclic AMP (Answer 15–41). The ball-and-stick are rotated 90° relative to each other to illustrate the similarities between the adenine ring and the fused ribose and phosphodiester rings.

15–45   False. It is the substrates for PKA, not PKA itself, that differ in different cell types.

15–46*

15–47

   A. If PKA were essential in hamster cells, it would have been impossible to isolate mutants that lack the enzyme, since they would not survive. Thus PKA is not essential to these hamster cells (nor is it essential in several other cell lines in which such mutants have been isolated). One should not conclude, however, that PKA is not essential for the organism. There are many examples of enzyme defects that have minimal consequences for cells in culture but that severely affect the intact organism.

   B. Mutations that eliminate the catalytic subunit would be unresponsive to high levels of cyclic AMP and therefore resistant to cyclic AMP. These mutations would be recessive, for in the presence of the wild-type catalytic subunit, cyclic AMP responsiveness would be restored. (It might seem that mutations that eliminate the regulatory subunits would be the same. Without the regulatory subunits, however, PKA would be active at all times, which is a lethal condition—excess PKA activity is the reason that these cells are killed by high levels of cyclic AMP. Thus, cell lines without regulatory subunits would not have been isolated in the first place.)

   Dominant mutations are somewhat more difficult to explain. In general, dominance indicates an altered activity rather than lack of activity. Dominant mutations have been found in both the regulatory and catalytic subunits. A possible dominant mutation in the regulatory subunit is one that increases its affinity for the catalytic subunit. Such a mutation might plausibly respond only to high levels of cyclic AMP (required to displace the tightly bound regulatory subunit), and it would be dominant because the mutant subunits would bind the catalytic subunits at the low cyclic AMP concentrations that would displace the normal regulatory subunits.

   Dominant mutations in the catalytic subunit are more difficult to explain. One possibility is a mutant catalytic subunit that binds its regulatory subunits more tightly, giving an altered responsiveness to cyclic AMP. Its dominance might be understood if a combination of a mutant catalytic subunit with a normal catalytic subunit rendered the heterodimer mutantlike in its binding to the regulatory subunits. If this were the case, an even mixture of mutant and normal catalytic subunits would be expected to have only one-quarter the normal PKA activity (only one-quarter of the catalytic dimers would have two wild-type subunits).

   C. These experimental results generally support the contention that all cyclic AMP effects are mediated through PKA, but they fall short of proving the point. Similar studies in a variety of other cell lines have also identified mutants solely in the PKA pathway. Furthermore, cells that completely lack PKA show none of the effects normally associated with cyclic AMP. Other roles for cyclic AMP have been identified; for example, olfactory neurons have cyclic AMP-gated cation channels.

   Reference: Gottesman MM (1985) Genetics of cyclic-AMP-dependent protein kinases. In Molecular Cell Genetics (MM Gottesman ed), pp 711–743. New York: Wiley.

15–48*

15–49   Both glycogen and triglyceride breakdown depend on cyclic AMP-dependent signaling pathways. By inhibiting cyclic AMP phosphodiesterase, caffeine increases cyclic AMP levels to promote glycogen breakdown in muscle and liver and to promote triglyceride breakdown in fat cells. The effects on glycogen are not so important as the effects on triglycerides. By breaking down triglycerides to release fatty acids earlier in the race, runners decrease their dependence on carbohydrate oxidation, thereby preserving carbohydrates for use throughout the race.

15–50* **References:** Byers D, Davis RL & Kiger JA (1981) Defect in cyclic AMP phosphodiesterase due to the *dunce* mutation of learning in *Drosophila melanogaster*. *Nature* 289, 79–81.

Chen C-N, Denome S & Davis RL (1986) Molecular analysis of cDNA clones and the corresponding genomic coding sequences of the *Drosophila dunce⁺* gene, the structural gene for cyclic AMP phosphodiesterase. *Proc. Natl. Acad. Sci. USA* 83, 9313–9317.

15–51

A. Observations 1, 2, and 3 are expected under both the proposed pathways for arachidonic acid production. Observations 4 and 5 distinguish between the pathways. If arachidonic acid arose by cleavage from the diacylglycerol produced along with IP_3, then its production should have paralleled that of IP_3 in both experiments; that is, arachidonic acid production should have been blocked by treatment with an inhibitor of phospholipase C (since no diacylglycerol would have been produced), and it should have been unaffected by pertussis toxin (since diacylglycerol production was unaffected).

B. Observations 1 and 2 indicate that arachidonic acid production is dependent on the α_1-adrenergic receptor. The stimulation of arachidonic acid production by GTPγS suggests that a G protein is involved in the pathway; the inhibition of arachidonic acid production by pertussis toxin confirms this. However, the toxin-sensitive G protein probably is not the same one that activates phospholipase C, since pertussis toxin does not affect phospholipase-C activity. The link between the G protein and phospholipase A_2 is not defined by these experiments.

Reference: Burch RM., Luini A & Axelrod J (1986) Phospholipase A_2 and phospholipase C are activated by distinct GTP-binding proteins in response to α_1-adrenergic stimulation in FRTL5 thyroid cells. *Proc. Natl. Acad. Sci. USA* 83, 7201–7205.

15–52* Figure 15–46.

Reference: Nishizuka Y (1983) Calcium, phospholipid turnover and transmembrane signaling. *Phil. Trans. R. Soc. Lond. (Biol.)* 302, 101–112.

15–53 IP_3-triggered Ca^{2+} responses are terminated in two ways: by adding or removing phosphates from IP_3 and by pumping Ca^{2+} outside of the cell or into the ER.

15–54*

15–55 EGTA, by chelating Ca^{2+}, would be expected to interfere with signaling pathways that use Ca^{2+} as a second messenger. Glucagon triggers glycogen breakdown in liver via a cyclic AMP pathway and thus would not be affected by EGTA. By contrast, vasopressin signals glycogen breakdown via a Ca^{2+} pathway and thus would be blocked by injection of EGTA.

15–56*

15–57

A. The difference in the behavior of myosin light-chain kinase when purified in the presence and absence of protease inhibitors suggests that the regulatory domain of the kinase is sensitive to cleavage by proteases in the cell extract. Activation of the kinase by proteolytic removal of the regulatory domain suggests that the domain in some way covers the active site. This regulatory 'flap' moves out of the way upon Ca^{2+}/calmodulin binding. This theme of a regulatory flap covering an active site is fairly common.

B. Upon entry into the cytosol, Ca^{2+} binds to calmodulin, causing a change in its conformation. The Ca^{2+}/calmodulin complex binds to the regulatory domain of myosin light-chain kinase, exposing the active site of the enzyme. The activated kinase then phosphorylates myosin light chains, altering their conformation so that they can bind actin and initiate contraction. As you might predict, contraction is terminated through the action of phosphatases that remove the phosphate from the myosin light chains.

Reference: Adelstein RS & Klee CB (1980) Smooth muscle myosin light-chain kinase. In Calcium and Cell Function (WY Cheung ed), Vol. 1, pp 167–182. New York: Academic Press.

15–58*

15–59

A. The complete G protein does not activate the K+ channels in the absence of acetylcholine presumably because, like other trimeric G proteins, the active portion is inhibited by one of the subunits. The ability of the $G_{\beta\gamma}$ subunit to open the K+ channel in the absence of acetylcholine and GTP suggests that it is the active portion of the G protein. This is different from the active component (G_{α}) of G-protein-linked receptors that activate adenylyl cyclase.

B. The opening of the K+ channel in the presence of GppNp and absence of acetylcholine may seem somewhat surprising, since the release of GDP and the binding GTP by G proteins normally are stimulated by an activated receptor. Even in the absence of an activated receptor, however, G proteins exchange their bound nucleotides with nucleotides in the cytoplasm. Exchange is slow, and any bound GTP is quickly hydrolyzed in the absence of an activated receptor, thereby keeping the channel closed. The K+ channels open slowly when GppNp is present because, each time a GDP is released and a GppNp is bound, the G protein is locked into an active form. Over the course of a minute, enough G protein is activated in this way to open the K+ channels in the absence of acetylcholine.

C. A simple scheme for the G-protein-mediated activation of K+ channels by acetylcholine is shown in Figure 15–47.

References: Logothetis DE, Kurachi Y, Galper J, Neer EJ & Clapham DE (1987) The βγ subunits of GTP-binding proteins activate the muscarinic K+ channel in heart. *Nature* 325, 321–326.

Reuveny E, Slesinger PA, Inglese J, Morales JM, Iniguez-Lluhi JA, Lefkowitz RJ, Bourne HR, Jan YN & Jan LY (1994) Activation of the cloned muscarinic potassium channel by G protein βγ subunits. *Nature* 370, 143–146.

15–60* **References:** Fung BK-K & Stryer L (1980) Photolyzed rhodopsin catalyzes the exchange of GTP for bound GDP in retinal rod outer segments. *Proc. Natl. Acad. Sci. USA* 77, 2500–2504.

Fung BK-K, Hurley JB & Stryer L (1981) Flow of information in the light-triggered cyclic nucleotide cascade of vision. *Proc. Natl. Acad. Sci. USA* 78, 152–156.

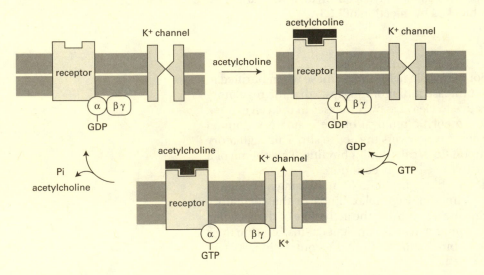

Figure 15–47 Diagram illustrating activation of K+ channels in heart by acetylcholine (Answer 15–59).

15–61

A. An inhibitor of cyclic GMP phosphodiesterase would prevent the reduction in cyclic GMP that normally occurs in response to light activation of rhodopsin. High levels of cyclic GMP would keep the Na^+ channels open, preventing the membrane hyperpolarization that is essential for the visual response.

B. A nonhydrolyzable analog of GTP would lead to prolonged activation of transducin in response to activated rhodopsin. Continued activation of transducin would keep cyclic GMP phosphodiesterase high, which would lead in turn to a protracted decrease in cyclic GMP, a prolonged hyperpolarization of the membrane, and an extended visual response.

C. An inhibitor of rhodopsin-specific kinase would prolong the visual response by increasing the signaling lifetime of the activated form of rhodopsin. Normally, rhodopsin-specific kinase adds a phosphate to the cytoplasmic tail of rhodopsin, inhibiting the interaction of activated rhodopsin with transducin.

15–62*

15–63 True. Nuclear receptors with a bound ligand bind to DNA sequences in the genome to activate (or inhibit) a specific gene; thus, there is a one-to-one correspondence between the signal and the response. By contrast, signaling pathways that involve enzymes or ion channels can significantly amplify a signal. One activated protein kinase, for example, can phosphorylate many molecules of its target.

15–64*

15–65 The β-adrenergic receptor is turned off directly by the conformational change that occurs when adrenaline is no longer bound. The G_α subunit becomes inactive when it hydrolyzes its attached GTP to GDP, which allows it to reassociate with the βγ subunits. Adenylyl cyclase becomes inactive as soon as G_α dissociates. Cyclic AMP is constantly being converted to AMP by cyclic AMP phosphodiesterase. In the absence of its continued synthesis, cyclic AMP quickly returns to its pre-stimulated level. At low concentrations, cyclic AMP dissociates from the regulatory subunits of PKA, which rebind the catalytic subunits to turn off PKA. In the absence of ongoing phosphorylation of phosphorylase kinase, a protein phosphatase removes the phosphates, turning off phosphorylase kinase. Similarly, a protein phosphatase quickly removes phosphates from glycogen phosphorylase, thereby turning it off.

SIGNALING THROUGH ENZYME-LINKED CELL-SURFACE RECEPTORS

15–66*

15–67 Ephrins and ephrin receptors are both membrane-bound proteins. An ephrin on one cell binds to an ephrin receptor on a second cell, triggering the ephrin receptor to initiate a signaling pathway in the second cell. Binding to the ephrin receptor also activates the ephrin so that it initiates a signaling pathway in the signaling cell. A ligand–receptor interaction that initiates signaling pathways in both interacting cells is called bidirectional signaling.

15–68* Figure 15–48.

15–69 False. Ligand binding usually causes a receptor tyrosine kinase to assemble into dimers, which activates the kinase domain. The receptors then phosphorylate themselves to initiate the intracellular signaling cascade. In some cases, the insulin receptor for example, the receptor exists as a dimer and ligand binding propagates a conformational change through the pair of membrane α helices.

15–70* **References:** Honegger AM, Kris RM, Ullrich A & Schlessinger J (1989) Evidence that autophosphorylation of solubilized receptors for epidermal growth factor is mediated by intermolecular cross-phosphorylation. *Proc. Natl. Acad. Sci. USA* 86, 925–929.

Honegger AM, Schmidt A, Ullrich A & Schlessinger J (1990) Evidence for epidermal growth factor (EGF)-induced intermolecular autophosphorylation of the EGF receptors in living cells. *Mol. Cell. Biol.* 10, 4035–4044.

15–71

A. This method could not be used as described for detecting proteins that bind to SH2 domains, which bind only to target proteins that carry a phosphotyrosine—a modification that doesn't occur normally in bacteria. One way to make the screen work would be to incubate the filters first with a protein tyrosine kinase and ATP. Alternatively, a protein tyrosine might be engineered into *E. coli* so that it could be turned on at the same time the cDNA library was expressed.

B. The main difference between protein–short-sequence interactions and the subunit–subunit interactions in multisubunit enzymes lies in their stability and reversibility. Both types of interaction depend largely on the total number and aggregate strength of the weak bonds involved in their formation. Large contact surfaces such as those found among subunits in multisubunit enzymes make for very stable structures, whereas most of the examples of short-sequence recognition are more transient, and in some cases conditional, as in the interaction of SH2 domains with phosphotyrosine-containing proteins.

Reference: Ren R, Mayer BJ, Cicchetti P & Baltimore D (1993) Identification of a ten-amino-acid proline-rich SH3 binding site. *Science* 259, 1157–1161.

15–72* **References:** Valius M & Kazlauskas A (1993) Phospholipase C-γ1 and phosphatidylinositol 3 kinase are the downstream mediators of the PDGF receptor's mitogenic signal. *Cell* 73, 321–334.

Valius M, Secrist JP & Kazlauskas A. (1995) The GTPase-activating protein of Ras suppresses platelet-derived growth factor β receptor signaling by silencing phospholipase C-γ1. *Mol. Cell. Biol.* 15, 3058–3071.

15–73 You would expect to see several differences. (1) You would expect a high background of Ras activity in the absence of an extracellular signal because Ras cannot be turned off efficiently. Since Ras activity depends on the balance between its binding to GTP and its GAP-enhanced hydrolysis of GTP, the balance would be somewhat more in favor of the GTP bound (active) form than normal. (2) As some Ras molecules will already be in their GTP-bound form, Ras activity in response to an extracellular signal would be greater than normal, but would saturate when all Ras molecules were converted to the GTP-bound form. (3) The response to signal would be less rapid because the signal-dependent increase in GTP-bound Ras would occur over an elevated background of preexisting GTP-bound Ras. (4) The response would be expected to be more prolonged than normal and to persist for a while even after the extracellular signal was removed because of the slower rate of conversion of GTP-bound Ras to its inactive GDP-bound form.

15–74*

15–75 In order for activation of Ras to depend on inactivation of a GAP, both the GAP and the GEF would need to be active in the absence of the signal. In this way the GEF would constantly load GTP onto Ras and the GAP would keep the concentration of Ras–GTP low by constantly inducing GTP hydrolysis to return Ras to its GDP-bound state. Under these conditions inactivation of the GAP would result in a rapid increase in Ras–GTP levels, allowing rapid signaling. Although this would be a perfectly effective way to regulate the levels of active Ras, it would be wasteful of energy. In order to keep Ras in its inactive state, GTP would be constantly hydrolyzed to GDP, which would then need to be reconverted to GTP (by ATP)—a drain on cellular energy metabolism. Regulation by activation of a GEF avoids this problem.

Although avoiding constant GTP hydrolysis is a rational explanation, eucaryotic cells are notoriously profligate in their energy expenditures. At several points in energy metabolism, for example, they operate so-called 'futile' cycles that hydrolyze ATP as a means for rapid regulation of the flux through metabolic pathways. Thus, it could be that constant hydrolysis of GTP by a Ras GEF and GAP would not unduly tax the cell's energy budget. Perhaps the cell's method of regulating Ras by controlling the activity of a GEF is simply an evolutionary happenstance.

15–76*

15–77 The very steep response curve for activation of MAPK converts it into a molecular switch. Thus, MAPK goes from inactive to active over a very narrow range of input stimulus. This kind of behavior keeps the cascade turned off below a threshold concentration of the input signal, yet delivers a maximum response once that threshold is exceeded.

Reference: Huang C-Y F & Ferrell JE (1996) Ultrasensitivity in the mitogen-activated protein kinase cascade. *Proc. Natl. Acad. Sci. USA* 93, 10078–10083.

15–78*

15–79 The analysis of individual oocytes shows clearly that the response to progesterone is all-or-none, with no oocytes having a partially activated MAP-kinase. Thus the graded response in the population results from an all-or-none response in individual oocytes, with different mixtures of fully mature or immature oocytes giving rise to intermediate levels of MAP-kinase activation (Figure 15–49). It is not so clear why individual oocytes respond differently to different concentrations of progesterone, although there is significant variability among oocytes in terms of age and size (and presumably in the number of progesterone receptors and the concentrations of components of the MAP-kinase cascade and downstream targets).

Whether a graded response in a population of cells indicates a graded response in each cell or a mixture of all-or-none responses is a question that arises in many contexts in biology.

Reference: Ferrell JE & Machleder EM (1998) The biochemical basis of an all-or-none cell fate switch in *Xenopus* oocytes. *Science* 280, 895–898.

15–80* Figure 15–50.

15–81 False. PI3-kinase phosphorylates inositol head groups at a position (number 3 on the inositol ring) that is not phosphorylated in IP_3. Phosphorylation at this site serves an entirely different function; it creates inositol head groups that can serve as docking sites for intracellular signaling proteins.

15–82* **Reference:** Toker A & Newton AC (2000) Akt/protein kinase B is regulated by autophosphorylation at the hypothetical PDK-2 site. *J. Biol. Chem.* 275, 8271–8274.

15–83

A. Many transcription factors are phosphorylated in response to cell stimulation: in some instances, this results in tighter binding to DNA, whereas in others it creates an acidic activation domain that promotes transcription. In this case phosphorylation seems to be necessary for DNA binding because anti-phosphotyrosine antibodies and phosphatase treatment inhibit DNA binding.

B. Free phosphotyrosine will bind to the SH2 domain of the transcription factor. If simple occupancy of the SH2 domain by a phosphotyrosine were all that was required, free phosphotyrosine would be expected to activate the transcription factor instead of inhibiting it. This implies that the SH2 domain must bind to the phosphotyrosine on the transcription factor. By interfering

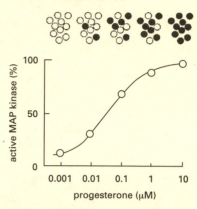

Figure 15–49 Graded or all-or-none responses in individual oocytes that give rise to a graded response in the population (Answer 15–79).

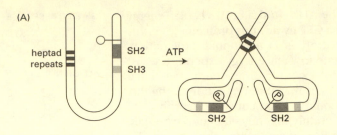

(A)

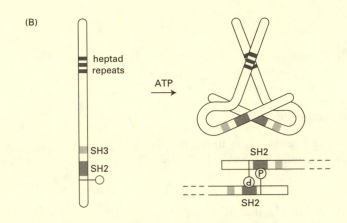

(B)

Figure 15–51 Two general ways for tyrosine phosphorylation to promote dimerization of the transcription factor (Answer 15–83).

with this interaction, free phosphotyrosine interferes with the factor's ability to bind to DNA.

C. Tyrosine phosphorylation of the transcription factor could promote its dimerization in two general ways. It could be that in the absence of phosphorylation the heptad repeats are masked by the tertiary structure of the protein and that phosphorylation and intramolecular binding to the SH2 domain causes a conformational change that exposes the heptad repeats and allows dimerization (Figure 15–51A). Alternatively, it could be that the heptad repeats do not promote a strong enough interaction for formation of a stable dimer and that phosphorylation and intermolecular binding to the SH2 domain is required (Figure 15–51B).

The dyad symmetry of the DNA sequence element suggests that it is composed of two half-sites for binding. The formation of the dimer allows the transcription factor to interact with both half-sites simultaneously, which greatly increases the strength of binding.

Reference: Sadowski HB, Shuai K, Darnell JE & Gilman MZ (1993) A common nuclear signal transduction pathway activated by growth factor and cytokine receptors. *Science* 261, 1739–1744.

15–84*

15–85 Replacing the extracellular domain of a receptorlike tyrosine phosphatase with the binding domain for a defined receptor (for example, the EGF receptor) converts the protein into a receptor for a known ligand (in this case, EGF). Since receptorlike tyrosine phosphatases are single-pass transmembrane proteins, they are likely to be regulated by dimerization, as the EGF receptor is. If this assumption is correct, then dimerization via the grafted EGF binding domain will bring the cytoplasmic domains of the hybrid protein together. If proximity is the triggering event, then it shouldn't matter how they are dimerized. For single-pass membrane receptors, making hybrids of this sort offers a general solution to studying their signaling pathways without first knowing the ligand involved.

15–86*

15–87 False. Atrial natriuretic peptides bind directly to a receptor guanylyl cyclase, which activates the cyclase to produce cyclic GMP, which then activates PGK.

15–88* **Reference:** Manson MD, Blank V, Brade G & Higgins CF (1986) Peptide chemotaxis in *E. coli* involves the Tap signal transducer and the dipeptide permease. *Nature* 321, 253–258.

15–89

A. The two cloned receptors, normal and truncated, both carry out signal transduction like the receptors in wild-type bacteria. Upon addition of aspartate, all three kinds of bacteria immediately suppress changes in direction of rotation. Thus the presence of the attractant (aspartate) in the medium is communicated to the flagella in all three kinds of bacteria.

B. The adaptive properties of bacteria containing the cloned receptors are very different from wild-type bacteria. Wild-type bacteria return to their normal rate of tumbling (reversal of direction of rotation) within 3 minutes. Bacteria with the cloned normal receptor return to the normal rate of tumbling only after about 50 minutes, and bacteria with the truncated receptor do not begin to tumble even after more than 3 hours. Thus bacteria with the cloned normal receptor adapt more slowly than wild-type bacteria, whereas bacteria with the cloned truncated receptor evidently do not adapt.

C. The alterations in adaptation in bacteria with the cloned receptors suggest differences in the methylation rates or extents. The inability of the truncated receptor to be methylated provides a molecular basis for the inability of bacteria to adapt to a high level of aspartate. The molecular basis for the difference between wild-type bacteria and bacteria with the cloned normal receptor is more subtle. The difference in time of adaptation between the two kinds of bacteria is about 17-fold (3 minutes versus 50 minutes), which is about the same as the difference in numbers of receptors per cell. Thus a reasonable explanation is that it takes the receptor methylase 15 times longer to methylate the more abundant normal receptor.

Reference: Russo AF & Koshland DE (1983) Separation of signal transduction and adaptation functions of the aspartate receptor in bacterial sensing. *Science* 220, 1016–1020.

SIGNALING PATHWAYS THAT DEPEND ON REGULATED PROTEOLYSIS

15–90*

15–91 The extracellular fragments of APP aggregate to form amyloid plaques outside the cells. The amyloid plaques are thought to interfere with nerve function, leading to the characteristic loss of mental acuity that is typical of Alzheimer's disease.

15–92* **Reference:** Aberle H, Bauer A, Stappert J, Kispert A & Kemler R (1997) β-Catenin is a target for the ubiquitin-proteasome pathway. *EMBO J.* 16, 3797–3804.

15–93 These results indicate that phosphorylation of β-catenin sensitizes it for degradation in proteasomes. If phosphorylation were irrelevant to degradation or if it protected against degradation, slower migrating, ubiquitylated forms of β-catenin should have been present in cell lines that were unable to phosphorylate β-catenin.

15–94* **Reference:** Winter CG, Wang B, Ballew A, Royou A, Karess R, Axelrod JD & Luo L (2001) *Drosophila* Rho-associated kinase (Drok) links Frizzled-mediated planar cell polarity signaling to the actin cytoskeleton. *Cell* 105, 81–91.

15–95 The *APC* gene is a tumor suppressor gene. Its normal function is to inhibit β-catenin by helping to hold it in the cytosol until a proper signal has been received. When both copies of the *APC* gene are inactivated, β-catenin is free to enter the nucleus in the absence of any signal, leading to uncontrolled stimulation of its target genes.

15–96* **Reference:** Porter JA, von Kessler DP, Ekker SC, Young KE, Lee JJ, Moses K & Beachy PA (1995) The product of *hedgehog* autoproteolytic cleavage active in local and long-range signaling. *Nature* 374, 363–366.

15–97 The N-terminus of the hedgehog precursor protein remains associated with the cells when cleaved naturally, but it is secreted when it is synthesized from the truncated construct. These data do not define the nature of the cell association. As a part of its cleavage mechanism, the N-terminal fragment could, for example, become associated with a component of the cell, either inside the cell or on the membrane; alternatively, it could be trapped in an intracellular compartment. The actual explanation is very surprising; the cleavage mechanism uses a membrane cholesterol molecule to complete the cleavage, leaving the N-terminal fragment attached to the membrane via a covalent linkage between glycine 257 and cholesterol.

Reference: Porter JA, von Kessler DP, Ekker SC, Young KE, Lee JJ, Moses K & Beachy PA (1995) The product of *hedgehog* autoproteolytic cleavage active in local and long-range signaling. *Nature* 374, 363–366.

15–98* **Reference:** Porter JA, von Kessler DP, Ekker SC, Young KE, Lee JJ, Moses K & Beachy PA (1995) The product of *hedgehog* autoproteolytic cleavage active in local and long-range signaling. *Nature* 374, 363–366.

15–99

1. The latent gene regulatory protein is attached to the membrane as part of the covalent structure of a transmembrane protein. When a valid signal is received the regulatory protein is cleaved from the membrane protein and enters the nucleus. Notch is an example.
2. The latent gene regulatory protein is actively degraded in the cytosol. When a valid signal is received, the protein is stabilized against degradation, allowing it to enter the nucleus. β-Catenin is an example.
3. The latent gene regulatory protein is anchored to a cytosolic structure and released in response to an appropriate signal. Cubitus interruptus is an example.
4. The latent gene regulatory protein is bound to a protein that masks its nuclear localization sequence. Upon receipt of an appropriate signal, the inhibitory protein is modified so that the nuclear localization sequence is exposed, allowing the gene regulatory protein to be transported into the nucleus. NF-κB is an example.

SIGNALING IN PLANTS

15–100*

15–101 If the basic mechanisms of cell communication arose in response to multicellularity, then fungi must have separated from the animal lineage after multicellularity evolved. This reasoning would suggest that unicellular fungi may have been derived from multicellular precursors. Not very long ago—and to great surprise—it was shown that *Saccharomyces cerevisiae* will form multicellular filamentous forms. Many members of the fungal kingdom have this ability, termed dimorphism, to switch between two morphological forms: a cellular form and a multicellular invasive form.

Reference: Madhuri HD & Fink GR (1998) The control of filamentous differentiation and virulence in fungi. *Trends Cell Biol.* 8, 348–353.

15–102*

15–103

A. The accepted explanation for the ability of an antisense RNA to block expression of the normal gene is that the two RNAs—the antisense RNA and the normal RNA—hybridize to make a double-stranded RNA that cannot be translated. This would effectively block synthesis of the ACC synthase enzyme and prevent formation of ethylene. But this may not be the true mechanism. In some plants a phenomenon called RIPing pairs duplicated sequences in meiosis and introduces mutations into both. Thus, it may be that the normal ACC synthase gene is inactivated in your transgenic tomatoes.

B. In all likelihood!

Reference: Oelier PW, Min-Wong L, Taylor LP, Pike DA, Theologis A (1991) Reversible inhibition of tomato fruit senescence by antisense RNA. *Science* 254, 437–439.

15–104*

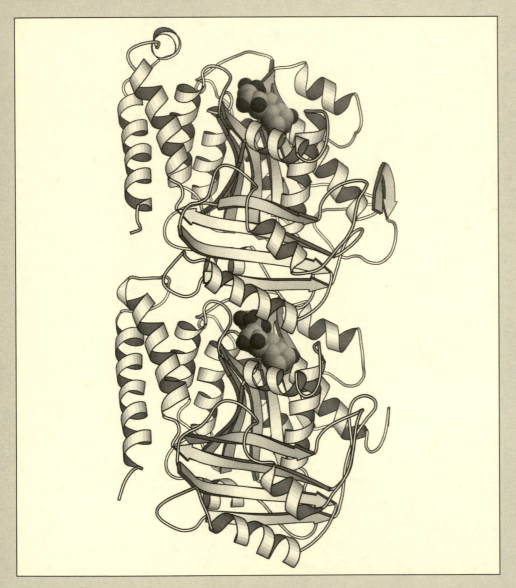

The tubulin dimer is the basic unit of microtubules. The β-subunit (with bound GDP) is above the α-subunit (containing GTP) in this picture. Coordinates from 1JFF determined by J Lowe, H Li, KH Downing & E Nogales *J. Mol. Biol.* 313:1045–1057, 2001.

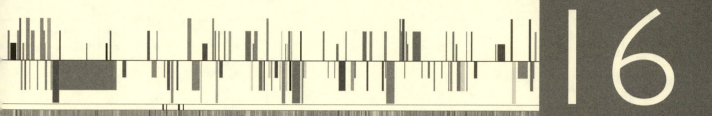

16

THE CYTOSKELETON

THE SELF-ASSEMBLY AND DYNAMIC STRUCTURE OF CYTOSKELETAL FILAMENTS

16–1 Intermediate filaments provide mechanical stability and resistance to shear stress. Microtubules determine the positions of membranous organelles and direct intracellular transport. Actin filaments determine the shape of the cell's surface and are necessary for whole-cell locomotion.

16–2*

16–3

A. Phase A corresponds to a lag phase (Figure 16–48A), during which actin monomers must assemble to form a nucleus for polymerization (thought to be a trimer of subunits). Formation of a nucleus (nucleation) is followed by rapid growth (phase B) as actin monomers are added to the ends of the growing filaments. At phase C, equilibrium is reached between the rate of addition of actin at the ends and its rate of release. Once equilibrium is reached, the concentration of free actin remains constant.

B. If the starting concentration of actin was doubled, the lag phase would be shorter, the growth phase would be more rapid (steeper), and the mass of polymer at equilibrium would be twice as great. The experimental curves generated at twice and half the initial actin concentrations illustrate these relationships (Figure 16–48A). The concentration of free actin monomers at equilibrium—the critical concentration (C_c)—would be the same regardless of the initial actin concentration. It can be estimated from the data in Figure 16–48A, as shown in Figure 16–48B.

Reference: Carlier M-F, Pantaloni D & Korn ED (1985) Polymerization of ADP- and ATP-actin under sonication and characteristics of the ATP-actin equilibrium polymer. *J. Biol. Chem.* 260, 6565–6571.

16–4* Figure 16–49.

(A) MEASUREMENTS

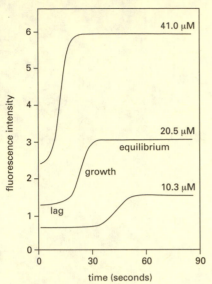

(B) CRITICAL CONCENTRATION

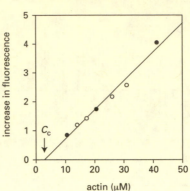

Figure 16–48 Analysis of actin polymerization (Answer 16–3). (A) Actin polymerization curves at three different actin concentrations, as indicated on the individual curves. (B) The critical concentration determined from many such experiments. The plateau values for increase in intensity of fluorescence (corrected for fluorescence of the monomers) are plotted against actin concentration. The *black circles* indicate the data shown in (A); *white circles* are additional data points from similar experiments. The critical concentration (C_c) is estimated by extrapolating to a value of zero increase in intensity.

16–5 Two tubulin dimers have a lower affinity for each other (because of a more limited number of interaction sites) than a tubulin dimer has for the end of a microtubule. At the end of an existing microtubule there are multiple possible interaction sites, both end-to-end as the tubulin dimers add to a protofilament and side-to-side as they bind to adjacent protofilaments in the microtubule lattice. Thus, to initiate a microtubule from scratch, enough tubulin dimers have to come together and remain bound to one another for long enough for other tubulin molecules to add to them. Only when several tubulin dimers have already assembled will the binding of the next subunit be favored.

16–6*

16–7 The heterotypic interactions between the protofilaments are likely to be weaker than the homotypic interactions between them. If the interactions between α-tubulin and β-tubulin were stronger than the homotypic interactions, the protofilaments would preferentially align so that heterotypic interactions, rather than homotypic ones, were maximized. If the two sets of interactions were the same strength, the arrangement of protofilaments might be mixed within the same microtubule, or two different types of microtubule—with protofilaments aligned either by homotypic or heterotypic lateral interactions—might be observed. You can imagine formation of the microtubule as building a sheet of protofilaments that curl into a microtubule by forming the seam (Figure 16–50).

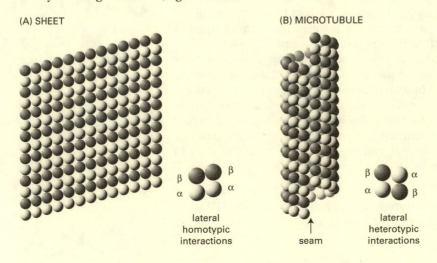

(A) SHEET

(B) MICROTUBULE

lateral homotypic interactions

seam

lateral heterotypic interactions

Figure 16–50 Interactions between protofilaments composed of αβ-tubulin dimers (Answer 16–7). (A) A sheet of protofilaments. An example of homotypic lateral interactions between αβ-tubulin dimers in the protofilaments is shown on the *right*. (B) A microtubule. An example of heterotypic lateral interactions between αβ-dimers at the seam is shown on the *right*.

Reference: Desai A & Mitchison TJ (1997) Microtubule polymerization dynamics. *Annu. Rev. Cell Dev. Biol.* 13, 83–117.

16–8* Figure 16–51.

References: Pollard TD, Blanchoin L & Mullins RD (2000) Molecular mechanisms controlling actin filament dynamics in nonmuscle cells. *Annu. Rev. Biophys. Biomol. Struct.* 29, 545–576.

Oosawa F (2001) A historical perspective of actin assembly and its interactions. *Res. Prob. Cell Diff.* 32, 9–21.

16–9 These experiments show that αβ-tubulin dimers are oriented with β-tubulin at the plus ends and α-tubulin at the minus ends. Because the GTP that is bound to the α-tubulin monomer is physically trapped at the dimer interface, it is never hydrolyzed or exchanged. By contrast, the GTP in β-tubulin is hydrolyzed and can be exchanged. Thus, when a microtubule is exposed to GTP-coated fluorescent beads, the GTP can bind to some of the β-tubulin subunits exposed at the end of the microtubule. Finding the fluorescent beads at the plus ends indicates that the αβ-tubulin dimer must be oriented with the β-tubulin monomer at the plus end. The presence of the beads only at one end, and not all along the microtubule, indicates that GTP can be exchanged only at the exposed ends.

Similarly, the presence of beads coated with antibodies specific for α-tubulin at the minus ends indicates that the αβ-tubulin dimer must be oriented with the α-tubulin monomer at the minus end. The presence of beads only at one end indicates that the peptide of α-tubulin with which the antibody reacts is buried at the interface between adjacent αβ-tubulin dimers, and thus is exposed only at the end.

References: Mitchison TJ (1993) Localization of an exchangeable GTP binding site at the plus end of microtubules. *Science* 261, 1044–1047.

Fan J, Griffiths AD, Lockhart A, Cross RA & Amos LA (1996) Microtubule minus ends can be labeled with a phage display antibody specific to α-tubulin. *J. Mol. Biol.* 259, 325–330.

Nogales E, Whittaker M, Milligan RA & Downing KH (1999) High-resolution model of the microtubule. *Cell* 96, 79–88.

16–10* **Reference:** Detrich WH, Parker SK, Williams RC, Nogales E & Downing KH (2000) Cold adaptation of microtubule assembly and dynamics. *J. Biol. Chem.* 275, 37038–37047.

16–11 Subunit 1 will add faster to the right end of the polymer than to the left end, subunit 2 will add to both ends at equal rates, and subunit 3 will add faster to the left end of the polymer. A difference in growth rates at the two ends reflects a change in conformation of the free subunit as it adds to the polymer. For example, subunit 1 can add to the right end of the polymer through an existing binding site (its pointed end) and change conformation later. To add to the left end, however, it must make the conformational change before it adds or during addition.

For the simple polymerization described here, both ends must grow or shrink; there is no concentration of subunit that can allow one end to grow while the other shrinks. This is because the conformations of the subunits at the two ends of the polymer are identical and they involve identical contacts. You could not tell from which end a free subunit derived. Thus, the ΔG for subunit loss, which determines the equilibrium constant for subunit association at an end, must be the same for both ends.

16–12* **Reference:** Horio T & Hotani H (1986) Visualization of the dynamic instability of individual microtubules by dark-field microscopy. *Nature* 321, 605–607.

16–13

A. An enzyme-catalyzed reaction reaches a plateau when the enzyme becomes saturated with substrate. Beyond that point an increase in substrate concentration cannot increase the rate of the reaction, because the enzyme is already working at maximum capacity. In contrast, growth of an actin filament does not saturate. Each time a monomer is added to the filament, a new site for addition of the next monomer is created. Addition of new monomers occurs through productive collisions with the end of the filament. The number of productive collisions increases linearly with concentration of actin monomers.

B. At concentration A, both ends would shrink. At concentration B, the minus end would shrink and the plus end would be unchanged. At concentration C, the plus end would grow and the minus end would shrink. At concentration D, the plus end would grow and the minus end would remain unchanged. At concentration E, both ends would grow, with the plus end growing faster than the minus end. The critical concentration is the concentration of actin at which an end neither shrinks nor grows. For plus ends the critical concentration is concentration B (0.12 µM); for the minus ends the critical concentration is concentration D (0.62 µM). At any concentration between these two critical concentrations the filament would exhibit treadmilling. At concentration C the plus end would grow at exactly the same rate as the minus end would shrink, giving treadmilling with no change in length of the filament.

References: Pollard TD, Blanchoin L & Mullins RD (2000) Molecular mechanisms controlling actin filament dynamics in nonmuscle cells. *Annu. Rev. Biophys. Biomol. Struct.* 29, 545–576.

Pollard TD (1986) Rate constants for the reactions of ATP- and ADP-actin with the ends of actin filaments. *J. Cell Biol.* 103, 2747–2754.

16–14* **Reference:** Chretien D, Fuller SD & Karsenti E (1995) Structure of growing microtubule ends: Two-dimensional sheets close into tubes at variable rates. *J. Cell Biol.* 117, 1311–1328.

16–15

A. The microtubule is shrinking because it has lost its GTP cap; that is, the tubulin subunits at its end are all in their GDP-bound form. GTP-loaded tubulin subunits from solution will still add to this end, but they will be short-lived—either because they hydrolyze their GTP or because they fall off as the microtubule rim around them disassembles. If, however, enough GTP-loaded subunits are added quickly enough to cover up the GDP-containing tubulin subunits at the microtubule end, then a new GTP cap can form and regrowth is favored.

B. The rate of addition of GTP-tubulin will be greater at higher tubulin concentrations. The frequency with which shrinking microtubules switch to the growing mode will therefore increase with increasing tubulin concentration. The consequence of this regulation is that the system is self-balancing. The more microtubules shrink (resulting in a higher concentration of the free tubulin), the more frequently microtubules will start to grow. As microtubules grow, the concentration of free tubulin will fall and the rate of GTP-tubulin addition will slow down. At some point GTP hydrolysis will catch up with new GTP-tubulin addition, the GTP cap will be destroyed, and the microtubule will switch to the shrinking mode.

C. If only GDP were present, microtubules would continue to shrink and eventually disappear, because tubulin dimers with GDP have very low affinity for each other and will not add stably to microtubules.

D. If a GTP analog that cannot be hydrolyzed is present, microtubules will continue to grow until all free tubulin subunits have been used up.

16–16* **Reference:** Detrich WH, Parker SK, Williams RC, Nogales E & Downing KH (2000) Cold adaptation of microtubule assembly and dynamics. *J. Biol. Chem.* 275, 37038–37047.

16–17 This simple purification procedure takes advantage of the properties of tubulin and microtubules. At 0°C microtubules dissociate into αβ-tubulin dimers, which remain in the supernatant when subjected to high centrifugal force. In the presence of GTP at 37°C the tubulin dimers polymerize into microtubules, which are large enough to pellet when subjected to high centrifugal force. This procedure purifies tubulin away from all other cell components. Large cell components are discarded each time the supernatant is saved; small cell components, including other proteins, are discarded each time the pellet is saved.

Reference: Sloboda RD, Dentler WL & Rosenbaum JL (1976) Microtubule-associated proteins and the stimulation of tubulin assembly *in vitro*. *Biochemistry* 15, 4497–4505.

16–18* **Reference:** Walker RA, Inoué S & Salmon ED (1989) Asymmetric behavior of severed microtubule ends after ultraviolet-microbeam irradiation of individual microtubules *in vitro*. *J. Cell Biol.* 108, 931–937.

16–19

A. Although the end points for polymerization and ATP hydrolysis were the same, the initial rate of ATP hydrolysis was less than the initial rate of polymerization. (Compare the slopes of the two curves in Figure 16–9 at short times.) At the time when all the actin was polymerized (about 30 seconds), less than half the ATP was hydrolyzed. It is the difference in initial rates that your advisor noticed, and, as he said, it proves that actin polymerization can occur in the absence of ATP hydrolysis.

B. Since the rate of polymerization is faster than the rate of ATP hydrolysis, newly added actin subunits must still retain bound ATP. Since the bound ATP is not hydrolyzed until some time after assembly, growing actin filaments have ATP 'caps.' Once an ATP-actin monomer has bound to a filament, the ATP can be hydrolyzed, giving rise to the bound ADP found interior to the ATP caps.

Reference: Carlier M-F, Pantaloni D & Korn ED (1984) Evidence for an ATP cap at the ends of actin filaments and its regulation of the F-actin steady state. *J. Biol. Chem.* 259, 9983–9986.

16–20*

16–21

A. The average time for a small molecule such as ATP to diffuse across a cell 10 μm (10^{-3} cm) in diameter is

$$t = x^2/2D$$
$$= (10^{-3} \text{ cm})^2/2 \, (5 \times 10^{-6} \text{ cm}^2/\text{sec})$$
$$= 0.1 \text{ sec}$$

Similarly, a protein molecule takes 1 second and a vesicle 10 seconds on average to travel 10 μm.

B. The diffusion of long, cytoskeletal filaments would be even slower than that of a membrane vesicle; hence it would take much longer to rearrange the cytoskeleton by diffusion. In addition to time, there's also the problem of length: polymerization allows filaments to be constructed to fit. Finally, if the long cytoskeletal elements were to rearrange by diffusion, they would become hopelessly entangled with one another.

16–22*

16–23 A few examples of the differences between bacteria and animal cells are listed below. This is by no means a complete list.

1. Animal cells are much larger, diversely shaped, and do not have a cell wall. Cytoskeletal elements are required to provide mechanical strength and shape in the absence of a cell wall.
2. Animal cells, and all other eucaryotic cells, have a nucleus that is shaped

and held in place by intermediate filaments; the nuclear lamins attached to the inner nuclear membrane support and shape the nuclear membrane, and a meshwork of intermediate filaments surrounds the nucleus and spans the cytosol.

3. Animal cells can move by a process that requires a change in cell shape. Actin filaments (and myosin motor proteins) are required for these activities.

4. Animal cells have a much larger genome than bacteria; this genome is fragmented into many chromosomes. For cell division, chromosomes need to be accurately distributed to the daughter cells, which requires the microtubules that form the mitotic spindle.

5. Animal cells have internal organelles. Their localization in the cell is dependent on motor proteins that move them along microtubules. A remarkable example is the long-distance travel of membrane-enclosed vesicles along microtubules in an axon, which can be up to a meter long in the case of the nerve cells that extend from your spinal cord to your feet.

16–24*

16–25 False. Intermediate filaments are found only in some metazoans, including vertebrates, nematodes, and snails. Even in these organisms intermediate filaments are not required in every cell type. The nuclear lamins, which are the ancestors of the intermediate filaments, form a meshwork of protein that lines the nuclear membrane; they are much more widely distributed among eucaryotes.

16–26*

16–27 Cells that migrate rapidly from one place to another, like amoebae (A) and sperm cells (E), do not in general need intermediate filaments in their cytoplasm, since they do not develop or sustain large tensile forces. Plant cells (F) are pushed and pulled by the forces of wind and water, but they resist these forces by means of their rigid cell walls, rather than by their cytoskeleton. Epithelial cells (B), smooth muscle cells (C), and the long axons of nerve cells (D) are all rich in cytoplasmic intermediate filaments, which prevent them from rupturing as they are stretched and compressed by the movements of surrounding tissues.

16–28* Figure 16–52.

References: Ho C-L, Martys JL, Mikhailov A, Gundersen GG & Liem RKH (1998) Novel features of intermediate filament dynamics revealed by green fluorescent protein chimeras. *J. Cell Sci.* 111, 1767–1778.

Chou Y-H, Helfand BT & Goldman RD (2001) New horizons in cytoskeletal dynamics: transport of intermediate filaments along microtubule tracks. *Curr. Opin. Cell Biol.* 13, 106–109.

16–29 In BHK-21 cells the entire vimentin network seems to depolymerize in preparation for mitosis and then reassembles afterwards (see Figure 16–10A). Note the absence of obvious filaments during the two phases of mitosis and their clear presence in the two daughter cells. By contrast, in PtK2 cells the vimentin network remains largely intact until late cytokinesis, when the portion of the network in the connecting cytoplasmic bridge is finally 'dissolved.' Thus, in these cells only a small portion of the vimentin network is disassembled during mitosis. It is unclear how these two quite different strategies are accomplished.

Reference: Yoon M, Moir RD, Prahlad V & Goldman RD (1998) Motile properties of vimentin intermediate filament networks in living cells. *J. Cell Biol.* 143, 147–157.

16–30* **References:** Ottaviano Y & Gerace L (1985) Phosphorylation of the nuclear lamins during interphase and mitosis. *J. Biol. Chem.* 260, 624–632.

Ward GE & Kirschner MW (1990) Identification of cell cycle-related phosphorylation sites on nuclear lamin C. *Cell* 61, 561–577.

Heald R & McKeon F (1990) Mutations of phosphorylation sites in lamin A that prevent nuclear lamina disassembly in mitosis. *Cell* 61, 579–589.

Peter M, Nakagawa J, Dorée M, Labbe JC & Nigg EA (1990) *In vitro* disassembly of the nuclear lamina and M phase-specific phosphorylation of lamins by cdc2 kinase. *Cell* 61, 591–602.

16–31   The disulfide bonds that cross-link keratin filaments in skin cells form after the cells have died. In the absence of cellular metabolism to maintain the reducing environment characteristic of a living cell, a dead cell's contents quickly become oxidized. It is in this postmortem environment that the keratin filaments become cross-linked by disulfide bonds.

16–32*   **References:** Herrmann H & Aebi U (2000) Intermediate filaments and their associates: multi-talented structural elements specifying cytoarchitecture and cytodynamics. *Curr. Opin. Cell Biol.* 12, 79–90.

Eliasson C, Sahlgren C, Berthold CH, Stakeberg J, Celis JE, Betsholtz C, Eriksson JE & Pekny M (1999) Intermediate filament protein partnership in astrocytes. *J. Biol. Chem.* 274, 23996–24006.

16–33

A. Figure 16–13 shows that cytochalasin B interferes with filament assembly by stopping actin polymerization at the plus end, which is normally the preferred end for addition of monomers. One plausible mechanism to explain this inhibition is that cytochalasin B binds to the plus end of the actin filament and physically blocks addition of new actin monomers.

   This mechanism can also account for the viscosity measurements. Since growth at the minus end is unaffected, the filaments continue to grow, but much more slowly. The slower growth rate explains the slower increase in viscosity in the presence of cytochalasin B. The lower viscosity at the plateau indicates that the actin filaments are shorter in the presence of cytochalasin B. The filaments are shorter when they are growing only from the minus ends because the critical concentration for assembly at the minus end is higher than the critical concentration for assembly at the plus end.

B. An actin filament normally grows at different rates at the plus and minus ends. This observation indicates that the monomer probably undergoes a conformational change upon addition to an actin filament. If all subunits, assembled and free, were identical in conformation, the rates of growth at the two ends should be the same (see Problem 16–11).

Reference: MacLean-Fletcher S & Pollard TD (1980) Mechanism of action of cytochalasin B on actin. *Cell* 20, 329–341.

16–34*   Figure 16–53.

Reference: Steinmetz MO, Stoffler D, Hoenger A, Bremer A & Aebi U (1997) Actin: From cell biology to atomic detail. *J. Struct. Biol.* 119, 295–320.

16–35

A. Phalloidin increases the growth rate of actin filaments by eliminating the off rate. Because the slopes of the lines in Figure 16–15A are identical, k_{on} is the same in the presence and absence of phalloidin. The y intercept ($-k_{off}$), however, is dramatically altered, from about -12 molecules/sec in the absence of phalloidin to 0 molecules/sec in its presence. These results suggest that the off rate is zero in the presence of phalloidin.

B. The results in Figure 16–15B confirm the interpretation in part A. Actin filaments made in the absence of phalloidin disassemble as expected when diluted in the absence of actin monomers. Filaments made in the presence of phalloidin, however, are rock-solid stable, as expected if the off rate were zero.

C. The critical concentration for actin assembly is the concentration of actin at which no growth occurs. In the absence of phalloidin this point occurs at about 1 μM. In the presence of phalloidin it occurs at an actin concentration of zero. This result is also consistent with phalloidin reducing the off rate to zero: at any concentration of actin the filament will grow.

D. Phalloidin interferes with actin assembly by binding to the filament to prevent dissociation of actin subunits. The requirement for a 1:1 molar mixture suggests that phalloidin is required stoichiometrically with actin, but does not tell you whether it binds to free monomers or to subunits in the filament. The stability of filaments prepared in the presence of phalloidin upon dilution indicates that phalloidin binds to the filaments. It is thought that phalloidin binds to actin subunits in a way that locks them in place.

Reference: Coluccio LM & Tilney LG (1984) Phalloidin enhances actin assembly by preventing monomer dissociation. *J. Cell Biol.* 99, 529–535.

16–36* **Reference:** Bubb MR, Spector I, Bershadsky AD & Korn ED (1995) Swinholide A is a microfilament disrupting marine toxin that stabilizes actin dimers and severs actin filaments. *J. Biol. Chem.* 270, 3463–3466.

16–37 Cell division depends on the ability of microtubules to polymerize and to depolymerize. During mitosis, cells first depolymerize most of their microtubules and then repolymerize them to form the mitotic spindle. Taxol-treated cells are prevented from depolymerizing their existing microtubules, and thus cannot form a mitotic spindle. Colchicine-treated cells cannot polymerize new microtubules, and thus are also prevented from forming a mitotic spindle. On a more subtle level, both drugs would block the dynamic instability of microtubules and thus would interfere with the workings of the mitotic spindle, even if one could be formed.

16–38* **References:** Stone JD, Peterson AP, Eyer J & Sickles DW (2000) Neurofilaments are nonessential elements of toxicant-induced reductions in fast axonal transport: pulse labeling in CNS neurons. *Neurotoxicology* 21, 447–457.

Sickles DW, Pearson JK, Beall A & Testino A (1994) Toxic axonal degeneration occurs independent of neurofilament accumulation. *J. Neurosci. Res.* 39, 347–354.

HOW CELLS REGULATE THEIR CYTOSKELETAL FILAMENTS

16–39 False. Although centrosomes, the major microtubule-organizing centers in almost all animal cells, do contain centrioles, a number of microtubule-organizing centers in plants, animals, and fungi do not. The common feature of all microtubule-organizing centers is an electron-dense matrix that usually contains γ-tubulin, which is used to nucleate microtubules.

16–40* **Reference:** Mitchison T & Kirschner M (1984) Microtubule assembly nucleated by isolated centrosomes. *Nature* 312, 232–237.

16–41
A. Microtubules assembled on flagellar axonemes are extensions of the microtubules already present in the axonemes. Therefore, the polarity of growth is fixed: the plus end of the axoneme will nucleate a microtubule that has its plus end free for addition of new subunits. The newly assembled microtubule, therefore, has its plus end pointing away from the axoneme and its minus end attached to the axoneme (Figure 16–54A).

B. The plus end of the microtubule must grow faster since microtubules with free plus ends (attached to the plus end of the axoneme) are longer than those with free minus ends (attached to the minus end of the axoneme).

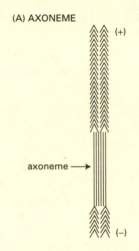

(A) AXONEME

axoneme →

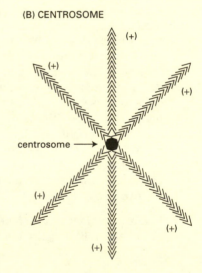

(B) CENTROSOME

centrosome →

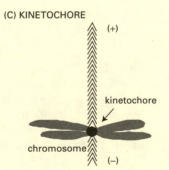

(C) KINETOCHORE

kinetochore

chromosome

Figure 16–54 Polarities of microtubules (Answer 16–41). (A) Nucleated on an axoneme. (B) Nucleated on a centrosome. (C) Nucleated on a kinetochore.

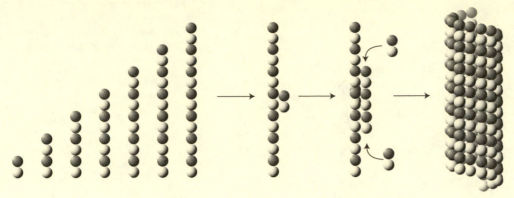

LINEAR GROWTH   LATERAL ASSOCIATION

Figure 16–55 Rapid addition of αβ-tubulin dimers to nucleation structure (Answer 16–43).

C. For axonemes, where the plus and minus ends can be distinguished, it is clear that the growth rate at the plus end is faster than at the minus end, since the microtubules attached at the plus end are longer than those attached to the minus end. It is this difference in growth rates that allows one to decide the polarity of growth nucleated by centrosomes and kinetochores.

Microtubules nucleated on centrosomes have a length that indicates their plus end is free. Thus, centrosomes nucleate microtubule growth by binding to the minus end of the microtubule (Figure 16–54B). Kinetochores, on the other hand, have a bimodal distribution of lengths, which suggests that some microtubules are attached by their minus ends (the longer ones) and some are attached by their plus ends (the shorter ones) (Figure 16–54C).

Reference: Mitchison T & Kirschner MW (1985) Properties of the kinetochore *in vitro*. I. Microtubule nucleation and tubulin binding. *J. Cell Biol.* 101, 755–765.

16–42* **Reference:** Leguy R, Melki R, Pantaloni D & Carlier M-F (2000) Monomeric γ-tubulin nucleates microtubules. *J. Biol. Chem.* 275, 21975–21980.

16–43   Once the first lateral association has occurred, the next αβ-dimer can bind much more readily because it is stabilized by both lateral and longitudinal contacts (Figure 16–55). The formation of a second protofilament stabilizes both protofilaments, allowing rapid addition of new αβ-tubulin dimers to form adjacent protofilaments and to extend existing ones. At some point the initial sheet of tubulin curls into a tube to form the microtubule.

Reference: Leguy R, Melki R, Pantaloni D & Carlier M-F (2000) Monomeric γ-tubulin nucleates microtubules. *J. Biol. Chem.* 275, 21975–21980.

16–44* Figure 16–56.

Reference: Leguy R, Melki R, Pantaloni D & Carlier M-F (2000) Monomeric γ-tubulin nucleates microtubules. *J. Biol. Chem.* 275, 21975–21980.

16–45   Whether γ-TuRC is present or not makes no difference to the lengths of the plus ends of the microtubules. However, γ-TuRC dramatically shifts the distribution of bright segments at the minus ends, suggesting that it nucleates growth at the minus ends. You might reasonably ask why there are any bright segments at all at the minus ends. Although 50% have very short, or non-existent bright segments, the remainder have a distribution of lengths that is not much different from that observed in the absence of γ-TuRC. Some of these microtubules may have spontaneously nucleated, γ-TuRC may have dissociated from some, or some microtubules may have been broken.

Reference: Zheng Y, Wong ML, Alberts B & Mitchison T (1995) Nucleation of microtubule assembly by a γ-tubulin-containing ring complex. *Nature* 378, 578–583.

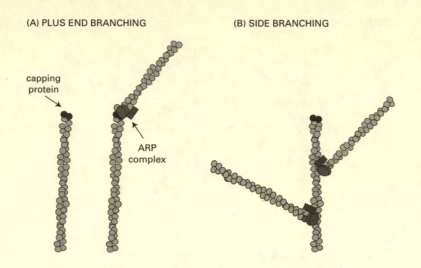

(A) PLUS END BRANCHING   (B) SIDE BRANCHING

capping protein

ARP complex

Figure 16–57 Expectations of two models for the branching of actin filaments induced by the binding of the ARP complex (Answer 16–49). (A) ARP-complex binding at the plus end. (B) ARP-complex binding to the side.

16–46*

16–47 Any actin-binding protein that stabilizes complexes of two or more actin monomers will facilitate the initiation (nucleation) of a new filament. The actin-binding proteins must not block the ends required for filament growth.

16–48*

16–49 The two alternatives make different predications about the kinds of structures that should be generated. As shown in Figure 16–57A, if the ARP complex binds to the plus ends, a capped actin filament might not be a substrate for ARP binding, in which case no branches would be seen. Alternatively, the ARP complex might bind to the capped structure, in which case a kinked filament would be generated. By contrast, if the ARP complex binds to the sides of the filament, then the plus end cap will be irrelevant and a typical branched structure will be generated (Figure 16–57B). Experiments of this kind support the idea that the ARP complex binds to the sides of actin filaments.

References: Amann KJ & Pollard TD (2001) The ARP2/3 complex nucleates actin filament branches from the sides of preexisting filaments. *Nature Cell Biol.* 3, 306–310.

Higgs HN & Pollard TD (2001) Regulation of actin filament network formation through ARP2/3 complex: Activation by a diverse array of proteins. *Annu. Rev. Biochem.* 70, 649–676.

16–50* **References:** Dramsi S & Cossart P (1998) Intracellular pathogens and the actin cytoskeleton. *Annu. Rev. Cell Dev. Biol.* 14, 137–166.

Welch MD, Rosenblatt J, Skoble J, Portnoy DA & Mitchison TJ (1998) Interaction of human Arp2/3 complex and the *Listeria monocytogenes* ActA protein in actin filament nucleation. *Science* 281, 105–108.

16–51 In cells most of the actin subunits are bound to thymosin, which locks actin into a form that cannot hydrolyze its bound ATP and cannot be added to either end of a filament. Thymosin reduces the concentration of free actin subunits to around the critical concentration. Actin subunits are recruited from this inactive pool by profilin, whose activity is regulated so that actin polymerization occurs when and where it is needed. The advantage of such an arrangement is that the cell can maintain a large pool of subunits for explosive growth at sites and times of its choosing.

16–52* **Reference:** Horio T & Hotani H (1986) Visualization of the dynamic instability of individual microtubules by dark-field microscopy. *Nature* 321, 605–607.

16–53   Cofilin preferentially binds to ADP-containing actin filaments, introducing strain into the filament, which makes it easier for the filament to be severed and for ADP–actin subunits to dissociate. Because polymerization is faster than ATP hydrolysis, the newly added subunits are resistant to depolymerization by cofilin. Thus, cofilin efficiently dismantles older filaments in the cell, ensuring that actin structures turn over rapidly.

16–54*

16–55   False. Most of the identified proteins that bind to the ends of actin filaments and microtubules cap the ends and prevent further polymerization. There are exceptions, however. For example, XMAP215 binds to the ends of microtubules in a way that stabilizes them without inhibiting continued polymerization, and catastrophins bind to ends to destabilize them and promote catastrophes.

16–56*

16–57
  A.   The observation that *Dictyostelium* with defects in α-actinin or gelation factor have normal motility and development, whereas the double mutant strains have profound problems, suggests that each protein alone can substitute for the functions normally provided by the other. Although α-actinin and gelation factor presumably serve somewhat different roles in *Dictyostelium*, they also provide a mutual backup system. This appears to be common in higher organisms. As reverse genetics is used to knock out more and more genes, it is becoming clear that many genes provide overlapping functions that can partially substitute for one another (see, for example, Problem 16–32).
  B.   The most peculiar observation in these experiments is that the amoebae of the doubly mutant strain move perfectly well, even though the two major actin-binding proteins are defective. This may mean that there are still other actin-binding proteins in *Dictyostelium* that function sufficiently well to allow amoebae to move, but not well enough to permit the multicellular slug to move properly.

  Reference: Witke W, Schleicher M & Noegel AA (1992) Redundancy in the microfilament system: abnormal development of *Dictyostelium* cells lacking two F-actin cross-linking proteins. *Cell* 68, 53–62.

16–58*

16–59   The signal cascade initially results in a massive influx of Ca^{2+} (B), which activates gelsolin (F) to cleave the existing short actin filaments into tiny fragments. With slower kinetics, the same signaling pathway causes a rise in PiP_2 (H) levels, which inactivates both gelsolin (F) and CapZ (C). The large number of exposed filament ends are then rapidly elongated by profilin (I)-mediated delivery of actin subunits to the ends. Some of these long actin filaments are bundled together by α-actinin (A) and fimbrin (E), and some are cross-linked into a gel by filamin (D). This allows the activated platelet to extend lamellipodia and filopodia to spread over the clot. Once the PiP_2 (H) signal subsides, CapZ (C) returns to the ends of the filaments, rendering them stable against depolymerization and locking the platelet into its spread form. Finally, the motor protein myosin II (G) uses ATP hydrolysis to slide the long filaments relative to one another, causing a contraction of the platelet, which pulls the edges of the wound together.

16–60*

16–61   These experimental results clearly support a role of N-WASP in the Cdc42-initiated polymerization of actin, and by implication, in the natural rearrangements of the actin cytoskeleton in cells as a result of Cdc42 activation. The inability of the mutant H208D to restore actin polymerization indicates that a direct interaction between Cdc42 and N-WASP is a critical feature of the activation pathway. The inability of the mutant Δcof to restore actin polymerization indicates that the cofilin domain of N-WASP is required for actin polymerization. The cofilin domain would allow N-WASP to bind to actin filaments, or perhaps to the actin subunits of the ARP complex, suggesting that such binding is essential for N-WASP-mediated actin polymerization.

Reference: Rohatgi R, Ma L, Miki H, Lopez M, Kirchhausen T, Takenawa T & Kirschner MW (1999) The interaction between N-WASP and the Arp2/3 complex links Cdc42-dependent signals to actin assembly. *Cell* 97, 221–231.

16–62* Figure 16–58.

Reference: Rohatgi R, Ma L, Miki H, Lopez M, Kirchhausen T, Takenawa T & Kirschner MW (1999) The interaction between N-WASP and the Arp2/3 complex links Cdc42-dependent signals to actin assembly. *Cell* 97, 221–231.

MOLECULAR MOTORS

16–63 Intermediate filaments have no polarity; their ends are indistinguishable. It would therefore be difficult for a hypothetical motor protein bound to the middle of the filament to sense a defined direction. Such a motor protein would be equally likely to attach to the filament facing in one direction as the other. The known molecular motors move in one direction along a filament of defined polarity.

16–64*

16–65

A. Motor proteins are unidirectional in their action; nearly all kinesins move toward the plus end of a microtubule and dyneins always move toward the minus end. Thus, if dynein molecules, for example, are attached to the coverslip, only those individual molecules that are correctly oriented relative to the microtubule that settles on them can attach to it, exert force, and propel it forward.

B. On a bed of dynein motors, microtubules will always move plus end first over the coverslip. The dynein motors 'walk' toward the minus end; thus, since the motors are fixed, the microtubule moves plus end first.

C. The protein on the coverslip is a plus end-directed motor. Since the bead, which marks the minus end of the microtubule is moving forward, the motor must be walking toward the opposite end—the plus end.

Reference: Fan J, Griffiths AD, Lockhart A, Cross RA & Amos LA (1996) Microtubule minus ends can be labeled with a phage display antibody specific to α-tubulin. *J. Mol. Biol.* 259, 325–330.

16–66* **Reference:** Fan J, Griffiths AD, Lockhart A, Cross RA & Amos LA (1996) Microtubule minus ends can be labeled with a phage display antibody specific to α-tubulin. *J. Mol. Biol.* 259, 325–330.

16–67

A. The differences in landing rates at low densities of two-headed and one-headed kinesins indicate that multiple one-headed kinesin motors are necessary to move a microtubule. At a high motor density the landing rates for both motors are about the same. The landing rate for two-headed kinesin declines linearly with density, whereas the landing rate for one-headed kinesin drops abruptly at lower densities. This behavior indicates that a single two-headed kinesin is sufficient to move a microtubule, but that several one-headed kinesins (4–6 according to the authors) are required. A one-headed kinesin can bind a microtubule, but when it lets go to take the next step the microtubule floats away. Thus, several one-headed kinesins are required so that some can hold onto the microtubule while others release and rebind.

B. Two heads are better than one. In principle, a single kinesin motor could move a vesicle for long distances along a microtubule track because it holds on with one 'hand,' while it releases and rebinds with the other. A one-headed motor would lose its way each time it released the microtubule to take a step.

Reference: Hancock WO & Howard J (1998) Processivity of the motor protein kinesin requires two heads. *J. Cell Biol.* 140, 1395–1405.

16–68* **Reference:** Svoboda K, Schmidt CF, Schnapp BJ & Block SM (1993) Direct observation of kinesin stepping by optical trapping interferometry. *Nature* 365, 721–727.

16–69

A. In each cycle the chemical free energy that drives the cycle is provided by hydrolysis of ATP. Although ATP hydrolysis is a common source of chemical free energy, it is not the only one. For example, the free energy in a Na^+ ion gradient usually drives active transport of sugars in animal cells, and GTP hydrolysis powers the movements of ribosomes during protein synthesis.

    The mechanical work accomplished during muscle contraction is the motion of actin thin filaments relative to myosin thick filaments. The mechanical work done during active transport of Ca^{2+} is the pumping of ions outside the cell against their concentration gradient.

B. Actin is bound tightly and then released in each cross-bridge cycle during muscle contraction; Ca^{2+} is bound tightly and then released during its active transport.

    In the diagram in Figure 16–32A, actin is tightly bound to myosin at each point where the two are in contact. The binding of ATP to the myosin head converts it to a weak binding form, allowing it to detach from actin. (Although each of these steps is shown separately in the diagram, the binding of ATP is thought to initiate a conformational change, which in turn reduces the affinity of myosin for actin, thereby promoting detachment of actin and the completion of the conformational change.)

    In the diagram in Figure 16–32B, Ca^{2+} is tightly bound to the transport protein when it is on the inside of the cell (upper drawing) but only weakly bound when it faces the outside of the cell (lower drawing). Although the tightness of binding is not immediately apparent in the diagrammatic representation, it follows from the concept of active transport. Since the pump transports Ca^{2+} against its concentration gradient, the pump must have a high affinity for Ca^{2+} on the inside of the cell (so that Ca^{2+} can be bound effectively at its low intracellular concentration) and a low affinity for Ca^{2+} on the outside of the cell (so that Ca^{2+} can be released effectively at its high external concentration).

C. In both cycles the 'power stroke' is the conformational change indicated on the right side of the cycles as drawn in Figure 16–32. The 'return stroke' in each case is the conformational change indicated on the left side of the drawings.

Reference: Eisenberg E & Hill TL (1985) Muscle contraction and free energy transduction in biological systems. *Science* 227, 999–1006.

16–70*

16–71 False. The centrosome, which establishes the principal array of micro-tubules in a cell, nucleates microtubule growth at the minus end. Thus, the plus ends of the microtubules are near the plasma membrane, and the minus ends are buried in the centrosome at the center of the cell. This orientation of the array requires that plus end-directed motors be used to transport cargo to the cell periphery and that minus end-directed motors be used for cargo delivery to the center of the cell.

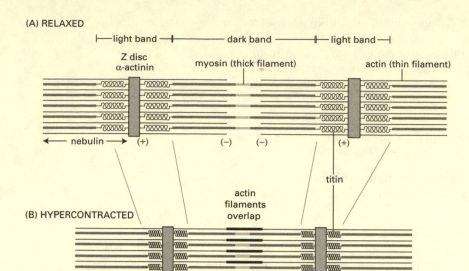

Figure 16–59 Schematic diagrams of electron micrographs in Figure 16–34 (Answer 16–75). (A) Relaxed muscle. (B) Hypercontracted muscle.

16–72* **Reference:** Reese EL & Haimo LT (2000) Dynein, dynactin, and kinesin II's interaction with microtubules is regulated during bidirectional organelle transport. *J. Cell Biol.* 151, 155–165.

16–73 Both filaments are composed of subunits of protein dimers that are held together by coiled-coil interactions. Moreover, in both cases, the dimers polymerize through their coiled-coil domains into filaments. Whereas intermediate filament dimers assemble head to head and create a filament that has no polarity, all myosin molecules in the same half of the myosin filament are oriented with their heads pointing in the same direction. This polarity is necessary for them to be able to develop a contractile force in muscle.

16–74*

16–75

A. The locations of the striated muscle components in the electron micrograph are illustrated schematically in Figure 16–59A. α-Actinin is a component of the Z disc; titin links the myosin II filaments to the Z disc; and nebulin binds along the length of each actin filament.

B. The micrograph in Figure 16–34B shows a hypercontracted muscle. The light band has entirely disappeared, and a new band, caused by the overlap of actin filaments, has appeared in the middle of the sarcomere. The schematic relationship of the two electron micrographs in Figure 16–34 is shown in Figure 16–59B.

16–76* **Reference:** Goldman YE, Hibberd MG, McCray JA & Trentham DR (1980) Relaxation of muscle fibers by photolysis of caged ATP. *Nature* 300, 701–705.

16–77 Sketches representing sarcomeres at each of the arrows in Figure 16–37 are shown in Figure 16–60. As illustrated in these pictures, the increase in tension with decreasing sarcomere length in segment I is due to increasing numbers of interactions between myosin heads and actin. In segment II, actin begins to overlap with the bare zone of myosin, yielding a plateau since the number of interacting myosin heads remains constant. In segment III, the actin filaments begin to overlap with each other, thereby interfering with the optimal interaction of actin and myosin and producing a decrease in tension. In segment IV, the spacing between the Z discs is less than the length of the myosin thick filaments, causing their deformation and a precipitous drop in muscle tension.

Reference: Gordon AM, Huxley AF & Julian FJ (1966) The variation in isometric tension with sarcomere length in vertebrate muscle fibres. *J. Physiol.* 184, 170–192.

16–78*

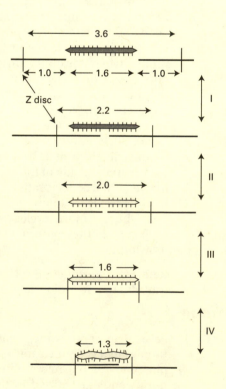

Figure 16–60 Schematic diagrams of sarcomeres at the points indicated by *arrows* in Figure 16–37 (Answer 16–77). Numbers refer to lengths in micrometers.

16–79   True. In resting muscle the troponin I–T complex pulls tropomyosin out of its normal binding groove in actin so that it interferes with the binding of myosin heads. An influx of Ca^{2+} causes troponin C to alter its conformation, which forces troponin I to release its hold on actin, allowing tropomyosin to slip back into its preferred position, thereby exposing binding sites for the myosin heads.

16–80*

16–81   ATP hydrolysis by the myosin motor domain is required for filament sliding in muscle contraction, and hydrolysis by the Ca^{2+}-ATPase is required to pump Ca^{2+} out of the cytosol.

16–82*

16–83   In the electron micrograph in Figure 16–39, the central pairs of microtubules are all oriented identically. The identical orientation suggests that some structural feature of the axoneme constrains the direction of bending; that is, that cilia are designed to bend in a particular way.

In fact, the orientation of the central pair of microtubules correlates with the direction of bending, which is always in a plane drawn between the two central microtubules. To make this relationship clearer, imagine that the axonemes shown in cross-section in Figure 16–39 extends straight up out of the page. The plane between the central pair of microtubules would also extend upward and run roughly from the top of the page to the bottom. The axonemes could bend toward the top or bottom of the page—staying within the plane—but not to the left or right, for example, which would break the plane.

If the microtubules of the central pair are aligned parallel to one another throughout the length of the cilium, then all bending in the power stroke and in the return stroke will be in one plane. It is not uncommon, however, for the central pair of microtubules to be twisted around one another, in which case the direction of bending rotates around the axis of the cilium as the bend propagates up the cilium. The consequence of this arrangement is a unidirectional power stroke (which depends only on the orientation of the central pair at the base of the cilium, where bending is initiated) and a helical return stroke as the bend moves to the tip of the cilium.

16–84*   **References:** Brokaw CJ, Luck DJL & Huang B (1982) Analysis of the movement of *Chlamydomonas* flagella: the function of the radial-spoke system is revealed by comparison of wild-type and mutant flagella. *J. Cell Biol.* 92, 722–732.

Gibbons IR (1981) Cilia and flagella of eukaryotes. *J. Cell Biol.* 91, 107s–124s.

16–85   One pattern of dynein activity that could account for the planar bending of an axoneme is depicted in Figure 16–61. The axonemes shown in this figure are oriented with their tips *below* the plane of the page. If the dynein arms on just the left half of the axoneme are active (arrows in Figure 16–61A), the cilium will bend upward toward the top of the page. This is difficult to imagine in three dimensions, but consider it step by step. First, the dynein arms push their neighbor doublets toward the *tip* of the axoneme, so the doublets are being pushed *below* the plane of the page. Second, the doublet at the *top* of the diagram in Figure 16–61A will be pushed the farthest below the page because its total displacement is the sum of incremental displacements produced by all four active dynein arms. Third, the doublet that moves the farthest defines the 'inside' of the bend (see Figure 16–41). Therefore, since the top doublet moves the farthest, the axoneme will bend upward (toward the top of the page) when the dynein arms on the left half of the axoneme are active.

The same reasoning argues that the axoneme will bend downward (toward the bottom of the page) if the dynein arms on the right half of the axoneme are active and the ones on the left half are passive (Figure 16–61B).

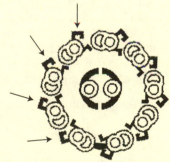

(A) UPWARD BEND

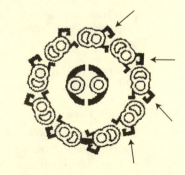

(B) DOWNWARD BEND

Figure 16–61 One possible pattern of dynein activity that could produce planar bending of an axoneme (Answer 16–85). (A) Upward bend. (B) Downward bend. *Arrows* indicate active dynein arms.

The actual pattern of dynein activity that gives rise to planar bending is not yet known. The two central singlet microtubules are natural candidates for regulatory elements: they are surrounded by nonidentical proteins; they contact different subsets of outer doublets; and they are linked (indirectly) to the two sets of dynein arms used in the model proposed above.

Reference: Satir P & Matsuoka T (1989) Splitting the ciliary axoneme: implications for a "switch-point" model of dynein arm activity in ciliary motion. *Cell Motility and the Cytoskeleton* 14, 345–358.

16–86*   Reference: Luck DJL (1984) Genetic and biochemical dissection of the eucaryotic flagellum. *J. Cell Biol.* 98, 789–794.

THE CYTOSKELETON AND CELL BEHAVIOR

16–87   During protrusion, cells extend actin-rich structures—filopodia, lamellipodia, or pseudopodia—in front of them. During attachment, the actin cytoskeleton in the extended structures makes connections with the substratum. During traction, contraction of the anchored actin cytoskeleton pulls the bulk of the cytoplasm forward.

16–88*

16–89   The minus ends of the growing actin filaments are anchored to the rest of the actin cytoskeleton, which supports the growing actin filaments and allows them to push on the membrane without simply sliding back into the cell's interior. The solution to the problem at the plus end is not so straightforward. Once the filament contacts the membrane, there would be no room for a new subunit to fit onto the end of the growing chain. It is thought that random thermal motions briefly expose the plus end of the filament, allowing a new subunit to be added. By taking advantage of these small windows of opportunity, actin polymerization acts as a ratchet to capture random thermal motions. It is unclear what motions the actin ratchet is capturing. It could be that membranes 'breathe' thermally, allowing polymerization. Alternatively, the actin filament may bend elastically, moving the plus end sufficiently to allow subunit addition.

16–90*   Reference: Tilney LG & Inoué S (1982) Acrosomal reaction of *Thyone* sperm. II. The kinetics and possible mechanism of acrosomal process elongation. *J. Cell Biol.* 93, 820–827.

16–91   Injection of activated Rac triggers actin polymerization over the entire membrane periphery, forming essentially one giant lamellipodium (see Figure 16–44C). Injection of activated Rho promotes the bundling of actin filaments with myosin II filaments to form stress fibers (see Figure 16–44B), which associate with other proteins at focal contacts. Injection of activated Cdc42 triggers actin polymerization and bundling to form filopodia (see Figure 16–44D).

Reference: Hall A (1998) Rho GTPases and the actin cytoskeleton. *Science* 279, 509–514.

16–92*

16–93   True. A site of bacterial infection is a source of bacterial proteins, some of which have retained the *N*-formylmethionine used for initiation of bacterial protein synthesis. As these proteins are degraded, *N*-formylated peptides are released that diffuse outward from the source, generating a gradient that can be sensed by neutrophils via membrane receptors. The binding of *N*-formylated peptides triggers changes in the cytoskeleton that allow the neutrophil to travel up the gradient to the site of infection.

16–94*

16–95 It would take a vesicle an average of 10^9 seconds—nearly 32 years—to diffuse the length of a 10 cm axon.

$$t = x^2/2D$$
$$= \frac{(10\text{ cm})^2}{2 \times (5 \times 10^{-8}\text{ cm}^2/\text{sec})}$$
$$= 10^9 \text{ second, or } 31.7 \text{ years}$$

16–96*

16–97 Kinesin motors use microtubules as tracks to deliver organelles and materials to nerve endings. The similar neuropathies that develop in mice and humans with only one functional copy of the gene for the kinesin motor KIF1B indicate that half the normal number of these motors cannot keep up with the needs of the nerves.

Reference: Zhao C, Takita J, Tanaka Y, Setou M, Nakagawa T, Takeda S, Yang HW, Terada S, Nakata T, Takei Y, Saito M, Tsuji S, Hayashi Y & Hirokawa N (2001) Charcot-Marie-Tooth disease type 2A caused by mutation in a microtubule motor KIF1Bβ. *Cell* 105, 587–597.

16–98* **References:** Yabe JT, Pimenta A & Shea TB (1999) Kinesin-mediated transport of neurofilament protein oligomers in growing axons. *J. Cell Sci.* 112, 3799–3814.

Shea TB & Yabe J (2000) Occam's razor slices through the mysteries of neurofilament axonal transport: Can it really be so simple? *Traffic* 1, 522–523.

16–99 Extension and contraction of this filopodium were regulated by the rate of actin polymerization. The rate of retrograde flow was constant at about –1 μm/min, whereas the rate of actin polymerization varied from 0 to 2 μm/min. The movements of the tip correlated with the rate of actin polymerization, extending when the polymerization rate was high and retracting when the polymerization rate was low. Observations on many individual filopodia support this general conclusion, although there were examples in which retrograde flow appeared to be regulated.

Reference: Mallavarapu A & Mitchison T (1999) Regulated actin cytoskeleton assembly at filopodium tips controls their extension and retraction. *J. Cell Biol.* 146, 1097–1106.

16–100* Figure 16–62.

References: Chant J & Herskowitz I (1991) Genetic control of bud site selection in yeast by a set of gene products that constitute a morphogenetic pathway. *Cell* 65, 1203–1212.

Chant J, Corrado K, Pringle JR & Herskowitz I (1991) Yeast BUD5, encoding a putative GDP-GTP exchange factor, is necessary for bud site selection and interacts with bud formation gene *BEM1*. *Cell* 65, 1213–1224.

Park H-O, Chant J & Herskowitz I (1993) *BUD2* encodes a GTPase-activating protein for Bud1/Rsr1 necessary for proper bud-site selection in yeast. *Nature* 365, 269–274.

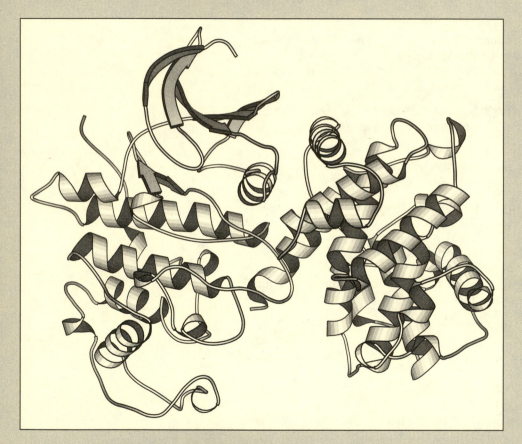

Cyclin A-CDK2 in its active form.
Coordinates from 1JST determined by
AA Russo, PD Jeffrey & NP Pavletich.
Nature Struct. Biol. 3:696–700, 1996.

17

THE CELL CYCLE AND PROGRAMMED CELL DEATH

AN OVERVIEW OF THE CELL CYCLE

17–1   No. Although a number of cells equivalent to an adult human is replaced about every three years, not all cells are replaced at the same rate. Blood cells and cells that line the gut are replaced at a high rate, whereas cells in most organs are replaced more slowly, and neurons are rarely replaced.

17–2*

17–3   True. The G_1 phase is the most critical time for cell growth. Its length can vary greatly depending on external conditions and extracellular signals from other cells.

17–4*

17–5   The constancy of the length of S phase in haploid versus diploid and diploid versus tetraploid organisms may not be so surprising. If one assumes that particular chromosomes and regions within chromosomes have a defined order of replication, then halving or doubling the number of chromosomes would not be expected to alter the schedule. Moreover, the ratio of genes encoding the replication machinery (DNA polymerases, helicases, initiation factors, etc.) to the amount of DNA would not change. By contrast, the DNA of different organisms might well be expected to have different ratios of critical genes to DNA content, which could account for the correlation seen in Table 17–1.

Reference: Prescott DM (1975) Reproduction of Eucaryotic Cells, pp 85–86. New York: Academic Press.

17–6*

17–7

A. The reciprocal temperature-shift experiments demonstrate that *cdc*101 is blocked at 37°C in the G_1 phase of the cell cycle. The first experiment shows that the mutational block precedes or coincides with the hydroxyurea block, or the cells would have divided when they were shifted to 20°C in the presence of hydroxyurea. The second experiment shows that the hydroxyurea block occurs after the mutational block because the cells divided once when they were shifted to 37°C in the absence of hydroxyurea. Together the two experiments indicate that the mutational block in *cdc*101 is in the G_1 phase of the cell cycle. In the first experiment at 37°C the *cdc*101 mutants accumulate in G_1; when shifted into hydroxyurea medium at 20°C, they move to S phase but are stopped there by the hydroxyurea block and thus do not divide. In the second experiment in the presence of hydroxyurea at 20°C the cells are blocked in S phase; when hydroxyurea is removed and the cells are shifted to 37°C, they progress normally through G_2 and M before they are blocked in G_1. Therefore, they undergo one round of cell division.

B. The results with *cdc*102 indicate that it is blocked at 37°C in S phase. The first experiment shows that the mutational block precedes or coincides with the hydroxyurea block, or the cells would have divided when they were shifted to 20°C. The second experiment shows that the hydroxyurea block precedes or coincides with the mutational block, or the cells would have divided when they were shifted to 37°C. Together the two experiments indicate that the mutational block and the hydroxyurea block coincide. Since hydroxyurea and the *cdc*102 mutation both affect the same phase of the cell cycle, the order of treatment makes no difference; the cells remain trapped in S phase and therefore do not divide.

Reference: Hartwell LH (1978) Cell division from a genetic perspective. *J. Cell Biol.* 77, 627–637.

17–8*   Figure 17–39.

Reference: Hartwell LH (1978) Cell division from a genetic perspective. *J. Cell Biol.* 77, 627–637.

17–9   To isolate the wild-type gene, you would first grow a population of the *cdc* mutant cells at 25°C. These cells would then be transfected with the cDNA library and grown at 37°C. Cells that receive a plasmid for a cDNA that does not correspond to the mutant gene will not be able to form a colony at the restrictive temperature. However, cells that receive the wild-type gene corresponding to the mutant gene will be able to progress through the cell cycle normally at the restrictive temperature and thus will form a colony. Plasmid DNA isolated from such colonies can be sequenced to determine the identity of the gene.

Among the cDNAs isolated in experiments like this one, it is not uncommon to find unexpected genes in addition to the one corresponding to the mutant gene. Occasionally, expression of a second gene at higher than normal levels, as usually occurs from plasmids, can compensate for the decreased function of the mutant gene. Oftentimes, characterization of such suppressor genes provides insight into the biological functions of both genes (see Problem 17–22).

17–10*

17–11   Since you examined 25,000 cells and found 3 in mitosis, the mitotic index is 3/25,000, which equals 0.00012. If mitosis is 30 minutes long and the frequency of cells in mitosis is 0.00012, then 30 minutes is 0.00012 of the length of the cell cycle. Thus, the cell cycle is 0.5/0.00012 = 16,600 hours in length, which is nearly two years.

17–12*

17–13
A. Only the cells that were in the S phase of their cell cycle—those that were making DNA—during the 30-minute labeling period contain any radioactive DNA.
B. Initially, mitotic cells contain no radioactive DNA because they were not engaged in DNA synthesis during the labeling period. It takes about 3 hours before the first labeled mitotic cells appear.

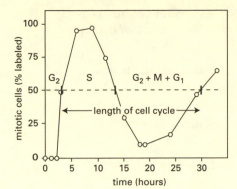

Figure 17–40 Lengths of phases of the cell cycle deduced from pulse labeling (Answer 17–13).

C. The initial rise of the curve corresponds to cells that were just finishing DNA replication when the radioactive thymidine was added. The curve rises to a peak that corresponds to times when all of the mitotic cells were in the S phase during the time of labeling. The labeled cells then exit mitosis, being replaced by unlabeled mitotic cells that were not yet in S phase during the labeling period. After 20 hours the curve starts rising again, because the labeled cells enter their second round of mitosis.

D. The ascending curve passes through 50% labeled mitoses at 3 hours, which corresponds to the length of the G_2 phase. The initial 3-hour lag before labeled mitotic cells appear corresponds to the time between the end of the S phase and the beginning of mitosis (Figure 17–40). The first labeled cells seen in mitosis were those that were just finishing S phase when the radioactive thymidine was added. The length of S phase can be estimated from the width of the first peak at 50% labeled mitoses, which is about 10.5 hours in this experiment (Figure 17–40). The overall length of the cell cycle is the time between the 50% points on the two ascending curves, which is about 27 hours (Figure 17–40). The total cell cycle minus G_2, S, and M is equal to G_1. Thus, G_1 is 13 hours long [27 – (3 + 0.5 + 10.5)].

Reference: Baserga R & Wiebel F (1969) The cell cycle of mammalian cells. *Intern. Rev. Exp. Pathol.* 7, 1–30.

17–14* **Reference:** Stanners CP & Till JE (1960) DNA synthesis in individual L-strain mouse cells. *Biochim. Biophys. Acta* 37, 406–419.

17–15

A. During the second thymidine block, all the cells will accumulate at the beginning of S phase, since they cannot synthesize DNA. Thus, upon release of the second block the synchronized population will begin S phase.

B. The first thymidine block halts all cells that are in S phase. Cells that are not in S phase traverse the cell cycle normally until they reach the beginning of S phase where they stop. Since G_2 + M + G_1 is only 15 hours long, the presence of thymidine for 18 hours should be sufficient for all the cells not originally in S phase to reach the beginning of S phase. Thus at the end of the first thymidine block the majority of the population is at the beginning of S phase, but the rest of the population is distributed throughout S phase. The release of the first thymidine block for 10 hours allows the entire population to move through S phase, but does not allow any of the population to reenter S phase. When the second thymidine block is applied, none of the cells are in S phase. Application of the second thymidine block for 16 hours allows the entire population to move through the cell cycle until they reach the beginning of S phase, where they accumulate.

In fact, in contrast to what is stated in the problem, it seems that a thymidine block does not totally stop DNA synthesis; rather the block slows it to a fraction of its normal rate. Thus a double thymidine block does not result in the entire population accumulating exactly at the G_1–S boundary. The population is actually distributed within the first bit of S phase.

References: Bootsma D, Budke L & Vos O (1964) Studies on synchronous division of tissue culture cells initiated by excess thymidine. *Exp. Cell Res.* 33, 301–309.

Bostock CJ, Prescott DM & Kirkpatrick JB (1971) An evaluation of the double thymidine block for synchronizing mammalian cells at the G_1-S border. *Exp. Cell Res.* 68, 163–168.

Rao PN & Johnson RT (1970) Mammalian cell fusion: I. Studies on the regulation of DNA synthesis and mitosis. *Nature* 225, 159–164.

Xeros N (1962) Deoxyriboside control and synchronization of mitosis. *Nature* 194, 682–683.

COMPONENTS OF THE CELL-CYCLE CONTROL SYSTEM

17–16* Figure 17–41.

17–17 The precise mechanism of the G_1 cell-size checkpoint is not yet known, but one can imagine in a general way how it might work. The amount of DNA in a G_1 cell is invariant; thus its concentration relative to that of cytosolic components steadily decreases as the cell gets bigger in G_1. The observation that the size of each particular type of cell is roughly proportional to its total amount of DNA suggests that the control mechanism depends on comparing the quantity of cytoplasmic components against the quantity of DNA, with the cell cycle being delayed until a threshold ratio of one or more cytoplasmic components to DNA mass is attained by the biosyntheses responsible for cell growth. If there were binding sites for such a component distributed along the DNA, then an appropriate size could be signaled by saturation of the DNA-binding sites (Figure 17–42).

17–18*

17–19

A. Since each transfer accomplishes a 20-fold dilution (50 nL/1000 nL), 10 transfers yield a dilution factor of 20^{10}, which is equal to 10^{13}. It is unreasonable for a molecule to have an undiminished biological effect over this range of dilution.

B. The appearance of MPF activity in the absence of protein synthesis suggests that an inactive precursor of MPF is being activated. In principle, activation could involve one of several kinds of posttranslational modifications such as protease cleavage or changes in protein phosphorylation. The final step of MPF activation is the dephosphorylation of its Cdc2 subunit.

C. In order for MPF to propagate its activated state through serial transfers, it must be able to activate itself. If it were a protease, for example, active MPF might activate its inactive precursor by cleavage, in the same way that trypsin-mediated cleavage of trypsinogen produces more trypsin. In the case of MPF, however, active MPF functions as a protein kinase that indirectly activates its inactive precursor by phosphorylation (see Problem 17–29). Note that it is not necessary that MPF activates itself; for example, it could activate another protein kinase, which in turn activates the precursor to MPF. Nevertheless, the principle is the same.

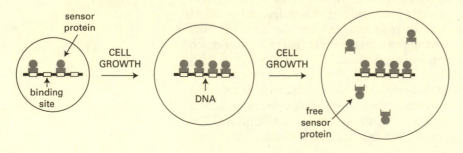

Figure 17–42 A hypothetical model for coordination of cell growth and cell-cycle progression (Answer 17–17). A sensor protein whose concentration increases during cell growth in G_1 is bound tightly to sites in the DNA until the sites are titrated at a particular degree of cell growth. Unbound sensor protein would signal adequate growth to permit entry into S phase.

D. Since there is no detectable MPF activity in an immature oocyte, MPF cannot be the source of the original activation event. Presumably, a protein synthesized in response to progesterone stimulation (therefore cycloheximide sensitive) is responsible, directly or indirectly, for the initial activation of MPF.

Reference: Wasserman WJ & Masui Y (1975) Effects of cycloheximide on a cytoplasmic factor initiating meiotic maturation in *Xenopus* oocytes. *Exp. Cell Res.* 91, 381–388.

17–20* **Reference:** Hunt T, Luca FC & Ruderman JV (1992) The requirement for protein synthesis and degradation, and the control of destruction of cyclins A and B in the meiotic and mitotic cell cycles of the clam embryo. *J. Cell Biol.* 116, 707–724.

17–21 At different transitions in the cell cycle in budding yeasts, the single Cdk (Cdk1) binds to different cyclins. These cyclins not only activate Cdk1, but also influence its target specificity. As a result, each cyclin–Cdk1 complex phosphorylates a different set of substrate proteins. Even though vertebrates use multiple Cdks, they use a variety of cyclins to target the cyclin–Cdk complexes to different substrates at different stages of the cell cycle.

17–22* **Reference:** Hadwiger JA, Wittenberg C, Richardson HE, de Barros Lopes M & Reed SI (1989) A family of cyclin homologs that control the G_1 phase in yeast. *Proc. Natl. Acad. Sci. USA* 86, 6255–6259.

17–23
A. When the promoter is turned on and Cln3 is expressed at high levels, many fewer cells are in G_1 and many more cells are present in S and G_2. Overexpression of Cln3 might reasonably be expected to increase the amount of active G_1-Cdk, thereby pushing cells through Start and committing them to DNA replication. If cells were constantly stimulated in this way, fewer would be expected in G_1 and more would be expected in S and G_2.
B. If the cells are being forced through G_1 more quickly than is normal, they will have less time to grow before cell division. As a result, the cells will be smaller than normal.

Reference: Richardson HE, Wittenberg C, Cross F & Reed SI (1989) An essential G_1 function for cyclin-like proteins in yeast. *Cell* 59, 1127–1133.

17–24*

17–25 You have isolated a gene with remarkable properties, which correspond precisely to those of the null allele and the *wee-ts* allele of the fission yeast *cdc2* gene, as discovered by Paul Nurse in 1974. Nurse had been very excited by his discovery of the *wee1* gene, which, when mutated, gave rise to small yeast cells that entered mitosis at a smaller-than-usual size compared to wild-type cells. To see if any other genes could be identified with similar properties, he set up a genetic screen based on looking at the mutants in the microscope, hunting for small cells. After identifying 49 new temperature-sensitive wee strains, all of which proved to be allelic to the original *wee1* gene, he was on the point of giving up when the 50th ts wee strain turned out not to map in the *wee1* locus. Instead, it corresponded to a different, but already known gene, *cdc2*. Because this gene could be mutated to inactivity or hyperactivity, Nurse concluded that it must encode a really important cell-cycle regulator. Many genes can be mutated so that they block cell-cycle progression, but this was the only one that could apparently be made more active. Time proved this conclusion to be absolutely correct.

Given that Wee1 is a protein kinase that phosphorylates Cdc2, one could well imagine that a mutation in tyrosine-15 of Cdc2, known to be the target of Wee1, could give rise to such a hyperactivated form of Cdc2. It could not be restrained by the normal signaling network that couples size to cell division. In principle, however, one could imagine that if it were possible to overproduce Wee1, or obtain hyperactive forms of Wee1 by mutation (perhaps by altering or deleting its own regulatory regions) then strains containing extra Wee1 activity would be unable, or be very slow, to enter mitosis. In actual fact, however, the mutations that give rise to *wee* alleles of *cdc2* map elsewhere in the molecule.

You might have proposed that mutations in a gene for an activator of cell division such as Cdc25 could be responsible for phenotypes. Indeed, loss of Cdc25 gives rise to cells that cannot enter mitosis and are large. If it were possible to hyperactivate Cdc25, then such strains ought to enter mitosis prematurely. No such mutations have ever been isolated in such screens, however.

Note that fission yeast continue to grow even though they cannot divide. Cdc2 does not control cell growth, although in normal cells it somehow responds to size, most likely through the regulatory network comprising *wee1*, *cdc25* and CDK inhibitors.

References: Nurse P (1975) Genetic control of cell size at cell division in yeast. *Nature* 256, 547–551.

Nurse P & Thuriaux P (1980) Regulatory genes controlling mitosis in the fission yeast *Schizosaccharomyces pombe*. *Genetics* 96, 627–637.

INTRACELLULAR CONTROL OF CELL-CYCLE EVENTS

17–26* **Reference:** Johnson RT & Rao PN (1971) Nucleo-cytoplasmic interactions in the achievement of nuclear synchrony in DNA synthesis and mitosis in multinucleate cells *Biol. Rev. Camb. Philos. Soc.* 46, 97–155.

17–27 S-Cdk could initiate the firing of replication origins—directly or indirectly— by activating origin-binding proteins. Such binding proteins might bind to pre-replicative complexes at different times in S phase as a result of their accessibility or surrounding DNA sequences. If the binding protein when bound to the pre-replicative complex created a binding site for S-Cdk (or an S-Cdk-activated protein kinase), then Cdc6 might be a target for phosphorylation only in those origins that have fired. Phosphorylation of Cdc6 leads to its ubiquitylation and destruction, thereby preventing re-replication. S-Cdk is also known to cause phosphorylation of Mcm proteins and their export from the nucleus. If S-Cdk maintained the phosphorylation of Cdc6 and Mcm proteins until it was inactivated at mitosis, re-replication would be effectively prevented.

17–28* Figure 17–43.

17–29

A. The state of phosphorylation of Wee1 and Cdc25 is the result of the balance between the protein kinase and protein phosphatase activities that regulate them. By inhibiting the protein phosphatases, okadaic acid causes Wee1 and Cdc25 to accumulate in their phosphorylated forms (Figure 17–44). Since this change activates M-Cdk, Wee1 and Cdc25 must have originally been present in the extract in their nonphosphorylated forms. Thus active Wee1 tyrosine kinase is nonphosphorylated, as is inactive Cdc25 tyrosine phosphatase (Figure 17–44). Knowing which forms are phosphorylated

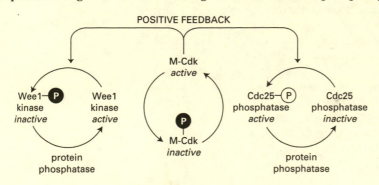

Figure 17–44 Control of M-Cdk activity by Wee1 kinase and Cdc25 phosphatase (Answer 17–29). Inhibitory phosphates are shown in *black*; activating phosphate is shown in *white*.

allows you to label the arrows that correspond to the kinases and phosphatases that control Wee1 and Cdc25 phosphorylation (Figure 17–44).

B. The protein kinases and phosphatases that control phosphorylation of Wee1 and Cdc25 must be specific for serine/threonine side chains because they are affected by okadaic acid, which inhibits only serine/threonine phosphatases.

C. Okadaic acid has no direct effect on Cdk1 phosphorylation because it is phosphorylated on a tyrosine side chains. Tyrosine phosphatases are unaffected by okadaic acid. The decrease in Cdk1 phosphorylation is a consequence of the change in activation of Wee1 kinase and Cdc25 phosphatase.

D. As soon as some active M-Cdk appears, it would begin to phosphorylate Wee1 and Cdc25, inactivating the kinase and activating the phosphatase. The resultant decrease in Wee1 kinase activity and increase in Cdc25 phosphatase activity would lead to dephosphorylation (and activation) of more M-Cdk. This in turn would further decrease the activity of Wee1 kinase and further increase the activity of Cdc25 phosphatase, leading to still more M-Cdk activity. Thus the initial appearance of a little M-Cdk activity would rapidly lead to its complete activation.

This sort of activation is referred to as a positive feedback loop. It is a common means of regulation when it is advantageous for a system to flip rapidly from one state to another without lingering in the intermediate states.

Reference: Kumagai A & Dunphy WG (1992) Regulation of the Cdc25 protein during the cell cycle in *Xenopus* extracts. *Cell* 70, 139–151.

17–30*

17–31 The doses of caffeine required to interfere with the DNA replication checkpoint mechanism are much higher than the amount imbibed by even the most excessive drinkers of coffee and colas. The concentration of caffeine in a cup of coffee is about 3.4 mM.

$$[\text{caffeine}] = \frac{100 \text{ mg}}{150 \text{ mL}} \times \frac{g}{1000 \text{ mg}} \times \frac{\text{mole}}{196 \text{ g}} \times \frac{1000 \text{ mL}}{\text{L}}$$
$$= 3.4 \times 10^{-3} \text{ M} = 3.4 \text{ mM}$$

Since the concentration in a cup is less than the 10 mM required to interfere with the DNA replication checkpoint mechanism, you can't really get a higher concentration by drinking it and diluting it in the water volume of the body. If you assume for the purposes of calculation, however, that the caffeine is not metabolized or excreted (but that all the liquid is), then you can ask how many cups of coffee would you need to drink (at 100 mg of caffeine per cup) to reach a concentration of 10 mM in 40 L of body water. You would need to drink 784 cups of coffee!

17–32* **Reference:** Holloway SL, Glotzer M, King RW & Murray AW (1993) Anaphase is initiated by proteolysis rather than by the inactivation of maturation-promoting factor. *Cell* 73, 1393–1402.

17–33

A. Sister chromatids behave as expected for wild-type cells: they are stuck together in small budded cells and separated into the mother and daughter cells at mitosis. The single spot of fluorescence in the small-budded cell in Figure 17–16 likely results from two sites of binding that are close together on the paired sister chromatids. The spot is brighter than the individual spots in cells that have two separated spots, and the size of the bud indicates that the site should have been replicated (by comparison with the size of the small-budded *scc1*[ts] cell with two spots).

B. In one of the small-budded cells from *scc1*[ts] it is clear from the presence of two spots of fluorescence that sister chromatids have already separated. (In the cell with the smaller bud and single spot of fluorescence the site of GFP binding may not have been replicated.) Although the sister chromatids are separated in the large-budded cells, they are abnormal because they have both remained in the mother cell.

C. Prematurely separated sister chromatids prevent the formation of a normal spindle apparatus and thus, prevent normal segregation of chromosomes into the mother and daughter cells. It is likely that they trigger the spindle-attachment checkpoint; that is, that unpaired sister chromatids behave as if they are unattached.

Reference: Michaelis C, Ciosk R & Nasmyth K (1997) Cohesins: chromosomal proteins that prevent premature separation of sister chromatids. *Cell* 91, 35–45.

17–34* **Reference:** Uhlmann F, Lottspeich F & Nasmyth K (1999) Sister-chromatid separation at anaphase onset is promoted by cleavage of the cohesin subunit of Scc1. *Nature* 400, 37–42.

17–35 Nocodazole arrests cells in M phase of the cell cycle. By preventing microtubule polymerization—hence spindle formation—nocodazole triggers the spindle-attachment checkpoint, which inhibits the APC ubiquitin ligase so that the metaphase-to-anaphase transition cannot occur.

17–36* **Reference:** Li R & Murray AW (1991) Feedback control of mitosis in budding yeast. *Cell* 66, 519–531.

17–37 M-phase cyclins are destroyed by proteasome digestion after they have been ubiquitylated by either Cdc20–APC or Hct1–APC. M-Cdk activity is required for activation of Cdc20–APC. The activity of Cdc20–APC rises after M-Cdk is activated and declines with a delay after the loss of M-Cdk activity, as the activating phosphates are removed by a phosphatase. M-Cdk also plays a direct role in controlling the activity of Hct1–APC by inhibiting it through phosphorylation. When the activity of M-Cdk begins to fall, the activity of Hct1–APC rises as the inhibitory phosphates are removed by a phosphatase.

17–38* **References:** Fung Y-KT, Murphree AL, T'Ang A, Qian J, Hinrichs SH & Benedict WF (1987) Structural evidence for the authenticity of the human retinoblastoma gene. *Science* 236, 1657–1661.

Knudson AG (1971) Mutation and cancer: statistical study of retinoblastoma. *Proc. Natl. Acad. Sci. USA* 68, 820–823.

17–39 Antiproliferative genes such as *Rb* encode proteins that stop the cell cycle. During normal cell division, these proteins must be turned off. If the proteins were overexpressed in all cells, it is likely that the machinery that keeps these proteins turned off would be overwhelmed, and cell division would stop. Thus, this cure for cancer might be successful, but the patient would be dead.

17–40*

17–41 The cells should be larger than normal because cell size would have to increase to generate sufficient Cln3 to occupy the extra binding sites. Only then would there be excess Cln3 to activate G_1-Cdk and trigger cell division.

17–42* **Reference:** Nurse P (1975) Genetic control of cell size at cell division in yeast. *Nature* 256, 547–551.

17–43 In order to generate more cells of the same size in the same amount of time, both the duration of the cell cycle and the rate of cell growth must be altered by overexpression of cyclin D and Cdk4. To account for the observed results, the cells must grow faster and divide more quickly (shorter cell cycle). To generate the same size cells, the increase in growth rate must be exactly balanced by the decrease in cell-cycle time. These results suggest that cyclin D and Cdk4 in some way coordinate growth rate and cycle time. (It is interesting to note that flies defective for both copies of Cdk4 have a slower growth rate and a longer cell-cycle time, but maintain normal sized cells.)

References: Meyer CA, Jacobs HW, Datar SA, Du W, Edgar BA & Lehner CF (2000) *Drosophila* Cdk4 is required for normal growth and is dispensable for cell cycle progression. *EMBO J.* 19, 4533–4542.

Datar SA, Jacobs HW, Flor A, de la Cruz A, Lehner CF & Edgar BA (2000) The *Drosophila* cyclin D-Cdk4 complex promotes cellular growth. *EMBO J.* 19, 4543–4554.

17–44* **References:** Meyer CA, Jacobs HW, Datar SA, Du W, Edgar BA & Lehner CF (2000) *Drosophila* Cdk4 is required for normal growth and is dispensable for cell cycle progression. *EMBO J.* 19, 4533–4542.

Datar SA, Jacobs HW, Flor A, de la Cruz A, Lehner CF & Edgar BA (2000) The *Drosophila* cyclin D-Cdk4 complex promotes cellular growth. *EMBO J.* 19, 4543–4554.

17–45

A. Radiation leads to DNA damage, which activates a DNA damage checkpoint (mediated by p53 and p21), which arrests the cell cycle until the DNA has been repaired.

B. In the absence of a functional DNA damage checkpoint, the cell will replicate the damaged DNA, introducing mutations into the genomes inherited by the daughter cells.

C. A checkpoint-deficient cell will be able to divide normally, but it will be prone to mutations, because some DNA damage always occurs as the result of natural processes (for example, by cosmic rays). The checkpoint mediated by p53 is mainly required as a safeguard against the devastating effects of DNA damage, but not for the natural progression of the cell cycle in undamaged cells.

D. Cell division is an ongoing process that does not cease upon reaching maturity. Blood cells, epithelial cells in the skin or lining the gut, and the cells of the immune system, for example, are being constantly produced by cell division to meet the body's needs; your body produces about 10^{11} new blood cells each day.

17–46*

17–47

A. Careful examination of the time-lapse pictures in Figure 17–23 shows that all the cells without buds arrested at the dumbbell stage, whereas all of the cells with buds formed viable colonies. The appearance of a bud corresponds with the beginning of S phase. Haploid cells that have partially or fully replicated their genomes are more resistant to x-ray-induced breaks because a break in one chromosome can be repaired by recombination with the intact sister chromatid. Haploid cells in G_1 are especially sensitive to breaks because they contain no second intact copy of the chromosome with which to recombine. After replication such a cell will contain two copies of the chromosome, but both will be broken at the same position. Thus even in G_2, a haploid cell that suffers a break in G_1 will not have an intact chromosome with which to repair itself by homologous recombination.

B. The observation that dumbbell-stage cells have a single nucleus and no spindle indicates that the cells are arrested in G_2 prior to mitosis. The presence of a bud indicates that the cells have already passed the G_1 checkpoint and such cells will complete S phase.

C. Half the wild-type cells temporarily arrest in the dumbbell stage while they wait for damage to be repaired. A cell-cycle checkpoint senses damaged DNA and halts the cell cycle until the damage is repaired. When repair is complete, the cells enter mitosis and divide to produce viable microcolonies. The nonviable cells died, either because they suffered damage too late to stop and divided with damaged chromosomes, or because they suffered so much damage that it could not be repaired.

D. *rad52* cells remain arrested at the dumbbell stage because they are incapable of repairing their damaged chromosomes. The continued signal from the damaged DNA prevents the cells from passing the mitotic entry checkpoint.

E. Very few *rad9* mutant cells arrest at the dumbbell stage because they are defective in their ability to sense DNA damage. In these cells the mitotic entry checkpoint does not function. Division in the absence of repair leads to haploid cells that have broken chromosomes and cells that are missing pieces of chromosomes. Both situations lead to nonviable cells. Only a small fraction of cells (30%) manages to repair their chromosomes in the absence of a checkpoint delay.

F.   If *rad9* cells are artificially delayed in mitosis, the number of viable cells increases. The artificial delay allows the cells time to repair their damaged chromosomes so that they then can complete mitosis with an intact genome. The important point is that *rad9* cells contain all the necessary enzymes required for DNA repair; they are defective only in sensing DNA damage.

References: Hartwell LH & Weinert TA (1989) Checkpoints: controls that ensure the order of cell cycle events. *Science* 246, 629–634.

Weinert TA & Hartwell LH (1988) The *rad9* gene controls the cell cycle response to DNA damage in *Saccharomyces cerevisiae*. *Science* 241, 317–322.

17–48* Figure 17–45.

Reference: Murray A & Hunt T (1993) The Cell Cycle: An Introduction, pp 143–144. New York: WH Freeman.

17–49

A.   Cells that cannot degrade M-phase cyclins would be unable to divide. The cells would enter mitosis but would not be able to exit.

B.   Cells that always expressed high levels of p21 would be unable to divide. The cells would arrest permanently in G_1 because their G_1/S-Cdk and their S-Cdk would be inactivated.

C.   Cells that cannot phosphorylate Rb would be unable to divide. The cells would not be able to activate the transcription of genes required for entry into S phase because the required regulatory proteins would be sequestered by unphosphorylated Rb.

PROGRAMMED CELL DEATH (APOPTOSIS)

17–50*

17–51   Because programmed cell death occurs on a large scale in both developing and adult tissues, it must not trigger alarm reactions that are normally associated with cell injury. In tissue injury, for example, signals are released that can cause a destructive inflammatory reaction. Moreover, the release of intracellular contents could elicit an immune response against molecules that are normally not encountered by the immune system. In normal development, such reactions would be self-defeating, even dangerous, if they occurred in response to programmed cell death.

17–52*

17–53   The plasma membrane of the cell that died by necrosis (see Figure 17–25A) is ruptured; several clear breaks are visible, for example, at 8, 9, and 12 o'clock. The cell's contents, mostly membranous and cytoskeletal debris, are seen spilling into the surroundings. The cytosol stains lightly, as most soluble components had been lost before the cell was fixed. By contrast, an intact membrane surrounds the cell that underwent apoptosis (see Figure 17–25B), and its cytosol is densely stained, indicating a normal concentration of cellular components. The cell's interior is remarkably different from a normal cell, however. Particularly characteristic are the large blobs that extrude from the nucleus, probably as the result of the breakdown of the nuclear lamina. The cytosol also contains many large, round membrane-enclosed vesicles of unknown origin that are not normally seen in healthy cells. The pictures visually confirm the notion that necrosis involves cell lysis, whereas cells undergoing apoptosis remain relatively intact until they are engulfed and digested inside a normal cell.

17–54* **Reference:** Siegel RM, Chan FK-M, Chun HJ & Lenardo MJ (2000) The multifaceted role of Fas signaling in immune cell homeostasis and auto-immunity. *Nature Immunol.* 1, 469–474.

17–55 Overexpression of a secreted protein that binds to Fas ligand would protect tumor cells from attack by killer lymphocytes. By binding to the Fas ligand on killer lymphocytes the secreted protein would prevent the Fas ligand from binding to Fas on tumor cells, thereby insulating them from death-inducing interactions with killer lymphocytes. Secreted proteins that bind to Fas ligand are commonly known as decoy receptors. They play a normal role in modulating the killing induced by interactions between Fas ligand and Fas. When tumor cells overproduce such decoy receptors, however, they subvert this normal mechanism into a cellular defense against immune-cytotoxic killing.

Reference: Pitti RM, Marsters SA, Lawrence DA, Roy M, Kischkel FC, Dowd P, Huang A, Donahue CJ, Sherwood SW, Baldwin DT, Godowski PJ, Wood WI, Gurney AL, Hillan KJ, Cohen RL, Goddard AD, Botstein D & Ashkenazi A (1998) Genomic amplification of a decoy receptor for Fas ligand in lung and colon cancer. *Nature* 396, 699–703.

17–56* **References:** Jacobson MD, Burne JF, King MP, Miyashita T, Reed JC & Raff MC (1993) Bcl-2 blocks apoptosis in cells lacking mitochondrial DNA. *Nature* 361, 365–369.

Liu X, Kim CN, Yang J, Jemmerson R & Wang X (1996) Induction of apoptotic program in cell-free extracts: Requirement for dATP and cytochrome *c*. *Cell* 86, 147–157.

17–57 *C. elegans* mutants that were defective for cytochrome *c* would not be viable. Cytochrome *c* is an essential component of the electron transport chain in mitochondria. Without it, no production of ATP by oxidative phosphory-lation would be possible and such a mutant organism could not survive. In mice, where reverse genetics is feasible, it is possible to knock out both copies of the gene for cytochrome *c*. These mice die *in utero*. Cells rescued from such embryos before they die, however, can be cultured under special conditions and studied. They are defective in cytochrome *c*-mediated apoptosis.

Reference: Ellis HM & Horvitz RH (1986) Genetic control of programmed cell death in the nematode *C. elegans*. *Cell* 44, 817–829.

17–58* **Reference:** Goldstein JC, Waterhouse NJ, Juin P, Evan GI & Green DR (2000) The coordinate release of cytochrome *c* during apoptosis is rapid, complete and kinetically invariant. *Nature Cell Biol.* 2, 156–162.

17–59 If the caspases that are activated downstream of cytochrome *c* release were required to accelerate cytochrome *c* release in order to yield the observed rapid response, the presence of the caspase inhibitor should slow the release of cytochrome *c*. Since the caspase inhibitor did not increase the time required for cytochrome *c* release, it is unlikely that a caspase-mediated, positive-feedback loop is involved for apoptosis induced by actinomycin D, staurosporine, or UV light.

Reference: Goldstein JC, Waterhouse NJ, Juin P, Evan GI & Green DR (2000) The coordinate release of cytochrome *c* during apoptosis is rapid, complete and kinetically invariant. *Nature Cell Biol.* 2, 156–162.

17–60* **Reference:** Wei MC, Zong W-X, Cheng EH-Y, Lindsten T, Panoutsakopoulou V, Ross AJ, Roth KA, MacGregor GR, Thompson CB & Korsmeyer SJ (2001) Proapoptotic BAX and BAK: A requisite gateway to mitochondrial dysfunc-tion and death. *Science* 292, 727–730.

17–61 Upon microinjection of cytochrome *c* both cell lines undergo apoptosis. The presence of cytochrome *c* in the cytosol is a signal for downstream events that lead to apoptosis. Cells that are defective for both Bax and Bak cannot release cytochrome *c* from mitochondria in response to upstream signals, but there is no defect in the downstream part of the pathway that is triggered by cytochrome *c*. Thus, microinjection bypasses the defects in the doubly defective cells, triggering apoptosis.

Reference: Wei MC, Zong W-X, Cheng EH-Y, Lindsten T, Panoutsakopoulou V, Ross AJ, Roth KA, MacGregor GR, Thompson CB & Korsmeyer SJ (2001) Proapoptotic BAX and BAK: A requisite gateway to mitochondrial dysfunction and death. *Science* 292, 727–730.

17–62* **Reference:** Li K, Li Y, Shelton JM, Richardson JA, Spencer E, Chen ZJ, Wang X & Williams RS (2000) Cytochrome *c* deficiency causes embryonic lethality and attenuates stress-induced apoptosis. *Cell* 101, 389–399.

17–63

A. All the treatments, except for Fas ligand, signal apoptosis at a point after Bid and before Bax and Bak. These results suggest that there is at least one other way to activate Bax and Bak in addition to activation via Bid. Fas ligand signals apoptosis through a pathway that does not involve Bid, Bax, or Bak.

B. It may be surprising that Fas ligand, which binds to Fas to activate caspase-8 cleavage of Bid (see Figure 17–29), still causes apoptosis in these deficient cells. Caspase-8, however, has additional targets for cleavage. In MEFs it can also trigger a caspase cascade directly, bypassing Bid, Bax, and Bak. In MEFs and other cells activation of caspase-8 may normally trigger both pathways as a way of ensuring rapid cell death.

Reference: Wei MC, Zong W-X, Cheng EH-Y, Lindsten T, Panoutsakopoulou V, Ross AJ, Roth KA, MacGregor GR, Thompson CB & Korsmeyer SJ (2001) Proapoptotic BAX and BAK: A requisite gateway to mitochondrial dysfunction and death. *Science* 292, 727–730.

17–64* **Reference:** Tournier C, Hess P, Yang DD, Xu J, Turner TK, Nimnual A, Bar-Sagi D, Jones SN, Flavell RA & Davis RJ (2000) Requirement of JNK for stress-induced activation of the cytochrome c-mediated death pathway. *Science* 288, 870–874.

17–65 Mice that are deficient for Apaf-1 or caspase-9 are defective for cytochrome *c*-dependent apoptosis. Apoptosis is a critical event in development, allowing excess brain cells to be weeded out. The extent of brain overgrowth and size of the cranial protrusions indicate that this weeding out process in the developing brain must be massive. The dramatic effects of the deficiencies of Apaf-1 and caspase-9 suggest that the cytochrome *c*-dependent apoptotic pathway must be critically important in brain development.

References: Cecconi F, Alvarez-Bolado G, Meyer BI, Roth KA & Gruss P (1998) Apaf1 (CED-4 homolog) regulates programmed cell death in mammalian development. *Cell* 94, 727–737.

Yoshida H, Kong Y-Y, Yoshida R, Elia AJ, Hakem A, Hakem R, Penninger JM & Mak TW (1998) Apaf1 is required for mitochondrial pathways of apoptosis and brain development. *Cell* 94, 739–750.

Kuida K, Haydar TF, Kuan C-Y, Gu Y, Taya C, Karasuyama H, Su MS-S, Rakic P & Flavell RA (1998) Reduced apoptosis and cytochrome *c*-mediated caspase activation in mice lacking caspase 9. *Cell* 94, 325–337.

Hakem R, Hakem A, Duncan GS, Henderson JT, Woo M, Soengas MS, Elia A, de la Pompa JL, Kagi D, Khoo W, Potter J, Yoshida R, Kaufman SA, Lowe SW, Penninger JM & Mak TW (1998) Differential requirement for caspase 9 in apoptotic pathways *in vivo*. *Cell* 94, 339–352.

17–66* **Reference:** Earnshaw WC, Martins LM & Kaufmann SH (1999) Mammalian caspases. *Annu. Rev. Biochem.* 68, 383–424.

17–67 These results show that alkylation of DNA is responsible for the apoptotic signal. MGMT-deficient and MGMT-overexpressing cells are equally sensitive to γ-irradiation, indicating that the apoptotic apparatus is intact in both cell lines. MGMT-overexpressing cells are much more resistant to MNNG treatment than are MGMT-deficient cells, suggesting that efficient removal of O^6-methylguanine lesions prevents apoptosis.

Reference: Meikrantz W, Bergom MA, Memisoglu A & Samson L (1998) O^6-alkylguanine DNA lesions trigger apoptosis. *Carcinogenesis* 19, 369–372.

17–68* **Reference:** D'Atri S, Tentori L, Lacal PM, Graziani G, Pagani E, Benincasa E, Zambruno G, Bonmassar E & Jiricny J (1998) Involvement of the mismatch repair system in temozolomide-induced apoptosis. *Mol. Pharmacol.* 54, 334–341.

EXTRACELLULAR CONTROL OF CELL DIVISION, CELL GROWTH, AND APOPTOSIS

17–69 Mitogens stimulate cell division, primarily by relieving intracellular negative controls that otherwise block progress through the cell cycle. Growth factors stimulate cell growth (an increase in cell mass) by promoting the synthesis of proteins and other macromolecules and by inhibiting their degradation. Survival factors promote cell survival by suppressing apoptosis.

17–70*

17–71

A. Most of the cells in an adult human are in this class, having withdrawn from the cell cycle into G_0. Liver cells, for example, remain quiescent for long periods, although they can grow and divide when the need arises.

B. Nerve cells grow as they extend axons over long distances, but do not divide. Fat cells can accumulate large quantities of triglyceride, which causes them to increase in size (although this is not properly growth, per se). Oocytes grow to become very large cells prior to fertilization.

C. This is the most rare category of cell, but the production of red blood cells is a good example. During production of red blood cells, precursor reticulocytes undergo five cell divisions with little increase in overall volume, ultimately generating very small red blood cells from a much larger precursor cell.

D. Most cells in the human body grow and divide actively at some point during development, until we become adults. Even in adults some cells continue to grow and divide; most notably, intestinal cells and hematopoietic cells, which must constantly renew the lining of the gut and the cells in the blood, respectively. Most other cells grow and divide often enough to balance cell death.

Reference: Dolznig H, Bartunek P, Nasmyth K, Mullner EW & Beug H (1995) Terminal differentiation of normal chicken erythroid progenitors: shortening of G_1 correlates with loss of D-cyclin/cdk4 expression and altered cell size control. *Cell Growth Differ.* 6, 1341–1352.

17–72* **Reference:** O'Keefe EJ & Pledger WJ (1983) A model of cell-cycle control: sequential events regulated by growth factors. *Mol. Cell. Endocrinol.* 31, 167–186.

17–73

A. Experiment 1 in Table 17–5 shows that the 170-kd protein has a high-affinity binding site for EGF and therefore is most likely the EGF receptor. The control experiment of incubating the membrane preparation in the presence of excess unlabeled EGF demonstrates that EGF binding to the 170-kd protein is specific, not random.

B. Experiments 2, 3, and 4 show that the 170-kd protein (the EGF receptor) becomes phosphorylated with radioactive phosphate in the presence of γ-^{32}P-ATP. Transfer of the phosphate from the γ position in ATP demonstrates that the EGF receptor is a substrate for a protein kinase. Since the labeling intensity of the 170-kd protein is increased in the presence of EGF, EGF stimulates the activity of the protein kinase.

C. Experiments 3 and 4 show that the EGF receptor can transfer phosphate from ATP to protein; thus, it is a protein kinase. Experiment 3 is less convincing than experiment 4. In experiment 3, although the antibody is specific for the EGF receptor, it is not unreasonable to question whether other proteins, perhaps including a protein kinase, might have been trapped within the antibody precipitate. In experiment 4 the EGF receptor is first separated by molecular weight from other proteins that might contaminate the antibody precipitate. Only in the unlikely event that the contaminating protein kinase was also a 170-kd protein would experiment 4 lead you astray.

D. With the caveat mentioned in part C, experiment 4 indicates convincingly that the EGF receptor is a substrate for its own protein kinase activity.

Reference: Cohen S, Ushiro H, Stocheck C & Chinkers M (1982) A native 170,000 epidermal growth factor receptor-kinase complex from shed plasma membrane vesicles. *J. Biol. Chem.* 257, 1523–1531.

17–74*

17–75 False. Serum-deprived cells continue through the current cell cycle until they reach the G_1 checkpoint where they enter G_0. Cells can only enter G_0 from G_1.

17–76* **Reference:** Baldin V, Lukas J, Marcote MJ, Pagano M & Draetta G (1993) Cyclin D1 is a nuclear protein required for cell cycle progression in G_1. *Genes Dev.* 7, 812–821.

17–77 Cancer cells have additional changes that typically disable cell-cycle checkpoints and apoptotic mechanisms. In the absence of these regulatory controls, which are fully operational in normal cells, overexpression of Myc drives cell growth and proliferation of cancer cells.

17–78*

17–79
A. As discussed in Problem 17–38, the absence of a shoulder on any of the three curves suggests that in all cases only a single event is needed to trigger tumor production in mice that are already expressing one or both oncogenes.

B. Although the rate of tumor production is much higher in mice with both oncogenes, activation of the cellular *Ras* gene cannot be a required event in the production of tumors in mice that are already expressing the MMTV-regulated *Myc* oncogene. Nor can activation of the cellular *Myc* gene be a required event in triggering tumor formation in mice that are already expressing the MMTV-regulated *Ras* oncogene. As indicated in part A, even when mice contain both *Myc* and *Ras*, some additional event is required to produce a tumor. If *Myc* plus *Ras* were sufficient for tumor formation, then all mice would develop tumors as soon as they passed through puberty.

C. The rate of tumor production in mice with both oncogenes is much higher than expected if the effects of the individual oncogenes were additive. Thus the two oncogenes together have a synergistic effect on the rate of tumor production. As argued in part B, however, activation of both oncogenes is not sufficient to generate a tumor. Thus, the two oncogenes acting together must open up a pathway to tumor production that can be triggered by any one of several low-frequency events or that can be triggered by one very common event. The nature of the activating events is unclear for any of these transgenic mice.

Reference: Sinn E, Muller W, Pattengale P, Tepler I, Wallace R & Leder P (1987) Coexpression of MMTV/v-Ha-*ras* and MMTV/c-*myc* genes in transgenic mice: synergistic action of oncogenes *in vivo*. *Cell* 49, 465–475.

17–80*

17–81 The gene for telomerase is turned off early in development and remains off in most cells in humans. Thereafter, each time a cell replicates its chromosomes, it fails to copy a short segment of the telomeric DNA at the very end of the chromosome (see Problems 5–50 and 5–51). As a result, the telomere becomes progressively shorter with each cell division; thus the length of the telomere is a rough gage of the number of times a cell has divided. When the telomere gets too short to function properly, it triggers a p53-dependent cell-cycle arrest.

17–82* **Reference:** Weinrich SL, Pruzan R, Ma L, Ouellette M, Tesmer VM, Holt SE, Bodnar AG, Lichtsteiner S, Kim NW, Trager JB, Taylor RD, Carlos R, Andrews WH, Wright WE, Shay JW, Harley CB & Morin GB (1997) Reconstitution of human telomerase with the template RNA component hTR and the catalytic protein subunit hTRT. *Nature Genet.* 17, 498–502.

17–83 False. Organism senescence (aging) is distinct from replicative senescence, which occurs in the absence of telomerase. Aging is thought to depend largely on progressive oxidative damage to macromolecules, as strategies that reduce metabolism, and thereby reduce the production of reactive oxygen species, can extend the lifespan of experimental animals.

17–84* **References:** Edgar BA & Lehner CF (1996) Developmental control of cell cycle regulators: A fly's perspective. *Science* 274, 1646–1652.

Prober DA & Edgar BA (2001) Growth regulation by oncogenes—new insights from model organisms. *Curr. Opin. Genet. Dev.* 11, 19–26.

Zecca M, Basler K & Struhl G (1995) Sequential organizing activities of engrailed, hedgehog and decapentaplegic in the *Drosophila* wing. *Development* 121, 2265–2278.

17–85 In alcoholism, liver cells proliferate because the organ is overburdened and becomes damaged by the large amounts of alcohol that have to be metabolized. This need for more liver cells activates the control mechanisms that normally regulate proliferation. Unless badly damaged, the liver will usually shrink back to a normal size after the patient stops drinking. In a liver tumor, in contrast, mutations abolish normal cell proliferation control, and as a result, cells divide and keep on dividing in an uncontrolled manner.

17–86* Table 17–7.

Reference: Folkman J & Moscona A (1978) Role of cell shape in growth control. *Nature* 273, 345–349.

Cortical spindles in an early *Drosophila* embryo. Divisions always occur on a plane parallel to the plasma membrane. Within this plane, spindles orient to maximize separation. Courtesy of Kristina Yu and William Sullivan.

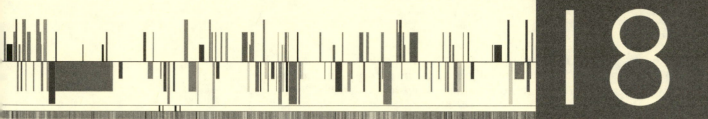

THE MECHANICS OF CELL DIVISION

AN OVERVIEW OF M PHASE

18–1 Cohesins must be present during S phase because it is only while DNA is being replicated that sister chromatids can be reliably identified. Once sister chromatids have separated, it is impossible for a nonspecific DNA-binding protein like cohesin to tell which chromosomes are sisters. And it would be virtually impossible for any protein to distinguish sister chromatids from homologous chromosomes. If sister chromatids are not kept together after their formation, they cannot be accurately segregated to the two daughter cells during mitosis.

Reference: Uhlmann F & Nasmyth K (1998) Cohesion between sister chromatids must be established during DNA replication. *Curr. Biol.* 8, 1095–1101.

18–2* **References:** Skibbens RV, Corson LB, Koshland D & Hieter P (1999) Ctf7 is essential for sister chromatid cohesion and links mitotic chromosome structure to the DNA replication machinery. *Genes Dev.* 13, 307–319.

Toth A, Ciosk R, Uhlmann F, Galova M, Schleiffer A & Nasmyth K (1999) Yeast cohesin complex requires a conserved protein, Eco1p(Ctf7), to establish cohesion between sister chromatids during DNA replication. *Genes Dev.* 13, 320–333.

18–3 In cells that fully condense their chromosomes at mitosis, as vertebrate cells do, most of the cohesin is released from the chromosomes at the start of mitosis, when condensins bind to drive condensation. With most of the cohesin out of the way, the condensins can coil individual sister chromatids into separate domains. The small amount of cohesin that remains is sufficient to hold sister chromatids together until anaphase, when the residual cohesins are degraded.

18–4* Figure 18–26.

References: Kimura K, Rybenkov VV, Crisona NJ, Hirano T, & Cozzarelli NR (1999) 13S condensin actively reconfigures DNA by introducing global positive writhe: Implications for chromosome condensation. *Cell* 98, 239–248.

Losada A & Hirano T (2001) Intermolecular DNA interactions stimulated by the cohesin complex *in vitro*: Implications for sister chromatid cohesion. *Curr. Biol.* 11, 268–272.

18–5 The separation of chromosomes and their distribution to daughter cells is accomplished by the bipolar mitotic spindle, which is composed of microtubules and a variety of microtubule-dependent motors. The division of an animal cell into daughter cells by cytokinesis is accomplished by the contractile ring, which is composed of actin and myosin filaments and is located just under the plasma membrane. As the ring constricts, it pulls the membrane inward, ultimately dividing the cell in two.

18–6*

18–7 Centrosome duplication is semi-conservative. The pair of centrioles in the centrosome separates, and each serves to nucleate synthesis of a new centriole. As a consequence, each new centrosome consists of one old and one new centriole. Thus, centrosome duplication is analogous to DNA replication: a new duplex consists of one old strand and one newly replicated strand.

Reference: Stearns T (2001) Centrosome duplication: A centriolar pas de deux. *Cell* 105, 417–420.

18–8* **References:** O'Connell KF, Caron C, Kopish KR, Hurd DD, Kemphues KJ, Li Y & White JG (2001) The *C. elegans zyg-1* gene encodes a regulator of centrosome duplication with distinct maternal and paternal roles in the embryo. *Cell* 105, 547–558.

Tokuyama Y, Horn HF, Kawamura K, Tarapore P & Fukasawa K (2001) Specific phosphorylation of nucleophosmin on Thr199 by cyclin-dependent kinase 2-cyclin E and its role in centrosome duplication. *J. Biol. Chem.* 276, 21529–21537.

18–9 The patterns of centriole duplication and splitting that account for the mercaptoethanol-induced abnormalities in cell division are diagrammed in Figure 18–27. In essence, treatment with mercaptoethanol allows the centrosome to split a second time without an intervening duplication of the centrioles (Figure 18–27A). (Centrosomes split the first time as the egg entered mitosis.) As a consequence, each of the spindle poles of the tetrapolar spindle has only one centriole rather than the normal pair of centrioles. Evidently, a centrosome with a single centriole is a perfectly adequate spindle pole.

The daughter cells from the four-way division of the egg receive a centrosome with a single centriole. During the next cell cycle the centriole is duplicated to form a normal centrosome with a centriole pair (Figure 18–27A). However, the usual form of the centrosome upon entry into mitosis has two centriole pairs instead of one. Thus the centrosome in the daughter cells looks like a single spindle pole and indeed forms a monopolar spindle (Figure 18–27B). Most commonly, cell division is aborted and the cell traverses the cell cycle again, allowing the centrosome to be duplicated a second time so that it possesses two centriole pairs (Figure 18–27B, top). This second duplication puts the centrosome cycle back in step with the cell cycle and

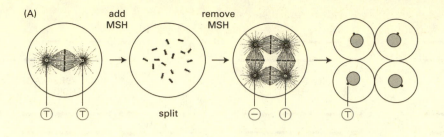

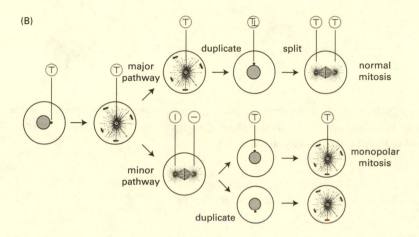

Figure 18–27 Duplication and splitting of centrosomes in mercaptoethanol-treated sea urchin eggs (Answer 18–9). (A) Mercaptoethanol-induced splitting of centrosomes. When MSH is removed, a tetrapolar spindle forms with each centriole serving as a spindle pole. (B) Abnormal cell divisions by cells that receive a single centriole. In most such cells (*upper pathway*) the centriole is duplicated twice—with an intervening aborted mitosis—to generate a normal pair of centrosomes and a normal bipolar spindle. In a few cells (*lower pathway*) the centrosome splits into single centrioles that form a bipolar spindle and allow cell division, but once again generate daughter cells with single centrioles.

subsequent cell divisions occur normally (Figure 18–27B, top). In the rarer cases in which the daughter cells form a bipolar spindle, the centrosomes split to form two centrosomes each with a single centriole (Figure 18–27B, bottom). Although the splitting allows cell division to occur, it presents the daughter cells with the same problem as the parent: a centrosome with a single centriole (Figure 18–27B, bottom). In this case the centrosome cycle remains out of step with the cell-division cycle.

Note that although the cell divisions can be restored to normal, eggs that undergo a four-way division develop abnormally because none of the cells receives a full complement of chromosomes.

References: Maizia D, Harris PJ & Bibring T (1960) The multiplicity of mitotic centers and the time-course of their duplication and separation. *J. Biophys. Biochem. Cytol.* 7, 1–20.

Sluder G & Rieder CL (1985) Centriole number and the reproductive capacity of spindle poles. *J. Cell Biol.* 100, 887–896.

18–10*

18–11 Prophase (see Figure 18–7E), prometaphase (see Figure 18–7D), metaphase (see Figure 18–7C), anaphase (see Figure 18–7A), telophase (see Figure 18–7F), and cytokinesis (see Figure 18–7B).

18–12*

18–13 The events occur in the following order: duplication of the centrosome (F), separation of centrosomes (J), condensation of chromosomes (D), breakdown of nuclear envelope (C), attachment of microtubules to chromosomes (B), alignment of chromosomes at the spindle equator (A), separation of sister chromatids (K), elongation of the spindle (G), re-formation of nuclear envelope (I), decondensation of chromosomes (E), and pinching of cell in two (H).

18–14* Figure 18–28.

References: Baltzer F (1967) Theodor Boveri: Life and Work of a Great Biologist. Berkeley: University of California Press.

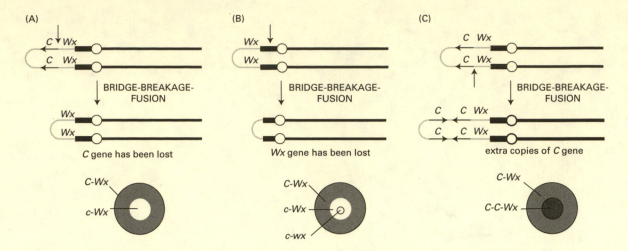

Figure 18–29 Formation of three different types of patches observed in speckled kernels (Answer 18–15). (A) Formation of a colorless, nonwaxy (*c-Wx*) spot. (B) Formation of a colorless, waxy (*c-wx*) spot inside a colorless, nonwaxy (*c-Wx*) spot formed as in (A). (C) Formation of an intensely colored, nonwaxy (*C-C-Wx*) spot. *Vertical arrows* pointing to the dicentric chromosomes show the positions of the breaks that lead to formation of the patches. In each case the *upper half* of the starting dicentric chromosome is arbitrarily shown to give rise to the product dicentric chromosome.

Boveri T (1902) Über mehrpolige Mitosen als mittel zur Analyse des Zellkerns. *Verh. d. phys.-med. Ges. Würzburg, N.F.* 35, 67–90. (Available in English translation, Foundations of Experimental Embryology, B Willier and J Oppenheimer eds. Englewood Cliffs, NJ: Prentice-Hall, 1964.)

Boveri T (1907) Zellenstudien VI: Die Entwicklung dispermer Seeigeleier. Ein Beitrag zur Befruchtungslehre und zur Theorie des Kernes. *Jenaische Zeitschr. Naturwissen.* 43, 1–292.

18–15

A. The formation of the three types of patches observed in speckled kernels is shown in Figure 18–29. The formation of a colorless, nonwaxy (*c-Wx*) patch results from a breakage that eliminates the dominant color (*C*) allele (Figure 18–29A). In the absence of the dominant allele the color of the patch is determined by the recessive colorless (*c*) allele on the normal chromosome (which is not shown in the figure).

   The formation of a colorless, waxy (*c-wx*) spot in a colorless, nonwaxy (*c-Wx*) patch is due to a second breakage event that eliminates the dominant non-waxy (*Wx*) allele (Figure 18–29B). In the absence of the dominant allele the spot is waxy (*wx*) due to the recessive allele on the normal chromosome (not shown).

   The formation of an intensely colored patch is due to a breakage event that leads to a dicentric chromosome with multiple copies of the dominant color allele (Figure 18–29C). Thus the genetic constitution of the intensely colored patch is *C-C-Wx*.

B. You would never expect to see a colored spot within a colorless patch because, once eliminated, the dominant color (*C*) allele cannot be regained by further bridge–breakage–fusion cycles.

C. You would expect to see colorless spots within an intensely colored patch because the dominant color (*C*) allele could be lost by subsequent bridge–breakage–fusion cycles.

   The demonstration by McClintock of bridge–breakage–fusion cycles in plants was one of the earliest indications that the broken ends of chromosomes are in some way 'sticky'—entirely different from natural chromosome ends. It is clear now that cells have an active repair pathway for joining broken DNA ends together as a defense against potentially lethal double-strand breaks. So long as breaks are rare, the correct ends are joined. But when multiple breaks are present, the wrong partners can be joined, leading to translocations or other genetic rearrangements. In humans such rearrangements are often associated with cancers.

Reference: McClintock B (1939) The behavior of successive nuclear divisions of a chromosome broken at meiosis. *Proc. Natl. Acad. Sci. USA* 25, 405–416.

18–16*

18–17   If the growth rate of microtubules is the same in mitotic and in interphase cells, their length is proportional to their lifetime. Thus, the average length of microtubules in mitosis is 1 μm (20 μm × 15 sec/300 sec). If microtubules are 20 times shorter on average, but are the same length in total, then there must be 20 times as many microtubules in mitosis. If there are 100 nucleation sites for microtubules in interphase (100 microtubules), then there must be 2000 microtubules (2000 nucleation sites) in mitosis. Since there are two centrosomes in mitotic cells, there would be 1000 nucleation sites per centrosome in mitosis.

18–18*

18–19   The two types of motors operate on the overlap microtubules. Plus-end-directed motors tend to push the spindle poles apart by decreasing the overlap, whereas minus-end-directed motors tend to pull the poles together by increasing the overlap (Figure 18–30).

18–20*   **Reference:** Hyman AA & Karsenti E (1996) Morphogenetic properties of microtubules and mitotic spindle assembly. *Cell* 84, 401–410.

18–21   True. Microtubules nucleated by the centrosomes grow outward toward the chromosomes in a highly dynamic process, alternately growing and shrinking. When they eventually attach to the kinetochore of a chromosome, they become stabilized and are referred to as kinetochore microtubules.

18–22*   **Reference:** Mitchison TJ & Kirschner MW (1985) Properties of the kinetochore *in vitro*. II. Microtubule capture and ATP-dependent translocation. *J. Cell. Biol.* 101, 766–777.

18–23

A.   Dicentric plasmids are stable in bacteria because bacteria use a completely different mechanism to segregate their chromosomes. A bacterial chromosome is polarized so that its single origin of replication (*oriC*) is located at one pole of the bacterium. As soon as *oriC* sequences are replicated, one copy is translocated to the opposite pole, ensuring that the daughter chromosomes lie on either side of the plane of fission that separates the bacterium into daughter cells. This mechanism of cell division makes bacteria indifferent to the presence of centromeres on the plasmid DNA.

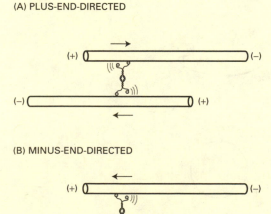

(A) PLUS-END-DIRECTED

(B) MINUS-END-DIRECTED

Figure 18–30 Operation of motors on the overlap microtubules in the spindle (Answer 18–19). (A) Plus-end-directed motors. (B) Minus-end-directed motors. Arrows indicate the direction of movement of the microtubules.

B. Dicentric plasmids are unstable in yeasts for the same reason that dicentric chromosomes are unstable in higher eucaryotes. If the two centromeres attach to opposite poles, the spindle apparatus can exert enough force on the DNA molecule to break its phosphodiester backbone. Roughly half the time a plasmid would be expected to orient itself on the spindle so that its two centromeres are attached to opposite poles. Thus there is a very high probability that a plasmid will be broken at each cell division, hence the instability.

C. Since monocentric plasmids are very stable, it seems most likely that the mechanism for deletion of centromeric sequences from dicentric plasmids relates to the breakage they suffer during mitosis. As illustrated in Figure 18–31, a circular plasmid must suffer two breaks to permit the centromeres to separate during mitosis. This breakage naturally separates the centromeres from one another onto linear fragments of the original plasmid. If the ends of a fragment join to make a circle, the resulting plasmid will contain a single centromeric sequence (Figure 18–31). Only those fragments that contain the yeast origin of replication (*ARS1*) and the selected marker (*TRP1*) can continue to grow in future generations.

This mechanism does not readily account for the loss of both centromeres; rather, it predicts that one centromere will be retained. Once the dicentric plasmid is reduced to a monocentric plasmid, it should be stable. The loss of both centromeres probably involves a process other than simple breakage. One likely possibility is that the broken ends are digested by exonucleases, which occasionally remove the remaining centromeric sequence before the fragment circularizes (Figure 18–31).

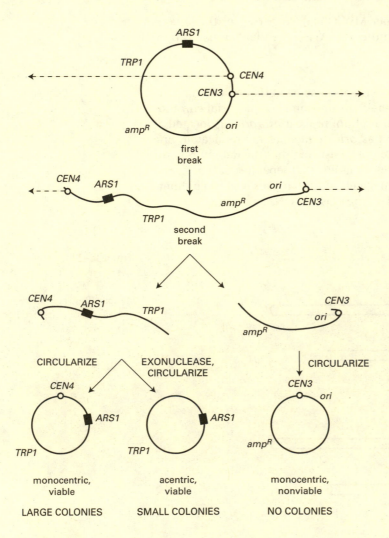

Figure 18–31 A mechanism for generating monocentric and acentric plasmids from a dicentric plasmid in yeast (Answer 18–23). *Dashed arrows* indicate the direction of pull toward the spindle poles. Viability refers to the ability of the plasmid to grow in yeast under selective conditions (which requires *ARS1* and *TRP1*).

Reference: Mann C & Davis RW (1983) Instability of dicentric plasmids in yeast. *Proc. Natl. Acad. Sci. USA* 80, 228–232.

18–24* **Reference:** Murray A & Szostak JW (1982) Pedigree analysis of plasmid segregation in yeast. *Cell* 34, 961–970.

18–25

A. If the second centromere were active, it would indeed destabilize the plasmid as described in Problem 18–23. In your experiments, however, growth on galactose keeps the introduced *CEN6* inactive by promoting transcription across it.

B. Transcription requires that RNA polymerase be able to unwind DNA and separate the two strands over a short distance as it moves along the DNA. The stable assembly of a kinetochore physically prevents unwinding of the DNA by RNA polymerase. This has ample precedent in the activity of some gene regulatory proteins, which bind tightly to DNA sequences in the path of RNA polymerase and thereby block its progression along the DNA.

   It is not so clear how transcription inactivates a kinetochore. Transcription apparently does not disturb the special arrangement of nucleosomes around centromeric DNA, and the majority of transcripts terminate at the border of the *CEN* sequence. It may be that the approach of the transcriptional apparatus close to the edge of the centromere destabilizes microtubule attachment just enough to kill the kinetochore.

C. Low fidelity of chromosome transmission can arise in several ways other than by disruption of kinetochore function. Mutations in tubulin genes and in microtubule-based motor proteins, which can interfere with attachment to kinetochores or with spindle function, also show elevated rates of chromosome loss. Mutations that affect DNA metabolism can also lead to chromosome loss by interfering with proper chromosome replication.

Reference: Doheny KF, Sorger PK, Hyman AA, Tugendreich S, Spencer F & Hieter P (1993) Identification of essential components of the *S. cerevisiae* kinetochore. *Cell* 73, 761–774.

18–26* **Reference:** Doheny KF, Sorger PK, Hyman AA, Tugendreich S, Spencer F & Hieter P (1993). Identification of essential components of the *S. cerevisiae* kinetochore. *Cell* 73, 761–774.

18–27 There are 46 human chromosomes, each with two kinetochores—one for each sister chromatid—thus, there are 92 kinetochores in a human cell at mitosis.

18–28* Table 18–3.

Reference: Bloom K (1993) The centromere frontier: kinetochore components, microtubule-based motility, and the CEN–value paradox. *Cell* 73, 621–624.

18–29 Both sister chromatids could end up in the same daughter cell for any of a number of reasons. If the microtubules or their connections with a kinetochore were to break during anaphase, both sister chromatids could be drawn to the same pole, hence the same daughter cell. If microtubules from the same spindle pole attached to both kinetochores, the chromosome would be pulled to the same pole. If the cohesins that link sister chromatids were not degraded, the pair of chromatids might be pulled to the same pole. If a chromosome never engaged microtubules, and was left out of the spindle, it would also end up in one daughter cell. Some of these errors in the mitotic process would be expected to engage a checkpoint mechanism—for example, the spindle-attachment checkpoint—and allow most such errors to be corrected, which is one reason why such errors are so rare.

   As a consequence of this error, one daughter cell would contain only one copy of all the genes carried on that chromosome and the other daughter cell would contain three copies. The altered gene dosage, leading to correspondingly changed amounts of the mRNAs and proteins produced, is often detrimental to the cell. In addition, there is the possibility that the cell with a single copy of the chromosome may be defective for a critical gene, a defect that was hidden by the presence of a second, good copy of the gene on the other chromosome.

18–30*

18–31   False. Equal and opposite forces on chromosomes would tend to position them at random locations between the poles. Some imbalance in the pulls or an additional force must act to position the chromosomes on the metaphase plate in the center of the spindle.

18–32*   **Reference:** Antonio C, Ferby I, Wilhelm H, Jones MJ, Karsenti E, Nebreda AR & Vernos I (2000) Xkid, a chromokinesin required for chromosome alignment on the metaphase plate. *Cell* 102, 425–435.

18–33   The separation of chromosomes is blocked in the presence of stable Xkid, even though the sister chromatids have been split apart. The continued function of Xkid, which constantly pushes the chromosomes away from the poles toward the equator, evidently is sufficient to oppose the normal force exerted by the shortening of the kinetochore microtubules. Note, however, that not all aspects of anaphase appear to be blocked by stable Xkid. The spindle, for example, has increased in size, suggesting that the poles have moved apart.

   Reference: Funabiki H & Murray AW (2000) The *Xenopus* chromokinesin Xkid is essential for metaphase chromosome alignment and must be degraded to allow anaphase chromosome movement. *Cell* 102, 411–424.

18–34*   **Reference:** de Saint Phalle B & Sullivan W (1998) Spindle assembly and mitosis without centrosomes in parthenogenetic *Sciara* embryos. *J. Cell Biol.* 141, 1383–1391.

18–35   Microtubules that are not directly connected to the centrosomes are held in place by minus-end-directed motors, which link them to the other microtubules and tend to 'walk' them toward the spindle pole.

   Reference: Compton DA (2000) Spindle assembly in animal cells. *Annu. Rev. Biochem.* 69, 95–114.

18–36*

18–37   In anaphase A the sister chromatids separate and the chromosomes move toward the poles by the shortening of the kinetochore microtubules. Anaphase A depends on the action of motors at the kinetochore. In anaphase B, which overlaps with anaphase A, the spindle poles move apart, separating the chromosomes still farther. Two sets of motors cooperate to accomplish spindle separation. Plus-end-directed motors acting on overlap microtubules push the spindle poles apart. This movement is accompanied by microtubule growth at their plus ends. Minus-end-directed motors that link the cell cortex to the astral microtubules act to pull the spindle poles apart.

18–38*

18–39   False. Kinetochore microtubules polymerize at their plus ends up to anaphase, at which point they begin to depolymerize. Prior to anaphase, kinetochore microtubules maintain a fairly constant length by treadmilling, with addition to the plus ends being balanced by removal at the minus ends. At anaphase, coincident with sister-chromatid separation, kinetochore microtubules begin to depolymerize at their plus ends as well, and therefore become shorter, moving the chromosomes toward the spindle poles.

18–40*

CYTOKINESIS

18–41 False. For most cells—typical cells—this statement is true. There are exceptions, however. Osteoclasts, for example, undergo mitosis without cytokinesis and become multinucleate.

18–42*

18–43

A. The chromosome-signaling hypothesis predicts that furrows will form only where chromosomes have been aligned. This prediction matches the result of the first division but not the result of the second division, where three furrows are formed instead of the two expected furrows.

B. The polar-relaxation hypothesis predicts that there should be two furrows at the first division instead of just one. In the toroidal egg the spindles should relax a ring of the cell surface on either side of the equatorial plane. As a consequence, there should be two regions where the cell surface furrows: at the equatorial plane and opposite it on the other side of the torus. The events of the second division match the expectations of the relaxation hypothesis.

C. The aster-stimulation hypothesis predicts that there should be a single furrow at the first cell division and three furrows at the second cell division, which matches the experimental observations. A single furrow is expected at the first division because the spindle fibers from the two poles interact only at the equatorial plane. At the second division, however, the spindle fibers interact not only at the two equatorial positions, but also at the position of the third furrow.

Reference: Rappaport R (1986) Establishment of the mechanism of cytokinesis in animal cells. *Int. Rev. Cytol.* 105, 245–281.

18–44*

18–45 The movement of chromosomes at anaphase depends on microtubules, not on actin or myosin. Injection of an antibody against myosin would therefore have no effect on chromosome movement during mitosis. Cytokinesis, on the other hand, depends on the assembly and contraction of a ring of actin and myosin filaments, which forms the cleavage furrow that splits the cell in two.

18–46*

18–47 Nocodazole treatment disassembles microtubules, preventing formation of a spindle. Because nuclei break down and chromosomes condense, these events must be independent of aster formation by centrosomes, for example, and of any other microtubule-dependent process. Mitosis finally arrests because the unattached chromosomes trigger a signal that engages the spindle-attachment checkpoint, which halts the cycle.

   Treatment with cytochalasin D does not affect mitosis because actin filaments are not involved in the process. Moreover, the disassembly of actin filaments does not trigger a cell-cycle arrest; the cell completes mitosis and forms a binucleate cell. Thus, there does not seem to be a cytokinesis checkpoint.

18–48* Figure 18–32.

Reference: Skop AR, Bergmann D, Mohler WA & White JG (2001) Completion of cytokinesis in *C. elegans* requires a brefeldin A-sensitive membrane accumulation at the cleavage furrow apex. *Curr. Biol.* 11, 735–746.

18–49 In the absence of the psychosine receptor the population of cells is distributed between 2N (unreplicated) cells and 4N (replicated) cells. In the presence of the receptor but in the absence of psychosine, there is the same distribution. When treated with psychosine, however, prominent peaks at 8N and 16N appear. FACS analysis cannot determine whether the increase

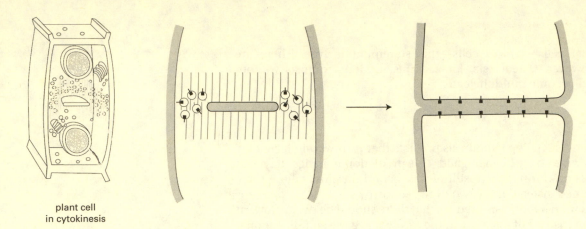

Figure 18–33 Fusion of Golgi-derived vesicles to form new cell wall (Answer 18–51).

plant cell
in cytokinesis

in DNA is due to multiple nuclei; it could arise by a total bypass of mitosis, leading to a single nucleus with several copies of the genome. When the cells in these experiments were examined microscopically, they were shown to contain multiple nuclei.

References: Im D-S, Heise CE, Nguyen T, O'Dowd BF & Lynch KR (2001) Identification of a molecular target of psychosine and its role in globoid cell formation. *J. Cell Biol.* 153, 429–434.

Mitchison T (2001) Psychosine, cytokinesis, and orphan receptors: Unexpected connections. *J. Cell Biol.* 153, F1–F3.

18–50* **References:** Nagata Y, Muro Y & Todokoro K (1997) Thrombopoietin-induced polyploidization of bone marrow megakaryocytes is due to a unique regulatory mechanism in late mitosis. *J. Cell Biol.* 139, 449–457.

Zimmet J & Ravid K (2000) Polyploidy: occurrence in nature, mechanisms, and significance for the megakaryocyte-platelet system. *Exp. Hematol.* 28, 3–16.

18–51 The membranes of the Golgi vesicles fuse to form the new plasma membranes of the two daughter cells. The interiors of the vesicles, which are filled with cell-wall material, become the new cell-wall matrix separating the two daughter cells. Proteins in the membranes of the Golgi vesicles thus become plasma membrane proteins. Those parts of the protein that were exposed to the lumen of the Golgi vesicle will end up exposed to the new cell wall (Figure 18–33).

References

Aberle H, Bauer A, Stappert J, Kispert A & Kemler R (1997) β-catenin is a target for the ubiquitin-proteasome pathway. *EMBO J.* 16, 3797–3804. (Problem 15–92)

Abrahams JP, Leslie AG, Lutter AGW & Walker JE (1994) Structure at 2.8 Å resolution of F1-ATPase from bovine heart mitochondria. *Nature* 370, 621–628. (Problem 14–18)

Abrami L, Simon M, Rousselet G, Berthonaud V, Buhler JM & Ripoche P (1994) Sequence and functional expression of an amphibian water channel, FA-CHIP: a new member of the MIP family. *Biochim. Biophys. Acta* 1192, 147–151. (Problem 11–3)

Adams GA & Rose JK (1985) Structural requirements of a membrane-spanning domain for protein anchoring and cell surface transport. *Cell* 41, 1007–1015. (Problem 13–32)

Adams KL, Song K, Roessler PG, Nugent JM, Doyle JL, Doyle JJ & Palmer JD (1999) Intracellular gene transfer in action: Dual transcription and multiple silencings of nuclear and mitochondrial *cox2* genes in legumes. *Proc. Natl. Acad. Sci. USA* 96, 13863–13868. (Problem 1–32)

Adelstein RS & Klee CB (1980) Smooth muscle myosin light-chain kinase. In Calcium and Cell Function (WY Cheung ed), Vol. 1, pp 167–182. New York: Academic Press. (Problem 15–57)

Alberts BM & Frey L (1970) T4 bacteriophage gene 32: a structural protein in the replication and recombination of DNA. *Nature* 227, 1313–1318. (Problem 5–24)

Allard WJ & Lienhard GE (1985) Monoclonal antibodies to the glucose transporter from human erythrocytes: identification of the transporter as a Mr = 55,000 protein. *J. Biol. Chem.* 260, 8668–8675. (Problem 11–9)

Amann KJ & Pollard TD (2001) The ARP2/3 complex nucleates actin filament branches from the sides of preexisting filaments. *Nat. Cell Biol.* 3, 306–310. (Problem 16–49)

Anderson JE (1993) Restriction endonucleases and modification methylases. *Curr. Opin. Struct. Biol.* 3, 24–30. (Problem 8–36)

Andreasen PH, Dreisig H & Kristiansen K (1987) Unusual ciliate-specific codons in *Tetrahymena* mRNAs are translated correctly in a rabbit reticulocyte lysate supplemented with a subcellular fraction from *Tetrahymena. Biochem. J.* 244, 331–335. (Problem 6–66)

Antonio C, Ferby I, Wilhelm H, Jones MJ, Karsenti E, Nebreda AR & Vernos I (2000) Xkid, a chromokinesin required for chromosome alignment on the metaphase plate. *Cell* 102, 425–435. (Problem 18–32)

Artandi SE, Chang S, Lee S-L, Alson S, Gottlieb GJ, Chin L & DePinho RA (2000) Telomere dysfunction promotes non-reciprocal translocations and epithelial cancers in mice. *Nature* 406, 641–645. (Problem 4–35)

Ashburner M, Chihara C, Meltzer P & Richards G (1973) Temporal control of puffing activity in polytene chromosomes. *Cold Spring Harbor Symp. Quant. Biol.* 38, 655–662. (Problem 15–24)

Bachmair A, Finley D & Varshavsky A (1986) *In vivo* half-life of a protein is a function of its amino-terminal residue. *Science* 234, 179–186. (Problems 6–78, 6–79)

Bachurski CJ, Theodorakis NG, Coulson RM & Cleveland DW (1994) An amino-terminal tetrapeptide specifies cotranslational degradation of β-tubulin but not α-tubulin mRNAs. *Mol. Cell Biol.* 14, 4076–4086. (Problem 7–80)

Baeuerle PA, & Baltimore D (1988) Activation of DNA-binding activity in an apparently cytoplasmic precursor of the NF-κB transcription factor. *Cell* 53, 211–217. (Problem 12–44)

Baldin V, Lukas J, Marcote MJ, Pagano M & Draetta G (1993) Cyclin D1 is a nuclear protein required for cell cycle progression in G_1. *Genes Dev.* 7, 812–821. (Problem 17–76)

Baltzer F (1967) Theodor Boveri: Life and Work of a Great Biologist. Berkeley: University of California Press. (Problem 18–14)

Bamberg E & Lauger P (1974) Temperature-dependent properties of gramicidin A channels. Biochim. Biophys. Acta 367, 127–133. (Problem 11–15)

Barnes G & Rine J (1985) Regulated expression of endonuclease EcoRI in *Saccharomyces cerevisiae*: nuclear entry and biological consequences. *Proc. Natl. Acad. Sci. USA* 82, 1354–1358. (Problems 12–25, 12–26)

Barron JT, Gu L & Parrillo JE (2000) NADH/NAD redox state of cytoplasmic glycolytic compartments in vascular smooth muscle. *Am. J. Physiol. Heart Circ. Physiol.* 279, H2872–H2878. (Problems 14–30, 14–31)

Bartel DP & Szostak JW (1993) Isolation of new ribozymes from a large pool of random sequences. *Science* 261, 1411–1418. (Problems 6–82, 6–84, 6–85, 6–86)

Baserga R & Wiebel F (1969) The cell cycle of mammalian cells. *Intern. Rev. Exp. Pathol.* 7, 1–30. (Problem 17–13)

Bender J & Kleckner N (1986) Genetic evidence that Tn10 transposes by a nonreplicative mechanism. *Cell* 45, 801–815. (Problem 5–82)

Benezra R, Davis RL, Lockshon D, Turner DL & Weintraub H (1990) The protein Id: a negative regulator of helix–loop–helix DNA binding proteins. *Cell* 61, 49–59. (Problem 7–19)

Bennett V & Stenbuck PJ (1979) The membrane attachment protein for spectrin is associated with band 3 in human erythrocyte membranes. *Nature* 280, 468–473. (Problem 10–44)

Bennett V & Stenbuck PJ (1980) Association between ankyrin and the cytoplasmic domain of band 3 isolated from the human erythrocyte membrane. *J. Biol. Chem.* 255, 6424–6432. (Problem 10–44)

Berg HC (1993) Random Walks in Biology, Expanded Edition, pp 5–6. Princeton, NJ: Princeton University Press. (Problems 2–65, 2–66)

Berg JM, Tymoczko JL & Stryer L (2002) Biochemistry, Fifth Edition, pp 436–437. New York: WH Freeman and Co. (Problem 2–78)

Berget SM, Berk AJ, Harrison T & Sharp PA (1977) Spliced segments at the 5′ termini of adenovirus-2 late mRNA: a role for heterogeneous nuclear RNA in mammalian cells. *Cold Spring Harbor Symp. Quant. Biol.* 42, 523–529. (Problem 6–23)

Bi X, Braunstein M, Shei G-J & Broach J (1999) The yeast *HML* I silencer defines a heterochromatin domain boundary by directional establishment of silencing. *Proc. Natl. Acad. Sci. USA* 96, 11934–11939. (Problem 7–45)

Bleil JD & Bretscher MS (1982) Transferrin receptor and its recycling in HeLa cells. *EMBO J.* 1, 351–355. (Problems 13–66, 13–67)

Bloom K (1993) The centromere frontier: kinetochore components, microtubule-based motility, and the CEN–value paradox. *Cell* 73, 621–624. (Problem 18–28)

Bloom KS & Carbon J (1982) Yeast centromere DNA is in a unique and highly ordered structure in chromosomes and small circular minichromosomes. *Cell* 29, 305–317. (Problem 4–60)

Boeke JD, Garfinkel DJ, Styles CA & Fink GR (1985) Ty elements transpose through an RNA intermediate. *Cell* 40, 491–500. (Problem 5–83)

Bogenhagen D & Clayton DA (1977) Mouse L cell mitochondrial DNA molecules are selected randomly for replication throughout the cell cycle. *Cell* 11, 719–727. (Problem 14–75)

Bonfanti L, Mironov Jr. AA, Martinez-Menarguez JA, Martella O, Fusella A, Baldassarre M, Buccione R, Geuze HJ, Mironov AA & Luini A (1998) Procollagen traverses the Golgi stack without leaving the lumen of cisternae: evidence for cisternal maturation. *Cell* 95, 993–1003. (Problem 13–40)

Bonner JT & Savage LJ (1947) Evidence for the formation of cell aggregates by chemotaxis in the development of the slime mold *Dictyostelium discoideum*. *J. Exp. Zool.* 106, 1–26. (Problem 15–5)

Bootsma D, Budke L & Vos O (1964) Studies on synchronous division of tissue culture cells initiated by excess thymidine. *Exp. Cell Res.* 33, 301–309. (Problem 17–15)

Bostock CJ, Prescott DM & Kirkpatrick JB (1971) An evaluation of the double thymidine block for synchronizing mammalian cells at the G_1-S border. *Exp. Cell Res.* 68, 163–168. (Problem 17–15)

Boveri T (1902) Über mehrpolige Mitosen als mittel zur Analyse des Zellkerns. *Verh. d. phys.-med. Ges. Würzburg, N.F.* 35, 67–90. (Available in English translation, Foundations of Experimental Embryology, B Willier and J Oppenheimer eds. Englewood Cliffs, NJ: Prentice-Hall, 1964.) (Problem 18–14)

Boveri T (1907) Zellenstudien VI: Die Entwicklung dispermer Seeigeleier. Ein Beitrag zur Befruchtungslehre und zur Theorie des Kernes. *Jenaische Zeitschr. Naturwissen.* 43, 1–292. (Problem 18–14)

Bretscher M (1972) Asymmetrical lipid bilayer structure for biological membranes. *Nature New Biol.* 236, 11–12. (Problem 10–11)

Brewer BJ & Fangman WL (1987) The localization of replication origins on ARS plasmids in *S. cerevisiae*. *Cell* 51, 463–471. (Problem 5–44)

Brinker A, Pfeifer G, Kerner MJ, Naylor DJ, Hartl FU & Hayer-Hartl M (2001) Dual function of protein confinement in chaperonin-assisted protein folding. *Cell* 107, 223–233. (Problem 6–72)

Brokaw CJ, Luck DJL & Huang B (1982) Analysis of the movement of *Chlamydomonas* flagella: the function of the radial-spoke system is revealed by comparison of wild-type and mutant flagella. *J. Cell Biol.* 92, 722–732. (Problem 16–84)

Brown CJ, Ballabio A, Rupert JL, Lafreniere RG, Grompe M, Tonlorenzi R & Willard HF (1993) A gene from the region of the human X inactivation center is expressed exclusively from the inactive X chromosome. *Nature* 349, 38–44. (Problem 7–59)

Brown DA & Rose JK (1992) Sorting of GPI-anchored proteins to glycolipid-enriched membrane subdomains during transport to the apical cell surface. *Cell* 68, 533–544. (Problem 10–18)

Brown MS & Goldstein JL (1979) Receptor-mediated endocytosis: insights from the lipoprotein receptor system. *Proc. Natl. Acad. Sci. USA* 76, 3330–3337. (Problems 13–59, 13–60)

Brown NR, Noble MEM, Lawrie AM, Morris MC, Tunnah P, Divita G, Johnson LN & Endicott JA (1999) Effects of phosphorylation of threonine 160 on cyclin-dependent kinase 2 structure and activity. *J. Biol. Chem.* 274, 8746–8756. (Problem 3–86)

Brutlag D & Kornberg A (1972) Enzymatic synthesis of deoxyribonucleic acid. 36. A proofreading function for the 3′ to 5′ exonuclease activity in deoxyribonucleic acid polymerases. *J. Biol. Chem.* 247, 241–248. (Problem 5–17)

Bubb MR, Spector I, Bershadsky AD & Korn ED (1995) Swinholide A is a microfilament disrupting marine toxin that stabilizes actin dimers and severs actin filaments. *J. Biol. Chem.* 270, 3463–3466. (Problem 16–36)

Burch RM, Luini A & Axelrod J (1986) Phospholipase A_2 and phospholipase C are activated by distinct GTP-binding proteins in response to α_1-adrenergic stimulation in FRTL5 thyroid cells. *Proc. Natl. Acad. Sci. USA* 83, 7201–7205. (Problem 15–51)

Byers D, Davis RL & Kiger JA (1981) Defect in cyclic AMP phosphodiesterase due to the *dunce* mutation of learning in *Drosophila melanogaster. Nature* 289, 79–81. (Problem 15–50)

Callan HG (1963) The nature of lampbrush chromosomes. *Int. Rev. Cytol.* 15, 1–34. (Problem 4–52)

Cantor CR & Schimmel PR (1980) Biophysical Chemistry, Part 1: The Conformation of Biological Macromolecules, pp 42–46. San Francisco: WH Freeman and Company. (Problem 2–35)

Cantor CR & Schimmel PR (1980) Biophysical Chemistry, pp 632–634. New York: WH Freeman and Company. (Problem 8–14)

Cantor CR & Schimmel PR (1980) Biophysical Chemistry, pp 674–675. New York: WH Freeman and Company. (Problems 8–17, 8–18)

Cantor CR & Schimmel PR (1980) Biophysical Chemistry, pp 944–945. New York: WH Freeman. (Problem 3–82)

Cantor CR & Schimmel PR (1980) Biophysical Chemistry, pp 954–956. New York: WH Freeman. (Problem 3–83)

Capaldi RA, Darley-Usmar V, Fuller S & Millet F (1982) Structural and functional features of the interaction of cytochrome *c* with complex III and cytochrome *c* oxidase. *FEBS Letters* 138, 1–7. (Problem 14–42)

Carlier M-F, Pantaloni D & Korn ED (1984) Evidence for an ATP cap at the ends of actin filaments and its regulation of the F-actin steady state. *J. Biol. Chem.* 259, 9983–9986. (Problem 16–19)

Carlier M-F, Pantaloni D & Korn ED (1985) Polymerization of ADP- and ATP-actin under sonication and characteristics of the STP-actin equilibrium polymer. *J. Biol. Chem.* 260, 6565–6571. (Problem 16–3)

Castagnola M (1998) Sensitive to the yoctomole limit. *Trends Biochem. Sci.* 23, 283. (Problem 8–29)

Cecconi F, Alvarez-Bolado G, Meyer BI, Roth KA & Gruss P (1998) Apaf1 (CED-4 homolog) regulates programmed cell death in mammalian development. *Cell* 94, 727–737. (Problem 17–65)

Cha TA & Alberts BM (1986) Studies of the DNA helicase-RNA primase unit from bacteriophage T4. A trinucleotide sequence on the DNA template starts RNA primer synthesis. *J. Biol. Chem.* 261, 7001–7010. (Problem 5–20)

Chamberlain JS, Gibbs RA, Ranier JE & Caskey CT (1990) Multiplex PCR for the diagnosis of Duchenne muscular dystrophy. In PCR Protocols (MA Innis, DH Gelfand, JJ Sninsky & TJ White, eds) pp 272–281. San Diego, CA: Academic Press. (Problem 8–60)

Chant J & Herskowitz I (1991) Genetic control of bud site selection in yeast by a set of gene products that constitute a morphogenetic pathway. *Cell* 65, 1203–1212. (Problem 16–100)

Chant J, Corrado K, Pringle JR & Herskowitz I (1991) Yeast BUD5, encoding a putative GDP-GTP exchange factor, is necessary for bud site selection and interacts with bud formation gene *BEM1. Cell* 65, 1213–1224. (Problem 16–100)

Chao S-H, Fujinaga K, Marion JE, Taube R, Sausville EA, Senderowicz AM, Peterlin BM & Price DH (2000) Flavopiridol inhibits P-TENb and blocks HIV-1 replication. *J. Biol. Chem.* 275, 28345–28348. (Problem 7–69)

Chapeville F, Lipmann F, von Ehrenstein G, Weisblum B, Ray, WJ & Benzer S (1962) On the role of soluble ribonucleic acid in coding for amino acids. *Proc. Natl. Acad. Sci. USA* 48, 1086–1092. (Problem 6–50)

Chardin P & McCormick F (1999) Brefeldin A: the advantage of being uncompetitive. *Cell* 97, 153–155. (Problem 13–11)

Chen C-N, Denome S & Davis RL (1986) Molecular analysis of cDNA clones and the corresponding genomic coding sequences of the *Drosophila dunce+* gene, the structural gene for cyclic AMP phosphodiesterase. *Proc. Natl. Acad. Sci. USA* 83, 9313–9317. (Problem 15–50)

Chen C-Y & Sarnow P (1995) Initiation of protein synthesis by the eucaryotic translation apparatus on circular RNAs. *Science* 268, 415–417. (Problem 7–87)

Chen S-H, Habib G, Yang CY, Gu ZW, Lee BR, Weng S-A, Silbermann SR, Cai S-J, Deslypere JP, Rosseneu M, Gotto AM, Li W-H & Chan L (1987) Apolipoprotein B-48 is the product of a messenger RNA with an organ-specific in-frame stop codon. *Science* 238, 363–366. (Problem 7–74)

Cheng C-HC & Chen L (1999) Evolution of an antifreeze protein. *Nature* 401, 443–444. (Problem 3–37)

Chou Y-H, Helfand BT & Goldman RD (2001) New horizons in cytoskeletal dynamics: transport of intermediate filaments along microtubule tracks. *Curr. Opin. Cell Biol.* 13, 106–109. (Problem 16–28)

Chrétien D, Fuller SD & Karsenti E (1995) Structure of growing microtubule ends: Two-dimensional sheets close into tubes at variable rates. *J. Cell Biol.* 117, 1311–1328. (Problem 16–14)

Cohen S, Ushiro H, Stocheck C & Chinkers M (1982) A native 170,000 epidermal growth factor receptor-kinase complex from shed plasma membrane vesicles. *J. Biol. Chem.* 257, 1523–1531. (Problem 17–73)

Coluccio LM & Tilney LG (1984) Phalloidin enhances actin assembly by preventing monomer dissociation. *J. Cell Biol.* 99, 529–535. (Problem 16–35)

Compton DA (2000) Spindle assembly in animal cells. *Annu. Rev. Biochem.* 69, 95–114. (Problem 18–35)

Conway L & Wickens M (1987) Analysis of mRNA 3′-end formation by modification interference: the only modifications which prevent processing lie in AAUAAA and the poly (A) site. *EMBO J.* 6, 4177–4184. (Problem 6–33)

Cox M & Lehman IR (1981) The polarity of the recA protein-mediated branch migration. *Proc. Natl. Acad. Sci. USA* 78, 6023–6027. (Problem 5–74)

Craigen WJ, Cook RG, Tate WP & Caskey CT (1985) Bacterial peptide chain release factors: conserved primary structure and possible frameshift regulation of release factor 2. *Proc. Natl. Acad. Sci. USA* 82, 3616–3620. (Problem 6–62)

Creighton TE (1993) Proteins, 2nd edn, pp 153–155. New York: WH Freeman. (Problem 3–1)

Creighton TE (1993) Proteins, 2nd edn, pp 288–289. New York: WH Freeman. (Problem 3–6)

Creighton TE (1993) Proteins, 2nd edn, pp 297–300. New York: WH Freeman. (Problem 3–4)

Crick F (1988) What Mad Pursuit: A Personal View of Scientific Discovery, p 109. New York: Basic Books, Inc. (Problem 6–1)

D'Atri S, Tentori L, Lacal PM, Graziani G, Pagani E, Benincasa E, Zambruno G, Bonmassar E & Jiricny J (1998) Involvement of the mismatch repair system in temozolomide-induced apoptosis. *Mol. Pharmacol.* 54, 334–341. (Problem 17–68)

Dammai V & Subramani S (2001) The human peroxisomal targeting signal receptor, Pex5p, is translocated into the peroxisomal matrix and recycled to the cytosol. *Cell* 105, 187–196. (Problem 12–73)

Datar SA, Jacobs HW, Flor A, de la Cruz A, Lehner CF & Edgar BA (2000) The *Drosophila* cyclin D-Cdk4 complex promotes cellular growth. *EMBO J.* 19, 4543–4554. (Problems 17–43, 17–44)

Davis RL, Weintraub H & Lassar AB (1987) Expression of a single transfected cDNA converts fibroblasts to myoblasts. *Cell* 51, 987–1000. (Problem 7–65)

de Saint Phalle B & Sullivan W (1998) Spindle assembly and mitosis without centrosomes in parthenogenetic *Sciara* embryos. *J. Cell Biol.* 141, 1383–1391. (Problem 18–34)

Deenen LLM & DeGier J (1974) Lipids of the red cell membrane. In The Red Blood Cell (D MacN Surgenor, ed), pp 147–211. New York: Academic Press. (Problem 10–11)

DeFranco D & Yamamoto KR (1986) Two different factors act separately or together to specify functionally distinct activities at a single transcriptional enhancer. *Mol. Cell. Biol.* 6, 993–1001. (Problem 15–22)

Delagado-Partin VM & Dalbey RE (1998) The proton motive force, acting on acidic residues, promotes translocation of amino-terminal domains of membrane proteins when the hydrophobicity of the translocation signal is low. *J. Biol. Chem.* 273, 9927–9934. (Problem 12–15)

Desai A & Mitchison TJ (1997) Microtubule polymerization dynamics. *Annu. Rev. Cell Dev. Biol.* 13, 83–117. (Problem 16–7)

Deshaies RJ & Schekman R (1987) A yeast mutant defective at an early stage in import of secretory protein precursors into the endoplasmic reticulum. *J. Cell Biol.* 105, 633–645. (Problem 12–84)

Detrich WH, Parker SK, Williams RC, Nogales E & Downing KH (2000) Cold adaptation of microtubule assembly and dynamics. *J. Biol. Chem.* 275, 37038–37047. (Problems 16–10, 16–16)

Deuerling E, Schulze-Specking A, Tomoyasu T, Mogk A & Bukau B (1999) Trigger factor and DnaK cooperate in folding of newly synthesized proteins. *Nature* 400, 693–696. (Problems 6–70, 6–71)

Dingwall C, Sharnick SV & Laskey RA (1982) A polypeptide domain that specifies migration of nucleoplasmin into the nucleus. *Cell* 30, 449–458. (Problem 12–23)

Dittmer F, Ulbrich EJ, Hafner A, Schmahl W, Meister T, Pohlmann R & von Figura K (1999) Alternative mechanisms for trafficking of lysosomal enzymes in mannose 6-phosphate receptor-deficient mice are cell type specific. *J. Cell Sci.* 112, 1591–1597. (Problem 13–51)

Dodson M, Dean FB, Bullock P, Echols H & Hurwitz J (1987) Unwinding of duplex DNA from the SV40 origin of replication by T-antigen. *Science* 238, 964–967. (Problem 5–40)

Doheny KF, Sorger PK, Hyman AA, Tugendreich S, Spencer F & Hieter P (1993) Identification of essential components of the *S. cerevisiae* kinetochore. *Cell* 73, 761–774. (Problems 18–25, 18–26)

Dolznig H, Bartunek P, Nasmyth K, Mullner EW & Beug H (1995) Terminal differentiation of normal chicken erythroid progenitors: shortening of G_1 correlates with loss of D-cyclin/cdk4 expression and altered cell size control. *Cell Growth Differ.* 6, 1341–1352. (Problem 17–71)

Donaldson JG, Finazzi D & Klausner RD (1992) Brefeldin A inhibits Golgi membrane-catalyzed exchange of guanine nucleotide into ARF protein. *Nature* 360, 350–352. (Problem 13–11)

Donzeau M, Káldi K, Adam A, Paschen S, Wanner G, Guiard B, Bauer MF, Neupert W & Brunner M (2000) Tim23 links the inner and outer mitochondrial membranes. *Cell* 101, 401–412. (Problem 12-59)

Dramsi S & Cossart P (1998) Intracellular pathogens and the actin cytoskeleton. *Annu. Rev. Cell Dev. Biol.* 14, 137–166. (Problem 16–50)

Dugaiczyk A, Woo SL, Lai EC, Mace Jr ML, McReynolds L & O'Malley BW (1978) The natural ovalbumin gene contains seven intervening sequences. *Nature* 274, 328–333. (Problem 6–24)

Dunham I et al. (1999) The DNA sequence of human chromosome 22. *Nature* 402, 489–495. (Problems 4–21, 4–22)

Dunn TM, Hahn S, Ogden S & Schleif RF (1984) An operator at –280 base pairs that is required for repression of *araBAD* operon promoter: addition of DNA helical turns between the operator and promoter cyclically hinders repression. *Proc. Natl. Acad. Sci. USA* 81, 5017–5020. (Problem 7–31)

Duysens LNM, Amesz J & Kamp BM (1961) Two photochemical systems in photosynthesis. *Nature* 190, 510–511. (Problem 14–63)

Earnshaw WC, Martins LM & Kaufmann SH (1999) Mammalian caspases. *Annu. Rev. Biochem.* 68, 383–424. (Problem 17–66)

Echols H (2001) Operators and Promoters: The Story of Molecular Biology and Its Creators, pp 46–47. Berkeley: University of California Press. (Problem 7–7)

Edelman P & Gallant J (1977) Mistranslation in *E. coli*. *Cell* 10, 131–137. (Problem 6–63)

Edgar BA & Lehner CF (1996) Developmental control of cell cycle regulators: A fly's perspective. *Science* 274, 1646–1652. (Problem 17–84)

Eilers M & Schatz G (1986) Binding of a specific ligand inhibits import of a purified precursor protein into mitochondria. *Nature* 322, 228–232. (Problem 12–56)

Eisenberg E & Hill TL (1985) Muscle contraction and free energy transduction in biological systems. *Science* 227, 999–1006. (Problem 16–69)

Eliasson C, Sahlgren C, Berthold CH, Stakeberg J, Celis JE, Betsholtz C, Eriksson JE & Pekny M (1999) Intermediate filament protein partnership in astrocytes. *J. Biol. Chem.* 274, 23996–24006. (Problem 16–32)

Elledge SJ, Mulligan JT, Ramer SW, Spottswood M & Davis RW (1991) λYES: a multifunctional cDNA expression vector for the isolation of genes by complementation of yeast and *Escherichia coli* mutations. *Proc. Natl. Acad. Sci. USA* 88, 1731–1735. (Problem 8–49)

Ellis HM & Horvitz RH (1986) Genetic control of programmed cell death in the nematode *C. elegans*. *Cell* 44, 817–829. (Problem 17–57)

Emerson R (1958) The quantum yield of photosynthesis. *Annu. Rev. Plant Physiol.* 9, 1–24. (Problem 14–61)

Fan J, Griffiths AD, Lockhart A, Cross RA & Amos LA (1996) Microtubule minus ends can be labeled with a phage display antibody specific to α-tubulin. *J. Mol. Biol.* 259, 325–330. (Problems 16–9, 16-65, 16-66)

Farrell PJ, Balkow K, Hunt T, Jackson RJ & Trachsel H (1977) Phosphorylation of initiation factor eIF-2 and the control of reticulocyte protein synthesis. *Cell* 11, 187–200. (Problem 7–85)

Faull RJ, Kovach NL, Harlan JM & Ginsberg MH (1993) Affinity modulation of integrin $\alpha_5\beta_1$: regulation of the functional response by soluble fibronectin. *J. Cell Biol.* 121, 155–162. (Problem 15–12)

Federman AD, Conklin BR, Schrader KA, Reed RR & Bourne HR (1992) Hormonal stimulation of adenylyl cyclase through G_i-protein $\beta\gamma$ subunits. *Nature* 356, 159–161. (Problem 15–44)

Ferrell JE & Machleder EM (1998) The biochemical basis of an all-or-none cell fate switch in *Xenopus* oocytes. *Science* 280, 895–898. (Problem 15–79)

Fersht A (1999) Structure and Mechanism in Protein Science, pp 360–361. New York: WH Freeman. (Problem 3–67)

Fersht A (1999) Structure and Mechanism in Protein Science, pp 575–600. New York: WH Freeman. (Problem 3–18)

Fields S & Song O (1989) A novel genetic system to detect protein–protein interactions. *Nature* 340, 245–246. (Problem 8–73)

Fijalkowska IJ, Jonczyk P, Tkaczyk MM, Bialoskorska M & Schaaper RM (1998) Unequal fidelity of leading strand and lagging strand DNA replication on the *Escherichia coli* chromosome. *Proc. Natl. Acad. Sci. USA* 95, 10020–10025. (Problem 5–29)

Flint AJ, Tiganis T, Barford D & Tonks NK (1997) Development of substrate-trapping mutants to identify physiological substrates of protein tyrosine phosphatases. *Proc. Natl. Acad. Sci. USA* 94, 1680–1685. (Problem 3–87)

Folkman J & Moscona A (1978) Role of cell shape in growth control. *Nature* 273, 345–349. (Problem 17–86)

Forbush B, Kok B & McGloin M (1971) Cooperation of charges in photosynthetic oxygen evolution II. Damping of flash yield, oscillation and deactivation. *Photochem. Photobiol.* 14, 307–321. (Problem 14–64)

Fornerod M, Ohno M, Yoshida M & Mattaj IW (1997) CRM1 is an export receptor for leucine-rich nuclear export signals. *Cell* 90, 1051–1060. (Problem 12–38)

Fox TD (1986) Nuclear gene products required for translation of specific mitochondrially coded mRNAs in yeast. *Trends Genet.* 2, 97–100. (Problem 14–89)

Foyer CH (1984) Photosynthesis, pp 176–195. New York: Wiley. (Problem 14–54)

Frayn KN (1996) Metabolic Regulation: A Human Perspective, p 179. London: Portland Press. (Problem 2–87)

Freeland FJ & Hurst LD (1998) The genetic code is one in a million. *J. Mol. Evol.* 47, 238–248. (Problem 1–5)

Friedberg EC, Walker GC & Siede W (1995) DNA repair and Mutagenesis, pp 92–103. New York: WH Freeman. (Problem 5–69)

Fries E, Gustafsson L & Peterson PA (1984) Four secretory proteins synthesized by hepatocytes are transported from the endoplasmic reticulum to Golgi complex at different rates. *EMBO J.* 3, 147–152. (Problem 13–71)

Fujiki Y (2000) Peroxisome biogenesis and peroxisome biogenesis disorders. *FEBS Let.* 476, 42–46. (Problem 12–72)

Funabiki H & Murray AW (2000) The *Xenopus* chromokinesin Xkid is essential for metaphase chromosome alignment and must be degraded to allow anaphase chromosome movement. *Cell* 102, 411–424. (Problem 18–33)

Fung BK-K & Stryer L (1980) Photolyzed rhodopsin catalyzes the exchange of GTP for bound GDP in retinal rod outer segments. *Proc. Natl. Acad. Sci. USA* 77, 2500–2504. (Problem 15–60)

Fung BK-K, Hurley JB & Stryer L (1981) Flow of information in the light-triggered cyclic nucleotide cascade of vision. *Proc. Natl. Acad. Sci. USA* 78, 152–156. (Problem 15–60)

Fung Y-KT, Murphree AL, T'Ang A, Qian J, Hinrichs SH & Benedict WF (1987) Structural evidence for the authenticity of the human retinoblastoma gene. *Science* 236, 1657–1661. (Problem 17–38)

Gall JG & Murphy C (1998) Assembly of lampbrush chromosomes from sperm chromatin. *Mol. Biol. Cell* 9:733–747. (Problem 4–54)

Garapin AC, Cami B, Roskam W, Kourilsky P, Le Pennec JP, Perrin F, Gerlinger P, Cochet M & Chambon P (1978) Electron microscopy and restriction enzyme mapping reveal additional intervening sequences in the chicken ovalbumin split gene. *Cell* 14, 629–639. (Problem 6–24)

Gartenberg MR & Crothers DM (1988) DNA sequence determinants of CAP-induced bending and protein binding affinity. *Nature* 333, 824–829. (Problem 7–11)

Gay DA, Yen TJ, Lau JTY & Cleveland DW (1987) Sequences that confer β-tubulin autoregulation through modulated mRNA stability reside within exon 1 of a β-tubulin mRNA. *Cell* 50, 671–679. (Problem 7–80)

Gelehrter TD & Collins FS (1990) Principles of Medical Genetics, p 80. Baltimore: Williams & Wilkins. (Problem 8–46)

Gerhart JH & Pardee AB (1963) The effect of the feedback inhibitor, CTP, on subunit interactions in aspartate transcarbamylase, *Cold Spring Harbor Symp. Quant. Biol.* 28, 491–496. (Problem 3–82)

Gibbons IR (1981) Cilia and flagella of eukaryotes. *J. Cell Biol.* 91, 107s–124s. (Problem 16–84)

Gilbert W & Müller-Hill B (1996) Isolation of the Lac Repressor. *Proc. Natl. Acad. Sci. USA* 56, 1891–1898. (Problem 3–47)

Glandsdorff N (1987) Biosynthesis of arginine and polyamines. In *Escherichia coli* and *Salmonella typhimurium*: Cellular and Molecular Biology (FC Neidhardt ed) pp 321–344. Washington DC: American Society for Microbiology. (Problem 7–24)

Godowski PJ, Rusconi S, Miesfeld R & Yamamoto KR (1987) Glucocorticoid receptor mutants that are constitutive activators of transcriptional enhancement. *Nature* 325, 365–368. (Problem 7–36)

Goldberg E (1983) Recognition, attachment, injection. In Bacteriophage T4 (CK Mathews, EM Kutter, G Mosig, PB Berget, eds), pp 32–39. Washington, DC: American Society for Microbiology. (Problem 3–53)

Goldberg J (1999) Structural and functional analysis of the ARF1–ARFGAP complex reveals a role for coatomer in GTP hydrolysis. *Cell* 96, 893–902. (Problem 13–12)

Goldman YE, Hibberd MG, McCray JA & Trentham DR (1980) Relaxation of muscle fibers by photolysis of caged ATP. *Nature* 300, 701–705. (Problem 16–76)

Goldstein JC, Waterhouse NJ, Juin P, Evan GI & Green DR (2000) The coordinate release of cytochrome *c* during apoptosis is rapid, complete and kinetically invariant. *Nat. Cell Biol.* 2, 156–162. (Problems 17–58, 17–59)

Gordon AM, Huxley AF & Julian FJ (1966) The variation in isometric tension with sarcomere length in vertebrate muscle fibres. *J. Physiol.* 184, 170–192. (Problem 16–77)

Görlich D, Prehn S, Laskey RA & Hartmann E (1994) Isolation of a protein that is essential for the first step of nuclear protein import. *Cell* 79, 767–776. (Problem 12–30)

Gottesman MM (1985) Genetics of cyclic-AMP-dependent protein kinases. In Molecular Cell Genetics (MM Gottesman ed), pp 711–743. New York: Wiley. (Problem 15–47)

Gottlieb TA, Beaudry G, Rizzolo L, Colman A, Rindler MJ, Adesnik M & Sabatini DD (1986) Secretion of endogenous and exogenous proteins from polarized MDCK monolayers. *Proc. Natl. Acad. Sci. USA* 83, 2100–2104. (Problem 13–78)

Goutte C & Johnson AD (1988) a1 protein alters the DNA-binding specificity of α2 repressor. *Cell* 52, 875–882. (Problem 7–49)

Guo W, Roth D, Walch-Solimena C & Novick P (1999) The exocyst is an effector for Sec4p, targeting secretory vesicles to sites of exocytosis. *EMBO J.* 18, 1071–1080. (Problem 13–21)

Hadwiger JA, Wittenberg C, Richardson HE, de Barros Lopes M & Reed SI (1989) A family of cyclin homologs that control the G_1 phase in yeast. *Proc. Natl. Acad. Sci. USA* 86, 6255–6259. (Problem 17–22)

Haigler HT, McKanna JA & Cohen S (1979) Rapid stimulation of pinocytosis in human A-431 carcinoma cells by epidermal growth factor. *J. Cell Biol.* 83, 82–90. (Problem 13–56)

Hakem R, Hakem A, Duncan GS, Henderson JT, Woo M, Soengas MS, Elia A, de la Pompa JL, Kagi D, Khoo W, Potter J, Yoshida R, Kaufman SA, Lowe SW, Penninger JM & Mak TW (1998) Differential requirement for caspase 9 in apoptotic pathways *in vivo*. *Cell* 94, 339–352. (Problem 17–65)

Hale SP, Auld DS, Schmidt E & Schimmel P (1997) Discrete determinants in transfer RNA for editing and aminoacylation. *Science* 276, 1250–1252. (Problem 6–52)

Hall A (1998) Rho GTPases and the actin cytoskeleton. *Science* 279, 509–514. (Problem 16–91)

Hancock WO & Howard J (1998) Processivity of the motor protein kinesin requires two heads. *J. Cell Biol.* 140, 1395–1405. (Problem 16–67)

Harada Y, Ohara O, Takatsuki A, Itoh H, Shimamoto N & Kinosita K (2001) Direct observation of DNA rotation during transcription by *Escherichia coli* RNA polymerase. *Nature* 409, 113–115. (Problem 6–17)

Harder T & Simons K (1997) Caveolae, DIGs, and the dynamics of sphingolipidcholesterol microdomains. *Curr. Opinion Cell Biol.* 9, 534–542. (Problem 10–17)

Harland RM & Laskey RA (1980) Regulated replication of DNA microinjected into eggs of *Xenopus laevis. Cell* 21, 761–771. (Problem 5–42)

Hartwell LH (1978) Cell division from a genetic perspective. *J. Cell Biol.* 77, 627–637. (Problems 17–7, 17–8)

Hartwell LH & Weinert TA (1989) Checkpoints: controls that ensure the order of cell cycle events. *Science* 246, 629–634. (Problem 17–47)

Hartwell LH, Hood L, Goldberg ML, Reynolds AE, Silver LM & Veres RC (2000) Genetics: From Genes to Genomes, pp 191–195. Boston, Massachusetts: The McGraw Hill Companies, Inc. (Problem 8–87)

Hartwell LH, Hood L, Goldberg ML, Reynolds AE, Silver LM & Veres RC (2000) Genetics: From Genes to Genomes, pp 217–218. New York: McGraw Hill. (Problem 5–3)

Hartwell LH, Hood L, Goldberg ML, Reynolds AE, Silver LM & Veres RC (2000) Genetics: From Genes to Genomes, pp 408–410. Boston, Massachusetts: The McGraw-Hill Companies. (Problem 7–66)

Heald R & McKeon F (1990) Mutations of phosphorylation sites in lamin A that prevent nuclear lamina disassembly in mitosis. *Cell* 61, 579–589. (Problem 16–30)

Hedjran F, Yeakley JM, Huh GS, Hynes RO & Rosenfeld MG (1997) Control of alternative pre-mRNA splicing by distributed repeats. *Proc. Natl. Acad. Sci. USA* 94, 12343–12347. (Problem 7–72)

Helms JB & Rothman JE (1992) Inhibition by brefeldin A of a Golgi membrane enzyme that catalyzes exchange of guanine nucleotide bound to ARF. *Nature* 360, 352–354. (Problem 13–11)

Hentze MW & Kühn LC (1996) Molecular control of vertebrate iron metabolism: mRNA-based regulatory circuits operated by iron, nitric oxide, and oxidative stress. *Proc. Natl. Acad. Sci. USA* 93, 8175–8182. (Problems 7–78, 7–79)

Herrmann H & Aebi U (2000) Intermediate filaments and their associates: multi-talented structural elements specifying cytoarchitecture and cytodynamics. *Curr. Opin. Cell Biol.* 12, 79–90. (Problem 16–32)

Hershey AD & Chase M (1952) Independent functions of viral protein and nucleic acid in growth of bacteriophage. *J. Gen. Physiol.* 36, 39–56. (Problem 4–8)

Hertel C, Muller P, Portenier M & Staehelin M (1983) Determination of the desensitization of beta-adrenergic receptors by [³H]CGP-12177. *Biochem. J.* 216, 669–674. (Problem 15–34)

Higgs HN & Pollard TD (2001) Regulation of actin filament network formation through ARP2/3 complex: Activation by a diverse array of proteins. *Annu. Rev. Biochem.* 70, 649–676. (Problem 16–49)

Higuchi R (1990) Recombinant PCR. In PCR Protocols (MA Innis, DH Gelfand, JJ Sninsky & TJ White, eds), pp 177–183. San Diego, CA: Academic Press. (Problem 8–69)

Hille B (1992) Ionic Channels of Excitable Membranes, 2nd edn, pp 1–20. Sunderland, MA: Sinauer. (Problem 11–46)

Hille B (1992) Ionic Channels of Excitable Membranes, 2nd edn, pp 23–58. Sunderland, MA: Sinauer. (Problem 11–43)

Hille B (1992) Ionic Channels of Excitable Membranes, 2nd edn, pp 315–336. Sunderland, MA: Sinauer. (Problem 11–56)

Ho C-L, Martys JL, Mikhailov A, Gundersen GG & Liem RKH (1998) Novel features of intermediate filament dynamics revealed by green fluorescent protein chimeras. *J. Cell Sci.* 111, 1767–1778. (Problem 16–28)

Holloway SL, Glotzer M, King RW & Murray AW (1993) Anaphase is initiated by proteolysis rather than by the inactivation of maturation-promoting factor. *Cell* 73, 1393–1402. (Problem 17–32)

Honegger AM, Kris RM, Ullrich A & Schlessinger J (1989) Evidence that autophosphorylation of solubilized receptors for epidermal growth factor is mediated by intermolecular cross-phosphorylation. *Proc. Natl. Acad. Sci. USA* 86, 925–929. (Problem 15–70)

Honegger AM, Schmidt A, Ullrich A & Schlessinger J (1990) Evidence for epidermal growth factor (EGF)-induced intermolecular autophosphorylation of the EGF receptors in living cells. *Mol. Cell. Biol.* 10, 4035–4044. (Problem 15–70)

Horio T & Hotani H (1986) Visualization of the dynamic instability of individual microtubules by dark-field microscopy. *Nature* 321, 605–607. (Problems 16–12, 16–52)

Horowitz S & Gorovsky MA (1985) An unusual genetic code in nuclear genes of *Tetrahymena. Proc. Natl. Acad. Sci. USA* 82, 2452–2455. (Problem 6–66)

Howard J (2001) Mechanics of Motor Proteins and the Cytoskeleton, pp 151–163. Sunderland, MA: Sinauer. (Problem 2–68)

Huang C-Y F & Ferrell JE (1996) Ultrasensitivity in the mitogen-activated protein kinase cascade. *Proc. Natl. Acad. Sci. USA* 93, 10078–10083. (Problem 15–77)

Huang S, Ratliff KS, Schwartz MP, Spenner JM & Matouschek A (1999) Mitochondria unfold precursor proteins by unraveling them from their N-termini. *Nature Struct. Biol.* 6, 1132–1138. (Problems 12–57, 12–58)

Huberman JA & Riggs AD (1968) On the mechanism of DNA replication in mammalian chromosomes. *J. Mol. Biol.* 32, 327–341. (Problem 5–37)

Huganir RL, Delcour AH, Greengard P & Hess GP (1986) Phosphorylation of the nicotine acetycholine receptor regulates its rate of desensitization. *Nature* 321, 774–776. (Problem 15–35)

Hunt T, Luca FC & Ruderman JV (1992) The requirement for protein synthesis and degradation, and the control of destruction of cyclins A and B in the meiotic and mitotic cell cycles of the clam embryo. *J. Cell Biol.* 116, 707–724. (Problem 17–20)

Hyman AA & Karsenti E (1996) Morphogenetic properties of microtubules and mitotic spindle assembly. *Cell* 84, 401–410. (Problem 18–20)

Im D-S, Heise CE, Nguyen T, O'Dowd BF & Lynch KR (2001) Identification of a molecular target of psychosine and its role in globoid cell formation. *J. Cell Biol.* 153, 429–434. (Problem 18–49)

Ingledew JW (1982) *Thiobacillus ferrooxidans*: the bioenergetics of an acidophilic chemolithotroph. *Biochim. Biophys. Acta* 683, 89–117. (Problems 14–24, 14–32)

Inman RB & Schnos M (1971) Structure of branch points in replicating DNA: presence of single-stranded connections in lambda DNA branch points. *J. Mol. Biol.* 56, 319–325. (Problem 5–12)

Jacks T & Varmus H (1985) Expression of the Rous sarcoma virus *pol* gene by ribosomal frameshifting. *Science* 230, 1237–1242. (Problem 6–62)

Jacobson MD, Burne JF, King MP, Miyashita T, Reed JC & Raff MC (1993) Bcl-2 blocks apoptosis in cells lacking mitochondrial DNA. *Nature* 361, 365–369. (Problem 17–56)

Jady B & Kiss T (2001) A small nucleolar guide RNA functions both in 2′-O-ribose methylation and pseudouridylation of the U5 spliceosomal RNA. *EMBO J.* 20, 541–551. (Problems 6–37, 6–38)

Jagendorf AT & Uribe E (1966) ATP formation caused by acid-base transition of spinach chloroplasts. *Proc. Natl. Acad. Sci. USA* 55, 170–177. (Problem 14–72)

Jain R, Rivera MC & Lake JA (1999) Horizontal gene transfer among genomes: The complexity hypothesis. *Proc. Natl. Acad. Sci. USA* 96, 3801–3806. (Problem 1–23)

Jensen RA, Thompson ME, Jetton TL, Szabo CI, van der Meer R, Helou B, Tronick SR, Page DL, King MC & Holt JT (1996) BRCA1 is secreted and exhibits properties of a granin. *Nature Genet.* 12, 303–308. (Problem 3–45)

Jia Z & Davies PL (2002) Antifreeze proteins: an unusual receptor–ligand interaction. *Trends Biochem. Sci.* 27, 101–106. (Problem 3–37)

Johnson KA (1993) Conformational coupling in DNA polymerase fidelity. *Annu. Rev. Biochem.* 62, 685–713. (Problem 5–28)

Johnson PJ, Kooter JM & Borst P (1987) Inactivation of transcription by UV irradiation of *T. brucei* provides evidence for a multicistronic transcription unit including a VSG gene. *Cell* 51, 273–281. (Problem 6–9)

Johnson RT & Rao PN (1971) Nucleo-cytoplasmic interactions in the achievement of nuclear synchrony in DNA synthesis and mitosis in multinucleate cells *Biol. Rev. Camb. Philos. Soc.* 46, 97–155. (Problem 17–26)

Jones KA (1997) Taking a new TAK on Tat transactivation. *Genes Dev.* 11, 2593–2599. (Problem 7–69)

Kabata H, Kurosawa O, Arai I, Washizu M, Margarson SA, Glass RE & Shimamoto N (1993) Visualization of single molecules of RNA polymerase sliding along DNA. *Science* 262, 1561–1563. (Problem 7–10)

Kantheti P, Qiao X, Diaz ME, Peden AA, Meyer GE, Carskadon SI, Kapfhamer D, Sufalko D, Robinson MS, Noebels JL & Burmeister M (1998) Mutation in AP-3 delta in the mocha mouse links endosomal transport to storage deficiency in platelets, melanosomes, and synaptic vesicles. *Neuron* 21, 111–122. (Problem 13–52)

Kaplan A, Achord DT & Sly WS (1977) Phosphohexosyl components of a lysosomal enzyme are recognized by pinocytosis receptors on human fibroblasts. *Proc. Natl. Acad. Sci. USA* 74, 2026–2030. (Problem 13–49)

Karrer KM & Gall JG (1976) The macronuclear ribosomal DNA of *Tetrahymena pyriformis* is a palindrome. *J. Mol. Biol.* 104, 421–453. (Problem 8–39)

Karzai AW, Susskind MM & Sauer RT (1999) SmpB, a unique RNA-binding protein essential for the peptide-tagging activity of SsrA (tmRNA). *EMBO J.* 18, 3793–3799. (Problems 3–48, 6–65)

Keilin D (1966) The History of Cell Respiration and Cytochrome. Cambridge, UK: Cambridge University Press. (Problem 14–35)

Kellems RE, Allison VF & Butow RA (1975) Cytoplasmic type 80S ribosomes associated with yeast mitochondria. IV. Attachment of ribosomes to the outer membrane of isolated mitochondria. *J. Cell Biol.* 65, 1–14. (Problem 12–47)

Kimura K & Hirano T (1997) ATP-dependent positive supercoiling of DNA by 13S condensin: A biochemical implication for chromosome condensation. *Cell* 90, 625–634. (Problem 4–67)

Kimura K, Rybenkov VV, Crisona NJ, Hirano T, & Cozzarelli NR (1999) 13S condensin actively reconfigures DNA by introducing global positive writhe: Implications for chromosome condensation. *Cell* 98, 239–248. (Problem 18–4)

Kinoshita N, Ghaedi K, Shimozawa N, Wanders RJA, Matsuzono Y, Imanaka T, Okumoto K, Suzuki Y, Kondo N & Fujiki Y (1998) Newly identified Chinese hamster ovary cell mutants are defective in biogenesis of peroxisomal membrane vesicles (peroxisome ghosts), representing a novel complementation group in mammals. *J. Biol. Chem.* 273, 24122–24130. (Problems 12–68, 12–71)

Kirchhausen T (2000) Clathrin. *Annu. Rev. Biochem.* 69, 699–727. (Problem 13–6, 13–7)

Kirschner MW & Schachman HK (1973) Local and gross conformational changes in aspartate transcarbamylase, *Biochemistry* 12, 2997–3004. (Problem 3–83)

Klemm RD, Austin RJ & Bell SP (1997) Coordinate binding of ATP and origin DNA regulates the ATPase activity of the origin recognition complex. *Cell* 88, 493–502. (Problem 5–47)

Knoop F (1905) Der Abbau aromatischer Fettsäuren im Tierkörper. *Beitr. Chem. Physiol.* 6, 150–162. (Problem 2–98)

Knudson AG (1971) Mutation and cancer: statistical study of retinoblastoma. *Proc. Natl. Acad. Sci. USA* 68, 820–823. (Problem 17–38)

Kobayashi T, Rein T & DePamphilis ML (1998) Identification of primary initiation sites for DNA replication in the hamster dihydrofolate reductase gene initiation zone. *Mol. Cell. Biol.* 18, 3266–3277. (Problem 5–48)

Koenig JH & Ikeda K (1999) Contribution of active zone subpopulation of vesicles to evoked and spontaneous release. *J. Neurophysiol.* 81, 1495–1505. (Problem 13–80)

Koleske AJ, Buratowski S, Nonet M & Young RA (1992) A novel transcription factor reveals a functional link between the RNA polymerase II CTD and TFIID. *Cell* 69, 883–894. (Problem 6–15)

Kondor-Koch C, Bravo R, Fuller SD, Cutler D & Garoff H (1985) Protein secretion in the polarized epithelial cell line MDCK. *Cell* 43, 297–306. (Problem 13–78)

Konkel DA, Maizel JV & Leder P (1979) The evolution and sequence comparison of two recently diverged mouse chromosomal β-globin genes. *Cell* 18, 865–873. (Problem 4–26)

Krause M & Hirsh D (1987) A trans-spliced leader sequence on actin mRNA in *C. elegans. Cell* 49, 753–761. (Problem 6–30)

Krebs HA & Johnson WA (1937) The role of citric acid in intermediate metabolism in animal tissues. *Enzymologia* 4, 148–156. (Problem 2–99)

Krings M, Stone A, Schmitz RW, Krainitzki H, Stoneking M & Pääbo S (1997) Neanderthal DNA sequences and the origin of modern humans. *Cell* 90, 19–30. (Problem 5–53)

Kristic RV (1997) Ultrastructure of the Mammalian Cell, p 207. Berlin, Germany: Springer-Verlag. (Problem 11–27)

Kudo N, Matsumori N, Taoka H, Fujiwara D, Schreiner EP, Wolff B, Yoshida M, & Horinouchi S (1999) Leptomycin B inactivates CRM1/exportin 1 by covalent modification at a cysteine residue in the central conserved region. *Proc. Natl. Acad. Sci. USA* 96, 9112–9117. (Problems 12–35, 12–36)

Kuhne T, Wieringa B, Reiser J & Weissmann C (1983) Evidence against a scanning model for RNA splicing. *EMBO J.* 2, 727–733. (Problem 6–29)

Kuida K, Haydar TF, Kuan C-Y, Gu Y, Taya C, Karasuyama H, Su MS-S, Rakic P & Flavell RA (1998) Reduced apoptosis and cytochrome c-mediated caspase activation in mice lacking caspase 9. *Cell* 94, 325–337. (Problem 17–65)

Kumagai A & Dunphy WG (1992) Regulation of the Cdc25 protein during the cell cycle in *Xenopus* extracts. *Cell* 70, 139–151. (Problem 17–29)

Kurjan I (1992) Pheromone response in yeast. *Annu. Rev. Biochem.* 61, 1097–1129. (Problem 15–40)

Kyte J (1995) Mechanism in Protein Chemistry, pp 175–177. New York: Garland Publishing. (Problem 3–47)

Kyte J (1995) Mechanism in Protein Chemistry, pp 207–208. New York: Garland Publishing. (Problem 3–66)

Landegren U, Kaiser R, Sanders J & Hood L (1988) A ligase-mediated gene detection technique. *Science* 241, 1077–1080. (Problem 8–43)

Lander ES et al (2001) Initial sequencing and analysis of the human genome. *Nature* 409, 860–921. (Problems 6–47, 4–19, 4–20, 4–21, 4–24, 4–28)

Larschan E & Winston F (2001) The *S. cerevisiae* SAGA complex functions *in vivo* as a coactivator for transcriptional activation by Gal4. *Genes Dev.* 15, 1946–1956. (Problem 7–41)

Lawlor DW (1987) Photosynthesis: Metabolism, Control, and Physiology. New York: Wiley. (Problem 14–61)

LeBowitz JH & McMacken R (1986) The *Escherichia coli* DnaB replication protein is a DNA helicase. *J. Biol. Chem.* 261, 4738–4748. (Problem 5–22)

Lee JT & Lu N (1999) Targeted mutagenesis of *Tsix* leads to nonrandom X inactivation. *Cell* 99, 47–57. (Problem 7–62)

Lee M-H & Schedl T (2001) Identification of *in vivo* mRNA targets of GLD-1, a maxi-KH motif containing protein required for *C. elegans* germ cell development. *Genes Dev* 15, 2408–2420. (Problem 7–84)

Leff SE, Evans RM & Rosenfeld MG (1987) Splice commitment dictates neuron-specific alternative RNA processing in calcitonin/CGRP gene expression. *Cell* 48, 517–524. (Problem 7–71)

Lefkowitz RJ, Limbird LE, Mukherjee C & Caron MG (1976) The β-adrenergic receptor and adenylate cyclase. *Biochim. Biophys. Acta* 457, 1–39. (Problem 15–36)

Leguy R, Melki R, Pantaloni D & Carlier M-F (2000) Monomeric γ-tubulin nucleates microtubules. *J. Biol. Chem.* 275, 21975–21980. (Problems 16–42, 16–43, 16–44)

Li K, Li Y, Shelton JM, Richardson JA, Spencer E, Chen ZJ, Wang X & Williams RS (2000) Cytochrome *c* deficiency causes embryonic lethality and attenuates stress-induced apoptosis. *Cell* 101, 389–399. (Problem 17–62)

Li L, Zhou J, James G, Heller-Harrison R, Czech MP & Olsen EN (1992) FGF inactivates myogenic helix–loop–helix proteins through phosphorylation of a conserved protein kinase C site in their DNA-binding domains. *Cell* 71, 1181–1194. (Problem 7–19)

Li R & Murray AW (1991) Feedback control of mitosis in budding yeast. *Cell* 66, 519–531. (Problem 17–36)

Li W-H (1997) Molecular Evolution, pp 228–230. Sinauer Associates, Inc. Sunderland MA. (Problems 1–36, 1–38)

Liebold EA & Munro HN (1988) Cytoplasmic protein binds *in vitro* to a highly conserved sequence in the 5′ untranslated region of ferritin heavy- and light-subunit mRNAs. *Proc. Natl. Acad. Sci. USA* 85, 2171–2175. (Problem 7–77)

Lindahl T, Demple B & Robins P (1982) Suicide inactivation of the *E. coli.* O^6-methylguanine-DNA methyltransferase. *EMBO J.* 1, 1359–1363. (Problem 5–64)

Lipmann F (1941) Metabolic generation and utilization of phosphate bond energy. *Adv. Enzymol.* 1, 99–162. (Problem 2–91)

Liu X, Kim CN, Yang J, Jemmerson R & Wang X (1996) Induction of apoptotic program in cell-free extracts: Requirement for dATP and cytochrome *c.* *Cell* 86, 147–157. (Problem 17–56)

Lo CW & Gilula NB (1979) Gap junctional communication in the preimplantation mouse embryo. *Cell* 18, 399–409. (Problem 15–15)

Lodish HF, Kong N, Snider M, & Strous GJAM (1983) Hepatoma secretory proteins migrate from rough endoplasmic reticulum to Golgi at characteristic rates. *Nature* 304, 80–83. (Problem 13–71)

Logothetis DE, Kurachi Y, Galper J, Neer EJ & Clapham DE (1987) The βγ subunits of GTP-binding proteins activate the muscarinic K⁺ channel in heart. *Nature* 325, 321–326. (Problem 15–59)

Losada A & Hirano T (2001) Intermolecular DNA interactions stimulated by the cohesin complex *in vitro*: Implications for sister chromatid cohesion. *Curr. Biol.* 11, 268–272. (Problem 18–4)

Luck DJL (1984) Genetic and biochemical dissection of the eucaryotic flagellum. *J. Cell Biol.* 98, 789–794. (Problem 16–86)

Luria SE & Delbruck M (1943) Mutations of bacteria from virus sensitivity to virus resistance. *Genetics* 28, 491–511. (Problem 5–2)

Maarse AC, Blom J, Grivell LA & Meijer M (1992) MPI1, an essential gene encoding a mitochondrial membrane protein, is possibly involved in protein import into yeast mitochondria. *EMBO J.* 11, 3619–3628. (Problem 12–51)

MacKinnon R, Aldrich RW & Lee AW (1993) Functional stoichiometry of *shaker* potassium channel inactivation. *Science* 262, 757–759. (Problem 11–39)

MacLean-Fletcher S & Pollard TD (1980) Mechanism of action of cytochalasin B on actin. *Cell* 20, 329–341. (Problem 16–33)

MacNicol A, Muslin AJ & Williams LT (1993) Raf-1 kinase is essential for early *Xenopus* development and mediates the induction of mesoderm by FGF. *Cell* 73, 571–583. (Problem 8–95)

Madhuri HD & Fink GR (1998) The control of filamentous differentiation and virulence in fungi. *Trends Cell Biol.* 8, 348–353. (Problem 15–101)

Madison-Antenucci S, Grams J & Hajduk SL (2002) Editing machines: the complexities of *Trypanosome* RNA editing. *Cell* 108, 435–438. (Problem 7–75)

Maizia D, Harris PJ & Bibring T (1960) The multiplicity of mitotic centers and the time-course of their duplication and separation. *J. Biophys. Biochem. Cytol.* 7, 1–20. (Problem 18–9)

Mallavarapu A & Mitchison TJ (1999) Regulated actin cytoskeleton assembly at filopodium tips controls their extension and retraction. *J. Cell Biol.* 146, 1097–1106. (Problem 16–99)

Mann C & Davis RW (1983) Instability of dicentric plasmids in yeast. *Proc. Natl. Acad. Sci. USA* 80, 228–232. (Problem 18–23)

Manoil C & Beckwith J (1986) A genetic approach to analyzing membrane protein topology. *Science* 233, 1403–1408. (Problem 12–88)

Manson MD, Blank V, Brade G & Higgins CF (1986) Peptide chemotaxis in *E. coli* involves the Tap signal transducer and the dipeptide permease. *Nature* 321, 253–258. (Problem 15–88)

Manson MD, Tedesco P, Berg HC, Harold FM & van der Drift C (1977) A protonmotive force drives bacteria flagella. *Proc. Natl. Acad. Sci. USA* 74, 3060–3064. (Problem 14–48)

Marks B, Stowell MHB, Vallis Y, Mills IG, Gibson A, Hopkins CR & McMahon HT. GTPase activity of dynamin and resulting conformation changes are essential for endocytosis. *Nature* 410, 231–235. (Problem 13–9)

Masaike T, Mitome N, Noji H, Muneyuki E, Yasuda R, Kinosita K & Yoshida M (2000) Rotation of F_1-ATPase and the hinge residues of the β subunit. *J. Exp. Biol.* 203, 1–8. (Problem 14–17)

Masutani C, Araki M, Yamada A, Kusumoto R, Nogimori T, Maekawa T, Iwai S & Hanaoka F (1999) Xeroderma pigmentosum variant (XP-V) correcting protein from HeLa cells has a thymine dimer bypass DNA polymerase activity. *EMBO J.* 18, 3491–3501. (Problem 5–62)

Matteoli M, Takei K, Perin MS, Südhof TC, & DeCamilli P (1992) Exo-endocytotic recycling of synaptic vesicles in developing processes of cultured hippocampal neurons. *J. Cell Biol.* 117, 849–861. (Problem 13–79)

McClintock B (1939) The behavior of successive nuclear divisions of a chromosome broken at meiosis. *Proc. Natl. Acad. Sci. USA* 25, 405–416. (Problem 18–15)

McKeon FD, Kirschner MW & Caput D (1986) Homologies in both primary and secondary structure between nuclear envelope and intermediate filament proteins. *Nature* 319, 463–468. (Problem 3–31)

McKnight SL & Kingsbury R (1982) Transcriptional control signals of a eukaryotic protein-coding gene. *Science* 217, 316–324. (Problem 6–10)

Meikrantz W, Bergom MA, Memisoglu A & Samson L (1998) O^6-Alkylguanine DNA lesions trigger apoptosis. *Carcinogenesis* 19, 369–372. (Problems 5–65, 17–67)

Meselson M & Stahl FW (1958) The replication of DNA in *Escherichia coli.* *Proc. Natl. Acad. Sci. USA* 44, 671–682. (Problem 8–14)

Meuse CW, Krueger S, Majkrzak CF, Dura JA, Fu J, Connor JT & Plant AL (1998) Hybrid bilayer membranes in air and water: infrared spectroscopy and neutron reflectivity studies. *Biophys. J.* 74, 1388–1398. (Problem 13–22)

Meyer CA, Jacobs HW, Datar SA, Du W, Edgar BA & Lehner CF (2000) *Drosophila* Cdk4 is required for normal growth and is dispensable for cell cycle progression. *EMBO J.* 19, 4533–4542. (Problem 17–43)

Meyer CA, Jacobs HW, Datar SA, Du W, Edgar BA & Lehner CF (2000) *Drosophila* Cdk4 is required for normal growth and is dispensable for cell cycle progression. *EMBO J.* 19, 4533–4542. (Problem 17–44)

Michaelis C, Ciosk R & Nasmyth K (1997) Cohesins: chromosomal proteins that prevent premature separation of sister chromatids. *Cell* 91, 35–45. (Problem 17–33)

Miller JH (1985) Mutagenic specificity of ultraviolet light. *J. Mol. Biol.* 182, 45–65. (Problem 5–61)

Milton RC, Milton SC & Kent SB (1992) Total chemical synthesis of a D-enzyme: the enantiomers of HIV-1 protease show reciprocal chiral substrate specificity. *Science* 256, 1445–1448. (Problem 3–65)

Minakami S, Suzuki C, Saito T & Yoshikawa H (1965) Studies on erythrocyte glycolysis. I. Determination of the glycolytic intermediates in human erythrocytes. *J. Biochem.* (*Tokyo*) 58, 543–550. (Problem 2–78)

Mitchell MB & Mitchell HK (1952) A case of "maternal" inheritance in *Neurospora crassa. Proc. Natl. Acad. Sci. USA* 38, 442–449. (Problem 14–86)

Mitchell MB, Mitchell HK & Tissieres A (1953) Mendelian and non-Mendelian factors affecting the cytochrome system in *Neurospora crassa. Proc. Natl. Acad. Sci. USA* 39, 605–613. (Problem 14–86)

Mitchison T & Kirschner MW (1985) Properties of the kinetochore *in vitro.* I. Microtubule nucleation and tubulin binding. *J. Cell Biol.* 101, 755–765. (Problem 16–40)

Mitchison T (2001) Psychosine, cytokinesis, and orphan receptors: Unexpected connections. *J. Cell Biol.* 153, F1–F3. (Problem 18–49)

Mitchison TJ (1993) Localization of an exchangeable GTP binding site at the plus end of microtubules. *Science* 261, 1044–1047. (Problem 16–9)

Mitchison TJ & Kirschner MW (1985) Properties of the kinetochore *in vitro.* II. Microtubule capture and ATP-dependent translocation. *J. Cell. Biol.* 101, 766–777. (Problem 18–22)

Mlynarczyk SK & Panning B (2000) X inactivation: *Tsix* and *Xist* as yin and yang. *Curr. Biol.* 10, R899–R903. (Problem 7–62)

Molecular Probes Handbook, 1999. (www.probes.com/handbook). (Problem 11–26)

Monod J (1947) The phenomenon of enzymatic adaptation. Growth Symposium XI:223–289. [Reprinted in Selected Papers in Molecular Biology by Jacques Monod (A Lwoff, A Ullmann, eds), pp 68–134, New York: Academic Press, 1947. (Problem 7–27)

Montoya J, Ojala D & Attardi G (1981) Distinctive features of the 5′-terminal sequences of the human mitochondrial mRNAs. *Nature* 290, 465–470. (Problem 14–83)

Moore MS & Blobel G (1993) The GTP-binding protein Ran/TC4 is required for protein import into the nucleus. *Nature* 365, 661–663. (Problem 12–30)

Moore MS & Blobel G (1994) Purification of a Ran-interacting protein that is required for protein import into the nucleus. *Proc. Natl. Acad. Sci. USA* 91, 10212–10216. (Problem 12–30)

Moore T & Haig D (1991) Genomic imprinting in mammalian development: a parental tug of war. *Trends Genet.* 7, 45–49. (Problem 7–67)

Morand OH, Allen LA, Zoeller RA & Raetz CR (1990) A rapid selection for animal cell mutants with defective peroxisomes. *Biochem. Biophys. Acta* 1034, 132–141. (Problem 12–70)

Moriyoshi K, Masu M, Ishii T, Shigemoto R, Mizuno N & Nakanishi S (1991) Molecular cloning and characterization of the rat NMDA receptor. *Nature* 354, 31–37. (Problem 11–64)

Mowry KL & Steitz JA (1987) Identification of human U7 snRNP as one of several factors involved in the 3′-end maturation of histone premessenger RNAs. *Science* 238, 1682–1687. (Problem 6–34)

Mueller-Storm HP, Sogo JM & Schaffner W (1989) An enhancer stimulates transcription in trans when attached to the promoter via a protein bridge. *Cell* 58, 767–777. (Problem 7–33)

Munro S & Pelham HR (1987) A C-terminal signal prevents secretion of luminal ER proteins. *Cell* 48, 899–907. (Problem 13–29)

Murray A & Hunt T (1993) The Cell Cycle: An Introduction, pp 143–144. New York: WH Freeman. (Problem 17–48)

Murray A & Szostak JW (1982) Pedigree analysis of plasmid segregation in yeast. *Cell* 34, 961–970. (Problem 18–24)

Murray EJ & Grosveld F (1987) Site-specific demethylation in the promoter of human γ-globin does not alleviate methylation mediated suppression. *EMBO J.* 6, 2329–2335. (Problem 7–64)

Nachury MV & Weis K (1999) The direction of transport through the nuclear pore can be inverted. *Proc. Natl. Acad. Sci. USA* 96, 9622–9627. (Problems 12–40, 12–41, 12–42)

Nagata K & Handa H (2000) Real-Time Analysis of Biomolecular Interactions: Applications of BIACORE. Tokyo, Japan: Springer. (Problem 8–81)

Nagata Y, Muro Y & Todokoro K (1997) Thrombopoietin-induced polyploidization of bone marrow megakaryocytes is due to a unique regulatory mechanism in late mitosis. *J. Cell Biol.* 139, 449–457. (Problem 18–50)

Nakamura TM, Morin GB, Chapman KB, Weinrich SL, Andrews WH, Lingner J, Harley CB & Cech TR (1997) Telomerase catalytic subunit homologs from fission yeast and human. *Science* 277, 955–959. (Problem 5–51)

Nathans J, Thomas D & Hogness DS (1986) Molecular genetics of human color vision: the genes encoding blue, green, and red pigments. *Science* 232, 193–202. (Problem 4–20)

Nathans J, Piantanida TP, Eddy RL, Shows TB & Hogness DS (1986) Molecular genetics of inherited variation in human color vision. *Science* 232, 203–210. (Problem 4–20)

Neupert W (1997) Protein import into mitochondria. *Annu. Rev. Biochem.* 66, 863–917. (Problem 12–48)

Newman M, Lunnen K, Wilson G, Greci J, Schildkraut I & Phillips SEV (1998) Crystal structure of restriction endonuclease BglI bound to its interrupted DNA recognition sequence. *EMBO J.* 17, 5466–5476. (Problem 8–36)

Nicholls DG & Ferguson SJ (1992) Bioenergetics 2, pp 65–104. London: Academic Press. (Problem 14–23)

Nicholls DG & Ferguson SJ (1992) Bioenergetics 2, pp 82–87. London: Academic Press. (Problem 14–38)

Nicholls DG & Ferguson SJ (1992) Bioenergetics 2, pp149–154. London: Academic Press. (Problem 14–47)

Nicholls DG & Ferguson SJ (1992) Bioenergetics 2, pp 215–218. London: Academic Press. (Problem 14–22)

Nichols BJ, Undermann C, Pelham HRB, Wickner WT & Haas A (1997) Homotypic vacuolar fusion mediated by t- and v-SNAREs. *Nature* 387, 199–202. (Problems 13–14, 13–27)

Ninfa AJ, Reitzer LJ & Magasanik B (1987) Initiation of transcription at the bacterial *glnAp2* promoter by purified *E. coli* components is facilitated by enhancers. *Cell* 50, 1039–1046. (Problem 7–30)

Nishizuka Y (1983) Calcium, phospholipid turnover and transmembrane signaling. *Phil. Trans. R. Soc. Lond. (Biol.)* 302, 101–112. (Problem 15–52)

Nogales E, Whittaker M, Milligan RA & Downing KH (1999) High-resolution model of the microtubule. *Cell* 96, 79–88. (Problem 16–9)

Noji H, Yasuda R, Yoshida M & Kinosita K (1997) Direct observation of the rotation of F$_1$-ATPase. *Nature* 386, 299–302. (Problem 14–18)

Norby JG (2000) The origin and the meaning of the little p in pH. *Trends Biochem. Sci.* 25, 36–37. (Problem 2–23)

Norman C, Runswick M, Pollock R & Treisman R (1988) Isolation and properties of cDNA clones encoding SRF, a transcription factor that binds to the *c-fos* serum response element. *Cell* 55, 989–1003. (Problem 7–20)

Nudler E, Mustaev A, Lukhtanov E, & Goldfarb A (1997) The RNA–DNA hybrid maintains the register of transcription by preventing backtracking of RNA polymerase. *Cell* 89, 33–41. (Problem 6–19)

Nugent JM & Palmer JD (1991) RNA-mediated transfer of the gene *coxII* from the mitochondrion to the nucleus during flowering plant evolution. *Cell* 66, 473–481. (Problem 1–33)

Nurse P (1975) Genetic control of cell size at cell division in yeast. *Nature* 256, 547–551. (Problems 17–25, 17–42)

Nurse P & Thuriaux P (1980) Regulatory genes controlling mistosis in the fission yeast *Schizosaccharomyces pombe*. *Genetics* 96, 627–637. (Problem 17–25)

O'Connell KF, Caron C, Kopish KR, Hurd DD, Kemphues KJ, Li Y & White JG (2001) The *C. elegans zyg-1* gene encodes a regulator of centrosome duplication with distinct maternal and paternal roles in the embryo. *Cell* 105, 547–558. (Problem 18–8)

O'Keefe EJ & Pledger WJ (1983) A model of cell-cycle control: sequential events regulated by growth factors. *Mol. Cell. Endocrinol.* 31, 167–186. (Problem 17–72)

O'Toole TE, Mandelman D, Forsyth J, Shattil SJ, Plow EF & Ginsberg MH (1991) Modulation of the affinity of integrin α$_{IIb}$β$_3$ (GPIIb-IIIa) by the cytoplasmic domain of α$_{IIb}$. *Science* 254, 845–847. (Problem 15–13)

Oeller PW, Min-Wong L, Taylor LP, Pike DA & Theologis A (1991) Reversible inhibition of tomato fruit senescence by antisense RNA. *Science* 254, 437–439. (Problem 15–103)

Ojala D, Montoya J & Attardi G (1981) tRNA punctuation model of RNA processing in human mitochondria. *Nature* 290, 470–474. (Problem 14–83)

Oka Y & Czech MP (1984) Photoaffinity labeling of insulin-sensitive hexose transporters in intact rat adipocytes: direct evidence that latent transporters become exposed to the extracellular space in response to insulin. *J. Biol. Chem.* 259, 8125–8133. (Problem 11–10)

Oosawa F (2001) A historical perspective of actin assembly and its interactions. *Res. Prob. Cell Diff.* 32, 9–21. (Problem 16–8)

Orci L, Glick BS & Rothman JE (1986) A new type of coated vesicular carrier that appears not to contain clathrin: its possible role in protein transport within the Golgi stack. *Cell* 46, 171–184. (Problem 13–72)

Orr-Weaver TL & Spradling AD (1986) *Drosophila* chorion gene amplification requires an upstream region regulating s18 transcription. *Mol. Cell. Biol.* 6, 4624–4633. (Problem 5–45)

Osinga KA, Swinkels BW, Gibson WC, Borst P, Veeneman GH, Van Boom JH, Michels PAM & Opperdoes FR (1985) Topogenesis of microbody enzymes: sequence comparison of the genes for the glycosomal (microbody) and cytosolic phosphoglycerate kinases of *Trypanosoma brucei*. *EMBO J.* 4, 3811–3817. (Problem 12–66)

Otero LJ, Devaux A & Standart N (2001) A 250-nucleotide UA-rich element in the 3′ untranslated region of *Xenopus laevis* Vg1 mRNA represses translation both *in vivo* and *in vitro*. *RNA* 7, 1753–1767. (Problem 7–86)

Ottaviano Y & Gerace L (1985) Phosphorylation of the nuclear lamins during interphase and mitosis. *J. Biol. Chem.* 260, 624–632. (Problem 16–30)

Ovchinnikov IV, Gotherstrom A, Romanova GP, Kharitonov VM, Liden K & Goodwin W (2000) Molecular analysis of Neanderthal DNA from the northern Caucasus. *Nature* 404, 490–493. (Problem 5–53)

Pace NR (2001) The universal nature of biochemistry. *Proc. Natl. Acad. Sci. USA* 98, 805–808. (Problem 1–1)

Park H-O, Chant J & Herskowitz I (1993) *BUD2* encodes a GTPase-activating protein for Bud1/Rsr1 necessary for proper bud-site selection in yeast. *Nature* 365, 269–274. (Problem 16–100)

Paschen SA, Rothbauer U, Kaldi K, Bauer MF, Neupert W & Grunner M (2000) The role of the TIM8-13 complex in the import of Tim23 into mitochondria. *EMBO J.* 19, 6392–6400. (Problem 12–59)

Patel LR, Curran T & Kerppola TK (1994) Energy transfer analysis of Fos–Jun dimerization and DNA binding. *Proc. Natl. Acad. Sci. USA* 91, 7360–7364. (Problem 7–18)

Pauling L (1948) Chemical achievement and hope for the future. *Am. Sci.* 36, 50–58. (Problem 3–66)

Payvar F, DeFranco D, Firestone GL, Edgar B, Wrange O, Okret S, Gustafsson J-A & Yamomoto KR (1983) Sequence-specific binding of glucocorticoid receptor to MMTV DNA at sites within and upstream of the transcribed region. *Cell* 35, 381–392. (Problem 15–23)

Pearse BMF, Smith CJ & Owen DJ (2000) clathrin coat construction in endocytosis. *Curr. Opin. Struc. Biol.* 10, 220–228. (Problems 13–5, 13–7)

Pelham HRB & Rothman JE (2000) The debate about transport in the Golgi—two sides of the same coin? *Cell* 102, 713–719. (Problem 13–41)

Penny GD, Kay GF, Sheardown SA, Rastan S & Brockdorff N (1996) Requirement for *Xist* in X chromosome inactivation. *Nature* 379, 131–137. (Problem 7–61)

Perara E, Rothman RE & Lingappa VR (1986) Uncoupling translocation from translation: implications for transport of proteins across membranes. *Science* 232, 348–352. (Problem 12–80)

Peter M, Nakagawa J, Dorée M, Labbe JC & Nigg EA (1990) *In vitro* disassembly of the nuclear lamina and M phase-specific phosphorylation of lamins by cdc2 kinase. *Cell* 61, 591–602. (Problem 16–30)

Pierschbacher MD & Ruoslahti E (1984) Cell attachment activity of fibronectin can be duplicated by small synthetic fragments of the molecule. *Nature* 309, 30–33. (Problem 3–52)

Pitti RM, Marsters SA, Lawrence DA, Roy M, Kischkel FC, Dowd P, Huang A, Donahue CJ, Sherwood SW, Baldwin DT, Godowski PJ, Wood WI, Gurney AL, Hillan KJ, Cohen RL, Goddard AD, Botstein D & Ashkenazi A (1998) Genomic amplification of a decoy receptor for Fas ligand in lung and colon cancer. *Nature* 396, 699–703. (Problem 17–55)

Pollard TD (1986) Rate constants for the reactions of ATP- and ADP-actin with the ends of actin filaments. *J. Cell Biol.* 103, 2747–2754. (Problem 16–13)

Pollard TD, Blanchoin L & Mullins RD (2000) Molecular mechanisms controlling actin filament dynamics in nonmuscle cells. *Annu. Rev. Biophys. Biomol. Struct.* 29, 545–576. (Problems 16–8, 16–13)

Pollock R & Treisman R (1990) A sensitive method for the determination of protein–DNA binding specificities. *Nucleic Acids Res.* 18, 6197–6204. (Problem 7–22)

Ponticelli AS, Schultz DW, Taylor AF & Smith GR (1985) Chi-dependent DNA strand cleavage by RecBCD enzyme. *Cell* 41, 145–151. (Problem 5–73)

Porter JA, von Kessler DP, Ekker SC, Young KE, Lee JJ, Moses K & Beachy PA (1995) The product of *hedgehog* autoproteolytic cleavage active in local and long-range signaling. *Nature* 374, 363–366. (Problems 15–96, 15–97, 15–98)

Potter H & Dressler D (1976) On the mechanism of genetic recombinaton: electron microscopic observation of recombination intermediates. *Proc. Natl. Acad. Sci. USA* 73, 3000–3004. (Problem 5–72)

Powell LM, Wallis SC, Pease RJ, Edwards YH, Knott TJ & Scott J (1987) A novel form of tissue-specific RNA processing produces apolipoprotein-B48 in intestine. *Cell* 50, 831–840. (Problem 7–74)

Prescott DM (1975) Reproduction of Eucaryotic Cells, pp 85–86. New York: Academic Press. (Problem 17–5)

Prober DA & Edgar BA (2001) Growth regulation by oncogenes—new insights from model organisms. *Curr. Opin. Genet. Dev.* 11, 19–26. (Problem 17–84)

Prunell A, Kornberg RD, Lutter L, Klug A, Levitt M & Crick FH (1979) Periodicity of deoxyribonuclease I digestion of chromatin. *Science* 204, 855–858. (Problem 4–38)

Ptashne M (1986) A Genetic Switch: Gene Control and Phage λ, p 114. Oxford, UK: Blackwell Scientific Press. (Problem 7–14)

Ptashne M & Gann A (2002) Genes & Signals, pp 75–76. Cold Spring Harbor, New York: Cold Spring Harbor Laboratory Press. (Problem 7–37)

Purdue PE, Allsop J, Isaya G, Rosenberg LE & Danpure CJ (1991) Mistargeting of peroxisomal L-alanine:glyoxylate aminotransferase to mitochondria in primary hyperoxaluria patients depends upon activation of a cryptic mitochondrial targeting sequence caused by a point mutation. *Proc. Natl. Acad. Sci. USA* 88, 10900–10904. (Problem 12–67)

Pytela R, Pierschbacher MD & Ruoslahti E (1985) Identification and isolation of a 140 kd cell surface glycoprotein with properties expected of a fibronectin receptor. *Cell* 40, 191–198. (Problem 3–52)

Racker E & Stoeckenius W (1974) Reconstitution of purple membrane vesicles catalysing light-driven proton uptake and adenosine triphosphate formation. *J. Biol. Chem.* 249, 662–663. (Problem 14–13)

Rao PN & Johnson RT (1970) Mammalian cell fusion: I. Studies on the regulation of DNA synthesis and mitosis. *Nature* 225, 159–164. (Problem 17–15)

Rappaport R (1986) Establishment of the mechanism of cytokinesis in animal cells. *Int. Rev. Cytol.* 105, 245–281. (Problem 18–43)

Rasko JEJ, Battini J-L, Kruglyak L, Cox DR & Miller AD (2000) Precise gene localization by phenotypic assay of radiation hybrid cells. *Proc. Natl. Acad. Sci. USA* 97, 7388–7392. (Problem 8–6)

Reese EL & Haimo LT (2000) Dynein, dynactin, and kinesin II's interaction with microtubules is regulated during bidirectional organelle transport. *J. Cell Biol.* 151, 155–165. (Problem 16–72)

Ren R, Mayer BJ, Cicchetti P & Baltimore D (1993) Identification of a ten-amino-acid proline-rich SH3 binding site. *Science* 259, 1157–1161. (Problem 15–71)

Reuveny E, Slesinger PA, Inglese J, Morales, JM, Iniguez-Lluhi JA, Lefkowitz RJ, Bourne HR, Jan YN & Jan LY (1994) Activation of the cloned muscarinic potassium channel by G protein βγ subunits. *Nature* 370, 143–146. (Problem 15–59)

Ribbeck K, Lipowsky G, Kent HM, Stewart M & Görlich D (1998) NTF2 mediates nuclear import of Ran. *EMBO J.* 17, 6587–6598. (Problem 12–31)

Richardson HE, Wittenberg C, Cross F & Reed SI (1989) An essential G_1 function for cyclin-like proteins in yeast. *Cell* 59, 1127–1133. (Problem 17–23)

Rief M, Gutel M, Oesterhelt F, Fernandez JM & Gaub HE (1997) Reversible folding of individual titin immunoglobulin domains by AFM. *Science* 276, 1109–1112. (Problem 3–32)

Roeder RG (1974) Multiple forms of deoxyribonucleic acid-dependent ribonucleic acid polymerase in *Xenopus laevis*. Isolation and partial characterization. *J. Biol. Chem.* 249, 241–248. (Problem 6–14)

Roger AJ, Svard SG, Tovar J, Clark CG, Smith MW, Gillin FD & Sogin ML (1998) A mitochondrial-like chaperonin 60 gene in *Giardia lamblia*: Evidence that diplomonads once harbored an endosymbiont related to the progenitor of mitochondria. *Proc. Natl. Acad. Sci. USA* 95, 229–234. (Problem 1–30)

Rohatgi R, Ma L, Miki H, Lopez M, Kirchhausen T, Takenawa T & Kirschner MW (1999) The interaction between N-WASP and the Arp2/3 complex links Cdc42-dependent signals to actin assembly. *Cell* 97, 221–231. (Problems 16–61, 16–62)

Rothman JE & Dawidowicz EA (1975) Asymmetric exchange of vesicle phospholipids catalyzed by the phosphatidylcholine exchange protein. Measurement of inside–outside transitions. *Biochemistry* 14, 2809–2816. (Problem 12–95)

Rothman JE, Miller RL & Urbani LJ (1984) Intercompartmental transport in the Golgi complex is a dissociative process: facile transfer of membrane protein between two Golgi populations. *J. Cell Biol.* 99, 260–271. (Problem 13–39)

Rousselet A, Guthmann C, Matricon J et al. (1976) Study of the transverse diffusion of spin labeled phospholipids in biological membranes: 1. Human red blood cells. *Biochim. Biophys. Acta* 426, 357–371. (Problems 10–8, 10–22)

Royzman I, Austin RJ, Bosco G, Bell SP and Orr-Weaver TL (1999) ORC localization in *Drosophila* follicle cells and the effects of mutations in *dEaF* and *dDP*. *Genes Dev.* 13, 827–840. (Problem 5–45)

Ruoslahti E & Pierschbacher MD (1986) Arg-Gly-Asp: a versatile cell recognition signal. *Cell* 44, 517–518. (Problem 3–52)

Russo AF & Koshland DE (1983) Separation of signal transduction and adaptation functions of the aspartate receptor in bacterial sensing. *Science* 220, 1016–1020. (Problem 15–89)

Sack RL, Brandes RW, Kendall AR & Lewy AJ (2000) Entrainment of free-running circadian rhythms by melatonin in blind people. *N. Engl. J. Med.* 343, 1114–1116. (Problem 7–53)

Sadowski HB, Shuai K, Darnell JE & Gilman MZ (1993) A common nuclear signal transduction pathway activated by growth factor and cytokine receptors. *Science* 261, 1739–1744. (Problem 15–83)

Safer B, Kemper W & Jagus R (1978) Identification of a 48S preinitiation complex in reticulocyte lysate. *J. Biol. Chem.* 253, 3384–3386. (Problem 6–67)

Sakmann B (1992) Elementary steps in synaptic transmission revealed by currents through single ion channels. *Science* 256, 503–512. (Problems 11–60, 11–61)

Satir P & Matsuoka T (1989) Splitting the ciliary axoneme: implications for a "switch-point" model of dynein arm activity in ciliary motion. *Cell Motil. Cyto.* 14, 345–358. (Problem 16–85)

Sawadogo M & Roeder RG (1985) Factors involved in specific transcription by human RNA polymerase II: analysis by a rapid and quantitative *in vitro* assay. *Proc. Natl. Acad. Sci. USA* 82, 4394–4398. (Problems 6–11, 6–12)

Sawadogo M & Roeder RG (1985) Interaction of a gene-specific transcription factor with the adenovirus major late promoter upstream of the TATA box region. *Cell* 43, 165–175. (Problem 7–35)

Scheidereit C, Geisse S, Westphal HM & Beato M (1983) The glucocorticoid receptor binds to defined nucleotide sequences near the promoter of mouse mammary tumor virus. *Nature* 304, 749–752. (Problem 15–23)

Schibler U (2000) Heartfelt enlightenment. *Nature* 404, 25–28. (Problem 7–54)

Schleif RF (1986) Genetics and Molecular Biology, Chap. 13. Reading, MA: Addison Wesley. (Problem 7–26)

Schubert U, Antón LC, Gibbs J, Norbury CC, Yewdell JW & Bennink JR (2000) Rapid degradation of a large fraction of newly synthesized proteins by proteasomes. *Nature* 404, 770–774. (Problem 6–75)

Schwoebel ED, Talcott B, Cushman I & Moore MS (1998) Ran-dependent signal-mediated nuclear import does not require GTP hydrolysis by Ran. *J. Biol. Chem.* 273, 35170–35175. (Problem 12–32)

Shah R, Cosstick R & West SC (1997) The RuvC protein dimer resolves Holliday junctions by a dual incision mechanism that involves base-specific contacts. *EMBO J.* 16, 1464–1472. (Problem 5–77)

Shea TB & Yabe J (2000) Occam's razor slices through the mysteries of neurofilament axonal transport: Can it really be so simple? *Traffic* 1, 522–523. (Problem 16–98)

Shibahara K & Stillman B (1999) Replication-dependent marking of DNA by PCNA facilitates CAF-1-coupled inheritance of chromatin. *Cell* 96, 575–585. (Problem 5–49)

Shimamoto N (1999) One-dimensional diffusion of proteins along DNA. *J. Biol. Chem.* 274, 15293–15296. (Problem 7–10)

Shtilerman M, Lorimer GH & Englander SW (1999) Chaperonin function: Folding by forced unfolding. *Science* 284, 822–825. (Problem 6–72)

Sickles DW, Pearson JK, Beall A & Testino A (1994) Toxic axonal degeneration occurs independent of neurofilament accumulation. *J. Neurosci. Res.* 39, 347–354. (Problem 16–38)

Siegel RM, Chan FK-M, Chun HJ & Lenardo MJ (2000) The multifaceted role of Fas signaling in immune cell homeostasis and autoimmunity. *Nat. Immunol.* 1, 469–474. (Problem 17–54)

Sinn E, Muller W, Pattengale P, Tepler I, Wallace R & Leder P (1987) Coexpression of MMTV/v-Ha-*ras* and MMTV/c-*myc* genes in transgenic mice: synergistic action of oncogenes *in vivo*. *Cell* 49, 465–475. (Problem 17–79)

Skibbens RV, Corson LB, Koshland D & Hieter P (1999) Ctf7 is essential for sister chromatid cohesion and links mitotic chromosome structure to the DNA replication machinery. *Genes Dev.* 13, 307–319. (Problem 18–2)

Skop AR, Bergmann D, Mohler WA & White JG (2001) Completion of cytokinesis in *C. elegans* requires a brefeldin A-sensitive membrane accumulation at the cleavage furrow apex. *Curr. Biol.* 11, 735–746. (Problem 18–48)

Sloboda RD, Dentler WL & Rosenbaum JL (1976) Microtubule-associated proteins and the stimulation of tubulin assembly *in vitro*. *Biochemistry* 15, 4497–4505. (Problem 16–17)

Sluder G & Rieder CL (1985) Centriole number and the reproductive capacity of spindle poles. *J. Cell Biol.* 100, 887–896. (Problem 18–9)

Smeekens S, Bauerle C, Hageman J, Keegstra K & Weisbeek P (1986) The role of the transit peptide in the routing of precursors toward different chloroplast compartments. *Cell* 46, 365–375. (Problem 12–64)

Smith HT, Ahmed AJ & Millet F (1981) Electrostatic interaction of cytochrome *c* with cytochrome c_1 and cytochrome oxidase. *J. Biol. Chem.* 256, 4984–4990. (Problem 14–42)

Söllner T, Whiteheart SW, Brunner M, Erdjument-Bromage H, Geromanos S, Tompst P & Rothman JE (1993) SNAP receptors implicated in vesicle targeting and fusion. *Nature* 362, 318–324. (Problem 13–19)

Spierer A & Spierer P (1984) Similar levels of polyteny in bands and interbands of *Drosophila* giant chromosomes. *Nature* 307, 176–178. (Problem 4–56)

Stanners CP & Till JE (1960) DNA synthesis in individual L-strain mouse cells. *Biochim. Biophys. Acta* 37, 406–419. (Problem 17–14)

Stanojevic D, Small S & Levine M (1991) Regulation of a segmentation stripe by overlapping activators and repressors in the *Drosophila* embryo. *Science* 254, 1385–1387. (Problem 7–43)

Stare FJ & Baumann CA (1936) The effect of fumarate on respiration. *Proc. Royal Soc. Lond. Ser.* B121, 338–357. (Problem 2–99)

Stearns T (2001) Centrosome duplication: A centriolar pas de deux. *Cell* 105, 417–420. (Problem 18–7)

Steinmetz MO, Stoffler D, Hoenger A, Bremer A & Aebi U (1997) Actin: From cell biology to atomic detail. *J. Struct. Biol.* 119, 295–320. (Problem 16–34)

Stern DB & Palmer JD (1984) Extensive and widespread homologies between mitochondrial DNA and chloroplast DNA in plants. *Proc. Natl. Acad. Sci. USA* 81, 1946–1950. (Problem 14–78)

Stone JD, Peterson AP, Eyer J & Sickles DW (2000) Neurofilaments are nonessential elements of toxicant-induced reductions in fast axonal transport: pulse labeling in CNS neurons. *Neurotoxicology* 21, 447–457. (Problem 16–38)

Svejstrup JQ, Li Yang, Fellows J, Gnatt A, Bjorklund S & Kornberg RD (1997) Evidence for a mediator cycle at the initiation of transcription. *Proc. Natl. Acad. Sci. USA* 94, 6075–6078. (Problem 6–15)

Svoboda K, Schmidt CF, Schnapp BJ & Block SM (1993) Direct observation of kinesin stepping by optical trapping interferometry. *Nature* 365, 721–727. (Problem 16–68)

Szent-Györgyi AV (1924) Über den mechanisms de Succin- und Paraphenylendiaminooxydation. Ein Betrag der Zellatmung. *Biochem Z.* 150, 141–149. (Problem 2–99)

Szent-Györgyi AV (1937) Nobel Lecture at www.nobel.se/medicine/laureates/1937/szent-gyorgyi-lecture.pdf. (Problem 2–99)

Szostak JW & Blackburn EH (1982) Cloning yeast telomeres on linear plasmid vectors. *Cell* 19, 245–255. (Problem 4–33)

Szostak JW, Bartel DP & Luisi L (2001) Synthesizing life. *Nature* 409, 387–390. (Problem 6–88)

Takeuchi Y, Porter CD, Strahan KM, Preece AF, Gustafsson K, Cosset FL, Weiss

RA & Collins MK (1996) Sensitization of cells and retroviruses to human serum by (α1-3) galactosyltransferase. *Nature* 379, 85–88. (Problem 13–35)

Tang W-J & Gilman AG (1991) Type-specific regulation of adenylyl cyclase by G protein $\beta\gamma$ subunits. *Science* 254, 1500–1503. (Problem 15–44)

Teasdale RD & Jackson MR (1996) Signal-mediated sorting of membrane proteins between the endoplasmic reticulum and the Golgi apparatus. *Annu. Rev. Cell Dev. Biol.* 12, 27–54. (Problem 13–30)

Teo I, Sedgwick B, Kilpatrick MW, McCarthy TV & Lindahl T (1986) The intracellular signal for induction of resistance to alkylating agents in *E. coli. Cell* 45, 315–324. (Problem 5–63)

Teter SA, Houry WA, Ang D, Tradler T, Rockabrand D, Fischer G, Blum P, Georgopoulos C & Hartl FU (1999) Polypeptide flux through bacterial Hsp70: DnaK cooperates with trigger factor in chaperoning nascent chains. *Cell* 97, 755–765. (Problems 6–70, 6–71)

Thomas JE, Smith M, Rubinfeld B, Gutowski M, Beckmann RP & Polakis P (1996) Subcellular localization and analysis of apparent 180-kDa and 220-kDa proteins of the breast cancer susceptibility gene, BRCA1. *J. Biol. Chem.* 271, 28630–28635. (Problem 3–45)

Thomas MJ, Platas AA & Hawley DK (1998) Transcriptional fidelity and proofreading by RNA polymerase II. *Cell* 93, 627–637. (Problem 6–4)

Thompson CM, Koleske AJ, Chao DM & Young RA (1993) A multisubunit complex associated with the RNA polymerase II CTD and TATA-binding protein in yeast. *Cell* 73, 1361–1375. (Problem 6–15)

Thompson NE & Burgess RR (1996) Immunoaffinity purification of RNA polymerase II and transcription factors using polyol-responsive monoclonal antibodies. *Methods Enzymol.* 274, 513–526. (Problem 3–51)

Thompson SR, Goodwin EB & Wickens M (2000) Rapid deadenylation and poly(A)-dependent translational repression mediated by the *Caenorhabditis elegans tra-2* 3′ untranslated region in *Xenopus* embryos. *Mol. Cell. Biol.* 20, 2129–2137. (Problem 7–83)

Tilney LG & Inoue S (1982) Acrosomal reaction of *Thyone* sperm. II. The kinetics and possible mechanism of acrosomal process elongation. *J. Cell Biol.* 93, 820–827. (Problem 16–90)

Ting AY, Kain KH, Klemke RL & Tsien RY (2001) Genetically encoded fluorescent reporters of protein tyrosine kinase activities in living cells. *Proc. Natl. Acad. Sci. USA* 98, 15003–15008. (Problem 8–71)

Toker A & Newton AC (2000) Akt/protein kinase B is regulated by autophosphorylation at the hypothetical PDK-2 site. *J. Biol. Chem.* 275, 8271–8274. (Problem 15–82)

Tokumaru H, Umayahara K, Pellegrini LL, Ishizuka T, Saisu H, Betz H, Augustine GJ & Abe T (2001) SNARE complex oligomerization by synaphin/complexin is essential for synaptic vesicle exocytosis. *Cell* 104, 421–432. (Problems 13–16, 13–17)

Tokuyama Y, Horn HF, Kawamura K, Tarapore P & Fukasawa K (2001) Specific phosphorylation of nucleophosmin on Thr199 by cyclin-dependent kinase 2-cyclin E and its role in centrosome duplication. *J. Biol. Chem.* 276, 21529–21537. (Problem 18–8)

Tonegawa S (1983) Somatic generation of antibody diversity. *Nature* 302, 575-581. (Problem 7–3)

Toth A, Ciosk R, Uhlmann F, Galova M, Schleiffer A & Nasmyth K (1999) Yeast cohesin complex requires a conserved protein, Eco1p(Ctf7), to establish cohesion between sister chromatids during DNA replication. *Genes Dev.* 13, 320–333. (Problem 18–2)

Tournier C, Hess P, Yang DD, Xu J, Turner TK, Nimnual A, Bar-Sagi D, Jones SN, Flavell RA & Davis RJ (2000) Requirement of JNK for stress-induced activation of the cytochrome c-mediated death pathway. *Science* 288, 870-874. (Problem 17–64)

Treisman R (1985) Transient accumulation of *c-fos* RNA following serum stimulation requires a conserved 5′ element and *c-fos* 3′ sequences. *Cell* 42, 889–902. (Problem 7–81)

Treisman R (1987) Identification and purification of a polypeptide that binds to the c-*fos* serum response element. *EMBO J.* 6, 2711–2717. (Problem 7–20)

Tsukamoto T, Yokota S & Fujiki Y (1990) Isolation of Chinese hamster ovary cell mutants defective in assembly of peroxisomes. *J. Cell. Biol.* 110, 651–660. (Problem 12–69)

Tugal T, Zou-Yang XH, Gavin K, Pappin D, Canas B, Kobayashi R, Hunt T & Stillman B (1998) The Orc4p and Orc5p subunits of the *Xenopus* and human origin recognition complex are related to Orc1p and Cdc6p. *J. Biol. Chem.* 273, 32421–32429. (Problem 8–72)

Tzagoloff A (1982) Mitochondria, pp 212–213. New York: Plenum Press. (Problem 14–22)

Uhlmann F & Nasmyth K (1998) Cohesion between sister chromatids must be established during DNA replication. *Curr. Biol.* 8, 1095–1101. (Problem 18–1)

Uhlmann F, Lottspeich F & Nasmyth K (1999) sister-chromatid separation at anaphase onset is promoted by cleavage of the cohesin subunit of Scc1. *Nature* 400, 37–42. (Problem 17–34)

Valius M, & Kazlauskas A (1993) Phospholipase C-γ1 and phosphatidylinositol 3 kinase are the downstream mediators of the PDGF receptor's mitogenic signal. *Cell* 73, 321–334. (Problem 15–72)

Valius M, Secrist JP & Kazlauskas A. (1995)The GTPase-activating protein of Ras suppresses platelet-derived growth factor β receptor signaling by silencing phospholipase C-γ1. *Mol. Cell. Biol.* 15, 3058–3071. (Problem 15–72)

van Arsdell SW & Weiner AM (1984) Human genes for U2 small nuclear RNA are tandemly repeated. *Mol. Cell. Biol.* 4, 492–499. (Problem 4–19)

Van Dyck E, Stasiak AJ, Stasiak A & West SC (1999) Binding of double-strand breaks in DNA by human Rad52 protein. *Nature* 398, 728–731. (Problem 5–67)

Van Meer G & Simons K (1986) The function of tight junctions in maintaining differences in lipid composition between the apical and the basolateral cell surface domains of MDCK cells. *EMBO J.* 5, 1455–1464. (Problem 10–52)

Varma R & Mayor S (1998) GPI-anchored proteins are organized in submicron domains at the cell surface. *Nature* 394, 798–801. (Problem 10–19)

Vidal M & Legrain P (1999) Yeast forward and reverse 'n'-hybrid systems. *Nucleic Acids Res.* 27, 919–929. (Problem 8–76)

Vojtek AB, Hollenberg SM & Cooper JA (1993) Mammalian Ras interacts directly with the serine/threonine kinase Raf. *Cell* 74, 205–214. (Problems 8–73, 8–74, 8–75)

Vollrath D, Nathans J & Davis RW (1988) Tandem array of human visual pigment genes at Xq28. *Science* 240, 1669–1671. (Problem 4–20)

Wahl GM, Padgett RA & Stark GR (1979) Gene amplification causes overproduction of the first three enzymes of UMP synthesis in *N*-(phosphonacetyl)-L-aspartate (PALA)–resistant hamster cells. *J. Biol. Chem.* 254, 8679–8689. (Problem 3–68)

Walker RA, Inoue S & Salmon ED (1989) Asymmetric behavior of severed microtubule ends after ultraviolet-microbeam irradiation of individual microtubules *in vitro*. *J. Cell Biol.* 108, 931–937. (Problem 16–18)

Walworth NC, Goud B, Kabcenell AK & Novick PJ (1989) Mutational analysis of *SEC4* suggests a cyclical mechanism for the regulation of vesicular traffic. *EMBO J.* 8, 1685–1693. (Problem 13–21)

Wang YH & Griffith J (1995) Expanded CTG triplet repeat blocks from the myotonic dystrophy gene create the strongest known natural nucleosome positioning elements. *Genomics* 25, 570–573. (Problem 4–44)

Ward GE & Kirschner MW (1990) Identification of cell cycle-related phosphorylation sites on nuclear lamin C. *Cell* 61, 561–577. (Problem 16–30)

Wasserman WJ & Masui Y (1975) Effects of cycloheximide on a cytoplasmic factor initiating meiotic maturation in *Xenopus* oocytes. *Exp. Cell Res.* 91, 381–388. (Problem 17–19)

Webb MR, Grubmeyer C, Penefsky HS & Trentham DR (1980) The stereochemical course of phosphoric residue transfer catalyzed by beef heart mitochondrial ATPase. *J. Biol. Chem.* 255, 11637–11639. (Problem 14–16)

Wei MC, Zong W-X, Cheng EH-Y, Lindsten T, Panoutsakopoulou V, Ross AJ, Roth KA, MacGregor GR, Thompson CB & Korsmeyer SJ (2001) Proapoptotic BAX and BAK: A requisite gateway to mitochondrial dysfunction and death. *Science* 292, 727–730. (Problems 17–60, 17–61, 17–63)

Weinert TA & Hartwell LH (1988) The *rad9* gene controls the cell cycle response to DNA damage in *Saccharomyces cerevisiae*. *Science* 241, 317–322. (Problem 17–47)

Weinrich SL, Pruzan R, Ma L, Ouellette M, Tesmer VM, Holt SE, Bodnar AG, Lichtsteiner S, Kim NW, Trager JB, Taylor RD, Carlos R, Andrews WH, Wright WE, Shay JW, Harley CB & Morin GB (1997) Reconstitution of human telomerase with the template RNA component hTR and the catalytic protein subunit hTRT. *Nat. Genet.* 17, 498–502. (Problem 17–82)

Weintraub H & Groudine M (1976) Chromosomal subunits in active genes have an altered conformation. *Science* 193, 848–856. (Problem 4–50)

Welch MD, Rosenblatt J, Skoble J, Portnoy DA & Mitchison TJ (1998) Interaction of human Arp2/3 complex and the *Listeria monocytogenes* ActA protein in actin filament nucleation. *Science* 281, 105–108. (Problem 16–50)

Whitehouse I, Flaus A, Cairns BR, White MF, Workman JL & Owen-Hughes T (1999) Nucleosome mobilization catalysed by the yeast SWI.SNF complex. *Nature* 400, 784–787. (Problem 4–47)

Whitmore D, Foulkes NS & Sassone-Corsi P (2000) Light acts directly on organs and cells in culture to set the vertebrate circadian clock. *Nature* 404, 87–91. (Problem 7–54)

Wickens M, Goodwin EB, Kimble J, Strickland S & Hentze M (2000) Translational control of developmental decisions. In Translational Control of Gene Expression (N Sonenberg, JWB Hershey, MB Mathews eds), pp 295–370. Cold Spring Harbor, NY: Cold Spring Harbor Laboratory Press. (Problem 7–82)

Wilmut I, Schnieke AE, McWhir J, Kind AJ & Campbell KHS (1997) Viable offspring derived from fetal and adult mammalian cells. *Nature* 385, 810–813. (Problem 7–2)

Wilson SM, Yip R, Swing DA, O'Sullivan TN, Zhang Y, Novak EK, Swank RT, Russell LB, Copeland NG & Jenkins NA (2000) A mutation in *Rab27* causes the vesicle transport defects observed in *ashen* mice. *Proc. Natl. Acad. Sci. USA* 97, 7933–7938. (Problem 13–53)

Wilson T & Treisman R (1988) Removal of poly(A) and consequent degradation of c-*fos* mRNA facilitated by 3′ AU-rich sequences. *Nature* 336, 396–399. (Problem 7–81)

Winter CG, Wang B, Ballew A, Royou A, Karess R, Axelrod JD & Luo L (2001) *Drosophila* Rho-associated kinase (Drok) links Frizzled-mediated planar cell polarity signaling to the actin cytoskeleton. *Cell* 105, 81–91. (Problem 15–94)

Witke W, Schleicher M & Noegel AA (1992) Redundancy in the microfilament system: abnormal development of *Dictyostelium* cells lacking two F-actin cross-linking proteins. *Cell* 68, 53–62. (Problem 16–57)

Wolff B, Sanglier J-J & Wang Y (1997) Leptomycin B is an inhibitor of nuclear export: inhibition of nucleo-cytoplasmic translocation of the human immunodeficiency virus type 1 (HIV-1) Rev protein and Rev-dependent mRNA. *Chem. Biol.* 4, 139–147. (Problem 12–33)

Wolffe AP & Brown DD (1986) DNA replication *in vitro* erases a *Xenopus* 5S RNA gene transcription complex. *Cell* 47, 217–227. (Problem 7–57)

Workman JL & Roeder RG (1987) Binding of transcription factor TFIID to the major late promoter during *in vitro* nucleosome assembly potentiates subsequent initiation by RNA polymerase II. *Cell* 51, 613–622. (Problem 7–38)

Wyrick JJ, Holstege FC, Jennings EG, Causton HC, Shore D, Grunstein M, Lander ES & Young RA (1999) Chromosomal landscape of nucleosome-dependent gene expression and silencing in yeast. *Nature* 402, 418–421. (Problem 4–59)

Xeros N (1962) Deoxyriboside control and synchronization of mitosis. *Nature* 194, 682–683. (Problem 17–15)

Xu T, Ashery U, Burgoyne RD & Neher E (1999) Early requirement for α-SNAP and NSF in the secretory cascade in chromaffin cells. *EMBO J.* 18, 3293–3304. (Problem 13–81)

Yabe JT, Pimenta A & Shea TB (1999) Kinesin-mediated transport of neurofilament protein oligomers in growing axons. *J. Cell Sci.* 112, 3799–3814. (Problem 16–98)

Yalow RS (1978) Radioimmunoassay: a probe for the fine structure of biologic systems. *Science* 200, 1236–1245. (Problem 15–6)

Yamano H, Tsurumi C, Gannon J & Hunt T (1998) The role of the destruction box and its neighbouring lysine residues in cyclin B for anaphase ubiquitin-dependent proteolysis in fission yeast: defining the D-box receptor. *EMBO J* 17, 5670–5678. (Problem 6–77)

Yanofsky C (2001) Advancing our knowledge in biochemistry, genetics, and microbiology through studies on tryptophan metabolism. *Annu. Rev. Biochem.* 70, 1–37. (Problem 2–105)

Yao M-C, Zhu S-G & Yao C-H (1985) Gene amplification in *Tetrahymena thermophila*: formation of extrachromosomal palindromic gene coding for rRNA. *Mol. Cell. Biol.* 5, 1260–1267. (Problem 8–39)

Ybe JA, Brodsky FM, Hofmann K, Lin K, Liu SH, Chen L, Earnest TN, Fletterick RJ & Hwang RK (1999) Clathrin self-assembly is mediated by a tandemly repeated superhelix. *Nature* 399, 371–375. (Problem 13–6)

Yen TJ, Machlin PS & Cleveland DW (1988) Autoregulated instability of β-tubulin mRNAs by recognition of the nascent amino terminus of β-tubulin. *Nature* 334, 580–585. (Problem 7–80)

Yoon M, Moir RD, Prahlad V & Goldman RD (1998) Motile properties of vimentin intermediate filament networks in living cells. *J. Cell Biol.* 143, 147–157. (Problem 16–29)

Yoshida H, Kong Y-Y, Yoshida R, Elia AJ, Hakem A, Hakem R, Penninger JM & Mak TW (1998) Apaf1 is required for mitochondrial pathways of apoptosis and brain development. *Cell* 94, 739–750. (Problem 17–65)

Zagotta WN, Hoshi T & Aldrich RW (1990) Restoration of inactivation in mutants of *shaker* potassium channels by a peptide derived from ShB. *Science* 250, 568–570. (Problem 11–38)

Zagouras P & Rose JK (1989) Carboxy-terminal SEKDEL sequences retard but do not retain two secretory proteins in the endoplasmic reticulum. *J. Cell Biol.* 109, 2633-2640. (Problem 13–31)

Zahringer J, Baliga BS & Munro HN (1976) Novel mechanism for translational control in regulation of ferritin synthesis by iron. *Proc. Natl. Acad Sci. USA* 73, 857–861. (Problem 7–77)

Zecca M, Basler K & Struhl G (1995) Sequential organizing activities of engrailed, hedgehog and decapentaplegic in the *Drosophila* wing. *Development* 121, 2265–2278. (Problem 17–84)

Zerangue N, Malan MJ, Fried SR, Dazin PF, Jan YN, Jan LY & Schwappach B (2001) analysis of endoplasmic reticulum trafficking signals by combinatorial screening in mammalian cells. *Proc. Natl. Acad. Sci. USA* 98, 2431–2436. (Problem 13–28)

Zhang X-F, Settleman J, Kyriakis JM, Takenchi-Suzuki E, Elledge SJ, Marshall MS, Bruder JT, Rapp UR & Avruch J (1993) Normal and oncogenic p21[ras] proteins bind to the amino-terminal domain of c-Raf-1. *Nature* 364, 308–313. (Problem 8–74, 8–75)

Zhao C, Takita J, Tanaka Y, Setou M, Nakagawa T, Takeda S, Yang HW, Terada S, Nakata T, Takei Y, Saito M, Tsuji S, Hayashi Y & Hirokawa N (2001) Charcot-Marie-Tooth disease type 2A caused by mutation in a microtubule motor KIF1Bβ. *Cell* 105, 587–597. (Problem 16–97)

Zhen L, Jiang S, Feng L, Bright NA, Peden AA, Seymour AB, Novak EK, Elliott R, Gorin MB, Robinson MS & Swank RT (1999) Abnormal expression and subcellular distribution of subunit proteins of the AP-3 adaptor complex lead to platelet storage pool deficiency in the *pearl* mouse. *Blood* 94, 146–155. (Problem 13–52)

Zheng Y, Wong ML, Alberts B & Mitchison T (1995) Nucleation of microtubule assembly by a γ-tubulin-containing ring complex. *Nature* 378, 578–583. (Problem 16–45)

Zhuang Y & Weiner AM (1986) A compensatory base change in U1 snRNA suppresses a 5′ splice site mutation. *Cell* 46, 827–835. (Problem 6–28)

Zieg J, Silverman M, Hilmen M & Simon M (1977) Recombinational switch for gene expression. *Science* 196, 170–172. (Problem 7–48)

Zimmet J & Ravid K (2000) Polyploidy: occurrence in nature, mechanisms, and significance for the megakaryocyte-platelet system. *Exp. Hematol.* 28, 3–16. (Problem 18–50)

Zinkel SS & Crothers DM (1987) DNA bend direction by phase sensitive detection. *Nature* 328, 178–181. (Problem 7–11)

Cited Researchers

Abe T (Problems 13–16, 13–17)
Aberle H (Problem 15–92)
Abrahams JP (Problem 14–18)
Abrami L (Problem 11–3)
Achord DT (Problem 13–49)
Adam A (Problem 12–59)
Adams GA (Problem 13–32)
Adams KL (Problem 1–32)
Adelstein RS (Problem 15–57)
Adesnik M (Problem 13–78)
Aebi U (Problems 16–32, 16–34)
Ahmed AJ (Problem 14–42)
Alberts BM (Problems 1–25, 5–20, 5–24, 16–45)
Aldrich RW (Problems 11–20, 11–38)
Allard WJ (Problem 11–9)
Allen LA (Problem 12–70)
Allison VF (Problem 12–47)
Allsop J (Problem 12–67)
Alson S (Problem 4–35)
Alvarez-Bolado G (Problem 17–65)
Amann KJ (Problem 16–49)
Amesz J (Problem 14–63)
Amos LA (Problems 16–9, 16–65, 16–66)
Anderson JE (Problem 8–36)
Andreasen PH (Problem 6–66)
Andrews WH (Problems 5–51, 17–82)
Ang D (Problems 6–70, 6–71)
Antón LC (Problem 6–75)
Antonio C (Problem 18-32)
Arai I (Problem 7–10)
Araki M (Problem 5–62)
Artandi SE (Problem 4–35)
Ashburner M(Problem 15–24)
Ashery U (Problem 13–81)
Ashkenazi A (Problem 17–55)
Attardi G (Problem 14–83)
Augustine GJ (Problems 13–16, 13–17)
Auld DS (Problem 6–52)
Austin RJ (Problems 5–46, 5–47)
Avruch J (Problems 8–74, 8–75)
Axelrod JD (Problems 15–51, 15–94)

Bachmair A (Problem 6–78)
Bachurski CJ (Problem 7–80)
Baeuerle PA (Problem 12–44)
Baldassarre M (Problem 13–40)
Baldin V (Problem 17–76)
Baldwin DT (Problem 17–55)
Baliga BS (Problem 7–77)
Balkow K (Problem 7–85)
Ballabio A (Problem 7–59)
Ballew A (Problem 15–94)
Baltimore D (Problems 12–44, 15–71)
Bamberg E (Problem 11–15)
Barford D (Problem 3–87)
Barnes G (Problems 12–25, 12–26)
Barron JT Gu L (Problem 14–31)
Bar-Sagi D (Problem 17–64)
Bartel DP (Problems 6–81, 6–84, 6–85, 6–86, 6–88)
Bartunek P (Problem 17–71)
Baserga R (Problem 17–13)
Basler K (Problem 17–84)
Battini J-L (Problem 8–6)
Bauer A (Problem 15–92)
Bauer MF (Problem 12–59)
Bauerle C (Problem 12–64)

Baumann CA (Problem 2–99)
Beachy PA (Problems 15–96, 15–97, 15–98)
Beall A (Problem 16–38)
Beato M (Problem 15–23)
Beaudry G (Problem 13–78)
Beckmann RP (Problem 3–45)
Beckwith J (Problem 12–88)
Bell SP (Problems 5–46, 5–47)
Bender J (Problem 5–82)
Benedict WF (Problem 17–38)
Benezra R (Problem 7–19)
Benincasa E (Problem 17–68)
Bennett V (Problem 10–44)
Bennink JR (Problem 6–75)
Benzer S (Problem 6–50)
Berg HC (Problems 2–65, 2–66, 14–48)
Berg JM (Problem 2–76)
Berget SM (Problem 6–23)
Bergmann D (Problem 18–48)
Bergom MA (Problems 5–65, 17–67)
Berk AJ (Problem 6–23)
Bershadsky AD (Problem 16–36)
Berthold CH (Problem 16–32)
Berthonaud V (Problem 11–3)
Betsholtz C (Problem 16–32)
Betz H (Problems 13–16, 13–17)
Beug H (Problem 17–71)
Bi X (Problem 7–45)
Bialoskorska M (Problem 5–29)
Bibring T (Problem 18–9)
Bjorklund S (Problem 6–15)
Blackburn EH (Problem 4–33)
Blanchoin L (Problems 16–8,16–13)
Blank V (Problem 15–88)
Bleil JD (Problems 13–66, 13–67)
Blobel G (Problem 12–30)
Block SM (Problem 16–68)
Blom J (Problem 12–51)
Bloom KS (Problems 4–60, 18–28)
Blum P (Problems 6–70, 6–71)
Bodnar AG (Problem 17–82)
Boeke JD (Problem 5–83)
Bogenhagen D (Problem 14–75)
Bonfanti L (Problem 13–40)
Bonmassar E (Problem 17–68)
Bonner JT (Problem 15–5)
Bootsma D (Problem 17–15)
Borst P (Problems 6–9, 12–66)
Bosco G (Problem 5–46)
Bostock CJ (Problem 17–15)
Botstein D (Problem 17–55)
Bourne HR (Problems 15–44, 15–59)
Boveri T (Problem 18–14)
Brade G (Problem 15–88)
Brandes RW (Problem 7–53)
Braunstein M (Problem 7–45)
Bravo R (Problem 13–78)
Bray D (Problem 1–25)
Bremer A (Problem 16–34)
Bretscher MS (Problems 10–11, 13–66, 13–67)
Brewer BJ 9 Problem 5–44)
Bright NA (Problem 13–52)
Brinker A (Problem 6–72)
Broach J (Problem 7–45)
Brockdorff N (Problem 7–61)
Brodsky FM (Problem 13–6)
Brokaw CJ (Problem 16–84)

Brown CJ (Problem 7–59)
Brown DA (Problem 10–18)
Brown DD (Problem 7–57)
Brown MS (Problems 13–59, 13–60)
Brown NR (Problem 3–86)
Bruder JT (Problems 8–74, 8–75)
Brunner M (Problems 12–59, 13–19)
Brutlag D (Problem 5–17)
Bubb MR (Problem 16–36)
Buccione R (Problem 13–40)
Budke L (Problem 17–15)
Buhler JM (Problem 11–3)
Bukau B (Problems 6–70, 6–71)
Bullock P (Problem 5–40)
Buratowski S (Problem 6–15)
Burch RM (Problem 15–51)
Burgess RR (Problem 3–51)
Burgoyne RD (Problem 13–81)
Burmeister M (Problem 13–52)
Burne JF (Problem 17–56)
Butow RA (Problem 12–47)
Byers D (Problem 15–50)

Cai S-J (Problem 7–74)
Cairns BR (Problem 4–47)
Callan HG (Problem 4–52)
Cami B (Problem 6–24)
Campbell KHS (Problem 7–2)
Canas B (Problem 8–72)
Cantor CR (Problems 2–35, 3–82, 3–83, 8–14, 8–17, 8–18)
Capaldi RA (Problem 14–42)
Caput D (Problem 3–31)
Carbon J (Problem 4–60)
Carlier M-F (Problems 16–3, 16–19, 16–42, 16–43, 16–44)
Carlos R (Problem 17–82)
Caron C (Problem 18–8)
Caron MG (Problem 15–35)
Carskadon SI (Problem 13–52)
Caskey CT (Problems 6–52, 8–60)
Castagnola M (Problem 8–29)
Causton HC (Problem 4–59)
Cecconi F (Problem 17–65)
Cech TR (Problem 5–51)
Celis JE (Problem 16–32)
Cha TA (Problem 5–20)
Chamberlain JS (Problem 8–60)
Chambon P (Problem 6–24)
Chan FK-M (Problem 17–54)
Chan L (Problem 7–74)
Chang S (Problem 4–35)
Chant J (Problem 16–100)
Chao DM (Problem 6–15)
Chao S-H (Problem 7–69)
Chapeville F (Problem 6–50)
Chapman KB (Problem 5–51)
Chardin P (Problem 13–11)
Chase M (Problem 4–8)
Chen C-N (Problem 15–50)
Chen C-Y (Problem 7–87)
Chen L (Problems 3–37, 13–6)
Chen S-H (Problem 7–74)
Chen ZJ (Problem 17–62)
Cheng C-H C (Problem 3–37)
Cheng EH-Y (Problems 17–60, 17–61, 17–63)

Chihara C (Problem 15–24)
Chin L (Problem 4–35)
Chinkers M (Problem 17–73)
Chou Y-H (Problem 16–28)
Chrétien D (Problem 16–14)
Chun HJ (Problem 17–54)
Cicchetti P (Problem 15–71)
Ciosk R (Problems 17–33, 18–2)
Clapham DE (Problem 15–59)
Clark CG (Problem 1–30)
Clayton DA (Problem 14–75)
Cleveland DW (Problem 7–80)
Cochet M (Problem 6–24)
Cohen RL (Problem 17–55)
Cohen S (Problems 13–56, 17–73)
Collins FS (Problem 8–46)
Collins MK (Problem 13–35)
Colman A (Problem 13–78)
Coluccio LM (Problem 16–35)
Compton DA (Problem 18–35)
Conklin BR (Problem 15–44)
Connor JT (Problem 13–22)
Conway L (Problem 6–33)
Cook RG (Problem 6–52)
Cooper JA (Problems 8–73, 8–74, 8–75)
Copeland NG (Problem 13–53)
Corrado K (Problem 16–100)
Corson LB (Problem 18–2)
Cossart P (Problem 16–50)
Cosset FL (Problem 13–35)
Cosstick R (Problem 5–77)
Coulson RM (Problem 7–80)
Cox DR (Problem 8–6)
Cox M (Problem 5–74)
Cozzarelli NR (Problem 18–4)
Craigen WJ (Problem 6–52)
Creighton TE (Problems 3–2, 3–4, 3–6)
Crick FH (Problems 4–38, 6–1)
Crisona NJ (Problem 18–4)
Cross F (Problem 17–23)
Cross RA (Problems 16–9, 16–65, 16–66)
Crothers DM (Problem 7–11)
Curran T (Problem 7–18)
Cushman I (Problem 12–32)
Cutler D (Problem 13–78)
Czech MP (Problems 7–19, 11–10)

D'Atri S (Problem 17–68)
Dalbey RE (Problem 12–15)
Dammai V (Problem 12–73)
Danpure CJ (Problem 12–67)
Darley-Usmar V (Problem 14–42)
Darnell JE (Problem 15–83)
Datar SA (Problems 17–43, 17–44)
Davies PL (Problem 3–37)
Davies RJ (Problem 17–64)
Davis RL (Problems 7–19, 7–65, 15–50)
Davis RW (Problems 4–20, 18–23, 8–49)
Dawidowicz EA (Problem 12–95)
Dazin PF (Problem 13–28)
de Barros Lopes M (Problem 17–22)
de la Cruz A (Problems 17–43, 17–44)
de la Pompa JL (Problem 17–65)
de Saint Phalle B (Problem 18–34)
Dean FB (Problem 5–40)
DeCamilli P (Problem 13–79)
Deenen LLM (Problem 10–11)
DeFranco D (Problems 15–22, 15–23)
DeGier J (Problem 10–11)
Delagado-Partin VM (Problem 12–15)
Delbruck M (Problem 5–2)
Delcour AH (Problem 15–35)
Demple B (Problem 5–64)
Denome S (Problem 15–50)
Dentler WL (Problem 16–17)
DePamphilis ML (Problem 5–48)
DePinho RA (Problem 4–35)
Desai A (Problem 16–7)
Deshaies RJ (Problem 12–84)
Deslypere JP (Problem 7–74)

Detrich WH (Problems 16–10, 16–16)
Deuerling E (Problems 6–70, 6–71)
Devaux A (Problem 7–86)
Diaz ME (Problem 13–52)
Dingwall C (Problem 12–23)
Dittmer F (Problem 13–51)
Divita G (Problem 3–86)
Dodson M (Problem 5–40)
Doheny KF (Problems 18–25, 18–26)
Dolznig H (Problem 17–71)
Donahue CJ (Problem 17–55)
Donaldson JG (Problem 13–11)
Donzeau M (Problem 12–59)
Doolittle R (Problems 1–34, 1–35)
Dorée M (Problem 16–30)
Dowd P (Problem 17–55)
Downing KH (Problems 16–9, 16–10, 16–16)
Doyle JJ (Problem 1–32)
Doyle JL (Problem 1–32)
Draetta G (Problem 17–76)
Dramsi S (Problem 16–50)
Dreisig H (Problem 6–66)
Dressler D (Problem 5–72)
Du W (Problems 17–43, 17–44)
Dugaiczyk A (Problems 5–73, 6–24)
Duncan GS (Problem 17–65)
Dunham I (Problems 4–21, 4–22)
Dunn TM (Problem 7–31)
Dunphy WG (Problem 17–29)
Dura JA (Problem 13–22)
Duysens LNM (Problem 14–63)

Earnest TN (Problem 13–6)
Earnshaw WC (Problem 17–66)
Echols H (Problem 5–40, 7–7)
Eddy RL (Problem 4–20)
Edelman P (Problem 6–63)
Edgar B (Problem 15–23)
Edgar BA (Problems 17–43, 17–44, 17–84)
Edwards YH (Problem 7–74)
Eilers M (Problem 12–56)
Eisenberg E (Problem 16–69)
Ekker SC (Problems 15–96, 15–97, 15–98)
Elia AJ (Problem 17–65)
Eliasson C (Problem 16–32)
Elledge SJ (Problems 8–49, 8–74, 8–75)
Elliott R (Problem 13–52)
Ellis HM (Problem 17–57)
Emerson R (Problem 14–61)
Englander SW (Problem 6–72)
Erdjument-Bromage H (Problem 13–19)
Eriksson JE (Problem 16–32)
Evan GI (Problems 17–58, 17–59)
Evans RM (Problem 7–71)
Eyer J (Problem 16–38)

Fan J (Problems 16–9, 16–65, 16–66)
Fangman WL (Problem 5–44)
Farrell PJ (Problem 7–85)
Faull RJ (Problem 15–12)
Federman AD (Problem 15–44)
Fellows J (Problem 6–15)
Feng L (Problem 13–52)
Ferby I (Problem 18-32)
Ferguson SJ (Problems 14–22, 14–23, 14–38, 14–47)
Fernandez JM (Problem 3–32)
Ferrell JE (Problems 15–77, 15–79)
Fersht A (Problems 3–18, 3–67)
Fields S (Problem 8–73)
Fijalkowska IJ (Problem 5–29)
Finazzi D (Problem 13–11)
Fink GR (Problems 5–83, 15–101)
Finley D (Problem 6–77)
Firestone GL (Problem 15–23)
Fischer G (Problems 6–70, 6–71)
Flaus A (Problem 4–47)
Flavell RA (Problems 17–64, 17–65)
Fletterick RJ (Problem 13–6)

Flint AJ (Problem 3–87)
Flor A (Problems 17–43, 17–44)
Folkman J (Problem 17–86)
Forbush B (Problem 14–64)
Fornerod M (Problem 12–38)
Forsyth J (Problem 15–13)
Foulkes NS (Problem 7–54)
Fox TD (Problem 14–89)
Foyer CH (Problem 14–54)
Frayn KN (Problem 2–87)
Freeland FJ (Problem 1–5)
Frey L (Problem 5–24)
Fried SR (Problem 13–28)
Friedberg EC (Problem 5–69)
Fries E (Problem 13–71)
Fu J (Problem 13–22)
Fujiki Y (Problems 12–68, 12–69, 12–71, 12–72)
Fujinaga K (Problem 7–69)
Fujiwara D (Problems 12–35, 12–36)
Fukasawa K (Problem 18–8)
Fuller SD (Problems 13–78, 14–42, 16–14)
Funabiki H (Problem 18–33)
Fung BK-K (Problem 15–60)
Fung Y-KT (Problem 17–38)
Fusella A (Problem 13–40)

Gall JG (Problems 4–54, 8–39)
Gallant J (Problem 6–63)
Galova M (Problem 18–2)
Galper J (Problem 15–59)
Gann A (Problem 7–37)
Gannon J (Problem 6–77)
Garapin AC (Problem 6–24)
Garfinkel DJ (Problem 5–83)
Garoff H (Problem 13–78)
Gartenberg MR (Problem 7–11)
Gaub HE (Problem 3–32)
Gavin K (Problem 8–72)
Gay DA (Problem 7–80)
Geisse S (Problem 15–23)
Gelehrter TD (Problem 8–46)
Georgopoulos C (Problems 6–70, 6–71)
Gerace L (Problem 16–30)
Gerhart JH (Problem 3–82)
Gerlinger P (Problem 6–24)
Geromanos S (Problem 13–19)
Geuze HJ (Problem 13–40)
Ghaedi K (Problems 12–68, 12–71)
Gibbons IR (Problem 16–84)
Gibbs J (Problem 6–75)
Gibbs PEM (Problem 5–73)
Gibbs RA (Problem 8–60)
Gibson A (Problem 13–9)
Gibson WC (Problem 12–66)
Gilbert W (Problem 3–47)
Gillin FD (Problem 1–30)
Gilman AG (Problem 15–44)
Gilman MZ (Problem 15–83)
Gilula NB (Problem 15–15)
Ginsberg MH (Problem 15–12, 15–13)
Glandsdorff N (Problem 7–24)
Glass RE (Problem 7–10)
Glick BS (Problem 13–72)
Glotzer M (Problem 17–32)
Gnatt A (Problem 6–15)
Goddard AD (Problem 17–55)
Godowski PJ (Problems 7–36, 17–55)
Goldberg E (Problem 3–53)
Goldberg J (Problem 13–12)
Goldberg ML (Problems 5–3, 7–66, 8–87)
Goldfarb A (Problem 6–17)
Goldman RD (Problems 16–28, 16–29)
Goldman YE (Problem 16–76)
Goldstein JC (Problems 17–58, 17–59)
Goldstein JL (Problems 13–59, 13–60)
Goodwin EB (Problems 7–82, 7–83)
Goodwin W (Problem 5–53)
Gordon AM (Problem 16–77)
Gorin MB (Problem 13–52)
Görlich D (Problems 12–30, 12–31)

Gorovsky MA (Problem 6–66)
Gottesman MM (Problem 15–47)
Gottlieb GJ (Problem 4–35)
Gottlieb TA (Problem 13–78)
Gotto AM (Problem 7–74)
Goud B (Problem 13–21)
Goutte C (Problem 7–49)
Grams J (Problem 7–75)
Graziani G (Problem 17–68)
Greci J (Problem 8–36)
Green DR (Problems 17–58, 17–59)
Greengard P (Problem 15–35)
Griffith J (Problem 4–44)
Griffiths AD (Problems 16–9, 16–65, 16–66)
Grivell LA (Problem 12–51)
Grompe M (Problem 7–59)
Grosveld F (Problem 7–64)
Groudine M (Problem 4–50)
Grubmeyer C (Problem 14–16)
Grunstein M (Problem 4–59)
Gruss P (Problem 17–65)
Gu Y (Problem 17–65)
Gu ZW (Problem 7–74)
Guiard B (Problem 12–59)
Gundersen GG (Problem 16–28)
Guo W (Problem 13–21)
Gurney AL (Problem 17–55)
Gustafsson J-A (Problem 15–23)
Gustafsson K (Problem 13–35)
Gustafsson L (Problem 13–71)
Gutel M (Problem 3–32)
Guthmann C (Problems 10–8, 10–22)
Gutowski M (Problem 3–45)

Haas A (Problems 13–14, 13–27)
Habib G (Problem 7–74)
Hadwiger JA (Problem 17–22)
Hafner A (Problem 13–51)
Hageman J (Problem 12–64)
Hahn S (Problem 7–31)
Haig D (Problem 7–67)
Haigler HT (Problem 13–56)
Haimo LT (Problem 16–72)
Hajduk SL (Problem 7–75)
Hakem A (Problem 17–65)
Hakem R (Problem 17–65)
Hale SP (Problem 6–52)
Hall A (Problem 16–91)
Hanaoka F (Problem 5–62)
Hancock WO (Problem 16–67)
Handa H (Problem 8–81)
Harada Y (Problem 6–17)
Harder T (Problem 10–17)
Harlan JM (Problem 15–12)
Harland RM (Problem 5–42)
Harley CB (Problems 5–51, 17–82)
Harold FM (Problem 14–48)
Harris PJ (Problem 18–9)
Harrison T (Problem 6–23)
Hartl FU (Problems 6–70, 6–71, 6–72)
Hartmann E (Problems 5–3, 7–66, 8–87, 12–30,
    17–7, 17–8, 17–47)
Hartwell LH (Problem 17–47)
Hawley DK (Problem 6–4)
Hayashi Y (Problem 16–97)
Haydar TF (Problem 17–65)
Hayer-Hartl M (Problem 6–72)
Heald R (Problem 16–30)
Hedjran F (Problem 7–72)
Heise CE (Problem 18–49)
Helfand BT (Problem 16–28)
Heller-Harrison R (Problem 7–19)
Helms JB(Problem 13–11)
Helou B (Problem 3–45)
Henderson JT (Problem 17–65)
Hentze MW (Problems 7–78, 7–79, 7–82)
Herrmann H (Problem 16–32)
Hershey AD (Problem 4–8)
Herskowitz I (Problem 16–100)
Hertel C (Problem 15–34)

Hess GP (Problem 15–35)
Hess P (Problem 17–64)
Hibberd MG (Problem 16–76)
Hieter P (Problems 18–2, 18–25, 18–26)
Higgins CF (Problem 15–88)
Higgs HN (Problem 16–49)
Higuchi R (Problem 8–69)
Hill TL (Problem 16–69)
Hillan KJ (Problem 17–55)
Hille B (Problems 11–43, 11–46, 11–56)
Hilmen M (Problem 7–48)
Hinrichs SH (Problem 17–38)
Hirano T (Problems 18–4, 4–67)
Hirokawa N (Problem 16–97)
Hirsh D (Problem 6–30)
Ho C-L (Problem 16–28)
Hoenger A (Problem 16–34)
Hofmann K (Problem 13–6)
Hogness DS (Problem 4–20)
Hollenberg SM (Problems 8–73, 8–74, 8–75)
Holloway SL (Problem 17–32)
Holstege FC (Problem 4–59)
Holt JT (Problem 3–45)
Holt SE (Problem 17–82)
Honegger AM (Problem 15–60)
Hood L (Problems 5–3, 7–66, 8–43, 8–87)
Hopkins CR (Problem 13–9)
Horinouchi S (Problems 12–35, 12–36)
Horio T (Problems 16–12, 16–52)
Horn HF (Problem 18–8)
Horowitz S (Problem 6–66)
Horvitz RH (Problem 17–57)
Hoshi T 9 Problem 11–38)
Houry WA (Problems 6–70, 6–71)
Howard J (Problems 2–68, 16–67)
Huang A (Problem 17–55)
Huang B (Problem 16–84)
Huang C-Y F (Problem 15–77)
Huang S (Problem 12–58)
Huberman JA (Problem 5–37)
Huganir RL (Problem 15–35)
Huh GS (Problem 7–72)
Hunt T (Problems 6–77, 7–85, 8–72, 17–20, 17–48)
Hurd DD (Problem 18–8)
Hurley JB (Problem 15–60)
Hurst LD (Problem 1–5)
Hurwitz J (Problem 5–40)
Huxley AF (Problem 16–77)
Hwang RK (Problem 13–6)
Hyman AA (Problems 18–20, 18–25, 18–26)
Hynes RO (Problem 7–72)

Ikeda K (Problem 13–80)
Im D-S (Problem 18–49)
Imanaka T (Problems 12–68, 12–71)
Ingledew JW (Problems 14–24, 14–32)
Inglese J (Problem 15–59)
Iniguez-Lluhi JA (Problem 15–59)
Inman RB (Problem 5–12)
Inoue S (Problems 16–17, 16–90)
Isaya G (Problem 12–67)
Ishii T (Problem 11–64)
Ishizuka T (Problems 13–16, 13–17)
Itoh H (Problem 6–17)
Iwai S (Problem 5–62)

Jacks T (Problem 6–52)
Jackson MR (Problem 13–30)
Jackson RJ (Problem 7–85)
Jacobs HW (Problems 17–43, 17–44)
Jacobson MD (Problem 17–56)
Jady B (Problem 6–37, 6–38)
Jagendorf AT (Problem 14–72)
Jagus R (Problem 6–68)
Jain R (Problem 1–23)
James G (Problem 7–19)
Jan LY (Problems 13–28, 15–59)
Jan YN (Problems 13–28, 15–59)

Jemmerson R (Problem 17–56)
Jenkins NA (Problem 13–53)
Jennings EG (Problem 4–59)
Jensen RA (Problem 3–45)
Jetton TL (Problem 3–45)
Jia Z (Problem 3–37)
Jiang S (Problem 13–52)
Jiricny J (Problem 17–68)
Johnson AD (Problems 1–25, 7–49)
Johnson KA (Problem 5–28)
Johnson LN (Problem 3–86)
Johnson PJ (Problem 6–9)
Johnson RT (Problems 17–15, 17–26)
Johnson WA (Problem 2–99)
Jonczyk P (Problem 5–29)
Jones KA (Problem 7–69)
Jones MJ (Problem 18–32)
Jones SN (Problem 17–64)
Juin P (Problems 17–58, 17–59)
Julian FJ (Problem 16–77)

Kabata H (Problem 7–10)
Kabcenell AK (Problem 13–21)
Kagi D (Problem 17–65)
Kain KH (Problem 8–71)
Kaiser R (Problem 8–43)
Káldi K (Problem 12–59)
Kamp BM (Problem 14–63)
Kantheti P(Problem 13–52)
Kapfhamer D (Problem 13–52)
Kaplan A (Problem 13–49)
Karasuyama H (Problem 17–65)
Karess R (Problem 15–94)
Karrer KM (Problem 8–39)
Karsenti E (Problems 16–24, 18–20, 18–32)
Karzai AW (Problems 3–48, 6–65)
Kaufman SA (Problem 17–65)
Kaufmann SH (Problem 17–66)
Kawamura K (Problem 18–8)
Kay GF (Problem 7–61)
Kazlauskas A (Problem 15–72)
Keegstra K (Problem 12–64)
Keilin D (Problem 14–35)
Kellems RE (Problem 12–47)
Kemler R (Problem 15–92)
Kemper W (Problem 6–68)
Kemphues KJ (Problem 18–8)
Kendall AR (Problem 7–53)
Kent HM (Problem 12–31)
Kent SB (Problem 3–65)
Kerner MJ (Problem 6–72)
Kerppola TK (Problem 7–18)
Khoo W (Problem 17–65)
Kiger JA (Problem 15–50)
Kilpatrick MW (Problem 5–63)
Kim CN (Problem 17–56)
Kim NW (Problem 17–82)
Kimble J (Problem 7–82)
Kimura K (Problems 4–67, 18–4)
Kind AJ (Problem 7–2)
King MC (Problem 3–45)
King MP (Problem 17–56)
King RW (Problem 17–32)
Kingsbury R (Problem 6–10)
Kinoshita N (Problems 12–68, 12–71)
Kinosita K (Problems 6–17, 14–17, 14–18)
Kirchhausen T (Problems 13–6, 13–7, 16–61,
    16–62)
Kirkpatrick JB (Problem 17–15)
Kirschner MW (Problems 3–31, 3–83, 16–30,
    16–41, 16–61, 16–62, 18–22)
Kischkel FC (Problem 17–55)
Kispert A (Problem 15–92)
Kiss T (Problems 6–37, 6–38)
Klausner RD (Problem 13–11)
Kleckner N (Problem 5–82)
Klee CB (Problem 15–57)
Klemke RL (Problem 8–71)
Klemm RD (Problem 5–47)
Klug A (Problem 4–38)

Mullins RD (Problems 16–8, 16–13)
Mullner EW (Problem 17–71)
Muneyuki E (Problem 14–17)
Munro HN (Problem 7–77)
Munro S (Problem 13–29)
Muro Y (Problem 18–50)
Murphree AL (Problem 17–38)
Murphy C (Problem 4–54)
Murray AW (Problems 17–32, 17–36, 17–48, 18–33, 18–24)
Murray EJ (Problem 7–64)
Muslin AJ (Problem 8–95)
Mustaev A (Problem 6–17)

Nachury MV (Problems 12–41, 12–42)
Nagata K (Problem 8–81)
Nagata Y (Problem 18–50)
Nakagawa J (Problem 16–30)
Nakagawa T (Problem 16–97)
Nakamura TM (Problem 5–51)
Nakanishi S (Problem 11–64)
Nakata T (Problem 16–97)
Nasmyth K (Problems 17–33, 17–34, 17–71, 18–1, 18–2)
Nathans J (Problem 4–20)
Naylor DJ (Problem 6–72)
Nebreda AR (Problem 18–32)
Neer EJ (Problem 15–59)
Neher E (Problem 13–81)
Neupert W (Problems 12–48, 12–59)
Newman M (Problem 8–36)
Newton AC (Problem 15–82)
Nguyen T (Problem 18–49)
Nicholls DG (Problems 14–22, 14–23, 14–38, 14–47)
Nichols BJ (Problem 13–14, 13–27)
Nigg EA (Problem 16–30)
Nimnual A (Problem 17–64)
Ninfa AJ (Problem 7–30)
Nishizuka Y (Problem 15–52)
Noble MEM (Problem 3–86)
Noebels JL (Problem 13–52)
Noegel AA (Problem 16–57)
Nogales E (Problems 16–9, 16–10, 16–16)
Nogimori T (Problem 5–62)
Noji H (Problem 14–17, 14–18)
Nonet M (Problem 6–15)
Norbury CC (Problem 6–75)
Norby JG (Problem 2–23)
Norman C (Problem 7–20)
Novak EK (Problem 13–52, 13–53)
Novick P (Problem 13–21)
Novick PJ (Problem 13–21)
Nudler E (Problem 6–19)
Nugent JM (Problems 1–32, 1–33)
Nurse P (Problem 17–25, 17–42)

O'Connell KF (Problem 18–8)
O'Dowd BF (Problem 18–49)
O'Keefe EJ (Problem 17–72)
O'Malley BW (Problem 6–24)
O'Sullivan TN (Problem 13–53)
O'Toole TE (Problem 15–13)
Oeller PW (Problem 15–103)
Oesterhelt F (Problem 3–32)
Ogden S (Problem 7–31)
Ohara O (Problem 6–17)
Ohno M (Problem 12–38)
Ojala D (Problem 14–83)
Oka Y (Problem 11–10)
Okret S (Problem 15–23)
Okumoto K (Problems 12–68, 12–71)
Olsen EN (Problem 7–19)
Oosawa F (Problem 16–8)
Opperdoes FR (Problem 12–66)
Orci L (Problem 13–72)
Orr-Weaver TL (Problems 5–45, 5–46)
Osinga KA (Problem 12–66)
Otero LJ (Problem 7–86)

Ottaviano Y (Problem 16–30)
Ouellette M (Problem 17–82)
Owen DJ (Problems 13–5, 13–7)
Owen-Hughes T (Problem 4–47)

Paäbo S (Problem 5–53)
Pace NR (Problem 1–1)
Padgett RA (Problem 3–68)
Pagani E (Problem 17–68)
Pagano M (Problem 17–76)
Page D (Problem 3–45)
Palmer JD (Problems 1–32, 1–33, 14–78)
Panning B (Problem 7–62)
Panoutsakopoulou V (Problems 17–60, 17–61, 16–63)
Pantaloni D (Problems 16–3, 16–19, 16–42, 16–43, 16–44)
Pappin D (Problem 8–72)
Pardee AB (Problem 3–82)
Park H-O (Problem 16–100)
Parker SK (Problems 16–10, 16–16)
Parrillo JE (Problem 14–31)
Paschen S (Problem 12–59)
Paschen SA (Problem 12–59)
Patel LR (Problem 7–18)
Pattengale P (Problem 17–79)
Pauling L (Problem 3–66)
Payvar F (Problem 15–23)
Pearse BMF (Problems 13–5, 13–7)
Pearson JK (Problem 16–38)
Pease RJ (Problem 7–74)
Peden AA (Problem 13–52)
Pekny M (Problem 16–32)
Pelham HRB (Problems 13–14, 13–27, 13–29, 13–41)
Pellegrini LL (Problem 13–16, 13–17)
Penefsky HS (Problem 14–16)
Penninger JM (Problem 17–65)
Penny GD (Problem 7–61)
Perara E (Problem 12–80)
Perin MS (Problem 13–79)
Perrin F (Problem 6–24)
Peter M (Problem 16–30)
Peterlin BM (Problem 7–69)
Peterson AP (Problem 16–38)
Peterson PA (Problem 13–71)
Pfeifer G (Problem 6–72)
Phillips SEV (Problem 8–36)
Piantanida TP (Problem 4–20)
Pierschbacher MD (Problem 3–52)
Pike DA (Problem 15–103)
Pimenta A (Problem 16–98)
Pitti RM (Problem 17–55)
Plant AL (Problem 13–22)
Platas AA (Problem 6–4)
Pledger WJ (Problem 17–72)
Plow EF (Problem 15–13)
Pohlmann R (Problem 13–51)
Polakis P (Problem 3–45)
Pollard TD (Problems 16–8, 16–13, 16–33, 16–49)
Pollock R (Problems 7–20, 7–22)
Ponticelli AS (Problem 5–73)
Portenier M (Problem 15–34)
Porter CD (Problem 13–35)
Porter JA (Problems 15–96, 15–97, 15–98)
Portnoy DA (Problem 16–50)
Potter H (Problem 5–72)
Potter J (Problem 17–65)
Powell LM (Problem 7–74)
Prahlad V (Problem 16–29)
Preece AF (Problem 13–35)
Prehn S (Problem 12–30)
Prescott DM (Problems 17–5, 17–15)
Price DH (Problem 7–69)
Pringle JR (Problem 16–100)
Prober DA (Problem 17–84)
Prunell A (Problem 4–38)
Pruzan R (Problem 17–82)
Ptashne M (Problems 7–14, 7–37))
Purdue PE (Problem 12–67)
Pytela R (Problem 3–52)

Qian J (Problem 17–38)
Qiao X (Problem 13–52)

Racker E (Problem 14–13)
Raetz CR (Problem 12–70)
Raff M (Problem 1–25)
Raff MC (Problem 17–56)
Rakic P (Problem 17–65)
Ramer SW (Problem 8–49)
Ranier JE (Problem 8–60)
Rao PN (Problems 17–15, 17–26)
Rapp UR (Problems 8–74, 8–75)
Rappaport R (Problem 18–43)
Rasko JEJ (Problem 8–6)
Rastan S (Problem 7–61)
Ratliff KS (Problem 12–58)
Ravid K (Problem 18–50)
Ray WJ (Problem 6–50)
Reed JC (Problem 17–56)
Reed RR (Problem 15–44)
Reed SI (Problems 17–22, 17–23)
Reese EL (Problem 16–72)
Rein T (Problem 5–48)
Reiser J (Problem 6–29)
Reitzer LJ (Problem 7–30)
Ren R (Problem 15–71)
Reuveny E (Problem 15–59)
Reynolds AE (Problems 5–3, 7–66, 8–87)
Ribbeck K (Problem 12–31)
Richards G (Problem 15–24)
Richardson HE (Problems 17–22, 17–23)
Richardson JA (Problem 17–62)
Rieder CL (Problem 18–9)
Rief M (Problem 3–32)
Riggs AD (Problem 5–37)
Rindler MJ (Problem 13–78)
Rine J (Problems 12–25, 12–26)
Ripoche P (Problem 11–3)
Rivera MC (Problem 1–23)
Rizzolo L (Problem 13–78)
Roberts K (Problem 1–25)
Robins P (Problem 5–64)
Robinson MS (Problem 13–52)
Rockabrand D (Problems 6–70, 6–71)
Roeder RG (Problems 6–11, 6–12, 6–14, 7–35, 7–38)
Roessler PG (Problem 1–32)
Roger AJ (Problem 1–30)
Rohatgi R (Problems 16–61, 16–62)
Rose JK (Problems 10–18, 13–31, 13–32)
Rosenbaum JL (Problem 16–17)
Rosenberg LE (Problem 12–67)
Rosenblatt J (Problem 16–50)
Rosenfeld MG (Problems 7–71, 7–72)
Roskam W (Problem 6–24)
Ross AJ (Problems 17–60, 17–61, 17–63)
Rosseneu M (Problem 7–74)
Roth D (Problem 13–21)
Roth KA (Problems 17–60, 17–61, 17–63)
Rothbauer U (Problem 12–59)
Rothman JE (Problems 12–95, 13–11, 13–19, 13–39, 13–41, 13–72)
Rothman RE (Problem 12–80)
Rousselet A (Problems 10–8, 10–22)
Rousselet G (Problem 11–3)
Roy M (Problem 17–55)
Royou A (Problem 15–94)
Royzman I (Problem 5–46)
Rubinfeld B (Problem 3–45)
Ruderman JV (Problem 17–20)
Ruffner DE (Problem 5–73)
Runswick M (Problem 7–20)
Ruoslahti E (Problem 3–52)
Rupert JL (Problem 7–59)
Rusconi S (Problem 7–36)
Russell LB (Problem 13–53)
Russo AF (Problem 15–89)
Rybenkov VV (Problem 18–4)

Turner TK (Problem 17–64)
Tymoczko JL (Problem 2–76)
Tzagoloff A (Problem 14–22)

Uhlmann F (Problems 17–34, 18–1, 18–2)
Ulbrich EJ (Problem 13–51)
Ullrich A (Problem 15–60)
Umayahara K (Problems 13–16, 13–17)
Undermann C (Problems 13–14, 13–27)
Urbani LJ (Problem 13–39)
Uribe E (Problem 14–72)
Ushiro H (Problem 17–73)

Valius M (Problem 15–72)
Vallis Y (Problem 13–9)
van Arsdell SW (Problem 4–19)
Van Boom JH (Problem 12–66)
van der Drift C (Problem 14–48)
van der Meer R (Problem 3–45)
Van Dyck E (Problem 5–67)
Van Meer G (Problem 10–52)
Varma R (Problem 10–19)
Varmus H (Problem 6–52)
Varshavsky (Problem 6–77)
Veeneman GH (Problem 12–66)
Veres RC (Problems 5–3, 7–66, 8–87)
Vernos I (Problem 18–32)
Vidal M (Problem 8–76)
Vojtek AB (Problems 8–73, 8–74, 8–75)
Vollrath D (Problem 4–20)
von Ehrenstein G (Problem 6–50)
von Figura K (Problem 13–51)
von Kessler DP (Problems 15–96, 15–97, 15–98)
Vos O (Problem 17–15)

Wahl GM (Problem 3–68)
Walch-Solimena C (Problem 13–21)
Walker GC (Problem 5–69)
Walker JE (Problem 14–18)
Walker RA (Problem 16–17)
Wallace R (Problem 17–79)
Wallis SC (Problem 7–74)
Walter P (Problem 1–25)
Walworth NC (Problem 13–21)
Wanders RJA (Problems 12–68, 12–71)
Wang B (Problem 15–94)
Wang Y (Problem 12–33)
Wang YH (Problem 4–44)
Wang X (Problems 17–56, 17–62)
Wanner G (Problem 12–59)

Ward GE (Problem 16–30)
Washizu M (Problem 7–10)
Wasserman WJ (Problem 17–19)
Waterhouse NJ (Problems 17–58, 17–59)
Webb MR (Problem 14–16)
Wei MC (Problems 17–60, 17–61, 17–63)
Weiner AM (Problems 4–19, 6–28)
Weinert TA (Problem 17–47)
Weinrich SL (Problems 5–51, 17–82)
Weintraub H (Problems 4–50, 7–19, 7–65)
Weis K (Problems 12–41, 12–42)
Weisbeek P (Problem 12–64)
Weisblum B (Problem 6–50)
Weiss RA (Problem 13–35)
Weissmann C (Problem 6–29)
Welch MD (Problem 16–50)
Weng S-A (Problem 7–74)
West SC (Problems 5–67, 5–77)
Westphal HM (Problem 15–23)
White JG (Problems 18–8, 18–48)
White MF (Problem 4–47)
Whiteheart SW (Problem 13–19)
Whitehouse I (Problem 4–47)
Whitmore D (Problem 7–54)
Whittaker M (Problem 16–9)
Wickens M (Problems 6–33, 7–82, 7–83)
Wickner WT (Problems 13–14, 13–27)
Wiebel F (Problem 17–13)
Wieringa B (Problem 6–29)
Wilhelm H (Problem 18–32)
Willard HF (Problem 7–59)
Williams LT (Problem 8–95)
Williams RC (Problems 16–10, 16–16)
Williams RS (Problem 17–62)
Wilmut I (Problem 7–2)
Wilson G (Problem 8–36)
Wilson SM (Problem 13–53)
Wilson T (Problem 7–81)
Winston F (Problem 7–41)
Winter CG (Problem 15–94)
Witke W (Problem 16–57)
Wittenberg C (Problems 17–22, 17–23)
Wolff B (Problems 12–33, 12–35, 12–36)
Wolffe AP (Problem 7–57)
Wong ML (Problem 16–45)
Woo M (Problem 17–65)
Woo SL (Problem 6–24)
Wood WI (Problem 17–55)
Workman JL (Problems 4–47, 7–38)
Workman JL (Problem)
Wrange O (Problem 15–23)
Wright WE (Problem 17–82)
Wyrick JJ (Problem 4–59)

Xeros N (Problem 17–15)
Xu J (Problem 17–64)
Xu T (Problem 13–81)

Yabe J (Problem 16–98)
Yabe JT (Problem 16–98)
Yalow RS (Problem 15–6)
Yamada A (Problem 5–62)
Yamamoto KR (Problems 7–36, 15–22, 15–23)
Yamano H (Problem 6–77)
Yang CY (Problem 7–74)
Yang DD (Problem 17–64)
Yang HW (Problem 16–97)
Yang J (Problem 17–56)
Yanofsky C (Problem 2–105)
Yao C-H (Problem 8–39)
Yao M-C (Problem 8–39)
Yasuda R (Problems 14–17, 14–18)
Ybe JA (Problem 13–6)
Yeakley JM (Problem 7–72)
Yen TJ (Problem 7–80)
Yewdell JW (Problem 6–75)
Yip R (Problem 13–53)
Yokota S (Problem 12–69)
Yoon M (Problem 16–29)
Yoshica H (Problem 17–65)
Yoshida M (Problems 12–35, 12–36, 12–38, 14–18)
Yoshida R (Problem 17–65)
Yoshikawa H (Problem 2–78)
Young KE (Problems 15–96, 15–97, 15–98)
Young RA (Problem 4–59, 6–15)

Zagotta WN (Problem 11–38)
Zagouras P (Problem 13–31)
Zahringer J (Problem 7–77)
Zambruno G (Problem 17–68)
Zecca M (Problem 17–84)
Zerangue N (Problem 13–28)
Zhang X-F (Problem 8–74, 8–75)
Zhang Y (Problem 13–53)
Zhao C (Problem 16–97)
Zhen L (Problem 13–52)
Zheng Y (Problem 16–45)
Zhou J (Problem 7–19)
Zhu S-G (Problem 8–39)
Zhuang Y (Problem 6–28)
Zieg J (Problem 7–48)
Zimmet J (Problem 18–50)
Zinkel SS (Problem 7–11)
Zoeller RA (Problem 12–70)
Zong W-X (Problems 17–60, 17–61, 17–63)
Zou-Yang XH (Problem 8–72)

Index

F

v-src 56–57
Onion-skin structure (DNA) 94
Oocyte 336
  see also Egg
  maturation 377–378
    Xenopus 375
    see also Microinjection experiments
Oogenesis, nematode gene control 178–179
Open reading frame 195
Operons 152–156
  lactose metabolism *see* Lactose *(lac)* operon
  trp operon 152, 154
Optical tweezers 362
ORC 94–95, 200–201, 380
  see also Replication origins
Organelles
  see also specific organelles
  compartmentalization 243–246
  genomes 308–313
    mitochondrial *see* Mitochondrial genome
  human 244
  inheritance 246, 312
  numbers 243
  transport mechanisms 244
Organic chemistry 11
Origin of replication *see* Replication origins
Origin recognition complex (ORC) 94–95, 200–201, 380
  see also Replication origins
Orthologous genes 4
Osmotic balance 225, 231
Ouabain, sodium–potassium pump effects 231
Overlap microtubules 406, 412
Ovotransferrin 56
Oxaloacetate 33
  regeneration 34
Oxidation reactions 24, 296
  palmitic acid 34
Oxidation–reduction (redox) reactions 24
  general principles 296–297
  lithotrophs 3
  standard redox potential (E_0') 296
  thermodynamics 296–298, 301, 307
Oxidation states 24
Oxidative phosphorylation
  ATP synthesis 290, 295
    uncoupling 301
  electron transport *see* Electron transport chains
Oxygen
  ATP regeneration 29
  cellular consumption 132
    cardiac muscle 295
    mitochondria 294, 299–300
  chemical properties 15
  citric acid cycle stoichiometry 34
  electron transport 290, 294
  hemoglobin binding affinity 56
  measurement 294
  photosynthesis 4, 306–307
Oxygen electrodes 294, 299

P

P9OH incorporation, peroxisomes 258
PAGE *see* Polyacrylamide-gel electrophoresis (PAGE)
Palindromic DNA sequence 151, 190
Palmitic acid 34
Paracoccus denitrificans, respiratory chain 310–311
Paracrine signaling 318
Parthenogenesis, mitosis 411
Pasteur, Louis 2, 290
Pasteur effect 290
Pasteur's experiment (spontaneous generation) 2
Patch clamp analysis
  inside-out 331
  membrane transport 228
  neuromuscular junction 238–239
  potassium channels 234, 235
    acetylcholine activation 331
  whole-cell 287
Paternal imprinting 172
Pauling, Linus 53–54
PCR *see* Polymerase chain reaction
PDGF *see* Platelet-derived growth factor
PDGF-R *see* Platelet-derived growth factor receptor
Pedigrees 172, 313, 407
Penile erection 320
Periodate 278

Peroxisomes
  alanine:glyoxylate aminotransferase (AGT) 257
  assay 257–258
  assembly 259
  catalase 257
  deficient cell selection 258
  distribution 256
  glycosomes 256–257
  membrane phospholipids 264
  P9OH incorporation 258
  primary hyperoxaluria type I (PH1) 257
  transmembrane transport 244, 256–259
    analysis 258–259
    cytosolic receptor 259
    folded proteins 259
    PEX2,5,6 genes 259
Pertussis toxin, GPCR effects 326, 327, 329
Petite *(pet)* mutants 313
PEX2,5,6 genes, peroxisome membrane transport 259
pH 15–20
  see also Acids; Bases; Buffers
  gradients 295
    see also Electrochemical proton gradients
  Henderson–Hasselbalch equation 16, 19
  intracellular 17, 229–230
    endosomes 283
    lysosomes 277, 278
    measurement 230–231
  ion-exchange chromatography 184–185
  isoelectric point 18
  mitochondria 290
  protein denaturation 38
  titration curves 16
Phage-display libraries 203–204
  panning 203
  reduced genetic code 204
Phagocytosis 280
Phalloidin, actin binding 352–353
Phase variation, DNA rearrangement 163
Phenotype 205
  genotype relationship 82
Phorbol esters 252, 328
Phosphate bonds, energetics 30–31, 89
Phosphate buffer 17
Phosphatidylcholine 214
  membrane exchange 264
Phosphatidylcholine exchange protein 264
Phosphatidylethanolamine 214
Phosphatidylinositol (PI), cell signaling 214
Phosphatidylinositol 3′-kinase 45–46, 334, 336–337
Phosphatidylinositol-dependent protein kinase (PDK1),
    PKB activation 337
Phosphatidylserine 214
  membrane redistribution 218
Phosphoanhydride bond 28
Phosphodiester bond 89
Phosphoenolpyruvate 30
Phosphoglucose isomerase, thermodynamics 26–27
2-Phosphoglycerate 30
3-Phosphoglycerate 304
  conversion to pyruvate 30, 31
  oxidation state 24
  structure 30
Phosphoglycerate kinase (PKG) 257
Phosphoglycerate mutase 30
Phosphoglycolate, enzyme inhibition 54
Phospholipase C
  neomycin inhibition 329
  PDGF receptor 334
Phospholipases
  membrane lysis 212, 214
  phospholipase C 329, 334
Phospholipid exchange proteins 264
Phospholipids
  see also specific lipids
  ER membrane 264
  ESR spectroscopy 213–214, 217–218
  membrane 214
    distribution 214, 217–218
    mitochondrial 264
    peroxisomes 264
    red blood cells 220
  phospholipase action 212, 214
  structure 22
  vesicle fusion 271
Phospholipid vesicles, synthetic 264
N-phosphonacetyl-L-aspartate (PALA) 54

Phosphorus ^{32}P isotope 12
  half-life 187
Phosphorylase kinase, signal integration 330
Phosphorylation 59–61
  see also Protein kinases
  ATP requirement 59
  autophosphorylation 333–335
  axonal transport role 371
  cell-cycle control 59
  histone tails 59, 72–73
  immunoblots 186–187
  intermediate filament disassembly 351–352
  myogenin 150
  nuclear import regulation 251, 252
  protein regulation 327–328
    coagulation cascade 329
    enzyme regulation 52, 59
    PDGF autophosphorylation 334–335
    receptor desensitization 324
    signaling pathways 322–324
    trk autophosphorylation 333–334
  proteosome sensitization 340
  RNA polymerase CTD 120
  substrate-level 30
Phosphotransfer reactions, stereochemistry 291–292
Phosphotyrosine phosphorylase 334
Photochemical reaction center 304
Photophosphorylation 306, 307
6-4 Photoproduct 100
Photoreactivation 104
Photoreceptors, rod cells 332
Photorespiration 304
  see also Carbon fixation
Photosynthesis 3, 23, 302–308
  see also Carbon fixation; Chloroplasts
  algal 305
  bacterial 290
    cyanobacteria 304
    purple sulfur bacteria 4
  cytochromes 306
  electron transport *see* Electron transport chains
  energetics 302–303
  oxygen evolution 4, 306–307
  photophosphorylation 306
    cyclic 307
    noncyclic 307
  photosystems 304–305, 306
  stoichiometry 4
  Z scheme 305–306
Photosystems 304–305, 306
  cooperation 305
  electron flow 306
  photosystem I 304, 305, 306
  photosystem II 305, 306
Phototransduction 331–332
Phototrophs 3
pH-sensitive fluorophores, SNARF-1 230–231
Phylogenetics, eucaryotes 4, 5, 342
Phylogenetic trees
  branching order (topology) 7, 8
  construction methods 7–8
  hemoglobin genes 7, 8, 9
  mitochondrial genes 6
Physics, life 2
PI3-kinase 45–46, 334, 336–337
Pigmentation
  control 363–364
  defects 279–280
Pinocytosis 280–281
pK values 16–20, 45
PKG 338
Planck's constant 302–303
Plant cells
  animal cells *versus* 5, 342
  characteristics 5
  chloroplasts *see* Chloroplasts
  cytokinesis 415
  energy conversion 302–308
    see also Photosynthesis
    efficiency 302–303
  expansion 342
  microtubules 342
  organellar genomes
    chloroplasts 308–313
    mitochondrial 6
  variegation 312, 404–405
Plant growth factors 342–343

Prefixes

Symbol	Name	Value		Symbol	Name	Value
d-	deci-	10^{-1}		Y-	yotta-	10^{24}
c-	centi-	10^{-2}		Z-	zetta-	10^{21}
m-	milli-	10^{-3}		E-	exa-	10^{18}
μ-	micro-	10^{-6}		P-	peta-	10^{15}
n-	nano-	10^{-9}		T-	tera-	10^{12}
p-	pico-	10^{-12}		G-	giga-	10^{9}
f-	femto-	10^{-15}		M-	mega-	10^{6}
a-	atto-	10^{-18}		k-	kilo-	10^{3}
z-	zepto-	10^{-21}		h-	hecto-	10^{2}
y-	yocto-	10^{-24}		da-	deca-	10^{1}

Geometric Formulas

Figure	Area	Surface Area	Volume
square	l^2		
circle	πr^2		
ellipse	$\pi r_1 r_2$		
cube		$6\,l^2$	l^3
cylinder		$2\,\pi rh + 2\,\pi r^2$	$\pi r^2 h$
sphere		$4\,\pi r^2$	$\frac{4}{3}\,\pi r^3$
cone			$\frac{1}{3}\,\pi r^2 h$

Radioactive Isotopes

Isotope	Emission	Half-life	Counting Efficiency[a]	Maximum Specific Activity[b]
^{4}C	beta	5730 years	96%	0.062 Ci/mmol
^{3}H	beta	12.3 years	65%	29 Ci/mmol
^{5}S	beta	87.4 days	97%	1490 Ci/mmol
^{25}I	gamma, auger, and conversion electrons	60.3 days	78%	2400 Ci/mmol
^{2}P	beta	14.3 days	100%	9120 Ci/mmol
^{31}I	beta and gamma	8.04 days	100%	16,100 Ci/mmol

[a] Maximum efficiency for an unquenched sample in a liquid scintillation counter. Most real samples are quenched to some extent.
[b] This value assumes one atom of radioisotope per molecule. If there are two radioactive atoms per molecule, the specific activity will be twice as great, and so on.